MW00807682

THEORY and PRACTICE

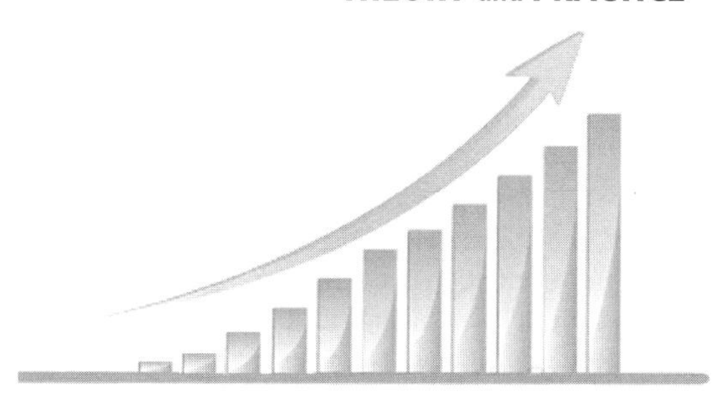

THE

KELLY CAPITAL GROWTH INVESTMENT CRITERION

World Scientific Handbook in Financial Economic Series

(ISSN: 2010-1732)

The Handbooks in Financial Economics (HIFE) are intended to be a definitive source for comprehensive and accessible information in the field of finance. Each individual volume in the series presents an accurate self-contained survey of a sub-field of finance, suitable for use by finance, economics and financial engineering professors and lecturers, professional researchers, investments, pension fund and insurance portfolio mangers, risk managers, graduate students and as a teaching supplement.

The HIFE series will broadly cover various areas of finance in a multi-handbook series. The HIFE series has its own web page that include detailed information such as the introductory chapter to each volume, an abstract of each chapter and biographies of editors. The series will be promoted by the publisher at major academic meetings and through other sources. There will be links with research articles in major journals.

The goal is to have a broad group of outstanding volumes in various areas of financial economics. The evidence is that acceptance of all the books is strengthened over time and by the presence of other strong volumes. Sales, citations, royalties and recognition tend to grow over time faster than the number of volumes published.

Published

Vol. 1 Stochastic Optimization Models in Finance (2006 Edition)
edited by William T. Ziemba & Raymond G. Vickson

Vol. 2 Efficiency of Racetrack Betting Markets (2008 Edition)
edited by Donald B. Hausch, Victor S. Y. Lo & William T. Ziemba

Vol. 3 The Kelly Capital Growth Investment Criterion: Theory and Practice
edited by Leonard C. MacLean, Edward O. Thorp & William T. Ziemba

World Scientific Handbook in Financial Economic Series — Vol. 3

THEORY and PRACTICE

THE KELLY CAPITAL GROWTH INVESTMENT CRITERION

Editors

Leonard C MacLean
Dalhousie University, USA

Edward O Thorp
University of California, Irvine, USA

William T Ziemba
Mathematical Institute, Oxford University, UK and University of British Columbia, Canada

NEW JERSEY • LONDON • SINGAPORE • BEIJING • SHANGHAI • HONG KONG • TAIPEI • CHENNAI

Published by

World Scientific Publishing Co. Pte. Ltd.
5 Toh Tuck Link, Singapore 596224
USA office: 27 Warren Street, Suite 401-402, Hackensack, NJ 07601
UK office: 57 Shelton Street, Covent Garden, London WC2H 9HE

Library of Congress Cataloging-in-Publication Data
The Kelly capital growth investment criterion : theory and practice / edited by Leonard C. MacLean, Edward O. Thorp, William T. Ziemba.
p. cm. -- (World Scientific handbook in financial economic series, 2010-1732 ; 3)
Includes bibliographical references.
ISBN-13: 978-9814293495
ISBN-10: 9814293490
ISBN-13: 978-9814293501
ISBN-10: 9814293504
1. Investments--Mathematical models. 2. Portfolio management--Mathematical models.
I. MacLean, L. C. (Leonard C.) II. Thorp, Edward O. III. Ziemba, W. T.
HG4515.2.K45 2010
332.63'2042--dc22

2010044902

British Library Cataloguing-in-Publication Data
A catalogue record for this book is available from the British Library.

Printed in Singapore.

To my wife Gwenolyn, for her patience and encouragement, and Dalhousie University for its constant support over many years
Leonard C. MacLean

To Vivian, with whom I've shared "the long run"
Edward O. Thorp

To Sandra for companionship, help, patience, and understanding over a long time and to the memory of Kelly criterion pioneeers, John L. Kelly, Henry A. Latane, Leo Breiman, and Kelly critic Paul A. Samuelson
William T. Ziemba

Contents

Preface

The modern development of the Kelly, or growth optimal, approach to allocating investments began with J.L. Kelly's seminal 1956 paper, over two hundred years after Daniel Bernoulli's 1738 introduction to the notion of logarithmic utility. Kelly's paper was followed by Latané's (1959) intuitive economic analysis and theoretical advances by Breiman (1960, 1961). Breiman showed that the Kelly maximization of expected utility with a logarithmic utility function also maximized the long run asymptotic growth of wealth while minimizing the expected time to reach arbitrarily large goals. Thorp (1962, 1966, 1969, 1971) pioneered the application of the Kelly criterion to actual gambling and investment. Ziemba and Vickson (1975) surveyed the literature presenting key papers, introductions and problems up to that time; for an update, see Ziemba and Vickson (2006). Algoet and Cover (1988) generalized the Breiman results to wider classes of assets and arbitrary ergodic market processes. MacLean, Ziemba and Blazenko (1992) show applications to a wide variety of sports and gambling events following Hausch, Ziemba and Rubinstein's (1981) application to racetrack betting.

Thorp (1960) suggested the term *Fortune's Formula* which later became the title of William Poundstone's 2005 book. This was in an abstract for a talk Thorp gave to the American Mathematical Society in January 1961, presenting his blackjack card counting discovery and his use of the Kelly approach to size bets in favorable situations. The term Kelly criterion appears to date from Thorp (1966) and is used in Thorp (1969)

Over the years both theory and practice have developed prolifically. The theory has been extended to managing portfolios of investments, results have been obtained for a broad range of distributional assumptions, the simultaneous management of assets and liabilities has been elaborated upon, and the various properties, advantages and disadvantages have been clarified.

We now have a fuller understanding of the tradeoff between risk and reward for fractional Kelly versus full Kelly and for Kelly subject to minimizing the underperformance of a benchmark or specified desired wealth path.

The theory has also benefitted from the practical experience of gamblers, traders, hedge fund managers and investors, especially from some of the greatest investors.

In this volume we present a selection of many of the most important papers from the now vast and growing literature on the subject. While we could not publish all the important papers, we feel that the main results appear here.

The volume is organized into six sections that cover the early ideas and contributions, classic papers and theories, relations to asset allocation including optimization with withdrawals, fractional Kelly wagering and its relations to benchmarks, assessing the good and bad properties of Kelly wagering, utility foundations and the use of Kelly type strategies by various investors including the greatest investors.

We thank our authors for their contributions, those who helped us with the editing and production especially Sandra Schwartz and our publisher, World Scientific, for their production and promotion of this volume. Special thanks go to Tom Cover for many helpful comments on earlier versions of the introductions and to Bryan Fitzgerald for valuable data.

Leonard C. MacLean
Edward O. Thorp
William T. Ziemba
March 2010

List of Contributors

Contributors	Chapters
Andrew R. Barron *Department of Statistics, Yale University, Connecticut, USA*	**13**
Bernhard K. Meister *Department of Physics, Renmin University of China, China*	**21**
Daniel Bernoulli *University of Basel, Switzerland*	**2**
David G. Luenberger *Stanford University, Stanford, USA*	**42**
Donald B. Hausch *School of Business, University of Wisconsin, Wisconsin, USA*	**46, 47, 48**
Eckhard Platen *School of Finance and Economics and Department of Mathematical Sciences, University of Technology, Sydney, Australia*	**28**
Edward O. Thorp *Edward O. Thorp and Associates, Newport Beach, CA, USA*	**6, 7, 36, 37, 38, 39, 54**
Erik Ordentlich *Hewlett Packard Labs, Palo Alto, California, USA*	**16**
George Blazenko *School of Business Administration, Simon Fraser University, British Columbia, Canada*	**24**

Contributors	Chapters
Harry M. Markowitz *IBM Thomas J. Watson Research Center,* *Yorktown Heights, New York, USA*	**35**
Henry A. Latané *Chapel Hill, University of North Carolina, USA*	**4**
Igor V. Evstigneev *Economics Department, School of Social Sciences,* *University of Manchester, UK*	**20, 29**
John L. Kelly Jr. *Bell Labs, New Jersey, USA*	**3**
John M. Mulvey *Princeton University, New Jersey, USA*	**50**
Klaus R. Schenk-Hoppé *Leeds University Business School and* *School of Mathematics,* *University of Leeds, UK*	**20, 29**
Leo Breiman *University of California, Los Angeles, USA*	**5**
Leonard C. MacLean *School of Business Administration,* *Dalhousie University, Halifax, Canada*	**19, 24, 25, 38, 39**
Mark Davis *Department of Mathematics,* *Imperial College London, London, UK*	**27**
Mark Finkelstein *University of California, Irvine, USA*	**17**
Markus Rudolf *WHU-Otto Beisheim Graduate School of Management,* *Dresdner Bank Chair of Finance, Germany*	**51**
Mehmet Bilgili *Alliance Bernstein, Equity Trading, New York, USA*	**50**

Contributors	Chapters
Michael Stutzer *Burridge Center for Securities Analysis and Valuation, Leeds School of Business, University of Colorado, Boulder, USA*	**43, 44**
Nils H. Hakansson *Walter A. Haas School of Business, University of California, Berkeley, USA*	**8, 9, 41, 49**
Paul A. Samuelson *Department of Economics, Massachusetts Institute of Technology, Cambridge, USA*	**31, 33, 34**
Paul H. Algoet *Boston University, Massachusetts, USA*	**14**
Rafael Sanegre *University of British Columbia, Canada*	**25**
Rachel E. S. Ziemba *Roubini Global Economics, London, UK*	**53**
Raymond G. Vickson *Professor Emeritus, University of Waterloo, Canada*	**32**
Robert M. Bell *Stanford University, Stanford, USA*	**12**
Robert R. Grauer *Simon Fraser University, Canada*	**49**
Robert Whitley *University of California, Irvine, USA*	**17, 37**
Richard Roll *Anderson School of Management, UCLA, Los Angeles, USA*	**10**
Sébastien Lleo *Department of Mathematics, Imperial College London, London, UK*	**27**

Contributors	Chapters
Sid Browne *Graduate School of Business, Columbia University, New York, USA*	**23, 26**
Taha M. Vural *Princeton University, New Jersey, USA*	**50**
Thomas M. Cover *Department of Statistics and Electrical Engineering, Stanford University, Stanford, USA*	**12, 13, 14, 15, 16**
Thorsten Hens *University of Zurich, Switzerland*	**20**
Vijay K. Chopra *Frank Russell Company*	**18**
William T. Ziemba *Professor Emeritus,* *University of British Columbia, Canada* *Visiting Professor,* *Oxford University and University of Reading, UK*	**18, 19, 24, 25, 32, 38, 39, 41, 46, 47, 48, 51, 52**
Yingdong Lv *Department of Physics,* *Renmin University of China, China*	**21**
Yonggan Zhao *School of Business, Dallhousie University, Halifax, Canada*	**25, 28**
Yuming Li *School of Business, California State University, Fullerton, USA*	**19**

Acknowledgements

We thank the following publishers and authors for permission to reproduce the articles listed below.

Advanced Applied Probability

Finkelstein, M. and R. Whitley (1981). Optimal strategies for repeated games. 13, 415–428.

Annals of Probability

Algoet, P. H. and T. M. Cover (1988). Asymptotic optimality and asymptotic equipartition properties of log-optimum investment. 16(2), 876–898.

Bell System Technical Journal

Kelly, Jr., J. R. (1956). A new interpretation of information rate. 35, 917–926.

Colloquia Mathematica Societatis Janos Bolyai

Thorp, E. O. and R. Whitley (1972). Concave utilities are distinguished by their optimal strategies. 813–830.

Econometrica

Bernoulli, D. (1954). Exposition of a new theory on the measurement of risk (translated by Louise Sommer). 22, 23–36.

Hakansson, N. H. (1970). Optimal investment and consumption strategies under risk for a class of utility functions. 38, 587–607.

Elsevier/North Holland

Hakansson, N. H. and W. T. Ziemba (1995). Capital growth theory. In R. A. Jarrow, V. Maksimovic, and W. T. Ziemba (Eds.), *Finance, Handbooks in OR & MS*, Vol. 9, 65–86.

Thorp, E. O. (2006). The Kelly criterion in blackjack sports betting and the stock market. In S. A. Zenios and W. T. Ziemba (Eds.), *Handbook of Asset and Liability Management*, Vol. 1, 387–428.

Review of the International Statistical Institute

Thorp, E. O. (1969). Optimal gambling systems for favorable games. 37(3), 273–293.

Springer

Lv, Y. and B. K. Meister (2009). Application of the Kelly criterion to Ornstein-Uhlenbeck processes. *Lecture Notes of the Institute for Computer Sciences*, 4, 1051–1062.

The Journal of Finance

Markowitz, H. M. (1976). Investment for the long run: New evidence for an old rule. 31(5), 1273–1286.

Roll, R. (1973). Evidence on the "growth-optimum" model. 28(3), 551–566.

Wiley

Thorp, E. O. (2008). Understanding the Kelly criterion. *Wilmott*, May and September.

Ziemba, R. E. S. and W. T. Ziemba (2007). Postscript: The Renaissance Medallion Fund. In *Scenarios for Risk Management and Global Investment Strategies*, 295–298.

World Scientific

Ziemba, W. T. and D. B. Hausch (2008). The Dr. Z betting system in England. In D. B. Hausch, V. Lo, and W. T. Ziemba (Eds.), *Efficiency of Racetrack Betting Markets*, 567–574.

We thank the following authors for permission to publish their new papers:

M. Bilgili	J. M. Mulvey
I. V. Evstigneev	E. Platen
M. H. A. Davis	K. R. Schenk-Hoppe
M. A. H. Dempster	M. Stutzer
T. Hens	E. O. Thorp
S. Lleo	T. M. Vural
L. C. Maclean	W. T. Ziemba

IEEE Transactions of Information Theory

Barron, A. R. and T. M. Cover (1988). A bound on the financial value of information. 34(5), 1097–1100.

Journal of Banking and Finance

Samuelson, P. A. (1979). Why we should not make mean log of wealth big though years to act are long. 3, 305–307.

Journal of Business

Grauer, R. R. and N. H. Hakansson (1986). A half century of returns on levered and unlevered portfolios of stocks, bonds and bills, with and without small stocks. 592, 287–318.

Hakansson, N. H. (1971). On optimal myopic portfolio policies, with and without serial correlation of yields. 44, 324–334.

Journal of Econometrics

Stutzer, M. (2003). Portfolio choice with endogenous utility: A large deviations approach. 116, 365–386.

Journal of Economic Dynamics and Control

Luenberger, D. G. (1993). A preference foundation for log mean-variance criteria in portfolio choice problems. 17, 887–906.

MacLean, L. C., R. Sanegre, Y. Zhao, and W. T. Ziemba (2004). Capital growth with security. 28(4), 937–954.

Rudolf, M. and W. T. Ziemba (2004). Intertemporal surplus management. 28, 975–990.

Journal of Political Economy

Latané, H. A. (1959). Criteria for choice among risky ventures. 67, 144–155.

Journal of Portfolio Management

Chopra, V. K. and W. T. Ziemba (1993). The effect of errors in means, variances and co-variances on optimal portfolio choice. 19, 6–11.

Ziemba, W. T. (2005). The symmetric downside-risk Sharpe ratio and the evaluation of great investors and speculators. 32(1), 108–122.

Mathematical Finance

Cover, T. M. (1991). Universal portfolios. 1(1), 1–29.

Mathematics of Operations Research

Bell, R. M. and T. M. Cover (1980). Competitive optimality of logarithmic investment. 5(2), 161–166.

Browne, S. (1997). Survival and growth with a liability: Optimal portfolio strategies in continuous time. 22(2), 468–493.

Ordentlich, E. and T. M. Cover (1998). The cost of achieving the best portfolio in hindsight. 23(4), 960–982.

Management Science

Browne, S. (2000). Risk-constrained dynamic active portfolio management. 46(9), 1188–1199.

Hausch, D. B., W. T. Ziemba, and M. E. Rubinstein (1981). Efficiency of the market for racetrack betting. 27, 1435–1452.

Hausch, D. B. and W. T. Ziemba (1985). Transactions costs, extent of inefficiencies, entries and multiple wagers in a racetrack betting model. 31, 381–394.

MacLean, L., W. T. Ziemba, and G. Blazenko (1992). Growth versus security in dynamic investment analysis. 38(11), 1562–1585.

Proceedings of the 4th Berkeley Symposium on Mathematical Statistics and Probability

Breiman, L. (1961). Optimal gambling systems for favorable games. 1, 63–68.

Proceedings of the Business and Economics Section of the American Statistical Association

Thorp, E. O. (1971). Portfolio choice and the Kelly criterion. 215–224.

Proceedings National Academy of Science

Samuelson, P. A. (1971). The “fallacy” of maximizing the geometric mean in long sequences of investing or gambling. 68, 2493–2496.

Quantitative Finance

MacLean, L. C., W. T. Ziemba, and Y. Li (2005). Time to wealth goals in capital accumulation. 5(4), 343–355.

Review of Economics and Statistics

Samuelson, P. A. (1969). Lifetime portfolio selection by dynamic stochastic programming. 51, 239–246.

Leonard MacLean at the 12th International Conference on Stochastic Programming, Dalhousie University, Halifax, Canada, August 19, 2010.

Vivian and Edward Thorp, Newport Beach, California, 2004.

Professor William T. Ziemba while visiting as a Christianson Fellow at St. Catherines College, Oxford University in 2003. Beginning in 2003 and yearly thereafter he has given a half-day lecture on the Kelly criterion theory and practice to the Masters students in the Mathematical Finance program at the Mathematical Institute of Oxford University.

Part I

The early ideas and contributions

1

Introduction to the Early Ideas and Contributions

We live in an age of instant contact through email, twitter, TV, and other modes of communication. Back in the 1700s, communication was slower and mail would take weeks or months between receipt and response. The first paper in this volume and arguably the first on log utility is Daniel Bernoulli's article, written in 1738 in Basel, Switzerland where he was professor of physics and philosophy. Bernoulli, at age 25, studied in Basel, went to St. Petersburg and then returned to Basel. He is a member of the famous family of Swiss mathematicians, who were known as the first to apply mathematical analysis to the movement of liquid bodies. His article on log utility and the St. Petersburg paradox reprinted here was translated by Dr. Louise Sommer of the American University with assistance from Karl Menger, mathematics professor at the Illinois Institute of Technology, and William J. Baumol, economics professor at Princeton University. The article was published in *Econometrica* in 1954.

A great paper often has only one new idea well developed with no major error. In his paper, Bernoulli develops two new ideas. The first idea is the development of declining marginal utility of wealth leading to logarithmic utility. His simple idea is that marginal utility should be proportional to current wealth. So upon integration, one has log utility. Bernoulli postulated monotone utility so that utility was increasing in wealth. Prior to this, it was assumed that decisions were made on an expected value or linear utility basis. The general idea of declining marginal utility or what we would later call "risk aversion" or "concavity" is crucial in modern decision theory.

The second idea is his contribution to the St. Petersburg paradox. This problem actually originates from Daniel Bernoulli's cousin, Nicolas Bernoulli, professor at the University of Basel. In 1708, he submitted five important problems to professor Pierre Montmort, one of which was the St. Petersburg paradox. The idea is to determine the expected value and what you would pay for the following gamble:

> A fair coin with $\frac{1}{2}$ probability of heads is repeatedly tossed until heads occurs, ending the game. The investor pays c dollars and receives in return 2^{k-1} with probability 2^{-k} for $k = 1, 2, \ldots$ should a head occur. Thus, after each succeeding loss, assuming a head does not appear, the bet is doubled to 2, 4, 8, ...etc. Clearly the expected value is $\frac{1}{2} + \frac{1}{2} + \frac{1}{2} + \ldots$ or infinity with linear utility.

Bell and Cover (1980) argue that the St. Petersburg gamble is attractive at any price c, but the investor wants less of it as $c \to \infty$. The proportion of the investor's wealth invested in the St. Petersburg gamble is always positive but decreases with the cost c as c increases. The rest of the wealth is in cash.

Bernoulli offers two solutions since he feels that this gamble is worth a lot less than infinity. In the first solution, he arbitrarily sets a limit to the utility of very large payoffs. Specifically, any amount over 10 million is assumed to be equal to 2^{24}. Under that assumption, the expected value is

$$\frac{1}{2}(1) + \frac{1}{4}(2) + \frac{1}{8}(4) + \cdots + \left(\frac{1}{2}\right)^{24}(2^{24}) + \left(\frac{1}{2}\right)^{25}(2^{24}) + \left(\frac{1}{2}\right)^{26}(2^{24}) + \cdots$$

$$= 12 + \text{the original } 1 = 13$$

If utility is $\sqrt{w}$, the expected value is

$$\frac{1}{2}\sqrt{1} + \frac{1}{4}\sqrt{2} + \frac{1}{8}\sqrt{4} + \ldots = \frac{1}{2-\sqrt{2}} \cong 2.9$$

When utility is log, as Bernoulli proposed, the expected value is

$$\frac{1}{2}\log 1 + \frac{1}{4}\log 2 + \frac{1}{8}\log 4 + \ldots = \log 2 = 0.69315$$

The use of a concave utility function does not eliminate the paradox. For example, the utility function $U(x) = x/\log(x + A)$, where $A > 2$ is a constant, is strictly concave, strictly increasing, and infinitely differentiable yet the expected value for the St. Petersburg gamble is $+\infty$.

As Menger (1934) pointed out, the log, the square root and many others, but not all, concave utility functions eliminate the original St. Petersburg paradox but it does not solve one where the payoffs grow faster than 2^n. So if log is the utility function, one creates a new paradox by having the payoffs increase at least as fast as log reduces them so one still has an infinite sum for the expected utility. With exponentially growing payoffs one has

$$\frac{1}{2}\log(e^1) + \frac{1}{4}\log(e^2) + \ldots = \infty$$

The super St. Petersburg paradox, in which even $E\log X = \infty$, is examined in Cover and Thomas (2006: 181, 182) where a satisfactory resolution is reached by looking at relative growth rates of wealth. Another solution to such paradoxes is to have bounded utility, for example, as Bernoulli suggested above 10 million. To solve the St. Petersburg paradox with exponentially growing payoffs, or any other growth rate, a second solution, in addition to that of bounding the utility function above, is simply to choose a utility function which, though unbounded, grows "sufficiently more slowly" than the inverse of the payoff function, e.g., like the log of the inverse function to the payoff function. The key is whether the valuation using a utility function is finite or not; if finite, the specific value does not matter since utilities are equivalent to within a positive linear transformation ($V = aU + b$, $a > 0$). So for any utility giving a finite result, there is an equivalent one that will give you any specified finite value as a result. Only the behavior of $U(x)$ as $x \to \infty$ matters and strict monotonicity is necessary for a paradox. For example, $U(x) = x$, $x \leqslant A$ will not produce a paradox, but the continuous concave utility function

$$U(x) = \frac{x}{2} + \frac{A}{2}, \quad x > A$$

will have a paradox. Samuelson (1977) provides an extensive survey of the paradox (see also Menger (1967) and Aase (2001)).

Kelly (1956) is given credit for the idea of using log utility in gambling and repeated investment problems, as such it is known as the Kelly criterion. Kelly's analyses use Bernoulli trials. Not only does he show that log is the utility function which maximizes the long run growth rate, but that this utility function is myopic in the sense that period by period maximization based only on current capital is optimal. Working at Bell Labs, Kelly was strongly influenced by information theorist Claude Shannon. Kelly defined the long run growth rate of the investor's fortune using

$$G = \lim_{N \to \infty} \log \frac{W_N}{W_0}$$

where W_0 is the initial wealth and W_N is the wealth after N trials in sequence. With Bernoulli trials, one wins $= +1$ with probability p and loses -1 with probability $q = 1 - p$. The wealth with M wins and $L = N - M$ loses is

$$W_N = (1 + f)^M (1 - f)^{N-M} W_0$$

where f is the fraction of wealth wagered on each of the N trials. Substituting this into G yields

$$\begin{aligned} G &= \lim_{N \to \infty} \left(\frac{M}{N} \log(1 + f) + \left(\frac{N - M}{N} \right) \log(1 - f) \right) \\ &= p \log(1 + f) + q \log(1 - f) = E \log W \end{aligned}$$

by the strong law of large numbers.

Maximizing G is equivalent to maximizing the expected value of the log of each period's wealth. The optimal wager for this is

$$f^* = p - q, \qquad p \geqslant q > 0$$

which is the expected gain per trial, or the edge. If there is no edge, the bet is zero.

If the payoff is $+B$ for a win and -1 for a loss, then the edge is $Bp - q$, the odds are B, and

$$f^* = \frac{Bp - q}{B} = \frac{edge}{odds}$$

Latané (1959) introduced log utility as an investment criterion to the finance world independent of Kelly's work. Focusing, like Kelly, on simple intuitive versions of the expected log criteria, he suggested that it had superior long run properties.

Breiman (1961), following his earlier intuitive paper Breiman (1960), established the basic mathematical properties of the expected log criterion in a rigorous fashion. He proves three basic asymptotic results in a general discrete time setting with intertemporally independent assets. Suppose in each period, N, there are K investment opportunities with returns per unit invested $X_{N1}, \ldots, X_{NK}$. Let $\Lambda = (\Lambda_1, \ldots, \Lambda_K)$ be the fraction of wealth invested in each asset. The wealth at the end of period N is

$$W_N = \left(\sum_{i=1}^{K} \Lambda_i X_{Ni} \right) W_{N-1}$$

Property 1. In each time period, two portfolio managers have the same family of investment opportunities with returns, X, and one uses a Λ^* which maximizes $E \log W_N$ whereas the other uses an *essentially different* strategy, Λ, so they differ infinitely often, that is

$$E \log W_N \Lambda^* - E \log W_N(\Lambda) \to \infty$$

Then

$$\lim_{N \to \infty} \frac{W_N(\Lambda^*)}{W_N(\Lambda)} \to \infty$$

So the wealth exceeds that with any other strategy by more and more as the horizon becomes more distant.

This generalizes the Kelly-Bernoulli trial setting to intertemporally independent and stationary returns.

Property 2. The expected time to reach a preassigned goal A is, asymptotically least as A increases with a strategy maximizing $E \log W_N$.

Property 3. Assuming a fixed opportunity set, there is a fixed fraction strategy that maximizes $E \log W_N$, which is independent of N.

Thorp (1969) discusses the general theory of optimal betting over time on favorable games or investments. Favorable games are those with a strategy such that

$$Prob \left[\lim_{N \to \infty} W_N > 0 \right] = 1$$

where W_N is the investor's capital after N trials. Thorp follows the footsteps of Kelly and Breiman, Bellman and Kalaba (1957), and Ferguson (1965), by discussing some favorable games such as blackjack, the side bet in Nevada-style baccarat (also called *chemin de fer*), roulette, the wheel of fortune, and the stock market. Once one has an edge, with positive expectation, Thorp outlines a general theory for optimal wagering on such games. For the stock market, Thorp discusses mispricings in warrants and how to hedge them with advantage and to choose the optimal amount to invest using the Kelly approach.

Markowitz arithmetic mean-variance (MV) efficiency applies to single period returns. For multiperiod returns we are interested in geometric or compound rates of return and the associated variance. Arithmetic MV-efficient portfolios are, in general, not geometric MV-efficient and conversely. The Kelly strategy is always geometric MV-efficient and has the highest growth rate of all such strategies. Any strategy which bets more is not geometric MV-efficient.

Thorp (1971) focuses on the theory of logarithmic utility as applied to portfolio selection and contrasts it with Markowitz mean-variance portfolio theory. He shows that the Kelly strategy is not necessarily arithmetic mean-variance efficient.

Hakansson (1970) presents arithmetic optimal strategies for the class of utility functions

$$\sum_{t=1}^{\infty} \alpha^{t-1} u(c_t), \qquad 0 < \alpha < 1$$

where c_t is the consumption in period t and either the relative $-cu''(c)/u'(c)$ or the absolute Arrow-Pratt risk aversion $u''(c)/u'(c)$ is a positive constant for all $c \geqslant 0$. This includes positive power $u(c) = c^\gamma$, $0 < \gamma < 1$, negative power $u(c) = -c^{-\gamma}$, $\gamma > 0$, logarithmic $u(c) = \log c$, and exponential $u(c) = -e^{-\gamma c}$, $\gamma > 0$.

Using this model, Hakansson is able to determine, in closed form, optimal consumption, investment and borrowing strategies. The investor has an initial capital position, which could be negative, and a known deterministic non-capital income stream, and an arbitrary number of possible investments that can be held long or shorted whose probability distributions satisfy an arbitrage free condition called *no easy money*. It means that no combination of risky assets can with probability one out return the constant rate of interest, that no combination of short sale investments exists where the probability is zero that a loss will exceed the lending rate of interest, and that no combination of short sale investments can guarantee against loss. It is assumed that there are no taxes or transaction costs, and that the joint distribution functions of the assets are known and stationary and have stochastically constant returns to scale. Double the investment provides double returns. The stationarity assumption can be generalized and the same results obtained.

The optimal investment strategies are independent of wealth, noncapital income, age and impatience to consume. Optimal consumption is linear and increasing in current wealth and in the present value of the noncapital income stream. The optimal asset mix depends only on the probability distribution of returns, the interest rate and the investor's one-period utility function of consumption.

Necessary and sufficient conditions for capital growth are derived. With logarithmic utility of consumption, the investor will always invest the capital available after consumption is paid so as to maximize the expected growth rate of capital plus the present value of the noncapital income stream.

Hakansson's definition of capital growth is asymptotic

$$\lim_{t\to\infty} Pr(W_t > W_1) = 1$$

When there is statistical independence period by period, this condition implies that

$$E(W_{t+1}|W_t) \geqslant W_t$$

but not the converse.

A necessary and sufficient condition for capital growth is $\alpha r > 1$ or that $E[\log W] > 0$. Here α is the time discount factor and r is the rate of interest. A sufficient but not necessary condition for growth is that there be a non-zero investment in at least one of the risky investments since then $Pr(W_{t+1} > W_t) > 0$ for all t.

A key property that aids multiperiod optimization calculations is myopia. Let $u(w)$ be a utility function of final wealth. The u is myopic if induced intermediate time utility functions of wealth are independent of the past and the future, in particular, they are independent of yields beyond the current period so they are positive linear transformations of $u(w)$. Hakansson (1971) shows that log utility induces myopia for general asset return distributions. Hakansson also clarifies earlier results of Mossin (1968) and Leland (1968) for power utility functions; for these one needs independence of period by period returns to induce myopia. For $u(w) = \log w$, for all time dependent markets one should, with myopia, maximize $E\{\log W_N|W_1, W_2, \ldots, W_{N-1}\}$, the conditional mean.

Mossin and Leland showed that with zero interest rates, myopia obtains for linear risk tolerance utility

$$u'(w)/u''(w) = a + bw$$

When interest rates are non-zero, there is myopia if these future interest rates are known. The linear risk tolerance family includes exponential, log and the negative and positive power utility families. All of these results are under the assumption of stochastically constant returns to scale, perfect liquidity and divisibility of the assets, no transaction costs, withdrawals, capital additions, taxes, and short sales.

Roll (1973) follows the finance literature starting with Latané (1959) rather than the Kelly (1956) literature. He investigates the relationship between the Kelly capital growth model, mean-variance and capital asset pricing (CAPM) analysis. Additional results on this topic are presented by Thorp (1969, 1971), who shows that Kelly weightings are not necessarily mean-variance efficient and Markowitz (1976), who argues that the log-optimal portfolio is a limiting mean-variance portfolio. MacLean, Ziemba and Blazenko (1992) argue that, from a growth versus security prospective, the Kelly portfolio is the most aggressive utility function that can be used and betting more leads to lower long run growth and less security. If estimation error is considered, the log utility is even riskier. Since, as Chopra and Ziemba (1993) show, with such low Arrow-Pratt absolute risk aversion (almost zero), errors in means are about 100 times errors in variances and co-variances in certainty equivalent value. Hence, obtaining accurate mean estimates is crucial to successful investing.

Roll (1973) observes the result proved by Breiman (1961) that growth-optimum "portfolios maximize the probability of exceeding a given level of wealth in a fixed time". Roll focuses on an empirical study of the implications for observed common stock returns of all investors selecting such a portfolio. Mean-variance type models have dominated finance theory and practice largely because of the relationship with the capital asset pricing model and its theoretical justification of index funds which beat about 75% of managers with less cost and effort. On the other hand, growth-optimal Kelly portfolios have been used not by ordinary investors but those, like Warren Buffett, who attempt to have superior returns. This may lead to less diversification, larger positions and more monthly losses, see Ziemba (2005).

For his test, Roll does not specify specific return distributions but rather relies on the fact that in each period the value of

$$\frac{1+\tilde{R}_{jt}}{1+R_{Ft}+\sum_i\left[\Lambda_{i,t-1}(\tilde{R}_{it}-R_{Ft})\right]}$$

expected by each individual is equal for all risky assets, where $\tilde{R}_{jt}$ is the return from asset j in period t, R_{Ft} is the risk free return in period t and Λ_{it} is the investment is asset i in period t. Under typical, rather strict, financial economics assumptions to get to the aggregate level, it is assumed that: (a) all investors hold identical probability beliefs or (b) that a "representative" investor exists with the invested proportions X^*s being equal to relative values of existing asset supplies. In either case, the denominator equals $1+R_{m.t}$, a "market return" defined as a value-weighted average of all individual asset returns. It is only approximately observable because no comprehensive value-weighted asset indices exist.

$$Z_{jt} \equiv \frac{1+R_{jt}}{1+R_{mt}}$$

will have the same mean for all securities. Then the return on stock j relative to the market will have the same mean for all securities. Such assumptions fly in the face of standard Kelly application which picks "the best stocks", and we see clearly the difference in philosophy from academic financial economics and other investment approaches. Using 1962–1969 data and rather clever ways of looking at the two models: growth-optimium and CAPM, Roll concludes that there is:

> "...a close correspondence was demonstrated between their qualitative implications. For example, both models imply that an asset's expected return will equal the risk-free interest rate if the covariance between the asset's return and the average return on all assets, $Cov(\tilde{R}_j, \tilde{R}_m)$, is zero. For most cases, the growth-optimum model also shares the Sharpe-Lintner implication that an asset's expected return will exceed the risk-free rate if and only if $Cov(\tilde{R}_j, \tilde{R}_m) > 0$.

There are, however, some cases of highly-skewed probability distributions where this implication does not follow for the growth-optimum model.

A close empirical correspondence between the two models was demonstrated for common stock returns. The procedure: (1) estimated returns and risk premia implied by the two models from time series; (2) calculated cross-sectional relations between estimated returns and risks; and (3) compared the cross sectional relations to the theoretical predictions of the two models. They could not be distinguished on an empirical basis. In every period, estimated corresponding coefficients of the two models were nearly equal; and indeed, they deviated much further from their theoretically anticipated levels than they deviated from each other."

The investment time horizon is, as Roll points out, important here since, for short intervals, the models are essentially equivalent after quadratic approximations and for long horizons, the normality of returns kicks in.

2

EXPOSITION OF A NEW THEORY ON THE MEASUREMENT OF RISK[1]

BY DANIEL BERNOULLI

§1. EVER SINCE mathematicians first began to study the measurement of risk there has been general agreement on the following proposition: *Expected values are computed by multiplying each possible gain by the number of ways in which it can occur, and then dividing the sum of these products by the total number of possible cases where, in this theory, the consideration of cases which are all of the same probability is insisted upon.* If this rule be accepted, what remains to be done within the framework of this theory amounts to the enumeration of all alternatives, their breakdown into equi-probable cases and, finally, their insertion into corresponding classifications.

§2. Proper examination of the numerous demonstrations of this proposition that have come forth indicates that they all rest upon one hypothesis: *since there is no reason to assume that of two persons encountering identical risks,*[2] *either*

[1] Translated from Latin into English by Dr. Louise Sommer, The American University, Washington, D. C., from "Specimen Theoriae Novae de Mensura Sortis," *Commentarii Academiae Scientiarum Imperialis Petropolitanae,* Tomus V [*Papers of the Imperial Academy of Sciences in Petersburg,* Vol. V], 1738, pp. 175–192. Professor Karl Menger, Illinois Institute of Technology has written footnotes 4, 9, 10, and 15.

EDITOR'S NOTE: In view of the frequency with which Bernoulli's famous paper has been referred to in recent economic discussion, it has been thought appropriate to make it more generally available by publishing this English version. In her translation Professor Sommer has sought, in so far as possible, to retain the eighteenth century spirit of the original. The mathematical notation and much of the punctuation are reproduced without change. References to some of the recent literature concerned with Bernoulli's theory are given at the end of the article.

TRANSLATOR'S NOTE: I highly appreciate the help of Karl Menger, Professor of Mathematics, Illinois Institute of Technology, a distinguished authority on the Bernoulli problem, who has read this translation and given me expert advice. I am also grateful to Mr. William J. Baumol, Professor of Economics, Princeton University, for his valuable assistance in interpreting Bernoulli's paper in the light of modern econometrics. I wish to thank also Mr. John H. Klingenfeld, Economist, U. S. Department of Labor, for his cooperation in the English rendition of this paper. The translation is based solely upon the original Latin text.

BIOGRAPHICAL NOTE: Daniel Bernoulli, a member of the famous Swiss family of distinguished mathematicians, was born in Groningen, January 29, 1700 and died in Basle, March 17, 1782. He studied mathematics and medical sciences at the University of Basle. In 1725 he accepted an invitation to the newly established academy in Petersburg, but returned to Basle in 1733 where he was appointed professor of physics and philosophy. Bernoulli was a member of the academies of Paris, Berlin, and Petersburg and the Royal Academy in London. He was the first to apply mathematical analysis to the problem of the movement of liquid bodies.

(On Bernoulli see: *Handwörterbuch der Naturwissenschaften,* second edition, 1931, pp. 800–801; "Die Basler Mathematiker Daniel Bernoulli und Leonhard Euler. Hundert Jahre nach ihrem Tode gefeiert von der Naturforschenden Gesellschaft," Basle, 1884 (Annex to part VII of the proceedings of this Society); and *Correspondance mathématique* . . ., edited by Paul Heinrich Fuss, 1843 containing letters written by Daniel Bernoulli to Leonhard Euler, Nicolaus Fuss, and C. Goldbach.)

[2] i.e., risky propositions (gambles). [Translator]

should expect to have his desires more closely fulfilled, the risks anticipated by each must be deemed equal in value. No characteristic of the persons themselves ought to be taken into consideration; only those matters should be weighed carefully that pertain to the terms of the risk. The relevant finding might then be made by the highest judges established by public authority. But really there is here no need for judgment but of deliberation, i.e., rules would be set up whereby anyone could estimate his prospects from any risky undertaking in light of one's specific financial circumstances.

§3. To make this clear it is perhaps advisable to consider the following example: Somehow a very poor fellow obtains a lottery ticket that will yield with equal probability either nothing or twenty thousand ducats. Will this man evaluate his chance of winning at ten thousand ducats? Would he not be ill-advised to sell this lottery ticket for nine thousand ducats? To me it seems that the answer is in the negative. On the other hand I am inclined to believe that a rich man would be ill-advised to refuse to buy the lottery ticket for nine thousand ducats. If I am not wrong then it seems clear that all men cannot use the same rule to evaluate the gamble. The rule established in §1 must, therefore, be discarded. But anyone who considers the problem with perspicacity and interest will ascertain that the concept of *value* which we have used in this rule may be defined in a way which renders the entire procedure universally acceptable without reservation. To do this the determination of the *value* of an item must not be based on its *price*, but rather on the *utility* it yields. The price of the item is dependent only on the thing itself and is equal for everyone; the utility, however, is dependent on the particular circumstances of the person making the estimate. Thus there is no doubt that a gain of one thousand ducats is more significant to a pauper than to a rich man though both gain the same amount.

§4. The discussion has now been developed to a point where anyone may proceed with the investigation by the mere paraphrasing of one and the same principle. However, since the hypothesis is entirely new, it may nevertheless require some elucidation. I have, therefore, decided to explain by example what I have explored. Meanwhile, let us use this as a fundamental rule: *If the utility of each possible profit expectation is multiplied by the number of ways in which it can occur, and we then divide the sum of these products by the total number of possible cases, a mean utility*[3] *[moral expectation] will be obtained, and the profit which corresponds to this utility will equal the value of the risk in question.*

§5. Thus it becomes evident that no valid measurement of the value of a risk can be obtained without consideration being given to its *utility*, that is to say, the utility of whatever gain accrues to the individual or, conversely, how much profit is required to yield a given utility. However it hardly seems plausible to make any precise generalizations since the utility of an item may change with circumstances. Thus, though a poor man generally obtains more utility than does a rich man from an equal gain, it is nevertheless conceivable, for example,

[3] Free translation of Bernoulli's "emolumentum medium," literally: "mean utility." [Translator]

that a rich prisoner who possesses two thousand ducats but needs two thousand ducats more to repurchase his freedom, will place a higher value on a gain of two thousand ducats than does another man who has less money than he. Though innumerable examples of this kind may be constructed, they represent exceedingly rare exceptions. We shall, therefore, do better to consider what usually happens, and in order to perceive the problem more correctly we shall assume that there is an imperceptibly small growth in the individual's wealth which proceeds continuously by infinitesimal increments. Now it is highly probable that *any increase in wealth, no matter how insignificant, will always result in an increase in utility which is inversely proportionate to the quantity of goods already possessed.* To explain this hypothesis it is necessary to define what is meant by the *quantity of goods.* By this expression I mean to connote food, clothing, all things which add to the conveniences of life, and even to luxury—anything that can contribute to the adequate satisfaction of any sort of want. There is then nobody who can be said to possess nothing at all in this sense unless he starves to death. For the great majority the most valuable portion of their possessions so defined will consist in their productive capacity, this term being taken to include even the beggar's talent: a man who is able to acquire ten ducats yearly by begging will scarcely be willing to accept a sum of fifty ducats on condition that he henceforth refrain from begging or otherwise trying to earn money. For he would have to live on this amount, and after he had spent it his existence must also come to an end. I doubt whether even those who do not possess a farthing and are burdened with financial obligations would be willing to free themselves of their debts or even to accept a still greater gift on such a condition. But if the beggar were to refuse such a contract unless immediately paid no less than one hundred ducats and the man pressed by creditors similarly demanded one thousand ducats, we might say that the former is possessed of wealth worth one hundred, and the latter of one thousand ducats, though in common parlance the former owns nothing and the latter less than nothing.

§6. Having stated this definition, I return to the statement made in the previous paragraph which maintained that, in the absence of the unusual, the *utility resulting from any small increase in wealth will be inversely proportionate to the quantity of goods previously possessed.* Considering the nature of man, it seems to me that the foregoing hypothesis is apt to be valid for many people to whom this sort of comparison can be applied. Only a few do not spend their entire yearly incomes. But, if among these, one has a fortune worth a hundred thousand ducats and another a fortune worth the same number of semi-ducats and if the former receives from it a yearly income of five thousand ducats while the latter obtains the same number of semi-ducats it is quite clear that to the former a ducat has exactly the same significance as a semi-ducat to the latter, and that, therefore, the gain of one ducat will have to the former no higher value than the gain of a semi-ducat to the latter. Accordingly, if each makes a gain of one ducat the latter receives twice as much utility from it, having been enriched by two semi-ducats. This argument applies to many other cases which, therefore, need not be discussed separately. The proposition is all the more valid for the majority of men who possess no fortune apart from their working capacity which is their only source of livelihood. True, there are men to whom one ducat means more than many ducats do to others who are less rich but more generous than they. But since we shall now concern ourselves only with one individual (in different states of affluence) distinctions of this sort do not concern us. The man who is emotionally less affected by a gain will support a loss with greater patience. Since, however, in special cases things can conceivably occur otherwise, I shall first deal with the most general case and then develop our special hypothesis in order thereby to satisfy everyone.

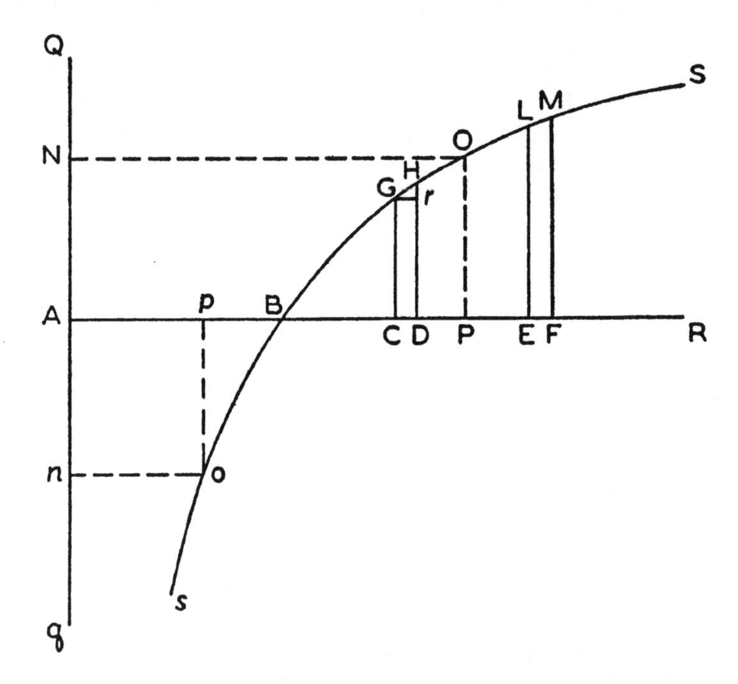

§7. Therefore, let AB represent the quantity of goods initially possessed. Then after extending AB, a curve $BGLS$ must be constructed, whose ordinates CG, DH, EL, FM, etc., designate *utilities* corresponding to the abscissas BC, BD, BE, BF, etc., designating gains in wealth. Further, let m, n, p, q, etc., be the numbers which indicate the number of ways in which gains in wealth BC, BD, BE, BF [misprinted in the original as CF], etc., can occur. Then (in accord with §4) the *moral* expectation of the risky proposition referred to is given by:

$$PO = \frac{m.CG + n.DH + p.EL + q.FM + \cdots}{m + n + p + q + \cdots}$$

Now, if we erect AQ perpendicular to AR, and on it measure off $AN = PO$, the straight line $NO - AB$ represents the gain which may properly be expected, or the value of the risky proposition in question. If we wish, further, to know how large a stake the individual should be willing to venture on this risky proposition, our curve must be extended in the opposite direction in such a way that the abscissa Bp now represents a loss and the ordinate po represents the corresponding decline in utility. Since in a fair game the disutility to be suffered by losing must be equal to the utility to be derived by winning, we must assume that $An = AN$, or $po = PO$. Thus Bp will indicate the stake more than which persons who consider their own pecuniary status should not venture.

COROLLARY I

§8. Until now scientists have usually rested their hypothesis on the assumption that all gains must be evaluated exclusively in terms of themselves, i.e., on the basis of their intrinsic qualities, and that these gains will always produce a *utility* directly proportionate to the gain. On this hypothesis the curve BS becomes a straight line. Now if we again have:

$$PO = \frac{m.CG + n.DH + p.EL + q.FM + \cdots}{m + n + p + q + \cdots},$$

and if, on both sides, the respective factors are introduced it follows that:

$$BP = \frac{m.BC + n.BD + p.BE + q.BF + \cdots}{m + n + p + q + \cdots},$$

which is in conformity with the usually accepted rule.

COROLLARY II

§9. If AB were infinitely great, even in proportion to BF, the greatest possible gain, the arc BM may be considered very like an infinitesimally small straight line. Again in this case the usual rule [for the evaluation of risky propositions] is applicable, and may continue to be considered approximately valid in games of insignificant moment.

§10. Having dealt with the problem in the most general way we turn now to the aforementioned particular hypothesis, which, indeed, deserves prior attention to all others. First of all the nature of curve sBS must be investigated under the conditions postulated in §7. Since on our hypothesis we must consider infinitesimally small gains, we shall take gains BC and BD to be nearly equal, so that their difference CD becomes infinitesimally small. If we draw Gr parallel to BR, then rH will represent the infinitesimally small gain in *utility* to a man whose fortune is AC and who obtains the small gain, CD. This *utility*, however, should be related not only to the tiny gain CD, to which it is, other things being equal, proportionate, but also to AC, the fortune previously owned to which it is inversely proportionate. We therefore set: $AC = x$, $CD = dx$, $CG = y$, $rH = dy$ and $AB = \alpha$; and if b designates some constant we obtain $dy = \frac{bdx}{x}$ or $y =$

$b \log \frac{x}{\alpha}$. The curve sBS is therefore a logarithmic curve, the subtangent[4] of which is everywhere b and whose asymptote is Qq.

§11. If we now compare this result with what has been said in paragraph 7, it will appear that: $PO = b \log AP/AB$, $CG = b \log AC/AB$, $DH = b \log AD/AB$ and so on; but since we have

$$PO = \frac{m.CG + n.DH + p.EL + q.FM + \cdots}{m + n + p + q + \cdots}$$

it follows that

$$b \log \frac{AP}{AB} = \left(mb \log \frac{AC}{AB} + nb \log \frac{AD}{AB} + pb \log \frac{AE}{AB} + qb \log \frac{AF}{AB} + \cdots\right) : (m + n + p + q + \cdots)$$

and therefore

$$AP = (AC^m.AD^n.AE^p.AF^q.\ \cdots)^{1/m+n+p+q+\cdots}$$

and if we subtract AB from this, the remaining magnitude, BP, will represent the value of the risky proposition in question.

§12. Thus the preceding paragraph suggests the following rule: *Any gain must be added to the fortune previously possessed, then this sum must be raised to the power given by the number of possible ways in which the gain may be obtained; these terms should then be multiplied together. Then of this product a root must be extracted the degree of which is given by the number of all possible cases, and finally the value of the initial possessions must be subtracted therefrom; what then remains indicates the value of the risky proposition in question.* This principle is essential for the measurement of the value of risky propositions in various cases. I would elaborate it into a complete theory as has been done with the traditional analysis, were it not that, despite its usefulness and originality, previous obligations do not permit me to undertake this task. I shall therefore, at this time, mention only the more significant points among those which have at first glance occurred to me.

[4] The tangent to the curve $y = b \log \frac{x}{\alpha}$ at the point $\left(x_0, \log \frac{x_0}{\alpha}\right)$ is the line $y - b \log \frac{x_0}{\alpha} = \frac{b}{x_0}(x - x_0)$. This tangent intersects the Y-axis ($x = 0$) at the point with the ordinate $b \log \frac{x_0}{\alpha} - b$. The point of contact of the tangent with the curve has the ordinate $b \log \frac{x_0}{\alpha}$. So also does the projection of this point on the Y-axis. The segment between the two points on the Y-axis that have been mentioned has the length b. That segment is the projection of the segment on the tangent between its intersection with the Y-axis and the point of contact. The length of this projection (which is b) is what Bernoulli here calls the "subtangent." Today, by the subtangent of the curve $y = f(x)$ at the point $(x_0, f(x_0))$ is meant the length of the segment on the X-axis (and not the Y-axis) between its intersection with the tangent and the projection of the point of contact. This length is $f(x_0)/f'(x_0)$. In the case of the logarithmic curve it equals $x_0 \log \frac{x_0}{\alpha}$.—Karl Menger.

§13. First, it appears that in many games, even those that are absolutely fair, both of the players may expect to suffer a loss; indeed this is Nature's admonition to avoid the dice altogether. . . . This follows from the concavity of curve sBS to BR. For in making the stake, Bp, equal to the expected gain, BP, it is clear that the disutility po which results from a loss will always exceed the expected gain in utility, PO. Although this result will be quite clear to the mathematician, I shall nevertheless explain it by example, so that it will be clear to everyone. Let us assume that of two players, both possessing one hundred ducats, each puts up half this sum as a stake in a game that offers the same probabilities to both players. Under this assumption each will then have fifty ducats plus the expectation of winning yet one hundred ducats more. However, the sum of the values of these two items amounts, by the rule of §12, to only $(50^1.150^1)^{\frac{1}{2}}$ or $\sqrt{50.150}$, i.e., less than eighty-seven ducats, so that, though the game be played under perfectly equal conditions for both, either will suffer an expected loss of more than thirteen ducats. We must strongly emphasize this truth, although it be self evident: the imprudence of a gambler will be the greater the larger the part of his fortune which he exposes to a game of chance. For this purpose we shall modify the previous example by assuming that one of the gamblers, before putting up his fifty ducat stake possessed two hundred ducats. This gambler suffers an expected loss of $200 - \sqrt{150.250}$, which is not much greater than six ducats.

§14. Since, therefore, everyone who bets any part of his fortune, however small, on a mathematically fair game of chance acts irrationally, it may be of interest to inquire how great an advantage the gambler must enjoy over his opponent in order to avoid any expected loss. Let us again consider a game which is as simple as possible, defined by two equiprobable outcomes one of which is favorable and the other unfavorable. Let us take a to be the gain to be won in case of a favorable outcome, and x to be the stake which is lost in the unfavorable case. If the initial quantity of goods possessed is α we have $AB = \alpha$; $BP = a$; $PO = b \log \frac{\alpha + a}{\alpha}$ (see §10), and since (by §7) $po = PO$ it follows by the nature of a logarithmic curve that $Bp = \frac{\alpha a}{\alpha + a}$. Since however Bp represents the stake x, we have $x = \frac{\alpha a}{\alpha + a}$ a magnitude which is always smaller than a, the expected gain. It also follows from this that a man who risks his entire fortune acts like a simpleton, however great may be the possible gain. No one will have difficulty in being persuaded of this if he has carefully examined our definitions given above. Moreover, this result sheds light on a statement which is universally accepted in practice: it may be reasonable for some individuals to invest in a doubtful enterprise and yet be unreasonable for others to do so.

§15. The procedure customarily employed by merchants in the insurance of commodities transported by sea seems to merit special attention. This may again be explained by an example. Suppose Caius,[5] a Petersburg merchant, has pur-

[5] Caius is a Roman name, used here in the sense of our "Mr. Jones." Caius is the older form; in the later Roman period it was spelled "Gaius." [Translator]

chased commodities in Amsterdam which he could sell for ten thousand rubles if he had them in Petersburg. He therefore orders them to be shipped there by sea, but is in doubt whether or not to insure them. He is well aware of the fact that at this time of year of one hundred ships which sail from Amsterdam to Petersburg, five are usually lost. However, there is no insurance available below the price of eight hundred rubles a cargo, an amount which he considers outrageously high. The question is, therefore, how much wealth must Caius possess apart from the goods under consideration in order that it be sensible for him to abstain from insuring them? If x represents his fortune, then this together with the value of the expectation of the safe arrival of his goods is given by $\sqrt[100]{(x + 10000)^{95}x^5} = \sqrt[20]{(x + 10000)^{19}x}$ in case he abstains. With insurance he will have a certain fortune of $x + 9200$. Equating these two magnitudes we get: $(x + 10000)^{19}x = (x + 9200)^{20}$ or, approximately, $x = 5043$. If, therefore, Caius, apart from the expectation of receiving his commodities, possesses an amount greater than 5043 rubles he will be right in not buying insurance. If, on the contrary, his wealth is less than this amount he should insure his cargo. And if the question be asked "What minimum fortune should be possessed by the man who offers to provide this insurance in order for him to be rational in doing so?" We must answer thus: let y be his fortune, then

$$\sqrt[20]{(y + 800)^{19}.(y - 9200)} = y$$

or approximately, $y = 14243$, a figure which is obtained from the foregoing without additional calculation. A man less wealthy than this would be foolish to provide the surety, but it makes sense for a wealthier man to do so. From this it is clear that the introduction of this sort of insurance has been so useful since it offers advantages to all persons concerned. Similarly, had Caius been able to obtain the insurance for six hundred rubles he would have been unwise to refuse it if he possessed less than 20478 rubles, but he would have acted much too cautiously had he insured his commodities at this rate when his fortune was greater than this amount. On the other hand a man would act unadvisedly if he were to offer to sponsor this insurance for six hundred rubles when he himself possesses less than 29878 rubles. However, he would be well advised to do so if he possessed more than that amount. But no one, however rich, would be managing his affairs properly if he individually undertook the insurance for less than five hundred rubles.

§16. Another rule which may prove useful can be derived from our theory. This is the rule that it is advisable to divide goods which are exposed to some danger into several portions rather than to risk them all together. Again I shall explain this more precisely by an example. Sempronius owns goods at home worth a total of 4000 ducats and in addition possesses 8000 ducats worth of commodities in foreign countries from where they can only be transported by sea. However, our daily experience teaches us that of ten ships one perishes. Under these conditions I maintain that if Sempronius trusted all his 8000 ducats of goods to one ship his expectation of the commodities is worth 6751 ducats. That is

$$\sqrt[10]{12000^9.4000^1} - 4000.$$

If, however, he were to trust equal portions of these commodities to two ships the value of his expectation would be

$$\sqrt[100]{12000^{81}.8000^{18}.4000} - 4000, \text{ i.e., } 7033 \text{ ducats.}$$

In this way the value of Sempronius' prospects of success will grow more favorable the smaller the proportion committed to each ship. However, his expectation will never rise in value above 7200 ducats. This counsel will be equally serviceable for those who invest their fortunes in foreign bills of exchange and other hazardous enterprises.

§17. I am forced to omit many novel remarks though these would clearly not be unserviceable. And, though a person who is fairly judicious by natural instinct might have realized and spontaneously applied much of what I have here explained, hardly anyone believed it possible to define these problems with the precision we have employed in our examples. Since all our propositions harmonize perfectly with experience it would be wrong to neglect them as abstractions resting upon precarious hypotheses. This is further confirmed by the following example which inspired these thoughts, and whose history is as follows: My most honorable cousin the celebrated *Nicolas Bernoulli*, Professor utriusque iuris[6] at the University of Basle, once submitted five problems to the highly distinguished[7] mathematician *Montmort*.[8] These problems are reproduced in the work *L'analyse sur les jeux de hazard* de *M. de Montmort*, p. 402. The last of these problems runs as follows: *Peter tosses a coin and continues to do so until it should land "heads" when it comes to the ground. He agrees to give Paul one ducat if he gets "heads" on the very first throw, two ducats if he gets it on the second, four if on the third, eight if on the fourth, and so on, so that with each additional throw the number of ducats he must pay is doubled. Suppose we seek to determine the value of Paul's expectation.* My aforementioned cousin discussed this problem in a letter to me asking for my opinion. Although the standard calculation shows[9] that the value of Paul's expectation is infinitely great, it has, he said, to be admitted that any fairly reasonable man would sell his chance, with great pleasure, for twenty ducats. The accepted method of calculation does, indeed, value Paul's prospects at infinity though no one would be willing to purchase it at a moderately high price.

[6] Faculties of law of continental European universities bestow up to the present time the title of a Doctor utriusque juris, which means Doctor of both systems of laws, the Roman and the canon law. [Translator]

[7] Cl., i.e., Vir Clarissimus, a title of respect. [Translator]

[8] Montmort, Pierre Rémond, de (1678–1719). The work referred to here is the then famous "Essai d'analyse sur les jeux de hazard," Paris, 1708. Appended to the second edition, published in 1713, is Montmort's correspondence with Jean and Nicolas Bernoulli referring to the problems of chance and probabilities. [Translator].

[9] The probability of heads turning up on the 1st throw is 1/2. Since in this case Paul receives one ducat, this probability contributes $1/2 \cdot 1 = 1/2$ ducats to his expectation. The probability of heads turning up on the 2nd throw is 1/4. Since in this case Paul receives 2 ducats, this possibility contributes $1/4 \cdot 2 = 1/2$ to his expectation. Similarly, for every integer n, the possibility of heads turning up on the n-th throw contributes $1/2^n \cdot 2^{n-1} = 1/2$ ducats to his expectation. Paul's total expectation is therefore $1/2 + 1/2 + \cdots + 1/2 + \cdots$, and that is infinite.—Karl Menger.

If, however, we apply our new rule to this problem we may see the solution and thus unravel the knot. The solution of the problem by our principles is as follows.

§18. The number of cases to be considered here is infinite: in one half of the cases the game will end at the first throw, in one quarter of the cases it will conclude at the second, in an eighth part of the cases with the third, in a sixteenth part with the fourth, and so on.[10] If we designate the number of cases through infinity by N it is clear that there are $\frac{1}{2}N$ cases in which Paul gains one ducat, $\frac{1}{4}N$ cases in which he gains two ducats, $\frac{1}{8}N$ in which he gains four, $\frac{1}{16}N$ in which he gains eight, and so on, ad infinitum. Let us represent Paul's fortune by α; the proposition in question will then be worth

$$\sqrt[N]{(\alpha+1)^{N/2}.(\alpha+2)^{N/4}.(\alpha+4)^{N/8}.(\alpha+8)^{N/16}\cdots} - \alpha$$
$$= \sqrt{(\alpha+1)}.\sqrt[4]{(\alpha+2)}.\sqrt[8]{(\alpha+4)}.\sqrt[16]{(\alpha+8)}\cdots - \alpha.$$

§19. From this formula which evaluates Paul's prospective gain it follows that this value will increase with the size of Paul's fortune and will never attain an infinite value unless Paul's wealth simultaneously becomes infinite. In addition we obtain the following corollaries. If Paul owned nothing at all the value of his expectation would be

$$\sqrt[2]{1}.\sqrt[4]{2}.\sqrt[8]{4}.\sqrt{8}\cdots$$

which amounts to two ducats, precisely. If he owned ten ducats his opportunity would be worth approximately three ducats; it would be worth approximately four if his wealth were one hundred, and six if he possessed one thousand. From this we can easily see what a tremendous fortune a man must own for it to make sense for him to purchase Paul's opportunity for twenty ducats. The amount which the buyer ought to pay for this proposition differs somewhat from the amount it would be worth to him were it already in his possession. Since, however, this difference is exceedingly small if α (Paul's fortune) is great,

[10] Since the number of cases is infinite, it is impossible to speak about one half of the cases, one quarter of the cases, etc., and the letter N in Bernoulli's argument is meaningless. However, Paul's expectation on the basis of Bernoulli's hypothesis concerning evaluation can be found by the same method by which, in footnote 9, Paul's classical expectation was determined. If Paul's fortune is α ducats, then, according to Bernoulli, he attributes to a gain of 2^{n-1} ducats the value $b \log \frac{\alpha + 2^{n-1}}{\alpha}$. If the probability of this gain is $1/2^n$, his expectation is $b/2^n \log \frac{\alpha + 2^{n-1}}{\alpha}$. Paul's expectation resulting from the game is therefore

$$\frac{b}{2}\log\frac{\alpha+1}{\alpha} + \frac{b}{4}\log\frac{\alpha+2}{\alpha} + \cdots + \frac{b}{2^n}\log\frac{\alpha+2^{n-1}}{\alpha} + \cdots$$
$$= b\log\left[(\alpha+1)^{1/2}(\alpha+2)^{1/4}\cdot\ldots\cdot(\alpha+2^{n-1})^{1/2^n}\cdot\ldots\right] - b\log\alpha.$$

What addition D to Paul's fortune has the same value for him? Clearly, $b \log \frac{\alpha + D}{\alpha}$ must equal the above sum. Therefore

$$D = (\alpha+1)^{1/2}(\alpha+2)^{1/4}\cdot\ldots\cdot(\alpha+2^{n-1})^{1/2^n}\cdot\ldots - \alpha.$$

—Karl Menger.

we can take them to be equal. If we designate the purchase price by x its value can be determined by means of the equation

$$\sqrt[2]{(\alpha + 1 - x)} . \sqrt[4]{(\alpha + 2 - x)} . \sqrt[8]{(\alpha + 4 - x)} . \sqrt[16]{(\alpha + 8 - x)} \cdots = \alpha$$

and if α is a large number this equation will be approximately satisfied by

$$x = \sqrt[2]{\alpha + 1} . \sqrt[4]{\alpha + 2} . \sqrt[8]{\alpha + 4} . \sqrt[16]{\alpha + 8} \cdots - \alpha.$$

After having read this paper to the Society[11] *I sent a copy to the aforementioned Mr. Nicolas Bernoulli, to obtain his opinion of my proposed solution to the difficulty he had indicated. In a letter to me written in* 1732 *he declared that he was in no way dissatisfied with my proposition on the evaluation of risky propositions when applied to the case of a man who is to evaluate his own prospects. However, he thinks that the case is different if a third person, somewhat in the position of a judge, is to evaluate the prospects of any participant in a game in accord with equity and justice. I myself have discussed this problem in* §2. *Then this distinguished scholar informed me that the celebrated mathematician, Cramer,*[12] *had developed a theory on the same subject several years before I produced my paper. Indeed I have found his theory so similar to mine that it seems miraculous that we independently reached such close agreement on this sort of subject. Therefore it seems worth quoting the words with which the celebrated Cramer himself first described his theory in his letter of* 1728 *to my cousin. His words are as follows:*[13]

"Perhaps I am mistaken, but I believe that I have solved the extraordinary" problem which you submitted to M. *de Montmort*, in your letter of September 9," 1713, (problem 5, page 402). For the sake of simplicity I shall assume that A" tosses a coin into the air and B commits himself to give A 1 ducat if, at the" first throw, the coin falls with its cross upward; 2 if it falls thus only at the" second throw, 4 if at the third throw, 8 if at the fourth throw, etc. The paradox" consists in the infinite sum which calculation yields as the equivalent which" A must pay to B. This seems absurd since no reasonable man would be willing" to pay 20 ducats as equivalent. You ask for an explanation of the discrepancy" between the mathematical calculation and the vulgar evaluation. I believe" that it results from the fact that, *in their theory*, mathematicians evaluate" money in proportion to its quantity while, *in practice*, people with common" sense evaluate money in proportion to the utility they can obtain from it. The" mathematical expectation is rendered infinite by the enormous amount which" I can win if the coin does not fall with its cross upward until rather late, perhaps" at the hundredth or thousandth throw. Now, as a matter of fact, if I reason" as a sensible man, this sum is worth no more to me, causes me no more pleasure"

[11] Bernoulli's paper had been submitted to the Imperial Academy of Sciences in Petersburg. [Translator]

[12] Cramer, Gabriel, famous mathematician, born in Geneva, Switzerland (1704–1752). [Translator]

[13] The following passage of the original text is in French. [Translator]

"and influences me no more to accept the game than does a sum amounting "only to ten or twenty million ducats. Let us suppose, therefore, that any "amount above 10 millions, or (for the sake of simplicity) above $2^{24} = 166777216$ "ducats be deemed by him equal in value to 2^{24} ducats or, better yet, that I "can never win more than that amount, no matter how long it takes before the "coin falls with its cross upward. In this case, my expectation is $\frac{1}{2}.1 + \frac{1}{4}.2 +$ "$\frac{1}{8}.4 \cdots + \frac{1}{2^{25}}.2^{24} + \frac{1}{2^{26}}.2^{24} + \frac{1}{2^{27}}.2^{24} + \cdots = \frac{1}{2} + \frac{1}{2} + \frac{1}{2} + \cdots$ "(24 times) $\cdots + \frac{1}{2} + \frac{1}{4} + \frac{1}{8} + \cdots = 12 + 1 = 13$. Thus, my moral ex-"pectation is reduced in value to 13 ducats and the equivalent to be paid for "it is similarly reduced—a result which seems much more reasonable than does "rendering it infinite."

Thus far[14] *the exposition is somewhat vague and subject to counter argument. If it, indeed, be true that the amount* 2^{25} *appears to us to be no greater than* 2^{24}, *no attention whatsoever should be paid to the amount that may be won after the twenty-fourth throw, since just before making the twenty-fifth throw I am certain to end up with no less than* $2^{24} - 1$,[15] *an amount that, according to this theory, may be considered equivalent to* 2^{24}. *Therefore it may be said correctly that my expectation is only worth twelve ducats, not thirteen. However, in view of the coincidence between the basic principle developed by the aforementioned author and my own, the foregoing is clearly not intended to be taken to invalidate that principle. I refer to the proposition that reasonable men should evaluate money in accord with the utility they derive therefrom. I state this to avoid leading anyone to judge that entire theory adversely. And this is exactly what Cl. C.*[16] *Cramer states, expressing in the following manner precisely what we would ourselves conclude. He continues thus:*[17]

"The equivalent can turn out to be smaller yet if we adopt some alternative "hypothesis on the moral value of wealth. For that which I have just assumed "is not entirely valid since, while it is true that 100 millions yield more satis-"faction than do 10 millions, they do not give ten times as much. If, for example, "we suppose the moral value of goods to be directly proportionate to the square "root of their mathematical quantities, e.g., that the satisfaction provided by "40000000 is double that provided by 10000000, my psychic expectation "becomes

$$"\tfrac{1}{2}\sqrt{1} + \tfrac{1}{4}\sqrt{2} + \tfrac{1}{8}\sqrt{4} + \tfrac{1}{16}\sqrt{8} + \cdots = \frac{1}{2 - \sqrt{2}}.$$

"However this magnitude is not the equivalent we seek, for this equivalent "need not be equal to my moral expectation but should rather be of such a "magnitude that the pain caused by its loss is equal to the moral expectation "of the pleasure I hope to derive from my gain. Therefore, the equivalent must,

[14] From here on the text is again translated from Latin. [Translator]

[15] This remark of Bernoulli's is obscure. Under the conditions of the game a gain of $2^{24} - 1$ ducats is impossible.—Karl Menger.

[16] To be translated as "the distinguished Gabriel." [Translator]

[17] Text continues in French. [Translator]

on our hypothesis, amount to $\left(\frac{1}{2 - \sqrt{2}}\right)^2 = \left(\frac{1}{6 - 4\sqrt{2}}\right) = 2.9 \cdots$, which" is consequently less than 3, truly a trifling amount, but nevertheless, I believe," closer than is 13 to the vulgar evaluation."

REFERENCES

There exists only one other translation of Bernoulli's paper:

Pringsheim, Alfred, *Die Grundlage der modernen Wertlehre: Daniel Bernoulli, Versuch einer neuen Theorie der Wertbestimmung von Glücksfällen* (Specimen Theoriae novae de Mensura Sortis). Aus dem Lateinischen übersetzt und mit Erläuterungen versehen von Alfred Pringsheim. Leipzig, Duncker und Humblot, 1896, Sammlung älterer und neuerer staatswissenschaftlicher Schriften des In- und Auslandes hrsg. von L. Brentano und E. Leser, No. 9.

For an early discussion of the Bernoulli problem, reference is made to

Malfatti, Gianfrancesco, "Esame critico di un problema di probabilita del Signor Daniele Bernoulli, e soluzione d'un altro problema analogo al Bernoulliano" in "*Memorie di Matematica e Fisica della Societa italiana*" Vol. I, Verona, 1782, pp. 768–824.

For more on the "St. Petersburg Paradox," including material on later discussions, see

Menger, Karl, "Das Unsicherheitsmoment in der Wertlehre. Betrachtungen im Anschluss an das sogenannte Petersburger Spiel," *Zeitschrift für Nationalökonomie*, Vol. 5, 1934.

This paper by Professor Menger, is the most extensive study on the literature of the problem, and the problem itself.

Recent interest in the Bernoulli hypothesis was aroused by its appearance in

von Neumann, John, and Oskar Morgenstern, *The Theory of Games and Economic Behavior*, second edition, Princeton: Princeton University Press, 1947, Ch. III and Appendix: "The Axiomatic Treatment of Utility."

Many contemporary references and a discussion of the utility maximization hypothesis are to be found in

Arrow, Kenneth J., "Alternative Approaches to the Theory of Choice in Risk-Taking Situations," ECONOMETRICA, Vol. 19, October, 1951.

More recent writings in the field include

Alchian, A. A., "The Meaning of Utility Measurement," *American Economic Review*, Vol. XLIII, March, 1953.

Friedman, M., and Savage, L. J., "The Expected Utility-Hypothesis and the Measurability of Utility," *Journal of Political Economy*, Vol. LX, December, 1952.

Herstein, I. N., and John Milnor, "An Axiomatic Approach to Measurable Utility," ECONOMETRICA, Vol. 21, April, 1953.

Marschak, J., "Why 'Should' Statisticians and Businessmen Maximize 'Moral Expectation'?", *Second Berkeley Symposium on Mathematical Statistics and Probability*, 1953.

Mosteller, Frederick, and Philip Nogee, "An Experimental Measurement of Utility," *Journal of Political Economy*, lix, 5, Oct., 1951.

Samuelson, Paul A., "Probability, Utility, and the Independence Axiom," ECONOMETRICA, Vol. 20, Oct. 1952.

Strotz, Robert H., "Cardinal Utility," *Papers and Proceedings of the Sixty-Fifth Annual Meeting of the American Economic Association, American Economic Review*, Vol. 43, May, 1953, and the comment by W. J. Baumol.

For dissenting views, see:

Allais, M., "Les Theories de la Psychologie du Risque de l'Ecole Americaine", *Revue d'Economie Politique*, Vol. 63, 1953.

——"Le Comportement de l'Homme Rationnel devant le Risque: Critique des Postu-

lats et Axiomes de l'Ecole Americaine," ECONOMETRICA, Oct., 1953
and
Edwards, Ward, "Probability-Preferences in Gambling," *The American Journal of Psychology*, Vol. 66, July, 1953.
Textbooks dealing with Bernoulli:
Anderson, Oskar, *Einführung in die mathematische Statistik*, Wien: J. Springer, 1935.
Davis, Harold, *The Theory of Econometrics*, Bloomington, Ind.: Principia Press, 1941.
Loria, Gino, *Storia delle Matematiche, dall'alba della civiltà al secolo XIX*, Second revised edition, Milan: U. Hopli, 1950.

3

A New Interpretation of Information Rate

By J. L. Kelly, jr.

If the input symbols to a communication channel represent the outcomes of a chance event on which bets are available at odds consistent with their probabilities (i.e., "fair" odds), a gambler can use the knowledge given him by the received symbols to cause his money to grow exponentially. The maximum exponential rate of growth of the gambler's capital is equal to the rate of transmission of information over the channel. This result is generalized to include the case of arbitrary odds.

Thus we find a situation in which the transmission rate is significant even though no coding is contemplated. Previously this quantity was given significance only by a theorem of Shannon's which asserted that, with suitable encoding, binary digits could be transmitted over the channel at this rate with an arbitrarily small probability of error.

INTRODUCTION

Shannon defines the rate of transmission over a noisy communication channel in terms of various probabilities.[1] This definition is given significance by a theorem which asserts that binary digits may be encoded and transmitted over the channel at this rate with arbitrarily small probability of error. Many workers in the field of communication theory have felt a desire to attach significance to the rate of transmission in cases where no coding was contemplated. Some have even proceeded on the assumption that such a significance did, in fact, exist. For example, in systems where no coding was desirable or even possible (such as radar), detectors have been designed by the criterion of maximum transmission rate or, what is the same thing, minimum equivocation. Without further analysis such a procedure is unjustified.

The problem then remains of attaching a value measure to a communication

[1] C.E. Shannon, A Mathematical Theory of Communication, B.S.T.J., **27**, pp. 379-423, 623-656, Oct., 1948.

system in which errors are being made at a non-negligible rate, i.e., where optimum coding is not being used. In its most general formulation this problem seems to have but one solution. A cost function must be defined on pairs of symbols which tells how bad it is to receive a certain symbol when a specified signal is transmitted. Furthermore, this cost function must be such that its expected value has significance, i.e., a system must be preferable to another if its average cost is less. The utility theory of Von Neumann[2] shows us one way to obtain such a cost function. Generally this cost function would depend on things external to the system and not on the probabilities which describe the system, so that its average value could not be identified with the rate as defined by Shannon.

The cost function approach is, of course, not limited to studies of communication systems, but can actually be used to analyze nearly any branch of human endeavor. The author believes that it is too general to shed any light on the specific problems of communication theory. The distinguishing feature of a communication system is that the ultimate receiver (thought of here as a person) is in a position to profit from any knowledge of the input symbols or even from a better estimate of their probabilities. A cost function, if it is supposed to apply to a communication system, must somehow reflect this feature. The point here is that an arbitrary combination of a statistical transducer (i.e., a channel) and a cost function does not necessarily constitute a communication system. In fact (not knowing the exact definition of a communication system on which the above statements are tacitly based) the author would not know how to test such an arbitrary combination to see if it were a communication system.

What can be done, however, is to take some real-life situation which seems to possess the essential features of a communication problem, and to analyze it without the introduction of an arbitrary cost function. The situation which will be chosen here is one in which a gambler uses knowledge of the received symbols of a communication channel in order to make profitable bets on the transmitted symbols.

THE GAMBLER WITH A PRIVATE WIRE

Let us consider a communication channel which is used to transmit the results of a chance situation before those results become common knowledge, so that a gambler may still place bets at the original odds. Consider first the case of a noiseless binary channel, which might be

[2]Von Neumann and Morgenstein, Theory of Games and Economic Behavior, Princeton Univ. Press, 2nd Edition, 1947.

used, for example, to transmit the results of a series of baseball games between two equally matched teams. The gambler could obtain even money bets even though he already knew the result of each game. The amount of money he could make would depend only on how much he chose to bet. How much would he bet? Probably all he had since he would win with certainty. In this case his capital would grow exponentially and after N bets he would have 2^N times his original bankroll. This exponential growth of capital is not uncommon in economics. In fact, if the binary digits in the above channel were arriving at the rate of one per week, the sequence of bets would have the value of an investment paying 100 per cent interest per week compounded weekly. We will make use of a quantity G called the exponential rate of growth of the gambler's capital, where

$$G = \lim_{N\to\infty} \frac{1}{N} \log \frac{V_N}{V_0}$$

where V_N is the gambler's capital after N bets, V_0 is his starting capital, and the logarithm is to the base two. In the above example $G = 1$.

Consider the case now of a noisy binary channel, where each transmitted symbol has probability, p, of error and q of correct transmission. Now the gambler could still bet his entire capital each time, and, in fact, this would maximize the expected value of his capital, $\langle V_N \rangle$, which in this case would be given by

$$\langle V_N \rangle = (2q)^N V_0$$

This would be little comfort, however, since when N was large he would probably be broke and, in fact, would be broke with probability one if he continued indefinitely. Let us, instead, assume that he bets a fraction, ℓ, of his capital each time. Then

$$V_N = (1+\ell)^W (1-\ell)^L V_0$$

where W and L are the number of wins and losses in the N bets. Then

$$\begin{aligned} G &= \lim_{N\to\infty} \left[\frac{W}{N} \log(1+\ell) + \frac{L}{N} \log(1-\ell) \right] \\ &= q\log(1+\ell) + p\log(1-\ell) \text{ with probability one} \end{aligned}$$

Let us maximize G with respect to ℓ. The maximum value with respect to the Y_i of a quantity of the form $Z = \sum X_i \log Y_i$, subject to the constraint $\sum Y_i = Y$, is obtained by putting

$$Y_i = \frac{Y}{X} X_i,$$

where $X = \sum X_i$. This may be shown directly from the convexity of the logarithm.

Thus we put

$$(1 + \ell) = 2q$$
$$(1 - \ell) = 2p$$

and

$$G_{max} = 1 + p \log p + q \log q$$
$$= R$$

which is the rate of transmission as defined by Shannon.

One might still argue that the gambler should bet all his money (make $\ell = 1$) in order to maximize his expected win after N times. It is surely true that if the game were to be stopped after N bets the answer to this question would depend on the relative values (to the gambler) of being broke or possessing a fortune. If we compare the fates of two gamblers, however, playing a nonterminating game, the one which uses the value ℓ found above will, with probability one, eventually get ahead and stay ahead of one using any other ℓ. At any rate, we will assume that the gambler will always bet so as to maximize G.

THE GENERAL CASE

Let us now consider the case in which the channel has several input symbols, not necessarily equally likely, which represent the outcome of chance events. We will use the following notation:

$p(s)$ the probability that the transmitted symbol is the s'th one.

$p(r/s)$ the conditional probability that the received symbol is the r'th on the hypothesis that the transmitted symbol is the s'th one.

$p(s, r)$ the joint probability of the s'th transmitted and r'th received symbol.

$q(r)$ received symbol probability.

$q(s/r)$ conditional probability of transmitted symbol on hypothesis of received symbol.

α_s the odds paid on the occurrence of the s'th transmitted symbol, i.e., α_s is the number of dollars returned for a one-dollar bet (including that one dollar).

$a(s/r)$ the fraction of the gambler's capital that he decides to bet on the occurrence of the s'th transmitted symbol *after* observing the r'th *received* symbol.

Only the case of independent transmitted symbols and noise will be considered. We will consider first the case of "fair" odds, i.e.,

$$\alpha_s = \frac{1}{p(s)}$$

In any sort of parimutuel betting there is a tendency for the odds to be fair (ignoring the "track take"). To see this first note that if there is no "track take"

$$\sum \frac{1}{\alpha_s} = 1$$

since all the money collected is paid out to the winner. Next note that if

$$\alpha_s > \frac{1}{p(s)}$$

for some s a bettor could insure a profit by making repeated bets on the sth outcome. The extra betting which would result would lower α_s. The same feedback mechanism probably takes place in more complicated betting situations, such as stock market speculation.

There is no loss in generality in assuming that

$$\sum_s a(s/r) = 1$$

i.e., the gambler bets his total capital regardless of the received symbol. Since

$$\sum \frac{1}{\alpha_s} = 1$$

he can effectively hold back money by placing canceling bets. Now

$$V_N = \prod_{r,s} [a(s/r)\alpha_s]^{W_{sr}} V_0$$

where W_{sr} is the number of times that the transmitted symbol is s and the received symbol is r.

$$\log \frac{V_n}{V_o} = \sum_{rs} W_{sr} \log \alpha_s a(s/r)$$
$$G = \lim_{N \to \infty} \frac{1}{N} \log \frac{V_N}{V_0} = \sum_{rs} p(s,r) \log \alpha_s a(s/r) \tag{1}$$

with probability one. Since

$$\alpha_s = \frac{1}{p(s)}$$

here

$$G = \sum_{rs} p(s,r) \log \frac{a(s/r)}{p(s)}$$
$$= \sum_{rs} p(s,r) \log a(s/r) + H(X)$$

where $H(X)$ is the source rate as defined by Shannon. The first term is maximized by putting

$$a(s/r) = \frac{p(s,r)}{\sum_k p(k,r)} = \frac{p(s,r)}{q(r)} = q(s/r)$$

Then $G_{max} = H(X) - H(X/Y)$, which is the rate of transmission defined by Shannon.

WHEN THE ODDS ARE NOT FAIR

Consider the case where there is no track take, i.e.,

$$\sum \frac{1}{\alpha_s} = 1$$

but where α_s is not necessarily

$$\frac{1}{p(s)}$$

It is still permissible to set $\sum_s a(s/r) = 1$ since the gambler can effectively hold back any amount of money by betting it in proportion to the $1/\alpha_s$. Equation (1) now can be written

$$G = \sum_{rs} p(s,r) \log a(s/r) + \sum_s p(s) \log \alpha_s.$$

G is still maximized by placing $a(s/r) = q(s/r)$ and

$$\begin{aligned} G_{max} &= -H(X/Y) + \sum_s p(s) \log \alpha_s \\ &= H(\alpha) - H(X/Y) \end{aligned}$$

where

$$H(\alpha) = \sum_s p(s) \log \alpha_s$$

Several interesting facts emerge here

(a) In this case G is maximized as before by putting $a(s/r) = q(s/r)$. That is, *the gambler ignores the posted odds* in placing his bets!

(b) Since the minimum value of $H(\alpha)$ subject to

$$\sum_s \frac{1}{\alpha_s} = 1$$

obtains when

$$\alpha_s = \frac{1}{p(s)}$$

and $H(X) = H(\alpha)$, any deviation from fair odds helps the gambler.

(c) Since the gambler's exponential gain would be $H(\alpha) - H(X)$ if he had no inside information, we can interpret $R = H(X) - H(X/Y)$ as the increase of G_{max} due to the communication channel. When there is no channel, i.e., $H(X/Y) = H(X)$, G_{max} is *minimized* (at zero) by setting

$$\alpha_s = \frac{1}{p_s}$$

This gives further meaning to the concept "fair odds."

WHEN THERE IS A "TRACK TAKE"

In the case there is a "track take" the situation is more complicated. It can no longer be assumed that $\sum_s a(s/r) = 1$. The gambler cannot make canceling bets since he loses a percentage to the track. Let $b_r = 1 - \sum_s a(s/r)$, i.e., the fraction not bet when the received symbol is the rth one. Then the quantity to be maximized is

$$G = \sum_{rs} p(s,r) \log[b_r + \alpha_s a(s/r)], \tag{2}$$

subject to the constraints

$$b_r + \sum_s a(s/r) = 1.$$

In maximizing (2) it is sufficient to maximize the terms involving a particular value of r and to do this separately for each value of r since both in (2) and in the associated constraints, terms involving different r's are independent. That is, we must maximize terms of the type

$$G_r = q(r) \sum_s q(s/r) \log[b_r + \alpha_s a(s/r)]$$

subject to the constraint

$$b_r + \sum_s a(s/r) = 1$$

Actually, each of these terms is the same form as that of the gambler's exponential gain where there is no channel

$$G = \sum_s p(s) \log[b + \alpha_s a(s)]. \tag{3}$$

We will maximize (3) and interpret the results either as a typical term in the general problem or as the total exponential gain in the case of no communication channel. Let us designate by λ the set of indices, s, for which $a(s) > 0$, and by λ' the set for which $a(s) = 0$. Now at the desired maximum

$$\begin{aligned}
\frac{\partial G}{\partial a(s)} &= \frac{p(s)\alpha_s}{b + a(s)\alpha_s} \log e = k \qquad \mathit{for}\ s \epsilon \lambda \\
\frac{\partial G}{\partial b} &= \sum_s \frac{p(s)}{b + a(s)\alpha_s} \log e = k \\
\frac{\partial G}{\partial a(s)} &= \frac{p(s)\alpha_s}{b} \log e \leqq k \qquad \mathit{for}\ s \epsilon \lambda'
\end{aligned}$$

where k is a constant. The equations yield

$$\begin{aligned}
k = \log e, \quad b = \tfrac{1-p}{1-\sigma} \\
a(s) = p(s) - \frac{b}{\alpha_s} \qquad \text{for } s \epsilon \lambda
\end{aligned}$$

where $p = \sum_\lambda p(s)$, $\sigma = \sum_\lambda (1/\alpha_s)$, and the inequalities yield

$$p(s)\alpha_s \leqq b = \frac{1-p}{1-\sigma} \qquad \text{for } s \epsilon \lambda'$$

We will see that the conditions

$$\begin{aligned}
&\sigma < 1 \\
&p(s)\alpha_s > \frac{1-p}{1-\sigma} \qquad \text{for } s \epsilon \lambda \\
&p(s)\alpha_s \leqq \frac{1-p}{1-\sigma} \qquad \text{for } s \epsilon \lambda'
\end{aligned}$$

completely determine λ.

If we permute indices so that

$$p(s)\alpha_s \geqq p(s+1)\alpha_{s+1}$$

then λ must consist of all $s \leqq t$ where t is a positive integer or zero. Consider how the fraction

$$F_t = \frac{1 - p_t}{1 - \sigma_t}$$

varies with t, where

$$p_t = \sum_1^t p(s), \quad \sigma_t = \sum_1^t \frac{1}{\alpha_s}; \qquad F_0 = 1$$

Now if $p(1)\alpha_1 < 1$, F_t increases with t until $\sigma_t \geqq 1$. In this case $t = 0$ satisfies the desired conditions and λ is empty. If $p(1)\alpha_1 > 1$ F_t decreases with t until $p(t+1)\alpha_{t+1} < F_t$ or $\sigma_t \geqq 1$. If the former occurs, i.e., $p(t+1)\alpha_{t+1} < F_t$, then $F_{t+1} > F_t$ and the fraction increases until $\sigma_t \geqq 1$. In any case the desired value of t is the one which gives F_t its minimum positive value, or if there is more than one such value of t, the smallest. The maximizing process may be summed up as follows:

(a) Permute indices so that $p(s)\alpha_s \geqq p(s+1)\alpha_{s+1}$

(b) Set b equal to the minimum positive value of

$$\frac{1 - p_t}{1 - \sigma_t} \quad \text{where} \quad p_t = \sum_1^t p(s), \quad \sigma_t = \sum_1^t \frac{1}{\alpha_s}$$

(c) Set $a(s) = p(s) - b/\alpha_s$ or zero, whichever is larger. (The $a(s)$ will sum to $1 - b$.)

The desired maximum G will then be

$$G_{max} = \sum_1^t p(s) \log p(s)\alpha_s + (1 - p_t) \log \frac{1 - p_t}{1 - \sigma_t}$$

where t is the smallest index which gives

$$\frac{1 - p_t}{1 - \sigma_t}$$

its minimum positive value.

It should be noted that if $p(s)\alpha_s < 1$ for all s no bets are placed, but if the largest $p(s)\alpha_s > 1$ some bets might be made for which $p(s)\alpha_s < 1$, i.e., the expected gain is negative. This violates the criterion of the classical gambler who never bets on such an event.

CONCLUSION

The gambler introduced here follows an essentially different criterion from the classical gambler. At every bet he maximizes the expected value of the logarithm of his capital. The reason has nothing to do with the value function which he attached to his money, but merely with the fact that it is the logarithm which is additive in repeated bets and to which the law of large numbers applies. Suppose the situation were different; for example, suppose the gambler's wife allowed him to bet one dollar each week but not to reinvest his winnings. He should then maximize his expectation (expected value of capital) on each bet. He would bet all his available capital (one dollar) on the event yielding the highest expectation. With probability one he would get ahead of anyone dividing his money differently.

It should be noted that we have only shown that our gambler's capital will surpass, with probability one, that of any gambler apportioning his money differently from ours but still in a fixed way for each received symbol, independent of time or past events. Theorems remain to be proved showing in what sense, if any, our strategy is superior to others involving $a(s/r)$ which are not constant.

Although the model adopted here is drawn from the real-life situation of gambling it is possible that it could apply to certain other economic situations. The essential requirements for the validity of the theory are the possibility of reinvestment of profits and the ability to control or vary the amount of money invested or bet in different categories. The "channel" of the theory might correspond to a real communication channel or simply to the totality of inside information available to the investor.

Let us summarize briefly the results of this paper. If a gambler places bets on the input symbol to a communication channel and bets his money in the same proportion each time a particular symbol is received, his capital will grow (or shrink) exponentially. If the odds are consistent with the probabilities of occurrence of the transmitted symbols (i.e., equal to their reciprocals), the maximum value of this exponential rate of growth will be equal to the rate of transmission of information. If the odds are not fair, i.e., not consistent with the transmitted symbol probabilities but consistent with some other set of probabilities, the maximum exponential rate of growth will be larger than it would have been with no channel by an amount equal to the rate of transmission of information. In case there is a "track take" similar results are obtained, but the formulae involved are more complex and have less direct information theoretic interpretations.

ACKNOWLEDGMENTS

I am indebted to R. E. Graham and C. E. Shannon for their assistance in the preparation of this paper.

4

CRITERIA FOR CHOICE AMONG RISKY VENTURES

HENRY ALLEN LATANÉ
Chapel Hill, North Carolina

THE SUBGOAL

THIS paper is concerned with the problem of how to make rational choices among strategies in situations involving uncertainty. Such choices can be expressed through payout matrices stated in terms of some measure of value to be maximized. These matrices show the probabilities of all relevant future occurrences and the payouts resulting from the combined effects of each possible strategy, on the one hand, and each relevant future occurrence, on the other.[1] All this information is needed to choose the proper strategy rationally. It would be impossible for a decision-maker to choose rationally among strategies if he disregarded either the probability of the relevant future occurrences or any of the possible payouts.

The problem of rational decision-making can be broken down into three steps: (1) deciding upon an objective and criteria for choosing among strategies; (2) filling out a payout matrix; and (3) choosing among available strategies on the basis of this matrix and the criteria. In real life the second step—deciding upon the size of the payout matrix, measured by the number of columns representing relevant future occurrences and rows representing available strategies, and filling in the matrix with reasonable estimates of payouts and probabilities—is by far the most difficult part of the decision-maker's job. This paper has little to say about these problems. It deals largely with the first step: the problem of setting up criteria for choosing among strategies on the basis of a filled-in payout matrix.

A hierarchy of goals and guides for reaching these goals is involved in rational choices among strategies. This hierarchy consists of (1) a goal; (2) a subgoal; and (3) a criterion for choosing among strategies to reach the subgoal, that is, a measure that must be maximized to attain the subgoal. The goal in rational decision-making is the maximization of some measure of value. Each decision is made for the sake of the difference it will make in terms of this

[1] Payout matrices are shown in Tables 1, 2, and 3. The payouts represent the possible final outcomes of choices among strategies. The matrices have single-valued payouts and probabilities. Many, if not all, decision problems can be reduced to such form. Consider first the probabilities. A subjective probability distribution of an imperfectly known underlying probability can be reduced to a subjective probability of the event itself. For example, suppose a gambler believes that there is a 0.5 probability that a coin is biased so that it always comes up tails and a 0.5 probability that it is unbiased. He has a subjective probability of 0.25 for heads and 0.75 for tails, and these probabilities would be used in his payout matrix as long as his probability beliefs remain unchanged. Consider next the payouts. In much discussion of decision theory the payouts are taken as given, with only the probabilities subject to uncertainty. However, in real life the sizes of the payouts often are as subject to uncertainty as are the probabilities. But, even when the payouts are uncertain a matrix filled in with single values can be constructed. If we have probability distributions of payouts for all specified occurrences (such as heads and tails in the toss of a coin), a payout matrix can be constructed listing each possible payout and the subjective probability of its occurrence. These large matrices often can be reduced to simple two-valued distributions of payouts without much loss of information.

objective. The decision-maker is confronted with a payout matrix expressed in terms of either a subjective utility measure such as utiles or an objective measure such as money or bushels of wheat. He wishes to choose the strategy that will give him the maximum payout. This is his goal. When some one strategy gives a higher payout than any other strategy in all relevant future occurrences, the goal itself enables the decision-maker to choose among strategies. He merely chooses the dominant strategy.

When there is no strategy superior to all the rest in all possible future occurrences, the decision-maker needs some other guide for making decisions, since the goal itself does not enable him to make his choice. This guide is here called the "subgoal." The need for a subgoal exists because the outcome of specific strategies is subject to probabilistic uncertainty. In utility theory the payout matrix is expressed in terms of some measure of subjective utility, say, utiles. Choice of the strategy that will give the maximum payout in utiles is the goal, and choice of the strategy with the maximum expected utility[2] is taken as the subgoal. Given a completely filled-in matrix, this subgoal can surely be reached. Whether or not the goal is reached depends on future occurrences, but, in any event, the subgoal of maximization of the expected value of the payouts expressed in utiles is logically related to the goal of maximization of the forthcoming payout also expressed in utiles.

In this paper a second subgoal is proposed for use when the choice is repetitious and has cumulative effects and when the goal is maximization of wealth at the end of a large number of choices. Under these conditions the choice of the strategy that has a greater probability (P') of leading to as much or more wealth than any other significantly different strategy at the end of a large number of choices also is a logical subgoal. The P' subgoal is not as general as the maximum expected utility subgoal. For example, it would not apply to unique choices. When a man is faced with a once-in-a-lifetime choice of risking his whole fortune and his life on a venture that will produce great rewards if successful, it does not help him to know that he is almost certain to be ruined if he takes such a risk often enough. The P' subgoal is not logically related to the goal in this case.[3] Such a man, however, could set up a payout matrix expressed in utiles and decide which course of action maximized his expected utility. Here the maximum expected utility subgoal is logically related to the goal even though the P' subgoal is not. The P' subgoal is less general but would seem to be more operational than the expected utility subgoal because of the difficulty of constructing a payout matrix expressed in terms of utiles, especially if the decision involves a firm or group of people.

The P' subgoal would seem to be particularly applicable to many business

[2] The expected utility of a strategy is computed by multiplying all possible payouts expressed in utiles by their respective probabilities and then summing the products.

[3] For certain utility functions and for certain repeated gambles, no amount of repetition justifies the rule that the gamble which is almost sure to bring the greatest wealth is the preferable one. For example, the P' subgoal is not appropriate for a decision-maker for whom the possibility of great gain, however small and diminishing, is more important than maximization of the probability of as much or more wealth than can be obtained by any other strategy in the long run. Such a decision-maker may adopt a course of action that is almost certain to result in less wealth in the long run. Whether or not his utility function is compatible with the specified goal of maximum long-run wealth is not at issue here.

decisions such as those involved in portfolio management. Wealth-holders have the option of holding their wealth in many different combinations of stocks, bonds, and cash. The allocation of wealth among these types of assets involves a series of choices extending over time. The fact that these choices are repetitive in nature with cumulative effects may be used as the key factor in defining a goal, a subgoal, and a criterion for choosing among portfolios.

The problem of choice among portfolios may be stated in terms of the payout matrix in Table 1. In this table p_j

TABLE 1

PAYOUT MATRIX FOR VARYING PORTFOLIOS

PORTFOLIO	RELEVANT FUTURE OCCURRENCES
	$1, \ldots, j, \ldots, k$
1	$a_{i1}, \ldots, a_{1j}, \ldots, a_{1k}$
.	
.	
i	$a_{i1}, \ldots, a_{ij}, \ldots, a_{ik}$
.	
.	
t	$a_{t1}, \ldots, a_{tj}, \ldots, a_{tk}$
Probability of occurrence	$p_1, \ldots, p_j, \ldots, p_k$

represents the probability of the jth occurrence, with $\Sigma p_j = 1$, and a_{ij} represents the return from the ith portfolio with $i = 1, \ldots, t$, if the jth occurrence takes place, with $j = 1, \ldots, k$. A return is the payout, including return of principal, per dollar of portfolio value per investment period (here called "year"). Returns cannot be negative, so that $a_i \geqq 0$. The portfolio manager is faced with such a payout matrix for n years and wants to choose in a rational manner one portfolio from all available portfolios in each of the n years.[4]

The goal of portfolio management is taken to be to select a portfolio so as to maximize wealth at the end of a period of years, assuming reinvestment of all returns.[5] Let W_i^n be the final value of $1.00 placed in portfolio i if returns are reinvested n times. Then the goal of portfolio management is taken to be to select the optimum portfolio so that $W_{opt}^n \geqq W_i^n$, with $i = 1, \ldots, t$. This goal cannot be used as a basis for choice among portfolios, since which portfolio will have the maximum W^n depends on future occurrences.

The subgoal proposed here is the choice of the portfolio that has a greater probability (P') of being as valuable or more valuable than any other significantly different portfolio at the end of n years, n being large. It is shown below

[4] The idea of maximizing wealth at the end of a large number of separate decisions based on the same payout matrix may appear unrealistic, but portfolio managers are continually being faced with choices having cumulative effects and involving approximately the same payouts and probabilities time after time. For example, year after year a portfolio manager may have probability beliefs such as: "I look for conditions in the next ten years to be very similar to those prevailing in 1926 through 1935. Bonds will yield about 4 per cent per annum during the whole period. Some day we are going to have a boom and a bust in the stock market, but I do not know which is going to come first." Choosing one portfolio to hold in each of the n years is not the same as choosing one portfolio at the beginning of the period to hold throughout the n years. For example, if the probability beliefs at the beginning of one year are such that the maximum P' allocation of the portfolio is 40 per cent in bonds and 60 per cent in stock and if these beliefs remain the same at the beginning of the next year, then the maximum P' allocation again will be 40 per cent in bonds and 60 per cent in stock at the beginning of the second year. If the relative prices of stocks and bonds have changed between the two dates, it will be necessary for the portfolio manager to make some sales and purchases in his portfolio to bring it into line with the desired proportions even if these proportions themselves have not changed.

[5] Few wealth-holders reinvest all returns, so the problem of maximizing wealth assuming no withdrawals is somethat unrealistic. However, this restriction can be modified. If withdrawals per unit of time are a fixed proportion of wealth (considered as interest, for example), they will not affect proper maximizing action. Whatever would maximize wealth, assuming no withdrawals, would maximize wealth, assuming proportionate withdrawals.

that the portfolio having a probability distribution of returns with the highest geometric mean, G, also has the greatest P'.

The central fact of this paper is a simple one: If the value of an asset, say, portfolio i, priced initially at $1.00 is believed to change after a year to, alternatively, a_{i1} or $a_{i2}, \ldots$, or a_{ik}, with respective probabilities $p_1, p_2, \ldots, p_k$, and if the proceeds are reinvested n times, then the final value of the investment, W_i^n, "converges in probability" to $G_i^n = a_{i1}^{p_1 n} \cdot a_{i2}^{p_2 n} \cdots a_{ik}^{p_k n}$. The probability that the absolute difference between W_i^n and G_i^n is smaller than any preassigned positive number will approach 1 as n increases indefinitely. In other words, the final return from $1.00 invested in portfolio i, assuming reinvestment of all annual returns for n years will converge in probability on G_i^n, the nth power of the geometric mean of the probability distribution of annual returns from that portfolio. This relationship is intuitively obvious, since the a_{i1} return will "tend" to occur np_1 times, the a_{ii} return will tend to occur np_i times, and so forth, if n is large. It can be proved rigorously by use of the law of large numbers applied to the logarithms of the annual returns.

Let n_j be the number of occurrences of the jth relevant future occurrence, with $\Sigma n_j = n$, with $j = 1, \ldots, k$, then $n_j/n \underset{\lim}{\rightarrow} p_j$ and $n_j/n \log a_{ij} \underset{\lim}{\rightarrow} p_j \log a_{ij}$ as $n \rightarrow \infty$. But $\log W_i = \Sigma n_j/n \log a_{ij}$, with $j = 1, \ldots, k$ and $\log G_i = \Sigma_{pj} \log a_{ij}$, so $\log W_i \underset{\lim}{\rightarrow} \log G_i$ and $W_i \underset{\lim}{\rightarrow} G_i$ as $n \rightarrow \infty$.[6] It follows from this that, if $G_i > G_j$, then the probability, P', that $W_i^n > W_j^n$ at the end of n years approaches 1 as n increases indefinitely. The portfolio with the highest G is almost certain to be more valuable than any other significantly different portfolio in the long run. For this reason G is accepted here as a rational criterion for choice among portfolios.

SUBGOALS AND SUBJECTIVE UTILITY

Rational choice among strategies is the ancient problem of the gambler who has the option to choose among bets. Classical writers on probability theory recommended that problems of this kind be solved by first computing the expected winnings (possibly negative) for each available bet and then choosing the bet with the highest mathematical expectation of winning. Since there was no reason to assume that, of two persons encountering identical risks, either should expect to have his desires more closely fulfilled, the classical writers thought that no characteristic of the risk-takers themselves ought to be taken into consideration; only those matters should be weighed carefully that pertain to the terms of the risk.[7] In 1738 Daniel Bernoulli in four short paragraphs demonstrated that the use of the mathematical expectation of winnings did not always apply and proposed instead that gamblers should evaluate bets on the basis of the mathematical expectation of the utilities of winnings.[8]

In terms of subgoals as defined in this study, Bernoulli showed that use of the

[6] The asymptotic quality of G is used in information theory as developed by Dr. Claude Shannon and was applied to a gambling situation by John Kelly in "A New Interpretation of Information Rate," *Bell System Technical Journal*, August, 1956, pp. 917–26. See also R. Bellman and R. Kalaba, "Dynamic Programming and Statistical Communication Theory," *Proceedings of the National Academy of Science*, XLIII (1957), 749–51. I had no knowledge of this work when I first proposed the P' subgoal at a Cowles Foundation Seminar in February, 1956.

[7] See Daniel Bernoulli, "Exposition of a New Theory on the Measurement of Risk," trans. Louise Sommer, *Econometrica*, XXII (January, 1954), 23.

[8] *Ibid.*, p. 24.

expected-value subgoal did not always lead to choices that seemed rational to him and proposed instead the use of the expected-utility subgoal. He used the following example:

Somehow a very poor fellow obtains a lottery ticket that will yield with equal probability either nothing or twenty thousand ducats. Will this man evaluate his chances of winning at ten thousand ducats? Would he not be ill-advised to sell this lottery ticket for nine thousand ducats? To me it seems that the answer is in the negative. On the other hand I am inclined to believe that a rich man would be ill-advised to refuse to buy the lottery ticket for nine thousand ducats. If I am not wrong then it seems clear that all men cannot use the same rule to evaluate the gamble.[9]

Bernoulli's example is somewhat aside from the daily business of living, but, when stripped of its gambling wrappings and expressed in terms of payouts and returns, it is seen to represent a major segment of economic decision-making. The hypothetical market price of the ticket, which has an equal probability of paying 20,000 ducats or nothing, is 9,000 ducats. Both the poor man and the rich man have the option either to hold the lottery ticket or to hold 9,000 ducats. Possible payouts range from 2.22 per ducat risked to 0. Payouts with ranges such as this—indeed, much greater ranges—are ordinary economic occurrences. The magnitude of the choice faced by the rich man is well within the range of ordinary business decisions, and the "poor man" today is continually faced with implicit or explicit decisions as serious as that faced by Bernoulli's lottery-ticket owner. He must decide whether to move to a new job, buy a new home, sign a second mortgage. He is continually offered the opportunity to undertake such risky ventures as purchasing his own truck, opening a restaurant, buying some uranium stock, some oil stock, or some investment shares. Some of these options may be highly advantageous, and he must choose some one course of action in each case. The effects of these choices are cumulative; that is, the decision-maker never comes back to exactly the same position he occupied before making his choice. The major difference between Bernoulli's problem and other choices among courses of action is that the ticket-holder's choice is clearly defined, while the other opportunities are usually ignored, or the choices are muddled.

[9] *Ibid.*

TABLE 2

PAYOUT MATRIX OF GAINS AND LOSSES

STRATEGY	FUTURE OCCURRENCE Ticket Wins	Ticket Loses	CRITERION A
a) Poor man:			
Hold ticket...	20	0	10
Sell ticket....	9	9	9
b) Rich man:			
Buy ticket...	11	−9	1
Not buy ticket	0	0	0
Probability of occurrence........	0.5	0.5	

Thus Bernoulli's example is representative of a wide class of choices. The decision-maker is being faced continually with such choices, and the outcome of each decision affects his entire future. In the following discussion this example is stated in payout matrices constructed to illustrate choices based on (*a*) classical mathematical expectation (the expected value subgoal); (*b*) Bernoulli's subjective utility (the expected-utility subgoal); and (*c*) the maximum chance (P') subgoal.

Table 2 shows the classical approach to choosing among risky ventures. The payout matrix, expressed in terms of thousands of ducats, shows the probability of the lottery ticket paying off or not and the net payout to the poor man and to the rich man for each of two courses of action. The classical writers would calculate the mathematical expectation, A, of the net payouts and choose that strategy which maximizes A. In this case they would recommend that the rich man buy the ticket for 9,000 ducats and that the poor man refuse to sell it at this price.

The mathematical expectation (that is, the arithmetic mean) of the probability distribution of payouts is, indeed, a good criterion when there are large numbers of independent trials. Even decision-makers who make repeated choices with cumulative effects (for example, the operators of roulette wheels and insurance companies) are rightly interested in this average when each risk is small in relation to total wealth. There is little or no conflict under these conditions between the use of the arithmetic mean as a criterion and the use of the geometric mean of the probability distribution of payouts per dollar of wealth as a criterion.[10] When a decision-maker can surely bet the same small amount on a large number of independent trials, he can maximize the expected value of his gain, and also the likelihood of having more gain than can be obtained by any other plan, by choosing that set of bets which gives him the greatest mathematically expected payout. For example, if Bernoulli's poor man had found 10,000 tickets involving 10,000 independent drawings, each with a payout equally likely to be 2 ducats or nothing, he clearly would be unwise to sell his block of tickets for 9,000 ducats. His winnings on 10,000 different trials would be almost certainly very close to 10,000 ducats, the mathematical expectation of the value of the set of tickets, and the advice of the classical writers would be sound.

Bernoulli used the lottery-ticket example to show that the expected values of the payouts are not good guides in making choices involving large risks. He proposed instead that the expected value of the utilities of the payouts be used as a criterion. He would fill in the payout matrix in Table 2 not with the money value of the gains and losses but with their utilities and then would use the mathematical expectation of these utilities as his criterion.

Whether or not particular payout matrices expressed in terms of subjective utility are realistic is not a problem here. But Bernoulli's procedure is very much at issue. He defines the "mean utility" of a course of action as the mathematical expectation of the probability distribution of the possible utilities from that course of action. He then states, with no discussion, that this mean utility can be used as a basis for valuing risks, that is, as a basis for choosing among courses of action. In other words, he explains why he expresses his profits (or losses) in terms of subjective utility, but he does not give any justification for maximizing the mathematical expectation of these utilities. Bernoulli's use of subjective utility has had wide recognition, and his use of mathematical expectation also has been widely adopted with little or no discussion.[11]

[10] When a gambler who has the choice of betting or not betting bets all his wealth on the toss of a fair coin with a payout of $3.00 per $1.00 bet if heads occur and nothing per $1.00 bet if tails occur, he is maximizing the expected value of the payout but not G. When he can bet only 1 per cent of his wealth, however, he will maximize both A and G by betting.

[11] Mathematical expectation now is used as a basis for defining utility. The present emphasis on the axiomatic approach to utility is largely derived from John von Neumann and Oskar Morgenstern, *Theory of Games and Economic Behavior* (rev. ed.; Princeton, N.J.: Princeton University Press, 1935). They say: "We have practically defined numerical utility as being that thing for which the calculus of mathematical expectations is legitimate" (p. 28).

Bernoulli's problem can also be solved by the use of the maximum-chance (P') subgoal. Table 3 shows the payout matrix of returns (that is, payouts, including return of principal, per dollar of wealth) for the poor man, who is assumed to have a wealth of 1,000 ducats aside from his lottery ticket, and for the rich man, who is assumed to have a wealth of 100,000 ducats. The arithmetic mean, A, of the probability distribution of returns is higher for the poor man when he holds the ticket and for the rich man when he buys the ticket. The geometric mean, G, of returns for the poor man is higher when the ticket is sold, however, and G for the rich man is higher when he buys the ticket.

TABLE 3

PAYOUT MATRIX OF RETURNS

Strategy	Future Occurrence: Ticket Wins	Ticket Loses	Criterion: A	G
a) Poor man:				
Hold ticket.	2.1	0.1	1.1	0.46
Sell ticket..	1.0	1.0	1.0	1.0
b) Rich man:				
Buy ticket.	1.11	0.91	1.01	1.005
Not buy ticket....	1.00	1.00	1.00	1.00
Probability of occurrence......	0.5	0.5		

Over a long enough period of time many economic choices involving returns of the same order of magnitude repeat themselves. Bernoulli's poor man may never find another lottery ticket, but he probably will have many options among courses of action with as wide, or wider, a range of returns. It is assumed here that both the rich man and the poor man will have many opportunities to risk the same proportions of their respective fortunes on approximately the same terms and that both men prefer more wealth to less wealth, everything else being equal. If these assumptions are valid, the maximization of P', the probability of having more wealth at the end of a long series of such choices than can be obtained by any other specified course of action, is a rational subgoal, and G is a rational criterion. The use of the maximum-chance subgoal results in courses of action for both the rich man and the poor man which seemed rational to Bernoulli.

The decision-maker who is interested in maximizing his wealth at the end of a long series of choices should ask himself how he would come out in the long run if he made the same choice on the same terms over and over again. It is not necessary for him to ask himself what his individual subjective utility of winning is. This is not to say that other goals, rather than the goal of maximum wealth at the end of a long series of choices, are irrational. Indeed, the use of subgoals based on the goal of maximum wealth often may be irrational. For example, the man who desperately needs $10.00 to escape a jail sentence and who has only $1.00 may well be justified in taking a gamble to get his money, even though this gamble would not stand the maximum-chance subgoal test. Even under these conditions, however, it would be useful for the man to know that he should not often act in such a manner, if he wants to build up his fortune so as to avoid similar predicaments in the future.

In his paper Bernoulli uses the expected utilities of the payouts as his criterion. He then reaches the conclusion that the utility resulting from any small increase in wealth usually is inversely proportional to the quantity of goods previously possessed.[12] Under these conditions the utilities of the returns vary as their logarithms, and the geometric mean, G, of the probability distributions

[12] This is generally credited with being the first use of a utility function.

of returns can be used as a criterion instead of expected utility. The arithmetic mean of the logarithms (utilities) of returns is maximized when G is maximized.[13]

Bernoulli gives a number of applications of his formula to gambling and to insurance. In each instance he is able to give a specific answer. He says that everyone who bets any part of his fortune on a mathematically fair game of chance is acting irrationally, and he then determines what odds a gambler with a specified fortune must obtain to break even in the long run. Most of his problems still are interesting in their own right, and many have a bearing on proper portfolio management. For instance, he demonstrates with numerical examples the advantages of diversification among equally risky ventures and between risky and safe assets.

Bernoulli's approach to the valuation of risky ventures is not contradictory to the maximum-chance (P') approach. Not only do the two approaches lead to the same conclusion when they both can be applied but they tend to support each other. Wealth-holders may be divided into two groups. The first group contains those to whom each risk is a unique event either because they do not expect it to recur or because they keep its effects entirely separate from the results of other risks. For example, the man who each year sets aside a small sum to bet on the races during his vacation, with the intention of "living it up" if he wins and writing it off to experience if he loses, presumably is not actuated by long-run profit-maximizing motives. The effects of each risk are kept separate. Analysis based on maximum chance has nothing to offer this first class of wealth-holders. The choice between profit and safety or expected return and variance is a matter of subjective utility. Bernoulli's assumption that the satisfaction derived from a small gain tends to vary in inverse proportion to the initial wealth may or may not be a shrewd guess.

The second class of wealth-holders includes those who expect to be faced repeatedly with risks of the same general type and magnitude. This group includes those making most business and portfolio decisions and hence is of great importance. It includes, specifically, all those who want to maximize the value of their portfolio at the end of n years, assuming reinvestment of all returns. Here there is a definite rule for choosing between risk and return, the P' subgoal, based on maximum-chance principles. This class may be subdivided further into (*a*) those who undertake only one risky venture at a time and (*b*) those who are able to diversify their risky ventures. Because so many economic phenomena, including yields on stocks, tend to fluctuate together over time, diversification among risky ventures cannot go as far toward eliminating risk as otherwise would be the case. Final choice among efficient portfolios for both groups, (*a*) and (*b*), is based on maximization of G, not because this maximizes subjective utility, but because it maximizes P'.

Bernoulli states that the wealth-holder should ask himself whether the added satisfaction associated with the expected gain justifies undertaking the risky venture. He bases an exact rule of behavior on his assumption as to how the added satisfaction varies with the size of the potential gain or loss in relation to the size of the portfolio. The rule may or

[13] As pointed out to me by Professor L. J. Savage (in correspondence), not only is the maximization of G the rule for maximum expected utility in connection with Bernoulli's function but (insofar as certain approximations are permissible) this same rule is approximately valid for all utility functions.

may not be empirically useful, but it is grounded on rather shaky evidence about the exact shape of the utility function. According to maximum-chance analysis, the wealth-holder or portfolio manager should ask himself how he can maximize his chances of getting as good or better return than can be obtained with any other specified plan, assuming that he risks the same proportion of his portfolio on the same terms over and over again. It turns out that the formula which enables the portfolio manager to answer the maximum-chance question is the same as that developed by Bernoulli on grounds of subjective utility.

In conclusion Bernoulli says:

> Though a person who is fairly judicious by natural instinct might have realized and spontaneously applied much of what I have here explained, hardly anyone believed it possible to define these problems with the precision we have employed in our examples. Since all of our propositions harmonize perfectly with experience it would be wrong to neglect them as abstractions resting upon precarious hypotheses.[14]

Professor Stigler, in a review article,[15] gives considerable space to Bernoulli's hypothesis about the slope of the wealth-holder's utility function, even though the major emphasis of the article is on utility not affected by probability. He mentions that Laplace and Marshall, among others, have accepted the law as a realistic guide. He also points out the similarity of Bernoulli's law to the Weber-Fechner psychological hypothesis that the just noticeable increment to any stimulus is proportional to the stimulus. Stigler says: "Bernoulli was right in seeking the explanation[16] in utility and he was wrong only in making a special assumption with respect to the slope of the utility curve for which there was no evidence and which he submitted to no tests."[17]

More recently Savage in a section on "Historical and Critical Comments on Utility" had this to say:

> Bernoulli went further than the law of diminishing marginal utility and suggested that the slope of utility as a function of wealth might, at least as a rule of thumb, be supposed, not only to decrease with, but to be inversely proportional to, the cash value of wealth. To this day, no other function has been suggested as a better prototype for Everyman's utility function. . . . Though it might be a reasonable approximation to a person's utility in a moderate range of wealth, it cannot be taken seriously over extreme ranges.[18]

INDIVIDUAL RISK PREFERENCE

As indicated in the previous section, Bernoulli took the following steps to develop his utility function and to justify diversification among risky ventures and between risk assets and safe assets. (1) He showed—subject to the implicit assumption about subgoals previously discussed—that the value of a risky venture to the individual wealth-holder is not the arithmetic mean of the probability distribution of returns (the mathematical expectation of returns) but may be taken to be the arithmetic mean of the probability distribution of the utilities of the returns. (2) He stated that, in the absence of the unusual, the gain in utility

[14] *Op. cit.*, p. 31.

[15] George J. Stigler, "The Development of Utility Theory," *Journal of Political Economy*, LVIII (1950), 373–77.

[16] Bernoulli is explaining the reason for the limited value of the game involved in the St. Petersburg paradox. This game is a type of risky venture with an infinitely large mathematically expected value but with an extremely small probability of winning.

[17] Stigler, *op. cit.*, p. 375.

[18] Leonard J. Savage, *The Foundations of Statistics* (New York: John Wiley & Sons, 1954), p. 94.

resulting from any small increase in wealth may be assumed to be inversely proportional to the quantity of goods previously possessed. (3) He developed a formula for calculating the utility of a risk asset to the individual wealth-holder using as a criterion the utility function developed in step 2. According to Bernoulli, the subjective utility of the wealth-holder's assets, including the risky venture, is measured by the geometric mean, *G*, of the probability distribution of payouts from such assets. (4) Using this formula, he was able to calculate exactly the utility of the wealth-holder's assets, including the risky venture, and to show that diversification among risky ventures increases the utility.[19]

Bernoulli's step 2 may be a reasonable assumption about utility,[20] but it is subject to so many qualifications and exceptions (it does not explain gambling, for example) that it has not been accepted as a suitable basis for erecting the superstructure of steps 3 and 4. The valuation of risky ventures has been left to individual risk preference without any criterion for deciding what this preference is likely to be. For example, Makower and Marschak present a hypothetical table in which an asset's marginal contribution is determined by adding together its contribution to "lucrativity" and safety measured in "lucrativity units" determined by the safety preference rate for a single individual.[21] These individual safety preference rates, in turn, are a matter of taste and must be accepted as given. Friedman and Savage build on Bernoulli's step 1 but modify step 2 by developing a doubly inflected curve comparing utility with income.[22]

Markowitz begins his analysis of portfolio selection by pointing out that "the portfolio with the maximum expected return is not necessarily the one with the minimum variance. There is a rate at which the investor can gain expected return by taking on variance, or reduce variance by giving up expected return."[23] He assumes that the investor considers, or should consider, expected return a desirable thing and variance of return an undesirable thing, and he defines an efficient portfolio as a portfolio with minimum variance for a given expected return or more and a maximum expected return for a given variance or less. He develops a method for selecting efficient portfolios from the set of all possible portfolios but does not give any basis for choice among the efficient portfolios except the individual's safety preference rate.

THE NEED FOR AN OBJECTIVE CRITERION

The difficulty of evaluating subjective risk preference and the need of an objective criterion are well indicated in the

[19] Bernoulli, *op. cit.*, pp. 24, 25, 28, 30.

[20] Cf. Alfred Marshall, *Principles of Economics* (8th ed.; New York: Macmillan Co., 1950), p. 135. Marshall says: "In accordance with a suggestion made by Daniel Bernoulli, we may regard the satisfaction which a person derives from his income as commencing when he has enough to support life, and afterwards as increasing by equal amounts with every equal successive percentage that is added to his income; and vice versa for loss of income."

See also Savage's comment quoted previously.

[21] Helen Makower and Jacob Marschak, "Assets, Prices and Marketing Theory," *Economica*, V (1938), 261–88. Reprinted in American Economic Association, *Readings in Price Theory* (Chicago: Richard D. Irwin, Inc., 1952), pp. 301–2.

[22] Milton Friedman and L. J. Savage, "The Utility Analysis of Choices Involving Risk," *Journal of Political Economy*, LVI (1948), 279–304.

[23] Harry Markowitz, "Portfolio Selection," *Journal of Finance*, VII (March, 1952), 79.

following quotation from a recent journal article dealing with selection of an optimum combination of crops for a farmer:

> The introduction of risk into an economic model of a firm and consequently into a linear programming model of a firm has been accomplished by describing risky outcomes as probability distributions and choosing from among alternate possible distributions by the expected utility hypothesis.
>
> Two basic weaknesses have appeared in applying this method of incorporating risk. One difficulty arises in choosing a value for the constant a, which in this case is some sort of risk aversion indicator, and is, to some degree, governed by the personal characteristics of the entrepreneur. A large value for a indicates that the entrepreneur places a great weight on the variance as a deciding factor and is consequently highly averse to risk, and vice versa. The estimation of such a constant to be used in a model is thus quite important; the wrong choice will invalidate any results obtained. The derivation of this constant is a delicate task beyond the scope of this paper.[24]

A major advantage of the criterion for choice among risky ventures developed in this paper is that it avoids the necessity for direct subjective determination of such factors as Marschak's "lucrativity units" or Freund's "risk aversion indicator." As Roy remarks, "A man who seeks advice about his actions will not be grateful for the suggestion that he maximize expected utility."[25]

The criteria for choice between risk and safety in portfolio management can be illustrated by assuming that a gambler has the choice of holding his money in cash or of betting on a gambling device which, with equal probability, will return $R-s$ on loss occasions and $R+s$ on gain occasions with an expected return of R per dollar played. The gambler's portfolio at any time consists of the proportion of his wealth held in cash plus the proportion bet on the gambling device. When the gambler bets none of his wealth, the expected return from his portfolio is 1, and the standard deviation of returns is 0. As the proportion bet increases, both the expected portfolio return and the standard deviation of returns increase. When he bets all his wealth, the expected portfolio return is R, and the expected standard deviation of returns is s. As long as R is greater than 1, and $R-s$ is less than 1, all possible combinations of the two assets in this range are efficient portfolios in that any one of the combinations gives the maximum possible expected return for some standard deviation or variance and the minimum standard deviation or variance for some expected return. Neither Marschak nor Friedman and Savage nor Markowitz would be able to help the gambler in choosing among these efficient portfolios beyond telling him that he should gamble heavily if he has a high preference for risk and should be very conservative in his betting if he has a high risk-aversion factor. In this paper an attempt is made to give the gambler (and wealth-holders, in general) an objective criterion for making this choice.

The wealth-holder who adopts the maximum-chance (P') subgoal can reach this subgoal by using the geometric mean, G, of the probability distribution of returns as his criterion and choose the strategy that has the probability distribution of returns with the highest G. Bernoulli also has shown that choice of that risky venture with the highest G is a rational choice (1) if maximization of the mathematical expectation of the utilities

[24] Rudolph J. Freund, "The Introduction of Risk into a Programming Model," *Econometrica*, XXIV (July, 1956), 253–63.

[25] A. D. Roy, "Safety First and the Holding of Assets," *Econometrica* XX (1952), 433.

of the payouts is a rational subgoal and (2) if the utility of a small gain or loss varies inversely with the amount of wealth already possessed.

Most economists recognize that the mathematical expectation and the variance of the probability distribution of returns and the chance of ruin are important to the wealth-holder—but they leave it to individual risk preference to balance one factor against the others. Since G depends on both the mathematical expectation and the variance of the probability distribution of returns, when G is maximized, there is no chance of ruin if the wealth-holder's probability beliefs are correct. Consequently, maximization of G falls within the generally accepted range of rational behavior. This is not to say that G is the only rational criterion for choice among strategies; it is to say, however, that it is a useful criterion in dealing with a broad range of problems.

5

OPTIMAL GAMBLING SYSTEMS FOR FAVORABLE GAMES

L. BREIMAN
UNIVERSITY OF CALIFORNIA, LOS ANGELES

1. Introduction

Assume that we are hardened and unscrupulous types with an infinitely wealthy friend. We induce him to match any bet we wish to make on the event that a coin biased in our favor will turn up heads. That is, at every toss we have probability $p > 1/2$ of doubling the amount of our bet. If we are clever, as well as unscrupulous, we soon begin to worry about how much of our available fortune to bet at every toss. Betting everything we have on heads on every toss will lead to almost certain bankruptcy. On the other hand, if we bet a small, but fixed, fraction (we assume throughout that money is infinitely divisible) of our available fortune at every toss, then the law of large numbers informs us that our fortune converges almost surely to plus infinity. What to do?

More generally, let X be a random variable taking values in the set $I = \{1, \cdots, s\}$ such that $P\{X = i\} = p_i$ and let there be a class $\mathcal{C}$ of subsets A_j of I, where $\mathcal{C} = \{A_1, \cdots, A_r\}$, with $\bigcup_j A_j = I$, together with positive numbers $(o_1, \cdots, o_r)$. We play this game by betting amounts $\beta_1, \cdots, \beta_r$ on the events $\{X \in A_j\}$ and if the event $\{X = i\}$ is realized, we receive back the amount $\sum_{i \in A_j} \beta_j o_j$ where the sum is over all j such that $i \in A_j$. We may assume that our entire fortune is distributed at every play over the betting sets $\mathcal{C}$, because the possibility of holding part of our fortune in reserve is realized by taking A_1, say, such that $A_1 = I$, and $o_1 = 1$. Let S_n be the fortune after n plays; we say that the game is *favorable* if there is a gambling strategy such that almost surely $S_n \to \infty$. We give in the next section a simple necessary and sufficient condition for a game to be favorable.

How much to bet on the various alternatives in a sequence of independent repetitions of a favorable game depends, of course, on what our goal utility is. There are two criterions, among the many possibilities, that seem pre-eminently reasonable. One is the minimal time requirement, that is, we fix an amount x we wish to win and inquire after that gambling strategy which will minimize the expected number of trials needed to win or exceed x. The other is a magnitude condition; we fix at n the number of trials we are going to play and examine the size of our fortune after the n plays.

This research was supported in part by the Office of Naval Research under Contract Nonr-222(53).

In this work, we are especially interested in the asymptotic point of view. We show that in the long run, from either of the two above criterions, there is one strategy Λ^* which is optimal. This strategy is found as that system of betting (essentially unique) which maximizes $E(\log S_n)$. The reason for this result is heuristically clear. Under reasonable betting systems S_n increases exponentially and maximizing $E(\log S_n)$ maximizes the rate of growth.

In the second section we investigate the nature of Λ^*. It is a conservative policy which consists in betting *fixed* fractions of the available fortune on the various A_j. For example, in the coin-tossing game Λ^* is: bet a fraction $p - q$ of our fortune on heads at every game. It is also, in general, a policy of diversification involving the placing of bets on many of the A_j rather than the single one with the largest expected return.

The minimal expected time property is covered in the third section. We show, by an examination of the excess in Wald's formula, that the desired fortune x becomes infinite, that the expected time under Λ^* to amass x becomes less than that under any other strategy.

Section four is involved with the magnitude problem. The content here is that Λ^* magnitudewise, does as well as any other strategy, and that if one picks a policy which in the long run does not become close to Λ^*, then we are asymptotically infinitely worse off.

Finally, in section five, we discuss the finite (nonasymptotic) case for the coin-tossing game. We have been unsuccessful in our efforts to find a strategy which minimizes the expected time for x fixed, but we state a conjecture which expresses a moderate faith in the simplicity of things. It is not difficult, however, to find a strategy which maximizes $P\{S_n \geqq x\}$ for fixed n, x and we state the results with only a scant indication of proof, and then launch into a comparison with the strategy Λ^* for large n.

The conclusion of these investigations is that the strategy Λ^* seems by all reasonable standards to be asymptotically best, and that, in the finite case, it is suboptimal in the sense of providing a uniformly good approximation to the optimal results.

Since completing this work we have been allowed to examine the most significant manuscript of L. Dubins and L. J. Savage [1], which will soon be published. Although gambling has been associated with probability since its birth, only quite recently has the question of gambling systems optimal with respect to some goal utility been investigated carefully. To the beautiful and deep results of Dubins and Savage, upon which work was commenced in 1956, must be given priority as the first to formulate systematically and solve the problems of optimal gambling strategies. We strongly recommend their work to every student of probability theory.

Although our original impetus came from a different source, and although their manuscript is almost wholly concerned with unfavorable and fair games, there are a few small areas of overlap which I should like to point out and acknowledge priority. Dubins and Savage did, of course, formulate the concept of a favorable game. For these games they considered the class of "fractionalizing strategies," which consist in betting a fixed fraction of one's fortune at every play, and noticed the interesting phenomenon that there was a critical fraction such that if one bets a fixed fraction less than this critical value, then $S_n \to \infty$ a.s. and if one bets a fixed fraction greater than this critical value, then $S_n \to 0$ a.s. In addition, our proposition 3 is an almost exact duplication of one of their theorems. In their work, also, will be found the solution to maximizing $P\{S_n \geqq x\}$ for an unfavorable game, and it is interesting to observe here the abrupt discontinuity in strategies as the game changes from unfavorable to favorable.

My original curiosity concerning favorable games dates from a paper of J. L. Kelly, Jr. [2] in which there is an intriguing interpretation of information theory from a gambling point of view. Finally, some of the last section, in problem and solution, is closely related to the theory of dynamic programming as originated by R. Bellman [3].

2. The nature of Λ^*

We introduce some notation. Let the outcome of the kth game be X_k and $R_n = (X_n, \cdots, X_1)$. Take the initial fortune S_0 to be unity, and S_n the fortune after n games. To specify a strategy Λ we specify for every n, the fractions $[\lambda_1^{(n+1)}, \cdots, \lambda_r^{(n+1)}] = \bar{\lambda}_{n+1}$ of our available fortune after the nth game, S_n, that we will bet on alternative $A_1, \cdots, A_r$ in the $(n+1)$st game. Hence

$$\sum_{j=1}^{r} \lambda_j^{(n+1)} = 1. \tag{2.1}$$

Note that $\bar{\lambda}_{n+1}$ may depend on R_n. Denote $\Lambda = (\bar{\lambda}_1, \bar{\lambda}_2, \cdots)$. Define the random variables V_n by

$$V_n = \sum_{\{j : i \in A_j\}} \lambda_j^{(n)} o_j, \qquad X_n = i, \tag{2.2}$$

so that $S_{n+1} = V_{n+1} S_n$. Let $W_n = \log V_n$, so we have

$$\log S_n = W_n + \cdots + W_1. \tag{2.3}$$

To define Λ^*, consider the set of vectors $\bar{\lambda} = (\lambda_1, \cdots, \lambda_r)$ with r nonnegative components such that $\lambda_1 + \cdots + \lambda_r = 1$ and define a function $W(\bar{\lambda})$ on this space $\mathfrak{F}$ by

$$W(\bar{\lambda}) = \sum_i p_i \log \Big(\sum_{i \in A_j} \lambda_j o_j \Big). \tag{2.4}$$

The function $W(\bar{\lambda})$ achieves its maximum on $\mathfrak{F}$ and we denote $W = \max_{\bar{\lambda} \in \mathfrak{F}} W(\bar{\lambda})$.

Proposition 1. *Let $\bar{\lambda}^{(1)}, \bar{\lambda}^{(2)}$ be in $\mathfrak{F}$ such that $W = W(\bar{\lambda}^{(1)}) = W(\bar{\lambda}^{(2)})$, then for all i, we have $\sum_{i \in A_j} \lambda_j^{(1)} o_j = \sum_{i \in A_j} \lambda_j^{(2)} o_j$.*

Proof. Let α, β be positive numbers such that $\alpha + \beta = 1$. Then if $\bar{\lambda} = \alpha \bar{\lambda}^{(1)} + \beta \bar{\lambda}^{(2)}$, we have $W(\bar{\lambda}) \leqq W$. But by the strict concavity of log

$$W(\bar{\lambda}) \geqq \alpha W(\bar{\lambda}^{(1)}) + \beta W(\bar{\lambda}^{(2)}) \tag{2.5}$$

with equality if and only if the conclusion of the proposition holds.

Now let $\bar{\lambda}^*$ be such that $W = W(\bar{\lambda}^*)$ and define Λ^* as $(\bar{\lambda}^*, \bar{\lambda}^*, \cdots)$. Although $\bar{\lambda}^*$ may not be unique, the random variables $W_1^*, W_2^*, \cdots$ arising from Λ^* are by proposition 1 uniquely defined, and form a sequence of independent, identically distributed random variables.

Questions of uniqueness and description of $\bar{\lambda}^*$ are complicated in the general case. But some insight into the type of strategy we get using Λ^* is afforded by

PROPOSITION 2. *Let the sets $A_1, \cdots, A_r$ be disjoint, then no matter what the odds o_j are, $\bar{\lambda}^*$ is given by $\lambda_j^* = P\{X \in A_j\}$.*

The proof is a simple computation and is omitted.

From now on we restrict attention to favorable games and give the following criterion.

PROPOSITION 3. *A game is favorable if and only if $W > 0$.*

PROOF. We have

$$\log S_n^* = \sum_1^n W_k^*. \tag{2.6}$$

If $W = EW_k^*$ is positive, then the strong law of large numbers yields $S_n^* \to \infty$ a.s. Conversely, if there is a strategy Λ such that $S_n \to \infty$ a.s. we use the result of section 4, which says that for any strategy Λ, $\lim_n S_n/S_n^*$ exists a.s. finite. Hence $S_n^* \to \infty$ a.s. and therefore $W \geqq 0$. Suppose $W = 0$, then the law of the iterated logarithm comes to our rescue and provides a contradiction to $S_n^* \to \infty$.

3. The asymptotic time minimization problem

For any strategy Λ and any number $x > 1$, define the random variable $T(x)$ by

$$T(x) = \{\text{smallest } n \text{ such that } S_n \geqq x\}, \tag{3.1}$$

and $T^*(x)$ the corresponding random variable using the strategy Λ^*. That is, $T(x)$ is the number of plays needed under Λ to amass or exceed the fortune x. This section is concerned with the proof of the following theorem.

THEOREM 1. *If the random variables $W_1^*, W_2^*, \cdots$ are nonlattice,*† *then for any strategy*

$$\lim_{x \to \infty} [ET(x) - ET^*(x)] = \frac{1}{W} \sum_1^\infty (W - EW_n) \tag{3.2}$$

and there is a constant α, independent of Λ and x such that

$$ET^*(x) - ET(x) \leqq \alpha. \tag{3.3}$$

Notice that the right side of (3.2) is always nonnegative and is zero only if Λ is equivalent to Λ^* in the sense that for every n, we have $W_n = W_n^*$. The reason for the restriction that W_n^* be nonlattice is fairly apparent. But as this restriction is on $\log V_n^*$ rather than on V_n^* itself, the common games with rational values of the odds o_j and probabilities p_i usually will be nonlattice. For instance, a little number-theoretic juggling proves that in the coin-tossing case the countable set of values of p for which W_n^* is lattice consists only of irrationals.*

The proof of the above theorem is long and will be carried out in a sequence of propositions. The heart is an asymptotic estimate of the excess in Wald's identity [4].

PROPOSITION 4. *Let $X_1, X_2, \cdots$ be a sequence of identically distributed, independent nonlattice random variables with $0 < EX_1 < \infty$. Let $Y_n = X_1 + \cdots + X_n$. For any real numbers x, ξ, with $\xi > 0$, let $F_x(\xi) = P\{\text{first } Y_n \geqq x \text{ is } < x + \xi\}$. Then there is a continuous distribution $G(\xi)$ such that for every value of ξ,*

$$\lim_{x\to\infty} F_x(\xi) = G(\xi). \tag{3.4}$$

PROOF. The above statement is contained in known results concerning the renewal theorem. If $X_1 > 0$ a.s. and has the distribution function F, it is known (see, for example, [5]) that $\lim_{x\to\infty} F_x(\xi) = (1/EX_1) \int_0^\xi [1 - F(t)]\,dt$. If X_1 is not positive, we use a device due to Blackwell [6]. Define the integer-valued random variables $n_1 < n_2 < \cdots$ by $n_1 = \{\text{first } n \text{ such that } X_1 + \cdots + X_n > 0\}$, $n_2 = \{\text{first } n \text{ such that } X_{n_1+1} + \cdots + X_n > 0\}$, and so forth. Then the random variables $X_1' = X_1 + \cdots + X_{n_1}$, $X_2' = X_{n_1+1} + \cdots + X_{n_2}$, $\cdots$ are independent, identically distributed, positive, and $EX_1' < \infty$ (see [6]). Letting $Y_n' = X_1' + \cdots + X_n'$, note that $P\{\text{first } Y_n \geqq x \text{ is } < x + \xi\} = P\{\text{first } Y_n' \geqq x \text{ is } < x + \xi\}$, which completes the proof.

We find it useful to transform this problem by defining for any strategy Λ, a random variable $N(y)$,

$$N(y) = \{\text{smallest } n \text{ such that } W_n + \cdots + W_1 \geqq y\} \tag{3.5}$$

with $N^*(y)$ the analogous thing for Λ^*. To prove (3.2) we need to prove

$$\lim_{y\to\infty} [EN(y) - EN^*(y)] = \frac{1}{W} \sum_1^\infty (W - EW_n), \tag{3.6}$$

and we use a result very close to Wald's identity.

PROPOSITION 5. *For any strategy Λ such that $S_n \to \infty$ a.s. and any y*

$$EN(y) = \frac{1}{W} E\left\{\sum_{k=1}^{N(y)} [W - E(W_k|R_{k-1})]\right\} + \frac{1}{W} E\left[\sum_{k=1}^{N(y)} W_k\right]. \tag{3.7}$$

PROOF. The above identity is derived in a very similar fashion to Doob's derivation [6] of Wald's identity. The difficult point is an integrability condition and we get around this by using, instead of the strategy Λ, a modification Λ_J which consists in using Λ for the first J plays and then switching to $\bar{\lambda}^*$. The condition $S_n \to \infty$ a.s. implies that none of the W_k may take on the value $-\infty$ and that $N(y)$ is well defined. Let $N_J(y)$ be the random variable analogous to $N(y)$ under Λ_J and $W_k^{(J)}$ to W_k. Define a sequence of random variables Z_n by

$$Z_n = \sum_1^n [W_k^{(J)} - E(W_k^{(J)}|R_{k-1})]. \tag{3.8}$$

This sequence is a martingale with $EZ_n = 0$. By Wald's identity, $EN_J(y) < \infty$ and it is seen that the conditions of the optional sampling theorem ([7], theorem 2.2–C_3) are validated with the conclusion that $EZ_{N_J} = 0$. Therefore

$$
\begin{aligned}
(3.9) \qquad WEN_J &= E\left[\sum_{k=1}^{N_J} W\right] \\
&= E\left\{\sum_{k=1}^{N_J} [W - E(W_k^{(J)}|R_{k-1})]\right\} + E\left[\sum_{k=1}^{N_J} W_k^{(J)}\right] \\
&= E\left\{\sum_{k=1}^{\min(N,J)} [W - E(W_k|R_{k-1})]\right\} + E\left[\sum_{k=1}^{N_J} W_k^{(J)}\right].
\end{aligned}
$$

The second term on the right satisfies

$$
(3.10) \qquad y \leqq E\left[\sum_{k=1}^{N_J} W_k^{(J)}\right] \leqq y + \alpha,
$$

where $\alpha = \max_j (\log o_j)$. Hence, if $EN = \infty$, then $\lim_J EN_J = \infty$, so that $E\{\sum_1^{N_J} [W - E(W_k|R_{k-1})]\} = \infty$ and (3.7) is degenerately true. Now assume that $EN < \infty$, and let $J \to \infty$. The first term on the right in (3.9) converges to $E\{\sum_1^N [W - E(W_k|R_{k-1})]\}$ monotonically. The random variables $\sum_1^{N_J} W_k^{(J)}$ converge a.s. to $\sum_1^N W_k$ and are bounded below and above by y and $y + \alpha$ so that the expectations converge. It remains to show that $\lim_J EN_J = EN$. Since

$$
(3.11) \qquad EN_J = \int_{\{N \leqq J\}} N\, dP + \int_{\{N > J\}} N_J\, dP,
$$

we need to show that the extreme right term converges to zero. Let

$$
(3.12) \qquad U_J = \sum_1^J W_k,\ N(U_J) = \{\text{first } n \text{ such that} \sum_{J+1}^{n+J} W_k^{(J)} \geqq y - U_J\}
$$

so that

$$
(3.13) \qquad \int_{\{N > J\}} N_J\, dP = JP(N > J) + \int_{\{N > J\}} N(U_J)\, dP.
$$

Since $EN < \infty$, we have $\lim_J JP\{N > J\} = 0$. We write the second term as $E\{E[N(U_J)|U_J]|N > J\}\, P\{N > J\}$. By Wald's identity,

$$
(3.14) \qquad E[N(U_J)|U_J] \leqq \frac{y - U_J + \alpha}{W}.
$$

On the other hand, since the most we can win at any play is α, the inequality

$$
(3.15) \qquad N \geqq \frac{y - U_J}{\alpha} + J
$$

holds on the set $\{N > J\}$. Putting together the pieces,

$$
(3.16) \qquad \int_{\{N > J\}} N(U_J)\, dP \leqq \frac{\alpha}{W} \int_{\{N > J\}} (N - J)\, dP + \frac{\alpha}{W} P(N > J).
$$

The right side converges to zero and the proposition is proven.

If we subtract from (3.7) the analogous result for Λ^* we get

$$
\begin{aligned}
(3.17) \quad & EN(y) - EN^*(y) \\
&= \frac{1}{W} E\left\{\sum_{k=1}^{N} [W - E(W_k|R_{k-1})]\right\} + \frac{1}{W} E\left[\sum_{k=1}^{N} W_k - \sum_{k=1}^{N^*} W_k^*\right].
\end{aligned}
$$

This last result establishes inequality (3.3) of the theorem. As we let $y \to \infty$, then $N(y) \to \infty$ a.s. and we see that

$$\lim_{y\to\infty} E\left\{\sum_{k=1}^{N} [W - E(W_k|R_{k-1})]\right\} = \sum_{k=1}^{\infty} (W - EW_k). \tag{3.18}$$

By proposition 4, the distribution F_y^* of $\sum_1^{N^*} W_k^* - y$ converges, as $y \to \infty$, to some continuous distribution F^* and we finish by proving that the distribution F_y of $\sum_1^N W_k - y$ also converges to F^*.

PROPOSITION 6. *Let* Y_n, ϵ_n *be two sequences of random variables such that* $Y_n \to \infty$, $Y_n + \epsilon_n \to \infty$ *a.s. If* Z is *any random variable, if* $\epsilon = \sup_{n\geqq 1} |\epsilon_n|$, *and if we define*

$$H_y(\xi) = P\{\textit{first } Y_n \geqq Z + y \textit{ is } < Z + y + \xi\}, \tag{3.19}$$

$$D_y(\xi) = P\{\textit{first } Y_n + \epsilon_n \geqq Z + y \textit{ is } < Z + y + \xi\}, \tag{3.20}$$

then for any $u > 0$,

(3.21)

$$F_{x+u}(\xi - 2u) - P\{\epsilon \geqq u\} \leqq D_y(\xi) \leqq D_y(\xi) \leqq H_{y-u}(\xi + 2u) + P\{\epsilon \geqq u\}.$$

PROOF.

(3.22)

$$\begin{aligned} D_y(\xi) &\leqq P\{\text{first } Y_n + \epsilon_n \geqq Z + y \text{ is } < Z + y + \xi, \epsilon < u\} + P\{\epsilon \geqq u\} \\ &\leqq P\{\text{first } Y_n > Z + y - u \text{ is } < Z + y + \xi + u, \epsilon < u\} + P\{\epsilon \geqq u\} \\ &\leqq H_{y-u}(\xi + 2u) + P\{\epsilon \geqq u\}. \end{aligned}$$

(3.23)

$$\begin{aligned} D_y(\xi) &\geqq P\{\text{first } Y_n + \epsilon_n \geqq Z + y \text{ is } < Z + y + \xi, \epsilon < u\} \\ &\geqq P\{\text{first } Y_n \geqq Z + y + u \text{ is } < Z + y + \xi - u, \epsilon < u\} \\ &\geqq H_{y+u}(\xi - 2u) - P\{\epsilon \geqq u\}. \end{aligned}$$

PROPOSITION 7. *Let* $X_1, X_2, \cdots$ *be a sequence of independent identically distributed nonlattice random variables,* $0 < EX_1 < \infty$, *with* $Y_n = X_1 + \cdots + X_n$. *If* Z *is any random variable independent of* $X_1, X_2, \cdots$, G *the limiting distribution of proposition* 4, *and*

$$F_{y,Z}(\xi) = P\{\textit{first } Y_n \geqq Z + y \textit{ is } < Z + y + \xi\}, \tag{3.24}$$

then $\lim_y F_{y,Z}(\xi) = G(\xi)$.

PROOF.

$$\begin{aligned} F_{y,Z}(\xi) &= E[P\{\text{first } Y_n \geqq Z + y \text{ is } < Z + y + \xi|Z\}] \\ &= E[F_{y+Z}(\xi)], \end{aligned} \tag{3.25}$$

where $F_y(\xi) = P\{\text{first } Y_n \geqq y \text{ is } < y + \xi\}$. But $\lim_y F_{y+Z}(\xi) = G(\xi)$ a.s. which, together with the boundedness of $F_{y+Z}(\xi)$, establishes the result.

We start putting things together with

PROPOSITION 8. *Let* $\sum_1^n W_k - \sum_1^n W_k^*$ *converge a.s. to an everywhere finite limit. If the* W_k^* *are nonlattice, if* $F_y(\xi)$ *is the distribution function for* $\sum_1^N W_k - y$, *then* $\lim_y F_y(\xi) = F^*(\xi)$.

PROOF. Fix m, let

$$(3.26)\qquad Z_m = -\sum_1^{m-1} W_k, \qquad \epsilon_{m,n} = \sum_m^n W_k - \sum_m^n W_k^*, \qquad \epsilon_m = \sup_n |\epsilon_{m,n}|,$$

and by assumption $\epsilon_m \to 0$ a.s. Now

$$(3.27)\qquad F_y(\xi) = P\{\text{first } \sum_1^n W_k \geqq y \text{ is } < y + \xi\}$$

$$= P\{\text{first } \left(\sum_m^n W_k^* + \epsilon_{m,n}\right) \geqq Z_m + y \text{ is } < Z_m + y + \xi\}.$$

If

$$(3.28)\qquad H_y(\xi) = P\{\text{first } \sum_m^n W_k^* \geqq Z_m + y \text{ is } < Z_m + y + \xi\},$$

then by proposition 6, for any $u > 0$,

$$(3.29)\qquad H_{y+u}(\xi - 2u) - P\{\epsilon_m \geqq u\} \leqq F_y(\xi) \leqq H_{y-u}(\xi + 2u) + P\{\epsilon_m \geqq u\}.$$

Letting $y \to \infty$ and applying proposition 7,

(3.30)

$$F^*(\xi - 2u) - P\{\epsilon_m \geqq u\} \leqq \varliminf_y F_y(\xi) \leqq \varlimsup_y F_y(\xi) \leqq F^*(\xi + 2u) + P\{\epsilon_m \geqq u\}.$$

Taking first $m \to \infty$ and then $u \to 0$ we get

$$(3.31)\qquad \varliminf_y F_y(\xi) = \varlimsup_y F_y(\xi) = F^*(\xi).$$

To finish the proof, we invoke theorems 2 and 3 of section 4. The content we use is that if $\sum_1^n W_k - \sum_1^n W_k^*$ does not converge a.s. to an everywhere finite limit, then $\sum_1^\infty [W - E(W_k|R_{k-1})] = +\infty$ on a set of positive probability. Therefore, if the conditions of propositions 5 and 8 are not validated, then by (3.17) both sides of (3.2) are infinite. Thus the theorem is proved.

3. Asymptotic magnitude problem

The main results of this section can be stated roughly as: asymptotically, S_n^* is as large as the S_n provided by any strategy Λ, and if Λ is not asymptotically close to Λ^*, then S_n^* is infinitely larger than S_n. The results are valid whether or not the games are favorable.

THEOREM 2. *Let* Λ *be any strategy leading to the fortune* S_n *after* n *plays. Then* $\lim_n S_n/S_n^*$ *exists a.s. and* $E(\lim_n S_n/S_n^*) \leqq 1$.

For the statement of theorem 3 we need

DEFINITION. Λ *is a nonterminating strategy if there are no values of* $\bar{\lambda}_n$ *such that* $\sum_{i \in A_l} \lambda_j^{(n)} o_j = 0$, *for any* n.

THEOREM 3. *If Λ is a nonterminating strategy, then almost surely*

$$\sum_1^\infty [W - E(W_k|R_{k-1})] = \infty \Leftrightarrow \lim_n \frac{S_n^*}{S_n} = \infty. \tag{4.1}$$

PROOFS. We present the theorems together as their proofs are similar and hinge on the martingale theorems. For every n

$$E\left(\frac{S_n}{S_n^*}\middle|R_{n-1}\right) = E\left(\frac{V_n}{V_n^*}\middle|R_{n-1}\right)\frac{S_{n-1}}{S_{n-1}^*}. \tag{4.2}$$

If we prove that $E(V_n/V_n^*|R_{n-1}) \leqq 1$ a.s., then S_n/S_n^* is a decreasing semi-martingale with $\lim_n S_n/S_n^*$ existing a.s. and

$$E \lim_n \frac{S_n}{S_n^*} \leqq E\frac{S_0}{S_0^*} = 1. \tag{4.3}$$

By the definition of Λ^*, for every $\epsilon > 0$,

$$E\{\log[(1-\epsilon)V_n^* + \epsilon V_n] - \log V_n^*|R_{n-1}\} \leqq 0. \tag{4.4}$$

Manipulating gives

$$\frac{1}{\epsilon}E\left[\log\left(1 + \frac{\epsilon}{1-\epsilon}\frac{V_n}{V_n^*}\right)\middle|R_{n-1}\right] \leqq \frac{1}{\epsilon}\log\frac{1}{1-\epsilon}. \tag{4.5}$$

By Fatou's lemma, as $\epsilon \to 0$

$$E\left(\frac{V_n}{V_n^*}\middle|R_{n-1}\right) = E\left[\underline{\lim}\left(\frac{1}{\epsilon}\log 1 + \frac{\epsilon}{1-\epsilon}\frac{V_n}{V_n^*}\right)\middle|R_{n-1}\right] \tag{4.6}$$

$$\leqq \underline{\lim}\frac{1}{\epsilon}\log\frac{1}{1-\epsilon} = 1.$$

Theorem 3 resembles a martingale theorem given by Doob ([6], pp. 323–324), but integrability conditions get in our way and force some deviousness. Fix a number $M > 0$ and take A to be the event $\{W - E(W_n|R_{n-1}) \geqq M \text{ i.o.}\}$. If $p = \min_i p_i$, then $E(W_n^* - W_n|R_{n-1}) \geqq M$ implies $P\{W_n^* - W_n \geqq M|R_{n-1}\} \geqq p$. By the conditional version of the Borel-Cantelli lemma ([7], p. 324), the set on which $\sum_1^\infty P\{W_n^* - W_n \geqq M|R_{n-1}\} = \infty$ and the set $\{W_n^* - W_n \geqq M \text{ i.o.}\}$ are a.s. the same. Therefore, a.s. on A, we have $W_n^* - W_n \geqq M$ i.o. and $\log(S_n^*/S_n) = \sum_1^n (W_k^* - W_k)$ cannot converge. We conclude that both sides of (4.1) diverge a.s. on A.

Starting with a strategy Λ, define an amended strategy Λ_M by: if $W - E(W_n|R_{n-1}) < M$, use Λ on the nth play, otherwise use Λ^* on the nth play. The random variables

$$U_n = \log\frac{S_n^*}{S_n^{(M)}} - \sum_1^n [W - E(W_k^{(M)}|R_{k-1})] \tag{4.7}$$

form a martingale sequence with

$$U_n - U_{n-1} = W_n^* - W_n^{(M)} - [W - E(W_n^{(M)}|R_{n-1})]. \tag{4.8}$$

For Λ_M, we have $E(W_n^* - W_n^{(M)}|R_{n-1}) < M$, leading to the inequalities,

$$\sup (W_n^* - W_n^{(M)}) \leqq \frac{M}{p}, \qquad U_n - U_{n-1} \leqq \frac{M}{p}. \tag{4.9}$$

On the other side, if

$$\alpha = \min_j \log \left(\sum_{i \in A_j} \lambda_j^* o_j \right), \qquad \beta = \max_j \log o_j, \tag{4.10}$$

then $U_n - U_{n-1} \geqq \alpha - \beta - M$. These bounds allow the use of a known martingale theorem ([7], pp. 319–320) to conclude that $\lim_n U_n$ exists a.s. whenever one of $\overline{\lim}\, U_n < \infty$, $\underline{\lim}\, U_n > -\infty$ is satisfied. This implies the statement

$$\lim \frac{S_n^*}{S_n^{(M)}} < \infty \Leftrightarrow \sum_1^\infty [W - E(W_k^{(M)}|R_{k-1})] < \infty. \tag{4.11}$$

However, on the complement of the set A the convergence or divergence of the above expressions involves the convergence or divergence of the corresponding quantities in (4.1) which proves the theorem.

COROLLARY 1. *If for some strategy* Λ, *we have* $\sum_1^\infty [W - E(W_k|R_{k-1})] = \infty$ *with probability* $\gamma > 0$, *then for every* $\epsilon > 0$, *there is a strategy* $\hat{\Lambda}$ *such that with probability at least* $\gamma - \epsilon$, $\overline{\lim}\, S_n/\hat{S}_n = 0$ *and except for a set of probability at most* ϵ, $\overline{\lim}\, S_n/\hat{S}_n \leqq 1$.

PROOF. Let E be the set on which $\lim S_n/S_n^* = 0$, with $P\{E\} = \gamma$. For any $\epsilon > 0$, for N sufficiently large, there is a set E_N, measurable with respect to the field generated by R_N such that $P\{E_N \Delta E\} < \epsilon$, where Δ denotes the symmetric set difference. Define $\hat{\Lambda}$ as follows: if $n < N$, use Λ, if R_n, with $n \geqq N$, is such that the first N outcomes $(X_1, \cdots, X_N)$ is not in E_N, use Λ, otherwise use Λ^*. On E_N, we have $\sum_1^\infty [W - E(\hat{W}_k|R_{k-1})] < \infty$, hence $\lim \hat{S}_n/S_n^* > 0$ so that $\lim S_n/\hat{S}_n = 0$ on $E_N \cap E$. Further, $P\{E_N \cap E\} \geqq P\{E\} - \epsilon = \gamma - \epsilon$. On the complement of E_N, we have $S_n = \hat{S}_n$, leading to $\overline{\lim}\, S_n/\hat{S}_n \leqq 1$, except for a set with probability at most ϵ.

5. Problems with finite goals in coin tossing

In this section we consider first the problem: fix an integer $n > 0$, and two numbers $y > x > 0$, find a strategy which maximizes $P\{S_n \geqq y|S_0 = x\}$. In this situation, then, only n plays of the game are allowed and we wish to maximize the probability of exceeding a certain return. We will also be interested in what happens as n, y become large. By changing the unit of money, note that

$$\sup P\{S_n \geqq y|S_0 = x\} = \sup P\left\{S_n \geqq 1|S_0 = \frac{x}{y}\right\}, \tag{5.1}$$

where the supremum is over all strategies. Thus, the problem reduces to the unit interval, and we may evidently translate back to the general case if we find an optimum strategy in the reduced case. Define, for $\xi \geqq 0$, $n \geqq 1$,

$$(5.2)\qquad \phi_n(\xi) = \begin{cases} \sup P\{S_n \geqq 1 | S_0 = \xi\}, & \xi < 1, \\ 1, & \xi \geqq 1, \end{cases}$$

and

$$(5.3)\qquad \phi_0(\xi) = \begin{cases} 0, & \xi < 1, \\ 1, & \xi \geqq 1. \end{cases}$$

In addition $\phi_n(\xi)$ satisfies

$$(5.4)\qquad \begin{aligned} \phi_n(\xi) &= \sup E[P\{S_n \geqq 1 | S_1, S_0\} | S_0 = \xi] \\ &\leqq \sup E[\phi_{n-1}(S_1) | S_0 = \xi] \\ &\leqq \sup_{0 \leqq z \leqq \xi} [p\phi_{n-1}(\xi + z) + q\phi_{n-1}(\xi - z)]. \end{aligned}$$

To find $\phi_n(\xi)$ and an optimal strategy, we define functions $\hat{\phi}_n(\xi)$ by

$$(5.5)\qquad \hat{\phi}_0(\xi) = \phi_0(\xi), \qquad \hat{\phi}_n(\xi) = \sup_{0 \leqq z \leqq \xi} [p\hat{\phi}_{n-1}(\xi + z) + q\hat{\phi}_{n-1}(\xi - z)]$$

having the property $\phi_n(\xi) \leqq \hat{\phi}_n(\xi)$, for all n, ξ. If we can find a strategy Λ such that under Λ we have $\hat{\phi}_n(\xi) = P\{S_n \geqq 1 | S_0 = \xi\}$, then, evidently, Λ is optimum, and $\hat{\phi}_n = \phi_n$. But, if for every $n \geqq 1$, and ξ there is a $z_n(\xi)$, with $0 \leqq z_n(\xi) \leqq \xi$, such that

$$(5.6)\qquad \hat{\phi}_n(\xi) = p\hat{\phi}_{n-1}[\xi + z_n(\xi)] + q\hat{\phi}_{n-1}[\xi - z_n(\xi)],$$

then we assert that the optimum strategy is Λ defined as: if there are m plays left and we have fortune ξ, bet the amount $z_m(\xi)$. Because, suppose that under Λ, for $n = 0, 1, \cdots, m$ we have $\hat{\phi}_n(\xi) = P\{S_n \geqq 1 | S_0 = \xi\}$, then

$$(5.7)\qquad \begin{aligned} P\{S_{m+1} \geqq 1 | S_0 = \xi\} &= E[P\{S_{m+1} \geqq 1 | S_1, S_0\} | S_0 = \xi] \\ &= E[\hat{\phi}_m(S_1) | S_0 = \xi] = \hat{\phi}_{m+1}(\xi). \end{aligned}$$

Hence, we need only solve recursively the functional equation (5.5) and then look for solutions of (5.6) in order to find an optimal strategy. We will not go through the complicated but straightforward computation of $\hat{\phi}_n(\xi)$. It can be described by dividing the unit interval into 2^n equal intervals $I_1, \cdots, I_{2n}$ such that $I_k = [k/2^n, (k + 1)/2^n]$. In tossing a coin with $P\{H\} = p$, rank the probabilities of the 2^n outcomes of n tosses in descending order $P_1 \geqq P_2 \geqq \cdots \geqq P_{2^n}$, that is, $P_1 = p^n$, $p_{2^n} = q^n$. Then, as shown in figure 1,

$$(5.8)\qquad \phi_n(\xi) = \sum_{j<k} P_j, \qquad \xi \in I_k,$$

Note that if $p > 1/2$, then $\lim_n \phi_n(\xi) = 1$, with $\xi > 0$; and in the limiting case $p = 1/2$, then $\lim_n \phi_n(\xi) = \xi$, with $\xi \leqq 1$, in agreement with the Dubins-Savage result [2].

There are many different optimum strategies, and we describe the one which seems simplest. Divide the unit interval into $n + 1$ subintervals $I_0^{(n)}, \cdots, I_n^{(n)}$, such that the length of $I_k^{(n)}$ is $2^{-n}\binom{n}{k}$ where the $\binom{n}{k}$ are binomial coefficients. On

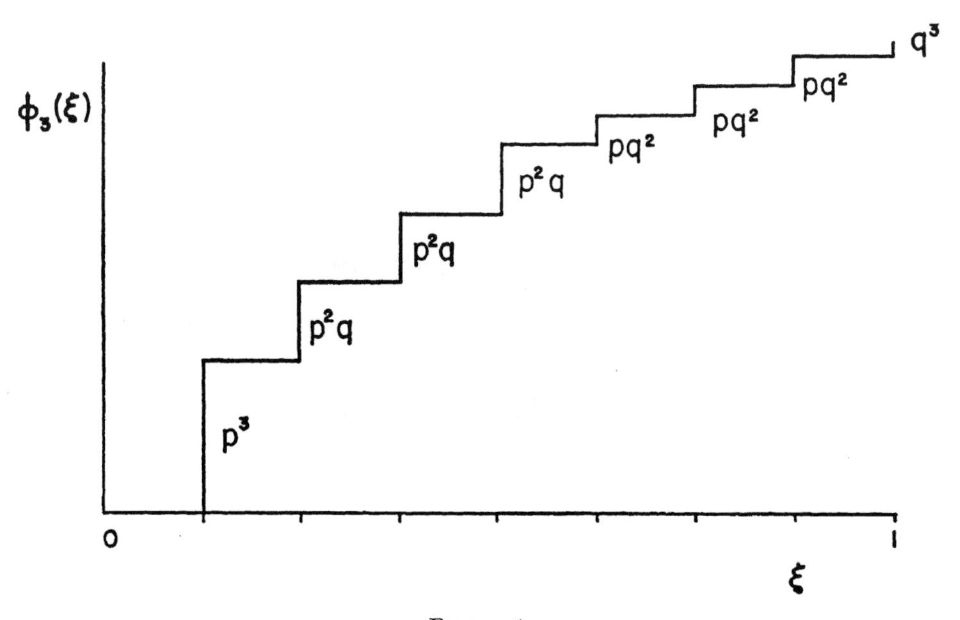

FIGURE 1

Graph of $\phi_n(\xi)$ for the case $n = 3$.

each $I_k^{(n)}$ as base, erect a 45°–45° isosceles triangle. Then the graph of $z_{n+1}(\xi)$ is formed by the sides of these triangles, as shown in figure 2. Roughly, this strategy calls for a preliminary "jockeying for position," with the preferred positions with m plays remaining being the midpoints of the intervals $I_k^{(m)}$. Notice that the endpoints of the intervals $\{I_k^{(n)}\}$ form the midpoints of the intervals $\{I_k^{(n-1)}\}$. So that if with n plays remaining we are at a midpoint of $\{I_k^{(n)}\}$, then at all remaining plays we will be at midpoints of the appropriate system of intervals. Very interestingly, this strategy is independent of the values of p so

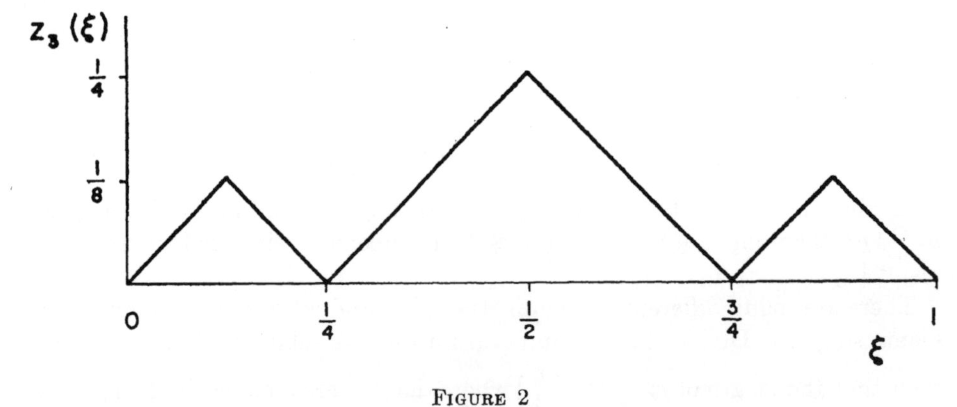

FIGURE 2

Graph of $z_{n+1}(\xi)$ for the case $n = 3$.

long as $p > 1/2$. The strategy Λ^* in this case is: bet a fraction $p - q$ of our fortune at every play. Let $\phi_n^*(\xi) = P\{S_n^* \geqq 1 | S_0 = \xi\}$. In light of the above remark, the following result is not without gratification.

THEOREM 4. $\lim_n \sup_\xi [\phi_n(\xi) - \phi_n^*(\xi)] = 0$.

PROOF. The proof is somewhat tedious, using the central limit theorem and tail estimates. However, some interesting properties of $\phi_n(\xi)$ will be discovered along the way. Let $P\{k|1/2\}$ be the probability of k or fewer tails in tossing a fair coin n times, $P\{k|p\}$ the probability of k or fewer tails in n tosses of a coin with $P\{H\} = p$. If $\xi = P\{k|1/2\} + 2^{-n}$, note that $\phi_n(\xi -) = P\{k|p\}$. Let $\sigma = \sqrt{pq}$, by the central limit theorem, if $\xi_{t,n} = P\{qn + t\sigma\sqrt{n}|1/2\} + 2^{-n}$, then

$$\lim_n \phi_n(\xi_{t,n} -) = \frac{1}{\sqrt{2\pi}} \int_{-\infty}^{t} e^{-x^2/2}\, dx, \tag{5.9}$$

uniformly in t. Thus, if we establish that

$$\lim_n \phi_n^*(\xi_{t,n}) = \frac{1}{\sqrt{2\pi}} \int_{-\infty}^{t} e^{-x^2/2}\, dx \tag{5.10}$$

uniformly for t in any bounded interval, then by the monotonicity of $\phi_n(\xi)$, $\phi_n^*(\xi)$, the theorem will follow.

By definition,

$$\begin{aligned} \phi_n^*(\xi) &= P\{W_1^* + \cdots + W_n^* \geqq 0 | W_0 \\ &= \log \xi\} = P\{W_1^* + \cdots + W_n^* \geqq -\log \xi\}, \end{aligned} \tag{5.11}$$

where the W_k^* are independent, and identically distributed with probabilities $P\{W_k^* = \log 2p\} = p$ and $P\{W_k^* = \log 2q\} = q$. Again using the central limit theorem, the problem reduces to showing that

$$\lim_n \frac{\log \xi_{n,t} + nEW_1^*}{\sqrt{n}\, \sigma(W_1^*)} = t \tag{5.12}$$

uniformly in any bounded interval. By a theorem on tail estimates [8], if $X_1, X_2, \cdots$ are independent random variables with $P\{X_k = 1\} = 1/2$ and $P\{X_k = 0\} = 1/2$, then

$$\log P\{X_1 + \cdots + X_n \geqq na\} = n\theta(a) + \mu(n, a) \log n, \tag{5.13}$$

where $\mu(n, a)$ is bounded for all n, with $1/2 + \delta \leqq a \leqq 1 - \delta$, and $\theta(a) = -a \log (2a) - (1 - a) \log [2(1 - a)]$. Now

$$\log \xi_{n,t} = \log [P\{X_1 + \cdots + X_n \geqq np - t\sigma\sqrt{n}\} + 2^{-n}] \tag{5.14}$$

so that the appropriate $a = p - t\sigma/\sqrt{n}$ with

$$\theta(a) = \theta(p) - \frac{t\sigma}{\sqrt{n}} \log \frac{q}{p} + O\left(\frac{1}{n}\right). \tag{5.15}$$

Since $\theta(p) > -\log 2$, we may ignore the 2^{-n} term and estimate

$$\log \xi_{n,t} = n\theta(p) - t\sigma\sqrt{n} \log \frac{q}{p} + O(\log n). \tag{5.16}$$

But $\theta(p) = -EW_1^*$, and the left-hand expression in (5.12) becomes

$$\frac{t\sigma \log \frac{p}{q}}{\sigma(W_1^*)} + O\left(\frac{\log n}{\sqrt{n}}\right). \tag{5.17}$$

Now the short computation resulting, $\sigma(W_1^*) = \sigma \log (p/q)$, completes the proof of the theorem.

There is one final problem we wish to discuss. Fix ξ, with $0 < \xi < 1$, and let

$$T(\xi) = E(\text{first } n \text{ with } S_n \geqq 1 | S_0 = \xi), \tag{5.18}$$

find the strategy which provides a minimum value of $T(\xi)$. We have not been able to solve this problem, but we hopefully conjecture that an optimal strategy is: there is a number ξ_0, with $0 < \xi_0 < 1$, such that if our fortune is less than ξ_0, we use Λ^*, and if our fortune is greater than or equal to ξ_0, we bet to 1, that is, we bet an amount such that, upon winning, our fortune would be unity.

REFERENCES

[1] L. Dubins and L. J. Savage, *How to Gamble if You Must* (tentative title), to be published.
[2] J. L. Kelly, Jr., "A new interpretation of information rate," *Bell System Tech. J.*, Vol. 35 (1956), pp. 917–926.
[3] R. Bellman, *Dynamic Programming*, Princeton, Princeton University Press, 1957.
[4] A. Wald, *Sequential Analysis*, New York, Wiley, 1947.
[5] E. B. Dynkin, "Limit theorems for sums of independent random quantities," *Izvestiia Akad. Nauk SSSR*, Vol. 19 (1955), pp. 247–266.
[6] D. Blackwell, "Extension of a renewal theorem," *Pacific J. Math.*, Vol. 3 (1953), pp. 315–320.
[7] J. L. Doob, *Stochastic Processes*, New York, Wiley, 1953.
[8] D. Blackwell and J. L. Hodges, Jr., "The probability in the extreme tail of a convolution," *Ann. Math. Statist.*, Vol. 30 (1959), pp. 1113–1120.

REVIEW OF THE INTERNATI
STATISTICAL INSTITUTI
Volume 37 : 3, 1969

6

OPTIMAL GAMBLING SYSTEMS FOR FAVORABLE GAMES[1]

by

E. O. Thorp

Mathematics Department, University of California at Irvine

INTRODUCTION

In the last decade it was found that the player may have the advantage in some games of chance. We shall see that blackjack, the side bet in Nevada-style Baccarat, roulette, and the wheel of fortune all may offer the player positive expectation. The stock market has many of the features of these games of chance [5]. It offers special situations with expected returns ranging above an annual rate of 25% [23].

Once the particular theory of a game has been used to identify favorable situations, we have the problem of how best to apportion our resources. Paralleling the discoveries of favorable situations in particular games, the outlines of a general mathematical theory for exploiting these opportunities has developed [2, 3, 10, 13].

We first describe the favorable games mentioned above, those being the ones with which the author is most familiar. Then we discuss the general mathematical theory, as it has developed thus far, and its application to these games. Detailed knowledge of particular games is not needed to follow the exposition. Each discussion of a favorable game in Part I motivates a concluding probabilistic summary of that game. These summaries suffice for the discussion in Part II so that a reader who has no interest in a particular game may skip directly to the summary.

References are provided for those who wish to explore particular games in detail. For the present, a favorable game means one in which there is a strategy such that $P(\lim S_n = \infty) > 0$ where S_n is the player's capital after n trials.

PART I. FAVORABLE GAMES

1. BLACKJACK

Blackjack, or twenty-one, is a card game played throughout the world. The casinos in Nevada currently realize an annual net profit of roughly eighty million dollars from the game. Taking a price/earnings ratio of 15 as typical for present day common stocks, the Nevada blackjack operation might be compared to a $ 1.2 billion corporation.

To begin the game a dealer randomly shuffles n decks of cards and players place their bets. (The value of n does not materially affect our discussion. It generally is 1, 2, or 4, and we shall use 1 throughout.) There are a maximum and a minimum allowed bet.

The minimum insures a positive probability of eventual ruin for the player who continues to bet. The maximum protects the casino from large adverse fluctuations and in particular prevents the game from being beaten by a martingale (e.g. doubling up), especially one starting with a massive bet. In fact, without a maximum, a casino

[1] The research for this paper was supported in part by the Air Force Office of Scientific Research through Grant AF-AFOSR 1113-66.
The paper is intended in large part to be an exposition for the general mathematical reader with some probability background, rather than for the expert.

in general be ruined a.s. (almost surely) by a player with
ıyer simply bets enough at each trial so that the casino is
l. A practical way for the player to have infinite resources
extend unlimited credit for the finite time it might be needed.
lealt after they have placed their bets. Each player then uses
ıtegy for improving his hand. Finally, the dealer plays out his hand according to a fixed strategy which does not allow skill, and bets are settled. In the case where play begins from one complete randomly shuffled deck, an approximate best strategy (i.e. one giving greatest expected return) was first given in 1956 [1].

Though the rules of blackjack vary slightly, the player following [1] typically has the tiny edge of + .10%. (The pessimistic figure of — .62% cited in [1] was erroneous and may have discouraged the authors from further analysis.) These mathematical results were in sharp contrast to the earlier and very different intuitive strategies generally recommended by card experts, and the associated player disadvantage of two or three per cent. We call the best strategy against a complete deck the *basic strategy*. Determined in 1965, it is almost identical with the strategy in [1] and it gives the player an edge of + 0.13% [22].

If the game were always dealt from a complete shuffled deck, we would have repeated independent trials. But for compelling practical reasons, the deck is not generally reshuffled after each round of play. Thus as successive rounds are played from a given deck, we have sampling without replacement and dependent trials. It is necessary to show the players most or all of the cards used on a given round of play before they place their bets for the next round. They can then use this knowledge of which cards have been played both to sharpen their strategy, and to more precisely estimate their edge. (The strategies for various card counting procedures, and their expectations, were determined directly from probability theory with the aid of computers. The results were reverified by independent Monte Carlo calculations.)

For a given card counting procedure and associated strategy, there is a probability distribution F_c describing the player's expectation on the next hand, provided c cards have been counted. As c increases, F_c spreads out. (This is a theorem, whose proof resembles that for the similar theorem in Baccarat, mentioned in [24], page 316). This spread in F_c can be exploited by placing large bets when the expectation is positive and small bets when it is negative. Part II indicates how best to do this.

If the basic strategy is always used, $E(F_c) = +0.13\%$, just as from a complete deck. But if an improved strategy, based on the card count, is used, $E(F_c)$ increases as c increases, approaching values of one to two per cent or more.

Ties, in which no money is won or lost, may be discounted. They occur about one tenth of the time. Most, but not all, of the other outcomes result in the player either winning or losing an amount equal to his original bet.

The conditional means $E(F_c \mid F_{c-k})$; $k = 1, 2, \ldots, c$, of the successive F_c are nondecreasing. The F_c are dependent; in particular when a deck "goes good", it tends to stay good.

Probabilistic summary

To a good first approximation, Blackjack is a coin toss where the probability p of success is selected independently on each trial from a known distribution F (which is a suitably weighted average of the F_c) and announced before each trial.

A more accurate model considers that the p's are dependent in short consecutive groups, corresponding to successive rounds of play from the same deck. Another more accurate observation is that insurance, naturals, doubling down, and pair splitting, each win or lose an amount different from the amount initially bet. We do not consider this more accurate model in part II because the improvement in results is slight and the increase in complexity is considerable.

2. BACCARAT

The terms Baccarat and Chemin de Fer are used, sometimes interchangeably, to refer to several closely related variants of what is essentially one card game. The game is currently popular in England and France, where it is sometimes played for unlimited stakes. It is also played in Nevada. The game-theoretic aspects of Baccarat have been discussed in [11, 14]. The Nevada game is analyzed in [24] which includes results of extensive computer calculations.

The studies of Baccarat show that the available bets generally offer an expectation on the order of —1 %. The use of mixed strategies, to the very limited extent that this is possible in some variants of the game, has but slight effect on the expectation. Despite the resemblances between Baccarat and Blackjack, the favorable situations detected by perfect card counting methods are not sufficient to make the game favorable. Thus Baccarat is not in general a favorable game.

The game as played in Nevada sometimes permits certain side bets. The minimum on the side bets was observed to be $ 5 to $ 20 and the maximum was $ 200. The bets either won nine times the amount bet or lost the amount bet. The game was played with eight well shuffled decks dealt from a dealing box, or shoe. Using the card counting techniques described in [24], the side bets were favorable about 20 % of the time. When they were favorable, the expectations ranged as high as + 100 %. The expectation initially was about — 5 % and as the number c of cards seen increased, the distribution F_c of expectations spread out ([24], page 316) as in Blackjack. In practice the betting methods discussed in part II, in which the bet increased with the expectation, doubled initial capital in twenty hours.

Unlike the Blackjack player, the Baccarat side bettor has no strategic decisions to make so $E(F_c)$ does not vary as c changes. When the expectation of the side bet falls below a certain value, it is best to make a "waiting" bet on one of the main bets. There are either two or four side bets, similar and dependent. How to apportion funds on the side bets is complicated by the fact that there are several of them. These complexities are treated in [24].

Probabilistic summary

When only one side bet is available, the pay-off for a one unit side bet is either + 9 or — 1. If p is the probability of success, we may suppose that p is selected independently from a known distribution F and announced before each trial. When several side bets are available, the situation is more complex. It illustrates the general setting of [3], page 65.

As in Blackjack, a more accurate model considers that the p's are dependent in consecutive groups, corresponding to successive rounds of play from the same ensemble of (eight) decks. It also considers the effect of waiting bets.

The situation here is more complex than in Blackjack. First, it is important to exploit any opportunities of making simultaneous bets on two or more favorable side bet situations. Second, the pay-off is never one to one.

276

3. ROULETTE

Roulette has long been the prototype of unbeatable gambling games. It is normally regarded as a repeated independent trials process which generates at each trial precisely one from a set of random numbers. In Monte Carlo these numbers are 0, 1, 2, . . . , 36. Players may wager on particular conventional subsets of random numbers (e.g. the first dozen, even, {27}, etc.), winning if the number which comes up is a number of the chosen subset. A player may wager on several subsets simultaneously, and each bet is settled without reference to the others. The expectation of each bet is negative (in Nevada generally — 5.26%, except for one worse bet, and in Monte Carlo — 1.35%.) Thus it has been long known that the classical laws of large numbers insure that the player will with probability one fall behind and stay behind, tending to lose in the long run at a rate close to the expectation of his bets.

Despite this, Henri Poincaré and Karl Pearson each examined roulette. Poincaré ([20], pages 69–70, pages 76–77; [21], pages 201–203; [9], pages 61–62) supposes that the uncertainty in initial conditions (e.g. the angular position and velocity of the ball and of the rotor at a given time) leads to a continuous probability density f in the ball's final position. He shows by an argument involving continuity only that if f has sufficient spread, then the finitely many final ball positions are to very high approximation equally likely.

Karl Pearson statistically analyzed certain published roulette data and found very significant patterns. In particular Pearson says, "If Monte Carlo roulette had gone on since the beginning of geological time on this earth, we should not have expected such an occurrence as this fortnight's play to have occurred *once* on the supposition that the game is one of chance." And again, "To sum up, then: Monte Carlo roulette . . . is. . . the most prodigious miracle of the nineteenth century." I've been told that it was later learned that the roulette data was supplied for a newspaper by journalists hired to sit at the wheel and record outcomes. The journalists instead simply made up numbers and submitted them. It was their personal bias that Pearson detected as statistically significant.

It brings to mind David Hume's essay *Of Miracles*: "No testimony is sufficient to establish a miracle, unless the testimony be of such a kind that its falsehood would be more miraculous than the fact that it endeavors to establish. . . . it is nothing strange . . . that men should lie in all ages."

Poincaré assumed a mechanically perfect roulette wheel. However, wheels sometimes have considerable bias due to mechanical imperfections. Some observed instances and their exploitation are discussed in detail in [25].

In Blackjack and Baccarat, we used the following fundamental principle: The payoff random variables, hence the favorability of a game to an optimal player, depend on the information set used to determine the optimal strategy. For instance, if used cards are ignored in Blackjack, then we simply have Bernoulli trials with $p = +0.13\%$. However, as more card counting information is employed, the distribution of p spreads out (has more structure), its expected value increases, and it can be more effectively exploited. The roulette system we now describe illustrates the use of an enlarged information set.

Play at roulette begins when the croupier launches the ball on a circular track which inclines towards the center so the ball will fall into the center when it slows down sufficiently. The center contains a rotor with a circle of congruent numbered pockets rotating in the opposite direction to the ball. The ball eventually slows and falls from its track on the stator, spiralling into the moving rotor and eventually coming to rest in a numbered pocket, the "winning number". Bets may generally be placed until the ball leaves its track. This is crucial for what follows.

A collaborator and I tried to use the mechanical perfection of the wheel – the very perfection needed to eliminate the bias method – to gain positive expectation. Our basic idea was to determine an initial position and velocity for the ball and rotor. We then hoped to predict the final position of the ball in much the same way that a planet's later position around the sun is predicted from initial conditions, hence the nickname "the Newtonian method".

The Newtonian method occurred to me in 1957, and by 1961 the work described here had been completed. Although the wheel of fortune device was mentioned in LIFE Magazine, March 27, 1964, pp. 80–91, we pointedly did not mention the roulette work there. However, we do so in [22], page 181–182. The Newtonian method is also mentioned in the significant book by R. A. Epstein, *The Theory of Gambling and Statistical Logic*, Academic Press, pp. 135–136, (1967).

The stator has metal deflectors placed to scatter the ball when it spirals down and the pockets are separated by vertical dividers ("frets") which also introduce scattering. These scatterings were measured and found to be far from sufficient to frustrate the Newtonian approach. However, there were additional sources of randomness which did frustrate this approach. (We never satisfactorily identified these causes and can only speculate – perhaps the causes included minute imperfections in track or ball or high sensitivity of the coefficient of friction to dirt or atmospheric humidity.)

We were led to a variation we called the quantum method. If a roulette wheel is tilted slightly the ball will not fall from a sector of the track on the "high" side. The effect is strong with a tilt of just 0.2°, which creates a forbidden zone of a quarter to a third of the wheel. The non-linear differential equation governing the ball's motion on the track is the equation for a pendulum which at first swings completely around its pivot, but is gradually slowed by air resistance. (It is illuminating to sketch the orbits of the equation, as indicated in [4], page 402, problem 3.) The experimental orbits of angle versus time could be plotted easily in the laboratory by taking a movie of the system in motion, along with a large electric clock whose hand swept out one revolution per second!

The existence of a forbidden zone partially quantizes the angle at which the ball can exit, and hence quantizes the final angular position of the ball on the rotor. The physics involved suggests that the quantization is in fact very sharp: Suppose the ball is going to exit beyond the low point of the tilted wheel. Then it must have been moving faster than a ball exiting at the low point, so it reaches its destination sooner. But it has also gone farther, and the two effects tend to cancel. They in fact cancel very well. A similar argument shows that balls which exit before the low point have been slower, hence later, offsetting the fact they have not gone as far. Observation verifies the conclusions of this heuristic argument.

The sharp quantization of ball final position, as a function of initial conditions, makes remarkably accurate prediction possible.

Using algorithms, it was possible by eye judgements alone to estimate the ball's final position three or four revolutions before exit (perhaps five to seven seconds before exit, which was ample time in which to bet) well enough to have a + 15% expectation on each of the five most favored numbers. A cigarette pack sized transistorized computer which we designed and built was able to predict up to eight revolutions in advance. The expectation in tests was + 44%.

One third of the Nevada roulette wheels which we observed had the desired tilt of at least 0.2°. The input to the computer consisted of four push-button hits: two when the 0 of the rotor crossed a fiducial mark during successive revolutions and two when the ball crossed a fiducial mark on successive revolutions. The decay constants of ball and rotor, approximately constant over the class of wheels observed, had been determined earlier by simple observations.

The ultimate weakness of the system was that the house could foil it by forbidding bets after the ball had been launched.

Probabilistic summary

Roulette on a slightly tilted wheel is repeated independent trials. At each trial the player may wager on one or more subsets of the finitely many elementary outcomes. A wager on a subset wins if and only if it contains the elementary outcome that occurs. There are subsets with expectations of 44%. Our procedure in practice was to bet on one of eight neighborhoods of five numbers. Thus the payoff for a bet of .2 units on each of five numbers was either —1 or + 6.2. The expectation of + 44% corresponds to a probability of success of .2. We remark that our knowledge of p increases with the sample size.

4. THE WHEEL OF FORTUNE

The wheel of fortune, featured in many Nevada casinos, is a six foot vertical wheel with horizontal equally spaced pegs in its rim. As the wheel spins, a rubber flapper strikes successive pegs, slowing the wheel. There are generally 48 to 54 spaces between the pegs, numbered with 1s, 2s, 5s, 10s, 20s, and two distinct 40s. A player betting a unit on one of these outcomes is paid that number of units if his outcome occurs. The wheel behaves to good approximation as though a constant increment of energy is lost each time a peg passes the flapper. Thus θ, the total angle of rotation, is proportional to the energy E, which equals $I\omega^2$, where I is the moment of inertia and ω is the angular velocity of the wheel.

In practice, a transistor timing device of match box size (a "spinoff" from the roulette technology) produced a faint click a chosen time after a push-button was hit. The button was hit when a specified 40 passed the flapper. The timer was set so the click was approximately when the second 40 reached the flapper. If it clicked after the second 40 reached the flapper, the wheel was "fast" and would go farther than average before stopping. If it clicked before the second 40 reached the flapper, the wheel was "slow".

For a given timer setting, a table was constructed empirically, giving the approximate final position of the wheel as a function of the number of spaces the second 40 was fast or slow when the click was heard.

In practice one could determine with certainty which of the two 40s could not occur. Thus, one could always bet on the "right" 40. On a wheel observed in the Riviera Hotel there were 50 numbers, including 22 ones, 14 twos, 7 fives, 3 tens, 2 twenties and 2 forties. Betting on the "right" 40 would win on average 80 units in 50 trials and lose 48, for an expectation of 32/50 or 64%.

Probabilistic summary

Ignoring obvious refinements, we have repeated independent trials with probability $p = 1/25$ of success at each trial, a payoff for a 1 unit bet of — 1 or + 40, and an expectation of + 64%.

5. THE STOCK MARKET

The stock market is a natural economic object for mathematical analysis because vast quantities of precise historical data are available in numerical form. There have been many attempts to mathematically predict future price behavior, using as a basis various subsets of the available information. Most notable are the attempts to predict future prices from past price behavior. These attempts have caused the view to be widespread in academic circles that, to first order, common stock prices are a random walk and changes in common stock prices are log normally distributed with a certain mean and standard deviation [5].

Practitioners hotly contest this view. Part of the dispute is caused by practitioners who are unwilling or unable to test their claims scientifically and part of it is due to the success of a few practitioners who use much more information than past price history alone. A recent study suggests strongly that "relative strength" in a price series is continued and, consequently, that past prices do have some value in predicting future prices [16].

Whether or not we can predict the future course of stock prices[1], there are investments in combinations of securities which can yield high expected return [23]. These investments involve convertible securities. A convertible security is one which, in some cases with the addition of money, is exchangeable (per share) for a certain number of shares of another security. Convertible securities include convertible bonds, convertible preferreds, stock options, stock rights, and warrants. There are several billion dollars worth of convertibles listed on the New York and American Stock Exchanges.

The analysis of other convertibles follows from the analysis of the common stock purchase warrant. We therefore restrict ourselves to these in our discussion, and shall refer to them simply as warrants.

A warrant is the right or option to buy a certain number of shares of common for a certain price, until a certain expiration date (warrants which do not expire are called perpetual). The terms ordinarily read: A warrants plus E dollars buy C shares until D date. To avoid normalization problems, we suppose $A = C = 1$. Then E is the "exercise price" of the warrant. The prices of warrant and common are related and it is this which allows successful investments. One observes: (1) The price W of the warrant should increase as the price S of the stock increases. (2) If $W + E < S$, warrants can be bought and common sold short, simultaneously. The warrants are then converted to common which is delivered against the short position. Neglecting commissions, a profit of $S - W - E$ per warrant results. The purchase of warrants tends to increase W and the sale of common tends to decrease S, until $W + E \geqq S$. Thus $W \geqq S - E$ normally holds. (3) The common has advantages over the warrant such as possible dividends, or voting rights, hence we also normally expect $W < S$.

Thus for practical purposes points (S, W) representing (nearly) simultaneous prices of a common stock and its warrant are confined to the part of the positive quadrant between the lines $W = S$ and $W = S - E$.

The prices W and S at a future time are random variables but they are related. As $E(S)$ increases we would expect, and past history verifies ([12, 23]), that $E(W)$ tends to increase. In fact the points (S, W) tend to lie on certain curves which depend

[1] The great mathematician Karl Friedrich Gauss was successful in the market but we have little knowledge of his methods. On a basic salary of 1000 thalers per year he left an estate in cash plus securities of 170,857 thalers ([7], page 237).

on several variables, most notably the time remaining until expiration of the warrant. Thus, although we may not know the price S of the common, or the price W of the warrant at a given future time, we do know that (W, S) is near one of these curves. The family of curves qualitatively resembles the family $W = (S^z + E^z)^{1/z} - E$, where $z = 1.3 + 5.3/T$ and T is the number of months remaining until expiration.

A historical study of expiring warrants (from here on we limit ourselves for convenience to warrants traded on the American Stock Exchange) suggests that during the last two years or so before expiration they tend to trade at prices which are much too high. For instance the average loss from buying each of a certain 11 listed warrants 18 months before expiration and holding until 2 months before expiration was 46.0%, an annual rate of 34.5% ([23], page 37). Thus selling warrants short seems to yield high expectation. However, it also happens to result in occasional large losses which, by the criterion of Part II, are extremely undesirable despite the high overall expectation. We can sharply reduce this high variance and yet retain a high expectation by using the so-called warrant hedge. The technique is to simultaneously sell short overpriced warrants and buy common in a fixed ratio (generally from one to three warrants will be shorted for each share of common bought). The position is held until just before expiration of the warrant (at which time the warrant sells at a "correct" price) and then it is liquidated.

Here is the rationale. We are mixing two investments with positive annual expectations of say 34.5% for the warrants and 10% for the common, resulting in an investment whose overall expectation must therefore be somewhere between these figures. (We suggest 10% for the common because this approximates the observed mean rate of return from common stocks during this century due to price appreciation plus dividends.) Buying the common leads to a gain when the common rises and a loss when it falls whereas shorting warrants leads to a gain when the common falls and leads to a loss only if the common rises substantially. Thus the risks tend to cancel out. In fact, the hedge generally yields a profit upon expiration of the warrant, for a wide range of prices of the common.

If we make assumptions about the probability distribution of the price of the common at expiration of the warrant, we get more precise information about the random variable representing the payoff from the hedge. Let the probability measure P with support $[0, \infty)$ describe the distribution of the stock price S_f at expiration. Then

$$E(S_f^n) = \int_0^\infty x^n \, dP(x).$$

Let S_0 be the present price and let E be the exercise price. Assume that $P(S_f \geqq S_0 + t) \geqq P(S_f \leqq S_0 - t)$ for each $t \geqq 0$, i.e. for any t, the chance of a price rise of at least t is no less than the chance of a price drop of at least t. This is a very weak assumption. Note that it does imply $E(S_f) \geqq S_0$.

Just before expiration $W_f \doteq 0$ if $S_f \leqq E$ and $W_f \doteq S_f - E$ if $S_f > E$. Thus the final gain from shorting a warrant at W_0 is W_0 if $S_f \leqq E$ and is $W_0 - S_f + E$ if $S_f > E$. The gain from buying a share of common at S_0 is, of course, $S_f - S_0$.

Hence if we assume one share of common is purchased at $.5E$ and one warrant is shorted at $.2E$, the final gain G_f is $S_f - .3E$ if $S_f \leqq E$ and $.7E$ if $S_f > E$. A standard measure-theoretic argument yields $E(G_f) \geqq .2E$. Using 100% margin, the percent profit is $E(G_f) / .7E > 28\%$. With 100% margin on the warrants and 70% margin on the common, it is at least $.2E / .55E > 36\%$. With 70% margin on each, it is at least $.2 / .49 > 40\%$, an annual rate of more than 20% if the warrant expires in two years.

It is interesting to calculate $E(G_f)$ by assuming that S_f is log normally distributed. Letting $s_f = S_f / E$, we thus assume that s_f has the density function
$f(x) = (x\sigma\sqrt{2\pi})^{-1} \exp\left[-(\ln x - \mu)^2 / 2\sigma^2\right]$, where μ and σ are parameters depending on the stock. We note $E(s_f) = \exp(\mu + \sigma^2/2)$.

If t is the time in months remaining until expiration (which is when s_f is realized), then we assume $\mu = \log s_0 + mt$ and $\sigma^2 = a^2 t$, where $S_0 = s_0 E$ is the present stock price and m and a are constants depending on the stock. Thus $E(s_f) = s_0 \exp[(m + a^2/2)t]$. A mean increase of 10% per year is approximated by setting $12(m + a^2/2) = .1$. If we estimate a^2 from past price changes we can solve for m.

Letting $w_f = W_f/E$, where W_f is the final warrant price, a calculation yields $E(w_f) = E(s_f)\,N(\mu/\sigma + \sigma) - N(\mu/\sigma)$, where N is the normal distribution. (Compare the equivalent expression from pp. 464–466 of [5].) Now suppose that $s_0 = .5$, that a has the realistic value of $.1 s_0$ or .05, and that $12(m + a^2/2) = .1$, whence $m = .085/12$. Then for $t = 24$ we have $\sigma = \sqrt{.06} = .245$ and $\mu = \log .5 + .17 = -.523$. This yields $E(w_f) = .0015$ and $E(s_f) = .61$, whence $E(G_f) = .20 + .11 = .31$. Thus the profit, with 70% margin on both warrant and common, is .31/.49 or 63.3% and the annual rate is 31.6%. Note that the warrant is virtually worthless!

Instead of selling one warrant short and buying one share of common, we can sell short w warrants and buy s shares of common. Neglecting commissions, which we do throughout for simplicity, the gain G at any point (S, W) is $s(S - S_0) - w(W - W_0)$. Thus the line $G = 0$, the zero profit line, is the line through (W_0, S_0) with positive slope $s/w = 1/m$. We call m the *mix*. Points below the zero profit line represent gain and points above it represent loss. If $1 < m < \infty$, the zero profit line intersects the S axis at $S_0 - m W_0$ and it intersects the line $W = S - E$ at $S = [m(W_0 + E) - S_0]/(m - 1)$; $W = (m W_0 + E - S_0)/(m - 1)$.

When the warrant expires the hedge position will yield a profit if S_f is between the S values of the two intersections and it will yield a loss if S_f is beyond the intersections. For instance, if $S_0 = .5E$ and $W_0 = .2E$, the choice $m = 2$ insures a final profit if $.1E < S_f < 1.9E$. Such safety is characteristic of the warrant hedge.

The final gain G_f is $s(S_f - S_0) + wW_0$ if $S_f \leqq E$ and it is $s(S_f - S_0) + w(W_0 + E - S_f)$ if $S_f > E$. Thus as a function of S_f it is an inverted "V" with apex above $S_f = E$. With 100% margin, the initial investment is $s S_0 + w W_0$ so the gain per unit invested is $g_f = G_f/(s S_0 + w W_0)$. With margin of α on the common and β on the warrants it is $g_f = G_f/(\alpha s S_0 + \beta w W_0)$.

We have assumed so far that a hedge position is held unchanged until expiration, then closed out. This static or "desert island" strategy is not optimal. In practice intermediate decisions in the spirit of dynamic programming lead to considerably superior dynamic strategies. The methods, technical details, and probabilistic summary are more complex so we defer the details for possible subsequent publication.

Probabilistic summary

The warrant hedge may offer high expectation with low risk. The gain per unit g_f is $g_f = [(S_f + S_0) + m W_0]/(\alpha S_0 + \beta m W_0)$ when $S_f \leqq E$ and $g_f = [(S_f - S_0) + m(W_0 + E - S_f)]/(\alpha S_0 + \beta m W_0)$ if $S_f > E$. The gain per unit depends only on the random variable S_f. This has an unknown distribution but it can be estimated. The other quantities are constants depending on circumstances.

282

PART II. A MATHEMATICAL THEORY FOR COMMITTING RESOURCES IN FAVORABLE GAMES

1. INTRODUCTION: COIN TOSSING

Suppose we are confronted with an infinitely rich adversary who will match all bets we make on repeated independent trials of a biased coin (whose two outcomes are "heads" and "tails".) Assume that we have finite capital X_0, that we bet B_i on the outcome of the ith trial, where X_i is our capital after the ith trial, and that the probability of heads is p, where $\frac{1}{2} < p < 1$. (This is approximately the situation in Nevada blackjack, except that the game is played with a "mix" of biased coins.) Our problem is to decide how much to bet at each trial. A classic criterion is to choose B_i so that our expected gain $E(X_i - X_{i-1})$ is maximized at each trial, which is equivalent to maximizing $E(X_n)$ for all n.

Define T_j by $T_j = 1$ if the jth trial results in success and $T_j = -1$ if the jth trial results in failure. Then $X_j = X_{j-1} + T_jB_j$, $j = 1, 2, \ldots$, and $X_n = X_o + \sum_{j=1}^{n} T_jB_j$.

We assume that T_j, X_j, and B_j are all random variables on a suitable sample space Ω. If, for example, B_j is a function of $X_0, X_1, \ldots, X_{j-1}$ as it is in the common gambling systems, e.g. Martingale, Labouchere, etc. (note that $B_k = | X_k - X_{k-1} |$ so we need not add the B_k, $k = 1, \ldots, j-1$), then we see by induction that B_j is a function of $X_0, T_1, T_2, \ldots$ Hence the underlying sample space can be taken to be the space of all sequences of successes and failures, with the usual product measure.

Suppose, more generally, that the player determined B_j by examining $X_0, \ldots, X_{j-1}$. and then "consulting" a chance device, e.g., a near-by roulette wheel. Then the sample space consisting of an infinite product of spaces, each of them a joint outcome of the roulette wheel and the latest trial, might be suitable. Such possibilities are included if we simply assume T_j, X_j and B_j are all random variables on *some* suitable sample space Ω.

When $B_j > X_{j-1}$, the player is betting more than he has. He is asking for credit. This is common in gambling casinos, in the stock market (buying on margin), in real estate (mortgages) and is not unrealistic.

When $B_j < 0$, the player is making a "negative" bet. To interpret this, we note that in our sequence of Bernoulli trials, or coin toss between *two* players, that what one wins, $X_j - X_0$, the other loses. To make a negative bet may be interpreted as "backing" the other side of the game, to taking the role of the "other" player.

In particular, the payoff B_jT_j from trial j may be written as $(-B_j)(-T_j)$. If $B_j \leqq 0$, then $-B_j \geqq 0$ and may be interpreted as a nonnegative bet by a player who succeeds when $-T_j = 1$, i.e., with probability q, and who fails with probability p. The $-T_j$ are independent so we have Bernoulli trials with success probabilities q, i.e., the other side of the game.

For simplicity we shall assume in what follows that $0 \leqq B_j \leqq X_{j-1}$, but we may wish at a future time to remove one or both of these limitations.

We also assume that B_j is independent of T_j, i.e., the amount bet on the jth outcome is independent of that outcome.

Definition: A betting strategy is a family $\{B_j\}$ such that $0 \leqq B_j \leqq X_{j-1}, j = 1, 2, \ldots$

Theorem 1: The betting strategies $B_j = X_{j-1}$ when $p > \frac{1}{2}$; $B_j = 0, p < \frac{1}{2}$; B_j arbitrary when $p = \frac{1}{2}$; are precisely the ones which maximize $E(X_j)$ for each j.

Proof: Since $X_n = X_o + \sum_{j=1}^{n} B_jT_j$, $E(X_n) = X_o + \sum_{j=1}^{n} E(B_jT_j) =$

283

$= X_0 + \sum_{j=1}^{n} (p-q) E(B_j)$. If $p - q = 0$, i.e., $p = \frac{1}{2}$, then $E(B_j)$ does not affect $E(X_n)$. If $p - q > 0$, i.e. $p > \frac{1}{2}$, then $E(B_j)$ should be maximized, i.e., $B_j = X_j$, to maximize $E(X_n)$. Similarly, if $p - q < 0$, i.e., $p < \frac{1}{2}$, then $E(B_j)$ should be minimized to maximize the jth term, i.e., $B_j = 0$. Clearly, these maxima are not attained with other choices for B_j. This establishes the theorem.

Remark. In the foregoing discussion, the Bernoulli trials and the T_j can be generalized, yielding a more general theorem. (The T_j become "payoff functions" that are not necessarily identically distributed; roulette is the classic example.) The particular case of blackjack is covered, for instance, by replacing p and q throughout by p_j and q_j for the respective probabilities that $T_j = 1$ or -1.

To maximize our expected gain we must bet our total resources at each trial. Thus if we lose once we are ruined, and the probability of this is $1 - p^n \rightarrow 1$ so maximizing expected gain is undesirable.

2. MINIMIZING THE PROBABILITY OF RUIN

Suppose instead that we play to minimize the probability of eventual ruin, where ruin occurs after the jth outcome if $X_j = 0$. If we impose no further restriction on B_j, then many strategies minimize the probability of ruin. For example, it suffices to choose $B_j < X_{j-1}/2$. The discreteness of money makes it realistic to assume $B_j \geqq C > 0$, where C is a non-zero constant. We further restrict ourselves to the subclass of strategies where B_j equals C whenever $0 < X_{j-1} < a$, $B_j = 0$ if $X_{j-1} \leqq 0$ or $X_{j-1} \geqq a$, and C divides both $a - z$ and z, where we have set $z = X_0$. This lets us use the gambler's ruin formulae ([8], page 314).

Consider the gambler's ruin situation: $X_0 = z$, $B_j = 1$ if $0 < X_{j-1} < a$, $B_j = 0$ if $X_{j-1} = 0$ or $X_{j-1} = a$, a and z are integers. Let r be a positive number (necessarily rational) such that zr and ar are integers. Let $R(r)$ be the ruin probability when z and a are replaced by zr and ar, respectively. This is equivalent to betting r^{-1} units when $0 < X_{j-1} < a$, in the original problem.

We have $R(r) = (\theta^{ar} - \theta^{zr}) / (\theta^{ar} - 1)$, where $0 < p \neq \frac{1}{2}$ and $\theta = q/p$.

Theorem 2: (a) If $1 > p > \frac{1}{2}$, $R(r)$ is a strictly decreasing function of r. (b) If $0 < p < \frac{1}{2}$, $R(r)$ is a strictly increasing function of r.

Proof: Follows from Lemma 3 below.

Part (a) of the Theorem says that in a favorable game, the chance of ruin is decreased by decreasing stakes. Note that for $p > \frac{1}{2}$, i.e., $\theta < 1$, $\lim_{r \rightarrow \infty} R(r) = 0$, hence by making stakes sufficiently small, the chance of ruin can be made arbitrarily small.

Lemma 3. Let $a > z > 0$, $x > 0$. If $0 < \theta < 1$, then $f(x) = (\theta^{zx} - \theta^{ax}) / (1 - \theta^{ax})$ is strictly decreasing as x increases, $x > 0$. If $\theta > 1$, $f(x)$ is strictly increasing as x increases, $x > 0$.

Proof: Elementary calculations which we omit.

Theorem 2(a) shows that, at least in the limited subclass of strategies to which it applies, we minimize ruin by making a minimum bet on each trial.

In fact, this holds for a broader class of strategies:

Theorem 3′: If $1 > p > \frac{1}{2}$, the strategy $B_j = 1$ if $1 \leqq z \leqq a-1$, $B_j = 0$ otherwise (*timid play*), uniquely minimizes the probability of ruin among the strategies where B_j is an integer satisfying $1 \leqq B_j \leqq \min(z, a-z)$ if $1 \leqq z \leqq a-1$, $B_j = 0$ otherwise.

Proof: We first show that if timid play is optimal, then it is uniquely so. Let q_z be the probability of ruin, starting from z, under timid play. To establish uniqueness it suffices to show

284

$$q_n < pq_{z+k} + q\, q_{z-k},\ 2 \leqq k \leqq a\text{–}2,\ z\text{–}k \geqq 0,\ z + k \leqq a. \qquad (1)$$

Using $q_z = (\theta^a\text{–}\theta^z) / (\theta^a\text{–}1)$ and simplifying (1), we find that it is equivalent to show

$$f(p) = p^{2k-1} + q^{2k-1} - p^{k-1}\, q^{k-1} > 0,\ 1/2 < p \leqq 1. \qquad (2)$$

This follows at once from the observations $f(\frac{1}{2}) = 0$, $f(1) = 1$, and $f'(p) > 0$, $\frac{1}{2} < p \leqq 1$.

To show that timid play is optimal, let $Q(z) = 1\text{–}q_z$, and adopt the terminology of [6]. Then $Q(z)$ is the probability of success, both for z in our original game, and for z/n in a normalized game where the possible fortunes are $F = \{ 0, 1/a, 2/a, \ldots, z/a, \ldots, 1 = a/a \}$, and the betting units and limits are $1/a$ as large as before.

The establishment of (1) shows $Q(z)$ is excessive. But obviously $u(z) \leqq Q(z) \leqq U(z)$ so by [6, Theorem 2.12.3], $Q(z) = U(z)$.

Thus timid play is the one and only strategy in our class of strategies which minimizes the probability of ruin.

Remark: In [6] it is shown that bold play is optimal but not necessarily unique when $p < \frac{1}{2}$ (pages 2, 87ff, 101ff). If there is also a legal upper limit to bets, there may be more than one optimal strategy; whether bold play is one of them seems to be unknown (page 4). Betting systems which minimize the probability of ruin in certain favorable games are also discussed in [10].

The strategy which minimizes ruin has the unsatisfactory consequence that it also minimizes our expected gain. Some strategy is called for which is intermediate between minimizing ruin (and expectation) and maximizing expectation (assuring ruin). A remarkable solution, in a certain sense very close to best possible, was proposed in [13].

3. THE KELLY CRITERION

Consider Bernoulli trials with $1 > p > \frac{1}{2}$ and $B_j = f X_{j-1}$, where $0 \leqq f \leqq 1$ is a constant. (This is sometimes called "fixed fraction" or "proportional" betting.) Let S_n and F_n be the number of successes and failures, respectively, in n trials. Then

$$X_n = X_0 (1 + f)^{S_n} (1 - f)^{F_n}.$$

Observe that $f = 0$ and $f = 1$ are uninteresting; we assume $0 < f < 1$. Note too that if $f < 1$, there is no chance that $X_n = 0$, ever. Hence ruin, in the sense of the gambler's ruin problem, cannot occur. We reinterpret "ruin" to mean that for each $\varepsilon > 0$, $\lim_n P[X_n \geqq \varepsilon] = 0$, and we shall see that this can occur. Note too that we are now assuming that capital is infinitely divisible. However, this assumption is not a serious problem in practical applications of the theory.

Remark: The min-max criterion of game theory is an inappropriate criterion in Bernoulli trials. If B_j is a positive integer for all j, the maximum loss, i.e. ruin, is always possible and all strategies have the same maximum possible loss, hence all are equivalent. If capital is infinitely divisible, ruin is as we redefined it, and we restrict ourselves to fixed fractions, then for an infinite series of trials the min-max criterion (suitably probabilitistically modified) considers all f with $1 \geqq f \geqq f_c$ equivalent and all f with $0 < f < f_c$ equivalent. It chooses the latter class. For a fixed number n of trials, smaller f are preferred over larger f. The criteria of minimizing ruin or of maximizing expectation likewise fail to make desirable distinctions between the fixed fraction strategies.

The quantity $\log [X_n / X_0]^{1/n} = (S_n / n) \log (1+f) + (F_n / n) \log (1-f)$ measures the rate of increase per trial. Since time is important it is plausible to in some sense maximize this. Kelly's choice [13] was to maximize $E \log [X_n / X_0]^{1/n} = p \log (1+f) + q \log (1-f) \equiv G(f)$, which we call the exponential rate of growth. The following theorems show the advantages of maximizing $G(f)$.

Theorem 4. If $1 > p > \frac{1}{2}$, $G(f)$ has a unique maximum at $f^* = p - q$, $0 < f^* < 1$, where $G(f^*) = p \log p + q \log q + \log 2 > 0$. There is a unique fraction $f_c > 0$ such that $G(f_c) = 0$, and f_c satisfies $f^* < f_c < 1$. Further, we have $G(f) > 0$, $0 < f < f_c$; $G(f) < 0, f > f_c$, with $G(f)$ strictly increasing, from 0 to $G(f^*)$, on $[0, f^*]$, and $G(f)$ strictly decreasing, from $G(f^*)$ to $-\infty$ on $[f^*, 1]$.

Theorem 5(a). If $G(f) > 0$, then $\lim_n X_n = \infty$ a.s., i.e., for each M, $P[\underline{\lim}_n X_n > M] = 1$.

(b) If $G(f) < 0$, then $\lim_n X_n = 0$ a.s., i.e., for each $\varepsilon > 0$, $P[\overline{\lim}_n X_n < \varepsilon] = 1$.

(c) If $G(f) = 0$, then $\overline{\lim}_n X_n = \infty$ a.s. and $\underline{\lim}_n X_n = 0$ a.s.

Thus for $0 < f < f_c$, the player's fortune will eventually permanently exceed any fixed bounds with probability one. For $f = f_c$ it will almost surely oscillate wildly between 0 and $+\infty$. If $f > f_c$, ruin is almost sure.

Proof: (a) By the Borel strong law ([17], page 19), $\lim_n \log [X_n/X_0]^{1/n} = G(f) > 0$ with probability 1. Hence, a.s., for $\omega \varepsilon \Omega$, where Ω is the space of all sequences of Bernoulli trials, there exists $N(\omega)$ such that for $n \geqq N(\omega)$,
$\log [X_n / X_0]^{1/n} \geqq G(f)/2 > 0$.
But then $X_n \geqq X_0 e^{n G(f)/2}$ for $n \geqq N(\omega)$ so $X_n \nearrow \infty$.

(b) The proof is similar to part (a).

(c) We use the fact that, given any M, $\overline{\lim}_n S_n \geqq np + M + 1$ and $\underline{\lim}_n S_n \leqq np - M - 1$. Then if $S_n \geqq np + M$, $\log [X_n / X_0]^{1/n} \geqq \frac{np + M}{n} \log (1+f) + \frac{n - (np+M)}{n} \log (1-f) = G(f) + \frac{M}{n} \log \frac{1+f}{1-f} = \frac{M}{n} \log \frac{1+f}{1-f}$, whence $X_n \geqq X_0 \left(\frac{1+f}{1-f}\right)^M$. Since $S_n \geqq np + M$ infinitely often, a.s., then $\overline{\lim}_n X_n \geqq X_0 \left(\frac{1+f}{1-f}\right)^M$ a.s. Since the right side may be chosen arbitrarily large, $\overline{\lim}_n X_n = \infty$ a.s.

The proof that $\underline{\lim}_n X_n = 0$ a.s. is similar.

Theorem 6: If $G(f_1) > G(f_2)$, then $\lim_n X_n(f_1)/X_n(f_2) = \infty$ a.s.

Proof: $\log [X_n(f_1) / X_0]^{1/n} - \log [X_n(f_2) / X_0)^{1/n}$
$= \log [X_n(f_1) / X_n(f_2)]^{1/n} = \frac{S_n}{n} \log \left(\frac{1+f_1}{1+f_2}\right) + \frac{F_n}{n} \log \left(\frac{1-f_1}{1-f_2}\right)$. Therefore, by the Borel strong law of large numbers, $\lim_n \log [X_n(f_1)/X_n(f_2)] \to G(f_1) - G(f_2) > 0$ with probability 1. Now proceed as in the proof of Theorem 5(a).

In particular, we see that if one player uses f^* and another, *betting on the same favor-*

able situation, uses any other fixed fraction strategy f, then $\lim_n X_n(f^*)/X_n(f) = \infty$ with probability 1. This is one of the important justifications of the criterion "bet to maximize $E \log X_n$."

Bellman and Kalaba ([2], pages 200–201) show that f^* not only maximizes $E \log X_n$ within the class of all fixed fraction betting strategies but in the class of "all" betting strategies.

This also is a consequence of the following theorem, part of which was suggested by a conversation with J. Holladay. Consider a series of independent trials in which the return on one unit bet on the ith outcome is the random variable Q_i. Then $X_n = \prod_{i=1}^{n} (X_i/X_{i-1})$ and $E \log X_n = \sum_{i=1}^{n} E \log (X_i/X_{i-1})$. We have $X_i = X_{i-1} + B_i Q_i$ and $X_i / X_{i-1} = 1 + (B_i / X_{i-1}) Q_i$. Thus each term is of the form $E \log (1 + F_i Q_i)$ where the random variable F_i depends only on the first i–1 trials, Q_i depends only on the ith trial, and hence F_i and Q_i are independent. We are free to choose the F_i to maximize $E \log (1 + F_i Q_i)$, subject to the constraint $0 \leqq F_i \leqq 1$.

Theorem 7: If for each i there is an f_i, $0 < f_i < 1$, such that $E \log (1 + f_i Q_i)$ is defined and positive, then for each i there is a number f_i^* such that $E \log (1 + F_i Q_i)$ attains its unique maximum for $F_i = f_i^*$ a.s. To avoid trivialities we assume $Q_i \neq 0$ a.s., each i.

Proof: It follows that the domain of definition of $E \log (1 + f_i Q_i)$ is an interval $[0, a_i)$ or $[0, a_i]$, where $a_i = \min(1, b_i)$ and $b_i = \sup \{f_i : f_i Q_i > 0 \text{ a.s.}\} > 0$. Since the second derivative with respect to f_i of $E \log (1 + f_i Q_i)$ is $-E(Q_i^2/(1 + f_i Q_i)^2)$, which is defined and negative, any maximum of $E \log (1 + f_i Q_i)$ is unique. The function is continuous on its domain so if it is defined at a_i, there is a maximum. If it is not defined at a_i, then $\lim_{f_i \uparrow a_i} E \log (1 + f_i Q_i) = -\infty$ so again there is a maximum.

By the independence of F_i and Q_i, we can consider $F_i(s_1)$ and $Q_i(s_2)$ as functions on a product measure space $S_1 \times S_2$. Then
$E \log (1 + F_i Q_i) = \int_{S_1} \int_{S_2} \log (1 + F_i(s_1) Q_i(s_2)) = E\, E \log (1 + F_i(s_1) Q_i)$
$\leqq E \log (1 + f_i^* Q_i)$ with equality if and only if $E \log (1 + F_i(s_1) Q_i) =$
$E \log (1 + f_i^* Q_i)$ a.s., which is equivalent to $f_i^* Q_i = F_i(s_1) Q_i$ a.s., and by the independence this means either $f_i^* = F_i$ a.s. or $Q_i = 0$ a.s. hence $f_i^* = F_i$ a.s., and the theorem is established.

We see in particular from the preceding theorem that for Bernoulli trials with success probability p_i on the ith trial and $1 > p_i > \frac{1}{2}$, $E \log X_n$ is maximized by simply choosing on each trial the fraction $f_i^* = p_i - q_i$ which maximizes $E \log (1 + f_i Q_i)$.

4. THE ADVANTAGES OF MAXIMIZING E LOG X_n

The desirability of maximizing $E \log X_n$ was established in a fairly general setting by Breiman [3]. Consider repeated independent trials with finitely many outcomes $I = \{1, \ldots, s\}$ for each trial. Let $P(i) = p_i$, $i = 1, \ldots, s$, and suppose that $\{A_1, \ldots, A_r\}$ is a collection of (betting) subsets of I, that each i is in some A_k, and that payoff odds o_k correspond to the A_k. We bet amounts $B_1, \ldots, B_r$ on the respective A_k and if outcome i occurs, we receive $\Sigma B_j o_j$ where the sum is over $\{j : i \in A_j\}$. We make the convention that $A_1 = I$ and $o_1 = 1$, which allows us to hold part of our fortune in reserve by simply betting it on A_1. We have, in effect, a generalized roulette game.

Roulette and the wheel of fortune, as described in Part I, are covered directly by Breiman's theory.

The theory easily extends to independent trials with finitely many outcomes which are not identically distributed but which are a mix of finitely many distinct distributions, each occurring on a given trial with specified probabilities. The theory so extended applies to Blackjack and Nevada Baccarat, as described in Part I.

Breiman calls a game (i.e., a series of such trials) favorable if there is a gambling strategy such that $X_n \to \infty$ a.s. Thus the infinite divisibility of capital is tacitly assumed. However, this is not a serious limitation of the theory. If the probability is "negligible" that the player's capital will at some time be "small," then the theory based on the assumption that capital is infinitely divisible applies to good approximation when the player's capital is discrete. This problem is considered for Nevada Baccarat in [24], pages 319 and 321.

Breiman establishes the following about strategies which maximize $E \log X_n$.

1. Allowing arbitrary strategies, there is a fixed fraction strategy $B_i = (f_1, \ldots, f_r)$ which maximizes $E \log X_n$.

2. If two players bet on the same game, one using a strategy Λ^* which maximizes $E \log X_n$ and the other using an "essentially different" strategy Λ, then $\lim_n X_n(\Lambda^*) / X_n(\Lambda) \to \infty$ a.s.

3. The expected time to reach a fixed preassigned goal x is, asymptotically as x increases, least with a strategy which maximizes $E \log X_n$.

Thus strategies which maximize $E \log X_n$ are (asymptotically) best by two reasonable criteria.

5. A STOCK MARKET EXAMPLE

Though in practice there are only finitely many outcomes of a bet in the stock market, it is technically convenient to approximate the finite distributions by discrete countably infinite distributions or by continuous distributions. In fact it is generally difficult not to do this. The additional hypotheses and difficulties which occur are, from the practical point of view, artificial consequences of the technique. Hence the new theory must preserve the conclusions of the finite theory so again we apportion our resources to maximize $E \log X_n$.

As a first example, consider the following stock market investment. It was the first to catch our interest, and was based on a tip from a company insider.

Suppose a certain stock now sells at 20 and that the anticipated price of the stock in one year is uniformly distributed on the interval [15, 35]. We first compute f^* and $G(f^*)$, assuming the stock is purchased and fully paid for now, and sold in one year. The purchase and selling fees have been included in the price. Thus, the outcome of this gamble, per unit bet, is described by $dF(s) = C_{(-\frac{1}{4}, \frac{3}{4})}(s)\, ds$, where F is the associated probability distribution and $C_A(s)$ is 1 for s in A and 0 for s not in A.

The mean m of F is $\frac{1}{4} > 0$. Also

$$G(f) = \int_{-\frac{1}{4}}^{\frac{3}{4}} \log_2(1+fs)\, ds, \quad G'(f) = \int_{-\frac{1}{4}}^{\frac{3}{4}} \frac{s \log_2 e}{1+fs}\, ds, \text{ and } \lim_{f \uparrow 4} G'(f) = -\infty.$$

Therefore Theorem 8 below applies and there is a unique f^* such that $0 < f^* < 4$ and $G'(f^*) = 0$. To obtain f^*, it suffices to solve

$$h(f) = 0 \text{ where } h(f) = fG'(f)/\log_2 e = \int_{-\frac{1}{4}}^{\frac{3}{4}} \frac{fs\, ds}{1+fs} = \int_{-\frac{1}{4}}^{\frac{3}{4}} ds - \int_{-\frac{1}{4}}^{\frac{3}{4}} \frac{ds}{1+fs} =$$

$$= 1 - (1/f) \log_e(1+fs) \Big|_{-\frac{1}{4}}^{\frac{3}{4}} \text{ which reduces to } 1 - (1/f) \log_e \frac{1+\frac{3}{4}f}{1-\frac{1}{4}f} = h(f).$$

Now $h(f)$ has the same sign and root as $G'(f)$ on (0,4). Since $h(3) = 1 - \frac{1}{3} \log_e 13 > 0$, $G'(f) > 0$ for $0 \leqq f \leqq 3$. Therefore $3 < f^* < 4$; calculation yields $f^* = 3.60^-$.

Thus if the maximum fraction of current capital which can be bet is 1, we should bet all our capital. However, if margin buying is allowed, we should (consistent with our ability to cover later) be willing to bet as much as possible, up to a fraction f^* which is 3.6 times our current capital.

The mathematical expectation for buying outright is 0.25 V_0 if buying on margin is excluded and 0.90 V_0 if unlimited buying on margin is permitted, and additional coverings can be made later as required, and we bet $f^* = 3.60$ of our current capital.

Integration yields
$G(f) = [(\log_2 e)/f]\{(1 + 3f/4)[\ln(1 + 3f/4) - 1] - (1 - f/4)[\ln(1 - f/4) - 1]\}$
from which we find $G(1) = 0.28$ and $G(3.60) = .59$.

Next, we compute f^* and $G(f^*)$, assuming that calls are purchased for 2 points per share. Thus the outcome of this gamble per unit bet is described by the probability distribution F with mass 5/20 at –1, and $dF(s) = 2/20$ if $-1 < s < 6.5$.

The mean $m = 1.8125 > 0$. Also $G(f) = (5/20) \log_2 (1-f) + (2/20) \int_{-1}^{6.5} \log_2 (1 + fs)\, ds$ and $G'(f) = \dfrac{-(5/20) \log_2 e}{1-f} + (2/20) \int_{-1}^{6.5} \dfrac{s \log_2 e}{1 + fs}\, ds$, from which it is clear that $\lim_{f \uparrow 1} G'(f) = -\infty$. Therefore, again by Theorem 8 below, there is a unique f^* such that $G'(f^*) = 0$ and $0 < f^* < 1$.

It suffices to solve $h(f) = 0$ where $h(f) = 20G'(f) / \log_2 e$
$= \dfrac{-5}{1-f} + \dfrac{2}{f}\left\{7.5 - \dfrac{1}{f} \log_e\left(\dfrac{1 + 6.5f}{1 - f}\right)\right\}$. We find $f^* = 0.57$. The mathematical expectation of the call purchase process is $1.8125 f^* V_0$ or about 1.03 V_0.

Integration yields
$G(f) = (\frac{1}{4}) \log_2 (1 - f) + (\ln_2 e / 10f)(1 + 6.5f)[\ln(1 + 6.5f) - 1] - (1 - f)[\ln(1 - f) - 1]$. We find $G(0.57) = 0.55$.

Thus we have the interesting result that the expectation from buying calls is higher than from buying on unlimited margin but that the growth coefficient is higher from buying on unlimited margin. Our criterion selects the latter investment.

For buying on margin $G(3^-) = .55$ so our criterion selects buying on margin if the margin requirement is less than $\frac{1}{3}^-$ and buying calls, if possible, if the margin requirement exceeds $\frac{1}{3}^-$.

In the preceding example we needed the following theorem to establish the uniqueness of f^*. We define $a = \sup\{t : F(-\infty, t) = 0\}$ and note that if $1 + fa > 0$ and the integral $G(f) = \int_a^\infty \log_2 (1 + fs)\, dF(s)$ is defined, then $G'(f) = \int_a^\infty \dfrac{s \log_2 e}{1 + fs} dF(s)$. (See, e.g. [17], page 126).

Theorem 8: The function $G'(f) = \int_a^\infty \dfrac{s \log_2 e}{1 + fs} dF(s)$ is monotone strictly decreasing on $[0, -1/a)$. If the mean $m = \int_a^\infty s\, dF(s) > 0$, then the equation $G'(f) = \int_a^\infty \dfrac{s \log_2 e}{1 + fs} dF(s) = 0$ has exactly one solution f^* in the interval $(0, -1/a)$ iff $\lim_{f \nearrow -1/a} G'(f) < 0$. In this event, $G(f)$ is monotonely strictly increasing for f in $[f^*, -1/a)$.

Proof: If $0 < f_1 < f_2 < -1/a$, $\dfrac{s}{1 + f_1 s} > \dfrac{s}{1 + f_2 s}$ $(0 \neq s \geqq a)$ so $G'(f_1) > G'(f_2)$.

From this and the right-hand continuity of $G'(f)$ at 0, G' is monotone strictly decreasing on $[0, -1/a)$. By hypothesis $G'(0) = m \log_2 e > 0$. Therefore, from the continuity of $G'(f)$ on $[0, -1/a)$, $G'(f)$ attains all values t on the interval $\lim_{f \nearrow -1/a} G'(f) < t \leqq G'(0)$ exactly once. Thus there is exactly one solution f^* in $(0, -1/a)$ iff $\lim_{f \nearrow -1/a} G'(f) < 0$.

The description of $G(f)$ is now evident.

6. WARRANT HEDGING

We next apply the criterion of maximizing $E \log X_n$ to the warrant hedge described in Part I. With the notation of Part I, and an assumed mix of 1, the gain X from a one unit bet is

$$X = (s_f - s_0 + w_0) / (\alpha s_0 + \beta w_0),\ s_f \leqq 1, \text{ and}$$
$$X = (w_0 + 1 - s_0) / (\alpha s_0 + \beta w_0),\ s_f > 1\ .$$

We wish to maximize the exponential rate of growth $G(f)$, given by $G(f) = E \log(1 + fX)$.

It can be shown that the situation is essentially the same as in Theorem 8 and that this depends on the a.s. boundedness of X; we have in fact a.s. $\sup X = (w_0 + 1 - s_0) / (\alpha s_0 + \beta w_0)$ and a.s. $\inf X = -(s_0 - w_0) / (\alpha s_0 + \beta w_0)$. Thus f^* can be computed when the mix is 1, though the details are tedious.

When the mix is greater than 1, more serious difficulties appear. The payoff function X has a.s. $\inf X = -\infty$ and a.s. $\sup X < \infty$. This means that, no matter what fraction $f > 0$ of our unit capital is bet, there is positive probability of losing at least the entire unit. Thus any bet is rejected! Yet this is unrealistic. We now find out what is wrong.

First, the assumption that arbitrarily large losses have positive probability of occurrence is not realistic. (a) The broker will automatically act to liquidate the position before the equity is lost. (b) The strategies for investing in hedges automatically lead to liquidating the position after the common is substantially above exercise price.

There is, then, a maximum imposed on X by practice but it is not easy in practice to specify this maximum. Further, this maximum will, in general, be a random variable (a.s. bounded, however) which is a function of the individual's investment strategy. It is not easy to determine the consequent probability distribution of s_f, yet this is required to calculate $E \log(1 + fX)$.

More generally, we might consider an individual's lifetime sequence of bets of various kinds. It is plausible to assume that $X_n = 0$ only upon the death of the individual, for although the individual may have no cash equity at a given instant, he does have a cash "worth", based on his future income, serendipity, etc., and this should be included in X_n. This is true even of a (Billie Sol Estes) bettor who loses more than he owns. The subtlety here, then, is that the accountant's figure for net assets (plus or minus) is not an accurate figure for X_n as X_n decreases below small positive amounts.

One can also object to X_n at death being assigned the value 0, by arguing that the chance of death in a time interval always has a small positive probability, thus making $E \log X_n = -\infty$ always. Also, individuals when choosing between two alternatives each involving a low probability of death generally do not meticulously select the safer alternative (e.g., air travel versus train travel). Thus death should really be treated as an event with a large but finite negative value.

Another common objection to $E \log X_n$ as a measure of "utility" is that, like all such measures which are not a.s. bounded, it allows the St. Petersburg paradox.

The foregoing objections to $E \log X_n$ only arise when we leave the case of finitely many outcomes. We say that these are artificial technical difficulties which can all be removed in the cases of practical importance. This may be tedious, as it is for the warrant hedge, so we defer such matters for a subsequent paper.

7. PORTFOLIO SELECTION USING E LOG X

The Breiman results were obtained for repeated independent trials with finitely many outcomes and finitely many ways to apportion our capital (amongst finitely many betting sets). The results extend, as we have remarked, to independent trials which are a mix of finitely many differently distributed trials (i.e., finitely many outcomes and betting sets) provided that as n tends to infinity, the number of trials with each distribution also tends to infinity.

There are significant real world situations, such as the selection and continuous revision of a portfolio of securities, to which this extended theory does not generally apply. A difficulty which we have already discussed is that it may be technically convenient to introduce continuously distributed and possibly unbounded payoffs, but now generalized to the apportionment of capital among a finite number of alternatives, rather than just betting a fraction on one alternative. Another problem is that the sequence of betting situations may change so that no two are ever the same. Further difficulties arise when we consider the possible dependence of trials. Still other problems appear when we consider that in the real world the spectrum of situations is changing continuously and that a potentially continuous portfolio revision is part of an optimal approach. (Actually, because of the transactions costs which occur in practice, portfolio revision is likely to occur in discrete steps.)

The extent to which Breiman's conclusions for the finite case can be generalized in these directions will be considered subsequently. For now we simply remark that the possible generalizations promise to be adequate for the real world problems of portfolio selection.

Assuming this to be the case, we shall see in the next section that economists and others now have for the first time an accurate guide for portfolio selection and revision.

8. THE KELLY CRITERION AND DEFICIENCIES IN THE MARKOWITZ THEORY OF PORTFOLIO SELECTION

How to apportion funds among investments has endlessly puzzled economists and decision-makers. The literature was noted for its lack of instruction in such matters. When Markowitz' work on portfolio selection appeared, first in articles and later in the monograph [18], it became the standard reference.

Markowitz considers situations in which there are r alternative and, in general correlated, investments, with the gain per unit invested of $X_1, \ldots, X_r$, respectively. (It is so much more dignified to call bets investments; we shall try to remember to do this in this section.) One of the investments is, of course, cash. The gain is given by $X_k = 0$ a.s.

To select a portfolio is to apportion our resources so that f_i is placed in the ith investment. Markowitz' basic idea is that a portfolio is better if it has higher expectation and at least as small a variance or if it has at least as great an expectation and has a lower variance. If two portfolios have the same expectation and variance, neither is preferable. As the f_i range over all possible admissible values, the set of portfolios is generated. Typically the assumptions on the f_i are $\Sigma f_i = 1$, and $f_i \geqq 0$ for $i = 1, \ldots, r$.

291

If a portfolio has the property that no other portfolio in the set is preferable, then it is called efficient. Markowitz says that the investor should always choose an efficient portfolio. Which efficient portfolio to choose depends on factors outside the theory, such as the investor's "needs".

The Markowitz theory has the obvious deficiency that if E_i and σ_i^2, $i = 1, 2$, are the expectation and variance of portfolios 1 and 2, then if $E_1 < E_2$ and $\sigma_1^2 < \sigma_2^2$, the theory cannot choose between the portfolios. Yet there are obvious instances where "everyone" will choose the second portfolio over the first, such as when $F_1(x) < F_2(x)$ for all x. Specifically, let X_1 be distributed uniformly on $[1, 3]$, let X_2 be uniformly distributed on $[10, 100]$ and let $X_3 = 0$ a.s. represent the possibility of holding some of our resources in cash. Suppose X_1, X_2, and X_3 are independent. Then $E\,\Sigma f_i X_i = 2f_1 + 55f_2$ and $\sigma^2\,\Sigma f_i X_i = \Sigma f_i^2 \sigma_i^2 = f_1^2/3 + 675f_2^2$. All cash, or $f_3 = 1$, is an efficient portfolio since this is the unique portfolio with zero expectation. The portfolio $f_2 = 1$ also is efficient since this is the unique portfolio with greatest expectation. There are, in fact, infinitely many efficient portfolios. (They lie on a curve in the f_1, f_2 plane connecting (0, 0) and (0, 1).) The theory doesn't tell us which is best, yet $f_2 = 1$ is clearly preferable to any alternative.

In the case where there are the two alternatives $X_1 = 0$ a.s. (cash) and X_2 with $E_2 > 0$ and $\sigma_2 > 0$, all portfolios are efficient and Markowitz' theory gives no information on which to choose. The Kelly criterion tells us to choose f_2 to maximize $E \log(1 + f_2X_2)$ and we know further from the theory of the Kelly criterion why this choice is good. As we have seen, repeated trials of such an investment with f_2 greater than the fraction f_c will lead to ruin a.s.

Remark: This incompleteness of Markowitz' theory is understandable since he only uses probability information about first and second moments. We note though that the examples he gives, and the real world applications, generally assume that more detailed structure is known. Hence, it is reasonable that the criterion $E \log X_n$, which does use higher moment information, can provide a sharper theory.

Next consider those two-point probability distributions with masses m_i located at x_i, $i = 1, 2$, and with mean and variance 1. These are indistinguishable by Markowitz' criterion. A calculation shows, however, that for X_1 defined by $x_1 = -1$, $x_2 = 3/2$, $m_1 = 1/5$, $m_2 = 4/5$, the optimal fraction f_1^* is $\frac{2}{3}$ and $G(f_1^*)$ is $-(1/5)\log 3 + (4/5) \log 2$. For X_2 defined by $x_1 = -2$, $x_2 = 4/3$, $m_1 = 1/10$, $m_2 = 9/10$, we have $f_2^* = \frac{3}{8}$ and $G(f_2^*) = -(1/10)\log 4 + (9/10)\log(3/2)$, which is smaller than $G(f_1^*)$. Hence if $X_{n,1}^*$ is the fortune after n repeated independent trials of an investor who invests f_1^* in X_1 at each trial and $X_{n,2}$ is the fortune after n trials of an investor who invests in any manner whatsoever in X_2 at each trial, we have $\lim X_{n,1}^*/X_{n,2} = \infty$ a.s.

As a final example, suppose we are to apportion our resources between the foregoing X_1 and X_2, which we now suppose to be independent, and cash, represented by X_3. We impose the constraints $f_i \geqq 0$, $i = 1, 2, 3$; $f_1 + f_2 + f_3 = 1$, and $f_1 + 2f_2 \leqq 1$. The latter constraint prevents investments where our losses exceed our total resources. (The analysis and conclusion are essentially the same without this constraint.) The admissible portfolios are represented by the closed triangular region of the positive quadrant bounded by the axes and the line $f_1 + 2f_2 = 1$.

We have $E\,\Sigma f_i X_i = f_1 + f_2$ and, because of the independence of X_1 and X_2, $\sigma^2\,\Sigma f_i X_i = f_1^2 + f_2^2$. The efficient portfolios are the points of the f_1, f_2 plane on the two closed line segments joining $(\frac{1}{3}, \frac{1}{3})$ to (0, 0) and to (1, 0).

The function $E \log(1 + f_1X_1 + f_2X_2) \equiv G(f_1, f_2)$ is given by $50\,G(f_1, f_2) =$

$36 \log(1 + 3f_1/2 + 4f_2/3) + 4 \log(1 + 3f_1/2 - 2f_2) + 9 \log(1 - f_1 + 4f_2/3) + \log(1 - f_1 - 2f_2)$. This function is undefined on the line joining (0, 1/3) and (1, 0). It is defined and continuous elsewhere on the triangle of portfolios and as (f_1, f_2) tends to the segment from this triangle, $G(f_1, f_2) \to -\infty$. It follows (by the continuity) that $G(f_1, f_2)$ attains an absolute maximum in the region of the triangle where it is defined. We also know that any such maximum is positive. It follows that, if an efficient portfolio maximizes $G(f_1, f_2)$, then it must be a portfolio from the interior of the segment joining (0, 0) and (1/3, 1/3). Hence the coordinates must simultaneously satisfy the equations $\partial G(f_1, f_2) / \partial f_1 = 0$ and $\partial G(f_1, f_2) / \partial f_2 = 0$. (We note that in repeated independent trials where the investor selects an efficient portfolio from the segment joining (1/3, 1/3) to (1, 0), he will be ruined with probability one.)

Setting $f_1 = f_2 = t$ in the equations $\partial G / \partial f_1 = 0$ and $\partial G / \partial f_2 = 0$ and attempting to solve simultaneously yields, upon elimination between the two equations of the last of the four fractions, the necessary condition $-2796 + 376t + 111t^2 = 0$. Since this is negative at $t = 0$ and $t = 1$, there are no roots in the interval $0 < t < 1/3$. Hence no efficient portfolio maximizes $G(f_1, f_2)$.

We conclude that if $X^*_{n,1}$ is the fortune after n trials of a player who bets to maximize $G(f_1, f_2)$ on each trial, and $X_{n,2}$ is the fortune of a player who chooses any efficient portfolio on each trial, then $\lim X^*_{n,1} / X_{n,2} = \infty$ a.s. Furthermore, the Kelly investor will reach a fixed goal x in less time, asymptotically as $x \to \infty$, than a Markowitz investor.

The Kelly criterion should replace the Markowitz criterion as the guide to portfolio selection.

REFERENCES

[1] Baldwin, Cantey, Maisel, and McDermott (1956). The optimum strategy in blackjack, *J. Amer. Statist. Assoc.*, 51, 429–439.

[2] Bellman, R., Kalaba, R. (Sept. 1957). On the role of dynamic programming in statistical communication theory. *IRE Trans. of the professional group on information theory*, IT-3 no. 3, 197–203.

[3] Breiman, L. (1961). Optimal gambling systems for favorable games. *Fourth Berkeley Symposium on probability and statistics*, I, 65–78.

[4] Coddington, E. A., Levinson, N. (1955). *Theory of Ordinary Differential Equations*. New York, McGraw-Hill.

[5] Cootner, P. H., editor (1964). *The Random Character of Stock Market Prices*. The M.I.T. press.

[6] Dubins, L., Savage, L. (1965). *How to Gamble if You Must*. New York, McGraw-Hill.

[7] Dunnington, G. Waldo (1955). *Carl Friedrich Gauss, Titan of Science*. New York, Hafner.

[8] Feller, W. (1957). *An Introduction to Probability Theory and Its Applications, Vol. I, Revised.* New York, Wiley.

[9] Feller, W. (1966). *An Introduction to Probability Theory and Its Applications, Vol. II*. New York, Wiley.

[10] Ferguson, T. S. (1965). Betting systems which minimize the probability of ruin. *J. Soc. Indust. Appl. Math.*, 13, no. 3.

[11] Foster, F. G. (1964). A computer technique for game-theoretic problems. I. Chemin-de-fer analyzed. *Comput. J.*, 1, 124–130.

[12] Kassouf, Sheen T. (1965). *A Theory and an Econometric Model for Common Stock Purchase Warrants*. Ph. D. Thesis, Columbia University, New York.

[13] Kelly, J. L. (1956). A new interpretation of information rate. *Bell System Technical Journal*, 35, 917–926.

[14] Kemeny, J. G., Snell, J. L. (1957). Game theoretic solution of baccarat, *Amer. Math. Monthly*, 114, no. 7, 465–9.

[15] Kendall, M. G., Murchland, J. D. Statistical aspects of the legality of gambling, *J. Roy. Statist. Soc. Ser. A.* To be published.

7

PORTFOLIO CHOICE AND THE KELLY CRITERION

Edward O. Thorp, University of California at Irvine*

1. Introduction. The Kelly (or capital growth) criterion is to maximize the expected value $E \log X$ of the logarithm of the wealth random variable X. Logarithmic utility has been widely discussed since Daniel Bernoulli introduced it about 1730 in connection with the Petersburg game [3, 28]. However, it was not until certain mathematical results were proven in a limited setting by Kelly in 1956 and then in much more general setting by Breiman in 1960 and 1961 that logarithmic utility was clearly distinguished by its properties from other utilities as a guide to portfolio selection. (See also [2, 4, 15], and the very significant paper of Hakansson [11].)

Suppose for each time period ($n = 1, 2, \cdots$) there are k investment opportunities with results per unit invested denoted by the family of random variables $X_{n,1}, X_{n,2}, \cdots, X_{n,k}$. Suppose also that these random variables have only finitely many distinct values, that for distinct n the families are independent of each other, and that the joint probability distributions of distinct families (as subscripted) are identical. Then Breiman's results imply that portfolio strategies Λ which maximize $E \log X_n$, where X_n is the wealth at the end of the n-th time period, have the following properties:

Property 1. (Maximizing $E \log X_n$ asymptotically maximizes the rate of asset growth.) If, for each time period, two portfolio managers have the same family of investment opportunities or investment universes, and one uses a strategy Λ^* maximizing $E \log X_n$ whereas the other uses an "essentially different" (i.e., $E \log X_n(\Lambda^*) - E \log X_n(\Lambda) \to \infty$) strategy Λ, then $\lim X_n(\Lambda^*)/X_n(\Lambda) \to \infty$ almost surely (a.s.).

Property 2. The expected time to reach a fixed preassigned goal x is, asymptotically as x increases, least with a strategy maximizing $E \log X_n$.

The qualification "essentially different" conceals subtleties which are not generally appreciated. For instance [11], which is close in method to this article, and whose conclusions we heartily endorse, contains numerous mathematically incorrect statements and several incorrect conclusions, mostly from overlooking the requirement "essentially different." We intend to present a detailed analysis elsewhere and only indicate the problem here: If $x_j = X_j/X_{j-1}$, then even though $E \log x_j > 0$ for all j it need not be the case that $P(\lim X_n = \infty) = 1$. In fact, we can have (just as in the case of Bernoulli trials and $E \log x_j = 0$; see [26]) $P(\lim\sup X_n = \infty) = 1$ and $P(\lim\inf X_n = 0) = 1$ (contrary to [11, p. 522, eq. (18)] and following assertions). Similarly, when $E \log x_j < 0$ for all j we can have these alternatives instead of $P(\lim X_n = 0) = 1$ (contrary to [11, p. 522, eq. (17)] and the following statements; footnote 1 is also incorrect.)

Note [6] that with the preceding assumptions, there is a fixed fraction strategy Λ which maximizes $E \log X_n$. A fixed fraction strategy is one in which the fraction of wealth $f_{n,j}$ allocated to investment $X_{n,j}$ is independent of n.

We emphasize that Breiman's results can be extended to cover many if not most of the more complicated situations which arise in real world portfolios. Specifically, the number and distribution of investments can vary with the time period, the random variables need not be finite or even discrete, and a certain amount of dependence can be introduced between the investment universes for different time periods. We have used such extensions in certain applications (e.g., [25; 26, p. 287]).

We consider almost surely having more wealth than if an "essentially different" strategy were followed, as the desirable objective for most institutional portfolio managers. (It also seems appropriate for wealthy families who wish mainly to accumulate and whose consumption expenses are only a small fraction of their total wealth.) Property 1 says that maximizing $E \log X_n$ is a recipe for approaching this goal asymptotically as n increases. This is our principal justification for selecting $E \log X$ as the guide to portfolio selection.

In any real application n is finite, the limit is not reached, and we have $P(X_n(\Lambda^*)/X_n(\Lambda) > 1 + M) = 1 - \epsilon(n, \Lambda, M)$ where $\epsilon \to 0$ as $n \to \infty$, $M > 0$ is given, Λ^* is the strategy which maximizes $E \log X_n$ and Λ is an "essentially different" strategy. Thus in any application it is important to have an idea of how rapidly $\epsilon \to 0$. Work needs to be done on this in order to reduce $E \log X$ to a guide that is useful (not merely valuable) for portfolio managers. Some illustrative examples for $n = 6$ appear in [11].

Property 2 shows us that maximizing $E \log X$ also is appropriate for individuals who have a set goal (e.g., to become a millionaire).

Appreciation of the compelling properties of the Kelly criterion may have been impeded by

Reprinted from the 1971 Business and Economics Statistics Section Proceedings
of the American Statistical Association

certain misunderstandings about it that persist in the literature of mathematical economics.

The first misunderstanding involves failure to distinguish among kinds of utility theories. We compare and contrast three types of utility theories: (1) descriptive, where data on observed behavior is fitted mathematically. Many different utility functions might be needed, corresponding to widely varying circumstances, cultures, or behavior types[1]; (2) predictive, which "explains" observed data: fits for observed data are deduced from hypotheses, with the hope future data will also be found to fit. Many different utility functions may be needed, corresponding to the many sets of hypotheses that may be put forward; (3) prescriptive (also called normative), which is a guide to behavior, i.e., a recipe for optimally achieving a stated goal. It is not necessarily either descriptive or predictive nor is it intended to be so.

We use logarithmic utility in this last way, and much of the misunderstanding of it comes from those who think it is being proposed as a descriptive or a predictive theory. The $E \log X$ theory is a prescription for allocating resources so as to (asymptotically) maximize the rate of growth of assets.

Another "objection" voiced by some economists to $E \log X$ and, in fact to all unbounded utility functions, is that it doesn't resolve the (generalized) Petersburg paradox. The rebuttal is blunt and pragmatic: The generalized Petersburg paradox does not arise in the real world because any one real world random variable is bounded (as is any finite collection). Thus in any real application the paradox does not arise.

To insist that a utility function resolve the paradox is an artificial requirement, certainly permissible, but obstructive and tangential to the goal of building a theory which is also a practical guide.

2. Samuelson's objections to logarithmic utility. Samuelson [21, pp. 245-6; 22, pp. 4-5] says that repeatedly authorities [5, 6, 14, 15, 30] "... have proposed a drastic simplification of the decision problem whenever T [the number of investment periods][2] is large.

Rule: Act in each period to maximize the geometric mean or the expected value of $\log x_t$.

The plausibility of such a procedure comes from the recognition of the following valid asymptotic result.

Theorem: Acting to maximize the geometric mean at every step will if the period is "sufficiently long," "almost certainly"[3] result in higher terminal wealth and terminal utility than from any other decision rule." ...

"From this indisputable fact, it is apparently tempting to believe in the truth of the following false corollary:

False corollary: If maximizing the geometric mean almost certainly leads to a better outcome, then the expected value utility of its outcomes exceeds that of any other rule, provided T is sufficiently large."

Samuelson then gives counterexamples to the corollary. We heartily agree that the corollary is false. In fact we had already shown this for one of the utilities Samuelson uses, for we noted [26] that in the case of Bernoulli trials with probability $1/2 < p < 1$ of success, one should commit a fraction $w = 1$ of his capital at each trial to maximize expected final gain $E X_n$ (page 283; the utility is $U(x) = x$) whereas to maximize $E \log X_n$ he should commit $w = 2p - 1$ of his capital at each trial (page 285, Theorem 4).

The statements which we have seen in print supporting this "false corollary" are by Latané [15, p. 151, fn. 13] as discussed in [21, p. 245, fn. 8], and Markowitz [16, pp. ix-x]. Latané may not have fully supported this corollary for he adds the qualifier "... (in so far as certain approximations are permissible)...".

That there were or are adherents of the "false corollary" seems puzzling in view of the following formulation. Consider a T stage investment process. At each stage we allocate our resources among the available investments. For each sequence A of allocations which we choose, there is a corresponding terminal probability distribution F_T^A of assets at the completion of stage T. For each utility function $U(\cdot)$, consider those allocations $A^*(U)$ which maximize the expected value of terminal utility $\int U(x)\, dF_T^A(x)$. Assume sufficient hypotheses on U and the set of F_T^A so that the integral is defined and that furthermore the maximizing allocation $A^*(U)$ exists. Then Samuelson says that $A^*(\log)$ is not in general $A^*(U)$ for other U. This seems intuitively evident.

Even more seems strongly plausible: that if U_1 and U_2 are inequivalent utilities then $\int U_1(x)\, dF_T^A(x)$ and $\int U_2(x)\, dF_T^A(x)$ will in general be maximized for different F_T^A. (Two utilities U_1 and U_2 are equivalent if and only if there are constants a and b such that $U_2(x) = aU_1(x) + b$, $a > 0$; otherwise U_1 and U_2 are inequivalent.) In this connection we have proved:

Theorem: Let U and V be utilities defined and differentiable on $(0, \infty)$, with $U'(x)$ and $V'(x)$ positive and strictly decreasing as x increases. Then if U and V are inequivalent, there is a one period investment setting such that U and V have distinct optimal strategies.[4]

All this is in the nature of an aside for Samuelson's correct criticism of the "false corollary" does not apply to our use of logarithmic utility. Our point of view is: if your goal is property 1 or property 2, then a recipe for achieving either goal is to maximize $E \log X$. These properties distinguish log from the prolixity of utility functions in the literature. Furthermore, we consider these goals appropriate for many (but not all) investors. Investors with other utilities, or with goals incompatible with logarithmic utility, will of course, find it inappropriate.

Property 1 implies that if Λ^* maximizes $E \log X_n(\Lambda)$ and Λ' is "essentially different," then $X_n(\Lambda^*)$ tends almost certainly to be better than $X_n(\Lambda')$ as $n \to \infty$. Samuelson [21, p. 246], apparently referring to this, says after refuting the "false corollary": "Moreover, as I showed elsewhere [20, p. 4], the ordering principle of selecting between two actions in terms of which has the greater probability of producing a higher result does not even possess the property of being transitive. ...we could have w^{***} better than w^{**}, and w^{**} better than w^*, and also have w^* better than w^{***}."

For some entertaining examples, see the discussion of non-transitive dice in [9]. (Consider the dice with equiprobable faces numbered as follows: $X = (3, 3, 3, 3, 3, 3)$; $Y = (4,4,4,4,1,1)$; $Z = (5,5,2,2,2,2)$. Then $P(Z>Y) = 5/9$, $P(Y>X) = 2/3$, $P(X>Z) = 2/3$.) What Samuelson does not tell us is that the property of producing a higher result almost certainly, as in property 1, is transitive. If we have $w^{***} > w^{**}$ almost certainly, and $w^{**} > w^*$ almost certainly, then we must have $w^{***} > w^*$ almost certainly.

One might object [20, p. 6] that in a real investment sequence the limit as $n \to \infty$ is not reached. Instead the process stops at some finite N. Thus we do not have $X_N(\Lambda^*) > X_n(\Lambda')$ almost certainly. Instead we have $P(X_n(\Lambda^*) > X_n(\Lambda')) = 1 - \epsilon_N$ where $\epsilon_N \to 0$ as $N \to \infty$, and transitivity can be shown to fail.

This is correct. But an approximate form of transitivity does hold: Let X, Y, Z be random variables with $P(X>Y) = 1 - \epsilon_1$, $P(Y>Z) = 1 - \epsilon_2$. Then $P(X>Z) \geqq 1 - (\epsilon_1 + \epsilon_2)$. To prove this, let A be the event $X>Y$, B be the event $Y>Z$, and C be the event $X>Z$. Then $P(A)+P(B) = P(A \cup B) + P(A \cap B) \leqq 1 + P(A \cap B)$. But $A \cap B \subset C$ so $P(C) \geqq P(A \cap B) \geqq P(A) + P(B) - 1$, i.e., $P(X>Z) \geqq 1 - (\epsilon_1 + \epsilon_2)$.

Thus our approach is not affected by the various Samuelson objections to the uses of logarithmic utility.

Markowitz [16, pp. ix-x] says "...in 1955-56, I concluded ... that the investor who is currently reinvesting everything for "the long run" should maximize the expected value of the logarithm of wealth." (This assertion seems to be regardless of the investor's utility and so indicates belief in the "false corollary.") Mossin [18] and Samuelson [20] "...have each shown that this conclusion is not true for a wide range of [utility] functions ... The fascinating Mossin-Samuelson result, combined with the straightforward arguments supporting the earlier conclusions, seemed paradoxical at first. I have since returned to the view of Chapter 6 (concluding that: for large T, the Mossin-Samuelson man acts absurdly, ..." Markowitz says here, in effect, that alternate utility functions (to log) are absurd. This position is unsubstantiated and unreasonable.

He continues "... like a player who would pay an unlimited amount for the St. Petersburg game ..." If you agree with us that the St. Petersburg game is not realizable and may be ignored when fashioning utility theories for the real world, then his continuation "... the terminal utility function must be bounded to avoid this absurdity; ..." does not follow.

Finally, Markowitz says "... and the argument in Chapter 6 applies when utility of terminal wealth is bounded." If he means by this that the "false corollary" holds if we restrict ourselves to bounded utility functions, then he is mistaken. Mossin [18] already showed that the optimal strategies for $\log x$ and x^γ/γ, $\gamma \neq 0$, are fixed fraction for these and only these utilities. Thus any bounded utility besides x^γ/γ, $\gamma<0$, will have optimal strategies which are not fixed fraction, hence not optimal for $\log x$. Samuelson [22] gives counterexamples which include the bounded utilities x^γ/γ, $\gamma < 0$. Since Mossin assumes U'' exists and our theorem only assumes that U' exists, it provides additional counterexamples.

3. An outline of the theory of logarithmic utility as applied to portfolio selection. The simplest case is Bernoulli trials with probability p of success, $0<p<1$. The unique strategy which maximizes $E \log X_n$ is to bet at trial n the fixed fraction $f^* = p - q$ of total current wealth X_{n-1} if $p > 1/2$ and to bet nothing otherwise.

To maximize $E \log X_n$ is equivalent to maximizing $E \log [X_n/X_0]^{1/n} \equiv G(f)$, which we call the (exponential) rate of growth (per time period). It turns out that for $p > 1/2$, $G(f)$ has a unique positive maximum at f^* and that there is a critical fraction f_c, $0 < f^* < f_c < 1$, such that $G(f_c) = 0$, $G(f)>0$ if $0 < f < f_c$, $G(f)<0$ if $f_c < f \leqq 1$ (we assume "no margin"; the case with margin is similar). If $f < f_c$, $X_n \to \infty$ a.s.; if $f = f_c$, $\lim \sup X_n = +\infty$ a.s., and $\lim \inf X_n = 0$ a.s.; if $f > f_c$, $\lim X_n = 0$ a.s. ("ruin").

Bernoulli trials exhibit many of the features

of the following more general case. Suppose we have at each trial $n = 1, 2, \cdots$ the k investment opportunities $X_{n,1}, X_{n,2}, \cdots, X_{n,k}$ and that the conditions of property 1, section 1 are satisfied. This means that the joint distributions of $\{X_{n,i_1}, X_{n,i_2}, \cdots, X_{n,i_j}\}$ are the same for all n, for each subset of indices $1 \leq i_1 < i_2 < \cdots < i_j \leq k$. Furthermore $\{X_{m,1}, \cdots, X_{m,k}\}$ and $\{X_{n,1}, \cdots, X_{n,k}\}$ are independent when $m \neq n$, and all random variables $X_{i,j}$ have only a finite number of distinct values. Thus we have in successive time periods repeated independent trials of "the same" investment universe.

Since Breiman has shown that there is for this case an optimal fixed fraction strategy $\Lambda^* = (f_1^*, \cdots, f_k^*)$, we will have an optimal strategy if we find a strategy which maximizes $E \log X_n$ in the class of fixed fraction strategies.

Let $\Lambda = (f_1, \cdots, f_k)$ be any fixed fraction strategy. We assume that $f_1 + \cdots + f_k \leq 1$ so there is no borrowing, or margin. The margin case is similar (the approach resembles [23]). Using the concavity of the logarithm, it is easy to show (see below) that the exponential rate of growth $E \log [X_n(\Lambda)/X_0]^{1/n} = G(f_1, \cdots, f_k)$ is a concave function of $(f_1, \cdots, f_k)$, just as in the Bernoulli trials case. The domain of $G(f)$ in the Bernoulli trials case was the interval $[0,1)$ with $G(f) \downarrow -\infty$ as $f \to 1$. The domain in the present instance is analogous. First, it is a subset of the k dimensional simplex $S_k = \{(f_1, \cdots, f_k): f_1 + \cdots + f_k \leq 1; f_1 \geq 0, \cdots, f_k \geq 0\}$.

To establish the analogy further, let $R_j = X_{n,j} - 1$, $j = 1, \cdots, k$, be the return per unit on the i-th investment opportunity at an arbitrary time period n. Let the range of R_j be $\{r_{j,1}; \cdots; r_{j,i_j}\}$ and let the probability of the outcome $[R_1 = r_{1,m_1}$ and $R_2 = r_{2,m_2}$ and $\cdots$ and $R_k = r_{k,m_k}]$ be $p_{m_1,m_2,\cdots,m_k}$. Then $E \log X_n/X_{n-1} = G(f_1, \cdots, f_k)$ $= \Sigma \{p_{m_1,\cdots,m_k} \log(1 + f_1 r_{1,m_1} + \cdots + f_k r_{k,m_k}):$ $1 \leq m_1 \leq i_1; \cdots; 1 \leq m_k \leq i_k\}$, from which the concavity of $G(f_1, \cdots, f_k)$ can be shown. Note that $G(f_1, \cdots, f_k)$ is defined if and only if $1 + f_1 r_{1,m_1} + \cdots + f_k r_{k,m_k} > 0$ for each set of indices $m_1, \cdots, m_k$. Thus the domain of $G(f_1, \cdots, f_k)$ is the intersection of all these open half-spaces with the k-dimensional simplex S_k. Note that the domain is convex and includes all of S_k in some neighborhood of the origin. Note too that the domain of G is all of S_k if (and only if) $R_j > -1$ for all j, i.e., if there is no probability of total loss on any investment. The domain of G includes the interior of S_k if $R_j \geq -1$. Both domains are particularly simple and most cases of interest are included.

If $f_1, \cdots, f_k$ are chosen so that $1 + f_1 r_{1,m_1} + \cdots + f_k r_{k,m_k} \leq 0$ for some $m_1, \cdots, m_k$, then $P(f_1 X_{n,1} + \cdots + f_k X_{n,k} \leq 0) = \varepsilon > 0$ for all n and ruin occurs with probability 1.

Computational procedures for finding an optimal fixed fraction strategy (generally unique in our present setting) are based on the theory of concave (dually, convex) functions [29] and will be presented elsewhere. (As Hakansson [11, p. 552] has noted, "... the computational aspects of the capital growth model are [presently] much less advanced" than for the Markowitz model.)

The theory may be extended to more general random variables and to dependence between different time periods. Most important, we may include the case where the investment universe changes with the time period, provided only that there be some mild regularity conditions on the $X_{i,j}$, such as that they be uniformly a.s. bounded and that they do not tend to 0 uniformly as $i \to \infty$. (See [15], and the generalization of the Bernoulli trials case as applied to blackjack in [26].) The techniques rely heavily on those used to generalize the law of large numbers.

Transactions costs, the use of margin, and the effect of taxes can be incorporated into the theory. Bellman's dynamic programming method is used here.

The general procedure for developing the theory into a practical tool imitates Markowitz [16]. Markowitz requires as inputs estimates of the expectations, standard deviations, and covariances of the $X_{i,j}$. We require joint probability distributions. This would seem to be a much more severe requirement, but in practice does not seem to be so [16, pp. 193-4, 198-201].

Among the actual inputs which Markowitz chose were (1) past history [16, ex., 8-20], (2) probability beliefs of analysts (pp. 26-33), and (3) models, most notably regression models, to predict future performance from past data (p. 33, pp. 99-100). In each instance one can get enough additional information to estimate $E \log (X_n/X_{n-1})$.

There are, however, two great difficulties which all theories of portfolio selection have, including ours and that of Markowitz. First, there seems to be no established method for generally predicting security prices which gives an edge of even a few per cent. The random walk is the best model for security prices today. (See [7, 10].)

The second difficulty is that for portfolios with many securities the volume of inputs called for is prohibitive: for 100 securities, Markowitz requires 100 expectations and 4950 covariances; and our theory requires somewhat more information. Although considerable attention has been given to finding condensed inputs that can be used instead, this aspect of portfolio theory still seems unsatisfactory.

In the fifth section we will show how both these difficulties were overcome in practice by an institutional investor. That investor, guided by the Kelly criterion, then outperformed for the year 1970 every one of the approximately 400 Mutual Funds listed by the S & P stock guide!

But first we relate our theory to that of Markowitz.

4. <u>Relation to the Markowitz theory; solution to problems therein.</u> The most widely used guide to portfolio selection today is probably the Markowitz theory. The basic idea is that a portfolio P_1 is superior to a portfolio P_2 if the expectation ("gain") is at least as great, i.e., $E(P_1) \geqq E(P_2)$ and the standard deviation ("risk") is no greater, i.e., $\sigma(P_1) \leqq \sigma(P_2)$, with at least one inequality. This partially orders the set $\mathcal{P}$ of portfolios. A portfolio such that no portfolio is superior (i.e., a maximal portfolio in the partial ordering) is called <u>efficient</u>. The goal of the portfolio manager is to determine the set of efficient portfolios, from which he then makes a choice based on his needs.

This is intuitively very appealing: It is based on standard quantities for the securities in the portfolio, namely expectation, standard deviation, and covariance (needed to compute the variance of the portfolio from that of the component securities). It also gives the portfolio manager "choice."

As Markowitz [16, Chapter 6] has pointed out, the optimal Kelly portfolio is approximately one of the Markowitz efficient portfolios under certain circumstances. If $E = E(P)$ and $R = P - 1$ is the return per unit of the portfolio P, let $\log P = \log(1+R) = \log((1+E) + (R-E))$. Expanding in Taylor's series about $1+E$ gives $\log P = \log(1+E) + (R-E)/(1+E) - (R-E)^2/2(1+E)^2 +$ higher order terms. Taking expectations and neglecting higher order terms gives $E \log P = \log(1+E) - \sigma^2(P)/2(1+E)^2$.

This leads to a simple pictorial relationship with the Markowitz theory. Consider the E-σ plane, and plot $(E(P), \sigma(P))$ for the efficient portfolios. The locus of efficient portfolios is a convex non-decreasing curve which includes its endpoints (Figure 1).

Then constant values of the growth rate

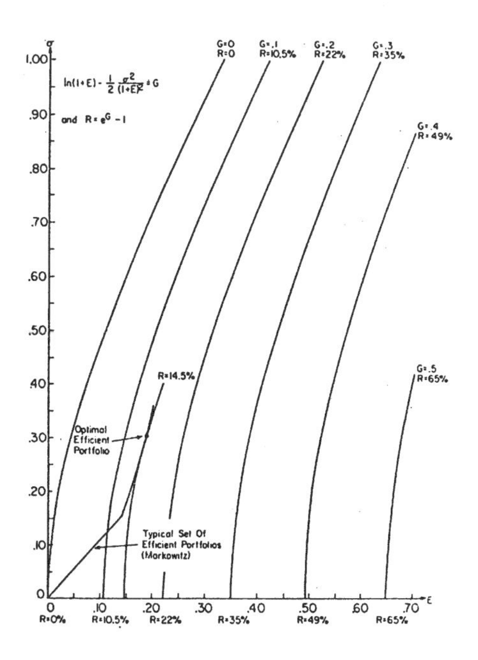

FIGURE 1. GROWTH RATE G (RETURN RATE R) IN THE E-σ PLANE ASSUMING THE VALIDITY OF THE POWER SERIES APPROXIMATION.

$G = E \log P$ approximately satisfy $G = \log(1+E) - \sigma^2(P)/2(1+E)^2$. This family of curves is illustrated in Figure 1 and the (efficient) portfolio which maximizes logarithmic utility is (approximately) the one which lies on the greatest G curve. Because of the convexity of the curve of efficient portfolios and the concavity of the G curves, the (E, σ) value where this occurs is unique.

The approximation to G breaks down badly in some significant practical settings, including that of the next section. But for portfolios with large numbers of "typical" securities, the approximation for G will generally provide an efficient portfolio which approximately maximizes asset growth. This solves the portfolio manager's problem of which efficient portfolio to choose. Also, if he repeatedly chooses his portfolio in successive time periods by this criterion he will tend to maximize the rate of growth of his assets, i.e., maximize "performance." We see also that in this instance the problem is reduced to that of finding the efficient portfolios plus the easy step of using Figure 1. Thus if the Markowitz theory can be applied in practice in this setting, so can our theory.

We have already remarked on the ambiguity of the set of efficient portfolios, and how our theory resolves them. To illustrate further that such ambiguity represents a defect in the Markowitz theory, let X_1 be uniformly distributed over $[1, 3]$, let X_2 be uniformly distributed over $[10, 100]$, let $\text{cor}(X_1, X_2) = 1$, and suppose these are the only securities. Then X_1 and X_2 are both efficient with $\sigma_1 < \sigma_2$ and $E_1 < E_2$ so Markowitz' theory does not choose between them. Yet "everyone" would choose X_2 over X_1 because the worst outcome with X_2 is far better than the best outcome from X_1. (We presented this example in [26]. Hakansson [11] presents further examples and an extended analysis. He formalizes the idea by introducing the notion of stochastic dominance: X stochastically dominates Y if $P(X \geq Y) = 1$ and $P(X > Y) > 0$. It is easy to prove the Lemma: An $E \log X$ optimal portfolio is never stochastically dominated. Thus our portfolio theory does not have this defect.)

There are investment universes $(X_1, \cdots, X_n)$ such that a unique portfolio P maximizes $E \log P$ yet P is not efficient in the sense of Markowitz. Hence choosing P in repeated independent trials will outperform any strategy limited to choosing efficient portfolios. In addition, the optimal Kelly strategy gives positive growth rate, yet some of the Markowitz-efficient strategies give negative growth rate and ruin after repeated trials. We gave such an example in [26] and another appears in [11]. See also [11, pp. 553-4] for further discussion of defects in the Markowitz model.

5. <u>The theory in action: results for a real institutional portfolio</u>. The elements of a practical profitable theory of convertible hedging were published in [27]. Thorp and Kassouf indicated an annualized return on investments of the order of 25% per year. Since then the theory has been greatly extended and refined with most of these new results thus far unpublished.

The historical data which has been used to develop the theory includes well over 100,000 observations of convertibles.

A convertible hedge transaction generally involves <u>two</u> securities, one of which is convertible into the other. Mathematical price relationships exist between pairs of such securities. When one of the pair is comparatively underpriced, a profitable convertible hedge may be set up by buying the relatively underpriced security and selling short an appropriate amount of the relatively overpriced security.

The purpose of selling short the overpriced security is to reduce the risk in the position. Typically, one sells short in a single hedge from 50% to 125% as much stock (in "share equivalents") as is held long. The exact proportions depend on the analysis of the specific situation; the quantity of stock sold short is selected to minimize risk. The risk (i.e., change in asset value with fluctuations in market prices) in a suitable convertible hedge should be much less than in the usual stock market long positions.

The securities involved in convertible hedges include common stock, convertible bonds, convertible preferreds, and common stock purchase warrants. Options such as puts, calls, and straddles may replace the convertible security. For this purpose, the options may be either written or purchased.

The theory of the convertible hedge is highly enough developed so that the probability characteristics of a single hedge can be worked out based on an assumption for the underlying distribution of the common. (Sometimes even this can almost be dispensed with! See [27, App. C].) A popular and plausible assumption is that the future price of the common is lognormally distributed about its' current price, with a trend and a variance proportional to the time. Plausible estimates of these parameters are readily obtained. Furthermore, it turns out that the return from the hedge is comparatively insensitive to changes in the estimates for these parameters.

Thus with convertible hedging we fulfill two important conditions for the practical application of our (or any other) theory of portfolio choice: (1) We have identified investment opportunities which are markedly superior to the usual ones. Compare the return rate of 20%-25% per year with the long term rate of 8% or so for listed common stocks. Further, it can be shown that the risks tend to be much less. (2) The probability inputs are available for computing $G(f_1, \cdots, f_n)$.

On November 3, 1969, a private institutional investor decided to commit all its resources to convertible hedging and to use the Kelly criterion to allocate its assets. The performance record appears in the Table.

The market period covered included one of the sharpest falling markets as well as one of the sharpest rising markets (up 50% in 11 months) since World War II. The gain was +16.3% for the year 1970, which outperformed all of the approximately 400 Mutual Funds Listed in the S & P stock guide. Unaudited figures show that gains were achieved during every single month.

The unusually low risk in the hedged positions is also indicated by the results for the 200 completed hedges. There were 190 winners, 6 break-evens, and 4 losses. The losses as a per cent of the long side of the specific investment

TABLE - PERFORMANCE RECORD

Date	Change To Date (%)	Elapsed Time (months)	Growth Rate To Date (%)++	Closing DJIA+	DJIA Chg. (%)++	Starting Even With DJIA+	Gain Over DJIA (%)
11- 3-69	0.0	0	---	855	0.0	855	0.0
12-31-69	+ 4.0	2	+26.8	800	- 6.3	889	+10.3
9- 1-70	+14.0	10	+17.0	758	-11.3	974	+25.3
12-31-70	+21.0	14	+17.7	839	- 1.8	1034	+22.8
6-30-71*	+39.9	20	+22.3	891	+ 4.2	1196	+35.7

Assets on 7-1-71 were 5.2 million.

* Preliminary-unaudited.

\+ DJIA = Dow Jones Industrial Average.

++ Compound growth rate, annualized.

ranged from 1% to 15%.

A characteristic of the Kelly criterion is that as risk decreases and expectation rises, the optimal fraction of assets to be invested in a single situation may become "large." On several occasions, the institution discussed above invested up to 30% of its assets in single hedge. Once it invested 150% of its assets in a single arbitrage. This characteristic of Kelly portfolio strategy is not part of the behavior of most portfolio managers.

To indicate the techniques and problems, we consider a simple portfolio with just one convertible hedge. We take as our example Kaufman and Broad common stock and warrants. A price history is indicated in Figure 2.

Price data shows that $W = .455S$ is a reasonable fit for $S \leqq 38$ and that $W = S - 21.67$ is a reasonable fit for $S \geqq 44$. Between $S = 38$ and $S = 44$ we have the line $W = .84S - 15.5$. For simplicity of calculation we replace this in our illustrative analysis by $W = .5S$ if $S \leqq 44$ and $W = S - 22$ if $S \geqq 44$. The lines are also indicated in Figure 2.

Past history at the time the hedge was instituted in late 1970 supported the fit for $S \leqq 38$. The conversion feature of the warrant ensured $W \geqq S - 21.67$ until the warrant expires. Thus $W = S - 21.67$ for $S \geqq 44$ underestimates the price of the warrant in this region. Extensive historical studies of warrants [12, 13, 24, 27] show that the past history fit would probably be maintained until about two years before expiration, i.e., until about March, 1972. Thus it is plausible to assume that for the next 1.3 years S may be roughly approximated by $W = .5S$ for $S \leqq 44$ and $W = S - 22$ for $S \geqq 44$.

Next we assume that S_t, the stock price at time $t > 0$ years after the hedge was initiated, is lognormally distributed with density

$f_{S_t}(x) = (x\sigma\sqrt{2\pi})^{-1} \exp[-(\log x - \mu)^2/2\sigma^2]$, hence mean $E(S_t) = \exp(\mu + \sigma^2/2)$ and standard deviation $\sigma(S_t) = E(S_t)\left(e^{\sigma^2} - 1\right)^{1/2}$. The functions $\mu \equiv \mu(t)$ and $\sigma \equiv \sigma(t)$ depend on the stock and on the time t. If t is the time in years until S_t is realized, it is plausible to assume $\mu(t) = \log S_0 + mt$ and $\sigma(t)^2 = a^2 t$, where S_0 is the present stock price and m and a are constants depending on the stock. For a detailed discussion, see [1, 19].

Then $E(S_t) = S_0 \exp[(m + a^2/2)t]$ and a mean increase of 10%/year is approximated by setting $m + a^2/2 = .1$. If we estimate a^2 from past price changes we can solve for m. In the case of Kaufman and Broad it is plausible to take $\sigma \doteq .45$ whence $a^2 = \sigma^2 \doteq .20$. This yields $m \doteq 0$. We then find $\sigma(S_t) \doteq .52 S_0$.

It is by no means established that the lognormal model is the appropriate one for stock price series [7, 10]. However, once we clarify certain general principles by working through our example on the basis of the lognormal model, it can be shown that the results are substantially unchanged by choosing instead any distribution that roughly fits observation!

For a time of one year, a computation shows the return $R(S)$ on the stock to be +10.5%, the return $R(W)$ on the warrant to be +34.8%, $\sigma(S) = .52$, $\sigma(W) = .92$, and the correlation coefficient $\mathrm{cor}(S, W) = .99$. The difference in $R(S)$ and $R(W)$ shows that the warrant is a much better buy than the common. Thus a hedge long warrants and short common has a substantial positive expectation. The value $\mathrm{cor}(S, W) = .99$ shows that a

FIGURE 2. PRICE HISTORY OF KAUFMAN AND BROAD COMMON, S, VERSUS THE WARRANTS, W.

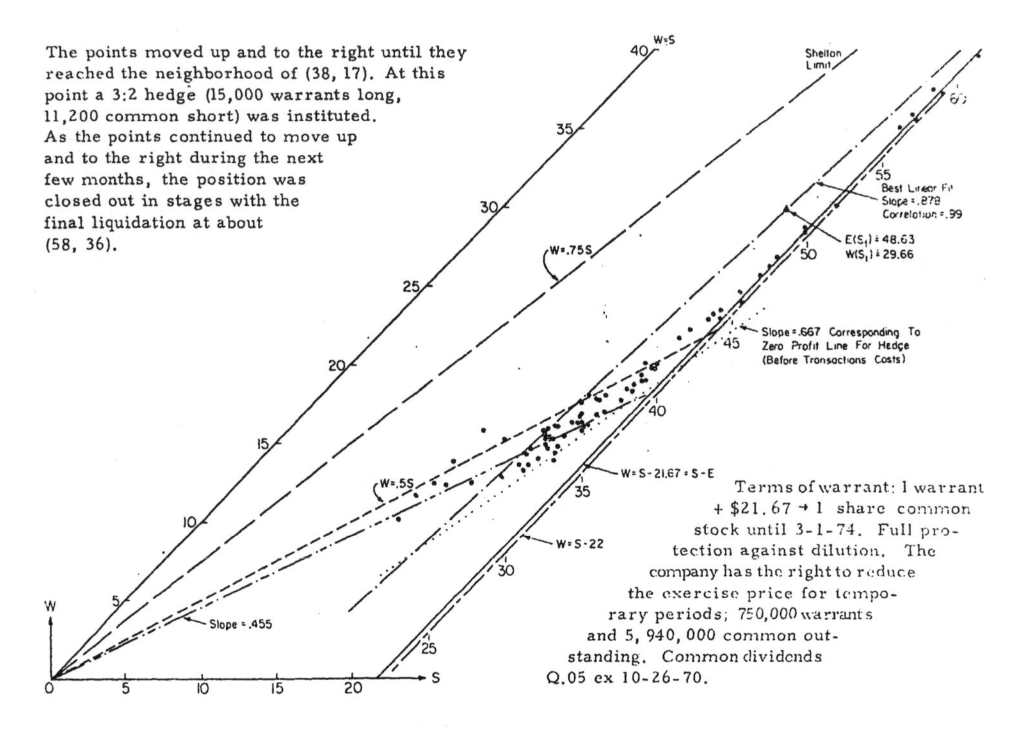
The points moved up and to the right until they reached the neighborhood of (38, 17). At this point a 3:2 hedge (15,000 warrants long, 11,200 common short) was instituted. As the points continued to move up and to the right during the next few months, the position was closed out in stages with the final liquidation at about (58, 36).
W=S
Shelton Limit
W=.75S
W=.5S
Slope = .455
Best Linear Fit Slope = .878 Correlation = .99
E(S1) = 48.63
W(S1) = 29.66
Slope = .667 Corresponding To Zero Profit Line For Hedge (Before Transactions Costs)
W = S - 21.67 = S - E
W = S - 22
W
S
0
5
10
15
20
25
30
35
40
45
50
55
Terms of warrant: 1 warrant + $21.67 → 1 share common stock until 3-1-74. Full protection against dilution. The company has the right to reduce the exercise price for temporary periods; 750,000 warrants and 5,940,000 common outstanding. Common dividends Q.05 ex 10-26-70.

hedge corresponding to the best linear fit of W to S has a standard deviation of approximately $(1-.99)^{1/2} = .1$ which suggests that $\sigma(P)$ for the optimal hedged portfolio is probably going to be close to .1. The high return and low risk for the hedge will remain, it can be shown, under wide variations in the choice of m and a.

To calculate the optimal mix of warrants long to common short we maximize $G(f_1, f_2) = E \log(1 + f_1 S + f_2 W)$. The detailed computational procedures are too lengthy and involved to be presented here. We plan to present them elsewhere.

Our institutional investor considered positions already held, some of which might have to be closed out to release assets, and also other current candidates for investment. The decision was made to short common and buy warrants in the ratio of three shares to four. The initial market value of the long side was about 14% of assets and for the short side about 20% of assets The net profit, in terms of the initial market value of the long side, was about 20% in six months. This resulted from a move in the common from about 40 to almost 60.

6. Concluding remarks. As remarked above, we do not propose logarithmic utility as descriptive of actual investment behavior, nor do we believe any one utility function could suffice. It would be of interest, however, to have empirical evidence showing areas of behavior which are characterized adequately by logarithmic utility. Neither do we intend logarithmic utility to be predictive; again, it would be of interest to know what it does predict.

We only propose the theory to be normative or prescriptive, and only for those institutions, groups, or individuals whose overriding current objective is maximization of the rate of asset growth. Those with a different "prime directive" may find that another utility function is a better guide.

We have found E log X to be valuable as a qualitative guide and suggest that this could be its most important use. Once familiarity with its properties is gained, many investment decisions can be guided by it without complex supporting calculations.

What sort of economic behavior can be expected from followers of E log X ? Insurance is "explained," i.e., even though it is a negative expectation investment for the insured and we assume both insurer and insured have the same probability information, it is often optimal for him (as well as for the insurance company) to insure [3]. It usually turns out that insurance against large losses is indicated and insurance against small losses is not. (Don't insure an old car for collision, take $200 deductible, not $25, etc.)

We find that if all parties to a security transaction are followers of E log X they will often find it mutually optimal to trade. This may be true whether the transactions be two party (no brokerage), or three party (brokerage), and whether or not the parties have the same probability information about the security involved, or even about the entire investment universe.

Maximizing logarithmic utility excludes portfolios which have positive probability of total loss of assets. Yet it can be argued that an impoverished follower of E log X might in some instances risk "everything." This agrees with some observed behavior, but is not what we might at first expect in view of the prohibition against positive probability of total loss. But consider each individual as a piece of capital equipment with an assignable monetary value. Then if he risks and loses all his cash assets, he hasn't really lost everything [3].

All of us behave as though death itself does not have infinite negative utility. Since the risk of death, although generally small, is ever present, a negative infinite utility for death would make all expected utilities negative infinite and utility theory meaningless. In the case of logarithmic utility as applied to the extended case of the (monetized) individual plus all his resources, death should be assigned a finite, though large and negative, utility. The value of this "death constant" is an additional arbitrary assumption for the enlarged theory of logarithmic utility.

In the case of investors who behave according to E log X (or other utilities unbounded below), it might be possible to discover their tacit "death constants."

Hakansson [11, p. 551] observes that logarithmic utility exhibits decreasing absolute risk aversion in agreement with deductions of Arrow and others on the qualities of "reasonable" utility functions. Hakansson says, "What the relative risk aversion index [given by $-xU''(x)/U'(x)$] would look like for a meaningful utility function is less clear ... In view of Arrow's conclusion that '... broadly speaking, the relative risk aversion must hover around 1, being, if anything, somewhat less for low wealths and somewhat higher for high wealths ...' the optimal growth model seems to be on safe ground." As he notes, for $U(x) = \log x$, the relative risk aversion is precisely 1. However, in both the extension to valuing the individual as capital equipment, and the further extension to include the death constant, we are led to $U(x) = \log(x+c)$ where c is positive. But then the relative risk aversion index is $x/(x+c)$ which behaves strikingly like Arrow's description. See also the discussion of $U(x) = \log(x+c)$ in [8, p. 103, p. 112].

Morgenstern [17] has forcefully observed that assets are random variables, not numbers, and that economic theory generally does not incorporate this. To replace assets by their expected utility in valuing companies, portfolios, property and the like, allows for comparisons when asset values are given as random variables. We think logarithmic utility will often be appropriate for such valuation.

I wish to thank James Bicksler for several stimulating and helpful conversations.

FOOTNOTES

*Edward O. Thorp is professor, Department of Mathematics, University of California at Irvine. This research was supported in part by the Air Force Office of Scientific Research under Grant AF-AFOSR 1870A. An expanded version of this paper will be submitted for publication elsewhere.

[1] Information on descriptive utility is sparse; how many writers on the subject have even been able to determine for us their own personal utility?

[2] Parenthetical explanation added since we have used n.

[3] "Almost certainly" and "almost surely" are synonymous.

[4] The proof of this theorem, and some further results obtained with R. Whitley, will appear elsewhere.

REFERENCES

[1] Ayres, Herbert F., "Risk Aversion in the Warrant Markets," S.M. Thesis, M.I.T., Industrial Management Review 5:1 (1963), 45-53. Reprinted in Cootner, pp. 479-505.

[2] Bellman, R. and Kalaba, R., "On the Role of Dynamic Programming in Statistical Communication Theory," IRE Transactions of the Professional Group on Information Theory, IT-3:3 (1957), 197-203.

[3] Bernoulli, Daniel, "Exposition of a New Theory on the Measurement of Risk," Econometrica, XXII (January 1954), 23-36, trans. Louise Sommer.

[4] Borch, Karl H., The Economics of Uncertainty, Princeton University Press, 1968.

[5] Breiman, Leo, "Investment Policies for Expanding Businesses Optimal in a Long Run Sense," Naval Research Logistics Quarterly, 7:4 (1960), 647-651.

[6] Breiman, Leo, "Optimal Gambling Systems for Favorable Games," Fourth Berkeley Symposium on Probability and Statistics, I, (1961), 65-78.

[7] Cootner, P. H., ed., The Random Character of Stock Market Prices, The M.I.T. Press, 1964.

[8] Freimer, Marshall and Gordon, Myron S., "Investment Behavior With Utility a Concave Function of Wealth," in K. Borch and J. Mossin, eds., Risk and Uncertainty, New York: St. Martin's Press, 1968, 94-115.

[9] Gardner, Martin, "Mathematical Games: The Paradox of the Non-Transitive Dice and the Elusive Principle of Indifference," Scientific American (December 1970), 110.

[10] Granger, Clive and Morgenstern, Oskar, Predictability of Stock Market Prices, Lexington, Massachusetts: D. C. Heath and Company, 1970.

[11] Hakansson, Nils, "Capital Growth and the Mean-Variance Approach to Portfolio Selection," Journal of Finance and Quantitative Analysis (January 1971), 517-557.

[12] Kassouf, Sheen T., "A Theory and an Econometric model for common stock purchase warrants," Thesis, Columbia University, 1965; New York: Analytic Publishers Company, 1965. A regression model statistical fit of normal price curves for warrants.. There are large systematic errors in the model due to faulty (strongly biased) regression techniques. The average mean square error in the fit is large. Thus, it is not safe to use the model in practice as a predictor of warrant prices. However, the model and the methodology are valuable as a first qualitative description of warrant behavior and as a guide to a more precise analysis.

[13] Kassouf, Sheen T., "An econometric model for option price with implications for investors' expectations and audacity," Econometrica, 37:4 (1969), 685-694. Based on the thesis. The variance of residuals is given as .248, or a standard deviation of about .50 in y, the normalized warrant price, and a standard error of about .34. The mid-range of y varies from 0 to .5 and is never greater, thus the caveat about not using the model for practical predictions!

[14] Kelly, J. L., "A New Interpretation of Information Rate," Bell System Technical Journal, 35 (1956), 917-926.

[15] Latané, Henry A., "Criteria for Choice Among Risky Ventures," Journal of Political Economy, 67 (1959), 144-155.

[16] Markowitz, H., Portfolio Selection, New York: John Wiley and Sons, Inc., 1959. See also the preface to the second printing of the Yale University Press, 1970 reprint.

[17] Morgenstern, Oskar, On the Accuracy of Economic Observations, 2nd ed. revised, Princeton University Press, 1963.

[18] Mossin, Jan, "Optimal Multiperiod Portfolio Policies," Journal of Business (April 1968).

[19] Osborne, M. F. M., "Brownian Motion in the Stock Market," Operations Research, 7 (1959), 145-173. Reprinted in Cootner, pp. 100-128.

[20] Samuelson, Paul A., "Risk and Uncertainty: A Fallacy of Large Numbers," Scientia, 6th Ser., 57th Yr., (April-May 1963).

[21] Samuelson, Paul A., "Lifetime Portfolio Selection by Dynamic Stochastic Programming," The Review of Economics and Statistics (August 1969), 239-246.

[22] Samuelson, Paul A., "The 'Fallacy' of Maximizing the Geometric Mean in Long Sequences of Investing or Gambling," Unpublished preliminary preprint, 1971.

[23] Schrock, Nicholas W., "The Theory of Asset Choice: Simultaneous Holding of Short and Long Positions in the Futures Market," Journal of Political Economy, 79:2 (1971), 270-293.

[24] Shelton, John P., "The Relation of the Price of a Warrant to Its Associated Common Stock," Financial Analysts Journal, 23:3 (1967). 143-151; and Financial Analysts Journal, 23:4 (1967), 88-99.

[25] Thorp, Edward, "A Winning Bet in Nevada Baccarat," Journal of the American Statistical Association, 61, Part I (1966), 313-328.

[26] Thorp, Edward, "Optimal Gambling Systems for Favorable Games," Review of the International Statistical Institute, 37:3 (1969), 273-293.

[27] Thorp, E. and Kassouf, S., Beat the Market, New York: Random House, 1967.

[28] Todhunter, I., A History of the Mathematical Theory of Probability, 1st ed., Cambridge, 1865, as reprinted by Chelsea, New York, 1965. (See pp. 213 ff. for details on Daniel Bernoulli's use of logarithmic utility.)

[29] Wagner, Harvey M., Principles of Operations Research, With Application to Managerial Decisions, New Jersey: Prentice-Hall, 1969.

[30] Williams, J. B., "Speculation and the Carry-over," Quarterly Journal of Economics, 50, (May 1936), 436-455.

ECONOMETRICA

VOLUME 38 September, 1970 NUMBER 5

8

OPTIMAL INVESTMENT AND CONSUMPTION STRATEGIES UNDER RISK FOR A CLASS OF UTILITY FUNCTIONS[1]

BY NILS H. HAKANSSON[2]

This paper develops a sequential model of the individual's economic decision problem under risk. On the basis of this model, optimal consumption, investment, and borrowing-lending strategies are obtained in closed form for a class of utility functions. For a subset of this class the optimal consumption strategy satisfies the permanent income hypothesis precisely. The optimal investment strategies have the property that the optimal mix of risky investments is independent of wealth, noncapital income, age, and impatience to consume. Necessary and sufficient conditions for long-run capital growth are also given.

1. INTRODUCTION AND SUMMARY

THIS PAPER presents a normative model of the individual's economic decision problem under risk. On the basis of this model, optimal consumption, investment, and borrowing-lending strategies are obtained in closed form for a class of utility functions. The model itself may be viewed as a formalization of Irving Fisher's model of the individual under risk, as presented in *The Theory of Interest* [**4**]; at the same time, it represents a generalization of Phelps' model of personal saving [**10**].

The various components of the decision problem are developed and assembled into a formal model in Section 2. The objective of the individual is postulated to be the maximization of expected utility from consumption over time. His resources are assumed to consist of an initial capital position (which may be negative) and a noncapital income stream which is known with certainty. The individual faces both financial opportunities (borrowing and lending) and an arbitrary number of productive investment opportunities. The returns from the productive opportunities are assumed to be random variables, whose probability distributions satisfy the "no-easy-money condition." The fundamental characteristic of the approach taken is that the portfolio composition decision, the financing decision, and the consumption decision are all analyzed simultaneously in *one* model. The vehicle of analysis is discrete-time dynamic programming.

In Section 3, optimal strategies are derived for the class of utility functions $\sum_{j=1}^{\infty} \alpha^{j-1} u(c_j)$, $0 < \alpha < 1$, where c_j is the amount of consumption in period j, such that either the relative risk aversion index, $-cu''(c)/u'(c)$, or the absolute risk

[1] This paper was presented at the winter meeting of the Econometric Society, San Francisco, California, December, 1966.

[2] This article is based on my dissertation which was submitted to the Graduate School of Business Administration of the University of California, Los Angeles, in June, 1966. I am greatly indebted to Professors George W. Brown (committee chairman), Jacob Marschak, and Jacques Drèze for many valuable suggestions and comments and to Professors Jack Hirshleifer, Leo Breiman, James Jackson, and Fred Weston for constructive criticisms. I am also grateful to the Ford Foundation for financial support over a three-year period.

aversion index, $-u''(c)/u'(c)$, is a positive constant for all $c \geqslant 0$, i.e., $u(c) = c^{\gamma}$, $0 < \gamma < 1$, $u(c) = -c^{-\gamma}$, $\gamma > 0$, $u(c) = \log c$, and $u(c) = -e^{-\gamma c}$, $\gamma > 0$.

Section 4 is devoted to a discussion of the properties of the optimal consumption strategies, which turn out to be linear and increasing in wealth and in the present value of the noncapital income stream. In three of the four models studied, the optimal consumption strategies precisely satisfy the properties specified by the consumption hypotheses of Modigliani and Brumberg [9] and of Friedman [5]. The effects of changes in impatience and in risk aversion on the optimal amount to consume are found to coincide with one's expectations. In response to changes in the "favorableness" of the investment opportunities, however, the four models exhibit an exceptionally diverse pattern with respect to consumption behavior.

The optimal investment strategies have the property that the optimal mix of risky (productive) investments in each model is *independent* of the individual's wealth, noncapital income stream, and impatience to consume. It is shown in Section 5 that the optimal mix depends in each case only on the probability distributions of the returns, the interest rate, and the individual's one-period utility function of consumption. This section also discusses the properties of the optimal lending and borrowing strategies, which are linear in wealth. Three of the models always call for borrowing when the individual is poor while the fourth model always calls for lending when he is sufficiently rich. The effect of differing borrowing and lending rates is also examined.

Necessary and sufficient conditions for capital growth are derived in Section 6. It is found that when the one-period utility function of consumption is logarithmic, the individual will always invest the capital available after the allotment to current consumption so as to maximize the expected growth rate of capital plus the present value of the noncapital income stream. Finally, Section 7 indicates how the preceding results are modified in the nonstationary case and under a finite horizon.

2. THE MODEL

In this section we shall combine the building blocks discussed in the previous section into a formal model. The following notation and assumptions will be employed:

c_j: amount of consumption in period j, where $c_j \geqslant 0$ (decision variable).

$U(c_1, c_2, c_3, \ldots)$: the utility function, defined over all possible consumption programs $(c_1, c_2, c_3, \ldots)$. The class of functions to be considered is that of the form

$$U(c_1, c_2, c_3, \ldots) = u(c_1) + \alpha U(c_2, c_3, c_4, \ldots) = \sum_{j=1}^{\infty} \alpha^{j-1} u(c_j), \quad 0 < \alpha < 1. \tag{1}$$

It is assumed that $u(c)$ is monotone increasing, twice differentiable, and strictly concave for $c \geqslant 0$. The objective in each case is to maximize $E[U(c_1, c_2, \ldots)]$, i.e., the expected utility derived from consumption over time.[3]

x_j: amount of capital (debt) on hand at decision point j (the beginning of the jth period) (state variable).

y: income received from noncapital sources at the *end* of each period, where $0 \leqslant y < \infty$.

[3] While we make use of the expected utility theorem, we assume that the von Neumann-Morgenstern postulates [12] have been modified in such a way as to permit unbounded utility functions.

M: the number of available investment opportunities.

S: the subset of investment opportunities which it is possible to sell short.

z_{ij}: amount invested in opportunity i, $i = 1, \ldots, M$, at the beginning of the jth period (decision variable).

$r - 1$: rate of interest, where $r > 1$.

β_i: transformation of each unit of capital invested in opportunity i in any period j (random variable); that is, if we invest an amount θ in i at the beginning of a period, we will obtain $\beta_i\theta$ at the end of that period (stochastically constant returns to scale, no transaction costs or taxes). The joint distribution functions of the β_i, $i = 1, \ldots, M$, are assumed to be known and independent with respect to time j. The $\{\beta_i\}$ have the following properties:

$$\beta_1 = r, \tag{2}$$

$$0 \leqslant \beta_i < \infty \qquad (i = 2, \ldots, M), \tag{3}$$

$$\Pr\left\{\sum_{i=2}^{M} (\beta_i - r)\theta_i < 0\right\} > 0, \tag{4}$$

for all finite θ_i such that $\theta_i \geqslant 0$ for all $i \notin S$ and $\theta_i \neq 0$ for at least one i.

$f_j(x_j)$: expected utility obtainable from consumption over all future time, evaluated at decision point j, when capital at that point is x_j and an optimal strategy is followed with respect to consumption and investment.

Y: present value at any decision point of the noncapital income stream capitalized at the rate of interest, i.e., $Y = y/(r - 1)$.

$\bar{v} \equiv (v_2, \ldots, v_M)$: a vector of real numbers.

$h(\bar{v}) \equiv E\left[u\left(\sum_{i=2}^{M} (\beta_i - r)v_i + r\right)\right]$.

k: maximum of $h(\bar{v})$ subject to (27) and (28) (see (26)).

$\bar{v}^*$: vector $\bar{v}$ which gives maximum k of $h(\bar{v})$ (see (26)).

$v^* \equiv \sum_{i=2}^{M} v_i^*$.

$c^*(x)$: an optimal consumption strategy.

$z_1^*(x)$: an optimal lending strategy.

$z_i^*(x)$: an optimal investment strategy for opportunity i, $i = 2, \ldots, M$.

$s_j \equiv x_j + Y$.

The limitations of utility functions of the form (1) are well known and need not be elaborated here. Condition (4) will be referred to as the "no-easy-money condition." In essence, this condition states (i) that no combination of productive investment opportunities exists which provides, with probability 1, a return at least as high as the (borrowing) rate of interest; (ii) that no combination of short sales exists in which the probability is zero that a loss will exceed the (lending) rate of interest; (iii) that no combination of productive investments made from the proceeds of any short sale can guarantee against loss. For these reasons, (4) may be viewed as a condition that the prices of the various assets in the market must satisfy in equilibrium.

Consumption and investment decisions are assumed to be made at the beginning of each period. The amount allocated to consumption is assumed to be spent immediately or, if spent gradually over the period, to be set aside in a nonearning account. We also assume that any debt incurred by the individual must at all times be fully secured, i.e., that the individual must be solvent at each decision point. In view of the "no-easy-money condition" (4), this implies that his (net) debt cannot exceed the present value, on the basis of the (borrowing) rate of interest, of his noncapital income stream at the end of any period.

We shall now identify the relation which determines the amount of capital (debt) on hand at each decision point in terms of the amount on hand at the previous decision point. This leads to the difference equation

$$x_{j+1} = rz_{1j} + \sum_{i=2}^{M} \beta_i z_{ij} + y \qquad (j = 1, 2, \ldots) \tag{5}$$

where

$$\sum_{i=1}^{M} z_{ij} = x_j - c_j \qquad (j = 1, 2, \ldots). \tag{6}$$

The first term of (5) represents the payment of the debt or the proceeds from savings, the second term the proceeds from productive investments, and the third term the noncapital income received. Combining (5) and (6) we obtain

$$x_{j+1} = \sum_{i=2}^{M} (\beta_1 - r)z_{ij} + r(x_j - c_j) + y \qquad (j = 1, 2, \ldots). \tag{7}$$

This is the difference equation, then, which governs the process we are about to study.

The definition of $f_j(x_j)$ may formally be written

$$f_j(x_j) \equiv \max E[U(c_j, c_{j+1}, c_{j+2}, \ldots)]|x_j. \tag{8}$$

From (1) we obtain, by the principle of optimality,[4] for all j,

$$f_j(x_j) = \max E[u(c_j) + \alpha\{\max E[U(c_{j+1}, c_{j+2}, \ldots)]|x_{j+1}\}]|x_j, \tag{9}$$

since we have assumed the $\{\beta_i\}$ to be independently distributed with respect to time j. By (8), (9) reduces to

$$f_j(x_j) = \max \{u(c_j) + \alpha E[f_{j+1}(x_{j+1})]\}, \quad \text{all } j. \tag{10}$$

Since by our assumptions we are faced with exactly the same problem at decision point $j + 1$ as when we are at decision point j, the time subscript may be dropped. Using (7), (10) then becomes

$$f(x) = \max_{c,\{z_i\}} \left\{ u(c) + \alpha E\left[f\left(\sum_{i=2}^{M} (\beta_i - r)z_i + r(x - c) + y \right) \right] \right\} \tag{11}$$

subject to

$$c \geqslant 0, \tag{12}$$

$$z_i \geqslant 0, \quad i \notin S, \tag{13}$$

and

$$\Pr\left\{ \sum_{i=2}^{M} (\beta_i - r)z_i + r(x - c) + y \geqslant -Y \right\} = 1 \tag{14}$$

at each decision point. Expression (14), of course, represents the solvency constraint.

[4] The principle of optimality states that an optimal strategy has the property that whatever the initial state and the initial decision, the remaining decisions must constitute an optimal strategy with regard to the state resulting from the first decision [2, p. 83].

For comparison, the model studied by Phelps [**10**] is given by the functional equation

$$f(x) = \max_{0 \leqslant c \leqslant x} \{u(c) + \alpha E[f(\beta(x - c) + y)]\}. \tag{15}$$

In this model, all capital not currently consumed obeys the transformation β, which is identically and independently distributed in each period. Since the amount invested, $x - c$, is determined once c is known, (15) has only one decision variable (c).[5]

Since x represents capital, $f(x)$ is clearly the utility of money at any decision point j. Instead of being assumed, as is generally the case, the utility function of money has in this model been induced from inputs which are more basic than the preferences for money itself. As (11) shows, $f(x)$ depends on the individual's preferences with respect to consumption, his noncapital income stream, the interest rate, and the available investment opportunities and their riskiness.

3. THE MAIN THEOREMS

We shall now give the solution to (11) for the class of one-period utility functions

$$u(c) = \frac{1}{\gamma} c^{\gamma}, \qquad 0 < \gamma < 1 \quad \text{(Model I)}; \tag{16}$$

$$u(c) = \frac{1}{\gamma} c^{\gamma}, \qquad \gamma < 0 \quad \text{(Model II)}; \tag{17}$$

$$u(c) = \log c, \qquad \text{(Model III)}; \tag{18}$$

$$u(c) = -e^{-\gamma c}, \qquad \gamma > 0 \quad \text{(Model IV)}. \tag{19}$$

[5] Phelps gives the solution to (15) for the utility functions $u(c) = c^{\gamma}$, $0 < \gamma < 1$, $u(c) = -c^{-\gamma}$, $\gamma < 0$, and for $u(c) = \log c$ when $y = 0$. Unfortunately, this solution is incorrect in the general case, i.e., whenever $y > 0$ *and* the distribution of β is nondegenerate. For example, when $u(c) = -c^{-\gamma}$, the solution is asserted to be, letting $\bar{\beta} \equiv E[\beta^{-\gamma}]$,

$$f(x) = -\left[\frac{(\alpha\bar{\beta})^{-1/(\gamma+1)}}{(\alpha\bar{\beta})^{-1/(\gamma+1)} - 1}\right]^{\gamma+1}\left[x + \frac{y}{\bar{\beta}^{-1/\gamma} - 1}\right]^{-\gamma}, \tag{15a}$$

$$c(x) = [1 - (\alpha\bar{\beta})^{1/(\gamma+1)}]\left[x + \frac{y}{\bar{\beta}^{-1/\gamma} - 1}\right], \tag{15b}$$

whenever $\alpha\bar{\beta} < 1$. But for this to be a solution, it would be necessary that one be able to write

$$E[(\beta(x - c) + y)^{-\gamma}] = E[\beta^{-\gamma}]\left[x - c + \frac{y}{E[\beta^{-\gamma}]^{-1/\gamma}}\right]^{-\gamma} \tag{15c}$$

which is clearly impossible unless the distribution of β is degenerate or $y = 0$ or both. The right side of (15c) may, of course, be regarded as a first-order approximation of the left side when the variance of β is small, but this negates the presence of uncertainty. In fact, the preceding solution holds even under certainty only when $\alpha\beta \geqslant 1$ and $x \geqslant [(\alpha\bar{\beta})^{-1/(\gamma+1)} - 1]y/(\bar{\beta} - 1)$, i.e., when $c(x)$ is less than or equal to x in *all* future periods.

It appears that an analytic solution to (15) does not exist when $y > 0$ and the distribution of β is nondegenerate. It is ironic, therefore, that when one generalizes Phelps' problem by introducing the possibility of *choice* among risky investment opportunities *and* the opportunity to borrow and lend (see (11)), an analytic solution does exist (as will be shown). It is the second of these generalizations which guarantees the solution in closed form.

Pratt [11] notes that (16)–(18) are the only monotone increasing and strictly concave utility functions for which the relative risk aversion index

$$(20) \qquad q^*(c) \equiv -\frac{u''(c)c}{u'(c)}$$

is a positive constant and that (19) is the only monotone increasing and strictly concave utility function for which the absolute risk aversion index

$$(21) \qquad q(c) \equiv -\frac{u''(c)}{u'(c)}$$

is a positive constant.[6]

THEOREM 1: *Let* $u(c), \alpha, y, r, \{\beta_i\}$, *and* Y *be defined as in Section 2. Then, whenever* $u(c)$ *is one of the functions* (16)–(18) *and* $ky < 1/\alpha$ *in Model I, a solution to* (11) *subject to* (12)–(14) *exists for* $x \geqslant -Y$ *and is given by*

$$(22) \qquad f(x) = Au(x + Y) + C,$$

$$(23) \qquad c^*(x) = B(x + Y),$$

$$(24) \qquad z_1^*(x) = (1 - B)(1 - v^*)(x + Y) - Y,$$

$$(25) \qquad z_i^*(x) = (1 - B)v_i^*(x + Y) \qquad (i = 2, \ldots, M)$$

where the constants v_i^* ($v^* \equiv \sum_{i=2}^{M} v_i^*$) *and* k *are given by*

$$(26) \qquad k \equiv E\left[u\left(\sum_{i=2}^{M} (\beta_i - r)v_i^* + r\right)\right] = \max_{\{v_i\}} E\left[u\left(\sum_{i=2}^{M} (\beta_i - r)v_i + r\right)\right],$$

subject to

$$(27) \qquad v_i \geqslant 0, \quad i \notin S,$$

and

$$(28) \qquad Pr\left\{\sum_{i=2}^{M} (\beta_i - r)v_i + r \geqslant 0\right\} = 1,$$

and the constants A, B, and C are given by

(i) *in the case of Models I–II,*

$$(29) \qquad A = (1 - (\alpha k y)^{1/(1-y)})^{y-1}, \quad B = 1 - (\alpha k y)^{1/(1-y)}, \quad C = 0;$$

[6] The underlying mathematical reason why solutions are obtained in closed form (Theorems 1 and 2) for the utility functions (16)–(19) is that these functions are also the only (monotone increasing and strictly concave utility function) solutions (see [8]) to the functional equations $u(xy) = v(x)w(y)$, $u(xy) = v(x) + w(y)$, $u(x + y) = v(x)w(y)$, and $u(x + y) = v(x) + w(y)$, which are known as the generalized Cauchy equations [1, p. 141].

(ii) *in the case of Model III,*

$$A = \frac{1}{1-\alpha},$$
$$B = 1 - \alpha, \tag{30}$$
$$C = \frac{1}{1-\alpha}\log(1-\alpha) + \frac{\alpha\log\alpha}{(1-\alpha)^2} + \frac{\alpha k}{(1-\alpha)^2}.$$

Furthermore, the solution is unique.

In proving this theorem, we shall make use of the following lemma and corollaries.

LEMMA: *Let $u(c)$, $\{\beta_i\}$, and r be defined as in Section 2 and let $\bar{v} \equiv (v_2, \ldots, v_M)$ be a vector of real numbers. Then the function*

$$h(v_2, v_3, \ldots, v_M) \equiv E\left[u\left(\sum_{i=2}^{M}(\beta_i - r)v_i + r\right)\right] \tag{31}$$

subject to the constraints

$$v_i \geqslant 0, \quad i \notin S, \tag{27}$$

and

$$Pr\left\{\sum_{i=2}^{M}(\beta_i - r)v_i + r \geqslant 0\right\} = 1, \tag{28}$$

has a maximum and the maximizing v_i ($\equiv v_i^*$) are finite and unique.

PROOF: Let D be the $(M-1)$-dimensional space defined by the set of points $\bar{v}$ which satisfy (27) and (28). We shall first prove that the set D is nonempty, closed, bounded, and convex, and that h is strictly concave on D.[7]

The nonemptiness of D follows trivially from the observation that $\bar{v}^0 \equiv (0, 0, \ldots, 0)$ is a member of D. By the boundedness of the β_i's and of r ((2) and (3)), there exists a neighborhood of $\bar{v}^0$ in relation to D. That is, there is a neighborhood of points $\bar{v}'$ such that

$$\Pr\left\{\sum_{i=2}^{M}(\beta_i - r)v_i' + r \geqslant 0\right\} = 1$$

where $v_i' \geqslant 0$ for all $i \notin S$.

Now consider the point $\bar{v}^\lambda \equiv \bar{v}^0 + \lambda\bar{v}' = \lambda\bar{v}'$ where $\lambda \geqslant 0$ and $\bar{v}'$ is one of the points in this neighborhood. Let $b(\bar{v})$ be the greatest lower bound on b such that

$$\Pr\left\{\sum_{i=2}^{M}(\beta_i - r)v_i < b\right\} > 0.$$

[7] The author gratefully acknowledges a debt to Professor George W. Brown for several valuable suggestions concerning the proof of the closure and the boundedness of D.

By the "no-easy-money condition" (4), $b(\bar{v}') \geqslant -r$ for $\bar{v}' \in D, b(\bar{v}^0) = 0$, and $b(\bar{v}) < 0$ for all $\bar{v} \neq \bar{v}^0$. Applying the "no-easy-money condition" with respect to the point $\bar{v}^\lambda$ and using the inequality

$$\Pr\left\{\sum_{i=2}^{M} (\beta_i - r)\lambda v_i < \lambda b\right\} > 0,$$

we obtain that $\lambda b(\bar{v}') = b(\lambda\bar{v}')$. But when $\lambda b(\bar{v}') < -r$, or $\lambda > -r/b(\bar{v}')$, the point $\bar{v}^\lambda$ cannot lie in D since $\lambda > -r/b(\bar{v}')$ implies that

$$\Pr\left\{\sum_{i=2}^{M} (\beta_i - r)\lambda v_i' + r \geqslant 0\right\} < 1.$$

Thus, $\lambda_0 \equiv -r/b(\bar{v}')$ is the greatest lower bound on λ such that $\bar{v}^\lambda \notin D$. Since $\lambda_0 b(\bar{v}') = -r$, $\bar{v}^{\lambda_0} \in D$ and is in fact the point farthest from $\bar{v}^0$ lying on the line through $\bar{v}^0$ and $\bar{v}'$ and belonging to D.

We shall only sketch the remainder of the proof establishing the closure and boundedness of D. Let $\bar{v} \neq \bar{v}^0$ be the limit of a sequence of points $\bar{v}^{(n)} \in D$. Since each point in the sequence belongs to D, $b(\bar{v}^{(n)}) \geqslant -r$ for all n. It can now be shown, by utilizing the fact that $\Sigma_{i=2}^{M} (\beta_i - r)\bar{v}_i$ is continuous at any $\bar{v} \neq \bar{v}^0$, uniformly with respect to the β_i's on any bounded set, that $\overline{\lim}_{n\to\infty} b(\bar{v}^{(n)}) \leqslant b(\bar{v})$, which implies that $\bar{v} \in D$. Consequently, D must be closed.

The boundedness of D is established as follows. Let S_R be the set of points $\bar{v}$ such that $|\bar{v}| = R > 0$. S_R is then clearly both closed and bounded. If $D' \equiv D \cap S_R$ is empty, the boundedness of D follows immediately. Let us therefore assume that D' is nonempty; in this case D' is also bounded and closed since D is closed and S_R is bounded and closed. If $\bar{v}$ is a limit point of the sequence $\langle \bar{v}^{(n)} \rangle$ such that $\bar{v}^{(n)} \in D'$, we must have that $\bar{v} \in D'$ since D' is closed. But $b(\bar{v}) < 0$ by the "no-easy-money condition" (4), since $\bar{v} \neq \bar{v}^0$ by assumption. Therefore, since we already have that $\overline{\lim}_{n\to\infty} b(\bar{v}^{(n)}) \leqslant b(\bar{v})$, 0 cannot be a limit point to the sequence $\langle b(\bar{v}^{(n)}) \rangle$, $\bar{v}^{(n)} \in D'$. Consequently, $b(\bar{v})$ for $\bar{v} \in D'$ is bounded away from zero, which implies that D must be bounded.

To prove convexity, let $\bar{v}''$ and $\bar{v}'''$ be two points in D. Then, for any $0 \leqslant \lambda \leqslant 1$,

$$\Pr\left\{\sum_{i=2}^{M} (\beta_i - r)\lambda v_i'' + \lambda r \geqslant 0\right\} = 1,$$

and

$$\Pr\left\{\sum_{i=2}^{M} (\beta_i - r)(1 - \lambda)v_i''' + (1 - \lambda)r \geqslant 0\right\} = 1,$$

which implies

$$\Pr\left\{\sum_{i=2}^{M} (\beta_i - r)(\lambda v_i'' + (1 - \lambda)v_i''') + r \geqslant 0\right\} = 1,$$

so that $\lambda\bar{v}'' + (1 - \lambda)\bar{v}''' \in D$. Thus, D is convex.

Let

$$\tilde{w}_n = \sum_{i=2}^{M} (\beta_i - r)v_i^n + r \qquad (n = 1, 2).$$

Then

$$h(\lambda \bar{v}^1 + (1 - \lambda)\bar{v}^2) = E[u(\lambda \tilde{w}_1 + (1 - \lambda)\tilde{w}_2)] \tag{32}$$

and

$$\lambda h(\bar{v}^1) + (1 - \lambda)h(\bar{v}^2) = \lambda E[u(\tilde{w}_1)] + (1 - \lambda)E[u(\tilde{w}_2)]. \tag{33}$$

For every pair of values $w_1 \neq w_2$ of the random variables $\tilde{w}_1$ and $\tilde{w}_2$ such that $\bar{v}^1$ and $\bar{v}^2 \in D$, we obtain, by the strict concavity of u,

$$u(\lambda w_1 + (1 - \lambda)w_2) > \lambda u(w_1) + (1 - \lambda)u(w_2), \quad 0 < \lambda < 1. \tag{34}$$

Consequently, (34) implies

$$E[u(\lambda \tilde{w}_1 + (1 - \lambda)\tilde{w}_2)] > \lambda E[u(\tilde{w}_1)] + (1 - \lambda)E[u(\tilde{w}_2)], \quad \bar{v}_1^1 \neq \bar{v}_2^2 \in D,$$
$$0 < \lambda < 1,$$

which, by (32) and (33), in turn implies that h is strictly concave on D.

Since our problem has now been shown to be one of maximizing a strictly concave function over a nonempty, closed, bounded, convex set, it follows directly that the function h has a maximum and that the v_i^* are finite and unique.

A number of corollaries obtain from this lemma which we shall also require in the proof of Theorem 1.

COROLLARY 1: *Let $u(c)$, $\{\beta_i\}$, and r be defined as in the Lemma. Moreover, let $u(c)$ be such that it has no lower bound. Then the v_i^* which maximize* (31) *subject to* (27) *and* (28) *are such that*

$$Pr\left\{\sum_{i=2}^{M} (\beta_i - r)v_i^* + r > 0\right\} = 1.$$

The proof is immediate from the observation that $h \to -\infty$ as the greatest lower bound on b such that Pr $\Sigma_{i=2}^{M} (\beta_i - r)v_i + r < b\} > 0$ approaches 0 from above.

COROLLARY 2: *Let $u(c)$, $\{\beta_i\}$, and r be defined as in the Lemma. Then the maximum of the function* (31) *subject to the constraints* (27) *and* (28) *is greater than or equal to $u(r)$.*

PROOF: When $v_i = 0$ for all i, which is always feasible, we obtain by (31) that $h = u(r)$.

COROLLARY 3: *Let $u(c)$, $\{\beta_i\}$, and r be defined as in the Lemma. Moreover, let $u(c)$ be such that $u(c) \leqslant b$. Then the vectors $\bar{v}$ which satisfy* (27) *and* (28) *are such that*

$$h(\bar{v}) \equiv E\left[u\left(\sum_{i=2}^{M} (\beta_i - r)v_i + r\right)\right] < b.$$

The proof is immediate from the observation that $u(c)$ is monotone increasing and that r, $\{\beta_i\}$, and the feasible v_i are bounded.

We are now ready to prove the theorem. The method of proof will be to verify that (22)–(25) is the (only) solution to (11).[8]

PROOF OF THEOREM 1 FOR MODELS I–II: Denote the right side of (11) by $T(x)$ upon inserting (22) for $f(x)$. This gives, for all decision points j,

$$T(x) = \max_{c,\{z_i\}} \left\{ \frac{1}{\gamma} c^{\gamma} + \alpha(1 - (\alpha k \gamma)^{1/(1-\gamma)})^{\gamma - 1} E\left[\frac{1}{\gamma} \left(\sum_{i=2}^{M} (\beta_i - r) z_i + r(x - c) + y + Y \right)^{\gamma} \right] \right\} \tag{35}$$

subject to

$$c \geqslant 0, \tag{12}$$

$$z_i \geqslant 0, \quad i \notin S, \tag{13}$$

and

$$\Pr\left\{ \sum_{i=2}^{M} (\beta_i - r) z_i + r(x - c) + y + (y/(r - 1)) \geqslant 0 \right\} = 1. \tag{14}$$

Since (14) may be written

$$\Pr\left\{ \sum_{i=2}^{M} (\beta_i - r) z_i + r(x + Y - c) \geqslant 0 \right\} = 1,$$

it follows from the "no-easy-money condition" (4) that (14) is satisfied if and only if *either*

$$s - c = 0 \tag{36}$$

and

$$z_i = 0 \qquad (i = 2, \ldots, M), \tag{37}$$

or

$$s - c > 0 \tag{38}$$

and

$$\Pr\left\{ \sum_{i=2}^{M} (\beta_i - r) z_i/(s - c) + r \geqslant 0 \right\} = 1, \tag{39}$$

where $s \equiv x + Y$.

Under feasibility with respect to (14), we then obtain

$$T(x) = \begin{cases} \max\left\{ \frac{1}{\gamma} s^{\gamma}, \bar{T}(x) \right\}, & 0 < \gamma < 1, \\ \max\{-\infty, \bar{T}(x)\}, & \gamma < 0, \end{cases} \tag{40}$$

[8] A proof based on the method of successive approximations may be found in [7].

where

(41) $$\bar{T}(x) = \sup_{c,(z_i)}\left\{\frac{1}{\gamma}c^\gamma + \alpha(1 - (\alpha k\gamma)^{1/(1-\gamma)})^{\gamma-1}(s-c)^\gamma \times E\left[\frac{1}{\gamma}\left(\sum_{i=2}^{M}(\beta_i - r)z_i/(s-c) + r\right)^\gamma\right]\right\}$$

subject to (12), (38), (39), and

(42) $$z_i/(s-c) \geqslant 0, \quad i \notin S,$$

since (42) is equivalent to (13) in view of (38). But by (31) the expectation factor in (41) may be written

(43) $$h(z_2/(s-c), \ldots, z_M/(s-c))$$

and (26), the Lemma, and Corollary 2 give

(44) $$k\gamma \geqslant r^\gamma > 0 \quad \text{(Model I)},$$

(45) $$k\gamma \leqslant r^\gamma < 1 \quad \text{(Model II)},$$

while (26), the Lemma, and Corollary 3 give

(46) $$k\gamma > 0 \qquad \text{(Model II)}.$$

Thus, $\partial\bar{T}/\partial h > 0$ always in Model II and in Model I whenever

(47) $$k\gamma < \frac{1}{\alpha}$$

under feasibility. When $k\gamma > 1/\alpha$ in Model I, $\bar{T}(x)$ does not exist; when $k\gamma = 1/\alpha$, (41) and (40) give $T(x) = (1/\gamma)s^\gamma \neq f(x)$. Consequently, it remains to consider the case when $\partial\bar{T}/\partial h > 0$.

Since the maximum of (43) subject to (42) and (39) is k by (26) and the Lemma, we obtain by the Lemma that the strategy

$$\frac{z_i}{s-c} = v_i^* \qquad (i = 2, \ldots, M)$$

or

(48) $$z_i^* = v_i^*(s-c) \qquad (i = 2, \ldots, M)$$

is optimal and unique for *every* c which satisfies (12) and (38) when (38) holds. It is clearly also optimal when (36) and (37) hold. Consequently, (40) reduces to

(49) $$T(x) = \max_{0 \leqslant c \leqslant s}\left\{\frac{1}{\gamma}c^\gamma + \alpha k(1 - (\alpha k\gamma)^{1/(1-\gamma)})^{\gamma-1}(s-c)^\gamma\right\}.$$

Since $u(c)$ is strictly concave and $u'(0) = \infty$ in Models I and II, $T(x)$ is strictly concave and differentiable with an "interior" unique solution $c^*(x)$ whenever

(50) $$\alpha k(1 - (\alpha k\gamma)^{1/(1-\gamma)})^{\gamma-1}\begin{cases} > 0 & \text{(Model I)}, \\ < 0 & \text{(Model II)}, \end{cases}$$

and $s \geqslant 0$. In this case, setting $dT/dc = 0$ and solving for c, we get,

$$c^{\gamma-1} - \alpha k\gamma(1 - (\alpha k\gamma)^{1/(1-\gamma)})^{\gamma-1}(s - c)^{\gamma-1} = 0$$

or

$$c^*(x) = (1 - (\alpha k\gamma)^{1/(1-\gamma)})(x + Y). \tag{51}$$

In Model I, (50) is satisfied whenever (44) and (47) hold; as noted earlier, no solution exists in Model I for those cases in which $k\gamma \geqslant 1/\alpha$. In Model II, (50) is always satisfied as seen from (45) and (46).

Inserting (51) in (48) we obtain

$$z_i^*(x) = (\alpha k\gamma)^{1/(1-\gamma)}v_i^*(x + Y) \qquad (i = 2, \ldots, M)$$

and (24) follows from (6) upon insertion of $c^*(x)$ and the $z_i^*(x)$. $T(x)$ now becomes, upon insertion of $c^*(x)$ in (49),

$$\begin{aligned} T(x) &= \frac{1}{\gamma}(1 - (\alpha k\gamma)^{1/(1-\gamma)})^{\gamma}s^{\gamma} + \alpha k(1 - (\alpha k\gamma)^{1/(1-\gamma)})^{\gamma-1}s^{\gamma}(\alpha k\gamma)^{\gamma/(1-\gamma)} \\ &= \frac{1}{\gamma}(1 - (\alpha k\gamma)^{1/(1-\gamma)})^{\gamma-1}s^{\gamma} \\ &= f(x) \end{aligned}$$

and the solution clearly exists for $s \geqslant 0$ or

$$x_j \geqslant -Y. \tag{52}$$

Since (52) is an induced constraint with respect to period $j - 1$, it remains to be verified that (52) is either redundant or not effective in period $j - 1$. Because (52) is already present in period $j - 1$ through (14), the induced constraint (52) is redundant, which completes the proof.

PROOF OF THEOREM 1 FOR MODEL III: Denote the right side of (11) $T(x)$ upon inserting (22) for $f(x)$. This gives, for all decision points j,

$$\begin{aligned} T(x) = \max_{c,\{z_i\}} \Bigg\{ \log c + \frac{\alpha}{1-\alpha} E\Bigg[\log \Bigg(\sum_{i=2}^{M} (\beta_i - r)z_i \\ + r(x - c) + y + Y \Bigg) \Bigg] + K \Bigg\} \end{aligned}$$

where

$$K \equiv \frac{\alpha}{1-\alpha}\log(1-\alpha) + \frac{\alpha^2 \log \alpha}{(1-\alpha)^2} + \frac{\alpha^2 k}{(1-\alpha)^2}$$

subject to (12), (13), and (14). By the reasoning for Models I and II, we obtain

$$T(x) = \max\{-\infty, \bar{T}(x)\} \tag{53}$$

where

(54) $$\bar{T}(x) = \sup_{c,\{z_i\}} \left\{ \log c + \frac{\alpha}{1-\alpha} \log (s - c) + \frac{\alpha}{1-\alpha} E\left[\log \left(\sum_{i=2}^{M} (\beta_i - r) z_i/(s - c) + r \right) \right] \right\} + K$$

subject to (12),

(38) $$s - c > 0,$$

(42) $$z_i/(s - c) \geqslant 0 \qquad (i = 2, \ldots, M),$$

and

(39) $$\Pr \left\{ \sum_{i=2}^{M} (\beta_i - r) z_i/(s - c) + r \geqslant 0 \right\} = 1.$$

By (31), the next to last term in (54) can be written

(55) $$\frac{\alpha}{1-\alpha} h(z_2/(s - c), \ldots, z_M/(s - c))$$

where $\partial \bar{T}/\partial h > 0$. Since the maximum of (55) subject to (42) and (39) is $(\alpha k/1 - \alpha)$ by (26) and the Lemma, we obtain from the Lemma that

(48) $$z_i^{*}(x) = v_i^{*}(x + Y - c) \qquad (i = 2, \ldots, M)$$

is optimal and unique for *every* c which satisfies (12) and (38). Thus, (53) reduces, in analogy with Models I and II, to

(56) $$T(x) = \max_{0 \leqslant c \leqslant s} \left\{ \log c + \frac{\alpha}{1-\alpha} \log (s - c) + \frac{\alpha k}{1-\alpha} + K \right\}$$

where $T(x)$ always exists since $0 < \alpha < 1$; furthermore, $T(x)$ is strictly concave and differentiable. Setting $\partial T/\partial c = 0$ we obtain

(57) $$c^*(x) = (1 - \alpha)(x + Y),$$

$$z_i^{*}(x) = \alpha v_i^{*}(x + Y) \qquad (i = 2, \ldots, M),$$

and (24), all unique. Inserting (57) into (56) gives

$$T(x) = \log (1 - \alpha) + \log s + \frac{\alpha}{1-\alpha} \log \alpha + \frac{\alpha}{1-\alpha} \log s + \frac{\alpha k}{1-\alpha} + \frac{\alpha}{1-\alpha} \log (1 - \alpha) + \frac{\alpha^2 \log \alpha}{(1-\alpha)^2} + \frac{\alpha^2 k}{(1-\alpha)^2}$$

$$= f(x).$$

Since $f(x)$ exists for $x_j \geqslant -Y$, which as an induced constraint with respect to period $j - 1$ is made redundant by (14) for that period, the proof is complete.

When $\gamma = 0$, the solution to (11) reduces to

$$f(x) = Au(x) + C,$$

$$c^*(x) = Bx,$$

$$z_1^*(x) = (1 - B)(1 - v^*)x,$$

$$z_i^*(x) = (1 - B)v_i^* x \qquad (i = 2, \ldots, M).$$

But then, letting $s \equiv x + Y$,

$$f(s) = Au(s) + C,$$

$$c^*(s) = Bs,$$

$$z_1^*(s) = (1 - B)(1 - v^*)s,$$

$$z_i^*(s) = (1 - B)v_i^* s \qquad (i = 2, \ldots, M).$$

As a result, except for $z_1^*(x + Y)$, the solution to the original problem is not altered when the individual, instead of receiving the noncapital income stream in installments, is given its present value Y in advance. Thus, instead of letting x be the state variable when there is a noncapital income, one could let $x + Y$ be the state variable (pretending there is no income), as long as Y is deducted from $z_1^*(x + Y)$.

Note that it is sufficient, though not necessary, for a solution *not* to exist in Model I that $r^{\gamma} \geqslant 1/\alpha$ (Corollary 2).

THEOREM 2: *Let α, $\{\beta_i\}$, r, γ, and Y be defined as in Section 2. Moreover, let $u(c) = -e^{-\gamma c}$ for $c \geqslant 0$ where $\gamma > 0$. Then a solution to* (11) *subject to* (12)–(14) *exists for $x \geqslant -Y + [r/(\gamma(r - 1)^2)] \log(-\alpha kr)$ and is given by*

$$f(x) = -\frac{r}{r - 1}(-\alpha kr)^{1/(r-1)} e^{-[\gamma(r-1)/r](x+Y)}, \tag{58}$$

$$c^*(x) = \frac{r - 1}{r}(x + Y) - \frac{1}{\gamma(r - 1)} \log(-\alpha kr), \tag{59}$$

$$z_1^*(x) = \frac{x}{r} - \frac{y}{r} + \frac{\log(-\alpha kr) - rv^*}{\gamma(r - 1)}, \tag{60}$$

$$z_i^*(x) = \frac{r}{\gamma(r - 1)} v_i^* \qquad (i = 2, \ldots, M), \tag{61}$$

where the constants k and v_i^* $(v^* \equiv \sum_{i=2}^{M} v_i^*)$ are given by

$$k \equiv E[-e^{-\sum_{i=2}^{M}(\beta_i - r)v_i^*}] = \max_{\{v_i\}} E[-e^{-\sum_{i=2}^{M}(\beta_i - r)v_i}] \quad \textit{subject to } (27) \tag{62}$$

provided that

$$\log(-\alpha kr) + b(\bar{v}^*) \geqslant 0 \tag{63}$$

where $b(\bar{v}^*)$ is the greatest lower bound on b such that

$$Pr\left\{\sum_{i=2}^{M}(\beta_i - r)v_i^* < b\right\} > 0$$

and $\bar{v}^* \equiv (v_2^*, \ldots, v_M^*)$. Moreover, the solution is unique.

Since the conditions under which Theorem 2 holds are quite restrictive, the reader is referred to [7] for the proof. Condition (63) insures that the individual's capital position x is nondecreasing over time with probability 1; it must hold for a solution to exist in closed form. The condition $\alpha r \geqslant 1$ is a necessary, but not sufficient, condition for (63) to be satisfied.[9]

4. PROPERTIES OF THE OPTIMAL CONSUMPTION STRATEGIES

In each of the four models we note that the optimal consumption function $c^*(x)$ is linear increasing in capital x and in noncapital income y. Whenever $y > 0$, positive consumption is called for even when the individual's net worth is negative, as long as it is greater than $-Y$ in Models I–III and greater than $-Y + [r/(\gamma(r-1)^2)] \log(-\alpha k r)$ in Model IV. Only *at* these end points would the individual consume nothing.

Since $x + Y$ may be viewed as permanent (normal) income and consumption is proportional $(0 < B < 1)$ to $x + Y$ in Models I–III, we see that the optimal consumption functions in these models satisfy the permanent (normal) income hypotheses precisely [9, 5, 3].

In each model, $c^*(x)$ is decreasing in α. Thus, the greater the individual's impatience $1 - \alpha$ is, the greater his present consumption would be. This, of course, is what we would expect.

By (20) and (21), the relative and absolute risk aversion indices of Models I–IV are as follows:

$$q^*(c) = 1 - \gamma \quad \text{(Models I–II)},$$

$$q^*(c) = 1 \quad \text{(Model III)},$$

$$q(c) = \gamma \quad \text{(Model IV)}.$$

[9] For example, when $u(c) = -e^{-.0001c}$, $\alpha = .99$, $y = \$10{,}000$, $r = 1.06$, $M = 2$, and β_2 assumes each of the values .96 and 1.17 with probability .5, a solution exists for $x \geqslant \$-22{,}986$. For selected capital positions, the optimal amounts to consume, lend, and invest in this case are as follows:

x	$c^*(x)$	$z_1^*(x)$	$z_2^*(x)$
\$ −22,986	0	\$−102,488	\$79,502
0	\$ 1,301	− 80,803	79,502
50,000	4,131	− 33,633	79,502
100,000	6,961	13,537	79,502
500,000	29,601	390,897	79,502
1,000,000	57,901	862,597	79,502

The maximum loss in each period from risky investment is \$3,180.

In Models I–II, we obtain

$$(64)\qquad \frac{\partial B}{\partial(1-\gamma)} = -(\alpha k\gamma)^{1/(1-\gamma)}\left\{\frac{[d(k\gamma)/d(1-\gamma)]}{k\gamma(1-\gamma)} - \frac{\log(\alpha k\gamma)}{(1-\gamma)^2}\right\}$$

where $d(k\gamma)/d(1-\gamma)$ is negative whenever $b(\bar{v}^*) \geqslant 1 - r$; otherwise the sign is ambiguous. Since $k\gamma > 0$ and $\alpha k\gamma < 1$, the sign of (64) is ambiguous in both cases; i.e., a change in relative risk aversion may either decrease or increase present consumption. In Model IV, on the other hand, $c^*(x)$ is increasing in γ; i.e., a more risk averse individual consumes more, ceterus paribus.

From (26) and (62) we observe that k is a natural measure of the "favorableness" of the investment opportunities. This is because k is a maximum determined by (the one-period utility function and) the distribution function (F); moreover, F is reflected in the solution only through k, and $f(x)$ is increasing in k. Let us examine the effect of k on the marginal propensities to consume out of capital and non-capital income.

Equation (29) gives

$$\frac{\partial B}{\partial k} = \frac{\alpha\gamma}{\gamma-1}(\alpha k\gamma)^{\gamma/(1-\gamma)}\begin{cases} < 0 & \text{(Model I)},\\ > 0 & \text{(Model II)}.\end{cases}$$

Thus, we find that the propensity to consume is *decreasing* in k in the case of Model I. This phenomenon can at least in part be attributed to the fact that the utility function is bounded from below but not from above; the loss from postponement of current consumption is small compared to the gain from the much higher rate of consumption thereby made possible later. In Model II, on the other hand, where the utility function has an upper bound but no lower bound, the optimal amount of present consumption is *increasing* in k, which seems more plausible from an intuitive standpoint.

In Model III, we observe from (30) the curious phenomenon that the optimal consumption strategy is independent of the investment opportunities in every respect. While the marginal propensity to consume is independent of k in Model IV also, the *level* of consumption in this case is an increasing function of k as is apparent from (59). We recall that the utility function in Model III is unbounded while that in Model IV is bounded both from below and from above. Thus, the class of utility functions we have examined implies an exceptionally rich pattern of consumption behavior with respect to the "favorableness" of the investment opportunities.

5. PROPERTIES OF THE OPTIMAL INVESTMENT AND BORROWING-LENDING STRATEGIES

The properties exhibited by the optimal investment strategies are in a sense the most interesting. Turning first to Model IV, we note that the portfolio of productive investments is constant, both in mix and amount, at all levels of wealth. The optimal portfolio is also independent of the noncapital income stream and the level of impatience $1 - \alpha$ possessed by the individual, as shown by (61) and (62).

Similarly, we find in Models I–III that, since for all $i, m > 1, z_i^*(x)/z_m^*(x) = v_i^*/v_m^*$ (which is a constant), the *mix* of risky investments is independent of wealth, noncapital income, and impatience to spend. In addition, the *size* of the total investment commitment in each period is proportional to $x + Y$. We also note that when $y = 0$, the ratio that the risky portfolio $\Sigma_{i=2}^{M} z_i^*(x)$ bears to the total portfolio $\Sigma_{i=1}^{M} z_i^*(x)$ is independent of wealth in each model.

In summary, then, we have the surprising result that the optimal mix of risky (productive) investments in each of Models I–IV is independent of the individual's wealth, noncapital income stream, and rate of impatience to consume; the optimal mix depends in each case only on the probability distributions of the returns, the interest rate, and the individual's one-period utility function of *consumption.*

In each case, we find that lending is linear in wealth. Turning first to Models I–III, we find that borrowing always takes place at the lower end of the wealth scale; (24) evaluated at $x = -Y$ gives $-Y < 0$ as the optimal amount to lend. From (24) we also find that $z_1^*(x)$ is increasing in x if and only if $1 - v^* > 0$ since $1 - B$ is always positive. As a result, the models always call for borrowing at least when the individual is poor; whenever $1 - v^* > 0$, they also always call for lending when he is sufficiently rich.

In Model IV, we observe that lending is always increasing in x. Thus, when an individual in this model becomes sufficiently wealthy, he will always become a lender. At the other extreme, when x is at the lower boundary point of the solution set, he will generally be a borrower, though not necessarily, since $z_1^*(x)$ evaluated at $x = -Y + [r/(\gamma(r-1)^2)] \log(-\alpha kr)$ gives

$$-Y + \frac{r \log(-\alpha kr)}{\gamma(r-1)^2} - \frac{rv^*}{\gamma(r-1)}$$

which may be either negative or positive.

We shall now consider the case when the lending rate differs from the borrowing rate as is usually the case in the real world. Let $r_B - 1$ and $r_L - 1$ denote the borrowing and lending rates, respectively, where $r_B > r_L$. Unfortunately, the sign of dv^*/dr is not readily determinable. However, since $f(x)$ is increasing in k, the analysis is straight-forward.[10]

[10] When $r_B > r_L$, the "no-easy-money condition" requires that the joint distribution function of $\beta_2, \ldots, \beta_M$ satisfies

$$\text{(4a)} \qquad \Pr\left\{\sum_{i=2}^{M} (\beta_i - r_B)\theta_i < 0\right\} > 0$$

for all finite numbers $\theta_i \geq 0$ such that $\theta_i > 0$ for at least one i;

$$\text{(4b)} \qquad \Pr\left\{\sum_{i \notin S} (\beta_i - r_L)\theta_i < 0\right\} > 0$$

for all finite numbers $\theta_i \leq 0$ such that $\theta_i < 0$ for at least one i; and

$$\text{(4c)} \qquad \Pr\left\{\sum_{\substack{i=2 \\ i \notin S^*}}^{M} \beta_i\theta_i - \sum_{k \in S^*} \beta_k\theta_k < 0\right\} > 0$$

for all finite numbers $\theta_i, \theta_k \geq 0$ and all $S^* \subseteq S$ such that

$$\sum_{\substack{i=2 \\ i \notin S^*}}^{M} \theta_i = \sum_{k \in S^*} \theta_k,$$

and $\theta_i > 0$ for at least one i. When $r_B = r_L$, 4(a)–4(c) reduce to (4).

Consider first Models I–III when noncapital income $y = 0$. In that case, it is apparent from (24) that when the individual is not in the trapping state (i.e., $x > -Y$), he either always borrows, always lends, or does neither, depending on whether $1 - v^*$ is negative, positive, or zero. Let k_L denote the maximum of (31) when the lending rate is used and the constraint

$$\text{(65)} \qquad \sum_{i=2}^{M} v_i \leqslant 1$$

is added to constraints (27) and (28). Since the set of vectors $\bar{v}$ which satisfy (65) is convex and includes $\bar{v} = (0, \ldots, 0)$, the Lemma still holds when (65) is added to the constraint set. Analogously, let k_B denote the maximum of (31) under the borrowing rate r_B subject to (27), (28), and

$$\text{(66)} \qquad \sum_{i=2}^{M} v_i \geqslant 1.$$

Again, the Lemma holds since the set of $\bar{v}$ satisfying (66) is convex and any $\bar{v}$ such that $\Sigma_{i=2}^{M} v_i = 1, v_i \geqslant 0$, for example, satisfies all constraints. Setting $k \equiv \max\{k_B, k_L\}$, Theorem 1 holds as before when $y = 0$.

When $y > 0$ in Models I–III and in the case of Model IV, no "simple" solution appears to exist when $r_B > r_L$.

6. THE BEHAVIOR OF CAPITAL

We shall now examine the behavior of capital implied by the optimal investment and consumption strategies of the different models. According to one school, capital growth is said to exist whenever

$$\text{(67)} \qquad E[x_{j+1}] > x_j \qquad (j = 1, 2, \ldots),$$

that is, capital growth is defined as expected growth [10]. We shall reject this measure since under this definition, as $j \to \infty$, x_j may approach a value less than x_1 with a probability which tends to 1. We shall instead define growth as asymptotic growth; that is, capital growth is said to exist if

$$\text{(68)} \qquad \lim_{j \to \infty} \Pr\{x_j > x_1\} = 1.$$

When the $>$ sign is replaced by the $\geqslant$ sign, we shall say that we have capital nondecline. If there is statistical independence with respect to j, (67) is implied by (68) but the converse does not hold, as noted.

Model IV will be considered first. From (63) it follows that nondecline of capital is always implied (in fact, the solution to the problem is contingent upon the condition that capital does not decrease, as pointed out earlier). It is readily seen that a sufficient, but not necessary, condition for growth is that there be a nonzero investment in at least one of the risky investment opportunities since in that case $\Pr\{x_{j+1} > x_j\} > 0, j = 1, 2, \ldots$, by (63). A necessary and sufficient condition for asymptotic capital growth is $\alpha r > 1$, which is readily verified by reference to (62), (63), and the foregoing statement.

Let us now turn to Models I–III and let, as before, $s_j \equiv x_j + Y$. From (7), (23), and (25) we now obtain

$$s_{j+1} = s_j(1 - B)\left[\sum_{i=2}^{M} (\beta_i - r)v_i^* + r\right] \tag{69}$$

$$= s_j W \qquad (j = 1, 2, \ldots)$$

where W is a random variable. By (28), $W \geqslant 0$. Attaching the subscript n to W for the purpose of period identification, we note that since

$$s_j = s_1 \prod_{n=1}^{j-1} W_n, \tag{70}$$

(70) verifies that

$$s_j \geqslant 0 \quad \text{for all } j \text{ whenever } s_1 \geqslant 0 \quad \text{(Models I–III)}.$$

Moreover, since $\Pr\{W > 0\} = 1$ in Models II and III by Corollary 1, it follows that

$$s_j > 0 \quad \text{whenever } s_1 > 0 \text{ for all finite } j \quad \text{(Models II–III)}. \tag{71}$$

From (70) we also observe that $s_j = 0$ whenever $s_k = 0$ for all $j > k$. Consequently, $x = -Y$ is a trapping state which, once entered, cannot be left. In this state, the optimal strategies in each case call for zero consumption, no productive investments, the borrowing of Y, and the payment of noncapital income y as interest on the debt. In Models II and III, it follows from (71) that the trapping state will never be reached in a finite number of time periods if initial capital is greater than $-Y$.

Equation (70) may be written

$$s_j = s_1 e^{\sum_{n=1}^{j-1} \log W_n}.$$

The random variable $\sum_{n=1}^{j-1} \log W_n$ is by the Central Limit Theorem asymptotically normally distributed; its mean is $(j - 1)E[\log W]$. By the law of large numbers,

$$\frac{\sum_{n=1}^{j-1} \log W_n}{j - 1} \to E[\log W] \quad \text{as } j \to \infty.$$

Thus, since $s_j > s_1$ if and only if $x_j > x_1$, it is necessary and sufficient for capital growth to exist that $E[\log W] > 0$.

It is clear that μ given by $\mu \equiv e^{E[\log W]}$ may be interpreted as the mean growth rate of capital. By (69), we obtain

$$E[\log W] = \log(1 - B) + E\left[\log\left\{\sum_{i=2}^{M} (\beta_i - r)v_i^* + r\right\}\right].$$

For Model III, this becomes, by (30) and (26),

$$E[\log W] = \log \alpha + \max_{\{v_i\}} E\left[\log\left\{\sum_{i=2}^{M} (\beta_i - r)v_i + r\right\}\right]$$

subject to (27) and (28). Thus, a person whose one-period utility function of consumption is logarithmic will always invest the capital available after the allotment to current consumption so as to maximize the mean growth rate of capital plus the present value of the noncapital income stream.

7. GENERALIZATIONS

We shall now generalize the preceding model to the nonstationary case. We then obtain, by the same approach as in the stationary case, for all j,

$$f_j(x_j) = \max_{c_j,\{z_{ij}\}} \left\{u(c_j) + \alpha_j E\left[f_{j+1}\left(\sum_{i=2}^{M_j} (\beta_{ij} - r_j)z_{ij} + r_j(x_j - c_j) + y_j\right)\right]\right\} \tag{72}$$

subject to

$$c_j \geqslant 0, \tag{73}$$

$$z_{ij} \geqslant 0, \quad i \notin S_j, \tag{74}$$

and

$$\Pr\{x_{j+1} \geqslant -Y_{j+1}\}, \tag{75}$$

where the patience factor α, the number of available investment opportunities M and S and their random returns $\beta_i - 1$, the interest rate r, and the noncapital income y may vary from period to period; this, of course, requires that they be time identified through subscript j. Time dependence on the part of any *one* of the preceding parameters also requires that $f(x)$ be subscripted.

As shown in [7], the solution to the nonstationary model is qualitatively the same as the solution to the stationary model.

In the case of a finite horizon, the problem again reduces to (72)–(75) with $f_{n+1}(x_{n+1}) \equiv 0$ if the horizon is at decision point $n + 1$. In this case, $f(x)$, x, c, z_i, and Y must clearly be time identified through subscript j even in the stationary model. Under a finite horizon, a solution always exists even for Model I. Again, the solution is qualitatively the same as in the infinite horizon case except that the constant of consumption proportionality B_j increases with time j, $B_n = 1$, and $z_{in}^* = 0$ for all i.[11]

University of California, Berkeley

Manuscript received September, 1966; *revision received January,* 1969.

[11] The implications of the results of the current paper with respect to the theory of the firm may be found in [6].

REFERENCES

[1] ACZÉL, J.: *Lectures on Functional Equations and Their Applications*. New York, Academic Press, 1966.

[2] BELLMAN, RICHARD: *Dynamic Programming*. Princeton, Princeton University Press, 1957.

[3] FARRELL, M. J.: "The New Theories of the Consumption Function," *Economic Journal*, December, 1959.

[4] FISHER, IRVING: *The Theory of Interest*. New York, MacMillan, 1930; reprinted, Augustus Kelley, 1965.

[5] FRIEDMAN, MILTON: *A Theory of the Consumption Function*. Princeton, Princeton University Press, 1957.

[6] HAKANSSON, NILS: "An Induced Theory of the Firm Under Risk: The Pure Mutual Fund," *Journal of Financial and Quantitative Analysis*, June 1970.

[7] ———: "Optimal Investment and Consumption Strategies for a Class of Utility Functions," Ph.D. Dissertation, University of California at Los Angeles, 1966; also, Working Paper No. 101, Western Management Science Institute, University of California at Los Angeles, June, 1966.

[8] ———: "Risk Disposition and the Separation Property in Portfolio Selection," *Journal of Financial and Quantitative Analysis*, December, 1969.

[9] MODIGLIANI, F., AND R. BRUMBERG: "Utility Analysis and the Consumption Function: An Interpretation of Cross-Section Data," *Post-Keynesian Economics* (ed. K. Kurihara), New Brunswick, Rutgers University Press, 1954.

[10] PHELPS, EDMUND: "The Accumulation of Risky Capital: A Sequential Utility Analysis," *Econometrica*, October, 1962.

[11] PRATT, JOHN: "Risk-Aversion in the Small and in the Large," *Econometrica*, January–April, 1964.

[12] VON NEUMANN, JOHN, and OSKAR MORGENSTERN: *Theory of Games and Economic Behavior*. Princeton University Press, 1947.

9

Reprinted from THE JOURNAL OF BUSINESS OF THE UNIVERSITY OF CHICAGO
Vol. 44, No. 3, July 1971

ON OPTIMAL MYOPIC PORTFOLIO POLICIES, WITH AND WITHOUT SERIAL CORRELATION OF YIELDS

NILS H. HAKANSSON

I. INTRODUCTION

In a recent paper, Mossin[1] attempts to isolate the class of utility functions of terminal wealth, $f(x)$, which, in the sequential portfolio problem, induces myopic utility functions of intermediate wealth positions. Induced utility functions of short-run wealth are said to be myopic whenever they are independent of yields beyond the current period; that is, they are positive linear transformations of $f(x)$. Mossin concludes (1) that the logarithmic function and the power functions induce completely myopic utility functions; (2) that, when the interest rate in each period is zero, all terminal wealth functions such that the risk tolerance index $-f'(x)/f''(x)$ is linear in x induce completely myopic utility functions of short-run wealth; (3) that, when interest rates are not zero, the last class of terminal wealth functions induces partially myopic utility functions (only future interest rates need be known); and (4) that all of the preceding is true whether the yields in the various periods are serially correlated or not. With the exception of the last assertion, the same conclusions are reached by Leland.[2] The purpose of this note is to show that the second and third statements are true only in a highly restricted sense even when yields are serially independent, and that, when investment yields in the various periods are statistically dependent, only the logarithmic function induces utility functions of short-run wealth which are myopic.

II. PRELIMINARIES

In this and the next three sections, the following notation will be employed:

x_j: amount of investment capital at decision point j (the beginning of the jth period);
M_j: number of investment opportunities available in period j;
S_j: the subset of investment opportunities which it is possible to sell short in period j;
$r_j - 1$: rate of interest in period j;
β_{ij}: proceeds per unit of capital invested in opportunity i, where $i = 2, \ldots, M_j$, in the jth period (random variable); that is, if we invest an amount θ in i at the beginning of the period, we will obtain $\beta_{ij}\theta$ at the end of that period;
z_{1j}: amount lent in period j (negative z_{1j} indicates borrowing) (decision variable);
z_{ij}: amount invested in opportunity i, $i = 2, \ldots, M_j$, at the beginning of the jth period (decision variable);
$f_j(x_j)$: utility of money at decision point j;
$z^*_{1j}(x_j)$: an optimal lending strategy at decision point j;
$z^*_{ij}(x_j)$: an optimal investment strategy for opportunity i, $i = 2, \ldots, M_j$, at decision point j.

$$F_j(y_2, y_3, \ldots, y_{M_j}) \equiv \Pr\{\beta_{2j} \le y_2, \beta_{3j} \le y_3, \ldots, \beta_{M_j j} \le y_{M_j}\}\ ; \quad \bar{z}_j \equiv (z_{2j}, \ldots, z_{M_j j})\ .$$

[1] Jan Mossin, "Optimal Multiperiod Portfolio Policies," *Journal of Business* 41 (April 1968): 215–29.

[2] Hayne Leland, "Dynamic Portfolio Theory" (Ph.D. thesis, Harvard University, 1968).

As most portfolio models do, we assume, in addition to stochastically constant returns to scale, perfect liquidity and divisibility of the assets at each (fixed) decision point, absence of transaction costs, withdrawals, capital additions, and taxes, and the opportunity to make short sales. Furthermore, we assume, until Section VI, that the yields in the various periods are stochastically independent.

Since the end-of-period capital position is given by the proceeds from current savings, or the negative of the repayment of current debt plus interest, plus the proceeds from current risky investments, we have

$$x_{j+1} = r_j z_{1j} + \sum_{i=2}^{M_j} \beta_{ij} z_{ij} \qquad j = 1, 2, \ldots, \qquad (1)$$

where

$$\sum_{i=1}^{M_j} z_{ij} = x_j \qquad j = 1, 2, \ldots. \qquad (2)$$

Combining (1) and (2) we obtain

$$x_{j+1} = \sum_{i=2}^{M_j} (\beta_{ij} - r_j) z_{ij} + r_j x_j \qquad j = 1, 2, \ldots. \qquad (3)$$

Let us now assume that $f_J(x_J)$ is given for some horizon J. Then, as Mossin shows, we may write, by the principle of optimality,[3]

$$f_j(x_j) = \max_{z_j \epsilon Z_j(x_j)} E[f_{j+1}(x_{j+1})] \qquad j = 1, \ldots, J-1 \qquad (4)$$

where $Z_j(x_j)$ is the set of feasible investments at decision point j given that capital is x_j. When there are two assets in each period (i.e., $M_j = 2$ for all j) and $Z_j(x_j) = \{z_{2j}: 0 \leq z_{2j} \leq x_j\}$, that is, borrowing and short sales are ruled out, Mossin concludes that $f_j(x_j) = a_j f_J(x_j) + b_j$ (where $a_j > 0$ and b_j are constants), $j = 1, \ldots, J-1$, that is, that the induced short-run utility functions at decision points $1, \ldots, J-1$ are completely myopic, if and only if (1) $f_J(x)$ is either logarithmic or a power function, or (2) $r_1 = r_2 = \ldots = r_{J-1} = 1$ and $f_J(x)$ is one of

$$f_J(x) = -e^{-\mu x}; \qquad (5)$$

$$f_J(x) = \log(x + \mu); \qquad (6)$$

$$f_J(x) = \frac{1}{\lambda - 1}(\lambda x + \mu)^{1-1/\lambda} \qquad \lambda \neq 0, \lambda \neq 1, \qquad (7)$$

where $\mu \neq 0$ and λ are constants. Note that λ and μ cannot both be negative.

While the first conclusion is beyond dispute, the second is incorrect, as are the conclusions concerning partial myopia in general, except in a severely restricted sense. We shall first demonstrate the assertion in the preceding case and then show that it also holds when borrowing and short sales are not ruled out.

III. AN EXAMPLE

Assume that $r_1 = r_2 = \ldots = r_{J-1} = 1$ and that there is only one risky opportunity (i.e., $M_j = 2$) in each period. Moreover, assume that the proceeds β_{2j} of

[3] Richard Bellman, *Dynamic Programming* (Princeton, N.J.: Princeton University Press, 1957).

this opportunity are

$$\beta_{2j} = \begin{cases} 0 \text{ with probability } 1/2 \\ 3 \text{ with probability } 1/2 \end{cases} \qquad \text{all } j \quad (8)$$

and that

$$f_J(x_J) = (x_J + d)^{1/2} \qquad d > 0\,. \quad (9)$$

(9) clearly belongs to the class (7). (4) and (3) now give

$$f_j(x_j) = \max_{0 \leq z_{2j} \leq x_j} E\{f_{j+1}[(\beta_{2j} - 1)z_{2j} + x_j]\} \qquad j = 1, \ldots, J-1\,, \quad (10)$$

where $f_J(x_J)$ is given by (9).

It is easily verified that

$$z^*_{2,J-1}(x_{J-1}) = \begin{cases} \text{does not exist} & x_{J-1} < 0 \\ x_{J-1} & 0 \leq x_{J-1} < d\,. \\ 1/2(x_{J-1} + d) & x_{J-1} \geq d \end{cases} \quad (11)$$

Thus,

$$f_{J-1}(x_{J-1}) = \begin{cases} 1/2(3x_{J-1} + d)^{1/2} + 1/2d^{1/2} & 0 \leq x_{J-1} < d \\ a_{J-1}(x_{J-1} + d)^{1/2} & x_{J-1} \geq d \end{cases}, \quad (12)$$

where

$$a_{J-1} = 1/2[(1/2)^{1/2} + 2^{1/2}]\,. \quad (13)$$

We now observe that $f_{J-1}(x)$ is a positive linear transformation of $f_J(x)$ *only* for $x \geq d$; for $x < d$, $f_{J-1}(x) < a_{J-1}f_J(x)$.

Proceeding with the solution to (10), we obtain

$$f_j(x_j) = \begin{cases} g_j(x_j) & 0 \leq x_j < b_j \\ a_j(x_j + d)^{1/2} & x_j \geq b_j \end{cases} \qquad j = 1, \ldots, J-1\,, \quad (14)$$

where a_j is a positive constant, $g_j(x_j) < a_j(x_j + d)^{1/2}$ for $0 \leq x_j < b_j$,

$$b_j = d + 2b_{j+1} \qquad (b_J = 0), j = 1, \ldots, J-1\,, \quad (15)$$

and

$$z^*_{2j}(x_j) = \begin{cases} h_j(x_j) & 0 \leq x_j < b_j \\ 1/2(x_j + d) & x_j \geq b_j \end{cases} \qquad j = 1, \ldots, J-1\,. \quad (16)$$

It is easily determined that $h_j(x_j)$ is highly irregular except for $j = J - 1$.

When $J = 11$ and $d = 1{,}000$, we obtain from (15) that $b_1 = 1.023$ million. Thus, when the horizon is ten periods distant, the optimal amount to invest in opportunity 2 is, in this example, proportional to $x_j + d$ only if initial wealth x_1 exceeds \$1 million by a substantial margin. Furthermore, while $f_j(x)$ is a positive linear transformation of $f_J(x)$ for $x \geq b_j$, $j = 1, \ldots, J-1$, it is not for $x < b_j$, that is, for $x_1 < 1.023$ million, $x_2 < 511{,}000$, etc., in the above example. Since the constant $b_j > 0$ depends on the distribution functions $F_j, \ldots, F_{J-1}$, the short-run utility functions induced by the terminal utility function (9) are clearly not myopic. In other words, to make an optimal decision at decision point j, not only F_j but $F_{j+1}, \ldots, F_{J-1}$ must be known.

IV. BORROWING AND SOLVENCY

The nonmyopic nature of the induced utility functions $f_1(x_1)$, $f_2(x_2), \ldots, f_{J-1}(x_{J-1})$ in the preceding example is clearly attributable to the constraint

$$0 \leq z_{2j} \leq x_j \qquad j = 1, \ldots, J-1\,, \quad (17)$$

which precludes borrowing and short sales. We shall now relax this constraint.

Case I.—(17) will first be replaced with

$$0 \leq z_{2j} \leq mx_j \qquad m > 1 \qquad j = 1, \ldots, J-1, \quad (18)$$

that is, $100/m$ is assumed to be the percentage margin requirement. The solution to (10) for $J-1$ now becomes

$$f_{J-1}(x_{J-1}) = \begin{cases} 1/2[(1-m)x_{J-1}+d]^{1/2} \\ \quad + 1/2[(2m+1)x_{J-1}+d]^{1/2} & 0 \leq x_{J-1} < \dfrac{d}{2m-1} \\ a_{J-1}f_J(x_{J-1}) & x_{J-1} \geq \dfrac{d}{2m-1} \end{cases}$$

$$z^*_{2,J-1}(x_{J-1}) = \begin{cases} mx_{J-1} & 0 \leq x_{J-1} < \dfrac{d}{2m-1} \\ 1/2(x_{J-1}+d) & x_{J-1} \geq \dfrac{d}{2m-1}, \end{cases} \quad (19)$$

and the total solution is represented by (14)–(16) with $b_j > 0, j = 1, \ldots, J-1$. Consequently, the optimal portfolio policy is nonmyopic in the case of constraint (18) also.

Case II.—Let us now introduce an absolute borrowing limit of m, that is, substitute

$$0 \leq z_{2j} \leq x_j + m \qquad m > 0 \qquad j = 1, \ldots, J-1 \quad (20)$$

for (17). When $m < d/2$ the solution to (9) is again given by (14)–(16) with $b_j > 0$, $j = 1, \ldots, J-1$. However, when $m \geq L = d/2$, the solution becomes

$$f_j(x_j) = a_j(x_j+d)^{1/2} \qquad j = 1, \ldots, J-1;$$

$$z^*_{1j}(x_j) = 1/2(x_j - d) \qquad j = 1, \ldots, J-1; \quad (21)$$

$$z^*_{2j}(x_j) = 1/2(x_j + d) \qquad j = 1, \ldots, J-1; \quad (22)$$

that is, the optimal investment policy would seem to be completely myopic on the basis of our assumptions. But L clearly depends on $F_{j+1}, \ldots, F_{J-1}$. Thus to know whether $m \geq L$, knowledge of future returns is necessary. Consequently, the optimal investment policy is not myopic in Case II either.

Let us now consider the realism of assumptions (18) and (20). With respect to (20), we observe from (21) that borrowing takes place, considering decision point $J-1$, only when $x_{J-1} < d$. By (3), (8), (21), and (22), we obtain

$$x_J = \begin{cases} 1/2(x_{J-1}-d) \text{ with probability } 1/2 \\ 2x_{J-1}+d \text{ with probability } 1/2\,. \end{cases}$$

Thus, the terminal wealth position has a 1/2 chance of being negative if and only if $x_{J-1} < d$, that is, when borrowing takes place. If the first event ($\beta_{2,J-1} = 0$) takes place and the investor declares bankruptcy at time J, the lender will stand to lose the entire loan of $|1/2(x_{J-1}-d)|$.

The point here is that it would be unreasonable for anyone to lend money to his investor when his optimal strategy calls for it; that is, m should be zero in (20)—which converts (20) to (17). In fact, (18) and (20) may be said to be inconsistent with the portfolio model itself. This is because a wealthier investor (one whose wealth exceeds d) with the same preferences and probability beliefs as a poorer one is, by (21), a possible lender to the poorer one whose wealth is less than d. But the model assumes that lending is safe, that is, that all loans are repaid with probability 1 while, as we have seen, the poorer investor may not be able to repay.

Case III.—A borrowing arrangement that is consistent with the assumed risklessness of lending is one which permits borrowing to the extent that ability to repay, that is, solvency, is guaranteed. Thus, a reasonable constraint on borrowing and short sales, with considerable intuitive appeal as well, is given by

$$\Pr\{x_{j+1} \geq 0\} = 1 \qquad j = 1, \ldots, J-1 . \quad (23)$$

When (23) is substituted for (17), the solution to (10) is the same as when (17) is used; that is, it is given by (14), (15), and (16). Thus, myopia is not optimal in this case either.

V. THE GENERAL CASE

It is readily verified that the conclusions of Sections III and IV are not changed if the number of risky investment opportunities is arbitrary. Moreover, the conclusions hold for all of the functions (5), (6), and (7) whenever $\mu > 0$, both with no borrowing and in each of Cases I–III. Finally, when $r_j \neq 1, j = 1, \ldots, J-1$, partial myopia, as defined by Mossin, is not optimal either in any of the preceding cases. It should be noted that a solution need not exist in Case III unless the "no-easy-money condition" holds.[4] A generalization of this condition (for the case when yields are serially correlated) is given in Section VI. In the most general version of Case III, the set $Z_j(x_j)$ in (4) is given by those $\hat{z}_j$ which satisfy

$$z_{ij} \geq 0 \qquad i \notin S_j \quad (24)$$

and (23).

When $\mu = 0$ in (6) and (7), [(5) is of no interest when $\mu \leq 0$], complete myopia is optimal in both Cases I and III but not in Case II, as is easily shown. The Mossin-Leland conclusions concerning complete myopia when $r_1 = r_2 = \ldots = r_{J-1} = 1$ and partial myopia do not apply in (6) and (7) when $\mu < 0$ either, except in Case III, as we shall demonstrate below. In doing so, we shall also show that, when $\mu \neq 0$, (5), (6), and (7) imply that the optimal investment policies at decision points 1, $\ldots, J-1$ are never myopic in the presence of explicit borrowing limits of any kind, with one exception.

When a solution to the portfolio problem at decision point $J-1$ exists in the presence of constraints (24) only, the optimal lending strategy $\hat{z}_{1,J-1}(x_{J-1})$ has the form $\hat{z}_{1,J-1}(x_{J-1}) = (1 - \lambda A_{J-1})x_{J-1} - A_{J-1}\mu$ in the case of (6) ($\lambda = 1$) and (7) and the form $\hat{z}_{1,J-1}(x_{J-1}) = x_{J-1} - B_{J-1}$ in the case of (5), where A_{J-1} and B_{J-1} are constants, generally positive,[5] which depend on F_{J-1}.[6]

Let us consider (6) and (7) when $\lambda,\mu > 0$. Since A_{J-1}, and hence, $-\hat{z}_{1,J-1}(x_{J-1})$,

[4] Nils Hakansson, "Optimal Investment and Consumption Strategies under Risk for a Class of Utility Functions," *Econometrica* 38 (September 1970): 587–607.

[5] Nonpositive A_{J-1} and B_{J-1} imply that total short sales exceed or equal total long investments.

[6] Nils Hakansson, "Risk Disposition and the Separation Property in Portfolio Selection," *Journal of Financial and Quantitative Analysis* 4 (December 1969): 401–16.

may be arbitrarily large, any finite borrowing limit has a chance to be binding. Consequently, for $f_{J-1}(x)$ to be a positive linear transformation of $f_J(x)$ in the presence of a borrowing limit, it must be a positive linear transformation of $f_J(x)$ *whether the borrowing limit is binding or not.* From Section IV, it is apparent that, when $M_j = 2$ and (17) holds, a necessary and sufficient condition for $f_{J-1}(x)$ to be a positive linear transformation of $f_J(x)$ is that $z^*_{2,J-1}(x_{J-1})$ has the form $z^*_{2,J-1}(x_{J-1}) = a_{2,J-1}(\lambda x_{J-1} + \mu)$, where $a_{2,J-1}$ is a nonnegative constant. When there is more than one risky asset and (24) holds, this condition generalizes to

$$z^*_{i,J-1}(x_{J-1}) = a_{i,J-1}(\lambda x_{J-1} + \mu) \qquad i = 2, \ldots, M_{J-1}, \tag{25}$$

where the $a_{i,J-1}$ are constants, nonnegative only for $i \notin S_{J-1}$.[7] It is now clear that the optimal investment strategy $\hat{z}^*_{J-1}$ will have the form (25) if and only if (1) the borrowing limit is not binding or (2) the borrowing limit has the form

$$-z_{1,J-1}\left(= \sum_{i=2}^{M_{J-1}} z_{i,J-1} - x_{J-1}\right) \leq (\lambda C_{J-1} - 1)x_{J-1} + C_{J-1}\mu \qquad \lambda C_{J-1} > 1, \quad \lambda, \mu > 0\,. \tag{26}$$

The latter assertion follows from (2) and the fact that this form of the borrowing limit does give the solution (25) for any F_{J-1}, as is easily verified; moreover, only (26) is capable of giving a solution of form (25) when the borrowing limit is binding. Since knowledge of whether any given borrowing limit is binding or not requires knowledge of F_{J-1}, it follows that $f_{J-1}(x)$ is myopic in the presence of a borrowing limit only if this limit has the form (26). By induction, $f_1(x), \ldots, f_{J-1}(x)$ are then myopic in the case of (6) and (7) for $\lambda, \mu > 0$ if and only if the borrowing limit in period j is given by

$$(\lambda C_j - 1)x_j + C_j\mu \qquad \lambda C_j > 1, \qquad \lambda, \mu > 0, \qquad j = 1, \ldots, J-1\,. \tag{27}$$

When $\mu < 0$ or $\lambda < 0$ in (6) and (7), any borrowing limit would, to be consistent with myopia, again have to have the form (27). But when $\mu < 0$, we must have $\lambda > 0$ and vice versa so that (27) cannot be nonnegative for all $x_j > 0$ for which borrowing may be desired, a basic requirement of any "true" borrowing limit. The situation in the case of function (5) is analogous. As a result, $f_1(x_1), \ldots, f_{J-1}(x_{J-1})$ can never be myopic for (5), (6), and (7) when $\lambda < 0$ or $\mu < 0$ in the presence of a borrowing limit.

Turning now to the solvency constraint (23), we obtain whenever a solution exists for (6) and (7) that the greatest lower bound on b such that $\Pr\{x_J < b\} > 0$, for any decision at decision point $J - 1$ which satisfies (24), is $K_{J-1}(\lambda x_{J-1} + \mu) + r\hat{z}_{1,J-1}(x_{J-1})$, where K_{J-1} is a constant which depends on F_{J-1}. Since $f'_J(-\mu/\lambda) = \infty$ for $\lambda > 0$, we obtain, letting $x_J \equiv K_{J-1}(\lambda x_{J-1} + \mu) + r\hat{z}_{1,J-1}(x_{J-1})$, that $\lambda x_J + \mu > 0$, which implies, since λ and μ cannot both be negative, $x_J > 0$ when $\mu \leq 0$. Thus the solvency constraint (23) is not binding when $\mu \leq 0$ but may be when $\mu > 0$. Consequently, the induced utility functions $f_1(x), \ldots, f_{J-1}(x)$ are myopic for the class (6) and (7) when $\mu < 0$ in the presence of (24) and the solvency constraint (23).

[7] Ibid.

In sum then, when $\mu \neq 0$ and interest rates are zero, the induced utility of wealth functions $f_1(x_1), \ldots, f_{J-1}(x_{J-1})$ are myopic in the presence of borrowing constraints if and only if the borrowing limits are of the form (27) and $f_J(x)$ is of the form (6) (7) with $\lambda,\mu > 0$; in the presence of the solvency constraint, $f_1(x_1), \ldots, f_{J-1}(x_{J-1})$ are myopic only if $f_J(x)$ has the form (6) or (7) with $\mu < 0$ (and hence $\lambda > 0$).

It should also be noted that, when $\mu < 0$, $f_J(x)$ is undefined for $x < |\mu/\lambda|$ and $f_j(x_j)$, $j = 1, \ldots, J-1$ is undefined for small x_j, a significant drawback. In addition, the relative risk aversion index $-xf_j''(x)/f_j'(x)$ is decreasing for these functions, whereas Arrow,[8] for example, suggests that plausible utility functions of money exhibit increasing relative risk aversion.

VI. SERIALLY CORRELATED YIELDS

We shall now consider the sequential investment problem when yields are serially correlated. In contrast to Mossin's assertion,[9] we shall find that the optimal investment policy is myopic in this case only for a small subset of the terminal utility functions which induce myopic short-run utility functions when returns are serially independent.

For simplicity, we assume that yields and the interest rate obey a Markov process. A distinction between risk due to general market forces, called the economy, and risk due to individual assets and periods is made. As a result, the assumptions and notation of Section II are modified as follows:

x_j: amount of investment capital at decision point j;
N_j: number of states of the economy at decision point j;
M_{jm}: number of investment opportunities available at decision point j, given that the economy is at state m at that time;
S_{jm}: the subset of investment opportunities which it is possible to sell short at decision point j, given that the economy is in state m at that time;
$r_{jm} - 1$: interest rate in period j, given that the economy is in state m at decision point j ($r_{jm} > 1$);
β_{ijmn}: proceeds at the end of period j, given that the economy is in state n at that time, per unit of investment in opportunity i, $i = 2, \ldots, M_{jm}$, at decision point j, given that the economy was in state m at that time;
p_{jmn}: probability that the economy makes a transition from state m to state n in period j

$$\left(p_{jmn} \geq 0, \sum_{n=1}^{N_{j+1}} p_{jmn} = 1\right);$$

z_{1jm}: amount lent in period j, given that the economy is in state m at decision point j (negative z_{1jm} indicate borrowing) (decision variable);
z_{ijm}: amount invested in opportunity i, $i = 2, \ldots, M_{jm}$, at decision point j, given that the economy is in state m at that point;
$f_{jm}(x_j)$: utility of money at decision point j, given that the economy is in state m at that time;
z^*_{1jm}: an optimal lending policy for state m at decision point j;
z^*_{ijm}: an optimal investment policy for state m at decision point j, $i = 2, \ldots, M_{jm}$.

$$\bar{z}_{jm} \equiv (z_{2jm}, \ldots, z_{M_{jm}jm}) ; \quad v_{ijm} \equiv \frac{z_{ijm}}{x_j} ; \quad i = 1, \ldots, M_{jm} .$$

$$\bar{v}_{jm} \equiv (v_{2jm}, \ldots, v_{M_{jm}jm}) ; \quad F_{jmn}(y_2, \ldots, y_{M_{jm}}) \equiv \Pr\{\beta_{2jmn} \leq y_2, \ldots, \beta_{M_{jm}jmn} \leq y_{M_{jm}}\} .$$

[8] Kenneth Arrow, *Aspects of the Theory of Risk-bearing* (Helsinki: Yrjö Jahnssonin Säätiö, 1965).

[9] Mossin, p. 222.

Clearly, v_{ijm} represents the proportion of capital x_j invested in opportunity i at decision point j, given that the economy is in state m at that point; thus

$$v_{1jm} = 1 - \sum_{i=2}^{M_{jm}} v_{ijm} .$$

It will be assumed that the joint distribution functions F_{jmn} are independent with respect to j. In addition, we postulate that the $\{\beta_{ijmn}\}$ satisfy the following conditions:

$$\Pr\{0 \leq \beta_{ijmn} < \infty\} = 1 , \qquad i = 2, \ldots, M_{jm} , \tag{28}$$

all j, m, and n

$$\Pr\left\{\sum_{i=2}^{M_{jm}} (\beta_{ijmn} - r_{jm})\theta_i < 0\right\} > 0 \tag{29}$$

for all j, all m, some n for which $p_{jmn} > 0$, and all finite θ_i such that $\theta_i \geq 0$ for all $i \notin S_{jm}$ and $\theta_i \neq 0$ for at least one i. (29) is a modification of the "no-easy-money-condition" for the case when the lending rate equals the interest rate.[10] This condition states that no combination of risky investment opportunities exists in any period which provides, with probability 1, a return at least as high as the (borrowing) rate of interest; no combination of short sales is available for which the probability is zero that a loss will exceed the (lending) rate of interest; and no combination of risky investments made from the proceeds of any combination of short sales can guarantee against loss. (29) may be viewed as a condition which the prices of all assets must satisfy in equilibrium.

(3) is now replaced by the conditional difference equations

$$x_{j+1}|mn = \sum_{i=2}^{M_{jm}} (\beta_{ijmn} - r_{jm})z_{ijm} + r_{jm}x_j \quad j = 1, \ldots, J-1, \text{ all } m, n , \tag{30}$$

and (4) becomes, for $j = 1, \ldots, J-1$ and all m,

$$f_{jm}(x_j) = \max_{z_{jm}} \sum_{n=1}^{N_{j+1}} p_{jmn}E[f_{j+1,n}(x_{j+1}|mn)] , \tag{31}$$

where $f_{Jm}(x_J)$ is given for all m, subject to

$$z_{ijm} \geq 0 \qquad i \notin S_{jm} , \tag{32}$$

and

$$\Pr\{x_{j+1}|mn \geq 0\} = 1 \qquad n = 1, \ldots, N_{j+1} . \tag{33}$$

VII. OPTIMAL MYOPIC POLICIES

On the basis of the finite yield assumption (28) and the "no-easy-money-condition" (29), we obtain the following:

Theorem.—Let r_{jm}, F_{jm}, and p_{jmn} be defined as in Section VI and let $u(x)$ be a monotone increasing and strictly concave function for all $x \geq 0$. Then the functions

$$h_{jm}(\bar{v}_{jm}) \equiv \sum_{n=1}^{N_{j+1}} p_{jmn}E\left\{u\left[\sum_{i=2}^{M_{jm}} (\beta_{ijm} - r_{jm})v_{ijm} + r_{jm}\right]\right\} , \tag{34}$$

[10] Hakansson, "Optimal Investment . . ." (see n. 4 above).

subject to

$$v_{ijm} \geq 0 \qquad i \notin S_{jm}\,; \quad (35)$$

and

$$\Pr\left\{\sum_{i=2}^{M_{jm}} (\beta_{ijmn} - r_{jm})v_{ijm} + r_{jm} \geq 0\right\} = 1 \quad n = 1, \ldots, N_{j+1} \quad (36)$$

have maxima for all $j = 1, \ldots, J - 1$ and all m. Moreover, the maximizing vectors, $\bar{v}^*_{jm}$, are finite and unique. The proof may be found in Hakansson (1968).[11]

Let us now assume that

$$f_{Jm}(x) = x^{1/2} \qquad \text{all } m\,; \quad (37)$$

and let k_{jm} denote the maximum of (34) subject to (35) and (36) when $u(x) = x^{1/2}$; that is,

$$k_{jm} \equiv \sum_{n=1}^{N_{j+1}} p_{jmn} E\left\{\left[\sum_{i=2}^{M_{jm}} (\beta_{ijmn} - r_{jm})v^*_{ijm} + r_{jm}\right]^{1/2}\right\} \quad j = 1, \ldots, J-1 \quad \text{all } m\,. \quad (38)$$

By the theorem, we know that k_{jm} exists.

Let us now determine $f_{J-1,m}(x_{J-1})$. From (31) we obtain for all m

$$f_{J-1,m}(x_{J-1}) = \max_{\bar{z}_{J-1,m}} \sum_{n=1}^{N_J} p_{J-1,mn} E[(x_J \mid mn)^{1/2}]\,, \quad (39)$$

subject to

$$z_{i,J-1,m} \geq 0 \qquad i \notin S_{J-1,m} \quad (40)$$

and

$$\Pr\left\{\sum_{i=2}^{M_{J-1,m}} (\beta_{i,J-1,mn} - r_{J-1,m})z_{i,J-1,m} + r_{J-1,m}x_{J-1} \geq 0\right\} = 1 \quad n = 1, \ldots, N_J\,. \quad (41)$$

By (29) and (41), $f_{J-1}(x_{J-1})$ does not exist for $x_{J-1} < 0$. For $x_{J-1} \geq 0$, (39) may be written, since (32) and (33) are equivalent to (35) and (36) when $x_{J-1} > 0$,

$$f_{J-1,m}(x_{J-1}) = x_{J-1}^{1/2} \max_{\bar{v}_{J-1,m}} \sum_{n=1}^{N_J} p_{J-1,mn} E\left\{\left[\sum_{i=2}^{M_{J-1,m}} (\beta_{i,J-1,mn} - r_{J-1,m})v_{i,J-1,m} + r_{J-1,m}\right]^{1/2}\right\}, \quad (42)$$

subject to

$$v_{i,J-1,m} \geq 0 \qquad i \notin S_{J-1,m} \quad (43)$$

and

$$\Pr\left\{\sum_{i=2}^{M_{J-1,m}} (\beta_{i,J-1,mn} - r_{J-1,m})v_{i,J-1,m} + r_{J-1,m} \geq 0\right\} = 1 \quad n = 1, \ldots, N_J\,. \quad (44)$$

By the theorem, we now obtain that $f_{J-1,m}(x_{J-1})$ exists for all m and $x_{J-1} \geq 0$ and is given by, using (38),

$$f_{J-1,m}(x) = k_{J-1,m}x^{1/2} \qquad \text{all } m\,. \quad (45)$$

[11] Nils Hakansson, "Optimal Entrepreneurial Decisions in a Completely Stochastic Environment," *Management Science: Theory* 17 (March 1971): 427–49.

By (31), the expression to be maximized at decision point $J-2$, given that the economy is in state m at that time, becomes

$$\sum_{n=1}^{M_{J-2,m}} p_{J-2,mn} k_{J-1\,n} E[(x_{J-1} | mn)^{1/2}] . \tag{46}$$

Since the constants $k_{J-1,n}$ will in general be different for different n, (46) and, therefore, the optimal portfolio $\bar{z}^*_{J-2,m}$, depend on the yields in period $J-1$. Thus the optimal investment policy is *not* myopic at decision point $J-2$; neither is it myopic at decision points $1, \ldots, J-3$, which is easily shown by induction.

The existence of positive constants $a_1, \ldots, a_{J-1}$ and of constants $b_{11}, \ldots, b_{J-1,N_{J-1}}$ such that

$$f_{jm}(x) = a_j f_{Jm}(x) + b_{jm} \qquad \text{all } m , \quad j = 1, \ldots, J-1 \tag{47}$$

are clearly both necessary and sufficient for myopia to be optimal in this model. As noted, (45) violates (47) for $j = J-1$ whenever $N_{J-1} > 1$. However, when

$$N_j = 1 , \quad j = 1, \ldots, J , \tag{48}$$

(47) is satisfied; but (48) also implies that yields are statistically independent in the various periods. This confirms Mossin's result that the optimal investment policy is myopic when returns are stochastically independent over time and the terminal utility function is $x^{1/2}$.

Let us now assume that $f_{Jm}(x)$ has the form

$$f_{Jm}(x) = \log x \qquad \text{all } m . \tag{49}$$

Letting $H_{J-1,m}$ denote the maximum of (34) subject to (35) and (36) when $u(x) = \log x$, that is,

$$H_{J-1,m} \equiv \sum_{n=1}^{N_J} p_{J-1,mn} E\left\{\log\left[\sum_{i=2}^{M_{J-1,m}} (\beta_{i,J-1,mn} - r_{J-1,m}) v^*_{i,J-1,m} + r_{J-1,m}\right]\right\} , \tag{50}$$

we obtain from (31)–(33), solving recursively,

$$f_{jm}(x) = \log x + b_{jm} \qquad \text{all } m \quad j = 1, \ldots, J-1 , \tag{51}$$

(where $b_{J-1,m} = H_{J-1,m}$, all m), which is consistent with (47). As a result, the induced utility functions

$$f_{11}(x), \ldots, f_{1N_1}(x), \ldots, f_{J-1\,1}(x), \ldots, f_{J-1,N_{J-1}}(x)$$

are myopic when the terminal utility function is logarithmic, both when yields are serially correlated and when they are not.

Just as in the case of (37), which is a special case of (7), the optimal investment policy is not myopic for *any* function (7) (whether $\mu = 0$ or not), nor for any function (5), when yields are serially dependent, as is easily verified. Since interest rates are assumed to be positive and state-dependent, myopia is not optimal for $\log (x + \mu)$, $\mu \neq 0$ either, or any other nonlogarithmic function, under serial dependence. Thus, when yields are serially correlated, only the logarithmic utility function of terminal wealth induces short-run utility of wealth functions which are myopic when yields in the various periods are nonindependent, contrary to Mossin's assertion.[12]

VIII. CONCLUDING REMARKS

In view of the difficulty of estimating future yields and their apparent serial correlation, the myopic property of the logarithmic utility functions is, of course, highly significant. However, this function also has other attractive properties.[13] Perhaps the most important of these is the property that maximization of the expected logarithm of end-of-period capital subject to (32) and (33) in each period also maximizes the expected growth rate of capital, whether returns are serially correlated[14] or not.[15]

As Mossin points out, the portfolio decision is in general not independent of the consumption decision. A realistic model of the investor's decision problem must, therefore, include consumption as a decision variable and a preference function for evaluating consumption programs. Consumption-investment models of this type have been developed by Hakansson, both when investment yields are serially correlated[16] and when they are not.[17]

[12] Mossin, p. 222.

[13] Some of the properties are reviewed in Nils Hakansson and Tien-Ching Liu, "Optimal Growth Portfolios When Yields Are Serially Correlated," *Review of Economics and Statistics* 52 (November 1970): 385–94.

[14] Ibid.

[15] Henry Latané, "Criteria for Choice among Risky Ventures," *Journal of Political Economy* 67 (April 1959): 144–55; and Leo Breiman, "Optimal Gambling Systems for Favorable Games," *Fourth Berkeley Symposium on Probability and Mathematical Statistics* (Berkeley: University of California Press, 1961).

[16] Hakansson, "Optimal Entrepreneurial Decisions . . ." (see n. 11 above).

[17] Hakansson, "Optimal Investment . . ." (n. 4 above); and "Optimal Investment and Consumption Strategies under Risk, an Uncertain Lifetime, and Insurance," *International Economic Review* 10 (October 1969): 443–66.

The Journal of FINANCE

Vol XXVIII JUNE 1973 No. 3

10

EVIDENCE ON THE "GROWTH-OPTIMUM" MODEL

Richard Roll*

I. Introduction

A portfolio owner may hope to maximize the long run growth rate of his real wealth. In 1959, Latané suggested maximum growth as an operational criterion for portfolio selection, contending that its (possible) suboptimality on theoretical grounds was practically unimportant and emphasizing Roy's [1952] warning that "A man who seeks advice about his actions will not be grateful for the suggestion that he maximize expected utility."

In the past ten years, however, little work on growth-maximization of portfolio value has appeared in the academic literature of finance or economics. The neglect was due to competing norms for asset selection, particularly to norms based on two-parameter, two-period portfolio models deriving from the work of Markowitz [1959], Tobin [1958], Sharpe [1964], and Lintner [1965]. These were developed into full theories of capital market equilibrium and the empirical evidence collected in their support seemed at least sufficient to justify continued research along the lines of relaxing assumptions and performing more tests on observed portfolio behavior.

Recently, Hakansson [1971] and Hakansson and Liu [1970] again brought the growth maximization criterion to our attention. Hakansson presented a persuasive theoretical argument that ". . . the mean-variance model [a special case of the aforementioned two-parameter model, was] severely compromised by the capital growth model in several significant respects."[1]

Most readers will find the following a significant respect: Given temporally independent returns, a number of mean-variance efficient portfolios can be shown to bring complete ruin after an infinite sequence of re-investments. It is true, of course, that such sequences may indeed be optimal from an expected

* Graduate School of Industrial Administration, Carnegie-Mellon University. The comments of Eugene Fama, Haim Levy, Robert Litzenberger and Myron Scholes are gratefully acknowledged. Remaining errors are due to the author alone.

This project was supported by the Ford Foundation which does not necessarily agree with the results and opinions.

1. Hakansson [1971, p. 517].

utility viewpoint, despite ultimate ruin. No mathematician can prove the contrary, especially when it is recognized that a typical investor's horizon will probably fall short of infinity and that he will likely consume a fraction of his assets each period. However, one of our basic concerns should not be with formal proof, but with practical and intuitively credible models of investor behavior. In this regard, "growth-optimum" portfolios possess appealing features even during a *finite* time span. For example, such portfolios maximize the probability of exceeding a given level of wealth within a fixed time.[2]

Hakansson also pointed out other features of the growth-optimum model that may be more or less appealing. It implies a logarithmic (in wealth) utility function, which displays decreasing absolute risk aversion, and it implies optimal decision rules that are myopic. As a further embellishment, Hakansson and Liu derived a "separation theorem" that holds the optimal sequence of investments to be independent of the sequence of wealth levels even when returns are stochastically *dependent* across time. (Myopia also holds with temporal dependence).

Although these are strong challenges to the practical superiority of two-parameter portfolio models, we should not abandon the latter too hurriedly. They have been successfully used in many empirical contexts and their competitors should be required to weather empirical examination; so the purpose of this paper is to report on some empirical tests of growth-optimum theory using common stock returns.

Briefly, the growth-optimum model receives mixed support. In some tests it performs extremely well while the results of other tests are puzzling. In comparison to the mean-variance model it also performs well but the test results are clouded by the close operational similarity of the two models.

II. A Test Statistic for the Growth-Optimum Model

The quantitative derivation of the growth-optimum rule will employ the following convenient

NOTATION:

n_j—number of shares purchased of security j initially

$p_{j,t}$—price per share of security j in period t.

$V_t \equiv \sum_{j=1}^{N} n_j p_{j,t}$—Value of a portfolio of N distinct securities in period t.

$X_{j,t} \equiv n_j p_{j,t} / \sum_j n_j p_{j,t}$—fraction of resources invested in security j in t.

$R_{j,t} \equiv \{[(p_{j,t} + D_{j,t})/p_{j,t-1}] - 1\}$—rate of return to security j from t − 1 to t.

$D_{j,t}$—per share dividend or coupon paid to security j between t − 1 and t.

E—Mathematical expectation.

2. Breiman [1961, section 5]. Hakansson seems to have erred slightly when he states that "Breiman has shown that if the objective is to achieve a certain level of capital as soon as possible, then the optimal-growth portfolio . . . minimizes the expected time to reach the given level," Hakansson [1971, p. 540]. Breiman conjectured that this was true but was unable to state a proof for a fixed level of wealth. As wealth grows indefinitely large, however, the limited expected minimum time is in fact achieved by the "growth-optimum" portfolio.

Since the rule for maximizing the expected growth of portfolio value is myopic, one only needs to optimize between successive periods. Thus, the growth-optimum rule is

$$\underset{X_{t-1}}{\text{maximize}} \quad E[\log_e(\tilde{V}_t/V_{t-1})] = E\{\log_e[\Sigma_i X_{i,t-1}(1+\tilde{R}_{i,t})]\} \tag{1}$$

subject to $\Sigma_i X_{i,t-1} = 1$. Negative values of X represent short sales which, by assumption, can be accomplished without penalty. Transaction costs are neglected.

Although the problem is easily solved without further qualifications, for historical comparison purposes it is worth assuming that the Nth asset, denoted F, is risk-free and that it receives the residual portion from the amounts invested in all risky assets; i.e.,

$$V_t/V_{t-1} = \sum_{j=1}^{N-1} X_{j,t-1}(1+R_{j,t}) + (1+R_{F,t})\left(1 - \sum_{j=1}^{N-1} X_{j,t-1}\right).$$

First-order conditions from problem (1) show that the investor's growth-optimum portfolio will be determined by proportions X* such that

$$E\left[\frac{\tilde{R}_{j,t} - R_{F,t}}{1 + R_{F,t} + \sum_{i=1}^{N-1} X^*_{i,t-1}(\tilde{R}_{i,t} - R_{F,t})}\right] = 0; \; j = 1, \ldots, N-1. \tag{2}$$

As illustrated by Hakansson [1971] and Breiman [1960], these optimal investment fractions generally will imply a diversified portfolio but their exact values cannot be determined without specifying the joint probability distribution of the R_i's.

To obtain a testable proposition from (2), however, it will not be necessary to specify that distribution. We can rely instead on the fact that in a given period, the value of

$$\frac{1+\tilde{R}_{j,t}}{1 + R_{F,t} + \sum_i [X_{i,t-1}(\tilde{R}_{i,t} - R_{F,t})]},$$

expected by each individual is equal for every risky security. Advancing to an aggregate level will require either of two traditional assumptions: (a) that all investors hold identical probability beliefs or (b) that a "representative" investor holds the expectations of equation (2) with the invested proportions (X*'s) being equal to relative values of existing asset supplies. In either case, the denominator of (2) is equal to $1 + R_{m,t}$, a "market return" defined as a value-weighted average of all individual asset returns. The approximately observable[3] variable

$$Z_{j,t} \equiv \frac{1 + R_{j,t}}{1 + R_{m,t}} \tag{3}$$

3. It is only approximately observable because no comprehensive value-weighted asset indexes exist.

will have the same mean for all securities. It will also be true, of course, that some unexplained variation may occur about these expectations so that the quantities $Z_{j,t}$ will not be exactly equal in every period. However, an appropriate test of the identity of expectations only requires that corresponding sample means be statistically equal; that is, the temporally averaged means

$$\bar{Z}_j = \sum_{t=1}^{T} \frac{1 + R_{j,t}}{1 + R_{m,t}} \Big/ T \tag{4}$$

must be insignificantly different across securities. Testing the equality of N sample means is an analysis of variance problem. Its application to New York and American Stock Exchange listed securities will be reported in the next section.

III. A Simple Test of the Growth-Optimum Model's Basic Validity

Data used in this section are rates of return obtained from the Wells Fargo rate-of-return tape prepared by M. Scholes. The tape contains daily price changes for all common stocks listed on the New York and American exchanges from June 2, 1962 through July 11, 1969. It is a condensation and thus a tractable version of the ISL Quarterly Historical Stock Price Tapes. The Standard & Poor's Composite Price Index (The "500") is used to obtain the "market return."

In the following test, returns were taken over weekly intervals. Thus, the statistic

$$Z_{j,t} = (1 + R_{j,t})/(1 + R_{m,t})$$

was calculated for stock j at the end of week t; where $1 + R_{j,t} = (p_{j,t} + D_{j,t})/p_{j,t-1}$ and $1 + R_{m,t}$ was similarly calculated using the S&P Composite Index. The null hypothesis requires the expected return ratios to be equal, $E(Z_{j,t}) = E(Z_{i,t})$, for all i, j, and t. Usually, this would be tested by one-way analysis of variance on the temporally-averaged means, $\bar{Z}_j = \frac{1}{T} \Sigma_t Z_{j,t}$ but simple analysis of variance procedures requires that all the $Z_{j,t}$'s be uncorrelated cross-sectionally and have equal variances under the null hypothesis. These assumptions are obviously too strong and are not required by the (null) growth-optimum hypothesis anyway. Furthermore, we have the evidence of many previous studies to confirm the existence of positive covariation between stock returns and returns on a market index. Some researchers (notably King, [1966]) have calculated directly a substantial co-movement among stock returns. Although the covariation between two individual stocks returns, say R_j and R_k, may be reduced in the return ratios (Z's) as the result of division by $1 + R_m$, it would be too audacious to assert a complete elimination. Furthermore, there is no *a priori* reason to suppose that $Var(\tilde{Z}_j) = Var(\tilde{Z}_k)$, as is required by a simple one-way of variance.

Fortunately, the Hotelling T^2 statistic[4] is available for precisely those cases

4. See Morrison [1967, pp. 117-124] or Graybill, [1961, pp. 205-206].

where independence among observations and equal variances do not hold. At the expense of considerable computer time, which is patiently borne by Carnegie-Mellon's undergraduates and the Nation's taxpayers, and is thus free to its current beneficiaries, this statistic was calculated for groups of 31 stocks selected alphabetically according to the procedure described below.

Hotelling's T^2 is computed as a quadratic form in the sample vector of differences between adjacent return ratios and the sample covariance matrix of those differences.[5] It is distributed as an F distribution with $k - 1$ and $N - k + 1$ degrees of freedom where N is the sample size (weeks) and k is the number of stocks in a group.[6] Since stocks are not always traded over coincidental calendar periods, and since coincidental observations are required in order to calculate sample covariances, the sample size N of each group was reduced to the number of weeks when all 31 stocks had been traded and recorded. At most, this was the number of weeks for the stock that had the minimum in its group and it was generally somewhat less. In fact, realizing in advance that some groups might be reduced to a very low number of coincidental observations, I decided to discard a stock if keeping it in the group meant reducing the second number of degrees of freedom, $N - k + 1$, below 30. When a stock was discarded, the next alphabetical one on the tape was added to the group. Of course this meant that a stock was then missing from the subsequent group and another had to be taken from the group following that and similarly to the end of the tape. Finally, 68 groups of 31 stocks remained for analysis and this number comprises all the stocks on the tape with sufficient coincidental observations for the test procedure. These 68 F statistics are depicted in Figure 1 and tabulated in Table 1.

The distribution of Figure 1 is stochastically below the expected null distribution. For example, using a Chi-square goodness-of-fit test with 17 classes to compare the F sampling distribution with the null distribution, the test statistic is about 70, which is far above the .005 level of significant difference between the two distributions. It should be emphasized, however, that only ***high*** F values reject the null growth-optimum hypothesis. Thus, growth-optimum theory is strongly, even too strongly, supported by this test of its basic validity.[7]

5. As further explanation, recall that $Z_{j,t} \equiv (1 + R_{j,t})/(1 + R_{m,t})$ is the ratio of return on stock j to the market return. For a given group of 31 stocks, the differences in return ratios are calculated as $y_{j,t} \equiv Z_{j,t} - Z_{j+1,t}$ for $j = 1, \ldots, 30$. The means of $y_{j,t}$ and covariances of $y_{j,t}$ and $y_{k,t}$ are then calculated over time. Hotelling's statistic is based on the sample quadratic form $\bar{y}'S^{-1}\bar{y}$ where $\bar{y}$ is the vector of sample mean differences and S is the sample covariance matrix of differences. Differences were calculated between adjacent alphabetic pairs but any other random scheme for selecting pairs would have been equally acceptable. Cf. Morrison [1967, pp. 135-138].

6. This is the rationale for using a group size of 31 stocks: since the number of degrees of freedom for the test is k—1, where k is the number of stocks, a group size of 31 is both large and makes tabular comparison easy. If the group size had been chosen larger, the second degrees of freedom parameter, N—k+1, (where N is the number of available time points), would be reduced to a low number. Thus, I thought k=31 would balance the two d.f. parameters and still leave them quite large.

7. Because the Hotelling test supports the null hypothesis *too* strongly, I decided to check several potential causes. One obvious possibility is the thick-tailed distributions of stock returns that have been pointed out by many researchers, (Cf. Fama [1965], Blume [1970]). The appropriate way to check this problem is to use a non-parametric analysis of variance, Friedman's multi-sample test, (Bradley, [1968, p. 127]). This was done for exactly the same sample of stocks grouped in

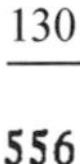

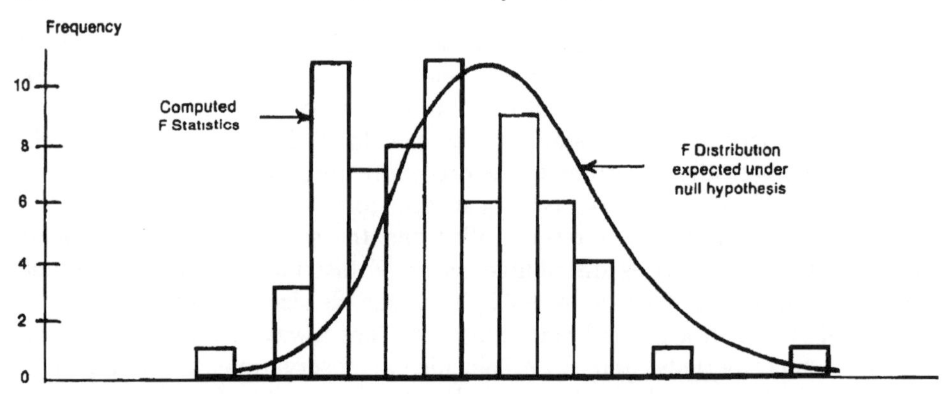

FIGURE 1
Distribution of F Statistics from Hotelling's T^2 Test of Equality Among Mean Return Ratios

IV. Relations between the Growth-Optimum and the Sharpe-Lintner Models

A. Theory

When one compares the first order conditions for the growth-optimum model

$$E\frac{1+\tilde{R}_j}{1+\tilde{R}_m} = (1+R_f).E\frac{1}{1+\tilde{R}_m}; \quad j=1,\ldots N-1 \tag{5}$$

with those for the Sharpe [1964]-Lintner [1965] model

$$\frac{E(1+\tilde{R}_j)}{E(1+\tilde{R}_m)} = \beta_j + (1-\beta_j)\frac{(1+R_f)}{E(1+\tilde{R}_m)}; \quad j=1,\ldots,N-1 \tag{6}$$

(where $\beta_j = \text{Cov}(\tilde{R}_j, \tilde{R}_m)/\text{Var}(\tilde{R}_m)$), some correspondence appears but it is rather difficult to evaluate fully just by inspection. For example, when the Sharpe-Lintner risk coefficient, β_j, is equal to zero, equation (6) becomes

$$\frac{E(1+\tilde{R}_j)}{E(1+\tilde{R}_m)} = \frac{1+R_f}{E(1+\tilde{R}_m)}.$$

But when $\beta_j = 0$, $\text{Cov}(\tilde{R}_j, \tilde{R}_m) = 0$, and the growth optimum condition expressed in equation (5) becomes

the same way. The results were identical to those obtained by using Hotelling's T^2, the growth-optimum model was too strongly supported.

A second possible misspecification is a deficiency in the market price index. The Standard & Poor's Composite Price Index, used in the preceding tests, is heavily-weighted in favor of a few stocks. Also, since it is essentially a "buy-and-hold" portfolio, weights of individual stocks change over time as relative prices change. Evans [1968] and Cheng and Deets [1971] have provided empirical evidence that such an index performs quite differently from a "fixed-investment proportion" or "rebalanced" index. Therefore, a rebalanced index composed of stocks on the tape was constructed and used in reporting the tests already done. Again, no difference was detected.

Thirdly, the possibility that an unrepresentative episode biased the entire sample period of seven years of weekly observations was checked by repeating the tests for annual sub-periods. There was no perceptible difference among the sub-periods or between the overall period and any sub-period. In every case the growth-optimum model was strongly supported.

Further details of all the tests in this footnote are available in an earlier working paper. (It was edited to conserve *Journal* space.)

TABLE 1
CALCULATED VALUES OF HOTELLING'S T^2 STATISTIC FROM COMMON STOCK MEAN RETURN RATIOS, 1962-1969

Group	Sample size (weeks)	T^2	Group	Sample size (weeks)	T^2
1	113	36.0	36	98	30.8
2	81	20.2	37	76	35.0
3	61	53.3	38	101	29.2
4	61	45.4	39	79	54.1
5	143	28.4	40	69	56.3
6	76	35.3	41	107	32.3
7	132	55.5	42	61	68.9
8	61	14.4	43	61	72.0
9	148	29.7	44	78	32.0
10	131	24.1	45	64	35.9
11	90	22.6	46	97	46.3
12	68	28.7	47	66	56.9
13	108	41.3	48	62	57.1
14	76	27.5	49	91	24.9
15	109	41.5	50	60	50.1
16	273	29.4	51	61	37.6
17	111	33.5	52	64	40.3
18	110	35.0	53	63	38.3
19	175	36.5	54	65	25.6
20	75	42.0	55	60	35.1
21	95	27.9	56	67	63.8
22	76	51.3	57	80	48.2
23	112	22.4	58	60	107.0
24	74	64.5	59	96	24.9
25	112	23.4	60	79	39.0
26	127	31.8	61	73	57.8
27	158	38.6	62	82	56.2
28	78	23.0	63	72	28.5
29	123	39.6	64	66	65.7
30	112	40.2	65	61	47.3
31	104	49.7	66	65	53.5
32	61	32.3	67	75	42.3
33	115	35.2	68	63	52.9
34	62	65.6			
35	70	30.6			

Note: The F statistic is calculated as $\frac{N-30}{(N-1)\,30} T^2$.

$$E(1+R_j)\,E\left(\frac{1}{1+\tilde{R}_m}\right) = (1+R_f)\,E\left(\frac{1}{1+\tilde{R}_m}\right).$$

Since the terms containing market returns cancel in both of the displayed equations just above, the growth optimum model provides a market diversification result which is well-known from Sharpe-Lintner theory; namely, a security whose portfolio risk is zero will sell at an expected return equal to the riskless rate.

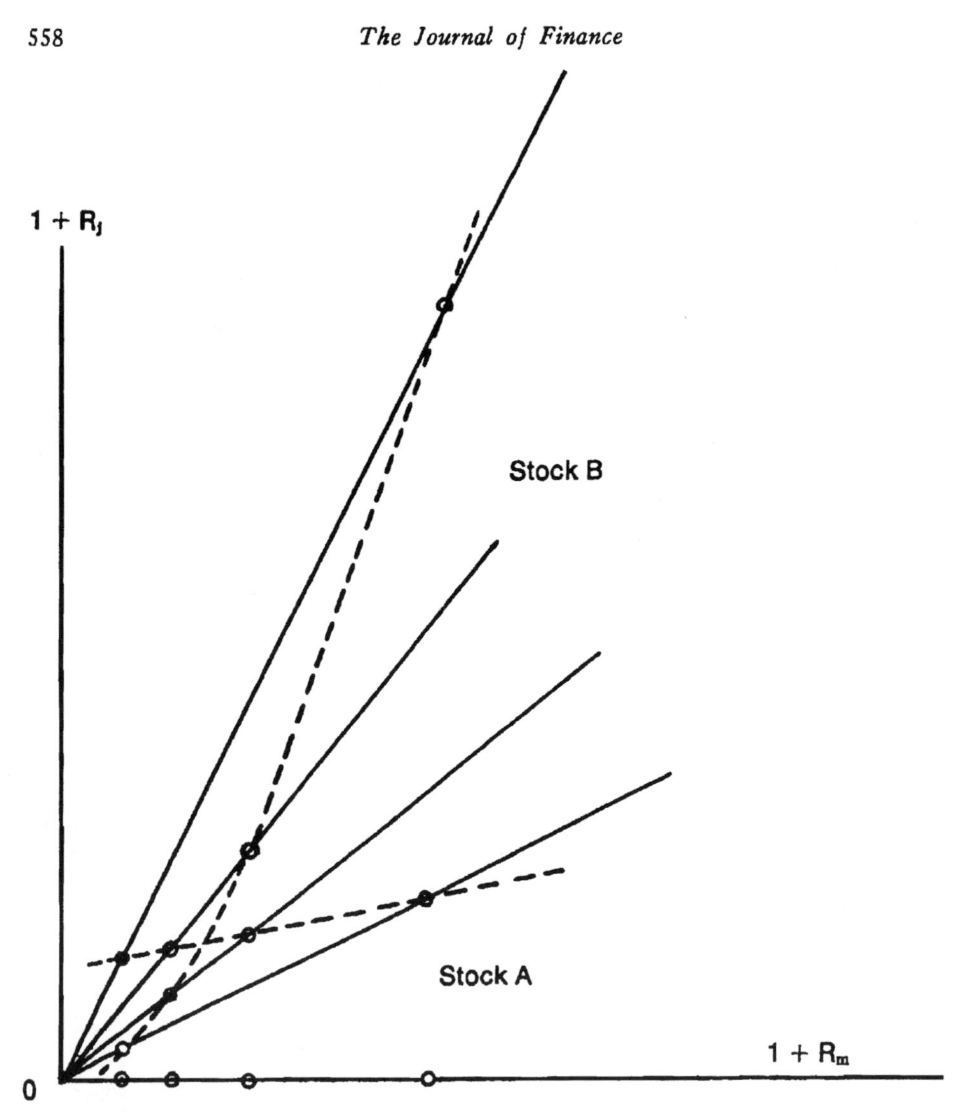

FIGURE 2
Stocks Satisfying the Growth-Optimum Model with Different Degrees of Market Response

It is *not* true that the growth-optimum model implies a constant β coefficient although it may seem to after a first glance at equation (6).[8] A simple example is sufficient to show the contrary. Figure 2 illustrates the *discrete* probability

8. Because (6) is

$$\frac{E(1+\tilde{R}_j)}{E(1+\tilde{R}_m)} = \frac{1+R_F}{E(1+\tilde{R}_m)} + \beta_j \left[1 - \frac{1+R_F}{E(1+\tilde{R}_m)}\right]$$

and (5) is

$$E\left(\frac{1+\tilde{R}_j}{1+\tilde{R}_m}\right) = E\left(\frac{1+\tilde{R}_F}{1+\tilde{R}_m}\right),$$

a reader not cautious about quotients and reciprocals of random variables might think that the latter model implies $\beta_j =$ zero, independent of j.

distributions of stocks A and B and of the market. In this example, only four equally likely returns are possible for the market and each stock can return corresponding amounts to those four. The growth-optimum model requires that

$$E\left(\frac{1+\tilde{R}_A}{1+\tilde{R}_m}\right)=E\left(\frac{1+\tilde{R}_B}{1+\tilde{R}_m}\right)$$

which in geometric terms implies that the expected *slopes* of rays from the origin through each possible point in the $1 + R_j$, $1 + R_m$ plane is equal for both stocks. This is clearly satisfied by the radially symmetric solid lines which pass through the points of possible occurrence in Figure 2. Nevertheless, the slopes of regression lines of R_j on R_m, indicated by dashes, are quite different. Stock A has low portfolio risk because its β is low, while stock B has much greater response to the market and thus higher risk. Both securities' returns are perfectly functionally related to the market return but it is already apparent that correlation per se will have the same relatively unimportant role in a growth-optimum as it has in the Sharpe-Lintner framework (i.e., the expected slopes of rays can be equal no matter what correlation occurs between a stock's return and the market's).

A close correspondence between the two models can be made more apparent by rearranging a few terms. The result (derived in the footnote[9]), is

$$E(\tilde{R}_j - R_F) = \left[\operatorname{Cov}\left(\tilde{R}_j, \frac{1}{1+\tilde{R}_m}\right)\Big/\operatorname{Cov}\left(\tilde{R}_m, \frac{1}{1+\tilde{R}_m}\right)\right] E(\tilde{R}_m - R_F); \tag{7}$$

which is very similar indeed to the Sharpe-Lintner equilibrium equation. In fact, since $\operatorname{Cov}\left(\tilde{R}_m, \frac{1}{1+\tilde{R}_m}\right)$ is negative, one is tempted to suggest that a second implication of Sharpe-Lintner theory is also satisfied by the growth-optimum model: namely, that security j's expected return will exceed the riskless return if and only if $\operatorname{Cov}(\tilde{R}_j, \tilde{R}_m)$ is positive. One would need to prove,

9. To obtain (7), note that equation (5) is equivalent to

$$\operatorname{Cov}[\tilde{R}_j, 1/(1+\tilde{R}_m)] = (1+R_F)K - E(1+\tilde{R}_j)K \tag{5a}$$

where $K \equiv E[1/(1+\tilde{R}_m)]$.

Multiplying both sides of (5a) by X_j, the proportion of wealth invested in security j, (or the fraction of aggregate economic wealth represented by security j),

$$\operatorname{Cov}[\sum_j X_j\tilde{R}_j, 1/(1+\tilde{R}_m)] = -K \sum_j [X_j E(\tilde{R}_j - R_F)]$$

and substituting for the definition of R_m, i.e., for $R_m \equiv R_F + \Sigma X_j(R_j - R_F)$,

$$\operatorname{Cov}[\tilde{R}_m - R_F(1-\sum_j X_j), 1/(1+\tilde{R}_m)] = -K\,E(\tilde{R}_m - R_F).$$

Since $R_F(1-\Sigma x_j)$ is a constant,

$$K = -\operatorname{Cov}[\tilde{R}_m, 1/(1+\tilde{R}_m)]/E(\tilde{R}_m - R_F).$$

Substituting this for K in (5a) provides equation (7).

however, that Cov $(R_j, R_m) > 0$ implies Cov $\left(R_j, \frac{1}{1+\tilde{R}_m}\right) < 0$ and this is definitely not true in general. Counterexamples can be displayed for highly-skewed distributions.[10]

B. *Testing Sharpe-Lintner vs. Growth-Optimum*

The two models will provide approximately equivalent empirical implications if certain restrictions are placed on the ranges of individual rates of return. To verify this, one only needs to note the following facts: (a) quadratic utility functions and homogenous anticipations will lead to the Sharpe-Lintner equilibrium result of equation (6) and (b) the logarithmic utility function implied by portfolio growth maximization can be approximated by a quadratic as

$$\log(1 + R_T) \approx R_T - 1/2\, R_T^2$$

provided that the portfolio's total return is restricted to less than 100 per cent per period.[11] It is indeed trivial to show that the two models are identical when this approximation to the logarithm is made. Thus, given the truth of one theory, we should not be surprised to find that an empirical test of the other supports it very well, especially when the observed rates of return used in testing fall predominantly near zero. Of course, this leads us to ask whether the strong empirical support for the growth-optimum model reported in the preceding section is really damnation for Sharpe-Lintner or just an accidental stroke of choosing a short time interval (one week) which guaranteed that returns were never observed far from zero.

An obvious way to test this is suggested by the logarithmic approximation. If growth-optimum theory appears to satisfy the data only because the logarithm approximates a quadratic when returns are near zero, one should choose a longer time interval for empirical testing so that many more large and small returns are observed.[12] This was done for both four week and twenty-six week periods with the same common stock data as used previously and the conclusions were identical to those for weekly periods already reported.

A more refined and direct test to discriminate between the two models can be based on equilibrium conditions of the two competing theories,

$$E(\tilde{R}_j - R_F) = \left\{ \text{Cov}\left[\tilde{R}_j, \frac{1}{1+\tilde{R}_m}\right] \Big/ \text{Cov}\left[\tilde{R}_m, \frac{1}{1+\tilde{R}_m}\right] \right\} E(\tilde{R}_m - R_f) \equiv \gamma_j \quad (8)$$

10. However, for at least one special asymmetric case, (lognormal distributions) the growth-optimum model does agree completely with the Sharpe-Lintner result that a security's expected return will be a linear function of systematic risk; i.e., of β_j. I am indebted to Robert Litzenberger for demonstrating this point.

11. And more than minus 100 per cent per period. Without short sales, this last restriction is presumably satisfied for common stocks by the existence of limited liability. Samuelson [1970] has derived a broader "fundamental approximation theorem" which shows the close match of mean-variance to *any* correct portfolio theory when the joint distribution of returns has a small dispersion.

12. This completely ignores the crucial question of investor horizon period that may have a significant effect on the *form* of the Sharpe-Lintner market model. Cf. Jensen [1969, pp. 186-191].

which is the growth-optimum condition (7), and

$$E(\tilde{R}_j - R_F) = \{Cov[\tilde{R}_j, \tilde{R}_m]/Var(\tilde{R}_m)\} E(\tilde{R}_m - R_F) \equiv \delta_j \quad (9)$$

which is the familiar Sharpe-Lintner condition. A suggested test procedure obtains the best estimates of all components of (8) and (9) from time series and then performs cross-sectional regression with those estimates.

A series of comparative tests of these two equations was conducted with the common stock returns mentioned previously. In each case, time series were used to calculate $\bar{R}_j$, the mean return, and $\hat{\varphi}_j$ or $\hat{\delta}_j$, the estimated risk measures implied by the growth-optimum or the Sharpe-Lintner model, respectively.[13] Then, cross-sectional computations were performed for the regression models

$$\bar{R}_j = \hat{a}_0 + \hat{a}_1\hat{\delta}_j \quad (10)$$

and

$$\bar{R}_j = \hat{b}_0 + \hat{b}_1\hat{\delta}_j. \quad (11)$$

Estimated coefficients from (10) and (11) should be compared to their theoretical counterparts: depending on which theory is correct, $\hat{a}_0$ or $\hat{b}_0$ should equal R_F, the average risk-free interest rate, and $\hat{a}_1$ or $\hat{b}_1$ should equal unity.

In the first test, all calculations were carried out with ***individual*** security returns during the same time period.[14] For example, 1192 separate values of $\bar{R}$, $\hat{\varphi}$, and $\hat{\delta}$ were obtained from weekly data covering the annual sub-period July, 1962 through June, 1963. Cross-sectional regressions using these 1192 estimates gave $\hat{a}_0 = 20.1$ and $\hat{b}_0 = 20.3$ per cent. The value of $\bar{R}_F$, as measured by the weekly average interest rate on short-term government debt obligations during the year, was only 3.27 per cent; so neither model satisfied its theoretical prediction very well in this particular annual sub-period.[15] Over all the seven years of available data, the estimated values of $\hat{a}_1$ and $\hat{b}_1$ were very significantly positive and they were scattered around unity in nice accord with their expected level. $\hat{a}_0$ and $\hat{b}_0$ were also reasonably close to $\bar{R}_F$, at least on average. As a distinguishing test of the two competing theories, however, these results failed miserably; for the estimates were very highly correlated between the models. The two competitive intercepts were practically identical in every

13. To be precise,

$$\hat{\varphi}_j \equiv \left[\hat{C}ov\left(R_j, \frac{1}{1+R_m} \right) \Big/ \hat{C}ov\left(R_m, \frac{1}{1+R_m} \right) \right] (\bar{R}_m - \bar{R}_F)$$

and

$$\hat{\delta}_j \equiv [\hat{C}ov(R_j,R_m)/\hat{V}ar(R_m)](\bar{R}_m - \bar{R}_F)$$

where ˆ indicates the sample analog of a population parameter, calculated from weekly observations over a specified period, and — indicates sample mean from the same period.

14. To save computation expense, only New York Exchange listed stocks with at least 30 weekly quotations during a year were included in the sample. The risk-free rate was measured by a weekly "average of short-term government debt obligations" taken from *Standard & Poor's Trade Statistics*. The market indexes used were: The S&P Composite Index and a rebalanced index constructed by weighting all NYSE Stock Returns equally each week. The results were very similar but only the results for the rebalanced index are quoted in the text.

15. To save space, only one annual sub-period is reported here but *all* the results are available from the author upon request.

period as were the two competitive slope coefficients. They deviated much more from theoretical predictions than from each other.

In order to sharpen the discriminatory resolution of the data, a different procedure[16] was necessary. The first problem to be alleviated was an extremely low explanatory power which was indicated by low R^2's (on the order of .05), for the cross-sectional models with individual stock returns. A technique to remedy this is to form portfolios of stocks and conduct cross-sectional tests on portfolios rather than on individual securities. A difficulty arises, however, because randomly-selected portfolios would have very similar values of the risk measures $\hat{\gamma}$ and $\hat{\delta}$. To create a cross-sectional spread in risk measures, portfolios must be selected on the basis of risk by grouping stocks with the lowest $\hat{\gamma}$'s or $\hat{\delta}$'s in one portfolio, stocks with higher $\hat{\gamma}$'s or $\hat{\delta}$'s in the next portfolio, and so on.

This procedure for forming portfolios makes another econometric problem obvious: In the cross-sectional regression, the explanatory variables contain errors which will tend to bias slope coefficient estimates toward zero.[17] To alleviate this problem somewhat, stocks were placed in portfolios based on their risk measures calculated from weekly data of a given year and then the mean portfolio return and risk measures for the portfolio as a whole were calculated from weekly data in the subsequent year. Cross-sectional models (10) and (11) were then estimated using these latter estimates and the results are given in Table 2 and Figure 3. To recapitulate, the procedure which resulted in the output of Table 2 and Figure 3 was as follows:

1. For each stock (j) in year t, the total risk premia $\hat{\gamma}_j$ and $\hat{\delta}_j$ were calculated from the time series of year t using the rebalanced market index (see note 7).
2. Stocks were ranked from smallest to largest $\hat{\gamma}_j$ and from smallest to largest $\hat{\delta}_j$.
3. Twenty portfolios were selected by assigning the lowest five per cent of ranked stocks to one portfolio, the next lowest five per cent to a second portfolio, etc. Thus, two sets of 20 portfolios each were formed; one set based on γ rankings and one set on δ rankings.
4. Each portfolio's mean return, $\bar{R}_p$, was calculated from time series <u>in year t + 1</u>. For each portfolio that had been formed on the basis of γ rankings, the growth-optimum risk premium $\hat{\gamma}_p$ was calculated from year t + 1 data. Similarly, the Sharpe-Lintner premium $\hat{\delta}_p$ was calculated from year t + 1 data for each portfolio that had been formed by δ rankings.
5. For each of six years (1963-64, 1964-65, . . . 1968-69), these calculated portfolio mean returns and risk measures are plotted in Figure 3 and regressions across portfolios are reported in Table 2.

In Figure 3, the scatters of mean portfolio returns versus the two competing risk measures are displayed for six different years. The solid lines mark

16. Blume [1970], Miller and Scholes [1972], Black, Jensen, and Scholes [1972], and Fama and MacBeth [1972], have originated and developed the procedures used here in their work with the empirical validity of the two-parameter portfolio model.

17. This is true, of course, for regressions using individual stocks as well as for those using portfolios.

TABLE 2
RELATIONS BETWEEN AVERAGE RETURNS AND RISK COEFFICIENTS OF 20 PORTFOLIOS SELECTED ON THE BASIS OF INDIVIDUAL RISK COEFFICIENTS CALCULATED ONE PERIOD EARLIER
(Rebalanced Index)

			Growth Optimum Model			Sharpe-Lintner Model			
Period (July through June)[a]	$\bar{R}_F$[e] % per annum	$\bar{R}_m$	$\hat{a}_0$ % per annum	$\hat{a}_1$[d]	R^2	$\hat{b}_0$ % per annum	$\hat{b}_1$[d]	R^2	Size of Portfolio[c] (No. of Stocks)
1963-1964	3.71	14.4	15.8 (6.29)	−.0817 (−.374)	.0077	15.8 (6.05)	−.0781 (.343)	.0065	59
1964-1965	3.81	12.6	12.4 (6.53)	.0195 (.0918)	.0005	12.5 (6.90)	.00604 (.0297)	.0001	58
1965-1966	4.34	21.0	−9.91 (−5.14)	1.85 (16.8)	.940	−9.68 (−5.62)	1.83 (18.6)	.951	62
1966-1967	4.61	29.4	−2.72 (−1.04)	1.27 (12.5)	.896	−3.02 (−1.14)	1.28 (12.4)	.895	63
1967-1968	5.15	29.2	21.7 (8.20)	.261 (2.43)	.247	22.2 (7.23)	.241 (1.93)	.172	62
1968-1969	5.36	−2.45	13.3 (3.55)	1.97 (4.24)	.500	13.2 (3.84)	1.96 (4.57)	.537	59

[a] The last date in 1969 was July 11. In other years the first date was the first Thursday in July, 196X, and the last date was the last Thursday in June, 196(X + 1).

[b] t-ratios are in parentheses.

[c] This is equal to the total number of stocks that have at least 30 available prices in periods $t-1$ and t, N, divided by 20. The remainder from $N/20$, $J = MOD(N,20)$, was distributed such that one extra stock was included in each of the first J portfolios.

[d] Means of sample values $\hat{a}_1 = .881$; $\hat{b}_1 = .873$; $\bar{R}_F = 4.50$; $\hat{a}_0 = 8.43$; $\hat{b}_0 = 8.50$.

[e] Mean of weekly observations of short-term government debt obligations, *Standard and Poor's Trade Statistics.*

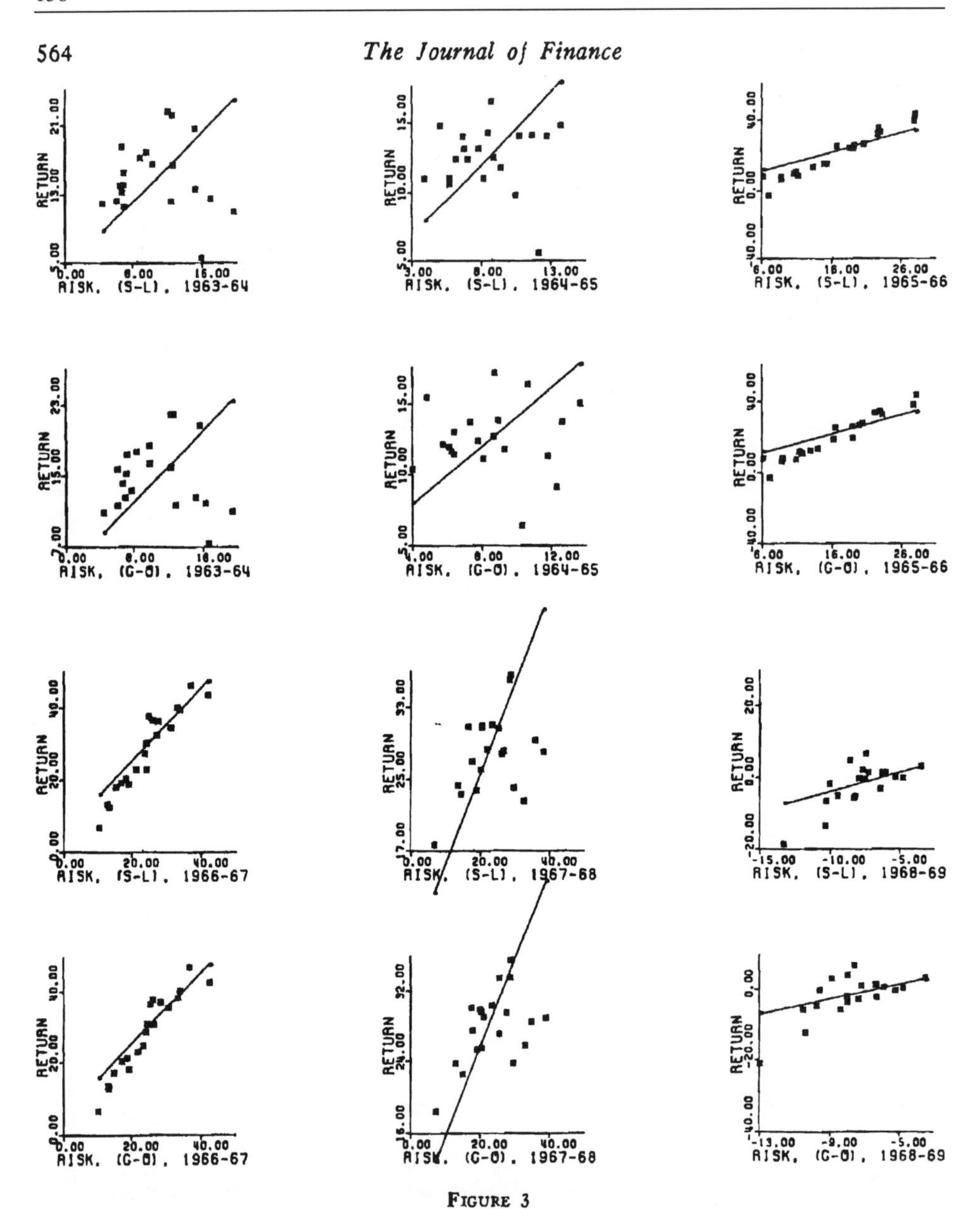

FIGURE 3
Mean Portfolio Returns and Estimated Portfolio Risk Premia, New York Exchange Stocks, 1962-1969, Rebalanced Index

* (S-L) Denotes Sharpe-Lintner Risk Premia and (G-O) denotes Growth-Optimum Premia.

the theoretical predictions, an intercept of $\bar{R}_F$ and a slope of unity.[18] Table 2 contains results from cross-sectional models (10) and (11) applied to these data.

On average across the six years, both the growth-optimum and the Sharpe-Lintner model seem to have excessive intercepts, ($\hat{a}_0$, $\hat{b}_0 > \bar{R}_F$), and deficient slopes, ($\hat{a}_1$, $\hat{b}_1 < 1$). The averages are given in footnote d of Table 2. These deviations from the anticipated can no doubt be attributed, at least in part,

18. Since the axes are scaled differently in each plot, the lines do not appear to have slopes of unity upon first examination. Note that the plotted lines are the theoretical predictions and are *not* the regression lines reported in Table 2.

to errors in the measurement of risk premia. In the first two years and in year 5 (1967-68) the wide scatter suggests that either the mean portfolio returns or the risk premia were measured inaccurately.

In the two years of best fit, 1965-66 and 1966-67, portfolios with low risk premia return less than anticipated while portfolios with high premia return more. This is unlikely to be just a sampling phenomena if the standard errors of $\hat{a}_1$ and $\hat{b}_1$ are credible. For example, the growth-optimum slope for 1965-66 is $\hat{a}_1 = 1.85$ which is nearly eight standard errors above unity.[19]

Perhaps the most striking characteristic of the two models is their very close relation over time. The slope coefficients and intercepts given in Table 2 vary widely across periods but are extremely close, between the two models, in each period. This is obvious from even a quick look at Figure 3.[20] Based on these results, one can only conclude that the two models are empirically identical. A qualification is in order, of course; assets whose returns are much more highly skewed (e.g., warrants), may permit a finer discriminatory test.

V. Summary and Conclusions

If investors wish to maximize the probability of achieving a given level of wealth within a fixed time, they should choose the "growth-optimum" portfolio; that is, the portfolio with highest expected rate of increase in value. This paper has examined the implications for observed common stock returns of all investors selecting such a portfolio.

Given some widely-used (and useful) aggregation assumptions, the growth-optimum model implies that the expected return ratio $E[(1 + \tilde{R}_j)/(1 + \tilde{R}_m)]$ is equal for all securities.[21] This implied equality of expected return ratios was utilized in analysis of variance tests with New York and American Exchange listed stocks from 1962-1969 in order to ascertain the basic validity of the growth-optimum model. The model was well-supported by the data.

The growth-optimum model was compared algebraically to Sharpe-Lintner theory, which is probably the most widely-used portfolio result in empirical work. A close correspondence was demonstrated between their qualitative implications. For example, both models imply that an asset's expected return will equal the risk-free interest rate if the covariance between the asset's return and the average return on all assets, $\text{Cov}(\tilde{R}_j, \tilde{R}_m)$, is zero. For most cases, the growth-optimum model also shares the Sharpe-Lintner implication that an asset's expected return will exceed the risk-free rate if and only if $\text{Cov}(R_j, R_m) > 0$. There are, however, some cases of highly-skewed probability distributions where this implication does not follow for the growth-optimum model.

A close empirical correspondence between the two models was demonstrated for common stock returns. The procedure (1) estimated returns and risk premia

19. For a more detailed discussion of this point, see Friend and Blume [1970].

20. The greatest difference between $\hat{a}_1$ and $\hat{b}_1$ is .02 which is only about 2.3 per cent of their average value. Between $\hat{a}_0$ $\hat{b}_0$, the greatest difference is about six per cent of their average value.

A close connection between the two models was previously implied by the work of Young and Trent [1969]. They showed that the geometric mean of portfolio returns was closely approximated by functions of the arithmetic mean and variance of returns. These functions were developed as approximations to the geometric mean. For accuracy, they require a minimal amount of skewness and are, therefore, analogous to the truncated (after two terms), Taylor series expansions of $\log_e(1+R)$.

21. R_j is the rate of return on security j and R_m is the rate of return on a portfolio of all assets.

implied by the two models from time series; (2) calculated cross-sectional relations between estimated returns and risks; and (3) compared the cross-sectional relations to the theoretical predictions of the two models. They could not be distinguished on an empirical basis. In every period, estimated corresponding coefficients of the two models were nearly equal; and indeed, they deviated much further from their theoretically anticipated levels than they deviated from each other.

REFERENCES

Fischer Black, Michael C. Jensen, and Myron Scholes. "The Capital Asset Pricing Model: Some Empirical Tests," in Jensen, ed., [1972].

Marshall E. Blume. "Portfolio Theory: A Step Towards Its Practical Application," *Journal of Business*, 43, (April, 1970), 152-173.

James V. Bradley. *Distribution-free Statistical Tests*, (Englewood Cliffs, N.J.: Prentice-Hall), 1968.

Leo Breiman. "Investment Policies for Expanding Businesses Optimal in a Long-Run Sense," *Naval Research Logistics Quarterly*, 7, (December, 1960), 647-651.

L. Breiman. "Optimal Gambling Systems for Favorable Games," *Proceedings of the Fourth Berkeley Symposium on Mathematical Statistics and Probability*, I (Berkeley, University of California Press), 1961, 65-78.

Pao L. Cheng and M. King Deets. "Portfolio Returns and the Random Walk Theory," *Journal of Finance*, 26 (March, 1971), 11-30.

John L. Evans. "The Random Walk Hypothesis, Portfolio Analysis and the Buy-and-Hold Criterion," *Journal of Financial and Quantitative Analysis*, 3 (September, 1968), 327-342.

Eugene F. Fama and James MacBeth. "Risk, Return and Equilibrium: Empirical Tests," (Working paper, University of Chicago, Graduate School of Business, February, 1972).

Eugene F. Fama. "The Behavior of Stock Market Prices," *Journal of Business*, 38 (January, 1965), 34-105.

Irwin Friend and Marshall Blume. "Measurement of Portfolio Performance Under Uncertainty," *American Economic Review*, 60, (September, 1970), 561-575.

Franklin A. Graybill. *An Introduction to Linear Statistical Models*, Vol. I, (New York: McGraw-Hill, 1961).

Nils H. Hakansson. "Capital Growth and the Mean-Variance Approach to Portfolio Selection," *Journal of Financial and Quantitative Analysis*, VI (January, 1971), 517-557.

Nils H. Hakansson and Tien-Ching Liu. "Optimal Growth Portfolios When Yields are Serially Correlated," *Review of Economics and Statistics*, 52 (November, 1970), 385-394.

Michael C. Jensen, ed., *Studies in The Theory of Capital Markets*, (New York, Praeger Publishing Co., 1972).

Michael C. Jensen. "Risk, The Pricing of Capital Assets, and the Evaluation of Investment Portfolios," *Journal of Business*, LXII (April, 1969), 167-247.

Benjamin F. King. "Market and Industry Factors in Stock Price Behavior," *Journal of Business*, 39 (January, 1966 supp.), 139-190.

Henry Allen Latané. "Criteria for Choice Among Risky Ventures," *Journal of Political Economy*, 67 (April, 1959), 144-155.

John Lintner. "The Valuation of Risk Assets and the Selection of Risky Investments in Stock Portfolios and Capital Budgets," *Review of Economics and Statistics*, 47 (February, 1965), 13-37.

Harry M. Markowitz. *Portfolio Selection: Efficient Diversification of Investments*, (New York: John Wiley & Sons, Inc.), 1959.

Merton H. Miller and Myron Scholes. "Rates of Return in Relation to Risk: A Re-examination of Some Recent Findings," in Jensen, ed., (1972).

Donald F. Morrison. *Multivariate Statistical Methods*, (New York: McGraw-Hill, 1967).

A. D. Roy. "Safety First and the Holding of Assets," *Econometrica*, 20 (July, 1952), 431-449.

P. A. Samuelson. "The Fundamental Approximation Theorem of Portfolio Analysis in Terms of Means, Variances and Higher Moments," *Review of Economic Studies*, 37 (October, 1970), 537-542.

William F. Sharpe. "Capital Asset Prices: A Theory of Market Equilibrium Under Conditions of Risk," *Journal of Finance*, 19 (September, 1964), 425-442.

J. Tobin. "Liquidity Preference as Behavior Toward Risk," *Review of Economic Studies*, 26 (February, 1958), 65-86.

William E. Young and Robert H. Trent. "Geometric Mean Approximations of Individual Security and Portfolio Performance," *Journal of Financial and Quantitative Analysis*, 4 (June, 1969), 179-199.

Part II

Classic papers and theories

11

Introduction to the Classic Papers and Theories

In this part of the book, we present papers that formalize, generalize, and extend the early results on the Kelly strategy. Although the early papers on the Kelly optimal growth strategy contained powerful results, the topic was not significant in mainstream financial economics. Unfortunately, despite lots of good evidence, it is still not a serious part of academic financial economics. Roll (1973), for example, shows that in his data set, capital growth and mean-variance portfolios are similar. Thorp (1971) had shown that the Kelly strategies were not necessarily mean-variance efficient and Markowitz (1976) argued that the Kelly strategy was the limiting mean-variance portfolio. These papers appear later in this book. Here we take up the work of those who extended the early results and begin to consider the good and bad properties of these strategies and the sensitivity of the results to data inputs and errors.

Bell and Cover (1980) consider the static one period behavior of two investors. They each can allocate among m stocks and have one dollar to invest. The winner has the most money after one period's returns of the portfolio. They show that the optimal one period policy is an appropriately randomized version $u\Lambda^*$ of the Kelly growth rate optimal portfolio Λ^*. Then for any other portfolio Λ, $Pr\{W(\Lambda) \geqslant W(u\Lambda^*\} \leqslant \frac{1}{2}$. So this establishes a good short run property in addition to the previously known good long term asymptotic Kelly criterion properties. In part IV, we review the good and bad properties of the $E \log$ Kelly capital growth criterion.

Algoet and Cover (1988) generalize the classic Breiman (1961) results to general dependent asset return distributions. Breiman assumed *iid* asset returns. So maximizing conditionally expected log in each period t given information up to t is optimal in that it maximizes the asymptotic growth rate of wealth for arbitrary asset returns. Algoet and Cover (1988) and Thorp (2006), in part VI, present the most general Kelly asymptotic optimality properties.

Cover (1991) presents an algorithm and theory for a universal portfolio that will perform as well as if the investor knew the realized distribution of the future asset returns. There are no assumption on these asset returns. The universal portfolio strategy is based on the past returns $x_i, \ldots, x_{t-1}$ up to the given period t, and will perform asymptotically as well as the best constant rebalanced portfolio based on foreknowledge of the sequence of price relatives. The universal portfolio in period 1 is uniform equally weighted over all the stocks, and the portfolio in period t is the performance weighted average of all constant rebalanced portfolios. Examples in Cover's paper show the exponential outperformance of the universal portfolio with respect to the assets used to construct this portfolio. Cover assumes zero transaction costs.

Ordentlich, E. and T. M. Cover (1998) greatly refine the statements in the 1991 paper and provide exact results for the minimax relative behavior of an investor's wealth with respect to the wealth of the universal investor.

When one sees the naturalness of growth optimality as a goal in investment and then finds the natural mathematics behind it (to maximize the geometric mean), it is tempting to think that the optimal strategy, which turns out to be Kelly gambling or, if you will, log optimal investment, has many other properties as well. Such correspondences certainly occur in mathematics. For example, numbers like π and e occur in hundreds of different contexts and are characterized by the set of all problems that give π and e as answers. π and e do not have just one defining property but many. Looking at it from this point of view, we ask for the properties of growth optimal investment beyond that of growth optimality itself.

The first question is what are the short run properties of growth optimal investment? Is Kelly also good for just one investment period and if so in what sense? Bell and Cover (1980) find the strategy that maximizes the probability of outperforming one's opponent during a single investment period. The optimal one period strategy is to choose a fairly randomized version of the portfolio that maximizes the expected logarithm of wealth. A generalization of this to other payoff functions appears in Bell and Cover (1988). All of these generalizations are achieved by randomized version of the log optimal strategy. Logarithms are nowhere in the statement of the problem. Is this a fortuitous coincidence or is it a property of the naturalness of the approach?

Next, we turn to Kelly and the value of side information. Kelly proves in his classic 1956 paper that the mutual information $I(X;Y)$ between a horse race market[1] X and side information Y gives an increase, $\triangle W = I(X;Y)$, in the growth rate of wealth in repeated investments. This is generalized in Barron and Cover (1988) to general stock market distributions. The mutual information is always an upper bound, $\triangle W \leqslant I$, on the increase in the growth rate of wealth, with equality if and only if the underlying stock market distribution is a horse race market.

A summary of the properties of Kelly gambling and growth optimal investment is given in Cover and Thomas (2006), Chapter 6 (Gambling and Data Compression) and Chapter 16 (Portfolio Theory).

Algoet and Cover (1988) use a simple sandwich argument to prove that the time average of the log wealth converges to a constant with probability one for any ergodic market. This, then, gives the proof of the famous Shannon MacMillan Breiman theorem as a special case in which the underlying market is a horse race, that is, there is one winner from all the entrants. This work also shows the asymptotic optimality of the investment scheme maximizing the conditional expected logarithm of wealth given the available past. These results are simplified and summarized in Cover and Thomas (2006, Chapter 16).

[1]By a horse race market we mean a situation where out of N entrants, only one wins.

The idea of a universal portfolio is to asymptotically achieve the same long-run growth rate as if one had known ahead of time the realized distribution of the stock market outcomes. This parallels universal data compression, where one wants a data compression scheme that works for any source of data, like music or voice or text, and compresses it to the limit that one could achieve if one knew ahead of time whether it was music or voice or text. The answer is Yes, there is such a scheme, and with a negligible cost of learning which washes out over time. Cover (1991) is the first in the series of three papers. Cover and Ordentlich (1996) and Ordentlich and Cover (1998) refine the statements in the 1991 paper. The latter gives exact results for the minimax relative behavior of an investor's wealth with respect to the wealth of the universal investor.

In summary, Kelly investing and its generalization to expected log optimal investing have good short run as well as long run properties. Furthermore, W_n/W_n^* is a martingale, where W_n^* is the wealth generated by Kelly investing, and W_n is the wealth generated by any other investment scheme. Moreover, there is an asymptotic equipartition result that says that $\frac{1}{n}\ln W_n^*$ converges to W, where W is the maximum expected log wealth conditioned on the infinite past. These are some of the natural properties of the Kelly criterion.

Finkelstein and Whitley (1981) generalize the Breiman (1960, 1961) results to *iid* assets that are not necessarily discretely distributed as Breiman assumed, but to arbitrary *iid* random variables with finite expectations. They then show how Breiman's basic results can be easily proved. For non *iid* assets the reader is referred to Algoet and Cover (1988) and Thorp (2006) in this book.

Chopra and Ziemba (1993) show how sensitive mean-variance optimization portfolios are to errors in the estimates of the input parameters. Mean-variance analysis assumes that all the means, variances and covariances are constant and known. In practice, they are estimated. Earlier studies by Kallberg and Ziemba (1981, 1984) showed that the errors in the means are the most important by far, being about ten times errors in variances, which are about twice as important as the co-variance errors. The loss is measured in certainty equivalent terms. Chopra and Ziemba (1993) refine the analysis and investigate the effect of the investor's risk aversion. They show that the relative importance of the mean errors increases as the risk aversion decreases. So instead of a 20 : 2 : 1 ratio of importance its more like 60 : 3 : 1 for low risk aversion and for extremely low risk aversion like log, it is well over 100, so the ratio is about 100 : 3 : 1. So Kelly bettors must accurately estimate their parameter values especially the means so as not to get into an overbet situation. Geyer and Ziemba (2008) show in a five year period asset-liability model that the sensitivity is huge in period 1, low in period 2 and by period 5, essentially non-existent.

MacLean, Ziemba and Li (2005) consider a dynamic investment problem where there are upper and lower limits on wealth. Instead of typical rebalancing at discrete fixed or variable points in time one rebalances when one of the limits is reached.

Assets are assumed to be lognormally distributed and the investor's goal is to minimize the expected time to reach the upper goal while maintaining a high probability of reaching that goal before falling to the lower wealth limit. The optimal strategy is fractional Kelly, being a blend of the Kelly expected log maximizing strategy and cash and is, under the log normal asset assumptions, equivalently represented by a negative power utility function. One has in this case the handy rule $f = \frac{1}{1-\alpha}$ where f is the Kelly fraction and $\alpha < 0$, is the parameter of the negative power utility function αw^{α}. By rebalancing when control limits are reached, the time to wealth goals approach provides greater control over downside risk and upside growth compared, for example, to an expected utility approach with fixed rebalancing times in an asset allocation problem with stocks, bonds and cash.

Evstigneev, Hens and Schenk-Hoppé (2009) survey current research in evolutionary finance focusing on the survival and stability properties of Kelly-type investment strategies. The approach to the study of dynamic wealth evolution follows Darwinian ideas of selection and mutation in N-person game theoretic markets. The result is an alternative to general equilibrium economic models. The setting is more broad than general equilibrium as well because only historical and current data influence the agents' behavior. No agreement about the future or coordinated behavior of the agents is required. The evolutionary model does not use unobservable agents characteristics such as subjective beliefs or utilities. Individual goals of investors are described in terms of properties such as survival with evolutionary stability holding almost surely rather than in terms of expected utility maximization. The Kelly rule insures the survival of those traders following this rule and yields global evolutionary stability.

Lv and Meister (2009) discuss the Kelly criterion in continuous time when the asset returns are multivariate Ornstein-Uhlenbeck mean-reverting processes assuming that there is a complete market. They develop the existence of the optimal self-financing trading strategy and the explicit form of the associated optimal investment fraction of the investor's current wealth.

MATHEMATICS OF OPERATIONS RESEARCH
Vol. 5, No. 2, May 1980
Printed in U.S.A.

12

COMPETITIVE OPTIMALITY OF LOGARITHMIC INVESTMENT*†

ROBERT M. BELL AND THOMAS M. COVER

Stanford University

Consider the two-person zero-sum game in which two investors are each allowed to invest in a market with stocks $(X_1, X_2, \dots, X_m) \sim F$, where $X_i \geqslant 0$. Each investor has one unit of capital. The goal is to achieve more money than one's opponent. Allowable portfolio strategies are random investment policies $\underline{B} \in \mathbb{R}^m, \underline{B} \geqslant \underline{0}, E\sum_{i=1}^m B_i = 1$. The payoff to player 1 for policy $\underline{B}_1$ vs. $\underline{B}_2$ is $P\{\underline{B}_1^t\underline{X} \geqslant \underline{B}_2^t\underline{X}\}$. The optimal policy is shown to be $\underline{B}^* = U\underline{b}^*$, where U is a random variable uniformly distributed on [0, 2], and $\underline{b}^*$ maximizes $E\ln \underline{b}^t\underline{X}$ over $\underline{b} \geqslant \underline{0}$, $\sum b_i = 1$.

Curiously, this competitively optimal investment policy $\underline{b}^*$ is the same policy that achieves the maximum possible growth rate of capital in repeated independent investments (Breiman (1961) and Kelly (1956)). Thus the immediate goal of outperforming another investor is perfectly compatible with maximizing the asymptotic rate of return.

1. Introduction. An investor is faced with a collection of stocks $(X_1, X_2, \dots, X_m)$ drawn according to some known joint distribution function F. We shall assume that stock values X_i are nonnegative. A *portfolio* is a vector $\underline{b} = (b_1, \dots, b_m)^t$, $b_i \geqslant 0, \sum b_i = 1$, with the interpretation that b_i is the proportion of capital allocated to stock i. The capital return S from investment portfolio $\underline{b}$ is

$$S = \underline{b}^t \underline{X} = \sum_{i=1}^{m} b_i X_i. \tag{1}$$

How should $\underline{b}$ be chosen? A currently accepted procedure is the efficient portfolio selection approach of Markowitz (1952, 1959). A portfolio $\underline{b}$ is said to be *efficient* if $(E\underline{b}^t\underline{X}, \operatorname{Var}\underline{b}^t\underline{X})$ is undominated. Criticisms of this approach are many. Only the first two moments are used in the analysis; there is no optimality of this procedure with respect to other obvious investment goals, and no choice procedure among the efficient portfolios is provided. (See Thorp (1971) and Samuelson (1969) for further comments.) Also, such a portfolio is not necessarily admissible [Hakansson (1971, p. 529), Thorp (1971, p. 20)] in the sense that it may be stochastically dominated by some other mixture S.

Another criterion for selecting $\underline{b}$, that of maximizing $E\ln S$, has been put forth by Kelly (1956) and Breiman (1961), and persuasively advocated by Thorp (1969, 1971, 1973). (Also see Latané (1959) and Williams (1936).) This portfolio is admissible, since it maximizes the expectation of a monotonic function of S. The resulting portfolio investment policy $\underline{b}^*$ has been demonstrated by Breiman to have the following properties:

P1. In repeated independent sequential investment, $\underline{b}^*$ maximizes $\liminf(1/n)\ln S_n$. Thus the asymptotic "interest rate" is maximized.

*Received January 19, 1979; revised September 1, 1979.
AMS 1970 subject classification. Primary 90D40. Secondary 90D15.
IAOR 1973 subject classification. Main: Investment. Cross reference: Games.
OR/MS Index 1978 subject classification. Primary 202 Finance, portfolio. Secondary 234 Games, noncooperative.
Key words. Competitive portfolio policy, two-person zero-sum games, logarithmic investment, maximum rate of return.
†This work was partially supported by National Science Foundation Grant ENG 76-03684 and JSEP N00014-C-0601.

0364-765X/80/0502/0161$01.25

P2. The time required to achieve a certain capital A is minimized by $\underline{b}^*$ (in a sense that can be made precise), in the limit as $A \to \infty$.

Yet $\underline{b}^*$ is not accepted in current economic practice. Perhaps one reason is that maximizing $E \ln S$ suggests that the investor has a logarithmic utility for money. However, the criticism of the choice of utility functions ignores the fact that maximizing $E \ln S$ is a consequence of the goals represented by properties P1 and P2, and has nothing to do with utility theory. (See Thorp (1971).) A thorough discussion of investment strategies and their relation to utility theory is developed in Arrow (1971).

We are not interested in utility theory in this paper insofar as utility theory deals with the consistency of subjective preferences. We wish instead to emphasize the objective aspects of portfolio selection, i.e., properties of optimal portfolios that have some objective appeal. In particular we wish to add another goal to the list P1, P2, namely that of outperforming another investor (or even of outperforming oneself with respect to what one could have done). If we can show that all three goals are uniquely achieved by a given policy, we are on our way to making an objective case for utility-independent optimality of the stated portfolio.

An objection to P1 and P2 and, by extension, to $\underline{b}^*$, held by Samuelson (1967, 1969) and others, is that not all investors are interested in long term goals. Samuelson (1969, p. 245) writes, "Our analysis enables us to dispel a fallacy that has been borrowed into portfolio theory from information theory of the Shannon type. Associated with independent discoveries by J.B. Williams (1936), John Kelly (1956), and H.A. Latané (1959) is the notion that if one is investing for many periods, the proper behavior is to maximize the geometric mean of return rather than the arithmetic mean. I believe this to be incorrect (except in the Bernoulli logarithmic case where it happens to be correct for reasons quite distinct from the Williams-Kelly-Latané reasoning) . . . . It is a mistake to think that, just because a w^{**} decision ends up with almost-certain probability to be better than a w^* decision, this implies that w^{**} must yield a better expected value of utility."

Another possible objection is that $\underline{b}^*$ may be too conservative, since it optimizes a concave (risk averse) function of the return S. One interpretation of "too conservative" could be that $\underline{b}^*$ will be outperformed (i.e., $\underline{b}'\underline{X} > \underline{b}^{*\prime}\underline{X}$) with high probability by a more ambitious policy $\underline{b}$. Alternatively, too conservative might mean that with substantial probability $\underline{b}^*$ will be outperformed by a large factor (i.e., $\underline{b}'\underline{X} > c\underline{b}^{*\prime}\underline{X}$, for some constant $c > 1$) by a more risky policy $\underline{b}$. Thus a reasonable goal for an individual investor or a mutual fund would be good short term competitive performance.

With the above objections in mind, we are led to the analysis of one-stage investments. Consider the two-person zero-sum game in which two investors seek portfolio policies that are competitively best in the sense that at least half the time one achieves more capital than one's opponent. Surprisingly, the game theoretic optimal strategy will be shown to be $U\underline{b}^*$ where U is an independent uniform $[0, 2]$ random variable, and $\underline{b}^*$ is the same log optimal policy as before. Furthermore, among non-randomized strategies, $\underline{b}^*$ is shown to be competitively best in the sense that it will not be beaten by very much very often. Thus the alleged conservatism of $\underline{b}^*$ must be established on other grounds; and the short term value of $U\underline{b}^*$ is established competitively.

In the next section, we shall argue for the naturalness of the random variable U in the competitive investment game. Theorem 1, establishing $U\underline{b}^*$ as the solution of the game, will be proved in §3.

2. A game-theoretic digression. Before proceeding, we establish the necessity for randomization in the competitive investment game.

Suppose 2 players each have 1 unit of capital. Their competitive positions are equal.

However, let us now suppose that player 2 has available to him any fair gamble ($EX = 1$, $X \geqslant 0$). By selecting the distribution of the gamble X judiciously, he can beat player 1 with probability $1 - \epsilon$. Simply let $P(X = 1/(1-\epsilon)) = 1 - \epsilon$, $P(X = 0) = \epsilon$. Then $P(X > 1) = 1 - \epsilon$. Therefore, player 1 must protect himself by randomizing his capital. This is a purely game theoretic maneuver and has nothing to do with maximizing investment return.

We now solve the following two-person zero-sum game. Let players 1 and 2 choose d.f.'s F and G, respectively, $\int x\,dF = \int y\,dG = 1$, $F(0^-) = G(0^-) = 0$. Assume $X \sim F$ and $Y \sim G$ are independently drawn. The freedom of choice we allow in the choice of F and G makes physical sense, since any capital distribution $F(x)$, $\int x\,dF(x) = 1$, $F(0^-) = 0$, is achievable from initial capital 1 by a sequential gambling scheme on fair coin tosses (Cover (1974)). The payoff to player 1 is

$$P(X \geqslant Y) = \int G\,dF. \tag{2}$$

LEMMA. *The value of this game is* $\frac{1}{2}$, *and the unique optimal strategies are*

$$F^*(t) = G^*(t) = \begin{cases} t/2, & 0 \leqslant t \leqslant 2, \\ 1, & t \geqslant 2. \end{cases} \tag{3}$$

PROOF. For F^* and for any G,

$$P(Y \geqslant X) = \int F^*\,dG = \int_0^\infty \min\{t/2, 1\}\,dG(t)$$

$$\leqslant \tfrac{1}{2} \int_0^\infty t\,dG(t) = \tfrac{1}{2}. \tag{4}$$

Thus F^* achieves $\frac{1}{2}$ against any G.

Uniqueness of the optimal distribution $F^*(x)$ is proved by assuming $P(Y \geqslant X) \leqslant \frac{1}{2}$ for (i) Y uniform $[0, 2]$, (ii) Y a two point distribution at 0 and a point $c \in [1, 2]$, and (iii) Y a two point distribution at $c \in [0, 1]$ and 2. Then (i)$\Rightarrow F^*(2) = 1$; (ii)$\Rightarrow F^*(c) \leqslant c/2$, $1 \leqslant c \leqslant 2$; and (iii)$\Rightarrow F^*(c) \leqslant c/2$, $0 \leqslant c \leqslant 1$. Since $\int t\,dF^*(t) = 1$, we see $F^*(t) = t/2$, $0 \leqslant t \leqslant 2$. The proof of the uniqueness of G^* follows by symmetry.

We see that a gambler must exchange his unit capital for a r.v. U uniformly distributed on $[0, 2]$ in order to protect himself. We mention parenthetically that one way to achieve this on a sequence of fair coin flips is to divide the one unit of initial capital into piles of size $(\frac{1}{2})^i$, $i = 1, 2, \ldots$, then bet the ith pile on the outcome of the ith coin flip. Letting $\omega_1, \omega_2, \ldots$ be i.i.d. Bernoulli $(1/2)$ r.v.'s, we have the return

$$S = \sum_{i=1}^{\infty} 2\omega_i 2^{-i} = \omega_1.\omega_2\omega_3\omega_4 \ldots$$

in binary, which is clearly uniformly distributed on $[0, 2]$.

3. The competitive investment game. Let $\mathfrak{B}$ be the set of all r.v.'s $\underline{B} = (B_1, B_2, \ldots, B_m)'$, $\underline{B} \geqslant 0$, a.e., $E\sum_{i=1}^m B_i = 1$. Note that the random investment policy $\underline{B}$ can be achieved by first exchanging the 1 unit initial capital for a fair random return W drawn according to the distribution of $\sum_{i=1}^m B_i$. Observe that $W \geqslant 0$, $EW = 1$. Then W is distributed across the stocks according to the conditional joint distribution of $(B_1, B_2, \ldots, B_m)$ given $\sum_{i=1}^m B_i = W$. The latter distribution can be performed on paper. Happily, the allowed conditional randomization is not necessary in the game theoretic optimal policy in Theorem 1 below.

Let the investment vector $\underline{X} = (X_1, X_2, \ldots, X_m)'$ be a r.v. with known distribution function $F(\underline{x})$. We assume that $\underline{X} \geqslant 0$, a.e. To eliminate degeneracy, we also assume that

$$-\infty < \sup_{\underline{b}} E \ln \underline{b}' \underline{X} < \infty.$$

Consider the two-person zero-sum game in which players 1 and 2 choose $\underline{B}^{(1)} \in \mathfrak{B}, \underline{B}^{(2)} \in \mathfrak{B}$, and player 1 receives payoff

$$P\{\underline{B}^{(1)\prime}\underline{X} \geqslant \underline{B}^{(2)\prime}\underline{X}\}. \tag{5}$$

It is assumed that $\underline{B}^{(1)}$, $\underline{B}^{(2)}$ and $\underline{X}$ are jointly independent.

THEOREM 1. *The solution for the competitive investment game is* $\underline{B}^* = U\underline{b}^*$, *where* U *is unif. on* $[0, 2]$, *independent of* $\underline{X}$, *and* $\underline{b}^*$ *maximizes* $E \ln \underline{b}'\underline{X}$. *The value of the game is* $\frac{1}{2}$.

PROOF. The Kuhn Tucker Theorem (1951) implies that the $\underline{b}^*$ maximizing $E \ln \sum_{i=1}^{m} b_i X_i$ subject to the constraint $\sum b_i = 1$, $b_i \geqslant 0$, satisfies

$$E \frac{X_i}{\sum b_j^* X_j} \begin{cases} = \lambda, & b_i^* > 0, \\ \leqslant \lambda, & b_i^* = 0,\ i = 1, 2, \ldots, m, \end{cases} \tag{6}$$

where λ is chosen so that $\sum b_i^* = 1$. But we see $\lambda = 1$, since

$$\begin{aligned} \lambda &= \textstyle\sum b_i^* \lambda = \sum b_i^* E X_i / (\sum b_j^* X_j) \\ &= E(\textstyle\sum b_i^* X_i)/(\sum b_j^* X_j) = 1. \end{aligned} \tag{7}$$

We now investigate the payoff of $\underline{B}^* = U\underline{b}^*$ against any other investment policy $\underline{B} \in \mathfrak{B}$:

$$\begin{aligned} P\{\underline{B}'\underline{X} \geqslant \underline{B}^{*\prime}\underline{X}\} &= P\{\underline{B}'\underline{X} \geqslant U\underline{b}^{*\prime}\underline{X}\} \\ &= P\{U \leqslant (\underline{B}'\underline{X})/(b^{*\prime}\underline{X})\} \leqslant 1/2 E((\underline{B}'\underline{X})/(\underline{b}^{*\prime}\underline{X})) \\ &= 1/2 \sum_{i=1}^{m} E B_i E(X_i/(\textstyle\sum b_i^* X_i)) \leqslant 1/2 \sum E B_i \lambda \\ &= \lambda/2 E \textstyle\sum B_i = \lambda/2 = 1/2. \end{aligned} \tag{8}$$

Thus $\underline{B}^* = U\underline{b}^*$ achieves the value of the game against any $\underline{B}$, and the proof is complete.

The above strategy $\underline{B}^*$ can be implemented by first exchanging the 1 unit initial capital for the fair gamble U, uniformly distributed over $[0, 2]$, then distributing U on the investments according to the solution $\underline{b}^*$ maximizing $E \ln \underline{b}'\underline{X}$.

This result can be generalized to show that $\underline{B}^* = U\underline{b}^*$ will not be beaten by very much very often:
0

COROLLARY 1. $P\{\underline{B}'\underline{X} \geqslant cU\underline{b}^{*\prime}\underline{X}\} \leqslant \frac{1}{2}c$, *for all* $\underline{B} \in \mathfrak{B}$, $c > 0$.

PROOF. $P\{cU \leqslant \underline{B}'\underline{X}/\underline{b}^{*\prime}\underline{X}\} \leqslant (\frac{1}{2}c)E(\underline{B}'\underline{X}/\underline{b}^{*\prime}\underline{X})$, and the proof proceeds as in Theorem 1.

Dropping the randomization U increases this probability by at most a factor of 2:

COROLLARY 2.

$$P(\underline{B}'\underline{X} \geqslant c\,\underline{b}^{*\prime}\underline{X}) \leqslant 1/c, \quad \textit{for all } \underline{B} \in \mathfrak{B},\ c > 0. \tag{9}$$

PROOF. By Markov's inequality and (8),

$$P(\underline{B}'\underline{X} \geqslant c\,\underline{b}^{*\prime}\underline{X}) \leqslant (1/c)E(\underline{B}'\underline{X}/\underline{b}^{*\prime}\underline{X}) \leqslant 1/c.$$

REMARK 1. This is the best that can be attained by any nonrandomized strategy, as can be seen from the discussion at the beginning of §2.

REMARK 2. Corollary 2 bears a strong resemblance to Markov's lemma, i.e., $Y \geqslant 0, EY = \mu \Rightarrow P(Y \geqslant c\mu) \leqslant 1/c, \forall c > 0$. This suggests that $\underline{b}^{*\prime}\underline{X}$ acts like the fixed amount of capital μ in Markov's lemma and that $\underline{b}^{*\prime}\underline{X}$ can be changed from a competitive standpoint only by *fair* randomization. Inequality (9) is true despite the fact that $E\underline{B}'\underline{X}$ may be greater than $E\underline{b}^{*\prime}\underline{X}$.

4. Example: The St. Petersburg paradox. In the St. Petersburg paradox, a gambler pays an entry fee c. He receives in return a random amount of capital X, where $P(X = 2^i) = 2^{-i}$, $i = 1, 2, \ldots, \infty$. Note that $EX = \infty$.

Suppose that a gambler has total initial capital S_0. He is allowed to receive 1 unit of St. Petersburg investment for each c units that he pays as an entry fee. Let him invest the amount bS_0, $0 \leqslant b \leqslant 1$, and retain $(1 - b)S_0$ in cash. Thus his return S is given by $S = S_0((1 - b) + (b/c)X)$. In the framework of the previous sections, the investment vector is $\underline{X} = (X_1, X_2)' = (1, X/c)'$.

Let $b^* \in [0, 1]$ maximize $E \ln S$. We calculate

$$\frac{dE \ln S}{db} = E \frac{-1 + X/c}{(1 - b) + (b/c)X}$$

$$= \sum_{i=1}^{\infty} \left(\frac{1}{2}\right)^i \frac{(2^i/c - 1)}{2^i b/c + (1 - b)}. \tag{10}$$

Letting $b = 1$, we see that $dE \ln S/db = 1 - (c/3)$, which is $\geqslant 0$ for $c \leqslant 3$. Thus $b^* = 1$, for $0 \leqslant c \leqslant 3$. For $c > 3$, the solution b^* to (10) tends monotonically to zero as the entry fee $c \to \infty$. Finally it can be seen that b^* and $\max E \ln S$ are always strictly positive.

Investing a proportion of capital b^* guarantees that

(1) The investor is acting in accordance with an investment policy maximizing $\liminf(1/n)\ln S_n$, regardless of whether or not the other investment opportunities are of the St. Petersburg form; and

(2) the investor investing Ub^* is competitively optimal in the St. Petersburg game.

Moreover, we see that all entry fees c are "fair." However, the proportion b^* of total capital invested varies as a function of c. Also, b^* is independent of the total initial capital S_0.

Finally, if the investment fee is low enough, i.e., $0 \leqslant c \leqslant 3$, then $b^* = 1$ and all of the capital is invested. This results in $S_n \sim S_0(4/c)^n$ in the sense that $(1/n)\log_2 S_n \to 2 - \log_2 c$, for $0 \leqslant c \leqslant 3$.

5. Conclusions. It should now be clear that the investment policy $\underline{b}^*$ achieving $\max E \ln \underline{b}'\underline{X}$ has good short run as well as good long run properties. In addition, $\underline{b}^*$ is admissible in the sense that no other policy $\underline{b}$ stochastically dominates $\underline{b}^*$.

We wish to comment on the use of $U\underline{b}^*$ (as opposed to $\underline{b}^*$ alone) in practice. We have seen in §2 that the use of U is a purely game theoretic protection against competition and has nothing to do with increasing a player's capital. Thus we feel that $\underline{b}^*$ alone is sufficient to achieve all reasonable competitive investment goals, and we do not choose to advocate the additional randomization U.

Finally, it is tantalizing that $\underline{b}^*$ arises as the solution to such dissimilar problems as maximizing $\liminf(1/n)\ln S_n$ and maximizing $P\{\underline{B}_1'\underline{X} > \underline{B}_2'\underline{X}\}$. The underlying reason for this coincidence will be investigated.

References

[1] Arrow, K. (1971). *Essays in the Theory of Risk-Bearing*. Markham Publishing Co., Chicago.

[2] Bicksler, J. and Thorp, E. (1973). The Capital Growth Model: An Empirical Investigation. *J. of Financial and Quantitative Analysis*. **VIII** 273–287.

[3] Breiman, L. (1961). Optimal Gambling Systems for Favorable Games. Fourth Berkeley Symposium. 1 65–78.
[4] Cover, T. (1974). Universal Gambling Schemes and the Complexity Measures of Kolmogorov and Chaitin. Stanford Statistics Dept. Tech. Report No. 12.
[5] Feller, W. (1950). *An Introduction to Probability Theory and Applications*. Vol. 1, second edition, 235–237.
[6] Hakansson, N. (1971). Capital Growth and the Mean-Variance Approach to Portfolio Selection. *J. of Financial and Quantitative Analysis*. **VI** 517–557.
[7] ——— and Liu, T. (1970). Optimal Growth Portfolios When Yields Are Serially Correlated. *Rev. Econom. Statist.* 385–394.
[8] Kelly, J. (1956). A New Interpretation of Information Rate. *Bell System Tech. J.* 917–926.
[9] Kuhn, H. and Tucker, A. (1951). Nonlinear Programming. Proc., Second Berkeley Symposium on Math. Stat. and Prob., University of California Press, Berkeley, Calif. 481–492.
[10] Latané, H. (1959). Criteria for Choice Among Risk Ventures. *J. of Political Economy*. **67** 144–155.
[11] Markowitz, H. (1952). Portfolio Selection. *The J. of Finance* **VII** 77–91.
[12] ———. (1959). *Portfolio Selection*. Wiley and Sons, Inc., New York.
[13] Samuelson, P. (1967). General Proof that Diversification Pays. *J. of Financial and Quantitative Analysis* **II** 1–13.
[14] ———. (1969). Lifetime Portfolio Selection by Dynamic Stochastic Programming. *Rev. Econom. Statist.* 239–246.
[15] Thorp, E. (1971). Portfolio Choice and the Kelly Criterion. *Business and Economics Statistics Proceedings, American Statistical Association*. 215–224.
[16] ———. (1969). Optimal Gambling Systems for Favorable Games. *Rev. Internat. Statist.* **37** 273–293.
[17] Williams, J. (1936). Speculation and the Carryover. *Quarterly J. of Economics*. **50** 436–455.

BELL: DEPARTMENT OF STATISTICS, STANFORD UNIVERSITY, STANFORD, CALIFORNIA 94305

COVER: DEPARTMENTS OF STATISTICS AND ELECTRICAL ENGINEERING, STANFORD UNIVERSITY, SEQUOIA HALL, ROOM 130, STANFORD, CALIFORNIA 94305

13

IEEE TRANSACTIONS ON INFORMATION THEORY, VOL. 34, NO. 5, SEPTEMBER 1988 1097

A Bound on the Financial Value of Information

ANDREW R. BARRON, MEMBER, IEEE, AND
THOMAS M. COVER, FELLOW, IEEE

Abstract—**It will be shown that each bit of information at most doubles the resulting wealth in the general stock market setup. This information bound on the growth of wealth is actually attained for certain probability distributions on the market investigated by Kelly. The bound will be shown to be a special case of the result that the increase in exponential growth of wealth achieved with true knowledge of the stock market distribution F over that achieved with incorrect knowledge G is bounded above by $D(F\|G)$, the entropy of F relative to G.**

I. Introduction

Let $X \geq 0$, $X \in \boldsymbol{R}^m$ denote a random stock market vector, with the interpretation that X_i is the ratio of the price of the ith stock at the end of an investment period to the price at the beginning. Let $B = \{b \in \boldsymbol{R}^m:\ b_i \geq 0, \Sigma_{i=1}^m b_i = 1\}$, be the set of all portfolios $\boldsymbol{b}$, where b_i is the proportion of wealth invested in the ith stock. The resulting wealth is

$$S = \sum_{i=1}^{m} b_i X_i = b^t X. \tag{1}$$

This is the wealth resulting from a unit investment allocated to the m stocks according to the portfolio $\boldsymbol{b}$.

II. Doubling Rate

Now let $F(x)$ be the probability distribution function of the stock vector X. We define the doubling rate $W(X)$ for the market by

$$W(X) = \max_{b \in B} \int \log b^t x \, dF(x). \tag{2}$$

The units for W are "doubles per investment." All logarithms in this correspondence are to the base 2. Let $b^* = b^*(F)$ denote a portfolio achieving $W(X)$. Note that $W(X)$ is a real number, a functional of F; the apparent dependence of W on X is for notational convenience.

Necessary and sufficient conditions for $\boldsymbol{b}$ to maximize $E \log b^t X$ are

$$E\frac{X_i}{b^t X} = 1, \qquad \text{for } b_i > 0$$
$$E\frac{X_i}{b^t X} \leq 1, \qquad \text{for } b_i = 0. \tag{3}$$

These are the Kuhn–Tucker conditions characterizing $b^*(F)$ (see Bell and Cover [3], Cover [4], and Finkelstein and Whitley [5]).

If current wealth is reallocated according to b^* in repeated independent investments against stock vectors $X_1, X_2, \cdots$ independent identically distributed (i.i.d.) according to $F(x)$, then the wealth S_n^* at time n is given by

$$S_n^* = \prod_{i=1}^{n} b^{*t} X_i. \tag{4}$$

The strong law of large numbers for products yields

$$(S_n^*)^{1/n} = 2^{(1/n)\Sigma_{i=1}^n \log b^{*t} X_i} \to 2^W, \tag{5}$$

with probability one. Moreover, no other portfolio achieves a higher exponent (Breiman [1]; Algoet and Cover [9]).

Now suppose side information Y is available. Here Y could be world events, the behavior of a correlated market, or past information on previous outcomes X. Again we define the maximum expected logarithm of the wealth, but this time we allow the portfolio $\boldsymbol{b}$ to depend on Y. Let the doubling rate for side information be

$$W(X|Y) = \max_{b(y)} \iint \log b^t(y) x \, dF(x, y) \tag{6}$$

and let $b^*(y) = b^*(F_{X|y})$ be the portfolio achieving $W(X|Y)$. It can be shown that $b^*(y)$ maximizes the conditional expected logarithm of the wealth $E\{\log b^t X | Y = y\}$.

In repeated investments against $X_1, X_2, \cdots, X_n$ where (X_i, Y_i) are i.i.d. $\sim F(x, y)$, and $b^*(Y_i)$ is the portfolio used at investment time i given side information Y_i, we have resulting wealth

$$S_n^{**} = \prod_{i=1}^{n} b^{*t}(Y_i) X_i \tag{7}$$

with asymptotic behavior

$$(S_n^{**})^{1/n} \to 2^{W(X|Y)} \tag{8}$$

with probability one. It follows that the ratio of wealth with side information to that without side information has limit

$$\left(\frac{S_n^{**}}{S_n^*}\right)^{1/n} \to 2^{W(X|Y) - W(X)} \tag{9}$$

with probability one.

Let the difference between the maximum expected logarithm of wealth with Y and without Y be

$$\Delta = W(X|Y) - W(X). \tag{10}$$

Thus Δ is the increment in doubling rate due to the side information Y. It is this difference that we wish to bound. As an example, if $\Delta = 1$ then the information Y yields an additional doubling of the capital in each investment period. Finally, we observe from (6) and (2) that $\Delta \geq 0$. Information never hurts. Kelly [6] identified Δ with the mutual information for a "horse-race" stock market, a result we will generalize here.

III. Mutual Information and Relative Entropy

The relative entropy (or Kullback Leibler information number) of probability distributions F and G is

$$D(F\|G) = \int \log(f/g) \, dF \tag{11}$$

where f and g are the respective densities with respect to any dominating measure. (*Note:* D is infinite if $g(x)$ is zero on a set of positive probability with respect to F.)

The relative entropy may be interpreted as the error exponent for the hypothesis test F versus G (Stein's lemma; see Chernoff [2]). Another interpretation of the relative entropy for a discrete random vector $X \sim P$ is that $D(P\|Q)$ is the expected increase in description length of the Shannon–Fano code based on the incorrect distribution Q.

Let X, Y be two random variables with joint distribution P_{XY}. The relative entropy between the conditional distribution $P_{X|Y}$

Manuscript received January 8, 1988; revised February 15, 1988. This work was supported in part by the National Science Foundation under a research fellowship and under Contract NCR-85-20136 A1 and in part by the Office of Naval Research under Contract N00014-86-K-06. This work was partially presented at the 6th Annual Symposium on Information Theory, Tashkent, USSR, September 1984.

A. R. Barron is with the Department of Statistics, University of Illinois, 725 South Wright Street, Champaign, IL 61282.
T. M. Cover is with the Departments of Electrical Engineering and Statistics, Stanford University, Durand, Room 121, Stanford, CA 94305.

IEEE Log Number 8824063.

0018-9448/88/0900-1097$01.00 ©1988 IEEE

and the marginal distribution P_X is the mutual information

$$I(X;Y) = \int D(P_{X|Y=y} \| P_X) P_Y(dy) = D(P_{XY} \| P_X P_Y). \quad (12)$$

Of the many alternative expressions for I, the most evocative is the identity

$$I(X;Y) = H(X) - H(X|Y) \quad (13)$$

where $H(X)$ is the entropy of X and $H(X|Y)$ is the conditional entropy. Thus I is the amount the entropy of X is decreased by knowledge of Y. One can compare (13) with (10) to see why a relationship between Δ and I might be expected.

The mutual information I can also be interpreted as the information rate achievable in communication over the communication channel $P(x,y)$. There is also an interpretation of $I(X;Y)$ in terms of efficient descriptions. Since $H(X)$ bits are required to describe the value of the random variable X (if X is discrete), and since $H(X|Y)$ bits are required to describe X given knowledge of Y, the decrement in the expected description length of X is given by $H(X) - H(X|Y) = I(X;Y)$.

In summary, the mutual information $I(X;Y)$ is 1) the decrease in the entropy of X when Y is made available, 2) the number of bits by which the expected description length of X is reduced by knowledge of Y, 3) The rate in bits at which Y can communicate with X by appropriate choice of Y, 4) the error exponent for the hypothesis test (X,Y) independent versus (X,Y) dependent.

IV. Portfolios Based on Incorrect Distributions

Suppose that it is believed that $X \sim G(x)$ when in fact $X \sim F(x)$. Thus the incorrect portfolio $b^*(G)$ is used instead of $b^*(F)$. The doubling rate associated with portfolio b and distribution F can be written

$$W(b,F) = \int \log b^t x \, dF(x) \quad (14)$$

with resulting growth of wealth

$$S_n \doteq 2^{nW(b,F)}. \quad (15)$$

The decrement in exponent from using $b^*(G)$ is

$$\Delta W(F,G) = W(b^*(F),F) - W(b^*(G),F). \quad (16)$$

The following theorem is central to our results.

Theorem 1:

$$0 \le \Delta W(F,G) \le D(F\|G). \quad (17)$$

Proof: The first inequality $0 \le \Delta$ follows by the optimality of $b^*(F)$ for the distribution F. The second inequality $\Delta \le D$ is shown to be a consequence of the optimality of $b^*(G)$ for the distribution G. Let F and G have densities f and g with respect to some dominating measure. The result $\Delta \le D$ is trivially true if $D(F\|G) = \infty$, so it is henceforth assumed that D is finite (whence $F \ll G$).

Let

$$S_1^* = b^{*t}(F)X \qquad S_2 = b^{*t}(G)X \quad (18)$$

be the wealth factors corresponding to the optimal portfolios with respect to F and G. From the Kuhn–Tucker conditions the wealth factor S_2 is strictly positive with probability one with respect to G (and with respect to F since $F \ll G$). It follows (again since $F \ll G$) that the set $A = \{x\colon S_2 > 0, f(x) > 0, g(x) > 0\}$ has probability one with respect to F. Then

$$\begin{aligned}\Delta W(F,G) &= \int_A \left(\log \frac{S_1^*}{S_2}\right) dF \\ &= \int_A \log\left(\frac{S_1^*}{S_2}\frac{g}{f}\frac{f}{g}\right) dF \\ &= \int_A \log\left(\frac{S_1^*}{S_2}\frac{g}{f}\right) dF + D(F\|G) \\ &\le \log \int_A \frac{S_1^*}{S_2}\, dG + D(F\|G) \\ &\le D(F\|G)\end{aligned} \quad (19)$$

where the first inequality follows from the concavity of the logarithm and the second from the Kuhn–Tucker conditions for the optimality of $b^*(G)$ for the distribution G.

We can improve Theorem 1 by normalizing X. Let $\tilde{F}$ denote the distribution of $X/\Sigma X_i$. We note that $E(\log b_1^t X / b_2^t X)$ depends on the distribution $F(x)$ only through the distribution of $X/\Sigma_{i=1}^m X_i \sim \tilde{F}$.

Corollary:

$$\Delta W(F,G) \le D(\tilde{F}\|\tilde{G}).$$

Remark: Another relationship between W and D is shown by Móri [13]. The doubling rate $W = W(b^*(F),F)$ is equal to the minimum of $D(F\|G)$ over all distributions G for which $E_G X_i \le 1$, for $i = 1,2,\cdots,m$.

V. The Information Bound for Side Information

We now ask how Δ and I are related for the stock market. We have

$$\Delta = E \log \frac{b^{*t}(Y)X}{b^{*t}X} \quad (20)$$

and

$$I = E \log \frac{f(X,Y)}{f(X)f(Y)} \quad (21)$$

where $(X,Y) \sim F(x,y)$. The first involves wealth and depends on the values X takes on. The second involves information and depends on X and Y only through the density $f(x,y)$. The following theorem establishes that the increment Δ in the doubling rate resulting from side information Y is less than or equal to the mutual information I.

Theorem 2:

$$0 \le \Delta \le I(X;Y). \quad (22)$$

Proof: For any y, let $P_{X|y}$ be the conditional distribution for X given that $Y = y$ and let P_X be the marginal distribution for X. Also let $b^{**} = b^*(P_{X|y})$. Apply Theorem 1, with $P_{X|y}$ and P_X in place of F and G, respectively, to obtain

$$0 \le E\left[\log \frac{b^{**t}X}{b^{*t}X} \middle| Y = y\right] \le D(P_{X|y}\|P_X). \quad (23)$$

Averaging with respect to the distribution of Y yields

$$0 \le \Delta \le I. \quad (24)$$

Remark: An alternative proof of this theorem, based on money ratio tests and Stein's lemma, appears in [7].

VI. Sequential Portfolio Estimation

Here we show that a good sequence of estimates of the true market distribution leads to asymptotically optimal growth rate of wealth. First we generalize Theorem 1 to handle the sequential setting.

IEEE TRANSACTIONS ON INFORMATION THEORY, VOL. 34, NO. 5, SEPTEMBER 1988 1099

Let $X_1, X_2, \cdots, X_n$ be a sequence of random stock vectors with joint probability distribution P^n. The log-optimal sequential strategy uses the portfolio $b_i^* = b^*(P_{X_i|x_1, x_2, \cdots, x_{i-1}})$, which maximizes the conditional expected value of $\log b^t X_i$ given that $X_1 = x_1, \cdots, X_{i-1} = x_{i-1}$. Suppose that instead of b^*, we use portfolios $\hat{b}_i = b^*(Q_{X_i|x_1, \cdots, x_{i-1}})$ which are optimal for an incorrect distribution Q^n for the sequence $X_1, \cdots, X_n$.

Let $P_{X_i|X_1, \cdots, X_{i-1}}$ and $Q_{X_i|X_1, \cdots, X_{i-1}}$ be the regular conditional distributions associated with P^n and Q^n, respectively. We compare the resulting wealth

$$\hat{S}_n = \prod_{i=1}^{n} \hat{b}_i^T X_i \tag{25}$$

with the wealth

$$S_n^* = \prod_{i=1}^{n} b_i^{*T} X_i. \tag{26}$$

Theorem 3:

$$0 \le E \log \frac{S_n^*}{\hat{S}_n} \le D(P^n \| Q^n). \tag{27}$$

Proof: Application of Theorem 1 shows that

$$0 \le E\left[\left(\log \frac{b_i^{*t} X_i}{\hat{b}_i^t X_i}\right) \middle| X_1, \cdots, X_{i-1}\right]$$
$$\le D(P_{X_i|X^{i-1}} \| Q_{X_i|X^{i-1}}). \tag{28}$$

Averaging with respect to the distribution of $X^{i-1} = (X_1, \cdots, X_{i-1})$ and then summing for $i = 1, 2, \cdots, n$ yields

$$0 \le \sum_{i=1}^{n} E\left(\log \frac{b_i^{*t} X_i}{\hat{b}_i^t X_i}\right)$$
$$\le \sum_{i=1}^{n} E D(P_{X_i|X^{i-1}} \| Q_{X_i|X^{i-1}})$$
$$= D(P^n \| Q^n) \tag{29}$$

by the chain rule, completing the proof.

Suppose $X_1, X_2, \cdots$ are independent with unknown density $p(x)$. Clearly, the optimal portfolio b^* does not depend on the time i or on the past. However, if $p(x)$ is unknown, a series of estimators of the distribution $\hat{P}_i(\cdot)$ corresponding to density estimators $\hat{p}_i(x)$ based on the past X^{i-1} may be used to obtain asymptotically optimal portfolios $\hat{b}_i = b^*(\hat{P}_i)$. It is often the case (see Barron [11], [12]) that there exists a sequence of estimators $\hat{P}_n$ converging to P in the sense that $ED(P\|\hat{P}_n) \to 0$, at least in the Cesaro sense, i.e.,

$$\lim_{n \to \infty} \frac{1}{n} \sum_{i=1}^{n} ED(P \| \hat{P}_i) = 0. \tag{30}$$

In this case Theorem 3 applies with $Q_{X_i|X^{i-1}}$ given by the estimator $\hat{P}_i(\cdot)$ to yield

$$\lim E \frac{1}{n} \log \frac{S_n^*}{\hat{S}_n} = 0. \tag{31}$$

It follows that the actual wealth $\hat{S}_n$ is close to the log-optimal wealth S_n^* as shown in the following theorem.

Theorem 4: Let $X_1, X_2, \cdots$ be i.i.d. $\sim P$. Let $\hat{P}_n$ be a sequence of estimators of the true distribution P such that

$$\frac{1}{n} \sum_{i=1}^{n} ED(P \| \hat{P}_i) \to 0 \tag{32}$$

and let

$$\hat{S}_n = \prod_{i=1}^{n} \hat{b}_i^t X_i \tag{33}$$

where

$$\hat{b}_i = b^*(\hat{P}_i). \tag{34}$$

Let

$$S_n^* = \prod_{i=1}^{n} b^{*t}(P) X_i \tag{35}$$

be the optimal wealth sequence. Then

$$\hat{S}_n = S_n^* 2^{n\,o(1)}, \tag{36}$$

where $o(1) \to 0$ in probability.

Consequently, if S_n^* has an exponential growth rate W^*, then $\hat{S}_n$ has the same asymptotic exponent.

Proof: To see that $\hat{S}_n / S_n^* = 2^{n\,o(1)}$ in probability, first observe that by Markov's inequality

$$P\left\{\frac{\hat{S}_n}{S_n^*} > 2^{n\epsilon}\right\} \le 2^{-n\epsilon} E \frac{\hat{S}_n}{S_n^*} \le 2^{-n\epsilon} \tag{37}$$

where the inequality $E(\hat{S}_n / S_n^*) \le 1$ follows from the Kuhn–Tucker conditions for the optimality of b^* (see Bell and Cover [8]).

On the other hand, using the notation $y^+ = \max\{0, y\}$, $y^- = \max\{0, -y\}$,

$$P\left\{\frac{S_n^*}{\hat{S}_n} > 2^{n\epsilon}\right\} = P\left\{\log \frac{S_n^*}{\hat{S}_n} > n\epsilon\right\}$$
$$\le \frac{1}{n\epsilon} E\left(\log \frac{S_n^*}{\hat{S}_n}\right)^+$$
$$\le \frac{1}{\epsilon}\left[\frac{1}{n} E \log \frac{S_n^*}{\hat{S}_n} + \frac{1}{n}\right] \tag{38}$$

where the first inequality follows from Markov's inequality and the second from

$$E(\log S_n^* / \hat{S}_n)^- = E \log \max\{\hat{S}_n / S_n^*, 1\}$$
$$\le E \log(1 + \hat{S}_n / S_n^*)$$
$$\le \log(1 + E(\hat{S}_n / S_n^*)) \le \log 2 = 1 \tag{39}$$

by the concavity of the logarithm and the Kuhn–Tucker conditions. Combining (31), (37) and (38), we have $\hat{S}_n / S_n^* = 2^{n\,o(1)}$ in probability, as claimed.

VII. Examples

We first give an example due to Kelly [6] in which $\Delta = I$. Here the stock market is a horse race, which, in the setup of (1), consists of a probability mass function $P\{X = O_i e_i\} = p_i$, $i = 1, 2, \cdots, m$, where e_i is a unit vector with a 1 in the ith place and 0's elsewhere, O_i equals the win odds (O_i for 1), and p_i is the probability that the ith horse wins the race.

Then

$$W(X) = \max_b E \log b^t X$$
$$= \max_b \sum_{i=1}^{m} p_i \log b_i O_i$$
$$= \sum p_i \log O_i - H(X) \tag{40}$$

where $H(X) = -\sum_{i=1}^{m} p_i \log p_i$. Also $b^* = p$, i.e., the optimal portfolio is to bet in proportion to the win probabilities, regardless of the odds.

For side information Y, where (X, Y) has a given distribution, a similar calculation yields

$$W(X|Y) = \sum p_i \log O_i - H(X|Y) \tag{41}$$

and

$$b_i^* = P(X = O_i e_i | y), \qquad i = 1, 2, \cdots, m.$$

Here the optimal portfolio is to bet in proportion to the conditional probabilities, given Y. Subtracting (40) from (41), we have

$$\Delta = W(X|Y) - W(X) = H(X) - H(X|Y) = I(X;Y). \tag{42}$$

Consequently, the information bound on Δ is tight.

Of course, it sometimes happens that the information Y about the market is useless for investment purposes. The next example has $\Delta = 0$, $I = 1$. Let $X = (1, 1/2)$ with probability 1/2, and $X = (1, 3/4)$ with probability 1/2. Let $Y = X$. An investment in the first stock always returns the investment, but an investment in the second stock may cut the investment capital to either 1/2 or 3/4 depending on the outcome X. It would be foolish to invest in the second stock, since the first stock dominates its performance. Thus $b^* = b^*(y) = (1, 0)$ for all y, and $\Delta = 0$. On the other hand, since the outcomes of X are equally likely, and $Y = X$, we see

$$\begin{aligned} I(X;Y) &= I(X;X) = H(X) - H(X|X) \\ &= H(X) = 1 \quad \text{bit}. \end{aligned} \tag{43}$$

Thus a bit of information is available, but $\Delta = 0$ and the growth rate is not improved.

VIII. Conclusion

We offer one final interpretation. Recall that $H(X) - H(X|Y) = I(X;Y)$ is the decrement in the expected description length of X due to the side information Y. Hence the inequality $\Delta \le I$ has the interpretation that the increment in the doubling rate of the market X is less than the decrement in the description rate of X.

IX. Acknowledgment

We would like to thank R. O. Duda for speculations that led to the statement of Theorem 2.

References

[1] L. Breiman, "Optimal gambling systems for favorable games," in *Proc. 4th Berkeley Symp.*, vol. 1, pp. 65–78, 1961.
[2] H. Chernoff, "Large-sample theory: Parametric case," *Ann. Math. Stat.*, pp. 1–22, 1956.
[3] R. Bell and T. Cover, "Competitive optimality of logarithmic investment," *Math. Operations Res.*, vol. 5, no. 2, pp. 161–166, 1980.
[4] T. Cover, "An algorithm for maximizing expected log investment return," *IEEE Trans. Inform. Theory*, vol. IT-30, no. 2, pp. 369–373, 1984.
[5] M. Finkelstein and R. Whitley, "Optimal strategies for repeated games," *Adv. Appl. Prob.*, vol. 13, pp. 415–428, 1981.
[6] J. Kelly, "New interpretation of information rate," *Bell Syst. Tech. J.*, vol. 35, pp. 917–926, 1956.
[7] T. Cover, "A bound on the monetary value of information," Stanford Statist. Dept. Tech. Rep. 52, 1984.
[8] R. Bell and T. Cover, "Game theoretic optimal portfolios," *Management Science*, vol. 34, no. 6, pp. 724–733, 1988.
[9] P. Algoet and T. Cover, "Asymptotic optimality and asymptotic equipartition properties of log-optimum investment," *Ann. Prob.*, vol. 16, no. 2, pp. 876–898, 1988.
[10] T. Cover and D. Gluss, "Empirical Bayes stock market portfolios," *Adv. Appl. Math.*, vol. 7, pp. 170–181, 1986.
[11] A. Barron, "Are Bayes rules consistent in information?" in *Open Problems in Communication and Computation*, T. Cover and B. Gopinath, Eds. New York: Springer-Verlag, 1987, pp. 85–91.
[12] A. Barron, "The exponential convergence of posterior probabilities with implications for Bayes estimators of density functions," submitted to *Ann. Statist.*, 1987.
[13] T. Móri, "*I*-divergence geometry of distributions and stochastic gains," in *Proc. 3rd Pannonian Symp. on Math. Stat.*, J. Mogyoródi, I. Vincze, W. Wertz, Eds., Visegrád, Hungary, 1982, pp. 231–238.

14

The Annals of Probability
1988, Vol. 16, No. 2, 876–898

ASYMPTOTIC OPTIMALITY AND ASYMPTOTIC EQUIPARTITION PROPERTIES OF LOG-OPTIMUM INVESTMENT

BY PAUL H. ALGOET[1] AND THOMAS M. COVER[2]

Boston University and Stanford University

We ask how an investor (with knowledge of the past) should distribute his funds over various investment opportunities to maximize the growth rate of his compounded capital. Breiman (1961) answered this question when the stock returns for successive periods are independent, identically distributed random vectors. We prove that maximizing conditionally expected log return given currently available information at each stage is asymptotically optimum, with no restrictions on the distribution of the market process.

If the market is stationary ergodic, then the maximum capital growth rate is shown to be a constant almost surely equal to the maximum expected log return given the infinite past. Indeed, log-optimum investment policies that at time n look at the n-past are sandwiched in asymptotic growth rate between policies that look at only the k-past and those that look at the infinite past, and the sandwich closes as $k \to \infty$.

1. Introduction. Suppose an investor starts with an initial fortune $S_0 = 1$. At the beginning of each period t (where t takes on discrete values $0, 1, \dots$), the current capital S_t is distributed over investment opportunities $j = 1, \dots, m$ according to some portfolio $b_t = (b_t^j)_{1 \le j \le m}$, a vector of nonnegative weights summing to 1. Let $X_t^j \ge 0$ denote the return per monetary unit allocated to stock j during period t, and $X_t = (X_t^j)_{1 \le j \le m}$ the vector of returns. The yield per unit invested according to portfolio b_t is the weighted average of the return ratios of the individual stocks, i.e., the inner product

$$(b_t, X_t) = \sum_{1 \le j \le m} b_t^j X_t^j. \tag{1}$$

Given that S_t units are invested at the beginning of period t, the total amount collected at the end of the period when the random outcome X_t is revealed is $S_{t+1} = S_t(b_t, X_t)$. This capital is redistributed at the beginning of the next round, and the compounded capital after n investment periods is

$$S_n = \prod_{0 \le t < n} (b_t, X_t). \tag{2}$$

Received September 1985; revised August 1987.

[1] Partially supported by Joint Services Electronics Program Grant DAAG 29-84-K-0047 and National Science Foundation Grant ECS-82-11568.

[2] Partially supported by National Science Foundation Grant ECS-82-11568 and DARPA contract N00039-84-C-0211.

AMS 1980 *subject classifications.* Primary 90A09, 94A15, 28D20; secondary 60F15, 60G40, 49A50.

Key words and phrases. Portfolio theory, gambling, expected log return, log-optimum portfolio, capital growth rate, ergodic stock market, asymptotic optimality principle, asymptotic equipartition property (AEP), Shannon–McMillan–Breiman theorem.

Portfolio b_t must be chosen on the basis of $\mathscr{F}_t$, a σ-field that embodies what is known at the beginning of period t. It obviously makes a difference whether decisions may depend on the history of an aggregate quantity like the Dow–Jones average, on detailed records of the past, or perhaps on inside information or help of a clairvoyant oracle. Our default assumption is that $\mathscr{F}_t = \sigma(X_0, \dots, X_{t-1})$ is the information contained in the past outcomes. We wish to distinguish an optimum strategy $\{b_t^*\}_{0 \le t < \infty}$ among all nonanticipating strategies $\{b_t\}_{0 \le t < \infty}$ such that b_t is $\mathscr{F}_t$-measurable for all $t \ge 0$.

We are dealing with a sequential version of the portfolio selection problem that has received much attention in the literature (not to speak of financial practice). Economic theory promotes the maximization of subjective expected utility as a guiding principle toward its solution, and this is certainly appropriate if the investor's preferences are sufficiently well elucidated so that they can be captured in a well-defined utility function. But subjective utilities are difficult to assess and many investors may prefer a less elusive and more objective criterion if there is some rationale for its use. The mean–variance analysis of Markowitz (1952, 1959) trades off expected return with risk as quantified by the standard deviation of the return. This approach is mathematically and computationally tractable, but it lacks generality [cf. Samuelson (1967, 1970)] and it fails to single out an optimum among the portfolios located on the efficient frontier. However, its economic foundation becomes more solid when cast in the form of the capital asset pricing model [cf. Sharpe (1985)]. Breiman (1960, 1961) considered a market with m stocks and independent, identically distributed discrete-valued return vectors $X_t = (X_t^j)_{1 \le j \le m}$, and proved asymptotic optimality of the portfolio b^* that attains the maximum expected log return $w^* = \sup_b E\{\log(b, X)\}$. Thorp (1971) exhibited certain optimality properties of the log return as a normative utility function, and Bell and Cover (1980, 1986) proved that log-optimum investment is also competitively optimum, from a game-theoretic point of view. Although some authors [e.g., Samuelson (1967, 1971)] have suggested that the log return should be considered just one among many possible utility functions, we hope to convince the reader of its more fundamental character.

We consider arbitrarily distributed outcomes $\{X_t\}$ and prove that maximizing the conditional expected log return given currently available information at each stage is optimum in the long run. A nonanticipating portfolio $b_t^* = b^*(X_0, \dots, X_{t-1})$ is called log-optimum (for period t) if it attains the maximum conditional expected log return

$$(3) \quad w_t^* = \mathbf{E}\{\log(b_t^*, X_t) | \mathscr{F}_t\} = \sup_{b = b(X_0, \dots, X_{t-1})} \mathbf{E}\{\log(b, X_t) | X_{t-1}, \dots, X_0\}.$$

Such b_t^* also attains the maximum (unconditional) expected log return

$$(4) \quad W_t^* = \mathbf{E}\{w_t^*\} = \mathbf{E}\{\log(b_t^*, X_t)\} = \sup_{b = b(X_0, \dots, X_{t-1})} \mathbf{E}\{\log(b, X_t)\}.$$

A log-optimum portfolio b_t^* always exists, and is unique if the conditional distribution of X_t given $\mathscr{F}_t$ has full support not confined to a hyperplane in $\mathscr{R}^m$. In any case, the return (b_t^*, X_t) is always uniquely defined, even if b_t^* is not.

The results of Breiman (1960, 1961) and Finkelstein and Whitley (1981) for independent, identically distributed $\{X_t\}$ are enhanced by the following theorem, which proves that b_t^* is optimum to first order in growth exponent.

THEOREM (Asymptotic optimality principle). *Let* $S_n^* = \prod_{0 \le t < n}(b_t^*, X_t)$ *and* $S_n = \prod_{0 \le t < n}(b_t, X_t)$, *respectively, denote the capital growth over n periods of investment according to the log-optimum strategy* $\{b_t^*\}_{0 \le t < \infty}$ *and a competing strategy* $\{b_t\}_{0 \le t < \infty}$. *Then* $\{S_n/S_n^*, \mathscr{F}_n\}_{0 \le n < \infty}$ *is a nonnegative supermartingale converging almost surely to a random variable Y with* $\mathbf{E}\{Y\} \le 1$, *and*

$$\limsup_n n^{-1}\log(S_n/S_n^*) \le 0 \quad a.s. \tag{5}$$

Thus $S_n < \exp(n\varepsilon)S_n^*$ *eventually for large n and arbitrary* $\varepsilon > 0$, *which means that no strategy can infinitely often exceed the log-optimum strategy by an amount that grows exponentially fast.*

The asymptotic optimality principle will be deduced from the Kuhn–Tucker conditions for log-optimality using Markov's inequality and the Borel–Cantelli lemma.

Now suppose $\{X_t\}_{-\infty < t < \infty}$ is a two-sided sequence of return vectors, and $\bar{b}_t^* = b^*(X_{-1}, \dots, X_{-t})$ is a log-optimum portfolio for period 0 based on the t-past $\bar{\mathscr{F}}_t = \sigma(X_{-1}, \dots, X_{-t})$. Portfolio $\bar{b}_t^*$ attains the maximum conditional expected log return for period 0 given $\bar{\mathscr{F}}_t$,

$$\bar{w}_t^* = \mathbf{E}\{\log(\bar{b}_t^*, X_0)|\bar{\mathscr{F}}_t\} = \sup_{b = b(X_{-1}, \dots, X_{-t})} \mathbf{E}\{\log(b, X_0)|X_{-1}, \dots, X_{-t}\}. \tag{6}$$

The maximum expected log return for period 0 given $\bar{\mathscr{F}}_t$ is given by

$$\bar{W}_t^* = \mathbf{E}\{\bar{w}_t^*\} = \mathbf{E}\{\log(\bar{b}_t^*, X_0)\} = \sup_{b = b(X_{-1}, \dots, X_{-t})} \mathbf{E}\{\log(b, X_0)\}. \tag{7}$$

The supremum is taken over a larger set of portfolios as t increases, so that $\bar{W}_t^*$ is monotonically increasing and $\{\bar{w}_t^*, \bar{\mathscr{F}}_t\}_{0 \le t < \infty}$ is a submartingale [strictly speaking only if all $\bar{W}_t^*$ are finite].

The information fields $\bar{\mathscr{F}}_t = \sigma(X_{-1}, \dots, X_{-t})$ increase to a limiting σ-field $\bar{\mathscr{F}}_\infty = \sigma(X_{-1}, X_{-2}, \dots)$. Any accumulation point of $\{\bar{b}_t^*\}$ is a log-optimum portfolio for period 0 based on $\bar{\mathscr{F}}_\infty$, and $\bar{b}_t^* = b^*(X_{-1}, \dots, X_{-t})$ almost surely converges to $\bar{b}_\infty^* = b^*(X_{-1}, X_{-2}, \dots)$ if the log-optimum portfolio for period 0 given $\bar{\mathscr{F}}_\infty$ is unique. Furthermore, $\bar{W}_t^*$ increases to the maximum expected log return given the infinite past,

$$\bar{W}_t^* \nearrow \bar{W}_\infty^* = \mathbf{E}\{\log(\bar{b}_\infty^*, X_0)\} = \sup_{b = b(X_{-1}, X_{-2}, \dots)} \mathbf{E}\{\log(b, X_0)\}. \tag{8}$$

We may use the expanded notation $W_t^* = W^*(X_t|X_{t-1}, \dots, X_0)$ and $\bar{W}_t^* = W^*(X_0|X_{-1}, \dots, X_{-t})$. Setting $\mathbf{E}\{\log S_n^*\} = W^*(X_0, \dots, X_{n-1})$ yields the chain rule

$$W^*(X_0, \dots, X_{n-1}) = \sum_{0 \le t < n} W^*(X_t|X_{t-1}, \dots, X_0). \tag{9}$$

If $\{X_t\}$ is stationary, then $\overline{W}_t^* = W^*(X_0|X_{-1},\ldots,X_{-t})$ is equal to $W_t^* = W^*(X_t|X_{t-1},\ldots,X_0)$ and these definitions are equivalent:

$$
\begin{aligned}
\overline{W}_\infty^* = W^*(X_0|X_{-1},X_{-2},\ldots) &= \lim_t \uparrow W^*(X_0|X_{-1},\ldots,X_{-t}) \\
&= \lim_t \uparrow W^*(X_t|X_{t-1},\ldots,X_0) \qquad (10)\\
&= \lim_n \uparrow n^{-1}W^*(X_0,X_1,\ldots,X_{n-1}).
\end{aligned}
$$

These identities for maximum capital growth rate generalize those for relative entropy rate in information theory. Indeed, suppose one stock will return m times the amount invested in it, whereas all other stocks return 0. Thus we must gamble against uniform odds, on the identity of the winning stock (indicated by the direction of X_t). Placing proportional bets, $b_t^j = \text{Prob}\{X_t^j \neq 0|X_{t-1},\ldots,X_0\}$ is log-optimum, and $W^*(X_t|X_{t-1},\ldots,X_0) = \log m - H^*(X_t|X_{t-1},\ldots,X_0)$, where $H^*(X_t|X_{t-1},\ldots,X_0)$ is the conditional entropy of X_t given $X_{t-1},\ldots,X_0$. Now $H^*(X_0|X_{-1},X_{-2},\ldots) = \lim_t \downarrow H^*(X_t|X_{t-1},\ldots,X_0)$ is the entropy rate of $\{X_t\}$, and

$$
W^*(X_0|X_{-1},X_{-2},\ldots) = \log m - H^*(X_0|X_{-1},X_{-2},\ldots). \qquad (11)
$$

The following AEP for log-optimum investment in a stationary ergodic market generalizes the Shannon–McMillan–Breiman theorem of information theory.

Theorem (Asymptotic equipartition property or AEP). *If $\{X_t\}$ is stationary ergodic, then $S_n^* = \prod_{0\le t<n}(b_t^*, X_t)$ grows exponentially fast with constant asymptotic rate almost surely equal to the maximum expected log return given the infinite past, i.e.,*

$$
n^{-1}\log S_n^* \to \overline{W}_\infty^* = W^*(X_0|X_{-1},X_{-2},\ldots) \quad a.s. \qquad (12)
$$

Equivalently, $S_n^* = \exp[n(\overline{W}_\infty^* + o(1))]$, where $o(1)\to 0$ a.s. The rate $\overline{W}_\infty^*$ is highest possible.

The AEP is an immediate consequence of the ergodic theorem if $\{X_t\}$ is finite order Markov. A sandwich argument and the asymptotic optimality principle will reduce the proof of the general case to applications of the ergodic theorem.

In the first half of the paper we discuss log-optimum investment for a single period. The Kuhn–Tucker conditions for log-optimality of a portfolio b^* are recalled in Section 2, and in Section 3 we examine log-optimum portfolio selections and the maximum expected log return as functions of the distribution P of the random outcome $X = (X^j)_{1\le j\le m}$ on $\mathscr{R}_+^m$. To simplify the analysis we use a divide-and-conquer approach. Namely, we consider the decomposition $X = (\beta, X)U$, where $\beta = (\beta^j)_{1\le j\le m}$ is a fixed reference portfolio and $U = X/(\beta, X)$ is the scaled outcome in the simplex $\mathscr{U} = \{u = (u^j)_{1\le j\le m} \in \mathscr{R}_+^m: (\beta, u) = 1\}$. The return (b, X) factors as $(\beta, X)(b, U)$, and the maximum expected log return $w^*(P) = \sup_b E_P\{\log(b, X)\}$ decomposes as the sum of a reference level $r(P) = E_P\{\log(\beta, X)\}$ that is affine in P and an extra term $w^*(Q) = \sup_b E_Q\{\log(b, U)\}$ that depends on P only through the marginal distribution Q of U. The term $w^*(Q)$ is nonnegative, bounded and continuous in Q when the space of probability measures on $\mathscr{U}$ is equipped with the weak topology, whereas the irregular term $r(P) = E_P\{\log(\beta, X)\}$ is irrelevant for portfolio selection.

We need the decomposition $w^*(P) = r(P) + w^*(Q)$ to show that the maximum conditional expected log return w_t^* is always attained by an $\mathscr{F}_t$-measurable portfolio b_t^*. Furthermore, the nonnegativity and lower semicontinuity of $w^*(Q)$ are essential in Section 4 when we argue that the maximum expected log return given the t-past converges to the maximum expected log return given the infinite past (i.e., $\overline{W}_t^* \nearrow \overline{W}_\infty^*$ at $t \to \infty$).

The asymptotic optimality principle is proved in Section 5, for an arbitrarily distributed sequence of return vectors. In Section 6 we argue that S_n^* has a well-defined growth rate if $\{X_t\}$ is stationary ergodic, and in Section 7 we examine whether the same is true if the market is stationary, or stationary in an asymptotic sense. Although the ergodic theorem is generally valid for asymptotically mean stationary processes (whose definition is recalled in Section 7), the AEP will hold for an asymptotically mean stationary market only if the investor can recover from transient losses before reaching the asymptotic regime. Finally, in Section 8 we specialize the investment game to gambling on the next outcome of a random process.

2. The Kuhn–Tucker conditions for log-optimality. When managing funds during a given investment period, an investor may diversify his risk by building a portfolio that includes several assets. The allocation of one unit of capital over elementary investment opportunities $j = 1, \dots, m$ is conveniently described by a vector of weights $b = (b^j)_{1 \le j \le m}$. The weights must be nonnegative (since no borrowing is allowed) and sum to 1. Thus a portfolio is a vector b in the unit simplex

$$\mathscr{B} = \left\{ b = (b^j)_{1 \le j \le m} \in \mathscr{R}_+^m \colon b^1 + \cdots + b^m = 1 \right\}. \tag{13}$$

Let $X^j \ge 0$ denote the return per monetary unit invested in stock j, and let $X = (X^j)_{1 \le j \le m}$ denote the vector of returns. Capital invested according to portfolio b will grow by the factor $(b, X) = \sum_{1 \le j \le m} b^j X^j$, that is, the weighted average of the per-unit returns of the individual stocks. Portfolio b must be selected at the beginning of the investment period, before the actual value of the random outcome X is revealed. However, the distribution of X on $\mathscr{R}_+^m$ is assumed to be known.

Let the expected log return of a portfolio b be denoted by

$$w(b) = E\{\log(b, X)\}. \tag{14}$$

We set $w(b) = -\infty$ if the expectation is not well defined in the usual sense.

DEFINITION. A portfolio b^* is called *log-optimum* if no competing portfolio b can improve the expected log return relative to b^*, i.e., if

$$E\left\{\log\left(\frac{(b, X)}{(b^*, X)}\right)\right\} \le 0, \quad \text{for all } b \in \mathscr{B}. \tag{15}$$

Every log-optimum portfolio b^* attains the maximum expected log return

$$w^* = \sup_{b \in \mathscr{B}} E\{\log(b, X)\}. \tag{16}$$

Conversely, if w^* is finite, then every portfolio b^* attaining $w^* = \sup_b w(b)$ is log-optimum. However, condition (15) may single out a unique log-optimum portfolio b^* even if $w(b)$ is infinite for all $b \in \mathscr{B}$.

We recall the Kuhn–Tucker conditions for log-optimality derived in Bell and Cover (1980). Let the expected score vector be defined for each portfolio b as

$$\alpha(b) = E\{X/(b, X)\}. \tag{17}$$

THEOREM 1. *Let $\alpha^* = \alpha(b^*)$ denote the expected score vector for portfolio b^*. Then b^* is log-optimum iff the Kuhn–Tucker conditions $\alpha^{*j} \le 1$ hold for all $1 \le j \le m$, or equivalently, iff*

$$(b, \alpha^*) = E\left\{\frac{(b, X)}{(b^*, X)}\right\} \le 1, \quad \textit{for all } b \in \mathscr{B}. \tag{18}$$

PROOF. For $b \in \mathscr{B}$ and $0 < \bar{\lambda} = 1 - \lambda < 1$ let $b_\lambda = \bar{\lambda} b^* + \lambda b$. Then

$$\frac{(b_\lambda, X)}{(b^*, X)} = \bar{\lambda} + \lambda \frac{(b, X)}{(b^*, X)} = 1 + \lambda Z, \quad \text{where } Z = \frac{(b, X)}{(b^*, X)} - 1.$$

Using a Taylor series expansion we obtain, for any $a > 0$,

$$\begin{aligned}\lambda Z \ge \log(1 + \lambda Z) &\ge \log(1 + \lambda(Z \wedge a)) \\ &= \lambda(Z \wedge a) - \tfrac{1}{2}\theta\lambda^2(Z \wedge a)^2 \quad (\text{for some } 0 < \theta < 1) \\ &\ge \lambda(Z \wedge a) - \tfrac{1}{2}\lambda^2 a^2.\end{aligned}$$

Choosing $a = a(\lambda)$ so that $a(\lambda) \to \infty$ and $\lambda a(\lambda) \to 0$ as $\lambda \searrow 0$, we see that $\lambda^{-1}E\{\log(1 + \lambda Z)\} \to E\{Z\}$ as $\lambda \searrow 0$. But $E\{Z\} = (b, \alpha^*) - 1$, so the right derivative at $\lambda = 0$ of $w(b_\lambda) = E\{\log(b_\lambda, X)\}$ is given by

$$\left.\frac{d}{d\lambda} w(b_\lambda)\right|_{\lambda = 0+} = \lim_{\lambda \searrow 0} \frac{E\{\log(1 + \lambda Z)\}}{\lambda} = E\{Z\} = (b, \alpha^*) - 1. \tag{19}$$

The Kuhn–Tucker conditions assert that b^* is log-optimum iff the directional derivative of the expected log return is nonpositive when moving from b^* to any competing portfolio b (in particular, when moving from b^* to any extreme point of $\mathscr{B}$). The infinitesimal conditions $dw(b_\lambda)/d\lambda|_{\lambda=0+} \le 0$ are necessary for log-optimality of b^*, and they are also sufficient because $w(b)$ is concave in b. □

The set B^* of log-optimum portfolios is never empty [cf. Cover (1984)]. In fact, let $\mathscr{L}$ denote the linear hull of the support of the distribution of X, that is, the smallest linear subspace of $\mathscr{R}^m$ such that $X \in \mathscr{L}$ with probability 1. Then $w(b)$ is strictly concave when restricted to $\mathscr{L}$ and constant along fibers perpendicular to $\mathscr{L}$. It follows that B^* is a polyhedral set (the intersection of $\mathscr{B}$ with a fiber orthogonal to $\mathscr{L}$), and the log-optimum portfolio b^* is unique if X has full support ($\mathscr{L} = \mathscr{R}^m$). The return (b^*, X) and log return $\log(b^*, X)$ are unambiguously defined, independent of the choice of log-optimum portfolio b^* in B^*.

3. Continuity and attainability of the maximum expected log return. We make explicit how various quantities depend on the distribution P of X on $\mathscr{R}_+^m$. Let $w(b, P) = E_P\{\log(b, X)\}$ denote the expected log return of portfolio b, $w^*(P) = \sup_b w(b, P)$ the maximum expected log return and $B^*(P)$ the set of log-optimum portfolios. It is clear that $w^*(P)$ is convex in P, since $w^*(P)$ is the supremum of functions $E_P\{\log(b, X)\}$ that are affine in P.

The direction of the return vector X embodies everything an investor needs to know in order to maximize the expected log return. To justify this claim, we choose a fixed reference portfolio $\beta = (\beta^j)_{1 \le j \le m}$ with $\beta^j > 0$ for all j, and we define the scaled return vector

$$U = u(X), \quad \text{where } u(x) = x/(\beta, x). \tag{20}$$

Thus U is obtained by projecting the return vector X on the simplex

$$\mathscr{U} = \{u = (u^j)_{1 \le j \le m} \in \mathscr{R}_+^m : (\beta, u) = 1\}. \tag{21}$$

If $X = 0$, then we set $U = u(0) = u_0$ for some arbitrary $u_0 \in \mathscr{U}$.

The distribution Q of $U = u(X)$ on $\mathscr{U}$ is obtained by integrating out the distribution P of X along rays through the origin. All mass accumulated along a ray is collected at the point where the ray crosses the simplex $\mathscr{U}$, except that mass found at $X = 0$ is transferred to $u(0) = u_0$. Thus Q is the image measure of P through $u: \mathscr{R}_+^m \to \mathscr{U}$, and for any Borel subset $A \subseteq \mathscr{U}$ we have

$$Q\{U \in A\} = P\{u(X) \in A\} = P(u^{-1}(A)). \tag{22}$$

Since $X = (\beta, X)u(X)$, the expected log return may be decomposed as the sum $E_P\{\log(b, X)\} = E_P\{\log(\beta, X)\} + E_P\{\log(b, u(X))\}$, or equivalently,

$$w(b, P) = r(P) + w(b, Q). \tag{23}$$

Here $r(P) = w(\beta, P)$ denotes the expected log return of the reference portfolio β. We interpret $r(P)$ as a reference level for the expected log return, since it is an inherent property of the market over which the investor has no control. Whereas $r(P) = E_P\{\log(\beta, X)\}$ is affine in P, it is also a very irregular function of P, possibly infinite or ill defined. Since our choice of b cannot affect its value, we shall subtract $r(P)$ from the expected log return $w(b, P)$. The remaining quantity $w(b, Q) = E_Q\{\log(b, U)\}$ depends on P only through the marginal distribution Q of the scaled outcome U, and represents the relative improvement in expected log return that results when portfolio b is chosen instead of β. The maximum expected log return can be expressed as the sum

$$(24) \qquad w^*(P) = r(P) + w^*(Q), \quad \text{where } w^*(Q) = \sup_{b \in \mathscr{B}} E_Q\{\log(b, U)\}.$$

Maximizing $w(b, P)$ or $w(b, Q)$ are equivalent operations, so that $B^*(P) = B^*(Q)$. Notice that $w^*(Q) = w^*(P) - r(P) \geq 0$, with equality iff the reference portfolio β is log-optimum.

It is an interesting fact that the maximum expected log return $w^*(Q)$ is always attained by some portfolio choice. However, we need a stronger result, namely, the existence of log-optimum portfolios $b^*(Q)$ that depend measurably on Q. To prove the existence of a measurable selection of log-optimum portfolios, we make use of topological properties, including compactness of $\mathscr{B}$ and upper semicontinuity of the expected log return $w(b, Q) = E_Q\{\log(b, U)\}$ in (b, Q).

The space $\mathscr{Q}$ of probability measures on the compact metric space $\mathscr{U}$ is compact and metrizable when equipped with the weak topology [that is the weakest topology on $\mathscr{Q}$ such that $Q \mapsto E_Q\{f(U)\}$ is continuous in $Q \in \mathscr{Q}$ for all bounded continuous functions $f\colon \mathscr{U} \to \mathscr{R}$]. Its Borel σ-field is the smallest σ-field on $\mathscr{Q}$ such that $A \mapsto Q(A)$ is measurable in Q for all Borel subsets $A \subseteq \mathscr{U}$.

THEOREM 2. *The maximum expect log return $w^*(Q) = \sup_{b \in \mathscr{B}} E_Q\{\log(b, U)\}$ is convex, bounded [between 0 and $\max_j(-\log \beta^j)$] and uniformly continuous when the space $\mathscr{Q}$ of probability measures on $\mathscr{U}$ is equipped with the weak topology. The set of log-optimum portfolios $B^*(Q)$ is a nonempty compact convex subset of $\mathscr{B}$ for every distribution Q on $\mathscr{U}$, and a log-optimum portfolio $b^*(Q) \in B^*(Q)$ can be selected for each $Q \in \mathscr{Q}$ so that $b^*(Q)$ is measurable in Q.*

PROOF. Clearly $w^*(Q)$ is convex in Q for the same reason that $w^*(P)$ is convex in P. We argue that $w^*(Q)$ is bounded below and lower semicontinuous on $\mathscr{Q}$, because (β, u) is bounded below on $\mathscr{U}$ and (b, u) is concave in $b \in \mathscr{B}$ and lower semicontinuous in $u \in \mathscr{U}$. We also prove that $w(b, Q)$ is bounded above and upper semicontinuous, using compactness of $\mathscr{B}$ and boundedness above and upper semicontinuity of (b, u) on $\mathscr{B} \times \mathscr{U}$. Boundedness and uniform continuity of $w^*(Q)$ and existence of a measurable selection of log-optimum portfolios $b^*(Q)$ will follow automatically.

First, we argue that $w^*(Q)$ is nonnegative and lower semicontinuous on $\mathscr{Q}$. For $0 \leq \lambda \leq 1$ and $b \in \mathscr{B}$, let $\bar{\lambda} = 1 - \lambda$, $b_\lambda = \bar{\lambda}\beta + \lambda b$, $\mathscr{B}_\lambda = \{b_\lambda\colon b \in \mathscr{B}\}$ and

$$(25) \qquad w_\lambda^*(Q) = \sup_{b \in \mathscr{B}_\lambda} E_Q\{\log(b, U)\} = \sup_{b \in \mathscr{B}} E_Q\{\log(b_\lambda, U)\}.$$

Observe that $w_\lambda^*(Q)$ is monotonically increasing in λ, since the supremum is taken over a larger set $\mathscr{B}_\lambda$ as λ increases. Furthermore, $\mathscr{B}_1 = \mathscr{B}$, $\mathscr{B}_0 = \{\beta\}$ and $(\beta, u) = 1$ for all $u \in \mathscr{U}$, so that

$$w^*(Q) = w_1^*(Q) \geq w_\lambda^*(Q) \geq w_0^*(Q) = E_Q\{\log(\beta, U)\} = 0.$$

If $\lambda < 1$, then $\log(b_\lambda, u)$ is bounded below (by $\log \bar{\lambda}$) and lower semicontinuous in u, so that $w(b_\lambda, Q) = E_Q\{\log(b_\lambda, U)\}$ and $w_\lambda^*(Q) = \sup_{b \in \mathscr{B}} w(b_\lambda, Q)$ are lower semicontinuous in Q. On the other hand, the inequality $(b_\lambda, u) \geq \lambda(b, u)$

implies that

$$w_\lambda^*(Q) \le w^*(Q) \le w_\lambda^*(Q) - \log \lambda,$$

and hence $w_\lambda^*(Q) \nearrow w^*(Q)$ as $\lambda \nearrow 1$. Since $w^*(Q)$ is the supremum of lower semicontinuous functions $w_\lambda^*(Q)$, it follows that $w^*(Q)$ is lower semicontinuous as well.

The expected log return $w(b, Q) = E_Q\{\log(b, U)\}$ is bounded above and upper semicontinuous on $\mathscr{B} \times \mathscr{Q}$, since (b, u) is bounded and upper semicontinuous on $\mathscr{B} \times \mathscr{U}$. Since $\mathscr{B}$ is compact, it follows [cf. Bertsekas and Shreve (1978), Proposition 7.33] that $w^*(Q) = \sup_{b \in \mathscr{B}} w(b, Q)$ is bounded above and upper semicontinuous on $\mathscr{Q}$. Furthermore, log-optimum portfolios $b^*(Q) \in B^*(Q)$ can be selected in a measurable fashion for all $Q \in \mathscr{Q}$ by the measurable selection theorem of Kuratowski and Ryll-Nardzewski (1961). The upper bound $w^*(Q) \le \max_j(-\log \beta^j)$ holds since $(b, u) \le \sum_j u^j \le \max_j(1/\beta^j)$ for all $b \in \mathscr{B}$ if u satisfies $(\beta, u) = 1$. □

It is impossible to select a portfolio $b^*(Q) \in B^*(Q)$ for all distributions Q on $\mathscr{U}$ so that $b^*(Q)$ is continuous in Q. However, if $\overline{Q}_n \to \overline{Q}_\infty$ and $\overline{b}_n^* \in B^*(\overline{Q}_n)$ for all n, then any accumulation point $\overline{b}_\infty^*$ of the sequence $\{\overline{b}_n^*\}$ is a point in $B^*(\overline{Q}_\infty)$. Furthermore, $(\overline{b}_n^*, U) \to (\overline{b}_\infty^*, U)$ almost surely under $\overline{Q}_\infty$. These continuity properties of the multivalued correspondence $Q \mapsto B^*(Q)$ follow from the following.

THEOREM 3. *The set $Gr(B^*) = \{(Q, b^*): b^* \in B^*(Q)\}$ is closed in $\mathscr{Q} \times \mathscr{B}$. Consequently, any selection of log-optimum portfolios $Q \mapsto b^*(Q) \in B^*(Q)$ is continuous at any $Q \in \mathscr{Q}$ such that $B^*(Q) = \{b^*(Q)\}$ is a singleton set.*

PROOF. Since $\mathscr{B}$ is compact, the theorem will follow from the following claim: If $\overline{Q}_n \to \overline{Q}_\infty$ in $\mathscr{Q}$, $\overline{b}_n^* \to \overline{b}_\infty^*$ in $\mathscr{B}$ and $\overline{b}_n^* \in B^*(\overline{Q}_n)$ for all n, then $\overline{b}_\infty^* \in B^*(\overline{Q}_\infty)$.

To prove the claim we consider the sequence of maximum expected log returns $w^*(\overline{Q}_n) = w(\overline{b}_n^*, \overline{Q}_n)$. It is clear that $w^*(\overline{Q}_n) \to w^*(\overline{Q}_\infty)$ since $\overline{Q}_n \to \overline{Q}_\infty$ in $\mathscr{Q}$ and $w^*(Q)$ is continuous in Q. On the other hand (see the proof of Theorem 2), $w(b, Q)$ is upper semicontinuous on $\mathscr{B} \times \mathscr{Q}$ and hence

$$\limsup_n w(\overline{b}_n^*, \overline{Q}_n) \le w\Big(\lim_n \overline{b}_n^*, \lim_n \overline{Q}_n\Big) = w(\overline{b}_\infty^*, \overline{Q}_\infty).$$

The claim $\overline{b}_\infty^* \in B^*(\overline{Q}_\infty)$ and Theorem 3 follow, since

$$w(\overline{b}_\infty^*, \overline{Q}_\infty) \ge \limsup_n w(\overline{b}_n^*, \overline{Q}_n) = \lim_n w^*(\overline{Q}_n) = w^*(\overline{Q}_\infty) = \sup_b w(b, \overline{Q}_\infty). \quad □$$

The maximum expected log return $w^*(P)$ is neither bounded nor continuous (for the weak topology) as P ranges over the space of probability measures on $\mathscr{R}_+^m$. But if the support of P is constrained to a closed subset $\mathscr{X}$ of $\mathscr{R}_+^m$, then $w^*(P)$ is lower semicontinuous and bounded below iff $\mathscr{X}$ is bounded away from 0, upper semicontinuous and bounded above iff $\mathscr{X}$ is bounded, and bounded and uniformly continuous iff $\mathscr{X}$ is bounded away from 0. In particular, if $\mathscr{X} = \mathscr{U}$ (i.e., if X is distributed on the simplex $\mathscr{U}$), then $P = Q$ and $w^*(P) = w^*(Q)$ is bounded and continuous.

Some of the conclusions of Theorems 2 and 3 continue to hold if the investor may distribute his funds over a countable set or even a separable metrizable space $\mathscr{A}$ of investment opportunities. Indeed, suppose every realization of the return X is a nonnegative lower semicontinuous function $x(a)$ on $\mathscr{A}$. (This is no restriction if $\mathscr{A}$ is a countable set with the discrete topology.) The average return $(b, x) = \int_{\mathscr{A}} x(a) b(da)$ is then well defined for every portfolio b [i.e., for every normalized measure $b(da)$ on the Borel σ-field of $\mathscr{A}$]. Further assume the existence of a reference portfolio β such that $(\beta, x) > 0$ is strictly positive for any return function $x(a)$ that is not identically 0. [Such β exists if $\mathscr{A}$ is locally compact, and, in particular, if $\mathscr{A}$ is countable.] If P and Q denote the distribution of X and $U = X/(\beta, X)$, then the maximum expected log return $w^*(P)$ admits the decomposition $r(P) + w^*(Q)$, and $w^*(Q)$ is nonnegative and lower semicontinuous by the argument presented in the proof of Theorem 2. If, moreover, $\mathscr{A}$ is compact and the return functions $x(a)$ are continuous and bounded by a fixed constant, then $w^*(Q)$ is bounded and continuous and a measurable selection of log-optimum portfolios $b^*(Q)$ exists by Theorem 2, and $Gr(B^*)$ is closed by Theorem 3.

4. Martingale properties. It will be shown that the maximum expected log return given increasing information fields tends to the maximum expected log return given the limiting σ-field. We assume that the random return vector $X(\omega) \in \mathscr{R}_+^m$ is defined on a perfect probability space $(\Omega, \mathscr{F}, \mathbf{P})$, so that X admits a regular conditional probability distribution given any sub-σ-field of $\mathscr{F}$. See Jiřina (1954) for a proof of this fact, and Ramachandran (1979) for a complete discussion of perfect measures.

THEOREM 4. *Suppose the random vector X is defined on a perfect probability space $(\Omega, \mathscr{F}, \mathbf{P})$, and $\{\bar{\mathscr{F}}_t\}_{0 \le t < \infty}$ is an increasing sequence of sub-σ-fields of $\mathscr{F}$ with limiting σ-field $\bar{\mathscr{F}}_\infty \subseteq \mathscr{F}$.*

(a) *If $\bar{\mathbf{P}}_t$ is a regular conditional probability distribution of X given $\bar{\mathscr{F}}_t$, then*

$$\bar{\mathbf{P}}_t \to \bar{\mathbf{P}}_\infty \quad \textit{weakly a.s.} \tag{26}$$

(b) *If $b^*(\cdot)$ is a measurable selector of log-optimum portfolios, then $\bar{b}_t^* = b^*(\bar{\mathbf{P}}_t)$ is an $\bar{\mathscr{F}}_t$-measurable portfolio attaining the maximum conditional expected log return given $\bar{\mathscr{F}}_t$. Moreover, $(\bar{b}_t^*, X) \to (\bar{b}_\infty^*, X)$ a.s., and hence*

$$\log(\bar{b}_t^*, X) \to \log(\bar{b}_\infty^*, X) \quad a.s. \tag{27}$$

If the log-optimum portfolio given $\bar{\mathscr{F}}_\infty$ is unique [$B^*(\bar{\mathbf{P}}_\infty) = \{\bar{b}_\infty^*\}$], then $\bar{b}_t^* \to \bar{b}_\infty^*$ a.s. as well.

(c) *If $w^*(\cdot)$ denotes the maximum expected log return function, then the maximum conditional expected log return given $\bar{\mathscr{F}}_t$ is given by*

$$\bar{w}_t^* = w^*(\bar{\mathbf{P}}_t) = \sup_{b \in \bar{\mathscr{F}}_t} \mathbf{E}\{\log(b, X) | \bar{\mathscr{F}}_t\} = \mathbf{E}\{\log(\bar{b}_t^*, X) | \bar{\mathscr{F}}_t\} \quad a.s. \tag{28}$$

Furthermore, $\{\bar{w}_t^*, \bar{\mathscr{F}}_t\}_{0 \le t \le \infty}$ *is a submartingale and*

$$\bar{w}_t^* \to \bar{w}_\infty^* \quad a.s. \left(\text{and in } L^1 \text{ if } \bar{W}_\infty^* < \infty\right). \tag{29}$$

(d) *The maximum expected log return given* $\bar{\mathscr{F}}_t$ *is given by*

$$\bar{W}_t^* = \mathbf{E}\{\bar{w}_t^*\} = \sup_{b \in \bar{\mathscr{F}}_t} \mathbf{E}\{\log(b, X)\} = \mathbf{E}\left\{\log\left(\bar{b}_t^*, X\right)\right\}. \tag{30}$$

Furthermore,

$$\bar{W}_t^* \nearrow \bar{W}_\infty^*, \quad as\ t \to \infty. \tag{31}$$

PROOF. Lévy's martingale convergence theorem for conditional expectations of a bounded continuous (or nonnegative measurable) function $f(x)$ states that

$$\int f\, d\bar{P}_t = \mathbf{E}\{f(X)|\bar{\mathscr{F}}_t\} \to \int f\, d\bar{P}_\infty = \mathbf{E}\{f(X)|\bar{\mathscr{F}}_\infty\} \quad \text{a.s.}$$

This proves (a), and assertion (b) follows in view of Theorem 3. Notice that $\bar{b}_t^* = b^*(\bar{P}_t)$ and $\bar{w}_t^* = w^*(\bar{P}_t)$ are $\bar{\mathscr{F}}_t$-measurable, since $\bar{P}_t$ is measurable on $(\Omega, \bar{\mathscr{F}}_t)$ and both $b^*(\cdot)$ and $w^*(\cdot)$ are measurable functions.

If $0 \le s \le t \le \infty$, then $\bar{\mathscr{F}}_s \subseteq \bar{\mathscr{F}}_t$, so that every $\bar{\mathscr{F}}_s$-measurable portfolio (including $\bar{b}_s^*$) is also $\bar{\mathscr{F}}_t$-measurable. It follows that

$$\mathbf{E}\left\{\log\left(\bar{b}_s^*, X\right)|\bar{\mathscr{F}}_t\right\} \le \bar{w}_t^* = \sup_{b \in \bar{\mathscr{F}}_t} \mathbf{E}\left\{\log(b, X)|\bar{\mathscr{F}}_t\right\}.$$

Taking $\bar{\mathscr{F}}_s$-conditional expectations proves that $\bar{w}_s^* = \mathbf{E}\{\log(\bar{b}_s^*, X)|\bar{\mathscr{F}}_s\} \le \mathbf{E}\{\bar{w}_t^*|\bar{\mathscr{F}}_s\}$, and hence $\{\bar{w}_t^*, \bar{\mathscr{F}}_t\}_{0 \le t \le \infty}$ is a submartingale. The maximum expected log returns $\bar{W}_t^* = \sup_{b \in \bar{\mathscr{F}}_t} \mathbf{E}\{\log(b, X)\}$ increase with t since the supremum is taken over larger and larger sets ($b \in \bar{\mathscr{F}}_s \Rightarrow b \in \bar{\mathscr{F}}_t$). More information does not hurt!

It remains to show that $\bar{W}_t^* \nearrow \bar{W}_\infty^*$ and $\bar{w}_t^* \to \bar{w}_\infty^*$ a.s. (and in L^1 if $\bar{W}_\infty^*$ is finite). For this purpose we choose a reference portfolio β (with $\beta^j > 0$ for all $1 \le j \le m$), and we recall the decomposition $w^*(P) = r(P) + w^*(Q)$ of the maximum expected log return into a reference level $r(P) = E_P\{\log(\beta, X)\}$ and a relative improvement $w^*(Q)$ that only depends on the distribution Q of the scaled return vector $U = u(X) = X/(\beta, X)$.

Let $\bar{Q}_t$ designate a regular conditional probability distribution of $U = u(X)$ given $\bar{\mathscr{F}}_t$, for $0 \le t \le \infty$. Then $\bar{Q}_t \to \bar{Q}_\infty$ weakly almost surely and $\{w^*(\bar{Q}_t), \bar{\mathscr{F}}_t\}_{0 \le t \le \infty}$ is a submartingale. Since $w^*(Q)$ is bounded and continuous in Q, it follows that $w^*(\bar{Q}_t) \to w^*(\bar{Q}_\infty)$ a.s. and in L^1, and $\mathbf{E}\{w^*(\bar{Q}_t)\} \nearrow \mathbf{E}\{w^*(\bar{Q}_\infty)\}$. The sequence $\{r(\bar{P}_t), \bar{\mathscr{F}}_t\}_{0 \le t \le \infty}$ [where $r(\bar{P}_t) = \mathbf{E}\{\log(\beta, X)|\bar{\mathscr{F}}_t\}$] is a martingale [at least if $\log(\beta, X)$ has finite expectation], and the martingale convergence theorem for conditional expectations asserts that

$$r\left(\bar{P}_t\right) = \mathbf{E}\{\log(\beta, X)|\bar{\mathscr{F}}_t\} \to r\left(\bar{P}_\infty\right) = \mathbf{E}\{\log(\beta, X)|\bar{\mathscr{F}}_\infty\} \quad \text{a.s.}$$

[and in L^1 if $\mathbf{E}\{\log(\beta, X)\}$ is finite]. Since $\bar{w}_t^* = r(\bar{P}_t) + w^*(\bar{Q}_t)$, we may conclude that $\{\bar{w}_t^*, \bar{\mathscr{F}}_t\}_{0 \le t < \infty}$ is a submartingale such that $\bar{w}_t^* \to \bar{w}_\infty^*$ a.s. (and in

L^1 if $\overline{W}_\infty^*$ is finite). The expectations satisfy

$$\overline{W}_t^* = \mathbf{E}\{\log(\beta, X)\} + \mathbf{E}\{w^*(\overline{Q}_t)\} \nearrow \overline{W}_\infty^* = \mathbf{E}\{\log(\beta, X)\} + \mathbf{E}\{w^*(\overline{Q}_\infty)\}.$$

□

The main conclusion of Theorem 4 is that no gap exists between $\lim_t \uparrow \overline{W}_t^*$ and $\overline{W}_\infty^*$. Thus the limit of the expectations $\overline{W}_t^* = \mathbf{E}\{\log(\bar{b}_t^*, X)\}$ coincides with $\overline{W}_\infty^* = \mathbf{E}\{\log(\bar{b}_\infty^*, X)\}$, which is the expectation of the limit $\log(\bar{b}_\infty^*, X) = \lim_t \log(\bar{b}_t^*, X)$.

We have shown that $\overline{w}_t^* \to \overline{w}_\infty^*$ a.s. and $\overline{W}_t^* \nearrow \overline{W}_\infty^*$, using boundedness and continuity of $w^*(Q)$. These convergence theorems also hold for a market with infinitely many investment opportunities, when $w^*(Q)$ is only nonnegative and lower semicontinuous. Indeed, $\{\overline{w}_t^*, \overline{\mathscr{F}}_t\}_{0 \le t \le \infty}$ is still a submartingale, so that $\overline{w}_t^* \le \mathbf{E}\{\overline{w}_\infty^* | \overline{\mathscr{F}}_t\}$ and $\lim_t \uparrow \mathbf{E}\{\overline{w}_t^*\} \le \mathbf{E}\{\overline{w}_\infty^*\}$ and hence, by Lévy's martingale convergence theorem for conditional expectations,

$$\limsup_t \overline{w}_t^* \le \lim_t \mathbf{E}\{\overline{w}_\infty^* | \overline{\mathscr{F}}_t\} = \mathbf{E}\{\overline{w}_\infty^* | \overline{\mathscr{F}}_\infty\} = \overline{w}_\infty^* \quad \text{a.s.} \tag{32}$$

Since $\{w^*(\overline{Q}_t), \overline{\mathscr{F}}_t\}_{0 \le t \le \infty}$ is a submartingale also, one similarly obtains

$$\begin{aligned} &\limsup_t w^*(\overline{Q}_t) \le w^*(\overline{Q}_\infty) \quad \text{a.s.} \\ \text{and} \quad &\lim_t \uparrow \mathbf{E}\{w^*(\overline{Q}_t)\} \le \mathbf{E}\{w^*(\overline{Q}_\infty)\}. \end{aligned} \tag{33}$$

But $\overline{Q}_t \to \overline{Q}_\infty$ weakly a.s. and $w^*(Q)$ is lower semicontinuous in Q, so that

$$\liminf_t w^*(\overline{Q}_t) \ge w^*(\overline{Q}_\infty) \quad \text{a.s.} \tag{34}$$

We conclude that $w^*(\overline{Q}_t) \to w^*(\overline{Q}_\infty)$ a.s. Since $w^*(Q)$ is also nonnegative Fatou's lemma implies that $\mathbf{E}\{w^*(\overline{Q}_t)\} \nearrow \mathbf{E}\{w^*(\overline{Q}_\infty)\}$. It follows that $\overline{w}_t^* \to \overline{w}_\infty^*$ a.s. and $\overline{W}_t^* \nearrow \overline{W}_\infty^*$, at least if $\mathbf{E}\{\log(\beta, X_0)\} > -\infty$ or $\sup_k \mathbf{E}\{w^*(\overline{Q}_k)\} < \infty$.

5. The asymptotic optimality principle. We now prove the asymptotic optimality principle for sequential log-optimum investment. The market is described by a sequence of return vectors $\{X_t\}_{0 \le t < \infty}$ defined on a perfect probability space $(\Omega, \mathscr{F}, \mathbf{P})$, and capital invested according to a portfolio b_t at the beginning of period t will grow by a factor (b_t, X_t) when the random outcome X_t is revealed at the end of that period. If the initial fortune is normalized to $S_0 = 1$, then the compounded capital S_n after n periods is given by

$$S_n = \prod_{0 \le t < n} (b_t, X_t). \tag{35}$$

The objective is to select portfolios b_t so as to maximize the capital growth rate $\liminf_n n^{-1} \log S_n$. Portfolio b_t must be selected on the basis of an information field $\mathscr{F}_t$ that embodies what is known at the beginning of period t. In other words, b_t must be $\mathscr{F}_t$-measurable ($b_t \in \mathscr{F}_t$, for short).

Let P_t denote a regular conditional probability distribution of X_t given $\mathscr{F}_t$, and let $b^*(\cdot)$ be a measurable selector of log-optimum portfolios. Then $b_t^* =$

$b^*(P_t)$ is an $\mathscr{F}_t$-measurable portfolio attaining the maximum conditional expected log return

(36) $$w_t^* = w^*(P_t) = \mathbf{E}\{\log(b_t^*, X_t)|\mathscr{F}_t\} = \sup_{b \in \mathscr{F}_t} \mathbf{E}\{\log(b, X_t)|\mathscr{F}_t\}.$$

The expectation of the log return $\log(b_t^*, X_t)$ and of its conditional expectation w_t^* are both equal to the maximum expected log return for period t given $\mathscr{F}_t$,

(37) $$W_t^* = \mathbf{E}\{w_t^*\} = \mathbf{E}\{\log(b_t^*, X_t)\} = \sup_{b \in \mathscr{F}_t} \mathbf{E}\{\log(b, X_t)\}.$$

We argue that $\{b_t^*\}_{0 \le t < \infty}$ is optimum in the long run.

THEOREM 5 (Asymptotic optimality principle). *Suppose the random outcomes $\{X_t\}_{0 \le t < \infty}$ are defined on a perfect probability space $(\Omega, \mathscr{F}, \mathbf{P})$, and $\{\mathscr{F}_t\}_{0 \le t < \infty}$ is an increasing sequence of sub-σ-fields of $\mathscr{F}$ such that $\sigma(X_0, \dots, X_{t-1}) \subseteq \mathscr{F}_t$ for all $0 \le t < \infty$. Let the compounded capital after n periods of investment according to the log-optimum strategy $\{b_t^*\}_{0 \le t < \infty}$ and some competing nonanticipating strategy $\{b_t\}_{0 \le t < \infty}$ be denoted by*

(38) $$S_n^* = \prod_{0 \le t < n} (b_t^*, X_t) \quad \text{and} \quad S_n = \prod_{0 \le t < n} (b_t, X_t).$$

Then $\{S_n/S_n^*, \mathscr{F}_n\}_{0 \le n < \infty}$ is a nonnegative supermartingale converging almost surely to be a random variable Y with $\mathbf{E}\{Y\} \le 1$. Furthermore, $\mathbf{E}\{S_n/S_n^*\} \le 1$ for all n, and

(39) $$\limsup_n n^{-1} \log\left(\frac{S_n}{S_n^*}\right) \le 0 \quad a.s.$$

PROOF. The log-optimum investor and his competitor start with equal fortunes, so that $S_0/S_0^* = 1$. The ratio $S_n/S_n^* = \prod_{0 \le t < n}(b_t, X_t)/(b_t^*, X_t)$ is $\mathscr{F}_n$-measurable, and the conditional log-optimality of b_n^* given $\mathscr{F}_n$ is equivalent to the Kuhn–Tucker condition

$$\mathbf{E}\left\{\frac{(b_n, X_n)}{(b_n^*, X_n)} \middle| \mathscr{F}_n\right\} \le 1.$$

It follows that

$$\mathbf{E}\left\{\frac{S_{n+1}}{S_{n+1}^*} \middle| \mathscr{F}_n\right\} = \mathbf{E}\left\{\frac{S_n}{S_n^*} \frac{(b_n, X_n)}{(b_n^*, X_n)} \middle| \mathscr{F}_n\right\} = \frac{S_n}{S_n^*} \mathbf{E}\left\{\frac{(b_n, X_n)}{(b_n^*, X_n)} \middle| \mathscr{F}_n\right\} \le \frac{S_n}{S_n^*}.$$

So $\{S_n/S_n^*, \mathscr{F}_n\}_{0 \le n < \infty}$ is a nonnegative supermartingale. Any nonnegative supermartingale converges almost surely, and the expectations decrease monotonically to a limit no smaller than the expectation of the limit, by Fatou's lemma. Thus S_n/S_n^* converges almost surely to a nonnegative random variable Y and

$$1 = \mathbf{E}\{S_0/S_0^*\} \ge \mathbf{E}\{S_n/S_n^*\} \ge \lim_n \downarrow \mathbf{E}\{S_n/S_n^*\} \ge \mathbf{E}\{Y\}.$$

Since $\mathbf{E}\{S_n/S_n^*\} \le 1$, it follows from the Markov inequality that, for $r_n > 0$,

$$\mathbf{P}\{S_n/S_n^* \ge r_n\} \le r_n^{-1}\mathbf{E}\{S_n/S_n^*\} \le r_n^{-1}.$$

If r_n increases sufficiently fast so that $\Sigma_n r_n^{-1} < \infty$, then

$$\sum_n \mathbf{P}\{S_n/S_n^* \ge r_n\} \le \sum_n r_n^{-1} < \infty,$$

and hence $S_n/S_n^* < r_n$ eventually for large n, by the Borel–Cantelli lemma. In particular, choosing $r_n = \exp(n\varepsilon)$ with $\varepsilon > 0$ proves that

$$\mathbf{P}\left\{n^{-1}\log(S_n/S_n^*) \ge \varepsilon \text{ infinitely often}\right\} = 0.$$

Since $\varepsilon > 0$ was arbitrary we may conclude that $\limsup_n n^{-1}\log(S_n/S_n^*) \le 0$ a.s. [This fact can be proved also by observing that S_n/S_n^* converges to a random variable Y with $\mathbf{E}\{Y\} \le 1$ and hence $0 \le Y < \infty$ a.s. Indeed, $S_n/S_n^* \le (1 + Y)$ for large n and hence $\limsup_n n^{-1}\log(S_n/S_n^*) \le \lim_n n^{-1}\log(1 + Y) = 0$ a.s.] □

Theorem 5 asserts that any alternative is dominated in the long run by the log-optimum strategy. Indeed, $\mathbf{E}\{S_n/S_n^*\} \le 1$ for all n, and the Borel–Cantelli lemma implies that $S_n/S_n^* < r_n$ eventually for any sequence $\{r_n\}$ such that $\Sigma_n r_n^{-1} < \infty$ (e.g., $r_n = n^{1+\varepsilon}$ or $r_n = e^{n\varepsilon}$). The maximal inequality for nonnegative supermartingales [cf. Neveu (1972), Proposition II-2-7, page 23] asserts that

$$\mathbf{P}\left\{\sup_n S_n/S_n^* \ge \lambda\right\} \le 1/\lambda. \tag{40}$$

Thus with probability at least $1 - 1/\lambda$, a competing investor will never outperform S_n^* by a factor greater than λ. The random variable $\sup_n S_n/S_n^*$ is finite almost surely, although its expectation is generally infinite. A game-theoretic sense in which S_n^* dominates S_n for games with payoff $\mathbf{E}\{\phi(S_n^{(1)}/S_n^{(2)})\}$ with ϕ increasing is given in Bell and Cover (1980, 1986).

The conclusions of Theorem 5 hold if $\{\mathscr{F}_t\}$ is an increasing sequence of information fields with $\sigma(U_0, \dots, U_{t-1}) \subseteq \mathscr{F}_t$ for all t. Indeed, $S_n/S_n^* = \prod_{0 \le t < n}(b_t, U_t)/(b_t^*, U_t)$ is completely determined by the history of the scaled outcomes $U_t = u(X_t)$.

6. The asymptotic equiparitition property. Breiman (1960, 1961) considered a market with outcomes $\{X_t\}$ that are independent and identically distributed according to an atomic measure, and he argued that repeated choice of the log-optimum portfolio b^* is optimum according to various criteria. In particular, the capital $S_n^* = \prod_{0 \le t < n}(b^*, X_t)$ will grow exponentially fast almost surely with limiting rate equal to the maximum expected log return $w^* = \sup_b E\{\log(b, X)\}$, by the strong law of large numbers,

$$n^{-1}\log S_n^* = n^{-1} \sum_{0 \le t < n} \log(b^*, X_t) \to w^* = E\{\log(b^*, X)\} \quad \text{a.s.} \tag{41}$$

We prove an asymptotic equipartition property for log-optimum investment in a market that is stationary ergodic. The successive outcomes $X_t(\omega) = X(T^t\omega)$ are defined in terms of a random vector $X(\omega) \in \mathscr{R}_+^m$ and an invertible measure-preserving and metrically transitive transformation T defined on a perfect probability space $(\Omega, \mathscr{F}, \mathbf{P})$. Since T is invertible, the returns can be embedded in a two-sided sequence $\{X_t\}_{-\infty<t<\infty}$.

Let b_t^* be a log-optimum portfolio for period t based on the t-past $\mathscr{F}_t = \sigma(X_0, \dots, X_{t-1})$, and let $\bar{b}_t^*$ be log-optimum for period 0 based on the shifted information field $\bar{\mathscr{F}}_t = T^t \mathscr{F}_t = \sigma(X_{-1}, \dots, X_{-t})$. Portfolios b_t^* and $\bar{b}_t^*$ attain the maximum conditional expected log returns $w_t^* = \sup_{b \in \mathscr{F}_t} \mathbf{E}\{\log(b, X_t) | \mathscr{F}_t\}$ and $\bar{w}_t^* = \sup_{b \in \bar{\mathscr{F}}_t} \mathbf{E}\{\log(b, X_0) | \bar{\mathscr{F}}_t\}$. We denote by $W_t^* = \mathbf{E}\{\log(b_t^*, X_t)\}$ and $\bar{W}_t^* = \mathbf{E}\{\log(\bar{b}_t^*, X_0)\}$ the maximum expected log returns. Then $\bar{W}_t^* = W^*(X_0 | X_{-1}, \dots, X_{-t})$ equals $W_t^* = W^*(X_t | X_{t-1}, \dots, X_0)$ by stationarity. If $\bar{b}_\infty^*$ is a log-optimum portfolio for period 0 based on the limiting σ-field $\bar{\mathscr{F}}_\infty = \sigma(X_{-1}, X_{-2}, \dots)$, then $W_t^* = \bar{W}_t^*$ increases monotonically to $\bar{W}_\infty^* = \mathbf{E}\{\log(\bar{b}_\infty^*, X_0)\}$. This limiting expectation is equal to the maximum expected log return given the infinite past, and is denoted by $\bar{W}_\infty^* = W^*(X_0 | X_{-1}, X_{-2}, \dots)$. It may be noted that $W^*(X_0 | X_{-1}, \dots, X_{-k})$ is the maximum expected log return given the infinite past under the stationary kth-order Markov process having the same $(k+1)$st-order marginal distribution as $\{X_t\}$.

Let $S_n^* = \prod_{0 \le t < n} (b_t^*, X_t)$ denote the capital growth over n periods of log-optimum investment. The AEP asserts that the time-averaged growth rate $n^{-1} \log S_n^*$ and its expectation $n^{-1} \mathbf{E}\{\log S_n^*\} = n^{-1} W^*(X_0, \dots, X_{n-1})$ converge to the same limit.

THEOREM 6 (Asymptotic equipartition property). *If the sequence of stock return vectors $\{X_t\}$ is stationary ergodic, then capital will grow exponentially fast under the log-optimum investment strategy, almost surely with constant asymptotic rate equal to the maximum expected log return given the infinite past*

$$n^{-1} \log S_n^* \to \bar{W}_\infty^* = W^*(X_0 | X_{-1}, X_{-2}, \dots) \quad a.s., \tag{42}$$

where

$$\begin{aligned} W^*(X_0 | X_{-1}, X_{-2}, \dots) &= \lim_t \uparrow W^*(X_0 | X_{-1}, \dots, X_{-t}) \\ &= \lim_t \uparrow W^*(X_t | X_{t-1}, \dots, X_0) \\ &= \lim_n \uparrow n^{-1} W^*(X_0, \dots, X_{n-1}). \end{aligned} \tag{43}$$

PROOF. One potential approach to establish the AEP for log-optimum investment is to invoke the extended ergodic theorem that was used by Breiman (1957/1960) to prove the AEP of information theory. This extension of the ergodic theorem asserts that

$$n^{-1} \log S_n^* = n^{-1} \sum_{0 \le t < n} \bar{\mathbf{w}}_t^*(T^t \omega) \to \bar{W}_\infty^* = \mathbf{E}\{\bar{\mathbf{w}}_\infty^*\} \quad \text{a.s. and in } L^1, \tag{44}$$

if $\bar{\mathbf{w}}_t^* = \log(\bar{b}_t^*, X_0)$ converges to $\bar{\mathbf{w}}_\infty^* = \log(\bar{b}_\infty^*, X_0)$ and $\{\bar{\mathbf{w}}_t^*\}_{0 \le t < \infty}$ is L^1-dominated. Theorem 4 asserts that $\bar{\mathbf{w}}_t^* \to \bar{\mathbf{w}}_\infty^*$ a.s., but it seems hard to check the integrability condition $\mathbf{E}\{\sup_t |\bar{w}_t^*|\} < \infty$. We shall instead reduce the AEP to direct applications of the ergodic theorem, using a sandwich argument.

The information field $\mathscr{F}_t = \sigma(X_0, \dots, X_{t-1})$ is approximated by a more refined σ-field $\mathscr{F}_t^{(\infty)}$ and by less refined σ-fields $\mathscr{F}_t^{(k)}$, defined for $0 \le k < \infty$ as follows:

$$\mathscr{F}_t^{(k)} = T^{-t}\overline{\mathscr{F}}_{t\wedge k} = \begin{cases} \sigma(X_0, \dots, X_{t-1}), & \text{if } 0 \le t < k, \\ \sigma(X_{t-k}, \dots, X_{t-1}), & \text{if } k \le t < \infty, \end{cases} \tag{45}$$

$$\mathscr{F}_t^{(\infty)} = T^{-t}\overline{\mathscr{F}}_\infty = \sigma(\dots, X_{-1}, X_0, \dots, X_{t-1}). \tag{46}$$

Let $b_t^{(k)}$ and $b_t^{(\infty)}$ denote log-optimum portfolios for period t based on the approximating σ-fields $\mathscr{F}_t^{(k)}$ and $\mathscr{F}_t^{(\infty)}$, and let the corresponding capital growths over n periods be denoted by

$$S_n^{(k)} = \sum_{0\le t<n} (b_t^{(k)}, X_t) \quad \text{and} \quad S_n^{(\infty)} = \prod_{0\le t<n} (b_t^{(\infty)}, X_t). \tag{47}$$

Thus $S_n^{(k)}$, S_n^* and $S_n^{(\infty)}$ denote the capital growth over n periods of log-optimum investment when the investor is allowed to look back at each stage, respectively, at the k-past (but not beyond period 0), up to time 0 and into the infinitely distant past.

Observe that $b_t^{(k)}(\omega) = b_k^*(T^{t-k}\omega)$ if $t \ge k$. Given the expansion

$$n^{-1}\log S_n^{(k)} = n^{-1}\log S_k^* + n^{-1} \sum_{k\le t<n} \log(b_t^{(k)}, X_t), \tag{48}$$

it follows from the ergodic theorem that

$$n^{-1}\log S_n^{(k)} \to W_k^* = \mathbf{E}\{\log(b_k^*, X_k)\} \quad \text{a.s.} \tag{49}$$

The sequence $\{\log(b_t^{(\infty)}, X_t)\}$ is stationary ergodic and $b_0^{(\infty)} = \overline{b}_\infty^*$, so that again by the ergodic theorem,

$$n^{-1}\log S_n^{(\infty)} = n^{-1} \sum_{0\le t<n} \log(b_t^{(\infty)}, X_t) \to \overline{W}_\infty^* = \mathbf{E}\{\log(\overline{b}_\infty^*, X_0)\} \quad \text{a.s.} \tag{50}$$

The log-optimum $\mathscr{F}_t^{(k)}$-measurable portfolio $b_t^{(k)}$ is $\mathscr{F}_t$-measurable since $\mathscr{F}_t^{(k)} \subseteq \mathscr{F}_t$, and the log-optimum $\mathscr{F}_t$-measurable portfolio b_t^* is $\mathscr{F}_t^{(\infty)}$-measurable since $\mathscr{F}_t \subseteq \mathscr{F}_t^{(\infty)}$. It follows from the asymptotic optimality principle that

$$\limsup_n n^{-1}\log\left(\frac{S_n^{(k)}}{S_n^*}\right) \le 0 \quad \text{and} \quad \limsup_n n^{-1}\log\left(\frac{S_n^*}{S_n^{(\infty)}}\right) \le 0 \quad \text{a.s.} \tag{51}$$

Thus we obtain the chain of asymptotic inequalities

$$\begin{aligned} W_k^* = \lim_n n^{-1}\log S_n^{(k)} &\le \liminf_n n^{-1}\log S_n^* \\ &\le \limsup_n n^{-1}\log S_n^* \le \lim_n n^{-1}\log S_n^{(\infty)} = \overline{W}_\infty^* \quad \text{a.s.} \end{aligned} \tag{52}$$

The AEP follows since $W_k^* = \overline{W}_k^* \nearrow \overline{W}_\infty^*$ with no gap as $k \to \infty$. □

The sandwich proof of the AEP remains valid if the log-optimum portfolios b_t^* are based on information fields $\mathscr{F}_t$ other than the history of past outcomes $\sigma(X_0, \dots, X_{t-1})$. However, $\{\mathscr{F}_t\}_{0\le t<\infty}$ must be monotonically increasing and the history of the scaled return vectors $\sigma(U_0, \dots, U_{t-1})$ must be contained in $\mathscr{F}_t$, so that the asymptotic optimality principle can be invoked. Monotonicity of $\{\mathscr{F}_t\}$ means that information available about the past should never be erased from memory. In addition, one must assume that the shifted fields $\bar{\mathscr{F}}_t = T^t\mathscr{F}_t$ are monotonically increasing to a limiting σ-field $\bar{\mathscr{F}}_\infty$, so that $\overline{W}_t^* \nearrow \overline{W}_\infty^*$ by Theorem 4. Monotonicity of $\{\bar{\mathscr{F}}_t\}$ means that later investors have an advantage in information when compared on common grounds, after shifting back to the reference period 0, where all face the same decision problem of selecting b_0.

Suppose in particular that side information $Y_t(\omega) = Y(T^t\omega)$ is revealed together with the return vector X_t at the end of period t. Then $\mathscr{F}_t = \sigma(X_0, Y_0, \dots, X_{t-1}, Y_{t-1})$ and $\bar{\mathscr{F}}_t = \sigma(X_{-t}, Y_{-t}, \dots, X_{-1}, Y_{-1})$ are monotonically increasing, and $n^{-1}\log S_n^* \to \overline{W}_\infty^*$ almost surely where $\overline{W}_\infty^* = W^*(X_0|X_{-1}, Y_{-1}, X_{-2}, Y_{-2}, \dots)$ is the maximum expected log return given the infinite past. The proof is identical to that of Theorem 6, except that $b_t^* = b^*(Q_t)$ and $\bar{b}_t^* = b^*(\overline{Q}_t)$ are now defined by applying a measurable selector of log-optimum portfolios $b^*(\cdot)$ to regular conditional probability distributions Q_t and $\overline{Q}_t$ of $U_t = u(X_t)$ given $\mathscr{F}_t$ and of $U_0 = u(X_0)$ given $\bar{\mathscr{F}}_t$.

The true log return $\log(b_t^*, X_t)$ will generally differ from the conditional expected log return $w_t^* = \mathbf{E}\{\log(b_t^*, X_t)|\mathscr{F}_t\}$. If conditional expected log returns were always exactly realized then the capital growth over n periods would be not S_n^* but rather

$$\tilde{S}_n^* = \prod_{0 \le t < n} \exp\left[\mathbf{E}\{\log(b_t^*, X_t)|\mathscr{F}_t\}\right]. \tag{53}$$

If $\tilde{S}_n = \prod_{0 \le t < n}\exp[\mathbf{E}\{\log(b_t, X_t)|\mathscr{F}_t\}]$ denotes the corresponding quantity under the competing strategy $\{b_t\}$, then $\tilde{S}_n \le \tilde{S}_n^*$ for all n, and hence

$$\limsup_n n^{-1}\log\left(\tilde{S}_n/\tilde{S}_n^*\right) \le 0 \quad \text{a.s.} \tag{54}$$

This may be called an asymptotic optimality principle for the hypothetical growth rate $\tilde{S}_n^*$. If the market is stationary ergodic, then an asymptotic equipartition property for $\tilde{S}_n^*$ can be proved as well, under certain integrability conditions. Let $L \log L$ designate the class of random variables $g(\omega)$ such that $\mathbf{E}\{|g|\log|g|\} < \infty$.

Theorem 7. *If the market is stationary ergodic and* $\mathbf{E}\{\log(\beta, X_0)|\bar{\mathscr{F}}_\infty\}$ *belongs to* $L \log L$, *then*

$$n^{-1}\log \tilde{S}_n^* \to \tilde{W}_\infty^* \quad a.s. \text{ and in } L^1. \tag{55}$$

Proof. Breiman's (1957/1960) extension of the ergodic theorem asserts that $n^{-1}\sum_{0 \le t < n} g_t(T^t\omega) \to \mathbf{E}\{g\}$ *a.s.* and in L^1 if $g_t \to g$ a.s. and $\mathbf{E}\{\sup_t|g_t|\} < \infty$. In particular, if $\{g_t, \mathscr{F}_t\}_{0 \le t < \infty}$ is a martingale or a nonnegative submartingale with limit g in $L \log L$, then the integrability condition $\mathbf{E}\{\sup_t|g_t|\} < \infty$ is satisfied. Indeed, Wiener's dominated ergodic theorem [cf. Chung (1974), example 7, page 355] asserts that

$$\mathbf{E}\left\{\sup_t|g_t|\right\} \le \frac{e}{e-1}\left[1 + \sup_t \mathbf{E}\{|g_t|\log^+|g_t|\}\right] \le \frac{e}{e-1}\left[1 + \mathbf{E}\{|g|\log^+|g|\}\right].$$

Consider the decomposition $\overline{w}_t^* = \bar{r}_t + w^*(\overline{Q}_t)$, where $\bar{r}_t = \mathbf{E}\{\log(\beta, X_0)|\overline{\mathscr{F}}_t\}$. Since $\{\bar{r}_t, \overline{\mathscr{F}}_t\}_{0 \le t < \infty}$ is a martingale with limit $\bar{r}_\infty = \mathbf{E}\{\log(\beta, X_0)|\overline{\mathscr{F}}_\infty\}$ in $L \log L$, Breiman's extended ergodic theorem implies that

$$n^{-1} \sum_{0 \le t < n} \mathbf{E}\{\log(\beta, X_t)|\mathscr{F}_t\} \to \mathbf{E}\{\log(\beta, X_0)\} \quad \text{a.s. and in } L^1.$$

Since $\{w^*(\overline{Q}_t)\}$ is bounded and $w^*(\overline{Q}_t) \to w^*(\overline{Q}_\infty)$ a.s., we also have

$$n^{-1} \sum_{0 \le t < n} \mathbf{E}\{\log(b_t^*, U_t)|\mathscr{F}_t\} \to \mathbf{E}\{w^*(\overline{Q}_\infty)\} \quad \text{a.s. and in } L^1.$$

By summation we may conclude that

$$n^{-1} \log \tilde{S}_n^* = n^{-1} \sum_{0 \le t < n} w_t^* \to \mathbf{E}\{\overline{w}_\infty^*\} = \overline{W}_\infty^* \quad \text{a.s. and in } L^1. \qquad \square$$

7. Stationary markets. We shall prove the AEP for markets that are stationary but not necessarily ergodic. A stationary market is a mixture of stationary ergodic modes [cf. Maitra (1977)], but no finite number of observations may suffice to exactly identify the (random) ergodic mode of $\{X_t\}$. However, log-optimum portfolios based on the t-past are better and better suited to the ergodic mode as t increases, and the log-optimum portfolio given the infinite past will be perfectly tailored because the ergodic mode is uniquely determined by the infinite past. It is therefore not surprising that S_n^* will grow with the same asymptotic rate as if the ergodic mode were known to begin with.

The AEP may hold even if the market is stationary in an asymptotic sense only. A dynamical system $(\Omega, \mathscr{F}, \mathbf{P}, T)$ asymptotically mean stationary (a.m.s.) if the Cesàro averages $n^{-1}\Sigma_{0 \le t < n}\mathbf{P}(T^{-t}F)$ converge for any event $F \in \mathscr{F}$. Setting the limit equal to $\overline{\mathbf{P}}(F)$ then defines a stationary (T-invariant) probability distribution $\overline{\mathbf{P}}$ on $(\Omega, \mathscr{F})$, and $\overline{\mathbf{P}}$ is perfect whenever $\mathbf{P}$ is. $\overline{\mathbf{P}}$ the stationary mean of $\mathbf{P}$, and expectations with respect to $\overline{\mathbf{P}}$ are denoted by $\overline{\mathbf{E}}\{\cdot\}$. The measures $\mathbf{P}$ and $\overline{\mathbf{P}}$ have the same restriction to the invariant σ-field $\mathscr{I} = \{F \in \mathscr{F}: T^{-1}F = F\}$, so that $\mathbf{E}\{\cdot|\mathscr{I}\} = \overline{\mathbf{E}}\{\cdot|\mathscr{I}\}$. See Gray and Kieffer (1980) for further discussion of asymptotically mean stationary measures, and Section 34.2 in Loève (1978) for a proof that the following strong law of large numbers holds for nonnegative measurable $g(\omega)$:

$$(56) \qquad n^{-1} \sum_{0 \le t < n} g(T^t\omega) \to \mathbf{E}\{g|\mathscr{I}\} = \overline{\mathbf{E}}\{g|\mathscr{I}\} \quad \text{a.s. } (\mathbf{P}) \text{ and a.s. } (\overline{\mathbf{P}}).$$

A market asymptotically mean stationary if the underlying dynamical system $(\Omega, \mathscr{F}, \mathbf{P}, T)$ is a.m.s. As before we assume that T is invertible, $\mathbf{P}$ is perfect, and $X_t(\omega) = X(T^t\omega)$ for some random vector $X(\omega) \in \mathscr{R}_+^m$. The AEP holds for an asymptotically mean stationary market, unless the investor goes broke after a few rounds and remains trapped in a state that is infinitely worse than any other. The investor should not be completely ruined by the time he reaches the asymptotic regime, so that he can recover from transient losses.

THEOREM 8. *Suppose the market is stationary, and $\bar{b}_\infty^*$ is a log-optimum portfolio for period* 0 *given the infinite past $\overline{\mathscr{F}}_\infty$. Then*

$$(57) \qquad n^{-1} \log S_n^* \to \mathbf{E}\{\log(\bar{b}_\infty^*, X_0)|\mathscr{I}\} \quad a.s.$$

The same conclusion holds if the market is asymptotically mean stationary and $\bar{b}_\infty^*$ *is log-optimum under the stationary mean, at least if* $n^{-1}\log S_{k_n}^* \to 0$ *a.s. for some sequence* $\{k_n\}$ *such that* $k_n \nearrow \infty$ *and* $k_n/n \to 0$.

PROOF. We consider the asymptotically mean stationary case. Recall that $S_n^* = \prod_{0 \le t < n}(b_t^*, X_t)$, where b_t^* is a log-optimum portfolio for period t based on the t-past $\mathscr{F}_t = \sigma(X_0, \dots, X_{t-1})$. Portfolio b_t^* is log-optimum with respect to the true distribution $\mathbf{P}$. Let $\bar{b}_k^*$ and $\bar{b}_\infty^*$ designate portfolios for period 0 that are log-optimum with respect to the stationary mean $\overline{\mathbf{P}}$, based on the shifted information field $\overline{\mathscr{F}}_k = \sigma(X_{-k}, \dots, X_{-1})$ and the limiting σ-field $\overline{\mathscr{F}}_\infty = \sigma(\dots, X_{-2}, X_{-1})$. If an investor selects log-optimum portfolios b_t^* during the first k periods $0 \le t < k$, and in later periods $t \ge k$ switches to suboptimum portfolios $\bar{b}_k^*(T^t\omega)$ (i.e., portfolios based on the k-past that are log-optimum with respect to the stationary mean $\overline{\mathbf{P}}$), then capital growth over n periods will be given by

$$S_n^{(k)} = \begin{cases} S_n^*, & \text{if } 0 \le n < k, \\ S_k^* \prod_{k \le t < n} \left(\bar{b}_k^*(T^t\omega), X_t\right), & \text{if } k \le n < \infty. \end{cases}$$

If the investor always selects the portfolio $\bar{b}_\infty^*(T^t\omega)$ that is log-optimum based on the infinite past with respect to the stationary mean $\overline{\mathbf{P}}$, then capital growth is given by

$$S_n^{(\infty)} = \prod_{0 \le t < n} \left(\bar{b}_\infty^*(T^t\omega), X_t\right).$$

It is clear that $\mathbf{E}\{S_n^{(k)}/S_n^*\} \le 1$ and $\overline{\mathbf{E}}\{S_n^*/S_n^{(\infty)}\} \le 1$, so that by Markov's inequality and the Borel–Cantelli lemma (cf. the proof of Theorem 5),

$$\limsup_n n^{-1}\log\left(\frac{S_n^{(k)}}{S_n^*}\right) \le 0 \quad \text{a.s. } (\mathbf{P}) \quad \text{and} \quad \limsup_n n^{-1}\log\left(\frac{S_n^*}{S_n^{(\infty)}}\right) \le 0 \quad \text{a.s. } (\overline{\mathbf{P}}).$$

The ergodic theorem for a.m.s. measures implies that

$$n^{-1}\log S_n^{(k)} \to \mathbf{E}\left\{\log\left(\bar{b}_k^*(T^k\omega), X_k\right)|\mathscr{I}\right\} = \overline{\mathbf{E}}\left\{\log\left(\bar{b}_k^*, X_0\right)|\mathscr{I}\right\} \quad \text{a.s. } (\mathbf{P})$$

and

$$n^{-1}\log S_n^{(\infty)} \to \overline{\mathbf{E}}\left\{\log\left(\bar{b}_\infty^*, X_0\right)|\mathscr{I}\right\} \quad \text{a.s. } (\overline{\mathbf{P}}).$$

Combining the previous results yields

$$\mathbf{E}\left\{\log\left(\bar{b}_k^*(T^k\omega), X_k\right)|\mathscr{I}\right\} = \overline{\mathbf{E}}\left\{\log\left(\bar{b}_k^*, X_0\right)|\mathscr{I}\right\} \le \liminf_n n^{-1}\log S_n^* \quad \text{a.s. } (\mathbf{P})$$

and

$$\limsup_n n^{-1}\log S_n^* \le \overline{\mathbf{E}}\left\{\log\left(\bar{b}_\infty^*, X_0\right)|\mathscr{I}\right\} \quad \text{a.s. } (\overline{\mathbf{P}}).$$

The last inequality also holds a.s. $(\mathbf{P})$ since both sides are invariant (the left-hand side by assumption). We obtain the chain of asymptotic inequalities

$$\begin{aligned}\overline{\mathbf{E}}\{\log(\bar{b}_k^*, X_0)|\mathscr{I}\} &\le \liminf_n n^{-1}\log S_n^* \\ &\le \limsup_n n^{-1}\log S_n^* \le \overline{\mathbf{E}}\{\log(\bar{b}_\infty^*, X_0)|\mathscr{I}\} \quad \text{a.s. } (\mathbf{P}).\end{aligned}$$

We claim that $\overline{\mathbf{E}}\{\log(\bar{b}_k^*, X_0)|\mathscr{I}\}$ is increasing in k. Indeed, if $k \le l$, then the event where $\overline{\mathbf{E}}\{\log(\bar{b}_k^*, X_0)|\mathscr{I}\}$ exceeds $\overline{\mathbf{E}}\{\log(\bar{b}_l^*, X_0)|\mathscr{I}\}$ must have zero probability, since conditioning on this invariant event and taking expectations would otherwise contradict the inequality $\overline{W}_k^* \le \overline{W}_l^*$. The expectations $\overline{W}_k^* = \mathbf{E}\{\log(\bar{b}_k^*, X_0)\}$ increase to $\overline{W}_\infty^* = \mathbf{E}\{\log(\bar{b}_\infty^*, X_0)\}$ as $k \to \infty$, so that by the monotone convergence theorem,

$$\overline{\mathbf{E}}\{\log(\bar{b}_k^*, X_0)|\mathscr{I}\} \nearrow \overline{\mathbf{E}}\{\log(\bar{b}_\infty^*, X_0)|\mathscr{I}\} \quad \text{a.s. } (\overline{\mathbf{P}}).$$

Convergence also holds a.s. ($\mathbf{P}$) since we are dealing with invariant random variables, and Theorem 8 follows since $\mathbf{E}\{\log(\bar{b}_\infty^*, X_0)|\mathscr{I}\} = \overline{\mathbf{E}}\{\log(\bar{b}_\infty^*, X_0)|\mathscr{I}\}$. □

8. Gambling as investment. We consider a market in which exactly one stock will yield a nonzero return, the jth stock with probability q^j. The random outcome X is then oriented along one of the coordinate axes of $\mathscr{R}^m$, and the scaled return U is an extreme point of the simplex $\mathscr{U} = \{u \in \mathscr{R}_+^m : (\beta, u) = 1\}$. As observed by Kelly (1956), investing in such a market is like gambling on the outcome of a horse race in which horse j has win probability q^j. Since one unit bet on horse j yields $U^j = 1/\beta^j$ if horse j wins, we have

$$w(b, Q) = E_Q\{\log(b, U)\} = \sum_{1\le j\le m} q^j \log(b^j/\beta^j) = D(q\|\beta) - D(q\|b).$$

The information divergence $D(q\|b) = \Sigma_{1\le j\le m} q^j \log(q^j/b^j)$ is nonnegative, and equal to zero iff $b = q$. It follows that the bet vector $b = q = (q^j)_{1\le j\le m}$ is the unique log-optimum portfolio. Thus the gambler should ignore the odds $1/\beta^j$ and place an amount on each horse j proportional to its win probability q^j. The maximum expected log of the scaled return $w^*(Q)$ is precisely the Kullback–Leibler divergence between the probability vector q and the reference portfolio β that defines the odds, i.e.,

$$w^*(Q) = D(q\|\beta) = \sum_{i\le j\le m} q^j \log(q^j/\beta^j). \tag{58}$$

Gambling on a set of m stocks out of which exactly one will yield a nonzero return is a most risky type of investment game. Least risky is a market whose return vector has a fixed direction, so that the stock(s) with highest return can be predicted with certainty. In general, we say that a distribution Q on $\mathscr{U}$ is less risky than another distribution Q', and write $Q \le Q'$, if there exists a dilation $\Gamma(du|\mu)$ of $\mathscr{U}$ such that $Q' = \Gamma Q$, i.e., $Q'(\cdot) = \int_{\mathscr{U}} \Gamma(\cdot|u)Q(du)$. [A dilation of $\mathscr{U}$ is a transition probability $\Gamma(du|\mu)$ from $\mathscr{U}$ to $\mathscr{U}$ such that μ is equal to the barycenter of $\Gamma(\cdot|\mu)$ for all $\mu \in \mathscr{U}$.] See Alfsen (1971) for more discussion of this so-called dilation or Choquet order on the space $\mathscr{Q}$ of probability measures on $\mathscr{U}$.

If $Q \le Q'$, then Q is less risky and more attractive than Q', in terms of expected log return. Indeed, $Q \le Q'$ iff $\int_{\mathscr{U}} \varphi(u) Q(du) \le \int_{\mathscr{U}} \varphi(u) Q'(du)$ for all lower semicontinuous convex $\varphi\colon \mathscr{U} \to (-\infty, \infty]$. Choosing $\varphi(u) = -\log(b, u)$ proves that $w(b, Q) = E_Q\{\log(b, U)\}$ is increasing in Choquet order, and taking suprema proves

THEOREM 9. *The maximum expected log return* $w^*(Q) = \sup_b E\{\log(b, U)\}$ *is monotonically decreasing in Choquet order on* $\mathscr{Q}$, *i.e.*,

$$\text{if } Q \le Q' \text{ in } \mathscr{Q}, \text{ then } w^*(Q) \ge w^*(Q'). \tag{59}$$

$\mathscr{U}$ is a Choquet simplex, so every distribution Q on $\mathscr{U}$ admits a barycenter $\mu(Q) \in \mathscr{U}$ and for every point $\mu \in \mathscr{U}$ there exists a unique probability measure π_μ on the set of extreme points of $\mathscr{U}$ that admits μ as barycenter. Two measures that are comparable in Choquet order have the same barycenter. The point mass δ_μ that is concentrated at μ is minimal and the measure π_μ on the extreme points of $\mathscr{U}$ is maximal with respect to Choquet order on $\mathscr{Q}$, among all distributions that admit the point $\mu \in \mathscr{U}$ as barycenter. Notice that $\pi_{\mu(Q)} = \Pi Q$, where $\Pi(du|\mu) = \pi_\mu(du)$ is the maximal dilation that sweeps all mass to the extreme points of $\mathscr{U}$; the minimal dilation is the identity kernel $\Delta(du|\mu) = \delta_\mu(du)$ that leaves all mass put.

Among all distributions Q on $\mathscr{U}$ with a given barycenter $\mu(Q) = \mu$, the most concentrated measure δ_μ is best and the most dilated measure π_μ is worst in terms of expected log return. Indeed, let $\mu = \mu(Q)$ and let q denote the probability vector proportional to μ [with components $q^j = \mu^j/(\sum_j \mu^j)$]. Since $\delta_\mu \le Q \le \pi_\mu$, Theorem 9 implies that

$$\max_j \log \mu^j = w^*(\delta_\mu) \ge w^*(Q), \tag{60}$$

$$w^*(Q) \ge w^*(\pi_\mu) = D(q\|\beta) = \sum_j q^j \log q^j. \tag{61}$$

The most natural choice for β is the uniform portfolio $(1/m)_{1 \le j \le m}$, which allocates an equal amount to each of the m stocks and whose yield (β, X) is the arithmetical average return $m^{-1}(X^1 + \cdots + X^m)$. Then $D(q\|\beta) = \log m - \mathscr{H}(q)$, where $\mathscr{H}(q) = -\sum_j q^j \log q^j$ is the Shannon entropy of the probability vector q. In general, one may interpret $h^*(Q) = \log m - w^*(Q)$ as the minimum loss of expected log return relative to the ideal reference level $r(P) + \log m$. When the chain of inequalities $0 \le w^*(\delta_\mu) \le w^*(Q) \le w^*(\pi_\mu) \le \log m$ is rewritten in terms of $h^*(Q)$, one obtains

$$0 \le \min_j(-\log q^j) = h^*(\delta_\mu) \le h^*(Q) = \log m - w^*(Q), \tag{62}$$

$$h^*(Q) \le h^*(\pi_\mu) = \mathscr{H}(q) = \sum_j q^j(-\log q^j). \tag{63}$$

If one starts with a point mass δ_μ located at $\mu \in \mathscr{U}$ and repeatedly dilates mass, then Q traces out a linearly $\le$-ordered chain of distributions all having barycenter $\mu(Q) = \mu$ in $\mathscr{U}$. Ultimately, one ends up with a measure π_μ, when all mass is swept to the corners (extreme points) of the simplex $\mathscr{U}$. Initially (when $Q = \delta_\mu$), one can place all bets on the stock(s) j for which the minimum information loss $(-\log q^j)$ is minimum, but in the end (when $Q = \pi_\mu$) one has to place proportional bets and concede an average loss equal to the Shannon entropy $\mathscr{H}(q) = \sum_{1 \le j \le m} q^j(-\log q^j)$.

Gambling on the next outcome of a horse race is a special type of investment game. Proportional betting is log-optimum, and the asymptotic optimality principle and asymptotic equipartition property can be formulated in a way that does not seem to involve a maximization, since the log-optimum strategy is explicitly known. The same is true for proportional betting on the next outcome of a random process with values in a Polish space. Indeed, let $p(x_0, \dots, x_{n-1})$ denote the marginal density with respect to some dominating measure of the first n outcomes of a random process $\{X_t\}$, and let $q(x_0, \dots, x_{n-1})$ denote the density under some alternative distribution. The likelihood ratio $q(X_0, \dots, X_{n-1})/p(X_0, \dots, X_{n-1})$ is then a nonnegative supermartingale converging almost surely to a random variable Y with $\mathbf{E}\{Y\} \le 1$, and

$$\limsup_n n^{-1} \log\left(\frac{q(X_0, \dots, X_{n-1})}{p(X_0, \dots, X_{n-1})}\right) \le 0 \quad \text{a.s.} \tag{64}$$

If, moreover, $\{X_t\}$ is stationary ergodic and densities are taken with respect to a Markovian reference measure, then $p(X_0, \dots, X_{n-1})$ will grow exponentially fast with constant limiting rate almost surely equal to the relative entropy rate of the true distribution with respect to the reference measure, i.e.,

$$n^{-1} \log p(X_0, \dots, X_{n-1}) \to \sup_n \mathbf{E}\{\log p(X_0, \dots, X_{n-1})\} \quad \text{a.s.} \tag{65}$$

See Barron (1985) and Orey (1985) for a proof of this generalized Shannon–McMillan–Breiman theorem using Breiman's extension of the ergodic theorem and Algoet and Cover (1988) for a sandwich proof.

Acknowledgments. We wish to acknowledge supporting discussions with John Gill and David Larson.

REFERENCES

ALFSEN, E. M. (1971). *Compact Convex Sets and Boundary Integrals*. Springer, New York.

ALGOET, P. H. and COVER, T. M. (1988). A sandwich proof of the Shannon–McMillan–Breiman theorem. *Ann. Probab.* **16** 899–909.

BARRON, A. R. (1985). The strong ergodic theorem for densities: Generalized Shannon–McMillan–Breiman theorem. *Ann. Probab.* **13** 1292–1303.

BELL, R. and COVER, T. M. (1980). Competitive optimality of logarithmic investment. *Math. Oper. Res.* **5** 161–166.

BELL, R. and COVER, T. M. (1986). Game-theoretic optimal portfolios. Preprint.

BERTSEKAS, D. P. and SHREVE, S. E. (1978). *Stochastic Optimal Control, the Discrete Time Case*. Academic, New York.

BREIMAN, L. (1957/1960). The individual ergodic theorem of information theory. *Ann. Math. Statist.* **28** 809–811. Correction **31** 809–810.

BREIMAN, L. (1960). Investment policies for expanding businesses optimal in a long run sense. *Naval Res. Logist. Quart.* **7** 647–651.

BREIMAN, L. (1961). Optimal gambling systems for favorable games. *Proc. Fourth Berkeley Symp. Math. Statist. Probab.* **1** 65–78. Univ. California Press.

CHUNG, K. L. (1974). *A Course in Probability Theory*, 2nd ed. Academic, New York.

COVER, T. M. (1984). An algorithm for maximizing expected log investment return. *IEEE Trans. Inform. Theory* **IT-30** 369–373.

FINKELSTEIN, M. and WHITLEY, R. (1981). Optimal strategies for repeated games. *Adv. in Appl. Probab.* **13** 415–428.

GRAY, R. M. and KIEFFER, J. C. (1980). Asymptotically mean stationary measures. *Ann. Probab.* **8** 962–973.

JIŘINA, M. (1954). Conditional probabilities on σ-algebras with countable basis. *Czechoslavak Math. J.* **4** 372–380. (English translation in *Amer. Math. Soc. Transl. Ser. 2* **2** (1962) 79–86.)

KELLY, J. L., JR. (1956). A new interpretation of information rate. *Bell System Tech. J.* **35** 917–926.

KURATOWSKI, K. and RYLL-NARDZEWSKI, C. (1961). A general theorem on selectors. *Bull. Acad. Polon. Sci.* **13** 397–403.

LOÈVE, M. (1978). *Probability Theory* **2**, 4th ed. Springer, New York.

MAITRA, A. (1977). Integral representations of integral measures. *Trans. Amer. Math. Soc.* **229** 209–225.

MARKOWITZ, H. M. (1952). Portfolio selection. *J. Finance* **7** 77–91.

MARKOWITZ, H. M. (1959). *Portfolio Selection.* Wiley, New York.

NEVEU, J. (1972). *Martingales à Temps Discret.* Masson, Paris.

OREY, S. (1985). On the Shannon-Perez-Moy theorem. *Contemp. Math.* **41** 319–327.

RAMACHANDRAN, D. (1979). *Perfect Measures I—Basic Theory* and *II—Special Topics. ISI Lecture Notes* **5**, **7**. Macmillan, New York.

SAMUELSON, P. (1967). General proof that diversification pays. *J. Financial and Quantitative Anal.* **2** 1–13.

SAMUELSON, P. (1970). The fundamental approximation theorem of portfolio analysis in terms of means, variances and higher moments. *Rev. Econom. Stud.* **37** 537–542.

SAMUELSON, P. (1971). The "fallacy" of maximizing the geometric mean in long sequences of investing or gambling. *Proc. Nat. Acad. Sci. U.S.A.* **68** 2493–2496.

SHARPE, W. F. (1985). *Investments*, 3rd ed. McGraw-Hill, New York.

THORP, E. O. (1971). Portfolio choice and the Kelly criterion. In *Stochastic Optimization Models in Finance* (W. T. Ziemba and R. G. Vickson, eds.) 599–619. Academic, New York.

COLLEGE OF ENGINEERING
BOSTON UNIVERSITY
110 CUMMINGTON STREET
BOSTON, MASSACHUSETTS 02215

DEPARTMENTS OF STATISTICS AND
ELECTRICAL ENGINEERING
STANFORD UNIVERSITY
STANFORD, CALIFORNIA 94305

Mathematical Finance, Vol. 1, No. 1 (January 1991), 1-29

15

UNIVERSAL PORTFOLIOS

THOMAS M. COVER[1]

Departments of Statistics and Electrical Engineering, Stanford University, Stanford, CA

We exhibit an algorithm for portfolio selection that asymptotically outperforms the best stock in the market. Let $\mathbf{x}_i = (x_{i1}, x_{i2}, \ldots, x_{im})^t$ denote the performance of the stock market on day i, where x_{ij} is the factor by which the jth stock increases on day i. Let $\mathbf{b}_i = (b_{i1}, b_{i2}, \ldots, b_{im})^t$, $b_{ij} \geq 0$, $\Sigma_j b_{ij} = 1$, denote the proportion b_{ij} of wealth invested in the jth stock on day i. Then $S_n = \Pi_{i=1}^n \mathbf{b}_i^t \mathbf{x}_i$ is the factor by which wealth is increased in n trading days. Consider as a goal the wealth $S_n^* = \max_{\mathbf{b}} \Pi_{i=1}^n \mathbf{b}^t \mathbf{x}_i$ that can be achieved by the best constant rebalanced portfolio chosen *after* the stock outcomes are revealed. It can be shown that S_n^* exceeds the best stock, the Dow Jones average, and the value line index at time n. In fact, S_n^* usually exceeds these quantities by an exponential factor. Let $\mathbf{x}_1, \mathbf{x}_2, \ldots,$ be an arbitrary sequence of market vectors. It will be shown that the nonanticipating sequence of portfolios $\hat{\mathbf{b}}_k = \int \mathbf{b} \Pi_{i=1}^{k-1} \mathbf{b}^t \mathbf{x}_i \, d\mathbf{b} / \int \Pi_{i=1}^{k-1} \mathbf{b}^t \mathbf{x}_i \, d\mathbf{b}$ yields wealth $\hat{S}_n = \Pi_{k=1}^n \hat{\mathbf{b}}_k^t \mathbf{x}_k$ such that $(1/n) \ln(S_n^*/\hat{S}_n) \to 0$, for every bounded sequence $\mathbf{x}_1, \mathbf{x}_2, \ldots,$ and, under mild conditions, achieves

$$\hat{S}_n \sim \frac{S_n^*(m-1)!(2\pi/n)^{(m-1)/2}}{|J_n|^{1/2}},$$

where J_n is an $(m-1) \times (m-1)$ sensitivity matrix. Thus this portfolio strategy has the same exponential rate of growth as the apparently unachievable S_n^*.

KEYWORDS: portfolio selection, robust trading strategies, performance weighting, rebalancing

1. INTRODUCTION

We consider a sequential portfolio selection procedure for investing in the stock market with the goal of performing as well as if we knew the empirical distribution of future market performance. Throughout the paper we are unwilling to make any statistical assumption about the behavior of the market. In particular, we allow for the possibility of market crashes such as those occurring in 1929 and 1987. We seek a robust procedure with respect to the arbitrary market sequences that occur in the real world.

We first investigate what a natural goal might be for the growth of wealth for arbitrary market sequences. For example, a natural goal might be to outperform the best buy-and-hold strategy, thus beating an investor who is given a look at a newspaper n days in the future.

We propose a more ambitious goal. To motivate this goal let us consider all constant rebalanced portfolio strategies. Let $\mathbf{x} = (x_1, x_2, \ldots, x_m)^t \geq 0$ denote a *stock market vector* for one investment period, where x_i is the *price relative* for

[1] I wish to thank Hal Stern for his invaluable contributions in obtaining and analyzing the stock market data. I also wish to thank the referee for helpful comments. This work was partially supported by NSF Grant NCR 89-14538.

Manuscript received January 1990; final revision received July 1990.

the ith stock—i.e., the ratio of closing to opening price for stock i. A *portfolio* $\mathbf{b} = (b_1, b_2, \ldots, b_m)^t$, $b_i \geq 0$, $\Sigma\, b_i = 1$, is the proportion of the current wealth invested in each of the m stocks. Thus $S = \mathbf{b}^t\mathbf{x} = \Sigma\, b_i x_i$, where $\mathbf{b}$ and $\mathbf{x}$ are considered to be column vectors, is the factor by which wealth increases in one investment period using portfolio $\mathbf{b}$.

Consider an arbitrary (nonrandom) sequence of stock vectors $\mathbf{x}_1, \mathbf{x}_2, \ldots, \mathbf{x}_n \in \mathbf{R}_+^m$. Here x_{ij} is the price relative of stock j on day i. A constant rebalanced portfolio strategy $\mathbf{b}$ achieves wealth

$$S_n(\mathbf{b}) = \prod_{i=1}^{n} \mathbf{b}^t\mathbf{x}_i, \tag{1.1}$$

where the initial wealth $S_0(\mathbf{b}) = 1$ is normalized to 1. Let

$$S_n^* = \max_{\mathbf{b}} S_n(\mathbf{b}) \tag{1.2}$$

denote the maximum wealth achievable on the given stock sequence maximized over all constant rebalanced portfolios. Our goal is to achieve S_n^*.

We will be able to show that there is a "universal" portfolio strategy $\hat{\mathbf{b}}_k$, where $\hat{\mathbf{b}}_k$ is based only on the past $\mathbf{x}_1, \mathbf{x}_2, \ldots, \mathbf{x}_{k-1}$, that will perform asymptotically as well as the best constant rebalanced portfolio based on foreknowledge of the sequence of price relatives. At first it may seem surprising that the portfolio $\hat{\mathbf{b}}_k$ should depend on the past, because the future has no relationship to the past. Indeed the stock sequence is arbitrary, and a malicious nature can structure future $\mathbf{x}_k$'s to take advantage of past beliefs as expressed in the portfolio $\hat{\mathbf{b}}_k$. Nonetheless the resulting wealth can be made to track S_n^*.

The proposed universal adaptive portfolio strategy is the performance weighted strategy specified by

$$\hat{\mathbf{b}}_1 = \left(\frac{1}{m}, \frac{1}{m}, \ldots, \frac{1}{m}\right), \qquad \hat{\mathbf{b}}_{k+1} = \frac{\int \mathbf{b} S_k(\mathbf{b})\, d\mathbf{b}}{\int S_k(\mathbf{b})\, d\mathbf{b}}, \tag{1.3}$$

where

$$S_k(\mathbf{b}) = \prod_{i=1}^{k} \mathbf{b}^t\mathbf{x}_i, \tag{1.4}$$

and the integration is over the set of $(m-1)$-dimensional portfolios

$$B = \left\{\mathbf{b} \in R^m \colon b_i \geq 0, \sum_{i=1}^{m} b_i = 1\right\}. \tag{1.5}$$

The wealth $\hat{S}_n$ resulting from the universal portfolio is given by

$$\hat{S}_n = \prod_{k=1}^{n} \hat{\mathbf{b}}_k^t \mathbf{x}_k. \tag{1.6}$$

Thus the initial universal portfolio $\hat{\mathbf{b}}_1$ is uniform over the stocks, and the portfolio $\hat{\mathbf{b}}_k$ at time k is the performance weighted average of all portfolios $\mathbf{b} \in B$. An approximate computation will be given in Section 8, and a generalization of this algorithm will be given in Section 9.

We will show that

$$(1.7) \qquad (1/n)\ln \hat{S}_n - (1/n)\ln S_n^* \to 0,$$

for arbitrary bounded stock sequences $\mathbf{x}_1, \mathbf{x}_2, \ldots$. Thus $\hat{S}_n$ and S_n^* have the same exponent to first order. A more refined analysis for two stocks shows

$$(1.8) \qquad \hat{S}_n \sim \sqrt{\frac{2\pi}{nJ_n}}\, S_n^*,$$

in a sense that will be made precise. It is difficult to summarize the behavior of $\hat{S}_n$ relative to S_n^* because of the arbitrariness of the sequence and the fact that we cannot assume a limiting distribution. For example, even the limit of $(1/n)\ln S_n^*$ cannot be assumed to exist.

The goal of uniformly achieving $S_n^*(\mathbf{x}_1, \mathbf{x}_2, \ldots, \mathbf{x}_n)$, as specified in (1.7), was partially achieved by Cover and Gluss (1986) for discrete-valued stock markets by using the theory of compound sequential Bayes decision rules developed in Robbins (1951), Hannan and Robbins (1955), and the game-theoretic approachability-excludability theory of Blackwell (1956a, b). Work on natural investment goals can be found in Samuelson (1967) and Arrow (1974). The vast theory of undominated portfolios in the mean-variance plane is exemplified in Markowitz (1952) and Sharpe (1963), while the theory of rebalanced portfolios for known underlying distributions is developed in Kelly (1956), Mossin (1968), Thorp (1971), Markowitz (1976), Hakansson (1979), Bell and Cover (1980, 1988), Cover and King (1978), Cover (1984), Barron and Cover (1988), and Algoet and Cover (1988). A spirited defense of utility theory and the incompatibility of utility theory with the asymptotic growth rate approach is made in Samuelson (1967, 1969, 1979) and Merton and Samuelson (1974).

We see the present work as a departure from the above model-based investment theories, whether they be based on utility theory or growth rate optimality. Here the goal $S_n^* = \max_{\mathbf{b}} \Pi_{i=1}^{n} \mathbf{b}^t \mathbf{x}_i$ depends solely on the data and does not depend upon underlying statistical assumptions. Moreover, Theorem 5.1, for example, provides a finite sample lower bound for the performance $\hat{S}_n$ of the universal portfolio with respect to S_n^*. Therefore the case for success rests almost entirely on the acceptance of S_n^* as a natural investment goal.

The performance of the universal portfolio is exhibited in Section 8, where numerous examples are given of $S_n(\mathbf{b})$, S_n^*, and $\hat{S}_n$ for various pairs of stocks. In general, volatile uncorrelated stocks lead to great gains of S_n^* and $\hat{S}_n$ over the best buy-and-hold strategy. However, ponderous stocks like IBM and Coca-Cola show only modest improvements.

2. ELEMENTARY PROPERTIES

We wish to show that the wealth $\hat{S}_n$ generated by the universal portfolio strategy $\hat{\mathbf{b}}_k$ exceeds the value line index and that $\hat{S}_n$ is invariant under permutations of the stock sequence $\mathbf{x}_1, \mathbf{x}_2, \ldots, \mathbf{x}_n$. We will use the notation

$$W(\mathbf{b}, F) = \int \ln \mathbf{b}^t\mathbf{x} \, dF(\mathbf{x}), \tag{2.1}$$

$$W^*(F) = \max_{\mathbf{b}} W(\mathbf{b}, F), \tag{2.2}$$

and we will denote by F_n the empirical distribution associated with $\mathbf{x}_1, \mathbf{x}_2, \ldots, \mathbf{x}_n$, where F_n places mass $1/n$ at each $\mathbf{x}_i$. In particular, we note that

$$S_n^* = \max_{\mathbf{b}} S_n(\mathbf{b}) = \max_{\mathbf{b}} \prod_{i=1}^{n} \mathbf{b}^t\mathbf{x}_i = e^{nW^*(F_n)}. \tag{2.3}$$

For purposes of comparison, we pay special attention to buy-and-hold strategies $\mathbf{b} = \mathbf{e}_j = (0, 0, \ldots, 0, 1, 0, \ldots, 0)$, where $\mathbf{e}_j$ is the jth basis vector. Note that

$$S_n(\mathbf{e}_j) = \prod_{k=1}^{n} \mathbf{e}_j^t\mathbf{x}_k = \prod_{k=1}^{n} x_{kj} \tag{2.4}$$

is the factor by which the jth stock increases in n investment periods. Thus $S_n(\mathbf{e}_j)$ is the result of the buy-and-hold strategy associated with the jth stock.

We now note some properties of the target wealth S_n^*:

PROPOSITION 2.1 (Target Exceeds Best Stock).

$$S_n^* \geq \max_{j=1,2,\ldots,m} S_n(\mathbf{e}_j). \tag{2.5}$$

Proof. S_n^* is a maximization of $S_n(\mathbf{b})$ over the simplex, while the right-hand side is a maximization over the vertices of the simplex. □

PROPOSITION 2.2 (Target Exceeds Value Line).

$$S_n^* \geq \left(\prod_{j=1}^{m} S_n(\mathbf{e}_j) \right)^{1/m}. \tag{2.6}$$

Proof. Each $S_n(\mathbf{e}_j)$ is $\leq S_n^*$. □

The next proposition shows that the target exceeds the DJIA.

PROPOSITION 2.3 (Target Exceeds Arithmetic Mean). *If* $\alpha_j \geq 0$, $\Sigma\, \alpha_j = 1$, *then*

$$S_n^* \geq \sum_{j=1}^{m} \alpha_j S_n(\mathbf{e}_j). \tag{2.7}$$

Proof.

$$S_n(\mathbf{e}_j) \le S_n^*, \qquad j = 1, 2, \ldots, m. \qquad \square \tag{2.8}$$

Thus S_n^* exceeds the arithmetic mean, the geometric mean, and the maximum of the component stocks. Finally, it follows by inspection that S_n^* does not depend on the order in which $\mathbf{x}_1, \mathbf{x}_2, \ldots, \mathbf{x}_n$ occur.

PROPOSITION 2.4 *$S_n^*(\mathbf{x}_1, \mathbf{x}_2, \ldots, \mathbf{x}_n)$ is invariant under permutations of the sequence $\mathbf{x}_1, \mathbf{x}_2, \ldots, \mathbf{x}_n$.*

Now recall the proposed portfolio algorithm in (1.3) with the resulting wealth

$$\hat{S}_n = \prod_{k=1}^{n} \hat{\mathbf{b}}_k^t \mathbf{x}_k. \tag{2.9}$$

It will be useful to recharacterize $\hat{S}_n$ in the following way.

LEMMA 2.5.

$$\hat{S}_n = \prod_{k=1}^{n} \hat{\mathbf{b}}_k^t \mathbf{x}_k = \int S_n(\mathbf{b})\, d\mathbf{b} \Big/ \int d\mathbf{b} \tag{2.10}$$

where

$$S_n(\mathbf{b}) = \prod_{i=1}^{n} \mathbf{b}^t \mathbf{x}_i. \tag{2.11}$$

Thus the wealth $\hat{S}_n$ resulting from the universal portfolio is the average of $S_n(\mathbf{b})$ over the simplex.

Proof. Note from (1.3) and (1.4) that

$$\hat{\mathbf{b}}_k^t \mathbf{x}_k = \int \mathbf{b}^t \mathbf{x}_k \prod_{i=1}^{k-1} \mathbf{b}^t \mathbf{x}_i\, d\mathbf{b} \Big/ \int \prod_{i=1}^{k-1} \mathbf{b}^t \mathbf{x}_i\, d\mathbf{b} \tag{2.12}$$

$$= \int \prod_{i=1}^{k} \mathbf{b}^t \mathbf{x}_i\, d\mathbf{b} \Big/ \int \prod_{i=1}^{k-1} \mathbf{b}^t \mathbf{x}_i\, d\mathbf{b}. \tag{2.13}$$

Thus the product in (2.9) telescopes into

$$\hat{S}_n = \prod_{k=1}^{n} \hat{\mathbf{b}}_k^t \mathbf{x}_k = \int \prod_{i=1}^{n} \mathbf{b}^t \mathbf{x}_i\, d\mathbf{b} \Big/ \int d\mathbf{b} = \int S_n(\mathbf{b})\, d\mathbf{b} \Big/ \int d\mathbf{b}. \qquad \square \tag{2.14}$$

We observe two properties of the wealth $\hat{S}_n$ achieved by the universal portfolio.

PROPOSITION 2.5 (Universal Portfolio Exceeds Value Line Index).

$$\hat{S}_n \geq \left(\prod_{j=1}^{m} S_n(\mathbf{e}_j)\right)^{1/m}. \tag{2.15}$$

Proof. Let F_n be the empirical cumulative distribution function induced by $\mathbf{x}_1, \mathbf{x}_2, \ldots, \mathbf{x}_n$. By two applications of Jensen's inequality and writing

$$\int S_n(\mathbf{b})\, d\mathbf{b} \Big/ \int d\mathbf{b} = E_{\mathbf{b}} S_n(\mathbf{b}), \tag{2.16}$$

we have

(2.17)

$$\begin{aligned} \hat{S}_n &= E_{\mathbf{b}} S_n(\mathbf{b}) = E_{\mathbf{b}} \exp\{nW(\mathbf{b}, F_n)\} \\ &\geq \exp\{nE_{\mathbf{b}} W(\mathbf{b}, F_n)\} = \exp\left\{nE_{\mathbf{b}} \int \ln \mathbf{b}^t\mathbf{x}\, dF_n(\mathbf{x})\right\} \\ &= \exp\left\{nE_{\mathbf{b}} \int \ln\left(\sum_{j=1}^{m} b_j \mathbf{e}_j^t\mathbf{x}\right) dF_n(\mathbf{x})\right\} \geq \exp\left\{nE_{\mathbf{b}} \sum_{j=1}^{m} b_j \int \ln(\mathbf{e}_j^t\mathbf{x})\, dF_n(\mathbf{x})\right\} \\ &= \exp\left\{n\left(\frac{1}{m}\sum \int \ln \mathbf{e}_j^t\mathbf{x}\, dF_n(\mathbf{x})\right)\right\} = \left(\prod_{j=1}^{m} S_n(\mathbf{e}_j)\right)^{1/m}. \end{aligned}$$

□

Thus the wealth induced by the proposed portfolio dominates the value line index for any stock sequence $\mathbf{x}_1, \mathbf{x}_2, \ldots, \mathbf{x}_n$, for all n.

Next, we observe that although $\hat{\mathbf{b}}_k$ depends on the order of the sequence $\mathbf{x}_1, \mathbf{x}_2, \ldots, \mathbf{x}_n$, the resulting wealth $\hat{S}_n = \Pi\, \hat{\mathbf{b}}_k^t \mathbf{x}_k$ does not.

PROPOSITION 2.6. *$\hat{S}_n$ is invariant under permutations of the sequence* $\mathbf{x}_1, \mathbf{x}_2, \ldots, \mathbf{x}_n$.

Proof. Since the integrand in

$$\hat{S}_n = \prod_{k=1}^{n} \hat{\mathbf{b}}_k^t \mathbf{x}_k = \int_B S_n(\mathbf{b})\, d\mathbf{b} \Big/ \int_B d\mathbf{b} = \int_B \prod_{i=1}^{n} \mathbf{b}^t\mathbf{x}_i\, d\mathbf{b} \Big/ \int_B d\mathbf{b} \tag{2.18}$$

is invariant under permutations, so is $\hat{S}_n$. □

This observation guarantees that the crash of 1929 will have no worse consequences for wealth $\hat{S}_n$ than if the bad days of that time had been sprinkled out among the good.

3. THE REASON THE PORTFOLIO WORKS

The main idea of the portfolio algorithm is quite simple. The idea is to give an amount $d\mathbf{b}/\int_B d\mathbf{b}$ to each portfolio manager indexed by rebalancing strategy $\mathbf{b}$, let him or her make $S_n(\mathbf{b}) = e^{nW(\mathbf{b}, F_n)}\, d\mathbf{b}$ at exponential rate $W(\mathbf{b}, F_n)$, and pool the wealth at the end. Of course, all dividing and repooling is done "on paper" at time k, resulting in $\hat{\mathbf{b}}_k$. Since the average of exponentials has, under suitable smoothness conditions, the same asymptotic exponential growth rate as the maximum, one achieves almost as much as the wealth S_n^* achieved by the best constant rebalanced portfolio. The trap to be avoided is to put a mass distribution on the market distributions $F(\mathbf{x})$. It seems that this cannot be done in a satisfactory way.

4. PRELIMINARIES

We now introduce definitions and conditions that will allow characterization of the behavior of $\hat{S}_n/S_n^*$. Let $F_n(\mathbf{x})$ denote the empirical probability mass function putting mass $1/n$ on each of the points $\mathbf{x}_1, \mathbf{x}_2, \ldots, \mathbf{x}_n \in \mathbf{R}_+^m$. Let the portfolio $\mathbf{b}^* = \mathbf{b}^*(F_n)$ achieve the maximum of $S_n(\mathbf{b}) = \Pi_{i=1}^n \mathbf{b}^t\mathbf{x}_i$. Equivalently, since $S_n(\mathbf{b}) = e^{nW(\mathbf{b}, F_n)}$, the portfolio $\mathbf{b}^*(F_n)$ achieves the maximum of $W(\mathbf{b}, F_n)$. Thus,

$$S_n^* = \max_{\mathbf{b} \in B} S_n(\mathbf{b}) = e^{nW^*(F_n)}. \tag{4.1}$$

DEFINITION. We shall say *all stocks are active* (at time n) if $(\mathbf{b}^*(F_n))_i > 0$, $i > 1, 2, \ldots, m$, for some $\mathbf{b}^*$ achieving $W^*(F_n)$. All stocks are *strictly active* if inequality is strict for all i and all $\mathbf{b}^*$ achieving $W^*(F_n)$.

DEFINITION. We shall say $\mathbf{x}_1, \mathbf{x}_2, \ldots, \mathbf{x}_n \in \mathbf{R}^m$ are of *full rank* if $\mathbf{x}_1, \mathbf{x}_2, \ldots, \mathbf{x}_n$ spans $\mathbf{R}^m$.

The condition of full rank is usually true for observed stock market sequences if n is somewhat larger than m, but the condition that all stocks be active often fails when certain stocks are dominated. The next definition measures the curvature of $S_n(\mathbf{b})$ about its maximum and accounts for the second-order behavior of $\hat{S}_n$ with respect to S_n^*.

DEFINITION. The *sensitivity matrix function* $J(\mathbf{b})$ of a market with respect to distribution $F(\mathbf{x})$, $\mathbf{x} \in \mathbf{R}_+^m$, is the $(m-1) \times (m-1)$ matrix defined by

$$J_{ij}(\mathbf{b}) = \int \frac{(x_i - x_m)(x_j - x_m)}{(\mathbf{b}^t\mathbf{x})^2}\, dF(\mathbf{x}), \qquad 1 \le i, j \le m-1. \tag{4.2}$$

The *sensitivity matrix* J^* is $J(\mathbf{b}^*)$, where $\mathbf{b}^* = \mathbf{b}^*(F)$ maximizes $W(\mathbf{b}, F)$.
We note that

$$J_{ij}^* = -\frac{\partial^2 W((\mathrm{b}_1^*, \mathrm{b}_2^*, \ldots, \mathrm{b}_{m-1}^*, 1 - \Sigma_{i=1}^{m-1} \mathrm{b}_i^*), F)}{\partial \mathrm{b}_i \partial \mathrm{b}_j}. \tag{4.3}$$

LEMMA 4.1. *J^* is nonnegative definite. It is positive definite if all stocks are strictly active.*

5. ANALYSIS FOR TWO ASSETS

We now wish to show that $\hat{S}_n/S_n^* \sim \sqrt{2\pi/nJ_n}$, where J_n is the curvature or volatility index. We show in detail that $\sqrt{2\pi/nJ_n}$ is an asymptotic lower bound on $\hat{S}_n/S_n^*$, and we develop explicit lower bounds on $\hat{S}_n/S_n^*$ for all n and any market sequence $\mathbf{x}_1, \ldots, \mathbf{x}_n$. We develop an upper bound by invoking strong conditions on the market sequence. Section 6 outlines the proof for m assets.

We investigate the behavior of $\hat{S}_n$ for $m = 2$ stocks. Consider the arbitrary stock vector sequence

$$\mathbf{x}_i = (x_{i1}, x_{i2}) \in \mathbf{R}_+^2, \qquad i = 1, 2, \ldots. \tag{5.1}$$

We now proceed to recast this two-variable problem in terms of a single variable. Since the portfolio choice requires the specification of one parameter, we write

$$\mathbf{b} = (\mathrm{b}, 1 - \mathrm{b}), \qquad 0 \le \mathrm{b} \le 1, \tag{5.2}$$

and rewrite $S_n(\mathbf{b})$ as

$$S_n(\mathrm{b}) = \prod_{i=1}^{n} (\mathrm{b}x_{i1} + (1 - \mathrm{b})x_{i2}), \qquad 0 \le \mathrm{b} \le 1. \tag{5.3}$$

Let

$$S_n^* = \max_{0 \le \mathrm{b} \le 1} S_n(\mathrm{b}), \tag{5.4}$$

and let b_n^* denote the value of b achieving this maximum. Section 8 contains examples.

The universal portfolio

$$\hat{\mathbf{b}}_k = (\hat{\mathrm{b}}_k, 1 - \hat{\mathrm{b}}_k) \tag{5.5}$$

is defined by

$$\hat{\mathrm{b}}_k = \int_0^1 \mathrm{b} S_k(\mathrm{b})\, d\mathrm{b} \bigg/ \int_0^1 S_k(\mathrm{b})\, d\mathrm{b} \tag{5.6}$$

and achieves wealth

$$\hat{S}_n = \prod_{i=1}^{n} (\hat{\mathrm{b}}_i x_{i1} + (1 - \hat{\mathrm{b}}_i)x_{i2}). \tag{5.7}$$

Let

$$W_n(\mathrm{b}) = \frac{1}{n} \ln S_n(\mathrm{b}) \tag{5.8}$$

$$= \frac{1}{n} \sum_{i=1}^{n} \ln(\mathrm{b}x_{i1} + (1 - \mathrm{b})x_{i2}) \tag{5.9}$$

$$= \int \ln(\mathrm{b}x_1 + (1 - \mathrm{b})x_2)\, dF_n(\mathbf{x}), \tag{5.10}$$

where $F_n(\mathbf{x})$ is the empirical cdf of $\{\mathbf{x}_i\}_{i=1}^n$. By Lemma 4.1, the wealth $\hat{S}_n$ achieved by the universal portfolio $\hat{\mathbf{b}}_k$ is given by

$$\hat{S}_n = \int_0^1 e^{nW_n(\mathrm{b})}\, d\mathrm{b}. \tag{5.11}$$

In order to characterize the behavior of $\hat{S}_n$ we define the following functions of the sequence $\mathbf{x}_1, \mathbf{x}_2, \ldots, \mathbf{x}_n$. Define the *relative range* τ_n of the sequence $\mathbf{x}_1, \mathbf{x}_2, \ldots, \mathbf{x}_n$ to be

$$\tau_n = 2^{1/3}\left(\frac{\max\{x_{ij}\}}{\min\{x_{ij}\}} - 1\right), \tag{5.12}$$

where the minimum and maximum are taken over $i = 1, 2, \ldots, n$; $j = 1, 2$. Let

$$J_n = \frac{1}{n}\sum_{i=1}^{n} \frac{(x_{i1} - x_{i2})^2}{(\mathrm{b}_n^* x_{i1} + (1 - \mathrm{b}_n^*)x_{i2})^2}, \tag{5.13}$$

where b_n^* maximizes $W_n(\mathrm{b})$. Let

$$W_n^* = \max_{0 \le \mathrm{b} \le 1} W_n(\mathrm{b}) = W_n(\mathrm{b}_n^*). \tag{5.14}$$

Thus τ_n corresponds to the relative range of the price relatives and J_n denotes the curvature of $\ln S_n(\mathrm{b})$ at the maximum.

THEOREM 5.1. *Let* $\mathbf{x}_1, \mathbf{x}_2, \ldots,$ *be an arbitrary sequence of stock vectors in* $\mathbf{R}_+^2$, *and let* $a_n = \min\{\mathrm{b}_n^*, 1 - \mathrm{b}_n^*, 3J_n/\tau_n^3\}$. *Then for any* $0 < \varepsilon < 1$, *and for any* n,

$$\frac{\hat{S}_n}{S_n^*} \ge \sqrt{\frac{2\pi}{nJ_n(1+\varepsilon)}} - \frac{2}{\varepsilon(1+\varepsilon)a_nJ_nn} \exp\left\{\frac{-\varepsilon^2(1+\varepsilon)a_nJ_nn}{2}\right\}. \tag{5.15}$$

REMARKS. This theorem says roughly that $\hat{S}_n/S_n^* \ge \sqrt{2\pi/nJ_n}$. So the universal wealth is within a factor of $C/\sqrt{n}$ of the (presumably) exponentially large S_n^*. It will turn out that every additional stock in the universal portfolio costs an additional factor of $1/\sqrt{n}$. But these factors become negligible to first order in exponent. It is important to mention that this theorem is a bound for each n. The bound holds for any stock sequence with bound a_n and volatility J_n.

Proof. We wish to bound $\hat{S}_n = \int_0^1 e^{nW_n(\mathrm{b})}\, d\mathrm{b}$. We expand $W_n(\mathrm{b})$ about the maximizing portfolio b_n^*, noting that $W_n(\mathrm{b})$ has different local properties for each n and indeed a different maximizing b_n^*. We have

$$W_n(\mathrm{b}) = W_n(\mathrm{b}_n^*) + (\mathrm{b} - \mathrm{b}_n^*)W_n'(\mathrm{b}_n^*) + \frac{(\mathrm{b} - \mathrm{b}_n^*)^2}{2} W_n''(\mathrm{b}_n^*) + \frac{(\mathrm{b} - \mathrm{b}_n^*)^3}{3!} W_n'''(\tilde{\mathrm{b}}_n), \tag{5.16}$$

where $\tilde{b}_n$ lies between b and b_n^*.

We now examine the terms.

(i) The first term is

$$W_n(b_n^*) = W^*(F_n) = \frac{1}{n} \log S_n^*, \tag{5.17}$$

where S_n^* is the target wealth at time n.

(ii) The second term is

$$W_n'(b_n^*) = \int \frac{x_1 - x_2}{\mathbf{b}_n^{*t}\mathbf{x}} \, dF_n(\mathbf{x}) \tag{5.18}$$

$$= 0, \qquad \text{if } 0 < b_n^* < 1,$$

by the optimality of b_n^*.

(iii) The third term is

$$W_n''(b_n^*) = -\int \frac{(x_1 - x_2)^2}{(\mathbf{b}_n^{*t}\mathbf{x})^2} \, dF_n(\mathbf{x}) = -J_n^*. \tag{5.19}$$

Thus, $W_n''(b_n^*) \geq 0$, with strict inequality if $0 < b_n^* < 1$ and $x_{i1} \neq x_{i2}$ for some time i. This term provides the constant in the second-order behavior of $\hat{S}_n$.

(iv) The fourth term is

$$W_n'''(\tilde{b}_n) = 2\int \frac{(x_1 - x_2)^3}{(\tilde{b}x_1 + (1 - \tilde{b})x_2)^3} \, dF_n(\mathbf{x}). \tag{5.20}$$

We have the bound

$$|W_n'''(\tilde{b}_n)| = \left| \int \frac{2(x_1 - x_2)^3}{(\tilde{b}_n x_1 + (1 - \tilde{b}_n)x_2)^3} \, dF_n(\mathbf{x}) \right| \tag{5.21}$$

$$\leq \tau_n^3, \qquad \text{for all } \tilde{b}_n \in [0, 1]. \tag{5.22}$$

Thus

$$S_n(b) \geq \exp\left(nW_n^* - \frac{n}{2}(b - b_n^*)^2 J_n - \frac{n|b - b_n^*|^3 \tau_n^3}{6} \right) \tag{5.23}$$

for $0 \leq b \leq 1$, where

$$J_n = \int \frac{(x_1 - x_2)^2}{(b_n^* x_1 + (1 - b_n^*)x_2)^2} \, dF_n(\mathbf{x}). \tag{5.24}$$

We now make the change of variable

$$u = \sqrt{n}\,(b - b_n^*), \tag{5.25}$$

where the new range of integration is

$$-\sqrt{n}\, b_n^* \leq u \leq \sqrt{n}\,(1 - b_n^*). \tag{5.26}$$

Then, noting $e^{nW_n^*} = S_n^*$, we have

$$\hat{S}_n = \int_0^1 S_n(\mathrm{b})\, d\mathrm{b} \tag{5.27}$$

$$\geq \frac{S_n^*}{\sqrt{n}} \int_{-\sqrt{n}\,\mathrm{b}_n^*}^{\sqrt{n}\,(1-\mathrm{b}_n^*)} \exp\left(-\frac{1}{2}u^2 J_n - \frac{1}{6\sqrt{n}}\,|u^3|\,\tau_n^3\right) du. \tag{5.28}$$

We wish to approximate this by the normal integral. To do so let $0 < \varepsilon \leq 1$ and note that

$$-\frac{1}{2}u^2 J_n - \frac{|u|^3 \tau_n^3}{6\sqrt{n}} \geq -\frac{1}{2}u^2 J_n(1+\varepsilon) \tag{5.29}$$

for

$$u \leq 3\varepsilon\sqrt{n}\, J_n/\tau_n^3. \tag{5.30}$$

Let Φ denote the cdf of the standard normal

$$\Phi(x) = \frac{1}{\sqrt{2\pi}} \int_{-\infty}^{x} e^{-u^2/2}\, du, \tag{5.31}$$

and let

$$a_n = \min\{\mathrm{b}_n^*,\ 1-\mathrm{b}_n^*,\ 3J_n/\tau_n^3\}. \tag{5.32}$$

Thus a_n is a measure of the degree to which $S_n(\mathrm{b})$ has a maximum of reasonable curvature within the unit interval. Then from (5.28), for any $0 < \varepsilon \leq 1$,

$$\frac{\sqrt{n}\,\hat{S}_n}{S_n^*} \geq \int_{-\sqrt{n}\,\mathrm{b}_n^*}^{\sqrt{n}(1-\mathrm{b}_n^*)} \exp\left(-\frac{1}{2}u^2 J_n - \frac{1}{6\sqrt{n}}\,|u|^3 \tau_n^3\right) du \tag{5.33}$$

$$\geq \int_{-\sqrt{n}\,a_n\varepsilon}^{\sqrt{n}\,a_n\varepsilon} \exp\left\{-\frac{1}{2}u^2 J_n(1+\varepsilon)\right\} du \tag{5.34}$$

$$= \int_{-\infty}^{\infty} \exp\left\{-\frac{1}{2}u^2 J_n(1+\varepsilon)\right\} du - 2\int_{-\infty}^{-\sqrt{n}\,a_n\varepsilon} \exp\left\{-\frac{1}{2}u^2 J_n(1+\varepsilon)\right\} du \tag{5.35}$$

$$= \sqrt{\frac{2\pi}{J_n(1+\varepsilon)}} - \sqrt{\frac{8\pi}{J_n(1+\varepsilon)}}\ \Phi\left(-\varepsilon a_n\sqrt{nJ_n(1+\varepsilon)}\right). \tag{5.36}$$

We use the inequality

$$\frac{1}{\sqrt{2\pi x^2}} \exp\left(-\frac{x^2}{2}\right)\left(1-\frac{1}{x^2}\right) < \Phi(-x) < \frac{1}{\sqrt{2\pi x^2}} \exp\left(-\frac{x^2}{2}\right), \tag{5.37}$$

for $x > 0$, to obtain the bound

$$\Phi\left(-\varepsilon a_n\sqrt{nJ_n(1+\varepsilon)}\right) < \frac{1}{\sqrt{2\pi\varepsilon^2 a_n^2 n J_n(1+\varepsilon)}} \exp\left\{-\frac{\varepsilon^2 a_n^2 n J_n(1+\varepsilon)}{2}\right\}. \tag{5.38}$$

Hence,

$$\text{(5.39)} \quad \frac{\sqrt{n}\,\hat{S}_n}{S_n^*} \geq \sqrt{\frac{2\pi}{J_n(1+\varepsilon)}} - \frac{2}{\varepsilon a_n J_n(1+\varepsilon)\sqrt{n}} \exp\left\{-\frac{\varepsilon^2 a_n^2 n J_n(1+\varepsilon)}{2}\right\}$$

for any $0 < \varepsilon \leq 1$, for all n, and all $\mathbf{x}_1, \mathbf{x}_2, \ldots$, which proves the theorem. □

The explicit bounds in Theorem 5.1 may be useful in practice, but a cleaner summary of performance is given in the following weaker theorem.

THEOREM 5.2. *Let* $\mathbf{x}_1, \mathbf{x}_2, \ldots$ *be a sequence of stock vectors in* $\mathbf{R}_+^2$ *and suppose* $\delta \leq \mathrm{b}_n^* \leq 1 - \delta$, $\tau_n \leq \tau < \infty$, *and* $J_n \geq J > 0$, *for a subsequence of times* $n_1, n_2, \ldots$. *Then*

$$\text{(5.40)} \quad \liminf_{n\to\infty} \frac{\hat{S}_n/S_n^*}{\sqrt{2\pi/nJ_n}} \geq 1,$$

along this subsequence.

Proof. The conditions of the theorem, together with Theorem 5.1 imply

$$\text{(5.41)} \quad \frac{\hat{S}_n/S_n^*}{\sqrt{2\pi/nJ_n}} \geq \sqrt{\frac{1}{1+\varepsilon_n}} - \frac{2\sqrt{J_n}}{\varepsilon_n\sqrt{2\pi nJ}\,\min\{\delta, 3J/\tau^3\}},$$

where τ is the bound ratio, and where we are free to choose $\varepsilon_n \in [0, 1]$ at each n. Noting that $J_n \leq \tau^2 < \infty$ and letting $\varepsilon_n = n^{-1/4}$ proves the theorem. □

We have just shown that $\hat{S}_n/S_n^*$ is as good as $\sqrt{2\pi/nJ_n}$. We now show that it is no better. For this we consider a subsequence of times such that $W_n(\mathrm{b})$ is approximately equal to some function $W(\mathrm{b})$, and we argue that upper bounds on $\int_0^1 e^{nW(\mathrm{b})}\,d\mathrm{b}$ suffice to limit the performance of the wealth $\hat{S}_n$. Toward that end, let us consider functions W such that

(5.42)
- (i) $W(\mathrm{b})$ is strictly concave on $[0, 1]$.
- (ii) $W'''(\mathrm{b})$ is bounded on $[0, 1]$.
- (iii) $W(\mathrm{b})$ achieves its maximum at $\mathrm{b}^* \in (0, 1)$.

We plan to pick out a subsequence of times such that $W_n(\mathrm{b}) = (1/n) \cdot \sum_{i=1}^n \ln \mathbf{b}^t\mathbf{x}_i$ approaches $W(\mathrm{b})$. We can expect such limit points from Arzelà's theorem on the compactness of equicontinuous functions on compact sets. Let b_n^* maximize $W_n(\mathrm{b})$. Let $\{n_i\}$ be a subsequence of times such that for $n = n_1, n_2, \ldots$,

(5.43)
- (i) $W_n(\mathrm{b}) \leq W(\mathrm{b}), \quad 0 \leq \mathrm{b} \leq 1.$
- (ii) $W_n''(\mathrm{b}_n^*) \to W''(\mathrm{b}^*)$.

Recall the notation $J_n = -W_n''(\mathrm{b}_n^*)$. The following theorem establishes the tightness of the lower bound in Theorem 5.2.

THEOREM 5.3. *For any* $\mathbf{x}_1, \mathbf{x}_2, \ldots \in \mathbf{R}_+^2$ *and for any subsequence of times* $n_1, n_2, \ldots$ *such that* $W_n(\mathrm{b})$ *satisfies conditions* (5.43) *for some* $W(\mathrm{b})$ *satisfying* (5.42), *we have*

(5.44) $$\hat{S}_n/S_n^* \sim \sqrt{2\pi/nJ_n}$$

along the subsequence.

Proof. The lower bound follows from Theorem 5.2. From Laplace's method of integration we have

(5.45) $$\int_0^1 e^{ng(u)}\,du \sim e^{ng(u^*)}\sqrt{\frac{2\pi}{n|g''(u^*)|}}$$

if g is three times differentiable with bounded third derivative, strictly concave, and the u^* maximizing $g(\cdot)$ is in the open interval (0, 1). Consequently,

(5.46) $$\hat{S}_n = \int_0^1 e^{nW_n(\mathrm{b})}\,d\mathrm{b} \le \int_0^1 e^{nW(\mathrm{b})}\,d\mathrm{b}$$
$$\sim e^{nW(\mathrm{b}^*)}\sqrt{\frac{2\pi}{n|W''(\mathrm{b}^*)|}} = S_n^*\sqrt{\frac{2\pi}{n|W''(\mathrm{b}^*)|}} \sim S_n^*\sqrt{\frac{2\pi}{nJ_n}},$$

and the theorem is proved. □

6. MAIN THEOREM

Here we prove the result for m assets under the assumption that all stocks are active and of full rank and $\mathbf{b}_n^*(F_n) \to \mathbf{b}^* \in \operatorname{int}(B)$. We discuss removing the conditions in Section 9. For example, lack of full rank reduces the dimension from m to m', as does the existence of inactive stocks. Finally, $\mathbf{b}_n^*(F_n)$ need not have a limit, in which case we can describe the behavior of $\hat{S}_n$ for convergent subsequences of $\mathbf{b}_n^*(F_n)$, as well as develop explicit bounds for all n.

From Lemma 2.1, we have

$$\hat{S}_n = \int_B S_n(\mathbf{b})\,d\mathbf{b} \Big/ \int_B d\mathbf{b},$$

where

$$S_k(\mathbf{b}) = \prod_{i=1}^{k} \mathbf{b}^t\mathbf{x}_i,$$

$$\hat{\mathbf{b}}_{k+1} = \int \mathbf{b}S_k(\mathbf{b})\,d\mathbf{b} \Big/ \int S_k(\mathbf{b})\,d\mathbf{b},$$

$$\hat{S}_n = \prod_{k=1}^{n} \hat{\mathbf{b}}_k^t\mathbf{x}_k.$$

A summary of the performance of $\hat{\mathbf{b}}_k$ is given by the following theorem.

THEOREM 6.1. *Suppose* $\mathbf{x}_1, \mathbf{x}_2, \ldots \in [a, c]^m$, $0 < a \le c < \infty$, *and at a subsequence of times* $n_1, n_2, \ldots$, $W_n(\mathbf{b}) \nearrow W(\mathbf{b})$ *for* $\mathbf{b} \in B$, $J_n^* \to J^*$, $\mathbf{b}_n^* \to \mathbf{b}^*$, *where $W(\mathbf{b})$ is strictly concave, the third partial derivatives of W are bounded on B, and $W(\mathbf{b})$ achieves its maximum at $\mathbf{b}^*$ in the interior of B. Then*

$$\frac{\hat{S}_n}{S_n^*} \sim \left(\sqrt{\frac{2\pi}{n}}\right)^{m-1} \frac{(m-1)!}{|J^*|^{1/2}} \tag{6.1}$$

in the sense that the ratio of the right- and left-hand sides converges to 1 *along the subsequence.*

Proof. (Outline) We define

$$C = \left\{(c_1, c_2, \ldots, c_{m-1}) \colon c_i \geq 0, \sum c_i \leq 1\right\} \tag{6.2}$$

and

$$S_n(\mathbf{c}) = \prod_{i=1}^{n} \mathbf{b}^t(\mathbf{c})\mathbf{x}_i, \qquad \mathbf{c} \in C, \tag{6.3}$$

where

$$\mathbf{b}(\mathbf{c}) = \left(c_1, c_2, \ldots, c_{m-1}, 1 - \sum_{i=1}^{m-1} c_i\right). \tag{6.4}$$

Note that

$$\mathrm{Vol}(C) = \int_C d\mathbf{c} = \frac{1}{(m-1)!}. \tag{6.5}$$

We shall prove only the lower bound associated with (6.1). From Lemma 2.1, the universal portfolio algorithm yields

$$\hat{S}_n = \int S_n(\mathbf{b})\, d\mathbf{b} \Big/ \int d\mathbf{b} = E_{\mathbf{b}} S_n(\mathbf{b}), \tag{6.6}$$

where $\mathbf{b}$ is uniformly distributed over the simplex B. Since a uniform distribution over B induces a uniform distribution over C, we have

$$\hat{S}_n = (m-1)! \int_C S_n(\mathbf{c})\, d\mathbf{c}. \tag{6.7}$$

We now expand $S_n(\mathbf{c})$ in a Taylor series about $\mathbf{c}^* = (b_1^*, \ldots, b_{m-1}^*)$, where $\mathbf{b}^*$ maximizes $W(\mathbf{b}, F_n)$. We drop the dependence of $\mathbf{b}^*$ on n for notational convenience. By assumption, $b_i^* > 0$, for all i. We have

$$S_n(\mathbf{c}) = e^{nW_n(\mathbf{c})}, \tag{6.8}$$

where

$$W_n(\mathbf{c}) = \frac{1}{n} \sum_{i=1}^{n} \ln \mathbf{b}^t\mathbf{x}_i = \int \ln \mathbf{b}^t\mathbf{x}\, dF_n(\mathbf{x}) \tag{6.9}$$

$$\triangleq E_{F_n} \ln \mathbf{b}^t\mathbf{X}$$

and

$$\mathbf{b} = \left(\mathbf{c},\ 1 - \sum c_i\right). \tag{6.10}$$

Expanding $W_n(\mathbf{c})$, we have

$$\begin{aligned} W_n(\mathbf{c}) &= W_n(\mathbf{c}^*) + (\mathbf{c} - \mathbf{c}^*)^t \nabla W_n(\mathbf{c}^*) - \frac{1}{2}(\mathbf{c} - \mathbf{c}^*)^t J_n^* (\mathbf{c} - \mathbf{c}^*) \\ &\quad + \frac{1}{6} \sum_{i,j,k} (c_i - c_i^*)(c_j - c_j^*)(c_k - c_k^*) E_{F_n} \\ &\qquad \cdot \frac{2(X_i - X_m)(X_j - X_m)(X_k - X_m)}{S^3(\tilde{\mathbf{c}})}, \end{aligned} \tag{6.11}$$

where $\tilde{\mathbf{c}} = \lambda \mathbf{c}^* + (1 - \lambda)\mathbf{c}$, for some $0 \le \lambda \le 1$, where λ may depend on $\mathbf{c}$, and

$$S(\mathbf{c}) = \sum_{i=1}^{m-1} c_i X_i + \left(1 - \sum_{i=1}^{m-1} c_i\right) X_m. \tag{6.12}$$

Here

$$\begin{aligned} S_n(\mathbf{c}) &= \prod_{i=1}^{n} \mathbf{b}^t(\mathbf{c}) \mathbf{x}_i, \\ \mathbf{b}(\mathbf{c}) &= \left(c_1, c_2, \ldots, 1 - \sum_{i=1}^{m-1} c_i\right), \\ W(\mathbf{c}) &= \int \ln\left(\sum_{i=1}^{m-1} c_i x_i + \left(1 - \sum_{1}^{m-1} c_i\right) x_k\right) dF_n(\mathbf{x}), \\ \frac{\partial W_n}{\partial c_i} &= \int \frac{x_i - x_m}{S(\mathbf{c})}\, dF_n(\mathbf{x}), \\ \frac{\partial^2 W_n}{\partial c_i\, \partial c_j} &= -\int \frac{(x_i - x_m)(x_j - x_m)}{S^2(\mathbf{c})}\, dF_n(\mathbf{x}), \end{aligned} \tag{6.13}$$

$$J_n^* = -\left[\frac{\partial^2 W_n(\mathbf{c}^*)}{\partial c_i\, \partial c_j}\right]. \tag{6.14}$$

The condition that all stocks be strictly active implies by Lemma 4.1 that $|J_n^*| > 0$, where $|\cdot|$ denotes determinant. We treat the terms one by one.

(i) By definition of $\mathbf{b}^*$,

$$W(\mathbf{c}^*) = W(\mathbf{b}^*, F_n) = W^*(F_n). \tag{6.15}$$

(ii) The second term is 0 because $\mathbf{b}^*$ is in the interior of B, $W_n(\mathbf{b})$ is differentiable, and $\mathbf{b}^*$ maximizes W_n. Thus,

$$\frac{\partial W_n(\mathbf{c}^*)}{\partial c_i} = E_{F_n}\frac{X_i - X_m}{\mathbf{b}^{*t}\mathbf{X}} \tag{6.16}$$

$$= 0, \qquad i = 1, 2, \ldots, m-1.$$

(iii) The third term is a positive definite quadratic form, where $J_n^* = J^*(\mathbf{b}^*(F_n))$.

(iv) For the fourth term,

(6.17)
$$\frac{1}{6}\sum_{i,j,k=1}^{m-1}(c_i - c_i^*)(c_j - c_j^*)(c_k - c_k^*)E_{F_n}\frac{2(X_i - X_m)(X_j - X_m)(X_k - X_m)}{S^3(\tilde{\mathbf{c}})},$$

we examine

$$E_{F_n}\frac{(X_i - X_m)(X_j - X_m)(X_k - X_m)}{S^3(\tilde{\mathbf{c}})}. \tag{6.18}$$

We note

$$S^3(\tilde{\mathbf{c}}) = (\tilde{\mathbf{b}}^t X)^3 \geq \left(\sum \tilde{b}_i a\right)^3 \geq a^3, \tag{6.19}$$

since $X_i \geq a$ for all i. Also since $X_i - X_m \leq 2c$, we have

$$-\frac{8c^3}{a^3} \leq E\frac{(X_i - X_m)(X_j - X_m)(X_k - X_m)}{S^3(\tilde{\mathbf{c}})} \leq \frac{8c^3}{a^3}. \tag{6.20}$$

We now make the change of variable $\mathbf{u} = \sqrt{n}(\mathbf{c} - \mathbf{c}^*)$, where we note the new range of integration $\mathbf{u} \in U = \sqrt{n}(C - \mathbf{c}^*)$. Thus

$$S_n(\mathbf{c}) = \exp\left\{nW_n(\mathbf{c}^*) - \frac{n}{2}(\mathbf{c} - \mathbf{c}^*)^t J_n^*(\mathbf{c} - \mathbf{c}^*) + \frac{n}{3}\sum\nolimits_3\right\} \tag{6.21}$$

$$= \exp\left(nW_n^* - \frac{1}{2}\mathbf{u}^t J_n^*\mathbf{u} + \frac{1}{3\sqrt{n}}\sum\nolimits_3\right),$$

where

$$\sum\nolimits_3 = \sum_{i,j,k=1}^{m-1} u_i u_j u_k E_{F_n}\left(\frac{(X_i - X_m)(X_j - X_m)(X_k - X_m)}{S^3(\tilde{\mathbf{c}})}\right). \tag{6.22}$$

Note that

$$\left[\sum\nolimits_3\right] \leq \left(\sum_{i=1}^{m-1}|u_i|\right)^3\frac{8c^3}{a^3}. \tag{6.23}$$

Observing

$$\sum|u_i| \leq \left(\sum u_i^2\right)^{1/2}\sqrt{m} \tag{6.24}$$

yields

$$S_n(\mathbf{c}) \geq \exp\left(nW_n^* - \frac{1}{2}\mathbf{u}^t J_n^* \mathbf{u} - \frac{m^{3/2}}{3\sqrt{n}} \frac{\|\mathbf{u}\|^3 8c^3}{a^3}\right). \tag{6.25}$$

The lower bound on $\hat{S}_n$ becomes

$$\text{(6.26)}\ \hat{S}_n = (m-1)! \int_{\mathbf{c}\in C} S_n(\mathbf{c})\, d\mathbf{c}$$

$$\geq (m-1)!\, S_n^* \int_{\mathbf{u}\in U} \exp\left(-\frac{1}{2}\mathbf{u}^t J_n^* \mathbf{u} - \frac{m^{3/2}}{3\sqrt{n}} \frac{\|\mathbf{u}\|^3 8c^3}{a^3}\right)\left(\frac{1}{\sqrt{n}}\right)^{m-1} d\mathbf{u},$$

which can now be bounded using the techniques in the two-stock proof. The upper bound follows from Laplace's method of integration, as in Theorem 5.3, from which the theorem follows. □

7. STOCHASTIC MARKETS

Another way to see the naturalness of the goal $S_n^* = e^{nW(\mathbf{b}^*(F_n), F_n)}$ is to consider random investment opportunities. Let $\mathbf{X}_1, \mathbf{X}_2, \ldots$ be independent identically distributed (i.i.d.) random vectors drawn according to $F(\mathbf{x})$, $\mathbf{x} \in \mathbf{R}^m$, where F is some known distribution function. Let $S_n(\mathbf{b}) = \Pi_{i=1}^n \mathbf{b}^t\mathbf{X}_i$ denote the wealth at time n resulting from an initial wealth $S_0 = 1$ and a reinvestment of assets according to portfolio $\mathbf{b}$ at each investment opportunity. Then

$$\text{(7.1)}\quad S_n(\mathbf{b}) = \prod_{i=1}^n \mathbf{b}^t\mathbf{X}_i = \exp\left(\sum_{i=1}^n \ln \mathbf{b}^t\mathbf{X}_i\right)$$

$$= \exp\{n(E \ln \mathbf{b}^t\mathbf{X} + o_p(1))\} = \exp\{n(W(\mathbf{b}, F) + o_p(1))\}$$

by the strong law of large numbers, where the random variable $o_p(1) \to 0$, a.e. We observe from the above that, to first order in the exponent, the growth rate of wealth $S_n(\mathbf{b})$ is determined by the expected log wealth

$$W(\mathbf{b}, F) = \int \ln \mathbf{b}^t\mathbf{x}\, dF(\mathbf{x}) \tag{7.2}$$

for portfolio $\mathbf{b}$ and stock distribution $F(\mathbf{x})$.

It follows for $\mathbf{X}_1, \mathbf{X}_2, \ldots$, i.i.d. $\sim F$ that $\mathbf{b}^*(F)$ achieves an exponential growth rate of wealth with exponent $W^*(F)$. Moreover Breiman (1961) establishes for i.i.d. stock vectors for any nonanticipating time-varying portfolio strategy with associated wealth sequence S_n that

$$\overline{\lim}\, \frac{1}{n} \ln S_n \leq W^*\{F\}, \quad \text{a.e.} \tag{7.3}$$

Finally, it follows from Breiman (1961), Finkelstein and Whitley (1981), Barron and Cover (1988), and Algoet and Cover (1988), in increasing levels of generality on the stochastic process, that $\lim_{n\to\infty} n^{-1} \ln S_n/S_n^* \le 0$, a.e., for every sequential portfolio. Thus $\mathbf{b}^*(F)$ is asymptotically optimal in this sense, and $W^*(F)$ is the highest possible exponent for the growth rate of wealth. Thus S_n^* is asymptotically optimal.

We omit the proof of the following.

THEOREM 7.1. *Let $\mathbf{X}_i$ be i.i.d. $\sim F(x)$. Let $\mathbf{b}^*(F)$ be unique and lie in the interior of B. Then the universal portfolio $\hat{\mathbf{b}}_k$ yields a wealth sequence $\hat{S}_n$ satisfying*

$$\frac{1}{n} \ln \hat{S}_n \to W^*(F), \quad \text{a.e.} \tag{7.4}$$

Thus, in the special case where the stocks are independent and identically distributed according to some unknown distribution F, the universal portfolio essentially learns F in the sense that the associated growth rate of wealth is equal to that achievable when F is known.

8. EXAMPLES

We now test the portfolio algorithm on real data. Consider, for example, Iroquois Brands Ltd. and Kin Ark Corp., two stocks chosen for their volatility listed on the New York Stock Exchange. During the 22-year period ending in 1985, Iroquois Brands Ltd. increased in price (adjusted in the usual manner for dividends) by a factor of 8.9151, while Kin Ark increased in price by a factor of 4.1276, as shown in Figure 8.1.

Prior knowledge (in 1963) of this information would have enabled an investor to buy and hold the best stock (Iroquois) and earn a 791% profit. However, a closer look at the time series reveals some cause for regret. Table 8.1 lists the performance of the constant rebalanced portfolios $\mathbf{b} = (b, 1-b)$. The graph of $S_n(b)$ is given in Figure 8.2. For example, reinvesting current wealth in the proportions $\mathbf{b} = (0.8, 0.2)$ at the start of each trading day would have resulted in an increase by a factor of 37.5. In fact, the best rebalanced portfolio for this 22-year period is $\mathbf{b}^* = (0.55, 0.45)$, yielding a factor $S_n^* = 73.619$. Here S_n^* is the target wealth (with respect to the coarse quantization of $B = [0, 1]$ we have chosen). The universal portfolio $\hat{\mathbf{b}}_k$ achieves a factor of $\hat{S}_n = 38.6727$. While $\hat{S}_n$ is short of the target, as it must be, $\hat{S}_n$ dominates the 8.9 and 4.1 factors of the constituent stocks. The daily performance of both stocks, the universal portfolio, and the target wealth are exhibited in Figure 8.3. The portfolio choice $\hat{b}_k$ as a function of time k is given in Figure 8.4.

To be explicit in the above analysis, we have quantized all integrals, resulting in the replacements of

$$S_n^* = \max_{b} S_n(b) \quad \text{by} \quad S_n^* = \max_{i=0,1,\ldots,20} S_n(i/20) \tag{8.1}$$

and

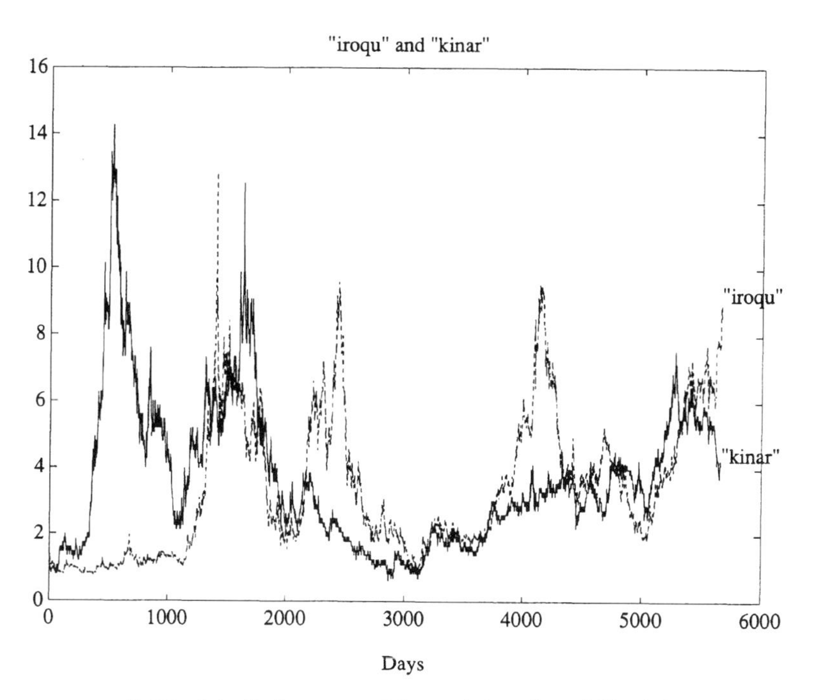

FIGURE 8.1. Performance of Iroquois brands and Kin Ark.

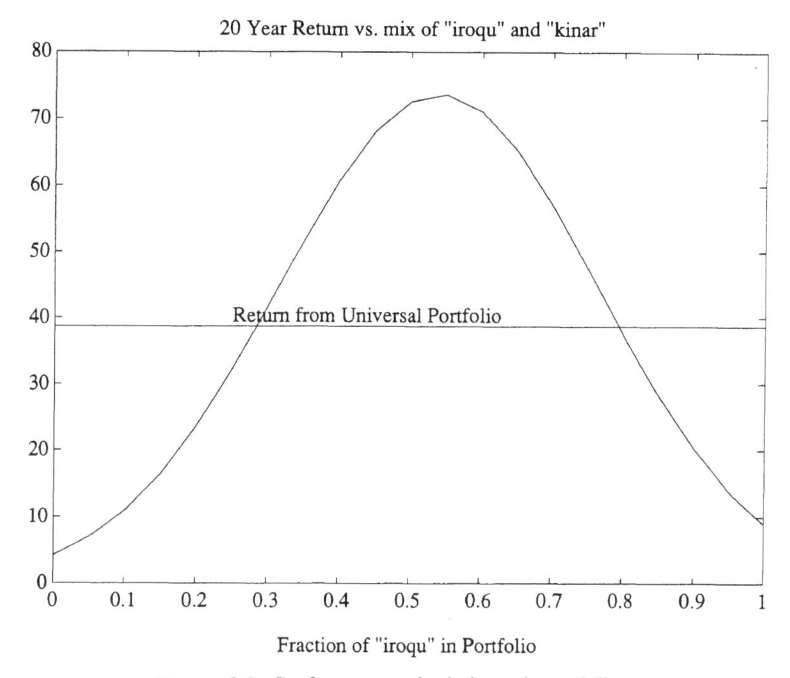

FIGURE 8.2. Performance of rebalanced portfolio.

TABLE 8.1
Iroquois Brands Ltd versus Kin Ark Corp

b	$S_n(\mathrm{b})$
1.00	8.9151
0.95	13.7712
0.90	20.2276
0.85	28.2560
0.80	37.5429
0.75	47.4513
0.70	57.0581
0.65	65.2793
0.60	71.0652
0.55	73.6190
0.50	72.5766
0.45	68.0915
0.40	60.7981
0.35	51.6645
0.30	41.7831
0.25	32.1593
0.20	23.5559
0.15	16.4196
0.10	10.8910
0.05	6.8737
0.00	4.1276

Target wealth: $S_n^* = 73.619$
Best rebalanced portfolio: $\mathrm{b}_n^* = 0.55$
Best constituent stock: 8.915
Universal wealth: $\hat{S}_n = 38.6727$

(8.2)

$$\hat{\mathrm{b}}_{k+1} = \int_0^1 \mathrm{b}S_k(\mathrm{b})\,d\mathrm{b} \Big/ \int_0^1 S_k(\mathrm{b})\,d\mathrm{b} \quad \text{by} \quad \hat{\mathrm{b}}_{k+1} = \sum_{i=0}^{20} \frac{i}{20} S_k\left(\frac{i}{20}\right) \Big/ \sum_{i=0}^{20} S_k\left(\frac{i}{20}\right).$$

The resulting wealth factor

$$\hat{S}_n = \prod_{k=1}^{n} \hat{\mathbf{b}}_k^t \mathbf{x}_k \tag{8.3}$$

is calculated using

$$\hat{\mathrm{b}}_k = \sum_{i=0}^{20} \frac{i}{20} S_k\left(\frac{i}{20}\right) \Big/ \sum_{i=0}^{20} S_k\left(\frac{i}{20}\right). \tag{8.4}$$

Telescoping still takes place under this quantization and it can be verified that $\hat{S}_n$ in (8.3) can be expressed in the equivalent form

$$\hat{S}_n = \frac{1}{21} \sum_{i=0}^{20} S_n\left(\frac{i}{20}\right). \tag{8.5}$$

Thus $\hat{S}_n$ is the arithmetic average of the wealths associated with the constant rebalanced portfolios.

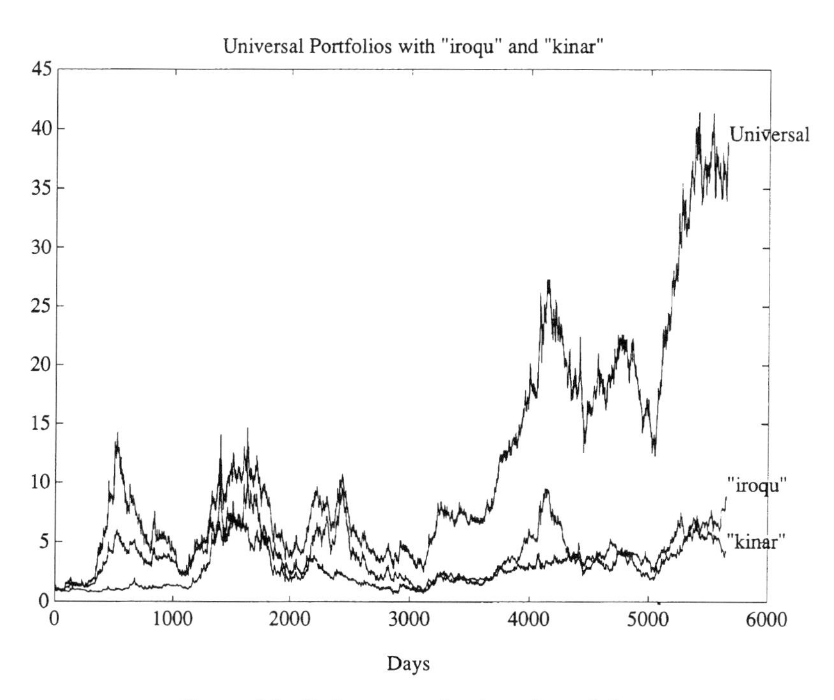

FIGURE 8.3. Performance of universal portfolio.

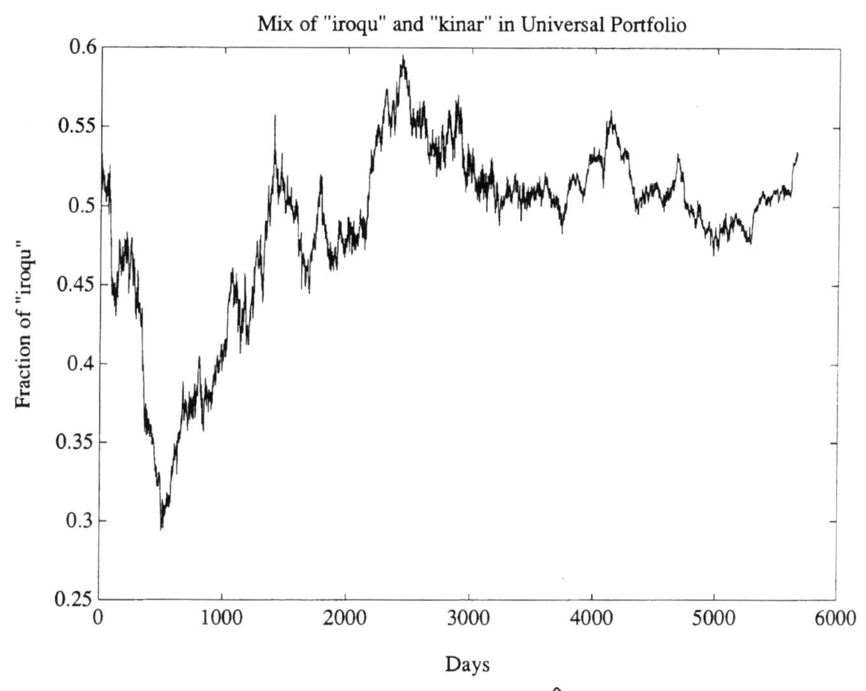

FIGURE 8.4. The portfolio $\hat{b}_k$.

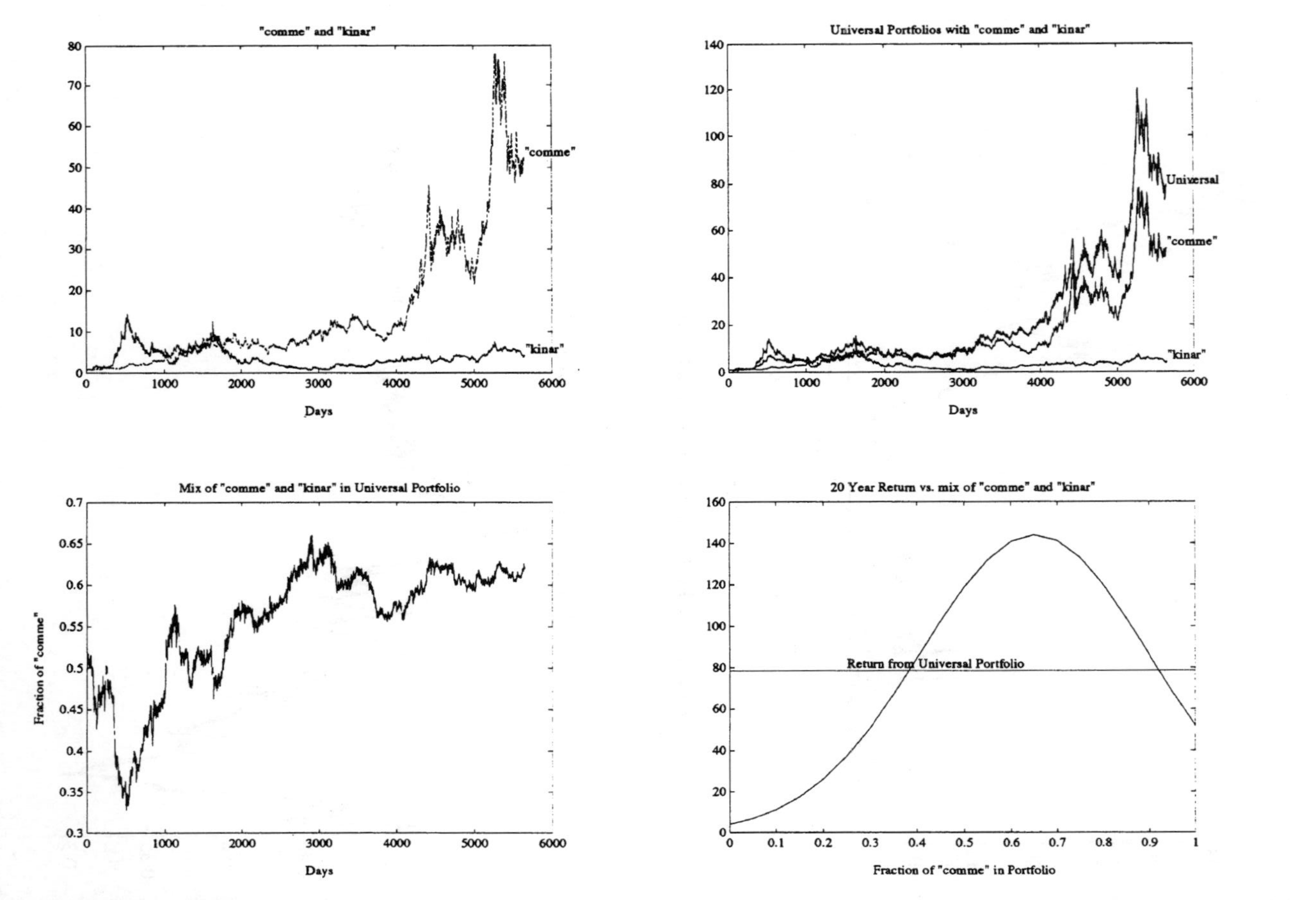

FIGURE 8.5. Commercial Metals and Kin Ark; Performance of Universal Portfolio; Universal Portfolio; Performance of Rebalanced Portfolio.

Finally, note the calculation of the portfolio $\hat{\mathbf{b}}_{n+1} = (\hat{b}_{n+1}, 1 - \hat{b}_{n+1})$ in this example. Merely compute the inner product of the b and $S_n(b)$ columns in Table 8.1 and divide by the sum of the $S_n(b)$ column to obtain $\hat{b}_{n+1}$. Note in particular that the universal portfolio $\hat{\mathbf{b}}_{n+1}$ is not equal to the log optimal portfolio $\mathbf{b}^*(F_n) = (0.55, 0.45)$ with respect to the empirical distribution of the past.

A similar analysis can be performed on Commercial Metals and Kin Ark over the same period. Here Commercial Metals increased by the factor 52.0203 and Kin Ark by the factor 4.1276 (Figure 8.5). It seems that an investor would not want any part of Kin Ark with an alternative like Commercial Metals available. Not so. The optimal constant rebalanced portfolio is $\mathbf{b}^* = (0.65, 0.35)$, and the universal portfolio achieves $\hat{S}_n = 78.4742$, outperforming each stock. See Table 8.2.

Next we put Commercial Metals (52.0203) up against Mei Corp (22.9160). Here $S_n^* = 102.95$ and $\hat{S}_n = 72.6289$, as shown in Figure 8.6 and Table 8.3. However, IBM and Coca-Cola show a lockstep performance, and, indeed, $\hat{S}_n$ barely outperforms them, as shown in Figure 8.7.

A final example crudely models buying on 50% margin. Suppose we have four investment choices each day: Commercial Metals, Kin Ark, and these same two stocks on 50% margin. Margin loans are settled daily at a 6% annual interest rate. The stock vector on the ith day is

$$\mathbf{x}_i = (x_i, 2x_i - 1 - r, y_i, 2y_i - 1 - r), \tag{8.6}$$

TABLE 8.2

Commercial Metals versus Kin Ark

b	$S_n(b)$
1.00	52.0203
0.95	68.2890
0.90	85.9255
0.85	103.6415
0.80	119.8472
0.75	132.8752
0.70	141.2588
0.65	144.0035
0.60	140.7803
0.55	131.9910
0.50	118.6854
0.45	102.3564
0.40	84.6655
0.35	67.1703
0.30	51.1127
0.25	37.3042
0.20	26.1131
0.15	17.5315
0.10	11.2883
0.05	6.9704
0.00	4.1276

Target wealth: $S_n^* = 144.0035$
Best rebalanced portfolio: $b_n^* = 0.65$
Best constituent stock: 52.0203
Universal wealth: $\hat{S}_n = 78.4742$

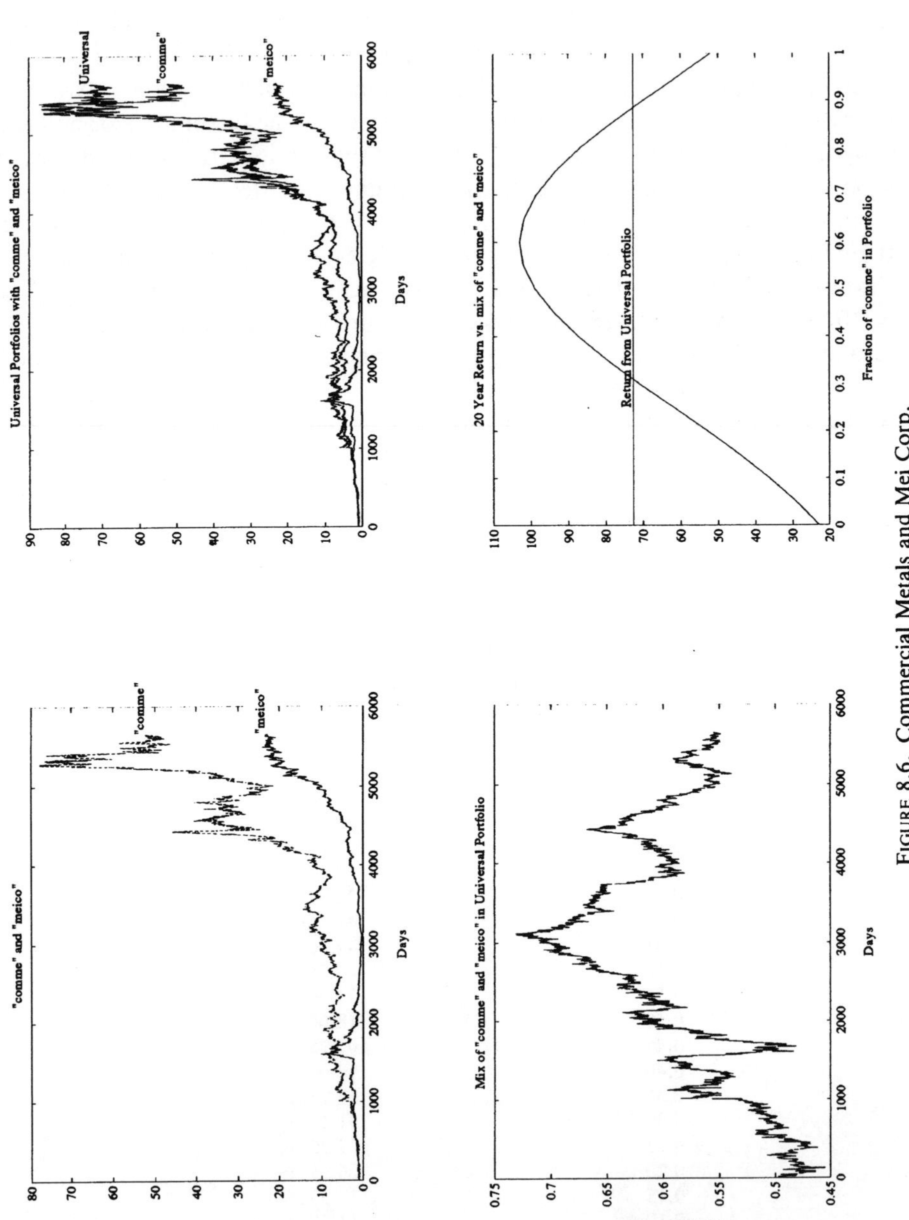

FIGURE 8.6. Commercial Metals and Mei Corp.

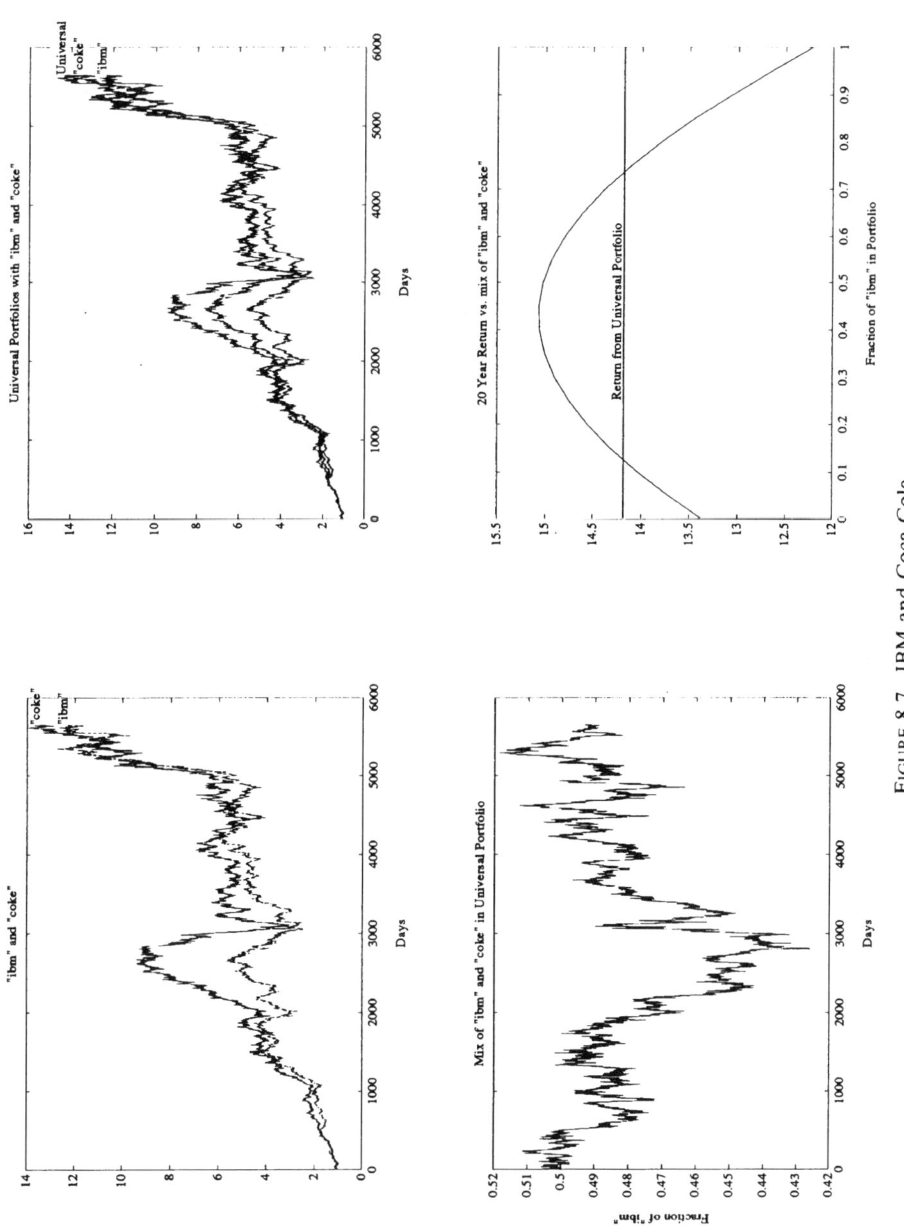

FIGURE 8.7. IBM and Coca-Cola.

TABLE 8.3

Commercial Metals versus Mei Corp

b	S_n(b)
1.00	52.0203
0.95	61.0165
0.90	70.0625
0.85	78.7602
0.80	86.6815
0.75	93.4026
0.70	98.5414
0.65	101.7927
0.60	102.9589
0.55	101.9691
0.50	98.8869
0.45	93.9033
0.40	87.3172
0.35	79.5057
0.30	70.8890
0.25	61.8932
0.20	52.9162
0.15	44.3012
0.10	36.3178
0.05	29.1538
0.00	22.9160

Target wealth: $S_n^* = 102.9589$
Best rebalanced portfolio: $b_n^* = 0.60$
Best constituent stock: 52.0203
Universal wealth: $\hat{S}_n = 72.6289$

$$r = 0.000233, \tag{8.7}$$

where x_i and y_i are the respective price relatives for Commercial Metals and Kin Ark on day i. Plunging on margin into Commercial Metals yields a factor 19.73, plunging into Kin Ark a factor of 0.0 (to four significant digits). Good as these stocks are, they cannot survive the down factors induced by the leverage. But a random sample of the simplex of portfolios listed in Table 8.4 reveals $\hat{S}_n = 98.4240$, while the optimal rebalanced portfolio $\mathbf{b}^* = (0.2, 0.5, 0.1, 0.2)$ results in a factor $\hat{S}_n^* = 262.4021$. Clearly 98.4 beats the factor of 78 achieved when margin is unavailable. Both factors exceed the performance 52.02 of the best stock.

We observe that $\hat{S}_n = 98.4$ exceeds the factor $\hat{S}_n = 78.47$ obtained for these stocks when margin is unavailable. This is borne out by the fact that $\mathbf{b}^*$ is positive in each component, calling for a small amount of leverage in the a posteriori optimal rebalanced portfolio.

9. THE GENERAL UNIVERSAL PORTFOLIO

If the best rebalanced portfolio $\mathbf{b}_n^*$ lies in the interior of a boundary k-face, then only k stocks are active in the best rebalanced portfolio. Thus we expect to obtain the previous bounds on $\hat{S}_n/S_n^*$ with m replaced by k. This is achieved if we start

TABLE 8.4

Two Stocks with Margin

Commercial metals	$52.0203 = \Pi_{i=1}^{n} x_i$
Commercial metals on margin	$19.7335 = \Pi_{i=1}^{n} (2x_i - 1 - r)$,
Kin Ark	$4.1276 = \Pi_{i=1}^{n} y_i$
Kin Ark on margin	$0.0000 = \Pi_{i=1}^{n} (2y_i - 1 - r)$
$r = 0.000233$/day = 6%/year	
$S_n^* = 262.4021$	$\mathbf{b}_n^* = (0.2, 0.5, 0.1, 0.2)$
Best constituent stock	52.0203
Wealth achieved by universal portfolio	$\hat{S}_n = 98.4240$

b	$S_n(\mathbf{b})$
(0.8, 0.2, 0.0, 0.0)	57.0535
(0.8, 0.1, 0.0, 0.1)	148.9951
(0.6, 0.1, 0.1, 0.2)	207.1143
(0.6, 0.0, 0.4, 0.0)	140.7803
(0.5, 0.0, 0.2, 0.3)	60.8358
(0.4, 0.0, 0.4, 0.2)	47.6074
(0.3, 0.5, 0.1, 0.1)	212.8928
(0.3, 0.4, 0.1, 0.2)	261.0452
(0.3, 0.2, 0.2, 0.3)	89.0330
(0.3, 0.1, 0.2, 0.4)	19.4840
(0.3, 0.0, 0.1, 0.6)	0.7700
(0.2, 0.7, 0.0, 0.1)	121.0142
(0.2, 0.2, 0.3, 0.3)	45.2562
(0.1, 0.8, 0.1, 0.0)	67.5882
(0.1, 0.5, 0.2, 0.2)	233.6328
(0.1, 0.4, 0.2, 0.3)	112.6695
(0.1, 0.3, 0.1, 0.5)	12.7702
(0.1, 0.2, 0.4, 0.3)	19.4840
(0.1, 0.1, 0.2, 0.6)	0.2354
(0.0, 0.5, 0.4, 0.1)	225.2524
(0.0, 0.4, 0.2, 0.4)	31.8076
(0.2, 0.5, 0.1, 0.2)	262.4021

with some mass on each face. To accomplish this, we let μ_S be the measure corresponding to the uniform distribution on $B(S) = \{\mathbf{b} \in R^m : \Sigma\, b_i = 1, b_i = 0, i \in S^c\}$, where $S \subseteq \{1, 2, \ldots, m\}$. Thus μ_S puts unit mass on the $|S|$-dimensional face of the portfolio simplex.

Let μ be the mixture of these measures given by

$$\mu = \frac{1}{2^m - 1} \sum \mu_S \tag{9.1}$$

where the sum is over all $S \neq \emptyset$, $S \subseteq \{1, 2, \ldots, m\}$. The generalized universal portfolio now becomes

$$\hat{\mathbf{b}}_{n+1} = \int \mathbf{b} S_n(\mathbf{b})\, \mu(d\mathbf{b}) \Big/ \int S_n(\mathbf{b})\, \mu(d\mathbf{b}) \tag{9.2}$$

with

$$S_n(\mathbf{b}) = \prod_{i=1}^{n} \mathbf{b}' \mathbf{x}_i, \qquad S_0(\mathbf{b}) = 1. \tag{9.3}$$

To state the results we define $J_n^{(k)}(F_n)$ to be the $k \times k$ sensitivity matrix with respect to the active stocks S, $|S| = k$, where S is the smallest set of stocks such that all optimal rebalanced portfolios $\mathbf{b}^*(F)$ are in the interior of $B(S)$. Then

$$\frac{\hat{S}_n}{S_n^*} \sim \frac{(k-1)!}{2^m - 1} \left(\frac{2\pi}{n}\right)^{k-1/2} \Big/ |J_n^{(k)}(F_n)|^{1/2} \tag{9.4}$$

will be the asymptotic behavior of $\hat{S}_n / S_n^*$.

10. CONCLUDING REMARKS

We now try to be sensible and ask how the universal portfolio works in practice. Of course, the examples are encouraging, as the universal portfolio outperforms the constituent stocks. However, we have ignored trading costs. In practice we would not trade daily, but only when the current empirical holdings were far enough from the recommended $\hat{\mathbf{b}}_k$. (A rule of thumb might be to trade only if the increase in W is greater than the logarithm of the normalized transaction costs.)

We are really interested in whether $\hat{S}_n$ will "take off," leaving the stocks behind. We first discuss the target wealth S_n^*. The best rebalanced portfolio $\mathbf{b}^*(F_n)$ based on prior knowledge of the stock sequence $\mathbf{x}_1, \mathbf{x}_2, \ldots, \mathbf{x}_n$ yields wealth $S_n^* = e^{nW_n^*}$. Now S_n^* grows exponentially fast to infinity under mild conditions. For example, if one of the constituent stocks is a risk-free asset with interest rate $r > 0$, then $W_n^* \geq \ln(1 + r) > 0$, for all n, and $S_n^* \geq (1 + r)^n \to \infty$. Since the universal portfolio yields

$$\hat{S}_n = \exp\left\{ n \left(W_n^* - O\left(\frac{\ln n}{n}\right)\right)\right\}, \tag{10.1}$$

it follows that $\hat{S}_n$ will tend to infinity, and $\hat{S}_n$ will have the same exponent as S_n^*, differing only in terms of order $(\ln n)/n$.

What state of affairs do we expect in the real world? Certainly we expect the stock sequence to be of full dimension m for n slightly greater than m. However, we do not expect all stocks to be active. But we do expect that two or more stocks will be active. This is important because it guarantees that the target growth rate W_n^* will be strictly greater than the growth rate of the constituent stocks. Consequently, we believe that the universal portfolio will achieve

$$\hat{S}_n / S_n(\mathbf{e}_i) \to \infty, \qquad i = 1, 2, \ldots, m,$$

exponentially fast, where $S_n(\mathbf{e}_i)$ is the wealth relative of the ith stock at time n.

However, n may need to be quite large before this exponential dominance manifests itself. In particular, we need n large enough that the difference in exponents between S_n^* and the stocks overcomes the $O((\ln n)/n)$ penalties incurred by universality. We conclude that $\hat{S}_n$ will leave the constituent stocks exponentially behind if there are at least two strictly active stocks in the best rebalanced portfolio.

REFERENCES

ALGOET, P., AND T. M. COVER (1988): "Asymptotic Optimality and Asymptotic Equipartition Properties of Log-Optimum Investment," *Ann. Probab.*, 16, 876-898.

ARROW, K. (1974): *Essays in the Theory of Risk Bearing.* Amsterdam North-Holland; New York: American Elsevier.

BARRON, A., AND T. M. COVER (1988): "A Bound on the Financial Value of Information," *IEEE Trans. Inform. Theory*, 34, 1097-1100.

BELL, R., AND T. M. COVER (1980): "Competitive Optimality of Logarithmic Investment," *Math. Oper. Res.*, 5, 161-166.

BELL, R., AND T. M. COVER (1988): "Game-Theoretic Optimal Portfolios," *Management Sci.*, 34, 724-733.

BLACKWELL, D. (1956a): "Controlled Random Walks," in *Proc. Int. Congress Math.* III, Amsterdam: North-Holland, 336-338.

BLACKWELL, D. (1956b): "An Analog of the Minimax Theorem for Vector Payoffs," *Pacific J. Math.* 6, 1-8.

BREIMAN, L. (1961): "Optimal Gambling Systems for Favorable Games." in *Fourth Berkeley Symp. on Mathematical Statistics and Probability*, 1, 65-78.

COVER, T. M. (1984): "An Algorithm for Maximizing Expected Log Investment Return," *IEEE Trans. Inform. Theory*, IT-30, 369-373.

COVER, T. M., AND D. GLUSS (1986): "Empirical Bayes Stock Market Portfolios," *Adv. Appl. Math.*, 7, 170-181. (A summary also appears in *Proceedings of Conference Honoring Herbert Robbins*, Springer-Verlag, 1986.)

COVER, T. M., AND R. KING (1978): "A Convergent Gambling Estimate of the Entropy of English," *IEEE Trans. Inform. Theory*, 24, 413-421.

FINKELSTEIN, M., AND R. WHITLEY (1981): "Optimal Strategies for Repeated Games," *Adv. Appl. Probab.*, 13, 415-428.

HAKANSSON, N. (1979): "A Characterization of Optimal Multiperiod Portfolio Policies," in *Portfolio Theory, 25 Years After: Essays in Honor of Harry Markowitz,* ed. E. Elton and M. Gruber. TIMS Studies in the Management Sciences, 11, Amsterdam; North-Holland, 169-177.

HANNAN, J. F., AND H. ROBBINS (1955): "Asymptotic Solutions of the Compound Decision Problem," *Ann. Math. Stat.*, 16, 37-51.

KELLY, J. L. (1956): "A New Interpretation of Information Rate," *Bell Systems Tech. J.*, 917-926.

MARKOWITZ, H. (1952): "Portfolio Selection," *J. Finance*, 8, 77-91.

MARKOWITZ, H. (1976): "Investment for the Long Run: New Evidence for an Old Rule," *J. Finance*, 31, 1273-1286.

MERTON, R. C., AND P. A. SAMUELSON (1974): "Fallacy of the Log-Normal Approximation to Optimal Portfolio Decision-Making over Many Periods," *J. Financial Econ.*, 1, 67-94.

MOSSIN, J. (1968): "Optimal Multiperiod Portfolio Policies," *J. Business*, 41, 215-229.

ROBBINS, H. (1951): "Asymptotically Subminimax Solutions of Compound Statistical Decision Problems," *Proc. Second Berkeley Symp. on Mathematical Statistics and Probability*, Berkeley: University of California Press.

SAMUELSON, P. A. (1967): "General Proof that Diversification Pays," *J. Financial Quant. Anal. II*, 1-13.

SAMUELSON, P. A. (1969): "Lifetime Portfolio Selection by Dynamic Stochastic Programming," *Rev. Econ. Statist.*, 239-246.

SAMUELSON, P. A. (1979): "Why We Should Not Make Mean Log of Wealth Big Though Years to Act Are Long," *J. Banking Finance*, 3, 305-307.

SHARPE, W. F. (1963); "A Simplified Model for Portfolio Analysis," *Management Sci.*, 9, 277-293.

THORP, E. (1971): "Portfolio Choice and the Kelly Criterion," *Business Econ. Statist. Proc. Amer. Stat. Assoc.*, 215-224.

MATHEMATICS OF OPERATIONS RESEARCH
Vol. 23, No. 4, November 1998
Printed in U.S.A.

16

THE COST OF ACHIEVING THE BEST PORTFOLIO IN HINDSIGHT

ERIK ORDENTLICH AND THOMAS M. COVER

For a market with m assets consider the minimum, over all possible sequences of asset prices through time n, of the ratio of the final wealth of a nonanticipating investment strategy to the wealth obtained by the best constant rebalanced portfolio computed in hindsight for that price sequence. We show that the maximum value of this ratio over all nonanticipating investment strategies is $V_n = [\Sigma\, 2^{-nH(n_1/n,\dots,n_m/n)}(n!/(n_1!\cdots n_m!))]^{-1}$, where $H(\cdot)$ is the Shannon entropy, and we specify a strategy achieving it. The optimal ratio V_n is shown to decrease only polynomially in n, indicating that the rate of return of the optimal strategy converges uniformly to that of the best constant rebalanced portfolio determined with full hindsight. We also relate this result to the pricing of a new derivative security which might be called the hindsight allocation option.

1. Introduction. Hindsight is not available when it is most useful. This is true in investing where hindsight into market performance makes obvious how one should have invested all along. In this paper we investigate the extent to which a nonanticipating investment strategy can achieve the performance of the best strategy determined in hindsight.

Obviously, with hindsight, the best investment strategy is to shift one's wealth daily into the asset with the largest percentage increase in price. Unfortunately, it is hopeless to match the performance of this strategy in any meaningful way, and therefore we must restrict the class of investment strategies over which the hindsight optimization is performed. Here we focus on the class of investment strategies called the constant rebalanced portfolios. A constant rebalanced portfolio rebalances the allocation of wealth among the available assets to the same proportions each day. Using all wealth to buy and hold a single asset is a special case. Therefore the best constant rebalanced portfolio, at the very least, outperforms the best asset.

In practice, one would expect the wealth achieved by the best constant rebalanced portfolio computed in hindsight to grow exponentially with a rate determined by asset price drift and volatility. Even if the prices of individual assets are going nowhere in the long run, short-term fluctuations in conjunction with constant rebalancing may lead to substantial profits. Furthermore, the best constant rebalanced portfolio will in all likelihood exponentially outperform any fixed constant rebalanced portfolio which includes buying and holding the best asset in hindsight.

The intuition that the best constant rebalanced portfolio is a good performance target is motivated by the well-known fact that if market returns are independent and identically distributed from one day to the next, the expected utility, for a wide range of utility functions including the log utility, is maximized by a constant rebalanced portfolio strategy. Additionally, "turnpike" theory (see Huberman and Ross 1983, Cox and Huang 1992, and references therein) finds an even broader class of utility functions for which, by virtue of their behavior at large wealths, constant rebalancing becomes optimal as the investment horizon tends to infinity. In all these settings, the optimal constant rebalanced

Received October 3, 1996; revised September 2, 1997; November 27, 1997.
AMS 1991 subject classification. Primary: 90A09.
OR/MS Index subject classification. Primary: Finance/Portfolio/Investment.
Key words. Portfolio selection, asset allocation, derivative security, optimal investment.

0364-765X/98/2304/0960/$05.00

portfolio depends on the underlying distribution, which is unknown in practice. Targeting the best constant rebalanced portfolio computed in hindsight for the actual market sequence is one way of dealing with this lack of information.

The question is: To what extent can a nonanticipating investment strategy perform as well as the best constant rebalanced portfolio determined in hindsight? We address this question from a *distribution-free*, worst-sequence perspective with no restrictions on asset price behavior. Asset prices can increase or decrease arbitrarily, even drop to zero. We assume no underlying randomness or probability distribution on asset price changes.

The analysis is best expressed in terms of a contest between an investor and nature. After the investor has selected a nonanticipating investment strategy, nature, with full knowledge of the investor's strategy (and its dependence on the past), selects that sequence of asset price changes which minimizes the ratio of the wealth achieved by the investor to the wealth achieved by the best constant rebalanced portfolio computed in hindsight for the selected sequence. The investor selects an investment strategy that maximizes the minimum ratio. In the main part of the paper we determine the optimum investment strategy and compute the max-min value of the ratio of wealths.

It may seem that such an analysis is overly pessimistic and risk averse since in reality there is no deliberate force trying to minimize investment returns. What is striking, however, is that if investment performance is measured in terms of rate of return or exponential growth rate per investment period, even this pessimistic point of view yields a favorable result. More specifically, the main result of this paper is the identification of an investment algorithm that achieves wealth $\hat{S}_n$ at time n that satisfies

$$\hat{S}_n \geq S_n^* \Big/ \sum_{\sum n_i = n} \binom{n}{n_1, \ldots, n_m} 2^{-nH(n_1/n, \ldots, n_m/n)} = S_n^* V_n, \tag{1}$$

for every market sequence, where S_n^* is the wealth achieved by the best constant rebalanced portfolio in hindsight, and $H(p_1, \ldots, p_m) = -\sum p_j \log p_j$ is the Shannon entropy function.

Since it can be shown that $V_n \sim \sqrt{2/(\pi n)}$ (for $m = 2$ assets), this factor, the price of universality, will not affect the exponential growth rate of wealth of $\hat{S}_n$ relative to S_n^*, i.e., $\lim \inf(1/n) \log(\hat{S}_n/S_n^*) \geq 0$. In other words, the rate of return achieved by the optimal strategy converges over time to that of the best constant rebalanced portfolio computed in hindsight, uniformly for every sequence of asset price changes. The bound (1) is the best possible; there are sequences of price changes that hold $\hat{S}_n/S_n^*$ to this bound for any nonanticipating investment strategy.

The problem of achieving the best portfolio in hindsight leads naturally to the consideration of a new derivative security which might be called the hindsight allocation option. The hindsight allocation option has a payoff at time n equal to S_n^*, the wealth earned by investing one dollar according to the best constant rebalanced portfolio (the best constant allocation of wealth) computed in hindsight for the observed stock and bond performance. This option might, for example, interest investors who are uncertain about how to allocate their wealth between stocks and bonds. By purchasing a hindsight allocation option, an investor achieves the performance of the best constant allocation of wealth determined with full knowledge of the actual market performance.

In §4 we argue that the max-min ratio computed above yields a tight upper bound on the price of this option. Specifically, Equation (1) suggests that $\hat{S}_n$ is an arbitrage opportunity if the option price is more than $1/V_n$. We compare this bound to the no-arbitrage option price for two well-known models of market behavior, the discrete time binomial lattice model and the continuous time geometric Wiener model. We consider only the simple case of a volatile stock and a bond with a constant rate of return. It is shown that the no-arbitrage prices for these restricted market models have essentially the same asymptotic $c\sqrt{n}$ behavior as the upper bound $1/V_n$. Different model parameter choices (volatility, interest rate) can yield more favorable constants c.

The pricing of the hindsight allocation option in the binomial and geometric Wiener models can also be thought of in terms of the max-min framework. The models can be viewed as constraints on nature's choice of asset price changes. The underlying distribution in the geometric Wiener model serves as a technical device for constraining the set of continuous asset price paths from which nature can choose. Because these markets are complete for the special case of one stock and one bond, the best constant rebalanced portfolio computed in hindsight can be hedged perfectly given a unique initial wealth. This wealth corresponds to the no-arbitrage price of the hindsight allocation option. Furthermore, the max-min ratio of wealths obtained by the investor and nature, when nature is constrained by these models, must be the reciprocal of this unique initial wealth.

Early work on universal portfolios (portfolio strategies performing uniformly well with respect to constant rebalanced portfolios) can be found in Cover and Gluss (1986), Larson (1986), Cover (1991), Merhav and Feder (1993), and Cover and Ordentlich (1996).

Cover and Gluss (1986) restrict daily returns to a finite set and provide an algorithm, based on the approachability-excludability theorem of Blackwell (1956a, 1956b), that achieves a wealth ratio $\hat{S}_n/S_n^* \geq e^{-c\sqrt{n}}$, for $m = 2$ stocks, where c is a positive constant. Larson (1986), also restricting daily returns to a finite set, uses a compound Bayes approach to achieve $\hat{S}_n/S_n^* = e^{-\delta n}$, for arbitrarily small $\delta > 0$. Cover (1991) defines a family of μ-weighted universal portfolios and uses Laplace's method of integration to show, for a bounded ratio of maximum to minimum daily asset returns, that $\hat{S}_n/S_n^* \geq c_n/n^{(m-1)/2}$ for m stocks, where c_n is the determinant of a certain sensitivity matrix measuring the empirical volatility of the price sequence. Merhav and Feder (1993) establish polynomial bounds on $\hat{S}_n/S_n^*$ under the same constraints.

The first individual sequence (worst-case) analysis of the universal portfolio of Cover (1991) is given in Cover and Ordentlich (1996), where it is shown that a Dirichlet$(\frac{1}{2})$ weighted universal portfolio achieves a worst case performance of $\hat{S}_n/S_n^* \geq c/n^{(m-1)/2}$. This analysis is also extended to investment with side information, with similar results. Jamshidian (1992) applies the universal portfolio of Cover (1991) (with μ uniform) to a geometric Wiener market, establishing the asymptotic behavior of $\hat{S}(t)/S^*(t)$, and showing $(1/t)\log \hat{S}(t)/S^*(t) \to 0$, for such markets.

The paper is organized as follows. Section 2 establishes notation and some basic definitions. The individual-sequence performance and game-theoretic analysis are established in §3. Section 4 contains the hindsight allocation option pricing analysis.

2. Notation and definitions. We represent the behavior of a market of m assets for n trading periods by a sequence of nonnegative, nonzero (at least one nonzero component) price-relative vectors $\mathbf{x}_1, \ldots, \mathbf{x}_n \in \mathbb{R}_+^m$. We refer to $\mathbf{x}^n = \mathbf{x}_1, \ldots, \mathbf{x}_n$ as the market sequence. The jth component of the ith vector denotes the ratio of closing to opening price of the jth asset for the ith trading period. Thus an investment in asset j on day i increases by a factor of x_{ij}.

Investment in the market is specified by a portfolio vector $\mathbf{b} = (b_1, \ldots, b_m)'$ with nonnegative entries summing to one. That is, $\mathbf{b} \in \mathcal{B}$, where

$$\mathcal{B} = \left\{ \mathbf{b} : \mathbf{b} \in \mathbb{R}_+^m, \sum_{j=1}^{m} b_j = 1 \right\}.$$

A portfolio vector $\mathbf{b}$ denotes the fraction of wealth invested in each of the m assets. An investment according to portfolio $\mathbf{b}_i$ on day i multiplies wealth by a factor of

$$\mathbf{b}_i^t \mathbf{x}_i = \sum_{j=1}^{m} b_{ij} x_{ij}.$$

A sequence of n investments according to portfolio choices $\mathbf{b}_1, \ldots, \mathbf{b}_n$ changes wealth by a factor of

$$\prod_{i=1}^{n} \mathbf{b}_i^t \mathbf{x}_i .$$

A constant rebalanced portfolio investment strategy uses the same portfolio $\mathbf{b}$ for each trading day. Assuming normalized initial wealth $S_0 = 1$, the final wealth will be

$$S_n(\mathbf{x}^n, \mathbf{b}) = \prod_{i=1}^{n} \mathbf{b}^t \mathbf{x}_i .$$

For a sequence of price-relatives $\mathbf{x}^n$ it is possible to compute the best constant rebalanced portfolio $\mathbf{b}^*$ as

$$\mathbf{b}^* = \arg\max_{\mathbf{b} \in \mathcal{B}} S_n(\mathbf{x}^n, \mathbf{b}),$$

which achieves a wealth factor of

$$S_n^*(\mathbf{x}^n) = \max_{\mathbf{b} \in \mathcal{B}} S_n(\mathbf{x}^n, \mathbf{b}).$$

The best constant rebalanced portfolio $\mathbf{b}^*$ depends on knowledge of market performance for time $1, 2, \ldots, n$; it is not a nonanticipating investment strategy.

This brings up the definition of a nonanticipating investment strategy.

DEFINITION 1. A nonanticipating investment strategy is a sequence of maps

$$\mathbf{b}_i : \mathbb{R}_+^{m(i-1)} \to \mathcal{B}, \quad i = 1, 2, \cdots$$

where

$$\mathbf{b}_i = \mathbf{b}_i(\mathbf{x}_1, \ldots, \mathbf{x}_{i-1})$$

is the portfolio used on day i given past market outcomes $\mathbf{x}^{i-1} = \mathbf{x}_1, \ldots, \mathbf{x}_{i-1}$.

3. Worst-case analysis. We now present the main result, a theorem characterizing the extent to which the best constant rebalanced portfolio computed in hindsight can be tracked in the worst case. Our analysis is best expressed in terms of a contest between an investor, who announces a nonanticipating investment strategy $\hat{\mathbf{b}}_i(\cdot)$, and nature, who, with full knowledge of the investor's strategy, selects a market sequence $\mathbf{x}^n = \mathbf{x}_1, \mathbf{x}_2, \ldots, \mathbf{x}_n$ to minimize the ratio of wealths $\hat{S}_n(\mathbf{x}^n)/S_n^*(\mathbf{x}^n)$, where $\hat{S}_n(\mathbf{x}^n)$ is the investor's wealth against sequence $\mathbf{x}^n$ and is given by

$$\hat{S}_n(\mathbf{x}^n) = \prod_{i=1}^{n} \hat{\mathbf{b}}_i^t(\mathbf{x}^{i-1}) \mathbf{x}_i .$$

Thus, nature attempts to induce poor performance on the part of the investor relative to the best constant rebalanced portfolio $\mathbf{b}^*$ computed with complete knowledge of $\mathbf{x}^n$. The investor, wishing to protect himself from this worst case, selects that nonanticipating investment strategy $\hat{\mathbf{b}}_i(\cdot)$ which maximizes the worst-case ratio of wealths.

THEOREM 1 (Max-min ratio). *For m assets and all n,*

$$\max_{\hat{\mathbf{b}}} \min_{\mathbf{x}^n} \frac{\hat{S}_n(\mathbf{x}^n)}{S_n^*(\mathbf{x}^n)} = V_n,$$

where

$$V_n = \left[\sum_{n_1 + \cdots + n_m = n} \binom{n}{n_1, \ldots, n_m} 2^{-nH(n_1/n, \ldots, n_m/n)} \right]^{-1}, \tag{2}$$

and

$$H(p_1, \ldots, p_m) = -\sum_{j=1}^{m} p_j \log p_j$$

is the Shannon entropy function.

REMARK. For $m = 2$, the value V_n is simply

$$V_n = \left(\sum_{k=0}^{n} \binom{n}{k} \left(\frac{k}{n}\right)^k \left(\frac{n-k}{n}\right)^{n-k} \right)^{-1}, \tag{3}$$

and it is shown in §3.2 that $2/\sqrt{n+1} \geq V_n \geq 1/(2\sqrt{n+1})$ for all n. Thus V_n behaves essentially like $1/\sqrt{n}$. For $m > 2$, $V_n \sim c(1/\sqrt{n})^{m-1}$.

REMARK. It is noted in §3.3 that

$$V_n \sim \frac{\Gamma\left(\frac{m}{2}\right)}{\sqrt{\pi}} \left(\frac{2}{n}\right)^{(m-1)/2},$$

in the sense that

$$\lim_{n \to \infty} \frac{V_n}{\Gamma\left(\frac{m}{2}\right) \Big/ \sqrt{\pi} \left(\frac{2}{n}\right)^{(m-1)/2}} = 1.$$

For $m = 2$ this reduces to $V_n \sim \sqrt{2/\pi}(1/\sqrt{n})$.

REMARK. The max-min optimal strategy for $m = 2$ will be specified in Equations (8)–(13). These equations are followed by an alternative definition of the optimal strategy in terms of extremal strategies.

We note that the negative logarithm of the max-min ratio of wealths given by Equation (2) also corresponds to the solution of a min-max pointwise redundancy problem in universal data compression theory. The pointwise redundancy problem was studied and solved by Shtarkov (1987) and earlier works referenced therein. A principal result of the present work, which is developed in greater information theoretic detail in Cover and Ordentlich (1996) and Ordentlich (1996), is that worst sequence market performance is bounded by worst sequence data compression.

The strategy achieving the maximum in Theorem 1, as developed in the proof below, depends on the horizon n. We note, however, that Cover and Ordentlich (1996) exhibits an infinite horizon investment strategy, the Dirichlet-weighted universal portfolio, denoted by $\hat{\mathbf{b}}^D(\cdot)$, which for $m = 2$ assets achieves a wealth ratio $\hat{S}_n^D(\mathbf{x}^n)/S_n^*(\mathbf{x}^n)$ satisfying

$$\min_{\mathbf{x}^n} \frac{\hat{S}_n^D(\mathbf{x}^n)}{S_n^*(\mathbf{x}^n)} \geq \frac{1}{\sqrt{2\pi}} V_n. \tag{4}$$

At time i, the Dirichlet-weighted universal portfolio investment strategy uses the portfolio

$$\hat{\mathbf{b}}_i^D = \hat{\mathbf{b}}_i^D(\mathbf{x}^{i-1}) = \frac{\int_{\mathcal{B}} \mathbf{b} S_{i-1}(\mathbf{b}, \mathbf{x}^{i-1}) d\mu(\mathbf{b})}{\int_{\mathcal{B}} S_{i-1}(\mathbf{b}, \mathbf{x}^{i-1}) d\mu(\mathbf{b})}, \quad i = 1, 2, \cdots \tag{5}$$

where

$$S_i(\mathbf{b}, \mathbf{x}^i) = S_i(\mathbf{b}) = \prod_{j=1}^{i} \mathbf{b}'\mathbf{x}_j, \quad \text{and} \quad S_0(\mathbf{b}, \mathbf{x}^0) = 1.$$

The measure μ on the portfolio simplex $\mathcal{B}$ is the Dirichlet(1/2, . . . , 1/2) prior with density

$$d\mu(\mathbf{b}) = \frac{\Gamma\left(\frac{m}{2}\right)}{\left[\Gamma\left(\frac{1}{2}\right)\right]^m} \left(1 - \sum_{j=1}^{m-1} b_j\right)^{-1/2} \prod_{j=1}^{m-1} b_j^{-1/2} d\mathbf{b},$$

$$\sum_{j=1}^{m-1} b_j \leq 1, \quad b_j \geq 0, \quad j = 1, \ldots, m-1,$$

where $\Gamma(\cdot)$ denotes the Gamma function. The running wealth factor achieved by the Dirichlet-weighted universal portfolio through each time n is

$$\hat{S}_n^D(\mathbf{x}^n) = \int_{\mathcal{B}} S_n(\mathbf{b}, \mathbf{x}^n) d\mu(\mathbf{b}).$$

Thus the max-min ratio can be achieved to within a factor of $\sqrt{2\pi}$ for all n by a single infinite horizon strategy. The bound (4) generalizes to $m > 2$ so that for each m, the worst-case wealth achieved by the Dirichlet-weighted universal portfolio is within a constant factor (independent of n) of V_n.

The significance of Theorem 1 can be appreciated by considering some naive choices for the optimum investor strategy $\hat{\mathbf{b}}$. Suppose, for $m = 2$ assets, that $\hat{\mathbf{b}}$ corresponds to investing half of the initial wealth in a buy-and-hold of asset 1 and the other half in a buy-and-hold of asset 2. In this case the first two portfolio choices are

$$\hat{\mathbf{b}}_1 = \left(\frac{1}{2}, \frac{1}{2}\right)' \quad \text{and} \quad \hat{\mathbf{b}}_2 = \left(\frac{x_{11}}{x_{11} + x_{12}}, \frac{x_{12}}{x_{11} + x_{12}}\right)'. \tag{6}$$

Since we are allowing nature to select arbitrary price-relative vector sequences, nature could set $\mathbf{x}_1 = (0, 2)'$ and $\mathbf{x}_2 = (2, 0)'$, in which case the investor using the split buy-and-hold strategy (6) goes broke after two days. On the other hand, for this two-day sequence, the best constant rebalanced portfolio is $\mathbf{b}^* = (1/2, 1/2)'$ and yields a wealth factor $S_2^*(\mathbf{x}_1, \mathbf{x}_2)$ of 1.

Suppose the investor instead opts to rebalance his wealth daily to the initial $(1/2, 1/2)$ proportions. Here $\hat{\mathbf{b}}_i$ is the constant rebalanced portfolio $\mathbf{b} = (1/2, 1/2)'$. If nature then chooses the sequence of price-relative vectors $\mathbf{x}^n = (2, 0)', (2, 0)', \ldots, (2, 0)'$ the investor earns a wealth factor $\hat{S}_n(\mathbf{x}^n) = 1$ while the best constant rebalanced portfolio $\mathbf{b}^* = (1, 0)'$ earns $S_n^*(\mathbf{x}^n) = 2^n$. The ratio $\hat{S}_n/S_n^*$ of these two wealths decreases exponentially in n while the max-min ratio V_n decreases only polynomially. In particular, the wealth achieved by the max-min optimal strategy is at least $2^n/(2\sqrt{n} + 1)$ for this sequence.

These two investment strategies are particularly naive. A more sophisticated scheme might start off with $\hat{\mathbf{b}}_1 = (1/2, 1/2)'$ and then use the best constant rebalanced portfolio for the observed past. This scheme, however, is also flawed, since if nature chooses $\mathbf{x}_1 = (1, 0)'$, the investor would use $\hat{\mathbf{b}}_2 = (1, 0)'$ the following day and then would go broke if nature set $\mathbf{x}_2 = (0, 1)'$. One might think of fixing this scheme by using a time varying mixture of the $(1/2, 1/2)$ portfolio and the best constant rebalanced portfolio for the past. However, this class of strategies also fails to achieve V_n.

We now proceed with the proof of Theorem 1. The following lemma is used. In the sequel we adopt the conventions that $a/0 = \infty$ if $a > 0$, and that $0/0 = 0$.

LEMMA 1. *If* $\alpha_1, \ldots, \alpha_n \geq 0$, $\beta_1, \ldots, \beta_n \geq 0$, *then*

$$\frac{\sum_{i=1}^n \alpha_i}{\sum_{i=1}^n \beta_i} \geq \min_j \frac{\alpha_j}{\beta_j}. \tag{7}$$

PROOF OF LEMMA 1. Let

$$J = \arg\min_j \frac{\alpha_j}{\beta_j}.$$

The lemma is trivially true if $\alpha_J = 0$ since the right side of (7) is zero. So assume $\alpha_J > 0$. Then, if $\beta_J = 0$ the lemma is true since both the left and right sides of (7) are infinity. Therefore assume $\alpha_J > 0$ and $\beta_J > 0$. Then

$$\frac{\sum_{j=1}^n \alpha_j}{\sum_{j=1}^n \beta_j} = \frac{\alpha_J\left(1 + \sum_{j \neq J} \frac{\alpha_j}{\alpha_J}\right)}{\beta_J\left(1 + \sum_{j \neq J} \frac{\beta_j}{\beta_J}\right)} \geq \frac{\alpha_J}{\beta_J}$$

because

$$\frac{\alpha_j}{\beta_j} \geq \frac{\alpha_J}{\beta_J}$$

which implies

$$\frac{\alpha_j}{\alpha_J} \geq \frac{\beta_j}{\beta_J}$$

for all j. □

PROOF OF THEOREM 1. For ease of exposition we prove the theorem for $m = 2$. The generalization of the argument to $m > 2$ is straightforward.

Thus, for the case of $m = 2$ we must show that

$$\max_{\hat{\mathbf{b}}} \min_{\mathbf{x}^n} \frac{\hat{S}_n(\mathbf{x}^n)}{S_n^*(\mathbf{x}^n)} = V_n,$$

where

$$V_n = \left(\sum_{k=0}^{n} \binom{n}{k} \left(\frac{k}{n}\right)^k \left(\frac{n-k}{n}\right)^{n-k} \right)^{-1}.$$

We prove that

$$\max_{\hat{\mathbf{b}}} \min_{\mathbf{x}^n} \frac{\hat{S}_n(\mathbf{x}^n)}{S_n^*(\mathbf{x}^n)} \geq V_n,$$

by explicitly specifying the max-min optimal strategy $\hat{\mathbf{b}}$. We define the strategy by keeping track of the indices of the terms in the product $\prod_{i=1}^n \hat{\mathbf{b}}_i^t \mathbf{x}_i$. For sequences $j^n \in \{1, 2\}^n$ let $n_1(j^n)$ and $n_2(j^n)$ denote, respectively, the number of 1's and the number of 2's in j^n. That is, if $j^n = (j_1, \ldots, j_n)$,

$$n_r(j^n) = \sum_{i=1}^{n} I(j_i = r), \tag{8}$$

where $I(\cdot)$ is the indicator function. Let

$$w(j^n) = V_n \left(\frac{n_1(j^n)}{n}\right)^{n_1(j^n)} \left(\frac{n_2(j^n)}{n}\right)^{n_2(j^n)}. \tag{9}$$

Then, since $\sum_{j^n \in \{1,2\}^n} w(j^n) = 1$, $w(j^n)$ is a probability measure on the set of sequences $j^n \in \{1, 2\}^n$. For $l < n$, let

$$w(j^l) = \sum_{j_{l+1},\ldots,j_n} w(j^l, j_{l+1}, \ldots, j_n) \tag{10}$$

be the marginal probability mass of $j_1, \ldots, j_l$. This marginal probability may also be denoted by $w(j^{l-1}, j_l)$. Finally, define the nonanticipating investment strategy $\hat{\mathbf{b}}_l = \hat{b}_{l1}, \hat{b}_{l2})^t$

$$\hat{b}_{l1}(\mathbf{x}^{l-1}) = \frac{\sum_{j^{l-1} \in \{1,2\}^{l-1}} w(j^{l-1}, 1) \prod_{i=1}^{l-1} x_{ij_i}}{\sum_{j^{l-1} \in \{1,2\}^{l-1}} w(j^{l-1}) \prod_{i=1}^{l-1} x_{ij_i}}, \tag{11}$$

and

$$(12)\qquad \hat{b}_{l2}(\mathbf{x}^{l-1}) = \frac{\sum_{j^{l-1}\in\{1,2\}^{l-1}} w(j^{l-1}, 2)\prod_{i=1}^{l-1} x_{ij_i}}{\sum_{j^{l-1}\in\{1,2\}^{l-1}} w(j^{l-1})\prod_{i=1}^{l-1} x_{ij_i}},$$

with

$$(13)\qquad \hat{b}_{11} = w(1) \quad \text{and} \quad \hat{b}_{12} = w(2).$$

An alternative characterization of the max-min optimal strategy, which turns out to be equivalent to the above, is as follows. Break the initial wealth into 2^n piles, one corresponding to each sequence j^n, where the fraction of initial wealth assigned to pile j^n is precisely $w(j^n)$ as given in (9). Now invest all the wealth in pile j^n in asset j_1 on day 1. From then on, for each day i, shift the entirety of the running wealth for this pile into asset j_i. Do this in parallel for each of the 2^n piles j^n. We refer to the strategy used to manage pile j^n as the extremal strategy corresponding to the sequence j^n.

The wealth factor achieved by the investor using (11) and (12) is

$$\hat{S}_n(\mathbf{x}^n) = \prod_{l=1}^{n} \hat{\mathbf{b}}_l^t \mathbf{x}_l$$

$$= \prod_{l=1}^{n} \frac{\left[\begin{array}{l}\sum_{j^{l-1}\in\{1,2\}^{l-1}} w(j^{l-1}, 1)x_{l1}\prod_{i=1}^{l-1} x_{ij_i} \\ \quad + \sum_{j^{l-1}\in\{1,2\}^{l-1}} w(j^{l-1}, 2)x_{l2}\prod_{i=1}^{l-1} x_{ij_i}\end{array}\right]}{\sum_{j^{l-1}\in\{1,2\}^{l-1}} w(j^{l-1})\prod_{i=1}^{l-1} x_{ij_i}}$$

$$= \prod_{l=1}^{n} \frac{\sum_{j^l\in\{1,2\}^l} w(j^l)\prod_{i=1}^{l} x_{ij_i}}{\sum_{j^{l-1}\in\{1,2\}^{l-1}} w(j^{l-1})\prod_{i=1}^{l-1} x_{ij_i}}$$

$$(14)\qquad = \sum_{j^n\in\{1,2\}^n} w(j^n)\prod_{i=1}^{n} x_{ij_i}$$

$$= V_n \sum_{k=0}^{k} \left(\frac{k}{n}\right)^k \left(\frac{n-k}{n}\right)^{n-k} X(k)$$

where

$$X(k) \triangleq \sum_{j^n: n_1(j^n)=k} \prod_{i=1}^{n} x_{ij_i},$$

and (14) follows from a telescoping of the product.

It is apparent from Equation (14) that the extremal strategy formulation of the max-min optimal strategy is equivalent to the portfolio formulation (8)–(13). The extremal strategies simply "pick off" the product of the price relatives corresponding to the sequence of assets with indices j^n. Equation (14) represents the sum of the wealths obtained by the extremal strategies operating in parallel.

Note that for $0 \le k \le n$,

$$(15)\qquad \left(\frac{k}{n}\right)^k \left(\frac{n-k}{n}\right)^{n-k} = \max_{0\le b\le 1} b^k(1-b)^{n-k}.$$

Also note that $S_n^*(\mathbf{x}^n)$ can be rewritten as

$$S_n^*(\mathbf{x}^n) = \prod_{i=1}^{n} \mathbf{b}^{*\prime}\mathbf{x}_i$$

$$(16) \qquad = \sum_{j^n \in \{1,2\}^n} \prod_{i=1}^{n} b_{j_i}^* x_{ij_i}$$

$$= \sum_{k=0}^{n} b^{*k}(1-b^*)^{n-k} X(k),$$

where $\mathbf{b}^* = (b^*, 1-b^*)'$ achieves the maximum in (16).

Therefore, for any market sequence $\mathbf{x}^n$, Lemma 1 and the above imply that

$$\frac{\hat{S}_n(\mathbf{x}^n)}{S_n^*(\mathbf{x}^n)} = \frac{V_n \sum_{k=0}^{n} \left(\frac{k}{n}\right)^k \left(\frac{n-k}{n}\right)^{n-k} X(k)}{\sum_{k=0}^{n} b^{*k}(1-b^*)^{n-k} X(k)}$$

$$(17) \qquad \geq V_n \min_{0 \leq k \leq n} \frac{\left(\frac{k}{n}\right)^k \left(\frac{n-k}{n}\right)^{n-k}}{b^{*k}(1-b^*)^{n-k}}$$

$$(18) \qquad \geq V_n,$$

where (17) follows from a combination of Lemma 1 and the cancellation of the sums of products of x_{ij_i}, and (18) follows from (15). Since the above holds for all sequences $\mathbf{x}^n$, we have shown that

$$(19) \qquad \max_{\hat{\mathbf{b}}} \min_{\mathbf{x}^n} \frac{\hat{S}_n(\mathbf{x}^n)}{S_n^*(\mathbf{x}^n)} \geq V_n.$$

To show equality in (19) we consider the following possibilities for $\mathbf{x}^n$. For each $j^n \in \{1,2\}^n$ define $\mathbf{x}^n(j^n) = \mathbf{x}_1(j_1), \ldots, \mathbf{x}_n(j_n)$, as

$$(20) \qquad \mathbf{x}_i(j_i) = \begin{cases} (1,0)' & \text{if } j_i = 1, \\ (0,1)' & \text{if } j_i = 2. \end{cases}$$

Let

$$\mathcal{K} = \{\mathbf{x}^n(j^n) : j^n \in \{1,2\}^n\}$$

be the set of such extremal sequences $\mathbf{x}^n$.

An important property shared by all nonanticipating investment strategies $\hat{\mathbf{b}}(\cdot)$ on the sequences (20) is that

$$(21) \qquad \sum_{\mathbf{x}^n \in \mathcal{K}} \hat{S}_n(\mathbf{x}^n) = 1.$$

Also note that, for $\mathbf{x}^n(j^n) \in \mathcal{K}$, the best constant rebalanced portfolio is easily verified to be

$$\mathbf{b}^*(\mathbf{x}^n(j^n)) = \frac{1}{n}(n_1(j^n), n_2(j^n))^t$$

so that

$$S_n^*(\mathbf{x}^n(j^n)) = \left(\frac{n_1(j^n)}{n}\right)^{n_1(j^n)} \left(\frac{n_2(j^n)}{n}\right)^{n_2(j^n)}$$
$$= \frac{w(j^n)}{V_n}.$$

Therefore

$$\sum_{\mathbf{x}^n \in \mathcal{K}} S_n^*(\mathbf{x}^n) = \frac{1}{V_n}.$$

Since the minimum is less than any average, we obtain equality in (19) from

$$\min_{\mathbf{x}^n \in \mathcal{K}} \frac{\hat{S}_n(\mathbf{x}^n)}{S_n^*(\mathbf{x}^n)} \leq \sum_{\tilde{\mathbf{x}}^n \in \mathcal{K}} \left(\frac{S_n^*(\tilde{\mathbf{x}}^n)}{\sum_{\mathbf{x}^n \in \mathcal{K}} S_n^*(\mathbf{x}^n)}\right) \frac{\hat{S}_n(\tilde{\mathbf{x}}^n)}{S_n^*(\tilde{\mathbf{x}}^n)} \tag{22}$$

$$= \sum_{\tilde{\mathbf{x}}^n \in \mathcal{K}} \frac{\hat{S}_n(\tilde{\mathbf{x}}^n)}{\sum_{\mathbf{x}^n \in \mathcal{K}} S_n^*(\mathbf{x}^n)} \tag{23}$$

$$= \frac{1}{\sum_{\mathbf{x}^n \in \mathcal{K}} S_n^*(\mathbf{x}^n)} \tag{24}$$

$$= V_n, \tag{25}$$

which holds for any $\hat{\mathbf{b}}$. Thus

$$\max_{\hat{\mathbf{b}}} \min_{\mathbf{x}^n} \frac{\hat{S}_n(\mathbf{x}^n)}{S_n^*(\mathbf{x}^n)} \leq V_n.$$

Combining this with (19) completes the proof of the theorem. □

Complexity. It appears from (11) and (12) that computing the max-min optimal portfolio requires keeping track of the products $\prod_{i=1}^{l-1} x_{ij_i}$ for each sequence j^{l-1}. This quickly becomes prohibitively complex, since the number of such sequences is exponentially increasing in l. Fortunately, a simplification can be made.

This follows from the observation that $w(j^n)$ defined in (9) depends on j^n only through its type $(n_1(j^n), n_2(j^n))$, the number of 1's and 2's. This implies that $w(j^{n-1}, j_n)$, for fixed j_n, is a function of j^{n-1} only through $(n_1(j^{n-1}), n_2(j^{n-1}))$. The same applies to $w(j^{n-1}) = \sum_{j_n} w(j^{n-1}, j_n)$. Thus, by induction $w(j^{l-1}, j_l)$ and $w(j^{l-1})$, for all l, are constant on j^{l-1} with the same type.

Using this fact, the numerator and denominator of (11) and (12) can be evaluated by grouping the products $\prod_{i=1}^{l-1} x_{ij_i}$ according to the type of j^{l-1}. More specifically, the numerator of (11), for example, can be written as

$$\sum_{j^{l-1}\in\{1,2\}^{l-1}} w(j^{l-1},1)\prod_{i=1}^{l-1} x_{ij_i} = \sum_{k=0}^{l-1} w'_{l-1}(k,1) \sum_{j^{l-1}:n_1(j^{l-1})=k}\prod_{i=1}^{l-1} x_{ij_i}$$

$$= \sum_{k=0}^{l-1} w'_{l-1}(k,1)X_{l-1}(k),$$

where $w'_{l-1}(k,1)$ equals $w(j^{l-1},1)$ when $n_1(j^{l-1})=k$ and

$$X_{l-1}(k) = \sum_{j^{l-1}:n_1(j^{l-1})=k}\prod_{i=1}^{l-1} x_{ij_i}.$$

The denominator can be rewritten in a similar way.

It is now clear that only the quantities $X_{l-1}(k)$ need be computed and stored instead of the exponentially many products $\prod_{i=1}^{l-1} x_{ij_i}$. The complexity of this is linear in l, since there are only l such quantities. The simple recursions

$$X_l(k) = x_{l1}X_{l-1}(k-1) + x_{l2}X_{l-1}(k)$$

$$X_l(0) = x_{l2}X_{l-1}(0)$$

$$X_l(l) = x_{l1}X_{l-1}(l-1)$$

suffice to update the $X_{l-1}(k)$.

The above generalizes in the obvious way to $m > 2$ assets resulting in a computational complexity growing like l^{m-1}. Therefore, the max-min optimal portfolio is, in fact, computationally feasible for moderate m.

3.1. Game-theoretic analysis. A full game-theoretic result can also be proved. Specifically, we imagine the same contest as above, except that mixed strategies are allowed. The payoff function is

$$A(\hat{\mathbf{b}},\mathbf{x}^n) = \frac{\hat{S}_n(\mathbf{x}^n)}{S_n^*(\mathbf{x}^n)}.$$

As before, the investor and nature respectively try to maximize and minimize the payoff. Let $\mathcal{G}_n$ denote the game when played with this payoff function.

A mixed strategy for the investor is a probability distribution $\mathcal{P}(\hat{\mathbf{b}})$ on the space of nonanticipating investment strategies, $\hat{\mathbf{b}} = (\hat{\mathbf{b}}_1, \hat{\mathbf{b}}_2(\mathbf{x}_1), \ldots, \hat{\mathbf{b}}_n(\mathbf{x}^{n-1}))$. Similarly, nature's mixed strategies are probability distributions on the space of price-relative sequences and will be denoted by $\mathcal{Q}(\mathbf{x}^n)$. The following theorem can then be proved.

THEOREM 2. *The value of the game $\mathcal{G}_n$ is*

$$\max_{\mathcal{P}(\hat{\mathbf{b}})}\min_{\mathcal{Q}(\mathbf{x}^n)} EA(\hat{\mathbf{b}},\mathbf{x}^n) = \min_{\mathcal{Q}(\mathbf{x}^n)}\max_{\mathcal{P}(\hat{\mathbf{b}})} EA(\hat{\mathbf{b}},\mathbf{x}^n) = V_n,$$

where V_n is given by (2). *Further, the investor's optimum strategy $\mathcal{P}^*$ is the pure strategy specified by* (8)–(13).

PROOF. We prove this for $m = 2$, the generalization being obvious. The pure strategy $\mathcal{P}^*$ is precisely the max-min optimal strategy (8)–(13) achieving the maximum in Theorem 1. Nature's optimum mixed strategy $\mathcal{Q}^*$ (for $m = 2$) consists of choosing sequences from

$$\mathcal{K} = \{\mathbf{x}^n(j^n) : j^n \in \{1, 2\}^n\}$$

according to the probability distribution $w(j^n)$ given by (9). The proof of

$$\min_{\mathcal{Q}(\mathbf{x}^n)} \max_{\mathcal{P}(\hat{\mathbf{b}})} EA(\hat{\mathbf{b}}, \mathbf{x}^n) \le V_n, \tag{26}$$

follows from Equations (22) through (25). The theorem follows from (26) and (19). □

The full game-theoretic analysis brings out a nice symmetry between the optimal investment strategy and nature's optimal strategy. The optimal investment strategy $\mathcal{P}^*$ is a pure strategy constructed from the distribution $w(j^n)$ on binary strings given by (9). Nature's optimal strategy, on the other hand, is to choose 0-1 price-relative vectors at random according to this same probability distribution.

This analysis generalizes to games with payoff

$$A_\phi(\hat{\mathbf{b}}, \mathbf{x}^n) = \phi\left(\frac{\hat{S}_n(\mathbf{x}^n)}{S_n^*(\mathbf{x}^n)}\right),$$

for which the following holds.

THEOREM 3. *For concave nondecreasing ϕ, the game $\mathcal{G}_n(\phi)$ with payoff $A_\phi(\hat{\mathbf{b}}, \mathbf{x}^n)$ has a value $V(\mathcal{G}_n(\phi))$ given by*

$$V(\mathcal{G}_n(\phi)) = \phi(V_n),$$

where V_n is given by (2) and the optimal strategies are the same as those for $\mathcal{G}_n$.

3.2. Bounds on V_n. We prove the following lemma for $m = 2$.

LEMMA 2. *For all n,*

$$\frac{1}{2\sqrt{n+1}} \le V_n \le \frac{2}{\sqrt{n+1}}.$$

PROOF. We first prove the lower bound. In Cover and Ordentlich (1996), a sequential portfolio selection strategy called the Dirichlet(1/2, . . . , 1/2) weighted universal portfolio was shown to achieve a wealth $\hat{S}_n^D(\mathbf{x}^n)$ satisfying

$$\min_{\mathbf{x}^n} \frac{\hat{S}_n^D(\mathbf{x}^n)}{S_n^*(\mathbf{x}^n)} \ge \frac{1}{2\sqrt{n+1}}.$$

Therefore

$$V_n = \max_{\hat{\mathbf{b}}} \min_{\mathbf{x}^n} \frac{\hat{S}_n(\mathbf{x}^n)}{S_n^*(\mathbf{x}^n)}$$

$$\geq \min_{\mathbf{x}^n} \frac{\hat{S}_n^D(\mathbf{x}^n)}{S_n^*(\mathbf{x}^n)}$$

$$\geq \frac{1}{2\sqrt{n+1}},$$

proving the lower bound on V_n.

We now establish the upper bound. Write $1/V_n$ as

$$\frac{1}{V_n} = \sum_{k=0}^{n} \binom{n}{k} \left(\frac{k}{n}\right)^k \left(\frac{n-k}{n}\right)^{n-k}$$

$$= \frac{\Gamma(n+1)}{n^n} \sum_{k=0}^{n} \frac{k^k (n-k)^{n-k}}{\Gamma(k+1)\Gamma(n-k+1)}, \tag{27}$$

where $\Gamma(x) = \int_0^\infty t^{x-1} e^{-t} dt$ is the Gamma function. If x is an integer, then $\Gamma(x+1) = x!$. In Marshall and Olkin (1979), it is shown that $(x_1, x_2) \mapsto (x_1^{x_1} x_2^{x_2})/(\Gamma(x_1+1)\Gamma(x_2+1))$ is Schur convex. This implies that under the constraint $x_1 + x_2 = n$, it is minimized by setting $x_1 = x_2 = n/2$. Therefore, each term in the summation (27) can be bounded from below by

$$\frac{k^k (n-k)^{n-k}}{\Gamma(k+1)\Gamma(n-k+1)} \geq \frac{\frac{n}{2}^{n/2} \frac{n}{2}^{n/2}}{\Gamma\left(\frac{n}{2}+1\right)\Gamma\left(\frac{n}{2}+1\right)}$$

to obtain

$$\frac{1}{V_n} \geq \frac{\Gamma(n+1)}{n^n} (n+1) \frac{\frac{n}{2}^{n/2} \frac{n}{2}^{n/2}}{\Gamma\left(\frac{n}{2}+1\right)\Gamma\left(\frac{n}{2}+1\right)}$$

$$= \frac{(n+1)\Gamma(n+1)}{2^n \Gamma^2\left(\frac{n}{2}+1\right)}.$$

The identity (see Rudin (1976))

$$\Gamma(n+1) = \frac{2^n}{\sqrt{\pi}} \Gamma\left(\frac{n+1}{2}\right) \Gamma\left(\frac{n}{2}+1\right), \tag{28}$$

can now be applied to obtain

$$\frac{1}{V_n} \geq \frac{(n+1)}{\sqrt{\pi}} \frac{\Gamma\left(\frac{n+1}{2}\right)}{\Gamma\left(\frac{n}{2}+1\right)}. \tag{29}$$

The log convexity of $\Gamma(x)$ (see Rudin 1976) now implies that

$$\Gamma\left(\frac{n}{2}+1\right) \leq \Gamma\left(\frac{n}{2}+\frac{1}{2}\right)^{1/2} \Gamma\left(\frac{n}{2}+\frac{3}{2}\right)^{1/2}$$

$$= \Gamma\left(\frac{n}{2}+\frac{1}{2}\right)\sqrt{\frac{n+1}{2}}, \tag{30}$$

where we have used the identity $\Gamma(x+1) = x\Gamma(x)$. Combining (29) and (30) we obtain

$$\frac{1}{V_n} \geq \frac{n+1}{\sqrt{\frac{n+1}{2}}} \frac{1}{\sqrt{\pi}}$$

$$= \sqrt{\frac{2(n+1)}{\pi}}$$

thereby proving that

$$V_n \leq \frac{2}{\sqrt{n+1}}$$

for all n. □

This bound can be generalized to $m > 2$ with the help of

$$\Gamma(n+1) = m^n \prod_{i=1}^{m} \frac{\Gamma\left(\frac{n+i}{m}\right)}{\Gamma\left(\frac{i}{m}\right)},$$

an extension of (28) to general m.

3.3. Asymptotics of V_n. The following lemma characterizes the asymptotic behavior of V_n for m stocks.

LEMMA 3. *For all m, V_n satisfies*

$$V_n \sim \frac{\Gamma\left(\frac{m}{2}\right)}{\sqrt{\pi}} \left(\frac{2}{n}\right)^{(m-1)/2} \tag{31}$$

in the sense that

$$\lim_{n\to\infty} \frac{V_n}{\Gamma\left(\frac{m}{2}\right) \Big/ \sqrt{\pi}\left(\frac{2}{n}\right)^{(m-1)/2}} = 1.$$

The quantity V_n arises in a variety of settings including the max-min data compression problem (see Shtarkov 1987), the distribution of the longest common subsequence between two random sequences (Karlin 1996), and bounds on the probability of undetected errors by linear codes (see Kløve 1995, Massey 1978, and Szpankowski 1995). Lemma 3 is proved in Shtarkov, Tjalkens, and Willems (1995) and an asymptotic expansion of V_n to arbitrary order is given in Szpankowski (1995, 1996). A direct proof of Lemma 3 based on a Riemann sum approximation is given in Ordentlich (1996).

In addition, Shtarkov (1987) obtains the bound

$$V_n \geq \left[\sum_{i=1}^{m} \binom{m}{i} \frac{\sqrt{\pi}}{\Gamma(i/2)} \left(\frac{n}{2}\right)^{(i-1)/2}\right]^{-1}$$

implying one half of the asymptotic behavior in Equation (31).

4. The hindsight allocation option. The results of the previous section motivate the analysis of the hindsight allocation option, a derivative security which pays $S_n^*(\mathbf{x}^n)$, the result of investing one dollar according to the best constant rebalanced portfolio computed in hindsight for the observed market behavior $\mathbf{x}^n$. Let

$$\bar{H}_n = \frac{1}{V_n}.$$

Certainly the price of the hindsight allocation option should be no higher than $\bar{H}_n$. This follows because $\bar{H}_n$ dollars invested in the nonanticipating strategy described in the proof of Theorem 1 is guaranteed to result in wealth at time n no less than $S_n^*(\mathbf{x}^n)$ for all market sequences $\mathbf{x}^n$. If the price of the hindsight allocation option were more than $\bar{H}_n$, selling the option and investing only $\bar{H}_n$ of the proceeds in the above strategy would be an arbitrage. Note that this argument assumes the existence of a riskless asset for investing the surplus.

Therefore, $\bar{H}_n$ is an upper bound on the price of the hindsight allocation option valid for any market model (with a risk free asset). Furthermore, while the return of the best constant rebalanced portfolio is expected to grow exponentially with n, the upper bound on the price of the hindsight allocation option $\bar{H}_n$ behaves like $\sqrt{n}$. This polynomial factor is exponentially negligible relative to S_n^*.

Is $\bar{H}_n$ a reasonable price for the hindsight allocation option? Probably not; the price should be lower. Pricing the option at $\bar{H}_n$ may be appropriate if no assumptions about market behavior can be made. This is the case in §3, where no restrictions are placed on nature's choice for the market behavior. Returns on assets can be arbitrarily high or low, even zero. Actual markets, however, are typically less volatile. We gain more insight into this issue by using established derivative security pricing theory to determine the no-arbitrage price of the hindsight allocation option for two much studied models of market behavior, the binomial lattice and continuous time geometric Brownian motion models.

4.1. Binomial lattice price. We consider a risky stock and a riskless bond. Accordingly, the price-relatives $\mathbf{x}_i$ are assumed to take on one of two values

$$\mathbf{x}_i \in \{(1+u, 1+r)^t, (1+d, 1+r)^t\}$$

with $r \geq 0$, $u > r > d$. The first component of $\mathbf{x}_i$ reflects the change in the price of the stock as measured by the ratio of closing to opening price. The second component indicates that the riskless bond compounds at an interest rate of r for each investment period. The parameters of the model are thus u, d, and r. If the stock price changes by a factor of $1 + u$ it has gone "(u)p"; if it changes by a factor of $1 + d$ it has gone "(d)own."

We will find that the no-arbitrage price H_n of the hindsight allocation option for this model is closely related to $\bar{H}_n$, the upper bound obtained in the previous sections. It will be apparent that for certain choices of d, u, and r, the upper bound $\bar{H}_n$ is essentially attained.

For a sequence of n price-relatives $\mathbf{x}^n = \mathbf{x}_1, \ldots, \mathbf{x}_n$, the wealth acquired by a constant rebalanced portfolio $\mathbf{b} = (b, 1-b)^t$ can be written as

$$S_n(\mathbf{b}) = [1 + r + b(u-r)]^k [1 + r + b(d-r)]^{n-k},$$

where k is the number of vectors $\mathbf{x}_i$ for which $x_{i1} = 1 + u$. Since $\log S_n(\mathbf{b})$ is concave in $\mathbf{b}$, the best constant rebalanced portfolio $\mathbf{b}^* = (b^*, 1 - b^*)^t$ is easily determined using calculus. For $0 < k < n$, define $\tilde{b}^*$ as the solution to

$$\frac{d \log S_n(\mathbf{b})}{db} = 0.$$

It is given by

$$\tilde{b}^* = \frac{(1+r)}{n}\left(\frac{k}{r-d} - \frac{n-k}{u-r}\right).$$

For $k = 0$, set $\tilde{b}^* = 0$, and for $k = n$, set $\tilde{b}^* = 1$. Then b^* is given by

$$b^* = \max(0, \min(1, \tilde{b}^*)).$$

We then obtain the wealth achieved by the best constant rebalanced portfolio as

$$S_n^*(\mathbf{x}^n) = [1 + r + b^*(u-r)]^k [1 + r + b^*(d-r)]^{n-k}$$

$$= \begin{cases} (1+r)^n & \text{if} \quad b^* = 0, \\ (1+u)^k (1+d)^{n-k} & \text{if} \quad b^* = 1, \\ [1 + r + \tilde{b}^*(u-r)]^k [1 + r + \tilde{b}^*(d-r)]^{n-k} & \text{if} \quad 0 < b^* < 1. \end{cases}$$

If $0 < b^* < 1$, the wealth achieved can be written more explicitly as

$$S_n^*(\mathbf{x}^n) = \left(1 + r + (1+r)\left[\frac{k}{n(r-d)} - \frac{n-k}{n(u-r)}\right](u-r)\right)^k$$

$$\cdot \left(1 + r + (1+r)\left[\frac{k}{n(r-d)} - \frac{n-k}{n(u-r)}\right](d-r)\right)^{n-k},$$

which simplifies to

$$S_n^*(\mathbf{x}^n) = (1+r)^n\left(\frac{k}{n}\right)^k\left(\frac{n-k}{n}\right)^{n-k}\left(\frac{u-d}{r-d}\right)^k\left(\frac{u-d}{u-r}\right)^{n-k}.$$

It is well known that for this model the no-arbitrage price P_n of any derivative security with payoff S_n at time n is given by

$$P_n = \frac{1}{(1+r)^n} E^Q(S_n),$$

where the expectation is taken with respect to Q, the so called equivalent martingale measure on asset price changes. The unique equivalent martingale measure for this market is a Bernoulli distribution on the sequence of "up" and "down" moves of the asset price with the probability of an "up" equal to $p_u = (r-d)/(u-d)$ and the "down" probability equal to $p_d = 1-(r-d)/(u-d) = (u-r)/(u-d)$.

We note that for the case of $0 < b^* < 1$

$$S_n^*(\mathbf{x}^n) = (1+r)^n\left(\frac{k}{n}\right)^k\left(\frac{n-k}{n}\right)^{n-k} p_u^{-k} p_d^{-(n-k)}$$

$$\triangleq S_n^*(k).$$

Therefore,

$$H_n = \frac{E^Q(S_n^*)}{(1+r)^n}$$

$$= \frac{1}{(1+r)^n}\sum_{k:0<b^*<1} S_n^*(k)\binom{n}{k}p_u^k p_d^{n-k} + \frac{1}{(1+r)^n}\sum_{k:b^*=0}(1+r)^n\binom{n}{k}p_u^k p_d^{n-k}$$

$$+ \frac{1}{(1+r)^n}\sum_{k:b^*=1}(1+u)^k(1+d)^{n-k}\binom{n}{k}p_u^k p_d^{n-k},$$

which simplifies to

$$H_n = \sum_{k:0<b^*<1}\binom{n}{k}\left(\frac{k}{n}\right)^k\left(\frac{n-k}{n}\right)^{n-k} + \sum_{k:b^*=0}\binom{n}{k}p_u^k p_d^{n-k}$$

$$+ \sum_{k:b^*=1}\binom{n}{k}\left(p_u\left(\frac{1+u}{1+r}\right)\right)^k\left(p_d\left(\frac{1+d}{1+r}\right)\right)^{n-k}.$$

The range of k such that $0 < b^* < 1$ is

$$p_u < \frac{k}{n} < p_u\left(\frac{u+1}{r+1}\right).$$

Thus

$$H_n = \sum_{p_u < k/n < p_u(u+1)/(r+1)} \binom{n}{k} \left(\frac{k}{n}\right)^k \left(\frac{n-k}{n}\right)^{n-k} + \sum_{k/n \le p_u} \binom{n}{k} p_u^k p_d^{n-k}$$

$$+ \sum_{k/n \ge p_u(u+1)/(r+1)} \binom{n}{k} \left(p_u\left(\frac{1+u}{1+r}\right)\right)^k \left(p_d\left(\frac{1+d}{1+r}\right)\right)^{n-k}.$$

It is useful to note that

$$p_u \frac{1+u}{1+r} + p_d \frac{1+d}{1+r} = 1.$$

This implies that

$$H_n \ge \sum_{p_u < k/n < p_u(u+1)/(r+1)} \binom{n}{k} \left(\frac{k}{n}\right)^k \left(\frac{n-k}{n}\right)^{n-k},$$

and

$$H_n \le \sum_{p_u < k/n < p_u(u+1)/(r+1)} \binom{n}{k} \left(\frac{k}{n}\right)^k \left(\frac{n-k}{n}\right)^{n-k} + 2.$$

Notice the similarities between these bounds and the expression for $V_n = 1/\bar{H}_n$ given by (3). It is possible to choose r, u, and d so that $p_u < 1/n$ and $p_u((u+1)/(r+1)) > (n-1)/n$, in which case the value of the hindsight allocation option is at least $\bar{H}_n - 2$.

In summary, the no-arbitrage price H_n of the hindsight allocation is given by

$$H_n = \sum_{p_u < k/n < p_u(u+1)/(r+1)} \binom{n}{k} \left(\frac{k}{n}\right)^k \left(\frac{n-k}{n}\right)^{n-k} + \sum_{k/n \le p_u} \binom{n}{k} p_u^k p_d^{n-k}$$

$$+ \sum_{k/n \ge p_u(u+1)/(r+1)} \binom{n}{k} \left(p_u\left(\frac{1+u}{1+r}\right)\right)^k \left(p_d\left(\frac{1+d}{1+r}\right)\right)^{n-k},$$

where the first summation comprises the bulk of the price for reasonable parameter values. The terms appearing in this sum are identical to those in the expression (3) for $V_n = 1/\bar{H}_n$. The number of such terms appearing in the sum depends on the parameter values. A Reimann sum approximation argument shows that for fixed parameter values $H_n \sim c\sqrt{n}$, where the constant c depends only on the parameters.

4.2. Geometric Brownian motion price. In this section, we give the price of the hindsight allocation option for the classical continuous time Black-Scholes market model with one stock and one bond. The stock price X_t follows a geometric Brownian motion and evolves according to the stochastic differential equation

$$dX_t = \mu X_t dt + \sigma X_t dB_t,$$

where μ and σ are constant, and B is a standard Brownian motion. Note that here X_t denotes a price, *not* a price-relative. The bond price β_t obeys

$$d\beta_t = \beta_t r dt$$

where r is constant and therefore

$$\beta_t = e^{rt}\beta_0.$$

Let $S_t(\mathbf{b})$ be the wealth obtained by investing one dollar at $t = 0$ in the constant rebalanced portfolio $\mathbf{b} = (b, 1-b)'$, where b is the proportion of wealth invested in the stock. Then $S_t(\mathbf{b})$ satisfies the stochastic differential equation

$$\frac{dS_t(\mathbf{b})}{S_t(\mathbf{b})} = b\frac{dX_t}{X_t} + (1-b)\frac{d\beta_t}{\beta_t}, \tag{32}$$

which can be solved to give

$$S_t(\mathbf{b}) = \exp\left(-\frac{b^2\sigma^2 t}{2} + b\left(\log\frac{X_t}{X_0} + \frac{\sigma^2 t}{2}\right) + (1-b)rt\right). \tag{33}$$

That this solves (32) can be verified directly using Ito's lemma (see Duffie 1996, Karatzas and Shreve 1991). Notice that, for fixed σ^2 and r, the wealth $S_t(\mathbf{b})$ depends on the stock price path only through the final price X_t.

The best constant rebalanced portfolio in hindsight at time T is obtained by maximizing the exponent of (33) for $t = T$ under the constraint that $0 \le b \le 1$. This results in

$$b_T^* = \max\left(0, \min\left(1, \frac{1}{2} + \frac{(1/T)\log(X_T/X_0) - r}{\sigma^2}\right)\right). \tag{34}$$

The wealth achieved by the best constant rebalanced portfolio is then obtained by evaluating (33) at $b = b_T^*$ resulting in

$$S_T^* = S_T(b_T^*) = \begin{cases} e^{rT} & \text{if } b_T^* = 0 \\ e^{(\sigma^2 T/2)b_T^{*2} + rT} & \text{if } 0 \le b_T^* \le 1 \\ \dfrac{X_T}{X_0} & \text{if } b_T^* \ge 1. \end{cases}$$

From the martingale approach to options pricing, the no-arbitrage price at $t = 0$ of the hindsight allocation option with duration T is given by

$$\begin{aligned} H_{0,T} &= \beta_0 E_Q \frac{S_T(\mathbf{b}_T^*)}{\beta_T} \\ &= e^{-rT} E_Q S_T(\mathbf{b}_T^*), \end{aligned} \tag{35}$$

where Q is the equivalent martingale measure or the unique (in this case) probability measure under which X_t/β_t is a martingale, and assuming that $S_T(\mathbf{b}_T^*)$ is integrable under Q, which it is.

It is well known (see Duffie 1996) that under the equivalent martingale measure Q the stock price X_t obeys

$$dX_t = rX_t dt + \sigma X_t dB_t.$$

This and Ito's lemma imply that under Q, the expression $\log(X_T/X_0)$ appearing in the exponent of (33) is normally distributed with mean $(r - (1/2)\sigma^2)T$ and variance $\sigma^2 T$. Therefore, the random variable

$$Y \triangleq \frac{\log(X_T/X_0) - (r - (1/2)\sigma^2)T}{\sqrt{\sigma^2 T}}$$

is standard normal. It can be rewritten as

$$Y = \sqrt{\sigma^2 T}\left(\frac{1}{2} + \frac{(1/T)\log(X_T/X_0) - r}{\sigma^2}\right), \tag{36}$$

so that, by equation (34),

$$\sqrt{\sigma^2 T}(\mathbf{b}_T^*) = \max(0, \min(\sqrt{\sigma^2 T}, Y)).$$

Equation (36) can be solved for X_T/X_0 resulting in

$$\frac{X_T}{X_0} = e^{Y\sqrt{\sigma^2 T} + (r - \sigma^2/2)T}.$$

The expectation (35) is then easily evaluated as

$$\begin{aligned} e^{-rT}E_Q[S_T(\mathbf{b}_T^*)] &= (E_Q[I(Y \le 0)] + E_Q[e^{Y^2/2} I(Y \in [0, \sqrt{\sigma^2 T}])] \\ &\quad + E_Q[e^{Y\sqrt{\sigma^2 T} - T\sigma^2/2} I(Y > \sqrt{\sigma^2 T})]). \end{aligned} \tag{37}$$

The first expectation is clearly equal to $\frac{1}{2}$, since Y is standard normal. The middle expectation is

$$\begin{aligned} E_Q[e^{Y^2/2} I(Y \in [0, \sqrt{\sigma^2 T}])] &= \frac{1}{\sqrt{2\pi}} \int_0^{\sqrt{\sigma^2 T}} e^{y^2/2} e^{-y^2/2}\, dy \\ &= \sqrt{\frac{\sigma^2 T}{2\pi}}. \end{aligned}$$

Finally, the third expectation is

$$\begin{aligned} E_Q[e^{Y\sqrt{\sigma^2 T} - T\sigma^2/2} I(Y > \sqrt{\sigma^2 T})] &= \frac{1}{\sqrt{2\pi}} \int_{\sqrt{\sigma^2 T}}^{\infty} e^{y\sqrt{\sigma^2 T} - T\sigma^2/2} e^{-y^2/2}\, dy \\ &= \frac{1}{\sqrt{2\pi}} \int_{\sqrt{\sigma^2 T}}^{\infty} e^{-(1/2)(y - \sqrt{\sigma^2 T})^2}\, dy \\ &= \frac{1}{2}. \end{aligned}$$

Thus (37) reduces to a surprisingly simple form. The no-arbitrage price $H_{0,T}$ of the hindsight allocation option is

$$H_{0,T} = 1 + \sqrt{\frac{\sigma^2 T}{2\pi}}.$$

The price is affinely increasing in the volatility σ and increases like the square root of the duration T. The dependence on duration matches the $\sqrt{n}$ growth of the discrete-time upper bound $\bar{H}_n$ and the binomial lattice price H_n. If the hindsight allocation option payoff is redefined to be $S_n^*(\mathbf{x}^n) - e^{rT}$ (the excess return of the best constant rebalanced portfolio beyond the return of the bond) then the price is simply $\sqrt{\sigma^2 T/(2\pi)}$. This can be thought of as a premium for volatility.

5. Conclusion. The worst sequence approach to the problem of achieving the best portfolio in hindsight leads to a favorable result: the max-min optimal portfolio strategy for m assets loses only $((m-1)/2)(\log n)/n$ in the rate of return in the worst case. This yields an asymptotically negligible difference in growth rate as the number of investment periods n grows to infinity. In practice we would expect even better performance, since real markets are less volatile than the max-min market identified here. This intuition is partially validated by the hindsight allocation pricing analysis for the binomial and geometric Wiener market models which indicates that the cost of achieving the best portfolio in hindsight depends monotonically on market volatility.

Acknowledgment. This work was supported by NSF grant NCR-9205663, JSEP contract DAAH04-94-G-0058, and ARPA contract JFBI-94-218-2. Portions of this paper were presented at CIFER 96, COLT 96, IMS 96.

References

Blackwell, D. (1956a). Controlled random walks. In *Proceedings of International Congress of Mathematics, vol. III*, pages 336–338, Amsterdam, North Holland.

——— (1956b). An analog of the minimax theorem for vector payoffs. *Pacific J. Mathe.* **6.**

Cover, T. M. (1991). Universal portfolios. *Math. Finance* **1**(1) 1–29.

———, D. Gluss (1986). Empirical Bayes stock market portfolios. *Adv. Appl. Math.* **7** 170–181.

———, E. Ordentlich (1996). Universal portfolios with side information. *IEEE Trans. Info. Theory* **42**(2).

Cox, J., C. Huang (1992). A continuous time portfolio turnpike theorem. *J. Economics Dynamics and Control* **16** 491–501.

Duffie, D. (1996). *Dynamic Asset Pricing Theory, Second Edition*. Princeton University Press, Princeton, New Jersey.

Huberman, G., S. Ross (1983). Portfolio turnpike theorem, risk aversion, and regularly varying functions. *Econometrica* **51**.

Jamshidian, F. (1992). Asymptotically optimal portfolios. *Math. Finance* **2**(2).

Karatzas, I., S. E. Shreve (1991). *Brownian Motion and Stochastic Calculus*. Graduate Texts in Mathematics. Springer-Verlag, second edition.

Karlin, S. (1996). Private communication.

Kløve, T. (1995). Bounds for the worst case probability of undetected error. *IEEE Trans. Info. Theory* **41**(1).

Larson, D. C. (1986). *Growth optimal trading strategies*. Ph.D. thesis, Stanford University, Stanford, California.

Marshall, A. W., I. Olkin (1979). Inequalities: Theory of Majorization and Its Applications, volume 143 of *Mathematics in Science and Engineering*. Academic Press, London.

Massey, J. (1978). Coding techniques for digital networks. In *Proceedings International Conference on Information Theory Systems*, Berlin, Germany.

Merhav, N., M. Feder (1993). Universal schemes for sequential decision from individual data sequences. *IEEE Trans. Info. Theory* **39**(4) 1280–1292.

Ordentlich, E. (1996). *Universal investment and universal data compression*. Ph.D. thesis, Stanford University, Stanford, California.

———, T. M. Cover (1996). On-line portfolio selection. In *Proceedings of Ninth Conference on Computational Learning Theory*, Desenzano del Garda, Italy.

Rudin, W. (1976). *Principles of Mathematical Analysis*. McGraw-Hill, third edition.

Shtarkov, Yu. M. (1987). Universal sequential coding of single messages. *Problems of Information Transmission* **23**(3), 3–17.

———, T. Tjalkens, F. M. Willems (1995). Multi-alphabet universal coding of memoryless sources. *Problems of Information Transmission* **31** 114–127.

Szpankowski, W. (1995). On asymptotics of certain sums arising in coding theory. *IEEE Trans. Info. Theory* **41**(6).

——— (1996). Some new sums arising in coding theory. Preprint.

E. Ordentlich: Hewlett-Packard Laboratories, 1501 Page Mill Road 3U-4, Palo Alto, California 94304-1126

T. M. Cover: Departments of Statistics and Electrical Engineering, Stanford University, Stanford, California 94305

17

Adv. Appl. Prob. 13, 415–428 (1981)
Printed in N. Ireland
0001-8678/81/020415-14$01.65

OPTIMAL STRATEGIES FOR REPEATED GAMES

MARK FINKELSTEIN* AND
ROBERT WHITLEY,* *University of California, Irvine*

Abstract

We extend the optimal strategy results of Kelly and Breiman and extend the class of random variables to which they apply from discrete to arbitrary random variables with expectations. Let F_n be the fortune obtained at the nth time period by using any given strategy and let F_n^* be the fortune obtained by using the Kelly–Breiman strategy. We show (Theorem 1(i)) that F_n/F_n^* is a supermartingale with $E(F_n/F_n^*) \leqq 1$ and, consequently, $E(\lim F_n/F_n^*) \leqq 1$. This establishes one sense in which the Kelly–Breiman strategy is optimal. However, this criterion for 'optimality' is blunted by our result (Theorem 1(ii)) that $E(F_n/F_n^*) = 1$ for many strategies differing from the Kelly–Breiman strategy. This ambiguity is resolved, to some extent, by our result (Theorem 2) that F_n^*/F_n is a submartingale with $E(F_n^*/F_n) \geqq 1$ and $E(\lim F_n^*/F_n) \geqq 1$; *and* $E(F_n^*/F_n) = 1$ if and only if at each time period j, $1 \leqq j \leqq n$, the strategies leading to F_n and F_n^* are 'the same'.

KELLY CRITERION; OPTIMAL STRATEGY; FAVORABLE GAME; OPTIMAL GAMBLING SYSTEM; PORTFOLIO SELECTION; CAPITAL GROWTH MODEL

1. Introduction

Suppose a gambler is given the opportunity to bet a fixed fraction γ of his (infinitely divisible) capital on successive flips of a biased coin: on each flip, with probability $p > \frac{1}{2}$ he wins an amount equal to his bet and with probability $q = 1 - p$ he loses his bet. What is a good choice for γ and why is it good?

This question is subtle because the obvious answer has an obvious flaw. The obvious answer is for the gambler to choose $\gamma = 1$ to maximize the expected value of his fortune. The obvious flaw is that he is then broke in n or fewer trials with probability $1 - p^n$, which tends to 1 as n tends to ∞.

A germinal answer was given by Kelly [10]: a gambler should choose $\gamma = p - q$ so as to maximize the expected value of the *log* of his fortune. He shows that a gambler who chooses $\gamma = p - q$ will 'with probability 1 eventually get ahead and stay ahead of one using any other value of γ' ([10], p. 920).

In an important paper Breiman [4] generalizes and considers strategies other

Received 23 April 1980; revision received 12 August 1980.
* Postal address: Department of Mathematics, University of California, Irvine, CA 92717, U.S.A.

than fixed-fraction strategies and generalizes the random variable as follows: Let X be a random variable taking values in $\{1, 2, \cdots, s\} = I$, $\mathscr{C}$ be a class $\{A_1, A_2, \cdots, A_r\}$ of subsets of I whose union is I, and $o_1, o_2, \cdots, o_r$ be positive numbers (odds). If for one round of betting a gambler bets fractional amounts $\beta_1, \beta_2, \cdots, \beta_r$ of his capital on the events $\{X \in A_1\}, \cdots, \{X \in A_r\}$, then when $X = i$ he gets a payoff of $\sum \beta_j o_j$ summed over those j with i in A_j. In this setting Breiman discusses several 'optimal' properties of the fixed-fraction strategy which chooses $\beta_1, \beta_2, \cdots, \beta_r$ so as to maximize the expected value of the log of the fortune and then bets these fractions on each trial, leading to the fortune F_n^* at the conclusion of the nth trial. He shows that if F_n is a fortune resulting from the use of any strategy, then $\lim F_n/F_n^*$ almost surely exists and $E(\lim F_n/F_n^*) \leqq 1$. In what follows we shall be concerned solely with magnitude results, like this asymptotic magnitude result of Breiman's, but the reader should be aware that under additional hypotheses Breiman also shows that $T(x)$, the time required to have a fortune exceeding x, has an expectation which is asymptotically minimized by the above fixed-fraction strategy.

The problem of how to apportion capital between various random variables is exactly the problem of portfolio selection, and so it is correct to suppose that these results on optimal allocation of capital are of considerable interest to economists, as Kelly recognized ([10], p. 926). He also prophetically realized that economists, familar with logarithmic utility, could easily misunderstand his result and think, incorrectly, that the choice of maximizing the expected value of the log of the fortune depended upon using logarithmic utility for money. For discussion see [15], p. 216 and [17]. An interesting concise discussion of the 'capital growth model of Kelly [10], Breiman [4], and Latané [11]' from an economic point of view can be found in [3].

A brief discussion of Kelly's proof will motivate his criterion and allow us to make an important conceptual distinction between his results and Breiman's. Suppose a gambler bets the fixed fraction γ of his capital at each toss of the p-coin. Kelly considers the exponential growth rate

$$G = \lim \log [(F_n/F_0)^{1/n}].$$

If our gambler has W wins and L losses in the first n trials, $F_n = (1+\gamma)^W(1-\gamma)^L F_0$, so

$$G = \lim \left(\frac{W}{n} \log(1+\gamma) + \frac{L}{n} \log(1-\gamma)\right) = p \log(1+\gamma) + q \log(1-\gamma),$$

by the law of large numbers. The growth rate G is maximized by $\gamma = p - q$, and if he uses another γ his G will be less and therefore eventually so will his fortune. A complication enters when we consider, as Kelly did not, strategies which are not fixed-fraction strategies. In that case we can have different strategies with the same G, e.g., use $\gamma = 1$ for the first 1000 trials and then use $\gamma = p - q$. This complication is intrinsic in the use of G and it has consequences which are quite serious for any application. For example, two strategies which at trial n give fortunes, respectively, of 1 and $\exp(\sqrt{n})$, both have $G = 1$! It is obviously unsatisfactory to regard these two strategies with the same G as 'the same', but it is done because using G makes it easy to extend the Kelly results to more general situations which involve more general random variables; using G the argument is a simple one employing either the law of large numbers or techniques which 'rely heavily on those used to generalize the law of large numbers' ([15], p. 218). Breiman understood the problems created by using G and so he considered F_n/F_n^*, not $(F_n/F_n^*)^{1/n}$. This is mathematically more difficult, but the results are more useful.

2. Definitions and lemmas

We shall consider situations with the property that at each time period a gambler can lose no more than the amount he invests, e.g., buying stock or betting on Las Vegas table games. Since there is a real limit to a gambler's liability, based on his total fortune, a broad interpretation of the phrase 'the amount he invests' will allow the inclusion of such situations as selling stock short or entering commodity futures contracts.

We suppose that there are a finite number of situations $1, 2, \cdots, N$ on which a gambler can bet various fractions of his (infinitely divisible) capital. The random variables $X_1, X_2, \cdots, X_N$ represent, respectively, the outcome of a unit bet on situations $1, 2, \cdots, N$. Because the loss can be no more than the investment, $X_k \geqq -1$ for $1 \leqq k \leqq N$. (Breiman considers the amount returned to the gambler after he has given up his bet in order to play, a real example of this sequence of events being betting on the horses. Here the amount the gambler gets back is $\geqq 0$, which corresponds to the amount he wins being $\geqq -1$). We further suppose, with no loss of applicability, that in all of what follows each X_k has an expectation, i.e., that $E(|X_k|)$ is finite. These will be the only restrictions on the random variables, and so we are considering a substantially larger class than those discrete random variables Breiman considers.

We also suppose that the gambler can repeatedly reinvest and change the proportion of the capital bet on the situations. The outcome at time j corresponds to the random variables $X_1^{(j)}, X_2^{(j)}, \cdots, X_N^{(j)}$. For each k, $1 \leqq k \leqq N$, the results of repeated betting of one unit on the kth situation is a sequence $X_k^{(1)}, X_k^{(2)}, \cdots, X_k^{(m)}, \cdots$ of independent random variables, each having the same distribution as X_k. In contrast to this independence, it is quite important for applications that $X_1, X_2, \cdots, X_N$ be allowed to be dependent.

A strategy for the game will be a sequence $\gamma^{(1)}, \cdots, \gamma^{(m)}, \cdots$ of vectors, $\gamma^{(m)} = (\gamma_1^{(m)}, \gamma_2^{(m)}, \cdots, \gamma_N^{(m)})$ giving the fractional amount $\gamma_k^{(m)}$ of the capital which at the mth bet is bet on the kth situation. Thus $\gamma_k^{(m)} \geqq 0$, $1 \leqq k \leqq N$, and $\sum_{k=1}^N \gamma_k^{(m)} \leqq 1$. We allow the possibility that $\gamma^{(m)}$ can depend, as a Borel-measurable function, on the past outcomes $X_1^{(1)}, \cdots, X_N^{(1)}$, $X_1^{(2)}, \cdots, X_N^{(2)}, \cdots, X_1^{(m-1)}, \cdots, X_N^{(m-1)}$. (Breiman includes the sure-thing bet $X_0 \equiv 1$, so that betting γ_0 on X_0 is the same thing as putting γ_0 aside; in this way his γ's always sum to 1. We shall not do this.)

Letting F_m be the fortune which is the result of m bets using $\gamma^{(1)}, \gamma^{(2)}, \cdots, \gamma^{(m)}$, and F_0 be the initial fortune,

$$F_m = F_0 \prod_{j=1}^m \left[1 + \sum_{k=1}^N \gamma_k^{(j)} X_k^{(j)}\right]. \tag{1}$$

To simplify the notation, let $X^{(j)} = (X_1^{(j)}, \cdots, X_N^{(j)})$ and denote the scalar product with $\gamma^{(j)} = (\gamma_1^{(j)}, \gamma_2^{(j)}, \cdots, \gamma_N^{(j)})$ by $\gamma^{(j)} \cdot X^{(j)}$, obtaining

$$F_m = F_0 \prod_{j=1}^m [1 + \gamma^{(j)} \cdot X^{(j)}]. \tag{2}$$

A *fixed-fraction strategy* is a strategy $\gamma^{(m)} = (\gamma_1, \gamma_2, \cdots, \gamma_N)$ which bets the same amount γ_k on situation k for all m. The result of using $\gamma = \gamma^{(j)}$, $1 \leqq j \leqq m$, for m bets is $F_m = F_0 \prod_{j=1}^N (1 + \gamma \cdot X^{(j)})$. We shall be particularly interested in 'the' fixed-fraction strategy $\gamma^* = (\gamma_1^*, \gamma_2^*, \cdots, \gamma_N^*)$ which maximizes $E(\log (F_m))$. In Lemma 3 we shall show that γ^* exists, and Lemma 1 shows in what sense it is unique. The strategy γ^* maximizes $E(\log (F_m))$ if and only if it maximizes the function

$$\phi(\gamma) = \phi(\gamma_1, \cdots, \gamma_N) = E(\log (1 + \sum \gamma_k X_k)) = E(\log (1 + \gamma \cdot X)) \tag{3}$$

over the domain

$$D = \{(\gamma_1, \cdots, \gamma_N) : \gamma_k \geqq 0, 1 \leqq k \leqq N, \sum \gamma_k \leqq 1\}. \tag{4}$$

Lemma 1.

(i) *The function* ϕ *of* (3) *is concave.*

(ii) *If* $\phi(a\alpha + (1-a)\beta) = a\phi(\alpha) + (1-a)\phi(\beta)$ *for* $0 < a < 1$ *with* $\phi(\alpha)$ *and* $\phi(\beta)$ *finite,* $\alpha \cdot X = \beta \cdot X$ *almost surely. In particular, if* $\alpha = (\alpha_1, \alpha_2, \cdots, \alpha_N)$ *and* $\beta = (\beta_1, \beta_2, \cdots, \beta_N)$ *both maximize* ϕ *over its domain D, then* $\sum \alpha_k X_k = \sum \beta_k X_k$ *a.s.*

(iii) *At* $\gamma = (\gamma_1, \cdots, \gamma_N)$ *in D with* $\sum \gamma_k < 1$, *the partial derivative* $\partial\phi/\partial\gamma_i$ *exists and equals* $E(X_i/(1 + \gamma \cdot X))$, $1 \leqq i \leqq N$.

Proof.

(i) The function $f(x) = \log (1 + x)$ is strictly concave on $(-1, \infty)$, and so for $x = (x_1, \cdots, x_N)$ a value of X, α and β in D, and $0 < a < 1$,

$$f(a\alpha \cdot x + (1-a)\beta \cdot x) \geqq af(\alpha \cdot x) + (1-a)f(\beta \cdot x), \tag{5}$$

an inequality which also holds if either $\alpha \cdot x$ or $\beta \cdot x$ is -1. Integrating (5) with respect to the probability measure P of the space on which X is defined,

$$\phi(a\alpha+(1-a)\beta)=\int f(a\alpha \cdot X+(1-a)\beta \cdot X)\, dP$$

$$\geqq \int (af(\alpha \cdot X)+(1-a)f(\beta \cdot X)\, dP = a\phi(\alpha)+(1-a)\phi(\beta).$$

(ii) Since ϕ is concave the set where it attains its max is convex. If $\phi(\alpha)=\phi(\beta)$ is a max, then for $0<a<1$, $\phi(a\alpha+(1-a)\beta)= a\phi(\alpha)+(1-a)\phi(\beta)$, or

$$\int (f(a \cdot \alpha X+(1-a)\beta \cdot X)-af(\alpha \cdot X)-(1-a)f(\beta \cdot X))\, dP=0. \tag{6}$$

From (5) and (6),

$$f(a\alpha \cdot X+(1-a)\beta \cdot X)=af(\alpha \cdot X)+(1-a)f(\beta \cdot X) \quad \text{a.s.} \tag{7}$$

Because f is strictly concave, $a\alpha \cdot X+(1-a)\beta \cdot X=\alpha \cdot X=\beta \cdot X$ at all values of X where f is finite. Both sides of (7) are $-\infty$ only at values of X where $\alpha \cdot X=\beta \cdot X=-1$. In any case, $\alpha \cdot X=\beta \cdot X$ almost surely.

(iii) Choose $\varepsilon>0$ so that $\sum \gamma_k<1-\varepsilon$. Then $1/|1+\gamma \cdot X| \leqq 1/\varepsilon$. The difference quotient for $\partial\phi/\partial\gamma_i$ is

$$\int \frac{\log\left(1+\sum\limits_{k\neq i} \gamma_k X_k+(\gamma_i+\Delta\gamma_i)X_i\right)-\log(1+\sum \gamma_k X_k)}{\Delta\gamma_i}\, dP. \tag{8}$$

Let $x=(x_1, \cdots, x_N)$ be a value of X and consider the function $g(\gamma_i)= \log(1+\sum_{k\neq i} \gamma_k x_k+\gamma_i x_i)$. By the mean value theorem,

$$\left|\frac{g(\gamma_i+\Delta\gamma_i)-g(\gamma_i)}{\Delta\gamma_i}\right|=\frac{|x_i|}{|1+\sum \gamma_k x_n+\xi_i x_i|},$$

$0<\xi_i<\Delta\gamma_i$. So the integrand in (8) is dominated by the L^1 function $|X_i|/\varepsilon$ for $\Delta\gamma_i$ small. Result (iii) follows from the Lebesgue dominated convergence theorem.

Here is a simple example which conceptually illustrates a practical use of the Kelly–Breiman criterion: maximize $E(\log F_m)$.

Example 1. Define two random variables X_1 and X_2 by flipping a fair coin: if heads, then $X_1=100$ and $X_2=-10$, if tails, then $X_1=-1$ and $X_2=1$. The payoff from X_1 is far superior to the payoff from X_2, but because X_1 and X_2 are (completely) correlated and have payoffs with opposite signs, the criterion will mix both in order to smooth out the rate of capital growth. A simple calculation shows that $\phi(\gamma)=E(\log(1+\gamma_1X_1+\gamma_2X_2))$ is maximized over D on the face $\gamma_1+\gamma_2=1$ at $\gamma_1^*\cong 0.54$ and $\gamma_2^*\cong 0.46$. (Lemma 3 will discuss the basic problem created by maxima occurring at non-interior points of D.) The extent to which the criterion will sacrifice expectation is surprising: $E(0.54X_1+0.46X_2)\cong 24.7$ vs. $E(X_1)=49.5$.

A fascinating example of the use of this criterion, in which the underlying idea is the same as this example, is in hedging a warrant against its stock as described in [15], pp. 220–222.

Example 2. For $\lambda>0$, let X have density $\lambda e^{-\lambda(x+1)}$ for $x\geqq -1$, and 0 for $x<-1$; an exponential shifted to allow losses. We shall show that there is a unique γ^*, $0\leqq\gamma^*<1$, which maximizes $\phi(\gamma)=E(\log(1+\gamma X))$, $0\leqq\gamma\leqq 1$; $\gamma^*=0$ iff $\lambda\geqq 1$.

By Lemma 1,

$$\phi'(\gamma)=-\int_{-1}^{\infty}\frac{x}{1+\gamma x}\lambda e^{-\lambda(x+1)}\,dx.$$

Further,

$$\phi''(\gamma)=\int_{-1}^{\infty}\frac{-x^2}{(1+\gamma x)^2}\lambda e^{-\lambda(x+1)}\,dx<0,$$

so ϕ is strictly concave on $[0,1)$. Since $\phi'(0)=E(X)=\lambda^{-1}-1$, the strict concavity of ϕ shows that $\gamma^*=0$ iff $\phi'(0)\leqq 0$.

It remains to show that for $0<\lambda<1$ there is a unique maximizing point γ^* with $0<\gamma^*<1$. By a change of variable, $\phi'(\gamma)=(\lambda/\gamma^2)e^a[g(a)-E1(a)]$, where $a=\lambda((1/\gamma)-1)$, $g(a)=e^{-a}/(a+\lambda)$, and the exponential integral $\mathrm{E1}(a)=\int_a^\infty e^{-t}/t\,dt$. Since $E1(0)=\infty$ and $g(0)=1/\lambda$, $\phi'(\gamma)<0$ for γ close to 1; as ϕ is strictly convex, there is thus a unique point $\gamma^*, 0<\gamma^*<1$, at which ϕ is maximized.

For future reference we note that it is not obvious that ϕ is continuous at 1; part of the computation involves an integration by parts and a change of variable to obtain $\phi(\gamma)=\log(1-\gamma)+e^aE1(a)$, a as above. The expansion $E1(a)=e^{-a}(-\log a-\gamma_0+o(a))$ ([1], p. 229), where $\gamma_0=0.577\cdots$ is Euler's constant, shows that $\lim_{\gamma\to 1}\phi(\gamma)=-\log(\lambda)-\gamma_0$.

In Example 2 $\gamma^*=0$ iff $E(X)\leqq 0$, i.e., a gambler bets on X only if it has positive expectation. This is a special case of a more general result. Breiman [4], p. 65, calls a game *favorable* if there is a strategy such that the associated fortune F_n tends almost surely to ∞ with n, and he shows that this condition is equivalent to $\phi(\gamma^*)$ being positive ([4], Proposition 3, p. 68). Lemma 2 establishes the equivalence with the intuitive Condition (iv).

Lemma 2. *The following are equivalent:*

(i) *There is a strategy with the associated fortune*

$$F_n \to \infty \quad a.s. \quad as \quad n \to \infty.$$

(ii) $F_n^* \to \infty$ *a.s.*

(iii) $\phi(\gamma^*) > 0$.

(iv) $E(X_i) > 0$ *for at least one* i, $1 \leqq i \leqq N$.

Proof.

(i) implies (ii). In Theorem 1 we show that F_n/F_n^* tends almost surely to a finite limit.

(ii) implies (iii). If $\phi(\gamma^*) = 0$, then $F_n^* = F_0$ for all n.

(iii) implies (iv). If $\gamma^* \cdot X = 0$, then $\phi(\gamma^*) = 0$. So $\gamma_i^* > 0$ for some i with X_i not identically 0. Fix all variables in ϕ but γ_i and set $\Psi(\gamma_i) = \phi(\gamma_1^*, \gamma_2^*, \cdots, \gamma_{i-1}^*, \gamma_i, \gamma_{i+1}^*, \cdots, \gamma_N^*)$, a concave function which has a positive max at γ_i^*. If $E(X_i) \leqq 0$, then $\Psi'(0) \leqq 0$, and Ψ has a local max at 0 and so has a global max there because it is concave. Hence $E(X_i) > 0$.

(iv) implies (i). Define Ψ by setting all the variables but γ_i equal to 0 in $\phi : \Psi(0, 0, \cdots, 0, \gamma_i, 0, \cdots, 0)$. As $E(X_i) > 0$, $\Psi'(0) > 0$ and so $\Psi(\gamma_i) > 0$ for γ_i close to 0. Thus $\phi(\gamma^*) > 0$. Since $\log(F_n^*/F_0) = \sum_1^n \log(1 + \gamma^* \cdot X^{(j)})$ and $E((\log F_n^*/F_0)/n) = \phi(\gamma^*) > 0$, the strong law of large numbers shows that $F_n^*/F_0 \to \infty$ almost surely.

Example 3. Let X_1 and X_2 be the coordinates of a point distributed uniformly on $[-1, b] \times [-1, b]$. Then $\phi(\gamma_1, \gamma_2) = E(\log(1 + \gamma_1 X_1 + \gamma_2 X_2))$ has a maximum at $\gamma^* = (\frac{1}{2}, \frac{1}{2})$ if $b \geqq \log(16) - 1$.

If (γ_1, γ_2) is a point where ϕ attains its max, then so is (γ_2, γ_1) by symmetry. Since ϕ is concave, $(\frac{1}{2})(\phi(\gamma_1, \gamma_2) + \phi(\gamma_2, \gamma_1)) \leqq \phi(\frac{1}{2}(\gamma_1 + \gamma_2), \frac{1}{2}(\gamma_1 + \gamma_2))$, and we may look for the maximum of ϕ along the diagonal (γ, γ), $0 \leqq \gamma \leqq \frac{1}{2}$. Then

$$\frac{d}{d\gamma} \phi(\gamma, \gamma) = \frac{1}{(b+1)^2} \int_{-1}^{b} \int_{-1}^{b} \frac{x_1 + x_2}{(1 + \gamma(x_1 + x_2))} dx_1 \, dx_2.$$

The second derivative is <0, and ϕ' is decreasing. A direct but tedious integration and calculation shows that

$$\lim_{\gamma \to \frac{1}{2}^-} \frac{d}{d\gamma} \phi(\gamma, \gamma) = 2\left(1 - \left(\frac{\log 16}{(b+1)}\right)\right).$$

Hence $\phi(\gamma, \gamma)$, which can be shown to be continuous on $[0, \frac{1}{2}]$, increases up to its max at $(\frac{1}{2}, \frac{1}{2})$ as long as $b \geqq \log(16) - 1 = 1.77 \cdots$.

It is interesting to compare this situation with betting on only one variable, say X_1. Then $\phi_1(\gamma_1) = E(\log(1 + \gamma_1 X_1))$ is continuous on $[0, 1]$. Continuity on $[0, 1)$ follows from Lemma 1 or inspection. Because of the singularity at

$\gamma_1 \cdot X_1 = -1$ in the integrand, continuity at 1 is not by inspection; one must compute $\phi_1(1) = \log(1+b) - 1$ and show it is equal to the limit. (The situation in two variables, which we dismissed with a word, is not easier.)

The function ϕ_1 is differentiable on $[0, 1)$ by inspection or Lemma 1. It is only differentiable at 1 in an extended sense; we can show that $\phi_1'(1) = -\infty$ and that $\lim_{\gamma \to 1} \phi_1'(\gamma) = -\infty$.

The function ϕ_1 has a unique maximum at a point $\gamma_1^*, 0 < \gamma_1^* < 1$. From Lemma 2, $\gamma_1^* = 0$ iff $E(X) = (b-1)/2 \leqq 0$. The existence of $\gamma_1^* < 1$ follows from $\phi_1'(0) = (b-1)/2$ and $\phi_1'(1) = -\infty$, the uniqueness from strict concavity.

The surprising fact is that for one variable X_1 a gambler does not bet all his fortune no matter what b is, but he does bet all his fortune on two independent copies of X_1 for b large enough.

The reader who has carried out the calculations of Examples 2 and 3 knows that, because of the possible singularity on $\sum \gamma_k = 1$, it is not clear that ϕ attains a maximum, and the differentiability of ϕ on the boundary $\sum \gamma_k = 1$ is even less clear.

Think of a continuous strictly increasing concave function f on $[0, 1]$ and redefine it at 1 so that its value there is less than $f(0)$. If this redefined function were $E(\log(1+\gamma X))$, then X would be a most interesting game with no Kelly–Breiman optimal strategy: with unit fortune, if a gambler bet an amount less than 1 he could always do better by betting slightly more, but betting all would be worst.

One result of Lemma 3 is that there is an optimal γ^* so no game can have the property discussed in the paragraph above. Another result of Lemma 3 is that ϕ is continuous, when finite. This is important because when we compute γ^*, a numerical calculation which will generally give γ^* to a certain number of decimals, we want to know that using this approximation to the exact γ^* will give close to optimal performance.

The other result is a substitute for differentiation when γ^* has $\sum \gamma_k^* = 1$, which allows us to derive the basic inequalities (9) and (10). Note that if all the random variables $X_1, X_2, \cdots, X_N$ are discrete, with a finite number of values, as they are in [4], then we can differentiate ϕ at γ^*: for then if $\gamma^* \cdot X$ equals -1 it does so with positive probability and $\phi(\gamma^*) = -\infty < \phi(0) = 0$, contrary to $\phi(\gamma^*)$ a maximum; thus ϕ is actually defined on a neighborhood of γ^* (which may extend outside D) and is differentiable as in Lemma 1. The problems which Lemma 3 resolves are those which arise from more general random variables.

Lemma 3.

(i) *The function ϕ is continuous where finite, and attains a maximum at a point γ^* in D.*

(ii) *Let* $K=\{k:\gamma_k^*\neq 0\}$. *For* k *in* K, $E(X_k/(1+\gamma^*\cdot X))$ *is finite and non-negative. If* k *belongs to* K, *then for all* m

$$E(X_k/(1+\gamma^*\cdot X))\geqq E(X_m/(1+\gamma^*\cdot X)). \tag{9}$$

If both k *and* m *belong to* K,

$$E(X_k/(1+\gamma^*\cdot X))=E(X_m/(1+\gamma^*\cdot X)). \tag{10}$$

Proof. Suppose that $\gamma^{(n)}$ converges to $\gamma^{(0)}$ with $\phi(\gamma^{(0)})$ finite. Denote the positive and negative parts of log by $\log^+$ and $\log^-$: $\log^+(x)=\log x$ if $x\geqq 1$, $\log^+(x)=0$ if $x\leqq 1$ and $\log^-(x)=-\log(x)$ if $x\leqq 1$, $\log^-(x)=0$ if $x\geqq 1$. Since $\log^+(1+\gamma\cdot X)\leqq|\gamma\cdot X|\leqq\max|X_k|$, ϕ is infinite only if it is $-\infty$. By the Lebesgue dominated convergence theorem $\lim\int\log^+(1+\gamma^{(n)}\cdot X)\,dP=\int\log^+(1+\gamma^{(0)}\cdot X)\,dP$, by Fatou's lemma $\liminf\int\log^-(1+\gamma^{(n)}\cdot X)\,dP\geqq\int\log^-(1+\gamma^{(0)}\cdot X)\,dP$, and putting these two facts together, $\limsup\phi(\gamma^{(n)})\leqq\phi(\gamma^{(0)})$. Thus ϕ is upper semicontinuous and therefore attains its maximum on the compact set $\{\gamma \text{ in } D:\phi(\gamma)\geqq 0\}$.

For $0<a<1$, $\phi(a\gamma+(1-a)\gamma^{(0)})\geqq a\phi(\gamma)+(1-a)\phi(\gamma^{(0)})$, and so $\liminf_{a\to 0}\phi(a\gamma+(1-a)\gamma^{(0)})\geqq\phi(\gamma^{(0)})$ if $\phi(\gamma)$ is finite, i.e., ϕ is continuous along lines directed towards $\gamma^{(0)}$ from points where ϕ is finite.

Suppose that $\gamma^{(1)}$ and $\gamma^{(2)}$ are in D with $\phi(\gamma^{(2)})$ finite, and let $a\in[0,1)$. Now

$$\begin{aligned}1+a\gamma^{(1)}\cdot X+(1-a)\gamma^{(2)}\cdot X&\geqq 1+a\textstyle\sum\gamma_k^{(1)}(-1)+(1-a)\gamma^{(2)}\cdot X\\&\geqq(1-a)+(1-a)\gamma^{(2)}\cdot X\\&=(1-a)(1+\gamma^{(2)}\cdot X).\end{aligned}$$

Since $\log^-(1+z)$ is a decreasing function of z, $\log^-(1+a\gamma^{(1)}\cdot X+(1-a)\gamma^{(2)}\cdot X)\leqq\log^-(1-a)(1+\gamma^{(2)}\cdot X)$ which equals $-\log(1-a)-\log(1+\gamma^{(2)}\cdot X)$ when $(1-a)(1+\gamma^{(2)}\cdot X)\leqq 1$, and equals 0 otherwise. Hence $\int\log^-(1+a\gamma^{(1)}\cdot X+(1-a)\gamma^{(2)}\cdot X)\,dP\leqq-\log(1-a)+\int\log^-(1+\gamma^{(2)}\cdot X)\,dP<\infty$, and thus $\phi(a\gamma^{(1)}+(1-a)\gamma^{(2)})$ is finite for $a\neq 1$.

Let a point γ be a given at which ϕ is finite. For $1\leqq k\leqq N$, let $e^{(k)}$ be the vector in D whose kth coordinate is 1, $e_i^{(k)}=\delta_{ik}$, $e^{(0)}=0$, and set $\gamma^{(k)}=ae^{(k)}+(1-a)\gamma$, a in $(0,1)$, for $0\leqq k\leqq N$. Note that D is the convex hull $\mathrm{co}(e^{(0)},e^{(1)},\cdots,e^{(N)})$. As we have seen above, $\phi(\gamma^{(k)})$ is finite and so ϕ is continuous on the line joining $\gamma^{(k)}$ to γ. Given $\varepsilon>0$, by choosing a small enough we have $|\phi(\gamma^{(k)})-\phi(\gamma)|\leqq\varepsilon$ for $0\leqq k\leqq N$. For any vector v in the convex hull $U=\mathrm{co}(\gamma^{(0)},\gamma^{(1)},\cdots,\gamma^{(N)})$, $v=\sum a_k\gamma^{(k)}$, $a_k\geqq 0$, $\sum a_k=1$, we have $\phi(v)\geqq\sum a_k\phi(\gamma^{(k)})\geqq\phi(\gamma)-\varepsilon$. The convex hull U is easily seen to have an interior (relative to D). Since ϕ is upper semicontinuous, $V=\{\alpha:\phi(\alpha)<\phi(\gamma)+\varepsilon\}$ is open. Therefore for v in the neighborhood $U^0\cap V$, $|\phi(v)-\phi(\gamma)|\leqq\varepsilon$ and ϕ is continuous at γ.

For γ in D with $\sum \gamma_i < 1$, and $0 \leqq a \leqq 1$, define $\Psi(a) = \phi(a\gamma + (1-a)\gamma^*)$. We have seen that Ψ is continuous on $[0, 1]$. Given $0 < \varepsilon < 1$, for a in $[\varepsilon, 1]$, $1 + (a\gamma + (1-a)\gamma^*) \cdot X \geqq \varepsilon(1 - \sum \gamma_i) > 0$. As in Lemma 1, $|(\gamma - \gamma^*) \cdot X/(1 + (a\gamma + (1-a)\gamma^*) \cdot X)|$ is bounded by the L^1 function $|(\gamma - \gamma^*) \cdot X/\varepsilon(1 - \sum \gamma_i)|$ and we may use the Lebesgue dominated convergence theorem to justify differentiating under the integral to obtain

$$\Psi'(a) = \int \frac{(\gamma - \gamma^*) \cdot X}{1 + (a\gamma + (1-a)\gamma^*) \cdot X} \, dP.$$

Since ε was arbitrary, this holds on $(0, 1]$.

If $\sum \gamma_k^* < 1$, then Lemma 1 establishes the fact that the expectations in (10) are 0. As in Lemma 2, if m is not in K, then fixing all the variables in ϕ but γ_m and considering that concave function with a max at 0 shows that $\partial\phi/\partial\gamma_m(\gamma^*) \leqq 0$ and (9) follows.

Now suppose that $\sum \gamma_k^* = 1$ and let $x = (x_1, \cdots, x_N)$ be some value of the random variable X in which not all the $x_k = -1$ for k in K. Note that the event $x_k = -1$ for all k in K has probability 0 since the integrand in the integral defining ϕ is $-\infty$ there and $\phi(\gamma^*)$ is finite.

For $0 \leqq a < 1$ and γ in D^0, define

$$f(a) = \log(1 + (a\gamma + (1-a)\gamma^*) \cdot x).$$

The function f is finite because x was so chosen, and is differentiable with

$$f'(a) = \frac{(\gamma - \gamma^*) \cdot x}{1 + (a\gamma + (1-a)\gamma^*) \cdot x}.$$

Because $f'' \leqq 0$, $f'(a)$ increases to $(\gamma - \gamma^*) \cdot x/(1 + \gamma^* \cdot x)$ as a decreases to 0. For $a \neq 0$, $(\gamma - \gamma^*) \cdot X/(1 + (a\gamma + (1-a)\gamma^*) \cdot X)$ in L^1 we may apply the B. Levi theorem to obtain

$$\lim_{a \downarrow 0} \Psi'(a) = \int \frac{(\gamma - \gamma^*) \cdot X}{1 + \gamma^* \cdot X} \, dP.$$

By the mean value theorem, $(\Psi(a) - \Psi(0))/a = \Psi'(b)$, $0 < b < a$. Then, because Ψ' is increasing, $\lim_{a\downarrow 0} \Psi'(a) = \lim_{a\downarrow 0} ((\Psi(a) - \Psi(0))/a)$. Since $\phi(\gamma^*)$ is a maximum the right-hand side is $\leqq 0$ and we obtain the basic

$$\int \frac{(\gamma - \gamma^*) \cdot X}{1 + \gamma^* \cdot X} \, dP \leqq 0. \tag{11}$$

For $0 < \varepsilon < 1$ and k in K, the choice $\gamma_j = \gamma_j^*$ for $j \neq k$ and $\gamma_k = \gamma_k^*(1 - \varepsilon)$ in (11) gives $-\int X_k/(1 + \gamma^* \cdot X) \, dP \leqq 0$, and $\int X_k/(1 + \gamma^* \cdot X) \, dP$ is non-negative and therefore finite.

For k in K and any m, and $0 < \varepsilon < \gamma_k^*$, the choice $\gamma_j = \gamma_j^*$ for j neither k nor

m, $\gamma_n = \gamma_n^* - \varepsilon$, $\gamma_m = \gamma_m^* + \varepsilon$ in (11) gives (9). Finally, (10) follows from (9) by symmetry.

3. Theorems

If F_n is a gambler's fortune at the nth time period, obtained by using some strategy $\gamma^{(1)}, \gamma^{(2)}, \cdots, \gamma^{(n)}, \cdots$, and F_n^* is the fortune obtained by using the fixed fraction strategy $\gamma^*, \gamma^*, \cdots, \gamma^*, \cdots$, then Breiman concludes ([4], Theorem 2, p. 72) that $\lim F_n/F_n^*$ almost surely exists and $E(\lim F_n/F_n^*) \leqq 1$. We extend these results in Theorem 1 below. Two comments are in order.

First, the advantage in using γ^*, as indicated by the fact that $E(\lim F_n/F_n^*) \leqq 1$, does not require passage to the limit. This was noted by Durham, for the case of two branching processes, in the proof of Theorem 1 of [6], p. 571. In fact, given that the limiting result is true and given a finite strategy $\gamma^{(1)}, \cdots, \gamma^{(m)}$, extend it by setting $\gamma^{(j)} = \gamma^*$ for $j > m$; then $F_n/F_n^* = F_m/F_m^*$ for $n \geqq m$ and $E(F_m/F_m^*) \leqq 1$. This should reassure the careful investor who wonders whether a strategy good in the long run may not be inferior in any practical number of trials—disregard of this point leads to an overevaluation of games of the type which produces the St. Petersburg paradox.

Second, by our analysis of ϕ, we are able to show in Theorem 1 that the presence of the expectation in $E(\lim F_n/F_n^*) \leqq 1$ raises serious problems in any superficial attempt to use this as an indication of the superiority of γ^*.

Theorem 1. *Let F_n be the fortune obtained by using a strategy $\gamma = \gamma^{(1)}, \cdots, \gamma^{(n)}$ for n repeated investment periods, and let F_n^* be the fortune obtained by using the fixed fraction strategy γ^*. Then*

(i) *F_n/F_n^* is a supermartingale with $E(F_n/F_n^*) \leqq 1$. Consequently, $\lim F_n/F_n^*$ exists almost surely as a finite number and $E(\lim F_n/F_n^*) \leqq 1$.*

(ii) *Suppose that γ bets only on those X_k with $\gamma_k^* > 0$, i.e., that $\gamma_l^{(j)} = 0$ if $l \notin K$, $1 \leqq l \leqq N$ and all j. If $\sum \gamma_k^* = 1$, then further suppose that $\sum \gamma_k^{(j)} = 1$ for all j. Then F_n/F_n^* is a martingale with $E(F_n/F_n^*) = 1$.*

Proof. Let m be given and let $\mathscr{C}_m$ be the sigma-algebra generated by $X_k^{(j)}$, $1 \leqq k \leqq N$, $1 \leqq j \leqq m$. Then

$$E\left(\frac{F_{m+1}}{F_{m+1}^*} \,\middle|\, \mathscr{C}_m\right) = E\left(\frac{1+\gamma^{(m+1)} \cdot X^{(m+1)}}{1+\gamma^* \cdot X^{(m+1)}} \cdot \frac{F_m}{F_m^*} \,\middle|\, \mathscr{C}_m\right).$$

Since F_m/F_m^* is $\mathscr{C}_m$-measurable, this equals

$$\frac{F_m}{F_m^*} E\left(\frac{1+\gamma^{(m+1)} \cdot X^{(m+1)}}{1+\gamma^* \cdot X} \,\middle|\, \mathscr{C}_m\right).$$

Because $\gamma^{(m+1)}$ is a strategy depending on the past values of the $X_k^{(j)}$, it too is

$\mathscr{C}_m$-measurable. That, together with the independence of $X^{(m+1)}$ and the previous $X^{(j)}$, shows this equals

$$\frac{F_m}{F_m^*} E\left(\frac{1}{1+\gamma^* \cdot X}\right) + \sum_{k=1}^{N} \gamma_k^{(m+1)} E\left(\frac{X_k^{(m+1)}}{1+\gamma^* \cdot X}\right).$$

Since $X^{(m+1)}$ has the same distribution as X, this can be written

$$\frac{F_m}{F_m^*}\left[E\left(\frac{1}{1+\gamma^* \cdot X}\right) + \sum_{k=1}^{N} \gamma_k^{(m+1)} E\left(\frac{X_k}{1+\gamma^* \cdot X}\right)\right] = A, \quad \text{say.} \tag{12}$$

By Lemma 3, $E(X_k/(1+\gamma^* \cdot X))$ equals a constant c for k in K and is $\leqq c$ for all k; this constant $c = 0$ if $\sum \gamma_k^* < 1$. Hence

$$A \leqq \frac{F_m}{F_n^*}\left[E\left(\frac{1}{1+\gamma^* \cdot X}\right) + \sum \gamma_k^* \cdot c\right] = \frac{F_m}{F_m^*} E\left[\frac{1+\gamma^* \cdot X}{1+\gamma^* \cdot X}\right] = \frac{F_m}{F_m^*}.$$

We have shown that F_n/F_n^* is a positive supermartingale, with $E(F_n/F_n^*) \leqq E(F_{n-1}/F_{n-1}^*) \leqq \cdots \leqq E(F_0/F_0) = 1$, and so by the supermartingale convergence theorem it converges almost surely to a finite limit. Using Fatou's lemma, $E(\lim F_n/F_n^*) \leqq \liminf E(F_n/F_n^*) \leqq 1$.

An examination of (12) shows that under the conditions of (ii), $E(F_{m+1}/F_{m+1}^* \mid \mathscr{C}_m) = F_m/F_m^*$, and (ii) follows.

The requirement in Theorem 1(ii) that if $\sum \gamma_n^* = 1$ then we must have $\sum \gamma_k^{(j)} = 1$, for all j, in order to be sure to get a martingale, is made clear by a one-variable example, Let $X \equiv 2$. Then $\phi(\gamma) = \log(1+2\gamma)$ is maximal at $\gamma^* = 1$. The gambler will do worse betting any amount less than 1, even though he still bets on the same random variable as γ^* does, the key observation being $\phi'(1) > 0$. In general, the 'partial derivatives' $E(X_k/(1+\gamma^* \cdot X))$, k in K, may be positive if $\sum \gamma_k^* = 1$, whereas they are all 0 if $\sum \gamma_k^* < 1$.

The surprising result of Theorem 1 is the broad conditions in (ii) under which $E(F_n/F_n^*) = 1$. To see what the surprise is, we shall superficially interpret Theorem 1(i): since 'on the average', and 'for large n', $F_n/F_n^* \leqq 1$, the gambler 'does better' with F_n^* than with F_n. But then Theorem 1(ii) tells us that if the gambler simply bets on the same variables as γ^* does, *but in any proportions at all*, and if γ^* bets all so does he, then $E(F_n/F_n^*) = 1$. So with the same intuitive interpretation as above, 'on the average' the gambler does the same with F_n as with F_n^*, so it really does not matter which strategy he uses! But we know that it does matter. For example, in a repeated biased-coin toss, if he plays a fixed fraction strategy betting an amount $\gamma \neq \gamma^* = p - q$, then almost surely $F_n/F_n^* \to 0$. *Yet we have* $E(F_n/F_n^*) = 1$. In general it will not help to look at $\lim F_n/F_n^*$. For example, if on the first flip of the coin he bets all his fortune, and from then on he bets $p - q$, $F_n/F_n^* > 1$ with probability p.

Theorem 2 will help our understanding of this situation by showing that F_n^* is the only denominator with $E(F_n/F_n^*) \leqq 1$ for *all* F_n; in fact $E(F_n^*/F_n) > 1$ if F_n does not come from a strategy equivalent to using γ^* repeatedly. The suspicious reader will note that this characterization of the sense in which γ^* is optimal contains an expectation. Anyone attempting to state intuitively the result of Theorem 2(i) in the form 'F_n^* is better than F_n because, on the average, $F_n^*/F_n \geqq 1$', should also be willing to apply the same interpretation to Theorem 1(ii) and conclude that often, '$F_n/F_n^* = 1$ on the average and so F_n and F_n^* are often the same after all'.

Theorem 2.

(i) F_n^*/F_n *is a submartingale with* $E(F_n^*/F_n) \geqq 1$. $\text{Lim}\, F_n^*/F_n$ *almost surely exists as an extended real number and* $E(\lim F_n^*/F_n) \geqq 1$.

(ii) $E(F_n^*/F_n) = 1$ *iff* $\gamma^{(1)}, \gamma^{(2)}, \cdots, \gamma^{(n)}$ *are all equivalent to* γ^*, *i.e., iff* $\gamma^{(j)} \cdot X = \gamma^* \cdot X$ *almost surely for almost all values of* $X^{(1)}, \cdots, X^{(j)}$ *(of which* $\gamma^{(j)}$ *is a function) for* $1 \leqq j \leqq n$.

Proof. By Theorem 1, the non-negative $\lim F_n/F_n^*$ almost surely exists, and so $\lim F_n^*/F_n$ almost surely exists as an extended real number.

As in Theorem 1,

$$E\left(\frac{F_{m+1}^*}{F_{m+1}} \,\middle|\, \mathscr{C}_m\right) = \frac{F_m^*}{F_m} E\left(\frac{1+\gamma^* \cdot X^{(m+1)}}{1+\gamma^{(m+1)} \cdot X^{(m+1)}} \,\middle|\, \mathscr{C}_m\right).$$

Suppose that $(X^{(1)}, \cdots, X^{(m)})$ takes on the value ω in R^{mN}, at which point $\gamma^{(m+1)}$ takes on the value $\gamma^{(m+1)}(\omega)$. Then

$$(13)\quad E\left(\frac{1+\gamma^* \cdot X^{(m+1)}}{1+\gamma^{m+1} \cdot X^{(m+1)}} \,\middle|_{(X^{(1)},\cdots,X^{(m)})=\omega}\right) = E\left(\frac{1+\gamma^* \cdot X^{(m+1)}}{1+\gamma^{(m+1)}(\omega) \cdot X^{(m+1)}}\right) = B, \quad \text{say,}$$

because $X^{(m+1)}$ is independent of the values of $(X^{(1)}, \cdots, X^{(m)})$ ([5], Corollary 4.38, p. 80). By Jensen's inequality,

$$(14)\quad B \geqq \exp\left(E(\log(1+\gamma^* \cdot X^{(m+1)})) - E(\log(1+\gamma^{(m+1)}(\omega) \cdot X^{(m+1)}))\right) = C, \quad \text{say.}$$

By Lemma 1, $C > 1$ unless $\gamma^* \cdot X = \gamma^{m+1}(\omega) \cdot X$ almost surely, in which case $C = 1$. Thus

$$(15)\qquad E\left(\frac{F_{m+1}^*}{F_{m+1}} \,\middle|\, \mathscr{C}_m\right) \geqq \frac{F_m^*}{F_m}$$

with equality holding iff $\gamma^* \cdot X = \gamma_{(\omega)}^{(m+1)} \cdot X$ almost surely for almost all values ω in the range of $(X^{(1)}, \cdots, X^{(m)})$. If $E(F_{m+1}^*/F_{m+1}) = 1$, then $1 = E(E(F_{m+1}^*/F_{m+1} \mid \mathscr{C}_m)) \geqq E(F_m^*/F_m) \geqq \cdots \geqq E(F_0/F_0) = 1$. By (15), equality holds in (15) and therefore $\gamma^* \cdot X = \gamma^{(m+1)}(\omega) \cdot X$ for almost all ω in the range of $(X^{(1)}, \cdots, X^{(m)})$. Continue for $m, m-1, \cdots, 1$, to obtain (ii).

Let $Y = \lim F_n^*/F_n$. By Theorem 1, $P(Y=0)=0$ and $E(1/Y) \leqq 1$. The function $g(x)=1/x$ is convex in $(0,\infty)$ and Jensen's inequality applies, to obtain $1 \geqq E(1/Y) \geqq 1/E(Y)$ which completes the proof.

References

[1] Abramowitz, M. and Stegun, A. (1965) *Handbook of Mathematical Functions*. National Bureau of Standards, Washington, D.C.

[2] Aucamp, D. (1975) Comment on "the random nature of stock market prices" authored by Barrett and Wright. *Operat. Res.* **23,** 587–591.

[3] Bicksler, J. and Thorp, E. (1973) The capital growth model: an empirical investigation. *J. Finance and Quant. Anal.* **VIII,** 273–287.

[4] Breiman, L. (1961) Optimal gambling systems for favorable games. *Proc. 4th Berkeley Symp. Math. Statist. Prob.* **1,** 65–78.

[5] Breiman, L. (1968) *Probability*. Addison-Wesley, New York.

[6] Durham, S. (1975) An optimal branching migration process. *J. Appl. Prob.* **12,** 569–573.

[7] Ferguson, T. (1965) Betting systems which minimize the probability of ruin. *J. SIAM* **13,** 795–818.

[8] Hewitt, E. and Stromberg, K. (1965) *Real and Abstract Analysis*. Springer-Verlag, New York.

[9] Hakansson, N. (1971) Capital growth and the mean-variance approach to portfolio selection. *J. Finance and Quant. Anal.* **VI,** 517–557.

[10] Kelly, J. (1956) A new interpretation of information rate. *Bell System Tech. J.* **35,** 917–926.

[11] Latané, H. (1959) Criteria for choice among risky ventures. *J. Political Econ.* **67,** 144–155.

[12] Latané, H., Tuttle, D., and James, C. (1975) *Security Analysis and Portfolio Management*. Wiley, New York.

[13] Olver, F. (1974) *Introduction to Asymptotics and Special Functions*. Academic Press, New York.

[14] Roberts, A. and Varberg, D. (1973) *Convex Functions*. Academic Press, New York.

[15] Thorp, E. (1971) Portfolio choice and the Kelly criterion. *Proc. Amer. Statist. Assoc, Business, Econ. and Stat. Section*, 215–224.

[16] Thorp, E. (1969) Optimal gambling systems for favorable games. *Rev. Internat. Statist. Inst.* **37,** 273–293.

[17] Thorp, E. and Whitley, R. (1972) Concave utilities are distinguished by their optimal strategies. *Coll. Math. Soc. Janos Bolya, European Meeting of Statisticians, Budapest (Hungary)*, 813–830.

18

The Effect of Errors in Means, Variances, and Covariances on Optimal Portfolio Choice*

Vijay K. Chopra and William T. Ziemba

Good mean forecasts are critical to the mean-variance framework

There is considerable literature on the strengths and limitations of mean-variance analysis. The basic theory and extensions of MV analysis are discussed in Markowitz [1987] and Ziemba & Vickson [1975]. Bawa, Brown & Klein [1979] and Michaud [1989] review some of its problems.

MV optimization is very sensitive to errors in the estimates of the inputs. Chopra [1993] shows that small changes in the input parameters can result in large changes in composition of the optimal portfolio. Best & Grauer [1991] present some empirical and theoretical results on the sensitivity of optimal portfolios to changes in means. This article examines the relative impact of estimation errors in means, variances, and covariances.

Kallberg & Ziemba [1984] examine the question of mis-specification in normally distributed portfolio selection problems. They discuss three areas of misspecification: the investor's utility function, the mean vector, and the covariance matrix of the return distribution.

They find that utility functions with similar levels of Arrow–Pratt absolute risk aversion result in similar optimal portfolios irrespective of the functional form of the utility[1]; Thus, mis-specification of the utility function is not a major concern because several different utility functions (quadratic, negative exponential, logarithmic, power) result in similar portfolio allocations for similar levels of risk aversion.

Misspecification of the parameters of the return distribution, however, does make a significant difference. Specifically, errors in means are at least ten times as important as errors in variances and covariances.

We show that it is important to distinguish between errors in variances and covariances. The relative impact of errors in means, variances, and covariances also depends on the investor's risk tolerance. For a risk tolerance of

*Reprinted, with permission, from *Journal of Portfolio Management*, 1993.

[1]For an investor with utility function U and wealth W, the Arrow–Pratt absolute risk aversion is ARA $= -U''(W)/U'(W)$. Friend and Blume [1975] show that investor behavior is consistent with decreasing ARA; that is, as investors' wealth increases, their aversion to a given risk decreases.

50, errors in means are about eleven times as important as errors in variances, a result similar to that of Kallberg & Ziemba.[2] Errors in variances are about twice as important as errors in covariances.

At higher risk tolerances, errors in means are even more important relative to errors in variances and covariances. At lower risk tolerances, the relative impact of errors in means, variances, and covariances is closer. Even though errors in means are more important than those in variances and covariances, the difference in importance diminishes with a decline in risk tolerance.

These results have an implication for allocation of resources according to the MV framework. The primary emphasis should be on obtaining superior estimates of means, followed by good estimates of variances. Estimates of covariances are the least important in terms of their influence on the optimal portfolio.

Theory

For a utility function U and gross returns $r - i$ (or return relatives) for assets $i = 1, 2, \ldots, N$, an investor's optimal portfolio is the solution to:

$$\text{maximize } Z(x) = E[U(W_0 \sum_{i=1}^{N} (r_i) x_i)]$$

$$\text{such that } x_i > 0, \quad \sum_{i=1}^{N} = 1,$$

where $Z(x)$ is the investor's expected utility of wealth, W_0 is the investor's initial wealth, the returns r_i have a distribution $F(r)$, and x_i are the portfolio weights that sum to one.

Assuming a negative exponential utility function $U(W) = -\exp(-aW)$ and a joint normal distribution of returns, the expected utility maximization problem is equivalent to the MV-optimization problem:

$$\text{maximize } Z(x) = \sum_{i=1}^{N} E[r_i] x_j - \frac{1}{t} \sum_{i=1}^{N} \sum_{j=1}^{N} x_i x_j E[\sigma_{ij}]$$

$$\text{such that } x_i > 0, \quad \sum_{i=1}^{N} x_i = 1,$$

[2]The risk tolerance reflects the investor's desired trade-off between extra return and extra risk (variance). It is the inverse slope of the investor's indifference curve in mean–variance space. The greater the risk tolerance, the more risk an investor is willing to take for a little extra return. Under fairly general input assumptions, a risk tolerance of 50 describes the typical portfolio allocations of large US pensions funds and other institutional investors. Risk tolerances of 25 and 75 characterize extremely conservative and aggressive investors, respectively.

EXHIBIT 1:
List of Ten Randomly Chosen DJIA Securities

1. Aluminum Co. of America
2. American Express Co.
3. Boeing Co.
4. Chevron Co.
5. Coca Cola Co.
6. E.I. Du Pont De Nemours & Co.
7. Minnesota Mining and Manufacturing Co.
8. Procter & Gamble Co.
9. Sears, Roebuck & Co.
10. United Technologies Co.

where $E[r_i]$ is the expected return for asset i, t is the risk tolerance of the investor, and $E[\sigma_{ij}]$ is the covariance between the returns on assets i and j.[3]

A natural question arises: How much worse off is the investor if the distribution of returns is estimated with an error? This is an important consideration because the future distribution of returns is unknown. Investors rely on limited data to estimate the parameters of the distribution, and estimation errors are unavoidable. Our investigation assumes that the distribution of returns is stationary over the sample period. If it is time-varying or nonstationary, the estimated parameters will be erroneous.

To measure how close one portfolio is to another, we compare the cash equivalent (CE) values of the two portfolios. The cash equivalent of a risky portfolio is the certain amount of cash that provides the same utility as the risky portfolio, that is, $U(\text{CE}) = Z(x)$ or $\text{CE} = U^{-1}[Z(x)]$ where, as defined before, $Z(x)$ is the expected utility of the risky portfolio.[4] The cash equivalent is an appropriate measure because it takes into account the investor's risk tolerance and the inherent uncertainty in returns, and it is independent of utility units. For a risk-free portfolio, the cash equivalent is equal to the certain return.

Given a set of asset parameters and the investors risk tolerance, a MV-optimal portfolio has the largest CE value of any portfolio of those assets. The

[3]Although the exponential utility function is convenient for deriving the MV problem with normally distributed returns, the MV framework is consistent with expected utility maximization for any concave utility function, assuming normality.

[4]For negative exponential utility, Freund [1956] shows that the expected utility of portfolio x is $Z(x) = 1 - \exp(-aE[x] + (a^2/2)\text{Var}[x])$, where $E[X]$ and $\text{Var}[x]$ are the expected return and variance of the portfolio. The cash equivalent is $\text{CE}_x = (1/a)\log(1 - Z(x))$. If returns are assumed to have a multivariate normal distribution, this is also the cash equivalent of an MV-optimal portfolio. See Dexter, Yu & Ziemba [1980] for more details.

percentage cash equivalent loss (CEL) from holding an arbitrary portfolio, x instead of an optimal portfolio o is

$$\text{CEL} = \frac{\text{CE}_o - \text{CE}_x}{\text{CE}_o}$$

where CE_o and CE_x are the cash equivalents of portfolio o and portfolio x respectively.

Data and Methodology

The data consist of monthly observations from January 1980 through December 1989 on ten randomly selected Dow Jones Industrial Average (DJIA) securities. We use the Center for Research in Security Prices (CRSP) database, having deleted one security (Allied–Signal, Inc.) because of lack of data prior to 1985. Each of the remaining twenty–nine securities had an equal probability of being chosen. The securities are listed in Exhibit 1.

MV optimization requires as inputs forecasts for: mean returns, variances, and covariances. We computed historical means ($\bar{r_i}$), variances (σ_{ii}), and covariances (σ_{ij}), and assumed that these are the 'true' values of these parameters. Thus, we assumed that $E[r_i] = \bar{r_i}$, $E[\sigma_{ii}] = \sigma_{ii}$, and $E[\sigma_{ij}] = \sigma_{ij}$. A base optimal portfolio allocation is computed on the basis of these parameters for a risk tolerance of 50 (equivalent to the parameter $a = 0.04$).

Our results are independent of the source of the inputs. Whether we use historical inputs or those based on a complete forecasting scheme, the results continue to hold as long as the inputs have errors.

Exhibit 2 gives the input parameters and the optimal base portfolio resulting from these inputs. To examine the influence of errors in parameter estimates, we change the true parameters slightly and compute the resulting optimal portfolio. This portfolio will be suboptimal for the investor because it is not based on the true input parameters.

Next we compute the cash equivalent values of the base portfolio and the new optimal portfolio. The percentage cash equivalent loss from holding the suboptimal portfolio instead of the true optimal portfolio measures the impact of errors in input parameters on investor utility.

To evaluate the impact of errors in means, we replaced the assumed true mean $\bar{r_i}$ for asset i by the approximation $\bar{r_i}(l + kz_i)$ where z_i has a standard normal distribution. The parameter k is varied from 0.05 through 0.20 in steps of 0.05 to examine the impact of errors of different sizes. Larger values of k represent larger errors in the estimates. The variances and covariances are left unchanged in this case to isolate the influence of errors in means.

The percentage cash equivalent loss from holding a portfolio that is optimal for approximate means $\bar{r_i}(1 + kz_i)$ but is suboptimal for the true means r, is

EXHIBIT 2:
Inputs to the Optimization and the Resulting Optimal Portfolio for a Risk Tolerance of 50 (January 1980–December 1989)

	Alcoa	Amex	Boeing	Chev.	Coke	Du Pont	MMM	P&G	Sears	U Tech
Means (% per month)	1.5617	1.9477	1.907	1.5801	2.1643	1.6010	1.4892	1.6248	1.4075	1.1537
Std. Dev. (% per month)	8.8308	8.4585	10.040	8.6215	5.988	6.8767	5.8162	5.6385	8.0047	8.212
Correlations										
Alcoa	1.0000									
Amex	0.3660	1.0000								
Boeing	0.3457	0.5379	1.0000							
Chev.	0.1606	0.2165	0.2218	1.0000						
Coke	0.2279	0.4986	0.4283	0.0569	1.0000					
Du Pont	0.5133	0.5823	0.4051	0.3609	0.3619	1.0000				
MMM	0.5203	0.5569	0.4492	0.2325	0.4811	0.6167	1.0000			
P&G	0.2176	0.4760	0.3867	0.2289	0.5952	0.4996	0.6037	1.0000		
Sears	0.3267	0.6517	0.4883	0.1726	0.4378	0.5811	0.5671	0.5012	1.0000	
U Tech	0.5101	0.5853	0.6569	0.3814	0.4368	0.5644	0.6032	0.4772	0.6039	1.0000
Optimal Port. Weights	0.0350	0.0082	0.0	0.1626	0.7940	0.0	0.0	0.0	0.00	0.00

then computed. This procedure is repeated with a new set of z values for a total of 100 iterations for each value of k.

To investigate the impact of errors in variances each variance forecast σ_{ii} was replaced by $\sigma_{ii}(1 + kZ_j)$. To isolate the influence of variance errors, the means and covariances are left unchanged.

Finally, the influence of errors in covariances is examined by replacing each covariance σ_{ij} $(i \neq j)$ by $\sigma_{ij} + kz_{ij}$ where z_{ij} has a standard normal distribution, while retaining the original means and variances. The procedure is repeated 100 times for each value of k, each time with a new set of z values, and the cash equivalent loss computed. The entire procedure is repeated for risk tolerances of 25 and 75 to examine how the results vary with investors' risk tolerance.

Results

Exhibit 3 shows the mean, minimum, and maximum cash equivalent loss over the 100 iterations for a risk tolerance of 50. Exhibit 4 plots the average CEL

EXHIBIT 3:
Cash Equivalent Loss (CEL) for Errors of Different Sizes

k (size of error)	Parameter with Error	Mean CEL	Min. CEL	Max. CEL
0.05	Means	0.66	0.01	5.05
0.05	Variances	0.05	0.00	0.34
0.05	Covariances	0.02	0.00	0.25
0.10	Means	2.45	0.01	15.61
0.05	Variances	0.22	0.00	1.39
0.10	Covariances	0.11	0.00	0.66
0.15	Means	5.12	0.15	24.35
0.15	Variances	0.55	0.00	3.35
0.15	Covariances	0.27	0.00	1.11
0.20	Means	10.16	0.17	36 09
0.20	Variances	0.90	0.01	4.16
0.20	Covariances	0.47	0.00	1.94

as a function of k. The CEL for errors in means is approximately eleven times that for errors in variances and over twenty times that for errors in covariances. Thus, it is important to distinguish between errors in variances and errors in covariances.[5] For example, for $k = 0.10$, the CEL is 2.45 for errors in means, 0.22 for errors in variances, and 0.11 for errors in covariances.

Our results on the relative importance of errors in means and variances are similar to those of Kallberg & Ziemba [1984]. They find that errors in means are approximately ten times as important as errors in variances and covariances considered together (they do not distinguish between variances and covariances).

Our results show that for a risk tolerance of 50 the importance of errors in covariances is only half as much as previously believed. Furthermore, the relative importance of errors in means, variances, and covariances depends upon the investor's risk tolerance.

Exhibit 5 shows the average ratio (averaged over errors of different sizes, k) of the CELs for errors in means, variances, and covariances. An investor with a high risk tolerance focuses on raising the expected return of the portfolio

[5]The result for covariances also applies to correlation coefficients, as the correlations differ from the covariances only by a scale factor equal to the product of two standard deviations.

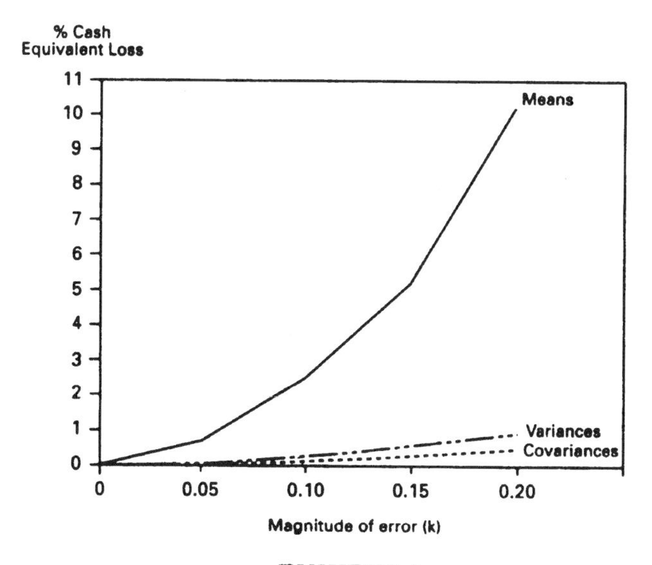

EXHIBIT 4
Mean percentage cash equivalent loss due to errors in inputs

EXHIBIT 5
Average Ratio of CELs for Errors in Means, Variances, and Covariances

Risk Tolerance	Errors in Means versus Variances	Errors in Means versus Covariances	Errors in Variances versus Covariances
25	3.22	5.38	1.67
50	1.98	22.50	2.05
75	21.42	56.84	2.68

and discounts the variance more relative to the expected return. To this investor, errors in expected returns are considerably more important than errors in variances and covariances. For an investor with a risk tolerance of 75, the average CEL for errors in means is over twenty–one times that for errors in variances and over fifty–six times that for errors in covariances.

Minimizing the variance of the portfolio is more important to an investor with a low risk tolerance than raising the expected return. To this investor, errors in means are somewhat less important than errors in variances and covariances. For an investor with a risk tolerance of 25, the average CEL for errors in expected returns is about three times that for errors in variances and about five times that for errors in covariances.

Most large institutional investors have a risk tolerance in the 40 to 60 range. Over that range, there is considerable difference in the relative importance of errors in means, variances, and covariances. Irrespective of the level of risk tolerance, errors in means are the most important, followed by errors in variances. Errors in covariances are the least important in terms of their influence on portfolio optimality.

Implications and Conclusions

Investors have limited resources available to spend on obtaining estimates of necessarily unknowable future parameters of risk and reward. This analysis indicates that the bulk of these resources should be spent on obtaining the best estimates of expected returns of the asset classes under consideration.

Sometimes, investors using the MV framework to allocate wealth among individual stocks set all the expected returns to zero (or a non-zero constant). This can lead to a better portfolio allocation because it is often very difficult to obtain good forecasts for expected returns. Using forecasts that do not accurately reflect the relative expected returns of different securities can substantially degrade MV performance.

In some cases it may be preferable to set all forecasts equal.[6] The optimization then focuses on minimizing portfolio variance and does not suffer from the error-in-means problem. In such cases it is important to have good estimates of variances and covariances for the securities, as MV optimizes only with respect to these characteristics.

Of course, if investors truly believe that they have superior estimates of the means, they should use them. In this case it may be acceptable to use historical values for variances and covariances.

For investors with moderate to high risk tolerance, the cash equivalent loss for errors in means is an order of magnitude greater than that for errors in variances or covariances. As variances and covariances do not much influence the optimal MV allocation (relative to the means), investors with moderate-to-high risk tolerance need not expend considerable resources to obtain better estimates of these parameters.

References

Bawa, Vijay S., Stephen J. Brown and Roger W. Klein (1979). 'Estimation Risk and Optimal Portfolio Choice.' *Studies in Bayesian Econometrics*, Bell Laboratories Series. North Holland.

[6]This approach is in the spirit of Stein estimation and is discussed in Chopra, Hensel, and Turner [1993]. As a practical matter, it should be used for assets that belong to the same asset class. e.g., equity indexes of different countries or stocks within a country. It would be inappropriate to apply it to financial instruments with very different characteristics; for example, stocks and T-bills.

Best, Michael J. and Robert R. Grauer (1991). 'On the Sensitivity of Means-Variance-Efficient Portfolios to Changes in Asset Means: Some Analytical and Computational Results.' *Review of Financial Studies* **4**, No. 2, 315–342.

Chopra, Vijay K. (1991). 'Mean-Variance Revisited: Near-Optimal Portfolios and Sensitivity to Input Variations.' *Russell Research Commentary*.

Chopra, Vijay K., Chris R. Hensel and Andrew L. Turner (1993). 'Massaging Mean-Variance Inputs: Returns from Alternative Global Investment Strategies in the 1980s.' *Management Science*, (July): 845–855.

Dexter, Albert S., Johnny N.W. Yu and William T. Ziemba (1980). 'Portfolio Selection in a Lognormal Market when the Investor has a Power Utility Function: Computational Results.' In *Proceedings of the International Conference on Stochastic Programming*, M.A.H. Dempster (ed.), Academic Press, 507–523.

Freund, Robert A. (1956). 'The Introduction of Risk into a Programming Model.' *Econometrica* **24** 253–263.

Freund, L. and M. Blume (1975). 'The Demand for Risky Assets.' *The American Economic Review*, December, 900–922.

Kallberg, Jarl G. and William T. Ziemba (1984). 'Mis-specification in Portfolio Selection Problems'. In *Risk and Capital*, G. Bamberg and K. Spremann (eds.), Lecture Notes in Econometrics and Mathematical Systems. Springer-Verlag.

Klein, Roger W, and Vijay S. Bawa (1976). 'The Effect of Estimation Risk on Optimal Portfolio Choice. *J. of Financial Economics* **3** (June), 215–231.

Markowitz, Harry M. (1987). *Mean-Variance Analysis in Portfolio Choice and Capital Markets*. Basil Blackwell.

Michaud. Richard O. (1989). 'The Markowitz Optimization Enigma: is 'Optimized' Optimal?' *Financial Analysts Journal* **45** (January–February), 31–42.

Ziemba, William T. and Raymond G. Vickson, eds. (1975) *Stochastic Optimization Models in Finance*. Academic Press.

Quantitative Finance, Vol. 5, No. 4, August 2005, 343–355

Routledge
Taylor & Francis Group

19

Time to wealth goals in capital accumulation

LEONARD C. MACLEAN*†, WILLIAM T. ZIEMBA‡§ and YUMING LI¶

†School of Business Administration, Dalhousie University, Halifax, NS, Canada B3H 1Z5
‡Sauder School of Business, University of British Columbia, Vancouver, BC, Canada V6T 1Z2
§Department of Finance, Sloan School of Management, 50 Memorial Drive E52-410, Massachusetts Institute of Technology, Cambridge, MA 02142-1347, USA
¶School of Business, California State University, Fullerton, CA 92834, USA

(*Received 26 January 2004; in final form 15 April 2005*)

This paper considers the problem of investment of capital in risky assets in a dynamic capital market in continuous time. The model controls risk, and in particular the risk associated with errors in the estimation of asset returns. The framework for investment risk is a geometric Brownian motion model for asset prices, with random rates of return. The information filtration process and the capital allocation decisions are considered separately. The filtration is based on a Bayesian model for asset prices, and an (empirical) Bayes estimator for current price dynamics is developed from the price history. Given the conditional price dynamics, investors allocate wealth to achieve their financial goals efficiently over time. The price updating and wealth reallocations occur when control limits on the wealth process are attained. A Bayesian fractional Kelly strategy is optimal at each rebalancing, assuming that the risky assets are jointly lognormal distributed. The strategy minimizes the expected time to the upper wealth limit while maintaining a high probability of reaching that goal before falling to a lower wealth limit. The fractional Kelly strategy is a blend of the log-optimal portfolio and cash and is equivalently represented by a negative power utility function, under the multivariate lognormal distribution assumption. By rebalancing when control limits are reached, the wealth goals approach provides greater control over downside risk and upside growth. The wealth goals approach with random rebalancing times is compared to the expected utility approach with fixed rebalancing times in an asset allocation problem involving stocks, bonds, and cash.

Keywords: Capital accumulation; Wealth goals; Investment of capital

1. Introduction

In capital accumulation under uncertainty, a decision-maker must determine how much capital to invest in riskless and risky investment opportunities at each point in time. The investment strategy yields a stream of capital over time, with investment decisions made so that the distribution of wealth has desirable properties. An investment strategy which has generated considerable interest is the growth optimal or Kelly strategy, where the expected logarithm of wealth is maximized (Kelly 1956). The wealth distribution of this strategy has many attractive characteristics (see e.g. Hakansson 1970, 1971, Markowitz 1976, Hakansson and Ziemba 1995). As Breiman (1960, 1961), and Algoet and Cover (1988) have shown, the Kelly strategy maximizes the long run expected rate of growth of capital and minimizes the expected time to reach a fixed level of wealth for sufficiently large goals under mild conditions. Researchers such as Thorp (1975), Hausch *et al.* (1981), Grauer and Hakansson (1986, 1987), and Mulvey and Vladimirow (1992) have used the optimal growth strategy to compute optimal portfolio weights in multi-asset and worldwide asset allocation problems.

The literature considers the stream of capital following from an investment policy from either a wealth or a time perspective. In most situations the expected utility of accumulated capital at a fixed point in time is analysed. There is particular interest in the logarithm of accumulated capital, and the corresponding rate of capital growth. Alternatively, the growth rate can be viewed as the time it takes for accumulated capital to reach wealth milestones.

The distribution of accumulated capital to a fixed point in time and the distribution of the first passage

*Corresponding author. Email: lmaclean@mgmt.dal.ca

Quantitative Finance
ISSN 1469–7688 print/ISSN 1469–7696 online © 2005 Taylor & Francis
http://www.tandf.co.uk/journals
DOI: 10.1080/14697680500149552

time to a fixed level of accumulated capital are variables controlled by the investment decisions. The Kelly strategy maximizes the expected logarithm of accumulated capital or the expected growth rate. However, the strategy is very aggressive. As Hausch and Ziemba (1985) and Clark and Ziemba (1987) have demonstrated, the optimal portfolio weights in the risky assets given by this strategy tend to be so large for favourable investments that the chances of losing a substantial portion of wealth are very high, particularly if the probability estimates are in error. In the time domain, the chances are high that the first passage to subsistence wealth occurs before achieving the established wealth goals.

When the investor is more risk averse, then this can be reflected in the utility function choice. If the utility function has an Arrow–Pratt risk aversion parameter, e.g. the constant relative risk aversion (CRRA) utility, then the value of this parameter captures risk tolerance; see Kallberg and Ziemba (1983).

Another approach is to define a measure of risk which depends on the investment decision. A standard measure of risk is volatility as defined by the variance of wealth at a point in time or the variance of the passage time to a wealth target. Mean-variance analysis of wealth has been widely used to determine investment strategies; see Markowitz (1952, 1987). In the time domain the mean-variance approach yields different strategies. However, the logarithm of wealth and the first passage time have consistent mean-variance properties; see Burkhardt (1998).

An alternative to variance is to use a *downside risk measure* (Breitmeyer *et al.* 1999). MacLean *et al.* (1992) considered, as risk measures, quantiles for wealth, log-wealth and first passage time in identifying investment strategies which achieve capital growth with a required level of security. Security is defined as controlling downside risk. Growth is traded for security with fractional Kelly strategies. In discrete time models with general return distributions this strategy is generally suboptimal, but it has attractive wealth/time distribution properties. See MacLean and Ziemba (1999) for extensions of this research. The emphasis in that trade-off work is the properties of the wealth process for a given fraction. For example, the probability of reaching upper wealth U before lower wealth L is calculated for various strategies. In this paper, the reverse problem of finding a strategy which achieves a specified probability (risk) is studied.

The most common downside risk measure is Value at Risk (VaR) (Jorion 1997). VaR has been studied extensively (Artzner *et al.* 1999, Gaivoronski and Pflug 2005). Basak and Shapiro (2001) consider VaR in a model with CRRA utility. Although VaR is an industry standard it has weaknesses in controlling risk. The most serious shortcoming is the insensitivity to very large losses which have small probability—the essence of risk. Measures based on lower partial moments (incomplete means) such as CvaR (Rockafeller and Uryasev 2000) and convex risk measures based on target violations (Carino and Ziemba 1998, Rockafeller and Ziemba 2000) attempt to deal with this problem.

Similar to mean-variance, a variety of mean-risk problems have been studied. Bi-criteria problems such as maximizing expected logarithm of wealth subject to a VaR constraint are consistent with stochastic dominance, the traditional concept for ordering wealth distributions. (Ogryczak and Ruszczynski 2002).

There is, however, another issue complicating the measurement of risk and return, which is the estimation of parameters defining the returns distribution. The value of the risk/return measures in practice are estimates of the true values, and the error of the estimates of measures is very sensitive to errors in the estimation of the returns distribution. Furthermore, the sensitivity to estimation errors of expected value measures (mean, CVaR) is much greater than quantile measures (median, VaR).

In this paper parameter estimation and risk control are considered in a model where the filtration and control processes are separate. A dynamic stochastic model for asset prices is presented in section 2. The model is a generalization to the multi-asset case of the random coefficients model of Browne and Whitt (1996). The inclusion of many assets facilitates the estimation of model coefficients since the correlation between assets is related to the intrinsic model structure. The existence of latent factors generating price movements is implied by the equations for price dynamics. Alternatively, price parameters could be related to observable state variables, as is the case in the model of Xia (2001). (See also Brennan 1998)

Given the estimated price dynamics, an investment decision is made to control the path of future wealth. In section 3, an approach to control is developed based upon wealth levels as stopping rules. In practice, an investment portfolio cannot be continuously rebalanced and a realistic approach is to reconsider the investment decision at regular discrete time intervals (Rogers 2000). At each rebalance time, with additional data and a change in wealth, price parameters are re-estimated and a new investment strategy is developed. The process is illustrated in figure 1. The most significant aspect of the rebalancing is the accuracy of the estimated returns distribution. If forecasts are accurate, then wealth will accumulate as expected. If prices are not as forecast over the next hold interval then unacceptable wealth levels may result. To protect against that outcome *wealth limits* can be placed on the wealth process. If the wealth trajectory is within the limits, the wealth process is under control. However, the asset pricing model and the investment decisions are reconsidered when a control limit is reached. So rebalancing occurs at a random time rather than at a given point in time. Intervention occurs when the wealth trajectory is not proceeding as expected and it can be concluded that the investment decision does not fit the true securities price process.

Whether rebalancing occurs at fixed or random times, at the rebalancing the investor:

1. generates *new estimates* for returns distributions;
2. establishes *next wealth limits* for random time rebalancing, or the next *hold time* for the fixed rebalance interval;

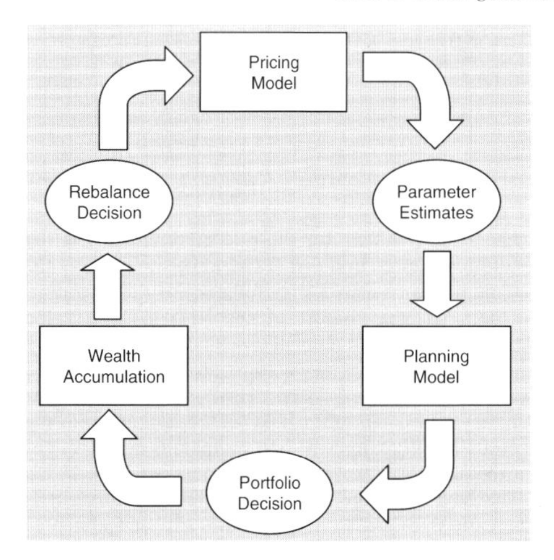

Figure 1. Dynamic investment process.

3. determines a *new investment strategy* based on the updated returns estimates, and control limits or hold time.

At a rebalance point, the decision on investment is based on accumulated wealth at a planning horizon. A number of bi-criteria decision models have been mentioned, but the model considered is maximization of expected utility at the horizon, with a negative power utility. It is significant that there exist wealth goals and an associated bi-criteria decision model which yields an identical investment strategy. Although demonstrated for the special case of negative power utility, this result is general, so that wealth goals reflect a wide range of preferences.

The wealth goals decision is put in the context of capital growth with security. So a strategy is determined which minimizes the expected time to the upper limit (*growth*) while maintaining a high probability of achieving the upper limit before falling to the lower limit (*security*). The optimal strategy for the power utility and growth–security models is shown to be *fractional Kelly*—a blend of the risk free (cash) strategy and the optimal growth strategy.

Although the negative power utility and wealth goals models are linked by the strategy, the approaches to risk are different. The expected utility model considers risk at the horizon, whereas the wealth goals model considers risk along the trajectory. In section 4, the wealth goals and power utility approaches are compared for the fundamental problem of investing in stocks, bonds, and cash over time. It is shown that there is an advantage from wealth goals, since rebalancing occurs at times indicated by the wealth trajectory, i.e. the wealth condition. This result is comparable to the superiority of condition based intervention in repairable systems (Aven and Jensen 1999). Appendix A contains proofs for propositions in the text.

2. Dynamic estimation of asset price distributions

In the dynamic investment process shown in figure 1, the returns on risky assets are re-estimated at the time of rebalancing a portfolio. The updated returns distributions are inputs to the investment decision models. In this section a linear pricing model is proposed where the estimation problem is separated from the control problem. The model is Bayesian, so that the updating of parameter estimates follows from Bayes theorem, and fits naturally into the dynamic investment process.

2.1. Price model

Suppose there are m risky assets, with $P_i(t)$ equal to the trading price of asset i at time t, $i=1,\ldots,m$, and a riskless asset with rate of return r at time t. The risky asset prices are assumed to have a joint log-normal distribution. Letting $Y_i(t)=\ell n P_i(t)$, $i=0,\ldots,m$, the price dynamics are defined by the stochastic differential equations

$$\begin{aligned} \mathrm{d}Y_0(t) &= r\,\mathrm{d}t, \\ \mathrm{d}Y_i(t) &= \alpha_i(t)\mathrm{d}t + \delta_i\,\mathrm{d}U_i, \quad i=1,\ldots,m, \end{aligned} \tag{1}$$

where $\mathrm{d}U_i$, $i=1,\ldots,m$, are independent standard Brownian motions. They represent the variation in price *specific* to each security. The other component of price variation is generated by the instantaneous mean rate of return $\alpha_i(t)$, $i=1,\ldots,m$. Assume that $\alpha_i(t)$ is a random variable with distribution defined by

$$\alpha_i(t) = \mu_i + \gamma_i Z_i(t), \quad i=1,\ldots,m. \tag{2}$$

The Z_i, $i=1,\ldots,m$, are correlated Gaussian variables, with ρ_{ij} being the correlation between Z_i and Z_j. So the rates of return are correlated, with the covariance between $\alpha_i(t)$ and $\alpha_j(t)$ given by $\gamma_i\gamma_j\rho_{ij}$.

The price process defined by (1) and (2) is a variation on the Merton model (1992) with the condition that the rates α_i, $i=1,\ldots,m$, are stochastic and the volatilities δ_i, $i=1,\ldots,m$, are not stochastic. The intention is to emphasize uncertainty in the mean since errors in estimating the mean have by far the greatest impact on portfolio decisions (Chopra and Ziemba 1993). The asset prices in the model are correlated, with the correlation generated by the relationship between the random rates of return on assets. Presumably those rates of return are related through some *common factors*. The factors may be *latent*, as is assumed in this paper, or they may be observable state variables as presented by Xia (2001), who considers the single risky asset case.

If the model for prices is correct then the error in forecasting prices arises from errors in parameter estimates. Error from incorrect modeling may be confounded with estimation error, so attention to the correct model in an application may be significant. There are generalizations of (1) and (2) which can be incorporated into the approach developed in this section. For example, the rates of return could be defined by stochastic differential equations: $d\alpha_i(t) = \mu_i dt + \gamma_i dq_i$, where dq_i are correlated Brownian motions, $i = 1, \ldots, m$. Then $\alpha_i(t) = \alpha_i(0) + \mu_i t + \gamma_i \sqrt{t} Z_i(t)$, and the rates are dynamic. The assumption in (2) simplifies the presentation, so that formulation is followed, but reference to the dynamic linear model will be made where appropriate. With regard to the stationary process (2), *the model assumption applies until the next rebalance point*. That is, the distribution for the random rates is the same between rebalance points, but the distribution at the next point may change. The model in (2) can be considered as an approximation between rebalance points. When used with control limits, where rebalancing occurs when the actual prices depart from model forecasts, this is a useful approach.

From (1) and (2) the relevant distributions for asset prices follow. Let $Y(t) = (Y_1(t), \ldots, Y_m(t))', \mu = (\mu_1, \ldots, \mu_m)'$, $\alpha = (\alpha_1, \ldots, \alpha_m)'$, $\Delta = \text{diag}(\delta_1^2, \ldots, \delta_m^2)$, and $\Gamma = (\gamma_{ij})$, where $\gamma_{ij} = \gamma_i \gamma_j \rho_{ij}$

$$\text{(Prior)} \quad \alpha \sim N(\mu, \Gamma), \tag{3}$$

$$\text{(Conditional)} \quad (Y(t)|\alpha, \Delta) \sim N(\alpha t, t\Delta), \tag{4}$$

$$\text{(Marginal)} \quad Y(t) \sim N(\mu(t), \Sigma(t)), \tag{5}$$

where $\mu(t) = t\mu$ and $\Sigma(t) = t^2\Gamma + t\Delta = \Gamma(t) + \Delta(t)$.

The Bayesian pricing model defined by (3)–(5) has an alternative structural model representation. The prior covariance has a decomposition $\Gamma = \Lambda\Lambda'$, where Λ is an $m \times l$ matrix with $l \leq m$. Consider the independent *latent factors* $F' = (F_1, \ldots, F_l)$, $F' \sim N(0, I)$, and the errors $\varepsilon'(t) = (\varepsilon_1(t), \ldots, \varepsilon_m(t))$, $\varepsilon'(t) \sim N(0, \Delta(t))$. Let $\Lambda(t) = t\Lambda$. Then the log prices at time t are

$$Y(t) = \mu(t) + \Lambda(t)F + \varepsilon(t). \tag{6}$$

From (6), $Y(t) \sim N(\mu(t), \Sigma(t))$, where $\Sigma(t) = \Lambda(t)\Lambda'(t) + \Delta(t)$. The relationship between prices is driven by underlying common factors which are unobserved. This manifests itself in rates of return, $\alpha = \mu + \Lambda F$, which are random.

The factor model in (6) has been the subject of numerous asset pricing studies, with the factors representing traded portfolio's in some cases (Campbell *et al.* 1997). Here, the factor model is used to define estimates for parameters in the inter-temporal pricing model of (1) and (2).

2.2. *Bayes estimation*

At the time an investor is rebalancing a portfolio, information is available on past realized securities prices. Consider the data available at time t, $\{Y(t), 0 \leq s \leq t\}$, and the corresponding filtration $F_t^Y = \sigma(Y(t), 0 \leq s \leq t)$, the σ-algebra generated by the process Y up to time t. Conditioned on the data, the distribution for the rate of return can be determined from Bayes theorem. That is,

$$\tilde{\alpha}(t) = \left(\alpha(t)|F_t^Y\right) \propto N(\hat{\alpha}(t), \hat{\Gamma}(t)), \tag{7}$$

where $\hat{\alpha}(t) = \mu(t) + (I - \Delta(t)\Sigma^{-1}(t))(\overline{Y}(t) - \mu(t))$, with

$$\overline{Y}(t) = \frac{1}{t} Y(t),$$

and

$$\hat{\Gamma}(t) = \frac{1}{t^2}\left(I - \Delta(t)\Sigma^{-1}(t)\right)\Delta(t).$$

The *Bayes estimate* for the rate of return $\alpha(t)$ is the conditional expectation $\hat{\alpha}(t) = E\tilde{\alpha}(t)$. This estimate is the minimum mean squared error forecast for the rate given the data. In the context of rebalancing, this is the planning value for α until the next decision time $t + \tau$ (when the stopping boundaries are reached or the fixed rebalance time occurs). Then, since no information is added in the hold interval, $E(\alpha(t)|F_t^Y) = E(E(\alpha(t)|F_{t+\tau}^Y)|F_t^Y)$, and the best estimate is $\hat{\alpha}(t)$ throughout the interval.

The Bayes estimate $\hat{\alpha}(t)$ has an appealing form. It is inherently dynamic, so that estimates can be updated as more information on prices is obtained. In particular, at the next rebalance time $t + \tau$, the posterior $\hat{\alpha}(t)$ becomes the next prior and Bayes theorem is used to update with the new information—$\overline{Y}(t + \tau)$. This approach is described in MacLean *et al.* (2004). There is another approach which emphasizes the current dynamics. The pricing model parameters (μ, Γ, Δ) are interpreted as values for the most recent rebalance cycle, and only data from that time interval is used in estimating the parameters. With control limits, the recent data contains information on a change in dynamics, and a change in prior parameters. The technique for implying the values for the prior parameters from data is empirical Bayes estimation.

To develop the estimation consider a rebalance interval $(0, t)$ and prices at the *discrete* times kt/n, $k = 0, \ldots, n$, with corresponding log prices $Y(k) := Y[(k/n)t]$. The change in log prices between times kt/n and $[(k+1)/n]t$ is $e(k) = Y(k) - Y(k-1)$, $k = 1, \ldots, n$. From the model equations

$$e(k) = \frac{t}{n}\alpha + \sqrt{\frac{t}{n}}\Delta^{1/2}Z, \quad k = 1, \ldots, n, \tag{8}$$

where $Z \sim N(0, I)$.

The increment vectors defined by (8) are independent and Gaussian for each k. The covariance matrix for each vector is

$$\Sigma_n(t) = \frac{t^2}{n^2}\Gamma + \frac{t}{n}\Delta = \Gamma_n(t) + \Delta_n(t),$$

and

$$\overline{Y}(t) = \frac{1}{n}\sum_{k=1}^{n} e(k)$$

Consider observed log prices on risky securities at the discrete time points: $\{Y_{ik}, i = 1, \ldots, m, k = 0, \ldots, n\}$. With $e_{ik} = Y_{ik} - Y_{i,k-1}$, the data on increments (first-order differences in log prices) are $\{e_{ik}, i = 1, \ldots, m, k = 1, \ldots, n\}$. Let the covariance matrix for the observed increments be S_m, the estimate of $\Sigma_n(t)$.

From the random effects model (8) the theoretical covariance is factored as $\Sigma_n(t) = \Gamma_n(t) + \Delta_n(t)$. If the rank $(\Gamma_n(t)) = l < m$, then the sample covariance matrix can be factored to produce estimates of $\Gamma_n(t)$ and $\Delta_n(t)$. In terms of the structural model for prices in (6), the assumption rank $(\Gamma_n(t)) = l < m$ is equivalent to the assumption *the number of latent factors l is strictly less than the number of risky assets m*.

Assuming $l < m$, a maximum likelihood factor analysis of S_{tn} with l factors will yield a *loading* matrix L_{tn} and a *specific error* matrix D_{tn}, where D_{tn} is diagonal. Since the number of factors is known, (L_{tn}, D_{tn}) is an efficient estimator of $(\Lambda_n(t), \Delta_n(t))$, with $\Lambda_n(t) = (t/n)\Lambda$ (Lawley and Maxwell 1971). Then $G_{tn} = L_{tn}L'_{tn}$ is an efficient estimator of $\Gamma_n(t)$.

The prior mean μ for the rates of return on securities in a rebalance interval needs to be estimated. If it is assumed that the rates have a *common prior mean*, i.e. $\mu_i = \mu, i = 1, \ldots, m$, then the overall mean $\bar{\bar{Y}}(t) = (1/I)\sum_{i=1}^{I} \bar{Y}_i(t)$ is an estimate of μ, where $\bar{Y}_i(t) = (1/t)Y_i(t)$. The assumption of a common value is consistent with a long run rate of return on securities. (Alternatively, securities in the same class could have a common mean, with differences between classes of securities.)

The discussion has identified estimates for all model parameters within a rebalance interval. The conditional distribution for securities prices can be estimated, and in particular the conditional mean rate of return for the next time interval can be forecast. Since the forecast is the Bayes formula with the prior parameters estimated from data it is an empirical Bayes estimate.

Proposition 1: *Let $Y_k, k = 0, \ldots, n$, be observations on the vector of log prices of risky assets at regular intervals of width t/n, with the statistics computed from first-order increments: S_{tn}, L_{tn}, D_{tn} and $\bar{\bar{Y}}_t$. Then $\hat{\Delta} = (n/t)D_{tn}$, $\hat{\Lambda} = (n/t)L_{tn}$ are estimates of the model parameters Δ and Λ, respectively. With $\hat{\Delta}(t) = t\hat{\Delta}, \hat{\Lambda}(t) = t\hat{\Lambda}$ and $S_t^* = \hat{\Lambda}(t)\hat{\Lambda}'(t) + \hat{\Delta}(t)$, an estimate for the conditional mean rate of return at time t is*

$$\hat{\hat{\alpha}}_n(t) = \bar{\bar{Y}}(t) + (I - \hat{\Delta}(t)S_t^*)(\bar{Y}(t) - \bar{\bar{Y}}(t)) \quad (9)$$

Furthermore, the limit as $n \to \infty$ of $\hat{\hat{\alpha}}_n(t)$ is $\hat{\alpha}(t)$.

The empirical Bayes estimate in (9) is a natural formulation given the proposed pricing model. If the model is correct then $\hat{\hat{\alpha}}_n(t)$ has smaller mean squared error than well known estimates for the rate of return, such as the maximum likelihood estimate and the James–Stein estimate. If the model is correct but the number of latent factors is unknown, then there is additional estimation error from estimating that number. However, results from similar models indicate such error is small; see MacLean and Weldon (1996). If a dynamic model for rates, $\alpha_i(t) = \alpha_i(0) + \mu_i t + \gamma_i\sqrt{t}Z_i(t), i = 1, \ldots, I$, is the correct formulation then the approach described above can be modified to obtain empirical Bayes estimates for parameters. The modification involves calculating *second-order differences* to obtain a random effects model with iid second-order increments.

With this estimation procedure, an essential component of the dynamic investment process is in place. The next component, the planning model, will have the forecasts from the estimated model as inputs.

3. Portfolio planning models

Assuming that at a rebalance point the data has been filtered to obtain estimates for the parameters, the decision on how much capital to allocate to the various assets is now considered. The objective is to control the path of accumulated capital generated by the decision and the unfolding asset prices. The decision will be developed from the estimated price parameters whereas capital will accumulate from the true process. Of course, if the estimates used in computing a strategy are substantially in error, then the trajectory of wealth will not proceed as anticipated, but the trajectory may still be under control. At the rebalancing time t, there are estimates for the conditional rate of return $\hat{\alpha}(t)$, and the volatility $\hat{\Delta}(t)$, based on the methods of section 2. The forecast dynamics for the price process, conditional on the estimates for parameters, are

$$\mathrm{d}Y(t) = \hat{\alpha}(t)\mathrm{d}t + \hat{\Delta}^{1/2}(t)\mathrm{d}\hat{Z}, \quad (10)$$

where the innovations process $\mathrm{d}\hat{Z}$ is standard Brownian motion.

An investment strategy is a vector process

$$\{(x_0(t), X(t)), t \geq 0\} = \{x_0(t), (x_1(t), \ldots, x_m(t)), t \geq 0\}, \quad (11)$$

where $\sum_{i=0}^{m} x_i(t) = 1$ for any t, with $x_0(t)$ the investment in the risk-free asset. The proportions of wealth invested in the m risky assets are unconstrained since $x_0(t)$ can always be chosen, with borrowing or lending, to satisfy the budget constraint.

The forecast change in wealth from an investment decision $X(t)$ is determined by the conditional price process (10). Suppose the investor at time t has wealth $W(t)$ and let $\hat{\phi}_i(t) = \hat{\alpha}_i(t) + (1/2)\hat{\delta}_i^2(t)$, $i = 1, \ldots, m$. Then the forecast for the instantaneous change in wealth is

$$\mathrm{d}W(t) = \left[\sum_{i=1}^{m} x_i(t)(\hat{\phi}_i(t) - r) + r\right] W(t)\mathrm{d}t + W(t)\sum_{i=1}^{m} x_i(t)\hat{\delta}_i(t)\mathrm{d}Z_i, \quad (12)$$

where $\mathrm{d}Z_i$ is standard Brownian motion.

There is a *growth condition* which will be required of any *feasible* investment strategy:

$$X_t = \left\{X(t)\left|\left[\sum_{i=1}^{m} x_i(t)(\hat{\phi}_i(t) - r) + r - \frac{1}{2}\sum_{i=1}^{m} x_i^2(t)\hat{\delta}_i^2(t)\right] \geq 0\right.\right\}.$$

With a fixed mix strategy over the hold interval satisfying the growth condition, the forecast wealth process $W(\tau)$, $\tau \geq t$, defined by (12) follows geometric Brownian motion such that $W(\tau) \geq 0$.

To determine an investment decision which controls wealth, two approaches are considered. First, a model which considers preferences for wealth at the planning horizon is defined. The standard approach is to maximize expected utility at a planning horizon. Because of uncertainty in the parameters for the asset pricing model, parameters are re-estimated and decisions revised at regular intervals in time up to the horizon. In the second approach, the preferences at the horizon are used to develop wealth goals (limits). Then a wealth goals model is defined, which has an equivalent investment strategy. The contrast from the approaches comes from rebalancing times. The wealth goals approach involves a random hold interval, with stopping/rebalancing occurring when specified upper or lower wealth goals are reached.

The matching of preferences in the two approaches does not mean that the same fixed mix strategy is used for both models. The rebalance times are different, and therefore there are different parameter estimates used in the strategy calculation.

The link between the planning models is stochastic dominance. Wealth W_1 dominates wealth W_2 iff $Eu(W_1)$ is greater than or equal to $Eu(W_2)$, with strict inequality for at least one utility function $u \in U$. Orders of stochastic dominance are determined by classes of utilities. If $U_k = \{u|(-1)^{j-1}u^{(j)} \geq 0, j = 1, \ldots, k\}$, so that $U_1 \supset U_2 \supset \ldots$, then U_1 is the class of monotone utilities, U_2 the class of concave monotone utilities,..., and U_∞ contains the CRRA utilities. Since the CRRA utilities are a limiting subset, they are a reasonable reference class.

There are alternative formulations of the various orders of stochastic dominance. In particular, first order is equivalent to dominance in the cumulative distribution of wealth. The wealth goals model will be defined on the distribution of wealth, and therefore is based on first-order dominance principles. However, it is a relaxation of first-order dominance in that it focuses on parts of the distribution in a bi-criteria problem.

3.1. *Expected utility strategy with fixed rebalance times*

Consider an investor whose objective is to maximize the expected utility of wealth at the end of a finite planning horizon T. To obtain an explicit solution we assume a constant relative risk aversion (CRRA) utility. The utility of wealth w is

$$U(w) = \frac{w^\beta}{\beta}, \quad \text{with } \beta < 1,\ \beta \neq 0.$$

At the current decision time, the parameters in the pricing model are estimated and an investment strategy is determined, conditional on those estimates. Consider the conditional CRRA wealth problem.

Definition 1: The conditional CRRA wealth problem at rebalance time t, with current wealth w_t, horizon T and estimated parameters $\hat{\alpha}(t), \hat{\Delta}(t)$, is to find the strategy $X(t) \in X_t$ which maximizes

$$E\left[\frac{W(T)^\beta}{\beta}\right],$$

where the forecast dynamics of wealth $W(t)$ are given by (12).

Proposition 2: *The optimal strategy for the conditional CRRA wealth problem is*

$$X^*(t) = \frac{1}{1-\beta}\hat{\Delta}^{-1}(t)(\hat{\phi}(t) - re). \tag{13}$$

As $\beta \to 0$ in definition 1 the utility is log and the efficient strategy is the benchmark optimal growth solution: $\tilde{X}(t) = \Delta^{-1}(\phi(t) - re)$. The fraction $f = 1/(1-\beta)$ captures the aversion to risk (Pratt 1964). When $0 < \beta < 1$, the positive power utility case, the optimal portfolio invests more than the benchmark, with the over-investment financed through borrowing (levered strategies). Although the strategy $X^\beta(t), 0 < \beta < 1$, is optimal, it is problematic from the perspective of the distribution of terminal wealth $W(T)$. Let F_β be the distribution function for terminal wealth following strategy $X^\beta(t)$, with F_0 the terminal wealth distribution following the benchmark strategy. Distributions can be compared with the first-order stochastic dominance relation: F_{β_1} dominates F_{β_2} iff $F_{\beta_1}(w) \leq F_{\beta_2}(w)$ for $w \in R$ with strict inequality for some w.

Proposition 3: *The terminal wealth distribution F_0 for the benchmark portfolio first-order stochastically dominates the distribution F_β for $0 < \beta < 1$.*

This result provides a rationale for a restriction $\beta \leq 0$ to the negative power and log utility functions, and the resulting class of fractional Kelly strategies $p\tilde{X}(t)$, where $p \leq 1$. First-order stochastic dominance is defined on the class of all monotone utilities, of which the CRRA utility is a small subset. Imposing the first-order dominance condition on the fraction from a CRRA utility model is a restriction, but other work, where a risk constraint such as VaR is included in the expected CRRA utility model, has shown that levered strategies are excluded (Basak and Shapiro 2001).

The solution in (13) is independent of current wealth w_t and the planning horizon T.

The decision is sensitive to estimation and modeling error, and at regular intervals of width s model parameters are re-estimated using the empirical Bayes methodology.

The actual investment decision at times $t, t+s, \ldots,$ would change with the new estimates.

3.2. *Wealth goals strategy and random rebalance times*

The uncertainty in returns on investment, whether from natural variability, variability in underlying factors, or errors in estimation, is the motivation for strategies that control risk. In the conditional CRRA utility problem the risk associated with the forecast wealth at the horizon is controlled with the risk aversion parameter β, with the estimation risk controlled by regularly updating estimates for parameters. Updates are appropriate when there is evidence that current estimates are substantially in error. This concept of updating based on the gap between expected and actual processes suggests another way to control the wealth process. That is, to set control limits (or wealth goals), as is the practice in process control. The wealth goals are action levels so that the investment portfolio is rebalanced if either of the goals is reached. With wealth w_t at time t, the lower and upper wealth goals $\underline{w}_t$ and $\overline{w}_t$ are set, where $\underline{w}_t < w_t < \overline{w}_t$. If the investment strategy $X(t)$ is chosen, then the strategy is followed until either $\underline{w}_t$ or $\overline{w}_t$ is reached. At that point the portfolio is rebalanced. So new estimates are generated, new goals are set and a new strategy is chosen. Between rebalancing points a fixed mix strategy is followed. Information on prices is obtained between rebalance points, but it is only used when the accumulated information on actual prices signals that the forecasts for price movements are substantially in error, that is, when a control limit is reached.

The choice of wealth goals is an important component of the control process. Wealth goals can be an expression of preferences as naturally as a utility function. Our approach is to first consider the form of an investment strategy for *given wealth goals*, assuming the wealth goals are feasible, i.e. admit an investment strategy. Then a methodology for specifying goals based on expected utility is developed.

The risk and return characteristics of the wealth process have traditionally guided the choice of strategy $X(t)$. Within the wealth goals context, therefore:

(i) the chance of hitting the lower wealth goal should be small. That is, the *downside risk* should be controlled;

(ii) the upper wealth goal, the preferred target, should be reached as quickly as possible. That is, the *growth rate* of capital should be maximized.

To formalize downside risk in the wealth goals approach, consider

$T_w(X(t), w_t|\hat{\alpha}(t), \hat{\Delta}(t))$ = *first passage time to wealth w starting with wealth w_t at time t and following investment strategy X(t), conditioned on the estimates for parameters at time t.*

Let the set of wealth goals be denoted by

$$B_t = \{b_t | b_t = (\underline{w}_t, \overline{w}_t), 0 < \underline{w}_t < w_t < \overline{w}_t\}.$$

For $b_t \in B_t$, downside risk for a strategy $X(t) \in X_t$ is measured by

$$\Re_{b_t}(X(t)) = \Pr\left[T_{\underline{w}}t\left(X(t), w_t|\alpha(t), \hat{\Delta}(t)\right) \leq T_{\overline{w}_t}\left(X(t), w_t|\hat{\alpha}(t), \hat{\Delta}(t)\right)\right]. \quad (14)$$

The risk is the probability the first passage time to $\underline{w}_t$ is smaller than the first passage time to $\overline{w}_t$.

Risk measures are associated with acceptance sets of capital accumulation paths (Artzner *et al.* 1999). Let $\Omega(X(t))$ be the set of all paths of $W(\tau), \tau \geq t$, and let $A(X(t))$ be the set of acceptable paths. If $A^c(X(t))$ is the complement, then the downside risk for $X(t)$ is $\Re(X(t)) = \sigma(A^c(X(t)))$. That is, the risk for a strategy is the probability measure of the unacceptable set of paths.

Desirable properties of risk measures have been considered in the literature. The following basic axioms for a *downside risk measure* are proposed in Breitmeyer *et al.* (1999).

A:	(non-negativity)	$\Re(X(t)) \geq 0$.
B:	(normalization)	If $A^c(X(t)) = \phi$, then $\Re(X(t)) = 0$.
C:	(downside focus)	If $A^c(X_1(t)) = A^c(X_2(t))$, then $\Re(X_1(t)) = \Re(X_2(t))$.
D:	(monotonicity)	If $A^c(X_1(t)) \subseteq A^c(X_2(t))$, then $\Re(X_1(t)) \leq \Re(X_2(t))$

Consider the measure in (14). From the growth condition defining feasible strategies, $X(t) \in X_t$, any path eventually reaches $\overline{w}_t$, and the *acceptance set* for the measure is defined as

$$A_{b_t}(X(t)) = \textit{paths of } W(\tau), \tau \geq t, \textit{ which reach } \overline{w}_t \textit{ before ever dropping to } \underline{w}_t.$$

Using this definition of acceptance sets, the risk measure defined in (14) satisfies axioms A–D.

With an appropriate risk measure defined in terms of wealth goals, risk can be controlled with a constraint on the measure. For control limits $b_t \in B_t$ and risk level $1-\gamma$, $0 \leq \gamma \leq 1$, feasible strategies $X(t) \in X_t$ are required to satisfy $\Re_{b_t}(X(t)) \leq 1-\gamma$.

The other characteristic of the wealth goals approach is the anticipation that the upper limit is reached as quickly as possible. With the expected time defined as

$$G_{b_t}(X(t)) = E\left\{T_{\overline{w}_t}(X(t), w_t|\hat{\alpha}(t), \hat{\Delta}(t))\right\}, \quad (15)$$

the objective is to minimize $G_{b_t}(X(t))$. Speed in the time domain is a growth measure. Suppose the random time to a fixed wealth goal $T_{\overline{w}_t}$ is mapped onto random wealth at a fixed time T by $W_T = \overline{w}_t \exp(T - T_{\overline{w}_t})$. Then, $E\{\ln W_T\} = [\ln \overline{w}_t + T] - E\{T_{\overline{w}_t}\}$. Minimizing the expected time to the wealth goal $\overline{w}_t$ is analogous to maximizing the expected log of wealth to the appropriate horizon T. The log optimal strategy maximizes the rate of growth of capital, so (15) is a capital growth criterion. With reference to first-order stochastic dominance, the expected log of wealth is equivalent to the median wealth, since wealth is lognormal.

With the risk-return characteristics specified by (14) and (15), a *Growth–Security* decision model can be defined. Security is the complement of risk, that is, $S_{b_t}(X(t)) = 1 - \Re_{b_t}(X(t))$.

Definition 2: For given control limits $b_t \in B_t$ and risk level $1-\gamma, 0 \leq \gamma \leq 1$, and estimates $\hat{\alpha}(t), \hat{\Delta}(t)$, the conditional Growth–Security strategy $X(t)$ is determined from the problem

$$\text{Minimize } \{G_{b_t}(X(t)) | S_{b_t}(X(t)) \geq \gamma, X(t) \in X_t\}. \quad (16)$$

Explicit expressions for the optimal solution to (16) will be developed for the geometric Brownian motion model. Consider the wealth process $W(\tau), \tau \geq t$, with forecast dynamics given by (12). Let

$$\psi_t(X(t)) = \sum_{i=1}^{m} x_i(t)(\hat{\phi}_i(t) - r) + r,$$

$$\sigma_t^2(X(t)) = \sum_{i=1}^{m} x_i^2(t)\delta_i^2(t), \varphi_t(X(t)) = 2\psi_t(X(t)) - \sigma_t^2(X(t)),$$

and

$$\theta_t(X(t)) = \frac{\varphi_t(X(t))}{\sigma_t^2(X(t))}.$$

The measures $G_{b_t}(X(t))$ and $S_{b_t}(X(t)) = 1 - \Re_{b_t}(X(t))$ for geometric Brownian motion are given by Karlin and Taylor (1981):

$$G_{b_t}(X(t)) = \frac{2\ln(\overline{w}_t/w_t)}{\varphi_t(X(t))}, \tag{17}$$

$$S_{b_t}(X(t)) = \frac{\left[1 - (\underline{w}_t/w_t)^{\theta_t(X(t))}\right]}{\left[1 - (\underline{w}_t/\overline{w}_t)^{\theta_t(X(t))}\right]}. \tag{18}$$

These expressions define growth and security in terms of the estimated parameters of the returns distribution and the control limits. Therefore, the growth–security strategy from (16) can be related to those inputs.

Proposition 4: *Suppose at rebalance time t, the investor has wealth w_t, parameter estimates $\hat{\alpha}(t), \hat{\Delta}(t)$, and the return on the risk-free asset is r. If control limits (wealth goals) are $b_t \in B_t$ and the risk level is $1 - \gamma$, then the conditional growth–security efficient strategy at time t is $X_{b_t}^*(t) = (x_{1b_t}^*(t), \ldots, x_{mb_t}^*(t))'$ with*

$$X_{b_t}^*(t) = p_t(w_t, b_t, \gamma)\hat{\Delta}^{-1}(t)(\hat{\phi}(t) - re), \tag{19}$$

where e is a vector of ones.

The expression in (19) reveals the approach to risk control with wealth goals. The component $\tilde{X}(t) = \hat{\Delta}^{-1}(t)(\hat{\phi}(t) - re)$ is the *optimal growth* portfolio (Merton 1992), or the *Kelly strategy* (Kelly 1956) or the benchmark portfolio (Basak and Shapiro 2001). The function $p_t(w_t, b_t, \gamma)$ defines a fraction invested in the risky assets that depends on current wealth, wealth goals, and risk level. Hence, the strategy which is growth–security efficient is *fractional Kelly*, as is the case for CRRA utility. An expression for the investment fraction follows from proposition 4.

Corollory 1: *For wealth goals $b_t \in B_t$, the fraction of wealth invested in risky assets at the rebalancing time t is* $\min\{1, p_t\}$ for

$$p_t(w_t, b_t, \gamma) = h_t \cdot H_t(\tilde{X}(t)) + \sqrt{[h_t \cdot H_t(\tilde{X}(t))]^2 + \frac{2rh_t}{\sigma_t^2(\tilde{X}(t))}}, \tag{20}$$

where

$$H_t(\tilde{X}(t)) = \frac{[\psi_t(\tilde{X}(t)) - r]}{\sigma_t^2(\tilde{X}(t))},$$

$$h_t = \frac{[\log w_t - \log \underline{w}_t]}{[\log w_t - \log \underline{w}_t - \log y^*]},$$

and y^* is the minimum positive root of $\gamma y^{c+1} - y + (1 - \gamma) = 0$ for

$$c_t = \frac{[\log \overline{w}_t - \log w_t]}{[\log w_t - \log \underline{w}_t]}.$$

The solutions to the conditional CRRA problem and the conditional G–S problem apply to one stage in the multistage investment process. Although the form of the solutions is fractional Kelly for both problems, there are differences. The CRRA solution is independent of current wealth and the planning horizon. The G–S solution depends on the current wealth and the horizon through the wealth goals. Both solutions depend on the estimates for price parameters.

3.3. Wealth goals

The dynamic investment process of figure 1 is almost complete. However, the specification of wealth goals in the G–S approach has not been addressed. The purpose in rebalancing is to develop a new investment policy when the current policy is not working. The method for determining goals in this analysis is based on preferences for wealth at the planning horizon as defined by the CRRA utility. A technique of *matched* strategies, which replicates the wealth preferences at the horizon, provides a fair comparison of the accumulated wealth for the contrasting models. That is, if expected utility is the standard, then performance of the G–S problem relative to that standard is studied. It is important to realize that the matching technique can be implemented with other standard problems. The wealth goals can be important milestones without reference to other problems.

Consider choices for β and $b_t = (\underline{w}_t, \overline{w}_t)$ which produce the same strategy, that is $p_t(w_t, b_t, \gamma) = 1/(1 - \beta)$. From the definition of $p_t(w_t, b_t, \gamma)$, there are potentially many combinations of goals which are linked to the same β. Since the expected utility fraction invested in the benchmark portfolio is independent of current wealth, whereas the process control fraction depends on the current wealth w_t and the probability γ, the values w_t and γ will be fixed, and the $\beta \to b_t$ link explored. Consider, then, the conditional β-class of goals, with $w_t = w_t^0, \gamma = \gamma^0$

$$B_t^0(\beta) = \left\{b_t | p_t(w_t^0, b_t, \gamma^0) = \frac{1}{1 - \beta}\right\}. \tag{21}$$

The set $B_t^0(\beta) \subseteq B_t$ defines wealth goals with the same Arrow–Pratt relative risk aversion given the current wealth w_t^0 and risk probability γ^0. Proposition 5 establishes that such goals actually exist.

Proposition 5: *At the rebalance time t consider current wealth w_t^0, risk probability γ^0 and risk aversion* *parameter β. Assume that the benchmark portfolio satisfies the growth condition, i.e.*

$$\frac{\psi_t(\tilde{X}(t))}{\sigma_t^2(\tilde{X}(t))} > \frac{1}{2}.$$

Then $B_t^0(\beta)$ is non-empty for $\beta < 0$.

Consider the following *matching heuristic* for setting wealth goals at any rebalancing time τ:

1. $\overline{w}_\tau^* = E(W^{EU}(\tau + s))$ is the expected wealth at time $\tau + s$ with the CRRA strategy

$$X^*(\tau) = \frac{1}{(1-\beta)} \tilde{X}(\tau).$$

2. $\underline{w}_\tau^* = \max\{\underline{w}_\tau | b_\tau \in B_\tau^0(\beta)\}$.

Setting goals at $b_\tau^* = (\underline{w}_\tau^*, \overline{w}_\tau^*)$ maintains the investment fraction at $p_\tau = 1/(1-\beta)$ for any rebalancing time τ, and links the goals to the holding time s.

4. Comparisons

The accumulated capital from the wealth goals with random rebalancing times, and expected utility with fixed rebalancing times is now analysed for the fundamental problem of allocating investment capital to stocks, bonds and cash over time. The matching heuristic approach is used to insure the problems are comparable, with the main difference being rebalancing time.

Daily prices for stocks and bonds are simulated using the dynamic pricing model, discretized to days. If $P_i(t)$ is the price for asset i on day t, where $i=0$ (cash), 1 (stock), and 2 (bonds), then consider the increments $e_i(t) = \ln P_i(t+1) - \ln P_i(t)$, $i = 0, 1, 2$. From the model, $e_i(t) = \alpha_i + \delta_i Z_i(t)$, where $Z_i(t)$ are independent Gaussian variables. For cash returns, $\alpha_0 = \ln(1+r)$. It is assumed that (α_1, α_2) are Gaussian variables dependent on a single latent state variable. So $\alpha_i = \mu + \lambda_i F, i = 1, 2$, with F standard Gaussian. There is a simple structure to model parameters in this single factor case. If the covariance for risky asset log prices is Σ, The model implies $\Sigma = \Lambda\Lambda' + \Delta$, where $\Lambda = (\lambda_1, \lambda_2)'$ and $\Delta = \mathrm{diag}(\delta_1^2, \delta_2^2)$. The factor solution is

$$\Lambda = (\sqrt{\rho}\sigma_{11}, \sqrt{\rho}\sigma_{22})',$$
$$\Delta = \mathrm{diag}((1-|\rho|)\sigma_{11}^2, (1-|\rho|)\sigma_{22}^2),$$
$$\rho = \frac{\sigma_{12}}{\sigma_{11}\sigma_{22}}.$$

If the true covariance is known, then model parameters are specified and daily prices can be simulated. Conversely, if a sample covariance matrix is determined from a sequence of prices then the parameter values can be estimated using these equations. The estimates are empirical Bayes, based on the methods in section 2.

For the simulation of asset prices, data on yearly asset returns corresponding to the S&P500, Solomon Brothers bond index, and U.S. T-bills for 1980–1990 were used to generate seed parameters. Statistics on the associated daily returns are in table 1. The corresponding statistics for log-rates were used as seeds.

Table 1. Daily rates of return.

	Stocks	Bonds	Cash
Mean	0.00050	0.00031	0.00019
Variance	0.00062	0.00035	0
Covariance	0.00014		

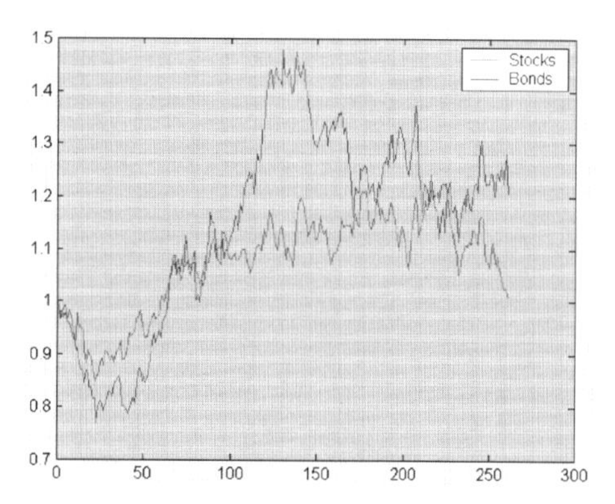

Figure 2. Daily prices for stocks and bonds.

Daily prices were generated for one year (260 trading days) and this was repeated 200 times. The one year horizon is appropriate for the simulation since it provides sufficient time for differentiation of capital accumulation from the PC and EU strategies. The contrasting approaches to selecting an investment strategy were applied to each trajectory of prices. Initial parameter estimates were taken from average daily returns for the prior year, 1979. For expected utility, rebalancing (revising estimates of price model parameters; calculating new strategy) was done every 30 days. For process control, rebalancing occurs when control limits are reached. The estimation is based on the covariance matrix. To increase stability the covariance analysed at the jth rebalancing at time t is

$$\tilde{S}_j = \frac{t}{260} S_j + \frac{1-t}{260} \tilde{S}_{j-1}.$$

As time increases the updating is weighted to the observed covariance S. The procedure for setting limits (wealth goals) is described in section 4, where the process control fraction is maintained at $p_\tau = 1/(1-\beta)$.

The expected utility and wealth goals (process control) methods are shown for the trajectory of prices in figure 2. This trajectory has a volatile return on stock and is a useful example.

The expected utility strategies and portfolio values for the setting $1 - \beta = 1.1$ are given in table 2. The rebalancing occurred at 30 day intervals. The strategy emphasizes stocks and is quite stable. The return on the portfolio at the end of the year was 24%.

Table 2. Expected utility performance.

	Strategy		
Time	Stocks	Bonds	Port value
1	0.4670	0.1465	1
31	0.7439	−0.1319	0.8943
61	0.8688	−0.1770	0.9930
91	0.8963	−0.2292	1.1287
121	0.9701	−0.4437	1.4089
151	0.9432	−0.4726	1.3114
181	0.9864	−0.4900	1.2164
211	1.0125	−0.4181	1.1508
241	0.9820	−0.3989	1.2178

Table 3. Wealth goals performance.

	Strategy			Wealth limits	
Time	Stocks	Bonds	Port value	Upper	Lower
1	0.4670	0.1465	1	1.0112	0.5765
64	0.5805	0.1140	1.0148	1.0281	0.5309
65	0.6346	0.1010	1.0379	1.0526	0.5212
66	0.6656	0.0910	1.0562	1.0717	0.5196
67	0.6744	0.0864	1.0787	1.0946	0.5301
74	0.6804	0.0898	1.1065	1.1229	0.5427
88	0.7106	0.0671	1.1433	1.1615	0.5385
107	0.7566	−0.0011	1.1644	1.1844	0.5231
108	0.7889	−0.0509	1.1910	1.2127	0.5177
115	0.7995	−0.0936	1.2375	1.2603	0.5351
118	0.8060	−0.1302	1.2771	1.3006	0.5514
120	0.8089	−0.1571	1.3387	1.3634	0.5779
123	0.8242	−0.1957	1.3640	1.3898	0.5798
130	0.7864	−0.2152	1.4043	1.4290	0.6234

The process control methodology was run on the same trajectory, with settings $1-\beta = 1.1, 1-\gamma = 0.03$. Those results are presented in table 3. There is initially more frequent rebalancing with wealth goals, because of the volatility. When prices settle down, the fixed mix strategy does not change. The return at the end of the year was 42%. It is noteworthy that the return was monotone and the advantage of the PC strategy increases with time.

The comparison of the expected utility and process control approaches is the *expected wealth ratio* at the end of the year. The wealth ratio for each trajectory is the accumulated wealth with the process control strategy divided by the accumulated wealth with the expected utility strategy.

Table 4 presents mean wealth ratios for values for the CRRA utility parameter β, and the process control risk probability $1-\gamma$. The levered strategies $(0<\beta<1)$ are included, although those strategies are not growth–security efficient.

The wealth goals (process control) approach has a substantial advantage over expected utility, demonstrating the effect of accounting for current wealth (at the control limit) in determining a strategy. Since the PC strategy is determined (matched) from the EU strategy, the expected wealth advantage of being less risk averse (decreasing $1-\beta$) is embedded in the PC strategy, but the PC advantage over EU grows since the volatility is controlled by

Table 4. Expected wealth ratios.

	$1-\gamma$				
$1-\beta$	0.01	0.02	0.03	0.04	0.05
1.15	1.291	1.290	1.289	1.265	1.264
1.10	1.322	1.321	1.320	1.298	1.293
1.05	1.365	1.364	1.359	1.340	1.337
1.00	1.410	1.408	1.403	1.384	1.381
0.95	1.460	1.459	1.457	1.443	1.433
0.90	1.523	1.522	1.520	1.507	1.491
0.85	1.608	1.608	1.607	1.602	1.584

the upper and lower limits. As the security requirement is relaxed (γ decreasing) the expected wealth advantage for PC decreases. This results from the specific construction of the strategy, where the lower control limit drops with decreased γ and given upper control limit.

The cases where $(1-\beta)<1$, the levered strategies, are high risk and are stochastically dominated as established in Proposition 4.

5. Conclusion

This paper considers the risk to accumulated capital in a multi-asset investment problem where the distribution of asset prices has random parameters. The basis for investment strategies is either process control, where wealth is controlled by upper and lower goals, or expected utility of terminal wealth at the horizon. In the process control approach, the portfolio is re-balanced when either wealth goal is reached.

At a re-balancing time the history of asset prices is used to update the conditional price distribution. An empirical Bayes estimator of distribution parameters is defined based on the observed correlation structure of asset prices. With the revised price distribution, new upper and lower wealth goals or the next re-balance time are selected and a new optimal strategy is derived. The process control strategy is growth–security efficient in the sense that it has maximum growth among strategies with a specified level of risk. Assuming a lognormal distribution for asset returns, the growth–security strategy is a blend of the risk-free portfolio and the optimal growth portfolio.

The wealth goals approach and the expected utility approach with constant relative risk aversion are compared. An equivalence between the wealth goals and the risk aversion parameter is developed. Although the alternative models have the same portfolio solution with the appropriate choices of parameter values, the wealth goals approach has the advantage of a direct control of downside risk and a re-balancing of the portfolio based on significant change in the price process and change in wealth. The benefit of intervening when the condition of wealth deviates from expectations rather than at regular intervals in time is clear. Re-balancing at specified points in time when wealth is proceeding as expected is analogous to *tampering* in process control, and typically introduces extra risk.

Acknowledgements

This work was presented at the Bachelier World Congress 2002, and seminars at The University of British Columbia and Dalhousie University. The authors wish to thank Yonggan Zhao for assistance with the computations. This research was supported by grants from the Natural Sciences and Engineering Research Council of Canada and the National Center of Competence in Research FINRISK, Swiss National Foundation.

Appendix A

This appendix contains proofs for the propositions in the paper.

Proposition 2: *The optimal strategy for the CRRA wealth problem is given by*

$$X^*(t) = \frac{1}{1-\beta}\hat{\Delta}^{-1}(t)(\hat{\phi}(t) - re). \tag{13}$$

Proof: The analysis is similar to theorem 6.2 in Janecek (1998). For investment strategy $X(t)$ the forecast dynamics of wealth are

$$dW(t) = \left[\sum_{i=1}^{m} x_i(t)(\hat{\phi}_i(t) - r) + r\right]W(t)\mathrm{d}t + W(t)\sum_{i=1}^{m} x_i(t)\hat{\delta}_i(t)\mathrm{d}Z_i.$$

With $X(\tau) = X(t)$ and $\hat{\phi}_i(\tau) = \hat{\phi}_i(t)$, $\tau \geq t$, based on information to time t, integration from t to T gives the forecast wealth at the horizon

$$\hat{W}(T-t) = w_t \exp\left\{\left[\sum_{i=1}^{m} x_i(t)(\hat{\phi}_i(t) - r) + r - \frac{1}{2}\sum_{i=1}^{m} x_i^2(t)\hat{\delta}_i^2(t)\right] \times (T-t) + (T-t)^{1/2}\sum_{i=1}^{m} x_i(t)\hat{\delta}_i(t)Z_i\right\},$$

where $Z_i \sim N(0,1)$. The expected utility of wealth is

$$E\left[\frac{\hat{W}^\beta(T-t)}{\beta}\right] = \frac{1}{\beta}\left[w_t^\beta \exp\left\{\left[\sum_{i=1}^{m} x_i(t)(\hat{\phi}_i(t) - r) + r - \frac{1}{2}\sum_{i=1}^{m} x_i^2(t)\hat{\delta}_i^2(t)\right](T-t)\beta\right\}. E\exp\left\{\beta(T-t)^{1/2}\sum_{i=1}^{m} x_i(t)\hat{\delta}_i(t)Z_i\right\}\right].$$

Consider the expression

$$E\exp\left\{\beta(T-t)^{1/2}\sum_{i=1}^{m} x_i(t)\hat{\delta}_i(t)Z_i\right\} = \exp\left\{\frac{1}{2}\beta^2(T-t)\sum_{i=1}^{m} x_i^2(t)\hat{\delta}_i^2(t)\right\}.$$

Then

$$E\left[\frac{\hat{W}^\beta(T-t)}{\beta}\right] = \frac{1}{\beta}\left[w_t^\beta \exp\left\{(T-t)\beta\left[\sum_{i=1}^{m} x_i(t)(\hat{\phi}_i(t) - r) + r - \frac{1}{2}(1-\beta)\sum_{i=1}^{m} x_i^2(t)\hat{\delta}_i^2(t)\right]\right\}\right].$$

The first-order conditions

$$\frac{\partial}{\partial x_i(t)} E\left(\frac{\hat{W}^\beta(T-t)}{\beta}\right) = 0$$

reduce to

$$(\hat{\phi}_i(t) - r) - (1-\beta)x_i^*(t)\hat{\delta}_i(t) = 0, \quad i = 1, \ldots, I,$$

and $X^*(t)$ is as stated. □

Proposition 3: *The terminal wealth distribution F_0 for the benchmark portfolio first-order stochastically dominates the distribution F_β* for $0 < \beta < 1$.

Proof: Let $\psi(\tilde{X})$ and $\sigma(\tilde{X})$ be the instantaneous mean rate of return and standard deviation, respectively, of the optimal growth (benchmark) portfolio. Let $f = 1/(1-\beta)$ and define

$$\psi(X^\beta) = \psi(f) = f(\psi(\tilde{X}) - r) + r), \sigma^2(f) = f^2\sigma^2(\tilde{X})$$

and

$$\phi(f) = 2\psi(f) - \sigma^2(f).$$

Under the assumptions of the model, terminal wealth $W(t+T)$ from the strategy X^β is log-normally distributed with

$$F_\beta(w) = N\left[\frac{\ln\{(w/w_t) - [\phi(f)/2]\}T}{\sigma(f)\sqrt{T}}\right],$$

where N is the Gaussian cumulative distribution.

Since $\sigma^2(f)$ is increasing in f and $\phi(f)$ is decreasing in f for $f > 1$ (optimality of benchmark), the statement in the proposition follows. □

Proposition 4: *Suppose at rebalance time t, the investor has wealth w_t, forecast dynamics of asset prices given the history $\{Y_s, 0 \leq s \leq t\}$, and the return on the risk-free asset is r. If control limits (wealth goals) are $b_t \in B_t$ and the risk level is $1-\gamma$, then the growth–security efficient strategy at time t is $X_{b_t}^*(t) = (x_{1b_t}^*(t), \ldots, x_{mb_t}^*(t))'$ with*

$$X_{b_t}^*(t) = p_t(w_t, b_t, \gamma)\hat{\Delta}^{-1}(t)(\hat{\phi}(t) - re), \tag{19}$$

where e is a vector of ones.

Proof: Let $\underline{w}_t = w_t/k$ and $\overline{w}_t = k^c w_t$, where $k > 1$ and $c > 0$. For simplicity the time variable t is deleted. From (16) and (17) the growth–security efficiency problem is

$$\max_x\left\{\varphi(X)\left|\frac{1-k^{-\theta(X)}}{1-k^{(c+1)\theta(X)}} \geq \gamma,\ X \text{ feasible}\right.\right\}.$$

Letting $y = k^{-\theta(X)}$, the constraint requires $\gamma y^{c+1} - y + (1-\gamma) \geq 0$. Consider the minimum positive root y^* of the equation $\gamma y^{c+1} - y + (1-\gamma) = 0$, where $y^* \leq 1$ since $y^* = 1$ is a root. Then $\gamma y^{c+1} - y + (1-\gamma) \geq 0$ for $y \leq y^*$ and the constraint is satisfied if $k^{-\theta(X)} \leq y^*$ or

$$\theta(X) \geq -\left(\frac{\log y^*}{\log k}\right) = q^* \geq 0,$$

where q^* depends upon $\underline{w}_t, \overline{w}_t$, and γ. But $\theta(X) \geq q^*$ if $2\psi(X) - (1-q^*)\sigma^2(X) \geq 0$. Hence, the efficiency problem is

$$\max_x \{2\psi(X) - \sigma^2(X) | 2\psi(X) - (1+q^*)\sigma^2(X) \geq 0,\ X \text{ feasible}\}.$$

If X^* is a solution to this problem then it also solves, for optimal multiplier λ^*, the Lagrangian problem

$$\max_x L(X, \lambda^*),$$

where

$$L(X, \lambda^*) = 2\psi(X) - \sigma^2(X) + \lambda^*[2\psi(X) - (1+q^*)\sigma^2(X)].$$

The first-order conditions $\nabla_x L(X^*, \lambda^*) = 0$ imply that the optimal X^* satisfies the linear system

$$2(1+\lambda^*)(\hat{\phi}_i - r) - 2(1+\lambda^*(1+q^*))[x_i \delta_i^2] = 0, \quad i = 1, \ldots, I.$$

The solution to this system is $X^* = p_t(w_t, \underline{w}_t, \overline{w}_t, \gamma) \times \Delta^{-1}(\hat{\phi}(t) - re)$ where

$$p_t(w_t, \underline{w}_t, \overline{w}_t, \gamma) = \frac{1+\lambda^*}{1+\lambda^*(1+q^*)}.$$

□

Corollory 1: *For wealth goals $b_t \in B_t$, the fraction of wealth invested in risky assets at the rebalancing time t is* $\min\{1, p_t\}$ for

$$p_t(w_t, b_t, \gamma) = h_t \cdot H_t(\tilde{X}(t)) + \sqrt{[h_t \cdot H_t(\tilde{X}(t))]^2 + \frac{2rh_t}{\sigma_t^2(\tilde{X}(t))}}, \tag{20}$$

where

$$H_t(\tilde{X}(t)) = \frac{[\psi_t(\tilde{X}(t)) - r]}{\sigma_t^2(\tilde{X}(t))},$$

$$h_t = \frac{[\log w_t - \log \underline{w}_t]}{[\log w_t - \log \underline{w}_t - \log y^*]},$$

and y^* is the minimum positive root of $\gamma y^{c+1} - y + (1-\gamma) = 0$ for

$$c_t = \frac{[\log \overline{w}_t - \log w_t]}{[\log w_t - \log \underline{w}_t]}.$$

Proof: Clearly, $p_t \leq 1$, with

$$p_t = \frac{1+\lambda^*}{1+\lambda^*(1+q^*)} = 1$$

if $\lambda^* = 0$, i.e. when the constraint is not binding. If the security constraint is binding then X^* is a solution to $k^{-\theta(X)} = y^*$. Let $\tilde{X} = \Delta^{-1}(\hat{\phi} - re)$ and $X^* = p\tilde{X}$. Then the equation becomes

$$\left[\left(1 - \frac{\log y^*}{\log k}\right)\sigma^2(\tilde{X})\right]p^2 - 2[\mu(\tilde{X}) - r]p - 2r = 0.$$

The statement in the corollary follows. □

Proposition 5: *At the rebalance time t consider current wealth w_t^0, risk probability γ^0 and risk aversion parameter β. Assume that the benchmark portfolio satisfies the growth condition, i.e.*

$$\frac{\psi_t(\tilde{X}(t))}{\sigma_t^2(\tilde{X}(t))} > \frac{1}{2}.$$

It follows that $B_t^0(\beta)$ is non-empty for $\beta < 0$

Proof: From corollary 1, $p_t(w_t, b_t, \gamma) = 1/(1-\beta)$ implies

$$h_t = \frac{\sigma^2(\tilde{X})}{2r(1-\beta)^2 + 2H(\tilde{X})\sigma^2(\tilde{X})(1-\beta)} = K > 0.$$

Since

$$h_t(\underline{w}_t, \overline{w}_t) = \frac{\log w_t - \log \underline{w}_t}{\log w_t - \log \underline{w}_t - \log y^*},$$

any $(\underline{w}_t, \overline{w}_t) \in B_t^0(\beta)$ must satisfy the equation $h(\underline{w}_t, \overline{w}_t) = K$. From the assumption $\beta < 0$ and $\psi/\sigma^2 > 1/2$ it follows that $0 < K < 1$. The solution y^* of the equation $\gamma y^{C+1} - y + (1-\gamma) = 0$ has the property $y^* \uparrow 1$ as $\overline{w}_t \downarrow w_t$, for fixed $\underline{w}_t^0 < w_t$. So there exists $(\underline{w}_t^0, \overline{w}_t^*)$ such that $h(\underline{w}_t^0, \overline{w}_t^*) > K$. Also for fixed $\overline{w}_t^*, h(\underline{w}_t, \overline{w}_t^*) \downarrow 0$ as $\underline{w}_t \uparrow w_t$. That is, there is a $\underline{w}_t^*$ with $h(\underline{w}_t^*, \overline{w}_t^*) = K$. □

References

Algoet, P.H. and Cover, T.M., Asymptotic optimality and asymptotic equipartiation properties of log-optimum investment. *Annals of Probability*, 1988, **16**, 876–898.

Artzner, P., Delbaen, F., Eber, J. and Heath, D., Coherent measures of risk. *Mathematical Finance*, 1999, **9**, 203–228.

Aven, T. and Jensen, U., *Stochastic Models in Reliability*, 1999 (Springer: New York).

Basak, S. and Shapiro, A., Value at Risk based risk management: optimal policies and asset prices. *Review of Financial Studies*, 2001, **14**, 371–405.

Breiman, L., Investment policies for expanding business optimal in a long-run sense. *Naval Research Logistics Quarterly*, 1960, **7**, 647–651.

Breiman, L., Optimal gambling systems for favorable games, in *Proceedings of the 4th Berkeley Symposium on Mathematics, Statistics and Probability*, Vol. 1, 1961, pp. 63–68.

Breitmeyer, C., Hakenes, H., Pfingsten, A. and Rechtien, C., Learing from poverty measurement: an axiomatic approach to measure downside risk. Working Paper, University of Muenster, Germany, 1999.

Brennan, M.J., The role of learning in dynamic portfolio decisions. *European Finance Review*, 1998, **I**, 295–306.

Browne, S. and Whitt, W., Portfolio choice and the Bayesian Kelly criterion. *Advances in Applied Probability*, 1996, **28**, 1145–1176.

Burkhardt, T., A mean-variance of first passage time approach to portfolio selection in a lognormal world, in *the VIII International Conference on Stochastic Programming*, Vancouver, August, 1998.

Campbell, J., Lo, A. and MacKinlay, C., *The Econometrics of Financial Markets*, 1997 (Princeton University Press: Princeton, NJ).

Carino, D.R. and Ziemba, W.T., Formulation of the Russell Yasuda Kasai Financial Planning Model. *Operations Research*, 1998, **46**, 450–462.

Chopra, V.K. and Ziemba, W.T., The effect of errors in means, variances, and covariances on optimal portfolio choice. *Journal of Portfolio Management*, 1993, **19**, 6–11.

Clark, R. and Ziemba, W.T., Playing the turn-of-the-year effect with index futures. *Operations Research*, 1987, **35**, 799–813.

Gaivoronski, A. and Pflug, G., Value-at-risk in portfolio optimization: properties and computational approach. *Journal of Risk*, 2005, **7**(2), 1–31.

Grauer, R.R. and Hakansson, N.H., A half century of returns on levered and unlevered portfolios of stocks, bonds and bills, with and without small stocks. *Journal of Business*, 1986, **59**, 287–318.

Grauer, R.R. and Hakansson, N.H., Gains from international diversification: 1968–85 returns on portfolios of stocks and bonds. *Journal of Finance*, 1987, **42**, 721–739.

Hakansson, N.H., Optimal investment and consumption strategies under risk for a class of utility functions. *Econometrica*, 1970, **38**, 587–607.

Hakansson, N.H., Capital growth and the mean-variance approach to portfolio selection. *Journal of Financial and Quantitative Analysis*, 1971, **6**, 517–557.

Hakansson, N.H. and Ziemba, W.T., Capital growth theory. In *Finance*, edited by R.A. Jarrow, V. Maksimovic and W.T. Ziemba, pp. 65–86, 1995 (North-Holland: Amsterdam).

Hausch, D.B. and Ziemba, W.T., Transactions costs, extent of inefficiencies, entries and multiple wagers in a racetrack betting model. *Management Science*, 1985, **31**, 381–392.

Hausch, D.B. Ziemba, W.T. and Rubinstein, M., Efficiency of the market for racetrack betting. *Management Science*, 1981, **27**, 1435–1452.

Janecek, K., Maximum growth strategies in gambling and investment. Unpublished MSc thesis, Charles University, Prague, 1998.

Jorion, P., *Value-at-Risk: The Benchmark for Controlling Market Risk*, 1997 (Irwin: Chicago).

Kallberg, J.G. and Ziemba, W.T., Comparison of alternative utility functions in portfolio selection problems. *Management Science*, 1983, **29**, 1257–1276.

Karlin, S. and Taylor, H.M., *A Second Course in Stochastic Processes*, 1981 (New York: Academic Press).

Kelly, J., A new interpretation of information rate. *Bell System Technology Journal*, 1956, **35**, 917–926.

Lawley, D.N. and Maxwell, A.E., *Factor Analysis as a Statistical Method*, 1971 (Butterworths: London).

MacLean, L.C., Sanegre, R., Zhao, Y. and Ziemba, W.T., Capital growth with security. *Journal of Economic Dynamics and Control*, 2004, **28**, 937–954.

MacLean, L.C. and Weldon, K.L., Estimating multivariate random effects without replication. *Commun. Statist. Theory Meth.*, 1996, **24**, 1447–1469.

MacLean, L.C. and Ziemba, W.T., Growth versus security tradeoffs in dynamic investment analysis. *Annals of Operations Research*, 1999, **85**, 193–225.

MacLean, L.C., Ziemba, W.T. and Blazenko, G., Growth versus security in dynamic investment analysis. *Management Science*, 1992, **38**, 1562–1585.

Markowitz, H.M., Portfolio Selection. *Journal of Finance*, 1952, **7**, 77–91.

Markowitz, H.M., Investment for the long run: new evidence for an old rule. *Journal of Finance*, 1976, **31**, 1273–1286.

Markowitz, H.M., *Mean-variance Analysis in Portfolio Choice and Capital Markets*, 1987 (Basil Blackwell: New York).

Merton, R.C., *Continuous Time Finance*, 2nd ed., 1992 (Blackwell: Malden, MA).

Mulvey, J.M. and Vladimirou, H., Stochastic network programming for financial planning problems. *Management Science*, 1992, **38**, 1642–1664.

Ogryczak, W. and Ruszczynski, A., Dual stochastic dominance and related mean risk models. *SIAM Journal on Optimization*, 2002, **13**, 60–78.

Pratt, J.W., Risk aversion in the small and in the large. *Econometrica*, 1964, **32**, 122–136.

Rockafeller, R.T. and Uryasev, S., Optimization of conditional value-at-risk. *Journal of Risk*, 2000, **2**, 21–41.

Rockafeller, R.T. and Ziemba, W., Modified risk measures and acceptance sets. Mimeo. University of British Columbia, 2000.

Rogers, L.C.G., The relaxed investor and parameter uncertainty. *Finance and Stochastics*, 2000, **5**, 131–154.

Thorp, E.O., Portfolio choice and the Kelly criterion. In *Stochastic Optimization Models in Finance*, edited by W.T. Ziemba and R.G. Vickson, pp. 599–619, 1975 (Academic Press: New York).

Xia, Y., Learning about predictability: the effects of parameter uncertainty on dynamic asset allocation. *Journal of Finance*, 2001, **56**, 205–246.

20

Survival and Evolutionary Stability of the Kelly Rule*

Igor V. Evstigneev
Economics Department, School of Social Sciences,
University of Manchester, United Kingdom
igor.evstigneev@manchester.ac.uk

Thorsten Hens
Swiss Banking Institute, University of Zurich, Switzerland
thens@isb.uzh.ch

Klaus Reiner Schenk-Hoppé
Leeds University Business School and School of Mathematics,
University of Leeds, United Kingdom
K.R.Schenk-Hoppe@leeds.ac.uk

Abstract

This chapter gives an overview of current research in evolutionary finance. We mainly focus on the survival and stability properties of investment strategies associated with the Kelly rule. Our approach to the study of the wealth dynamics of investment strategies is inspired by Darwinian ideas on selection and mutation. The goal of this research is to develop an evolutionary framework for practical investment advice.

1 Introduction

The principal objective of our evolutionary approach to the study of financial market dynamics is the development and analysis of models that constitute a plausible alternative to the conventional general equilibrium framework. Our aim is to provide a framework that is suitable for delivering practical investment advice.

The equilibrium concept most commonly used in financial economics is due to Radner (1972). This equilibrium notion involves the plans and price expectations

*Financial support by the Finance Market Fund, Norway (project Stability of Financial Markets: An Evolutionary Approach) and the National Center of Competence in Research "Financial Valuation and Risk Management", Switzerland (project Behavioural and Evolutionary Finance) is gratefully acknowledged. This chapter was written during KRSH's visit to the Department of Finance and Management Science at the Norwegian School of Economics and Business Administration in August 2009. We are grateful to Terje Lensberg, Edward O. Thorp and William T. Ziemba for their helpful comments.

of agents as well as market prices. A well-known drawback of that framework is the necessity of agents to have *perfect foresight* (rational expectations) in order to establish an equilibrium (see the discussion in Laffont (1989) and Dubey *et al.* (1987)) market participants have to agree on the future prices for each of the possible future realizations of the states of the world (without knowing which particular state will be realized).

Our evolutionary model differs radically from that approach: only historical observations and the current state of the world influence the agents' behavior; no agreement about the future market structure is required and no coordinated actions by the agents are assumed. From a practical finance perspective, the important distinction between our approach and the conventional general equilibrium paradigm lies in the data required to formulate the model. The evolutionary model does not use agents' characteristics that are unobservable (such as individual utilities or subjective beliefs). We describe aims of investors in terms of properties (survival, evolutionary stability, etc.) holding almost surely, rather than in terms of the maximization of expected utilities. We consider this robust modeling approach as the basis for developing a new generation of dynamic equilibrium models that could be used for practical quantitative recommendations applicable in the financial industry.

The general approach underlying this direction of work is to apply evolutionary dynamics — mutation and selection — to the analysis of the long-run performance of investment strategies. A stock market is understood as a heterogeneous population of frequently interacting portfolio rules in competition for market capital. The ultimate goal is to build a *Darwinian theory* of portfolio selection. Evolutionary ideas have a long history in the social sciences going back to Malthus, who played an inspirational role for Darwin (for a review of the subject see, e.g., Hodgeson (1993). A more recent stage of development of these ideas began in the 1950s with the publications of Alchian (1950) and others. An important role in this line of work has been played by the interdisciplinary research conducted in the 1980s and 1990s under the auspices of the Santa Fe Institute in New Mexico, USA — see, e.g., Arthur *et al.* (1997), Farmer and Lo (1999), LeBaron *et al.* (1999), Blume and Easley (1992), and Blume and Durlauf (2005).

Our model also revives the literature on stochastic dynamic games going back to Shapley (1953). The framework also shares some conceptual features with the *games of survival* pioneered by Milnor and Shapley (1957) and the dynamic market game in Shubik and Whitt (1973). The main difference is that we use the notion of survival strategies rather than that of Nash equilibrium. In a finance context, the focus on survival is an advantage (over those approaches invoking expectations) because it is a property holding almost surely and not requiring discounted or undiscounted utility.

This survey is based on the research carried out by Amir *et al.* (2005, 2008, 2009), Evstigneev *et al.* (2002, 2006, 2008) and Hens and Schenk-Hoppé (2005).

A general survey of evolutionary finance is provided by Evstigneev *et al.* (2009). The issue of survival of traders in Radner's setting has been studied by Blume and Easley (1992); see also the surveys Blume and Easley (2008, 2009).

Section 2 explains in details the model, Section 3 discusses its dynamics and defines the concepts of survival and evolutionary stability, Section 4 surveys the results obtained in the literature, and Section 5 concludes.

2 Model

Consider a market in which $K \geq 2$ *assets* are traded. Each asset $k = 1, 2, ..., K$ pays dividends at dates $t = 1, 2, ...$ The non-negative payoff $D_{t,k}(s^t) \geq 0$ depends on the history $s^t = (s_1, ..., s_t)$ of states of the world up to date t. The states are random factors modeled in terms of an exogenous stochastic process $s_1, s_2, ...$, where s_t is a random element of a measurable space S_t. We assume functions to be measurable throughout the following. At each date (and in each random situation), at least one asset is assumed to pay a strictly positive dividend:

$$\sum_{k=1}^{K} D_{t,k}(s^t) > 0 \text{ for all } t, s^t \,. \tag{1}$$

The total net supply of asset k is equal to the random amount $V_{t,k}(s^t) > 0$ (a constant $V_{0,k} > 0$ at the initial time $t = 0$). The vector of market prices is $p_t = (p_{t,1}, ..., p_{t,K}) \in R_+^K$, where $p_{t,k}$ is the price per unit. We assume $V_{t,k}(s^t)/V_{t-1,k}(s^{t-1}) \geq \gamma > 0$ for $t \geq 0$ and all s^t. This condition is satisfied, e.g., if the asset supply is constant or increasing; the supply however cannot decrease at an ever-increasing rate.

There are $N \geq 2$ *investors*, or *traders*, acting in the market. A *portfolio* of investor i at date $t = 0, 1, ...$ is specified by a vector $\theta_t^i = (\theta_{t,1}^i, ..., \theta_{t,K}^i) \in R_+^K$ where $\theta_{t,k}^i$ is the number of units of asset k held in the portfolio. The value of investor i's portfolio at date t is $\langle p_t, \theta_t^i \rangle = \sum_{k=1}^K p_{t,k}\theta_{t,k}^i$. The *state of the market* at each date t is characterized by the set $(p_t, \theta_t^1, ..., \theta_t^N)$ consisting of the price vector and all traders' portfolios.

Investors $i = 1, 2, ..., N$ enter the market at date $t = 0$ with endowments $w_0^i > 0$ that form their initial budgets. At date $t \geq 1$, investor i's budget is $\langle D_t(s^t) + p_t, \theta_{t-1}^i \rangle$, where $D_t(s^t) = (D_{t,1}(s^t), ..., D_{t,K}(s^t))$. This budget consists of two components linked to the portfolio θ_{t-1}^i: the dividends $\langle D_t(s^t), \theta_{t-1}^i \rangle$ received and the market value $\langle p_t, \theta_{t-1}^i \rangle$ expressed in terms of the current prices p_t.

A fraction α of the budget is invested in assets. We will assume that the *investment rate* $\alpha \in (0, 1)$ is a fixed number, the same for all the traders. The remaining amount of wealth is withdrawn from the market through a *tax rate* $1 - \alpha$ (or, if you dislike taxes, through a *consumption rate* $1 - \alpha$). The assumption of a common $1 - \alpha$ for all the investors is quite natural when interpreted as a tax rate. As a consumption rate it might seem a restrictive condition, however, it is indispensable since we focus on the analysis of the comparative performance of trading strategies in the long run. Without this assumption, an analysis of this kind does not make sense: a seemingly worse long-run performance of an investment strategy might be simply due to a higher consumption rate.

For each $t \geq 0$, every trader $i = 1, 2, ..., N$ selects a vector of *portfolio weights* $\lambda_t^i = (\lambda_{t,1}^i, ..., \lambda_{t,K}^i)$ according to which he/she plans to distribute the available budget between assets. Vectors λ_t^i belong to the unit simplex $\Delta^K = \{(a_1, ..., a_K) \geq 0 : a_1 + ... + a_K = 1\}$. In game-theoretic terms, the vectors λ_t^i represent the investors' *actions.* The portfolio weights at each date $t \geq 0$ are selected by the N investors simultaneously and independently—the investors participate in a simultaneous-move N-person dynamic game.

The investors' actions can depend (for $t \geq 1$) on the history s^t of the states of the world and the *history of the game* $(p^{t-1}, \theta^{t-1}, \lambda^{t-1})$, where $p^{t-1} = (p_0, ..., p_{t-1})$ is the history of asset price vectors and

$$\theta^{t-1} = (\theta_0, \theta_1, ..., \theta_{t-1}),\ \theta_l = (\theta_l^1, ..., \theta_l^N)$$

$$\lambda^{t-1} = (\lambda_0, \lambda_1, ..., \lambda_{t-1}),\ \lambda_l = (\lambda_l^1, ..., \lambda_l^N)$$

are the sets of vectors describing the portfolios and the portfolio weights of all the traders at all the dates up to $t - 1$.

The history of the game contains information about the *market history* — the sequence $(p_0, \theta_0), ..., (p_{t-1}, \theta_{t-1})$ of the states of the market — and about the actions λ_l^i of all the investors $i = 1, ..., N$ at all the dates $l = 0, ..., t - 1$. A vector $\Lambda_0^i \in \Delta^K$ and a sequence of measurable functions with values in Δ^K

$$\Lambda_t^i(s^t, p^{t-1}, \theta^{t-1},\ \lambda^{t-1}),\ t = 1, 2, ...$$

form an *investment strategy* Λ^i of trader i, specifying a *portfolio rule* according to which trader i selects portfolio weights at each date $t \geq 0$. This is a general game-theoretic definition of a strategy, assuming full information about the history of the game, including the players' previous actions, and the knowledge of all the past and present states of the world. Among general investment strategies, we will distinguish those for which Λ_t^i depends only on s^t and not on the market history. We will call such strategies *basic.* Note that basic strategies are in general not simple (constant).

To complete the description of the market, it remains to define a dynamic equilibrium for the assets. Suppose that at date 0, each investor i has selected some portfolio weights $\lambda_0^i = (\lambda_{0,1}^i, ..., \lambda_{0,K}^i) \in \Delta^K$. Then the amount invested in asset k by trader i is $\alpha\lambda_{0,k}^i w_0^i$ and the total amount invested in asset k is $\alpha \sum_{i=1}^N \lambda_{0,k}^i w_0^i$. It is assumed that the market is always in equilibrium (asset supply is equal to asset demand), which makes it possible to determine the equilibrium price $p_{0,k}$ of each asset k from the equation

$$p_{0,k} V_{0,k} = \alpha \sum_{i=1}^{N} \lambda_{0,k}^i w_0^i \tag{2}$$

The left-hand side is the total market value $p_{0,k}V_{0,k}$ of the supply of asset k while the right-hand side represents the total wealth invested in asset k by all the investors. Equilibrium ensures that both sides are equal, i.e., (2) holds. The portfolio weights λ_0^i chosen by the traders at date 0 determine their portfolios θ_0^i at date 0 by

$$\theta_{0,k}^i = \frac{\alpha \lambda_{0,k}^i w_0^i}{p_{0,k}} \tag{3}$$

i.e., the current market value $p_{0,k}\theta_{0,k}^i$ of the kth position of investor i's portfolio is equal to the fraction $\lambda_{0,k}^i$ of the investor's investment budget αw_0^i.

Consider now the situation at any date $t \geq 1$. Suppose all investors have chosen their portfolio weights $\lambda_t^i = (\lambda_{t,1}^i, ..., \lambda_{t,K}^i)$. Then the equilibrium of asset supply and demand determines the market-clearing prices of assets $k = 1, ..., K$ through the relation

$$p_{t,k}V_{t,k} = \alpha \sum_{i=1}^{N} \lambda_{t,k}^i \langle D_t(s^t) + p_t, \theta_{t-1}^i \rangle \tag{4}$$

The investment budgets $\alpha\langle D_t(s^t) + p_t, \theta_{t-1}^i\rangle$ of the traders $i = 1, 2, ..., N$ are distributed between assets in the proportions $\lambda_{t,k}^i$, so that the kth position of the trader i's portfolio θ_t^i is

$$\theta_{t,k}^i = \frac{\alpha \lambda_{t,k}^i \langle D_t(s^t) + p_t, \theta_{t-1}^i \rangle}{p_{t,k}} \tag{5}$$

The price vector p_t is determined implicitly as the solution to the system of equations (4). The existence and uniqueness of a non-negative vector p_t solving (4) (for any s^t and any feasible θ_{t-1}^i and λ_t^i) is proved in Amir *et al.* (2009, Proposition 1).

For a given strategy profile $(\Lambda^1, ..., \Lambda^N)$ of investors and their endowments $w_0^1, ..., w_0^N$, a path of the market game is generated by setting

$$\lambda_0^i = \Lambda_0^i \tag{6}$$

$$\lambda_t^i = \Lambda_t^i(s^t, p^{t-1}, \theta^{t-1}, \lambda^{t-1}) \tag{7}$$

for all $t = 1, 2, ...,$ and $i = 1, ..., N$; and by defining p_t and θ_t^i recursively according to equations (2)–(5). The random dynamical system (Arnold 1998) described above defines step by step the vectors of portfolio weights $\lambda_t^i(s^t)$, the equilibrium prices $p_t(s^t)$ and the investors' portfolios $\theta_t^i(s^t)$ as measurable vector functions of s^t for each moment of time $t \geq 0$ (for $t = 0$ these vectors are constant). Thus, we obtain a random path of the game

$$(p_t(s^t); \theta_t^1(s^t), ..., \theta_t^N(s^t); \lambda_t^1(s^t), ..., \lambda_t^N(s^t)) \tag{8}$$

as a vector stochastic process in $R_+^K \times R_+^{KN} \times R_+^{KN}$.

The above description is correct only if the price of each asset is strictly positive. Strategy profiles which guarantee this property at each period will be called *admissible*. Admissibility is satisfied, e.g., if the strategy profile contains an investor who uses a strategy with strictly positive portfolio weights. We will deal only with such strategy profiles from now on and, therefore, have well-definedness of the random dynamical system. Then (using induction) the equilibrium path all the portfolios $\theta_t^i = (\theta_{t,1}^i, ..., \theta_{t,K}^i)$ are non-zero and the wealth

$$w_t^i = \langle D_t + p_t, \theta_{t-1}^i \rangle \quad (9)$$

is strictly positive.

Equation (5) implies $\sum_{i=1}^N \theta_{t,k}^i = V_{t,k}$, i.e., market clearing for every asset k and each date $t \geq 1$ (analogously for $t = 0$). For every equilibrium state of the market one therefore has that markets clear, $p_t > 0$ and $\theta_t^i \neq 0$ for all i.

The description of a financial market is game-theoretic: the market dynamics is formulated as a simultaneous-move N-person dynamic game. Less general versions of the model can be motivated by invoking Marshall's (1949) principle of temporary equilibrium (Evstigneev *et al.*, 2008) or evolutionary dynamics (Evstigneev *et al.*, 2006).

3 Dynamics and Stability

This section provides details on the dynamics of the model and defines notions of survival and evolutionary stability. An explicit formulation of the dynamics for the investors' wealth as well as their relative wealth is given.

3.1 *Dynamics and the role of basic strategies*

Let $(\Lambda^1, ..., \Lambda^N)$ be an admissible strategy profile of the investors. Consider the path (8) of the random dynamical system generated by this strategy profile and the given initial budgets. Let $w_t^i > 0$ denote the investor i's wealth at date $t \geq 0$. For $t \geq 1$, $w_t^i = w_t^i(s^t)$ is given by formula (9) while each investor i's endowment, w_0^i, is a constant.

When studying the dynamics of this wealth process for a *fixed* strategy profile, it is sufficient to consider the class of basic strategies, i.e., those strategies for which Λ_t^i depends only on the history of states until time t, s^t, and not on the market history. This assertion holds because any sequence of vectors $w_t = (w_t^1, ..., w_t^N)$ of wealth generated by some strategy profile $(\Lambda^1, ..., \Lambda^N)$ can be generated by a strategy profile $(\lambda_t^1(s^t), ..., \lambda_t^N(s^t))$ consisting of basic portfolio rules. The corresponding vector functions $\lambda_t^i(s^t)$ can be defined recursively by (6) and (7), using (2)–(5).

This observation is very useful when studying the dynamic properties (such as survival or evolutionary stability) for a basic strategy — as we will do here. However this 'reduction' of the model to basic strategies has its limitations when dealing with the characteristics of a general (non-basic) strategy which competes in markets with different opponents. Such a strategy will in general take on different basic strategies, dependent on the pool of competitors — it is therefore not basic.

The procedure described above gives the system of equations

$$w_{t+1}^i = \sum_{k=1}^{K} \left[\alpha \frac{1}{V_{t+1,k}} \langle \lambda_{t+1,k}, w_{t+1} \rangle + D_{t+1}(s_{t+1}) \right] V_{t,k} \frac{\lambda_{t,k}^i w_t^i}{\langle \lambda_{t,k}, w_t \rangle} \tag{10}$$

$i = 1, ..., N$. This dynamic on the space $\{w \in R^N \mid w \geq 0 \text{ and } w \neq 0\}$ is well-defined under the assumptions imposed in Section 2 (e.g., Amir *et al.* (2009) or Section 4.1 in Evstigneev *et al.* (2008) which applies with minor modifications).

Our analysis will focus on the long-run behavior of the *relative wealth* or the *market shares* $r_t^i = w_t^i / \sum_{j=1}^{N} w_t^j$ of the traders. Both the stability and survival of portfolio rules will be studied in this framework. An explicit random dynamical system can be derived for the vector r_t as follows.

Assume that the asset supply changes over time at the same rate $\gamma > 0$:

$$V_{t,k} = \gamma^t V_k \tag{11}$$

where $V_k > 0$ $(k = 1, 2, ..., K)$ are the initial amounts of the assets. In the case of real dividend-paying assets — involving long-term investments in the real economy (e.g., real estate, transportation, media, infrastructure, etc.) — the above assumption means that the economic system under consideration is on a *balanced growth path*. Define the *relative dividends* of the assets $k = 1, ..., K$ by $R_t = (R_{t,1}, ..., R_{t,K})$ where

$$R_{t,k} = R_{t,k}(s^t) = \frac{D_{t,k}(s^t) V_k}{\sum_{m=1}^{K} D_{t,m}(s^t) V_m} \tag{12}$$

for $t \geq 1$. Further, assume that $\alpha < \gamma$ and define

$$\rho = \alpha/\gamma, \text{ and } \rho_t = \rho^{t-1}(1 - \rho)$$

One finds the dynamic for the vector of market shares:

$$r_{t+1}^i = \sum_{k=1}^{K} [\rho \langle \lambda_{t+1,k}, r_{t+1} \rangle + (1 - \rho) R_{t+1,k}] \frac{\lambda_{t,k}^i r_t^i}{\langle \lambda_{t,k}, r_t \rangle} \tag{13}$$

with $i = 1, ..., N$. This equation can be written in explicit form (e.g., Evstigneev *et al.*, 2009):

$$r_{t+1} = (1 - \rho) \left[\text{Id} - \rho \Theta_t \lambda_{t+1} \right]^{-1} \Theta_t R_{t+1} \tag{14}$$

where $\lambda_t^T = (\lambda_{t,1}^T, ..., \lambda_{t,K}^T) \in R^{N \times K}$ is the matrix of portfolio weights and $\Theta_t \in R^{N \times K}$ is the matrix of portfolios given by $\Theta_{t,k}^i = \lambda_{t,k}^i r_t^i / \langle \lambda_{t,k}, r_t \rangle$.

Setting $\rho = 0$ in (14) one obtains a model with short-lived assets akin to parimutuel betting markets, see e.g. Evstigneev *et al.* (2009).

3.2 *Survival and extinction of portfolio rules*

We study the dynamics of wealth shares from an evolutionary perspective. Our focus is on the questions of "survival and extinction" of portfolio rules.

A portfolio rule $\lambda^i = (\lambda_t^i(s^t))$ (or the investor i using it) *survives* with probability one if $\inf_{t\geq 0} r_t^i > 0$ almost surely. This means that for almost all realizations of the process of states of the world $s_1, s_2, ...$, the market share of the first investor is bounded away from zero by a strictly positive random constant. We say that λ^i *becomes extinct* with probability one if $\lim_{t\to\infty} r_t^i = 0$ almost surely. Survival and extinction can be defined for general non-basic investment strategies without any changes. An investment strategy Λ is called a *survival strategy* if the investor using it survives with probability one regardless of what other strategies are present in the market. In terms of the wealth process w_t^i, $i = 1, 2, ..., N$, no investor's wealth can grow asymptotically faster than the wealth of investors who use survival strategies. In this sense, all survival strategies are competitive.

A portfolio rule $\lambda = (\lambda_t(s^t))$ is called *globally evolutionarily stable* if the following condition holds. Suppose, in a group of investors $i = 1, 2, ..., J$ $(1 \leq J < N)$, all use the portfolio rule λ, while all the others, $i = J+1, ..., N$ use portfolio rules $\hat{\lambda}^i$ distinct from λ. Then those investors who belong to the former group $(i = 1, ..., J)$ survive with probability one, whereas those who belong to the latter $(i = J + 1, ..., N)$ become extinct with probability one (cf. Evstigneev *et al.*, 2008)). An analogous concept of *local* evolutionary stability can be defined. In that definition of stability, the initial market share $r_0^{J+1} + ... + r_0^N$ of the group of investors who use strategies $\hat{\lambda}^i$ distinct from λ is supposed to be small enough (cf. Evstigneev *et al.*, 2006).

4 Results

The findings on the long-run outcome of the dynamics of the above game fall in two categories, survival and evolutionary stability. The study of the first is carried out in the general model while the latter requires placing restrictions on the set of admissible strategies. This is simply caused by the extreme generality of the asset market game introduced above. For instance, if one strategy is constant while some other strategy always imitates the decision made in the previous period, both end up with the same portfolio weights. However, it can be shown that mimicking strategies are not survival strategies.

The central role in our analysis is played by a (generalized) Kelly rule. Define the investment strategy Λ^* with the vectors of portfolio weights $\lambda_t^*(s^t)$ by

$$\lambda_{t,k}^* = E_t \sum_{l=1}^{\infty} \rho_l R_{t+l,k} \tag{15}$$

where $E_t(\cdot) = E(\cdot|s^t)$ is the conditional expectation given s^t (unconditional expectation $E(\cdot)$ if $t = 0$).

The portfolio rule specified by (15) prescribes to distribute wealth across assets in accordance with the proportions of the expected flow of their discounted future relative dividends. The discount rate $\rho_{t+1}/\rho_t = \rho$ is equal to the investment rate α divided by the growth rate γ.

The portfolio weights of the strategy Λ^* generally depend on time t and the history of states of the world s^t, but do not depend on the history of the game $(p^{t-1}, \theta^{t-1}, \lambda^{t-1})$ — the strategy is basic. The strategy Λ^* is a generalization of the Kelly portfolio rule of 'betting your beliefs' playing an important role in capital growth theory — see Kelly (1956), Breiman (1961), Thorp (1971), Algoet and Cover (1988), and Hakansson and Ziemba (1995).

4.1 *Survival of the Kelly rule*

This section collects the results on the survival of the Kelly rule in different specifications of the model. Survival of the Kelly rule is ensured for the case of general investment strategies in both stock and betting markets.

Amir *et al.* (2009, Theorem 1) shows that *the Kelly rule* (15) *is a survival strategy.* The only additional assumption needed for this result is that for all k and t, $E_t R_{t+1,k}(s_{t+1}) > \delta$ almost surely for some constant $\delta > 0$. This results extends the findings in Amir *et al.* (2008) to the stock market case (long-lived dividend-paying assets, $\rho > 0$). In Amir *et al.* (2008), betting markets ($\rho = 0$) are studied in which the assets are short-lived and are reissued at each date. The Kelly rule in this setting is obtained from (15) by using the discount rates $\rho_1 = 1$ and $\rho_l = 0$ for $l > 1$. One has

$$\lambda^*_{t,k} = E_t R_{t+1,k} \tag{16}$$

for $k = 1, ..., K$. Under the assumption $E \ln E_t R_{t+1,k}(s^{t+1}) > -\infty$ (which ensures strict positivity of all $\lambda^*_{t,k}$), Amir *et al.* (2008, Theorem 1) asserts that $\lambda^*_{t,k}$ defined in (16) is a survival strategy. The existence of a non-basic survival strategy is an open problem.

In both cases, the Kelly rule is asymptotic unique among all basic survival strategies. One has that $\sum_{t=0}^{\infty} ||\lambda^*_t - \lambda_t||^2 < \infty$ almost surely for any basic survival strategy $\Lambda = (\lambda_t)$ (see Amir *et al.* (2008, Theorem 2) and Amir *et al.* (2009, Theorem 2)).

4.2 *Evolutionary stability of the Kelly rule*

This section reports on properties of the Kelly rule that are stronger than 'merely' ensuring survival. These features however only surface when restricting the set of admissible strategies (as well as the type of randomness driving the dividend process). The main results are that evolutionary stability with iid asset payments holds when the set of admissible strategies is restricted to constant strategies; and that local evolutionary stability hold is strategies and states are Markov. These theoretical findings are complemented by simulation studies of markets in which the pool of investment strategies changes over time through a process of mutation.

Assume that there are finitely many states of the world $s \in S$, states $s_1, s_2, ...$ are independent identically distributed (iid) with $P\{s_t = s\} > 0$ for each $s \in S$ and that the relative dividends $R_{t,k}(s^t) = R_k(s_t)$ depend only on the current state s_t and do not explicitly depend on t. In addition to (1), we assume that $ER_k(s_t) > 0$ for $k = 1, ..., K$. Under these conditions, (15) takes the special form

$$\lambda^*_{t,k} = \sum \rho_t ER_k(s_t) = ER_k(s_t) \tag{17}$$

which means that the strategy Λ^* is formed by the sequence of constant vectors $(ER_1(s_t), ..., ER_K(s_t))$ (independent of t and s^t). Note that in this special case, the formula for Λ^* does not involve the investment rate ρ. In this case, the 'beliefs' at each date t are concerned simply with the expected relative dividends (which do not depend on t). We call an investment strategy Λ *simple*, or *fixed-mix*, if all the portfolio weights are constant. The investment strategy (17) is simple and completely mixed (i.e., all components are strictly positive).

Suppose there are no redundant assets, i.e., the functions $R_1(s), ..., R_K(s)$ are linearly independent. Evstigneev *et al.* (2008), Theorem 1, shows that the Kelly rule defined in (17) is globally evolutionarily stable in any market in which all the portfolio rules are simple. This result demonstrates that the Kelly rule Λ^* has good properties beside ensuring survival: the group of investors following this rule outperform all other simple strategies and dominate the market. These investors eventually gather the total market wealth, while those who use simple strategies distinct from Λ^* become extinct. The related result for short-lived assets ($\rho = 0$) is proved in Evstigneev *et al.* (2002) and Amir *et al.* (2005). The latter deals with a more general setting in which the state follows a Markov process and the relative asset payoffs at time $t + 1$ depend on s_t and s_{t+1}.

Results on the local evolutionary stability of the Kelly rule are obtained in Hens and Schenk-Hoppé (2005) (dealing with short-lived assets) and Evstigneev *et al.* (2006) (considering long-lived assets and basic Markovian strategies). The Kelly rule is the only portfolio rule that is locally evolutionary stable — and, thus, the only candidate for a global evolutionary stable strategy. The advantage of considering the local dynamics is that stability can be characterized by growth rates (of the linearized dynamics) that do have explicit representations. It is therefore straightforward to verify whether particular portfolio rules co-exist or whether selection can occur through extinction of some portfolio rules.

In a strict sense, all of the above papers deal with the issue of selection rather than the process of mutation that creates novel behavior. The question whether the Kelly rule *emerges* in a market where traders adapt their strategies is investigated in Lensberg and Schenk-Hoppé (2007). They combine the evolutionary model with traders whose decisions are made by genetic programs. In that setting, the strategies and the dynamics of wealth shares co-evolve in a process of market interaction and tournament selection of strategies (in which poor strategies are replaced by novel rules). Their numerical analysis shows that the Kelly rule evolves in the long-run but prices converge much faster than the individual strategies to those prescribed by the Kelly rule. Successful traders, interestingly, invest according to a *fractional Kelly rule* rather than its pure form. This approach drastically reduces the volatility of investment returns, which helps to avoid deletion by the tournament selection process. The optimality properties of the fractional Kelly rule for practical investment are discussed in MacLean *et al.* (1992).

5 Conclusion

This survey presents in detail an evolutionary model of financial markets. The model is described as a simultaneous-move N-person dynamic game. Restricting the space of strategies to those depending only on the history of states (rather than the entire information consisting of the market history and revealed strategies of competing traders), one obtains the earlier, behavioral models evolutionary finance. The main findings of the survival and evolutionary stability properties of investment strategies are discussed. It turns out that the Kelly rule (appropriately defined to fit the framework under consideration) ensures the survival of the traders following this portfolio rule. The Kelly rule brings the additional benefit of (global) evolutionary stability.

References

Alchian, A. (1950). Uncertainty, evolution and economic theory. *Journal of Political Economy*, 58, 211–221.

Algoet, P. H. and T. M. Cover (1988). Asymptotic optimality and asymptotic equipartition properties of log-optimum investment. *Annals of Probability*, 16, 876–898.

Amir, R., I. V. Evstigneev, T. Hens and K. R. Schenk-Hoppé (2005). Market selection and survival of investment strategies. *Journal of Mathematical Economics*, 41, 105–122.

Amir, R., I. V. Evstigneev, T. Hens and L. Xu (2009). Evolutionary finance and dynamic games. Working Paper No. 581. NCCR "Financial Valuation and Risk Management", Switzerland, August 2009.

Amir, R., I. V. Evstigneev and K. R. Schenk-Hoppé (2008). Asset market games of survival. Working Paper No. 505. NCCR Financial Valuation and Risk Management, Switzerland, Revised version, June 2010.

Arnold, L. (1998). *Random Dynamical Systems*. US: Springer.

Arthur, W. B., J. H. Holland, B. LeBaron, R. G. Palmer and P. Taylor (1997). Asset pricing under endogenous expectations in an artificial stock market. In *The Economy as an Evolving Complex System II* (eds. Arthur, W. B., Durlauf, S. and Lane, D.), pp. 15–44. MA: Addison Wesley.

Blume, L. and S. Durlauf, (eds.) (2005). *The Economy as an Evolving Complex System III*. New York: Oxford University Press.

Blume, L. and D. Easley (1992). Evolution and market behavior. *Journal of Economic Theory*, 58, 9–40.

Blume, L. and D. Easley (2008). Market competition and selection. In *The New Palgrave Dictionary of Economics*, Vol. 5 (eds. Blume, L. and Durlauf, S. N.), pp. 296–300. UK: Macmillan.

Blume, L. and D. Easley (2009). Market selection and asset pricing. In *Handbook of Financial Markets: Dynamics and Evolution* (eds. Hens, T. and Schenk-Hoppé, K. R.), Chapter 7, pp. 403–437. Amsterdam: North-Holland.

Breiman, L. (1961). Optimal gambling systems for favorable games. In *Proceedings of the Fourth Berkeley Symposium on Mathematical Statistics and Probability, Vol. 1* (ed. Neyman, J.), pp. 65-78. US: University of California Press.

Dubey, P., J. Geanakoplos and M. Shubik (1987). The revelation of information in strategic market games: A critique of rational expectations equilibrium. *Journal of Mathematical Economics*, 16, 105–137.

Evstigneev, I. V., T. Hens and K. R. Schenk-Hoppé (2002). Market selection of financial trading strategies: Global stability. *Mathematical Finance*, 12, 329–339.

Evstigneev, I. V., T. Hens and K. R. Schenk-Hoppé (2006). Evolutionary stable stock markets. *Economic Theory*, 27, 449–468.

Evstigneev, I. V., T. Hens and K. R. Schenk-Hoppé (2008). Globally evolutionarily stable portfolio rules. *Journal of Economic Theory*, 140, 197–228.

Evstigneev, I. V., T. Hens and K. R. Schenk-Hoppé (2009). Evolutionary Finance. In *Handbook of Financial Markets: Dynamics and Evolution* (eds. Hens, T. and Schenk-Hoppé, K. R.), Chapter 9, pp. 507–566. Amsterdam: North-Holland.

Farmer, J. D. and A. W. Lo (1999). Frontiers of finance: Evolution and efficient markets. *Proceedings of the National Academy of Sciences*, 96, 9991–9992.

Hakansson, N. H. and W. T. Ziemba (1995). Capital growth theory. In *Handbooks in Operations Research and Management Science, Vol. 9, Finance* (eds. Jarrow, R. A., Maksimovic, V. and Ziemba, W. T.), Chapter 3, pp. 65–86. Amsterdam: North-Holland.

Hens, T. and K. R. Schenk-Hoppé (2005). Evolutionary stability of portfolio rules in incomplete markets. *Journal of Mathematical Economics*, 41, 43–66.

Hodgeson, G. M. (1993). *Economics and Evolution: Bringing Life Back into Economics.* US: Blackwell.

Kelly, J. (1956). A new interpretation of information rate. *Bell System Technical Journal*, 35, 917–926.

Laffont, J.-J. (1989). *The Economics of Uncertainty and Information.* MA: MIT Press.

LeBaron, B., W. B. Arthur and R. Palmer (1999). Time series properties of an artificial stock market. *Journal of Economic Dynamics and Control*, 23, 1487–1516.

Lensberg, T. and K. R. Schenk-Hoppé (2007). On the evolution of investment strategies and the Kelly rule – A Darwinian approach. *Review of Finance*, 11, 25–50.

MacLean, L. C., W. T. Ziemba and G. Blazenko (1992). Growth versus security in dynamic investment analysis. *Management Science*, 38, 1562–1585.

Marshall, A. (1949). *Principles of Economics* (8th edition). UK: Macmillan.

Milnor, J. and L. S. Shapley (1957). On games of survival. In *Contributions to the Theory of Games III* (eds. Dresher, M., Tucker, A. W. and Wolfe, P.) Annals of Mathematics Studies 39, pp. 15–45. US: Princeton University Press.

Radner, R. (1972). Existence of equilibrium of plans, prices, and price expectations in a sequence of markets. *Econometrica*, 40, 289–303.

Shapley, L. S. (1953). Stochastic games. *Proceedings of the National Academy of Sciences of the USA*, 39, 1095–1100.

Shubik, M. and W. Whitt (1973). Fiat money in an economy with one durable good and no credit. In: *Topics in Differential Games* (ed. Blaquiere, A.), pp. 401–448. Amsterdam: North-Holland.

Thorp, E. O. (1971). Portfolio choice and the Kelly criterion. In *Business and Economics Statistics Section, Proceedings of the American Statistical Association*, pp. 215–224.

21

Application of the Kelly Criterion to Ornstein-Uhlenbeck Processes

Yingdong Lv & Bernhard K. Meister

Department of Physics, Renmin University of China
Email:lyd08250@163.com & b_meister@ruc.edu.cn

Abstract. In this paper, we study the Kelly criterion in the continuous time framework building on the work of E.O. Thorp and others. The existence of an optimal strategy is proven in a general setting and the corresponding optimal wealth process is found. A simple formula is provided for calculating the optimal portfolio for a set of price processes satisfying some simple conditions. Properties of the optimal investment strategy for assets governed by multiple Ornstein-Uhlenbeck processes are studied. The paper ends with a short discussion of the implications of these ideas for financial markets.

Keywords: utility function; optimal investment strategy; self-financing; complete market; risk-neutral measure; Brownian motion; Ornstein-Uhlenbeck.

1 Introduction

The Kelly Criterion [1], [2] was initially introduced in 1956 to find the optimal betting amount in games with fixed known odds, and was later extended to the field of financial investments by E. O. Thorp and others. The strategy maximizes the entropy and with probability one outperforms any other strategy asymptotically [3]. This approach was recently further developed by Kargin [4], who applied the criterion to a mean-reverting asset process under liquidity and credit constraints.

The Kelly Criterion tells us that the optimal betting fraction is given by p-q, if a gambler is faced with a bet, where the probability to double the money is p and to lose the initial stake is q ($p>q$). The optimal betting fraction maximizes the expected log wealth. The question, why investors should choose to maximize the log wealth, has a simple answer: according to Breiman's theorem [3], it gives the asymptotically optimal pay-out and dominates any other strategy.

In this paper, we start by extending the original idea to the general continuous time framework with n correlated assets. Our task is to find the optimal self-financing trading strategy. We will prove that if the market is complete, this optimal self-financing trading strategy always exists. A limited number of applications are discussed in the context of Ornstein-Uhlenbeck processes.

The paper is organized as follows. In section 2 we review the standard assumptions and prove the optimization theorem. The theorem covers both the existence of the optimal trading strategy and the explicit form of the associated optimal investment fraction. In section 3 we apply the theory to a market of n correlated assets given by

Ornstein-Uhlenbeck mean-reverting processes. The optimal investment strategy is calculated for some representative examples. In the last section we put the results into the financial context and describe some open problems.

2 General Theory

In the first section, we will cover the assumptions and the theoretical framework. In 2.2 we will provide the main result, which is contained in Theorem 1.

2.1 Basic Assumptions and other Preliminaries

We will use the standard notation and conventions of financial mathematics. In section 2 the basic assumption is that the market is complete and frictionless [5]. Let us further assume that we consider all the processes in the finite time interval 0 to the terminal time T. There exists a probability space $(\Omega,\ \mathrm{P}_T, \mathcal{F}_T)$, on which all of the random variables are constructed, where Ω is the sample space, $\mathcal{F}_T$ is a σ-algebra which denotes the information accumulated up to time T and P_T is the spot probability measure [5]. The filtration $\mathcal{F}_t$, $t \in [0,T]$, represents the information accumulated up to time t. The sub-probability space $(\Omega,\ \mathrm{P}_t, \mathcal{F}_t)$ is introduced at time t, where P_t is the restriction of P_T on the filtration $\mathcal{F}_t$.

We assume there are $n+1$ investable assets in the market including the wealth process B_t, representing a saving account with value 1 at the initial time 0. We assume B_t follows

$$dB_t = B_t r_t dt \quad , \tag{1}$$

where r_t is the short term rate at time t.

The other n assets in the market are denoted by $S_i(t), t \in [0,T], i = 1,2,...,n$, and we define a n-dimensional vector by $\mathbf{S}_t = (S_1(t), S_2(t),...,S_n(t))^T$, where' ^{T} 'represents the transposition of a matrix. Let us define the relative assets price process by $\tilde{\mathbf{S}}_t = \mathbf{S}_t B_t^{-1}$. Let $\phi_0(t)$ denotes the number of units of B_t an investor holds at time t and $\phi_i(t), t \in [0,T], i = 1,2,...,n$, denotes the number of units of the ith asset an investor holds at time t. In addition, the n-dimensional vector ϕ_t is defined as $\phi_t = (\phi_1(t), \phi_2(t),...,\phi_n(t))^T$. $V_t(\psi)$ is the total value of the portfolio $\psi_t = (\phi_0(t), \phi_t)$. So we have

$$V_t(\psi) = \phi_0(t) B_t + \phi_t \cdot \mathbf{S}_t \quad , \tag{2}$$

where $\phi_t \cdot S_t$ is the inner product of two vectors.

Application of the Kelly Criterion to Ornstein-Uhlenbeck Processes

Definition 1 *A self-financing trading strategy* $\psi_t = (\phi_0(t), \phi_t)$ *is a strategy that satisfies:*

$$dV_t(\psi) = \phi_0(t) dB_t + \phi_t \cdot d\mathbf{S}_t, \quad \forall t \in [0, T] . \tag{3}$$

We assume $V_0(\psi) = 1$.

Definition 2 *A self-financing trading strategy* $\psi_t = (\phi_0(t), \phi_t)$ *is said to be admissible if and only if*

$$V_t(\psi) \geq 0, \ \mathrm{P_T} \ \text{a.s.} \quad \forall t \in [0, T] . \tag{4}$$

$U(x), x \geq 0$, is defined to be a concave function representing the utility of wealth. Here concaveness means

$$U((1-p)x_1 + px_2) \geq (1-p)U(x_1) + pU(x_2), \forall x_2 \geq x_1 \geq 0 \text{ and } 0 \leq p \leq 1 . \tag{5}$$

Further it is assumed that $U(x)$ has a first order derivative for $\forall x \in (0, +\infty)$. The first order derivative at $x = 0$ can be either finite or infinite, and the first order derivative of $U(x)$, $x \geq 0$, is a strictly decreasing function of x with $\lim_{x \to +\infty} U'(x) = 0$. If $U'(0) = +\infty$, let $I(x), x \geq 0$, be the inverse function of $U'(x)$ with $I(0) = +\infty$ and $I(+\infty) = 0$. For $U'(0) = b > 0$, we denote by $I_b(x), x \in [0, b]$, the inverse function of $U'(x)$, with $I(0) = +\infty$. In this case, we define $I(x)$ as

$$I(x) = \begin{cases} I_b(x), & x \in [0, b] \\ 0 \ , & x \in (b, +\infty) \end{cases} . \tag{6}$$

Let us denote by $\mathscr{D}$ the class of all of the admissible self-financing trading strategies. We say a self-financing trading strategy $\psi^* \in \mathscr{D}$ is the optimal trading strategy, if and only if

$$E_{\mathrm{P_T}}\left[U\left(V_T\left(\psi^*\right)\right)\right] \geq E_{\mathrm{P_T}}\left[U\left(V_T(\psi)\right)\right], \quad \forall \psi \in \mathscr{D} . \tag{7}$$

Our task is to find an optimal $\psi^* \in \mathscr{D}$, which satisfies eq.7.

2.2 The Optimal Strategy

To find the optimal strategy, we will first need to introduce the following lemma..

Lemma 1 *The function* $I(x)$, $x \in [0, +\infty)$ *satisfies the following inequality:*

$$U(I(y)) - yI(y) \geq U(c) - yc, \quad \forall y, c \in [0, +\infty) . \tag{8}$$

Proof:
If $I(y)=c$, then eq.8 is obviously satisfied. If $I(y)>c$, then the average growth rate of $U(x)$ from c to $I(y)$ should be larger than the first order derivative of $U(x)$ at $I(y)$, which is given by $U'(I(y))$, since the first order derivative of $U(x)$ is a strictly decreasing function of x.
The average growth rate of $U(x)$ from c to $I(y)$ is

$$\frac{U(I(y))-U(c)}{I(y)-c} .$$

This yields the following inequality

$$\frac{U(I(y))-U(c)}{I(y)-c} \geq U'(I(y)) = y$$

$$\Rightarrow U(I(y)) - yI(y) \geq U(c) - yc .$$

An almost identical argument can be applied in the case $I(y)<c$. □

Let us define $\tilde{P}_T$ as the martingale of the market and $Z_t = \frac{dP_t}{d\tilde{P}_t}$. Then $(Z_t, \mathcal{F}_t)$ is a $\tilde{P}_T$ martingale [5]. Define $\eta_t^* = y_t Z_t^{-1} B_t^{-1}$, $V_t^* = I(\eta_t^*)$, where we assume y_t is a deterministic function of t and is defined in such a way that $\tilde{V}_t^* = V_t^* B_t^{-1}$ is a $\tilde{P}_T$ martingale. So y_t solves the equation

$$\tilde{E}\left[B_t^{-1} I\left(y_t Z_t^{-1} B_t^{-1}\right)\right] = 1 . \qquad \textbf{(9)}$$

The deterministic property of y_t seems contrived, but is necessary for the proof of Proposition 1. As we shall see in section 3, in the case of the log utility function the deterministic function y_t indeed exists and is a constant.

Proposition 1 $V_T^* = I(\eta_T^*)$ *satisfies the inequality given by eq.7.*

Proof:
Let $V_T(\psi)$, $\forall \psi \in \mathcal{D}$, be the wealth process corresponding to a special trading strategy ψ, then

$$E_{P_T}\left[U(V_T^*)\right] - E_{P_T}\left[U(V_T(\psi))\right]$$

$$= E_{P_T}\left[\left(U(I(\eta_T^*)) - \eta_T^* I(\eta_T^*)\right) - \left(U(V_T(\psi)) - \eta_T^* V_T(\psi)\right)\right] + E_{P_T}\left[\eta_T^*\left(V_T^* - V_T(\psi)\right)\right] \qquad \textbf{(10)}$$

Application of the Kelly Criterion to Ornstein-Uhlenbeck Processes

According to lemma 1, the first term of the right hand side of eq.10 is positive. The second term is equal to zero,

$$E_{P_T}\left[\eta_T^*\left(V_T^*-V_T(\psi)\right)\right]=\tilde{E}\left[Z_T\eta_T^*\left(V_T^*-V_T(\psi)\right)\right]=y_T\tilde{E}\left[\tilde{V}_T^*-\tilde{V}_T(\psi)\right]=0 \quad . \quad \textbf{(11)}$$

The last equality of the above equation is deduced from the fact that both $\tilde{V}_t^*$ and $\tilde{V}_t(\psi)$ are martingales under the martingale measure.
Combining eq.10 and eq.11, we directly get

$$E_{P_T}\left[U\left(V_T^*\right)\right]\geq E_{P_T}\left[U\left(V_T(\psi)\right)\right]$$

□

Proposition 1 only state the fact that $I\left(\eta_t^*\right)$ satisfies eq.7. It doesn't necessarily mean that $I\left(\eta_t^*\right)$ is the optimal wealth process. We will prove in the following theorem that $I\left(\eta_t^*\right)$ is in fact the optimal wealth process.

Theorem 1 *Given a concave utility function* $U(x)$, *there exists an optimal self-financing trading strategy* ψ^*, *such that for each time* $t\in[0,T]$, *the wealth process* $V_t\left(\psi^*\right)$ *of this strategy satisfies*

$$E_{P_T}\left[U\left(V_t\left(\psi^*\right)\right)\right]\geq E_{P_T}\left[U\left(V_t(\psi)\right)\right],\quad \forall\psi\in\mathscr{D} \quad .$$

And the optimal wealth process is given by: $V_t\left(\psi^*\right)=I\left(\eta_t^*\right),\ t\in[0,T]$.

Proof:

Define $\tilde{V}_t^*$ to be $B_t^{-1}I\left(\eta_t^*\right)$. For $t=T$, we have $V_T^*=B_T\tilde{V}^*=I\left(\eta_T^*\right)$, which represents a general contingent claim in the market. Since the market is complete, the contingent claim V_T^* is attainable. This means there exists a self-financing trading strategy ψ^* such that $\tilde{V}_T^*=V_T\left(\psi^*\right)B_t^{-1}$, where $V_t\left(\psi^*\right)$ is the wealth process of this self-financing trading strategy. So the relative wealth process $\tilde{V}_t\left(\psi^*\right)=V_t\left(\psi^*\right)B_t^{-1}$ is a martingale under the martingale measure $\tilde{P}_T$. $\tilde{V}_t^*$ is also a martingale under the martingale measure $\tilde{P}_T$, and we have

$$V_t^*=B_t\tilde{E}\left[\tilde{V}_T^*\mid\mathscr{F}_t\right]=B_t\tilde{E}\left[\tilde{V}_T\left(\psi^*\right)\mid\mathscr{F}_t\right]=V_t\left(\psi^*\right),\quad \forall t\in[0,T] \quad . \quad \textbf{(12)}$$

Eq.12 shows that ψ^* is a self-financing trading strategy and replicates the optimal wealth process $\tilde{V}_t^*=B_t^{-1}I\left(\eta_t^*\right)$. From the combination of V_t^*, satisfying eq.7 for any time t before T, and eq.12, we can see that $V_t(\psi^*)$ also satisfies eq.7. This proves that the strategy ψ^* is both self-financing and optimal. □

It follows from Proposition 1 and Theorem 1 that the existence of a self-financing trading strategy $\psi^* \in \mathscr{D}$, where the total wealth at a fixed time T is consistent with eq.7, implies that the wealth process $V_t(\psi^*)$ satisfies

$$E_{P_T}\left[U\left(V_t\left(\psi^*\right)\right)\right] \geq E_{P_T}\left[U\left(V_t\left(\psi\right)\right)\right], \quad \forall \psi \in \mathscr{D} .$$

Therefore, an optimal trading strategy for a fixed time T will be optimal for any time before T. It follows further that an optimal trading strategy is only based on the information up to time t. The optimal trading strategy ψ_t^* is an adapted process with respect to the filtration $\{\mathscr{F}_t, t \in [0, +\infty)\}$. In the next section, we will apply the theorem to the case of a financial market containing a saving account and n correlated assets, whose price processes follow Ornstein- Uhlenbeck mean-reverting processes.

3 Implications for Ornstein-Uhlenbeck Processes

In this section we set $r_t = r$ and $S_i(t) = \exp(x_i(t))$, $t \in [0, T]$, $i = 1, 2, ..., n$. Each $x_i(t)$ is governed by

$$dx_i(t) = \left[a_i - b_i x_i(t)\right]dt + \sum_{j=1}^{n} \sigma_{i,j} dW_t^j, \quad i, j = 1, 2, ..., n \quad .$$

where a_i is some fixed real number, $b_i > 0$ is some nonnegative real number and $\sigma_{i,j}$ are constants. $\mathbf{W}_t = \left(W_t^1, W_t^2, ..., W_t^n\right)^T$ is a standard n-dimensional Brownian motion.

Define $\mathbf{a}$ to be the vector $(a_1, a_2, ..., a_n)^T$ (' T ' is the transposed of a matrix), $\mathbf{b}$ to be the $n \times n$ matrix of the form $\mathbf{b} = \begin{cases} \mathbf{b}_{i,j} = b_i, & i = j \\ \mathbf{b}_{i,j} = 0, & i \neq j \end{cases}$, and $\boldsymbol{\sigma}$ to be the matrix $\boldsymbol{\sigma}_{i,j} = \sigma_{i,j}$, for $1 \leq i, j \leq n$. The matrix $\boldsymbol{\sigma}$ has a non-zero determinant. Then the dynamic equation of $\mathbf{x}_t = \left(x_1(t), x_2(t), ..., x_n(t)\right)^T$ can be expressed as

$$d\mathbf{x}_t = \left[\mathbf{a} - \mathbf{b}\mathbf{x}_t\right]dt + \boldsymbol{\sigma} d\mathbf{W}_t \quad . \tag{13}$$

Let $\mathbf{S}_t = \left(S_1(t), S_2(t), ..., S_n(t)\right)^T$, $\tilde{\mathbf{S}}_t = B_t^{-1}\mathbf{S}_t$.

According to Ito's lemma, the dynamic of $\mathbf{S}_t$ is

Application of the Kelly Criterion to Ornstein-Uhlenbeck Processes

$$dS_i(t) = S_i(t)\mu_i(t)dt + S_i(t)\sum_{j=1}^{n}\sigma_{i,j}dW_t^j \ , i = 1,2,\ldots,n \ .$$

where $\mu_i(t) = a_i - b_i \log(S_i(t)) + \frac{1}{2}\|\boldsymbol{\sigma}_i\|^2$ and $\boldsymbol{\sigma}_i = (\sigma_{i,1}, \sigma_{i,2}, \ldots, \sigma_{i,n})^T$.

Define

$$\frac{d\mathrm{P}_T}{d\tilde{\mathrm{P}}_T} = \exp\left(\int_0^T \boldsymbol{\theta}_u \cdot d\tilde{\mathbf{W}}_u - \frac{1}{2}\int_0^T \|\boldsymbol{\theta}_u\|^2 du\right) , \quad \textbf{(14)}$$

where $\boldsymbol{\theta}_u = (\theta_1(u), \theta_2(u), \ldots, \theta_n(u))^T$ is a n-dimensional adapted stochastic process and $\|\boldsymbol{\theta}_u\| = \sqrt{\theta_1^2(u) + \ldots + \theta_n^2(u)}$ is the Euclidean vector norm, and $\tilde{\mathbf{W}}_t$ a Girsanov transformed Brownian motion, i.e. $\mathbf{W}_t = \tilde{\mathbf{W}}_t - \int_0^t \boldsymbol{\theta}_u du$.

Then under $\tilde{\mathrm{P}}_T$, $\tilde{\mathbf{S}}_t$ it follows

$$d\tilde{S}_i(t) = \tilde{S}_i(t)\left[\mu_i(t) - r - \sum_{j=1}^{n}\sigma_{i,j}\theta_j(t)\right]dt + \tilde{S}_i(t)\sum_{j=1}^{n}\sigma_{i,j}d\tilde{W}_t^j \ .$$

Define $c_i(t) = \mu_i(t) - r$ and in vector form $\mathbf{c}_t = (c_1(t), c_2(t), \ldots, c_n(t))^T$. If $\boldsymbol{\theta}_t$ solves

$$\boldsymbol{\sigma\theta}_t = \mathbf{c}_t \Rightarrow \boldsymbol{\theta}_t = \boldsymbol{\sigma}^{-1}\mathbf{c}_t \ , \quad \textbf{(15)}$$

then under $\tilde{\mathrm{P}}_T$ it follows that

$$d\tilde{\mathbf{S}}_t = \tilde{\mathcal{S}}_t \boldsymbol{\sigma} d\tilde{\mathbf{W}}_t \ , \quad \textbf{(16)}$$

where the matrix $\tilde{\mathcal{S}}_t$ is defined as: $\tilde{\mathcal{S}}_t = \begin{cases} \tilde{\mathcal{S}}_{i,j}(t) = \tilde{S}_i(t), & i = j \\ \tilde{\mathcal{S}}_{i,j}(t) = 0 & , i \neq j \end{cases}$. $\tilde{\mathbf{S}}_t$ is a martingale under $\tilde{\mathrm{P}}_T$.

To apply theorem 1, we need first to prove the completeness of the market price processes under consideration. The next lemma tells us that indeed the market is complete. The proof is given in an earlier presentation [7].

Lemma 2 *The mean-reverting market given above is complete.*

In case $U(x) = \log(x)$, we find $I(x) = 1/x$. Using eq.9, we can show $y_t = 1$. Then the optimal discounted wealth process is

$$\tilde{V}_t^* = Z_t = \exp\left(\int_0^t \boldsymbol{\theta}_u \cdot d\tilde{\mathbf{W}}_u - \frac{1}{2}\int_0^t \|\boldsymbol{\theta}_u\|^2 du\right).$$

Now we are in a position to derive a general result for Ornstein-Uhlenbeck processes.

Application of the Kelly Criterion to Ornstein-Uhlenbeck Processes

Theorem 2 *The optimal trading strategy* $\psi_t^* = \left(\phi_0^*(t), \phi_t^*\right)$ *is given by:*

$$\phi_0^*(t) = B_t^{-1} V_t^* \left(1 - B_t^{-1} \boldsymbol{\theta}_t^T \boldsymbol{\lambda}_t \mathbf{S}_t\right), \phi_i^*(t) = B_t^{-1} V_t^* \sum_{j=1}^{n} \theta_j(t) \lambda_{j,i}(t), \ i = 1, 2, ..., n \ . \quad \textbf{(17)}$$

where $\boldsymbol{\lambda}_t = \left(\mathscr{S}_t \tilde{\boldsymbol{\sigma}}\right)^{-1}$.

Proof:
First, we can show immediately

$$V_t^* = \phi_t^* \cdot \mathbf{S}_t + \phi_0^*(t) B_t \ ,$$

where V_t^* is the optimal wealth process given by

$$V_t^* = B_t \exp\left(\int_0^t \boldsymbol{\theta}_u \cdot d\tilde{\mathbf{W}}_u - \frac{1}{2} \int_0^t \|\boldsymbol{\theta}_u\|^2 du \right) .$$

Using Ito's lemma for $\tilde{V}_t^*$, we get

$$d\tilde{V}_t^* = dZ_t = \tilde{V}_t^* \boldsymbol{\theta}_t^T d\tilde{\mathbf{W}}_t \ .$$

From eq.16 we know that

$$d\tilde{\mathbf{W}}_t = \boldsymbol{\lambda}_t d\tilde{\mathbf{S}}_t \ .$$

Combining the above two equations, we have

$$d\tilde{V}_t^* = \tilde{V}_t^* \boldsymbol{\theta}_t^T \boldsymbol{\lambda}_t d\tilde{\mathbf{S}}_t = B_t^{-1} \tilde{V}_t^* \left(\boldsymbol{\theta}_t^T \boldsymbol{\lambda}_t d\mathbf{S}_t - r \boldsymbol{\theta}_t^T \boldsymbol{\lambda}_t \mathbf{S}_t dt \right) \ , \quad \textbf{(18)}$$

and

$$d\tilde{V}_t^* = B_t^{-1} dV_t^* - r B_t^{-1} V_t^* dt \ . \quad \textbf{(19)}$$

Combining eq.18 and eq.19, we arrive at

$$dV_t^* = V_t^* \left[B_t^{-1} \boldsymbol{\theta}_t^T \boldsymbol{\lambda}_t d\mathbf{S}_t + B_t^{-1} \left(1 - B_t^{-1} \boldsymbol{\theta}_t^T \boldsymbol{\lambda}_t \mathbf{S}_t\right) dB_t \right] = \phi_t^* \cdot d\mathbf{S}_t + \phi_0^*(t) dB_t \ . \quad \textbf{(20)}$$

Eq.20 shows directly that $\psi_t^* = \left(\phi_0^*(t), \phi_t^*\right)$ given by eq.17 is the optimal self-financing trading strategy. □

The optimal fraction vector $\mathbf{f}_t^* = \left(f_1^*(t), f_2^*(t), ..., f_n^*(t)\right)^T$ is composed of the individual $f_i^*(t)$, e.g. $\phi_i^*(t) S_i^*(t) / V_t^*$. By simple calculations based on Theorem 2, we have

$$\mathbf{f}_t^* = \mathbf{R}^{-1} \mathbf{c}_t \ , \quad \textbf{(21)}$$

where $\mathbf{R} = \boldsymbol{\sigma}\boldsymbol{\sigma}^T$ is a symmetric matrix and called correlation matrix. We will show in a separate paper that the matrix **R** denotes the correlations of the yield rates, e.g. the correlation of the ith and j^{th} assets is a deterministic function of $\boldsymbol{\sigma}_i \cdot \boldsymbol{\sigma}_j$. If the standard inverse of the volatility matrix does not exist, then one can resort to the generalized Moore-Penrose inverse to obtain a related result for the optimal investment fractions in markets without arbitrage.

Another derivation of the optimal fraction can be based on the function

$$\mathrm{F}(\mathbf{x}) = \mathbf{c}_t^T \mathbf{x} - \frac{1}{2}\mathbf{x}^T \mathbf{R}\mathbf{x}, \qquad \forall \mathbf{x} \in \mathrm{R}^n \quad , \tag{22}$$

linked to the mean-variance approach, since the optimal fraction given by eq.21 is the maximum of the function F. This indicates the close relationship between the utility maximization and the mean-variance method.

In the special case where the market is composed of only one stock, by eq.21, the optimal fraction is $f_t^* = (\mu_t - r)/\sigma^2$, where $\mu_t = a - b\log(S(t)) + 0.5\sigma^2$. Fig. 1 shows a sample path for the stock process and the associated optimal investment fraction and wealth process. As an aside, if one assumes zero interest rates, than the sensitivity of the optimal fraction to a percentage estimation error in the drift μ is twice the negative of a similar error in σ, e.g. a *1%* overestimation in volatility has approximately the same impact as an underestimation of the drift by *2%*.

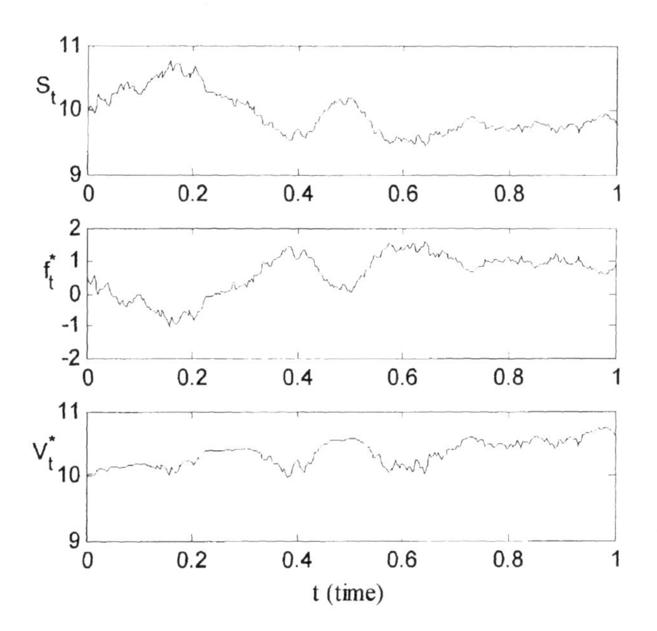

Fig. 1. Simulation of the stock price process, the corresponding optimal strategy f_t^*, and the wealth process V_t^* with parameters a=0.5, b=0.2, σ=0.1, r=0.03, S_0=10 and V_0=10.

Besides the sensitivity to estimation errors, it would be interesting to analyze the impact of the correlation matrix on the optimal trading strategy. Positive correlation has a tendency to reduce the number of 'independent' assets and forces investors to reduce leverage, e.g. the sum of the absolute values of the investment fractions is smaller.

Next, we study a special case where the assets have local correlations. The different assets only correlate to the neighboring assets but have no correlation to the rest assets. Let's set the risk-free rate to zero and the volatility matrix to be

$$\boldsymbol{\sigma} = \begin{bmatrix} \sigma & \sigma & 0 & \cdots & 0 \\ 0 & \sigma & \sigma & \ddots & \vdots \\ 0 & 0 & \ddots & \ddots & 0 \\ \vdots & \ddots & \ddots & \ddots & \sigma \\ 0 & 0 & 0 & \cdots & \sigma \end{bmatrix}_{n\times n} , n \geq 2 \; .$$

In this case, we will only study the large time limit. Let us denote by $\overline{f_S^*}(\infty)$

$$\overline{f_S^*}(\infty) := \lim_{t\to+\infty} E_{\mathrm{P}_T}\left[\sum_{i=1}^{n} f_i^*(t)\right] \; . \qquad \textbf{(23)}$$

After some simple manipulations we get

$$\overline{f_S^*}(\infty) = \begin{cases} \dfrac{n+1}{4} & \text{for n odd} \\ \dfrac{n}{4} & \text{for n even} \end{cases} \; . \qquad \textbf{(24)}$$

For a fixed odd integer n, the limit of the expected total fraction is $(n+1)/4$, which is identical to the value for the next even number.

Now, let us investigate another correlation structure where the assets have global correlations. As a simple example the volatility matrix is chosen as

$$\boldsymbol{\sigma} = \sigma \begin{bmatrix} 1 & 0 & 0 & \cdots & 0 \\ 1 & 1 & 0 & \cdots & 0 \\ 1 & 1 & \ddots & \ddots & \vdots \\ \vdots & \vdots & \ddots & 1 & 0 \\ 1 & 1 & \cdots & 1 & 1 \end{bmatrix}_{n\times n} , \; n \geq 2 \; ,$$

and the corresponding inverse matrix is

$$\boldsymbol{\sigma}^{-1} = \sigma^{-1} \begin{bmatrix} 1 & 0 & 0 & \cdots & 0 \\ -1 & 1 & 0 & \cdots & 0 \\ 0 & -1 & \ddots & \ddots & \vdots \\ \vdots & \vdots & \ddots & 1 & 0 \\ 0 & 0 & \cdots & -1 & 1 \end{bmatrix}_{n\times n} \; .$$

Application of the Kelly Criterion to Ornstein-Uhlenbeck Processes

Thus the optimal fraction $f_i^*(t)$ is given by

$$f_i^*(t)=\begin{cases} c_1(t)-(c_2(t)-c_1(t)) & i=1 \\ (c_i(t)-c_{i-1}(t))-(c_{i+1}(t)-c_i(t)) & i=2,3,\ldots,(n-1) \\ c_n(t)-c_{n-1}(t) & i=n \end{cases}. \tag{25}$$

The total optimal fraction is

$$f^*(t)=\sum_{i=1}^{n} f_i^*(t)=c_1(t)=\frac{\mu_1(t)-r}{\sigma^2}. \tag{26}$$

The total fraction here is equal to the optimal fraction in another market containing only the first asset. This surprising result is partly due to the fact that in the multi-dimensional case the investment fractions are likely to have both positive and negative signs. The expected total optimal fraction is

$$\bar{f}^*(t)=\mathrm{E}\left[f^*(t)\right]=\bar{c}_1(t)=\frac{\left[a_1-b_1\log(S_1(0))\right]e^{-b_1 t}+\frac{1}{2}\sigma^2-r}{\sigma^2}. \tag{27}$$

As time approaches infinity, the following limit is reached:

$$\bar{f}^*(\infty)=\lim_{t\to+\infty}\bar{f}^*(t)=\begin{cases} \dfrac{a_1+\frac{1}{2}\sigma^2-r}{\sigma^2} & \text{for } b_1=0 \\ \dfrac{\frac{1}{2}\sigma^2-r}{\sigma^2} & \text{for } b_1\neq 0 \end{cases}. \tag{28}$$

For additional examples, in particular in higher dimensions, we refer the reader to [7].

4 Conclusions

In the earlier sections we presented a discussion of the Kelly criterion in the continuous-time framework. The main theorem shows that in a complete market there exists an optimal self-financing trading strategy that maximizes the logarithmic utility function. The optimal investment fractions were explicitly calculated.

One general implication of the Kelly's criterion is maybe worth mentioning. It follows from Breiman's theorem [2], which shows that a logarithmic utility maximizer outperforms with probability one in the long run any substantially different trading strategy. This theorem has surprising consequences, for example it has spanned a smallish field called 'evolutionary finance' [8]. According to evolutionary finance 'natural selection' should favor agents with log utility. Such agents maximize the growth rate of their wealth with probability one, and thus dominate eventually the market. The stark claim is that either the investor maximizes utility or is marginalized. The authors are doubtful, if such a strong claim is justified, since only in the long time limit does the utility maximizer almost surely outperform. In the real world, where one has multiple independent agents and frequent paradigm shifts, maybe an even more aggressive strategy is warranted. Being 'overinvested' can be 'superior' (lower utility, but higher winning probability) in the short term. Even in the medium term the log maximizer has difficulties to outperform, if many independent agents exist. This could be a partial explanation for the regular crisis in financial markets, e.g. investors, who seek a short-term competitive advantage, invest over-aggressively. This is on top of the significant inherent volatility of utility maximization. A proper understanding of the impact of the Kelly criterion on the optimal behavior of individual agents is the precursor to consistent multi-agent modeling.

As a speculative aside, maybe utility maximization has a role in the study of the punctuated equilibrium observed in the evolutionary history of the earth, since utility maximization could provide a potential explanation without necessarily having to resort to external causes like asteroid impacts or volcanic eruptions for rare widespread extinction events.

A further interesting aside is to relate Kelly's criteria to market bubbles. A bubble is sometimes defined as a self-reinforcing dynamic associated with an excessive increase in asset prices followed by a sudden collapse. One simple and generic way to achieve a bubble is to find a reinforcing process that reduces the volatility σ, which would increase the optimal leverage as perceived by the investors, since the optimal Kelly fraction is given by $(\mu_t - r)/\sigma^2$. Next a description of a possible self-reinforcing process: In a number of investment strategies investors are short volatility. This selling of implied volatility can lead to a decrease of implied volatility due to supply and demand imbalances, which leads on the one hand to an increase in the optimal leverage and on the other hand to mark-to-market profits for the short volatility positions. As volatility decreases the optimal position size, e.g. leverage, increases, putting additional pressure on implied volatilities. This process continues until the friction associated with obtaining additional leverage stops the process. This 'virtuous circle' is then replaced by a 'vicious circle', since the extremal point is unstable, as implied volatility increases and leverage decreases. Here we do not discuss realized and implied volatility separately, since they are positively correlated and for the qualitative description presented here a dampening process for either type of volatility is sufficient. The analysis sketched out above can be extended to show that many bubbles, i.e. credit and stock market bubble, are driven by changes in the optimal leverage ratio as derived from the Kelly criterion.

Due the space limitations, we are not able to provide even a brief description of the application of the result in the area of statistical arbitrage. The present discussions are based on the continuous time framework, but realistic markets have an inherent discreteness. Furthermore there are different types of frictions, e.g. transaction cost, bid-offer spreads and liquidity constraints, which impose portfolio readjustment frequency restrictions. Not all of those influences are small and can be neglected. In an earlier presentation [7] correction terms for reducing the investment fractions were explicitly calculated. It would be of interest to give a comprehensive analysis of the impact of the different types of frictions for statistical arbitrage strategies. This will be done by the authors in a separate paper.

Application of the Kelly Criterion to Ornstein-Uhlenbeck Processes

In conclusion this article gives a quantitative insight into the trade-off between risk and return as diversification opportunities are added, correlation structure changed, and other constraints modified.

Acknowledgements. YL expresses his thanks to one of his dearest friends, Huiqing Chen, for her kind encouragement and support. BKM acknowledges support from the NSF of China and multiple informative discussions with DC Brody and O Peters.

References

1. Kelly, J.L.: A New Interpretation of Information Rate. Bell System Technical Journal. 35, 917--926 (1956)
2. Thorp, E.O.: The Kelly Criterion in Blackjack, Sports Betting, and the Stock Market. The 10th International Conference on Gambling and Risk Taking (1997)
3. Breiman, L.: Optimal Gambling Systems for Favorable Games. Jerzy Neyman, Proceedings of the Berkeley Symposium on Mathematical Statistics and Probability. 1, 65-78 (1961)
4. Kargin, V.: Optimal Convergence Trading , arXiv: math.OC/0302104 (2003)
5. Musiela, M., Rutkowski, M.: Martingale Methods in Financial Modelling. Springer, New York (1997)
6. Karatzas, I., Shreve, S.: Brownian Motion and Stochastic Calculus. Springer, Heidelberg (1998)
7. Lv, Y., Meister, B.K., Peters, O.: Implications of the Kelly Criterion for Multiple Ornstein-Uhlenbeck Processes, Bachelier Finance Society Fifth World Congress (2008)
8. Hakansson, N.H.: On Optimal Myopic Portfolio Policies, with and without Serial Correlation of Yields , Journal of Business 44,324-34 (1971)

Part III

The relationship of Kelly to asset allocation

22

Introduction to the Relationship of Kelly Optimization to Asset Allocation

The Kelly growth optimum approach is an attractive formula for investing. If the formula is robust and can be adapted to include realistic constraints on investing, then the practicality of the method is clear. The various papers here discuss these issues including liabilities, fractional Kelly, benchmarks, fixed mix strategies, and volatility induced wealth growth.

The first paper discusses asset allocations with withdrawals. Browne (1997) analyzes the optimal behavior of an investor (or pension fund manager) who withdraws funds continuously at a fixed rate per unit time. The investor can allocate funds to a given number of risky assets whose prices follow geometric Brownian motion and a risk free asset with a known constant rate of return. He shows that there are safe and risky zones with the latter being a region with a positive probability of ruin and the former where there is no chance of bankruptcy with those asset allocations. It is not possible to reach the safe region with probability one before bankruptcy or to determine an optimal policy to reach an upper level before a given lower level. But the safe region can be reached (optimized) with $\in$-optimality. Brown then finds the optimal growth policy for safe zone investors, which is a $E \log$ (Kelly) maximizing policy where the investor invests a constant proportion of wealth above a stochastic target floor in a form of constant proportions portfolio insurance.

The next two papers discuss fractional Kelly strategies. MacLean, Ziemba and Blazenko (1992) discuss growth versus security tradeoffs in dynamic investment analysis. We know that the $E \log$ maximizing strategies have the highest asymptotic long term growth rates. This approach is combined with the maximal security literature of Ferguson (1965), Epstein (1977), and Feller (1962). They propose three growth measures and three security measures. The growth measures are:

(1) The mean accumulation of wealth at the end of time t, that is how much do we expect to have, on average, after t periods.
(2) The mean exponential growth rate and its limit as $t \to \infty$.
(3) The mean first passage time to reach the set $[U, \infty)$, that is, how long on average does it take the investor to reach a specific level of wealth.

The measures of security are:

(1) The probability that the investor will have a specific accumulated wealth at time t, that is what are the chances of reaching a given target in a fixed amount of time.

(2) The probability that the investor's wealth is above a specified path.
(3) The probability of reaching a goal U which is higher than the initial wealth y_0 before falling to a wealth level L below y_0.

MacLean *et al.* study growth-security efficiency analgously to mean-variance efficiency as well as the weaker conditions of growth-security monotonicity. This latter tradeoff is called effective. They show that fractional Kelly strategies trace out effective growth-security tradeoffs. Later in MacLean, Ziemba, and Li (2005), it is shown that these fractionally Kelly strategies are efficient, if the assets are log-normally distributed, so they maximize growth for any given security level. The analysis leads to simple two dimensional graphs that compare growth rates with the probability of achieving a higher goal before falling to a lower level. Applications are made to blackjack, horseracing, lotto games, and futures trading. This range of applications runs the full range of asset weights from very small to very large percentages of the investor's wealth.

MacLean, Sanegre, Zhao, and Ziemba (MSZZ) (2004) provide an approach to calculate the optimal fractional Kelly weights at discrete intervals of time. The idea is to maximize the expected log subject to being above a specified wealth path with high probability. This is a value at risk type of constraint and leads to a model where a modified Kelly strategy can be computed at the discrete intervals of the planning horizon. This strategy is not necessarily fractional Kelly. In general, as risk rises, the expected return will fall so there is an *effective* tradeoff. With geometric Brownian motion in continuous time, fractional Kelly is optimal for the value at risk (VaR) and conditional value at risk (CVaR) where returns in the VaR tails are lineraly penalized constraints. The algorithm proposed uses a disjunctive form for the probabilistic constraints which identifies an outer problem of choosing an optimal set of scenarios and an inner (conditional) problem of finding the optimal decisions for a given discrete scenario set. The multiperiod inner problem is composed of a sequence of conditional one period problems. The theory is illustrated for the dynamic allocation of wealth in stocks, bonds and cash.

There are two ways to model the application, maximize the expected log of final wealth subject to

(1) wealth being above the specified wealth path with high probability; and
(2) a secured annual drawdown with a drawdown amount and a security level both exogenously specified.

The VaR model only controls loss at the horizon but not at intermediate decision points. The more stringent risk control constraint in the drawdown model considers the loss in each period.

At low levels of risk control, the full Kelly strategy is optimal. As the risk control requirements are raised, the strategy becomes more conservative, especially when close to the planning horizon. Assets are assumed to be log-normally distributed with geometric random walk but the computational procedure is general and applies to general price distributions.

In an extension, MacLean, Zhao, and Ziemba (2009) add one more feature to the MSZZ model, namely that when the predetermined exogenous path is violated, there is a convex penalty for there shortfall violations (see Mulvey, Bilgili, and Vural (2010) in part VI for an application along these lines).

The next three papers discuss the impact of benchmarks. Browne (2000) considers a dynamic portfolio management problem where the objective concerns the tradeoff of performance goals and the risk of a shortfall below a benchmark target. Return is measured by the expected time to reach investment goals relative to the benchmark. This return is maximized subject to the risk constraint in terms of the probability of shortfall relative to the benchmark. The setting is Merton's (1971) continuous time framework.

Davis and Lleo (2009) extend the definition of fractional Kelly strategies to the situation in which the investor's objective is to outperform a benchmark. These benchmarked fractional Kelly strategies are efficient portfolios even when the asset returns are not log-normally distributed. They find the benchmarked fractional Kelly strategies for various types of benchmarks such as the S&P500 and the Salomon Smith Barney World Government Bond Index. They also determine the relations between the investor's risk aversion and the given benchmarks. The setting is a stochastic control model in continuous time. They develop, following Merton, two and three mutual fund theorems. This means that the optimal weights over n assets can be found by optimally weighting the two or three mutual funds which are linear combinations of the original n assets, where $n > 3$, see Rudolf and Ziemba (2004) for a four mutual fund theory with liabilities in part VI; see also Davis and Lleo (2008a,b). In addition to single index benchmarks, there can be composite benchmarks plus alpha or composite weighted multiple index benchmarks with or without an alpha.

The Davis and Lleo two fund theorem for the case without and with benchmarks is that the assets can be split between the full Kelly portfolio and a correction portfolio C with the weights dependent on the investor's risk sensitivity. C is an intertemporal hedging portfolio, which can be the risk free asset under certain assumptions. They develop a three mutual fund theorem for the case of benchmarks with risky and risk free assets. Formulas are developed for the optimal equity portfolio weightings in the various mutual funds.

Any portfolio can be expressed as a linear combination of a mutual fund, an index fund and a long-short hedge fund with risky and risk free asset allocations. Again, the weights are dependent on the risk sensitivity. The risk sensitivity varies from zero (the full Kelly expected log case) to infinity where the investor replicates the benchmark with an index fund. With moderate risk sensitivity there will be a blend of all three asset funds.

Platen (2010) discusses the benchmark approach for dynamic investing and pricing. A numeraire portfolio exists which is strictly positive that makes all benchmarked nonnegative portfolios downward trending or trendless. The benchmark portfolio is the Kelly $E\log$ maximizing portfolio which cannot be outperformed by another long only portfolio. The benchmark portfolio can be used for pricing and computing conditional expectations.

The final paper in this part deals with fixed-mix and volatility induced wealth growth. Dempster, Evstigneev, and Schenk-Hoppé (2009) discuss the literature concerning the growth of wealth over time through volatility using fixed-mix strategies. These self-financing strategies rebalance the portfolio at each decision point to keep constant the proportions of wealth invested in various assets. Through volatility, the portfolio will gain wealth even with assets that do not have strictly positive means. This is a generalization of the buy-low sell-high investment strategy. But they show that there is much more to the story than this implies. The analysis does not need utility assumptions and compares the fixed-mix rebalancing with buy and hold strategies. The key assumption is that the period by period returns of the various assets are stationary. Dempster, Evstigneev, and Schenk-Hoppé discuss some counter-intuitive examples such as growth with assets having zero means. They discuss the effects of transaction costs on their results and determine conditions for growth to occur with sufficiently small transaction costs. They also show that the conjecture, *the higher the volatility, the higher the induced growth rate* is not true in general. The key idea is to rebalance to an arbitrary fixed mix portfolio.

MATHEMATICS OF OPERATIONS RESEARCH
Vol 22, No 2, May 1997
Printed in U.S.A

23

SURVIVAL AND GROWTH WITH A LIABILITY: OPTIMAL PORTFOLIO STRATEGIES IN CONTINUOUS TIME

SID BROWNE

We study the optimal behavior of an investor who is forced to withdraw funds continuously at a fixed rate per unit time (e.g., to pay for a liability, to consume, or to pay dividends). The investor is allowed to invest in any or all of a given number of risky stocks, whose prices follow geometric Brownian motion, as well as in a riskless asset which has a constant rate of return. The fact that the withdrawal is continuously enforced, regardless of the wealth level, ensures that there is a region where there is a positive probability of ruin. In the complementary region ruin can be avoided with certainty. Call the former region the danger-zone and the latter region the safe-region. We first consider the problem of maximizing the probability that the safe-region is reached before bankruptcy, which we call the survival problem. While we show, among other results, that an optimal policy does not exist for this problem, we are able to construct explicit ϵ-optimal policies, for any $\epsilon > 0$. In the safe-region, where ultimate survival is assured, we turn our attention to growth. Among other results, we find the *optimal growth policy* for the investor, i.e., the policy which reaches another (higher valued) goal as quickly as possible. Other variants of both the survival problem as well as the growth problem are also discussed. Our results for the latter are intimately related to the theory of Constant Proportions Portfolio Insurance.

1. Introduction. The problem considered here is to solve for the optimal investment decision of an investor who must withdraw funds (e.g., to pay for some liability or to consume) continuously at a given rate per unit time. Income can be obtained only from investment in any of $n + 1$ assets: n risky stocks, and a bond with a deterministic constant return. The objectives considered here relate solely to what can be termed "goal problems," in that we assume the investor is interested in reaching some given values of wealth (called goals) with as high a probability as possible and/or as quickly as possible.

The fact that the investor must continuously withdraw funds at a fixed rate introduces a new difficulty that was not present in the previous studies of objectives related to reaching goals quickly (cf. Heath et al. 1987). Specifically the forced withdrawals ensure that at certain levels of wealth, there is a positive probability of going bankrupt, and thus the investor is forced to invest in the risky stocks to avoid ruin. In this paper we address the two basic problems faced by such an investor: how to survive, and how to grow. The *survival* problem turns out to be somewhat tricky, in that we prove that no fully optimal policy exists. Nevertheless, we are able to construct ϵ-optimal policies, for any $\epsilon > 0$. The *growth* problem is answered completely, once the survival aspect is clarified.

While this model is directly applicable to the workings of certain economic enterprises, such as a pension fund manager with fixed expenses that must be paid continuously (regardless of the level of wealth in the fund), our results are also related to investment strategies that are referred to as *Constant Proportion Portfolio*

Received January 11, 1995; revised October 3, 1995; July 20, 1996.
AMS 1991 subject classification. Primary: 90A09, 60H10; Secondary: 93E20, 60G40, 60J60.
OR/MS Index 1978 subject classification. Primary: Finance/Portfolio; Secondary: Optimal control/Stochastic.
Key words. Stochastic control, portfolio theory, diffusions, martingales, bankruptcy, optimal gambling, Hamilton-Jacobi-Bellman equations, portfolio insurance.

0364-765X/97/2202/0468/$05.00

Insurance (CPPI). In fact a related model was used as the economic justification of CPPI in Black and Perold (1992), where both the theory and application of such strategies is described. In Black and Perold (1992), optimal strategies were obtained for the objective of maximizing utility of consumption, for a very specific utility function, subject to a minimum consumption constraint. However, the analysis and policies of Black and Perold (1992) are relevant only when initial wealth is in a particular region (specifically, when initial wealth is above the "floor"), wherein for that policy, there is no possibility of bankruptcy. Black and Perold (1992) did not address the fact that for the model described there, ruin, or bankruptcy, is a very real possibility when initial wealth is in the complementary region (below the "floor").

Here we focus on the objectives of survival and growth, which are intrinsic objective criteria that are independent of any specific individual utility function. As such, our results for both aspects of the problem will therefore complement the results of Black and Perold (1992) (as well as the more recent related work of Dybvig 1995). Firstly, the survival problem has not been addressed before for this model (although see Majumdar and Radner 1991 and Roy 1995), and secondly, the optimal growth policies we obtain provides another objective justification for the use of the CPPI policies prescribed in Black and Perold (1992), since for this problem we get similar policies as those obtained there.

The remainder of the paper is organized as follows: In the next section, we will describe the model in greater detail, and prove a general theorem in stochastic control from which all our subsequent results will follow. To facilitate the exposition, we will at first consider the case where there is only one risky stock and where the withdrawal rate is constant per unit time. It turns out that the state space (for wealth) can be divided into two regions, which we will call the "danger-zone" and the "safe-region." In the latter region, the investor need never face the possibility of ruin, and so we can concentrate purely on the growth aspects of the investor. (The aforementioned studies of Black and Perold 1992 and Dybvig 1995 considered only this region in their analyses of the maximization of utility from consumption problem.) In the former region, ruin, or bankruptcy, is a possibility (hence the term "danger-zone") and therefore we first concentrate on passing from the danger-zone into the safe-region. This is the *survival problem* and it is completely analyzed in §3. In particular, two problems are considered, *maximizing the probability of reaching the "safe-region" before going bankrupt*, and *minimizing the discounted penalty that must be paid upon reaching bankruptcy*. It is the former problem that does not admit an optimal policy, although we are able to explicitly construct an ϵ-optimal policy. The latter problem does admit an optimal policy, which we find explicitly. The structure of both policies are quite similar, in that they both essentially invest a (different) proportional amount of the distance to the safe-region. In §4 we consider the *growth problem* in the safe-region. We define growth as reaching a given (high) level of wealth as quickly as possible. Two related problems are then solved completely. First, we find the policy that *minimizes the expected time to the* (good) *goal*, and then we find the policy that *maximizes the expected discounted reward of getting to the goal*. Our resulting optimal growth policies turn out to be quite similar to the CPPI policies obtained for a different problem by Black and Perold (1992), in that they invest a (different) proportional amount of the distance from the danger-zone. Extensions to the multiple asset case as well as the case of a wealth-dependent withdrawal rate are discussed in §§5 and 6.

All the control problems considered in this paper are special cases of a particular general control problem that is solved in Theorem 2.1 in §2 below. For this problem we use the Hamilton-Jacobi-Bellman (HJB) equations of stochastic control (see, e.g., Fleming and Rishel 1975, or Krylov 1980) to obtain a candidate optimal policy in terms of a candidate value function and this value function is then in turn given as the solution to a particular nonlinear Dirichlet problem. These candidate values are then verified and rigorously proved to be optimal by the *martingale optimality principle* (see §V.15 in Rogers and Williams 1987, or §2 in Davis and Norman 1990). The resulting nonlinear differential equations are then solved in turn for each of the problems considered below, yielding the optimal solutions in explicit form.

2. The model and continuous-time stochastic control. Without loss of generality, we assume that there is only one risky stock available for investment (e.g., a mutual fund), whose price at time t will be denoted by P_t. (Extension to the multidimensional case (for a *complete market*) is quite straightforward, and since the excess notation required adds little to the understanding, we will simply outline how to obtain the results for the multidimensional case in a later section.) As is quite standard (see, e.g., Merton 1971, 1990, Davis and Norman 1990, Black and Perold 1992, Grossman and Zhou 1993, Pliska 1986), we will assume that the *price* process of the risky stock follows a *geometric Brownian motion*, i.e., P_t satisfies the stochastic differential equation

$$dP_t = \mu P_t \, dt + \sigma P_t \, dW_t \tag{1}$$

where μ and σ are positive constants and $\{W_t: t \geq 0\}$ is a standard Brownian motion defined on the complete probability space $(\Omega, \mathscr{F}, P)$, where $\{\mathscr{F}_t\}$ is the P-augmentation of the natural filtration $\mathscr{F}_t^W := \sigma\{W_s; 0 \leq s \leq t\}$. (Thus the instantaneous *return* on the risky stock, dP_t/P_t, is a *linear* Brownian motion.)

The other investment opportunity is a bond, whose price at time t is denoted by B_t. We will assume that

$$dB_t = rB_t \, dt \tag{2}$$

where $r > 0$. To avoid triviality, we assume $\mu > r$.

We assume, for now, that the investor must withdraw funds continuously at a constant rate, say $c > 0$ per unit time, regardless of the level of wealth. (This would be applicable for example if the investor faces a constant liability to which \$$c$ must be paid continuously.) In a later section we generalize this to a case where the withdrawals are wealth-dependent.

Let f_t denote the *total amount of money invested in the risky stock at time t under an investment policy f*. An investment policy f is *admissible* if $\{f_t, t \geq 0\}$ is a measurable, $\{\mathscr{F}_t\}$-adapted process for which $\int_0^T f_t^2 \, dt < \infty$, a.s., for every $T < \infty$. Let $\mathscr{G}$ denote the set of admissible policies.

For each admissible control process $f \in \mathscr{G}$, let $\{X_t^f, t \geq 0\}$ denote the associated wealth process, i.e., X_t^f is the wealth of the investor at time t, if he follows policy f. Since any amount not invested in the risky stock is held in the bond, this process then evolves as

$$\begin{aligned} dX_t^f &= f_t \frac{dP_t}{P_t} + \left(X_t^f - f_t\right)\frac{dB_t}{B_t} - c\,dt \\ &= \left[rX_t^f + f_t(\mu - r) - c\right] dt + f_t \sigma \, dW_t \end{aligned} \tag{3}$$

upon substituting from (1) and (2).

Thus, for Markov control processes f, and functions $\Psi(t,x) \in \mathscr{C}^{1,2}$, the generator of the wealth process is

$$\mathscr{A}^f\Psi(t,x) = \Psi_t + [f_t(\mu - r) + rx - c]\Psi_x + \frac{1}{2}f_t^2\sigma^2\Psi_{xx}. \tag{4}$$

We will put no constraints on the control f_t (other than admissibility). In particular, we will allow $f_t < 0$, as well as $f_t > X_t^f$. In the first instance, the company is selling the stock short, while in the second instance it is borrowing money to invest long in the stock. (While we do allow shortselling, it turns out that none of our optimal policies will ever in fact do this.)

What will differentiate our model and results from previous work are the objectives considered and the fact that here the withdrawal rate c is constant, and not a decision variable.

The usual portfolio and asset allocation problems considered in the financial economics literature deal with an investor whose wealth also evolves according to a stochastic differential equation as in (3), where instead of being constant, c is now a control variable as well, i.e., the consumption function $c_t = c(X_t^f)$. For a specific utility function $u(\cdot)$, the investor's objective is then to maximize the expected utility of consumption and terminal wealth over some finite horizon, i.e., for $T > 0$, and some "bequest function" $\Psi(\cdot)$, the investor wishes to solve

$$\sup_{f,c} E_x\left(\int_0^T e^{-\lambda t}u(c_t)\,dt + e^{-\lambda T}\Psi(X_T^f)\right), \tag{5}$$

for some discount factor $\lambda \geq 0$. Alternatively, the investor may wish to solve the discounted infinite horizon problem

$$\sup_{f,c} E_x\int_0^\infty e^{-\lambda t}u(c_t)\,dt. \tag{6}$$

In both of these cases, since the *process* $\{c_t\}$ is usually assumed to be completely controllable, it is clear that for certain utility functions at least, ruin need never occur, since we may simply stop consuming at some level. Alternatively, as is the case when the utility function is of the form $u(c) = c^{1-R}/(1-R)$ for some $R < 1$, or $u(c) = \ln(c)$, the resulting optimal policy takes both investment f_t and consumption c_t to be proportional to wealth, i.e., $f_t = \pi_1(t)X_t$, $c_t = \pi_2(t)X_t$, which in turns makes the optimal wealth process into a geometric Brownian motion, and thus the origin becomes an inaccessible barrier. Classical accounts of such (and more sophisticated) problems are discussed in Merton (1971, 1990) and Davis and Norman (1990) among others.

Optimal investment decisions with *constraints on consumption* have also been considered in the literature previously. Most relevant to our model is the literature on *constant proportion portfolio insurance* (CPPI), as introduced in Black and Perold (1992), where the resulting policy is to invest a *constant proportion of the excess of wealth over a given constant floor*. (As its name suggests, *portfolio insurance* can be loosely considered any trading and investment strategy that ensures that the value of a portfolio never decrease below some limit. Alternative approaches to portfolio insurance using options and other techniques are described in e.g., Luskin 1988.) Black and Perold (1992) introduced this policy as the solution to the discounted infinite horizon problem of (6) *subject to the constraint that* $c_t \geq c_{\min}$, where $c_{\min}$ is some given constant. The specific utility function considered there was

$$u(c) = \begin{cases} c^{1-R}/(1-R) & \text{for } c \geq c^*, \\ K_1 - K_2 c & \text{for } c \leq c^*, \end{cases}$$

where c^* is a given constant, $R \leq 1$ and K_1, K_2 are constants chosen to ensure $u(\cdot)$ continuous throughout. While others (e.g., Dybvig 1995) have raised some technical questions about the analysis in Black and Perold (1992), more relevant to our point of view is the fact that this model (and the resulting optimal policy) allows for the possibility of *ruin*, or bankruptcy, if wealth is initially below the given floor. This possibility was never addressed in Black and Perold (1992).

In this paper we do not concentrate on the usual utility maximization problems of (5) and (6). Rather, here we are concerned with the *objective* problems of survival and growth. In particular, we first study the problem of how the investor (whose wealth evolves according to (3)) should invest to maximize the probability that the investor survives forever (which turns out to be related to *maximizing the probability of achieving a given fixed fortune before going bankrupt*), as well as the problem of how the investor should invest so as *to minimize the time until a given level of wealth has been achieved.* The former problem is called the *survival problem*, and is discussed in §3. The latter is called the *growth problem* and is the content of §4. Related problems have been studied in general under the label of "goal problems" in the works of Pestien and Sudderth (1985, 1988), Heath et al. (1987) and Orey et al. (1987). The survival problem for some specific related models were studied in Browne (1995) and Majumdar and Radner (1991). The former treated an "incomplete market" model, where the withdrawals are not fixed but rather follow a stochastic process, and the latter treated a model with forced constant consumption but without the possibility of investing in a risk free asset.

Recently, in order to provide a consumption based economic justification for the interesting portfolio strategies introduced in Grossman and Zhou (1993) (where the optimal policy invests a constant proportion of wealth over a *stochastic* floor), Dybvig (1995) considered the consumption-investment problem of (6) with the constraint that *consumption never decrease*, i.e., that $c_t > c_s$, for all $t \geq s$, with $c_0 > 0$. Thus consumption is forced in his model as well. He considered utility functions of the form $u(c) = c^{1-R}/(1-R)$ and $u(c) = \ln(c)$. However he only considered the problem in the *feasible region*, where initial wealth X_0, satisfies $X_0 > c_0/r$, and so for which ruin need not occur. Dybvig (1995) did not consider the case when $X_0 > c_0/r$, and hence where ruin is possibility, and so our results on this problem in §3 will complement his analysis as well. Since in this paper our objectives deals solely with the achievement of particular goals associated with *wealth*, it is clear that if there is a constraint on consumption as in the models of Black and Perold (1992) and Dybvig (1995), we should always set consumption at the minimum level, which in both cases is a constant ($c_{\min}$ in Black and Perold 1992 and c_0 in Dybvig 1995). This is consistent with the model we analyze here, where we will (at least at first) take consumption as a fixed constant c per unit time. This implies that at least for some values of wealth, the origin is accessible, and thus ruin is in fact a possibility.

In the next section we consider the problem of how to invest in order to survive. However, before we study that problem, we need a preliminary result from control theory that will provide the basis of all our future results.

2.1. Optimal control. The problems of survival and growth considered in this paper are all special cases of (Dirichlet-type) optimal control problems of the following form: For each admissible control process $\{f_t, t \geq 0\}$, let

$$\tau_z^f := \inf\{t > 0: X_t^f = z\}$$

denote the first hitting time to the point z of the associated wealth process $\{X_t^f\}$ of (3), under policy f. For given numbers (l, u) with $l < X_0 < u$, let $\tau^f := \min\{\tau_l^f, \tau_u^f\}$ denote the first escape time from the interval (l, u).

For a given nonnegative continuous function $\lambda(x) \geq 0$, a given real bounded continuous function $g(x)$, and a function $h(x)$ given for $x = l, x = u$, let $\nu^f(x)$ be defined by

$$(7) \quad \nu^f(x) = E_x\left(\int_0^{\tau^f} g(X_t^f)\exp\left\{-\int_0^t \lambda(X_s^f)\,ds\right\}dt + h(X_{\tau^f}^f)\exp\left\{-\int_0^{\tau^f}\lambda(X_s^f)\,ds\right\}\right)$$

with

$$\nu(x) = \sup_{f \in \mathcal{G}} \nu^f(x) \quad \text{and} \quad f_\nu^*(x) = \arg\sup_{f \in \mathcal{G}} \nu^f(x).$$

We note at the outset that we are only interested in controls (and initial values x) for which $\nu^f(x) < \infty$.

As a matter of notation, we note first that here, and throughout the remainder of the paper, the parameter γ will be defined by

$$(8) \qquad \gamma := \frac{1}{2}\left(\frac{\mu - r}{\sigma}\right)^2.$$

THEOREM 2.1. *Suppose that $w(x)$: $(l, u) \to (-\infty, \infty)$ is a $\mathcal{C}^2$ function that is the concave increasing (i.e., $w_x > 0$ and $w_{xx} < 0$) solution to the nonlinear Dirichlet problem*

$$(9) \qquad (rx - c)w_x(x) - \gamma\frac{w_x^2(x)}{w_{xx}(x)} + g(x) - \lambda(x)w(x) = 0, \quad \textit{for } l < x < u,$$

with

$$(10) \qquad w(l) = h(l) \quad \textit{and} \quad w(u) = h(u),$$

and satisfies the conditions:

(i) *$w_x^2(x)/w_{xx}(x)$ is bounded for all x in (l, u);*

(ii) *there exists an integrable random variable Y such that for all $t \geq 0$, $w(X_t^f) \geq Y$;*

(iii) *$w_x(x)/w_{xx}(x)$ is locally Lipschitz continuous.*

Then $w(x)$ is the optimal value function, i.e., $w(x) = \nu(x)$, and moreover the optimal control, f_ν^*, can then be written as

$$(11) \qquad f_\nu^*(x) = -\left(\frac{\mu - r}{\sigma^2}\right)\frac{w_x(x)}{w_{xx}(x)}, \quad \textit{for } l < x < u.$$

PROOF. The appropriate HJB optimality equation of dynamic programming for maximizing $\nu^f(x)$ of (7) over control policies f_t, to be solved for a function ν is $\sup_{f \in \mathcal{G}}\{\mathcal{A}^f\nu + g - \lambda\nu\} = 0$, subject to the Dirichlet boundary conditions $\nu(l) = h(l)$ and $\nu(u) = h(u)$ (cf. Theorem 1.4.5 of Krylov 1980). Since $\nu(x)$ is independent of time, the generator of (4) shows that this is equivalent to

$$\sup_{f \in \mathcal{G}} \left\{ (f(\mu - r) + rx - c)\nu_x + \frac{1}{2} f^2 \sigma^2 \nu_{xx} + g - \lambda\nu \right\} = 0. \tag{12}$$

Assuming now that (12) admits a classical solution with $\nu_x > 0$ and $\nu_{xx} < 0$ (see, e.g., Fleming and Soner 1993), we may then use standard calculus to optimize with respect to f in (12) to obtain the maximizer $f_\nu^* = -((\mu - r)/\sigma^2)\nu_x/\nu_{xx}$ (compare with (11)). When this $f_\nu^*(x)$ is then substituted back into (12) and the resulting equation is simplified, we obtain the nonlinear Dirichlet problem of (9) (with $\nu = w$).

It remains only to verify that the policy f_ν^* is indeed optimal. The aforementioned theorem in Krylov (1980) does not apply here, since in particular the degeneracy condition (Krylov 1980, page 23) is not met. We will use instead the martingale optimality principle, which entails finding an appropriate functional which is a uniformly integrable martingale under the (candidate) optimal policy, but a *supermartingale* under any other admissible policy, with respect to the filtration $\mathcal{F}_t$ (see Rogers and Williams 1987, Davis and Norman 1990).

To that end, let $\Lambda^f(s, t) := \int_s^t \lambda(X_v^f)\, dv$, and define the process

$$M(t, X_t^f) := e^{-\Lambda^f(0,t)} w(X_t^f) + \int_0^t e^{-\Lambda^f(0,s)} g(X_s^f)\, ds, \quad \text{for } 0 \le t \le \tau^f, \tag{13}$$

where w is the concave increasing solution to (9).

Optimality of f_ν^* of (11) is then a direct consequence of the following lemma.

LEMMA 2.2. *For any admissible policy f, and $M(t, \cdot)$ as defined in* (13), *we have*

$$E(M(t \wedge \tau^f, X_{t \wedge \tau^f}^f)) \le M(0, X_0) = w(x), \tag{14}$$

with equality holding if and only if $f = f_\nu^*$, where f_ν^* is the policy given in (11). *Moreover, under policy f_ν^*, the process $\{M(t \wedge \tau^f, X_{t \wedge \tau^f}^*)\}$ is a uniformly integrable martingale.*

PROOF. Applying Ito's formula to $M(t, X_t^f)$ of (13) using (3) shows that for $0 \le s \le t \le \tau^f$

$$M(t, X_t^f) = M(s, X_s^f) + \int_s^t e^{-\Lambda^f(s,v)} Q(f_v; X_v^f)\, dv + \int_s^t e^{-\Lambda^f(s,v)} \sigma f_v w_x(X_v^f)\, dW_v \tag{15}$$

where $Q(z; y)$ denotes the quadratic (in z) defined by

$$Q(z; y) := z^2 \left[\frac{1}{2} \sigma^2 w_{xx}(y) \right] + z[(\mu - r) w_x(y)]$$
$$+ (ry - c) w_x(y) + g(y) - \lambda(y) w(y).$$

Recognize now that since $Q_{zz}(z; y) = \sigma^2 w_{xx}(y) < 0$, we always have $Q(z; y) \le 0$, and the maximum is achieved at the value

$$z^*(y) := -\left(\frac{\mu - r}{\sigma^2} \right) \frac{w_x(y)}{w_{xx}(y)}$$

with corresponding maximal value

$$Q(z^*; y) = (ry - c) w_x(y) - \gamma \frac{w_x(y)^2}{w_{xx}(y)} + g(y) - \lambda(y) w(y) \equiv 0$$

where the final equality follows from (9). Therefore the second term in the r.h.s. of (15) is always less than or equal to 0. Moreover (15) shows that we have

$$\int_0^{t\wedge\tau^f} e^{-\Lambda^f(0,v)}\sigma f_v w_x(X_v^f)\,dW_v$$

$$= M(t\wedge\tau^f, X^f_{t\wedge\tau^f}) - w(x) - \int_0^{t\wedge\tau^f} e^{-\Lambda^f(0,v)}Q(f_v; X_v^f)\,dv$$

$$\geq M(t\wedge\tau^f, X^f_{t\wedge\tau^f}) - w(x).$$

Thus, by (ii) we see that the stochastic integral term in (15) is a local martingale that is in fact a *supermartingale*. Hence, taking expectations on (15), with $s = 0$, therefore shows that

$$(16)\quad E\big(M(t\wedge\tau^f, X^f_{t\wedge\tau^f})\big) \leq w(x) + E\left(\int_0^{t\wedge\tau^f} e^{-\Lambda^f(0,v)}Q(f_v; X_v^f)\,dv\right)$$

$$\leq w(x) + E\left(\int_0^{t\wedge\tau^f} e^{-\Lambda^f(0,v)}\left[\sup_{f_v} Q(f_v; X_v^f)\right]dv\right)$$

$$= w(x)$$

with the equality in (16) being achieved at the policy f_v^*.

Thus we have established (14).

Note that under the policy f_v^* of (11), the wealth process X^* satisfies the stochastic differential equation

$$(17)\qquad dX_t^* = \left[\left(rX_t^* - c - 2\gamma\frac{w_x(X_t^*)}{w_{xx}(X_t^*)}\right)dt - \sqrt{2\gamma}\,\frac{w_x(X_t^*)}{w_{xx}(X_t^*)}dW_t\right]I_{\{t\leq\tau^*\}}$$

where $\tau^* := \tau^{f_v^*}$. By (iii) this equation admits a unique strong solution (Karatzas and Shreve 1988, Theorem 5.2.5).

Furthermore note that under the (optimal) policy, f_v^*, we have, for all $0\leq s\leq t\leq \tau^*$,

$$(18)\quad M(t, X_t^*) = M(s, X_s^*) - \sqrt{2\gamma}\int_s^t \exp\left\{-\int_s^v \lambda(X_\rho^*)\,d\rho\right\}\frac{w_x^2(X_v^*)}{w_{xx}(X_v^*)}dW_v$$

which by (i) above is seen to be a uniformly integrable martingale. This completes the proof of the theorem. □

We now return to the survival problem.

3. Maximizing survival. We consider in this section two objectives related to maximizing the survival of the investor. First we consider the problem of *minimizing the probability of ruin* which is related to the problem of *maximizing the probability of reaching a particular given upper level of wealth before a given lower level*. We will show that an optimal strategy for this latter problem does not exist, although exploiting the solution to a related solvable problem will allow us to explicitly construct ϵ-optimal ones. Next we consider the related objective of *minimizing the expected discounted penalty of ruin*, which is equivalent to *minimizing the expected discounted time to* *bankruptcy*. This problem does admit an optimal solution and we find it explicitly. The structure of the (optimal) survival policies obtained in this section are similar in that they all invest a fixed fraction of the positive distance of wealth to a particular goal.

3.1. Minimizing the probability of ruin. The evolutionary equation (3) exhibits clearly that under policy f, the wealth process is a diffusion with drift function m and diffusion coefficient function v given respectively by

$$m(f,x,t) = f_t(\mu - r) + rx - c, \qquad v(f,x,t) = f_t^2\sigma^2. \tag{19}$$

Thus for any admissible control $f < \infty$ there is a region (in X space) where there is a positive probability of *bankruptcy*. This is due to the fact that while the variance of the wealth process is completely controllable, as is apparent from (19), the *drift* is not completely controllable due to the fact that $c > 0$, and hence the drift can be negative at certain wealth levels. This feature differentiates this model from those usually studied in the investment literature (e.g., Merton 1971, 1990, Pliska 1986, Davis and Norman 1990), with Majumdar and Radner (1991) being a notable exception. (For results on an "incomplete market" model where the variance, as well as the drift, is also not completely controllable, see Browne 1995.) Specifically, let a denote the bankruptcy level or point, with corresponding "bankruptcy time" (or *ruin time*) τ_a^f, where $0 \leq a < X_0$. One survival objective is then to choose an investment policy which minimizes the probability of ruin, i.e., one which minimizes $P(\tau_a^f < \infty)$, or equivalently, maximizes $P(\tau_a^f = \infty)$ (see, e.g., Majumdar and Radner 1991, Browne 1995, Roy 1995).

Clearly this objective is meaningless for $X_t^f \geq c/r$. To see this directly, consider the case where the wealth level is $x > c/r$. We may then choose a policy which puts all wealth into the bond, and then under this policy the probability of bankruptcy is 0. Specifically, if we take $f = 0$ for $x > c/r$, (3) shows that the wealth will then follow the deterministic differential equation $dX_t = (rX_t - c)\,dt$, $X_0 = x > c/r$, which exhibits exponential growth and for which $P(\tau_{x-\epsilon} = \infty) = 1$, for all $\epsilon > 0$. Thus *the survival problem is interesting and relevant only in the region* $a < x < c/r$, which we will call the "danger-zone." This is of course due to the fact that $c/r = c\int_0^\infty e^{-rt}\,dt$ is the amount that is needed to be invested in the perpetual bond to pay off the forced withdrawals forever. Since the investor need never face the possibility of ruin for $x > c/r$, we will call the region $(c/r, \infty)$ the "safe-region."

Our objective in this section therefore is to determine a strategy that *maximizes the probability of hitting the safe-region or "safe point,"* c/r, *prior to the "bankruptcy point,"* a, *when initial wealth is in the danger-zone*, i.e., $a < x < c/r$. As noted above, we will show that an optimal policy for this problem *does not exist*, necessitating the construction of an ϵ-optimal strategy.

A somewhat related survival problem with constant withdrawals was studied in Majumdar and Radner (1991) in a different setting, although without a risk-free investment, and hence without a safe-region. Moreover, their results are *not applicable* to our case since here $\inf_f v^2(f,x,t) = 0$, which violates the conditions of the model in Majumdar and Radner (1991). As we shall see, it is in fact precisely this fact that negates the existence of an optimal policy for our problem. A related survival model which allows for investment in a risk-free asset, but where the "withdrawals" are assumed to follow another (possibly dependent). Brownian motion with drift, was treated in Browne (1995). Since the Brownian motion is unbounded, there was no safe-region in Browne (1995) either. A discrete-time model with constant withdrawals that does allow for a risk-free investment was treated in Roy (1995), but with no borrowing allowed and bounded support for the return on the risky asset.

To show explicitly why no policy obtains optimality for the model treated here, and how we may construct ϵ-optimal strategies, we will first consider the following problem: for any point b in the danger-zone, i.e., with $a < x < b < c/r$, we will find the optimal policy to maximize the probability of hitting b before a. For b strictly less than c/r, an optimal policy *does exist* for this problem, and we will identify it in the following theorem. To that end let

$$V(x\colon a,b) = \sup_{f\in\mathcal{G}} P_x(\tau_a^f > \tau_b^f), \quad \text{and let} \quad f_V^* = \arg\sup_{f\in\mathcal{G}} P_x(\tau_a^f > \tau_b^f).$$

THEOREM 3.1. *The optimal policy is to invest, at each wealth level $a < x < b$, the state dependent amount*

$$f_V^*(x) = \frac{2r}{\mu - r}\left(\frac{c}{r} - x\right). \tag{20}$$

The optimal value function is

$$V(x\colon a,b) = \frac{(c - ra)^{\gamma/r+1} - (c - rx)^{\gamma/r+1}}{(c - ra)^{\gamma/r+1} - (c - rb)^{\gamma/r+1}}, \quad \textit{for } a \le x \le b, \tag{21}$$

where γ is defined by (8).

REMARK 3.1. Note that the policy of (20) invests less as the wealth gets closer to the goal b. In fact, it invests a *constant proportion of the distance to the "safe point" c/r, regardless of the value of the goal b, and the bankruptcy point a*. It is interesting to observe that while here this constant proportion is *independent of the underlying diffusion parameter σ^2*, this does not hold when there are multiple risky stocks in which to invest in (see §5 below). The constant proportion is greater (less) than 1 as $\mu/r <$ ($>$) 3. Thus it is interesting to observe that as the wealth gets closer to the bankruptcy point, a, the optimal policy does *not* "panic" and start investing an enormous amount, rather the optimal policy stays calm and invests at most $f_V^*(a) = 2(c - ra)/(\mu - r)$. The investor does get increasingly more cautious as his wealth gets closer to the goal b. (This behavior should be compared with the "timid" vs. "bold" play in the discrete-time problems considered in the classic book of Dubins and Savage 1965. See also Majumdar and Radner 1991 and Roy 1995.)

Observe further that the investor is borrowing money to invest in the stock only when $x < 2c/(\mu + r)$ but not when $2c/(\mu + r) < x < c/r$. This can be seen by observing directly that in the former case $f_V^*(x) > x$ and in the latter case $f_V^*(x) < x$. (The fact that $2c/(\mu + r) < c/r$ follows from the assumption that $\mu > r$.)

PROOF. While we could prove Theorem 3.1 from a more general theorem in Pestien and Sudderth (1985) (see also Pestien and Sudderth 1988) which we will discuss later (see Remark 3.4 below), recognize that this is simply a special case of the control problem solved in Theorem 2.1 for $l = a, u = b$ with $\lambda = 0$, $g = 0$ and $h(b) = 1$, $h(a) = 0$. As such the nonlinear Dirichlet problem of (9) for the optimal value function V becomes in this case

$$(rx - c)V_x - \gamma\frac{V_x^2}{V_{xx}} = 0, \quad \text{for } a < x < b \tag{22}$$

subject to the (Dirichlet) boundary conditions $V(a) = 0, V(b) = 1$.

The general solution to the second-order nonlinear ordinary differential equation of (22) is $K_1 - K_2(c - rx)^{\gamma/r+1}$, where K_1, K_2 are arbitrary constants which will be determined from the boundary conditions. The boundary condition $V(a) = 0$ determines that $K_1 = K_2(c - ra)^{\gamma/r+1}$, and the boundary condition $V(b) = 1$ then determines K_2, which then leads directly to the function $V(x)$ given in (21). It is clear that this function V is in $\mathscr{C}^2$ and does in fact satisfy $V_x > 0$ and $V_{xx} < 0$, and moreover satisfies conditions (i), (ii) and (iii) of Theorem 2.1 on the interval (a, b). (Condition (ii) is trivially met since V is bounded on (a, b).) As such V is indeed the optimal value function and the associated optimal control function f_V^* of (20) is then obtained by substituting the function V of (21) for w in (11). □

Note that under policy f_V^*, the wealth process, say X_t^*, satisfies the stochastic differential equation

$$(23) \qquad dX_t^* = (c - rX_t^*)\,dt + \frac{2\sigma}{\mu - r}(c - rX_t^*)\,dW_t, \quad \text{for } t \le T^*,$$

where $T^* = \min\{\tau_a^*, \tau_b^*\}$, and $\tau_z^* = \inf\{t > 0\colon X_t^* = z\}$. This is obtained by placing the control (20) into the evolutionary equation (3). Equation (23) defines a linear stochastic differential equation, i.e., X^* is a time-homogeneous diffusion on (a, b) with drift function $\mu_*(x) = c - rx$, and diffusion coefficient function $\sigma^2{}_*(x) = ((2\sigma/(\mu - r))(c - rx))^2 \equiv (2/\gamma)(c - rx)^2$. As such its *scale* function is defined by

(24)

$$S^*(x) = \int^x \exp\left\{-\int^y \frac{2\mu_*(u)}{\sigma_*^2(u)}du\right\} dy \equiv -(\gamma + r)^{-1}(c - rx)^{\gamma/r+1}, \quad \text{for } a \le x \le b$$

where $\gamma = \frac{1}{2}((\mu - r)/\sigma)^2$. For this process therefore,

$$P_x(\tau_a^* > \tau_b^*) = \frac{S^*(x) - S^*(a)}{S^*(b) - S^*(a)} \equiv \frac{(c - ra)^{\gamma/r+1} - (c - rx)^{\gamma/r+1}}{(c - ra)^{\gamma/r+1} - (c - rb)^{\gamma/r+1}},$$

which of course agrees with (21). Thus the process $\{S^*(X_t^*)\}$ is a diffusion in *natural scale*, and is therefore a (uniformly integrable) martingale with respect to the filtration $\mathscr{F}_t$ (as is the optimal value function), i.e., $E(S^*(X_t^*)|\mathscr{F}_s) = S^*(X_s^*)$ for $0 \le s \le t \le \tau^*$, where $\tau^* := \min\{\tau_a^*, \tau_b^*\}$. Note further that the scale function $S^*(x)$ of (24), is increasing in x (although *negative*) for $0 \le x < c/r$.

3.1.1. Inaccessibility of the safe-region under f_V^* and ϵ-optimal strategies. While we have found a policy that maximizes the probability of reaching any $b < c/r$ before any $a < b$, it is important to realize that if we extend b to c/r, then this policy *will never achieve the safe point c/r with positive probability in finite time*. We can of course extend the function displayed in (21) to the point c/r to get

$$(25) \quad V(x\colon a, c/r) = P_x(\tau_a^* > \tau_{c/r}^*)$$

$$= \frac{S^*(x) - S^*(a)}{S^*(c/r) - S^*(a)} \equiv 1 - \left(\frac{c - rx}{c - ra}\right)^{\gamma/r+1} < \infty, \quad \text{for } a < x \le c/r$$

which shows than in fact c/r is an *attracting barrier* for the process X^*. However, it is an *unattainable* barrier. (See §15.6 in Karlin and Taylor (1981) for a discussion of the boundary classification terminology used here.) To verify this, first recall that if we let $s^*(x) = dS^*(x)/dx$ denote the *scale density* of the diffusion X^*, then its *speed density* is given by

$$m^*(x) = \left(\sigma_*^2(x)s^*(x)\right)^{-1} \equiv \left(\frac{2}{\gamma}(c-rx)^2(c-rx)^{\gamma/r}\right)^{-1} \tag{26}$$

$$= \frac{\gamma}{2}(c-rx)^{-(\gamma/r+2)}$$

and it is well known then that

$$E_x\left(\min\{\tau_a^*, \tau_{c/r}^*\}\right) < \infty \quad \text{if and only if} \quad \int_x^{c/r}[S^*(c/r) - S^*(y)]m^*(y)\,dy < \infty.$$

However, it can be seen from (24) and (26) that the latter quantity is

$$\int_x^{c/r}[S^*(c/r) - S^*(y)]m^*(y)\,dy = \frac{\gamma}{2(\gamma+r)}\int_x^{c/r}\frac{1}{c-ry}dy = \infty,$$

and thus we see that while f_V^* minimizes the probability of hitting the ruin point a, and so is in fact optimal for the problem of $\min_f P_x(\tau_a^f = \infty)$, it does so in a way which makes the upper goal, c/r, unattainable in finite expected time. In fact, under f_V^* we have $\tau_{c/r}^* = \infty$ a.s., and thus *no optimal policy exists* for the problem of maximizing the probability of reaching the safe-region prior to bankruptcy!

Intuitively, what's going on is that as the wealth gets closer to the boundary of the safe-region, c/r, the investor gets increasingly more cautious so as not to forfeit his chances of getting there. This of course entails investing less and less, but in continuous-time, where the wealth is infinitely divisible, this just means eventually investing (close to) nothing. However while this in turn does in effect shut off the drift of the resulting wealth process (see (23)), it also shuts off the variance, and some positive variance is needed to cross over the c/r-barrier from the danger-zone into the safe-region. This is not supplied by the policy described above, which essentially tells the investor that the best he can hope to do (i.e., with maximal probability) is to try to get pulled into an asymptote that is drifting toward c/r.

In terms of our Theorem 2.1, it is clear that V is no longer concave increasing for $x > c/r$ (i.e., for $x > c/r$, we have $V_x(x) > 0$ and $V_{xx}(x) < 0$), and thus Theorem 2.1 is not valid for any $u > c/r$.

REMARK 3.3. This difficulty disappears if $r = 0$, since if there is no risk-free investment, the investor always faces a positive probability of ruin and the only way to survive is to always invest in the risky stock. To see this, note that letting $a = 0$ and taking limits as $r \downarrow 0$ (so that the "safe point" goes to infinity, i.e., when $r = 0$, there is always a positive probability of bankruptcy) shows that the value function, $V(x\colon 0, c/r)$, then goes to an exponential, i.e., as $r \downarrow 0$,

$$V(x\colon 0, c/r) \to 1 - \exp\left\{-\frac{\mu}{2\sigma^2 c}x\right\} \tag{27}$$

and for this case the (unconstrained) optimal control to minimize the probability of ruin is to always invest the *fixed constant* $2c/\mu$. (This model then becomes a degenerate special case of Browne 1995.) In this case the optimal wealth process follows a *linear Brownian motion* with drift c and diffusion coefficient $2c\sigma/\mu$, for which the probability of ruin is the exponential (27). Ferguson (1965) conjectured that an ordinary investor (in discrete-time and space) can asymptotically *minimize the probability of ruin* by *maximizing the exponential utility of terminal wealth, for some risk aversion parameter*. It is interesting to observe that for this model the conjecture turns out to be true. To verify this, one would have to solve the finite-horizon utility maximization problem for the utility function $u(x) = \delta - \eta \exp\{-2cx/\mu\}$, with arbitrary $\eta > 0$ and δ. Since this problem is then essentially a special case of the (Cauchy) problem considered in §3 of Browne (1995), we refer the reader there for further details. If we impose the constraint that the investor is not allowed to borrow, then it can be shown (Browne 1995, Theorem 3) that the optimal control in this case is $f^* = \max\{x, 2c/\mu\}$, whereby the investor must invest all his wealth in the risky stock when wealth is below the critical level $2c/\mu$. In this case the value function is no longer concave below $2c/\mu$. Such extremal behavior (or "bold" play, ala Dubins and Savage 1965) and nonconcavity of the value function below a threshold is also a feature of the optimal policies in the related survival models studied in Majumdar and Radner (1991) and Roy (1995), where borrowing is not allowed.

REMARK 3.4. This inaccessibility and the resulting nonexistence of an optimal policy can be best understood in the context of the more general "goal" problem: Consider a controlled diffusion $\{Y_t^f\}$ on the interval (a, b) satisfying

$$dY_t^f = m(f, y)\,dt + v(f, y)\,dW_t,$$

with the objective of determining a control to maximize the probability of hitting b before a. Let $\Psi(x)$ denote the optimal value function for this problem, i.e., $\Psi(x) = \sup_{f \in \mathcal{G}} P_x(\tau_a^f > \tau_b^f)$ with optimal control $\psi(x) = \arg\sup_{f \in \mathcal{G}} P_x(\tau_a^f > \tau_b^f)$. This problem was first studied by Pestien and Sudderth (1985, 1988), who showed—using a different formulation—that

$$\psi(x) = \arg\sup_f \left\{ \frac{m(f, x)}{v^2(f, x)} \right\}, \tag{28}$$

and indeed our $f_V^*(x)$ of (20) can be obtained from maximizing m/v^2 for m and v^2 in (19). However as noted in Pestien and Sudderth (1985, 1988), this is the case only when $\inf_x v_*^2(f, x) > 0$, where $\psi = m_*/v^2{}_*$.

These results can be obtained from somewhat simpler methods (albeit with some lesser generality) then those used in Pestien and Sudderth (1985, 1988) as follows: $\Psi(x)$ must satisfy the HJB equation

$$\sup_f \left\{ m(f, x)\Psi_x + \frac{1}{2} v^2(f, x)\Psi_{xx} \right\} \equiv \sup_f \left\{ \left[\frac{m(f, x)}{v^2(f, x)} \Psi_x + \frac{1}{2}\Psi_{xx} \right] \cdot v^2(f, x) \right\} = 0, \tag{29}$$

subject to the Dirichlet boundary conditions $\Psi(a) = 0$, $\Psi(b) = 1$.

If Ψ is a *classical solution* to the HJB equation (29), then we must have $\Psi_x > 0$ and $\Psi_{xx} < 0$. Therefore as long as $v^2(f, x) > 0$, it is clear from (29) that the maximum of (29) occurs at the maximum of m/v^2, which by (28) is denoted by ψ. If we now let $\rho(x) = \sup_f\{m(f, x)/v^2(f, x)\}$, i.e., $\rho(x) = m(\psi(x), x)/v^2(\psi(x), x)$, then the solution to (29) subject to the Dirichlet conditions is simply $\Psi(x) = \int_a^x s(z)\,dz / \int_a^b s(z)\,dz$, where $s(z) = \exp\{-2\int^z p(y)\,dy\}$, with which our value function (21) of course agrees, for $b < c/r$.

However, it is also clear from (29) that for $v^2(f, x) = 0$, the HJB equation *need not hold*, and therefore, no policy is in general optimal when this is the case, which is precisely what is happening here for $b = c/r$ (see also Example 4.1 in Pestien and Sudderth 1988).

For more details on the general problem from a different perspective, we refer the reader to the fundamental papers of Pestien and Sudderth (1985, 1988). We now return to the problem of determining a 'good' strategy for crossing the c/r barrier.

An ϵ-optimal strategy. As we have just seen, the inaccessibility of c/r is due to the fact that f_V^* dictates an investment policy that causes the drift and variance of the resulting wealth process to go to zero as the c/r barrier is approached from below. A practical way around this difficulty is to modify f_V^* as follows:

Let f_δ^* denote the (suboptimal) policy which agrees with f_V^* below the point $c/r - \delta$, and then above it invests κ in the risky stock until the c/r barrier is crossed, i.e.,

$$f_\delta^*(x) = \begin{cases} f_V^*(x) & \text{for } x \le c/r - \delta, \\ \kappa & \text{for } x > c/r - \delta. \end{cases}$$

Now $V(x_0, a, c/r)$ as given in (25) is an *upper bound* on the probability of escaping the interval $(a, c/r)$ into the safe-region starting from an initial wealth level $x_0 < c/r$ (see Krylov 1980, page 5). Without loss of generality, we may take $a = 0$ here. Thus for any $\epsilon > 0$, and initial wealth $x_0 < c/r$, the best we can do is find a policy which gives

$$V(x_0 : 0, c/r) - \epsilon = 1 - \left(1 - \frac{rx_0}{c}\right)^{\gamma/r+1} - \epsilon \tag{30}$$

as its value. Therefore for any given ϵ, and initial wealth $x_0 < c/r$, we need to find $\delta = \delta(x_0, \epsilon)$ and $\kappa = \kappa(x_0, \epsilon)$ which will achieve the value (30). To keep the drift and diffusion parameters continuous, we must take $\kappa = (2r/(\mu - r))\delta(x_0, \epsilon)$, and so $f_\delta(x) = (2r/(\mu - r))\max\{c/r - x, \delta\}$, which then gives a corresponding wealth process X^δ which has the (continuous) drift function $\mu_\delta(x)$ and diffusion function $\sigma_\delta^2(x)$ given by $\mu_\delta(x) = \max\{c - rx, 2r\delta + rx - c\}$, and $\sigma_\delta^2(x) = \max\{2(c - rx)^2/\gamma, 2r^2\delta^2/\gamma\}$.

The scale *density* for this new process, defined by

$$s_\delta(y) = \exp\left\{-\int^y \frac{2\mu_\delta(z)}{\sigma_\delta^2(z)}dz\right\}$$

can then be written as

$$s_\delta(y) = \begin{cases} (c - ry)^{\gamma/r} & \text{for } y \le c/r - \delta, \\ (r\delta)^{\gamma/r}\, e^{\gamma/(2r)}\sqrt{2\pi}\,\phi\left(\dfrac{y + 2\delta - c/r}{\sqrt{r\delta^2/\gamma}}\right) & \text{for } y \ge c/r - \delta, \end{cases}$$

where ϕ denotes the standard normal p.d.f.

The probability of reaching the safe-region from initial wealth $x_0 < c/r$ under this policy is therefore

$$V_\delta(x_0: 0, c/r) = \frac{\int_0^{x_0} s_\delta(y)\,dy}{\int_0^{c/r} s_\delta(y)\,dy} \equiv V(x_0: 0, c/r)\left(1 + \delta^{\gamma/r+1} H(\gamma, r, c)\right)^{-1}$$

where V is as in (25) and H is given by

$$H(\gamma, r, c) = (1 + \gamma/r)(r/c)^{\gamma/r+1}\, e^{\gamma/(2r)}\sqrt{2\pi r/\gamma}\left[\Phi\left(2\sqrt{\gamma/r}\right) - \Phi\left(\sqrt{\gamma/r}\right)\right],$$

where Φ denotes the standard normal c.d.f. Setting $V_\delta = V - \epsilon$, and then solving for δ therefore gives

$$\delta(x_0, \epsilon) = \left(\frac{\epsilon}{H(\gamma, r, c)\left[V(x_0: 0, c/r) - \epsilon\right]}\right)^{r/(\gamma+r)}. \tag{31}$$

Therefore, for the particular $\delta(x_0, \epsilon)$ given in (31), the policy f_δ^* is within ϵ of optimality. Since we chose $a = 0$ here purely for notational convenience, we summarize this in the following theorem for the case with an arbitrary bankruptcy point a, with $0 \le a < x_0$.

THEOREM 3.2. *The policy* f_δ^*, *given by*

(32)

$$f_\delta^*(x) = \begin{cases} f_V^*(x) & \text{for } a < x \le c/r - \delta, \\ \dfrac{2r}{\mu - r}\left[\epsilon/\left(H(\gamma, r, c)\left[V(x_0: a, c/r) - \epsilon\right]\right)\right]^{r/(\gamma+r)} & \text{for } x \ge c/r - \delta, \end{cases}$$

is an ϵ*-optimal policy for maximizing the probability of crossing the* c/r *barrier before hitting the point* a, *starting from an initial wealth level* x_0, *where* $a < x_0 < c/r$, *and* $V(\cdot: a, c/r)$ *is the function given by* (25).

3.2. Minimizing discounted penalty of bankruptcy. Suppose now that instead of minimizing the *probability* of ruin, we are instead interested in choosing a policy that *maximizes the time until bankruptcy*, in some sense. Obviously, this problem is nontrivial only in the danger-zone $a < x < c/r$, which is the case considered here. Maximizing the *expected* time until bankruptcy is a trivial problem, since there are any number of policies under which the expected time to bankruptcy is in fact infinite. In particular the ϵ-optimal policy described above gives a positive probability of reaching the c/r barrier, and since the safe-region $(x \ge c/r)$ is absorbing, it therefore gives an infinite expected time to ruin. Thus we need to look at other criteria. Here we will consider the objective of minimizing the expected *discounted* time until bankruptcy (a related problem without forced withdrawals was treated in Dutta 1994 in a different framework, and in an incomplete market in Browne 1995). In particular, suppose there is a large penalty, say M, that must be paid if and when the ruin point a is hit. If there is a (constant) discount rate $\lambda > 0$, then the amount due upon hitting this point is therefore $Me^{-\lambda\tau_a^f}$, and we would like to find a policy that minimizes the expected value of this penalty. Clearly, this policy is equivalent to the policy that *minimizes* $E_x(e^{-\lambda\tau_a^f})$.

To that end, let $F(x) = \inf_{f \in \mathcal{G}} E_x(e^{-\lambda\tau_a^f})$, and let f_F^* denote the associated optimal policy, i.e., $f_F^* = \arg\inf_{f \in \mathcal{G}} E_x(e^{-\lambda\tau_a^f})$. For reasons that will become clear soon, define the constants η^+ and D by

$$\eta^+ = \eta^+(\lambda) = \frac{1}{2r}\left[(r+\gamma+\lambda) + \sqrt{D}\right], \tag{33}$$

$$D = D(\lambda) = (\gamma + \lambda - r)^2 + 4r\gamma, \tag{34}$$

where γ is defined by (8). The optimal policy and optimal value function for this problem is then given in the following theorem.

THEOREM 3.3. *The optimal control is*

$$f_F^*(x) = \frac{\mu - r}{\sigma^2(\eta^+ - 1)}\left(\frac{c}{r} - x\right), \quad \text{for } a < x < c/r, \tag{35}$$

and the optimal value function is

$$F(x) = \left(\frac{c - rx}{c - ra}\right)^{\eta^+}, \quad \text{for } a \le x \le c/r. \tag{36}$$

REMARK 3.5. Note that $\eta^+ > 1$, and that $F(a) = 1$, $F(c/r) = 0$, and $F(x)$ is monotonically decreasing on the interval $(a, c/r)$, as is the optimal policy f_F^*, which once again invests a *constant proportion of the distance to the goal*. Observe too that we are therefore once again faced with the problem that under this policy, the safe-point c/r is inaccessible. However, in this case it is indeed the unique optimal policy. The condition $F(a) = 1$ is by construction, but the fact that $F(c/r) = 0$ is determined by the optimality equation itself, i.e., optimality determines that the c/r barrier is inaccessible. The intuition behind this is that this policy—although it never allows the fortune to cross the c/r-barrier—does indeed minimize the expected *discounted* time until ruin. The best one can do in this case is to get trapped in an asymptote approaching c/r, which this policy tries to do. Any additional investment near the c/r-barrier (such as in the ϵ-optimal strategy of the previous problem) allows a greater possibility of hitting a, thus increasing the value of $E_x(e^{-\lambda\tau_a})$.

REMARK 3.6. As a consistency check, note too that when we substitute the control f_F^* of (35) back into the evolutionary equation (3), we obtain a wealth process, say X_t^λ, that satisfies the stochastic differential equation

(37)

$$dX_t^\lambda = \left[\frac{2\gamma}{(\eta^+ - 1)r} - 1\right](c - rX_t^\lambda)\,dt + \frac{\mu - r}{\sigma(\eta^+ - 1)r}(c - rX_t^\lambda)\,dW_t, \quad \text{for } t < T^\lambda,$$

where $T^\lambda = \min\{\tau_a^\lambda, \tau_{c/r}^\lambda\}$, where $\tau_z^\lambda = \inf\{t > 0: X_t^\lambda = z\}$.

For this process, it is well known that the Laplace transform of τ_a^λ evaluated at the point λ, say $L(x\colon \lambda) = E_x(e^{-\lambda\tau_a^\lambda})$, is the unique solution of the Dirichlet problem

$$\left[\frac{2\gamma}{(\eta^+ - 1)r} - 1\right](c - rx)L_x + \frac{1}{2}\left(\frac{\mu - r}{\sigma(\eta^+ - 1)r}(c - rx)\right)^2 L_{xx} - \lambda L = 0,$$

with $L(a\colon \lambda) = 1$ and $L(c/r\colon \lambda) = 0$. It can be easily checked that we do in fact have $F(x) \equiv L(x\colon \lambda)$. It should be noted that τ_a^λ is a defective random variable, with $E_x(\tau_a^\lambda) = \infty$, as can be seen from the fact that

$$L(x\colon 0) = \left(\frac{c - rx}{c - ra}\right)^{\gamma/r + 1} \equiv 1 - V(x\colon a, c/r),$$

where $V(x: \cdot, \cdot)$ is the function defined by (21). This of course is due to the fact that c/r essentially acts as an absorbing barrier, and it can be hit with positive probability (albeit in infinite time). Specifically, as $\lambda \downarrow 0$, it is clear that $\eta^+(\lambda) \to \eta^+(0) \equiv \gamma/r + 1$, and thus $F(x)$ converges (uniformly in x) to the probability that the bankruptcy point a is hit before the safe point c/r, which implies that the control $f_F^*(x)$ converges (uniformly in x) to the control $f_V^*(x)$ of (20), i.e. as $\lambda \downarrow 0$:

$$F(x) \to 1 - V(x: a, c/r) \quad \text{and} \quad f_F^*(x) \to \frac{2}{\mu - r}(c - rx) \equiv f_V^*(x).$$

Note also that for $\lambda > 0$, we have $f_V^* > f_U^*$, which is of course consistent with the fact that a "bolder" strategy maximizes the probability of survival, while a "timid" strategy maximizes expected playing time (for subfair games).

PROOF. Theorem 2.1 is again relevant, however since Theorem 2.1 deals with the maximization problem, recognize that $F = -\sup_f\{-E_x(e^{-\lambda\tau_a^f})\}$. We can now apply Theorem 2.1 to $\tilde{F} := -F$ with $\lambda(x) = \lambda$, $g = 0$, $h(a) = -1$. Reverting back to F, we then see that the nonlinear Dirichlet problem of (9) for F becomes then:

$$(rx - c)F_x - \gamma\frac{F_x^2}{F_{xx}} - \lambda F = 0, \quad \text{for } a < x < c/r, \tag{38}$$

subject to the Dirichlet boundary condition $F(a) = 1$, where $\gamma = \frac{1}{2}((\mu - r)/\sigma)^2$. Observe of course that we now require $F_x < 0$ and $F_{xx} > 0$.

The nonlinear second-order ordinary differential equation in (38) admits the two solutions $C(c - rx)^{\eta^+}$, and $K(c - rx)^{\eta^-}$, where C and K are constants to be determined from the boundary condition, and where η^+, η^- are the roots to the quadratic equation $\tilde{Q}(\eta) = 0$, where

$$\tilde{Q}(\eta) = \eta^2 r - \eta(\gamma + \lambda + r) + \lambda. \tag{39}$$

To determine which (if any) of these two solutions are appropriate we need to examine these roots in greater detail. The discriminant of (39) is the constant D of (34) which is clearly positive, and thus the two roots are real and, for $\lambda > 0$, distinct. In particular

$$\eta^+ = \frac{1}{2r}\left[(r + \gamma + \lambda) + \sqrt{D}\right] \quad \text{and} \quad \eta^- = \frac{1}{2r}\left[(r + \gamma + \lambda) - \sqrt{D}\right]. \tag{40}$$

Since $\eta^+\eta^- = \lambda/r > 0$, both roots are of the same sign, and since $\eta^+ > 0$, they are both positive. The boundary condition $F(a) = 1$ determines the constants C, K as $C = (c - ra)^{-\eta^+}$ and $K = (c - ra)^{-\eta^-}$ and so clearly $C > 0$ and $K > 0$, and therefore, $F_x < 0$ for both solutions. However it is easy to check the roots in (40) to see that $\eta^+ > 1$, while $\eta^- < 1$, and so $F_{xx} > 0$ only for the root η^+. Thus we find that the (unique) solution to the (38) that satisfies $F(a) = 1$ and $F_x < 0$ $F_{xx} > 0$ is given by the function $F(x)$ defined in (36). Moreover, it is a simple matter to check that conditions (i), (ii) and (iii) of Theorem 2.1 are indeed met for F (F is bounded on (a, b)), and so we may conclude that F is optimal. The associated optimal control function, f_F^* of (35), is then obtained by placing F (or $\tilde{F} = -F$) into (11). □

REMARK 3.7. An alternative proof of Theorem 3.4 can be constructed by modifying the arguments in Orey et al. (1987), who treat the converse problem of maximizing discounted time to a goal, to deal with the minimization problem treated here.

The evolutionary equation (3) would have to be reparameterized by taking $f_t = \pi_t \cdot (c/r - X_t^f)$, for admissible control processes π, and then applying the results of Orey et al. (1987) to the further transformed process $Y_t^\pi = \ln[(c - rX_t^\pi)/(c - rb)]$.

Note that since the wealth process, say X_t^λ, under the policy f_F^*, satisfies the stochastic differential equation (37), we can apply Ito's formula to the function $F(\cdot)$ given by (36) to show, after simplification, that

(41)

$$dF(X_t^\lambda) = F(X_t^\lambda)\left[\frac{(\eta^+)^2 r - \eta^+(r+\gamma)}{\eta^+ - 1}dt - \sqrt{2\gamma}\frac{\eta^+}{\eta^+ - 1}dW_t\right], \quad \text{for } 0 < t < T^\lambda.$$

The quadratic of (39) then shows that $(\eta^+)^2 r - \eta^+(r+\gamma) = \lambda(\eta^+ - 1)$, and thus substituting this into the r.h.s. of (41), and then solving the resulting (linear) stochastic differential equation gives

$$F(X_t^\lambda) = F(X_0)\exp\left\{\left(\lambda - \gamma\left(\frac{\eta^+}{\eta^+ - 1}\right)^2\right)t - \sqrt{2\gamma}\frac{\eta^+}{\eta^+ - 1}W_t\right\}, \quad \text{for } t \le T^\lambda,$$

which shows that the value function $F(\cdot)$ operating on the process X_t^λ is a *geometric Brownian motion* on the interval $(0, 1)$, for $a \le X_t^\lambda \le c/r$.

Unfortunately, as noted above, this policy, while optimal for the stated problem, *will never cross the c/r barrier into the safe-region*, and thus the investor should utilize a policy similar to the ϵ-optimal policy described earlier to get into the safe-region. Since this can be achieved at relatively little cost, we will assume for the sequel that the investor does in fact invest in a way that will allow a positive probability of getting into the safe-region. When (if) the safe-region is achieved, the investor no longer faces the problem of bankruptcy, and should then be concerned with other optimality criteria. We consider two such criteria in the next section.

4. Optimal growth policies, in the safe-region. Suppose now that we have survived, i.e., we have achieved a level $x > c/r$. As noted earlier it is clear that in this region there need never be a possibility of ruin, and therefore the investor who has achieved this safe-region will be interested in criteria other than survival. In particular, we assume here that in this region the investor is interested in *growth*, by which we mean achieving a high level of wealth as quickly as possible. Suppose therefore that there is now some target goal, which we will denote again by b with $b > x$, which the investor wants to get to (e.g., to pay out dividends) as quickly as possible. In this section we consider two related aspects of this problem. First we consider the problem of *minimizing the expected time to the goal*, and then we consider the related problem of *maximizing the expected discounted reward of achieving the goal.* In both cases, the optimal strategies are interesting generalizations of the *Kelly criterion* that has been studied in discrete-time in Kelly (1956), Breiman (1961) and Thorp (1969), and in continuous-time in Pestien and Sudderth (1985) and Heath et al. (1987). (See also Theorem 6.5 in Merton 1990, where it is called the *growth-optimum strategy*. For Bayesian versions of both the discrete and continuous-time Kelly criterion, see Browne and Whitt 1996.) Such policies dictate investing a *constant multiple* of the wealth in the risky stock. Here our policies invest a constant multiple of the *excess wealth* over the boundary c/r, in the risky stock. This will make the c/r boundary inaccessible from above, ensuring that the investor will stay in the safe-region forever, almost surely.

4.1. Minimizing the time to a goal. To formalize this, let $X_0 = x$, but now with $c/r < x < b$. For $\tau_b^f := \inf\{t > 0: X_t^f = b\}$, let

$$G(x) = \inf_{f \in \mathcal{G}} E_x(\tau_b^f), \quad \text{and let} \quad f_G^* = \arg\inf_{f \in \mathcal{G}} E_x(\tau_b^f).$$

THEOREM 4.1. *For the problem of minimizing the expected time to the goal b, the optimal policy is to invest*

$$f_G^*(x) = \frac{\mu - r}{\sigma^2}\left(x - \frac{c}{r}\right), \quad \textit{for } c/r < x < b. \tag{42}$$

The optimal value function is

$$G(x) = \frac{1}{r + \gamma}\ln\left(\frac{rb - c}{rx - c}\right), \quad \textit{for } c/r < x \leq b. \tag{43}$$

REMARK 4.1. Note that the proportion $(\mu - r)/\sigma^2$ in (42) is the same proportion as in the ordinary continuous-time Kelly criterion (or optimal growth policy) (see Heath et al. 1987, Merton 1990, Browne and Whitt 1996). However in our policy f_G^*, this proportion operates only on *the excess wealth over the boundary* c/r. Under this policy therefore, the lower boundary, c/r is inaccessible. It is quite interesting to observe that this policy is independent of the goal b. This is quite remarkable, since while it was to be expected a priori that the optimal policy should look something like (42) near the point c/r, which ensures that c/r is inaccessible from above, it is not clear why one should expect such behavior to continue throughout even when the wealth is far away from c/r. Nevertheless, it appears that the best one can do is to simply put c/r into the safe asset, and leave it there forever, continuously compounding at rate r. This is the endowment which will finance the withdrawal at the constant rate c forever. (Recall, $c/r = c\int_0^\infty e^{-rt}\,dt$.) Once this is done, the optimal policy then plays the best *ordinary* optimal growth game with the remainder of the wealth, $x - c/r$. This policy is quite similar to the policy prescribed in Proposition 11 of Black and Perold (1992) as a form of CPPI (see also Dybvig 1995). Thus, we have shown that CPPI has another optimality property associated with it, namely that of *optimal growth*.

PROOF. Since here we are minimizing expected time, we could apply Theorem 2.1 to $\tilde{G}(x) = \sup_f\{-E_x(\tau_b^f)\}$, with $g(x) = -1$, $\lambda = 0$, $h(b) = 0$. Recognizing that $G = -\tilde{G}$, it is then seen that in terms of G, Theorem 2.1 now requires $G_x < 0$ and $G_{xx} > 0$, and that the nonlinear Dirichlet problem of (9) specializes to

$$(rx - c)G_x - \gamma\frac{G_x^2}{G_{xx}} + 1 = 0, \quad \text{for } c/r < x < b, \tag{44}$$

subject to the boundary condition $G(b) = 0$. It is readily verified that the function G of (43) satisfies this, and that moreover for this function we have $G_x < 0$ and $G_{xx} > 0$ for all $c/r < x < b$. The control function $f_G^*(x)$ of (42) is obtained by substituting G of (43) for ν in (11). However, note that while it is easy to see that conditions (i) and (iii) of Theorem 2.1 are satisfied by G, it is also clear that G is *unbounded* on $(c/r, b)$, since $G(x) \to \infty$ as $x \downarrow c/r$. Thus it is doubtful that condition (ii) of

Theorem 2.1 holds for this case. Nevertheless, we will show that Theorem 4.1 holds and f_G^* is indeed the optimal policy, however the final proof of this awaits the development in §4.2, and we will complete the proof there after Lemma 4.3. □

Note that when we substitute the control f_G^* of (42) back into the evolutionary equation (3), we obtain an (optimal) wealth process, say X^b, that satisfies

$$(45)\quad dX_t^b = (r+2\gamma)\left(X_t^b - \frac{c}{r}\right)dt + \sqrt{2\gamma}\left(X_t^b - \frac{c}{r}\right)dW_t, \quad \text{for } 0 < t < \tau_b^*,$$

where $\tau_b^* := \inf\{t > 0\colon X_t^b = b\}$, which is again a linear stochastic differential equation. (It is clear from this that c/r is in fact an inaccessible *lower* boundary for X^b.) The solution to (45) is

$$X_t^b = \left(X_0^b - \frac{c}{r}\right)\exp\{(r+\gamma)t + \sqrt{2\gamma}W_t\} + \frac{c}{r}, \quad \text{for } 0 \le t < \tau_b^*,$$

from which it follows that

$$(46)\qquad G(X_t^b) = G(X_0) - t - \frac{\sqrt{2\gamma}}{r+\gamma}W_t, \quad \text{for } 0 \le t < \tau_b^*,$$

i.e., under the (optimal) policy f_G^*, the process $\{G(X_t^b) - G(X_0)\}$, follows a simple Brownian motion on $(0,\infty)$ with a drift coefficient equal to -1. (From this it is easy to recover the value function (43) from (46) by evaluating the expected value of (46) at $t = \tau_b^*$ using the fact that $G(b) = 0$, which then gives $E_x(\tau_b^*) = G(x)$.)

REMARK 4.2. The "minimal time to a goal" problem for the case $c = 0$ was first solved in the fundamental paper of Heath et al. (1987) without direct recourse to HJB methods (see also Schäl 1993). The result in that case is simply (42) with $c = 0$ (see §4 in Heath et al. 1987). Merton (1990, Theorem 6.5), also obtained this policy via another, rather complicated, argument. Since the proof given here holds too for the case $c = 0$, our results also provide an alternative and complementary proof for that case to the ones in Heath et al. (1987) and Merton (1990).

In fact, it is possible to apply the results of Heath et al. (1987) to construct a different proof of Theorem 4.1. First one would need to reparameterize the wealth equation (3) by taking $f_t = \pi_t \cdot (X_t^f - c/r)$, and then applying results of Heath et al. (1987) to the further transformed process $Y_t^\pi = \ln[(rX_t^\pi - c)/(rb - c)]$. However the results in Heath et al. (1987) are specific to the case where the controls must lie on a given constant set that is independent of the current wealth, while the approach here, based on the HJB methods of Theorem 2.1 could be modified to allow for a state dependent opportunity set.

4.2. Maximizing expected discounted reward of achieving the goal. Suppose now that instead of minimizing the expected time to the goal b, we are instead interested in maximizing $E_x(e^{-\lambda\tau_b^f})$, for $c/r < x \le b$. To that end let

$$U(x) = \sup_{f\in\mathcal{G}} E_x(e^{-\lambda\tau_b^f}), \quad \text{and let} \quad f_U^*(x) = \arg\sup_{f\in\mathcal{G}} E_x(e^{-\lambda\tau_b^f}).$$

As we show in the following theorem, the optimal policy for this problem also invests a (different) constant proportion of the excess wealth above the c/r barrier, and is hence another version of the CPPI strategy as in Black and Perold (1992).

THEOREM 4.2. *The optimal control is*

$$f_U^*(x) = \frac{\mu - r}{\sigma^2(1-\eta^-)}\left(x - \frac{c}{r}\right), \quad \textit{for } c/r < x < b, \tag{47}$$

and the optimal value function is

$$U(x) = \left(\frac{rx - c}{rb - c}\right)^{\eta^-}, \quad \textit{for } c/r \le x \le b, \tag{48}$$

where $\eta^- \equiv \eta^-(\lambda)$ *was defined previously in* (40).

REMARK 4.3. Recall that η^- is the root that satisfies $0 \le \eta^- < 1$ to the quadratic equation $\tilde{Q}(\eta) = 0$, where $\tilde{Q}(\cdot)$ is given in (39). Note that $U(b) = 1, U(c/r) = 0$, with $U(x)$ monotonically increasing on $(c/r, b)$. As was the case earlier in §3, the fact that $U(b) = 1$ is by construction, but it is optimality that causes $U(c/r) = 0$, and hence makes the danger-zone inaccessible from the safe-region.

PROOF. The proof is essentially the same as for Theorem 3.3. Specifically, here Theorem 2.1 applies directly with $u = b$, $\lambda(x) = \lambda > 0$, $g = 0$ and $h(b) = 1$. Thus the nonlinear Dirichlet problem of (9) for this case specializes to

$$(rx - c)U_x - \gamma\frac{U_x^2}{U_{xx}} - \lambda U = 0, \quad \text{for } c/r < x < b, \tag{49}$$

subject to the boundary condition $U(b) = 1$. Since we require $U_x > 0$ and $U_{xx} < 0$, it is clear the solution of interest here involves the smaller root, η^-, to the quadratic $\tilde{Q}(\eta) = 0$ (see (39)), since $\eta^- < 1$. The control function $f_U^*(x)$ of (47) is then obtained by substituting U of (48) for ν into (11). Finally, it is easy to check that U of (48) satisfies conditions (i), (ii) and (iii) of Theorem 2.1, and we may therefore conclude that f_U^* is indeed optimal. □

It is interesting to observe that when we place the control f_U^* back into the evolutionary equation (3), we find that the resulting optimal wealth process, say $\tilde{X}_t^\lambda$, satisfies the stochastic differential equation

(50)

$$d\tilde{X}_t^\lambda = \left[\frac{2\gamma}{(1-\eta^-)r} + 1\right]\left(r\tilde{X}_t^\lambda - c\right)dt + \frac{\sqrt{2\gamma}}{(1-\eta^-)r}\left(r\tilde{X}_t^\lambda - c\right)dW_t, \quad \text{for } t < \tau_b^\lambda$$

where $\tau_b^\lambda = \inf\{t > 0: \tilde{X}_t^\lambda = b\}$. An application now of Ito's formula to the function $U(\cdot)$ of (48) using (50) (and (39)) gives

$$U(\tilde{X}_t^\lambda) = U(X_0)\exp\left\{\left(\lambda - \gamma\left(\frac{\eta^-}{1-\eta^-}\right)^2\right)t + \sqrt{2\gamma}\frac{\eta^-}{1-\eta^-}W_t\right\}, \quad \text{for } t \le \tau_b^\lambda,$$

which shows that the value function $U(\cdot)$ operating on the process $\tilde{X}_t^\lambda$ is a *geometric Brownian motion* on the interval $(0, 1)$, for $c/r < \tilde{X}_t^\lambda < b$.

REMARK 4.4. Orey, Pestien and Sudderth (1987), using different methods, studied some general goal problems with a similar objective as that considered here, and as a particular example study a version of our problem with $r = c = 0$ (Orey Pestien and Sudderth 1987, page 1258). An alternative proof of Theorem 4.3 can therefore be constructed by using the results of Orey et al. (1987) using the transformation and reparameterization described above in Remark 4.2.

We may now use the results of Theorem 4.2 to complete the proof of Theorem 4.1. However, we first need the following lemma, which is of independent interest since it is applicable to more general processes than those considered here (for related results, see Schäl 1993, §4).

LEMMA 4.3. *Suppose that for every* $\lambda > 0$, *we have*

$$\nu(x;\lambda) = \inf_f E_x\left(\frac{1-e^{-\lambda\tau^f}}{\lambda}\right), \quad \textit{with optimal control } f^*(x;\lambda),$$

with $\nu(x;\lambda) < \infty$, $\lim_{\lambda\downarrow 0}\nu(x;\lambda) = \nu(x;0) < \infty$, *and* $\lim_{\lambda\downarrow 0} f^*(x;\lambda) = f(x;0)$.
Then

$$\lim_{\lambda\downarrow 0}\inf_f E_x\left(\frac{1-e^{-\lambda\tau^f}}{\lambda}\right) = \inf_f E_x\left(\lim_{\lambda\downarrow 0}\frac{1-e^{-\lambda\tau^f}}{\lambda}\right) \equiv \inf_f E_x(\tau^f), \tag{51}$$

with $\inf_f E_x(\tau^f) = \nu(x;0)$ *and with optimal control* $f(x;0)$.

PROOF. It is the first equality in (51) that needs to be established since the second is just an identity. To proceed, it is obvious that $\lambda^{-1}[1-e^{-\lambda\tau^f}] \le \tau^f$ for all $\lambda \ge 0$, and hence $E_x(\lambda^{-1}[1-e^{-\lambda\tau^f}]) \le E_x(\tau^f)$, as well as $\inf_f E_x(\lambda^{-1}[1-e^{-\lambda\tau^f}]) \le \inf_f E_x(\tau^f)$. Since the r.h.s. of this inequality is independent of the parameter λ, it follows that we may take limits on λ to get

$$\lim_{\lambda\downarrow 0}\inf_f E_x\left(\lambda^{-1}\left[1-e^{-\lambda\tau^f}\right]\right) \le \inf_f E_x(\tau^f). \tag{52}$$

For notational convenience now, let f_λ^* denote the policy $f^*(\cdot;\lambda)$, and for any policy f, let $\tau[f] = \tau^f$. Note that under this notation, we may write $\nu(x;\lambda) = E_x(\lambda^{-1}[1 - e^{-\lambda\tau[f_\lambda^*]}])$.

To go the other way now, suppose that there is an admissible policy, say $\tilde{f}$, such that $\tau[f_\lambda^*] \to^{\text{as}} \tau[\tilde{f}]$ as $\lambda\downarrow 0$. Then it is clear that

$$\inf_f E_x(\tau^f) \le E_x\left(\tau\left[\tilde{f}\right]\right) = E_x\left(\lim_{\lambda\downarrow 0}\lambda^{-1}[1-e^{-\lambda\tau[f_\lambda^*]}]\right).$$

An application of Fatou's lemma then shows that

$$E_x\left(\lim_{\lambda\downarrow 0}\lambda^{-1}[1-e^{-\lambda\tau[f_\lambda^*]}]\right) \le \lim_{\lambda\downarrow 0}E_x\left(\lambda^{-1}[1-e^{-\lambda\tau[f_\lambda^*]}]\right),$$

and since $E_x(\lambda^{-1}[1-e^{-\lambda\tau[f_\lambda^*]}]) \equiv \inf_f E_x(\lambda^{-1}[1-e^{-\lambda\tau^f}])$, we in turn conclude that

$$\inf_f E_x(\tau^f) \le \lim_{\lambda\downarrow 0}\inf_f E_x\left(\lambda^{-1}\left[1-e^{-\lambda\tau^f}\right]\right). \tag{53}$$

The inequalities (52) and (53) yield (51). □

COMPLETION OF PROOF OF THEOREM 4.1. Observe first that $\eta^-(\lambda) \to 0$ as $\lambda\downarrow 0$, and that therefore $f_U^* \to f_G^*$ as $\lambda\downarrow 0$, where f_U^* and f_G^* are given by (47) and (42).

Note further that for any $c/r < x < b$, we have

$$\lim_{\lambda \downarrow 0} \frac{1 - U(x)}{\lambda} = G(x),$$

where U and G are given by (48) and (43). Finally, since $\eta^{-}(\lambda) \to 0$ as $\lambda \downarrow 0$, we have $X^{b}_{t \wedge \tau_b} \to^{a.s.} \bar{X}^{\lambda}_{t \wedge \tau^{\lambda}_b}$ as $\lambda \downarrow 0$, where X^{b}_{t} and $\bar{X}^{\lambda}_{t}$ are defined by (45) and (50), from which it is clear that $\tau^{\lambda}_{b} \to^{a.s.} \tau_b$ as $\lambda \downarrow 0$. Therefore, Lemma 4.3 may be applied directly to Theorem 4.2 to deduce Theorem 4.1. □

5. The multiple asset case. As promised earlier, here we show how all of our previous results extend in a very straightforward way to the case with multiple risky stocks. The model here is that of a complete market (as in, e.g., Karatzas and Shreve 1988) where there are n risky assets generated by n independent Brownian motions. The prices of these stocks evolve as

$$dP_i(t) = P_i(t)\left[\mu_i\, dt + \sum_{j=1}^{n} \sigma_{ij}\, dW_t^{(j)}\right], \qquad i = 1, \dots, n, \tag{54}$$

while the riskless asset, B_t, still evolves as $dB_t = rB_t\, dt$. The wealth of the investor therefore evolves as

$$dX_t^f = \left[rX_t^f - c + \sum_{i=1}^{n} f_i(\mu_i - r)\right] + \sum_{i=1}^{n}\sum_{j=1}^{n} f_i \sigma_{ij}\, dW_t^{(j)}, \tag{55}$$

where now f_i denotes the total amount of money invested in the ith stock.

If we introduce now the matrix $\boldsymbol{\sigma} = (\sigma)_{ij}$, and the (column) vectors $\boldsymbol{\mu} = (\mu_1, \dots, \mu_n)^T, \mathbf{f} = (f_1, \dots, f_n)^T$, and then set $\mathbf{A} = \boldsymbol{\sigma}\boldsymbol{\sigma}^T$, we may write the generator of the (one-dimensional) wealth process, for functions $\Psi(x) \in \mathscr{C}^2$ as

$$\mathscr{A}^{\mathbf{f}}\Psi(x) = (\mathbf{f}^T(\boldsymbol{\mu} - r\mathbf{1}) + rx - c)\Psi_x + \frac{1}{2}\mathbf{f}^T\mathbf{A}\mathbf{f}\Psi_{xx}, \tag{56}$$

where $\mathbf{1}$ denotes a vector of 1's. The assumption of completeness implies that $\mathbf{A}^{-1}$ exists, and thus all our results will go through exactly as before. In particular, if an optimal value function for a specific problem is denoted by $\nu(x)$, the optimal control vector is $\mathbf{f}^*_\nu(x)$ where

$$\mathbf{f}^*_\nu(x) = -\mathbf{A}^{-1}(\boldsymbol{\mu} - r\mathbf{1})\frac{\nu_x}{\nu_{xx}}. \tag{57}$$

The differential equations ((22), (38), (44) and (49)), and hence the value functions ((21), (36), (43) and (48)), all remain the same except for the fact that now the scalar γ is evaluated as

$$\gamma = \frac{1}{2}(\boldsymbol{\mu} - r\mathbf{1})^T\mathbf{A}^{-1}(\boldsymbol{\mu} - r\mathbf{1}). \tag{58}$$

It is interesting to note that for the problem of maximizing the probability of reaching b before a when b is in the danger-zone, considered in §3.1, the optimal policy now *does depend on the variances and covariances* of the risky assets, since instead of (20), in the multiple asset case we now get

$$\mathbf{f}_V^*(x) = \mathbf{A}^{-1}(\boldsymbol{\mu} - r\mathbf{1})\frac{r}{\gamma}\left(\frac{c}{r} - x\right). \tag{59}$$

The ϵ-optimal policy of §3 needs to be modified, but the extension is straightforward and we leave the details for the reader. For reference, we note further that if we define the vector $\mathbf{K}$ by $\mathbf{K} := \mathbf{A}^{-1}(\boldsymbol{\mu} - r\mathbf{1})$, then the optimal controls (35), (42) and (47) of §§3.2, 4.1 and 4.2 become, respectively

$$\mathbf{f}_F^*(x) = \mathbf{K}(\eta^+ - 1)^{-1}\left(\frac{c}{r} - x\right), \qquad \mathbf{f}_G^*(x) = \mathbf{K}\left(x - \frac{c}{r}\right), \tag{60}$$

$$\mathbf{f}_U^*(x) = \mathbf{K}(1 - \eta^-)^{-1}\left(x - \frac{c}{r}\right).$$

6. Linear withdrawal rate. In this section we show how all of our previous results and analysis for the case of forced withdrawals at the *constant* rate $c > 0$ can be generalized to the case where there is a *wealth-dependent* withdrawal rate, $c(x)$ where

$$c(x) = c + \theta x.$$

Here we will only consider the case where $0 \le \theta < r$. For notational ease, we will consider again only the case with one risky stock. The generalization to the multiple stock case as in the previous section is very straightforward, and so we leave the details for the reader.

For this case the evolutionary equation (3) becomes

$$\begin{aligned} dX_t^f &= f_t\frac{dP_t}{P_t} + \left(X_t^f - f_t\right)\frac{dB_t}{B_t} - \left(c + \theta X_t^f\right)dt \\ &= \left[(r - \theta)X_t^f + f_t(\mu - r) - c\right]dt + f_t\sigma\, dW_t. \end{aligned} \tag{61}$$

If we now define $\tilde{r} := r - \theta > 0$, then for Markov control processes f, and $\Psi \in \mathscr{C}^2$ the generator of the wealth process is

$$\mathscr{A}^f\Psi(x) = [f(\mu - r) + \tilde{r}x - c]\Psi_x + \frac{1}{2}f^2\sigma^2\Psi_{xx}. \tag{62}$$

The parameter $\tilde{r}$ is simply the *adjusted (risk-free) compounding rate.* Essentially, nothing really changes except for the fact that the danger-zone is now the region $x < c/\tilde{r}$, and the safe-region is its complement. The differential equations (22), (38), (44) and (49) all remain the same except for the fact that we must replace $rx - c$ with $\tilde{r}x - c$. The parameter γ in all those equations, as well as here, is still defined as in (8), i.e., $\gamma = \frac{1}{2}((\mu - r)/\sigma)^2$, where r is the *standard* interest rate. Thus, the previous analysis will go through with relatively little change, and so we will only point out the essential differences. In particular, the structure of the policies remain the same, in that the optimal *survival* policies of §3 invest a fixed proportion of the distance to the (new) safe-region barrier, $c/\tilde{r}$, while the optimal *growth* policies of §4 invest a fixed proportion of the excess of wealth over the barrier.

6.1. Survival problems in the danger-zone. The analysis of §§3.1 and 3.2 can be repeated almost verbatim. What changes is that now for $a < x < b < c/\tilde{r}$, the value function $V(x; a, b)$ of (21) becomes instead

$$V(x: a, b) = \frac{(c - \tilde{r}a)^{\gamma/\tilde{r}+1} - (c - \tilde{r}x)^{\gamma/\tilde{r}+1}}{(c - \tilde{r}a)^{\gamma/\tilde{r}+1} - (c - \tilde{r}b)^{\gamma/\tilde{r}+1}}, \quad \text{for } a \le x \le b. \tag{63}$$

Since (11) still holds, the resulting optimal control becomes, instead of (20),

$$f_V^*(x) = \frac{2\tilde{r}}{\mu - r}\left(\frac{c}{\tilde{r}} - x\right). \tag{64}$$

For the discounted problem of §3.2 (as well as for the discounted problem of §4.2), the quadratic $\tilde{Q}(\cdot)$ of (39) changes to $\tilde{Q}(\eta) = \eta^2\tilde{r} - \eta(\gamma + \lambda + \tilde{r}) + \lambda$, and thus the two (real) roots to $\tilde{Q}(\eta) = 0$, denoted by $\tilde{\eta}^+$ and $\tilde{\eta}^-$, become, instead of (40),

$$\tilde{\eta}^{+,-} = \frac{1}{2\tilde{r}}\left[(\tilde{r} + \gamma + \lambda) \pm \sqrt{(\gamma + \lambda + \tilde{r})^2 - 4\tilde{r}\lambda}\right]. \tag{65}$$

It is easy to check that once again, we have $0 \le \tilde{\eta}^- < 1 < \tilde{\eta}^+$, and so the optimal value function (36) and the optimal control function (35) become, respectively

$$F(x) = \left(\frac{c - \tilde{r}x}{c - \tilde{r}a}\right)^{\tilde{\eta}^+}, \qquad f_F^*(x) = \frac{\mu - r}{\sigma^2(\tilde{\eta}^+ - 1)}\left(\frac{c}{\tilde{r}} - x\right). \tag{66}$$

6.2. Growth policies in the safe-region. Once again, the analysis is almost identical to that in §§4.1 and 4.2. The value and the optimal control functions for the minimal expected time to the goal problem, (43) and (42) are replaced respectively by

$$G(x) = \frac{1}{\tilde{r} + \gamma}\ln\left(\frac{\tilde{r}b - c}{\tilde{r}x - c}\right), \qquad f_G^*(x) = \frac{\mu - r}{\sigma^2}\left(x - \frac{c}{\tilde{r}}\right), \quad \text{for } c/\tilde{r} < x < b. \tag{67}$$

Note that the optimal (Kelly) proportion of the excess wealth invested in the stock, $(\mu - r)/\sigma^2$, is unchanged in this case.

Similarly for the discounted problem considered in §4.2, the value function (48) and optimal policy (47) become

$$U(x) = \left(\frac{\tilde{r}x - c}{\tilde{r}b - c}\right)^{\tilde{\eta}^-}, \qquad f_U^*(x) = \frac{\mu - r}{\sigma^2(1 - \tilde{\eta}^-)}\left(x - \frac{c}{\tilde{r}}\right). \tag{68}$$

The case where $\theta > r$, and hence with $\tilde{r} < 0$, introduces new difficulties that will be discussed elsewhere.

Acknowledgments. The author is most grateful to Professors Ioannis Karatzas of Columbia University and Bill Sudderth of the University of Minnesota for very helpful discussions. In particular, the proof of Lemma 4.3 is due to Ioannis Karatzas. He is also thankful to two anonymous referees for very helpful comments and suggestions and for bringing the references Schäl (1993), Dutta (1994) and Roy (1995) to his attention.

References

Black, F., A. F. Perold (1992). Theory of constant proportion portfolio insurance. *Jour. Econ. Dyn. and Cntrl.* **16** 403–426.

Breiman, L. (1961). Optimal gambling systems for favorable games *Fourth Berkeley Symp. Math. Stat. and Prob.* **1** 65–78.

Browne, S. (1995). Optimal investment policies for a firm with a random risk process: Exponential utility and minimizing the probability of ruin. *Math. Oper. Res.* **20** 937–958.

____, W. Whitt (1996). Portfolio choice and the Bayesian Kelly criterion. *Adv. Appl. Prob.* **28** 1145–1176.

Davis, M. H. A., A. R. Norman (1990). Portfolio selection with transactions costs, *Math. Oper. Res.* **15** 676–713.

Dubins, L. E., L. J. Savage (1965). *How to Gamble If You Must: Inequalities for Stochastic Processes*, McGraw-Hill, New York.

Dutta, P. (1994). Bankruptcy and expected utility maximization. *Jour. Econ. Dyn. and Cntrl.* **18** 539–560.

Dybvig, P. H. (1995). Dusenberry's ratcheting of consumption: Optimal dynamic consumption and investment given intolerance for any decline in standard of living. *Review of Economic Studies* **62** 287–313.

Ferguson, T. (1965). Betting systems which minimize the probability of ruin. *J. SIAM* **13** 795–818.

Fleming, W. H., R. W. Rishel (1975). *Deterministic and Stochastic Optimal Control*, Springer-Verlag, New York.

____, H. M. Soner (1993). *Controlled Markov Processes and Viscosity Solutions*, Springer-Verlag, New York.

Grossman, S. J., Z. Zhou (1993). Optimal investment strategies for controlling drawdowns. *Math. Fin.* **3** 241–276.

Heath, D., S. Orey, V. Pestien, W. Sudderth (1987). Minimizing or maximizing the expected time to reach zero. *SIAM J. Contr. and Opt.* **25** 195–205.

Karatzas, I., S. Shreve (1988). *Brownian Motion and Stochastic Calculus*, Springer-Verlag, New York.

Karlin, S., H. M. Taylor (1981). *A Second Course on Stochastic Processes*, Academic, New York.

Kelly, J. (1956). A new interpretation of information rate. *Bell Sys. Tech. J.* **35** 917–926.

Krylov, N. V. (1980). *Controlled Diffusion Processes*, Springer-Verlag, New York.

Luskin, D. L. (Editor) (1988). *Portfolio Insurance: A Guide to Dynamic Hedging*, Wiley, New York.

Majumdar, M., R. Radner (1991). Linear models of economic survival under production uncertainty. *Econ. Theory* **1** 13–30.

Merton, R. (1971). Optimum consumption and portfolio rules in a continuous time model. *J. Econ. Theory* **3** 373–413.

____ (1990). *Continuous Time Finance*, Blackwell, Massachusetts.

Orey, S., V. Pestien, W. Sudderth (1987). Reaching zero rapidly. *SIAM J. Cont. and Opt.* **25** 1253–1265.

Pestien, V. C., W. Sudderth (1985). Continuous-time red and black: How to control a diffusion to a goal. *Math. Oper. Res.* **10** 599–611.

____, ____ (1988). Continuous-time casino problems. *Math. Oper. Res.* **13** 364–376.

Pliska, S. R. (1986). A stochastic calculus model of continuous trading: Optimal portfolios. *Math. Oper. Res.* **11** 371–382.

Rogers, L. C. G., D. Williams (1987). *Diffusions, Markov Processes, and Martingales, Vol. 2*, Wiley, New York.

Roy, S. (1995). Theory of dynamic portfolio choice for survival under uncertainty. *Math. Soc. Sci.* **30** 171–194.

Schäl, M. (1993). On hitting times for jump-diffusion processes with past dependent local characteristics. *Stoch. Proc. Appl.* **47** 131–142.

Thorp, E. O. (1969). Optimal gambling systems for favorable games. *Rev. Int. Stat. Inst.* **37** 273–292.

S. Browne: Graduate School of Business, Columbia University, 402 Uris Hall, New York, NY 10027; e-mail: sb30@columbia.edu

MANAGEMENT SCIENCE
Vol. 38, No. 11, November 1992
Printed in U.S.A.

24

GROWTH VERSUS SECURITY IN DYNAMIC INVESTMENT ANALYSIS*

L. C. MacLEAN, W. T. ZIEMBA AND G. BLAZENKO
School of Business Administration, Dalhousie University, Halifax, Nova Scotia, Canada B3H 1Z5
Faculty of Commerce, University of British Columbia, Vancouver, British Columbia, Canada V6T 1Z2
School of Business Administration, Simon Fraser University, Burnaby, British Columbia, Canada V5A 1S6

This paper concerns the problem of optimal dynamic choice in discrete time for an investor. In each period the investor is faced with one or more risky investments. The maximization of the expected logarithm of the period by period wealth, referred to as the Kelly criterion, is a very desirable investment procedure. It has many attractive properties, such as maximizing the asymptotic rate of growth of the investor's fortune. On the other hand, instead of focusing on maximal growth, one can develop strategies based on maximum security. For example, one can minimize the ruin probability subject to making a positive return or compute a confidence level of increasing the investor's initial fortune to a given final wealth goal. This paper is concerned with methods to combine these two approaches. We derive computational formulas for a variety of growth and security measures. Utilizing fractional Kelly strategies, we can develop a complete tradeoff of growth versus security. The theory is applicable to favorable investment situations such as blackjack, horseracing, lotto games, index and commodity futures and options trading. The results provide insight into how one should properly invest in these situations.

(CAPITAL ACCUMULATION; FRACTIONAL KELLY STRATEGIES; EFFECTIVE GROWTH-SECURITY TRADEOFF; BLACKJACK; HORSERACING; LOTTO GAMES; TURN OF THE YEAR EFFECT)

This paper develops an approach to the analysis of risky investment problems for practical use by individuals. The idea is to provide simple-to-understand two-dimensional graphs that provide essential information for intelligent investment choice. Although the situation studied is multiperiod, the approach is couched in a growth versus security fashion akin to the static Markowitz mean-variance portfolio selection tradeoff. The approach is a marriage of the capital growth literature of Kelly (1956), Breiman (1961), Thorp (1966, 1975), Hakansson (1971, 1979), Algoet and Cover (1988) and others, which emphasizes maximal growth, with the maximal security literature of Ferguson (1965), Epstein (1977), and Feller (1962). In §1, we formulate a general investment model and describe three growth measures and three security measures. The measures all relate to single-valued aggregates of the investment results over multiple periods such as the mean first passage time to a particular wealth level and the probability of doubling one's wealth before halving it. In §2, we model the investment process as a random walk. Then, using familiar procedures from probability theory, we can easily generate computable quantities which are usually close approximations to the measures of interest. To generate tradeoffs of growth versus security one can utilize fractional Kelly strategies as discussed in §3. A simple but powerful result is that a complete trade-off of growth versus security for the most interesting growth and security measures is implementable simply by choosing various fractional Kelly strategies. Theoretical justification for the fractional Kelly strategies in a multiperiod context can be made using a continuous time approach and log normality assumptions (see Li et al. 1990 and Wu and Ziemba 1990).

0025-1909/92/3811/1562$01.25

Application of the theory to four favorable investment situations is made in §4. In each case the basic game or investment situation is unfavorable to the typical or average player. However, systems have been developed that beat the game in the sense that they have positive expected value. The question then remains how large should the wagers be and how confident is one that particular goals will be achieved. The various games, blackjack, horseracing, lotto games and the turn of the year effect differ markedly in their character. The size of the wagers vary from over half of one's fortune to less than one millionth of the fortune. The graphs that are outputted for each of these applications show how to trade off risk and return and provide crucial insight into how one should invest intelligently in these situations.

1. The Basic Investment Problem

An investor has initial wealth $y_0 \in \mathbb{R}$ and is facing n risky investments in periods 1, 2, $\ldots, t, \ldots$. The return on investments follows a stochastic process defined on the probability space (Ω, B, P) with corresponding product spaces (Ω^t, B^t, P^t). Given the realization history $\omega^{t-1} \in \Omega^{t-1}$, the investor's capital at the *beginning* of period t is $Y_{t-1}(\omega^{t-1})$. The investment in period t in each opportunity i, $i = 1, \ldots, n$, is $X_{it}(\omega^{t-1})$, $X_{it}(\omega^{t-1}) \le Y_{t-1}(\omega^{t-1})$. The investment decisions in terms of proportions are $X_{it}(\omega^{t-1}) = p_{it}(\omega^{t-1})Y_{t-1}(\omega^{t-1})$ for $i = 1, \ldots, n$, and $p_t(\omega^{t-1}) = (p_{0t}(\omega^{t-1}), \ldots, p_{nt}(\omega^{t-1}))$, where $\sum_{i=0}^{n} p_{it}(\omega^t) = 1$ and $p_{0t}(\omega^{t-1})Y_{t-1}(\omega^{t-1})$ is the investment at time t in riskless cash-like instruments. In this general form the investment strategy $p_t(\omega^{t-1})$ depends on time and the history Ω^{t-1} of the investment process. The net return per unit of capital invested in i, $i = 1, \ldots, n$, given the outcome $\omega_t \in \Omega$, in period t is given by $K_{it}(\omega_t)$, $i = 1, \ldots, n$, $t = 1, \ldots$. The return on the risk-free asset is given by $K_{0t} = 0$. It is assumed that $K_{it}(\omega_i)$ is defined by the multiplicative model

$$K_{it}(\omega_t) = \alpha_i(t)E_i(\omega_t),$$

where $\alpha_i(t) > 0$ is the average time path and $E_i(\omega_i)$ is an independent error term. Special cases result when $\alpha_i(t) = \alpha_i$ or $\alpha_i(t) = \alpha_i^t$ for $t = 1, \ldots$. The error terms in this model are not autocorrelated. A more general approach would be to consider the conditional return at time t given the history ω^{t-1}, denoted by $K_{it}(\omega_t | \omega^{t-1})$. A Bayesian analysis with such a model is theoretically tractable, but it is not as suited to computations as the multiplicative model.

An investment environment is said to be favorable if $EK_i(\omega) > 0$ for some i. We are only concerned with favorable environments. The total return on investment is $R_{it}(\omega_i) = 1 + K_{it}(\omega_t)$, and the investment decisions are $p = (p_1, p_2(\omega^1), \ldots, p_t(\omega^{t-1}), \cdots)$. Then the accumulated wealth at the end of period t is

$$Y_t(p, \omega^t) = y_0 \prod_{s=1}^{t} \left(\sum_{i=0}^{n} R_{is}(\omega_s) p_{is}(\omega^{s-1}) \right), \qquad \omega^t \in \Omega^t, \qquad t = 1, \ldots. \tag{1}$$

For the stochastic capital accumulation process $\{Y_t(p)\}_{t=1}^{\infty}$ we are interested in characterizing the accumulation paths as the investment decisions vary. Consider the following measures of growth and security.

1.1. *Measures of Growth*

G1. $\mu_t(y_0, p) = EY_t(p)$. This is the mean accumulation of wealth at the end of time t. That is, how much do we expect to have, on average, after t periods.

G2. $\phi_t(y_0, p) = E \ln (Y_t(p)^{1/t})$. This is the mean exponential growth rate over t periods. It is a measure of how fast the investor is accumulating wealth. We also consider the long-run growth rate $\phi(p) = \lim_{t \to \infty} \phi_t(y_0, p)$.

G3. $\eta(y_0, p) = E\tau_{\{Y(p) \geq U\}}$. This is the mean first passage time τ to reach the set $[U, \infty)$. That is, how long on average does it take the investor to reach a specific level of wealth. For example, how long must the investor wait before he is a millionaire.

1.2. *Measures of Security*

S1. $\gamma_t(y_0, p) = \Pr[Y_t(p) \geq b_t]$. This is the probability that the investor will have a specific accumulated wealth $b_t \in \mathbb{R}$ at time t. That is, what are his chances of reaching a given target in a fixed amount of time. For example, what chance does the investor have of accumulating $200,000 one hundred days from now?

S2. $\alpha(y_0, p) = \Pr[Y_t(p) \geq b_t, t = 1, \cdots]$. This is the probability that the investor's wealth is above a specified path. Our concern for an investor is that his wealth will *fall back* too much in any period. He may want protection against losing more than, say, 10% of current wealth at any point in time.

S3. $\beta(y_0, p) = \Pr[\tau_{\{Y(p) \geq U\}} < \tau_{\{Y(p) \geq L\}}]$. This is the probability of reaching a goal U which is higher than his initial wealth y_0, before falling to a wealth level L which is less than y_0. For example, if $U = 2y_0$ and $L = y_0/2$, this is the probability of doubling before halving.

The rationale for the measures selected is to *profile* the growth and security dimensions of the accumulation process. In each case there is a natural pairing of the measures: (μ_t, γ_t) for accumulated wealth at a point in time; (ϕ, α) for the behavior of growth paths; and (η, β) for first passage to terminal or stopping states. The choice of profile (pair of measures) is largely a function of personal preference and problem context. It is expected that the information contained in each of the profiles would be useful in selecting an investment strategy.

The usual criterion for evaluating a decision rule is $t\phi_t(y_0, p)$, the expected log of accumulated wealth. When $K_{it}(\omega_t) = \alpha_i(t)E_i(\omega_t)$, $i = 1, \ldots, n$, $t = 1, \ldots$, then the log optimal strategy is *proportional*, $\bar{p}_{it}(\omega^{t-1}) = \bar{p}_{it}$. Furthermore, if $\bar{p}_i$, $i = 1, \ldots, n$, solves the one-period problem

$$\max \left\{ E \ln \left(1 + \sum_{t=1}^{n} E_i(\omega) p_i \right) \middle| \sum_{i=1}^{n} p_i \leq 1, \quad p_i \geq 0, \quad i = 1 \cdots, n \right\},$$

then $\bar{p}_{it} = \inf\{\alpha_i^{-1}(t)\bar{p}_i, 1\}$ for $i = 1, \ldots, n$, $t = 1, \ldots$. In the case where $\alpha_i(t) = \alpha_i$, $i = 1, \ldots, n$, the log optimal strategy is a fixed fraction (independent of time), and is referred to as the Kelly (1956) strategy.

Considering the other growth measures we find that the Kelly strategy (i) maximizes the long run exponential growth rate $\phi(y_0, p)$, and (ii) minimizes the expected time to reach large goals $\eta(y_0, p)$ (Breiman 1961, Algoet and Cover 1988). The main properties of the Kelly strategy are summarized in Table 1. The performance of the Kelly strategy on the security measures is not so favorable. As with any fixed fraction rule, the Kelly strategy never risks ruin, but in general it entails a considerable risk of losing a substantial portion of wealth.

In terms of security an optimal policy is often to keep virtually all wealth in the riskless asset (Eithier 1987). We will let $p_0 = (1, 0, \ldots, 0)$ be the extreme security strategy, where all assets are held in *cash*.

Our purpose is to monitor the growth and security measures as the investment decisions vary with the goal of balancing the measures to obtain a strategy which performs well on both the growth and security dimensions. For a particular (growth, security) combination, the complete set of values is given by the *graph* (see Figure 1)

$$U_i = \{(G_i(p), S_i(p)) \mid p \text{ is a feasible investment strategy}\}.$$

TABLE 1—PART 1
Main Properties for the Kelly Criterion

Good/Bad	Property	Reference
G	Maximizing $E \log X$ asymptotically maximizes the rate of asset growth.	Breiman (1961), Algoet and Cover (1988)
G	The expected time to reach a preassigned goal is asymptotically as X increases least with a strategy maximizing $E \log X_N$.	Breiman (1961), Algoet and Cover (1988)
G	Maximizing median $\log X$.	Ethier (1987)
B	False Property: If maximizing $E \log X_N$ almost certainly leads to a better outcome then the expected utility of its outcome exceeds that of any other rule provided n is sufficiently large. Counterexample: $u(x) = X$, $\frac{1}{2} < p < 1$, Bernoulli trials, $\hat{f} = 1$ maximizes $EU(x)$ but $f^* = 2p - 1 < 1$ maximizes $E \log X_N$.	Thorp (1975)
G	The $E \log X$ bettor never risks ruin.	Hakansson and Miller (1975)
B	If the $E \log X$ bettor wins then loses or loses then wins he is behind. The order of win and loss is immaterial for one, two, . . . sets of trials.	$(1 + \gamma)(1 - \gamma)X_0 = (1 - \gamma^2)X_0 < 0$
G	The absolute amount bet is monotone in wealth.	$(\partial E \log X)/\partial W_0 > 0$
B	The bets are extremely large when the wager is favorable and the risk is very low.	Roughly the optimal wager is proportional to the edge divided by the odds. Hence for low risk situations and corresponding low odds the wager can be extremely large. For one such example, see Ziemba and Hausch (1985, pp. 159–160). There in a \$3 million race the optimal fractional wager on a 3–5 shot was 64%.

The most obvious points in U_i to identify are *efficient* points given by $(G_i(p_\alpha), S_i(p_\alpha)) \in U_i$ with

$$G_i(p_\alpha) = \max \{G_i(p) \mid S_i(p) \geq \alpha, p \text{ feasible}\}.$$

Alternatively the efficient points are $(G_i(p_\beta), S_i(p_\beta))$ with

$$S_i(p_\beta) = \max \{S_i(p) \mid G_i(p) \geq \beta, p \text{ feasible}\}.$$

Given the required level of security (or growth) the efficient points are undominated. The set of efficient points constitutes the *efficient frontier*.

Two key efficient points are the unconstrained optimal growth and optimal security points, respectively $(G_i(\bar{p}), S_i(\bar{p}))$ with $G_i(\bar{p}) = \max \{G_i(p) \mid p \text{ feasible}\}$ and $(G_i(p_0), S_i(p_0))$ with $S_i(p_0) = \max \{S_i(p) \mid p \text{ feasible}\}$. The optimal growth strategy is the Kelly strategy $\bar{p}$ and the optimal security strategy is the cash strategy p_0. These are both fixed fraction strategies. Other strategies yielding points on the efficient frontier would not typically be fixed fraction, but rather would be dynamic strategies, depending on current wealth and/or time. Gottlieb (1985) has derived such a strategy for the criteria of maximizing mean first passage time to a goal (G_3), subject to a high probability of reaching the goal (S_3).

TABLE 1—PART 2

Main Properties for the Kelly Criterion

Good/Bad	Property	Reference
B	One overbets when the problem data is uncertain.	Betting more than the optimal Kelly wager is dominated in a growth-security sense. Hence if the problem data provides probabilities, edges and odds that may be in error, then the suggested wager will be too large. This property is discussed in and largely motivates this paper.
B	The total amount wagered swamps the winnings—that is, there is much "churning."	Ethier and Tavaré (1983) and Griffin (1985) show that the Expected Gain/E Bet is arbitrarily small and converges to zero in a Bernoulli game where one wins the expected fraction p of games.
B	The unweighted average rate of return converges to half the arithmetic rate of return.	Related to property 5 this indicates that you do not seem to win as much as you expect; see Ethier and Tavare (1983) and Griffin (1985).
G	The $E \log X$ bettor is never behind any other bettor on average in 1, 2, . . . trials.	Finkelstein and Whitley (1981)
G	The $E \log X$ bettor has an optimal myopic policy. He does not have to consider prior to subsequent investment opportunities.	This is a crucially important result for practical use. Hakansson (1971b) proved that the myopic policy obtains for dependent investments with the log utility function. For independent investments and power utility a myopic policy is optimal.
G	The chance that an $E \log X$ wagerer will be ahead of any other wagerer after the first play is at least 50%.	Bell and Cover (1980)
G	Simulation studies show that the $E \log X$ bettor's fortune pulls way ahead of other strategies' wealth for reasonably-sized sequences of investments.	Ziemba and Hausch (1985)

Rather than trying to move from optimal growth to optimal security along the efficient frontier, we will consider the path generated by convex combinations of the optimal growth and optimal security strategies, namely

$$p(\lambda) = \lambda \bar{p} + (1 - \lambda) p_0, \qquad 0 \leq \lambda \leq 1.$$

Clearly $p(\lambda)$ is a fixed fraction strategy and we will call it *fractional Kelly* since it blends part of the Kelly with the cash strategy. The fractional Kelly strategy is n-dimensional, specifying investments in all opportunities, and the key parameter is λ, which we can consider as a *tradeoff* index.

The path traced by the fractional Kelly strategies is

$$U_i^0 = \{(G_i(p(\lambda)), S_i(p(\lambda))) | 0 \leq \lambda \leq 1\}.$$

The fractional Kelly path is illustrated in Figure 1. This path has attractive properties. First, it is easily *computable*, as shown in the next section. When λ varies from 0 to 1 we continuously trade growth for security. Although this tradeoff is not always efficient, it is *effective* in that we see a *monotone increase in security as we decrease growth*. However, the rate of change in growth or security with respect to λ is not constant and thus the Kelly path is curvilinear. In fact, MacLean and Ziemba (1990) established that the Kelly path is above the straight line (secant) joining the optimal growth and optimal security points. These properties are discussed in §3.

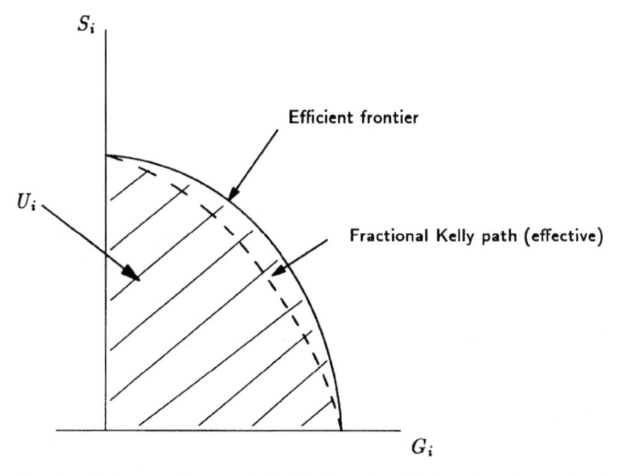

FIGURE 1. Graph of (Growth, Security) Profile Showing Efficient Frontier and Effective Path.

2. Computation of Measures

If we select a strategy which specifies our investment in the various opportunities and we stick with that strategy, then each of the measures described in §2 is easily computable. We will now develop the computational formulae. Since most of our discussion focuses on fractional Kelly strategies, we work with them. However, more general fixed strategies could be used.

Consider the fractional Kelly strategy $p(\lambda)$ and use the transformation

$$Z_t(p(\lambda), \omega^t) = \ln Y_t(p(\lambda), \omega^t).$$

From (1),

$$Z_t(p(\lambda), \omega^t) = z_0 + \sum_{s=1}^{t} \ln\left(\sum_{i=0}^{n} R_i(\omega_s)p_i(\lambda)\right) = Z_{t-1}(p(\lambda), \omega^{t-1}) + J(p(\lambda), \omega_t),$$

where $J(p(\lambda), \omega) = \ln(\sum_{i=0}^{n} R_i(\omega)p_i(\lambda))$ is a stationary jump process and $z_0 = \ln y_0$. Hence, Z_t, $t = 1, \ldots,$ is a random walk and we can evaluate the various measures of interest using this process. An alternative approach to evaluating these measures is presented in Ethier (1987) where the discrete time (and maybe discrete state) process $Y_t(p(\lambda))$ is approximated by geometric Brownian motion.

The mean wealth accumulation measure G1 is

$$\mu_t(y_0, p(\lambda)) = EY_t(p(\lambda)) = y_0 E\prod_{s=1}^{t}\left(\sum_{i=0}^{n} R_i(\omega_s)p_i(\lambda)\right) = y_0\left(\sum_{i=0}^{n} \bar{R}_i p_i(\lambda)\right)^t, \quad (2)$$

where $\bar{R}_i = ER_i(w)$, $i = 0, \ldots, n$.

For the mean growth rate measure G2

$$\phi_t(y_0, p(\lambda)) = E\ln(Y_t(p(\lambda))^{1/t}) = 1/tEZ_t(p(\lambda))$$

$$= E\ln\left(\sum_{i=0}^{n} R_i(\omega)p_i(\lambda)\right) + \frac{1}{t}z_0 = EJ(p(\lambda), \omega) + \frac{1}{t}z_0, \quad (3)$$

which is an easy calculation to make.

For measure S1 assume the process

$$Z_t(p(\lambda), \omega^t) = z_0 + \sum_{s=1}^{t} J(p(\lambda), \omega_s)$$

has an approximate normal distribution. Using this approximation yields

$$\gamma_t(y_0, p(\lambda)) = 1 - F\left(\frac{b_t^* - E(Z_t(p(\lambda)))}{\sigma(Z_t(p(\lambda)))}\right), \tag{4}$$

where F is the cumulative for the standard normal, $b_t^* = \ln b_t$, and $E(Z_t(p(\lambda)))$, $\sigma(Z_t(p(\lambda)))$ are the mean and standard deviation respectively.

Measure S2 is given by

$$\alpha(y_0, p(\lambda)) = \Pr[Z_t(p(\lambda)) \geq b_t^*, t = 1, \cdots] \tag{5}$$

$$= \prod_{t=1}^{\infty} \Pr[Z_t(p(\lambda)) \geq b_t^* | Z_s(p(\lambda)) \geq b_s^*, s = 1, \ldots, t-1]$$

$$= \prod_{t=1}^{\infty} \alpha_t(z_0, p(\lambda)). \tag{6}$$

Then with $h_t(z_t | Z_s \geq b_s^*, s = 1, \ldots, t-1)$ the conditional density

$$\alpha_t = \int_{z_t \geq b_t^*} h_t(z_t | Z_s \geq b_s^*, s = 1, \ldots, t-1) dz_t$$

$$= \int_{z_t \geq b_t^*} \left\{ \int_{z_t - x \geq b_{t-1}^*} h_{t-1}(z_t - x | Z_s \geq b_s^*, s = 1, \ldots, t-1) \pi(x) dx \right\} dz_t,$$

where $\pi(x)$ is the density for $J(p(\lambda))$,

$$= \int_{z_t \geq b_t^*} \left\{ \int_{x \leq z_t - b_{t-1}^*} \frac{h_{t-1}(z_t - x | Z_s \geq b_s^*, s = 1, \ldots, t-2)}{\alpha_{t-2}} \pi(x) dx \right\} dz_t. \tag{7}$$

Equation (7) provides a sequential procedure to compute measure S2, requiring only the distribution from the previous stage and the jump probabilities.

For the other measures assume that the random variable $J(p(\lambda))$ has finite support given by the interval $[J_m(p(\lambda)), J_M(p(\lambda))]$. Consider measure S3

$$\beta(y_0, p(\lambda)) = \Pr[\tau_{\{Y(p(\lambda)) \geq U\}} < \tau_{\{Y(p(\lambda)) \leq L\}} | y_0].$$

Transforming to an additive process with $z = \ln y$ and letting

$$\beta(z, p(\lambda)) = \Pr[\tau_{\{Z(p(\lambda)) \geq \ln U\}} < \tau_{\{Z(p(\lambda)) \leq \ln L\}} | z]$$

yields

$$\beta(z_0, p(\lambda)) = E\{\beta(z_0 + J(p(\lambda)), p(\lambda))\}. \tag{8}$$

The expectation in (8) is over the random variable $J(p(\lambda))$. We must solve (8) for β subject to the boundary conditions

$$\beta(z, p(\lambda)) = 0 \quad \text{if} \quad z \leq \ln L \tag{8a}$$

$$\beta(z, p(\lambda)) = 1 \quad \text{if} \quad z \geq \ln U. \tag{8b}$$

A solution to (8) is of the form $\beta(z, p(\lambda)) = \sum_{k=1}^{s} A_k \theta_k^z$, where A_k are chosen to satisfy (8a, 8b) and the roots $\theta_1, \ldots, \theta_s$ satisfy the equation $E\theta^{J(p(\lambda))} = 1$. It is easy to show that there are two positive roots $\theta_1 = 1$ and $\theta_2 = \theta$, where $\theta_2 < 1$ or $\theta_2 > 1$ depending on whether $EJ(p(\lambda)) > 0$ or $EJ(p(\lambda)) < 0$ respectively. Similar to the methods discussed in Feller (1962, pp. 330–334), we take the minimum $J_m(p(\lambda)) < 0$ and maximum

TABLE 2

Computational Formulas for Growth Measures G1–G3 *and Security Measures* S1–S3

G1	$\mu_t(Y_0, p) = Y_0\left(\sum_{i=0}^{n} \bar{R}_i p_i\right)^t$
G2	$\phi(p) = E \ln\left(\sum_{i=0}^{n} R_i(\omega) p_i\right)$
G3	$\bar{\eta}(Z_0, p) = \frac{\theta^{Z_0}}{2EJ(p)}\left(\frac{U_*(p)}{\theta^{U_*(p)}} + \frac{U^*(p)}{\theta^{U^*(p)}}\right) - \frac{Z_0}{EJ(p)}$
S1	$\bar{\gamma}(Z_0, p) = 1 - F\left(\frac{b_t^* - Z_0 - tEJ(p)}{\sqrt{t}\sigma(J(p))}\right)$
S2	$\alpha(Z_0, p) = \prod_{t=1}^{\infty} \alpha_t(Z_0, p),$
	$\alpha_t(Z_0, p) = \sum_{Z_t \geq b_t^*} \left[\sum_{J_i \geq Z_t - b_{t-1}^*} \frac{G_{t-1}(Z_t - J_i(p))}{\alpha_{t-1}(Z_0, p)} \cdot \pi_i\right]$
S3	$\bar{\beta}(Z_0, p) = \frac{(\theta^{Z_0} - \theta^{U^*})(\theta^{U^*} - \theta^{L^*(p)}) + (\theta^{Z_0} - \theta^{L^*(p)})(\theta^{U^*(p)} - \theta^{L^*})}{2(\theta^{U^*(p)} - \theta^{L^*})(\theta^{U^*} - \theta^{L^*(p)})}$

$J_M(p(\lambda)) > 0$ jumps and find bounds on the solution by solving a simple system with the extreme boundary conditions satisfied as equalities. For the upper bound,

$$\beta(\ln L + J_m(p(\lambda)), p(\lambda)) = 0 \quad \text{and} \quad \beta(\ln U, p(\lambda)) = 1.$$

For the lower bound,

$$\beta(\ln L, p(\lambda)) = 0 \quad \text{and} \quad \beta(\ln U + J_M(p(\lambda)), p(\lambda)) = 1.$$

Let $L^* = \ln L$, $L^*(p(\lambda)) = \ln L + J_m(p(\lambda))$, $U^* = \ln U$, and $U^*(p(\lambda)) = \ln U + J_M(p(\lambda))$.

The bounds for measure S3 are then

$$\frac{\theta^{z_0} - \theta^{L^*}}{\theta^{U^*(p(\lambda))} - \theta^{L^*}} \leq \beta(y_0, p(\lambda)) \leq \frac{\theta^{z_0} - \theta^{L^*(p(\lambda))}}{\theta^{U^*} - \theta^{L^*(p(\lambda))}}. \tag{9}$$

For measure G3,

$$\eta(y_0, p(\lambda)) = E\tau_{\{Y(p(\lambda)) \geq U | y_0\}} = E\tau_{\{Z(p(\lambda)) \geq \ln U | z_0\}},$$

we use the recursive equation

$$\eta(z_0, p(\lambda)) = E\{\eta(z_0 + J(p(\lambda)), p(\lambda))\} + 1 \tag{10}$$

subject to the boundary condition

$$\eta(z, p(\lambda)) = 0 \quad \text{if} \quad z \geq \ln U. \tag{10a}$$

One solution to (10) is $\eta^*(z, p(\lambda)) = -z/(EJ(p(\lambda)))$. Any other solution can be written as $\eta(z, p(\lambda)) = \eta^*(z, p(\lambda)) + \Delta(z, p(\lambda))$ where $\Delta(z, p(\lambda))$ satisfies the system

$$\Delta(z, p(\lambda)) = E\{\Delta(z + J(p(\lambda)), p(\lambda))\}. \tag{10b}$$

Solutions of (10b) have the form $\Delta(z, p(\lambda)) = \sum A_k \theta_k^z$. Solutions of (10) are then

$$\eta(z, p) = \sum A_k \theta_k^z - z/(EJ(p(\lambda)))$$

where the A_k are chosen to satisfy the boundary conditions (10a). With $U_*(p(\lambda))$

$= \ln U - J_m(p(\lambda))$ we can solve the extreme boundary conditions as equalities, namely $\eta(U^*(p(\lambda)), p(\lambda)) = 0$ and $\eta(U_*(p(\lambda)), p(\lambda)) = 0$ yields the following bounds on G3:

$$\left[\frac{U_*(p(\lambda))}{EJ(p(\lambda))}\right]\left(\frac{\theta^{z_0}}{\theta^{U_*(p(\lambda))}}\right) - \frac{z_0}{EJ(p(\lambda))} \le \eta(y_0, p(\lambda))$$

$$\le \left[\frac{U^*(p(\lambda))}{EJ(p(\lambda))}\right]\left(\frac{\theta^{z_0}}{\theta^{U^*(p(\lambda))}}\right) - \frac{z_0}{EJ(p(\lambda))}. \quad (11)$$

The bounds in (9) and (11) are quite tight when the jumps are small relative to the initial wealth z_0.

Table 2 summarizes the formulas we can use to compute the six growth and security measures. For G3 and S3 we have bounds and the formulas given provide midpoint estimates of these bounds. A similar table appears in Ethier (1987) for the geometric Brownian motion model.

3. Effective Growth-security Tradeoff

The fact that the various growth and security measures can be easily evaluated for fractional Kelly strategies provides an opportunity to interactively investigate the performance of strategies for various measures and scenarios. The ability to evaluate various performance measures and determine a preferred strategy will depend on an *effective tradeoff* between complementary measures of growth and security. A tradeoff is effective if the loss of performance in one dimension (say growth) is compensated by a gain in the other dimension (security). In this section we establish that the tradeoff using fractional Kelly strategies is effective for each of the profile combinations. In calculating the rates of change for the various measures we will see that the tradeoff between growth and security is not constant for all λ, $0 \le \lambda \le 1$.

(i) *Rate Profile*: (ϕ, α).

LEMMA 3.1. *Suppose $p(\lambda)$ is a fractional Kelly strategy and $\bar{p}$ is the Kelly strategy. Then*

$$\frac{\partial \phi(y_0, p(\lambda))}{\partial \lambda} > 0, \qquad 0 \le \lambda \le 1.$$

PROOF. Consider the optimal growth strategy given by $\bar{p} = (\bar{p}_0, \ldots, \bar{p}_n)$, and the optimal security strategy $p_0 = (1, 0, \ldots, 0)$.

Then

$$p(\lambda) = (\lambda\bar{p}_0 + (1-\lambda), \ldots, \lambda\bar{p}_n) = (p_0(\lambda), \ldots, p_n(\lambda)) \qquad \text{and}$$

$$\frac{d}{d\lambda} p_0(\lambda) < 0, \qquad \frac{d}{d\lambda} p_i(\lambda) > 0, \qquad i = 1, \ldots, n.$$

We have the optimal growth problem

$$\max \phi(p(\lambda)) = \max E \ln\left(\sum_{i=0}^{n} R_{it}(\omega) p_i(\lambda)\right).$$

Then

$$\frac{\partial}{\partial p_0} \phi(p(\lambda)) = E\left(\frac{1 - \sum_{i=1}^n R_{it}(\omega)}{\sum_{i=0}^n R_{it}(\omega)p_i(\lambda)}\right) \qquad \text{and}$$

$$\frac{\partial^2}{\partial p_0^2} \phi(p(\lambda)) = -E\left[\frac{1 - \sum_{i=1}^n R_{it}(\omega)}{\sum_{i=0}^n R_{it}(\omega)p_i(\lambda)}\right]^2 < 0.$$

So $\phi(p(\lambda))$ is concave in p_0 and with $\bar{p}_0$ such that $\partial\phi(p(\lambda))/\partial p_0 = 0$, we have

$$\frac{\partial}{\partial p_0}\phi(p(\lambda)) < 0, \qquad p_0 < \bar{p}_0.$$

Therefore

$$\frac{d}{d\lambda}\phi(p(\lambda)) = \frac{dp_0}{d\lambda}\left(\frac{\partial}{\partial p_0}\phi(p(\lambda))\right) > 0 \qquad \text{and } \phi(p(\lambda)) \text{ is increasing in } \lambda. \qquad \square$$

LEMMA 3.2. *Let* $p(\lambda)$ *be a fractional Kelly strategy and* $\ln a < 0$ *be a given fallback rate. Then*

$$\frac{\partial}{\partial\lambda}\alpha(a, p(\lambda)) \le 0.$$

PROOF. We have for the security measure

$$\begin{aligned}
\alpha(a, p(\lambda)) &= \text{Prob}\,[Y_t(p(\lambda)) \ge y_0 a^t, t = 1, \cdots] \\
&= \prod_{t \ge 1} \text{Prob}\,[Y_t(p(\lambda)) \ge a^t \,|\, Y_s(p(\lambda)) \ge a^s, s = 1, \ldots, t-1] \\
&= \prod_{t \ge 1} \alpha_t(a, p(\lambda)).
\end{aligned}$$

Along any path $1, y_1, \ldots, y_{t-1}$ with $y_s \ge a^s$, $s = 1, \ldots, t-1$.

$$\alpha_t(a, p(\lambda)) = \text{Prob}\left[\ln\left(\sum R_{jt}(\omega)p_j(\lambda)\right) > \ln a - \frac{1}{t-1}\sum_{s=1}^{t-1} z_s\right],$$

where $z_s = \ln y_s$ and $\ln a - (t-1)^{-1}\sum_{s=1}^{t-1} z_s < 0$. With $b = \ln a - (t-1)^{-1}\sum z_s$ and $R_t p(\lambda) = \sum R_{jt}(\omega)p_j(\lambda)$ we have $\alpha_t(\lambda) \equiv$

$$\begin{aligned}
\alpha_t(a, p(\lambda)) &= \text{Prob}\,[\ln(R_t p(\lambda)) \ge b] \\
&= \text{Prob}\,[\ln(\lambda R_t \bar{p} + (1-\lambda)R_t p_0) \ge b] \\
&= \text{Prob}\,[\lambda \ln(R_t \bar{p}) + (1-\lambda)\ln(R_t p_0) + \Delta \ge b] \\
&= \text{Prob}\left[\ln(R_t \bar{p}) \ge \frac{b - \Delta}{\lambda}\right] = 1 - F_t(x(\lambda)),
\end{aligned}$$

where $x(\lambda) = (b - \Delta)/\lambda$ and F_t is the distribution function for the optimal growth process $\ln(R_t p_0)$.

Then $dx(\lambda)/d\lambda > 0$ and F nondecreasing imply $d\alpha_t(\lambda)/d\lambda \le 0$. Hence $d\alpha(a, p(\lambda))/d\lambda \le 0$. $\square$

Thus along the fractional Kelly path growth is increasing and security is decreasing and an *effective* tradeoff is possible.

(ii) *Wealth Profile*: (μ_t, γ_t).

LEMMA 3.3. *If* $p(\lambda)$ *is a fractional Kelly strategy then*

$$\frac{d}{d\lambda}\mu_t(p(\lambda)) > 0, \qquad 0 \le \lambda \le 1.$$

PROOF. We have

$$\frac{\partial}{\partial p_0}\mu_t(p(\lambda)) = E\frac{\partial}{\partial p_0}Y_t(p(\lambda)) = E\frac{\partial}{\partial p_0}\left[y_0\prod_{s=1}^{t}\left(\sum_{j=1}^{n}R_j(\omega_s)p_j(\lambda)\right)\right]$$

$$= E\left[y_0\sum_{r=1}^{t}\frac{\partial}{\partial p_0}\left(\sum_{j=0}^{n}R_j(\omega_r)p_j(\lambda)\right)\prod_{s\neq r}\left(\sum_{j=0}^{n}R_j(\omega_s)p_j(\lambda)\right)\right]$$

$$= -ty_0E[K(\omega_t)Y_{t-1}(\omega^{t-1})] = -ty_0(EK)(EY_{t-1}) < 0.$$

Then

$$\frac{d}{d\lambda}\mu_t(p(\lambda)) = \frac{dp_0(\lambda)}{d\lambda}\frac{\partial}{\partial p_0}\mu_t(p(\lambda)) > 0. \qquad \square$$

LEMMA 3.4. *Suppose $p(\lambda)$ is a fractional Kelly strategy and consider the growth rate $\alpha = 1$. Then*

$$\frac{d}{d\lambda}\gamma_t(y_0, p(\lambda)) < 0 \qquad \text{for} \qquad 0 \leq \lambda < 1.$$

PROOF. We have

$$\gamma_t(y_0, p(\lambda)) = \text{Prob}\,[Y_t(p(\lambda)) \geq y_0a^t] = \text{Prob}\left[\sum_{s=1}^{t}\ln R_sp(\lambda) \geq t\ln a\right]$$

$$= \text{Prob}\left[\sum \ln(\lambda R_s\bar{p} + (1-\lambda)R_sp_0) \geq 0\right]$$

$$= \text{Prob}\left[\lambda\sum\ln(R_s\bar{p}) + \sum\Delta_s \geq 0\right] \qquad \text{where} \qquad \Delta_s > 0$$

$$= \text{Prob}\left[\sum\ln(R_s\bar{p}) \geq \frac{-\sum\Delta_s}{\lambda}\right]$$

$$= 1 - F^t(x_t(\lambda)),$$

where $x_t(\lambda) = (-\sum\Delta_s)/\lambda$ and F^t is the distribution function for $\sum_{s=1}^{t}\ln(R_s\bar{p})$. Then $(d/d\lambda)x(\lambda) > 0$ and F^t nondecreasing gives the desired result. $\square$

Hence for the wealth profile an *effective* tradeoff is possible along the fractional Kelly path.

(iii) *Stopping Rule Profile*: (η^{-1}, β). Our analysis of the stopping rule profile focuses on the computational formulas $\bar{\eta}(p(\lambda))$ and $\bar{\beta}(p(\lambda))$. Since we have upper and lower bounds for the true values, the accuracy of the approximation is easy to determine in a particular application. Our choice of strategy is based on the profile graph and that requires $\bar{\eta}$ and $\bar{\beta}$.

The formulas for $\bar{\eta}$ and $\bar{\beta}$ are based on the random walk $Z_t(p(\lambda))$ with jump process $J(p(\lambda))$. The process $J(p(\lambda))$ is *flexible* at the fractional Kelly strategy $p(\lambda)$ if the condition

$$\frac{L'^{*}(p(\lambda))\theta^{L^{*}(p(\lambda))}}{L^{*}(p(\lambda))\theta^{L^{*}(p(\lambda))}} < \frac{EJ'(p(\lambda))\theta^{J(p(\lambda))}}{EJ(p(\lambda))\theta^{J(p(\lambda))}} < \frac{U'^{*}(p(\lambda))\theta^{U^{*}(p(\lambda))}}{U^{*}(p(\lambda))\theta^{U^{*}(p(\lambda))}}$$

holds, where $\theta < 1$ is a positive root of the equation $E\theta(p(\lambda))^{J(p(\lambda))} = 1$, and prime denotes the derivative w.r.t. λ.

LEMMA 3.5. *Suppose $p(\lambda)$ is a fractional Kelly strategy and consider the wealth process $Y_t(p(\lambda))$, $t = 1, \cdots$ and the absorbing states $U = ky_0$, $L = k^{-1}y_0$, $k > 1$. If $J(p(\lambda))$ is flexible at $p(\lambda)$ then for $0 \leq \lambda \leq 1$*

$$\text{(i)} \qquad \frac{\partial}{\partial\lambda}\,\bar{\eta}(y_0, p(\lambda)) < 0,$$

$$\text{(ii)} \qquad \frac{\partial}{\partial\lambda}\,\bar{\beta}(y_0, p(\lambda)) < 0.$$

PROOF. (i) For the mean first passage time consider the lower bound

$$\eta^2(p(\lambda)) = \frac{U_*(p(\lambda))}{EJ(p(\lambda))\theta^{U_*(p(\lambda))}},$$

where without loss of generality assume $y_0 = 1$ ($z_0 = 0$). Then

$$\frac{\partial}{\partial p_0}\,\eta^2(p(\lambda)) = \frac{N^2(p(\lambda))}{D^2(p(\lambda))},$$

where $\quad D^2(p(\lambda)) = [EJ(p(\lambda))\theta^{U_*(p(\lambda))}]^2 \quad$ and

$$\begin{aligned} N^2(p(\lambda)) = [EJ(p(\lambda))\theta^{U_*(p(\lambda))}J'_m(p(\lambda)) &- U_*(p(\lambda))EJ'(p(\lambda))\theta^{U_*(p(\lambda))} \\ &- U_*(p(\lambda))EJ(p(\lambda))\theta^{U_*(p(\lambda))}J'_m(p(\lambda))\ln\theta \\ &- U_*(p(\lambda))EJ(p(\lambda))\theta^{U_*(p(\lambda))}U_*(p(\lambda))\theta'/\theta], \end{aligned}$$

where the prime denotes the first derivative w.r.t. p_0. Since $EJ'(p(\lambda)) < 0$ and $EJ(p(\lambda)) > 0$, $J'_m(p(\lambda)) > 0$, $\theta' < 0$, we get all terms in $N^2(p(\lambda))$ positive and $(\partial/\partial p_0)\eta^2(p(\lambda)) > 0$. In the same way for the upper bound $\eta^1(p(\lambda))$ we get $(\partial/\partial p_0)\eta^1(p(\lambda)) > 0$. With $(d/d\lambda)p_0(\lambda) < 0$ the average of the bounds yields (1).

(ii) For the first passage probability we utilize the upper bound

$$\beta^1(p(\lambda)) = \frac{1 - \theta^{L^*(p(\lambda))}}{\theta^{U^*} - \theta^{L^*(p(\lambda))}}.$$

Then the numerator of $(\partial/\partial p_0)\beta^1(p(\lambda))$ is

$$N(p(\lambda)) = (\theta^{U^*} - 1)\frac{\partial}{\partial p_0}\theta^{L^*(p(\lambda))} - (1 - \theta^{L^*(p(\lambda))})\frac{\partial}{\partial p_0}\theta^{U^*}.$$

We have $E\theta(p(\lambda))^{J(p(\lambda))} = 1$ and $(\partial/\partial p_0)E\theta(p(\lambda))^{J(p(\lambda))} = 0$ from which

$$\begin{aligned} \frac{\partial}{\partial p_0}\theta(p(\lambda)) &= -\theta(p(\lambda))\ln\theta(p(\lambda))[EJ'(p(\lambda))\theta^{J(p(\lambda))}/EJ(p(\lambda))\theta^{J(p(\lambda))}] \\ &= (-\theta(p(\lambda))\ln\theta(p(\lambda)))\psi(p(\lambda)). \end{aligned}$$

Then

$$\begin{aligned} \frac{\partial}{\partial p_0}\theta^{L^*(p(\lambda))} &= \theta^{L^*(p(\lambda))}\left(\frac{L^*(p(\lambda))}{\theta(p(\lambda))}\frac{\partial}{\partial p_0}\theta + \ln\theta\,\frac{\partial}{\partial p_0}L^*(p(\lambda))\right) \\ &= -L^*(p(\lambda))\theta^{L^*(p(\lambda))}\psi(p(\lambda))\ln\theta + L'^*(p(\lambda))\theta^{L^*(p(\lambda))}\ln\theta, \end{aligned}$$

and

$$\frac{\partial}{\partial p_0}\theta^{U^*} = U^*\theta^{U^*-1}\frac{\partial}{\partial p_0}\theta = -U^*\theta^{U^*}\psi(p(\lambda))\ln\theta.$$

Substituting into $N(p(\lambda))$ with a little algebra yields

$$N(p(\lambda)) = -(\theta^{U^*} - 1)L^*(p(\lambda))\theta^{L^*(p(\lambda))}\psi(p(\lambda)) \ln \theta + (1 - \theta^{L^*(p(\lambda))})U^*\theta^{U^*}\psi(p(\lambda)) \ln \theta - (\theta^{U^*} - 1)L'^*(p(\lambda))\theta^{L^*(p(\lambda))} \ln \theta.$$

With $U = ky_0$ and $L = k^{-1}y_0$ it is easy to show that

$$\frac{U^*\theta^{U^*}}{1-\theta^{U^*}} < \frac{L(p(\lambda))\theta^{L^*(p(\lambda))}}{1-\theta^{L^*(p(\lambda))}}$$

and therefore

$$A = (1 - \theta^{L^*(p(\lambda))})U^*\theta^{u^*} > -(\theta^{U^*} - 1)L^*(p(\lambda))\theta^{L^*(p(\lambda))} = B.$$

We have, then, substituting B for A

$$N(p(\lambda)) > 2(1 - \theta^{U^*}) \ln \theta(L^*(p(\lambda))\theta^{L^*(p(\lambda))}\psi(p(\lambda)) + L'^*(p(\lambda))\theta^{L^*(p(\lambda))}) > 0$$

since J flexible at $p(\lambda)$ implies

$$L^*(p(\lambda))\theta^{L^*(p(\lambda))}\psi(p(\lambda)) + L'^*(p(\lambda))\theta^{L^*(p(\lambda))} < 0.$$

So

$$\frac{\partial}{\partial p_0}\beta^1 > 0.$$

With the same result for the lower bound and with $(d/d\lambda)p(\lambda) < 0$ we get (ii). □

The proof of the above lemma depends on the flexibility condition but not the requirement that the strategy be fractional Kelly. The significance of the fractional Kelly strategies is that the flexibility condition is always satisfied if it is satisfied for the Kelly strategy.

For the stopping rule profile $(\bar{\eta}^{-1}, \bar{\beta})$ we have along the fractional Kelly path growth increasing and security decreasing. The inverse of expected stopping time is a measure of growth since faster growth would imply getting to the target U sooner.

In summary, for each of the profile types we have the same type of behavior. The complementary measures, growth and security, move in opposite directions as we change the blend between the Kelly and cash strategies. This enables a continuous tradeoff between growth and security.

TRADEOFF THEOREM. *Suppose $p(\lambda)$ is a fractional Kelly strategy. Then for each profile there is an effective tradeoff between growth and security by varying λ, the tradeoff index.*

Additional theoretical results concerning these tradeoffs are presented in MacLean and Ziemba (1990).

Friedman (1982) seems to have first considered fractional Kelly strategies in the context of blackjack. His results are along the lines discussed in §4.1 below. Other references on fractional Kelly strategies are Gottlieb (1984); Ethier (1987); Li, MacLean and Ziemba (1990); and Wu and Ziemba (1990).

The tradeoff theorem in this section establishes the use of growth and security measures in determining an investment strategy. There are many possible tradeoffs depending upon the choice of measures and the parameters specified in these measures. Rather than further considering the preferred measures and tradeoff in theory, we will apply the concepts to some real world examples and explore the choice of investment decisions in practice.

4. Applications

The methods described in previous sections are now applied to a variety of well-known problems in gambling and investment. Rather than concentrating on the results in the tradeoff theorem, we will explore a variety of growth and security measures to see how interactively a satisfactory decision is achieved from both perspectives. In each case we emphasize a pair of measures (growth, security). There are a few points we should make about our approach to these applications. Satisfaction with a particular investment strategy will depend on many factors. The measures of performance we have described are helpful, but they are functions of *inputs* such as initial wealth, return distribution, and performance standards. We must vary those inputs in appropriate ways to see a decision rule under different scenarios. In cases where inputs are soft we may need many diverse scenarios.

Throughout the theory and examples we use fixed strategies. This is not to imply that the particular fixed strategy is used forever, but rather a statement that if all inputs remain constant we can predict the results from the fixed strategy. Of course, the inputs are dynamic and we would regularly (continuously) update the strategy and measures of performance based on new input data. This process is facilitated by easily computable measures.

4.1. *Blackjack*: (ϕ, β)

The game of blackjack seems to have evolved from several related card games in the 19th century. It became popular in World War I and has since reached enormous popularity. It is played by millions of people in casinos around the world. Billions of dollars are lost each year by people playing the game. A relatively small number of professionals and advanced amateurs, using various methods such as card counting, are able to beat the game. The object is to reach, or be close to, twenty-one with two or more cards. Scores above twenty-one are said to bust or lose. Cards two to ten are worth their face value, Jacks, Queens, and Kings are worth ten points and Aces are worth one or eleven at the player's choice. The game is called blackjack because an Ace and a ten-valued card was paid three for two and an additional bonus accrued if the two cards were the Ace of Spades and the Jack of Spades or Clubs. While this extra bonus has been dropped by current casinos, the name has stuck. Dealers normally play a fixed strategy of hit until

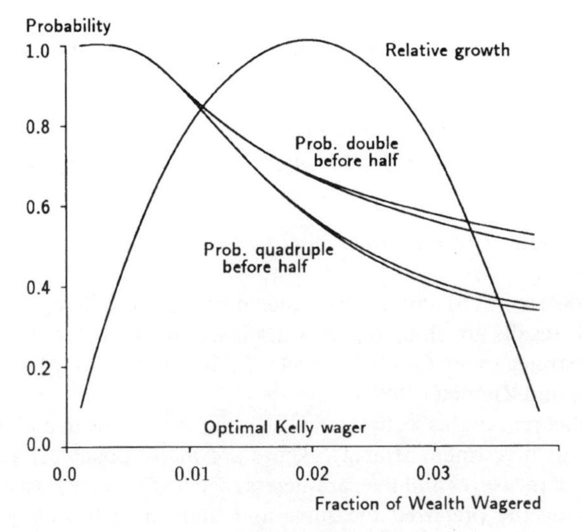

FIGURE 2. Probability of Doubling and Quadrupling before Halving and Relative Growth Rates Versus Fraction of Wealth Wagered for Blackjack (2% Advantage, $p = 0.51$ and $q = 0.49$).

TABLE 3

Growth Rates Versus Probability of Doubling before Halving for Blackjack

	λ		$\bar{\beta}(0, p)$ = P[Doubling before Halving]		$\phi(p)$ = Relative Growth Rate
Range for Blackjack Teams	0.1		0.999		0.19
	0.2		0.998		0.36
	0.3	↑	0.98	↑	0.51
	0.4	SAFER	0.94	LESS GROWTH	0.64
	0.5		0.89		0.75
	0.6	RISKIER	0.83	MORE GROWTH	0.84
	0.7	↓	0.78	↓	0.91
	0.8		0.74		0.96
	0.9		0.70		0.99
	1.0	KELLY	0.67		1.00
	1.5		0.56		0.75
Overkill → Too Risky	2.0		0.50		0.00

a seventeen is reached and then stay. A variation is whether or not a soft seventeen (an ace with cards totalling six) is hit. It is slightly better for the player if the dealer stands on soft seventeen. The house has an edge of 2–10% against typical players. For example, the strategy mimicking the dealer loses about 8% because the player must hit first and busts about 28% of the time ($0.28^2 \cong 0.08$).

In general, the edge for a successful card counter varies from about −10% to +10% depending upon the favorability of the deck. By wagering more in favorable situations and less or nothing when the deck is unfavorable, an average edge weighted by the size of the bet of about 2% is reasonable. Hence, an approximation that will provide us insight into the long-run behavior of a player's fortune is to simply assume that the game is a Bernoulli trial with a probability of success $\pi = 0.51$ and probability of loss $1 - \pi = 0.49$.

We then have $\Omega = \{0, 1\}$, $K(0, p) = p$ with probability π, and $K(1, p) = -p$ with probability $1 - \pi$. The mean growth rate is

$$E \ln (1 + K(\omega, p)) = \pi \ln (1 + p) + (1 - \pi) \ln (1 - p).$$

Simple calculus gives the optimal fixed fraction strategy: $p^* = 2\pi - 1$ if $EK > 0$; $p^* = 0$ if $EK \leq 0$. (This optimal strategy may be interpreted as the edge divided by the odds (1–1 in this case).) In general, for win or lose for two outcome situations where the size of the wager does not influence the odds, the same intuitive formula holds. Hence with a 2% edge betting on a 10–1 shot, the optimal wager is 0.2% of one's fortune.) The growth rate of the investor's fortune is

$$\phi(p) = \pi \ln (1 + p) + (1 - \pi) \ln (1 - p)$$

and this is shown in Figure 2. It is nearly symmetrical around $p^* = 0.02$. Security measure S3 is also displayed in Figure 2, in terms of the probability of doubling or quadrupling before halving. The bounds, from (5), are fairly sharp. Since the growth rate and the security are both decreasing for $p > p^*$, it follows that it is never advisable to wager more than p^*. However, one may wish to trade off lower growth for more security using a fractional Kelly strategy. Table 3 illustrates the relationship between the fraction λ and growth and security. For example, a drop from $p = 0.02$ to 0.01 for a 0.5 fractional Kelly strategy, drops the growth rate by 25%, but increases the chance of doubling before halving from 67% to 89%.

In the blackjack example we see how the additional information provided by the security measure $\beta(y_0, p)$ was valuable in reaching a final investment decision. In a flexible or adaptive decision environment, competing criteria would be balanced to achieve a satisfactory path of accumulated wealth. With this approach professional blackjack teams typically use a fractional Kelly wagering strategy with the fraction $\lambda = 0.2$ to 0.8. See Gottlieb (1985) for discussion including the use of adaptive strategies.

4.2. *Horseracing*: (μ_t, γ_t)

Suppose we have n horses entered in a race. Only the first three finishers have positive return to the bettor. For the remaining positions, you lose your bet. Then

$$\Omega = \{(1, 2, 3), \ldots, (i, j, k), \ldots, (n-2, n-1, n)\}$$

is the set of all outcomes with probability π_{ijk}. We wager the fractions p_{i1}, p_{i2}, p_{i3} of our fortune Y_0 on horse i to win, place, or show, respectively. One collects on a win bet only when the horse is first, on a place bet when the horse is first or second, and on a show bet when the horse is first, second, or third. The order of finish does not matter for place and show bets. The bettors, wagering on a particular horse, share the net pool in proportion to the amount wagered, once the original amount of the bets are refunded and the winning horses share the resulting profits. Let p be the $n \times 3$ matrix of wager fractions, where $\sum_{i=1}^{n} \sum_{j=1}^{3} p_{ij} \leq 1$. The return function for a particular (i, j, k) outcome is

$$K((i,j,k), p) = (QW - w_i)\frac{p_{i1}}{w_i} + \frac{(QP - P_i - P_j)}{2}\left(\frac{p_{i2}}{P_i} + \frac{p_{j2}}{P_j}\right)$$
$$+ \left(\frac{QS - S_i - S_j - S_k}{3}\right)\left(\frac{p_{i3}}{S_i} + \frac{p_{j3}}{S_j} + \frac{p_{k3}}{S_k}\right)$$
$$- \left(\sum_{l \neq i} p_{l1} + \sum_{l \neq i,j} p_{l2} + \sum_{l \neq i,j,k} p_{l3}\right),$$

where $Q = 1 -$ the track take, W_i, P_j and S_k are the total amounts bet to win, place, and show on the various horses, respectively, and $W = \sum W_i$, $P = \sum P_j$, $S = \sum S_k$.

As an example we will consider a simplified scheme where a wager is made on a single horse per race and there are five races occurring simultaneously across tracks (Hausch and Ziemba 1990). The required information on the races is given below. Each of the horses has an expected return on a \$1 bet of \$1.14, so this is a very favorable situation.

Race	Horse	Probability of Collecting on Wager	Odds
1	A	0.570	1–1
2	B	0.380	2–1
3	C	0.285	3–1
4	D	0.228	4–1
5	E	0.190	5–1

The Kelly strategy is

$$p^* = (p_0^*, p_1^*, p_2^*, p_3^*, p_4^*, p_5^*) = (0.8529, 0.0140, 0.0210, 0.0141, 0.0700, 0.0280).$$

So 1.4% of your fortune is bet on horse A in race 1. Using the growth measure $\mu_t(p)$ = mean accumulation of wealth, and the security measure $\gamma_t(p)$ = probability that accumulated wealth will exceed b_t, the Kelly strategy was compared to half Kelly and the fixed fraction strategies which bet 0.01, 0.05 and 0.10 respectively, distributed across

TABLE 4
$\gamma_t(Y_0, p)$ = Prob $Y_t(p)$ *Will Exceed* b_t (t = 700, Y_0 = 1000)

	Strategy – p				
			Proportional		
b_t	Kelly	0.5 Kelly	1%	5%	10%
500	0.940	0.992	1.000	0.915	0.518
1,000	0.892	0.961	0.980	0.856	0.517
5,000	0.713	0.691	0.061	0.631	0.358
10,000	0.603	0.445	0.003	0.512	0.304
100,000	0.231	0.028	0.000	0.148	0.141
$\mu_t(p)$*	$17,497	$8,670	$2,394	$10,464	$1,176

* The median is used here since $Y_t(p)$ is approximately log normal (skewed) (Ethier 1987).

horses in the proportions (0.1, 0.3, 0.3, 0.2, 0.1). These fixed fraction strategies are not fractional Kelly.

The results for $\mu_t(p)$ and $\gamma_t(p)$ are given in Table 4. The Kelly strategy with a 6% betting fraction and the 5% fixed fraction are similar, both having favorable growth and security. The smaller betting fractions provide slightly more security but have much less growth. Since the 10% strategy is beyond the Kelly fraction it has less growth and less security. An interesting point is that the total fraction bet is more significant than the distribution of the bets across the horses.

Implicit in the above example is the ability to identify races where there is a substantial edge in the bettor's favor. There has been considerable research into that question. Hausch et al. (1981) demonstrated the existence of anomalies in the place and show market. At thoroughbred racetracks about 2–4 profitable wagers with an edge of 10% or more exist on an average day. The profitable wagers occur mainly because: (1) the public has a distaste for the high probability-low payoff wagers that occur on short priced horses to place and show, and (2) the public's inability to properly evaluate the worth of place and show wagers because of their complexity—for example, in a ten-horse race there are 72 possible show finishes, each with a different payoff and chance of occurrence. In Hausch et al. (1981) and more fully in Hausch and Ziemba (1985) equations are developed that approximate the expected return and optimal Kelly wagers based on minimal amounts of data to make the edge operational in the limited time available at the track.

Ziemba and Hausch (1987) implement and discuss these ideas and explore various applications, extensions, simulations and results. A laymen's article discussing this appears in the May 1989 issue of *OMNI*. A survey of the academic literature on racetrack and other forms of sports betting is in Hausch and Ziemba (1992).

4.3. *Lotto Games*: (η, β)

Lotteries have been played since before the birth of Christ. Organizations and governments have long realized the enormous profits that can be made, based on the greed and hopes of the players. Lotteries tend to go through various periods of government control and when excessive abuses occur, they have often been shut down or outlawed, only to resurface later. Since 1964, there has been an unprecedented growth in lottery games in the United States and Canada. Current yearly sales are over thirty billion dollars, while net profits to various governmental bodies are well over ten billion dollars. Since the prizes are very large, the public cares little that the expected return per dollar wagered is only 40–50 cents. Typically, half the money goes to prizes, a sixth to expenses, and a third to profits. With such a low payback, it is exceedingly difficult to win at these games

TABLE 5

Lotto Games

Prizes	Probability of Winning	Case A Prize	Case A Contribution to Expected Value	Case B Prize	Case B Contribution to Expected Value
Jackpot	1/13,983,816	$6 M	42.9	$10 M	71.5
Bonus, 5/6^{+}	1/2,330,636	$0.8	34.3	$1.2 M	51.5
5/6	1/55,492	M	9.0	$10,000	18.0
4/6	1/1,032	$5,000	14.5	$250	24.2
3/6	1/57	$150	17.6	$10	17.5
		$10			
			118.1		182.7
Edge			18.1%		82.7%
Optimal Kelly Bet			0.000,000,11		0.000,000,65
Optimal Number of Tickets Purchased per Draw with $10 M Bankroll			11		65

and the chances of winning any prize at all, let alone one of the big prizes, is very small. Is it possible to beat such a game with a scientific system? The only hope of winning is to wager on unpopular numbers in pari-mutuel lotto games. A lotto game is based on a small selection of numbers chosen by the players. Typically, games have forty to forty-nine numbers and you must choose five or six numbers. Management then draws the winning numbers and prizes typically accrue to those with three, four or more correctly matching numbers. By pari-mutuel, it is meant that the net pools for the various prizes are shared by those with winning tickets. Hence, if a small number of people, with so-called unpopular numbers, win a prize, it will be larger than if more people with popular numbers are sharing. Ziemba et al. (1986) investigated the Canadian 6/49, which played across the country, and other regional games. They found that numbers ending in nines and zeros and high numbers tended to be unpopular. Collections of unpopular numbers have a slight edge, a dollar wagered was worth more than a dollar on average, and this edge became fairly substantial when there was a large carryover. With a large carryover, that is, a jackpot pool that is steadily growing because the jackpot has not been won lately, the edge can be as high as 100% or more. Despite this promise, investors may still lose because of the very reasons that Chernoff found, essentially reverberation to the mean and gamblers ruin. Still, with such substantial edges, it is interesting to see how an investor might do playing such numbers over a long period of time.

Two realistic cases are developed in Table 5. Case A corresponds to the situation when the investor wagers only when there is a medium-sized carryover and the numbers that are drawn are quite unpopular. Case B corresponds to the situation where there is a huge carryover and the numbers drawn are the most unpopular. About one draw in every five to ten is similar to Case A and one draw in every twenty or more corresponds to Case B. Also, we have been generous in the suggested prizes. In short, we are giving the unpopular number system an at least fair chance to see if it has any hope of being a winning system in the long run. These cases correspond to the Canadian situation of paying the lotto winnings in cash up front tax free. The US situation of paying the lotto winnings over twenty years and taxing these winnings yields prizes with expected returns about one-third of those in Canada.

First we observe that the optimal Kelly wagers are miniscule. This is not surprising since for Case A, 77.2 cents of the $1.18 expected value and for Case B, $1.23 of the

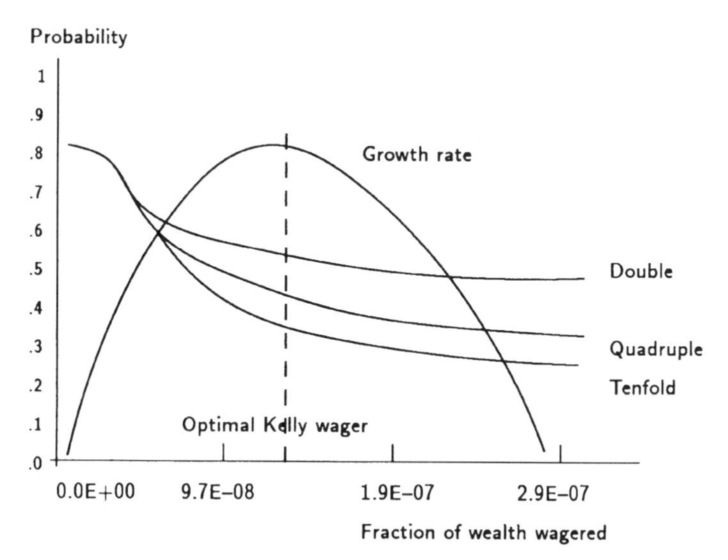

FIGURE 3. Probability of Doubling, Quadrupling and Tenfolding before Halving for Lotto 6/49, Case A.

\$1.83 expected value, respectively, is composed of less than a one in a million chance of winning the jackpot or the bonus prize. One needs a bankroll of \$1 million to justify even one \$1 ticket for Case A and over \$150,000 for Case B. If one had a bankroll of \$10 million, the optimal Kelly wagers are just 11 and 65 \$1 tickets respectively. Figures 3 and 4 show the chance that the investor will double, quadruple, or tenfold his fortune before it is halved, using Kelly and fractional Kelly strategies for Cases A and B respectively. For the Kelly strategies, these chances are 0.4 to 0.6 for Case A and 0.55 to 0.80 for Case B. With fractional Kelly strategies in the range of 0.00000004 and 0.00000025 or less, the chance of tenfolding one's initial fortune before halving it is 95% or more with Cases A and B respectively. This is encouraging, but it takes an average 294 and 55 billion

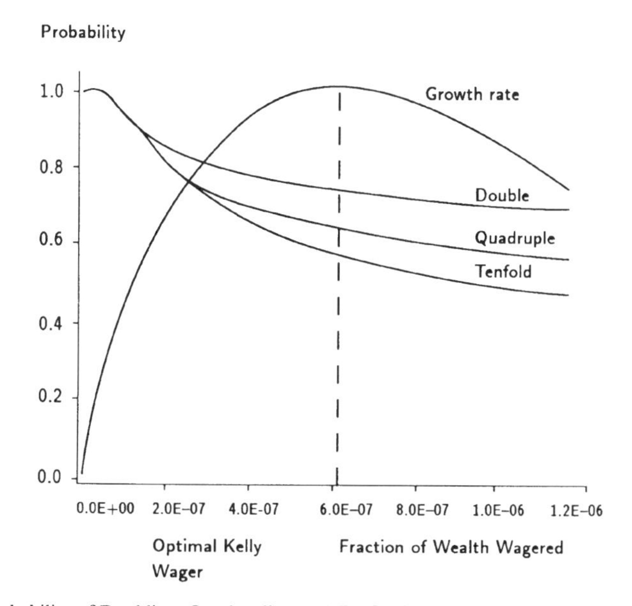

FIGURE 4. Probability of Doubling, Quadrupling and Tenfolding before Halving for Lotto 6/49, Case B.

 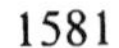

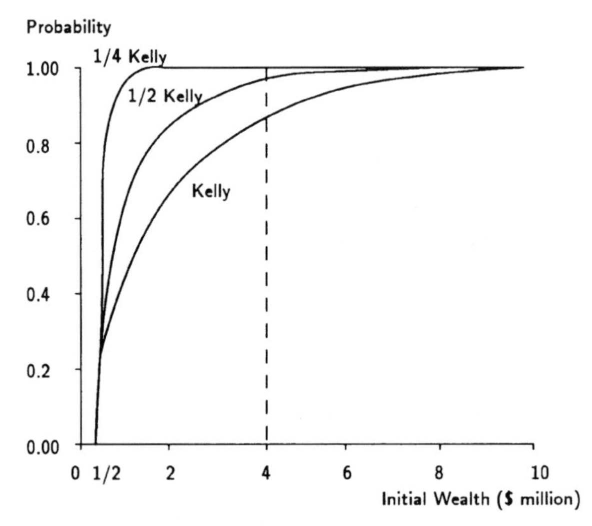

FIGURE 5. Probability of Reaching the Goal of \$10 Million before Falling to \$$\frac{1}{2}$ Million with Various Initial Wealth Levels for Kelly, $\frac{1}{2}$ Kelly and $\frac{1}{4}$ Kelly Wagering Strategies for Case A.

years, respectively, to achieve this goal. These calculations assume that there are 100 draws per year.

Figure 5 shows the probability of reaching \$10 million before falling to \$$\frac{1}{2}$ million for various initial wealth levels in the range \$$\frac{1}{2}$–\$10 million for Cases A and B with full Kelly, half Kelly and quarter Kelly wagering strategies. The results are encouraging to the millionaire lotto player especially with the smaller wagers. For example, starting with \$1 million there is over a 95% chance of achieving this goal with the quarter Kelly strategy for Cases A and B. Again, however, the time needed to do this is very long being 914 million years for Case A and 482 million years for Case B. More generally we have that Case A with full Kelly will take 22 million years, half Kelly 384 million years and quarter Kelly 915 million years. Case B with full Kelly will take 2.5 million years, half Kelly

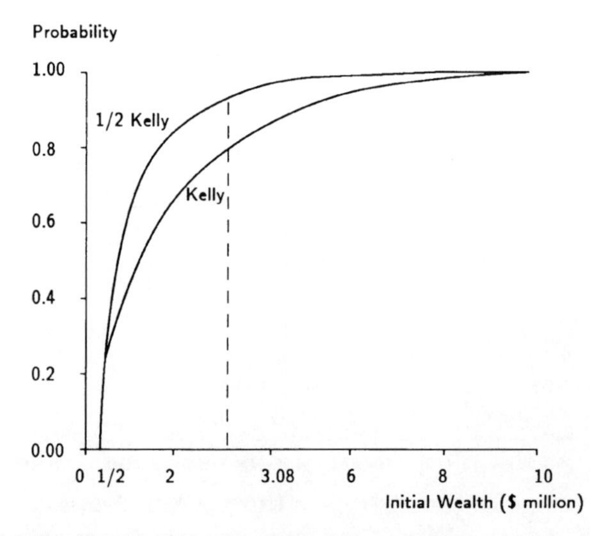

FIGURE 6. Probability of Reaching \$10 million before Falling to \$25,000 with Various Initial Wealth Levels for Kelly and $\frac{1}{2}$ Kelly Wagering Strategies for Case B.

19.3 million years and quarter Kelly 482 million years. It will take a lot less time to merely double one's fortune rather than tenfolding it, but it is still millions of years. For Case A it will take 4.6, 2.6 and 82.3 million years for full, half, and quarter Kelly, respectively. For Case B it will take 0.792, 2.6 and 12.7 million years for full, half and quarter Kelly.

Finally, we investigate the situation for the nonmillionaire wishing to become one. First, our aspiring gambler must pool his funds with some colleagues to get enough bankroll to proceed since at least $150,000 is needed for Case B and $1 million for Case A. Such a tactic is legal in Canada and in fact highly encouraged by the lottery corporations who supply legal forms for such an arrangement. For Case A, our player needs a pool of $1 million even if the group wagers only $1 per draw. Hence the situation is well modeled by Figure 5. Our aspiring millionaire "puts up" $100,000 along with nine other friends for the $1 million bankroll and when they reach $10 million each share is worth $1 million. The pool must play full Kelly and has a chance of success of nearly 50% before disbanding if they lose half their stake. Each participant does not need to put up the whole $100,000 at the start. Indeed, the cash outflow is easy to bankroll, namely 10 cents per week per participant. However, to have a 50% chance of reaching the $1 million goal each participant (and his heirs) must have $50,000 at risk. On average it will take 22 million years to achieve the goal.

The situation is improved for Case B players. First, the bankroll needed is about $154,000 since 65 tickets are purchased per draw for a $10 million wealth level. Suppose our aspiring nouveau riche is satisfied with $500,000 and is willing to put all but $25,000/2 or $12,500 of the $154,000 at risk. With one partner he can play half Kelly strategy and buy one ticket per Case B type draw. Figure 6 indicates that the probability of success is about 0.95. On average with initial wealth of $308,000 and full Kelly it will take $\frac{3}{4}$ million years to achieve this goal. With half Kelly it will take 2.7 million years and with quarter Kelly it will take 300 million years.

The conclusion then seems to be: millionaires who play lotto games can enhance their dynasties' long-run wealth provided their wagers are sufficiently small and made only when carryovers are reasonably large; and it is not possible for non-already-rich people, except in pooled syndicates, to use the unpopular numbers in a scientific way to beat the lotto and have confidence of becoming rich; moreover these aspiring millionaires are most likely going to be residing in a cemetery when their distant heir finally reaches the goal.

4.4. *Playing the Turn of the Year Effect with Index Futures*: (ϕ, β)

Ibbotson Associates (1986) have considered the actual returns received from investments in US assets with different levels of risk during the period 1926–1985. While small stocks outperformed common stocks by more than 4 to 1, in terms of the cumulative wealth levels, their advantage is totally in the last two decades and most of the gains are in the 1974+ bull market. These returns are before taxes so that the net return after taxes adjusted for inflation for the "riskless" investments and bonds may well be negative for many investors.

One way to invest in this anomaly in light of the Roll (1983) and Ritter (1988) results is to hold long positions in a small stock index and short positions in large stock indices, because the transaction costs are less than a tenth of that of trading the corresponding basket of securities. During the time of this study the March Value Line index was a geometric average of the prices of about 1,700 securities and emphasizes the small stocks while the S&P 500 is a value weighted index of 500 large stocks. Hence the VL/S&P special makes are long in small stocks and short in big stocks at the end of the year. Each point change in the index spread is worth $500. By January 15, the biggest gains are over and the risks increase. On average, the spread drops 0.92 points in this period with a

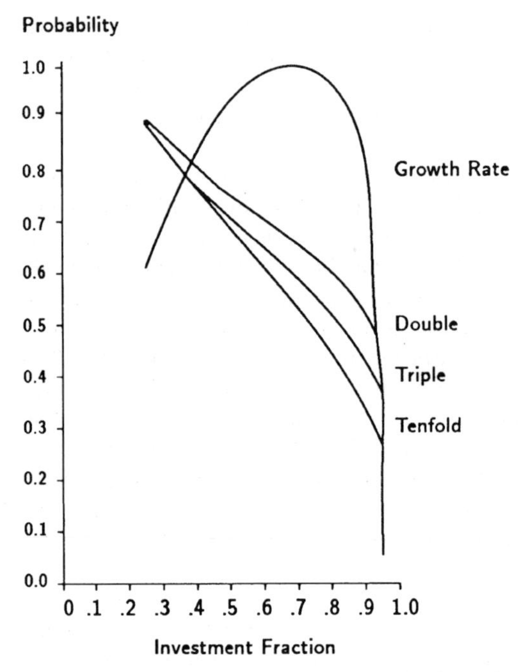

FIGURE 7. Turn of Year Effect: Relative Growth Rate and the Probability of Doubling, Tripling and Tenfolding before Halving for Various Fractional Kelly Strategies.

high variance. The projected gain from a successful trade is 0–5 points and averages 2.85 points or $1,342.50 per spread, assuming a commission of 1.5 × $55. On average, the December 15 to (−1) day gain on the spread, is 0.57 points. However, it was 1.05 in 1985 and 3.15 in 1986 which may reflect the fact that with the thin trading in the VL index, the market can be moved with a reasonably small number of players, who are learning about the success of this trade, i.e. the basis was bid up anticipating the January move. The average standard deviation of the VL/S&P spread was about 3.0. With a mean of 2.85 the following is an approximate return distribution for the trade:

Gain	7	6	5	4	3	2	1	0	−1
Probability	0.007	0.024	0.070	0.146	0.217	0.229	0.171	0.091	0.045

The optimal Kelly investment based on the return distribution is a shocking 74% of one's fortune! Such high wagers are typical for profitable situations with a small probability of loss. Given the uncertainty of the estimates involved and the volatility and margin requirements of the exchanges a much smaller wager is suggested.

Figure 7 displays the probability of doubling, tripling and tenfolding one's fortune before losing half of it, as well as the growth rates, for various fractional Kelly strategies. At fractional strategies of 25% or less the probability of tenfolding one's fortune before halving it exceeds 90% with a growth rate in excess of 50% of the maximal growth rate. Figure 8 gives the probability of reaching the distant goal of $10 million before ruining for Kelly, half Kelly and quarter Kelly strategies with wealth levels in the range of $0–$10 million. The results indicate that the quarter Kelly strategy seems very safe with a 99+% chance of achieving this goal.

These concepts were used in a $100,000 speculative account for a client of CARI Ltd., a Canadian investment management company. Five VL/S&P spreads were purchased to approximate a slightly less than 25% fractional Kelly strategy. Watching the market carefully, these were bought on December 17, 1986 at a spread of −22.18 which was

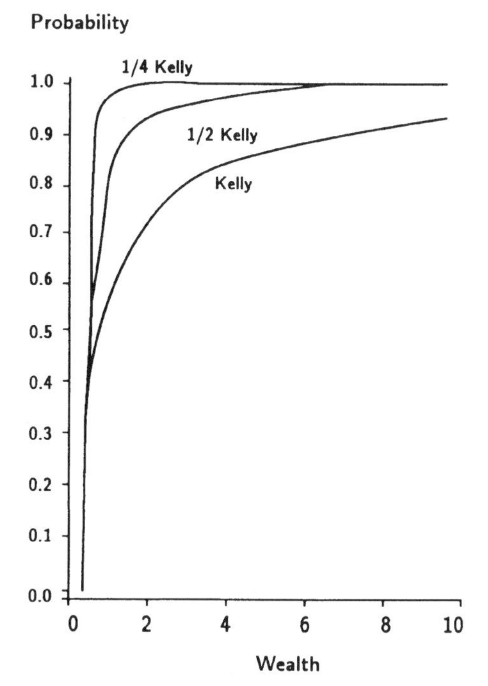

FIGURE 8. Turn of Year Effect: Probability of Reaching \$10 Million before Ruin for Kelly, $\frac{1}{2}$ Kelly and $\frac{1}{4}$ Kelly Strategies.

very close to the minimum that the spread traded at around December 15. The spread continued to gain and we cashed out at −16.47 on the 14th for a gain of 5.55 points per contract or \$14,278.50 after transactions costs.

A more detailed discussion of many of the issues in this section appears in Clark and Ziemba (1987), which was updated by Van der Cruyssen and Ziemba (1992a, b).[1]

[1] This research was partially supported by Natural Sciences and Engineering Research Council of Canada grants 5-87147 and A-3152 and by the U.S. National Science Foundation. Without implicating them we would like to thank T. Cover, S. Zenios, and the referees for helpful comments on earlier drafts of this paper.

References

ALGOET, P. H. AND T. M. COVER, "Asymptotic Optimality and Asymptotic Equipartition Properties at Log-optimum Investment," *Ann. Prob.*, 16 (1988), 876–898.

BELL, R. M., AND T. M. COVER, "Competitive Optimality of Logarithmic Investment," *Math. Oper. Res.*, 5, 2 (1980), 161–166.

BREIMAN, L., "Optimal Gambling System for Favorable Games," *Proc. 4th Berkeley Symposium on Math. Statistics and Prob.*, 1 (1961), 63–68.

CHERNOFF, H., "An Analysis of the Massachusetts Numbers Game," Technical Report No. 23, Department of Math., Massachusetts Institute of Technology, Cambridge, MA, 1980 shortened version published in *Math. Intelligence*, 3 (1981), 166–172.

CLARK, R. AND W. T. ZIEMBA, "Playing the Turn of the Year with Index Futures," *Oper. Res.*, 35 (1987), 799–813.

EPSTEIN, R. A., *The Theory of Gambling and Statistical Logic*, 2nd ed., Academic Press, New York, 1977.

ETHIER, S. N., "The Proportional Bettor's Fortune," *Proc. Seventh International Conf. on Gambling and Risk Táking*, Department of Economics, University of Nevada, Reno, NV, 1987.

———, AND S. TAVARÉ, "The Proportional Bettor's Return on Investment," *J. Appl. Prob.*, 20, 563–573.

FELLER, W., *An Introduction to Probability Theory and its Applications*, Vol. 1, 2d ed., John Wiley & Sons, New York, 1962.

FERGUSON, T. S., "Betting Systems which Minimize the Probability of Ruin," *J. Soc. Appl. Math.*, 13 (1965), 795–818.

FINKELSTEIN, M. AND R. WHITLEY, "Optimal Strategies for Repeated Games," *Adv. Appl. Prob.*, 13 (1981), 415–428.

FRIEDMAN, J., "Using the Kelly Criterion to Select Optimal Blackjack Bets," Mimeo, Stanford University, 1982.

GOTTLIEB, G., "An Optimal Betting Strategy for Repeated Games," Mimeo, New York University, 1984.

———, "An Analystic Derivation of Blackjack Win Rates," *Oper. Res.*, 33 (1985), 971–988.

GRIFFIN, P., "Different Measures of Win Rate for Optimal Proportional Betting," *Management Sci.*, 30, 12 (1985), 1540–1547.

HAKANSSON, N. H. "Capital Growth and the Mean-Variance Approach to Portfolio Selection," *J. Fin. Quant. Anal.*, 6 (1971a), 517–557.

———, "On Optimal Myopic Portfolio Policies With and Without Serial Correlation," *J. Business*, 44 (1971b), 324–334.

———, "Capital Growth and the Mean-Variance Approach to Portfolio Selection," *J. Fin. Quant. Anal.*, 6 (1971c), 517–557.

———, "Optimal Multi-Period Portfolio Policies," *Portfolio Theory, TIMS Studies in the Management Sciences*, Vol. 11, E. Elton and M. J. Gruber, Eds., North-Holland, Amsterdam, 1979.

——— AND B. L. MILLER, "Compound-Return Mean-Variance Efficient Portfolios Never Risk Ruin," *Management Sci.*, 22, 4 (1975), 391–400.

HAUSCH, D. AND W. T. ZIEMBA, "Transactions Costs, Extent of Inefficiencies, Entries and Multiple Wagers in a Racetrack Betting Model," *Management Sci.*, 31, 4 (1985), 381–392.

——— AND ———, "Arbitrage Strategies for Cross-Track Betting on Major Horse Races," *J. Business*, 63, 1 (1990), 61–78.

——— AND ———, "Efficiency of Sports and Lottery Betting Markets," In *New Palgrave Dictionary on Money and Finance*, J. Eatwell, P. Newman and M. Milgatel, Eds., Macmillan, London, 1992.

———, ——— AND M. RUBINSTEIN, "Efficiency of the Market for Racetrack Betting," *Management Sci.*, 27 (1981), 1435–1452.

IBBOTSON ASSOCIATES, *Stocks, Bonds, Bills and Inflation: Market Results for* 1926–1985, Ibbotson Associates, Chicago, IL, 1986.

KELLY, J., "A New Interpretation of Information Rate," *Bell System Technology J.*, 35 (1956), 917–926.

LI, Y., L. MACLEAN AND W. T. ZIEMBA, "Security Aspects of Optimal Growth Models with Minimum Expected Time Criteria," Mimeo, University of British Columbia, British Columbia, Canada, 1990.

MACLEAN, L. AND W. T. ZIEMBA, "Growth-Security Profiles in Capital Accumulation Under Risk," *Ann. Oper. Res.*, 31 (1990), 501–509.

RITTER, J. R., "The Buying and Selling Behavior of Individual Investors at the Turn of the Year: Evidence of Price Pressure Effects," *J. Finance*, 43 (1988), 701–719.

ROLL, R., "Was ist das? The Turn of the Year Effect and the Return Premia of Small Firms," *J. Portfolio Management*, 10 (1983), 18–28.

THORP, E. O., *Beat the Dealer*, 2d ed., Random House, New York, 1966.

———, "Portfolio Choice and the Kelly Criterion," In *Stochastic Optimization Models in Finance*, W. T. Ziemba and R. G. Vickson, Eds., Academic Press, New York, 1975.

VAN DER CRUYSSEN, B. AND W. T. ZIEMBA, "The Turn of the Year Effect," *Interfaces*, forthcoming (1992a).

——— AND ———, "Investing in the Turn of the Year Effect in the Futures Markets," *Interfaces*, forthcoming, (1992b).

WU, M. G. H. AND W. T. ZIEMBA, "Growth Versus Security Tradeoffs in Dynamic Investment Analysis," Mimeo, University of British Columbia, British Columbia, Canada, 1990.

ZIEMBA, W. T., "Security Market Inefficiencies: Strategies for Making Excess Profits in the Stock Market," TIMS Workshop on Investing in the Stock Market, 1986.

———, S. L. BRUMELLE, A. GAUTIER AND S. L. SCHWARTZ, *Dr. Z's 6/49 Lotto Guidebook*, Dr. Z Investments Inc., Vancouver, 1986.

——— AND D. B. HAUSCH, *Dr. Z's Beat the Racetrack*, William Morrow, New York (revised and expanded second edition of Ziemba-Hausch 1984), 1987.

——— AND ———, *Betting at the Racetrack*, Dr. Z Investments, Los Angeles, 1985.

Available online at www.sciencedirect.com

ELSEVIER Journal of Economic Dynamics & Control 28 (2004) 937–954

JOURNAL OF Economic Dynamics & Control

www.elsevier.com/locate/econbase

25

Capital growth with security

Leonard C. MacLean[a], Rafael Sanegre[b], Yonggan Zhao[c], William T. Ziemba[d,*]

[a]*School of Business Administration, Dalhousie University, Halifax, NS, Canada B3H3J5*
[b]*Meta4, Edf. Roma, Rozabella, 8, 28230 Las Rozas, Madrid, Spain*
[c]*Nanyang Business School, Nanyang Technological University, Singapore 639798, Singapore*
[d]*Faculty of Commerce and Business Administration, The University of British Columbia, Vancouver, BC, Canada V6T 1Z2*

Abstract

This paper discusses the allocation of capital over time with several risky assets. The capital growth log utility approach is used with conditions requiring that specific goals are achieved with high probability. The stochastic optimization model uses a disjunctive form for the probabilistic constraints, which identifies an outer problem of choosing an optimal set of scenarios, and an inner (conditional) problem of finding the optimal investment decisions for a given scenarios set. The multiperiod inner problem is composed of a sequence of conditional one period problems. The theory is illustrated for the dynamic allocation of wealth in stocks, bonds and cash equivalents.

JEL classification: C61; D92; F47; G11

Keywords: Capital growth; Drawdown constraint; Growth and security; Kelly strategy; Scenario generation; Value at risk

1. Introduction

The problem of capital accumulation under uncertainty has occupied an important place in the theory of financial economics. Given a set of risky investment opportunities, a decision maker must choose how much of available capital to invest in each asset at each point in time. When the criterion for selecting an investment policy is maximizing the expected value of the logarithm of accumulated capital, the resulting policy,

* Corresponding author. Tel.: 604-261-1343; fax: 604-263-9752.
E-mail addresses: lmaclean@mgmt.dal.c (L.C. MacLean), rafaelsll@meta4.com (R. Sanegre), aygzhao@ntu.edu.sg (Y. Zhao), ziemba@interchange.ubc.ca (W.T. Ziemba).

0165-1889/03/$ - see front matter © 2003 Elsevier B.V. All rights reserved.
doi:10.1016/S0165-1889(03)00056-3

known as the Kelly or capital growth criterion, has many desirable properties. The maximum expected logarithm strategy asymptotically maximizes the log run expected growth rate of capital (Kelly, 1956; Breiman, 1961). Moreover, the optimal policy is myopic and period-by-period optimization can be used to compute the optimal decisions (Hakansson, 1972). Breiman (1961) has shown that the expected time to reach asymptotically large wealth levels is minimized by this strategy. A theoretical exposition of the properties of the capital growth strategy in the intertemporally independent and weakly dependent cases appears in Algoet and Cover (1988). Rotando and Thorp (1992) apply the Kelly strategy to long-term investment in the U.S. stock market and demonstrate some of the benefits and liabilities of that strategy. Thorp (1998) updates the 1992 paper with additional discussions on blackjack and sports betting. Additional discussions appear in Aucamp (1993), Ethier and Tavarè (1983), Finkelstein and Whitley (1981), Markowitz (1976), and Rubinstein (1977, 1991). MacLean et al. (1992) discuss a theory of growth versus security using fractional Kelly strategies which are convex combinations of cash and the Kelly fraction and apply this to several speculative investment applications; see also MacLean and Ziemba (1991, 1999). MacLean et al. (2000) show that the fractional Kelly strategies lie on a growth-security efficient frontier if the assets are lognormally distributed. Without lognormality, the tradeoff of growth versus security is monotone but not necessarily efficient. Hakansson and Ziemba (1995) review the capital growth literature and various applications.

The optimality properties of the Kelly strategy are related to expected values, either of log wealth or first passage times. But the fraction of wealth invested may be unacceptably large because the Arrow–Pratt risk aversion index is the reciprocal of wealth and is essentially zero for reasonable wealth levels. Furthermore, if uncertainty in the return on investment is considerable then the probability of wealth becoming negligible at some point is high with the capital growth strategy. [1] This downside risk of investment strategies has led to a growing interest in risk control.

The traditional approach to risk control has been to include the variance of wealth in the decision problem, and solve for mean-variance efficient strategies (Markowitz, 1959). More recently Value-at-Risk-based risk management has emerged as the standard (Jorion, 1997). The VaR is the floor below which wealth can fall in a specified time interval, with a prespecified small probability.

Investment models can be formulated in discrete or continuous time. In continuous time the dynamics of asset prices are usually defined by geometric Brownian motion, with the asset prices having log-normal distributions (Merton, 1969, 1992). A continuous time model with a VaR constraint is studied by Basak and Shaprio (2001), where the VaR risk managers' strategies are contrasted with portfolio insurance. The discrete time analogue to the Brownian motion model is a random walk. Computational approaches to obtaining optimal growth strategies in discrete time without assumptions

[1] For example, Ziemba and Hausch (1986) ran a simulation where an investor has initial wealth \$1000 and makes 700 independent wagers, all of which have an expected value of \$1.14 per \$1 wagered and all of which have not small probabilities of winning. In 166 of 1000 replications the Kelly bettor had a final fortune of \$1,000,000 or more. However, the minimum final wealth was only \$18.

on asset price distributions are presented by Cover (1991); see also Helmbold et al. (1996).

In this paper the computation of strategies for investment in discrete time which achieve maximal capital growth subject to a VaR constraint is considered. The emphasis is on defining a multistage stochastic programming problem where the constraints are identified by a selection of scenarios sampled from the space of potential financial market outcomes. Using the VaR constraint, the outcomes are classified as critical and noncritical, with feasible investment strategies maintaining the critical outcomes at a small percentage. An algorithm for classifying scenarios is developed and illustrated with the optimal trade-off of cash, bonds and stocks over time.

2. Capital accumulation model

Consider an investor with initial wealth W_0 and the opportunity to invest in m risky securities. The following assumptions are made about capital markets: no transactions costs; no taxes; infinite divisibility of assets; assets have limited liability; borrowing and lending are allowed at the same rate; and short selling is permitted. These conditions will be referred to as the market assumptions.

The trading price of security i at time t is $P_i(t)$, $i=1,\ldots,m$. In discrete time the rate of return on a unit of capital invested in security i at time t is

$$\frac{P_i(t+1)}{P_i(t)} = R_i(t), \quad i=1,\ldots,m. \tag{1}$$

It is assumed that the returns follow a log-normal distribution, so consider $Y_i(t) = \ln P_i(t)$, $i=1,\ldots,m$. Whereas $P_i(t)$ is a geometric process, $Y_i(t)$ is an arithmetic random walk with increments $Z_i(t) = \ln R_i(t)$, $i=1,\ldots,m$, with $Z_i(t)$ having a normal distribution. Therefore, the increments can be represented as a random effects linear model

$$Z_i(t) = \pi_i + \delta_i \varepsilon_i(t), \quad i=1,\ldots,m, \tag{2}$$

where $\varepsilon(t)^\top = (\varepsilon_1(t),\ldots,\varepsilon_m(t))$ is $N(0,I)$ and $\pi^\top = (\pi_1,\ldots,\pi_3)$ is $N(\mu,\Gamma)$, with $\mu^\top = (\mu_1,\ldots,\mu_m)$ and $\Gamma = (\gamma_{ij})$. So, γ_{ij} is the covariance between π_i and π_j, the random expected rates of return of securities i and j, respectively.

From the arithmetic random walk with normal increments the conditional distribution of log-prices at time t, given π and $\Delta = diag(\delta_1^2,\ldots,\delta_m^2)$ is $(Y(t)|\pi,\Delta) \sim N(\pi t, t\Delta)$. The marginal distribution of log-prices is $Y(t) \sim N(\mu t, \Sigma(t))$, where $\Sigma(t) = t\Delta + t^2\Gamma$.

Although this log-normal model for securities prices is specialized, the random rates of return provide the flexibility required to match the theoretical prices to observations. This discrete time model is the analogue to the geometric Brownian models in continuous time. The dynamics of price movements are clear from the distribution parameters. The underlying parameters (μ,Γ,Δ) generate the securities prices. If these parameters are known or estimated then the price distributions can be specified. Parameter estimation is discussed in Section 6.

From the initial values the forward price process evolves as a random walk with intertemporally independent increments defined by (2). Consider the rate of returns process $R_a(t)^\top = (R_1(t), \ldots, R_m(t))$, $t = 1, \ldots, T$, and denote the multivariate rate of returns distribution at time t by F_t. For the stochastic process $R_a(t)^\top = (R_1(t), \ldots, R_m(t))$, a trajectory or realization of the data process is associated with an outcome ω in the sample space Ω of all outcomes (trajectories). The distributions $F_1, \ldots, F_T$ generate a probability measure P on Ω and the associated probability space (Ω, B, P). The sample space can be represented as $\Omega = \Omega_1 \times \cdots \times \Omega_T$, with $\omega_t \in \Omega_t$, the data for time t. The information available to the investor at time t is the data on the past, and is represented by the filtration $B_0 := \{\emptyset, \Omega\} \subset B_1 \subset \cdots \subset B_r := B$, where $B_t := \sigma(\omega^t)$ is the σ-field generated by the history ω^t of the data process ω to time t. So the stochastic process is adapted to $\{B_t; t = 1, \ldots, T\}$, the augmented filtration generated by ω.

In addition to the risky securities defined on $(\Omega, B, \{B_t\}, P)$ there is a riskless asset with rate of return $R_0(t) = 1 + r(t)$. Let $R(t)^\top = (R_0(t), R_a(t)^\top)$.

An investment decision at time t is the proportion of wealth to allocate to each asset, given by

$$X(t)^\top = (X_0(t), \ldots, X_m(t)). \tag{3}$$

It is assumed that $X(t)$ can depend on the data history ω^t but not on unknown future returns, so it is B_t predictable. The budget constraint at each time requires $\sum_{i=0}^m X(t) = 1$. The proportions of wealth invested in risky assets are unconstrained, since the proportion invested in the risk-free asset can always be chosen (with borrowing or lending) to satisfy the budget constraint.

An investment strategy is an $m+1$ vector process, $X = \{X(t),\ t = 1, \ldots, T\}$, where T is the planning horizon. With initial wealth W_0, rate of returns process R and investment process X, the capital accumulated to time t is

$$W(t) = W_0 \prod_{s=1}^{t} R(s)^\top X(s), \quad t = 1, \ldots, T. \tag{4}$$

The paths of the capital accumulation process are controlled by the investment strategy, and the investor selects a strategy based on anticipated performance as indicated by measures for growth and security (risk).

3. Downside risk control

For the geometric growth process of capital accumulation in (4) a natural performance measure is the geometric mean $E[W(T)^{1/T}] = E[(W_0 \prod_{t=1}^T R(t)^\top X(t))^{1/T}]$. Since

$$W(T)^{1/T} = W_0 \left(\exp\left\{ \frac{1}{T} \sum_{t=1}^{T} \ln(R(t)^\top X(t)) \right\} \right),$$

$G(X) = \frac{1}{T} \sum_{t=1}^T \ln(R(t)^\top X(t))$ is the growth rate of capital. The geometric mean is maximized by the optimal expected growth rate strategy from

$$\max_X E[G(X)] = \frac{1}{T} \sum_{t=1}^{T} E[\ln(R(t)^\top X(t))]. \tag{5}$$

The growth optimal strategy X^* from (5) is called the Kelly (1956) strategy. This strategy has been studied extensively. For advanced proofs of optimality properties with minimal assumptions see Algoet and Cover (1988).

The Kelly strategy is very risky. Although it provides optimum growth in the long run it is possible to experience negative growth in the medium term and in any period experience a substantial loss of capital (drawdown). Discussion of these properties appears in Table 1 of MacLean and Ziemba (1999).

Because of the volatility of financial markets it is prudent (and frequently it is a legal requirement) to include downside risk control in the decisions on investment strategy. To put risk measurement in context, consider the following definition, an adaptation of one provided in Breitmeyer et al. (1999).

Definition 1 (Downside risk measure). Consider a wealth process $\{W(t),\ t=1,\dots,T\}$ with corresponding distributions $\{H_t,\ t=1,\dots,T\}$. Let $q \in \mathfrak{R}$ be an arbitrary number which partitions wealth trajectories into acceptable and unacceptable sets, denoted by C and $\bar{C}$, respectively. If V is the set of probability distributions for the wealth process, a downside risk measure ϕ is a function $\phi: V \times \mathfrak{R} \to \mathfrak{R}$ satisfying the axioms

1. (non-negativity): $\phi(H,q) \geqslant 0$,
2. (normalization): If $H(\omega) = 0$ for all $\omega \in C$, then $\phi(H,q) = 0$,
3. (downside focus): If H and $\bar{H}$ are distributions over wealth trajectories such that $H(\omega) = \bar{H}(\omega)$, for $\omega \in C$, then $\phi(H,q) = \phi(\bar{H},q)$.

There are additional axioms which consider properties such as consistency, continuity and invariance (Breitmeyer et al., 1999; Artzner et al., 1999), but the properties in our definition are basic.

The standard measure for downside risk in the financial industry is VaR (Jorion, 1997). It is defined as *the loss*, which is exceeded with some given probability α, over a given time horizon. The intention is to control VaR, so a prespecified minimum value w is given and the value at risk with probability α must exceed w. Equivalently, the probability that wealth exceeds w at the VaR horizon is at least $1-\alpha$.

Although VaR is widely used, it has some undesirable properties (see Artzner et al., 1999; Basak and Shaprio, 2001). Other proposed measures are the period-by-period drawdown (Grossman and Zhou, 1993) and the incomplete mean (Basak and Shaprio, 2001). The drawdown used in this paper considers the potential fraction of wealth lost at each period. The incomplete mean is the partial expected value in the lower α percentile of the wealth distribution.

Definition 2 (Risk measures). Consider a wealth process $\{W(t),\ t=1,\dots,T\}$ with corresponding distributions $\{H_t,\ t=1,\dots,T\}$.

1. The value at risk measure for horizon T and wealth value w is

$$\phi_1(H,w) = \Pr[W(T) \geqslant w] = 1 - H_T(w).$$

942 L.C. MacLean et al. / Journal of Economic Dynamics & Control 28 (2004) 937–954

2. The drawdown measure for decay fraction $b \in [0,1]$ is

$$\phi_2(H,b) = \Pr[W(t+1) \geqslant bW(t), t = 1,\ldots,T].$$

3. The incomplete mean measure for horizon T and specified percentile α is

$$\phi_3(H,\alpha) = \int_0^{\omega_\alpha} w \, \mathrm{d}H_T(w),$$

where $\Pr[W(T) \leqslant w_\alpha] = \alpha$.

Each of these measures satisfies the basic axioms for downside risk. $\phi_1(H,w_\alpha)$ and $\phi_3(H,\alpha)$ have the same unacceptable sets, and ϕ_3 is a measure of the expected loss on the unacceptable set. With respect to ϕ_2 it is possible to link b to the value w in ϕ_1.

Let $b(w) = (w/W_0)^{1/T}$, and see that the unacceptable set for ϕ_1 is contained in the unacceptable set for ϕ_2. Therefore $\phi_2(H,b(w)) \leqslant \phi_1(H,w)$. So ϕ_2 and ϕ_3 are more stringent risk measures than ϕ_1.

The purpose in defining a risk measure is to develop an investment strategy which achieves capital growth while controlling for risk. Subsequent discussion of risk will concentrate on the VaR measure ϕ_1. The approach followed is easily adapted to other risk measures.

Consider the general growth with security problem

$$\sup_X \{E[G(X)] | \phi_1(H(X),w) \geqslant 1-\alpha\}, \tag{6}$$

where $H(X) = (H_1(X),\ldots,H_T(X))$ is the distribution over wealth generated by investment strategy $X = \{X(t), t = 1,\ldots,T\}, w$ is a prespecified wealth floor and $1-\alpha$ is a confidence level. It is assumed that the measurability conditions previously discussed for the return process and the investment process are imposed.

Let $w^* = w/W_0$. Problem (6) can be written more explicitly in terms of rate of returns as (CCP):

$$\sup_X \left\{ \sum_{t=1}^{T} E \ln(R(t)^\top X(t)) \middle| \Pr\left[\sum_{t=1}^{T} \ln(R(t)^\top X(t)) \geqslant \ln w^* \right] \geqslant 1-\alpha \right\}. \tag{7}$$

The rate of return process $R = \{R(t), t = 1,\ldots,T\}$ and the investment process $X = \{X(t), t = 1,\ldots,T\}$ are defined on $(\Omega, B, \{B_t\}, P)$. For a given trajectory $\omega \in \Omega$ let the associated return path be $R(\omega) = \{R(\omega,t),\ t = 1,\ldots,T\}$ and the investment path be $X(\omega) = \{X(\omega,t), t = 1,\ldots,T\}$. The risk measure refers to acceptable and unacceptable sets of wealth trajectories. Consider sets of measure $1-\alpha$ in the probability space, given by $B_\alpha = \{A | P(A) \geqslant 1-\alpha\}$. There are associated sets of wealth trajectories for A and its complement $\bar{A}$. An equivalent formulation to (7) based on trajectories

L.C. MacLean et al. / Journal of Economic Dynamics & Control 28 (2004) 937–954 943

is (DP)

$$\sup_{A \in B_\alpha} \left[\sup_X \left\{ \sum_{t=1}^{T} E \ln(R(t)^\top X(t)) \middle| \sum_{t=1}^{T} \ln(R(\omega,t))^\top X(\omega,t) \geqslant \ln \omega^*, \omega \in A \right\} \right]. \quad (8)$$

The disjunctive formulation in (8) defines a sequence of stochastic convex dynamic programming problems, Rockafeller and Wets (1978) referred to as *inner problems*, with a constraint for each $\omega \in A$.

Another reformulation of (7), found by introducing a weighing variable $\theta(\omega)$, $\omega \in \Omega$, with $0 \leqslant \theta(\omega) \leqslant 1$, is (NCP):

$$\sup_{(x,\theta)} \left\{ \sum_{t=1}^{T} E \ln(R(t)^\top X(t)) \middle| \theta(\omega) \left[\sum_{t=1}^{T} \ln(R(t)^\top X(t)) \geqslant \ln \omega^* \right] \right.$$

$$\left. \geqslant 0, \; E[\theta(\omega)] \geqslant 1 - \alpha \right\}. \quad (9)$$

At the optimal solution to (9) the weighing variable is an indicator function. That is, if (X^*, θ^*) is optimal then there is a set A such that $\theta^*(\omega) = 1$ for $\omega \in A$, $\theta^*(\omega) = 0$ for $\omega \in \bar{A}$. Hence, for the optimal investment strategy: (CCP) $\Leftrightarrow$ (DP) $\Leftrightarrow$ (NCP).

The growth with security problems (7)–(9) provide a framework for individual portfolio choice, where risk is controlled at a specific level. In multidimensional financial markets it is common to identify the underlying sources of systematic risk, and to define a small set of generating portfolios or mutual funds. These funds serve an intermediary role for fund managers to create products which satisfy investor preferences. It is important that the growth with security problem works within this intermediation theory.

To understand the generating portfolios (mutual funds) recall for the price process it is assumed there exist $p < m$ independent latent factors $U^\top = (U_1, \ldots, U_p)$, $U_i \sim N(0,1)$, $i = 1, \ldots, p$, so that the log-prices are represented as

$$Y(t) = \mu t + t \Lambda U + \sqrt{t}\xi, \quad (10)$$

where $\mu^\top = (\mu_1, \ldots, \mu_m)$, $\lambda = (\lambda_{ij})$ for loadings λ_{ij}, $i = 1, \ldots, m$ and $j = 1, \ldots, p$, and the covariance $Cov(\xi) = diag(\delta_1^2, \ldots, \delta_p^2)$. Assume that $E[\xi] = 0$, and U and ξ are independent. $Y(t)$ has the distribution $N(\mu t, \Sigma(t))$, where $\Sigma(t) = t\Delta + t^2\Gamma$ and $\Gamma = \Lambda\Lambda^\top$. The factor model (10) and the associated conditions are referred to as the *structural model assumptions*.

In this structure imposed on the price process, the factors $U^\top = (U_1, \ldots, U_p)$ are the standardized log-prices for the mutual funds.

Theorem 1 (Financial intermediation). *Suppose there exist m risky securities and a risk-free asset satisfying the market and structural model assumptions. Then there exists a risk-free and p < m risky mutual funds, so that growth with security investors as defined by* (7) *are indifferent between choosing portfolios from among the original securities and choosing portfolios from the mutual funds.*

944 L.C. MacLean et al. / Journal of Economic Dynamics & Control 28 (2004) 937–954

Proof. Consider the return in period t, $R(t)^\top X(t)=\sum_{i=0}^m R_i(t)X_i(t)$. From the model for securities prices $R_i(t)=\exp(\mu_i+\Lambda_i^\top U+\delta_i\varepsilon_i)$, where Λ_i is the ith row of the $m\times p$ loading matrix Λ in (10). Let $A_0=(a_{ij})$, where $a_{ij}=\lambda_{ij}^2/(\sigma_i^2-\delta_i^2)$, and

$$A=\begin{pmatrix} 1 & 0 \\ 0 & A_0^\top \end{pmatrix}.$$

Define $q(t)=A\cdot X(t)$. From (10), $\sum_{j=1}^p \lambda_{ij}^2=\sigma_i^2-\delta_i^2$ and

$$\begin{aligned}\sum_{j=0}^{p} q_j(t) &= X_0(t)+\sum_{j=1}^{p}\sum_{i=1}^{m}\left(\frac{\lambda_{ij}^2}{(\sigma_i^2-\delta_i^2)}X_i(t)\right)\\ &= X_0(t)+\sum_{i=1}^{m}\left(\sum_{j=1}^{p}\frac{\lambda_{ij}^2}{(\sigma_i^2-\delta_i^2)}X_i(t)\right)\\ &= \sum_{i=0}^{m} X_i(t)=1.\end{aligned}$$

With A^-, the generalized inverse of A, let $X(t)=A^-q(t)$. Hence, the return in period t is $R(t)^\top X(t)=(R(t)^\top A^-)q(t)=M(t)^\top q(t)=\sum_{j=0}^p M_j(t)q_j(t)$, with $M_j(t)$ the return on risky mutual fund j, $j=1,\ldots,p$ and $M_0(t)=1+r(t)$. Since the returns in each period are the same for the mutual funds and the original securities, the statement in the theorem holds. □

The investment process $X=\{X(t),t=1,\ldots,T\}$ is defined on the space of trajectories $(\Omega,B,\{B_t\},P)$, where $X(t)$ is B_t predictable. An important property of the investment process is path independence (Cox and Leland, 2000), so that it depends only on the reinvested return on the securities, not on the whole price history. This property is satisfied by the growth with security investor.

Theorem 2 (Path independence). *The optimal growth with security investment strategy X^* is path independent, i.e. $X(t)^*$ depends on the level of wealth at time t, w_t, but not on the particular path to achieve that wealth.*

Proof. Consider the growth with security problem in the disjunctive form

$$\text{DP}:\quad \sup_{A\in B_\alpha}\left[\sup_{X}\left\{\sum_{t=1}^{T} E\ln(R(t)^\top X(t))\,\middle|\,\sum_{t=1}^{T}\ln(R(\omega,t)^\top X(\omega,t))\geqslant \ln w^*,\omega\in A\right\}\right].$$

If I_A is the indicator function for the set A and $\lambda(\omega)\geqslant 0$ is a multiplier so that $\lambda\in L(\Omega,B,P)$, the space of Lebesgue integrable functions on Ω, then a Lagrangian for (DP) is

$$\mathscr{L}(X,\lambda,A) = E\left[\sum_{t=1}^{T} \ln(R(\omega,t)^\top X(\omega,t)) \right.$$
$$\left. + I_A\lambda(\omega)\left(\sum_{t=1}^{T} \ln(R(\omega,t)^\top X(\omega,t)) - \ln w^*\right)\right]. \tag{11}$$

If a solution exists for (DP) then there exist elements (A^*, λ^*) so that the solution to (DP) is given by $\sup_X \mathscr{L}(X, \lambda^*, A^*)$. The Lagrangian is

$$\mathscr{L}(X,\lambda^*,A^*) = \sum_{t=1}^{T} E[(1 + I_{A^*}\lambda^*)\ln(R(t)^\top X(t))].$$

Consider $\eta_0^* = E[1 + I_{A^*}\lambda^*]$ and $\eta_t^* = E_{T-t}(1 + I_{A^*}\lambda^*)$, where E_{T-t} is the expectation over the data process from time $t+1$ to time T. Then, $\eta_t^* = \eta_{t-1}^* \cdot \beta_t^*$, where β_t^* depends only on the additional data in period t. Since $\ln \eta(\omega^t) = \sum_{s=1}^{t} m_s(\omega_s)$, where $m_s(\omega_s) = E_{t-s}m(\omega^t) - E_{t-s+1}m(\omega^t)$, then

$$\sup_X \sum_{t=1}^{T} E[(1 + I_{A^*})\ln(R(t)^\top X(t))] = \sup_X \sum_{t=1}^{T} E[\eta_{t-1}^* \cdot \beta_t^* \ln(R(t)^\top X(t))]$$
$$= \sup_X \sum_{t=1}^{T} E[\beta_t^* \cdot E_{t-1}[\eta_{t-1}^* \ln(R(t)^\top X(t))]],$$

where E_{t-1} is the expectation with respect to data for periods 1 to $t-1$. With $\bar{\eta}_{t-1}^* = E_{t-1}\eta_{t-1}^*$,

$$E_{t-1}[\eta_{t-1}^* \ln(R(t)^\top X(t))] = \bar{\eta}_{t-1}^* E_{t-1}\left[\frac{\eta_{t-1}^*}{\bar{\eta}_{t-1}^*}\ln(R(t)^\top X(t))\right]$$
$$\leqslant \bar{\eta}_{t-1}^* \ln\left(E_{t-1}\left[R(t)^\top \frac{\eta_{t-1}^*}{\bar{\eta}_{t-1}^*} X(t)\right]\right)$$

from Jensen's inequality and B_{t-1} measurability of η_{t-1}^* and $X(t)$. But $R_i(t)=\exp(Z_i(t))$, where the increments $Z_i(t)$ are intertemporally independent. With $\tilde{X}(t) = E_{t-1}[(\eta_{t-1}^*/\bar{\eta}_{t-1}^*)X(t)]$, the problem becomes

$$\sup_X \sum_{t=1}^{T} \bar{\eta}_{t-1}^* E[\beta_t^* \ln(R(t)^\top \tilde{X}(t))],$$

where $\tilde{X}(t)$ does not depend on the path ω^{t-1}, $t = 1,\ldots,T$. □

The path independence property is within the context of planning for T periods with a given distribution over returns. As described in Section 6, at the start of the next T period plan the history of prices is used to update the distribution over returns.

946 L.C. MacLean et al. / Journal of Economic Dynamics & Control 28 (2004) 937–954

So decisions depend on trajectories through the estimation (filtration) process, which is separated from the optimization.

4. Scenario selection

The growth with security problem in its disjunctive form (8) or equivalent Lagrangian form (11) involves complicated multivariate integration. Some form of discrete approximation to distributions for securities prices is required to proceed with computation.

Consider the distributions F_t on the returns $R_a(t)$, $t = 1,\dots,T$. Assume that in $\Re^m$ a grid is constructured with rectangles so that the probability mass from F_t in each rectangle is equal. Sample a point from each rectangle, generating the empirical distribution F_t^n. Let the space of trajectories corresponding to the empirical distribution F_t^n, $t=1,\dots,T$, be $\Omega^{(n)}$, and the corresponding probability space be $(\Omega^{(n)}, B^{(n)}, P^{(n)})$. So $\Omega^{(n)} = \{\omega_1,\dots,\omega_I\}$ and $P(\omega_i) = 1/I$.

With this lattice structure, a discrete approximation to the growth with security problem is (DP_n):

$$\sup_{A^{(n)}\in B^{(n)}} \left\{ \sup_{X^n} \left\{ E\sum_{t=1}^{T} \ln(R^n(t)^\top X^n(t)) \middle| \sum_{t=1}^{T} \ln(R^n(\omega,t)^\top X^n(\omega,t) \right.\right.$$

$$\left.\left. \geqslant \ln, w^*, \omega \in A^{(n)} \right\}\right\}. \tag{12}$$

The inner problem and the set selection problem are now discrete. For a given A and the corresponding $A^{(n)}$ the convergence of the solution of the discrete inner problem to the solution of the continuous inner problem has been established and bounds on the error for given n developed (Pflug, 1999). The identification of the optimal set $A^{(n)}$, where $P^{(n)}(A^{(n)}) \geqslant 1-\alpha$, is required. Assume that the specified target w^* at the horizon T is such that problem (12) has a solution for $A^{(n)} = \Omega^{(n)}$. The process for identifying the optimal set of scenarios is *backward elimination* (Culioli, 1996). If k is the largest integer such that $k/I \leqslant \alpha$, then up to k constraints corresponding to scenarios in $\Omega^{(n)}$ get eliminated and the remaining constraints yield the optimal $A^{(n)}$. The elimination is one constraint at a time starting from $\Omega^{(n)}$ and working backward. In describing the elimination procedure the notation $A^{(n)}(i_j|i_{j-1},\dots,i_1)$ refers to a set of scenarios in the family of sets $A_j^{(n)}(i_{j-1},\dots,i_1)$ defined by the $j-1$ elimination steps, where scenarios $\omega_{i_1},\dots,\omega_{i_{j-1}}$ have been eliminated sequentially. The basis for the algorithm is the inner problem

$$P(A^{(n)}):\quad \sup_{X^n} \left\{ \sum_{t=1}^{T} E\ln(R^n(t)^\top X^n(t)) \middle| \sum_{t=1}^{T} \ln(R^n(\omega,t)^\top X^n(\omega,t)) \geqslant \ln w^*, \omega \in A^{(n)} \right\}.$$

The algorithm proceeds as follows:

[1] Set $j=0$ and $A_o^{(n)} = \{\Omega^{(n)}\}$.

[2] (Solution step) For each $A^{(n)}(i_j|\cdot) \in A_j^{(n)}(\cdot)$, solve the problem $P(A^{(n)}(i_j|\cdot))$, denoting the optimal (*strategy*, *value*) by $(X(A^{(n)}(i_j|\cdot)), G^*(A^{(n)}(i_j|\cdot)))$, and the set of scenarios corresponding to the active constraints by $C_{j+i}(A^{(n)}(i_j|\cdot))$. If $j = k$ or $C_{j+1}(A^{(n)}(i_j|\cdot)) = \emptyset$, then designate the problem $P(A^{(n)}(i_j|\cdot))$ as k-reduced and go to step [4].

[3] (Reduction step) For each $A^{(n)}(i_j|\cdot) \in A_j^{(n)}(\cdot)$, generate a family of reduced sets, indexed by i_{j+1},

$$A_{j+1}^{(n)}(i_j, \ldots, i_1) = \{A^{(n)}(i_{j+1}|\cdot)|A^{(n)}(i_{j+1}|\cdot) = A^{(n)}(i_j|\cdot)/\{\omega\}, \omega \in C_{j+1}(A^{(n)}(i_j|\cdot))\}.$$

Return to step [2] with j updated to $j + 1$.

[4] From the set of all k-reduced problems select the set $A^{(n)}$ with the maximum optimal value $G^*(A^{(n)})$. The optimal strategy for the growth with security problem (DP_n) is $X^*(A^{(n)})$.

The reduction step in the elimination algorithm is founded on the following simple result which establishes the existence of candidate scenarios for elimination.

Theorem 3 (Scenario elimination). *Assume there exists an "optimal" set $A^{(n)}$ for the problem (DP_n) and consider the j step reduced problem $P(A^{(n)}(i_j|\cdot))$ where j active constraints have been sequentially eliminated. That is, the corresponding scenarios are in $\Omega^{(n)}/A^{(n)}$. If $C_{j+1}^{(n)}(A^{(n)}(i_j|\cdot)) \neq \emptyset$ are the scenarios corresponding to active constraints in the solution of the reduced problem, then*

$$C_{j+1}(A^{(n)}(i_j|\cdot)) \cap [\Omega^{(n)}/A^{(n)}] \neq \emptyset.$$

Proof. It suffices to consider the first elimination. In the solution to the problem with all constraints, $P(\Omega^{(n)})$, there are sets of scenarios $C_1(\Omega^{(n)})$ and $\Omega^{(n)}/C_1(\Omega^{(n)})$ corresponding to active and inactive constraints, respectively, where $C_1(\Omega^{(n)}) \neq \emptyset$. Since the inactive constraints exceed the goal $\ln w^*$

$$C_1(\Omega^{(n)}) \subseteq A^{(n)} \quad \text{and} \quad C_1(\Omega^{(n)}) \cap [\Omega^{(n)}/A^{(n)}] \neq \emptyset.$$

The procedure for identifying $A^{(n)}$ has a tree structure for reduced problems. If, for example, the number of active constraints in each reduced problem is $r \geqslant 2$, then the k eliminations require solving $\sum_{j=1}^{k} r^j$ problems. This could be unmanageable. A simple heuristic is to solve one reduced problem at each stage corresponding to eliminating the active constraint in the previous stage with the best result (highest expected log wealth). With this heuristic method at most rk problems are solved. The heuristic is equivalent to the exact method if there is at most one active constraint at each stage.

5. Application to the fundamental problem of asset allocation over time

The computation of growth with security strategies is now illustrated with the determination of optimal fractions over time in cash, bonds and stocks. Table 1 has information on the yearly asset returns of the S&P500, the Salomon Brothers Bond index and U.S. T-bills for 1980–1990 with data from Data Resources, Inc. Without

Table 1
Yearly rate of return on assets relative to cash (%)

Parameter	Stocks	Bonds	Cash
Mean: μ	108.75	103.75	100
Standard deviation: σ	12.36	5.97	0
Correlation: ρ	0.32		

Table 2
Rates of return scenarios

Scenarios	Stocks	Bonds	Cash	Probability
1	95.00	101.50	100	0.25
2	106.50	110.00	100	0.25
3	108.50	96.50	100	0.25
4	125.00	107.00	100	0.25

loss of generality cash returns are set to one in each period and the mean returns for other assets are adjusted for this shift. The standard deviation for cash is small and is set to 0 for convenience.

A simple grid was constructed from the assumed lognormal distribution for stocks and bonds by partitioning $\Re^2$ at the centroid along the principal axes. A sample point was selected from each quadrant with the goal of approximating the parameter values. The sample points are shown in Table 2.

The planning horizon is $T = 3$, so that there are 64 scenarios each with probability 1/64. Problems are solved with the VaR constraint and then for comparison, with the stronger drawdown constraint.

(i) VaR control with $w^* = a$.
Consider the problem

$$\max_X \left\{ E \sum_{t=1}^{3} \ln\left(R(t)^\top X(t)\right) \middle| Pr\left[\sum_{t=1}^{3} \ln(R(t)^\top X(t)) \geqslant 3 \ln a\right] \geqslant 1 - \alpha \right\}.$$

With initial wealth $W(0) = 1$, the value at risk is a^3.

Table 3 presents the optimal investment decisions and optimal growth rate for several values of a, the secured average annual growth rate and $1 - \alpha$, the security level.

The heuristic was used to determine A, the set of scenarios for the security constraint. Since only a single constraint was active at each stage the solution is optimal. The mean return structure for stocks is quite favorable in this example, as is typical over long horizons (see Keim and Ziemba, 2000), and the Kelly strategy, not surprisingly, is to invest all capital in stock most of the time. It is only when security requirements are high that some capital is in bonds. As the requirements increase the fraction invested in the more secure bonds increases. The three-period investment decisions become more

Table 3
Growth with secured rate

Secured growth rate a	Secured level $1-\alpha$	Period 1			2			3			Optimal growth rate (%)
		Stocks	Bonds	Cash	Stocks	Bonds	Cash	Stocks	Bonds	Cash	
0.95	0	1	0	0	1	0	0	1	0	0	23.7
	0.85	1	0	0	1	0	0	1	0	0	23.7
	0.9	1	0	0	1	0	0	1	0	0	23.7
	0.95	1	0	0	1	0	0	1	0	0	23.7
	0.99	1	0	0	0.492	0.508	0	0.492	0.508	0	19.6
0.97	0	1	0	0	1	0	0	1	0	0	23.7
	0.85	1	0	0	1	0	0	1	0	0	23.7
	0.9	1	0	0	1	0	0	1	0	0	23.7
	0.95	1	0	0	1	0	0	1	0	0	23.7
	0.99	1	0	0	0.333	0.667	0	0.333	0.667	0	18.2
0.99	0	1	0	0	1	0	0	1	0	0	23.7
	0.85	1	0	0	1	0	0	1	0	0	23.7
	0.9	1	0	0	1	0	0	1	0	0	23.7
	0.95	1	0	0	0.867	0.133	0	0.867	0.133	0	19.4
	0.99	0.456	0.544	0	0.27	0.73	0	0.27	0.73	0	12.7
0.995	0	1	0	0	1	0	0	1	0	0	23.7
	0.85	1	0	0	0.996	0.004	0	0.996	0.004	0	23.7
	0.9	1	0	0	0.996	0.004	0	0.996	0.004	0	23.7
	0.95	1	0	0	0.511	0.489	0	0.442	0.558	0	19.4
	0.99	0.27	0.73	0	0.219	0.59	0.191	0.218	0.59	0.192	12.7
0.999	0	1	0	0	1	0	0	1	0	0	23.7
	0.85	1	0	0	0.956	0.044	0	0.956	0.044	0	23.4
	0.9	1	0	0	0.956	0.044	0	0.956	0.044	0	23.4
	0.95	1	0	0	0.381	0.619	0	0.51	0.49	0	19.1
	0.99	0.27	0.73	0	0.008	0.02	0.972	0.008	0.02	0.972	5.27

conservative as the horizon approaches. Although this example is simplified the patterns observed illustrate the effect of security constraints on decisions and growth.

(ii) Secured annual drawdown: b.

The VaR condition only controls loss at the horizon. At intermediate times the investor could experience substantial loss, and in practice be unable to continue. The more stringent risk control constraint, referred to as drawdown, considers the loss in each period. Consider, then, the problem

$$\max_X \left\{ E\sum_{t=1}^{3} \ln(R(t)^\top X(t)) \,\middle|\, Pr[\ln(R(t)^\top X(t)) \geqslant \ln b, t=1,2,3] \geqslant 1-\alpha \right\}.$$

950 L.C. MacLean et al. / Journal of Economic Dynamics & Control 28 (2004) 937–954

Table 4
Growth with secured maximum drawdown

Drawdown b	Secured level $1-\alpha$	Period 1 Stocks	Period 1 Bonds	Period 1 Cash	Period 2 Stocks	Period 2 Bonds	Period 2 Cash	Period 3 Stocks	Period 3 Bonds	Period 3 Cash	Optimal growth rate (%)
0.96	0	1	0	0	1	0	0	1	0	0	23.7
	50	1	0	0	1	0	0	0.846	0.154	0	23.1
	75	1	0	0	0.846	0.154	0	0.846	0.154	0	22.5
	100	0.846	0.154	0	0.846	0.154	0	0.846	0.154	0	21.9
0.97	0	1	0	0	1	0	0	1	0	0	23.7
	50	1	0	0	1	0	0	0.692	0.308	0	22.5
	75	1	0	0	0.692	0.308	0	0.692	0.308	0	21.3
	100	0.692	0.308	0	0.692	0.308	0	0.692	0.308	0	20.1
0.98	0	1	0	0	1	0	0	1	0	0	23.7
	50	1	0	0	1	0	0	0.538	0.462	0	21.2
	75	1	0	0	0.538	0.462	0	0.538	0.462	0	18.6
	100	0.538	0.462	0	0.538	0.462	0	0.538	0.462	0	16.1
0.99	0	1	0	0	1	0	0	1	0	0	23.7
	50	1	0	0	1	0	0	0.385	0.615	0	21.2
	75	1	0	0	0.385	0.615	0	0.385	0.615	0	18.6
	100	0.385	0.615	0	0.385	0.615	0	0.385	0.615	0	16.1
0.999	0	1	0	0	1	0	0	1	0	0	23.7
	50	1	0	0	1	0	0	0.105	0.284	0.611	17.7
	75	1	0	0	0.105	0.284	0.611	0.105	0.284	0.611	11.8
	100	0.105	0.284	0.611	0.105	0.284	0.611	0.105	0.284	0.611	5.84

The simple form of the constraint follows from the arithmetic random walk $\ln W(t)$, where

$$Pr[W(t+1) \geqslant bW(t),\ t=0,1,2] = Pr[\ln W(t+1) - \ln W(t) \geqslant \ln b,\ t=0,1,2]$$
$$= Pr[\ln (R(t)^{\top} X(t)) \geqslant \ln b,\ t=1,2,3].$$

In Table 4 the optimal investment decisions and growth rate for several values of b, the drawdown and $1-\alpha$, the security level are presented. The heuristic is used in determining scenarios in the solution. The security levels are different in the table since constraints are active at different probability levels in this discretized problem.

As with the VaR constraint, investment in the more secure bonds and cash increases as the drawdown rate and/or the security level increases. Also the strategy is more conservative as the horizon approaches. For similar requirements (compare $a=0.97$, $1-\alpha=0.85$ and $b=0.97$, $1-\alpha=0.75$), the drawdown condition is more stringent, with the Kelly strategy (all stock) optimal for VaR constraint, but the drawdown constraint requires substantial investment in bonds in the second and third periods. In general, consideration of drawdown requires a heavier investment in secure assets and at an earlier time point. It is not a feature of this aggregate example, but both the VaR and drawdown constraints are insensitive to large losses, which occur with small probability. Control of that effect would require the lower partial mean violations condition or a model with a convex risk measure that penalizes more and more as larger constraint violations occur, see e.g. Cariño and Ziemba (1998). These results can be compared with those of Grauer and Hakansson (1997) who do calculations with the standard capital growth-Kelly model and Brennan and Schwartz (1998) who use a Merton, continuous time model with the instantaneous mean returns dependent upon fundamental factors. These models have hair trigger type behavior that is very sensitive to small changes in mean values (see Chopra and Ziemba, 1993).

6. Portfolio rebalancing

The approach to investment planning with downside risk control is to develop an optimal strategy over a T period planning horizon using projections of the multivariate returns distributions on securities. Although securities prices are dynamic, the changes are generated from a pricing model with seed parameters (μ, Γ, Δ).

It is anticipated that the planning horizon is short, and the values of the seed parameters will be reconsidered at the end of the horizon. An important feature of the proposed pricing model is the ability to revise the seed parameters using data collected during the planning period.

Consider the observations $\{Y(1), \ldots, Y(T)\}$. Since $(Y(T)|\pi) \sim N(\pi T, T\Delta)$ and $\pi \sim N(\mu, \Gamma)$, from Bayes Theorem

$$(\pi(T)|Y(T)) \sim N(\pi_T, \Gamma_T), \tag{13}$$

where

$$\pi_T = \mu + (I - \Delta_T \Sigma_T^{-1})(\bar{Y}_T - \mu),$$

$$\Gamma_T = \frac{1}{T^2}(I - \Delta_T \Sigma_T^{-1})\Delta_T,$$

$$\bar{Y}_T = \frac{1}{T} Y(T),$$

$$\Delta_T = T\Delta$$

and

$$\Sigma_T = T\Delta + T^2 \Gamma. \tag{14}$$

Furthermore, from the increments $Z_i(t) = Y_i(t+1) - Y_i(t)$, $t = 1, \ldots, T-1$, the covariance matrix S_T can be computed. With the number of factors (mutual funds) set at $p < m$, perform a factor analysis on S_T obtaining a loading matrix L_T and the specific variance matrix $D_T = diag(D_1^2, \ldots, D_m^2)$. Let $S_T^* = L_T L_T^\top + D_T$. Then S_T^* is an estimate of Σ_T, and D_T is an estimate of Δ_T. If there is a common mean, $\mu^\top = (\mu, \ldots, \mu)$

and $\bar{\bar{Y}}_T = \frac{1}{m}\sum_{i=1}^{m} \bar{Y}_i(T)$, then $\bar{\bar{Y}}_T$ is an estimate of μ. All parameters in (13) and (14) are now estimated and the rate of return distribution for the next T period planning cycle can be specified.

The revision of the rate of return distributions produces a rebalanced portfolio in the next planning cycle in light of new information on securities prices. The formula for π_T in (14) displays reversion to the grand mean. Considering the impact of errors in estimating mean values (Chopra and Ziemba, 1993), the improved estimates lead to more reliable investment decisions.

An alternative to blending the mean estimate with a grand mean would be to blend the mean with the prior Bayes estimate. That is, for successive planning periods $\{1,\dots,T_1,T_1+1,\dots,T_1+T_2\}$ the revised estimate for the mean rate of return is $\pi_{T_2} = \pi_{T_1} + (I - \Delta_{T_2}\Sigma_{T_2}^{-1})(\bar{Y}_{T_2} - \pi_{T_1})$. This approach provides smoothed estimates where the full history of prices is considered with the past being weighted in the manner of exponential smoothing.

7. Conclusion

This paper considers the problem of investment in risky securities with the objective of achieving maximal capital growth while controlling for downside risk. Working in discrete time, a geometric random walk model for asset prices was developed. The model has two important features. The increments in the random walk have a Bayes framework, so that the asset prices depend on hyper parameters. In addition the correlation in the asset price distributions was related to a structural model and thereby the hyper parameters were identified from data.

A variety of risk measures were defined and corresponding capital growth with security problems were presented. The emphasis was on the Value at Risk, but control of period-by-period drawdown was also considered in the application.

An algorithm for the computation of growth with security strategies was presented for the problems using discrete approximations to the distributions on asset returns. The computational procedure is general and applies to any price distributions, although it was presented in the context of the geometric random walk and log normal asset prices.

The methods were applied to an example where investment capital is allocated to stocks, bonds and cash over time. At low levels of risk control the capital growth or Kelly strategy is optimal. As the risk control requirements are tightened the strategy becomes more conservative, particularly close to the planning horizon. The solved problems are discrete in time and state, and in contrast to the continuous time lognormal model (MacLean et al., 2000) a fractional Kelly strategy is generally not optimal.

Acknowledgements

This research was partially supported by The Natural Sciences and Engineering Research Council of Canada. It was presented at the 16th International Symposium of

Mathematical Programming, Lausanne, Switzerland, August 1997, the 21st Meeting of the EURO Working Group on Financial Modelling, Venice, Italy, October 1997, the 9th International Conference on Stochastic Programming, Berlin, August 2001 and the Bachelier Finance Society, Crete, June 2002.

References

Algoet, P., Cover, T., 1988. Asymptotic optimality and asymptotic equipartition properties of log-optimum investments. Annals of Probability 16, 876–898.

Artzner, P., Delbaen, F, Eber, J., Heath, D., 1999. Coherent measures of risk. Mathematical Finance 9, 203–228.

Aucamp, D., 1993. On the extensive number of plays to achieve superior performance with the geometric mean strategy. Management Science 39, 1163–1172.

Basak, S., Shaprio, A., 2001. Value-at-risk based risk management: optimal policies and asset prices. Review of Financial Studies 14, 371–405.

Breiman, L., 1961. Optimal gambling system for favorable games. Proceedings of the Fourth Berkeley Symposium on Mathematical Statistics and Probability, Vol. 1, pp. 63–68.

Breitmeyer, C., Hakenes, H., Pfingsten, A., Rechtien, C., 1999. Learning from poverty measurement: an axiomatic approach to measure downside risk. Working Paper, University of Muenster, Germany.

Brennan, M., Schwartz, E., 1998. The use of treasury bill futures in strategic asset and liability modelling. In: Ziemba, W.T., Mulvey, J.M. (Eds.), Worldwide Asset and Liability Modelling. Cambridge University Press, Cambridge, pp. 205–228.

Cariño, D., Ziemba, W.T., 1998. Formulation of Russell–Yasudda Kasai financial planning model. Operations Research 46, 433–449.

Chopra, V., Ziemba, W.T., 1993. The effect of errors in mean, variance and co-variance estimates on optimal portfolio choice. Journal of Portfolio Management Winter 19, 6–11.

Cover, T., 1991. Universal portfolios. Mathematical Finance 1, 1–29.

Cox, J., Leland, H., 2000. On dynamic investment strategies. Journal of Economic Dynamics and Control 24, 1859–1880.

Culioli, J.C., 1996. Problèmes de minimax partial. Rapport Interne CAS, Ecole des Mines de Paris.

Ethier, S.N., Tavarè, 1983. The proportional bettor's return on investment. Journal of Applied Probability 20, 563–573.

Finkelstein, M., Whitley, R., 1981. Optimal strategies for repeated games. Advanced Applied Probability 13, 415–428.

Grauer, R.R., Hakansson, N.H., 1997. On naive approaches to timing the market: the empirical probability assessment approach with an inflation adapter. In: Ziemba, W.T., Mulvey, J.M. (Eds.), Worldwide Asset and Liability Modelling. Cambridge University Press, Cambridge, pp. 149–181.

Grossman, S.J., Zhou, Z., 1993. Optimal investment strategies for controlling drawdowns. Mathematical Finance 3, 241–276.

Hakansson, N.H., 1972. On optimal myopic portfolio policies with and without serial correlation. Journal of Business 44, 324–334.

Hakansson, N.H., Ziemba, W.T., 1995. Capital growth theory. In: Jarrow, R.A., Maksimovic, V., Ziemba, W.T. (Eds.), Finance Handbook. North-Holland, Amsterdam, pp. 123–144.

Helmbold, D., Schapire, R., Singer, Y., Warmuth, M., 1996. On-line portfolio selection using multiplicative updates. In: Machine Learning. Proceedings of the 13th International Conference, pp. 243–251.

Jorion, P., 1997. Value-at-Risk: The New Benchmark for Controlling Market Risk. Irwin, Chicago.

Keim, D., Ziemba, W.T. (Eds.), 2000. Security Market Imperfections in Worldwide Equity Markets. Cambridge University Press, Cambridge.

Kelly, J., 1956. A new interpretation of information rate. Bell System Technology Journal 35, 917–926.

MacLean, L., Ziemba, W.T., 1991. Growth-security profiles in capital accumulation under risk. Annal of Operations Research 31, 501–510.

MacLean, L., Ziemba, W.T., 1999. Growth versus security tradeoffs in dynamic investment analysis. Annals of Operations Research 85, 193–227.
MacLean, L., Ziemba, W.T., Blazenko, G., 1992. Growth versus security in dynamic investment analysis. Management Science 38, 1562–1585.
MacLean, L., Ziemba, W.T., Li, Y., 2000. Time to wealth goals in capital accumulation and the optimal trade-off of growth versus security. Working Paper, School of Business Administration, Dalhousie University.
Markowitz, H., 1959. Portfolio Selection. Yale University Press, New Haven, CT.
Markowitz, H., 1976. Investment for the long run: New evidence for an old rule. Journal of Finance 31, 1273–1286.
Merton, R.C., 1969. Lifetime portfolio selection under uncertainty: the continuous time case. Review of Economics and Statistics 51, 247–259.
Merton, R.C., 1992. Continuous-Time Finance, 2nd Edition. Blackwell Publishers, Cambridge, MA.
Pflug, G., 1999. Stochastic programs and statistical data. Annals of Operations Research 85, 59–79.
Rockafeller, T., Wets, R.J.B., 1978. The optimal recourse problem in discrete time: L'-multipliers for inequality constraints. SIAM Journal of Control and Optimization 16, 16–36.
Rotando, L.M., Thorp, E.O., 1992. The Kelly criterion and the stock market. The American Mathematical Monthly December, 992–1032.
Rubinstein, M., 1977. The strong case for log as the premier model for financial modeling. In: Levy, H., Sarnet, M. (Eds.), Financial Decisions Under Uncertainty. Academic Press, New York.
Rubinstein, M., 1991. Continuously rebalanced investment strategies. Journal of Portfolio Management 17, 78–81.
Thorp, E.O., 1998. The Kelly criterion in blackjack, sports betting, and the stock market. Mimeo.
Ziemba, W.T., Hausch, D., 1986. Betting at the Racetrack. Dr. Z Investments Inc., Los Angeles, CA.

26

Risk-Constrained Dynamic Active Portfolio Management

Sid Browne
Goldman, Sachs and Company, Firmwide Risk Management, 10 Hanover Square, New York, New York 10005, and Graduate School of Business, Columbia University, New York, New York 10027
sid.browne@gs.com • sb30@columbia.edu

Active portfolio management is concerned with objectives related to the outperformance of the return of a target benchmark portfolio. In this paper, we consider a dynamic active portfolio management problem where the objective is related to the tradeoff between the achievement of performance goals and the risk of a shortfall. Specifically, we consider an objective that relates the probability of achieving a given performance objective to the time it takes to achieve the objective. This allows a new direct quantitative analysis of the risk/return tradeoff, with risk defined directly in terms of probability of shortfall relative to the benchmark, and return defined in terms of the expected time to reach investment goals relative to the benchmark. The resulting optimal policy is a state-dependent policy that provides new insights. As a special case, our analysis includes the case where the investor wants to minimize the expected time until a given performance goal is reached subject to a constraint on the shortfall probability.
(*Portfolio Theory; Benchmarking; Active Portfolio Management; Stochastic Control*)

1. Introduction

In this paper we analyze an optimal dynamic portfolio and asset allocation policy for an investor who is concerned about the performance of a portfolio relative to the performance of a given benchmark. We take as our setting the standard continuous-time framework pioneered by Merton (1971) and others. Portfolio problems where the objective is to exceed the performance of a selected target benchmark is sometimes referred to as *active* portfolio management, whereas *passive* portfolio management just tries to *track* a benchmark, see, for example, Sharpe et al. (1995). Many professional and institutional investors in fact follow this benchmarking procedure: For example, many mutual funds take the Standard and Poors (S&P) 500 Index as a benchmark; commodity funds seek to beat the Goldman Sachs Commodity Index; bond funds try to beat the Lehman Brothers Bond Index, etc. Moreover, benchmarking is not specific to professional investors, as many ordinary investors implicitly follow a benchmarking procedure, for example, by trying to beat inflation, exchange rates, or other indices. In other applications, such as pension funds, the benchmark might be a liability. See Litterman and Winkelman (1996) for more detail on these and other benchmarks. For a treatment of active portfolio management in a static setting, see Grinold and Kahn (1995).

This paper extends the earlier analysis in Browne (1999a) of active portfolio management problems with objectives related to the achievement of relative performance goals and shortfalls (see Browne 2000 for related stochastic differential games). Specifically, Browne (1999a) considered a general problem in an incomplete market where the benchmark was only partially correlated with the active investor's investment opportunities and with investment objectives related to the achievement of investment goals and shortfalls relative to the benchmark. The specific objectives that were explicitly solved for there include:

Management Science © 2000 INFORMS
Vol. 46, No. 9, September 2000 pp. 1188–1199

0025-1909/00/4609/1188$05.00
1526-5501 electronic ISSN

maximizing the probability that the investor's wealth achieves a certain performance goal relative to the benchmark before falling below it to a predetermined shortfall; minimizing the expected time to reach the performance goal; and maximizing the expected reward obtained upon reaching the goal, as well as minimizing the expected penalty paid upon falling to the shortfall level. The corresponding optimal policies obtained there are all constant proportion, or constant mix, portfolio allocation strategies, whereby the portfolio is continuously rebalanced so as to always keep a constant proportion of wealth in the various asset classes, regardless of the level of wealth. (Observe that if the proportion associated with an asset class is positive, then this rebalancing requires selling an asset when its price rises relative to the other prices, and conversely, buying the asset when its price drops relative to the others.) It is well-known that such policies have a variety of optimality properties associated with them for the ordinary portfolio problem (see, e.g., Merton 1990 or Browne 1998 for surveys) and are widely used in asset allocation practice (see Perold and Sharpe 1988 and Black and Perold 1992). Nevertheless, some investors object to using constant proportion strategies in that they dictate the same strategy for every wealth level, while their individual intuition would suggest otherwise. In this paper we address some of these issues for the complete market case where the investor is allowed to invest in all the individual components of the benchmark. From an analytic point of view, the problem addressed here is solvable only in the complete market case, which is a somewhat more restrictive setting than that of Browne (1999a). However, it is in fact the complete market case that is of most interest to active portfolio practitioners. Our analysis allows us to extend the domain of goal/shortfall-related objectives with known explicit solutions to a case that allows for a very interesting and intuitive state-dependent optimal policy. (A continuous-time active portfolio management problem with a finite-horizon probability-maximizing objective in a complete market setting was studied in Browne (1999b, 1999c). The optimal portfolio policy in that case turns out to be intimately related to hedging strategies for certain options, and as such is *both* time and state dependent.)

An outline of the remainder of the paper, and a summary of our main results are as follows: In the next section, we provide a description of the model and the problems studied. For the objectives considered here, the relevant state variable is the *ratio* of the investor's wealth to the benchmark. We then state for reference a general theorem in stochastic control for our model, which contains the specific goal-related objectives considered in the sequel as a special case. The upshot of this theorem is that it shows how the optimal value function and associated optimal control function for a general control problem can be obtained as the solution to a particular nonlinear Dirichlet problem. The theorem is a special case of the more general result in Browne (1999a), and so it is stated without proof. Because the specific goal-related problems considered in the sequel are special cases, we need only identify and then solve the appropriate nonlinear Dirichlet problem.

In §3 we apply the theorem to show that the ordinary optimal-growth portfolio policy for the case of an investor without a benchmark is once again optimal in our extended model, in that regardless of the underlying benchmark, the ordinary optimal-growth strategy will minimize the expected time until that benchmark is outperformed by any given percentage. While this result is not new, it is quite important for the sequel. In particular, it highlights the rather disturbing property of this optimal-growth policy in that it yields no insight for the portfolio manager as to how the benchmark affects the investment decision, because for this objective the benchmark is in fact irrelevant. Moreover, as we show below, not only is the policy independent of the benchmark, but the *probability* that the active manager using the optimal-growth policy reaches an investment goal before falling to a shortfall level relative to the benchmark is also *independent* of the benchmark as well as any other parameters of the assets. Given these disturbing results, we move on in §4 to consider a fractional objective that relates the time to beat the benchmark to the probability of a shortfall relative to the benchmark. For this objective, the optimal strategy is no longer a constant proportion, but rather a state-dependent amount that modulates the amount invested in the risky assets in inverse proportion to wealth. As a special case, our results are then applied to the objective of minimizing the expected time to reach the goal subject to a constraint on the shortfall probability. This objective is motivated by the interesting gambling model of Gottlieb (1985).

2. The Model

The model under consideration here consists of $k+1$ underlying processes: k (correlated) risky assets or stocks $S^{(1)},\ldots,S^{(k)}$ and a riskless asset B called a bond. The investor may invest in the risky stocks and the bond, whose price processes will be denoted, respectively, by $\{S_t^{(i)}, t\geq 0\}_{i=1}^k$ and $\{B_t, t\geq 0\}$.

The probabilistic setting is as follows: We are given a filtered probability space $(\Omega, \mathcal{F}, \{\mathcal{F}_t\}, \mathrm{P})$, supporting k independent standard Brownian motions, $(W^{(1)},\ldots,W^{(k)})$, where $\mathcal{F}_t$ is the P-augmentation of the natural filtration $\mathcal{F}_t^W = \sigma\{W_s^{(1)}, W_s^{(2)},\ldots,W_s^{(k)}; 0\leq s\leq t\}$ (see e.g., Duffie 1996 for a brief review of the relevant terminology).

It is assumed that these k Brownian motions generate the prices of the k risky stocks. Specifically, following Merton (1971) and many others, we will assume that the risky stock prices are correlated geometric Brownian motions, i.e., $S_t^{(i)}$ satisfies the stochastic differential equation

$$dS_t^{(i)} = \mu_i S_t^{(i)} dt + \sum_{j=1}^k \sigma_{ij} S_t^{(i)} dW_t^{(j)}, \quad \text{for } i=1,\ldots,k, \quad (1)$$

where $\{\mu_i : i=1,\ldots,k\}$ and $\{\sigma_{ij} : i,j=1,\ldots,k\}$ are constants. The price of the risk-free asset is assumed to evolve according to

$$dB_t = r B_t\, dt, \quad (2)$$

where $r\geq 0$. We assume that $\mu_i > r$ for all $i=1,\ldots,k$.

An investment policy is a (column) vector control process $f=\{f_t : t\geq 0\}$ in R^k with individual components $f_t^{(i)}$, $i=1,\ldots,k$, where $f_t^{(i)}$ is the fraction (we use f for fraction) or proportion of the investor's wealth invested in the risky stock i at time t, for $i=1,\ldots,k$, with the remainder invested in the risk-free bond. It is assumed that $\{f_t, t\geq 0\}$ is a suitable, admissible $\mathcal{F}_t$-adapted control process, i.e., f_t is a nonanticipative function that satisfies $\int_0^T f_t' f_t dt < \infty$ a.s. for every $T<\infty$. We place no other restrictions on f, for example, we allow $\sum_{i=1}^k f_t^{(i)} \geq 1$, whereby the investor is leveraged and has borrowed to purchase the stocks, as well as $f_t^{(i)} < 0$, whereby the investor is selling stock i short.

Let X_t^f denote the *wealth* of the investor at time t under policy f, with $X_0 = x$. Since any amount not invested in the risky stock is held in the bond, this process then evolves as

$$\begin{aligned} dX_t^f &= X_t^f\left(\sum_{i=1}^k f_t^{(i)} \frac{dS_t^{(i)}}{S_t^{(i)}}\right) + X_t^f\left(1-\sum_{i=1}^k f_t^{(i)}\right)\frac{dB_t}{B_t} \\ &= X_t^f\left(r + \sum_{i=1}^k f_t^{(i)}(\mu_i - r)\right)dt \\ &\quad + X_t^f \sum_{i=1}^k\sum_{j=1}^k f_t^{(i)}\sigma_{ij}\, dW_t^{(j)} \end{aligned} \quad (3)$$

upon substituting from (1) and (2). This is the wealth equation first studied by Merton (1971).

If we introduce now the matrix $\sigma = (\sigma)_{ij}$ and the column vectors $\mu = (\mu_1,\ldots,\mu_k)'$, $\mathbf{1} = (1,\ldots,1)'$, and $W_t = (W_t^{(1)},\ldots,W_t^{(k)})'$, we can rewrite the wealth process of (3) as

$$dX_t^f = X_t^f[(r + f_t'(\mu - r\mathbf{1}))dt + f_t'\sigma\, dW_t]. \quad (4)$$

For the sequel, we will also need the matrix $\Sigma = \sigma\sigma'$. It is assumed for the sequel that the square matrix σ is of full rank, hence σ^{-1} (and Σ^{-1}) exists.

2.1. The Benchmark Portfolio

As described above, our interest lies in determining investment strategies that are optimal relative to the performance of a benchmark. The benchmark we work with here is the wealth associated with another portfolio strategy $\pi = (\pi(1),\ldots,\pi(k))'$ where $\pi(i)$ denotes the fraction of the benchmark wealth invested in the i-th stock. Accordingly, the benchmark portfolio evolves similarly to (4), as

$$dX_t^\pi = X_t^\pi[(r + \pi'(\mu - r\mathbf{1}))dt + \pi'\sigma\, dW_t]. \quad (5)$$

For example, if $\pi(i)=0$ for each $i=1,\ldots,k$, then the benchmark is simply "cash," which is the relevant benchmark in a variety of situations (see Litterman and Winkelman 1996). Alternatively, if $\pi(i)=1$, with $\pi(j)=0$ for all $j\neq i$, then the benchmark is just the i-th stock or asset. We note that while the problems studied in this paper can in fact be treated with more general benchmarks (and in more general settings), here we only consider the constant coefficients case for analytical and economic simplicity.

2.2. Optimal Growth

Let π^* be the constant vector defined by

$$\pi^* = \Sigma^{-1}(\mu - r\mathbf{1}). \tag{6}$$

The vector π^* plays a fundamental role in the theory of finance (see Merton 1990, Ch. 6) and will also play a fundamental role in the sequel. Following Merton (1990), we refer to the vector π^* as the *optimal-growth* portfolio strategy. The reason for this is the policy $f_t = \pi^*$, for all t, has many optimality properties associated with it in an ordinary portfolio setting (where there is no benchmark) that are relevant for growth-related objectives. In particular, for an investor whose wealth evolves according to (4), and who is not concerned with performance relative to any benchmark, (i) π^* maximizes the expected logarithm of terminal wealth, for any fixed terminal time T, hence (ii) π^* maximizes the (actual and expected) rate at which wealth compounds. More interesting, perhaps, and certainly more relevant to our concerns here is the property (iii): π^* *minimizes the expected time* until any given level of wealth is achieved (needless to say, so long as that level is greater than the initial wealth). Merton (1990, Ch. 6) contains a comprehensive review of these properties (see also Browne 1998 for further optimality properties). Given these results, it is not surprising that the policy π^* has extended optimality properties in our benchmark-based model as well, as we show below. Most relevant to our concerns is the fact that indeed π^* is the policy that minimizes the expected time until the benchmark portfolio strategy is beaten by any given percentage (see Corollary 1 below). The fact that this holds for *any* benchmark, however, severely limits the applicability of this result in that it does not provide any insight for the active portfolio manager as to the role the benchmark plays in the investment decision, and as such might not be a reasonable objective for an active portfolio manager.

We note also that the ratio of the wealth process of any (admissible) portfolio strategy to the wealth process determined by the optimal growth strategy is a supermartingale. This fact has important consequences for pricing contingent claims, as described, for example, in Merton (1990, Ch. 6).

2.3. Active Portfolio Management

There are of course many possible objectives related to outperforming a benchmark. Here, as in Browne (1999a), we consider objectives related solely to the achievement of relative performance goals and shortfalls or drawdowns. Specifically, for numbers l, u with $lX_0^\pi < X_0^f < uX_0^\pi$, we say that performance goal u is reached (relative to the benchmark asset allocation strategy π) if $X_t^f = uX_t^\pi$, for some $t > 0$, and that performance shortfall level l occurs if $X_t^f = lX_t^\pi$ for some $t > 0$. The active portfolio management problems considered for an *incomplete market* (i.e., where there are more sources of risk than there are traded securities) in Browne (1999a) are: (i) maximizing the probability that performance goal u is reached before shortfall l occurs; (ii) minimizing the expected time until the performance goal u is reached; (iii) maximizing the expected time until shortfall l is reached; (iv) maximizing the expected discounted reward obtained upon achieving goal u; and (v) minimizing the expected discounted penalty paid upon falling to shortfall level l. Among other scenarios, these objectives are relevant to institutional money managers, whose performance is typically judged by the return on their managed portfolio relative to the return of a benchmark. Browne (1999a) showed that the optimal strategy for each of these objectives was in fact a *constant proportions* asset allocation strategy. While constant proportion strategies are optimal in a variety of other settings as well, many professional investors object to them on the grounds that they do not take into account the wealth level of the investor.

In this paper we consider an extended fractional objective that relates the probability in Objective (i) to the expected time in Objective (ii). It turns out that for this objective the optimal strategy is no longer constant, but rather state dependent. In particular, the policy is hyperbolic in the state variable, where the relevant state variable is the *ratio of the wealth process to the benchmark*. We use standard techniques of stochastic control theory (e.g., Krylov 1980) to establish our results since this ratio is a controlled diffusion.

In particular, because X_t^f is a controlled geometric Brownian motion, and X_t^π is another geometric Brownian motion, it follows directly that the ratio process, Z^f, where $Z_t^f = X_t^f / X_t^\pi$, is also a controlled geometric Brownian motion. Specifically, a direct application to Ito's formula gives:

PROPOSITION 1. *For X_t^f, X_t^π defined by (4) and (5), let Z_t^f be defined by $Z_t^f = X_t^f / X_t^\pi$. Then, using the definition of the vector π^* as given in (6), we have*

$$dZ_t^f = Z_t^f (f_t - \pi)' \Sigma (\pi^* - \pi)\, dt + Z_t^f (f_t - \pi)' \sigma\, dW_t. \quad (7)$$

Alternatively, in integral form we have

$$Z_t^f = Z_0 \exp\left\{ \int_0^t (f_s - \pi)' \Sigma \left(\pi^* - \tfrac{1}{2}(f_s + \pi)\right) ds + \int_0^t (f_s - \pi)' \Sigma (f_s - \pi)\, dW_s \right\}. \quad (8)$$

Next, we provide a general theorem in stochastic optimal control for the process $\{Z_t^f, t \geq 0\}$ of (7) that covers the specific problems treated here as special cases.

2.4. Optimal Control

The active portfolio management problems considered in this paper are special cases of optimal control problems of the following (Dirichlet-type) form: For the ratio process $\{Z_t^f, t \geq 0\}$ given by (7), let

$$\tau_x^f = \inf\{t > 0 : Z_t^f = x\} \quad (9)$$

denote the first hitting time to the point x under a specific policy $f = \{f_t, t \geq 0\}$. For given numbers l, u, with $l < Z_0 < u$, let $\tau^f = \min\{\tau_l^f, \tau_u^f\}$ denote the first escape time from the interval (l, u), under this policy f.

For a given real bounded continuous function $g(z)$ and a function $h(z)$ given for $z = l, z = u$, with $h(u) < \infty$, let $\nu^f(z)$ be the reward function under policy f, defined by

$$\nu^f(z) = E_z\left(\int_0^{\tau^f} g(Z_t^f)\, dt + h(Z_{\tau^f}^f) \right), \quad (10)$$

with

$$\nu(z) = \sup_{f \in \mathcal{G}} \nu^f(z), \quad \text{and} \quad f_\nu^*(z) = \arg\sup_{f \in \mathcal{G}} \nu^f(z) \quad (11)$$

denoting, respectively, the optimal value function and associated optimal control function, where $\mathcal{G}$ denotes the set of admissible controls. (Here and in the sequel, we use the notations $P_z(\cdot)$ and $E_z(\cdot)$ as shorthand for $P(\cdot | Z_0 = z)$ and $E(\cdot | Z_0 = z)$.) We note at the outset that we are only interested in controls (and initial values z) for which $\nu^f(z) < \infty$.

REMARK 1. Observe that the reward functional in (10) is sufficiently general to cover a variety of goal-related objectives. For example, the probability of beating the benchmark before being beaten by it, following a given strategy $\{f_t\}$, i.e., $P_z(\tau_u^f < \tau_l^f)$, is a special case with $g(\cdot) = 0$, $h(u) = 1$ and $h(l) = 0$. Similarly, by taking $g(\cdot) = 1$ and $h(u) = 0 = h(l)$, we obtain $E_z(\tau^f)$. Related optimal control problems have been treated previously in various forms for a variety of models. In particular see Pestien and Sudderth (1985), Heath et al. (1987), and Browne (1995, 1997, 1999a). Related stochastic differential games are treated in Browne (2000).

As a matter of notation, we note first that here, and throughout the remainder of the paper, the parameter γ will be defined by

$$\gamma = \gamma(\pi) = (\pi^* - \pi)' \Sigma (\pi^* - \pi)/2, \quad (12)$$

where π^* is the optimal-growth policy of (6) and π is the benchmark under consideration.

The following theorem, which is a special case of the more general Theorem 1 in Browne (1999a), shows that the optimal value function is the solution to a particular nonlinear ordinary differential equation with Dirichlet boundary conditions, and that the optimal policy is given in terms of the first two derivatives of this solution.

THEOREM 1. *Suppose that $w(z)$ is twice continuously differentiable with the first two derivatives given by w_z and w_{zz}, and is the strictly concave increasing (i.e., $w_z > 0$ and $w_{zz} < 0$) solution to the nonlinear Dirichlet problem*

$$-\gamma \frac{w_z^2(z)}{w_{zz}(z)} + g(z) = 0, \quad \text{for } l < z < u, \quad (13)$$

with

$$w(l) = h(l), \quad \text{and} \quad w(u) = h(u), \quad (14)$$

and satisfies the following three conditions:

(i) $(w_z^2(z)/w_{zz}(z))$ is bounded for all z in (l, u);

(ii) for every $t \geq 0$, and every admissible policy f, we have

$$E\int_0^t (Z_s^f w_z(Z_s^f))^2\, ds < \infty; \text{ and} \tag{15}$$

(iii) $(w_z(z)/w_{zz}(z))$ is locally Lipschitz-continuous. Then $w(z)$ is the optimal value function, i.e., $w(z) = v(z)$, and, moreover, the optimal control vector, f_v^*, can then be written as

$$f_v^*(z) = \pi - (\pi^* - \pi)\left(\frac{w_z(z)}{zw_{zz}(z)}\right), \tag{16}$$

where π^* is the vector defined in (6).

The utility of Theorem 1 for our purposes is that for various choices of the functions $g(\cdot)$ and $h(\cdot)$, it addresses the objective problems discussed earlier. Moreover, it shows that for each of these problems, all we need do is solve the ordinary differential Equation (13) and then take the appropriate derivatives to determine the optimal control by (16). Conditions (i), (ii), and (iii) are just technical conditions that ensure integrability of certain functionals, which in turn ensure optimality. We will not discuss them further here, but the interested reader should see Browne (1999a). Conditions (i) and (iii) are easy to check, and while Condition (ii) seems potentially hard to verify, for the cases considered here it is in fact easy, as demonstrated below.

REMARK 2. Observe that the representation of the optimal control vector $f_v^*(z)$ of (16) demonstrates that the optimal portfolio strategy consists of two distinct parts: (i) the *tracking* component π and (ii) the *active* component, $-(\pi^* - \pi)w_z/(zw_{zz})$. Because w is increasing and concave in z, the active component associated with asset i is positive if $\pi^*(i) > \pi(i)$, and negative if $\pi^*(i) < \pi(i)$. That is, the active manager will invest more heavily in asset i than the benchmark if the benchmark is underinvested in asset i relative to the vector π^*, and vice versa. The extent to which this occurs depends on the specifics of the value function $w(z)$.

REMARK 3. As noted earlier, Theorem 1 is a special case of a more general result in Browne 1999a, and as such we do not provide a formal proof. However, to provide some insight into the result, observe that the Hamilton–Jacobi–Bellman (HJB) optimality equation of dynamic programming for maximizing $v^f(z)$ of (10) over control policies f, to be solved for an optimal value function v, is

$$\sup_f\{(f-\pi)'\,\Sigma(\pi^*-\pi)zv_z + (f-\pi)'\,\Sigma(f-\pi)z^2v_{zz} + g\} = 0, \tag{17}$$

subject to the Dirichlet boundary conditions $v(l) = h(l)$ and $v(u) = h(u)$ (see, e.g., Krylov 1980, Theorem 1.4.5, or Fleming and Soner 1993, §IV.5).

Assuming now that (17) admits a classical solution with $v_z > 0$ and $v_{zz} < 0$, we may then use standard calculus to optimize with respect to f in (17) to obtain the optimal control function $f_v^*(x)$ of (16), with $v = w$. When (16) is then substituted back into (17) and simplified, we obtain the nonlinear Dirichlet problem of (13) (with $v = w$). To complete the proof now, one only needs to verify that the solution to the HJB equation is indeed the optimal value function, and hence that the policy f_v^* is indeed optimal. A verification argument based on the martingale optimality principle given in Browne (1999a) covers the case at hand, provided that Conditions (i), (ii), and (iii) hold. A stochastic differential game-theoretic version of the verification argument and martingale principle appears in Browne (2000).

3. Minimizing the Expected Time to Beat the Benchmark

3.1. Optimality of π^*

We can use Theorem 1 to show that, as claimed earlier, the ordinary optimal growth portfolio policy, π^* of (6), is indeed also optimal for minimizing the expected time to beat the benchmark by any predetermined amount, *regardless of the underlying benchmark strategy* π. Indeed, we state this formally in the following corollary to Theorem 1.

COROLLARY 1. *Let $G^*(z) = \inf_f E_z(\tau_u^f)$ with optimizer $f^*(z) = \arg\inf_f E_z(\tau_u^f)$. Then for any $\pi \neq \pi^*$, and γ as defined in (12), we have*

$$G^*(z) = \frac{1}{\gamma}\ln\left(\frac{u}{z}\right), \quad \text{with } f^*(z) = \pi^*, \text{ for all } z \leq u. \tag{18}$$

PROOF. Observe first that while Theorem 1 is stated in terms of a maximization problem, it obviously contains the minimization case, as we can apply

Theorem 1 to $\tilde{G}(z) = \sup_f\{-E_z(\tau_u^f)\}$, and then recognize that $G^* = -\tilde{G}$. As such, Theorem 1 applied with $g(z) = 1$ and $h(u) = 0$ shows that G^* must solve the ordinary differential equation

$$-\gamma \frac{G_z^2(z)}{G_{zz}(z)} + 1 = 0, \tag{19}$$

together with the boundary condition $G^*(u) = 0$. Moreover, G^* must be *convex decreasing* (since it is the solution to a minimization problem). It is easy to substitute the claimed values from (18) into (19) to verify that in fact that is the case. Furthermore, we have $G_z^*/zG_{zz}^* = -1$, and as such (16) of Theorem 1 shows that the optimal control for this case reduces to π^*.

It remains to verify whether the Conditions (i) (ii) and (iii) of Theorem 1 hold: It is clear that (i) and (iii) hold. Condition (ii) is seen to hold for this case since we have $dG^*(z)/dz = -1/(z\gamma)$, and as such Requirement (15) reduces here to $\int_0^t \gamma^{-2} ds < \infty$, which holds trivially. □

Remark 4. We have just shown that the ordinary optimal growth policy, π^*, minimizes the expected time to a goal in the presence of a benchmark. To see that this same policy maximizes logarithmic utility of the ratio for the investor, simply observe that $\sup_f\{E[\ln(Z_T^f)]\} = \sup_f\{E[\ln(X_T^f)]\} - E[\ln(X_T^\pi)]$.

3.2. Properties of the Growth-Optimal Ratio

When π^* is substituted back into (7), we obtain the following stochastic differential equation for the ratio of the growth-optimal wealth to the benchmark

$$dZ_t(\pi^*, \pi) = Z_t(\pi^*, \pi)[\gamma dt + (\pi^* - \pi)'\sigma dW_t], \tag{20}$$

which implies that under π^*, this ratio process is the geometric Brownian motion given by

$$Z_t(\pi^*, \pi) = Z_0 \exp\{\gamma t + (\pi^* - \pi)'\sigma W_t\}. \tag{21}$$

While we did not consider a lower shortfall barrier, l, in the development above proving optimality of the standard growth-optimal portfolio policy, it is of importance in many applications to consider one, since many investors indeed are interested in avoiding substantial shortfalls. In the following proposition we give two fundamental results for the wealth process for an investor following the optimal-growth strategy π^*: (i) the *probability* that the investment goal u is reached before a shortfall of size l occurs; and (ii) the expected time to escape the interval (l, u) (which is not the same as the expected time to the goal).

Proposition 2. *For the process $Z_t(\pi^*, \pi)$ of (21), let τ denote the first escape time from the interval (l, u), and let $\theta(z:l,u)$ denote the probability of "successful" escape, i.e., $\tau = \inf\{t : Z_t(\pi^*, \pi) \notin (l, u)\}$, and $\theta(z:l,u) = P_z(Z_\tau(\pi^*, \pi) = u)$. Then $\theta(z)$ is given by*

$$\theta(z:l,u) = \frac{u}{z}\left(\frac{z-l}{u-l}\right). \tag{22}$$

Also, the expected time of first escape from the interval is given by

$$E_z(\tau(\pi^*, \pi)) = \gamma^{-1}\left[\theta(z:l,u)\ln\left(\frac{u}{l}\right) - \ln\left(\frac{z}{l}\right)\right]. \tag{23}$$

Thus, we see that while π^* is the policy under which any given investment goal will be reached in minimal expected time, and hence in a sense maximizes expected return, it does so with a risk of a shortfall of size l occurring with probability $1 - \theta(z:l,u)$, where $\theta(z:l,u)$ is the probability given in (22).

Remark 5. It is important to note that the probability $\theta(z:l,u)$ is *independent of any of the underlying parameters* associated with the underlying model. In particular, this probability is independent of the *benchmark* policy π. (For example, the probability of the ratio doubling ($u = 2z$) before being halved ($l = 0.5z$) is always $\frac{2}{3}$.) This limits the usefulness of the optimal-growth policy in the active portfolio management setting, since it provides no guide to the manager in how to choose a benchmark. Of course, the expected time to escape the interval (l, u), given by (23), does depend on the underlying parameters, but only through the parameter γ.

Remark 6. Observe that for $l = 0$, we obtain $\theta(z:0,u) = 1$, and the expected escape time from the interval $E_z(\tau(\pi^*, \pi))$ reduces, as it should, to the optimal expected first passage time to the upper barrier $E_z(\tau_u(\pi^*, \pi))$. Of course, for $l > 0$, the expected hitting time of the optimal ratio process to the lower shortfall level, $E_z(\tau_l(\pi^*, \pi))$, is infinite due to the fact that the drift in (20) is positive.

REMARK 7. The probability in (22) and the expected hitting time in (23) can be established directly via a variety of different ways. Most directly we have the following lemma for geometric Brownian motion:

LEMMA 1. *Let X_t denote a geometric Brownian motion that satisfies*

$$dX_t = mX_t dt + \sqrt{2s}\, X_t dW_t, \quad \text{with } X_0 = x \qquad (24)$$

and let $\tau_z = \inf\{t > 0 : X_t = z\}$. *For* $0 \le a \le x \le b$, *define now* $\tau = \min\{\tau_a, \tau_b\}$, *and*

$$K(x) = P_x(\tau_b < \tau_a) \equiv P_x(\tau = \tau_b) \qquad (25)$$

$$H(x) = E_x(\tau). \qquad (26)$$

Then for $m \neq s$ *we have*

$$K(x) = \frac{x^v - a^v}{b^v - a^v}, \quad \text{where } v = 1 - \frac{m}{s}, \qquad (27)$$

$$H(x) = \frac{1}{m - s}\left(K(x)\ln\left(\frac{b}{a}\right) - \ln\left(\frac{x}{a}\right)\right), \qquad (28)$$

while for $m = s$ *we have*

$$K(x) = \frac{\ln(x/a)}{\ln(b/a)}, \qquad (29)$$

$$H(x) = \frac{1}{2m}\left([(\ln b)^2 - (\ln a)^2]\frac{\ln(x/a)}{\ln(b/a)} - [(\ln x)^2 - (\ln a)^2]\right). \qquad (30)$$

PROOF. Recognize first that $\{X_t\}$ is a diffusion process with drift function $\mu(x) = mx$ and diffusion function $\sigma^2(x) = 2sx$. Therefore, if follows from elementary results about one-dimensional diffusions that K and H are the unique solutions to the respective (Dirichlet) problems:

$$mxK_x + sx^2K_{xx} = 0: \quad K(a) = 0,\ K(b) = 1 \qquad (31)$$

$$mxH_x + sx^2H_{xx} + 1 = 0: \quad H(a) = 0,\ H(b) = 0 \qquad (32)$$

(see e.g., Karlin and Taylor 1981, pp. 192–193). The general solution to the second order differential equation in the left-hand side of (31), for $m \neq s$, is $K(x) = C_1 x^v + C_2$, and the boundary conditions determine the constants C_1, C_2, as in (27). For $m = s$, the general solution is $C_1 \ln x + C_2$, and the boundary conditions give (29).

The general solution to the left-hand side of (32) is $C_1 + C_2 x^v - (m - s)^{-1}\ln x$ for $m \neq s$, and from the boundary conditions we determine C_1 and C_2, giving

$$\frac{1}{m - s}\left(\frac{1}{b^v - a^v}\right)[-(b^v - a^v)\ln x + (x^v - a^v)\ln b + (b^v - x^v)\ln a],$$

which is equivalent to (28) above when we simplify using the definition of $K(x)$. For $m = s$, the general solution of (32) is $C_1 - (\frac{1}{2}m)(\ln x)^2 + C_2 \ln x$, which gives (30) after applying the boundary conditions. □

Using this lemma for our purposes, we first note that (20) is distributionally equivalent to a diffusion that evolves according to the stochastic differential equation

$$dX_t = X_t(2\gamma dt + \sqrt{2\gamma}\, dW_t),$$

where W is an independent one-dimensional Brownian motion. As such, identify $m = 2\gamma$, $s = \gamma$, and hence $v = -1$, and then using $b = u$ and $a = l$, (22) and (23) follow upon substitution. □

4. Risk-Related Objectives

4.1. Objective Function and Optimal Policy

The results of the previous section indicate that the optimal growth strategy, π^*, regardless of its many optimality properties in the ordinary portfolio setting, may not be appropriate for an active portfolio manager who cares about downside risk as well as upside growth relative to a benchmark.

As such, in this section we treat an objective that is perhaps more appropriate in that it allows the active portfolio manager to incorporate the shortfall probability directly in the risk/return tradeoff relative to expected growth. In particular, we consider a linear tradeoff between the shortfall probability and the expected time to get to the surplus level. More specifically, we now treat the following objective: For given nonnegative constants α and β, let

$$V^*(z) = \sup_f\{\alpha P_z(Z^f_{\tau^f} = u) - \beta E_z(\tau^f)\}. \qquad (33)$$

As we will show, the optimal dynamic portfolio strategy for this objective is no longer a constant asset allocation strategy. Rather, as we will establish below, the optimal strategy hyperbolically modulates the fractions invested in the risky assets by the level of the ratio process.

THEOREM 2. *Let $V^*(z)$ denote the optimal value function in (33), and let $f^*_V(z)$ denote the associated optimal control function. Then, $V^*(z)$ is given by*

$$V^*(z) = \alpha \ln\left(\frac{z+b}{l+b}\right) \Big/ \ln\left(\frac{u+b}{l+b}\right), \quad \text{for } l \le z \le u, \qquad (34)$$

where the scalar $b = b(\alpha, \beta : u, l, \gamma)$ is given by

$$b = \frac{ue^{-\gamma\alpha/\beta} - l}{1 - e^{-\gamma\alpha/\beta}}. \qquad (35)$$

The optimal portfolio policy, $f^*_V(z)$, is given by

$$f^*_V(z) = \pi - (\pi^* - \pi)\left(1 + \frac{b}{z}\right) \equiv \pi^* + (\pi^* - \pi)\frac{b}{z}, \qquad (36)$$

where b is given by (35), π^* is the ordinary optimal growth vector given earlier in (6),

$$\pi^* = \Sigma^{-1}(\mu - r\mathbf{1}),$$

and π is the benchmark strategy.

REMARK 8. Observe that the portfolio strategy $f^*_V(z)$ is a state-dependent policy that is inversely modulated by the level of the ratio process Z. The final representation in (36) shows that the policy is composed of two parts: First it just uses the optimal-growth policy π^*, and then multiplies the difference between the optimal-growth policy π^* and the tracking portfolio π by the correction term b/z. The sign of the correction factor is determined by the sign of b. Some direct manipulations on (35) shows that the sign of b is the sign of

$$\frac{1}{\gamma}\ln\left(\frac{u}{l}\right) - \frac{\alpha}{\beta}.$$

Comparison now with (18) reveals that we can write this quantity as $G^*(l) - \alpha/\beta$, where $G^*(l)$ is the minimal possible expected time to get from the shortfall level l to the surplus goal u. Thus, b is positive (negative) if the ratio α/β is less (more) than this minimal expected time.

Observe further that if $b > 0$, then the active manager invests more heavily in the i-th stock than does π^* so long as the benchmark is underinvested in that stock relative to the optimal-growth policy, i.e., so long as $\pi^*(i) > \pi(i)$.

Finally, note that (35) shows that we must always have $b \ge -l$.

PROOF. Theorem 1 applies directly to this case with $g(z) = -\beta$, with $h(u) = \alpha$ and $h(l) = 0$. As such, we require that V^* be the concave increasing solution to the nonlinear Dirichlet problem:

$$-\gamma\frac{V_z^2}{V_{zz}} - \beta = 0 \quad \text{for} \quad l < z < u, \qquad (37)$$

and that V^* satisfy the boundary conditions $V^*(u) = \alpha$ and $V^*(l) = 0$.

The general form of the solution to the nonlinear ordinary differential equation in (37) is of the form $V(z) = (\beta/\gamma)\ln(z + C_1) + C_2$, where C_1 and C_2 are arbitrary constants which we can choose to match the boundary conditions. Observe that this function is concave. The boundary condition $V^*(u) = \alpha$ determines that $C_2 = \alpha - (\beta/g)\ln(u + C_1)$, and the boundary condition $V^*(l) = 0$ determines that $C_1 = b$, where b is the constant in (35). As such, the value function is given explicitly by

$$V^*(z) = \alpha + \frac{\beta}{\gamma}\ln\left[(z+b)\frac{(1 - e^{-\gamma\alpha/\beta})}{u - l}\right]. \qquad (38)$$

Observe now that we can invert the relation in (35) to write $\beta/\gamma = \alpha/\ln[(u+b)/(l+b)]$, which when placed into (38) and then simplified, gives the value function in the form that it is given in (34).

For this value function, the conditions given in Theorem 1: (i) and (iii) are seen to hold directly, and (ii) can be established by first noting that $(zw_z)^2 = (\beta/\gamma)^2(z/[z+b])^2$, and then by using the fact that since $z/[u+b] \le z/[z+b] \le z/[l+b]$, we have

$$E\int_0^t (Z_s^f w_z(Z_s^f))^2 ds = \left(\frac{\beta}{\gamma}\right)^2 E\int_0^t \left(\frac{Z_s^f}{Z_s^f+b}\right)^2 ds$$

$$< \left(\frac{\beta/\gamma}{l+b}\right)^2 \int_0^t E(Z_s^f)^2 ds < \infty,$$

where the final inequality holds by the assumption of admissibility.

Because we now have an explicit form for the optimal policy, we may place it into (16) to obtain the optimal control vector given by $f_V^*(z)$ of (36). □

4.2. The Optimal Process

Observe that when we place the optimal control $f_V^*(Z_t)$ of (36) back into (7), we obtain an optimal-wealth process $Z(f_V^*, \pi)$ which we will denote by Z^*, that follows the stochastic differential equation, using the definition of γ from (12),

$$dZ_t^* = 2\gamma(Z_t^*+b)\,dt + (Z_t^*+b)(\pi^*-\pi)'\sigma dW_t. \quad (39)$$

The unique strong solution to (39) is given by

$$Z_t^* = (Z_0+b)\exp\{\gamma t + (\pi^*-\pi)'\sigma W_t\} - b. \quad (40)$$

Comparison with the results on the optimal-growth policy of the last section shows that we can write this in terms of the optimal-growth ratio as

$$Z_t^* = \left(1+\frac{b}{Z_0}\right) Z_t(\pi^*, \pi) - b, \quad (41)$$

where $Z_t(\pi^*, \pi)$ is the ratio of the ordinary optimal-growth wealth to the benchmark, as given in (21).

In the next proposition we list the following two properties of the optimal ratio process Z^*: (i) the probability of reaching the upper surplus goal u before the lower shortfall level l; and (ii) the expected time it takes to escape from the interval (l, u). Recall that the scalar b depends on the benchmark through the parameter γ, and that we always have $b \ge -l$.

PROPOSITION 3. *Let $\{Z_t^*, t \ge 0\}$ denote the optimal-wealth process associated with the control function $f_V^*(z)$ of (36), and let τ^* denote the associated first escape time from the interval (l, u), i.e., $\tau^* = \inf\{t : Z_t^* \notin (l, u)\}$. Also, let $\phi(z : b, l, u)$ denote the probability of successful escape from the interval, i.e., $\phi(z : b, l, u) = P_z(Z_{\tau^*}^* = u)$. Then*

$$\phi(z : b, l, u) = \frac{(z-l)(u+b)}{(z+b)(u-l)}. \quad (42)$$

The expected time of the first escape is given by

$$E_z(\tau^*) = \frac{1}{\gamma}\left[\phi(z : b, l, u)\ln\left(\frac{u+b}{l+b}\right) - \ln\left(\frac{z+b}{l+b}\right)\right]. \quad (43)$$

REMARK 9. Observe that the probability in (42) is now dependent on the benchmark policy π through the parameter b. Note further that as intuition would suggest, the probability ϕ in (42) is larger than the associated probability for the optimal-growth strategy, θ obtained earlier in (22), if $b<0$, and smaller if $b>0$.

PROOF. The results above can be established directly from the earlier Lemma 1 applied to the process Z_t^*+b, because as (41) exhibits, Z_t^*+b is a geometric Brownian motion, with initial state Z_0+b. In fact, (41) implies that Z_t^*+b is distributionally equivalent to a multiple of the optimal-growth ratio described in the last section, i.e., $Z_t^*+b \stackrel{d}{=} (1+b/Z_0)Z_t(\pi^*, \pi)$. As such, we have

$$\phi(z : b, l, u) \equiv P_z(Z_{\tau^*}^* = u) = P_z(Z_{\tau^*}^* + b = u+b)$$

$$= P_z\left(\left(1+\frac{b}{z}\right)Z_\tau(\pi^*, \pi) = u+b\right) \quad (44)$$

where in the latter τ denotes the first escape time of the process $(1+b/z)Z_t(\pi^*, \pi)$ from the interval $(l+b, u+b)$. As such, the results of the previous section allows us to evaluate this latter probability in terms of the function $\theta(\cdot)$ defined earlier in (22). Specifically, the argument in (44) shows that

$$\phi(z : b, l, u) = \theta\left(z : \frac{l+b}{1+b/z}, \frac{u+b}{1+b/z}\right)$$

$$\equiv \theta(z+b : l+b, u+b), \quad (45)$$

and indeed (42) is obtained when we substitute appropriately into (22).

The optimal expected hitting time, $E_z(\tau^*)$ of (43) is derived directly from the fact that under the optimal policy f_V^*, the value function is

$$V^*(z) = \alpha P_z(Z_{\tau^*}^* = u) - \beta E_z(\tau^*),$$

and therefore

$$E_z(\tau^*) = \frac{1}{\beta}[\alpha P_z(Z^*_{\tau^*} = u) - V^*(z)]. \quad (46)$$

Substituting now for $P_z(Z^*_{\tau^*} = u)$ from (42) and for $V^*(z)$ from (34), and using $1/\beta = \ln[(u+b)/(l+b)]/(\alpha\gamma)$ gives (43). □

4.3. Risk-Constrained Minimal Time

The results of the previous section can now be applied directly to the active portfolio management case where the shortfall probability is prespecified. Specifically, suppose that the shortfall probability is prespecified to the active manager to be no more than $1 - p$, where p is a given number between 0 and 1, i.e., the active manager is told that he must have $P_z(Z^f_{\tau^f} = l) \leq 1 - p$, or equivalently, that he must have $P_z(Z^f_{\tau^f} = u) \geq p$. The risk-constrained active portfolio management problem is now to minimize the expected time to beat the benchmark subject to a constraint on the shortfall probability, specifically, to find the strategy $\{f^*_t, t \geq 0\}$ that minimizes $E(\tau^f)$ subject to $P(Z^f_{\tau^f} = u) \geq p$, where p is a given number in $(0, 1)$. This is now related to the gambling problem first solved in Gottlieb (1985). Following Gottlieb, we observe first that the dual of the risk-constrained active portfolio management problem is to maximize the probability that $P(Z^f_{\tau^f} = u)$ subject to a constraint on $E(\tau^f)$. Moreover, observe that should a solution exist, then the constraint will be met at equality, and so we would have $P(Z^f_{\tau^f} = u) = p$.

Let us write the solution to the dual problem, should it exist, as

$$\Psi(z) = \sup_f [P_z(Z^f_{\tau^f} = u) - \beta E_z \tau^f], \quad (47)$$

where β is now the value of a Lagrangian multiplier.

The control problem in (47) is a special case of the problem treated above with $\alpha = 1$, and as such, from (34) we know that the solution is given by

$$\Psi(z) = \Psi(z) = \ln\left(\frac{z + \tilde{b}}{l + \tilde{b}}\right) \Big/ \ln\left(\frac{u + \tilde{b}}{l + \tilde{b}}\right), \quad (48)$$

where $\tilde{b}$ is the value of b in (35) evaluated at $\alpha = 1$, i.e., $\tilde{b} = b(1, \beta)$. The value of β, the Lagrangian multiplier, will be determined from the risk constraint $P(Z^f_{\tau^f} = u) = p$.

We also know from (36) that the associated risk-constrained optimal portfolio strategy is given by

$$f^*(z) = \pi^* + (\pi^* - \pi)\frac{\tilde{b}}{z}. \quad (49)$$

Observe now that we may invert the identity for b in (35) in this case to write the unknown β in terms of the unknown $\tilde{b}$ as

$$\beta = \gamma \left[\ln\left(\frac{u + \tilde{b}}{l + \tilde{b}}\right)\right]^{-1}. \quad (50)$$

Because we require $\beta > 0$, this implies that we require $\tilde{b} \geq -l$ (we always need $b \geq -l$, to ensure that the probability $\phi(z : \cdot, \cdot)$ does not exceed 1, i.e., to keep $\phi \leq 1$).

To determine the value of $\tilde{b}$, we can use the risk constraint evaluated at the initial time 0, i.e., *set*

$$\phi(Z_0; \tilde{b}) = p.$$

Using (42) with $b = \tilde{b}$, this gives us

$$\frac{(Z_0 - l)(u + \tilde{b})}{(Z_0 + \tilde{b})(u - l)} = p,$$

which in turn now can be solved for $\tilde{b}$, giving

$$\tilde{b} = \tilde{b}(p, Z_0, u, l) = \frac{pZ_0(u - l) - u(Z_0 - l)}{Z_0 - l - p(u - l)}. \quad (51)$$

Remark 9. Observe that

$$\inf_{\tilde{b} > -l} \phi(Z_0 : \tilde{b}) = \frac{Z_0 - l}{u - l}$$

and as such, the risk-constrained problem is feasible only for an initial probability level p that satisfies

$$p > \frac{Z_0 - l}{u - l}.$$

Observe too that for $p = 1$, $\tilde{b}$ reduces to $\tilde{b} = -l$, which makes the lower barrier l unattainable as in many "portfolio insurance" models (see Browne 1997). The insurance level $\tilde{b}$ in (51) is positive for values of p satisfying

$$\frac{Z_0 - l}{u - l} < p < \frac{u}{Z_0}\left(\frac{Z_0 - l}{u - l}\right)$$

and $\tilde{b}$ is negative for larger values in the region

$$p > \frac{u}{Z_0}\left(\frac{Z_0 - l}{u - l}\right) \equiv \theta(Z_0),$$

where $\theta(Z_0)$ is the initial probability that the optimal growth wealth/ratio hits u before l, as given in (22). Thus, as intuition suggests, to have a higher ''success'' probability than the optimal-growth strategy, the active portfolio manager must take less risk and invest less (since $\tilde{b} < 0$) than the ordinary optimal-growth investor.

The optimal expected hitting time, $E_z(\tau^*)$ can be obtained directly as

$$E_z(\tau^*) = \frac{1}{\gamma}\left[\frac{z-l}{z+\tilde{b}}\left(\frac{u+\tilde{b}}{u-l}\right)\ln\left(\frac{u+\tilde{b}}{l+\tilde{b}}\right) - \ln\left(\frac{z+\tilde{b}}{l+\tilde{b}}\right)\right]. \tag{52}$$

5. Conclusions

We have studied a goal-related objective for the problem of outperforming a benchmark. The objective relates the time to outperform the benchmark to the shortfall probability. The resulting optimal policy is state-dependent in an intuitive way not captured by previous studies where the optimal policy was of a constant proportions type. Moreover, this optimal policy directly addresses and alleviates one of the undesirable features, in the benchmark outperformance problem, of the ordinary optimal-growth policy, whose shortfall probability was shown to be independent of the benchmark and any other model-specific parameter. As a special case of this objective, we studied the problem of minimizing the expected first passage time subject to a constrained probability of successful escape.

References

Black, F., A. F. Perold. 1992. Theory of constant proportion portfolio insurance. *J. Econom. Dynam. and Control* **16** 403–426.

Browne, S. 1995. Optimal investment policies for a firm with a random risk process: Exponential utility and minimizing the probability of ruin. *Math. Oper. Res.* **20** 937–958.

——. 1997. Survival and growth with a fixed liability: Optimal portfolios in continuous time. *Math. Oper. Res.* **22** 468–493.

——. 1998. The return on investment from proportional portfolio strategies. *Adv. Appl. Probab.* **30**(1) 216–238.

——. 1999a. Beating a moving target: Optimal portfolio strategies for outperforming a stochastic benchmark. *Finance and Stochastics* **3** 275–294.

——. 1999b. Reaching goals by a deadline: Digital options and continuous-time active portfolio management. *Adv. Appl. Probab.* **31** 551–577.

——. 1999c. The risk and reward of minimizing shortfall probability. *J. Portfolio Management* **25**(4) 76–85.

——. 2000 Stochastic differential portfolio games. *J. Appl. Probab.* In press, **37**(1).

Duffie, D. 1996. *Dynamic Asset Pricing Theory*, 2nd ed. Princeton University Press, Princeton, NJ.

Fleming, W. H., H. M. Soner. 1993. *Controlled Markov Processes and Viscosity Solutions*. Springer-Verlag, New York.

Gottlieb, G. 1985. An optimal betting strategy for repeated games. *J. Appl. Probab.* **22** 787–795.

Grinold, R. C., R. N. Kahn. 1995. *Active Portfolio Management*. Irwin, Probus, IL.

Heath, D., S. Orey, V. Pestien, W. Sudderth. 1987. Minimizing or maximizing the expected time to reach zero. *SIAM J. Control and Optim.* **25**(1) 195–205.

Karlin, S., H. M. Taylor. 1981. *A Second Course in Stochastic Processes*. Academic, New York.

Krylov, N. V. 1980. *Controlled Diffusion Processes*. Springer-Verlag, New York.

Litterman, R., K. Winkelmann. 1996. Managing market exposure. *J. Portfolio Management* **22**(4) 32–48.

Merton, R. 1971. Optimum consumption and portfolio rules in a continuous time model. *J. Econom. Theory* **3** 373–413.

——. 1990. *Continuous Time Finance*. Blackwell, MA.

Perold, A. F., W. F. Sharpe. 1988. Dynamic strategies for asset allocation. *Financial Anal. J.* **44**(1) 16–27.

Pestien, V. C., W. D. Sudderth. 1985. Continuous-time red and black: How to control a diffusion to a goal. *Math. Oper. Res.* **10**(4) 599–611.

Sharpe, W. F., G. F. Alexander, J. V. Bailey. 1995. *Investments*, 5th ed. Prentice Hall, NJ.

Accepted by Paul Glasserman; received April 1, 1999. This paper was with the authors 5 months for 1 revision.

27

Fractional Kelly Strategies for Benchmarked Asset Management

Mark Davis
Department of Mathematics, Imperial College London,
London SW7 2AZ, England
mark.davis@imperial.ac.uk

Sébastien Lleo
Department of Mathematics, Imperial College London,
London SW7 2AZ, England
sebastien.lleo@imperial.ac.uk

Abstract

In this paper, we extend the definition of fractional Kelly strategies to the case where the investor's objective is to outperform an investment benchmark. These benchmarked fractional Kelly strategies are efficient portfolios even when asset returns are not lognormally distributed. We deduce the benchmarked fractional Kelly strategies for various types of benchmarks and explore the interconnection between an investor's risk-aversion and the appropriateness of their investment benchmarks.

1 Introduction

Classically, a fractional Kelly strategy with fraction f consists in investing a proportion f of one's wealth in the Kelly criterion, or log utility, optimal portfolio and a proportion $1 - f$ in the risk-free asset. Fractional Kelly strategies play an important role in active investment management in the asset only case, that is when the investor's objective is to maximize the terminal value of his/her wealth, without any benchmark to track or liability to pay (see for example Thorp (2006) and Ziemba (2003) for discussion and additional references). In this chapter, we analyze the role of fractional Kelly strategies in the asset allocation of a benchmarked investor, that is an investor whose objective is to outperform a given investment benchmarks, such as the S&P 500 or the Salomon Smith Barney World Government Bond Index.

The argument developed in this chapter builds on and applies the stochastic control-based model proposed by Davis and Lleo (2008a). Their methodology is founded on the theory of risk-sensitive stochastic control, rather than on the classical stochastic control theory-based approach proposed by Merton (e.g., Merton (1969,

1971, 1992)). In the asset only case, this choice has the benefit of being consistent with both the Merton model and with the mean-variance analysis while allowing for the explicit inclusion of underlying valuation factors and while admitting a simpler analytical solution than the Merton model. An added advantage of this model is the relative ease with which the asset only problem as formalized by Bielecki and Pliska (1999) can be extended to include a benchmark, as in Davis and Lleo (2008a), or a liability, as in Davis and Lleo (2008b). In in these paper, it is seen that fractional Kelly strategies as defined above are not necessarily optimal, with the notable exception of the Merton model where they arise naturally as a consequence of the assumption that asset prices are lognormally distributed. As a result, to be able to interpret the solution to our benchmarked asset allocation problem in terms of fractional Kelly strategies and therefore guarantee their optimality, we will find it necessary to expand slightly their definition.

In Section 2, we introduce risk-sensitive asset management from the perspective of an asset only investor. The interpretation of the resulting optimal asset allocation formula is then formalized in Section 3 both as a Mutual Fund Theorem and in terms of a redefinition of fractional Kelly strategies. Section 4 presents a comparison of the classical Merton model with the risk-sensitive asset management approach from the perspective of fractional Kelly strategies. Then, in Section 5, we present the analytical solution to the risk-sensitive benchmarked asset allocation problem, before interpreting this solution in a Mutual Fund theorem and in terms of benchmarked fractional Kelly strategies in Section 6. Finally, Section 7 contains a number of case studies showing applications of these ideas to specific types of benchmarks and to Kelly criterion investors.

2 Risk Sensitive Asset Management

2.1 *Risk sensitive control*

Risk-sensitive control is most simply defined as a generalization of classical stochastic control in which the degree of risk aversion or risk tolerance of the optimizing agent is explicitly parameterized in the objective criterion and influences directly the outcome of the optimization. Risk sensitive control was introduced by Jacobson (1973) and has been developed by many authors, notably Whittle (1990) and Bensoussan and van Schuppen (1985) before being applied to finance by Lefebvre and Montulet (1994) and to asset management by Bielecki and Pliska (1999).

While in classical stochastic control the objective of the decision maker is to maximize $\mathbf{E}[F]$, the expected value of some performance criterion F, in risk-sensitive control the decision maker's objective is to select a control policy $h(t)$ maximizing the criterion

$$J(t,x,h;\theta) := -\frac{1}{\theta}\ln \mathbf{E}\left[e^{-\theta F(t,x,h)}\right] \tag{1}$$

where

- t and x are the time and the state variable;
- F is a reward function;
- the risk sensitivity $\theta \in (-1, 0) \cup (0, \infty)$ represents the decision maker's degree of risk aversion.

A formal Taylor expansion of J around $\theta = 0$ evidences the vital role played by the risk sensitivity:

$$J(x, t, h; \theta) = \mathbf{E}\left[F(t, x, h)\right] - \frac{\theta}{2}\mathbf{Var}\left[F(t, x, h)\right] + O(\theta^2) \tag{2}$$

- $\theta \to 0$ corresponds to the "risk-null" case and to classical stochastic control;
- when $\theta < 0$, we have the "risk-seeking" case that is a maximization of the expectation of a convex decreasing function of $F(t, x, h)$;
- finally, $\theta > 0$ is the "risk-averse" case that is a minimization of the expectation of a convex increasing function of $F(t, x, h)$.

To summarize, risk-sensitive control differs from traditional stochastic control in that it explicitly models the risk-aversion of the decision maker as an integral part of the control framework, rather than importing it in the problem via an externally defined utility function.

2.2 *Risk sensitive asset management*

Bielecki and Pliska (1999) pioneered the application of risk-sensitive control to asset management. They proposed to take the logarithm of the investor's wealth V as the reward function, i.e.

$$F(t, x, h) = \ln V(t, x, h) \tag{3}$$

The natural interpretation of this choice is that the investor's objective is to maximize the risk-sensitive (log) return of the his/her portfolio. With this choice of reward function, the control criterion is

$$J(t, x, h; \theta) := -\frac{1}{\theta} \ln \mathbf{E}\left[e^{-\theta \ln V(t,x,h)}\right] \tag{4}$$

and interpret the expectation

$$\mathbf{E}\left[e^{-\theta \ln V(t,x,h)}\right] = \mathbf{E}\left[V(t, x, h)^{-\theta}\right] =: U_\theta(V_t) \tag{5}$$

as the expected utility of time t wealth under the power utility (HARA) function. The investor's objective is to maximize the utility of terminal wealth. The Taylor expansion becomes

$$J(t, x; \theta) = \mathbf{E}\left[\ln V(t, x, h)\right] - \frac{\theta}{2}\mathbf{Var}\left[\ln V(t, x, h)\right] + O(\theta^2) \tag{6}$$

Ignoring higher order terms, we recover the mean-variance optimization criterion and the log utility or Kelly criterion portfolio in the limit as $\theta \to 0$.

2.3 *The risk sensitive asset management model*

2.3.1 *Asset and factor dynamics*

Embedding the investor's risk-sensitivity in the control criterion provides more leeway in the specification of the asset market than would be obtained in the classical stochastic control of the Merton approach. Bielecki and Pliska (1999), in particular, propose a factor model in which the prices of the m risky assets follow a SDE of the form

$$\frac{dS_i(t)}{S_i(t)} = (a + AX(t))_i dt + \sum_{k=1}^{n+m} \sigma_{ik} dW_k(t)$$
$$S_i(0) = s_i, \quad i = 1, \dots, m \tag{7}$$

where $W(t)$ is a $N := n + m$-dimensional Brownian motion and the market parameters a, A, $\Sigma := [\sigma_{ij}]$, $i = 1, \dots, m$, $j = 1, \dots, N$ are matrices of appropriate dimensions. To avoid redundancy, we assume $\Sigma\Sigma' > 0$. To these m risky securities, we add a money market asset with dynamics

$$\frac{dS_0(t)}{S_0(t)} = (a_0 + A_0' X(t))\, dt, \qquad S_0(0) = s_0 \tag{8}$$

Finally, the asset prices drift depends on n valuation factors modelled as affine stochastic processes with constant diffusion

$$dX(t) = (b + BX(t))dt + \Lambda dW(t), \qquad X(0) = x \tag{9}$$

where $X(t)$ is the $\mathbb{R}^n$-valued factor process with components $X_j(t)$ and the parameters b, B, $\Lambda := [\Lambda_{ij}]$, $i = 1, \dots, n$, $j = 1, \dots, N$ are matrices of appropriate dimensions. These valuation factors must be specified, but they could include macroeconomic, microeconomic or abstract statistical variables.

Under these conditions, the logarithm of the investor's wealth is given by the SDE

$$\ln V(t) = \ln v + \int_0^t (a_0 + A_0' X(s)) + h(s)' \left(\hat{a} + \hat{A} X(s)\right) ds$$
$$-\frac{1}{2} \int_0^t h(s)' \Sigma\Sigma' h(s) ds + \int_0^t h(s)' \Sigma dW(s) \tag{10}$$

where $V(0) = v$, h is the m-dimensional vector of portfolio weights and we used the notation $\hat{a} := a - a_0 \mathbf{1}$, $\hat{A} := A - \mathbf{1} A_0'$ and $\mathbf{1}$ is a n-element column vector with all entries set to 1.

The equation for V solely depends on the valuation factors (the state process) and is independent of the asset prices. This implies that the effective dimension of the risk-sensitive control problem will be n, the number of factors, rather than m, the number of risky assets. The limited impact of the number of assets is significant since for practical applications we would typically use only a few factors (possibly 3 to 5) to parametrize a large cohort of assets and asset classes (possibly several dozens). The risk-sensitive asset management model is therefore particularly efficient from a computational perspective.

2.3.2 *The associated linear exponential-of-quadratic Gaussian control problem*

The next step in the analysis is due to Kuroda and Nagai (2002) who ingeniously observed that under an appropriately chosen change of probability measure (via the Girsanov theorem), the risk-sensitive criterion can be expressed as

$$I(v, x; h; t, T) = \ln v - \frac{1}{\theta} \ln \mathbf{E}_h^\theta \left[\exp \left\{ \theta \int_t^T g(X_s, h(s); \theta) ds \right\} \right] \tag{11}$$

where the expectation $\mathbf{E}_h^\theta [\cdot]$ is taken with respect to a newly-defined measure $\mathbb{P}_h^\theta$ depending on the investment strategy h, the functional g is

$$g(x, h; \theta) = \frac{1}{2} (\theta + 1) h'\Sigma\Sigma' h - a_0 - A_0' x - h'(\hat{a} + \hat{A}x) \tag{12}$$

and the factor dynamics under the new measure $\mathbb{P}_h^\theta$ is

$$dX_s = (b + BX_s - \theta\Lambda\Sigma' h(s))\, ds + \Lambda dW_s^\theta \tag{13}$$

(see Davis and Lleo (2008a) for details).

In this formulation, the problem is a standard Linear Exponential-of-Quadratic Gaussian (LEQG) control problem which can be solved exactly, up to the resolution of a system of Riccati equations. But before solving this control problem and deriving the optimal asset allocation, we will first develop some intuition by considering the simple case in which the security and factor risk are uncorrelated, i.e., when $\Lambda\Sigma' = 0$.

2.3.3 *Special case: Uncorrelated assets and factors*

When $\Lambda\Sigma' = 0$, security risk and factor risk are uncorrelated and the evolution of X_t under the measure $\mathbb{P}_h^\theta$ given in equation (13) simplifies to

$$dX_s = (b + BX_s)\, ds + \Lambda dW_s^\theta \tag{14}$$

The evolution of the state is therefore independent of the control variable h and, as a result, the control problem can be solved through a pointwise maximisation of the auxiliary criterion function $I(v, x; h; t, T)$.

In this case, the optimal control h^* is simply the maximizer of the function $g(x; h; t, T)$ given by

$$h^* = \frac{1}{\theta + 1} (\Sigma\Sigma')^{-1} \left(\hat{a} + \hat{A}x \right) \tag{15}$$

which represents a position of $\frac{1}{\theta+1}$ in the Kelly criterion portfolio.

Let $\Phi(t, x)$ be the value function corresponding to the auxiliary criterion function $I(v, x; h; t, T)$. Substituting the value of h^* in the equation for g yields

$$\begin{aligned} \Phi(t, x) &= \sup_{h \in \mathcal{A}(T)} I(v, x; h; t, T) \\ &= -\frac{1}{\theta} \ln \mathbf{E}_h^\theta \left[\exp \left\{ \theta \int_0^{T-t} g(x, h^*(s); t, T; \theta) ds \right\} v^{-\theta} \right] \end{aligned} \tag{16}$$

The PDE for Φ can now be obtained directly via an exponential transformation and an application of Feynman-Kac.

2.3.4 *The general case*

In the general case, the value function Φ for the auxiliary criterion function $I(v,x;h;t,T)$, defined as

$$\Phi(t,x) = \sup_{\mathcal{A}(T)} I(v,x;h;t,T) \tag{17}$$

satisfies the Hamilton-Jacobi-Bellman Partial Differential Equation (HJB PDE)

$$\frac{\partial \Phi}{\partial t} + \sup_{h\in\mathbb{R}^m} L_t^h \Phi(X(t)) = 0 \tag{18}$$

where

$$\begin{aligned} L_t^h \Phi(t,x) &= (b + Bx - \theta \Lambda \Sigma' h(s))' D\Phi + \frac{1}{2}\mathrm{tr}\left(\Lambda\Lambda' D^2\Phi\right) \\ &\quad - \frac{\theta}{2}(D\Phi)'\Lambda\Lambda' D\Phi - g(x,h;\theta) \end{aligned} \tag{19}$$

where $D\Phi$ is the gradient vector defined as $D\Phi := \left(\frac{\partial \Phi}{\partial x_1}, \dots, \frac{\partial \Phi}{\partial x_i}, \dots, \frac{\partial \Phi}{\partial x_n}\right)'$ and $D^2\Phi$ is the Hessian matrix defined as $D^2\Phi := \left[\frac{\partial^2 \Phi}{\partial x_i x_j}\right]$, $i,j = 1,\dots,n$. The value function Φ satisfies the terminal condition $\Phi(T,x) = \ln v$.

Solving the optimization problem gives the optimal investment policy $h^*(t)$

$$h^*(t) = \frac{1}{\theta+1}\left(\Sigma\Sigma'\right)^{-1}\left[\hat{a} + \hat{A}X(t) - \theta\Sigma\Lambda' D\Phi(t,X(t))\right] \tag{20}$$

Moreover, the solution of the PDE is of the form

$$\Phi(t,x) = x'Q(t)x + x'q(t) + k(t) \tag{21}$$

where $Q(t)$ solves a n-dimensional matrix Riccati equation and $q(t)$ solves a n-dimensional linear ordinary differential equation depending on Q (see Kuroda and Nagai (2002) for details).

3 Fractional Kelly Strategies in the Risk-Sensitive Asset Management Model

3.1 *A mutual fund theorem*

Theorem 1 (Mutual Fund Theorem (Davis and Lleo, 2008a)). *Any portfolio can be expressed as a linear combination of investments into two "mutual funds" with respective risky asset allocations:*

$$\begin{aligned} h^K(t) &= (\Sigma\Sigma')^{-1}\left(\hat{a} + \hat{A}X(t)\right) \\ h^C(t) &= -(\Sigma\Sigma')^{-1}\Sigma\Lambda'\left(q(t) + Q(t)X(t)\right) \end{aligned} \tag{22}$$

and respective allocation to the money market account given by

$$\begin{aligned} h_0^K(t) &= 1 - \mathbf{1}'(\Sigma\Sigma')^{-1}\left(\hat{a} + \hat{A}X(t)\right) \\ h_0^C(t) &= 1 + \mathbf{1}'(\Sigma\Sigma')^{-1}\Sigma\Lambda'\left(q(t) + Q(t)X(t)\right) \end{aligned} \tag{23}$$

Moreover, if an investor has a risk sensitivity θ, then the respective weights of each mutual fund in the investor's portfolio equal $\frac{1}{\theta+1}$ and $\frac{\theta}{\theta+1}$, respectively.

The main implication of this theorem is that the allocation between the two funds is a sole function of the investor's risk sensitivity θ. As $\theta \to 0$, the investor's wealth gets invested in the Kelly criterion portfolio (portfolio K). On the other hand, as $\theta \to \infty$, the investor's wealth gets invested in the correction portfolio C. The investment strategy of this portfolio can be interpreted as a large position in the short-term rate and a set of positions trading on the comovement of assets and valuation factors. In financial economics, portfolio C is referred to as the 'intertemporal hedging term'.

When we assume that there are no underlying valuation factors, the risky securities follow geometric Brownian motions with drift vector μ and the money market account becomes the risk-free asset (i.e., $a_0 = r$ and $A_0 = 0$). In this case, $\Sigma\Lambda' = 0$ and we can then easily see that fund C is fully invested in the risk-free asset. As a result, we recover Merton's Mutual Fund Theorem for m risky assets and a risk-free asset (e.g., Theorem 15.1, p. 489 in Merton (1992)).

3.2 *Fractional Kelly strategies in the risk-sensitive asset management model*

Fractional Kelly strategies arise naturally in the Merton investment model, since Merton's Mutual Fund Theorems guarantees that the optimal investment strategy can be split in an allocation to the risk-free asset and an allocation to a mutual fund investing in the Kelly criterion portfolio. However, the optimality of fractional Kelly strategies is the exception rather than the rule: in the Merton model, fractional Kelly strategies are optimal as a result of the assumption that asset prices are lognormally distributed. In the factor-based risk-sensitive asset management model, we cannot expect either that the 'classical' definition of fractional Kelly strategies to yield optimal or near optimal asset allocations as soon as $\theta \neq 0$.

To address this difficulty, Davis and Lleo (2008a) proposed a generalization of the concept of fractional Kelly strategy based on the findings expressed in the Mutual Fund Theorem 1. Rather than regarding the fractional Kelly strategy as a split between the Kelly portfolio and the short-term rate, Davis and Lleo propose to define it as a split between the Kelly portfolio and the portfolio C, as defined in the Mutual Fund Theorem 1. In this case, the Kelly fraction, which represents the proportion of wealth invested in the Kelly portfolio, is inversely proportional to the investor's risk sensitivity and is equal to $\frac{1}{\theta+1}$.

This redefinition of fractional Kelly strategies has two important consequences:

- The fractional Kelly portfolios are always optimal portfolios;
- In the lognormal case (i.e. when $n = 0$), the generalized definition of fractional Kelly strategies reverts to the 'classical' definition. This can be verified from the fact that in the lognormal case, the Mutual Fund Theorem 1 simplifies into Merton's Mutual Fund Theorem for m risky assets and a risk-free asset.

4 Fractional Kelly Strategies, Risk-Sensitive Asset Management and the Merton Model with Power Utility

4.1 *Objective*

In this section, we go further in our analysis by drawing a comparison between risk-sensitive asset management and the Merton model from the perspective of fractional Kelly strategies.

For comparison purpose, we consider:

- a fractional Kelly strategy with fraction f;
- the Merton model maximizing utility of terminal wealth, with the utility function chosen to be the homogeneous power utility or hyperbolic absolute risk aversion (HARA) function; and
- the diffusion risk-sensitive asset management model without any underlying valuation factors.

In order to perform a comparison between the risk-sensitive asset allocation model and the original Merton model, we assume that the market is comprised of m risky assets and that the investor's objective is to find an optimal m-dimensional portfolio allocation vector $h(t)$, where $h_i(t)$ represents the proportion of the portfolio invested in risky security i. We also assume that there are no valuation factors and hence $n = 0$. We will show that the treatment of fractional Kelly strategies is comparable in both cases up to a sign difference in the risk-aversion/risk-sensitive coefficient.

4.2 *A brief review of the Merton model*

In the Merton model, the objective of an investor is to maximize the utility of terminal wealth represented by the criterion:

$$I(t, s, h; \theta) := \mathbf{E}\left[U(V(t, s, h))\right] \tag{24}$$

where

- $V(t, x, h)$ is the investor's wealth at time t in response to a securities market with price vector s and an investment policy h;
- U is the homogeneous power utility or hyperbolic absolute risk aversion (HARA) function, defined as

$$U(z) = \frac{z^\gamma}{\gamma} \tag{25}$$

  where $\gamma \in (-\infty, 0) \cup (0, 1)$ is the risk-aversion coefficient.

The dynamics of the prices of the m risky assets follows a geometric Brownian motion of the form

$$\frac{dS_i(t)}{S_i(t)} = \mu_i dt + \sum_{k=1}^{m} \sigma_{ik} dW_k(t), \qquad S_i(0) = s_i, \quad i = 1, \ldots, m \tag{26}$$

where $W(t)$ is a m-dimensional Brownian motion. The price of the money market asset satisfies

$$\frac{dS_0(t)}{S_0(t)} = rdt, \qquad S_0(0) = s_0 \tag{27}$$

In this setting, the optimal asset allocation is given by

$$h^*(t) = \frac{1}{1-\gamma}(\Sigma\Sigma')^{-1}\hat{\mu} \tag{28}$$

with $\hat{\mu} := \mu - r\mathbf{1}$ where $\mathbf{1}$ is the m-element unit vector and Σ is the diffusion matrix, defined as $\Sigma = [\sigma_{ij}]$.

4.3 *Observations and conclusions*

From a fractional Kelly perspective, the two models are strikingly similar. Taking the case of an investor allocating a fraction f of his/her wealth to the Kelly portfolio, we see that

- in the Merton model, such strategy would be optimal for an investor with a level of risk aversion γ equal to

$$\gamma = 1 - \frac{1}{f} \tag{29}$$

- in the risk-sensitive model, such strategy would be optimal for an investor with a level of risk-sensitivity θ equal to

$$\theta = \frac{1}{f} - 1 \tag{30}$$

Looking solely at the Kelly component of the optimal investment policy, this would imply that

$$\theta = -\gamma \tag{31}$$

This similarity is in not surprising. Indeed, when we restrict the risk-sensitive approach to a 0 factor model, we get the same optimal asset allocation as in the Merton model with homogeneous power utility, but for the fact that γ has been replaced by $-\theta$, i.e.

$$h^*(t) = \frac{1}{\theta+1}(\Sigma\Sigma')^{-1}\hat{\mu} \tag{32}$$

This observation is confirmed by both the range of θ and γ and by the functional form of the utility function associated with the risk-sensitive approach.

These findings are summarized in Table 1.

Key Points:

- The situation between the Merton model with homogeneous power utility and the risk-sensitive asset management model is parallel, and we can use the guideline correspondence $\theta = -\gamma = 1/f - 1$ to link them with the fractional Kelly approach;

Table 1 Comparison of the Merton Model and of the Risk-Sensitive Asset Management Model from a Kelly Perspective

	Risk-Sensitive Asset Management	Merton model with homogeneous power utility
Risk-sensitive parameter/Risk aversion coefficient	$\theta \in (-1, 0) \cup (0, \infty)$	$\gamma \in (-\infty, 0) \cup (0, 1)$
Range of the risk-sensitive parameter/risk aversion coefficient for risk-averse investors	$(0, \infty)$	$(-\infty, 0)$
Form of Utility Function	$U(z) = z^{-\theta}$	$U(z) = \frac{z^\gamma}{\gamma}$
Optimal asset allocation $h^*(t)$	$\frac{1}{\theta+1}(\Sigma\Sigma')^{-1}(\hat{\mu}(X(t)) + \Sigma\Lambda' D\Phi)$	$\frac{1}{1-\gamma}(\Sigma\Sigma')^{-1}\hat{\mu}$
Value of the risk aversion coefficient/risk-sensitive parameter corresponding to a Kelly fraction f	$1/f - 1$	$1 - 1/f$
Range of Kelly fractions for risk-averse investors	$(0, 1)$	$(0, 1)$

- The risk-sensitive approach is therefore consistent with the Merton model with homogeneous power utility;
- In the case when there are no factors (i.e., $n = 0$), the optimal asset allocation obtained in the risk-sensitive approach reverts to that associated with the Merton model;
- In the general case ($n > 0$), the definition of fractional Kelly strategies in the risk-sensitive asset management model can be extended as proposed above while remaining consistent with the 'classical' definition of fractional Kelly strategies as a split bewteen the Kelly portfolio and the risk-free asset.

5 Adding an Investment Benchmark: Risk Sensitive Benchmarked Asset Management Model

So far, we have introduced risk sensitive asset management in the classical asset only context of an investor attempting to maximize the terminal utility of his/her wealth. We will now examine the related benchmarked asset management problem in which the investor selects an asset allocation to outperform a given investment benchmark.

In the benchmarked case, Davis and Lleo (2008a) propose that the reward function $F(t, x; h)$ be defined as the (log) excess return of the investor's portfolio over the return of the benchmark, i.e.

$$F(t,x,h) := \ln \frac{V(t,x,h)}{L(t,x,h)} \tag{33}$$

where L is the level of the benchmark.

Furthermore, the dynamics of the benchmark is modelled by the SDE:

$$\frac{dL(t)}{L(t)} = (c + C'X(t))dt + \varsigma' dW(t), \qquad L(0) = l \tag{34}$$

where C is a scalar constant, C is a n-element column vector, and ς is a N-element column vector.

This formulation is wide enough to encompass a multitude of situations such as:

- *the single benchmark case*, where the benchmark is, e.g. an equity index such as the S&P500 or the FTSE 100.
- *the single benchmark plus alpha*, where, for example, a hedge fund has for benchmark a target based on a short-term interest rate plus alpha.
- *the composite benchmark case*, e.g., a benchmark constituted of 5% cash, 35% Citigroup World Government Bond Index, 25% S&P 500 and 35% MSCI EAFE.
- *the composite benchmark plus alpha* that is a combination of the previous two cases.

By Itô's lemma, the log of the excess return in response to a strategy h is

$$\begin{aligned} F(t,x;h) &= \ln \frac{v}{l} + \int_0^t d\ln V(s) - \int_0^t d\ln L(s) \\ &= \ln \frac{v}{l} + \int_0^t \left(a_0 + A_0'X(s) + h(s)'\left(\hat{a} + \hat{A}X(s)\right)\right) ds \\ &\quad - \frac{1}{2}\int_0^t h(s)'\Sigma\Sigma' h(s) ds + \int_0^t h(s)'\Sigma dW(s) \\ &\quad - \int_0^t (c + C'X(s))ds + \frac{1}{2}\int_0^t \varsigma'\varsigma ds \\ &\quad - \int_0^t \varsigma' dW(s) \end{aligned} \tag{35}$$

$$F(0,x;h) = f_0 := \ln \frac{v}{l}$$

Following an appropriate change of measure, the criterion function can be expressed as

$$I(f_0, x; h; t, T) = \ln f_0 - \frac{1}{\theta} \ln \mathbf{E}_h^{\theta}\left[\exp\left\{\theta \int_0^{T-t} g(X_s, h(s); \theta) ds\right\}\right] \tag{36}$$

where

$$g(x, h; \theta) = \frac{1}{2}(\theta + 1) h'\Sigma\Sigma' h - a_0 - A_0' x - h'(\hat{a} + \hat{A}x) \\ - \theta h'\Sigma\varsigma + (c + C'x) + \frac{1}{2}(\theta - 1)\varsigma'\varsigma \tag{37}$$

Once again, the control problem simplifies into a LEQG problem.

The value function Φ for the auxiliary criterion function $I(f_0, x; h; t, T)$. Then Φ is defined as

$$\Phi(t, x) = \sup_{\mathcal{A}(T)} I(f_0, x; h; t, T) \tag{38}$$

and it satisfies the HJB PDE

$$\frac{\partial \Phi}{\partial t} + \sup_{h \in \mathbb{R}^m} L_t^h \Phi = 0 \tag{39}$$

where

$$L_t^h \Phi = (b + Bx - \theta\Lambda(\Sigma' h - \varsigma))' D\Phi + \frac{1}{2}\mathrm{tr}\left(\Lambda\Lambda' D^2 \Phi\right) \\ - \frac{\theta}{2}(D\Phi)'\Lambda\Lambda' D\Phi - g(x, h; \theta) \tag{40}$$

Solving the optimization problem gives the optimal investment policy $h^*(t)$

$$h^* = \frac{1}{\theta + 1}(\Sigma\Sigma')^{-1}\left(\hat{a} + \hat{A}x - \theta\Sigma\Lambda' D\Phi + \theta\Sigma\varsigma\right) \tag{41}$$

The solution of the PDE is still of the form

$$\Phi(t, x) = x'Q(t)x + x'q(t) + k(t) \tag{42}$$

where $Q(t)$ solves a n-dimensional matrix Riccati equation and $q(t)$ solves a n-dimensional linear ordinary differential equation.

6 What About the Kelly Criterion?

6.1 *A mutual fund theorem*

In the benchmarked case, we will follow the same logic as in the asset only and rewrite the optimal asset allocation as an allocation between two funds. We will then use this result to define benchmarked optimal fractional Kelly strategies, that is fractional Kelly strategies, which are also optimal investment policies in the benchmarked asset management model. All we will need to do after that is to verify that this new definition is consistent with the definition we gave earlier in the asset only case, and thus, with the 'classical' definition of fractional Kelly strategies in the limit as our model converges to the Merton model.

The following benchmarked mutual fund theorem is due to Davis and Lleo (2008a):

Theorem 2 (Benchmarked Mutual Fund Theorem). *Given a time t and a state vector $X(t)$, any portfolio can be expressed as a linear combination of investments into two "mutual funds" with respective risky asset allocations*

$$\begin{aligned} h^K(t) &= (\Sigma\Sigma')^{-1}\left(\hat{a} + \hat{A}X(t)\right) \\ h^C(t) &= (\Sigma\Sigma')^{-1}\left[\Sigma\varsigma - \Sigma\Lambda'\left(q(t) + Q(t)X(t)\right)\right] \end{aligned} \tag{43}$$

and respective allocation to the money market account given by

$$\begin{aligned} h_0^K(t) &= 1 - \mathbf{1}'(\Sigma\Sigma')^{-1}\left(\hat{a} + \hat{A}X(t)\right) \\ h_0^C(t) &= 1 - \mathbf{1}'(\Sigma\Sigma')^{-1}\left[\Sigma\varsigma - \Sigma\Lambda'\left(q(t) + Q(t)X(t)\right)\right] \end{aligned} \tag{44}$$

Moreover, if an investor has a risk sensitivity θ, then the respective weights of each mutual fund in the investor's portfolio equal $\frac{1}{\theta+1}$ and $\frac{\theta}{\theta+1}$, respectively.

There are two main differences between Theorems 1 and 2: the definition of portfolio C and the role played by the risk sensitivity θ. In the asset only case of Theorem 1, portfolio C is comprised of the money market asset and of a strategy trading the comovement of assets and valuation factors. In the benchmark case of Theorem 2, portfolio C still includes an allocation to the money market asset and the asset-factor co-movement strategy, but it also contains an allocation to a strategy designed to replicate the risk profile of the benchmark. Indeed, the term $\hat{u} := (\Sigma\Sigma')^{-1}\Sigma\varsigma$ represents an unbiased estimator of a linear relationship between asset risks and benchmark risk $\varsigma = \Sigma' u$. In financial economics, the interpretation of $\hat{u}$ is as the vector of asset systematic exposure, or "betas", computed with respect to the benchmark.

Hence, when θ is low, the investor will take more active risk by investing larger amounts into the log-utility or Kelly portfolio. On the other hand, when θ is high, the investor will divert most of his/her wealth to the correction fund, which is dominated by the term $(\Sigma\Sigma')^{-1}\Sigma\varsigma$ and designed to track the index. In short, while an investor with $\theta = 0$ is a Kelly criterion investor, an investor with extremely high θ is a passive investor.

This first observation leads us to reconsider the role and definition of the risk sensitive parameter θ. Indeed, while in the asset only case, θ represents the sensitivity of an investor to total risk, in the benchmark case, θ corresponds to the investor's sensitivity to active risk. To some extent, in the benchmark case the investor already takes the benchmark risk as granted. The main unknown is therefore how much additional risk the investor is willing to take in order to outperform the benchmark. This amount of risk is directly quantified by the risk sensitive parameter θ.

6.2 *Fractional Kelly strategies revisited*

In the benchmark case, as in the asset only case, we can expand the classical definition of fractional Kelly strategies by defining them as a split between the Kelly portfolio K and the benchmark-tracking portfolio C, as per the Mutual Fund Theorem 2 and which is includes a benchmark replicating strategy.

We can quickly check that the extended definition in the benchmarked case is consistent with the definition given earlier in the asset only case by setting the benchmark to 0, which de facto implies the absence of a benchmark. We see then that the definitions of fund C given in Theorems 2 and 1 are identical: the two theorems now coincide. As expected, the asset only problem can be viewed as a special case of the benchmarked allocation.

Thus, if we set $n = 0$ so that no valuation factor is considered and set the benchmark to 0, the risk-sensitive benchmarked asset management model is the Merton model and the definition of optimal benchmarked fractional Kelly strategies reverts to the classical definition of fractional Kelly strategies.

7 Case Studies

We will now apply the model to study some specific benchmark structures and assess the appropriateness of benchmarks for Kelly criterion investors.

7.1 *Benchmark as a portfolio of traded assets*

First, we revisit two examples considered by Davis and Lleo (2008a) in which the benchmark is a portfolio of traded assets respectively with or without the money market asset.

7.1.1 *Benchmark as a portfolio of traded assets and the money market asset*

Developing the ideas from the previous paragraph, we consider a benchmark whose dynamics is given by

$$\begin{aligned}\frac{dL_t}{L_t} &= (a_0 + A_0'X(t)) + \nu'(t)(\hat{a} + \hat{A}X(t))dt + \nu'(t)\Sigma dW_t \\ &= \left[(1-\nu'\mathbf{1})\,a_0 + \nu'a + ((1-\nu'\mathbf{1})A_0' + \nu'A)\,X(t)\right]dt \\ &\quad + \nu'(t)\Sigma dW_t \end{aligned} \tag{45}$$

where ν is an m-element allocation vector satisfying the budget equation

$$\mathbf{1}'\nu = 1 - h_0^L \tag{46}$$

and h_0^L is the allocation left in the money market account. The process $L(t)$ represents the (log) return of a constant proportion portfolio with risky allocation vector ν and the remainder (i.e., $1 - \mathbf{1}'\nu$) invested in the money market account.

Corollary 3. *(Fund Separation Theorem with a Constant Proportion Benchmark (II)). Given a time t and a state vector $X(t)$, any portfolio can be expressed as* *a linear combination of investments into a "mutual fund", an index fund and a "long-short hedge fund" with respective risky asset allocations*

$$\begin{aligned} h^K(t) &= (\Sigma\Sigma')^{-1}\left(\hat{a} + \hat{A}X(t)\right) \\ h^I(t) &= \nu \\ h^H(t) &= -(\Sigma\Sigma')^{-1}\Sigma\Lambda'\left(q(t) + Q(t)X(t)\right) \end{aligned} \tag{47}$$

and respective allocation to the money market account given by

$$\begin{aligned} h_0^K(t) &= 1 - \mathbf{1}'(\Sigma\Sigma')^{-1}\left(\hat{a} + \hat{A}X(t)\right) \\ h_0^I(t) &= 1 - \mathbf{1}'\nu \\ h_0^H(t) &= 0 \end{aligned} \tag{48}$$

Moreover, if an investor has a risk sensitivity θ, then the respective weights of each fund in the investor's portfolio equal $\frac{1}{\theta+1}$, $\frac{\theta}{\theta+1}$ and $\frac{\theta}{\theta+1}$, respectively.

We can interpret this Corollary in terms of optimal fractional Kelly strategies in a similar way as we did for Corollary 4. The second component of the strategy, the correction fund C, can again be split into:

1. a portfolio I with risky asset allocation given by ν which is designed to strictly replicate the risk exposure of the index;
2. a "long-short hedge fund" H with risky asset allocation given by $h^H(t)$ and whose sole purpose is to trade the comovement of assets and factors.

Here again, we could show that H is a zero net weight strategy, i.e.

$$1'h^H = 0 \tag{49}$$

7.1.2 *Benchmark as a portfolio of traded assets only*

We assume that the benchmark follows a constant proportion strategy invested in a combination of traded assets. Its dynamics is given by the equation

$$\frac{dL_t}{L_t} = \nu'(a + AX(t))dt + \nu'\Sigma dW_t \tag{50}$$

where ν is a m-element allocation vector satisfying the budget equation

$$\mathbf{1}'\nu = 1 \tag{51}$$

In this setting, Corollary 3 can be expressed as:

Corollary 4. *(Fund Separation Theorem with a Constant Proportion Benchmark (I)). Given a time t and a state vector $X(t)$, any portfolio can be expressed as* *a linear combination of investments into a "mutual fund", an index fund and a "long-short hedge fund" with respective risky asset allocations*

$$\begin{aligned} h^K(t) &= (\Sigma\Sigma')^{-1}\left(\hat{a} + \hat{A}X(t)\right) \\ h^I(t) &= \nu \\ h^H(t) &= -(\Sigma\Sigma')^{-1}\Sigma\Lambda'\left(q(t) + Q(t)X(t)\right) \end{aligned} \tag{52}$$

and respective allocation to the money market account given by

$$\begin{aligned} h_0^K(t) &= 1 - \mathbf{1}'(\Sigma\Sigma')^{-1}\left(\hat{a} + \hat{A}X(t)\right) \\ h_0^I(t) &= 0 \\ h_0^H(t) &= \mathbf{1}'(\Sigma\Sigma')^{-1}\Sigma\Lambda'\left(q(t) + Q(t)X(t)\right) \end{aligned} \tag{53}$$

Moreover, if an investor has a risk sensitivity θ, then the respective weights of each mutual fund in the investor's portfolio equal $\frac{1}{\theta+1}$, $\frac{\theta}{\theta+1}$ and $\frac{\theta}{\theta+1}$, respectively.

From the perspective of optimal fractional Kelly strategies, the implication of this corollary is that the second component of the strategy, the correction fund C, can now be explicitly split into two sub portfolios:

1. a portfolio I with asset allocation given by ν which is designed to strictly replicate the risk exposure of the index. Observe that the linear estimator $\hat{u} := (\Sigma\Sigma')^{-1}\Sigma\varsigma$ has vanished and been replaced by the actual asset allocation ν: since the benchmark is now comprised of traded assets it can be replicated by a direct investment in the appropriate asset allocation and does not require any further estimation;
2. a "long-short hedge fund" H with risky asset allocation given by

$$-(\Sigma\Sigma')^{-1}\Sigma\Lambda'\left(q(t) + Q(t)X(t)\right) \tag{54}$$

and whose sole purpose is to trade the comovement of assets and factors.

The "long-short hedge fund" H is particularly interesting because it has zero net weight in the sense that the sum of all the long positions in the portfolio is exactly matched by the sum of short positions. As a result, fund H satisfies the budget equation

$$1'h^H + h_0^H = 0 \tag{55}$$

This "long-short hedge fund" can therefore be viewed as a macro-oriented overlay strategy within the asset allocation.

7.2 *Benchmark with alpha target*

We now consider two new examples involving a benchmark based on an interest rate plus some measure of risk-adjusted excess return known as alpha.

7.2.1 *Money market rate plus alpha benchmark*

Money market rate plus some predetermined alpha has been adopted as a benchmark by a large number of hedge funds. What are the implications of this choice in terms of the optimal fractional Kelly strategies?

The dynamics of such a money market rate plus alpha benchmark can be expressed as

$$\frac{dL_t}{L_t} = (a_0 + A_0 X(t))dt + \alpha dt \tag{56}$$

where α represents the instantaneous level of risk-adjusted excess return, or alpha, required of the fund manager.

The optimal asset allocation in this case is

$$h^* = \frac{1}{\theta+1}(\Sigma\Sigma')^{-1}\left(\hat{a} + (\hat{A} - \theta\Sigma\Lambda' Q)x - \theta\Sigma\Lambda' q\right) \tag{57}$$

which is simply the asset only asset allocation as given in (20). As a result, the conclusion of the Mutual Fund Theorem 1 hold and optimal fractional Kelly strategies can be defined accordingly.

So, what has happened to the benchmark? In our risk-sensitive framework, the benchmark is tracked through its risk profile rather than its return profile. Since a money market plus alpha benchmark does not have any direct risk, there cannot be any comovement between the assets and the benchmark. Thus, the benchmark risk profile cannot be replicated, and as a result, the asset allocation is established without any regard to the benchmark.

The money market plus alpha benchmark will have an impact on the value function Φ. Indeed, in the benchmarked case, the auxiliary criterion function $I(f_0, x; h; t, T)$ reflects excess return over the benchmark, rather than total return as in the asset only case. The value function for the benchmarked problem is therefore consistent with excess return rather than total return, resulting in different value functions for the benchmark problem and for the asset only case.

The main conclusion from this case study is that from the perspective of a rational risk-sensitive investor with no investment constraints, money market plus alpha benchmarks results in the same investment strategy as not having any benchmark at all. To some extent, in a risk-sensitive setting, money market plus alpha is a benchmark-free benchmark. In particular, and somehow counter intuitively, the choice a higher value for alpha does not result in a the design of a riskier asset allocation. In fact, the value of alpha has no impact at all!

7.2.2 *Bill rate or bond yield plus alpha benchmark*

The situation would however be slightly different if the benchmark was for example 3 month Treasury bill rate plus alpha or a 5 year Treasury note yield plus alpha. To generalize slightly, the benchmark dynamics could be expressed as an extension of the model considered above in the "Benchmark as a Portfolio of Traded Assets Only" case (in Section 7.1.2) [1]:

$$\frac{dL_t}{L_t} = \nu'(a + AX(t))dt + \alpha dt + \nu'\Sigma dW_t \tag{58}$$

where ν is a m-element allocation vector satisfying the budget equation

$$\mathbf{1}'\nu = 1 \tag{59}$$

As expected, the level of alpha does not influence the optimal asset allocation and the conclusions stemming from Corollary 4 are still valid in this case.

7.2.3 *Solving the alpha puzzle*

Why is the alpha not producing any impact on the asset allocation? The reason is that the risk-sensitive benchmarked asset management model does not penalize for a non-achievement of the benchmark return. As we have seen above, risk sensitivity implies that the risk of the asset portfolio relative to the benchmark matters, but not the expected return. Since an alpha target is in essence a "pure return" target, it does not change the behaviour of the risk sensitive investor.

Should the investor be penalized for a non-achievement of the benchmark return? Not in our opinion. For an investor who decides of their own asset allocation, the benchmark is a measure of acceptable risk rather than an minimum return objective. Indeed, the portfolio optimization process will always select, for a given risk sensitivity, the asset allocation which maximizes the relative return of the asset portfolio with respect to its benchmark. The issue of an alpha target is irrelevant in this problem.

The situation might however be different if the investor hires a manager to perform the asset allocation. The alpha imperative may in this case be understood as a control imposed by the investor to force the asset manager to select an optimal (or near optimal) asset allocation. The investor should still not be penalized for a non achievement of the benchmark, but it is conceivable that the investor may want to penalize the manager for non achievement of the benchmark return. Here, the penalization would occur at the level of the fee earned by the manager and not at the level of the investor's terminal wealth. This would result in a completely different control problem in which we would need to take the perspective of the manager whose objective is to allocate the investor's assets in order to maximize the management fee received and subject to a non-achievement penalty.

[1]This slight generalization is required in the event the benchmark is based on a bond. Indeed, in the risk-sensitive model, zero coupon bonds are the representatives of the fixed income asset class in the investment universe. Any coupon bond must therefore be "recreated" as a linear combination of zero coupon bonds. In the event, the benchmark is a Treasury bill, then we have the degenerate case in which ν is a vector with the element corresponding to the Treasury bill set to 1 and all other elements equal to 0

7.3 *Alpha-omega targets*

An alternative to the pure alpha target is an alpha-omega target, where omega represents the variability of alpha (see Grinold and Kahn (1999) for a view of the role of alpha and omega in active management). This approach implictly recognises that alpha generation is not only uncertain but also risky. The introduction of the omega terms changes only slightly the asset allocation problem. The dynamics of a benchmark with alpha - omega targets can be modelled as

$$\begin{aligned}\frac{dL(t)}{L(t)} &= [(c + C'X(t))dt + \varsigma' dW(t)] + [\alpha dt + \omega' dW(t)] \\ &= (c + \alpha + C'X(t))dt + (\varsigma + \omega)' dW(t), \qquad L(0) = l \end{aligned} \tag{60}$$

where the scalar constant c, the n-element column vector C, and ς is a N-element column vector are related to the benchmark itself and the scalar α and vector ω refer to the excess return demanded by the investor. The last element of ς is set to 0 while the first $n + m$ element of the vector ω are set to 0. This condition ensures that the variability of alpha is uncorrelated with the factors and assets, as active asset management theory suggests.

7.3.1 *The optimal investment policy*

Extending the reasoning in Davis and Lleo (2008a), we would find that the optimal investment policy h^* is

$$h^* = \frac{1}{\theta + 1}(\Sigma\Sigma')^{-1}\left(\hat{a} + \hat{A}x - \theta\Sigma\Lambda' D\Phi + \theta\Sigma(\varsigma + \omega)\right) \tag{61}$$

The solution of the PDE is still of the form

$$\Phi(t, x) = x'Q(t)x + x'q(t) + k(t) \tag{62}$$

where $Q(t)$ solves a n-dimensional matrix Riccati equation and $q(t)$ solves a n-dimensional linear ordinary differential equation. Specifically,

$$\dot{Q}(t) - Q(t)K_0Q(t) + K_1'Q(t) + Q(t)K_1 + \frac{1}{\theta + 1}\hat{A}'(\Sigma\Sigma')^{-1}\hat{A} = 0 \tag{63}$$

for $t \in [0, T]$, with *terminal* condition $Q(T) = 0$ and with

$$K_0 = \theta\left[\Lambda\left(I - \frac{\theta}{\theta + 1}\Sigma'(\Sigma\Sigma')^{-1}\Sigma\right)\Lambda'\right] \tag{64}$$

$$K_1 = B - \frac{\theta}{\theta + 1}\Lambda\Sigma'(\Sigma\Sigma')^{-1}\hat{A} \tag{65}$$

$$\begin{aligned}&\dot{q}(t) + (K_1' - Q(t)K_0)\,q(t) + Q(t)b + \theta Q'(t)\Lambda(\varsigma + \omega) + A_0 - C \\ &\quad + \frac{1}{\theta + 1}\left(2\hat{A}' - \theta Q'(t)\Lambda\Sigma'\right)(\Sigma\Sigma')^{-1}\left(\hat{a} + \theta\Sigma(\varsigma + \omega)\right) \\ &= 0\end{aligned} \tag{66}$$

with *terminal* condition $q(T) = 0$.

$$k(s) = f_0 + \int_s^T l(t)dt \tag{67}$$

for $0 \leq s \leq T$ and where

$$\begin{aligned}
l(t) &= \frac{1}{2}\mathrm{tr}\left(\Lambda\Lambda' Q(t)\right) - \frac{\theta}{2}q'(t)\Lambda\Lambda' q(t) + b'q(t) \\
&+ \frac{1}{\theta+1}\hat{a}'(\Sigma\Sigma')^{-1}\hat{a} + \frac{1}{\theta+1}\theta^2 q'(t)\Lambda\Sigma'(\Sigma\Sigma')^{-1}\Sigma\Lambda' q(t) \\
&- \frac{\theta}{\theta+1}q'(t)\Lambda\Sigma'(\Sigma\Sigma')^{-1}\hat{a} - \frac{2\theta^2}{\theta+1}q'(t)\Lambda\Sigma'(\Sigma\Sigma')^{-1}\Sigma\gamma \\
&+ \theta(\varsigma+\omega)'\Lambda' q(t) - \frac{1}{2}(\theta-1)(\varsigma+\omega)'(\varsigma+\omega) + \frac{\theta}{\theta+1}\hat{a}'(\Sigma\Sigma')^{-1}\Sigma(\varsigma+\omega) \\
&+ \frac{1}{\theta+1}\theta^2\gamma'\Sigma(\Sigma\Sigma')^{-1}\Sigma(\varsigma+\omega) + a_0 - (c+\alpha)
\end{aligned} \tag{68}$$

7.3.2 *Mutual fund theorem*

We can now restate the optimal asset allocation in terms of the following Mutual Fund Theorem:

Theorem 5 (Alpha-Omega Benchmarked Mutual Fund Theorem). *Given a time t and a state vector $X(t)$, any portfolio can be expressed as a linear combination of investments into two "mutual funds" with respective risky asset allocations*

$$\begin{aligned}
h^K(t) &= (\Sigma\Sigma')^{-1}\left(\hat{a} + \hat{A}X(t)\right) \\
h^C(t) &= (\Sigma\Sigma')^{-1}\left[\Sigma(\varsigma+\omega) - \Sigma\Lambda'\left(q(t) + Q(t)X(t)\right)\right]
\end{aligned} \tag{69}$$

and respective allocation to the money market account given by

$$\begin{aligned}
h_0^K(t) &= 1 - \mathbf{1}'(\Sigma\Sigma')^{-1}\left(\hat{a} + \hat{A}X(t)\right) \\
h_0^C(t) &= 1 - \mathbf{1}'(\Sigma\Sigma')^{-1}\left[\Sigma(\varsigma+\omega) - \Sigma\Lambda'\left(q(t) + Q(t)X(t)\right)\right]
\end{aligned} \tag{70}$$

Moreover, if an investor has a risk sensitivity θ, then the respective weights of each mutual fund in the investor's portfolio equal $\frac{1}{\theta+1}$ and $\frac{\theta}{\theta+1}$, respectively.

Proof. This Corollary can be proved in a similar fashion to Theorem 3 in Davis and Lleo (2008a). □

7.3.3 *Kelly strategies*

In terms of Kelly strategies, this Mutual Fund Theorem represents a split between a fraction $\frac{1}{\theta+1}$ invested in the Kelly portfolio K, and a fraction $\frac{\theta}{\theta+1}$ invested in the correction fund C. As in Theorem 2, portfolio C includes an allocation to the money market asset and the asset-factor co-movement strategy and an allocation to a strategy designed to replicate the risk profile of the benchmark. In addition, the allocation to portfolio C features a new term related to the variability of alpha, defined as $\hat{u} := (\Sigma\Sigma')^{-1}\Sigma\omega$ and which represents an unbiased estimator of a linear relationship between asset risks and alpha risk $\omega = \Sigma' u$.

When θ is high, the investor will divert most of the wealth to the correction fund in order to replicate the risk profile of the benchmark and the appropriate level of active risk given by ω. On the other hand, when θ is low, the investor will take more active risk by investing larger amounts into the Kelly portfolio and will not attempt to stay within the bounds of active risk imposed by ω.

7.4 *Benchmarks and Kelly investors: No benchmark for buffett!*

Our last application of the idea of benchmarked fractional Kelly strategies takes the form of an anecdote concerning the appropriateness of benchmarks for Kelly investors. A few years ago, a controversy shook the North American investment industry: what would be a proper investment benchmark to assess the performance of legendary investor Warren Buffettt's Berkshire Hathaway? Although this question may appear, at first glance, trivial, it is in fact the reflection of a wider concern shared by asset managers and investors alike on what constitutes an appropriate investment benchmark. This question has received an increasing deal of attention in the past two decades and it is considered so important in the investment management industry that the Research Foundation of the CFA Institute (then called AIMR), a leading professional association in the investment management industry, has recently devoted a monograph to the question (see Siegel (2003) for more details).

Although we do not intend on entering the benchmark design and specification debate[2], the results we derived reveal a new dimension to the problem: the impact of the investment benchmark on a fund manager's investment strategy depends to a significant extent on the risk-aversion of the manager or investor. Indeed, we have shown that in the risk-sensitive benchmarked asset management model, the importance of the benchmark in the investment decision increases as the risk aversion of the investor increases. In fact, the risk-sensitivity θ is a direct indication of the amount of active risk that an investor is willing to take. Thus, the exercise consisting in setting an investment benchmark becomes increasingly irrelevant as risk-aversion gets nearer to 0: Kelly criterion investors invest in the log-utility portfolio and have no regards for a benchmark.

Before setting a benchmark, it is imperative to know the investor's risk preferences as this will influences his/her investment style. A very rough correspondence would be to identify extremely high risk aversion investors to passive portfolio managers whose mandate is to track an index, low risk aversion investors to purely active managers, whose mandate is to generate the best possible risk-adjusted return in

[2] We have, after all, conveniently assumed throughout that an appropriate care had been taken beforehand to select the benchmark.

a specific market, and medium risk-aversion investors to so called "core plus" or "index plus" managers who track closely and index while trying to generate some incremental extra return or "alpha". The only time these three broad categories of investors will coincide is when their benchmark is the Kelly portfolio.

Going back to our original question, Warren Buffet focuses on long-term growth maximization rather than the avoidance of short-term losses (see Ziemba (2005) for a discussion of Berkshire Hathaway's risk-adjusted performance and Thorp (2006) for a Kelly criterion perspective). In this sense, Mr. Buffet behaves similarly to a Kelly criterion bettor. The RSBAM model then demonstrates that benchmarks are irrelevant to Mr. Buffet's investment strategy.

As a concluding note, since at least 1995, the Berkshire Hathaway annual report features a table presenting a proxy for the performance of the Berkshire Hathaway share against the return of the S&P500 going back to 1965. Warren Buffet, however, does not recognize the S&P500 as his benchmark. He made it clear that this comparative table had been included in the annual reports at the request of investors but that he personally sees little value in it.

8 Conclusion

Historically, Kelly and fractional Kelly strategies have played an important part in asset-only investment management. But the essential role of fractional Kelly strategies does not stop at this level: it also extends to benchmarks and even to asset and libaility management (see Davis and Lleo (2008b)). In benchmarked asset allocation problems, fractional Kelly strategies highlight the fundamental split between a purely active growth maximizing strategy (the Kelly criterion portfolio) and pure benchmark replication. Moreover, the Kelly fraction, which represent the fraction of one's wealth invested in the Kelly portfolio, is a sole function of the investor's degree of risk-sensitivity.

This has profound implications. A Kelly investor, with 0 risk-sensitivity, will invest solely in the purely active portfolio and as a result, benchmarks are irrelevant to Kelly investors. On the other end of the spectrum, investors with extremely large risk-sensitivity will tend to opt for full replication of the benchmark: they are passive investors. Somewhere in the middle, we find investors with an average risk-sensitivity, who adopt core-plus strategies: a basic replication of the benchmark with some departure in order to generate excess returns. The only time these three broad categories of investors will coincide is when their benchmark is the Kelly portfolio.

References

A. Bensoussan and J. H. van Schuppen (1985). Optimal control of partially observable stochastic systems with an exponential-of-integral performance index. *SIAM Journal on Control and Optimization*, 23(4), 599–613.

Bielecki, T. R. and S. R. Pliska (1999). Risk-sensitive dynamic asset management. *Applied Mathematics and Optimization*, 39, 337–360.

Davis, M. H. A. and S. Lleo (2008a). Risk-sensitive benchmarked asset management. *Quantitative Finance*, 8(4), 415–426.

Davis, M. H. A. and S. Lleo (2008b). A risk sensitive asset and liability management model. *Working Paper.*

Grinold, R. and R. Kahn (1999). *Active Portfolio Management: A Quantative Approach for Producing Superior Returns and Selecting Superior Money Managers.* New York: McGraw-Hill.

Jacobson, D. H. (1973). Optimal stochastic linear systems with exponential criteria and their relation to deterministic differential games. *IEEE Transactions on Automatic Control*, 18(2), 114–131.

Kuroda, K. and H. Nagai (2002). Risk-sensitive portfolio optimization on infinite time horizon. *Stochastics and Stochastics Reports*, 73, 309–331.

Lefebvre, M. and P. Montulet (1994). Risk-sensitive optimal investment policy. *International Journal of Systems Science*, 22, 183–192.

Merton, R. C. (1969). Lifetime portfolio selection under uncertainty: The continuous time case. *Review of Economics and Statitsics*, 51, 247–257.

Merton, R. C. (1971). Optimal consumption and portfolio rules in a continuous-time model. *Journal of Economic Theory*, 3, 373–4113.

Merton, R. C. (1992). *Continuous-Time Finance.* US: Blackwell Publishers.

Siegel, L. (2003). *Benchmarks and Investment Management.* The Research Foundation of AIMR.

Thorp, E. (2006). The Kelly criterion in blackjack, sports betting and the stock market. In S. A. Zenios and W. T. Ziemba, editors, *Handbook of Asset and Liability Management, Volume 1*, Chapter 9, pp. 385–428. Amersdam: North Holland.

P. Whittle (1990). *Risk Sensitive Optimal Control.* New York: John Wiley & Sons.

Ziemba, W. T. (2003). *The Stochastic Programming Approach to Asset, Liability, and Wealth Management.* Research Foundation Publication. CFA Institute.

Ziemba, W. T. (2005). The symmetric downside-risk sharpe ratio. *Journal of Portfolio Management*, 32(1), 108–122.

28

A Benchmark Approach to Investing and Pricing

Eckhard Platen
University of Technology Sydney,
School of Finance & Economics and Department of Mathematical Sciences,
PO Box 123, Broadway, NSW, 2007, Australia

Abstract

This paper introduces a general market modeling framework, the benchmark approach, which assumes the existence of the *numéraire* portfolio. This is the strictly positive portfolio that when used as benchmark makes all benchmarked non-negative portfolios supermartingales, that is intuitively speaking downward trending or trendless. It can be shown to equal the Kelly portfolio, which maximizes expected logarithmic utility. In several ways, the Kelly or *numéraire* portfolio is the "best" performing portfolio and cannot be outperformed systematically by any other non-negative portfolio. Its use in pricing as *numéraire* leads directly to the real world pricing formula, which employs the real world probability when calculating conditional expectations. In a large regular financial market, the Kelly portfolio is shown to be approximated by well-diversified portfolios.

JEL Classification: G10, G13

1991 *Mathematics Subject Classification:* primary 90A12; secondary 60G30, 62P20.

Keywords and phrases: Kelly portfolio, real world pricing, *numéraire* portfolio, strong arbitrage, diversification.

1 Introduction

The classical asset pricing theories, as developed in Debreu (1959), Sharpe (1964), Lintner (1965), Merton (1973a,b), Ross (1976), Harrison and Kreps (1979), Constantinides (1992), and Cochrane (2001) represent forms of relative pricing, since the existence of an equivalent risk neutral probability measure is typically requested. However, relative pricing ignores in the long run the presence of the equity premium. This paper presents an extension of classical risk neutral pricing in a general modeling framework, the benchmark approach, where a form of absolute pricing naturally emerges, see Platen (2002) Platen (2002, 2006) and Platen and Heath (2006). The benchmark represents here the "best" performing, strictly positive, tradable portfolio, which turns out to be the *Kelly portfolio*, see Kelly (1956). Most results we present can be stated in a model independent manner. The benchmark approach only requires the existence of the benchmark, playing the role of the *numéraire portfolio*, originally discovered in Long (1990). The benchmark approach with the Kelly portfolio as benchmark is directly applicable in portfolio optimization and covers a wider modeling world than classical theories allow.

We will see that non-negative portfolios, when denominated in units of the benchmark are supermartingales. This supermartingale property has been already observed for particular settings, for instance, in Long (1990), Bajeux-Besnainou and Portait (1997), Becherer (2001), Platen (2002), Bühlmann and Platen (2003), Platen and Heath (2006), Karatzas and Kardaras (2007), and Kardaras and Platen (2008b). For the purpose of pricing, the inverse of the benchmark plays the role of the stochastic discount factor in the language of Cochrane (2001). Within this paper, there will be no major additional assumption made beyond the request on the existence of a *numéraire* portfolio. A series of fundamental results follows from this assumption by a few basic arguments. Some of these describe "best" performance properties of the Kelly portfolio which underline its fundamental importance in investing.

In Platen and Heath (2006), the benchmark approach has been described for jump-diffusion markets. Probably the most striking feature of the rich benchmark framework is the possible co-existence of several self-financing portfolios that may perfectly replicate one and the same payoff. The presence of different replicating portfolios is not consistent with the classical Law of One Price. The proposed *Law of the Minimal Price* identifies for a given contingent claim the corresponding minimal replicating portfolio process, which characterizes also in an incomplete market the economically correct price process. The Law of the Minimal Price yields directly the *real world pricing formula*, where the expectation is taken with respect to the real world probability measure and the Kelly portfolio appears as the *numéraire*. No change of probability measure is performed under real world pricing, which extends the classical risk neutral approach. Consequently, the request on the existence of an equivalent risk neutral probability measure is here avoided and a much wider modeling world is available. The real world pricing formula generalizes the risk neutral pricing formula as well as the actuarial pricing formula. These formulae represent the central pricing rules in their respective streams of literature and nothing else is here requested than the existence of the Kelly portfolio. The fact that the Law of One Price does not hold does not create strong arbitrage opportunities in the sense of this paper. We will see that there is no economic reason to exclude any weaker form of arbitrage as under the classical no-arbitrage approach.

A Diversification Theorem will be derived along the lines of Platen (2005). It states that any diversified portfolio in a regular financial market approximates the Kelly portfolio. This makes the benchmark approach rather practical and allows one to interpret a well-diversified portfolio as proxy for the Kelly portfolio. For further results on the benchmark approach the reader is referred to Platen and Heath (2006).

The remainder of the paper is organized as follows: It introduces the *numéraire* portfolio in Section 2. Section 3 derives various manifestations of "best" performance for the Kelly portfolio. The Law of the Minimal Price is proposed in Section 4. In Section 5 we discuss the concept of real world pricing. Section 6 introduces a strong form of arbitrage. Finally, Section 7 provides a version of the Diversification Theorem.

2 Benchmark Approach

Along the lines of Platen (2002) (Platen, 2002, 2006) and Platen and Heath (2006), we consider a general financial market in continuous time with d risky, non-negative, primary securities, $d \in \{1, 2, \ldots\}$. These securities could be, for instance, shares, currencies or other traded securities. Denote by S_t^j the value of the corresponding j^{th} *primary security account*, $j \in \{0, 1, \ldots, d\}$, at time $t \geq 0$. This non-negative account holds units of the jth primary security together with all dividends or interest payments reinvested. The 0th primary security account S_t^0 denotes the value of the locally riskless *savings account* at time $t \geq 0$. The dynamics of the primary security accounts need not be specified when formulating the main statements.

The market participants can form self-financing portfolios with primary security accounts as constituents. A portfolio value S_t^δ at time t is described by the number δ_t^j of units held in the jth primary security account S_t^j for all $j \in \{0, 1, \ldots, d\}$, $t \geq 0$. For simplicity, assume that the units of the primary security accounts are perfectly divisible, and that for all $t \in [0, \infty)$ the values $\delta_t^0, \delta_t^1, \ldots, \delta_t^d$, for any given *strategy* $\delta = \{\delta_t = (\delta_t^0, \delta_t^1, \ldots, \delta_t^d)^\top, t \geq 0\}$, depend only on information available at time t. The portfolio value at this time is given by the sum

$$S_t^\delta = \sum_{j=0}^{d} \delta_t^j \, S_t^j$$

We consider only self-financing portfolios where changes in their value are only due to changes in the values of the primary security accounts. We neglect any market frictions or liquidity effects.

By $\mathcal{V}_x^+$ denote the set of all strictly positive, finite, self-financing portfolios with initial capital $x > 0$. The benchmark approach employs a very special strictly positive portfolio as benchmark, which we denote by $S^{\delta*} \in \mathcal{V}_x^+$. Later it will become apparent that it is the *Kelly portfolio*, see Kelly (1956), which is in several ways the "best" performing strictly positive tradable portfolio. On the other hand, it will turn out that it is also the natural *numéraire* , see Long (1990), for pricing any type of claim when employing the real world probability for calculating the respective conditional expectation in the resulting pricing formula.

Let $E_t(X)$ denote the conditional expectation under the real world probability measure P given the information available at time $t \geq 0$.

Definition 2.1. *For given $x > 0$, a strictly positive, finite, self-financing portfolio $S^{\delta_*} \in \mathcal{V}_x^+$ is called a numéraire portfolio if all non-negative portfolios S^δ, when denominated in units of S^{δ_*}, are supermartingales.*

The notion of a *numéraire* portfolio was originally introduced by Long (1990) in a rather special setting. Later it was generalized in Bajeux-Besnainou and Portait (1997) and Becherer (2001). These authors worked under classical no-arbitrage assumptions that are consistent with the existence of an equivalent risk neutral probability measure, see Delbaen and Schachermayer (1998). More recently, Platen (2002), Bühlmann and Platen (2003), Platen and Heath (2006), and Platen (2006) emphasized that in a more general setting one still obtains a viable financial market model, as long as a *numéraire* portfolio exists. Also Fernholz and Karatzas (2005), Karatzas and Kardaras (2007), and Kardaras and Platen (2008b) consider financial market models beyond the classical no-arbitrage framework.

To provide a basis, let us formulate the only major assumption of the paper:

Assumption 2.2. *For given $x > 0$, there exists a numéraire portfolio $S^{\delta_*} \in \mathcal{V}_x^+$.*

This assumption is satisfied for a wide range of financial market models used in practice. For instance, in Platen and Heath (2006) it has been verified for jump-diffusion markets. Karatzas and Kardaras (2007) and Kardaras and Platen (2008b) confirm the validity of Assumption 2.2 for a wide range of semi-martingale markets.

Now, under the benchmark approach we choose the *numéraire* portfolio as benchmark. The benchmarked value $\hat{S}_t^\delta$ of a portfolio S^δ is given by the ratio

$$\hat{S}_t^\delta = \frac{S_t^\delta}{S_t^{\delta_*}}$$

for all $t \geq 0$. Definition 2.1 leads by Assumption 2.2 directly to the following conclusion:

Corollary 2.3. *The benchmarked value $\hat{S}_t^\delta$ of any non-negative portfolio S^δ satisfies the supermartingale property*

$$\hat{S}_t^\delta \geq E_t\left(\hat{S}_s^\delta\right) \tag{2.1}$$

for all $0 \leq t \leq s < \infty$.

Consequently, the currently observed benchmarked value of a non-negative portfolio is always greater than or equal to its expected future benchmarked value. This means intuitively, if there were any trend in a benchmarked non-negative portfolio, then this trend could only point downward.

The supermartingale property (2.1) is the fundamental property of a financial market. For instance, it yields easily the uniqueness of the benchmark by the following argument: Consider two strictly positive portfolios that are supposed to be *numéraire* portfolios. According to Corollary 2.3, the first portfolio, when expressed in units of the second one, must satisfy the supermartingale property (2.1). By the same argument, the second portfolio, when expressed in units of the first one, must also satisfy the supermartingale property. Consequently, by Jensen's inequality the portfolios have to be identical, and the value process $S^{\delta *} \in \mathcal{V}_x^+$ of a *numéraire* portfolio is unique. Note that the stated uniqueness does not imply that the number of units invested has to be unique, which is due to potential redundancies in primary security accounts.

3 Best Performance of the Benchmark

In this section we list several manifestations of the fact that our benchmark $S^{\delta *}$ is the "best" performing, strictly positive, tradable portfolio and equals the Kelly portfolio:

3.1 *Numéraire portfolio*

First, the definition of the *numéraire* portfolio $S^{\delta *}$ itself, given by Definition 2.1, expresses the fact that this portfolio performs "best" in the sense that the expected returns of benchmarked non-negative portfolios never become strictly positive.

3.2 *Kelly portfolio*

As a second manifestation of "best" performance, we derive the *growth optimality* of the benchmark, which identifies it very generally as the Kelly portfolio, a fact that has been documented for special models in the literature, for instance, in Long (1990). The *expected growth* $g^{\delta}_{t,h}$ of a strictly positive portfolio S^{δ} over the time period $(t, t+h]$ for $t, h > 0$ is given by the conditional expectation

$$g^{\delta}_{t,h} = E_t\left(\ln\left(A^{\delta}_{t,h}\right)\right)$$

of the logarithm of the portfolio ratio

$$A^{\delta}_{t,h} = \frac{S^{\delta}_{t+h}}{S^{\delta}_t}$$

To identify the strictly positive portfolio that maximizes the expected growth, let us perturb at time $t \geq 0$, the investment in a given strictly positive portfolio $S^{\underline{\delta}} \in \mathcal{V}_x^+$, $x > 0$, by some small fraction $\varepsilon \in (0, \frac{1}{2})$ of some non-negative portfolio S^{δ}. For analyzing the changes in the expected growth of the perturbed portfolio S^{δ_ε} define the *derivative of expected growth* in the direction of S^{δ} as the right hand limit

$$\left.\frac{\partial g^{\delta_\varepsilon}_{t,h}}{\partial \varepsilon}\right|_{\varepsilon=0} = \lim_{\varepsilon \to 0} \frac{1}{\varepsilon}\left(g^{\delta_\varepsilon}_{t,h} - g^{\underline{\delta}}_{t,h}\right) \tag{3.1}$$

for $t, h \geq 0$. Obviously, if the portfolio that maximizes expected growth coincides in (3.1) with the portfolio $S^{\underline{\delta}}$, then the resulting derivative of expected growth will never be greater than zero for all non-negative portfolios S^{δ}. This leads to the following definition of *growth optimality*:

Definition 3.1. A strictly positive portfolio $S^{\underline{\delta}}$ is called *growth optimal* if the corresponding derivative of expected growth is less than or equal to zero for all non-negative portfolios S^{δ}, that is,

$$\left.\frac{\partial g^{\delta_\varepsilon}_{t,h}}{\partial \varepsilon}\right|_{\varepsilon=0} \leq 0$$

for all t, $h \geq 0$.

Note that this definition is different to the classical characterization of the *Kelly portfolio* or *growth optimal portfolio.* In the literature, it is typically based on the maximization of expected logarithmic utility from terminal wealth, as used in Kelly (1956) and later also employed in a stream of literature, including Latané (1959), Breiman (1960), Hakansson (1971), Merton (1973a), Roll (1973) and Markowitz (1976), among many others. It is clear from Definition 3.1 and the standard log-utility definition of the Kelly portfolio, see for instance Kelly (1956) and Thorp (1972), that the above growth optimal portfolio equals the Kelly portfolio. The following result provides a convenient method for the identification of the benchmark or *numéraire* portfolio in a given investment universe by searching for its Kelly portfolio.

Theorem 3.2. *The numéraire portfolio is growth optimal.*

Proof: For $\varepsilon \in (0, \frac{1}{2})$, two consecutive times t and $t + h$ with $h > 0$, and a non-negative portfolio S^{δ}, with $S^{\delta}_t > 0$, consider the perturbed portfolio S^{δ_ε} under the choice $S^{\underline{\delta}}_t = S^{\delta *}_t$ in (3.1), yielding a portfolio ratio $A^{\delta_\varepsilon}_{t,h} = \varepsilon\, A^{\delta}_{t,h} - (1-\varepsilon)\, A^{\delta *}_{t,h} > 0$. One then obtains by the well-known inequality $\ln(x) \leq x - 1$ for $x \geq 0$, the relations

$$G^{\delta_\varepsilon}_{t,h} = \frac{1}{\varepsilon} \ln\left(\frac{A^{\delta_\varepsilon}_{t,h}}{A^{\delta *}_{t,h}}\right) \leq \frac{1}{\varepsilon}\left(\frac{A^{\delta_\varepsilon}_{t,h}}{A^{\delta *}_{t,h}} - 1\right) = \frac{A^{\delta}_{t,h}}{A^{\delta *}_{t,h}} - 1 \tag{3.2}$$

and

$$G^{\delta_\varepsilon}_{t,h} = -\frac{1}{\varepsilon} \ln\left(\frac{A^{\delta *}_{t,h}}{A^{\delta_\varepsilon}_{t,h}}\right) \geq -\frac{1}{\varepsilon}\left(\frac{A^{\delta *}_{t,h}}{A^{\delta_\varepsilon}_{t,h}} - 1\right) = \frac{A^{\delta}_{t,h} - A^{\delta *}_{t,h}}{A^{\delta_\varepsilon}_{t,h}} \tag{3.3}$$

Since $A^{\delta_\varepsilon}_{t,h} > 0$ one obtains from (3.3) for $A^{\delta}_{t,h} - A^{\delta *}_{t,h} \geq 0$ the inequality

$$G^{\delta_\varepsilon}_{t,h} \geq 0 \tag{3.4}$$

and for $A^{\delta}_{t,h} - A^{\delta *}_{t,h} < 0$, because of $\varepsilon \in (0, \frac{1}{2})$ and $A^{\delta}_{t,h} \geq 0$, the relation

$$G^{\delta_\varepsilon}_{t,h} \geq -\frac{A^{\delta *}_{t,h}}{A^{\delta_\varepsilon}_{t,h}} = -\frac{1}{1 - \varepsilon + \varepsilon \frac{A^{\delta}_{t,h}}{A^{\delta *}_{t,h}}} \geq -\frac{1}{1-\varepsilon} \geq -2 \tag{3.5}$$

Summarizing (3.2)–(3.5) yields the upper and lower bounds

$$-2 \leq G^{\delta_\varepsilon}_{t,h} \leq \frac{A^{\delta}_{t,h}}{A^{\delta *}_{t,h}} - 1 \tag{3.6}$$

where by Definition 2.1 one has

$$E_t\left(\frac{A^{\delta}_{t,h}}{A^{\delta *}_{t,h}}\right) \leq 1 \tag{3.7}$$

By using (3.6) and (3.7) it follows by the Dominated Convergence Theorem, see Shiryaev (1984), that

$$\begin{aligned}
\left.\frac{\partial g^{\delta_\varepsilon}_{t,h}}{\partial \varepsilon}\right|_{\varepsilon=0} &= \lim_{\varepsilon \to 0+} E_t\left(G^{\delta_\varepsilon}_{t,h}\right) = E_t\left(\lim_{\varepsilon \to 0+} G^{\delta_\varepsilon}_{t,h}\right) \\
&= E_t\left(\frac{\partial}{\partial \varepsilon} \ln\left(\frac{A^{\delta_\varepsilon}_{t,h}}{A^{\delta *}_{t,h}}\right)\bigg|_{\varepsilon=0}\right) = E_t\left(\frac{A^{\delta}_{t,h}}{A^{\delta *}_{t,h}}\right) - 1
\end{aligned}$$

This proves by Definitions 2.1 and 3.1 that the *numéraire* portfolio $S^{\delta *}$ is growth optimal. □

3.3 *Long term growth*

To formulate a third manifestation of "best" performance for the benchmark, define the *long term growth rate* g^{δ} of a strictly positive portfolio $S^{\delta} \in \mathcal{V}^{+}_{x}$ as the upper limit

$$g^{\delta} = \limsup_{t \to \infty} \frac{1}{t} \ln\left(\frac{S^{\delta}_{t}}{S^{\delta}_{0}}\right) \tag{3.8}$$

The long term growth rate (3.8) is defined pathwise almost surely and does not involve any expectation. By exploiting the supermartingale property (2.1), the following fascinating property of the Kelly portfolio will be shown very generally below.

Theorem 3.3. *The numéraire portfolio $S^{\delta *} \in \mathcal{V}^{+}_{x}$ achieves the maximum long term growth rate. This means, when compared with any other strictly positive portfolio $S^{\delta} \in \mathcal{V}^{+}_{x}$, one has*

$$g^{\delta} \leq g^{\delta *} \tag{3.9}$$

Proof: Similar as in Karatzas and Shreve (1998) consider a strictly positive portfolio $S^\delta \in \mathcal{V}_x^+$, $x > 0$, with the same initial capital as the *numéraire* portfolio, that is, $S_0^\delta = S_0^{\delta *} = x > 0$. By Corollary 2.3, we can use the following maximal inequality, derived in Doob (1953), where for any $k \in \{1, 2, \ldots\}$ and $\varepsilon \in (0, 1)$ one has

$$\exp\{\varepsilon\, k\}\, P\left(\sup_{k \le t < \infty} \hat{S}_t^\delta > \exp\{\varepsilon\, k\}\right) \le E_0\left(\hat{S}_k^\delta\right) \le \hat{S}_0^\delta = 1$$

One finds for fixed $\varepsilon \in (0, 1)$ that

$$\sum_{k=1}^{\infty} P\left(\sup_{k \le t < \infty} \ln\left(\hat{S}_t^\delta\right) > \varepsilon\, k\right) \le \sum_{k=1}^{\infty} \exp\{-\varepsilon\, k\} < \infty$$

By the Borel-Cantelli lemma, see Shiryaev (1984), there exists a random variable k_ε such that for all $k \ge k_\varepsilon$ and $t \ge k$ it holds that

$$\ln\left(\hat{S}_t^\delta\right) \le \varepsilon\, k \le \varepsilon\, t$$

Therefore, it follows for all $k > k_\varepsilon$ the estimate

$$\sup_{t \ge k} \frac{1}{t} \ln\left(\hat{S}_t^\delta\right) \le \varepsilon$$

which implies that

$$\limsup_{t \to \infty} \frac{1}{t} \ln\left(\frac{S_t^\delta}{S_0^\delta}\right) \le \limsup_{t \to \infty} \frac{1}{t} \ln\left(\frac{S_t^{\delta *}}{S_0^{\delta *}}\right) + \varepsilon \tag{3.10}$$

Since the inequality (3.10) holds for all $\varepsilon \in (0, 1)$ one obtains with (3.8) the relation (3.9). □

According to Theorem 3.3, the trajectory of the Kelly portfolio outperforms in the long run those of all other strictly positive portfolios that start with the same initial capital. This property is independent of any model choice and, therefore, very robust. An investor, who is aiming in the long run for the highest possible wealth, has to invest her or his total tradable wealth into the Kelly portfolio.

3.4 *Systematic outperformance*

Over short and medium time horizons, almost any strictly positive portfolio can generate larger returns than those exhibited by the Kelly portfolio. However, the fourth manifestation of "best" performance will show that such short term outperformance cannot be achieved systematically. To formulate a corresponding statement properly, we will employ the following definition:

Definition 3.4. *A non-negative portfolio S^δ systematically outperforms a strictly positive portfolio $S^{\tilde{\delta}}$ if*

(i) *both portfolios start with the same initial capital $S_{t_0}^\delta = S_{t_0}^{\tilde{\delta}}$;*

(ii) *at a later time t the portfolio value* S_t^δ *is at least equal to* $S_t^{\tilde{\delta}}$, *that is* $P(S_t^\delta \geq S_t^{\tilde{\delta}}) = 1$, *and*

(iii) *the probability for* S_t^δ *being strictly greater than* $S_t^{\tilde{\delta}}$ *is strictly positive so that* $P\left(S_t^\delta > S_t^{\tilde{\delta}}\right) > 0$.

Systematic outperformance of one portfolio by another one is possible under the benchmark approach, see Platen and Heath (2006). The above notion of systematic outperformance was introduced in Platen (2004) and was motivated by the supermartingale property (2.1). It relates in some sense to the notion of relative arbitrage, later studied in Fernholz and Karatzas (2005), and also to the notion of a maximal element earlier employed in Delbaen and Schachermayer (1998). The following result presents a fourth manifestation of "best" performance of the benchmark:

Theorem 3.5. *The numéraire portfolio cannot be systematically outperformed by any non-negative portfolio.*

Proof: Consider a non-negative portfolio S^δ with benchmarked value $\hat{S}_t^\delta = 1$ at a given time $t \geq 0$, where $\hat{S}_s^\delta \geq 1$ almost surely at some later time $s \in [t, \infty)$. Then it follows by the supermartingale property given in Corollary 2.3 that

$$0 \geq E_t\left(\hat{S}_s^\delta - \hat{S}_t^\delta\right) = E_t\left(\hat{S}_s^\delta - 1\right)$$

Since one has $\hat{S}_s^\delta \geq 1$ almost surely and $E_t(\hat{S}_s^\delta) \leq 1$, it can only follow that $\hat{S}_s^\delta = 1$. This means that one has at time s the equality $S_s^\delta = S_s^{\delta *}$ almost surely. Therefore, according to Definition 3.4 the portfolio S^δ does not systematically outperform the *numéraire* portfolio. □

As a consequence of Theorem 3.5, one can conclude that in the given very general modeling setting, no active fund manager can systematically outperform the benchmark. On the other hand, if the market portfolio is not the *numéraire* portfolio, which is most likely the case, and a fund approximates well the Kelly portfolio, then this fund will not be outperformed systematically by the market portfolio or any other significantly different portfolio. In the long run its path will by property (3.8) beat the path of the market portfolio almost surely.

There are further manifestations of "best" performance of the Kelly portfolio. For instance, in Kardaras and Platen (2008a), it is shown that the Kelly portfolio minimizes some expected market time to reach a given wealth level. Results in this direction can be found, for instance, in Browne (1998).

4 The Law of the Minimal Price

The fundamental supermartingale property (2.1) ensures that the maximum expected return of a benchmarked non-negative portfolio can at most equal zero. In the case when it equals zero for a given benchmarked price process for all time instances, then the current benchmarked value of the price is always the best forecast of its future benchmarked values. In this case, one has equality in relation (2.1) and we call such a price process *fair*. A benchmarked fair price process forms a martingale. In general, not all primary security accounts and portfolios need to be fair under the benchmark approach. This is the key property that differentiates this approach from the classical approaches. For instance, there may exist benchmarked portfolios that form local martingales but are not true martingales, as is extensively discussed in Platen and Heath (2006). Furthermore, note that non-negative fair price processes are somehow minimal, as we shall see below.

It may be puzzling to some readers that discounting with another discount factor than the savings account ought to lead to "fair" prices or could be particularly meaningful. Below we will give a valid reason why the Kelly portfolio is the "universal currency" that should be used for discounting when pricing under the real world probability. It stems from the fact that the Kelly portfolio represents the best performing portfolio and, thus, the natural *numéraire* for valuation when the real world probability measure for calculating expectations.

The following *Law of the Minimal Price* substitutes the widely postulated classical Law of One Price, which no longer holds in our general setting:

Theorem 4.1. *Law of the Minimal Price. If a fair portfolio process replicates a given non-negative payoff at a given maturity date, then this portfolio represents the minimal replicating portfolio among all non-negative portfolios that replicate this payoff.*

Proof: A stochastic process $\hat{S}^{\delta}$, which satisfies relation (2.1) is a supermartingale, see Shiryaev (1984). When equality holds in (2.1) then the process is fair and its benchmarked value process is a martingale. Within a family of non-negative supermartingales with the same value at a given future payoff date, it is the martingale among these supermartingales which attains almost surely the minimal possible value at all times before the maturity date. This fundamental fact about the optimality of martingales in a family of supermartingales proves directly Theorem 4.1. □

The existence of unfair price processes under the benchmark approach creates new realistic effects that can be modeled and are not captured by any classical no-arbitrage framework. For a given replicable payoff, it follows by Theorem 4.1 that the corresponding fair replicating portfolio describes the least expensive hedge portfolio. This is also the economically correct price process in a competitive market. We emphasize that the *Law of the Minimal Price* generates a unique price system for contracts and derivatives under the benchmark approach which only relies on the existence of a *numéraire* portfolio. As shown in Platen (2009), pricing purely based on hedging or classical no-arbitrage arguments, as employed under the classical Arbitrage Pricing Theory, see Ross (1976), can lead in our general setting to significantly more expensive prices than provided by fair price processes.

5 Real World Pricing

Define a *contingent claim* H_T as a non-negative payoff delivered at maturity $T \in (0, \infty)$, which is expressed in units of the domestic currency and has finite expectation

$$E_0\left(\frac{H_T}{S_T^{\delta_*}}\right) < \infty. \tag{5.1}$$

By the *Law of the Minimal Price* one can identify the corresponding fair price process via the following *real world pricing formula:*

Corollary 5.1. *If for a contingent claim H_T, $T \in (0, \infty)$, there exists a fair portfolio S^{δ_H} that replicates this claim at maturity T such that $H_T = S_T^{\delta_H}$, then its minimal replicating price at time $t \in [0, T]$ is given by the real world pricing formula*

$$S_t^{\delta_H} = S_t^{\delta_*}\, E_t\left(\frac{H_T}{S_T^{\delta_*}}\right) \tag{5.2}$$

Relation (5.2) is called the real world pricing formula because it involves the conditional expectation E_t with respect to the real world probability measure. This formula can be interpreted in the sense of Cochrane (2001) as a pricing formula that uses the stochastic discount factor $(S_t^{\delta_*})^{-1}$. Also the closely related use of the corresponding state price density, pricing kernel and deflator are consistent with the real world pricing formula (5.2) under appropriate assumptions. However, all the classical pricing approaches exclude classical arbitrage, which is equivalent to the existence of an equivalent risk neutral probability measure, see Delbaen and Schachermayer (1998). In this manner, they postulate the *Law of One Price*, see for instance Ingersoll (1987), Long (1990), Constantinides (1992), Duffie (2001), and Cochrane (2001) for classical no-arbitrage settings. The benchmark approach with its real world pricing concept does not require a risk neutral probability measure to exist or equivalent classical no-arbitrage constraints and goes beyond the *Law of One Price.* The real world pricing formula (5.2) only requires the existence of the *numéraire* portfolio and the finiteness of the expectation in (5.1). No measure transformation needs to be applied.

An important special case of the real world pricing formula (5.2) arises when H_T is independent of $S_T^{\delta_*}$. In this case one obtains the *actuarial pricing formula*

$$S_t^{\delta_H} = P(t,T)\, E_t(H_T) \tag{5.3}$$

with the fair zero coupon bond price

$$P(t,T) = S_t^{\delta_*}\, E_t\left((S_T^{\delta_*})^{-1}\right) \tag{5.4}$$

that pays one monetary unit at maturity T. The fair zero coupon bond price $P(t,T)$ provides the discount factor in (5.3) in a way as it has been used intuitively by actuaries for obtaining the, so called, net present value of H_T. The real world pricing formula provides a rigorous and general derivation of the actuarial pricing formula that has been in use for centuries.

Let us now derive risk neutral pricing as a special case of real world pricing. For this purpose we rewrite for $t = 0$ the real world pricing formula (5.2) in the form

$$S_0^{\delta_H} = E_0\left(\Lambda_T \frac{S_0^0}{S_T^0} H_T\right) \tag{5.5}$$

while employing the benchmarked normalized savings account $\Lambda_T = \frac{\hat{S}_T^0}{\hat{S}_0^0}$. By the supermartingale property (2.1) of the normalized benchmarked savings account process $\Lambda = \{\Lambda_t = \frac{\hat{S}_t^0}{\hat{S}_0^0}, t \geq 0\}$ we have $1 = \Lambda_0 \geq E_0(\Lambda_T)$. Together with equation (5.5) this yields the inequality

$$S_0^{\delta_H} \leq \frac{E_0\left(\Lambda_T \frac{S_0^0}{S_T^0} H_T\right)}{E_0(\Lambda_T)} \tag{5.6}$$

If the benchmarked savings account is not a martingale, then equality does not hold in relation (5.6). To ensure equality in (5.6), one needs to impose the strong assumption that the savings account is fair, that is, intuitively its benchmarked value has no trend and is a martingale. In this particular case, the expression on the right hand side of (5.6) can be interpreted by Bayes' formula as the conditional expectation of the discounted contingent claim under the, in this case existing, equivalent risk neutral probability measure Q with Radon-Nikodym derivative $\Lambda_T = \frac{dQ}{dP}$. Only in this case when Λ is a martingale the relation (5.6) yields, in general, for a contingent claim H_T the classical risk neutral pricing formula

$$S_0^{\delta_H} = E_0^Q\left(\frac{S_0^0}{S_T^0} H_T\right)$$

see, for instance, Harrison and Kreps (1979) or Karatzas and Shreve (1998). Here E_0^Q denotes the conditional expectation at time $t = 0$ under the equivalent risk neutral probability measure Q. By inequality (5.6) it follows that the fair derivative price is never more expensive than the price obtained under some formal application of the standard risk neutral pricing rule.

Finally, it shall be noted for not perfectly hedgable contingent claims that utility indifference pricing, in the sense of Davis (1997), leads in the case of incomplete jump-diffusion markets again to the real world pricing formula (5.2), see Platen and Heath (2006). Under real world pricing, the hedgable part of the claim is replicated via the minimal possible hedge portfolio and the benchmarked unhedgable part remains untouched. In some sense, the real world pricing formula provides the least squares projection of a given unhedgable benchmarked contingent claim into the set of current benchmarked prices. The benchmarked hedge error has zero mean and its variance is minimized. This generalizes also the notion of local-risk minimization for pricing in incomplete markets, as advocated in Föllmer and Schweizer (1991), Hofmann et al. (1992), and Schweizer (1995). In practice, when benchmarked hedge errors can be diversified in the large book of a well diversified bank or insurance company, then the total benchmarked hedge error vanishes by the *Law of Large Numbers* asymptotically from the bank's trading book and the market becomes asymptotically complete from this perspective. This shows that in practice real world pricing makes perfect sense. Any systematically more expensive pricing would make an institution less competitive. On the other hand, any lower prices would make it unsustainable.

6 Strong Arbitrage

Since the benchmark approach goes significantly beyond the classical no-arbitrage world, it is of importance to clarify the potential existence of arbitrage opportunities. In the literature, there exist many different mathematical definitions of arbitrage, and one has to ensure that the given modeling framework is economically viable. Obviously, arbitrage opportunities can only be exploited by market participants. These have to use their portfolios of total tradable wealth when trying to exploit potential arbitrage opportunities. Due to the established legal concept of limited liability, only non-negative total tradable wealth processes have to be considered when studying the exploitation of potential arbitrage. Therefore, any realistic arbitrage concept has to focus on non-negative, self-financing portfolios.

An obvious, strong form of arbitrage arises when a market participant can generate strictly positive wealth from zero initial capital. This leads to the following definition of *strong arbitrage*, which was introduced in Platen (2002) motivated by the supermartingale property (2.1):

Definition 6.1. *A non-negative portfolio S^δ is a strong arbitrage if it starts with zero initial capital, that is $S_0^\delta = 0$, and generates strictly positive wealth with strictly positive probability at a later time $t \in (0, \infty)$, that is, $P(S_t^\delta > 0) > 0$.*

The exclusion of the above form of arbitrage has been independently argued for on purely economic grounds by Loewenstein and Willard (2000). Important is that if only strong arbitrage is excluded, then weaker forms of arbitrage may still exist. However, this does not harm the economic viability of the market model. For instance, there may exist, so-called, *free snacks* and *cheap thrills*, in the sense of Loewenstein and Willard (2000), which usually represent situations of systematic outperformance in the sense of Definition 3.4. Also *free lunches with vanishing risk*, as excluded in Delbaen and Schachermayer (1998), can exist without creating complications from an economic point of view. Furthermore, by looking at any of the weaker forms of arbitrage one realizes that these cannot be exploited in practice without providing adequate collateral, see Platen and Heath (2006). This request, however, makes the corresponding theoretical notions of weaker forms of arbitrage questionable from their practical relevance. By exploiting the supermartingale property (2.1), the following result can be established:

Theorem 6.2. *There does not exist any non-negative portfolio that is a strong arbitrage.*

Proof: For a non-negative portfolio S^δ, which starts with zero initial capital, it follows by the supermartingale property given in Corollary 2.3 that

$$0 = S_0^\delta = x\,\hat{S}_0^\delta \geq x\,E_0\left(\hat{S}_t^\delta\right) = x\,E(\hat{S}_t^\delta) \geq 0$$

for $t \geq 0$, where $E(\cdot)$ denotes expectation under the real world probability. By the nonnegativity of S_t^δ and the strict positivity of $S_t^{\delta*}$, the event $S_t^\delta > 0$ can only have zero probability, that is

$$P\left(S_t^\delta > 0\right) = 0$$

This leads to the conclusion that S_t^δ equals zero for all $t \geq 0$, which proves by Definition 6.1 the Theorem 6.2. □

Theorem 6.2 states that strong arbitrage is automatically excluded in the given general benchmark framework. Therefore, different to the classical no-arbitrage approaches, pricing by excluding strong arbitrage does not make any sense. Instead, one should use real world pricing which is economically and also theoretically meaningful, as we have seen.

7 Diversification

To conclude the paper, we consider the practical problem of identifying or approximating the Kelly portfolio of a given market. We will indicate how to construct proxies for the Kelly portfolio that can be used for portfolio optimization and valuation under the benchmark approach. For simplicity, let us consider a continuous financial market with its benchmarked j^{th} primary security account value $\hat{S}_t^j$ at time t satisfying the driftless stochastic differential equation (SDE)

$$d\hat{S}_t^j = \hat{S}_t^j \sum_{k=1}^{j} \sigma_t^{j,k}\, dW_t^k \tag{7.1}$$

for $t \in [0,\infty)$, with $\hat{S}_0^j > 0$, see Platen and Heath (2006). The more general setting of jump diffusion markets has been considered in Platen (2005) and forthcoming work will generalize the results presented below, avoiding any particular assumptions. In (7.1) $W^k = \{W_t^k,\, t \in [0,\infty)\}$ denotes an independent standard Wiener process, $k \in \{1, 2, \ldots\}$. A benchmarked portfolio $\hat{S}^{\delta_d}$ with fractions $\pi_{\delta_d,t}^j$ invested at time

t in the jth primary security account, where $j \in \{1,2,\ldots,d\}$ with $d \in \{1,2,\ldots\}$, satisfies the SDE

$$d\hat{S}_t^{\delta_d} = \hat{S}_t^{\delta_d} \sum_{j=1}^{d} \pi_{\delta_d,t}^{j} \sum_{k=1}^{j} \sigma_t^{j,k}\, dW_t^k \tag{7.2}$$

for $t \in [0,\infty)$, with $\hat{S}_0^{\delta_d} > 0$.

It is clear that the benchmarked Kelly portfolio $\hat{S}_t^*$ equals the constant one and its diffusion coefficients vanish. This leads us to define a sequence $(\hat{S}^{\delta_d})_{d\in\{1,2,\ldots\}}$ of *benchmarked approximate Kelly portfolios* as a sequence of strictly positive portfolios such that for all $\varepsilon > 0$ we have

$$\lim_{d\to\infty} P\left(\sum_{k=1}^{d}\left(\sum_{j=1}^{d} \pi_{\delta_d,t}^{j}\,\sigma_t^{j,k}\right)^2 \geq \varepsilon\right) = 0 \tag{7.3}$$

for $t \in [0,\infty)$.

Furthermore, a sequence $(\hat{S}^{\delta_d})_{d\in\{1,2,\ldots\}}$ of benchmarked portfolios is called a sequence of benchmarked *diversified portfolios* if some constants $K_1, K_2 \in (0,\infty)$ and $K_3 \in \{1,2,\ldots\}$ exist independently of d, such that for $d \in \{K_3, K_3+1,\ldots\}$ each fraction $\pi_{\delta_d,t}^{j}$ is bounded in the form

$$0 \leq \pi_{\delta_d,t}^{j} \leq \frac{K_2}{d^{\frac{1}{2}+K_1}} \tag{7.4}$$

almost surely for all $j \in \{1,2,\ldots,d\}$ and $t \in [0,\infty)$.

If most benchmarked primary security accounts would have the same driving Wiener process, then it may become difficult to form a benchmarked portfolio with vanishing volatility. To avoid such a situation we assume that the given continuous financial market is *regular*, which means that for all $t \in [0,\infty)$ and $k \in \{1,2,\ldots\}$ there exists a finite adapted stochastic process $C = \{C_t,\, t \in [0,\infty)\}$ such that

$$\left(\sum_{j=1}^{\infty} \left|\sigma_t^{j,k}\right|\right)^2 \leq C_t \tag{7.5}$$

almost surely.

This allows us to prove, similar as in Platen (2005), and Platen and Heath (2006), the following *Diversification Theorem*:

Theorem 7.1. *In a regular continuous financial market each sequence of diversified portfolios is a sequence of approximate Kelly portfolios.*

Proof: For a sequence of diversified portfolios it follows from (7.4) and (7.5) that for each $d \in \{1, 2, \ldots\}$ one has

$$\sum_{k=1}^{d} \left(\sum_{j=1}^{d} \pi_{\delta_d,t}^{j} \sigma_t^{j,k} \right)^2 \leq \sum_{k=1}^{d} \left(\sum_{j=1}^{d} \left| \pi_{\delta_d,t}^{j} \right| \left| \sigma_t^{j,k} \right| \right)^2$$

$$\leq \sum_{k=1}^{d} \frac{(K_2)^2}{d^{1+2K_1}} \left(\sum_{j=1}^{d} \left| \sigma_t^{j,k} \right| \right)^2$$

$$\leq \frac{(K_2)^2 C_t}{d^{2K_1}}.$$

Consequently, we have for all $\varepsilon > 0$ that

$$\lim_{d \to \infty} P \left(\sum_{k=1}^{d} \left(\sum_{j=1}^{d} \pi_{\delta_d,t}^{j} \sigma_t^{j,k} \right)^2 \geq \varepsilon \right) = 0,$$

which proves (7.3). □

8 Conclusion

A general financial modeling and pricing framework, the benchmark approach, has been presented, which only assumes the existence of the *numéraire* portfolio. It turns out that this portfolio coincides with the Kelly portfolio, which is in several ways the "best" performing strictly positive portfolio. It can be used as benchmark in the traditional sense of portfolio optimization but also as *numéraire* in derivative pricing. Under the benchmark approach, the classical *Law of One Price* does generally not hold. It has been replaced by the *Law of the Minimal Price*, according to which the minimal replicating price process for a given contingent claim is trendless when expressed in units of the Kelly portfolio. By exploiting this fact, the real world pricing concept emerges with the Kelly portfolio as *numéraire* and the real world probability measure as pricing measure. Real world pricing turns out to be the natural pricing concept for nonhedgable contingent claims in an incomplete market. Certain weak forms of arbitrage may exist under the benchmark approach that are excluded under the classical no-arbitrage approach. However, this does not harm the economic viability of the resulting general modeling framework. It has been shown that diversified portfolios approximate asymptotically in a regular market the Kelly portfolio, as the number of constituents increases.

References

Bajeux-Besnainou, I. and R. Portait (1997). The numeraire portfolio: A new perspective on financial theory. *The European Journal of Finance*, 3, 291–309.

Becherer, D. (2001). The numeraire portfolio for unbounded semimartingales. *Finance and Stochastics*, 5, 327–341.

Breiman, L. (1960). Investment policies for expanding business optimal in a long run sense. *Naval Research Logistics Quarterly*, 7(4), 647–651.

Browne, S. (1998). The return on investment from proportional portfolio strategies. *Advances in Applied Probability*, 30(1), 216–238.

Bühlmann, H. and E. Platen (2003). A discrete time benchmark approach for insurance and finance. *ASTIN Bulletin*, 33(2), 153–172.

Cochrane, J. H. (2001). *Asset Pricing*. US: Princeton University Press.

Constantinides, G. M. (1992). A theory of the nominal structure of interest rates. *Review Financial Studies*, 5, 531–552.

Davis, M. H. A. (1997). Option pricing in incomplete markets. In M. A. H. Dempster and S. R. Pliska (Eds.), *Mathematics of Derivative Securities*, pp. 227–254. UK: Cambridge University Press.

Debreu, G. (1959). *Theory of Value*. New York: John Wiley.

Delbaen, F. and W. Schachermayer (1998). The fundamental theorem of asset pricing for unbounded stochastic processes. *Mathematische Annalen*, 312, 215–250.

Doob, J. L. (1953). *Stochastic Processes*. New York: John Wiley.

Duffie, D. (2001). *Dynamic Asset Pricing Theory* (3rd ed.). US: Princeton University Press.

Fernholz, E. R. and I. Karatzas (2005). Relative arbitrage in volatility-stabilized markets. *Annals of Finance*, 1(2), 149–177.

Föllmer, H. and M. Schweizer (1991). Hedging of contingent claims under incomplete information. In M. H. A. Davis and R. J. Elliott (Eds.), *Applied Stochastic Analysis*, Vol. 5 of *Stochastics Monograph*, pp. 389–414. New York: Gordon and Breach.

Hakansson, N. H. (1971). Capital growth and the mean-variance approach to portfolio selection. *Journal of Financial and Quantitative Analysis*, 6(1), 517–557.

Harrison, J. M. and D. M. Kreps (1979). Martingale and arbitrage in multiperiod securities markets. *Journal of Economic Theory*, 20, 381–408.

Hofmann, N., E. Platen, and M. Schweizer (1992). Option pricing under incompleteness and stochastic volatility. *Mathematical Finance*, 2(3), 153–187.

Ingersoll, J. E. (1987). *Theory of Financial Decision Making*. Studies in Financial Economics. Rowman and Littlefield.

Karatzas, I. and C. Kardaras (2007). The numeraire portfolio in semimartingale financial models. *Finance Stochastics*, 11(4), 447–493.

Karatzas, I. and S. E. Shreve (1998). *Methods of Mathematical Finance*, Volume 39 of *Appl. Math.* Springer.

Kardaras, C. and E. Platen (2008a). Minimizing the expected market time to reach a certain wealth level. Technical report, University of Technology, Sydney. QFRC Research Paper 230, to appear in *SIAM J. Financial Mathematics*.

Kardaras, C. and E. Platen (2008b). On financial markets where only buy-and-hold trading is possible. Technical report, University of Technology, Sydney. QFRC Research Paper 213.

Kelly, J. R. (1956). A new interpretation of information rate. *Bell Syst. Techn. J.*, 35, 917–926.

Latané, H. (1959). Criteria for choice among risky ventures. *J. Political Economy*, 38, 145–155.

Lintner, J. (1965). The valuation of risk assets and the selection of risky investments in stock portfolios and capital budgets. *Rev. Econom. Statist.*, 47, 13–37.

Loewenstein, M. and G. A. Willard (2000). Local martingales, arbitrage, and viability: Free snacks and cheap thrills. *Econometric Theory*, 16(1), 135–161.

Long, J. B. (1990). The numeraire portfolio. *J. Financial Economics*, 26, 29–69.

Markowitz, H. (1976). Investment for the long run: New evidence for an old rule. *J. Finance*, 31(5), 1273–1286.

Merton, R. C. (1973a). An intertemporal capital asset pricing model. *Econometrica*, 41, 867–888.

Merton, R. C. (1973b). Theory of rational option pricing. *Bell J. Econ. Management Sci.*, 4, 141–183.

Platen, E. (2002). Arbitrage in continuous complete markets. *Adv. in Appl. Probab.*, 34(3), 540–558.

Platen, E. (2004). A benchmark framework for risk management. In *Stochastic Processes and Applications to Mathematical Finance*, pp. 305–335. Proceedings of the Ritsumeikan Intern. Symposium: World Scientific.

Platen, E. (2005). Diversified portfolios with jumps in a benchmark framework. *Asia-Pacific Financial Markets*, 11(1), 1–22.

Platen, E. (2006). A benchmark approach to finance. *Math. Finance*, 16(1), 131–151.

Platen, E. (2009). Real world pricing of long term contracts. Technical report, University of Technology, Sydney. QFRC Research Paper.

Platen, E. and D. Heath (2006). *A Benchmark Approach to Quantitative Finance*. New York: Springer.

Roll, R. (1973). Evidence on the "Growth-Optimum" model. *J. Finance*, 28(3), 551–566.

Ross, S. A. (1976). The arbitrage theory of capital asset pricing. *J. Economic Theory*, 13, 341–360.

Schweizer, M. (1995). On the minimal martingale measure and the Föllmer-Schweizer decomposition. *Stochastic Anal. Appl.*, 13, 573–599.

Sharpe, W. F. (1964). Capital asset prices: A theory of market equilibrium under conditions of risk. *J. Finance*, 19, 425–442.

Shiryaev, A. N. (1984). *Probability*. New York: Springer.

Thorp, E. O. (1972). Portfolio choice and the Kelly criterion. In *Proceedings of the 1971 Business and Economics Section of the American Statistical Association*, Volume 21, pp. 5–224.

29

Growing Wealth with Fixed-Mix Strategies*

Michael A. H. Dempster
Centre for Financial Research, University of Cambridge, United Kingdom
mahd2@cam.ac.uk

Igor V. Evstigneev
Economics Department, School of Social Sciences,
University of Manchester, United Kingdom
igor.evstigneev@manchester.ac.uk

Klaus Reiner Schenk-Hoppé
Leeds University Business School and School of Mathematics,
University of Leeds, United Kingdom
K.R.Schenk-Hoppe@leeds.ac.uk

Abstract

This chapter surveys theoretical research on the long-term performance of fixed-mix investment strategies. These self-financing strategies rebalance the portfolio over time so as to keep constant the proportions of wealth invested in various assets. The main result is that wealth can be grown from volatility. Our findings demonstrate the benefits of active portfolio management and the potential of financial engineering.

1 Introduction

Investment advice is usually based on some optimality principle such as the maximization of expected utility or the growth rate. The present chapter takes a broader view by studying generic features of an investment style based on constant proportions, or fixed-mix, strategies. These self-financing strategies aim to maintain fixed proportions between the value of portfolio positions by trading in the market at

*Financial support by the Finance Market Fund, Norway (Stability of Financial Markets: An Evolutionary Approach) and the National Center of Competence in Research Financial Valuation and Risk Management, Switzerland (Behavioural and Evolutionary Finance) is gratefully acknowledged. This chapter was written during KRSH's visit to the Department of Finance and Management Science at the Norwegian School of Economics and Business Administration in June 2009. We are grateful to editors Edward O. Thorp and William T. Ziemba for their many helpful comments.

specific points in time. It is an active portfolio management that rebalances positions by selling assets, whose portfolio value exceed the given benchmark to finance the purchase of those assets with a too low weight in the portfolio.

The significance of constant proportions strategies for investment science was established in Kelly (1956)'s work on information theory and its application to betting markets. A detailed account is provided in Part I of this volume. Kelly's research was inspired by Claude Shannon's lectures on investment problems in which the founder of the mathematical theory of information outlined his pioneering ideas in the field of investment science (though he never published in it); see Cover (1998) for the history of this approach.

The optimality property of the Kelly rule, which maximizes the growth rate of capital in the case of independent and identically distributed returns on investment, motivated the research on the log-optimum investment principle, Breiman (1961), Algoet and Cover (1988), MacLean, Ziemba and Blazenko (1992), Hakansson and Ziemba (1995), Browne and Whitt (1996), Cover (1991), Li (1998), and others. Constant proportions strategies have been studied in many different frameworks, e.g., Browne (1998), Luenberger (1998), Aurell *et al.* (2000) and Aurell and Muratore-Ginanneschi (2000), Fernholz (2002), Fernholz and Karatzas (2005), Fernholz *et al.* (2005), Evstigneev, Hens and Schenk-Hoppé (2009a, 2009b), Kuhn and Luenberger (forthcoming). This approach has proved quite successful in practical finance as witnessed by a considerable body of empirical literature, see Thorp (1971), Perold and Sharpe (1988), Ziemba and Mulvey (1998), Mulvey (2001, 2009), Dries *et al.* (2002), Dempster *et al.* (2003), Mulvey *et al.* (2007), and others as well as Part VI in this volume.

In contrast to this literature on investment, we will ignore questions of optimality of trading strategies and will not make use of expected utility or related concepts. Our goal is to analyze the performance of *arbitrary* fixed-mix (constant proportions) strategies in markets with different characteristics. We are only interested in 'generic' properties of (not necessarily optimal) fixed-mix strategies and their performance relative to buy-and-hold strategies.[1]

This survey is based on the authors' research, Evstigneev and Schenk-Hoppé (2002) and Dempster, Evstigneev, and Schenk-Hoppé (2003, 2007, 2008). In these papers, we study the wealth dynamics of investors employing fixed-mix strategies. Very general (and mostly counter-intuitive) results on fixed-mix strategies are obtained under the assumption that markets will exhibit some degree of stationarity either of (relative) prices or returns. This assumption of stationarity of asset returns is widely accepted in financial theory (and, thus, the basis for practical investment advice) allowing, as it does, expected exponential price growth and mean reversion, volatility clustering and very general intertemporal dependence, such as long memory effects, of returns. Stationarity of returns for instance is a salient feature of the Cox-Ross-Rubinstein binomial asset pricing model.

[1]Of course it is well-known that optimally-chosen rebalancing strategies perform at least as well as any buy-and-hold portfolio, see Part II in this volume.

The remainder of this chapter is organized as follows. Section 1.1 invites the readers to test their intuition while Section 1.2 provides a very simple illustrative example. The general theory of constant proportions strategies in stationary markets is covered in Section 2. A generalization to fixed-mix strategies as well as an application to currency market models is given in Section 3. The most intriguing case of stock markets with stationary returns, covered in Section 4, delivers counter-intuitive results that even experienced researchers find puzzling. Section 5 discusses several prominent explanations for volatility-induced growth and provides an interpretation that agrees with our findings. Section 6 concludes.

1.1 *A test of your intuition*

There is barely any 'better' result than one that is (completely) counter-intuitive. But before one is tempted to make such a claim on one's own findings, it seems appropriate to test the audience's intuition. In the following, we ask a few questions to which the reader has to promptly guess the answer. Many scientists have been asked these questions casually in private, at seminars, and at conferences. The following insights were first reported in Dempster, Evstigneev, and Schenk-Hoppé (2007). Non-technically minded readers can skip to the next section which provides the informal discussion of an intriguing example.

Stationary markets: puzzles and misconceptions. Consider a market with K assets whose price process $p_t = (p_t^1, \ldots, p_t^K)$ is ergodic and stationary. (It is assumed that prices are log-integrable.) The assumption of stationarity of asset prices, perhaps after some detrending, seems plausible when modeling currency markets where 'prices' are determined by exchange rates of all the currencies with respect to some selected reference currency.

Question 1. Suppose vectors of asset prices $p_t = (p_t^1, \ldots, p_t^K)$ fluctuate randomly, forming a stationary stochastic process (assume even that the vectors p_t are iid (independent identically distributed)). Consider a fixed-mix self-financing investment strategy prescribing rebalancing one's portfolio at each of the dates $t = 1, 2, \ldots$ so as to keep equal investment proportions of wealth in all the assets. What is the tendency of the portfolio value in the long run, as $t \to \infty$? Will the value: (a) decrease; (b) increase; or (c) fluctuate randomly, converging (in some sense) to a stationary process?

The audience of our respondents was quite broad and professional, but practically *nobody* succeeded in guessing the correct answer, which is (b). Among those with a firm view, *nearly all* selected (c). There were also a couple of respondents who decided to bet on (a). Common intuition suggests that if the market is stationary, then the portfolio value for a constant proportions strategy must converge in one sense or another to a stationary process. The usual intuitive argument in support of this conjecture appeals to the self-financing property. The self-financing constraint seems to exclude possibilities of unbounded growth. This argument is also substantiated by the fact that in the deterministic case both the prices and the portfolio value are constant. This way of reasoning makes the answer (c) to the above question more plausible a priori than the others.

It might seem surprising that the wrong guess (c) has been put forward even by those who have known about examples of volatility pumping for a long time. The reason for this might lie in the non-traditional character of the setting where not only the asset returns but the prices themselves are stationary. Moreover, the phenomenon of volatility-induced growth is more paradoxical in the case of stationary prices, where growth emerges "from nothing". In the conventional setting of stationary returns, volatility serves as the cause of an acceleration of growth, rather than its emergence from prices with zero growth rates.

A potentially promising attempt to understand the correct answer to Question 1 might be to refer to the concept of *arbitrage.* Getting something from nothing as a result of an arbitrage opportunity seems to be similar to the emergence of growth in a stationary setting where there are no obvious sources for growth. As long as we deal with an infinite time horizon, we would have to consider some kind of *asymptotic* arbitrage. All known concepts of this kind[2] however are much weaker than what we would need in the present context. According to our results, growth is exponentially fast, unbounded wealth is achieved with probability one, and the effect of growth is demonstrated for specific (constant proportions) strategies. None of these properties can be directly deduced from asymptotic arbitrage.

Thus, there are no convincing arguments showing that volatility-induced growth in stationary markets can be derived from, or explained by, asymptotic arbitrage over an infinite time horizon. But what can be said about relations between stationarity and arbitrage over finite time intervals? As is known, there are no arbitrage opportunities (over a finite time horizon) if and only if there exists an equivalent martingale measure. A stationary process can be viewed as an 'antipodal concept' to the notion of a martingale. This might lead to the conjecture that in a stationary market arbitrage is a typical situation. Is this true or not? Formally, the question can be stated as follows.

Question 2. Suppose vectors of asset prices $p_t = (p_t^1, \ldots, p_t^K)$ form a stationary stochastic process (assume even that the vectors p_t are iid) Furthermore, suppose the first asset $k = 1$ is riskless with constant price $p_t^1 = 1$. The market is frictionless and there are no portfolio constraints (in particular, short selling is allowed). Does this market have arbitrage opportunities over a finite time horizon?

An arbitrage opportunity over a fixed time horizon is understood in the conventional sense: the existence of a trading strategy that does not require any investment

[2]E.g. Ross (1976), Huberman (1982), Kabanov and Kramkov (1994), and Klein and Schachermayer (1996).

Table 1 Net return of the two assets in the market

	Asset 1	Asset 2
Heads	+50%	−55%
Tails	−40%	+100%

at the initial time, is self-financing, does not incur a loss at the terminal time, and makes a gain with strictly positive probability. Again, the answer to this question is practically never guessed immediately. The correct answer depends on whether the distribution of the price vector of the risky assets is continuous or discrete. For example, if $(p_t^2, \dots, p_t^K)$ takes on a finite number of values, then an arbitrage opportunity exists. But if its distribution is continuous, there are no arbitrage opportunities. For details see Evstigneev and Kapoor (2006).

If your intuition led you astray, the remainder of this chapter will help to offer insights into these counter-intuitive results. Even if your intuition led you to the correct answers, you might be interested in the precise formulation of the results and the ideas behind their proofs.

1.2 *An illustrative example*

The potential of constant proportions investment strategies for financial growth can be demonstrated by means of the following example which seems simple enough to be discussed at high-school level. A related example is presented in Dempster, Evstigneev, and Schenk-Hoppé (2008) (see also Luenberger, 1998, Chapter 15). A more demanding illustration, though with a sound empirical background, is given in Ziemba (2008).

Two investment opportunities available at every point in time: 0, 1, 2.... The realized returns of the assets between two points in time are determined by flipping a fair coin (only one!). Think of the outcome (heads/tails) as a economy-wide event that effects the two investments in different ways. The net returns are defined in Table 1.

Investors who buy and hold the first asset will see their wealth decline (exponentially fast) over time. The grow rate is negative:

$$g_1 := 0.5 \ln(1.5) + 0.5 \ln(0.6) \approx -0.05268 < 0$$

Investment in only the second asset does not yield any better performance because the growth rate is negative as well and equal to g_1:

$$g_2 := 0.5 \ln(0.45) + 0.5 \ln(2.0) = g_1 < 0$$

Now consider an investor with a constant proportions strategy. Suppose the investor divides his wealth 50 : 50 between the two assets. After each change in the value of the two portfolio positions, the investor rebalances his portfolio by trading assets to restore the 50 : 50 split of wealth. The growth rate of his wealth is

$$0.5 \ln(0.5 \cdot 1.5 + 0.5 \cdot 0.45) + 0.5 \ln(0.5 \cdot 0.6 + 0.5 \cdot 2.0) \approx +0.1185 > 0$$

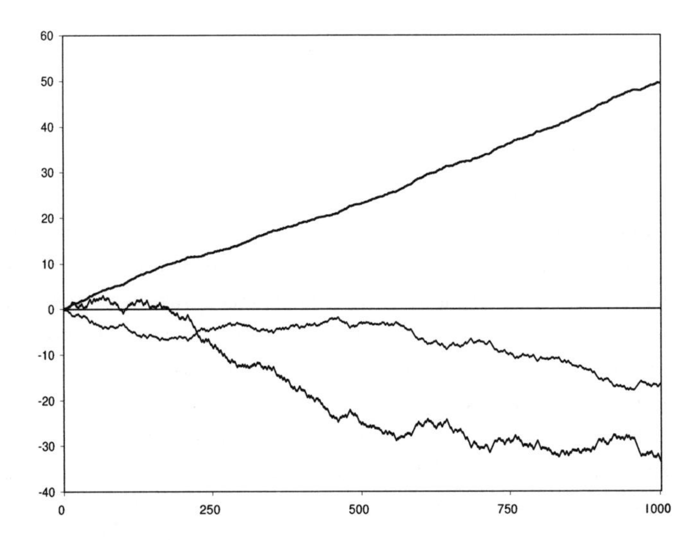

Figure 1 Wealth dynamics of the two passive investors (both decreasing) and the active investor with 50 : 50 constant proportions strategy (increasing). Initial wealth is 1.0. Logarithmic scale on y-axis.

Both passive investors (holding only one asset) will see their wealth halve in less than 14 periods on average. In contrast, the investor following a constant proportions strategies with proportions 50 : 50 will, on average, double his wealth more often than every 6th period. A typical realization of the wealth dynamics is depicted in Figure 1.

The active investor's choice of the proportions is arbitrary. Indeed, qany constant proportions investment strategy holding both assets will generate growth in excess of either buy-and-hold strategy. The 50 : 50 proportions create positive growth while strategies with proportions close to 100 : 0 (or 0 : 100) entail growth at negative rate but one that is higher than $g^1 = g^2$.

Examples, examples, examples. The above example is just one of many. Indeed most descriptions (and analyses) of the phenomenon of generating financial growth from volatility are restricted to examples involving specialized models. Since the deduction of general principles from examples is an unreliable route if a thorough mathematical analysis is feasible, it might come as no surprise that several terms have been coined to describe this phenomenon and many different explanations for its origins have been put forward. In the remainder of this chapter, we will present mathematical results on the growth-volatility nexus in general (discrete-time) models and attempt to de-mystify its origins and explanations.

1.3 *Notation*

All models discussed in this chapter share the basic setup. The definitions and concepts are listed and discussed here. An investor observes prices and takes actions only at particular points in time. These time periods are denoted $t = 0, 1, 2, \ldots$. The price dynamics are driven by random factors. These factors are described by a stochastic process s_t with $t = 0, \pm 1, \pm 2, \ldots$. The process takes values in a measurable space S. The realization of the random parameter s_t corresponds to the state of the world at time t. Denote by P the probability measure induced by the stochastic process s_t, $t = 0, \pm 1, \pm 2, \ldots$, on the space of its paths. There are $K \geq 2$ assets traded in the market. Asset prices $p_t = (p_t^1, \ldots, p_t^K) > 0$, at the time periods $t = 0, 1, 2, \ldots$ are described as a sequence of strictly positive random vectors with values in the K-dimensional linear space R^K. We will assume that the price vector p_t depends on the history of the process s_t up to time t:

$$p_t = p_t(s^t) \in R^K, \quad s^t = (\ldots, s_{t-1}, s_t)$$

It is supposed, without further mentioning, that all functions of s^t will be measurable.

A market is called *stationary* if the process s_t is stationary and the price vectors p_t do not explicitly depend on t, i.e., $p_t = p(s^t)$. We will further assume that the process s_t is ergodic and that $E|\ln p^k(s^t)| < \infty$ for all $k = 1, \ldots, K$.

A stochastic process $\xi_1, \xi_2, \ldots$ is *stationary* if, for any $m = 0, 1, 2, \ldots$ and any measurable function $\phi(x_0, x_1, \ldots, x_m)$, the distribution of the random variable $\phi_t := \phi(\xi_t, \xi_{t+1}, \ldots, \xi_{t+m})$ $(t = 0, 1, \ldots)$ does not depend on t. According to this definition, all probabilistic characteristics of the process ξ_t are time-invariant. If ξ_t is stationary, then for any measurable function ϕ for which $E|\phi(\xi_t, \xi_{t+1}, \ldots, \xi_{t+m})| < \infty$, the averages

$$\frac{\phi_1 + \cdots + \phi_t}{t} \tag{1}$$

converge almost surely (a.s.) as $t \to \infty$ by Birkhoff's ergodic theorem — see, e.g., Billingsley (1965). If the limit of all averages of the form (1) is non-random (equal to a constant a.s.), then the process ξ_t is called *ergodic*. In this case, the above limit is equal a.s. to the expectation $E\phi_t$, which does not depend on t by virtue of stationarity of ξ_t.

The exponential *growth rate* of a stochastic process $\xi_1, \xi_2, \ldots$ with $\xi_t > 0$ is defined by

$$\lim_{t \to \infty} \frac{1}{t} \ln \xi_t \tag{2}$$

provided the limit exists. Suppose the process ξ is stationary and ergodic with $E \ln \xi_m < \infty$, then its growth rate zero because $t^{-1} \ln \xi_t \to 0$ a.s. by Proposition 4.1.3 in Arnold (1998). One also has

$$\frac{1}{t} \ln(\xi_1 \cdot \ldots \cdot \xi_t) = \frac{\ln \xi_1 + \cdots + \ln \xi_t}{t} \to E \ln \xi_m = const. \ (a.s.) \tag{3}$$

This property is frequently used below.

At each time period t, the investor chooses a portfolio $h_t(s^t) = (h_t^1(s^t), \ldots, h_t^K(s^t))$ with a non-negative number of units of asset k, $h_t^k(s^t) \geq 0$. Short selling of assets is ruled out in our model by this assumption of non-negativity. A sequence $h_t(s^t)$, $t = 0, 1, 2, \ldots$, of portfolios is called a trading strategy. The value of a portfolio h_t at time t is $p_t h_t = \sum_k p_t^k h_t^k$.

A trading strategy $H = (h_0, h_1, \ldots)$ with *initial wealth* $w_0 > 0$ is *self-financing* if $p_0 \, h_0 = w_0$ and

$$p_t(s^t)\, h_t(s^t) \leq p_t(s^t)\, h_{t-1}(s^{t-1}), \quad t = 1, 2, \ldots \quad \text{(a.s.)} \tag{4}$$

The inequalities in (4) are supposed to hold almost surely with respect to the probability measure P. The market value of the portfolio after trade does not exceed the vale of the yesterday's portfolio at current prices. This budget constraint restricts the investor's choice.

A *balanced* trading strategy H is of the form

$$h_t(s^t) = \gamma(s^1) \cdot \ldots \cdot \gamma(s^t)\, \tilde{h}(s^t), \quad t = 1, 2, \ldots \tag{5}$$

with a scalar-valued function $\gamma(\cdot) > 0$ and a vector function $\tilde{h}(\cdot) \geq 0$. If $t = 0$, we assume that $h_0(s^0) = \tilde{h}(s^0)$. It is assumed that $\ln \gamma(s^t)$ and $\ln |\tilde{h}(s^t)|$ are integrable with respect to the measure P, i.e., $E|\ln \gamma(s^t)|$ and $E|\ln |\tilde{h}(s^t)||$ are finite. Define $|h| = \sum_k |h^k|$ for a vector $h = (h^k)$.

These strategies are called balanced because they are of balanced growth: (5) implies that all proportions between the amounts of different assets in the portfolio

$$\frac{h_t^j(s^t)}{h_t^k(s^t)} = \frac{\tilde{h}^j(s^t)}{\tilde{h}^k(s^t)}, \quad j \neq k \tag{6}$$

form stationary stochastic processes. The random growth rate of the amount of each asset $k = 1, \ldots, K$, in the portfolio

$$\frac{h_t^k(s^t)}{h_{t-1}^k(s^{t-1})} = \gamma(s^t)\, \frac{\tilde{h}^k(s^t)}{\tilde{h}^k(s^{t-1})}$$

is a stationary process. A balanced strategy is self-financing (4) if and only if

$$\gamma(s^t)\, p(s^t)\, \tilde{h}(s^t) \leq p(s^t)\, \tilde{h}(s^{t-1}) \quad \text{(a.s.)} \tag{7}$$

Stationarity implies that if (7) holds for some t, it holds for all t.

Suppose the vector of asset prices is stationary. Then every non-negative vector function $\tilde{h}(s^t)$ with $E|\ln |\tilde{h}(s^t)|| < \infty$ defines a self-financing balanced strategy (5) by

$$\gamma(s^t) := \frac{p(s^t)\, \tilde{h}(s^{t-1})}{p(s^t)\, \tilde{h}(s^t)} \tag{8}$$

since $E|\ln p^k(s^t)| < \infty$, and relations (4) and (7) hold as equalities.

Balanced trading strategies have the property that the growth rate of wealth of any investor employing it is completely determined by the expected value of γ. Proposition 1 in Evstigneev and Schenk-Hoppé (2002) states that for any balanced trading strategy (5)

$$\lim_{t\to\infty} \frac{1}{t} \ln(p(s^t)\, h_t(s^t)) = \lim_{t\to\infty} \frac{1}{t} \ln |h_t(s^t)| = E \ln \gamma(s^0) \quad (a.s.) \tag{9}$$

This result shows strict positivity of $E \ln \gamma(s^0) \equiv E \ln \gamma(s^t)$ implies exponential growth of wealth, i.e., $p(s^t)\, h_t(s^t) \to \infty$ a.s. exponentially fast.

2 Asset Markets with Stationary Prices

We first discuss the simplest model in which one can analyze generic properties of fixed-mix strategies: a market in which prices are stationary processes. It turns out that any constant proportions strategy produces growth, though in this stationary market the growth rate of each asset price is zero and, therefore, buy-and-hold strategies do not yield positive growth. This results might seem, at the first glance, counter-intuitive. This section is based on Evstigneev and Schenk-Hoppé (2002). All notations are introduced in Section 1.3.

2.1 *The model*

A constant proportions strategy in a market with K assets is characterized by a vector $\lambda = (\lambda_1, \ldots, \lambda_K)$ in the set

$$\Delta = \left\{ (\lambda_1, \ldots, \lambda_K) \in R^K : \lambda_k > 0, \sum_{k=1}^{K} \lambda_k = 1 \right\} \tag{10}$$

To avoid pathological cases, we assume strict positivity of all components. Proportional investment rules of this kind are sometimes termed *completely mixed.*

The trading strategy h_t is a *constant proportions strategy* if

$$p_t^k\, h_t^k = \lambda_k\, p_t\, h_{t-1} \tag{11}$$

for all $t = 1, 2, \ldots$ and $k = 1, \ldots, K$. The investor rebalances the portfolio in every period in time by investing the constant share λ_k of the wealth $p_t\, h_{t-1}$ into the kth asset. The investor's wealth at the beginning of period t is determined by evaluating the portfolio h_{t-1} (bought in the previous period) at the current price system p_t^k. We assume that all the coordinates of h_t are non-negative: short sales are ruled out.

Constant proportions strategies are self-financing because (11) implies

$$p_t\, h_t = p_t\, h_{t-1}$$

Given a vector λ, the strategy is uniquely defined by its initial portfolio $h_0(s^0)$. Recursive application of (11) determines $h_t(s^t)$ for every $t = 1, 2, \ldots$.

In the study of the growth rate of wealth, the initial portfolio does not matter, see Evstigneev and Schenk-Hoppé (2002, pp. 568). Indeed if there are two constant proportions strategies h_t and $\hat{h}_t$ both generated by the same $\lambda \in \Delta$ but with different (non-zero) initial portfolios, then there exist $\underline{c}, \overline{c} > 0$ such that $\underline{c} h_t^k \leq \hat{h}_t^k \leq \overline{c} h_t^k$ for all k and all $t \geq 1$. Obviously, $\lim t^{-1} \ln |h_t| > 0$ a.s. if and only if $\lim t^{-1} \ln |\hat{h}_t| > 0$ a.s. (and exponential growth is at the same speed).

2.2 *The growth rate*

Constant proportions strategies generate balanced portfolios. From the representation (5), the growth rate of the investors wealth is obtained as the expected value $E \ln \gamma(s^0)$. The notion of balanced portfolios goes back, though in a somewhat different form, to Radner (1971)'s study of stochastic generalizations of the von Neumann economic growth model. Arnold, Evstigneev, and Gundlach (1999) study this concept of balanced paths in a general setting. The application to financial problems is recent.

The central result on the growth rate of constant proportions strategies in markets with stationary asset prices is obtained by showing that they can be written as balanced strategies and that $E \ln \gamma(s^0) \geq 0$. To have a strictly positive growth rate, the price process $p(s^t)$ must be non-degenerate.

We impose the following assumption:

(**A**) With strictly positive probability, the random variable

$$p^k(s^t)/p^k(s^{t-1})$$

is not constant with respect to $k = 1, 2, \ldots, K$, i.e., there exist m and n (that might depend on s^t) for which

$$\frac{p^m(s^t)}{p^m(s^{t-1})} \neq \frac{p^n(s^t)}{p^n(s^{t-1})} \tag{12}$$

All results in this Handbook section will be proved under this assumption, though it will be convenient to repeat this condition in different disguises–each best suited for the particular application.

Under assumption (**A**), one has the following result, see Evstigneev and Schenk-Hoppé (2002, Theorem 1).

Theorem 1. *Fix any* $\lambda = (\lambda_1, \ldots, \lambda_K) \in \Delta$, *and let* w_0 *be a strictly positive number. Then there exists a vector function* $h_0(s^0) \geq 0$ *such that the constant proportions strategy* h_t *generated by* λ *and* h_0 *is a balanced strategy with initial wealth* w_0, *and we have*

$$\lim_{t\to\infty} \frac{1}{t} \ln p_t h_t = \lim_{t\to\infty} \frac{1}{t} \ln |h_t| > 0 \quad (a.s.) \tag{13}$$

The construction of the vector function $h_0(s^0) \geq 0$ is as follows. Suppose the proportions $\lambda \in \Delta$ and initial wealth $w_0 > 0$ are given. To show that h_t is a balanced trading strategy one needs to define a vector function $\tilde{h}(\cdot) \geq 0$ and a scalar-valued function $\gamma(\cdot) > 0$ such that h_t coincides with the strategy defined in (5) (and that $h_t(s^0) = \tilde{h}(s^0)$). Define

$$\tilde{h}(s^t) = \left(\frac{\lambda_1 w_0}{p^1(s^t)}, \ldots, \frac{\lambda_K w_0}{p^K(s^t)} \right) \tag{14}$$

and

$$\gamma(s^t) = \frac{p(s^t)\,\tilde{h}(s^{t-1})}{w_0} \left[= \sum_{k=1}^{K} \lambda_k \frac{p^k(s^t)}{p^k(s^{t-1})} \right] \tag{15}$$

This is a balanced trading strategy which coincides with h_t recursively defined by (11).

Strictly positive growth ($E \ln \gamma(s^t) > 0$) holds because Jensen's inequality (applied to the probability measure λ_k on the set $\{1, \ldots, K\}$) implies that

$$\ln \sum_{k=1}^{K} \lambda_k \frac{p^k(s^t)}{p^k(s^{t-1})} \geq \sum_{k=1}^{K} \lambda_k \ln \frac{p^k(s^t)}{p^k(s^{t-1})} \tag{16}$$

with strict inequality on a set of positive probability by assumption (12), which ensures

$$E \ln \gamma(s^t) > \sum_{k=1}^{K} \lambda_k\, E \ln \frac{p^k(s^t)}{p^k(s^{t-1})} = 0$$

Strict positivity of $E \ln \gamma(s^t)$ means that wealth tends to infinity at an exponential rate. If the non-degeneracy condition (**A**) is not satisfied, the market is essentially deterministic and all prices are constant. In this case, all strategies give zero growth. Indeed, (12) is a very weak requirement that is satisfied in virtually every market.

The result presented in Theorem 1 holds under (sufficiently) small proportional transaction costs, see Evstigneev and Schenk-Hoppé (2002, Theorem 2).

2.3 *Interpretation*

Our analysis shows that constant proportions strategies provide growth in a market in which buy-and-hold strategies do not deliver any. The intuition behind this result is that constant proportions strategies 'exploit' the persistent fluctuation of prices. When keeping a fixed fraction of wealth invested in each asset, a change in prices leads the investor to sell those assets that are expensive relative to the other assets and to purchase relatively cheap assets. The stationarity of prices implies that this portfolio rule yields a strictly positive expected rate of growth, despite the fact that each asset price has growth rate zero. The notion of an asset being 'cheap' resp. 'expensive' makes sense when prices are stationary. If the current price is below the expected value, then the tomorrow's price has a higher-than-average chance of being larger than today's price. This asset is cheap today. In other words, if prices are stationary, there is *reversion to the median* (which is also true for price ratios). Rebalancing strategies, on average, buy low and sell high. In the model with stationary returns (rather than prices), accepting this interpretation would be falling victim to the gambler's fallacy, see Section 4.

The result highlights the benefit of financial engineering. The implementation of a constant proportions strategy requires to invest actively in the available assets by rebalancing at discrete time periods — the growth is financially engineered by making use of Jensen's inequality.

3 Fixed-Mix Strategies in Stationary Markets

The model with constant proportions strategies can be generalized to accommodate the transfer of fractions between different portfolio positions. An application to currency markets is provided. In this model, the exchange rates fluctuate randomly in time as stationary stochastic processes. This aspect of our analysis is inspired, in particular, by recent work of Kabanov (1999) and Kabanov and Stricker (2001). This section follows Dempster, Evstigneev, and Schenk-Hoppé (2003). Again, the main result holds under small proportional transaction costs.

3.1 *The model*

A fixed-mix strategy is determined by a (non-random) matrix λ_{kj}, $k, j = 1, \ldots, K$, such that

$$\lambda_{kj} > 0\,, \quad \sum_{k=1}^{K} \lambda_{kj} = 1 \tag{17}$$

In each time period, this strategy prescribes the transfer of a fixed share $\lambda_{kj} > 0$ of the jth position of the portfolio to the kth position $(k, j \in \{1, \ldots, K\})$. The dynamics of the wealth invested in the portfolio positions are given by

$$p_t^k h_t^k = \sum_{j=1}^{K} \lambda_{kj} p_t^j h_{t-1}^j \tag{18}$$

For any matrix λ_{kj} satisfying (17)), a strategy H is called a *fixed-mix strategy associated with the matrix* $\lambda = (\lambda_{kj})$, if (18) holds for all k, t and s^t.

Any fixed-mix strategy is self-financing: $p_t h_t = p_t h_{t-1}$. As in the previous section, we are interested in the asymptotic behavior of the portfolio h_t (and its market value) of a trader employing a fixed-mix investment rule.

In the deterministic case, the analysis of the long-term dynamics is simple. The price $p_t = p = (p^1, \ldots, p^k) > 0$ is a constant vector. The corresponding process h_t defined by (18) is deterministic and will always converge to a steady state. This claim follows from results on positive matrices, Kemeny and Snell (1960). This result, combined with the assumption of stationarity, might seem to rule out unbounded growth and could lead to the (wrong) conjecture of the convergence of h_t to a stationary distribution in the stochastic case.

The case studied in Section 2 corresponds to the situation in which λ_{kj} does not depend on j:

$$\lambda_{kj} = \lambda_k \text{ with } \lambda_k > 0\,, \quad \text{and} \quad \lambda_1 + \cdots + \lambda_K = 1 \tag{19}$$

because (18) reduces to (11).

3.2 *Currency markets*

The foreign exchange market is arguably the real-world example closest to a market in which prices are stationary. We show to formulate the wealth dynamics of a currency trader with a fixed-mix strategy in the above model.

Currencies $k = 1, 2, \ldots, K$ are traded in a frictionless market. The exchange rates $\pi_t^{kj} = \pi^{kj}(s^t) > 0$ fluctuate randomly in time. The real number π_t^{kj} denotes the amount of currency k which can be purchased by selling one unit of currency j at time t. We assume absence of arbitrage at each point in time t, i.e.z the exchange rates must satisfy

$$\pi_t^{kj} = \pi_t^{km}\, \pi_t^{mj} \tag{20}$$

for all k, m and j.

Assume the trader follows any fixed-mix strategy (17) by dividing the holdings $h_{t-1}^j \geq 0$ of currency j purchased at time $t-1$ according to the proportions $\lambda_{kj} > 0$, $k = 1, \ldots, K$ at time t. The amount $\lambda_{kj} h_{t-1}^j$ is exchanged into currency k. After execution of all these transactions, the amount of currency k obtained at time t is equal to

$$h_t^k = \sum_{j=1}^{K} \lambda_{kj}\, \pi_t^{kj}\, h_{t-1}^j \tag{21}$$

The dynamics (21) can be written in the form (18). Take currency 1 as a numeraire and define

$$p_t^k = \pi_t^{1k}$$

The relation (20) implies $\pi_t^{kj} = 1/\pi_t^{jk}$ and $\pi_t^{jj} = 1$. Therefore, $\pi_t^{kj} = p_t^j/p_t^k$. Multiplying (21) by π_t^{1k} and using these relations, yields the formulation (18) of the wealth dynamics.

3.3 *The growth rate*

The analysis of the long-term growth of an investor following a fixed-mix strategy is conceptually identical to that of constant proportions strategies. First, one shows that there is an initial portfolio such that the fixed-mix strategy leads to a balanced portfolio. Then, one proves that the growth rate is independent of the initial portfolio. Finally, one shows that this growth rate is strictly positive under a non-degeneracy condition. The mathematical tools required in the proof however are very different. The case of fixed-mix strategies requires considerably more advanced methods.

As in Section 2 we assume that the market is stationary and the state of the world s_t is ergodic. The price process $p_t = p(s^t)$ satisfies $E|\ln p^k(s^t)| < \infty$ for all $k = 1, \ldots, K$. The growth rate of each asset price is zero because $t^{-1} \ln p_t \to 0$ a.s. and, therefore, buy-and-hold strategies do not yield positive growth.

We assume non-degeneracy of the price process $p(s^t)$:

(**B**) The vector $\bar{p}(s^t) = (\bar{p}^1(s^t), \ldots, \bar{p}^K(s^t))$ of normalized prices

$$\bar{p}^j(s^t) := \frac{p^j(s^t)}{\sum_m p^m(s^t)} \quad j \in \{1, \ldots, K\}$$

is not constant a.s. with respect to s^t.

This assumption says that there is no constant vector c for which $\bar{p}(s^t) = c$ almost surely. By virtue of stationarity of (s_t), condition (**B**) holds for all t if is satisfied for some t. If s_t is ergodic, the condition (**B**) is equivalent to the following requirement (Dempster, Evstigneev, and Schenk-Hoppé, 2003, pp. 271): With positive probability, the ratios $p^k(s^t)/p^k(s^{t-1})$ are not constant with respect to k. This is condition (**A**).

The main result on the performance of an investor employing any fixed-mix strategy is as follows, Dempster, Evstigneev, and Schenk-Hoppé (2003, Theorem 1).

Theorem 2. *Let $h_t(s^t)$, $t \geq 0$, be a fixed-mix strategy associated with the matrix $\lambda = (\lambda_{kj})$ satisfying (17). For each $k \in \{1, 2, \ldots, K\}$, the limit*

$$\lim_{t \to \infty} \frac{1}{t} \ln h_t^k \tag{22}$$

exists and is strictly positive almost surely. Furthermore, this limit does not depend on k, and

$$\lim_{t \to \infty} \frac{1}{t} \ln h_t^k = \lim_{t \to \infty} \frac{1}{t} \ln p_t h_t > 0 \quad \text{(a.s.)} \tag{23}$$

The result ensures that the wealth of a fixed-mix investor tends to infinity at an exponential rate. All portfolio positions and the investor's wealth grow at the same positive exponential rate. This finding generalizes the results in Section 2 to the general case of fixed-mix investment strategies.

The proof of Theorem 2 rests on the observation that one can define a balanced fixed-mix strategy for any given matrix λ satisfying (17). Recall that a trading strategy H is balanced if $h_t(s^t) = \gamma(s^1) \ldots \gamma(s^t)\, \tilde{h}(s^t)$ (a.s.) for $t \geq 1$. For $t = 0$, we assume $h_0(s^0) = \tilde{h}(s^0)$. The function $\gamma(\cdot) > 0$ is scalar-valued and $\tilde{h}(\cdot) > 0$ is a vector function such that $E|\ln \gamma(s^t)| < \infty$ and $|\tilde{h}(s^t)| = 1$. The norm $|h|$ of a vector $h = (h^k)$ is defined as $\sum_k |h^k|$. The assumption $|\tilde{h}(s^t)| = 1$, which was not imposed in (5), will be satisfied automatically in the present case.

For balanced strategies, all the proportions between the amounts of different assets in the portfolio and the random growth rate of the amount of each asset in the portfolio are stationary stochastic processes. To show existence of a balanced fixed-mix strategy associated with the matrix $\lambda = (\lambda_{kj})$, one needs to show there are appropriate functions $\gamma(\cdot)$ and $\tilde{h}(\cdot)$. Indeed, Theorem 2 in Dempster, Evstigneev, and Schenk-Hoppé (2003) ensures that for each matrix λ satisfying (17), there is a unique balanced fixed-mix strategy with $|\tilde{h}(s^t)| = 1$.

The construction can be sketched as follows. Denote by $A_t = A(s^t) = (A_{kj}(s^t))$ the positive random $K \times K$ matrix defined by

$$A_{kj}(s^t) = \lambda_{kj} \frac{p^j(s^t)}{p^k(s^t)} \tag{24}$$

We have $E|\ln A_{kj}(s^t)| < \infty$. With this definition, the fixed-mix strategy H can be represented as

$$h_t(s^t) = A(s^t)A(s^{t-1}) \cdots A(s^1)h_0(s^0) \tag{25}$$

Stationarity of (s_t) implies that functions $\gamma(\cdot)$ and $\tilde{x}(\cdot)$ (satisfying $E|\ln \gamma(s^t)| < \infty$ and $|\tilde{x}(s^t)| = 1$) generate a balanced λ-strategy if and only if

$$\gamma(s^t)\tilde{x}(s^t) = A(s^t)\tilde{x}(s^{t-1}) \text{ (a.s.)} \tag{26}$$

The existence of a solution to this equation follows from a stochastic version of the Perron–Frobenius theorem, Theorem A.1 in Dempster, Evstigneev, and Schenk-Hoppé (2003). The problem (26) cannot be solved by applying the conventional Perron–Frobenius theorem because the vector $\tilde{x}(s^t)$ on the left-hand side does not coincide with the vector $\tilde{x}(s^{t-1})$ on the right-hand side as the function $\tilde{x}(s^t)$ is obtained by 'time-shifting' $\tilde{x}(s^{t-1})$.

The independence of the growth rate from the initial portfolio is proved by couching the one-step forward portfolio between multiples of the portfolio $\tilde{x}(s^1)$, analogous to the outline in Section 2. All have the same growth rate.

Finally, the strict positivity of the fixed-mix strategies' growth rate is asserted by proving that $E \ln \gamma(s^t) > 0$. This can be seen as follows. Denote by $(\gamma(\cdot), \tilde{x}(\cdot))$ the balanced strategy corresponding to the fixed mix strategy λ. The relation (26) implies

$$\gamma(s^t)\tilde{x}^k(s^t) = \sum_{j=1}^{K} \lambda_{kj} \frac{p^j(s^t)}{p^k(s^t)} \tilde{x}^j(s^{t-1}) \quad k \in \{1, \ldots, K\} \tag{27}$$

The Perron–Frobenius theorem yields existence of a vector $r = (r_1, \ldots, r_K) > 0$ with

$$r_k = \sum_{j=1}^{K} \lambda_{kj} r_j \quad k \in \{1, \ldots, K\}$$

Put $\beta_{kj} = r_k^{-1} \lambda_{kj} r_j$. One finds that

$$\gamma(s^t) = \sum_{j=1}^{K} \beta_{kj} \frac{p^j(s^t)}{p^j(s^{t-1})} \frac{p^j(s^{t-1})\tilde{x}^j(s^{t-1}) r_k}{p^k(s^t)\tilde{x}^k(s^t) r_j} \quad k \in \{1, \ldots, K\} \tag{28}$$

From this representation of $\gamma(s^t)$ one can conclude (using Jensen's inequality and several other arguments, see Dempster, Evstigneev and Schenk-Hoppé, 2003, Section 3) that $E \ln \gamma(s^t) > 0$ under assumption (**B**).

3.4 *Price processes with trend*

The concept of a stationary market, where asset prices p_t fluctuate as stationary stochastic processes, is an idealization. The assumption that only the relative proportions p_t^j / p_t^k are stationary seems more realistic. Following Dempster, Evstigneev, and Schenk-Hoppé (2003), let us assume that the prices are of the form

$$p_t = \xi_t \hat{p}_t$$

where $\hat{p}_t = \hat{p}(s^t)$ is a process satisfying the assumptions we previously imposed on p_t, and $\xi_t = \xi_t(s^t) > 0$ is *any* sequence of strictly positive random variables. The factors ξ_t represent the dynamics of a price index. The normalized prices $\hat{p}_t$ are free of this trend.

The above analysis can directly be applied to this more general price process. The portfolio value $p_t h_t$ satisfies

$$\frac{1}{t} \ln p_t h_t = \frac{1}{t} \ln \xi_t + \frac{1}{t} \ln \hat{p}_t h_t$$

The growth rate of $p_t h_t$ is determined by that of ξ_t and $\hat{p}_t h_t$.

Under assumption (**B**), the process $\hat{p}_t h_t$ grows exponentially fast almost surely. Consequently, if the price index ξ_t grows at an exogenous exponential rate r, then the investor's wealth $p_t h_t$ will grow almost surely at a rate r' *strictly greater* than r.

4 Stock Markets with Stationary Returns

The observation that rebalancing a portfolio by following any constant proportions or, more generally, any fixed-mix strategy leads to a strictly higher growth rate of wealth than any buy-and-hold portfolio has been confirmed in markets with stationary prices. After seeing the results and following the intuition that a reversion to the mean is a powerful source of capital growth, one might wonder about the case in which returns (rather than prices) are stationary processes. Indeed, as explained in Section 1.1, there is no general agreement that the previous results hold for stationary returns. This is quite sensible because the any notion of cheap (or expensive) assets based on the argument that reversion to the long run trend renders an asset cheap after a series of low returns would mean to fall victim to the gambler's fallacy. This lack of a simple intuition for the wealth dynamics of constant proportions strategies in markets where returns (not prices) are stationary demands a thorough analysis of this case. How can one rest without having gained an understanding of this problem? The presentation follows Dempster, Evstigneev, and Schenk-Hoppé (2007), which also contains a proof of the main result under small (proportional) transaction costs.

4.1 *The model*

Denote the (gross) return on asset k between time $t-1$ and t by

$$R_t^k := \frac{p_t^k}{p_{t-1}^k} \quad k = 1, 2, \ldots, K, \ t \geq 1 \tag{29}$$

Let $R_t := (R_t^1, \ldots, R_t^K)$. We impose the assumption:

(**R**) The vector stochastic process R_t, $t = 1, 2, \ldots,$ is stationary and ergodic. The expected values $E|\ln R_t^k|$, $k = 1, 2, \ldots, K$, are finite.

The typical example of a stationary ergodic process is a sequence of iid (independent identically distributed) random variables. To avoid misunderstandings, we emphasize that Brownian motion and a random walk are *not* stationary.

The asset price at time t and the initial price are related as $p_t^k = p_0^k R_1^k \cdot \ldots \cdot R_t^k$, where the random sequence R_t^k is stationary by (**R**). This assumption on the structure of the price process is a fundamental hypothesis in finance. Moreover, it is quite often assumed that the random variables $R_t^k, t = 1, 2, \ldots$ are independent, i.e., the price process p_t^k forms a *geometric random walk.* This postulate, which is much stronger than the hypothesis of stationarity of R_t^k, lies at the heart of the classical theory of asset pricing (Black, Scholes, Merton), see e.g. Luenberger (1998).

Birkhoff's ergodic theorem implies

$$\lim_{t \to \infty} \frac{1}{t} \ln p_t^k = \lim_{t \to \infty} \frac{1}{t} \sum_{n=1}^{t} \ln R_n^k = E \ln R_t^k \quad \text{(a.s.)} \tag{30}$$

for each $k = 1, 2, \ldots, K$. This means that the price of each asset k has almost surely a well-defined and finite (asymptotic, exponential) *growth rate*, which turns out to be equal a.s. to the expectation $\rho_k := E \ln R_t^k$, the *drift* of this asset's price. The drift can be positive, zero or negative. It does not depend on t in view of the stationarity of R_t.

Consider an investor following a constant proportions strategy with proportions (10). Fix any vector $\lambda \in \Delta$. Given an initial portfolio $h_0 > 0$, the (self-financing) trading strategy H is defined recursively by

$$h_t^k = \lambda_k p_t h_{t-1} / p_t^k \quad k = 1, 2, \ldots, K, \ t \geq 1 \tag{31}$$

This definition is equivalent to (11). Our aim is to study the asymptotic behavior of the portfolio value $V_t = p_t h_t$ as $t \to \infty$, i.e. the limit $\lim_{t \to \infty} t^{-1} \ln(V_t)$ describing the (exponential) growth rate of the strategy.

4.2 *The growth rate*

We impose the following non-degeneracy condition:

(**C**) With strictly positive probability,

$$\frac{p_t^k(s^t)}{p_t^m(s^t)} \neq \frac{p_{t-1}^k(s^{t-1})}{p_{t-1}^m(s^{t-1})} \quad \text{for some } 1 \le k, m \le K \text{ and } t \ge 1$$

This condition is a very mild assumption on the existence of volatility of the price process. Condition (**C**) does not hold if and only if the relative prices of the assets are constant in time (a.s.), i.e., if with probability one, the ratio p_t^k/p_t^m of the prices of any two assets k and m does not depend on t. This condition is also equivalent to (**B**), see Dempster, Evstigneev, and Schenk-Hoppé (2003, pp. 271).

The condition (**C**) on asset prices has an equivalent formulation for asset returns:

(**D**) For some $t \ge 1$ (and hence, by virtue of stationarity, for each $t \ge 1$), the probability

$$P\{R_t^k \neq R_t^m \text{ for some } 1 \le k, m \le K\}$$

is strictly positive.

Equivalence can bee seen as follows. Note that $p_t^k/p_t^m \neq p_{t-1}^k/p_{t-1}^m$ if and only if $p_t^k/p_{t-1}^k \neq p_t^m/p_{t-1}^m$, i.e. $R_t^k \neq R_t^m$. Denote by δ_t the random variable that is equal to 1 if the event $\{R_t^k \neq R_t^m$ for some $1 \le k, m \le K\}$ occurs and 0 otherwise. Condition (**C**) means that $P\{\max_{t\ge1} \delta_t = 1\} > 0$, while (**D**) states that, for some t (and hence for each t), $P\{\delta_t = 1\} > 0$. The latter property is equivalent to the former because $\{\max_{t\ge1} \delta_t = 1\} = \cup_{t=1}^{\infty}\{\delta_t = 1\}$.

We can now state the main result on the growth of wealth of investors following constant proportions strategies.

Theorem 3. *Fix any* $\lambda \in \Delta$.

(i) The growth rate of the constant proportions strategy is almost surely equal to a constant which is strictly greater than $\sum_k \lambda_k \rho_k$, *where* ρ_k *is the drift of the price of asset* k.

(ii) Suppose all the assets have the same drift (therefore, almost surely the same asymptotic growth rate), i.e., $E \ln R_t^k = \rho$ *for each* $k = 1, \ldots, K$ *with some real number* ρ. *Then the growth rate of the constant proportions strategy is almost surely strictly greater than the growth rate of each individual asset.*

In Theorem 3, assertion (ii) immediately follows from (i). The result (ii) shows that any completely mixed constant proportions strategy grows at a rate strictly greater than ρ, the growth rate of each particular asset. The growth of the investor's wealth is only driven by the volatility of the price process. This result seems to contradict conventional finance theory which usually regards the volatility of asset prices as an impediment to financial growth. In the present context, volatility serves as an endogenous source of its acceleration. Theorem 3 (ii) asserts the validity of the conclusion drawn in the illustrative example discussed in Section 1.2. Indeed, the constant proportions strategy $(.5, .5)$ (as well as any other completely mixed constant proportions strategy) yields a higher growth rate than the price of each asset.

The first part of Theorem 3 places a floor under the constant proportions strategy's growth rate. When asset prices grow at different rates, there is no general result that constant proportions strategy grow faster than any (completely diversified) buy-and-hold strategy. The latter grows at rate $\max_k \rho_k$ which is not strictly dominated by $\sum_k \lambda_k \rho_k$. The growth-optimal constant proportions strategy however will always grow at least as fast as any buy-and-hold in the model.

The proof of the above result is surprisingly simple and can be presented with success to an audience with knowledge of only elementary probability and calculus: Fix any vector $\lambda \in \Delta$. The random wealth dynamics of the corresponding constant proportions strategy (31) are given by

$$\begin{aligned} V_t = p_t h_t &= \sum_{k=1}^{K} p_t^k h_{t-1}^k = \sum_{k=1}^{K} \frac{p_t^k}{p_{t-1}^k} p_{t-1}^k h_{t-1}^k = \sum_{k=1}^{K} \frac{p_t^k}{p_{t-1}^k} \lambda_k p_{t-1} h_{t-1} \\ &= \left[\sum_{k=1}^{K} R_t^k \lambda_k \right] V_{t-1} = (R_t \lambda) V_{t-1} \end{aligned} \tag{32}$$

for each $t \geq 1$. This implies

$$V_t = (R_t \lambda) \cdot \ldots \cdot (R_1 \lambda) V_0 \tag{33}$$

for all $t \geq 1$. The ergodic theorem ensures

$$\lim_{t \to \infty} \frac{1}{t} \ln V_t = \lim_{t \to \infty} \frac{1}{t} \sum_{n=1}^{t} \ln(R_n \lambda) = E \ln(R_t \lambda) \quad \text{(a.s.)} \tag{34}$$

It remains to show that $E \ln(R_t \lambda) > \sum_{k=1}^{K} \lambda_k \rho_k$ under the condition (**D**). Indeed, Jensen's inequality and (**D**) ensure the relation

$$\ln \sum_{k=1}^{K} R_t^k \lambda_k > \sum_{k=1}^{K} \lambda_k (\ln R_t^k)$$

with strictly positive probability, while the non-strict inequality holds always. Consequently,

$$E \ln(R_t \lambda) > \sum_{k=1}^{K} \lambda_k E(\ln R_t^k) = \sum_{k=1}^{K} \lambda_k \rho_k \tag{35}$$

This proves Theorem 3.

4.3 *Interpretation*

The rigorous proof of the presence of volatility-induced growth in markets where asset returns are stationary leaves open the question on the intuition behind this result.

It seems any explanation one can give is nothing but a repetition of the mathematical reasoning behind this result: If $R_t^1, \ldots, R_t^K$ are the random returns of the K assets, then the asymptotic growth rates of these assets are $E \ln R_t^k$, while the asymptotic growth rate of a constant proportions strategy is $E \ln(\sum_k \lambda_k R_t^k)$, which is strictly greater than $\sum_k \lambda_k E \ln(R_t^k)$ by Jensen's inequality because the logarithmic function is strictly concave. The proof in Section 2 confirms that Jensen's inequality is the central tool used in the proof. Any explanation not based on this fact would be flawed.

It is well-known that a linear combination of assets can produce non-linearity in a portfolio's characteristics. Indeed, this feature drives mean-variance portfolio choice, cf. Luenberger (1998). In the present model, the rebalancing of the portfolio so as to maintain constant proportions causes a non-linear effect in the portfolio's growth rate with the feature that $E \ln(\sum_k \lambda_k R_t^k) - \sum_k \lambda_k E \ln(R_t^k) > 0$ for any $\lambda = (\lambda_1, \ldots, \lambda_K) \in \Delta$ provided assumption (**C**) holds.

Other attempts to appeal to intuition are discusses (and rejected) in the next section.

An interesting problem is the question under which (general) assumptions $E \ln(\sum_k \lambda_k R_t^k) > 0$ even if $\max_k \rho_k \leq 0$ (or, less restrictive, $\min_k \rho_k \leq 0$). The example in Section 1.2 possesses this property. In that example, $\rho_1 = \rho_2 < 0$, but $E \ln(\lambda_1 R_t^1 + \lambda_2 R_t^2) > 0$ for the constant proportions strategy $\lambda = (.5, .5)$. But if λ is close to $(1, 0)$ or $(0, 1)$ the growth rate of wealth becomes negative as well. A partial answer to this 'black swan' question is provided in Section 5. We discuss a case in which the addition of a slower growing asset always enhances the growth of a constant proportions strategy and raises the growth rate over that of any buy-and-hold strategy.

5 Myths and Misconceptions

The phenomenon of volatility-induced growth is simply counter-intuitive as confirmed by our test in Section 1.1. Providing a common-sense explanation for the mathematical results can therefore not be a simple task. In this section, we discuss some of the more prominent suggestions for intuition put forward in the literature.

5.1 *Volatility pumping*

The term 'volatility pumping' appears to have been first used by Luenberger (1998). His suggestion is that constant proportions strategies force the investor to 'buy low and sell high'—the common sense dictum of stock market trading. Those assets whose prices have risen from the last rebalance date will be overweighted in the portfolio, and their holdings must be reduced to meet the required proportions and to be replaced in part by assets whose prices have fallen and whose holdings must therefore be increased. This behavior leads to growth if asset returns exhibit some stationarity properties.

In Section 2.3, we argued that this explanation is correct for stationary prices. In the present case, accepting this explanation would mean to fall victim to the gambler's fallacy. This explanation is not valid because when the price follows a geometric random walk, the set of its values is generally unbounded and for every value there is a smaller as well as a larger value. Suppose returns are iid, then 'high' and 'low' do not have any meaning. Reversion to the long run trend (as postulated by the ergodic theorem) is not an explanation either: arguing that the longer the run of black numbers, the higher the odds of red numbers at the next spin of the roulette wheel is the gambler's fallacy. When returns are determined by the flip of a coin, an asset's upside and downside potential does not change over time. Such an asset is not cheap or expensive at any point in time.

5.2 *The importance of constancy*

A more substantial lacuna in the above reasoning is that it does not reflect the assumption of *constancy* of investment proportions. This leads to the question: what will happen if the portfolio is rebalanced so as to sell *all* those assets that gain value and buy only those ones which lose it? (This should lead to an even higher growth rate — provided volatility could be 'pumped'.)

Assume, for example, that there are two assets, the price p_t^1 of the first (riskless) is always 1, and the price p_t^2 of the second (risky) follows a geometric random walk, so that the gross return on it can be either 2 or 1/2 with equal probabilities. Suppose the investor sells the second asset and invests all wealth in the first if the price p_t^2 goes up and performs the converse operation, betting all wealth on the risky asset, if p_t^2 goes down. Then the sequence $\lambda_t = (\lambda_{1,t}, \lambda_{2,t})$ of the vectors of investment proportions will be iid with values $(0, 1)$ and $(1, 0)$ taken on with equal probabilities. Furthermore, λ_{t-1} will be independent of R_t. By virtue of (34), the growth rate of the portfolio value for this strategy is equal to $E \ln(R_t \lambda_{t-1}) = [\ln(0 \cdot 1 + 1 \cdot 2) + \ln(0 \cdot 1 + 1 \cdot \frac{1}{2}) + \ln(1 \cdot 1 + 0 \cdot 2) + \ln(1 \cdot 1 + 0 \cdot \frac{1}{2})]/4 = 0$, which is the same as the growth rate of each of the two assets $k = 1, 2$. But this growth rate is strictly lower than that of any completely mixed constant proportions strategy.

5.3 *Energy-interpretation of volatility*

Our results highlight the importance of volatility because fixed-mix strategies do not produce growth beyond buy-and-hold portfolios in any of the above models when prices or returns are constant. It is tempting to see volatility as a source of energy that can be tapped to generate growth.

Indeed, Luenberger (1998) presents an intriguing continuous-time example which supports this view. (A closely related observation is made in Fernholz and Shay (1982).) There are K assets whose prices follow independent (but identical) geometric Brownian motions. The constant proportions strategy with equal weights on all assets has a higher growth rate than any asset price. The remarkable feature however is that increasing the number of assets K leads to a higher growth rate and –at the same time– to a reduction in the volatility of the return. It would seem as if fixed-mix strategies turn the classical return-volatility tradeoff on its head.

The framework for this example is the well-known continuous-time model developed by Merton and others, in which the price processes S_t^k, $t \geq 0$, of two assets $k = 1, 2$ are supposed to be solutions to the stochastic differential equations $dS_t^k/S_t^k = \mu_k dt + \sigma_k dW_t^k$, where the W_t^k are independent (standard) Wiener processes and $S_0^k = 1$. As is well-known, these equations admit explicit solutions $S_t^k = \exp[\mu_k t - (\sigma_k^2/2)t + \sigma_k W_t^k]$. Given some $\theta \in (0, 1)$, the value V_t of the constant proportions portfolio prescribing investing the proportions θ and $1-\theta$ of wealth into assets $k = 1, 2$ is the solution to the equation

$$dV_t/V_t = [\theta\mu_1 + (1-\theta)\mu_2]dt + \theta\sigma_1 dW_t^1 + (1-\theta)\sigma_2 dW_t^2$$

Equivalently, V_t can be represented as the solution to the equation $dV_t/V_t = \bar{\mu}dt + \bar{\sigma}dW_t$, where $\bar{\mu} := \theta\mu_1 + (1-\theta)\mu_2$, $\bar{\sigma}^2 := (\theta\sigma_1)^2 + [(1-\theta)\sigma_2]^2$ and W_t is a standard Wiener process. Thus, $V_t = \exp[\bar{\mu}t - (\bar{\sigma}^2/2)t + \bar{\sigma}W_t]$, and so the growth rate and the volatility of the portfolio value process V_t are given by $\bar{\mu} - (\bar{\sigma}^2/2)$ and $\bar{\sigma}$. In particular, if $\mu_1 = \mu_2 = \mu$ and $\sigma_1 = \sigma_2 = \sigma$, then the growth rate and the volatility of V_t are equal to

$$\mu - (\bar{\sigma}^2/2) \quad \text{and} \quad \bar{\sigma} = \sigma\sqrt{\theta^2 + (1-\theta)^2} < \sigma \tag{36}$$

while for each individual asset, the growth rate and the volatility are $\mu - (\sigma^2/2)$ and σ, respectively.

Thus, in this example, the use of a constant proportions strategy prescribing investing in a mixture of two assets leads (due to diversification) to an increase of the growth rate and to a simultaneous decrease of the volatility. When looking at the expressions in (36), the temptation arises even to say that the volatility reduction is the *cause* of volatility-induced growth. Indeed, the growth rate $\mu - (\bar{\sigma}^2/2)$ is greater than the growth rate $\mu - (\sigma^2/2)$ because $\bar{\sigma} < \sigma$. This suggests speculation along the following lines. Volatility is something like energy. When constructing a mixed portfolio, it converts into growth and therefore decreases. The greater the volatility reduction, the higher the growth acceleration.

5.4 *A counter-example*

Do the observations made in the previous section have any grounds in the general case, or do they have a justification only in the above example? To formalize this question and answer it, let us return to the discrete-time framework. Suppose there are two assets with iid vectors of returns $R_t = (R_t^1, R_t^2)$. Let $(\xi, \eta) := (R_1^1, R_1^2)$ and assume, to avoid technicalities, that the random vector (ξ, η) takes on a finite number of values and is strictly positive. The value V_t of the portfolio generated by a fixed-mix strategy with proportions x and $1 - x$ $(0 < x < 1)$ is computed according to the formula

$$V_t = V_1 \prod_{n=2}^{t} [xR_n^1 + (1-x)R_n^2], \; t \geq 2$$

see (33). The growth rate of this process and its volatility are given, respectively, by the expectation $E \ln \zeta_x$ and the standard deviation $\sqrt{Var \ln \zeta_x}$ of the random variable $\ln \zeta_x$, where $\zeta_x := x\xi + (1-x)\eta$. We know from the above analysis that the growth rate increases when mixing assets with the same growth rate. What can be said about volatility? Specifically, we consider the following question.

Question 3. (a) Suppose $Var \ln \xi = Var \ln \eta$. Is it true that $Var \ln[x\xi + (1-x)\eta] \leq Var \ln \xi$ when $x \in (0,1)$? (b) More generally, is it true that $Var \ln[x\xi + (1-x)\eta] \leq \max(Var \ln \xi, Var \ln \eta)$ for $x \in (0,1)$?

Query (b) asks whether the logarithmic variance is a quasi-convex functional. Questions (a) and (b) can also be stated for volatility defined as the square root of logarithmic variance. They will have the same answers because the square root is a strictly monotone function. Positive answers to these questions would substantiate the above conjecture of volatility reduction — negative, refute it.

In general (without additional assumptions on ξ and η), the above questions 3(a) and 3(b) have *negative* answers. To show this, consider two iid random variables U and V with values 1 and $a > 0$ realized with equal probabilities. Consider the function

$$f(y) := Var \ln[yU + (1-y)V], \; y \in [0,1] \tag{37}$$

By evaluating the first and the second derivatives of this function at $y = 1/2$, one can show the following. There exist some numbers $0 < a_- < 1$ and $a_+ > 1$ such that the function $f(y)$ attains its minimum at the point $y = 1/2$ when a belongs to the closed interval $[a_-, a_+]$ and it has a local maximum (!) at $y = 1/2$ when a does not belong to this interval. The numbers a_- and a_+ are given by

$$a_\pm = 2e^4 - 1 \pm \sqrt{(2e^4-1)^2 - 1}$$

where $a_- \approx 0.0046$ and $a_+ \approx 216.388$. If $a \in [a_-, a_+]$, the function $f(y)$ is convex, but if $a \notin [a_-, a_+]$, its graph has the shape illustrated in Figure 2.

Fix any a for which the graph of $f(y)$ looks like the one depicted in Figure 2. Consider any number $y_0 < 1/2$ which is greater than the smallest local minimum of $f(y)$ and define $\xi := y_0 U + (1-y_0)V$ and $\eta := y_0 V + (1-y_0)U$. (U and V may be interpreted as "factors" on which the returns ξ and η on the two assets depend.) Then $Var \ln[(\xi+\eta)/2] > Var \ln \xi = Var \ln \eta$, which yields a negative answer both

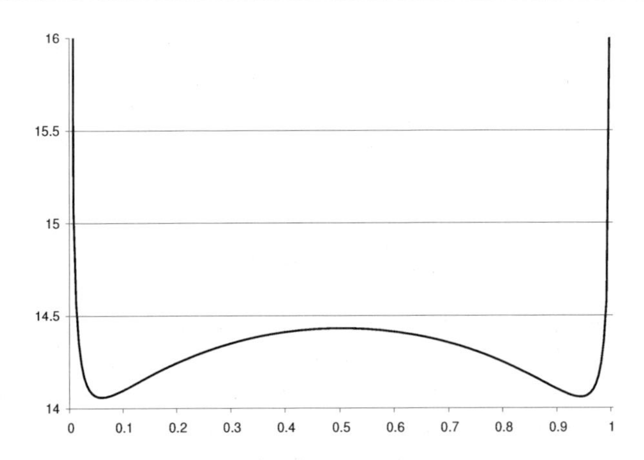

Figure 2 Graph of the function $f(y)$ in equation (37) for $a = 10^4$.

to (a) and (b). In this example, ξ and η are dependent. It would be of interest to investigate questions (a) and (b) for general independent random variables ξ and η. It can be shown that the answer to (b) is positive if one of the variables ξ and η is constant. But even in this case the function $Var \ln[x\xi + (1-x)\eta]$ is not necessarily convex: it may have an inflection point in $(0,1)$, which can be easily shown by examples involving two-valued random variables.

Thus, it can happen that a fixed-mix portfolio may have a greater volatility than each of the assets from which it has been constructed. Consequently, the above conjecture and the 'energy interpretation' of volatility are generally not valid. It is interesting, however, to find additional conditions under which assertions regarding volatility reduction hold true. In this connection, we can assert the following fact, see Theorem 4 in Dempster, Evstigneev, and Schenk-Hoppé (2007).

Theorem 4. *Let U and V be independent random variables bounded above and below by strictly positive constants. If U is not constant, then one has that $Var \ln[yU + (1-y)V] < Var \ln U$ for all $y \in (0,1)$ sufficiently close to* 1.

Volatility can be regarded as a quantitative measure of instability of the portfolio value. The above result shows that small independent noise can reduce volatility. This result is akin to a number of known facts about *noise-induced stability*, e.g., Abbott (2001) and Mielke (2000). An analysis of links between the topic of the present work and results about stability under random noise might constitute an interesting theme for further research.

5.5 *Growth under transaction costs*

In this section, we consider an example (a binomial model) in which quantitative estimates for the size of the transaction costs needed for the validity of the result on volatility-induced growth can be provided. Suppose that there are two assets $k = 1, 2$: one riskless and one risky. The price of the former is constant and equal to 1. The price of the latter follows a geometric random walk. It can either jump up by $u > 1$ or down by u^{-1} with equal probabilities. Thus, both security prices have growth rate zero.

Suppose the investor pursues the constant proportions strategy prescribing to keep 50% of wealth in each of the securities. There are no transaction costs for buying and selling the riskless asset, but there is a transaction cost rate for buying and selling the risky asset of $\varepsilon \in [0, 1)$. Assume the investor's portfolio at time $t-1$ contains v units of cash; then the value of the risky position of the portfolio must be also equal to v. At time t, the riskless position of the portfolio will remain the same, and the value of the risky position will become either uv or $u^{-1}v$ with equal probability. In the former case, the investor rebalances his/her portfolio by selling an amount of the risky asset worth A so that

$$v + (1 - \varepsilon)A = vu - A \tag{38}$$

By selling an amount of the risky asset of value A in the current prices, the investor receives $(1 - \varepsilon)A$, and this sum of cash is added to the riskless position of the portfolio. After rebalancing, the values of both portfolio positions must be equal, which is expressed in (38). From (38) we obtain $A = v(u-1)(2-\varepsilon)^{-1}$. The positions of the new (rebalanced) portfolio, measured in terms of their current values, are equal to $v + (1 - \varepsilon)A = v[1 + (1 - \varepsilon)(2 - \varepsilon)^{-1}(u - 1)]$. In the latter case (when the value of the risky position becomes $u^{-1}v$), the investor buys some amount of the risky asset worth B, for which the amount of cash $(1 + \varepsilon)B$ is needed, so that

$$v - (1 + \varepsilon)B = u^{-1}v + B$$

From this, we find $-B = v(u^{-1} - 1)(2 + \varepsilon)^{-1}$, and so $v - (1 + \varepsilon)B = v[1 + (1 + \varepsilon)(2 + \varepsilon)^{-1}(u^{-1} - 1)]$.

Thus, the portfolio value at each time t is equal to its value at time $t - 1$ multiplied by the random variable ξ such that $P\{\xi = g'\} = P\{\xi = g''\} = 1/2$, where $g' := 1+(1+\varepsilon)(2+\varepsilon)^{-1}(u^{-1}-1)$ and $g'' := 1+(1-\varepsilon)(2-\varepsilon)^{-1}(u-1)$. Consequently, the asymptotic growth rate of the portfolio value, $E \ln \xi = (1/2)(\ln g' + \ln g'')$, is equal to $(1/2) \ln \phi(\varepsilon, u)$, where

$$\phi(\varepsilon, u) := \left[1 + (1 + \varepsilon)\frac{u^{-1} - 1}{2 + \varepsilon}\right]\left[1 + (1 - \varepsilon)\frac{u - 1}{2 - \varepsilon}\right]$$

We have $E \ln \xi > 0$, i.e., the phenomenon of volatility induced growth takes place, if $\phi(\varepsilon, u) > 1$. For $\varepsilon \in [0, 1)$, this inequality turns out to be equivalent to the following very simple relation: $0 \le \varepsilon < (u - 1)(u + 1)^{-1}$. Thus, given $u > 1$, the asymptotic growth rate of the fixed-mix strategy under consideration is greater than zero if the transaction cost rate ε is less than $\varepsilon^*(u) := (u - 1)(u + 1)^{-1}$. We call $\varepsilon^*(u)$ the *threshold transaction cost rate.* Volatility-induced growth takes place—in the present example, where the portfolio is rebalanced in every one period—when $0 \le \varepsilon < \varepsilon^*(u)$.

The volatility σ of the risky asset under consideration (the standard deviation of its logarithmic return) is equal to $\ln u$. In the above considerations, we assumed that σ—or equivalently, u—is fixed, and we examined $\phi(\varepsilon, u)$ as a function of ε. Let us now examine $\phi(\varepsilon, u)$ as a function of u when the transaction cost rate ε is fixed and strictly positive. For the derivative of $\phi(\varepsilon, u)$ with respect to u, we have

$$\phi'_u(\varepsilon, u) = \frac{1+\varepsilon}{4-\varepsilon^2}\left[\frac{1-\varepsilon}{1+\varepsilon} - u^{-2}\right]$$

If $u = 1$, then $\phi'_u(\varepsilon, 1) < 0$. Thus, if the volatility of the risky asset is small, the performance of the constant proportions strategy at hand is *worse* than the performance of each individual asset. This fact *refutes the possible conjecture that 'the higher the volatility, the higher the induced growth rate'*. Further, the derivative $\phi'_u(\varepsilon, u)$ is negative when $u \in [0, u_*(\varepsilon))$, where $u_*(\varepsilon) := (1-\varepsilon)^{-1/2}(1+\varepsilon)^{1/2}$. For $u = u_*(\varepsilon)$, the asymptotic growth rate of the constant proportions strategy at hand attains its minimum value. For those values of u which are greater than $u_*(\varepsilon)$, the growth rate increases, and it tends to infinity as $u \to \infty$. Thus, although the assertion 'the greater the volatility, the greater the induced growth rate' is not valid always, it is valid (in the present example) under the additional assumption that the volatility is large enough.

6 Conclusion

This chapter surveys the authors' recent theoretical work on the phenomenon of volatility-induced growth. We study the performance of fixed-mix strategies in markets where asset prices or returns are stationary ergodic. We have established the surprising result that these strategies generate portfolio growth rates in excess of the individual asset growth rates, provided some volatility is present. As a consequence, even if the growth rates of the individual securities all have mean zero, the value of a fixed-mix portfolio tends to infinity with probability one.

By contrast with the twenty five years in which the effects of 'volatility pumping' have been investigated in the literature by example, our results are quite general. They are obtained under assumptions that accommodate virtually all the empirical market return properties discussed in the literature. We have in this paper also dispelled the notion that the demonstrated acceleration of portfolio growth is simply a matter of 'buying low and selling high' or stems from some 'volatility–energy' link.

The example of Section 5.2 shows that our result depends critically on rebalancing to an arbitrary *fixed* mix of portfolio proportions. Any such mix defines the relative magnitudes of individual asset returns realized from volatility effects. This observation and our analysis of links between growth, arbitrage, and noise-induced stability suggest that financial growth driven by volatility is a subtle and delicate phenomenon.

A major obstacle in practical applications of our results is the fact that investment is not over an infinite period. This point has been forcefully argued by

Samuelson (1979), but its validity is not as clear cut as one may be led to believe, see MacLean, Zhao, and Ziemba (2006), Ziemba (2008) and the discussion in Part IV in this volume. Other issues are the size of transaction costs in real markets. Though our findings do hold under sufficiently small proportional transaction costs, this issue can only be resolved empirically. Another intriguing question is how much stationarity is present in real markets. All of these questions are left for future research.

References

Abbott, D. (2001). Overview: Unsolved problems of noise and fluctuations. Chaos 11, 526–538.

Algoet, P. H. and T. M. Cover (1988). Asymptotic optimality and asymptotic equipartition properties of log-optimum investment. *Annals of Probability*, 16, 876–898.

Arnold, L. (1998). *Random Dynamical Systems.* New York: Springer.

Arnold, L., I. V. Evstigneev and V. M. Gundlach (1999). Convex-valued random dynamical systems: A variational principle for equilibrium states. *Random Operators and Stochastic Equations*, 7, 23–28.

Aurell, E., R. Baviera, O. Hammarlid, M. Serva and A. Vulpiani (2000). A general methodology to price and hedge derivatives in incomplete markets. *International Journal of Theoretical and Applied Finance*, 3, 1–24.

Aurell, E. and P. Muratore-Ginanneschi (2000). Financial friction and multiplicative Markov market games. *International Journal of Theoretical and Applied Finance*, 3, 501–510.

Billingsley, P. (1965). *Ergodic Theory and Information.* New York: John Wiley.

Breiman, L. (1961). Optimal gambling systems for favorable games. In *Proceedings of the Fourth Berkeley Symposium on Mathematical Statistics and Probability, Vol. 1* (ed. Neyman, J.), pp. 65–78. LA: University of California Press.

Browne, S. (1998). The return on investment from proportional portfolio strategies. *Advances in Applied Probability*, 30, 216–238.

Browne, S. and W. Whitt (1996). Portfolio choice and the Bayesian Kelly criterion. *Advances in Applied Probability*, 28, 1145–1176.

Cover, T. M. (1991). Universal portfolios. *Mathematical Finance*, 1, 1–29.

Cover, T. M. (1998). Shannon and investment. IEEE Information Theory Society Newsletter, Summer 1998, Special Golden Jubilee Issue, pp. 10–11.

Dempster, M. A. H., I. V. Evstigneev and K. R. Schenk-Hoppé (2003). Exponential growth of fixed-mix strategies in stationary asset markets. *Finance & Stochastics*, 7, 263–276.

Dempster, M. A. H., I. V. Evstigneev and K. R. Schenk-Hoppé (2007). Volatility-induced financial growth. *Quantitative Finance*, 7, 151–160.

Dempster, M. A. H., I. V. Evstigneev and K. R. Schenk-Hoppé (2008). Financial Markets: The joy of volatility. *Quantitative Finance*, 8, 1–3.

Dempster, M. A. H., M. Germano, E. A. Medova and M. Villaverde (2003). Global asset liability management. *British Actuarial Journal*, 9, 137–195.

Dries, D., A. Ilhan, J. Mulvey, K. D. Simsek and R. Sircar (2002). Trend-following hedge funds and multi-period asset allocation. *Quantitative Finance*, 2, 354–361.

Evstigneev, I.V., T. Hens and K. R. Schenk-Hoppé (2009a). Evolutionary Finance. In *Handbook of Financial Markets: Dynamics and Evolution* (eds. Hens, T. and Schenk-Hoppé, K. R.), Chapter 9, pp. 507–566. Amsterdam: North-Holland.

Evstigneev, I. V., T. Hens and K. R. Schenk-Hoppé (2009b). Evolutionary stability of investment rules.

Evstigneev, I. V. and D. Kapoor (2006). Arbitrage in stationary markets. Discussion Paper 0608, School of Economic Studies, University of Manchester.

Evstigneev, I. V. and K. R. Schenk-Hoppé (2002). From rags to riches: On constant proportions investment strategies. *International Journal of Theoretical and Applied Finance*, 5, 563–573.

Fernholz, R. and B. Shay (1982). Stochastic portfolio theory and stock market equilibrium. *Journal of Finance*, 37, 615–624.

Fernholz, E. R. (2002). Stochastic Portfolio Theory. New York: Springer.

Fernholz, E. R. and I. Karatzas (2005). Relative arbitrage in volatility-stabilized markets. *Annals of Finance*, 1, 149–177.

Fernholz, E. R., I. Karatzas and C. Kardaras (2005). Diversity and arbitrage in equity markets. *Finance & Stochastics*, 9, 1–27.

Hakansson, N. H. and W. T. Ziemba (1995). Capital Growth Theory. In *Handbooks in Operations Research and Management Science, Vol. 9, Finance* (eds. Jarrow, R. A., Maksimovic, V. and Ziemba, W. T.), Chapter 3, pp. 65–86. Amsterdam: North-Holland.

Huberman, G. (1982). A simple approach to Arbitrage Pricing Theory. *Journal of Economic Theory*, 28, 183–191.

Kabanov, Yu. M. (1999). Hedging and liquidation under transaction costs in currency markets. *Finance & Stochastics*, 3, 237–248.

Kabanov, Yu. M. and D. A. Kramkov (1994). Large financial markets: Asymptotic arbitrage and contiguity. *Theory of Probability and its Applications*, 39, 222–228.

Kabanov, Yu. M. and Ch. Stricker (2001). The Harrison–Pliska arbitrage pricing theorem under transaction costs. *Journal of Mathematical Economics*, 35, 185–196.

Kelly, J. (1956). A new interpretation of information rate. *Bell System Technical Journal*, 35, 917–926.

Kemeny, J. G. and J. L. Snell (1960). *Finite Markov Chains.* New York: Van Nostrand.

Klein, I. and W. Schachermayer (1996). Asymptotic arbitrage in non-complete large financial markets. *Theory of Probability and its Applications*, 41, 927–934.

Kuhn, D. and D. G. Luenberger (2010). Analysis of the rebalancing frequency in log-optimal portfolio selection. *Quantitative Finance.* (Forth coming).

Li, Y. (1998). Growth-security investment strategy for long and short runs. *Management Science*, 39, 915–924.

Luenberger, D. G. (1998). *Investment Science.* UK: Oxford University Press.

MacLean, L. C., Y. Zhao and W. T. Ziemba (2006). Dynamic portfolio selection with process control. *Journal of Banking and Finance*, 30, 317–339.

MacLean, L. C., W. T. Ziemba and G. Blazenko (1992). Growth versus security in dynamic investment analysis. *Management Science*, 38, 1562–1585.

Mielke, A. (2000). Noise induced stability in fluctuating bistable potentials. *Physical Review Letters*, 84, 818–821.

Mulvey, J. M. (2001). Multi-period stochastic optimization models for long-term investors. In M. Avellaneda (Ed.), *Quantitative Analysis in Financial Markets, Vol. 3*, pp. 66–85. Singapore: World Scientific.

Mulvey, J. M. (2009). Applying Kelly and related strategies during turbulent markets.

Mulvey, J. M., C. Ural and Z. Zhang (2007). Improving performance for long-term investors: Wide diversification, leverage, and overlay strategies. *Quantitative Finance*, 7, 175–187.

Perold, A. F. and W. F. Sharpe (1988). Dynamic strategies for asset allocation. *Financial Analysts Journal*, 44, 16–27.

Radner, R. (1971). Balanced stochastic growth at the maximum rate. In G. Bruckman and W. Weber (Eds.), *Contributions to the von Neumann Growth Model (Zeitschrift für Nationalökonomie, Suppl. 1)*, pp. 39–62. New York: Springer.

Ross, S. A. (1976). The arbitrage theory of asset pricing. *Journal of Economic Theory*, 13, 341–360.

Samuelson, P. A. (1979). Why we should not make mean log of wealth big though years to act are long. *Journal of Banking and Finance*, 3, 307–309.

Thorp, E. O. (1971). Portfolio choice and the Kelly criterion. In *Business and Economics Statistics Section, Proceedings of the American Statistical Association*, pp. 215–224. Reprinted in: *Stochastic Optimization Models in Finance* (eds. Ziemba, W. T. and Vickson, R. G.), pp. 599–619. US: Academic Press (1975, reprinted in 2006).

Ziemba, W. T. (2008). The Kelly Criterion. *The Actuary*, September issue, Features: Investment.

Ziemba, W. T. and J. M. Mulvey (eds.) (1998). *Worldwide Asset and Liability Modeling.* UK: Cambridge University Press.

Part IV

Critics and assessing the good and bad properties of Kelly

30

Introduction to the Good and Bad Properties of Kelly

Multiperiod lifetime investment-savings optimization dates at least to Ramsey (1928). Phelps (1962) extended the model to include uncertainty while maximizing expected utility of lifetime consumption by choosing between consumption and investment in a single risky asset using an additive utility function. He obtained explicit solutions for a constant member of the isolastic utility class. Samuelson (1969) and Merton (1969) in companion articles develop, following Ramsey (1928) and Phelps (1962), in both discrete-time and continuous time, lifetime portfolio selection models where the objective function is the discounted sum of concave functions of period by period consumption. Samuelson solves the case when there are interior maxima, and shows that for isoelastic period by period utility functions $u'(C) = C^{\delta-1}$, $\delta < 1$, the optimal portfolio decisions are independent of current wealth at each stage and independent of all consumption-savings decisions with a stationary optimal policy to invest a fixed proportion of current wealth in each period. Ziemba and Vickson (2010) review this literature and point to some queries regarding the validity of the interior maxima as discussed in problems in Ziemba and Vickson (1975, 2006).

For log utility, $\delta \to 0$, Samuelson (1969) showed that the optimal decision splits into two independent parts with the Ramsey savings problem independent of the lifetime portfolio selection problem. Log utility frequently splits the problem in this way, see Rudolf and Ziemba (2004) reprinted in part VI of this book for a similar splitting in a multiperiod pension model.

Samuelson (1971) focuses on the third point of his 1979 paper that follows. He states the following result which has broad agreement.

Theorem: If one acts to maximize the geometric mean at each step, and if the period is "sufficiently long", "almost certainly" higher terminal wealth and terminal utility [for a log utility investor][1] will result than from any other [essentially different] decision rule.

Samuelson points out the following:

False Corollary: If maximizing the geometric mean almost certainly leads to a better outcome, then the expected utility of its outcomes exceeds that of any other rule, provided T is sufficiently large.

The editors of this book agree with Samuelson that the $E\log$ maximization is for the special utility function $u(w) = \log w$ and not for other utility functions.

[1] [] indicate phrase added by the editors

Samuelson presents one example that illustrates this. Samuelson's criticisms are with the pure theory and not in conflict with any of the conclusions of this book, much of which is summarized in the papers in this part of the book. But he ends with the observation that his critical remarks

> "...do not deny that this criterion, arbitrary as it is, still avoids some of the even greater arbitrariness of conventional mean-variance analysis. Its essential defect is that it attempts to replace the pair of "asymptotically sufficient parameters" $E \log X_i$, Variance($\log X_i$) by the first of these alone, thereby gratuitously ruling out arbitrary γ in the family $u(x) = \frac{x^\gamma}{\gamma}$ in favor of $u(x) = \log x$".

Luenberger (1993), reprinted in part V, provides a careful analysis of these two parameters. The paper on Kelly simulations in part IV by MacLean *et al.* also considers both parameters in an analysis of proportional investment strategies.

Samuelson' paper (1979) is written in a style, with one syllable words, that is hard to read, but when you cut through the language he makes a few points:

(1) Those who follow the rule maximization of mean log of wealth will with higher and higher probability have more wealth in the long run than those that use an [essentially different] strategy. This is agreed and is one of the basic results due to Breiman and others.
(2) Some of those who have favorable asset returns period by period and maximize the expected log of wealth can lose a lot and in fact almost all of their wealth. This is agreed and examples of this are in the simulation paper by MacLean, Thorp, Zhao and Ziemba (MTZZ) (2010) which appears in this part of the book. While the Kelly expected log criterion faced with favorable positive expectation bets will provide very high returns much of the time, in a small percent of the time there will be huge losses. In one example in Ziemba and Hausch (1986), which is redone in MTZZ with a 14% advantage on each play you can lose 98% of your initial wealth. Of course, with fairly high probability, you can achieve a 10- or 100- fold increase in your initial wealth. This is, of course, the pure theory and in actual applications by hedge funds and other investors, financial engineering methods may be used to limit losses should they occur.
(3) Latane (1959, 1978) seems to have claimed (although even that is debatable) that the expected log maximizing strategy is in some sense better than a strategy based on some other utility function, this is not discussed or claimed by those involved with this book. The expected log maximizing strategy is about long run wealth maximizing most of the time. The only utility functions used are log in general for full Kelly and negative power in the special case with log normally distributed assets. The papers in this part of the book discuss these and other points further.

Markowitz (1976) argues that when one traces out the set of mean-variance (MV) efficient portfolios, one passes through a portfolio which gives approximately the maximum $E \log X$. Thorp (1969) gives an example of an asset distribution such that the Kelly log optimal portfolio is not on the efficient frontier, so the approximation is not exact. Markowitz argues that this Kelly portfolio should be considered the upper limit for conservative choice among E,V efficient portfolios, since portfolios with higher (arithmetic) mean give greater short-run variability with less return in the long run. A real investor might prefer a lower mean and variance giving up return in the long run for stability in the short run. Markowitz (1952) conjectured and Young and Trent (1969) proved that

$$E \log(X_i) \approx \log E - \frac{1}{2}\left\{\frac{V}{E^2}\right\}$$

for a wide class of ex post distributions of portfolio returns, where X = gross return, $E = EX$ and $V = VarX$.

Thorp (2008) provides an introduction to the Kelly criterion and discusses a number of interesting topics about it. His paper and the following ones on simulations of Kelly returns and good and bad properties show what Kelly strategies do and do not do. Thorp begins with how he coined the term "Fortune's Formula" in 1960 in his blackjack application, where he developed favorable card counting strategies that have had a huge impact across the world. He discusses various examples and paradoxes and the issue of bet concentration, that is, limited diversification. Experts with good asset situations can plunge heavily into a few assets as Warren Buffett has done. Thorp's treatment and Ziemba's paper on great investors in part VI of this book provide clues that Buffett acts like a Kelly bettor. Buffett himself argues that amateurs are likely better off diversifying into index funds, after all, they beat about 75% of the professional managers in net returns with essentially no work. Thorp cautions the reader against common errors in the application of Kelly betting such as the failure to consider all investments, risk tolerance misjudgments, forgetting how aggressive Kelly betting is in the short run, overbetting because of data errors, forgetting about the impact of bad scenario black swans and forgetting that most of the superior Kelly betting properties are for the very long run. He presents some results on fractional Kelly wagering that supplement the papers in part III of this book. He discusses the subtle concept of "essentially different" that confounds even some top experts. He discusses some subtleties such as strategies that can beat Kelly strategies. An important topic is the use of Kelly strategies by great investors trying to multiply their fortunes, akin to the Kelly property that it will get you to a wealth goal faster than any other essentially different strategy for asymptotically high goals. Thorp shows the relationship between fractional Kelly strategies and Markowitz mean-variance strategies with the full Kelly being the limiting portfolio on this surface supplementing the Markowitz paper discussed above.

Thorp also discusses and deals with Paul Samuelson's criticisms of Kelly investment strategies and his discussion supplements the three Samuelson papers and the last papers in this part. The editors of this book agree that even when you make a long sequence of very good bets, you can, in fact, lose and the loss can be a lot. But most of the time you win a lot. Thorp mentions two papers he co-authored in the 1970s which show that different utility functions indeed have different optimal solutions. That, as fully discussed in the Thorp and Whitley's (1972) paper reprinted here, deals with the false notion that Kelly proponents think that log optimal strategies are optimal for other concave utility functions, which notion is neither claimed nor true.

Thorp's paper concludes with a discussion of Proebsting's paradox which shows you can make a series of favorable bets and be asymptotically ruined. But the paradox needs a series of correlated bets made at different points in time to show this result. So it does not contradict the property that betting full Kelly or any fixed fraction less than full Kelly leads to exponential growth assuming the bets are independent (as in Breiman's paper in part I or have weak dependence as in Cover's papers in part II or Thorp's paper in part VI).

One approach to successfully model black swans is to use a scenario optimization stochastic programming model where you include the possibility of an extreme event, specifying its consequences but not actually indicating the nature of the event (see Geyer and Ziemba (2008) for the application to the Siemens Austria Pension Fund). Correlations change as the scenario sets move from normal conditions to volatile to crash which include the black swans (see also Ziemba (2003) for additional applications of this approach). This is, of course, fully applicable to Kelly betting and will lower the wager.

Thorp and Whitley (1972) show, for *interesting* concave utilities, that different utility functions, namely those which differ by more than a positive linear transformation, have distinctly different sets of optimal strategies. Moreover, for the case of cash and a two valued risky investment, the optimal strategies are unique. This means that if two utility functions have the same set of optimal strategies for each such investment, then these two utility functions are equivalent, that is, they are positive linear transformations of one another. Hence, given any concave utility different from log there are investment settings in which the two utilities have different optimal strategies. Thorp and Whitley show that utility functions that are close to one another may have very different optimal solutions. Kallberg and Ziemba (1983) show that two utility functions with the same Rubinstein risk aversion,[2] under the assumption of normality, will have the same sets of optimal solutions. Thorp and Whitley show that for two continuous non-decreasing utility functions that are not equivalent, there is a one-period two security investment setting such that these two utility functions have distinct optimal strategies if either (a) the two utility

[2]The Rubinstein risk aversion $-\frac{E_{\xi}u''(\xi' x)}{E_{\xi}u'(\xi' x)}w_0$ is a constant.

functions are either concave or convex, or (b) they have a second derivative that exists except possibly for a set of isolated points.

Short term Kelly and even fractional Kelly strategies are very risky since the Arrow-Pratt risk aversion index is very low. MacLean, Thorp, Zhao and Ziemba (2010) present three simple investment situations to simulate the behavior of full Kelly and fractional Kelly over medium time horizons. These simulations extend and redo over longer horizons with more scenarios earlier studies by Bicksler and Thorp (1973) and Ziemba and Hausch (1986). The results show that:

(1) The great superiority of full Kelly and close to full Kelly strategies over medium term horizons (40 to 700 periods) with very large gains a large fraction of the time.
(2) The short term performance of Kelly and high fractional Kelly strategies is very risky.
(3) There is a well defined and consistent tradeoff of growth versus security as a function of the bet size determined by the various strategies.
(4) No matter how favorable the investment opportunities are or how long the finite horizon is, a sequence of bad scenarios can lead to poor final wealth outcomes with a loss of most of the investor's initial capital.

The calculations show the final wealth distributions including possible huge gains and huge losses and show the fraction of the time these are likely to occur. Mean-variance tradeoffs are shown versus the Kelly fraction. The mean is highest for full Kelly and decreases for lower and higher Kelly fractions. The variance increases with the Kelly fraction confirming that full Kelly is very volatile. The minimum and maximum wealth levels are shown versus the Kelly fraction. The wealth accumulated from the full Kelly strategy does not stochastically dominate fractional Kelly strategies. To lower risk, one must lower the Kelly fraction. The results showed that for reasonable horizons with three different experiments, that the full Kelly and fractional Kelly strategies are not merely long run approaches. Proper use over short and medium time horizons can yield good wealth goals while protecting against drawdowns most of the time.

The Kelly criterion has many good long-term properties and some favorable short-term properties. MacLean, Thorp, and Ziemba (2010) discuss the good and bad properties of the Kelly criterion. The main advantage is the long run growth rate maximization. That is an asymptotic mathematical result. Also the medium term simulation results show that most of the time the Kelly bettor has a lot higher final wealth than that from other strategies. But for a small percentage of the time, despite many periods of play with very favorable investments, there can be huge losses. Besides this, the main disadvantage of the Kelly criterion is that it is very risky short term and the near zero Arrow-Pratt risk aversion index typically leads to wild swings in the wealth path. But the Kelly strategy has favorable short run competitive optimality compared to other strategies. Moreover, the extreme sensitivity of $E\log$ to errors in the mean signals that mean estimates must be carefully and conservatively estimated to avoid overbetting which can lead to large portfolio losses. Fractional Kelly strategies such as half Kelly, moderate this but they can still have large losses. There is a tradeoff of lower growth for more security as the Kelly fraction is reduced and more is invested in riskless cash. Also it may take a long time for the Kelly bettor to dominate an essentially different strategy with high probability. No one disagrees with the Samuelson objection that $E\log$ maximizing policies are optimal for only log utility functions. Thorp and Whitley's (1972, 1974) result shows that this "objection" is simply a universal fact, true for all utilities. None the less, the long run properties indicate very high final wealth most of the time for Kelly and fractional Kelly bettors.

31

LIFETIME PORTFOLIO SELECTION BY DYNAMIC STOCHASTIC PROGRAMMING

Paul A. Samuelson *

Introduction

MOST analyses of portfolio selection, whether they are of the Markowitz-Tobin mean-variance or of more general type, maximize over one period.[1] I shall here formulate and solve a many-period generalization, corresponding to lifetime planning of consumption and investment decisions. For simplicity of exposition I shall confine my explicit discussion to special and easy cases that suffice to illustrate the general principles involved.

As an example of topics that can be investigated within the framework of the present model, consider the question of a "businessman risk" kind of investment. In the literature of finance, one often reads; "Security A should be avoided by widows as too risky, but is highly suitable as a businessman's risk." What is involved in this distinction? Many things.

First, the "businessman" is more affluent than the widow; and being further removed from the threat of falling below some subsistence level, he has a high propensity to embrace variance for the sake of better yield.

Second, he can look forward to a high salary in the future; and with so high a present discounted value of wealth, it is only prudent for him to put more into common stocks compared to his present tangible wealth, borrowing if necessary for the purpose, or accomplishing the same thing by selecting volatile stocks that widows shun.

Third, being still in the prime of life, the businessman can "recoup" any present losses in the future. The widow or retired man nearing life's end has no such "second or n^{th} chance."

Fourth (and apparently related to the last point), since the businessman will be investing for so many periods, "the law of averages will even out for him," and he can afford to act almost as if he were not subject to diminishing marginal utility.

What are we to make of these arguments? It will be realized that the first could be purely a one-period argument. Arrow, Pratt, and others[2] have shown that any investor who faces a range of wealth in which the elasticity of his marginal utility schedule is great will have high risk tolerance; and most writers seem to believe that the elasticity is at its highest for rich — but not ultra-rich! — people. Since the present model has no new insight to offer in connection with statical risk tolerance, I shall ignore the first point here and confine almost all my attention to utility functions with the same relative risk aversion at all levels of wealth. Is it then still true that lifetime considerations justify the concept of a businessman's risk in his prime of life?

Point two above does justify leveraged investment financed by borrowing against future earnings. But it does not really involve any increase in relative risk-taking once we have related what is at risk to the proper larger base. (Admittedly, if market imperfections make loans difficult or costly, recourse to volatile, "leveraged" securities may be a rational procedure.)

The fourth point can easily involve the innumerable fallacies connected with the "law of large numbers." I have commented elsewhere[3] on the mistaken notion that multiplying the same kind of risk leads to cancellation rather

* Aid from the National Science Foundation is gratefully acknowledged. Robert C. Merton has provided me with much stimulus; and in a companion paper in this issue of the REVIEW he is tackling the much harder problem of optimal control in the presence of continuous-time stochastic variation. I owe thanks also to Stanley Fischer.

[1] See for example Harry Markowitz [5]; James Tobin [14], Paul A. Samuelson [10]; Paul A. Samuelson and Robert C. Merton [13]. See, however, James Tobin [15], for a pioneering treatment of the multi-period portfolio problem; and Jan Mossin [7] which overlaps with the present analysis in showing how to solve the basic dynamic stochastic program recursively by working backward from the end in the Bellman fashion, and which proves the theorem that portfolio proportions will be invariant only if the marginal utility function is iso-elastic.

[2] See K. Arrow [1]; J. Pratt [9]; P. A. Samuelson and R. C. Merton [13].

[3] P. A. Samuelson [11].

than augmentation of risk. I.e., insuring many ships adds to risk (but only as $\sqrt{n}$); hence, only by insuring more ships and by *also* subdividing those risks among more people is risk on each brought down (in ratio $1/\sqrt{n}$).

However, before writing this paper, I had thought that points three and four could be reformulated so as to give a valid demonstration of businessman's risk, my thought being that investing for each period is akin to agreeing to take a $1/n^{\text{th}}$ interest in insuring n independent ships.

The present lifetime model reveals that investing for many periods does not *itself* introduce extra tolerance for riskiness at early, or any, stages of life.

Basic Assumptions

The familiar Ramsey model may be used as a point of departure. Let an individual maximize

$$\int_0^T e^{-\rho t} U[C(t)]\,dt \tag{1}$$

subject to initial wealth W_0 that can always be invested for an exogeneously-given certain rate of yield r; or subject to the constraint

$$C(t) = rW(t) - \dot{W}(t) \tag{2}$$

If there is no bequest at death, terminal wealth is zero.

This leads to the standard calculus-of-variations problem

$$J = \underset{\{W(t)\}}{\text{Max}} \int_0^T e^{-\rho t} U[rW - \dot{W}]\,dt \tag{3}$$

This can be easily related [4] to a discrete-time formulation

$$\text{Max} \sum_{t=0}^{T} (1+\rho)^{-t} U[C_t] \tag{4}$$

subject to

$$C_t = W_t - \frac{W_{t+1}}{1+r} \tag{5}$$

or,

$$\underset{\{W_t\}}{\text{Max}} \sum_{t=0}^{T} (1+\rho)^{-t} U\left[W_t - \frac{W_{t+1}}{1+r}\right] \tag{6}$$

[4] See P. A. Samuelson [12], p. 273 for an exposition of discrete-time analogues to calculus-of-variations models. Note: here I assume that consumption, C_t, takes place at the beginning rather than at the end of the period. This change alters slightly the appearance of the equilibrium conditions, but not their substance.

for prescribed (W_0, W_{T+1}). Differentiating partially with respect to each W_t in turn, we derive recursion conditions for a regular interior maximum

$$\frac{(1+\rho)}{1+r} U'\left[W_{t-1} - \frac{W_t}{1+r}\right] = U'\left[W_t - \frac{W_{t+1}}{1+r}\right] \tag{7}$$

If U is concave, solving these second-order difference equations with boundary conditions (W_0, W_{T+1}) will suffice to give us an optimal lifetime consumption-investment program.

Since there has thus far been one asset, and that a safe one, the time has come to introduce a stochastically-risky alternative asset and to face up to a portfolio problem. Let us postulate the existence, alongside of the safe asset that makes \$1 invested in it at time t return to you at the end of the period \$1$(1+r)$, a risk asset that makes \$1 invested in, at time t, return to you after one period \$1Z_t, where Z_t is a random variable subject to the probability distribution

$$\text{Prob}\{Z_t \leqq z\} = P(z). \qquad z \geqq 0 \tag{8}$$

Hence, $Z_{t+1} - 1$ is the percentage "yield" of each outcome. The most general probability distribution is admissible: i.e., a probability density over continuous z's, or finite positive probabilities at discrete values of z. Also I shall usually assume independence between yields at different times so that $P(z_0, z_1, \ldots, z_t, \ldots, z_T) = P(z_t)P(z_1)\ldots P(z_T)$.

For simplicity, the reader might care to deal with the easy case

$$\text{Prob}\{Z = \lambda\} = 1/2 = \text{Prob}\{Z = \lambda^{-1}\}, \qquad \lambda > 1 \tag{9}$$

In order that risk averters with concave utility should not shun this risk asset when maximizing the expected value of their portfolio, λ must be large enough so that the expected value of the risk asset exceeds that of the safe asset, i.e.,

$$\frac{1}{2}\lambda + \frac{1}{2}\lambda^{-1} > 1 + r, \text{ or}$$

$$\lambda > 1 + r + \sqrt{2r + r^2}.$$

Thus, for $\lambda = 1.4$, the risk asset has a mean yield of 0.057, which is greater than a safe asset's certain yield of $r = .04$.

At each instant of time, what will be the optimal fraction, w_t, that you should put in the risky asset, with $1 - w_t$ going into the safe asset? Once these optimal portfolio fractions are known, the constraint of (5) must be written

$$C_t = \left[W_t - \frac{W_{t+1}}{[(1-w_t)(1+r) + w_t Z_t]} \right]. \tag{10}$$

Now we use (10) instead of (4), and recognizing the stochastic nature of our problem, specify that we maximize the expected value of total utility over time. This gives us the stochastic generalizations of (4) and (5) or (6)

$$\underset{\{C_t, w_t\}}{\text{Max}} E \sum_{t=0}^{T} (1+\rho)^{-t} U[C_t] \tag{11}$$

subject to

$$C_t = \left[W_t - \frac{W_{t+1}}{(1+r)(1-w_t) + w_t Z_t} \right]$$

W_0 given, W_{T+1} prescribed.

If there is no bequeathing of wealth at death, presumably $W_{T+1} = 0$. Alternatively, we could replace a prescribed W_{T+1} by a final bequest function added to (11), of the form $B(W_{T+1})$, and with W_{T+1} a free decision variable to be chosen so as to maximize (11) + $B(W_{T+1})$. For the most part, I shall consider $C_T = W_T$ and $W_{T+1} = 0$.

In (11), E stands for the "expected value of," so that, for example,

$$E\, Z_t = \int_0^\infty z_t dP(z_t) .$$

In our simple case of (9),

$$EZ_t = \frac{1}{2}\lambda + \frac{1}{2}\lambda^{-1}.$$

Equation (11) is our basic stochastic programming problem that needs to be solved simultaneously for optimal saving-consumption and portfolio-selection decisions over time.

Before proceeding to solve this problem, reference may be made to similar problems that seem to have been dealt with explicitly in the economics literature. First, there is the valuable paper by Phelps on the Ramsey problem in which capital's yield is a prescribed random variable. This corresponds, in my notation, to the $\{w_t\}$ strategy being frozen at some fractional level, there being no portfolio selection problem. (My analysis could be amplified to consider Phelps'[5] wage income, and even in the stochastic form that he cites Martin Beckmann as having analyzed.) More recently, Levhari and Srinivasan [4] have also treated the Phelps problem for $T = \infty$ by means of the Bellman functional equations of dynamic programming, and have indicated a proof that concavity of U is sufficient for a maximum. Then, there is Professor Mirrlees' important work on the Ramsey problem with Harrod-neutral technological change as a random variable.[6] Our problems become equivalent if I replace $W_t - W_{t+1}[(1+r)(1-w_t) + w_t Z_t]^{-1}$ in (10) by $A_t f(W_t/A_t) - nW_t - (W_{t+1} - W_t)$ let technical change be governed by the probability distribution

$$\text{Prob}\,\{A_t \leqq A_{t-1}Z\} = P(Z);$$

reinterpret my W_t to be Mirrlees' per capita capital, K_t/L_t, where L_t is growing at the natural rate of growth n; and posit that $A_t f(W_t/A_t)$ is a homogeneous first degree, concave, neoclassical production function in terms of capital and efficiency-units of labor.

It should be remarked that I am confirming myself here to regular interior maxima, and not going into the Kuhn-Tucker inequalities that easily handle boundary maxima.

Solution of the Problem

The meaning of our basic problem

$$J_T(W_0) = \underset{\{C_t, w_t\}}{\text{Max}}\; E \sum_{t=0}^{T} (1+\rho)^{-t} U[C_t] \tag{11}$$

subject to $C_t = W_t - W_{t+1}[(1-w_t)(1+r) + w_t Z_t]^{-1}$ is not easy to grasp. I act now at $t = 0$ to select C_0 and w_0, knowing W_0 but not yet knowing how Z_0 will turn out. I must act now, knowing that one period later, knowledge of Z_0's outcome will be known and that W_1 will then be known. Depending upon knowledge of W_1, a new decision will be made for C_1 and w_1. Now I can only guess what that decision will be.

As so often is the case in dynamic programming, it helps to begin at the end of the planning period. This brings us to the well-known

[5] E. S. Phelps [8].

[6] J. A. Mirrlees [6]. I have converted his treatment into a discrete-time version. Robert Merton's companion paper throws light on Mirrlees' Brownian-motion model for A_t.

one-period portfolio problem. In our terms, this becomes

$$J_1(W_{T-1}) = \underset{\{C_{T-1},w_{T-1}\}}{\text{Max}} U[C_{T-1}] + E(1+\rho)^{-1}U[(W_{T-1} - C_{T-1})\{(1-w_{T-1})(1+r) + w_{T-1}Z_{T-1}\}]. \quad (12)$$

Here the expected value operator E operates only on the random variable of the next period since current consumption C_{T-1} is known once we have made our decision. Writing the second term as $EF(Z_T)$, this becomes

$$EF(Z_T) = \int_0^\infty F(Z_T)dP(Z_T|Z_{T-1},Z_{T-2},\ldots,Z_0)$$

$$= \int_0^\infty F(Z_T)dP(Z_T),$$ by our independence postulate.

In the general case, at a later stage of decision making, say $t = T-1$, knowledge will be available of the outcomes of earlier random variables, $Z_{t-2},\ldots$; since these might be relevant to the distribution of subsequent random variables, conditional probabilities of the form $P(Z_{T-1}|Z_{T-2},\ldots)$ are thus involved. However, in cases like the present one, where independence of distributions is posited, conditional probabilities can be dispensed within favor of simple distributions.

Note that in (12) we have substituted for C_T its value as given by the constraint in (11) or (10).

To determine this optimum (C_{T-1}, w_{T-1}), we differentiate with respect to each separately, to get

$$0 = U'[C_{T-1}] - (1+\rho)^{-1}EU'[C_T]\{(1-w_{T-1})(1+r) + w_{T-1}Z_{T-1}\} \quad (12')$$

$$0 = EU'[C_T](W_{T-1} - C_{T-1})(Z_{T-1} - 1 - r)$$
$$= \int_0^\infty U'[(W_{T-1} - C_{T-1})\{(1-w_{T-1}(1+r) - w_{T-1}Z_{T-1}\}](W_{T-1} - C_{T-1})(Z_{T-1} - 1 - r)dP(Z_{T-1}) \quad (12'')$$

Solving these simultaneously, we get our optimal decisions (C^*_{T-1}, w^*_{T-1}) as functions of initial wealth W_{T-1} alone. Note that if somehow C^*_{T-1} were known, (12″) would by itself be the familiar one-period portfolio optimality condition, and could trivially be rewritten to handle any number of alternative assets.

Substituting (C^*_{T-1}, w^*_{T-1}) into the expression to be maximized gives us $J_1(W_{T-1})$ explicitly. From the equations in (12), we can, by standard calculus methods, relate the derivatives of U to those of J, namely, by the envelope relation

$$J_1'(W_{T-1}) = U'[C_{T-1}]. \quad (13)$$

Now that we know $J_1[W_{T-1}]$, it is easy to determine optimal behavior one period earlier, namely by

$$J_2(W_{T-2}) = \underset{\{C_{T-2},w_{T-2}\}}{\text{Max}} U[C_{T-2}] + E(1+\rho)^{-1}J_1[(W_{T-2} - C_{T-2})\{(1-w_{T-2})(1+r) + w_{T-2}Z_{T-2}\}]. \quad (14)$$

Differentiating (14) just as we did (11) gives the following equations like those of (12)

$$0 = U'[C_{T-2}] - (1+\rho)^{-1}EJ_1'[W_{T-2}]\{(1-w_{T-2})(1+r) + w_{T-2}Z_{T-2}\} \quad (15')$$

$$0 = EJ_1'[W_{T-1}](W_{T-2} - C_{T-2})(Z_{T-2} - 1 - r)$$
$$= \int_0^\infty J_1'[(W_{T-2} - C_{T-2})\{(1-w_{T-2})(1+r) + w_{T-2}Z_{T-2}\}](W_{T-2} - C_{T-2})(Z_{T-2} - 1 - r)\,dP(Z_{T-2}). \quad (15'')$$

These equations, which could by (13) be related to $U'[C_{T-1}]$, can be solved simultaneously to determine optimal (C^*_{T-2}, w^*_{T-2}) and $J_2(W_{T-2})$.

Continuing recursively in this way for $T-3$, $T-4, \ldots, 2, 1, 0$, we finally have our problem solved. The general recursive optimality equations can be written as

$$\begin{cases} 0 = U'[C_0] - (1+\rho)^{-1}EJ'_{T-1}[W_0]\{(1-w_0)(1+r) + w_0Z_0\} \\ 0 = EJ'_{T-1}[W_1](W_0 - C_0)(Z_0 - 1 - r) \end{cases}$$

$$\cdots\cdots\cdots$$

$$0 = U'[C_{T-1}] - (1+\rho)^{-1}EJ'_{T-t}[W_t]\{(1-w_{t-1})(1+r) + w_{t-1}Z_{t-1}\} \quad (16')$$

$$0 = EJ'_{T-t}[W_{t-1} - C_{t-1})(Z_{t-1} - 1 - r), \quad (t = 1,\ldots,T-1). \quad (16'')$$

In (16′), of course, the proper substitutions must be made and the E operators must be over the proper probability distributions. Solving (16″) at any stage will give the optimal decision rules for consumption-saving and for portfolio selection, in the form

$$C^*_t = f[W_t; Z_{t-1},\ldots,Z_0]$$
$$= f_{T-t}[W_t]$$ if the Z's are independently distributed

$$w^*_t = g[W_t; Z_{t-1}, \ldots, Z_0]$$
$$= g_{T-t}[W_t] \text{ if the } Z\text{'s are independently distributed.}$$

Our problem is now solved for every case but the important case of infinite-time horizon. For well-behaved cases, one can simply let $T \to \infty$ in the above formulas. Or, as often happens, the infinite case may be the easiest of all to solve, since for it $C^*_t = f(W_t)$, $w^*_t = g(W_t)$, independently of time and both these unknown functions can be deduced as solutions to the following functional equations:

$$0 = U'[f(W)] - (1+\rho)^{-1} \int_0^\infty U'[(W - f(W))\{(1+r) - g(W)(Z-1-r)\}][(1+r) - g(W)(Z-1-r)]dP(Z) \quad (17')$$

$$0 = \int_0^\infty U'[\{W - f(W)\}\{1 + r - g(W)(Z-1-r)\}][Z-1-r]dP(Z) \quad (17'')$$

Equation (17′), by itself with $g(W)$ pretended to be known, would be equivalent to equation (13) of Levhari and Srinivasan [4, p. f]. In deriving (17′)–(17″), I have utilized the envelope relation of my (13), which is equivalent to Levhari and Srinivasan's equation (12) [4, p. 5].

Bernoulli and Isoelastic Cases

To apply our results, let us consider the interesting Bernoulli case where $U = \log C$. This does not have the bounded utility that Arrow [1] and many writers have convinced themselves is desirable for an axiom system. Since I do not believe that Karl Menger paradoxes of the generalized St. Petersburg type hold any terrors for the economist, I have no particular interest in boundedness of utility and consider $\log C$ to be interesting and admissible. For this case, we have, from (12),

$$J_1(W) = \underset{\{C,w\}}{\text{Max}} \log C + E(1+\rho)^{-1} \log[(W-C)\{(1-w)(1+r) + wZ\}]$$
$$= \underset{\{C\}}{\text{Max}} \log C + (1+\rho)^{-1}\{\log[W-C] + \underset{\{w\}}{\text{Max}} \int_0^\infty \log[(1-w)(1+r) + wZ]dP(Z)\} \quad (18)$$

Hence, equations (12) and (16′)–(16″) split into two independent parts and the Ramsey-Phelps saving problem becomes quite independent of the lifetime portfolio selection problem. Now we have

$$0 = (1/C) - (1+\rho)^{-1}(W-C)^{-1} \text{ or}$$
$$C_{T-1} = (1+\rho)(2+\rho)^{-1}W_{T-1} \quad (19')$$

$$0 = \int_0^\infty (Z-1-r)[(1-w)(1+r) + wZ]^{-1} dP(Z) \text{ or}$$
$$w_{T-1} = w^* \text{ independently of } W_{T-1}. \quad (19'')$$

These independence results, of the C_{T-1} and w_{T-1} decisions and of the dependence of w_{T-1} on W_{T-1}, hold for all U functions with isoelastic marginal utility. I.e., (16′) and (16″) become decomposable conditions for all

$$U(C) = 1/\gamma\, C^\gamma, \qquad \gamma < 1 \quad (20)$$

as well as for $U(C) = \log C$, corresponding by L'Hôpital's rule to $\gamma = 0$.

To see this, write (12) or (18) as

$$J_1(W) = \underset{\{C,w\}}{\text{Max}} \frac{C^\gamma}{\gamma} + (1+\rho)^{-1}\frac{(W-C)^\gamma}{\gamma} \int_0^\infty [(1-w)(1+r) + wZ]^\gamma dP(Z)$$
$$= \underset{\{C\}}{\text{Max}} \frac{C^\gamma}{\gamma} + (1+\rho)^{-1}\frac{(W-C)^\gamma}{\gamma} \times \underset{w}{\text{Max}} \int_0^\infty [(1-w)(1+r) + wZ]^\gamma dp(Z). \quad (21)$$

Hence, (12″) or (15″) or (16″) becomes

$$\int_0^\infty [(1-w)(1+r) + wZ]^{\gamma-1}(Z-r-1)\, dP(Z) = 0, \quad (22'')$$

which defines optimal w^* and gives

$$\underset{\{w\}}{\text{Max}} \int_0^\infty [(1-w)(1+r) + wZ]^\gamma dP(Z)$$
$$= \int_0^\infty [(1-w^*)(1+r) + w^*Z]^\gamma dP(Z)$$
$$= [1 + r^*]^\gamma, \text{ for short.}$$

Here, r^* is the subjective or util-prob mean return of the portfolio, where diminishing marginal utility has been taken into account.[7] To get optimal consumption-saving, differentiate (21) to get the new form of (12′), (15′), or (16′)

[7] See Samuelson and Merton for the util-prob concept [13].

$$0 = C^{\gamma-1} - (1+\rho)^{-1}(1+r^*)^{\gamma}(W-C)^{\gamma-1}. \quad (22')$$

Solving, we have the consumption decision rule

$$C^*_{T-1} = \frac{a_1}{1+a_1} W_{T-1} \quad (23)$$

where

$$a_1 = [(1+r^*)^{\gamma}/(1+\rho)]^{1/\gamma-1}. \quad (24)$$

Hence, by substitution, we find

$$J_1(W_{T-1}) = b_1 W^{\gamma}_{T-1}/\gamma \quad (25)$$

where

$$b_1 = a_1^{\gamma}(1+a_1)^{-\gamma} + (1+\rho)^{-1}(1+r^*)^{\gamma}(1+a_1)^{-\gamma}. \quad (26)$$

Thus, $J_1(\cdot)$ is of the same elasticity form as $U(\cdot)$ was. Evaluating indeterminate forms for $\gamma = 0$, we find J_1 to be of log form if U was.

Now, by mathematical induction, it is easy to show that this isoelastic property must also hold for $J_2(W_{T-2})$, $J_3(W_{T-3}), \ldots,$ since, whenever it holds for $J_n(W_{T-n})$ it is deducible that it holds for $J_{n+1}(W_{T-n-1})$. Hence, at every stage, solving the general equations (16′) and (16″), they decompose into two parts in the case of isoelastic utility. Hence,

Theorem:

For isoelastic marginal utility functions, $U'(C) = C^{\gamma-1}$, $\gamma < 1$, the optimal portfolio decision is independent of wealth at each stage and independent of all consumption-saving decisions, leading to a constant w^*, the solution to

$$0 = \int_0^\infty [(1-w)(1+r)+wZ]^{\gamma-1}(Z-1-r)dP(Z).$$

Then optimal consumption decisions at each stage are, for a no-bequest model, of the form

$$C^*_{T-i} = c_i W_{T-i}$$

where one can deduce the recursion relations

$$c_1 = \frac{a_1}{1+w_1},$$

$$a_1 = [(1+\rho)/(1+r^*)^{\gamma}]^{1/1-\gamma}$$

$$(1+r^*)^{\gamma} = \int_0^\infty [(1-w^*)(1+r) + w^* Z]^{\gamma} dP(Z)$$

$$c_i = \frac{a_1 c_{i-1}}{1+a_1 c_{i-1}}$$

$$= \frac{a^i_1}{1+a_1+a^2_1+\ldots+a^i_1} < c_{i-1}$$

$$= \frac{a^i_1(a_1-1)}{a^{i+1}_1-1}, \qquad a_1 \neq 1$$

$$= \frac{1}{1+i}, \qquad a_1 = 1.$$

In the limiting case, as $\gamma \to 0$ and we have Bernoulli's logarithmic function, $a_1 = (1+\rho)$, independent of r^*, and all saving propensities depend on subjective time preference ρ only, being independent of technological investment opportunities (except to the degree that W_t will itself definitely depend on those opportunities).

We can interpret $1+r^*$ as kind of a "risk-corrected" mean yield; and behavior of a long-lived man depends critically on whether $(1+r^*)^{\gamma} \gtrless (1+\rho)$, corresponding to $a_1 \lessgtr 1$.

(i) For $(1+r^*)^{\gamma} = (1+\rho)$, one plans always to consume at a uniform rate, dividing current W_{T-i} evenly by remaining life, $1/(1+i)$. If young enough, one saves on the average; in the familiar "hump saving" fashion, one dissaves later as the end comes sufficiently close into sight.

(ii) For $(1+r^*)^{\gamma} > (1+\rho)$, $a_1 < 1$, and investment opportunities are, so to speak, so tempting compared to psychological time preference that one consumes nothing at the beginning of a long-long life, i.e., rigorously

$$\lim_{i\to\infty} c_i = 0, \qquad a_1 < 1$$

and again hump saving must take place. For $(1+r^*)^{\gamma} > (1+\rho)$, the *perpetual* lifetime problem, with $T = \infty$, is divergent and ill-defined, i.e., $J_i(W) \to \infty$ as $i \to \infty$. For $\gamma \leq 0$ and $\rho > 0$, this case cannot arise.

(iii) For $(1+r^*)^{\gamma} < (1+\rho)$, $a_1 > 1$, consumption at very early ages drops only to a limiting positive fraction (rather than zero), namely

$$\lim_{i\to\infty} c_i = 1 - 1/a_1 < 1, \ a_1 > 1.$$

Now whether there will be, on the average, initial hump saving depends upon the size of $r^* - c_\infty$, or whether

$$r^* - 1 - \frac{(1+r^*)^{\gamma/1-\gamma}}{(1+\rho)^{1/1-\gamma}} > 0.$$

This ends the *Theorem*. Although many of the results depend upon the no-bequest assumption, $W_{T+1} = 0$, as Merton's companion paper shows (p. 247, this *Review*) we can easily generalize to the cases where a bequest function $B_T(W_{T+1})$ is added to $\Sigma^T_0 (1+\rho)^{-t}U(C_t)$. If B_T is itself of isoelastic form,

$$B_T \equiv b_T(W_{T+1})^{\gamma}/\gamma,$$

the algebra is little changed. Also, the same comparative statics put forward in Merton's continuous-time case will be applicable here, e.g., the Bernoulli $\gamma = 0$ case is a watershed between cases where thrift is enhanced by riskiness rather than reduced; etc.

Since proof of the theorem is straightforward, I skip all details except to indicate how the recursion relations for c_i and b_i are derived, namely from the identities

$$\begin{aligned} b_{i+1}W^\gamma/\gamma &= J_{i+1}(W) \\ &= \underset{C}{\text{Max}}\,\{C^\gamma/\gamma \\ &\quad + b_i(1+r^*)^\gamma(1+\rho)^{-1}(W-C)^\gamma/\gamma\} \\ &= \{c^\gamma_{i+1} + b_i(1+r^*)^\gamma \\ &\quad (1+\rho)^{-1}(1-c_{i+1})^\gamma\}\,W^\gamma/\gamma \end{aligned}$$

and the optimality condition

$$\begin{aligned} 0 &= C^{\gamma-1} - b_i(1+r^*)^\gamma(1+\rho)^{-1}(W-C)^{\gamma-1} \\ &= (c_{i+1}W)^{\gamma-1} - b_i(1+r^*)^\gamma(1+\rho)^{-1} \\ &\quad (1-c_{i+1})^{\gamma-1}W^{\gamma-1}, \end{aligned}$$

which defines c_{i+1} in terms of b_i.

What if we relax the assumption of isoelastic marginal utility functions? Then w_{T-j} becomes a function of W_{T-j-1} (and, of course, of r, ρ, and a functional of the probability distribution P). Now the Phelps-Ramsey optimal stochastic saving decisions do interact with the optimal portfolio decisions, and these have to be arrived at by simultaneous solution of the nondecomposable equations (16′) and (16″).

What if we have more than one alternative asset to safe cash? Then merely interpret Z_t as a (column) vector of returns $(Z^2{}_t, Z^3{}_t, \ldots)$ on the respective risky assets; also interpret w_t as a (row) vector $(w^2{}_t, w^3{}_t, \ldots)$, interpret $P(Z)$ as vector notation for

$$\begin{aligned} &\text{Prob}\,\{Z^2{}_t \leqq Z^2, Z^3{}_t \leqq Z^3, \ldots\} \\ &= P(Z^2, Z^3, \ldots) = P(Z), \end{aligned}$$

interpret all integrals of the form $\int G(Z)dP(Z)$ as multiple integrals $\int G(Z^2, Z^3, \ldots)dP(Z^2, Z^3, \ldots)$. Then (16″) becomes a vector-set of equations, one for each component of the vector Z_t, and these can be solved simultaneously for the unknown w_t vector.

If there are many consumption items, we can handle the general problem by giving a similar vector interpretation to C_t.

Thus, the most general portfolio lifetime problem is handled by our equations or obvious extensions thereof.

Conclusion

We have now come full circle. Our model denies the validity of the concept of businessman's risk; for isoelastic marginal utilities, in your prime of life you have the same relative risk-tolerance as toward the end of life! The "chance to recoup" and tendency for the law of large numbers to operate in the case of repeated investments is not relevant. (Note: if the elasticity of marginal utility, $-U'(W)/WU''(W)$, rises empirically with wealth, and if the capital market is imperfect as far as lending and borrowing against future earnings is concerned, then it seems to me to be likely that a doctor of age 35–50 might rationally have his highest consumption then, and certainly show greatest risk tolerance then — in other words be open to a "businessman's risk." But not in the frictionless isoelastic model!)

As usual, one expects w^* and risk tolerance to be higher with algebraically large γ. One expects C_t to be higher late in life when r and r^* is high relative to ρ. As in a one-period model, one expects any increase in "riskiness" of Z_t, for the same mean, to decrease w^*. One expects a similar increase in riskiness to lower or raise consumption depending upon whether marginal utility is greater or less than unity in its elasticity.[8]

Our analysis enables us to dispel a fallacy that has been borrowed into portfolio theory from information theory of the Shannon type. Associated with independent discoveries by J. B. Williams [16], John Kelly [2], and H. A. Latané [3] is the notion that if one is investing for many periods, the proper behavior is to maximize the *geometric* mean of return rather than the arithmetic mean. I believe this to be incorrect (except in the Bernoulli logarithmic case where it happens [9] to be correct for reasons

[8] See Merton's cited companion paper in this issue, for explicit discussion of the comparative statical shifts of (16)'s C^*_t and w^*_t functions as the parameters (ρ, γ, r, r^*, and $P(Z)$ or $P(Z_t, \ldots)$ or $B(W_T)$ functions change. The same results hold in the discrete-and-continuous-time models.

[9] See Latané [3, p. 151] for explicit recognition of this point. I find somewhat mystifying his footnote there which says, "As pointed out to me by Professor L. J. Savage (in correspondence), not only is the maximization of G [the geometric mean] the rule for maximum expected utility in connection with Bernoulli's function but (in so far as certain approximations are permissible) this same rule is approximately valid for all utility functions." [Latané, p. 151, n.13.] The geometric mean criterion is definitely too conservative to maximize an isoelastic utility function corresponding to positive γ in my equation (20), and it is definitely too daring to maximize expected utility when $\gamma < 0$. Professor Savage has informed me recently that his 1969 position differs from the view attributed to him in 1959.

quite distinct from the Williams-Kelly-Latané reasoning).

These writers must have in mind reasoning that goes something like the following: If one maximizes for a distant goal, investing and reinvesting (all one's proceeds) many times on the way, then the probability becomes great that with a portfolio that maximizes the geometric mean at each stage you will end up with a larger terminal wealth than with any other decision strategy.

This is indeed a valid consequence of the central limit theorem as applied to the additive logarithms of portfolio outcomes. (I.e., maximizing the geometric mean is the same thing as maximizing the arithmetic mean of the logarithm of outcome at each stage; if at each stage, we get a mean log of $m^{**} > m^{*}$, then after a large number of stages we will have $m^{**}T >> m^{*}T$, and the properly normalized probabilities will cluster around a higher value.)

There is nothing wrong with the logical deduction from premise to theorem. But the implicit premise is faulty to begin with, as I have shown elsewhere in another connection [Samuelson, 10, p. 3]. It is a mistake to think that, just because a w^{**} decision ends up with almost-certain probability to be better than a w^{*} decision, this implies that w^{**} must yield a better expected value of utility. Our analysis for marginal utility with elasticity differing from that of Bernoulli provides an effective counter example, if indeed a counter example is needed to refute a gratuitous assertion. Moreover, as I showed elsewhere, the ordering principle of selecting between two actions in terms of which has the greater probability of producing a higher result does not even possess the property of being transitive.[10] By that principle, we could have w^{***} better than w^{**}, and w^{**} better than w^{*}, and also have w^{*} better than w^{***}.

[10] See Samuelson [11].

REFERENCES

[1] Arrow, K. J., "Aspects of the Theory of Risk-Bearing" (Helsinki, Finland: Yrjö Jahnssonin Säätiö, 1965).

[2] Kelly, J., "A New Interpretation of Information Rate," *Bell System Technical Journal* (Aug. 1956), 917–926.

[3] Latané, H. A., "Criteria for Choice Among Risky Ventures," *Journal of Political Economy* 67 (Apr. 1959), 144–155.

[4] Levhari, D. and T. N. Srinivasan, "Optimal Savings Under Uncertainty," Institute for Mathematical Studies in the Social Sciences, Technical Report No. 8, Stanford University, Dec. 1967.

[5] Markowitz, H., *Portfolio Selection: Efficient Diversification of Investment* (New York: John Wiley & Sons, 1959).

[6] Mirrlees, J. A., "Optimum Accumulation Under Uncertainty," Dec. 1965, unpublished.

[7] Mossin, J., "Optimal Multiperiod Portfolio Policies," *Journal of Business* 41, 2 (Apr. 1968), 215–229.

[8] Phelps, E. S., "The Accumulation of Risky Capital: A Sequential Utility Analysis," *Econometrica* 30, 4 (1962), 729–743.

[9] Pratt, J., "Risk Aversion in the Small and in the Large," *Econometrica* 32 (Jan. 1964).

[10] Samuelson, P. A., "General Proof that Diversification Pays," *Journal of Financial and Quantitative Analysis* II (Mar. 1967), 1–13.

[11] ———, "Risk and Uncertainty: A Fallacy of Large Numbers," *Scientia*, 6th Series, 57th year (April-May, 1963).

[12] ———, "A Turnpike Refutation of the Golden Rule in a Welfare Maximizing Many-Year Plan," Essay XIV *Essays on the Theory of Optimal Economic Growth*, Karl Shell (ed.) (Cambridge, Mass.: MIT Press, 1967).

[13] ———, and R. C. Merton, "A Complete Model of Warrant Pricing that Maximizes Utility," *Industrial Management Review* (in press).

[14] Tobin, J., "Liquidity Preference as Behavior Towards Risk," *Review of Economic Studies*, XXV, 67, Feb. 1958, 65–86.

[15] ———, "The Theory of Portfolio Selection," *The Theory of Interest Rates*, F. H. Hahn and F. P. R. Brechling (eds.) (London: Macmillan, 1965).

[16] Williams, J. B., "Speculation and the Carryover," *Quarterly Journal of Economics* 50 (May 1936), 436–455.

32

Models of Optimal Capital Accumulation and Portfolio Selection and the Capital Growth Criterion*

William T. Ziemba

Professor Emeritus, University of British Columbia, Vancouver, BC
Visiting Professor, Mathematical Institute, Oxford University, UK
ICMA Centre, University of Reading, UK
University of Bergamo, Italy

Raymond G. Vickson

Professor Emeritus, University of Waterloo

Abstract

This edited and updated excerpt from the Ziemba and Vickson (1975, 2006) volume discusses various aspects of the history of optimal capital accumulation and the capital growth criterion, and supplements the introduction to Part IV of this volume.

1 Optimal Capital Accumulation

Phelps (1962) extended the Ramsey (1928) model of lifetime saving to include uncertainty. In Phelps' problem, the choice in each period was between consumption and investment in a single risky asset, with the objective being the maximization of expected utility of life time consumption. Phelps assumed that the utility function was additive in each period's utility for consumption, and he obtained explicit solutions when each utility function was the same and a member of the isoelastic marginal utility family. The papers reprinted in Ziemba and Vickson (1975, 2006) (ZV) and in this volume generalize and extend the work of Phelps in several directions: Portfolio choice is included in Samuelson (1969) and Hakansson (1970, 1971a) papers, and more general utility functions are treated in Neave (1971b), which is in ZV. These papers also constitute generalizations of the multiperiod consumption-investment papers of Part IV.[1]

Neave (1971b) treats the multiperiod consumption-investment problem for a preference structure given by a sum of utility functions of consumption in different periods, including a utility for terminal bequests. Future utilities are discounted by

*Reprinted, edited and updated from Ziemba and Vickson (1975, 2006).

[1]See Zabel (1973) for an interesting extension of the Phelps-Samuelson-Hakansson model to include proportional transaction costs.

an "impatience" factor, compounded in time. The portfolio problem is neglected, with the only decisions at any time being the amount of wealth to be consumed, the non-negative remainder being invested in a single risky asset. In each period, terminal wealth consists of a certain, fixed income plus gross return on investment, and this terminal wealth becomes initial wealth for the following period, from which consumption and investment decisions are again to be made, and so forth. Neave is concerned with conditions under which risk-aversion properties are preserved in the induced utility functions for wealth. He demonstrates the important result that the property of non-increasing Arrow-Pratt absolute risk aversion is preserved under rather general conditions. For intertemporally independent risky returns, if the single-period utilities for consumption and the utility for terminal bequests exhibit non-increasing absolute risk aversion, then all the induced utility functions for wealth also have this property. Mathematically, this result arises from the fact that convex combinations of functions having decreasing absolute risk aversion also have this property (see Pratt, 1964, reprinted in ZV). Thus, the decreasing absolute risk-aversion property is preserved under the expected-value operation, and is further preserved under maximization. The paper specifically includes the possibility of boundary as well as interior solutions in the presentation.

Neave (1971b) also considers the behavior of the Arrow-Pratt relative risk-aversion index of induced utility for wealth. He presents sufficient conditions under which the property of non-decreasing relative risk aversion is preserved. Since this property is not, in general, preserved under convex combinations, results can be obtained only under special conditions. Unfortunately, these conditions are not easily interpreted, and involve the specific values of the optimal decision variables as well as the properties of the utility functions. In general, however, non-decreasing relative risk aversion obtains for sufficiently large values of wealth. In ZV's Exercise V-ME-6, the reader is asked to attempt to widen Neave's classes of utility functions that generate desirable risk-aversion properties for the induced utility functions. Neave also relates properties of the relative risk-aversion measures to wealth elasticity of demand for risky investment. Arrow (1971) performed a similar study in the context of choice between a risk-free and a risky asset, with no consumption. In this case, the choice is between consumption and risky investment, with no risk-free asset. Neave shows that if the utilities for consumption and terminal wealth in any period exhibit constant relative risk aversion equal to unity, then the wealth elasticity for risky investment in that period is also equal to unity. He further shows that the wealth elasticity of risky investment exceeds unity if and only if the relative risk-aversion index of expected utility for wealth is less than or equal to that of utility for consumption in the relevant period.

Samuelson (1969) studies the optimal consumption-investment problem for an investor whose utility for consumption over time is a discounted sum of single-period utilities, with the latter being constant over time and exhibiting constant relative risk aversion (power-law functions or logarithmic functions). Samuelson assumes that the investor possesses in period t an initial wealth Wt which can be consumed or invested between two assets. One of the assets is safe, with constant known rate of return r per period, while the other asset is risky, with known probability distribution of return Zt in period t. The Zt are assumed to be intertemporally independent and identically distributed. Samuelson thus generalizes Phelps' model to include portfolio choice as well as consumption.

Terminal wealth $Wt + 1$ in period t consists of gross return on investment, and becomes initial wealth governing the consumption-investment decisions in period $t + 1$. The problem is to determine the fraction of wealth to be consumed and the optimal mix between riskless and risky investment in each period (given an initial wealth wo), so as to maximize expected utility of lifetime consumption. The usual backward induction procedure of dynamic programming yields a sequence of induced single-period utility functions Ut for terminal wealth in period t, with optimal consumption and investment in period t being determined by maximizing expected utility of consumption plus terminal wealth. Assuming interior maxima in each period, a recursive set of first-order conditions is derived for the optimal consumption-investment program.

The explicit form of the optimal solution is derived for the special case of utility functions having constant relative risk aversion. Samuelson shows that the optimal portfolio decision is independent of time, of wealth, and of the consumption decision at each stage. This optimal portfolio is identical with the optimal portfolio resulting from the investment of one dollar in the two assets so as to maximize expected utility ofgross return, using the instantaneous utility for consumption as the "utility for wealth" function in the optimization procedure. Furthermore, in each period, the optimal amount to consume is a known fraction of initial wealth in the period, with the optimal fraction being dependent only on the subjective discount factor for consumption over time, the return on the risk-free asset, the probability distribution ofrisky return, and the relative risk-aversion index. For logarithmic utilities, the optimal consumption fraction further simplifies to a function of the subjective discount factor alone, and is independent of the asset returns.

Samuelson utilizes his solutions to analyze optimal consumption and investment behavior as a function of time of life. As observed above, the optimal mix of riskless versus risky investment is (for constant relative risk aversion) completely independent of time of life. Samuelson relates the optimal consumption over time to the asset returns and the subjective discount factor, and discusses conditions under which the consumer-investor will save during early life and dissave in old age.

Throughout the paper, Samuelson assumes the validity of interior maxima in the optimization problems. The validity of such an assumption in the multiperiod portfolio problem was seriously questioned in Hakansson (1970, 1971a), and in ZV-Exercises including IV-CR-3, 5, and 6. When consumption decisions are included, as they are in the present case, the question of boundary solutions is presumably even more important than it was in the pure investment case. The reader may wish to ponder whether these considerations can materially affect Samuelson's conclusions. ZV-Exercise V-CR-16 shows that a stationary policy is not optimal in a multiperiod mean-variance model even under strong independence assumptions.

Hakansson (1970) studies a generalization of the problem treated by Samuelson (1969). Hakansson's assumptions regarding utility for consumption are identical to Samuelson's, except for an assumed infinite lifetime. The investment possibilities facing Hakansson's investor are, however, more general. The individual is assumed to possess a (possibly negative) initial capital position plus a non-capital steady income stream which is known with certainty. In each period, borrowing and lending at a known, time-independent rate of interest, and investment in a number of risky assets with known joint distribution of returns, are possible. The risky asset returns are assumed to be independent identically distributed (iid) over time. It is further assumed that a subset of the risky assets may be sold short. In each period, initial wealth to be consumed and invested consists of the certain income plus gross return on the previous period's investment. The objective is maximization of the expected utility of consumption over an infinite lifetime.

Hakansson introduces two important and apparently realistic constraints on the risky returns and the investor's decision variables. The capital market is assumed to impose a so-called no-easy-money restriction on the risky asset returns. This restriction ensures that no mix of risky assets exists, which provides with probability 1 a return exceeding the risk-free rate, and that no mix of long and short sales exists, which guarantees against loss in excess of the riskfree lending rate. The consumption-investment decision variables are assumed to satisfy the so-called solvency constraint. This requires that the investor always remain solvent with probability 1, i.e., that in any period t, the investor's initial capital position plus the capitalized value of the future income stream be non-negative with certainty.

Hakansson uses the familiar Bellman backward induction procedure of dynamic programming to set up a functional equation for the optimal return function (i.e., the maximum expected utility of present and future consumption, given any value of initial wealth). Since the risky returns are iid and the lifetime infinite, the functional equation obtained is stationary over time: The optimal decision variables in period t depend only on initial wealth (in period t), on the single-period asset returns, and on the relative risk-aversion index of consumption. Hakansson's treatment of the infinite horizon dynamic programming problem is intuitively plausible, but not wholly satisfactory from a rigorous standpoint. (The results are correct, however.) First, he assumes that the optimal return function exists and satisfies the functional equation; in a totally rigorous development, these facts need proof. Next, he verifies the form of the optimal solution by direct substitution, that is, by showing that the assumed form of solution actually solves the functional equation. Here the question of uniqueness arises, and is not treated satisfactorily in the development. One way of treating the problem rigorously would be to deal first with the finite horizon problem (as Samuelson does) and subsequently go to the infinite horizon limit. At each stage in the finite horizon backward induction procedure, one would have a functional equation similar to Hakansson's (but rigorously justified). As shown in the paper, the decision variable feasibility region defined by the solvency constraint is compact and convex, and (as in Hakansson's proof of his lemma, and its corollaries) the functional equation has a finite, unique solution. By considering the limiting behavior of the optimal solution and optimal return function, the given infinite horizon results could be justified. For the case of a power-law utility function unbounded from above, Hakansson needs a restriction on the utility function parameters, the optimal risky asset returns, and the subjective discount factor, in order to obtain a solution. In a limiting argument as outlined above, such a restriction would be required in order to obtain a finite return function in the limit of infinite horizon.

For constant relative risk-aversion utilities, the form of the optimal solution is (i) a fixed fraction of initial wealth plus capitalized future income is consumed in each period; and (ii) a fixed proportion of initial wealth plus capitalized future income is invested in each of the risky assets, with the optimal asset proportions being independent of time, initial wealth, capitalized income stream, and impatience to consume. The remainder of initial wealth is invested in the risk-free asset. For logarithmic utilities, the optimal consumption fraction is completely independent of all asset returns, and is a function only of the impatience factor. Hakansson also considers the interesting case of constant absolute risk aversion (exponential utilities), and states without proof the form of the optimal policy under special restrictions. The conditions required for the solution and the form of the optimal policy are more difficult to interpret than the corresponding results for the constant relative risk-aversion case. The reasons for this different behavior in the two classes of utilities may be understood as follows. For a power-law (or logarithmic) utility, expected utility of terminal wealth decomposes into a product (or sum) of utility of initial wealth and expected utility of rate of return on one dollar of investment (since terminal wealth equals initial wealth times rate of return). Such a factorization does not obtain for exponential utilities. See ZV Exercise IV-CR-3 for a clarification of this point in a simple example.

A number of conclusions regarding consumption and investment behavior are derived as corollaries to the optimal solution. As stated previously, the optimal level of consumption is a fixed fraction of initial wealth plus capitalized future income. Furthermore, the optimal amount of consumption increases with increasing impatience to consume, as expected. The optimal solution is such that the individual borrows when poor and generally lends when rich. Additional behavioral implications of the solutions are discussed in the paper.

In the Samuelson-Hakansson additive utility models of this section, the individual's consumption versus saving and investment behavior may vary as a function of time of life, but attitudes toward risk do not. In these models, the same risky portfolios are chosen in youth and in old age. In ZV-Exercise V-ME-21, a multiplicative utility model, based on Pye (1972), is developed. In this model, risk aversion in portfolio selection increases or decreases with age according as risk tolerance is greater or less than that of the logarithmic utility function. Logarithmic utility is a member of both the additive and multiplicative families. In the multiplicative family, risk aversion in portfolio selection depends on impatience to consume as well as on age. The multiplicative family thus allows greater richness in portfolio selection behavior over time. On the other hand, consumption behavior over time is simpler in the multiplicative model: Given the impatience factor and the length of remaining life, the propensity to consume is identical for all members of the multiplicative family, and is independent of present and future investment opportunities. Furthermore, in the infinite horizon limit, portfolio selection behavior changes over time, but consumption behavior does not: A constant fraction of wealth is consumed in every period.

To summarize, these papers show that an optimal lifetime consumption-investment program is simple to calculate for additive utilities belonging to the constant relative risk-aversion family. The optimal portfolio remains unchanged throughout life for asset returns which are iid over time. Consumption in any period is always proportional to wealth (including capitalized future income), the proportionality factor being dependent in general on impatience to consume, on the capital market, and possibly also on time of life. The optimal portfolio and the optimal consumption fraction are easily determined by solving a static, one-period, pure portfolio problem using utility for consumption as the appropriate utility for wealth function. Effects due to age-dependent risk aversion can be incorporated through a multiplicative form of utility function. In this case, the optimal consumption program may be calculated (in the absence of exogenous income, at least) without solving any "optimization" problem and without reference to the capital market. Optimal portfolio choice depends on time, however, and must be continually recalculated in successive periods. In both the additive and multiplicative cases, a useful form of myopia obtains: In each period, past or future realized random outcomes need not be known in choosing the optimal portfolio, and past outcomes affect present consumption decisions only by fixing the available wealth.

Some of the conclusions would remain unchanged under a slight weakening of hypotheses. In particular, the assumption of identically distributed asset returns over time may be dropped (while retaining independence). For additive utilities having constant relative risk aversion, the optimal amount of consumption will still be proportional to wealth. The relative risk aversion of induced utility for wealth will remain constant over time. The optimal asset proportions at any time t will be independent of wealth, consumption, and past or future asset returns, and will depend only on the nature of the capital market at time t. Calculation of the optimal consumption fraction at time t will, however, generally require the solution of all "future" consumption and portfolio problems, and will thus be much more complex than in the iid case. Hakansson (1971b) studied a significant generalization of his model where the individual's impatience is variable, his lifetime is stochastic, and the capital assets (both risk free and risky) change in time. The individual consumes, borrows (or lends), invests in risky assets, and purchases life insurance. For additive utilities having constant relative risk aversion, solutions are obtained for the optimal lifetime consumption, investment, and insurance-buying program. For logarithmic utilities, Miller (1974ab) analyzed the optimal consumption strategies over an infinite lifetime with a stochastic (possible nonstationary) income stream and a single, safe asset. Rentz (1971, 1972, 1973) has analyzed optimal consumption and portfolio policies with life insurance, for stochastic lifetimes and changing family size. Additional extensions are in Hakansson (1969), Long (1972), and Neave (1971a).

2 The Capital Growth Criterion

The papers by Breiman and Thorp are concerned with Kelly's capital growth criterion for long-term portfolio growth. The criterion states that in each period one should allocate funds to investments so that the expected logarithm of wealth is maximized. Hence, the investor behaves in a myopic fashion using the stationary logarithmic utility function and the current distribution of wealth.

Kelly (1956) supposed that the maximization of the exponential rate of growth of wealth was a very desirable investment criterion. Mathematically the criterion is the limit as time t goes to infinity of the logarithm of period t's wealth relative to initial wealth divided by t. If one considers Bernoulli investments in which a fixed fraction of each period's present wealth is invested in a specific favorable double-or-nothing gamble, then the criterion is easily shown to be equivalent to the maximization of the expected logarithm of wealth. ZV Exercise V-CR-18 illustrates the calculations involved and related elementary properties of Kelly's criterion.

The desirability of Kelly's criterion was further enhanced when Breiman [Breiman (1961) and his 1960 article reprinted here] showed that the expected log strategy produced a sequence of decisions that had two additional desirable properties. First, if in each period two investors have the same investment opportunities and one uses the Kelly criterion while the other uses an essentially different strategy (i.e., the limiting difference in expected logs is infinite), then in the limit the former investor will have infinite times as much wealth as the latter investor with probability 1. Second, the expected time to reach a pre-assigned goal is asymptotically (as the goal increases) least with the expected log strategy. Latane (1959) provided an earlier intuitive justification of the first property.

Breiman (1961) established these results for discrete intertemporally independent identically distributed random variables. In this case, as in the Bernoulli case, a stationary policy is optimal. In ZV Exercise V-ME-22, the reader is asked to develop these and related results for the Bernoulli case. The main technical tool involved in the proof is the Borel strong law of large numbers, which states that with probability 1 the limiting ratio of the number of successes to trials equals the Bernoulli probability of a success. Breiman's paper reprinted here proves the first result for positive bounded random variables. Results from the theory of martingales form the basis of proof of his conclusions. ZV Exercise V-CR-19 provides the reader with some elementary background on martingales. The reader is asked in ZV Exercise V-CR-20 to consider some questions concerning Breiman's analysis.

The proof of the results in Breiman's paper involves difficult and advanced mathematical tools. ZV Exercise V-ME-23 shows how similar results may be proved in a simple fashion using the Chebychev inequality. The crucial assumption is that the variance of the relative one-period gain be finite in every investment period. Breiman's assumption that the random returns are finite and bounded away from zero is sufficient but not necessary for the satisfaction of this assumption. ZV Exercise V-CR-22 presents an extension of Breiman's model to allow for more general constraint sets and value functions. The development extends his Theorem 1 and indicates that no strategy has higher expected return than the expected log strategy in any period. The model applies to common investment circumstances that include borrowing, transactions and brokerage costs, taxes, possibilities. of short sales, and so on. In ZV Exercise V-ME-24, the reader is invited to consider the relationship between properties 1 and 2, to attempt to verify the validity of property 2 for Breiman's model reprinted here, and to consider the discrete asset allocation case.

Thorp's paper provides a lucid expository treatment of the Kelly criterion and Breiman's results. He also discusses some relationships between the max expected log approach and Markowitz's mean-variance approach. In addition, he points out some of the misconceptions concerning the Kelly criterion, the most notable being the fact that decisions that maximize the expected log of wealth do not necessarily maximize expected utility of terminal wealth for arbitrarily large time horizons. The basic fallacy is that points that maximize expected log of wealth do not generally maximize the expected utility of wealth if an investor has nonlogarithmic utility function; see ZV Exercise V-ME-25 for one such example and Thorp and Whitley (1973) for a general analysis. See Markowitz (1976) for a refutation, in a limited sense, of the fallacy if the investor's utility function is bounded. For some enlightening discussion of this and other fallacies in dynamic stochastic investment analysis see Merton and Samuelson (1974) and ZV Exercise V-ME-26. Miller (1974b) shows how one can avoid the fallacy altogether by utilizing what is called the utility of an infinite capital sequence criterion. Under this criterion, the investor's utility is assumed to depend only on the wealth levels in one or more periods infinitely distant from the present; that is, capital is accumulated for its own sake, namely its prestige. In this formulation, the time and expectation limit operators are reversed (from the conventional formulation) and hence the improper limit exchange that yields the fallacy does not need to be made. One disadvantage of this formulation is that the admissible utility functions generally are variants of the unconventional form: limit infimum of the utility of period t's wealth. The reader is invited in ZV Exercise V-CR-21 to consider some questions concerning Thorp's paper. In ZV Exercise V-ME-27, the reader is asked to determine whether good decisions obtained from other utility functions have a property "similar" to the expected log strategy when they produce infinitely more expected utility.

For additional discussion and results concerning the Kelly criterion the reader may consult Aucamp (1971), Breiman (1961), Dubins and Savage (1965), Goldman (1974), Hakansson (1971a,b,d), Hakansson and Miller (1972), Jen (1971, 1972), Latane (1959, 1972), Markowitz (1976), Roll (1972), Samuelson (1971), Merton and Samuelson (1974), Thorp (1969), Young and Trent (1969), and Ziemba (1972b). For an elementary presentation of the theory of martingales the reader may consult Doob (1971). More advanced material may be found in the work of Breiman (1968), Burrill (1972), and Chow *et al.* (1971).

The highly technical Merton paper discusses the optimal consumption technical investment problem in continuous time. Because of its heavy reliance on stochastic differential equations and stochastic optimal control theory, the paper may be quite intimidating upon first reading. To make the paper intelligible to readers unfamiliar with these mathematical concepts, we deviate from our previous policy of keeping formalism out of the introduction: Appendix A contains a brief, intuitive introduction to stochastic differential equations and stochastic control theory. Although not rigorous, the arguments are, hopefully, sufficiently plausible as to enable the reader to understand Merton's paper with relative ease.

Merton assumes the investor's utility for lifetime consumption to be an integral of utilities for instantaneous consumption over time. This is a natural, continuous-time version of the additive utility assumption in discrete time. The general formalism of the optimal control problem is outlined first for utilities that vary arbitrarily in time and is specialized in later sections to the pure "impatience" case, as in the Samuelson and Hakansson discrete time models. Most of Merton's explicit solutions pertain to utility functions of the hyperbolic absolute risk aversion (HARA) class, that is, to the class where the reciprocal of the Arrow-Pratt absolute risk-aversion index is linear in wealth. This class contains as special cases utilities having constant absolute or relative risk aversion. The individual must decide at each point in time how to allocate his existing wealth to between consumption and investment in a number of financial assets. The risky asset returns are governed by a known Markov process. The objective is maximization of expected utility of lifetime consumption, including terminal bequests.

In the spirit of dynamic programming, there exists at each time t a derived utility function for wealth w, namely, the maximum expected utility of future lifetime consumption given that wealth is w at time t. The instantaneous consumption-investment problem requires the maximization of utility of consumption and terminal wealth at time t, or rather, at an "infinitesimal" time later than t. For risky asset returns governed by a stationary log-normal process (see Appendix A), terminal wealth at the end of the "infinitesimal" time span is almost deterministic. The expected utility maximization thus involves only means and variances of risky returns, as in the Samuelson paper in Part III, Chapter 1. The portfolio selection aspect of the continuous-time consumption-investment problem is thus solved exactly using mean-variance analysis. As in the Lintner and Ziemba papers in Part III, this implies that the optimal portfolio possesses the mutual fund separation property: All investors will choose a linear combination of two composite assets, which are, moreover, independent of individual preferences. If there exists a risk-free asset, it may be chosen as one of the mutual funds, and all portfolios reduce to a mix of this risk-free asset and a single composite risky asset which is the same for all investors. The returns on the mutual fund are also governed by a stationary log-normal process. Note that such a result is only true in continuous time.

In discrete time, a non-negative linear combination of log-normally distributed random returns is not log-normally distributed. Through the use of a continuous-time formulation, the consumption-investment problem is thus reduced to a two-asset model with consumption. Using this reduction, Merton obtains explicit solutions for the optimal consumption-investment program for utility functions of the HARA class. As in the papers of Part V, the optimal amount of consumption and the optimal amount of investment are linear functions of wealth. Furthermore, when a steady noncapital income stream is included, this remains true with "wealth" reinterpreted as present wealth plus future income (capitalized at the risk-free rate of return).

Merton discusses the effects on optimal consumption and investment due to the possibility of large but rare random events. Specifically, among such rare but significant events which he discusses are: (1) a bond which is otherwise risk free, may suddenly become worthless, or (2) the investor dies. The modeling of such processes in continuous time is achieved through the use of Poisson differential equations (see Appendix A). Merton derives the optimal consumption-investment policy for choice between a log-normally distributed common stock and a "suddenly worthless" bond, for utilities having constant relative risk aversion. As before, the optimal amount of consumption is linear in wealth. The optimal investment in the common stock is an increasing function ofthe probability of bond default, as expected. Merton also discusses the effect of uncertain lifetime, using a stationary Poisson process to model the arrival of the consumer's death. He shows that the optimal consumption-investment problem in this case (with no bequests) is identical to that of an infinite-lifetime model, with the consumer's utility discounted by a subjective rate of time preference equal to the reciprocal of life expectancy.

There are a number of other topics covered in the paper, which we mention only briefly. These pertain to the important and usually neglected question of nonstationary random asset returns. Merton presents solutions for the optimal consumption-investment problem in three such nonstationary models. In the first model, he assumes there exists an asymptotic, deterministic price curve pet), toward which the stochastic asset prices tend (in expected value) in the distant future. In the second model, he assumes that the risky return is given by a log-normal distribution whose mean is itself a random process of a special form. Finally, in the third model, he assumes that the risky return is given by a stationary log-normal process whose parameters are unknown, but must be estimated from past behavior of the random variable. Explicit solutions for these three models are obtained in a two-asset setting, for utility functions having constant absolute risk aversion. ZV Exercise V-ME-28 considers a deterministic continuous-time consumption-investment model. If utility is intertemporally additive, then it is possible to develop explicit optimality criteria that yield an algorithm for constructing an optimal policy.

References

Arrow, K. J. (1971). *Essays in the Theory of Risk Bearing.* Chicago: Markham Publishing.

Aucamp, D. C. (1971). A new theory of optimal investment. Working Paper, Southern Illinois University.

Breiman, L. (1961). Optimal gambling system for favorable games. *Proceedings of the 4th Berkeley Symposium on Mathematical Statistics and Probability*, 1, 63–8.

Breiman, L. (1968). *Probability.* MA: Addison-Wesley, Reading.

Burrill, C. W. (1972). *Measure, Integration and Probability.* NY: McGraw Hill.

Chow, Y. S. (1971). *Great Expectations: The Theory of Optimal Stopping.* MA: Houghton, Boston.

Doob, J. L. (1971). What is a Martingale? *AMM*, 78, 451–462.

Dubins, L. and L. Savage (1965). *How to Gamble if You Must.* NY: McGraw-Hill.

Goldman, M. B. (1974). A negative report on the "near optimality" of the max-expected log policy and appled to bounded utilities for long-lived programs. *Journal of Financial Economics*, 1, 97–103.

Hakansson, N. H. (1969). Optimal investment and consumption strategies under risk, an uncertain lifetime, and insurance. *International Economic Review*, 10, 443–466.

Hakansson, N. H. (1970). Optimal investment and consumption strategies under risk for a class of utility functions. *Econometrica*, 38, 587–607.

Hakansson, N. H. (1971a). Capital growth and the mean-variance approach to portfolio selection. *JFQA*, 6, 517–557.

Hakansson, N. H. (1971b). Mean-variance analysis of average compound returns. Working Paper, University of California, Berkeley.

Hakansson, N. H. (1971c). On optimal myopic portfolio policies, with and without serial correlation of yields. *The Journal of Business*, 44, 324–334.

Hakansson, N. H. (1971d). Optimal entrepreneurial decisions in a completely stochastic environment. *Management Science*, 17, 427–449.

Hakansson, N. H. and B. L. Miller (1972). Compound-return mean-variance efficient portfolios never riskruin. Working Paper No. 8, Research Program in Finance, Graduate School of Business Administration, University of California, Berkeley.

Jen, F. (1971). Multi-period portfolio strategies. Working Paper No. 108. State University of New York, Buffalo.

Jen, F. (1972). Criteria in multi-period portfolio decisions. Working Paper No. 131. State University of New York, Buffalo.

Kelly, Jr., J. R. (1956). A new interpretation of the information rate. *Bell System Technical Journal*, 35, 917–926.

Latané, H. (1959). Criteria for choice among risky ventures. *Journal of Political Economy*, 67, 144–155.

Latané, H. A. (1972). An optimum growth portfolio selection model. In G. Szego and K. Shell (Eds.), *Mathematical Methods in Investment and Finance*, pp. 336–343. Amsterdam: NorthHolland.

Lintner, J. (1975). The valuation of risk assets and the selection of risky investments in stock portfolios and capital budgets. In W. T. Ziemba and R. G. Vickson (Eds.), *Stochastic Optimization Models in Finance*, pp. 131–155. NY: Academic Press.

Long, Jr, J. B. (1972). Consumption-investment decisions and equilibrium in the securities market. In M. C. Jensen (Ed.), *Studies in the Theory of Capital Markets*, pp. 146–222. New York: Praeger.

Markowitz, H. M. (1976). Investment for the long run: New evidence for an old rule. *Journal of Finance*, 31(5), 1273–1286.

Merton, R. C. and P. A. Samuelson (1974). Fallacy of the log-normal approximation to optimal portfolio decision-making over many periods. *Journal of Financial Economics*, 1, 67–94.

Miller, B. L. (1974a). Optimal consumption with a stochastic income stream. *Econometrica*, 42, 253–266.

Miller, B. L. (1974b). Optimal portfolio decision making where the horizon is infinite. Working Paper, Western Management Science Institute, University of California, Los Angeles.

Neave, E. H. (1971a). Multiperiod consumption-investment decisions and risk preference. Working paper, Queen's University, Kingston, Ontario, Canada.

Neave, E. H. (1971b). Optimal consumption-investment decisions and discrete-time dynamic programming. *Journal of Economic Theory*, 3, 40–53.

Phelps, E. S. (1962). The accumulation of risky capital: A sequential utility analysis. *Econometrica*, 30, 729–743.

Pratt, J. W. (1964). Risk aversion in the small and the large. *Econometrica*, 32, 122–136.

Pye, G. (1971). Minimax policies for selling an asset and dollar averaging. *Management Science*, 17, 379–393.

Ramsey, F. P. (1928). A mathematical theory of saving. *Economic Journal*, 38, 543–559.

Rentz, W. F. (1971). Optimal consumption and portfolio policies. Unpublished dissertation, University of Rochester, Rochester, New York.

Rentz, W. F. (1972). Optimal multi-period consumption and portfolio policies with life insurance. Working Paper No. 72-36, Graduate School of Business University of Texas, Austin.

Rentz, W. F. (1973). The family's optimal multiperiod consumption, term insurance and portfolio policies under risk. Working Paper No. 73-20. Graduate School of Business, University of Texas, Austin.

Roll, R. (1972). Evidence on the 'growth-optimum' model. *Journal of Finance*, 28, 551–566.

Samuelson, P. A. (1969). Lifetime portfolio selection by dynamic stochastic programming. *Review of Economics and Statistics*, 51, 239–246.

Samuelson, P. A. (1971). The fallacy of maximizing the geometric mean in long sequences of investing or gambling. *Proceedings National Academy of Science*, 68, 2493–2496.

Thorp, E. O. (1969). Optimal gambling systems for favorable games. *Review of the International Statistical Institute*, 37(3), 273–293.

Thorp, E. O. and R. Whitley (1973). Concave utilities are distinguished by their optimal strategies. Working Paper, Mathematics Department, University of California, Irvine.

Young, W. E. and R. H. Trent (1969). Geometric mean approximations of individual security and portfolio performance. *JFQA*, 4, 179–199.

Zabel, E. (1973). Consumer choice, portfolio decisions and transaction costs. *Econometrica*, 44, 321–335.

Ziemba, W. T. (1972). Note on optimal growth portfolios when yields are serially correlated. *JFQA*, 7, 1995–2000.

Ziemba, W. T. and R. G. Vickson (Eds.) (1975). *Stochastic Optimization Models in Finance.* NY: Academic Press.

Ziemba, W. T. and R. G. Vickson (Eds.) (2006). *Stochastic Optimization Models in Finance*, (2 ed.). Singapore: World Scientific.

33

Proc. Nat. Acad. Sci. USA
Vol. 68, No. 10, pp. 2493–2496, October 1971

The "Fallacy" of Maximizing the Geometric Mean in Long Sequences of Investing or Gambling

(maximum geometric mean strategy/uniform strategies/asymptotically sufficient parameters)

PAUL A. SAMUELSON

Department of Economics, Massachusetts Institute of Technology, Cambridge, Mass. 02139

ABSTRACT **Because the outcomes of repeated investments or gambles involve products of variables, authorities have repeatedly been tempted to the belief that, in a long sequence, maximization of the expected value of terminal utility can be achieved or well-approximated by a strategy of maximizing at each stage the geometric mean of outcome (or its equivalent, the expected value of the logarithm of principal plus return). The law of large numbers or of the central limit theorem as applied to the logs can validate the conclusion that a maximum-geometric-mean strategy does indeed make it "virtually certain" that, in a "long" sequence, one will end with a higher terminal wealth and utility. However, this does not imply the false corollary that the geometric-mean strategy is optimal for any finite number of periods, however long, or that it becomes asymptotically a good approximation. As a trivial counter-example, it is shown that for utility proportional to x^γ/γ, whenever $\gamma \neq 0$, the geometric strategy is suboptimal for all T and never a good approximation. For utility bounded above, as when $\gamma < 0$, the same conclusion holds. If utility is bounded above and finite at zero wealth, *no* uniform strategy can be optimal, even though it can be that the best uniform strategy will be that of the maximum geometric mean. However, asymptotically the same level of utility can be reached by an infinity of nearby uniform strategies. The true optimum in the bounded case involves nonuniform strategies, usually being more risky than the geometric-mean maximizer's strategy at low wealths and less risky at high wealths. The novel criterion of maximizing the expected average compound return, which asymptotically leads to maximizing of geometric mean, is shown to be arbitrary.**

BACKGROUND

Suppose one begins with initial wealth, X_0, and after a series of decisions one is left with terminal wealth, X_T, subject to a conditional probability distribution

$$\text{prob}\,\{X_T \leqslant X|X_0 = A\} = P_T(X,A) \tag{1}$$

Then an "expected utility" maximizer, by definition, will choose his decisions to

$$\max E\{u(X_T)\} = \max \int_{-\infty}^{\infty} u(X)P_T(dX,A), \tag{2}$$

where E stands for the "expected value" and where $u(x)$ is a specified utility function that is unique except for arbitrary scale and origin parameters, b and a, in $a + bu(x)$, $b > 0$.

In a portfolio, or gambling situation, at each period one can make investments or bets proportional to wealth at the beginning of that period, X_t, so that the outcome of wealth for the next period is the random variable

$$X_{t+1} = X_t\{w_1Y_1 + \ldots + w_nY_n\},\ \sum_1^n w_j = 1 \tag{3}$$

where the w's are the proportions decided upon for investment in the different securities (or gambling games). The returns per dollar invested in each alternative, respectively, are subject to known probability distributions.

$$\text{prob}\,\{Y_1 \leq y_1,\ldots,Y_n \leq y_n\} = F(y_1,\ldots,y_n), \tag{4}$$

where for simplicity the vector outcomes at any t are assumed each to be independent of outcomes at any other time periods, and to remain the same distribution over all time periods. Usually Y_t is restricted to being nonnegative to avoid bankruptcy. A complete decision involves selecting over the interval of time $t = 1,2,\ldots,T$, all the vectors $[w_j(t)]$, which will be nonnegative if short-selling and bankruptcy are ruled out.

In particular, if the same strategy is followed at all times, so that $w_j(t) \equiv w_j$, the variables become

$$\begin{aligned} X_1 &= X_0x_1,\ X_2 = X_1x_2 = X_0x_1x_2 \\ X_t &= X_{t-1}x_t = X_0x_1x_2\ldots x_t \end{aligned} \tag{5}$$

and all x_t are independently distributed according to the same distribution, which we may write as

$$\text{prob}\,\{x_t \leqslant x\} = \Pi(x). \tag{6}$$

Of course, $\Pi(x)$ will depend on the w strategy chosen, and is short for $\Pi(x;\,w_1,\ldots,w_n)$ in this particular case, and on the assumed $F(y_1,\ldots,y_n)$ function.

It can be easily shown that

$$\begin{aligned} \text{prob}\,\{X_t \leq X|X_0 = Z\} &= P_t(X,Z) \\ P_t(X,Z) &= P_t(X/Z,1) \\ &= P_t(X/Z) \text{ for short} \end{aligned} \tag{7}$$

and

$$\begin{aligned} P_1(x) &\equiv \Pi(x) \\ P_2(x) &= \int_{-\infty}^{\infty} P_1(x/s)dP_1(s) \\ &\ldots\ldots\ldots\ldots \\ P_T(x) &= \int_{-\infty}^{\infty} P_{T-1}(x/s)dP_1(s) \end{aligned} \tag{8}$$

EXACT SOLUTIONS

For general $u(.)$ functions, a different decision-vector $[wl(t)]$ is called for at each intermediate time period, $t = 1,2,\ldots,T-1$. This is tedious, but both inevitable and feasible. It is well known that for one, and only one, family of utility functions, namely

$$\begin{aligned} u(x;\gamma) &= x^{\gamma}/\gamma \qquad \gamma \neq 0 \\ &= \log x \qquad \gamma = 0 \end{aligned} \tag{9}$$

it is optimal to use the same repeated strategy. For this case the common optimal strategy is that of a $T = 1$ period problem, namely

$$X_0^{\gamma} \max_{w_i} Ex^{\gamma}/\gamma = X_0^{\gamma} \max_{w_i} \int_0^{\infty} (x^{\gamma}/\gamma) dP_1(x; w_1,\ldots,w_n) \tag{10}$$

The $\log x$ case is included in this formulation, since as $\gamma \to 0$, the indeterminate form is easily evaluated.

PROPOSED CRITERION

Often in probability problems, as the number of variables becomes large, $T \to \infty$. In this case, certain asymptotic simplifications become feasible. Repeatedly, authorities (1–3) have proposed a drastic simplification of the decision problem whenever T is large.

Rule. Act in each period to maximize the geometric mean or the expected value of $\log x_t$.

The plausibility of such a procedure comes from recognition of the following valid asymptotic result.

Theorem. If one acts to maximize the geometric mean at every step, if the period is "sufficiently long," "almost certainly" higher terminal wealth and terminal utility will result than from any other decision rule.

To prove this obvious truth, one need only apply the central-limit theorem, or even the weaker law of large numbers, to the sum of independent variables

$$\log X_T = \log X_0 + \sum_1^T \log x_t \tag{11}$$

We may note the following fact about $P_T(x; \max \text{g.m.}) = Q_T^*(x)$ and $P_T(x; \text{other rule}) = Q_T(x)$.

$$Q_T^*(x) < Q_T(x) \qquad \text{for } T > M(x) \tag{12}$$

The crucial point is that M is a function of x that is unbounded in x.

From this indisputable fact, it is tempting to believe in the truth of the following false corollary:

False Corollary. If maximizing the geometric mean almost certainly leads to a better outcome, then the expected utility of its outcomes exceeds that of any other rule, provided T is sufficiently large.

The temptation to error is compounded by the consideration that both distributions approach asymptotically log normal distributions.

$$\begin{aligned} \lim_{T\to\infty} Q_T^*(x) &\cong \frac{1}{\sqrt{2\pi}\sigma^* T^{1/2}} \int_{-\infty}^{\log x} \left\{\exp -\frac{1}{2}[s-\mu^* T]^2/(\sigma^*)^2 T\right\} ds = L_T^*(x) \\ \lim_{T\to\infty} Q_T(x) &\cong \frac{1}{\sqrt{2\pi}\sigma T^{1/2}} \int_{-\infty}^{\log x} \left\{\exp -\frac{1}{2}[s-\mu T]^2/\sigma^2 T\right\} ds = L_T(x) \end{aligned} \tag{13}$$

Since $\mu^* > \mu$ by hypothesis, it follows at once that, for T large enough $L_T^*(x)$ can be made smaller than $L_T(x)$ for *any* x. From this it is thought, apparently, that one can validly deduce that $\int_0^{\infty} u(x) dQ_T^* > \int_0^{\infty} u(x) dQ_T(x)$.

FALSITY OF COROLLARY

A single example can show that the needed corollary is not generally valid. Suppose $\gamma = 1$, and one acts, in the fashion recommended by Pascal, to maximize expected money wealth itself. Let the gambler–investor face a choice between investing completely in safe cash, Y_1, or completely in a "security" that yields for each dollar invested, \$2.70 with probability 1/2 or only \$0.30 with probability 1/2. To maximize the geometric mean, one must stick only to cash, since $[(2.7)(.3)]^{1/2} = .9 < 1$. But, Pascal will always put all his wealth into the risky gamble.

Isn't he a fool? If he wins and loses an equal number of bets, and in the long run that will be his median position, he ends up with

$$(2.7)^T(.3)^T = (.9)^T \to 0 \text{ as } T \to \infty \tag{14}$$

"Almost certainly," he will be "virtually ruined" in a "long enough" sequence of play.

No, Pascal is not a fool according to *his* criterion. In those rare long sequences (and remember *all* sequences are *finite*, albeit very large) when he does experience relatively many wins, he makes more than enough to compensate him, according to the max EX_T criterion, for the more frequent times when he is ruined.

Actually,

$$\begin{aligned} E\{X_T\} &= E\left\{X_0 \prod_1^T x_t\right\} \\ &= X_0 \prod_1^n E\{x_t\}, \text{ for independent variates} \qquad (15) \\ &= X_0\{Ex_1\}^T \\ &= X_0 1.5^T > X_0 1^T \text{ for } all\ T \end{aligned}$$

But, you may say, Pascal is foolish to court ruin just for a large money gain. A dollar he wins is surely worth less in utility than the dollar he loses. Very well, as with eighteenth-century writers on the St. Petersburg Paradox, let us assume a concave utility-function, $u(x)$, with $u''(x) < 0$. Let us now test the false corollary for $u(x) = x^{\gamma}/\gamma$, $1 > \gamma \neq 0$. Let the decision according to the geometric mean rule lead to

$$E\{\log x_1^*\} > E \log x_1 \tag{16}$$

But let the alternative decision produce

$$E\{x_1^{\gamma}/\gamma\} > E\{x_1^{*\gamma}/\gamma\} \tag{17}$$

Then, as before, except for the changed value of the γ exponent

$$E\{X_T^{\gamma}/\gamma\} > E\{X_T^{*\gamma}/\gamma\}, \tag{18}$$

since

$$E\{X_T^\gamma/\gamma\} = X_0^\gamma \prod_1^T E\{x_t^\gamma/\gamma\}$$
$$= X_0^\gamma[E\{x_1^\gamma/\gamma\}]^T \quad (19)$$
$$> X^\gamma[E\{x_1^{*\gamma}/\gamma\}]^T = EX_T^{*\gamma}/\gamma$$

This strong inequality holds for all T, however large. Thus, the false corollary is seen to be invalid in general, even for concave utilities.

BOUNDED UTILITIES

Most geometric-mean maximizers are convinced by this reasoning (**5**). But not all. Thus, Markowitz, in the new 1971 preface to the reissue of his classic work on portfolio analysis (**4**), says that "boundedness" of the utility function will save the geometric-mean rule. For $\gamma < 0$, $u = x^\gamma/\gamma$ will be bounded from above, and a run of favorable gains will not bring the decision maker a utility gain of more than a finite amount. But, as the case $\gamma = -1$ shows, that cannot save the false corollary. For in all cases when $\gamma < 0$, the false rule leads to *over*-riskiness, just as for $\gamma > 0$ it leads to *under*-riskines. Those few times, and they will happen for all T, however large, with a positive (albeit diminishing probability), whenever the false rule brings you closer to zero terminal wealth (or ruin), the unboundedness of u as $x \to 0$ and $u \to -\infty$ puts a prohibitive penalty against the false rule.

What about the case where utility is bounded above and is finite at $X_T = 0$ or ruin? It is easy to show that *no* uniform decision rule can be optimal for such a case, where

$$-\infty < u(0) \leqslant u(X) \leqslant M < \infty \quad (20)$$

But suppose we choose among all suboptimal uniform strategies that have the property of "limited liability," so that each x_t is confined to the range of nonnegative numbers. Then the following theorem, which seems to contain the germ of truth the geometric mean maximizers are groping for, is valid.

Theorem. For $u(x)$ bounded above and below in the range of nonnegative numbers, the uniform rule of maximizing the geometric mean $E\{\log X_T^*\}$, will asymptotically outperform any other uniform strategy's result, $E\{u(X_T)\}$, in the following sense

$$E\{u(X_T^*)\} > E\{u(X_T)\},\ T > \bar{T}(X_0) \quad (21)$$

The limited worth of this theorem is weakened a bit further by the consideration that an infinite number of blends of a suboptimal strategy with the geometric mean rule, of the form $vx^* + (1 - v)x$, v sufficiently small and positive, will do negligibly worse as $T \to \infty$.

Thus for all X_0,

$$\lim_{T\to\infty} E\{u(X_T^*)\} = u(0) \text{ or } M = \sup_X u(X), \quad (22)$$

depending upon whether

$$E(\log x_1^*) \leqslant 0 \text{ or } > 0$$

For a range of v's, the same utility level will be approached as $T \to \infty$.

OPTIMAL NONUNIFORM STRATEGY

To illustrate that no uniform strategy can be optimal when utility is bounded, consider the well-known case of $u = -e^{-bx}$.

Suppose X_t can be invested at each stage into Z_t dollars of a risky Y_2 such that $E\{Y_2\} > 1$, or into $X_t - Z_t$ dollars of safe cash. Under "limited liability," if the lowest Y_2 could get were r, Z_t could not exceed $X_t/(1 - r)$.

First, disregard limited liability and permit negative X_T. Then a well-known result of Pfansangl (**6**) shows that at every stage, it is optimal to set $Z_t = Z^*$, where

$$\max_z \int_{-\infty}^{\infty} -\exp[-b(X-Z) - bY_2Z]dF_2(Y_2)$$
$$= -[\exp - b(X - Z^*)] \int_{-\infty}^{\infty} \exp - [bY_2Z^*]dF_2(Y_2) \quad (23)$$

Then optimally

$$X_T = X_0 - TZ^* + (Y_2^{(1)} + Y_2^{(2)} + \dots + Y_2^{(T)})Z^* \quad (24)$$

As $T \to \infty$, X_T approaches a normal distribution, not a log-normal distribution as in the case of uniform strategies. However, the utility level itself will approach a log-normal distribution.

$$U_t \equiv \exp - b[X_0 - TZ^*] \prod_1^T (\exp - bY_2^{(t)}Z^*)$$
$$\lim_{T\to\infty} E\{U_T\} = 0 = M \quad (25)$$

For this option of cash or a risky asset with positive return, it will necessarily be the case that

$$\max_w E\{\log[w + (1 - w)Y_2]\} > 1$$
$$\lim_{T\to\infty} E\{U_T^*\} = 0 = M$$

Nonetheless, at every wealth level, above or below a critical number, the optimal policy will generally differ from that of maximum geometric mean.

When we reintroduce limited liability in the problem and never permit an investor to take a position that could leave him bankrupt with negative wealth, the optimal strategy at each X_t will be to put min (X_t, Z^*) into the risky asset and an exact solution becomes more tedious. But clearly the optimal strategy is nonuniform and will outperform any and all uniform strategies, including that of the geometric-mean maximizer.

MAXIMIZING EXPECTED AVERAGE COMPOUND GROWTH

Hakansson (**7**) has presented an analysis with a bearing on geometric-mean maximization by the long-run investor. Defining the rate of return in any period as x_t and the average compound rate of return as $(x_1x_2 \dots x_T^{1/T})$, one can propose as a criterion of portfolio selection maximizing the expected value of this magnitude. After the conventional scaling-factor T is introduced, this gives

$$\max_{w_i} TX_0E(x_1x_2 \dots x_T)^{1/T} = X_0\{\max_{w_i} E[x_j^{1/T}/(1/T)]\}^T \quad (26)$$

This problem we have already met in (19) for $\gamma = 1/T$. For finite T, this new criterion leads to slightly more risk-taking than does geometric-mean or expected-logarithm maximizing: for $T = 1$, it leads to Pascal's maximizing of expected money gain; for $T = 2$, it leads to the eighteenth century square-root utility function proposed by Cramer to resolve the St. Petersburg Paradox.

However, as $T \to \infty$ and $\gamma = 1/T \to 0$,

$$\lim_{\gamma \to 0} x^\gamma/\gamma = \log x \tag{27}$$

and we asymptotically approach the geometric-mean maximizing.

Thus, one wedded psychologically to a utility function $-x^{-1}$ will find the new criterion leads to rash investing. Example: modify the numerical example of (14) above so that 2.7 and .3 are replaced by 2.4 and .6. Because the geometric mean of these numbers equals $1.2 > 1$, none in cash is better for such an investor than is all in cash. But, since the harmonic means of these numbers equals $.96 < 1$, our hypothesized investor would prefer to satisfy his own pyschological tastes and choose to invest all in cash rather than none in cash—no matter how great T is and in the full recognition that he is violating the new criterion. The few times that following that criterion leads him to comparative losses are important enough in his eyes to scare him off from use of that criterion.

Indeed, if commissions were literally zero, then no matter how short were T in years, the number of transaction periods would become indefinitely large: Hence, with $\gamma = 0$, the novel criterion would lead to geometric-mean maximization, not just asymptotically for long-lived investors, but for any T. To be sure, as one shortens the time period between transactions, my assertion of independence of probabilities between periods might become unrealistic. This opens a Pandora's Box of difficulties. Fortuitously, the utility function $\log x$ is the one case that is least complicated to handle when probabilities are intertemporally dependent. This makes $\log x$ an attractive candidate for Santa Claus examples in textbooks, but will not endear it to anyone whose psychological tastes deviate significantly from $\log x$. (For what it is worth, I may mention that I do not fall into that category, but that does not affect the logic of the problem.)

These remarks critical of the criterion of maximum expected average compound growth do not deny that this criterion, arbitrary as it is, still avoids some of the even greater arbitrariness of conventional mean-variance analysis. Its essential defect is that it attempts to replace the pair of "asymptotically sufficient parameters" $[E\{\log x_t\}, \text{Variance}\{\log x_t\}]$ by the first of these alone, thereby gratuitously ruling out arbitrary γ in the family $u(x) = x^\gamma/\gamma$ in favor of $u(x) = \log x$. This diagnosis can be substantiated by the valuable discussion in the cited Hakansson paper of the efficiency properties of the pair [E{average–compound–return}, Variance{average–compound–return}], which are asymptotically surrogates for the above sufficient parameters.

Financial aid from the National Science Foundation and editorial assistance from Mrs. Jillian Pappas are gratefully acknowledged. I have benefited from conversations with H. M. Markowitz, H. A. Latané, and L. J. Savage, but cannot claim that they would hold my views. N. H. Hakansson has explicitly warned that the purpose of his paper was not to favor maximizing the expected average–compound-return criterion.

1. Williams, J. B., "Speculation and Carryover," *Quarterly Journal of Economics*, **50**, 436–455 (1936).
2. Kelley, J. L., Jr., "A New Interpretation of Information Rate," *Bell System Technical Journal*, 917–926 (1956).
3. Latané, H. A., "Criteria for Choice Among Risky Ventures," *Journal of Political Economy*, **67**, 144–155 (1956); Kelley, J. L., Jr., and L. Breiman, "Investment Policies for Expanding Business Optimal in a Long-Run Sense," *Naval Research Logistics Quarterly*, **7** (4), 647–651 (1960); Breiman, L., "Optimal Gambling Systems for Favorable Games," ed. J. Neyman, *Proceedings of the Fourth Berkeley Symposium on Mathematical Statistics and Probability* (University of California Press, Berkeley, Calif., 1961).
4. Markowitz, H. M., *Portfolio Selection. Efficient Diversification of Investments* (John Wiley & Sons, New York, 1959), Ch. 6.
5. Samuelson, P. A., "Lifetime Portfolio Selection by Dynamic Stochastic Programming," *Review of Economics and Statistics*, **51**, 239–246 (1969).
6. Pfanzangl, J., "A General Theory of Measurement-Applications to Utility," *Naval Research Logistics Quarterly*, **6**, 283–294 (1959).
7. Hakansson, N. H., "Multi-period Mean-variance Analysis: Toward a General Theory of Portfolio Choice," *Journal of Finance*, **26**, 4 (September, 1971).

Journal of Banking and Finance 3 (1979) 305–307. © North-Holland Publishing Company

34

WHY WE SHOULD NOT MAKE MEAN LOG OF WEALTH BIG THOUGH YEARS TO ACT ARE LONG

Paul A. SAMUELSON*

Massachusetts Institute of Technology, Cambridge, MA 02139, USA

> He who acts in N plays to make his mean log of wealth as big as it can be made will, with odds that go to one as N soars, beat me who acts to meet my own tastes for risk.

Who doubts *that*? What we do doubt[1] is that *it* should make us change our views on gains and losses – should taint our tastes for risk.

To be clear is to be found out. Know that life is not a game with net stake of one when you beat your twin, and with net stake of nought when you do not. A win of ten is not the same as a win of two. Nor is a loss of two the same as a loss of three. *How much* you win by counts. *How much* you lose by counts.

As soon as we see *this* clear truth, we are back to our own tastes for risk. Mean log of wealth then bores those of us with tastes for risk not real near to one odd (thin!) point on the line of *all* the tastes for risk – and this holds for each N, *with N as big as you like.*

Why then do some still think they should want to make mean log of wealth big? They nod. They feel 'That way I *must* end up with more. More sure beats less'. But they err. What they do not see is this:

> When you lose – and you *sure can* lose – with N large, you can lose real big. Q.E.D.

Long since, in Samuelson (1963, p. 4), I had to prove what is not hard to grasp:

> If it does not pay to do an act once, it will not pay to do it twice, thrice,..., or at all.

*I owe thanks for aid to NSF Grant 750-4053-A01-SOC.

[1]Cf. the views of Ophir (1978, 1979) and Latané (1978).

Can we bring the dead rule back to life by what it tells us about the mean growth rate? No. Here's why not.

> For large N, when you act at each turn to make the mean of log of wealth big, you will make your *mean growth rate* big in this sense:
>
> As N grows large, the odds go to one that my mean growth rate (per turn) will end up real close to a rate less than that which you (with big odds) *end up* close to.

Who doubts *that* truth? But *it* does not rule out this clear truth.

> For N as large as one likes, your growth rate can well (and at times must) turn out to be less than mine – and turn out so much less that my tastes for risk will force me to shun your mode of play. To make N large will *not* (say it again, *not*) make me change my mind so as to tempt me to your mode of play. Q.E.D.

No doubt some will say: 'I'm not sure of my taste for risk. I lack a rule to act on. So I grasp at one that at least ends doubt: better to act to make the odds big that I win than to be left in doubt?' Not so. There is more than one rule to end doubt. Why pick on one odd one? Why not try to come a bit more close to that which is not clear but which you ought to try to make more clear?

No need to say more.[2] I've made my point.[3] And, save for the last word, have done so in prose of but one syllable.

[2]We should spare the dead. When a chap has said he now doubts that '...this same rule [of max of mean of log of wealth] is approximately valid for all utility functions [...insofar as certain approximations are permissible...]...', we should take him at his word and free his shade of all guilt. For a live friend to still say: 'given the qualifications it seems to me that this [above quoted] statement of Savage is very difficult to refute', as the French say, gives one to cry. Those key words are false when we make them clear. When we don't make them clear, there is nought to talk about (to say Yes or No to). As the French say too, it is a case of put up or... For more on this, see Latané (1959, p. 151; 1978, p. 397) and Samuelson (1959, p. 245).

[3]Let me tie down one loose end. Look at this Odd Rule:

> From Acts $A(N)$ and $B(N)$ pick Act $A(N)$ if, for their two end wealths, $W_A(N)$ and $W_B(N)$, with odds of more than one half (or more than $1-\varepsilon_N$, $0<\varepsilon_N\ll 1$), $W_A(N)>W_B(N)$.

This Odd Rule is odd since it can put you in this Fix:

> You may well pick $A(N)$ from $A(N)$ and $B(N)$, and pick $B(N)$ from $B(N)$ and $C(N)$, and yet still pick $C(N)$ from $A(N)$ and $C(N)$.

References

Latané, H.A., 1959, Criteria for choice among risky ventures, Journal of Political Economy 67, 144–155.
Latané, H.A., 1978, The geometric-mean principle revisited: A reply, Journal of Banking and Finance 2, 395–398.
Ophir, T., 1978, The geometric-mean principle revisited, Journal of Banking and Finance 2, 103–107.
Ophir, T., 1979, The geometric-mean principle revisited: A reply to a reply, Journal of Banking and Finance, this issue.
Samuelson, P.A., 1963. Risk and uncertainty: A fallacy of large numbers, Scientia 1–6. Reproduced in: 1966, Collected scientific papers of Paul A. Samuelson – I (MIT Press, Cambridge, MA) 153–158.
Samuelson, P.A., 1969, Lifetime portfolio selection, Review of Economics and Statistics 51, 239–246. Reproduced in: 1972, Collected scientific papers of Paul A. Samuelson – II (MIT Press, Cambridge, MA) 883 890 (cf. p. 889).
Thorp, E., 1971, Portfolio choice and the Kelley criterion. Reproduced in: W.T. Ziemba and R.G. Vickson, eds., 1975, Stochastic optimization models in finance (Academic Press, New York) 599–619.

This Fix *can* come for *all* N, as large as we choose to make N. It can do so though the truth of Thorp (1971, p. 603) holds,

$$\operatorname*{plim}_{N\to\infty} W_A(N) = \bar{W}_A > \bar{W}_B = \operatorname*{plim}_{N\to\infty} W_B(N), \qquad \bar{W}_B > \bar{W}_C = \operatorname*{plim}_{N\to\infty} W_C(W),$$

rules out

$$\operatorname*{plim}_{N\to\infty} W_C(N) > \operatorname*{plim}_{N\to\infty} W_A(N).$$

That is so since $\bar{W}_A > \bar{W}_B$ and $\bar{W}_B > \bar{W}_C$ means $\bar{W}_A > \bar{W}_C$.

But I'd not said: 'When you act to make mean of log of wealth large, you could get in the Fix'. To see why you can't, we need only note that

$$\text{mean of log of } W_A(N) = L_A(N) > L_B(N) = \text{mean of log of } W_B(N),$$

and

$$\text{mean of log of } W_B(N) = L_B(N) > L_C(N) = \text{mean of log of } W_C(N),$$

rules out

$$L_C(N) > L_A(N),$$

and it does so for *all* N, as when N is as small as one.

Some goals are strange, but need not be as bad as the odd rules some seek to base them on.

The Journal of FINANCE

VOL. XXXI DECEMBER 1976 NO. 5

35

INVESTMENT FOR THE LONG RUN: NEW EVIDENCE FOR AN OLD RULE

HARRY M. MARKOWITZ*

I. BACKGROUND

"INVESTMENT FOR THE LONG RUN," as defined by Kelly [7], Latané [8] [9], Markowitz [10], and Breiman [1] [2], is concerned with a hypothetical investor who neither consumes nor deposits new cash into his portfolio, but reinvests his portfolio each period to achieve maximum growth of wealth over the indefinitely long run. (The hypothetical investor is assumed to be not subject to taxes, commissions, illiquidities and indivisibilities.) In the long run, thus defined, a penny invested at 6.01% is better—eventually becomes and stays greater—than a million dollars invested at 6%.

When returns are random, the consensus of the aforementioned authors is that the investor for the long run should invest each period so as to maximize the expected value of the logarithm of (1 + single period return). The early arguments for this "maximum-expected-log" (MEL) rule are most easily illustrated if we assume independent draws from the same probability distribution each period. Starting with a wealth of W_0, after T periods the player's wealth is

$$W_T = W_0 \cdot \prod_{t=1}^{T} (1 + r_t) \tag{1}$$

where r_t is the return on the portfolio in period t. Thus

$$\log(W_T / W_0) = \sum_{t=1}^{T} \log(1 + r_t) \tag{2}$$

If $\log(1 + r)$ has a finite mean and variance, the weak law of large numbers assues us that for any $\epsilon > 0$

$$\text{Prob}\left(\left|\frac{1}{T} \cdot \log(W_T / W_0) - E\log(1 + r)\right| > \epsilon\right) \to 0 \tag{3}$$

* IBM Thomas J. Watson Research Center, Yorktown Heights, New York.

and the strong law[1] assures us that

$$\lim_{T\to\infty} \frac{1}{T}\cdot\log(W_T/W_0)=E\log(1+r) \tag{4}$$

with probability 1.0. Thus if $E\log(1+r)$ for portfolio A exceeds that for portfolio B, then the weak law assures us that, for sufficiently large T, portfolio A has a probability as close to unity as you please of doing better than B when time $=T$; and the strong law assures us that

$$W_T^A/W_T^B\to\infty \qquad \text{with probability one.} \tag{5}$$

Some authors have argued that the strategy which is optimal for the player for the long run is also a good rule for some or all real investors. My own interest in the subject stems from a different source. In tracing out the set of mean, variance (E, V) efficient portfolios one passes through a portfolio which gives approximately maximum $E\log(1+r)$.[2] I argued that this "Kelly-Latané" point should be considered the upper limit for conservative choice among E, V efficient portfolios, since portfolios with higher (arithmetic) mean give greater short-run variability with less return in the long run. A real investor might, however, perfer a smaller mean and variance, giving up return in the long run for stability in the short run.

Samuelson [14] and [15] objected to MEL as the solution to the problem posed in [1], [2], [7], [8], [9], [10]. Samuelson's objection may be illustrated as follows: suppose again that the same probability distributions of returns are available in each of T periods, $t=1,2,\ldots,T$. (Samuelson has also treated the case in which t is continuous; but his objections are asserted as well for the original discussion of discrete time. The latter, discrete time, analysis is the subject of the present paper.) Assume that the utility associated with a play of a game is

$$U=W_T^\alpha/\alpha \qquad \alpha\neq 0 \tag{6}$$

where W_T is final wealth. Samuelson shows that, in order to maximize expected utility for the game as a whole, the same portfolio should be chosen each period. This always chosen portfolio is the one which maximizes single period

$$EU=E(1+r)^\alpha/\alpha. \tag{7}$$

Furthermore, if EU_T^0 is the expected return provided by this strategy for a T

1. In most cases the early literature on investment for the long run used the weak law of large numbers. The results in Breiman [1], however, specialize to a strong law of large numbers in the particular case of unchanging probability distributions. See also the Doob [4] reference cited by Breiman.

2. Markowitz [10] Chapters 6 and 13 conjectures, and Young and Trent [16] confirm that

$$E\log(1+r)\approx\log(1+E)-\tfrac{1}{2}\cdot(V/(1+E)^2)$$

for a wide class of actual ex post distributions of annual portfolio returns.

period game, and EU_T^L is that provided by MEL, usually we will have

$$EU_T^0/EU_T^L \to \infty \qquad \text{as } T \to \infty \tag{8}$$

Thus, despite (3), (4) and (5), MEL does not appear to be asymptotically optimal for this apparently reasonable class of games.

Von Newman and Morgenstern [17] have directly and indirectly persuaded many, including Samuelson and myself, that, subject to certain caveats, the expected utility maxim is the correct criterion for rational choice among risky alternatives. Thus if it were true that the laws of large numbers implied the general superiority of MEL, but utility analysis contradicted this conclusion, I am among those who would accept the conclusions of utility analysis as the final authority. But not every model involving "expected utility" is a valid formalization of the subject purported to be analyzed. In particular I will argue that, on closer examination, utility analysis supports rather than contradicts MEL as a quite general solution to the problem of investment for the long run.

II. The Sequence of Games

It is important to note that (8) is a result concerning a *sequence of games*. For fixed T, say $T=100, EU=EW_{100}^{\alpha}/\alpha$ is the expected utility (associated with a particular strategy) of a game involving precisely 100 periods. For $T=101, EW_{101}^{\alpha}/\alpha$ is the expected utility of a game lasting precisely 101 periods; and so on for $T=102, 103, \ldots$.

That (8) is a statement about a sequence of games may be seen either from the statement of the problem or from the method of solution. In Samuelson's formulation W_T is *final* wealth—wealth at the *end* of the game. If we let T vary (as in "$T\to\infty$") we are talking about games of varying length.

Viewed differently, imagine computing the solution by dynamic programming starting from the last period and working in the direction of earlier periods. (Here we may ignore the fact that essentially the same solution reemerges in each step of the present dynamic program. Our problem here is not how to compute a solution economically, but what problem is being solved). If we allow our dynamic programming computer to run backwards in time for 100 periods, we arrive at the optimum first move, and the expected utility for the game as a whole given any initial W_0, for a game that is to last 100 moves. If we allow the computer to continue for an additional 100 periods we arrive at the optimum first move, and the expected utility for the game as a whole given any initial W_0, for a game that is to last for 200 moves; and so on for $T=201, 202, \ldots$.

In particular, equation (8) is not a proposition about a single game that lasts forever. This particular point will be seen most clearly later in the paper when we formalize the utility analysis of unending games.

To explore the asymptotic optimality of MEL, we will need some notation concerning sequences of games in general. Let $T_1 < T_2 < T_3 \cdots$ be a sequence of strictly increasing positive integers. In this paper[3] we will denote by $G_1, G_2, G_3 \ldots$ a

3. A somewhat different, but equivalent, notation was used in [11].

sequence of games, where the ith game lasts T_i moves. (In case the reader feels uncomfortable with the notion of a sequence of games, as did at least one of our colleges who read [11], perhaps the following remarks may help. The notion of a sequence of games is similar to the notion of a sequence of numbers, or a sequence of functions, or a sequence of probability distributions. In each case there is a first object (i.e., a first number or function or distribution or game) which we may denote as G_1; a second object (number, function, distribution, game) which we may denote by G_2; etc.).

In general we will not necessarily assume that the same opportunities are available in each of the T_i periods of the game G_i. We will always assume that—as part of the rules that govern G_i—the game G_i is to last exactly T_i periods, and that the investor is to reinvest his entire wealth (without commissions, etc.) in each of the T_i periods. Beyond this, specific assumptions are made in specific analyses.

In addition to a sequence of games, we shall speak of a sequence of strategies $s_1, s_2, s_3, \ldots$ where s_i is a strategy (i.e., a complete rule of action) which is valid for (may be followed in) the game G_i. By convention, we treat the utility function as part of the specification of the rules of the game. The rules of G_i and the strategy s_i together imply an expected utility to playing that game in that manner.

III. Alternate Sequence-of-Games Formalizations

Let g equal the rate of return achieved during a play of the game G_i; i.e., writing T for T_i:

$$W_T = W_0 \cdot (1+g)^T \tag{9}$$

or

$$g = (W_T / W_0)^{1/T} - 1. \tag{10}$$

In the Samuelson sequence of games, here denoted by $G_1, G_2, G_3, \ldots$, the utility function of each game G_i was assumed to be

$$U = f(W_T) = W_T^{\alpha} / \alpha. \tag{11}$$

We can imagine another sequence of games—call them $H_1, H_2, H_3, \ldots$—which have the same number of moves and the same opportunities per move as $G_1, G_2, G_3, \ldots$, respectively, but have a different utility function. Specifically imagine that the utility associated with a play of each game H_i is

$$U = V(g). \tag{11a}$$

for some increasing function of g. For a fixed game of length $T = T_i$, we can always find a function $V(g)$ which gives the same rankings of strategies as does some specific $f(W_T)$. For example, for fixed T (11) associates the same U to each possible play as does

$$U = V(g) = W_0^{\alpha} \cdot (1+g)^{\alpha T} / \alpha. \tag{11b}$$

Thus for a given T it is of no consequence whether we assume that utility is a function of final wealth W_T or of rate of return g.

On the other hand, the assumption that some utility function $V(g)$ remains constant in a sequence of games, as in $H_1, H_2, H_3, \ldots$ has quite different consequences than the assumption that some utility function $f(W_T)$ remains constant as in $G_1, G_2, \ldots$. Markowitz [11] shows that if $V(g)$ is continuous then

$$EV_T^L / EV_T^0 \to 1 \qquad \text{as } T \to \infty \tag{12a}$$

where EV_T^L is the expected utility provided by the MEL strategy for the game H_i, and EV_T^0 is the expected utility provided by the optimum strategy (if such an optimum exists[4]); and if $V(g)$ is discontinuous then

$$EV_T^L / EV_T^0 \geqslant 1 - \epsilon - \delta \tag{12b}$$

where δ is the largest jump in V at a point of discontinuity, and $\epsilon \to 0$ as $T \to \infty$. (12a) and (12b) do not require the assumption that the same probability distributions are available each period. It is sufficient to assume that the return r is bounded by two extremes $\underline{r}$ and $\bar{r}$:

$$-1 < \underline{r} \leqslant r \leqslant \bar{r} < \infty \tag{13}$$

e.g., the investor is assumed to not lose more than, say, 99.99% nor make more than a million percent on any one move in any play of any game of the sequence. Note also that $V(g)$ is not required to be concave, nor strictly increasing nor differentiable; but of course it is allowed to be such.

Thus under quite general assumptions, if $V(g)$ is continuous MEL is asymptotically optimal in the sense of 12a. If $V(g)$ has small discontinuities, then MEL may possibly fail to be asymptotically optimal by small amounts as in 12b. These results are in contrast to (8), derived on the assumption of constant $U = f(W_T)$.

In [11] I argued that the assumption of constant $V(g)$ in a sequence of games is a more reasonable formalization of "investment for the long run" than is the assumption of constant $U(W_T)$. Given the basic assumptions of utility analysis, the choice between constant $V(g)$ and constant $U(W_T)$ is equivalent to deciding which of two types of questions would be more reasonable to ask (or determine from revealed preferences) of a rational player who invests for the long run in the sense under discussion.

Example of question of type I: what probability would make you indifferent between (a) a strategy which yields 6% with certainty in the long run: and (b) a strategy with a probability α of yielding 9% in the long run versus a probability of $1-\alpha$ of yielding 3% in the long run.

Example of question of type II: if your initial wealth is \$10,000.00, what

4. The assumptions of [11] do not necessarily imply that an optimum strategy exists. In any case (12a) and (12b) apply to any "other" strategy such that

$$EV_T^0 \geqslant EV_T^L \qquad \text{for all } T.$$

probability β would make you indifferent between (a) a strategy which yields \$20,000 with certainty in the long run, versus (b) a strategy which yields \$25,000 with probability β and \$15,000 with probability $(1-\beta)$ in the long run.

Question I has meaning if constant $V(g)$ is assumed; question II if constant $U(W_T)$ is assumed. It seemed to me (and still does) that preferences among probability distributions involving, e.g., 3%, 6% or 9% return in the indefinitely long run are more reasonable to assume than preferences among probability distributions involving a final wealth of, e.g., \$10,000, \$15,000 or \$20,000 in the long run. I will not try to further argue the case for constant $V(g)$ as opposed to constant $U(W_T)$ at this point, other than to encourage the reader to ask himself questions of type I and type II to judge.

In [11] I also argued that even if we were to assume constant $U(W_T)$ rather than constant $V(g)$, we would have to assume that U was bounded (from above and below) in order to avoid paradoxes like those of Bernoulli [3] and Menger [13]. I then show that MEL is asymptotically optional for bounded $U(W_T)$. Merton and Samuelson [12] and Goldman [6] object to my definition of asymptotic optimality, although it is essentially the same as the criteria by which we judge, e.g. a statistic to be asymptotically efficient. Merton and Samuelson proposed, and Goldman adopted, an alternative criterion in terms of the "bribe" required to make a given strategy as good as the optimum strategy. But this bribe criteria seems to me unacceptable, since it violates a basic tenant of game theory—that the normalized form of a game (as described in [17]) is all that is needed for the comparison of strategies. It is not possible to infer the Samuelson-Merton-Goldman bribe from the normalized form of a game. Strategies Ia and Ib in game I may have the same expected utilities, respectively, as strategies IIa and IIb in game II; but a different bribe may be required to make Ia indifferent to Ib than is required to make IIa indifferent to IIb. Strategies IIIa and IIIb in a third game (not necessarily an investment game) may have the same pair of expected utilities as Ia and Ib in game I, or IIa and IIb in game II, but the notion of a bribe may have no meaning whatsoever in game III.[5] Thus unless we are prepared to reject the equivalence between the normalized and extensive form of a game in evaluating strategies, we must reject the Merton-Samuelson-Goldman bribe as part of a precise, formal definition of asymptotic optimality.

5. For example, suppose that strategy (a) has $EU_a=0$ and strategy (b) has $EU_b=\frac{1}{2}$. What bribe will make (a) as good as (b)? Consider the answer, e.g., for one period games I and II in which (a) accepts $W=\frac{1}{2}$ with certainty and (b) elects a 50-50 chance of $W=0$ versus $W=1$. In (I) suppose

$$U=\begin{cases} 0 & \text{for} \quad W<\frac{1}{2} \\ 10^*(W-\frac{1}{2}) & \text{for} \quad 0.5<W\leqslant 0.6 \\ 1 & \qquad W>0.6 \end{cases}$$

while in II suppose

$$U=\begin{cases} 0 & \text{for} \quad W<0.9 \\ 10^*(W-0.9) & \qquad 0.9<W<1.0 \\ 1 & \qquad W>1.0 \end{cases}$$

In game I, (a) requires a bribe of 0.05; in game II (a) requires 0.45.

IV. UNENDING GAMES

Even if we agree that a player playing a fixed finite game should maximize expected utility, we cannot determine whether MEL is asymptotically optimal for a given sequence of games $\{G_i\}$ unless we can agree on criteria for asymptotic optimality. What is needed is either "metacriteria" regarding how to choose criteria of asymptotic optimality, or else an alternate method of analyzing the desirability of strategies for the long run. This section presents such an alternate method, namely the utility analysis of unending games.

Consider a game G_∞ which is like one of the games G_i described above with this one exception: the game G_∞ never terminates. Instead of having a first move, a second move, and so on through a T_ith move, we have an unending sequence of moves. As with a game G_i, a strategy for a G_∞ is a rule specifying the choice of portfolio at each time t as a function of the information available at that time. The only difference is that now the rule is defined for each positive integer $t=1,2,3,\ldots$ rather than only for $1 \leqslant t \leqslant T$.

Given a particular game G_∞ and a strategy (s), a play of the game involves an infinite sequence of "spins of the wheel" and results in an infinite sequence of wealths at each time:

$$(W_0, W_1, W_2, \ldots, W_t, \ldots) \tag{14}$$

where W_0 is initial wealth, and

$$W_t = W_{t-1}^*(1+\text{return at time } t) \tag{15}$$

as in G_i.

The reader should find it no more unthinkable to imagine an infinite sequence of spins than to imagine drawing a uniformly distributed random variable. For example, if the same wheel is to be spun each time in an unending game, and if this wheel has ten equally probable stopping points, which we may label 0 through 9, then the infinite decimal expansion of a uniform [0, 1] random variable may be taken as the infinite sequence of random stopping points of the wheel.[6] If the wheel has sixteen stopping points, then the hexadecimal expansion of the random number may be used. In either case the infinite sequence of wealths $(W_0, W_1, W_2, \ldots)$ is implied by the rules of the game, the player's strategy, and the uniform random number drawn.

In general, a given G_∞ and a given strategy imply a probability distribution of wealth-sequences $(W_0, W_1, W_2, \ldots)$.

Since G_∞ has no "last period", we cannot speak of "final wealth". We can, however, assume that the player has preferences among alternate wealth-sequences: e.g., he may prefer the sequence of passbook entries provided by a savings account which compounds his money at 6%, starting with W_0, to one that compounds it at 3%. Given any two sequences:

$$W^a = (W_0, W_1^a, W_2^a, \ldots)$$

6. The fact that some numbers have two decimal expansions, like 0.4999... versus 0.5000..., may be resolved in any manner without effect on the analysis; since such numbers occur with zero probability.

and

$$W^b = (W_0, W_1^b, W_2^b, \ldots)$$

we may assume that the player either prefers W^a to W^b, or W^b to W^a or is indifferent. Further, we may assume that given a choice between any two probability distributions among sequences of wealth

$$\Pr_A(W_0, W_1, W_2, \ldots)$$

versus

$$\Pr_B(W_0, W_1, W_2, \ldots)$$

he either prefers probability distribution A to B, or B to A, or is indifferent between the two probability distributions.

We shall not only assume that the player has such preferences, but also that he maximizes expected utility. In other words, we assume that he attaches a (finite) number

$$U(W_0, W_1, W_2, \ldots)$$

to each sequence of wealths, and chooses among alternate strategies so as to maximize EU.

The only additional assumption we make about the utility function $U(\ldots)$, is this:

If the sequence $W^a = (W_0, W_1^a, W_2^a, \ldots)$ eventually pulls even with, and then stays even with or ahead of the sequence

$$W^b = (W_0, W_1^b, W_2^b, \ldots)$$

then W^a is at least as good as W^b; i.e., if there exists a T such that

$$W_t^a \geqslant W_t^b \qquad \text{for } t \geqslant T \tag{16}$$

then $U(W^a) \geqslant U(W^b)$. This assumption expresses the basic notion that, in the sense that we have used the terms throughout this controversy, if player A eventually gets and stays ahead of player B (or at least stays even with him) then player A has done at least as well as player B "in the long run".

At first it may seem appropriate to make a stronger assumption that if W_t^a eventually pulls ahead of W_t^b, and stays ahead, then the sequence W^a is preferable to the sequence W^b. In other words, if there is a T such that

$$W_t^a > W_t^b \qquad \text{for } t \geqslant T \tag{16a}$$

then

$$U(W^a) > U(W^b).$$

As shown in the footnote[7], this stronger assumption is too strong in that no utility function $U(W_0, W_1, W_2, \ldots)$ can have this property. Utility functions can however have the weaker requirement in (16).

The analysis of unending games is particularly easy if we assume that the same opportunities are available at each move, and that we only consider strategies which select the same probability distribution of returns each period. We shall make these assumptions at this point. Later we will summarize more general results derived in the appendix to this paper.

Without further loss of generality we will confine our discussion to just two strategies, namely, MEL and any other strategy, and will consider when the expected utility supplied by MEL is at least as great as that supplied by the other strategy. Letting W_t^L and W_t^0 be the wealth at t for a particular play of the game using MEL or the other strategy, respectively,

$$U(W_0, W_1^L, W_2^L, \ldots) \geqslant U(W_0, W_1^0, W_2^0, \ldots) \tag{17}$$

is implied if there is a T such that

$$W_t^L \geqslant W_t^0 \qquad \text{for } t \geqslant T. \tag{18}$$

7. If U orders all sequences $W=(W_0, W_1, W_2, \ldots)$ in a manner consistent with (16a), then in particular it orders sequences of the form

$$\begin{cases} W_0 - \text{given} \\ W_1 - \text{any positive number} \\ W_t = (1+\alpha)\cdot W_{t-1} \qquad \text{for } t \geqslant 2;\ \alpha > -1. \end{cases} \tag{N.1}$$

Since this family of sequences depends only on W_1 and α, we may here write

$$V(W_1, \alpha) = U(W_0, W_1, W_2, \ldots). \tag{N.2}$$

Then (16a) requires

$$V(W_1^A, \alpha^A) > V(W_1^B, \alpha^B) \qquad \text{if either } \alpha^A > \alpha^B \text{ or}$$

$$\alpha^A = \alpha^B \qquad \text{and} \qquad W_1^A > W_1^B. \tag{N.3}$$

For any α let

$$U_{\text{low}}(\alpha) = GLB\ V(W_1, \alpha) \tag{N.4}$$

$$U^{\text{hi}}(\alpha) = LUB\ V(W_1, \alpha).$$

Then (N.3) implies

$$U_{\text{low}}(\alpha) < U^{\text{hi}}(\alpha) \qquad \text{for every } \alpha \tag{N.5a}$$

as well as

$$U_{\text{low}}(\alpha^A) > U^{\text{hi}}(\alpha^B) \qquad \text{if } \alpha^A > \alpha^B. \tag{N.5b}$$

But (N.5b) implies that we can have $U_{\text{low}}(\alpha) < U^{\text{hi}}(\alpha)$ for at most a countable number of values of α, since at most a countable number of values of α can have $U^{\text{hi}}(\alpha) - U_{\text{low}}(\alpha) > 1/N$ for $N = 1, 2, 3, \ldots$. But this contradicts (N.5a).

Equation (2) implies that we have $W_t^L \geqslant W_t^0$ if and only if

$$\frac{1}{t}\sum_{i=1}^{t}\log(1+r_i^L) \geqslant \frac{1}{t}\sum_{i=1}^{t}\log(1+r_i^0). \tag{19}$$

Thus (18) will hold in any play of the game in which there exists a T such that

$$\frac{1}{t}\sum_{i=1}^{t}\log(1+r_i^L) > \frac{1}{t}\sum_{i=1}^{t}\log(1+r_i^0) \qquad \text{for all } t > T. \tag{20}$$

Or, if we let

$$y_i = \log(1+r_i) \tag{21}$$

(20) may be written as

$$\frac{1}{t}\sum_{i=1}^{t} y_i^L > \frac{1}{t}\sum_{i=1}^{t} y_i^0 \qquad \text{for } t > T. \tag{22}$$

Under the present simplified assumptions

$$Ey^L > Ey^0 \tag{23}$$

by definition of MEL. But for random variables $y_1, y_2, \ldots$ with identical distributions and with (finite) expected value μ, we have

$$\lim_{t\to\infty}\frac{1}{t}\sum_{i=1}^{t}(y_i-\mu)=0 \tag{24}$$

except for a set of probability measure zero. In other words

$$\lim_{t\to\infty}\frac{1}{t}\sum_{i=1}^{t} y_i=\mu \tag{25}$$

except for a set of sequences which have (in total) zero probability of occurrence (c.f. the strong law of large numbers in [4] or [5]). But (23) and (25) imply (as a simple corollary of the definition of the limit of sequence) that there exists T such that

$$\frac{1}{t}\sum_{i=1}^{t} y_i^L > \frac{1}{t}\sum_{i=1}^{t} y_i^0 \qquad \text{for } t \geqslant T \tag{26}$$

except on a set of probability zero; hence (17) holds except on a set of measure zero. Since

$$EU = \int U(W_0, W_1, \ldots)\, dP(W_0, W_1, \ldots) \tag{27}$$

is not affected by arbitrarily changing the value of U on a set of measure zero, we have

$$EU(W_0, W_1^L, W_2^L, \ldots) \geqslant EU(W_0, W_1^0, W_2^0, \ldots). \tag{28}$$

Thus, given our simplifying assumption of an unchanging probability distribution of returns for a given strategy, the superiority of MEL follows quite generally.

The case in which opportunities change from period to period and, whether or not opportunities change, strategies may select different distributions at different times, is treated in the appendix. It is shown there that if a certain continuity condition holds, then MEL is optimal quite generally. If this continuity condition does not hold, however, then there can exist games for which MEL is not optimal.

In this respect the results for the unending game are similar to those for the sequence of games with constant $V(g)$. In the latter case we found that MEL was asymptotically optimal for the sequence of games if $V(g)$ was continuous, but could fail to be so if $V(g)$ was discontinuous. In the case of the unending game, the theorem is not concerned with asymptotic optimality in a sequence of games, but optimality for a single game. Given a particular continuity condition, MEL is the optimum strategy.

V. Conclusions

The analysis of investment for the long run in terms of the weak law of large numbers, Breiman's strong law analysis, and the utility analysis of unending games presented here each imply the superiority of MEL under broad assumptions for the hypothetical investor of [1], [2], [7], [8], [9], [10]. The acceptance or rejection of a similar conclusion for the sequence-of-games formalization depends on the definition of asymptotic optimality. For example, if constant $V(g)$ rather than constant $U(W_T)$ is assumed, as this writer believed plausible on *a priori* grounds, then the conclusion of the asymptotic analysis is approximately the same (even in terms of where MEL fails) as those of the unending game.

I conclude, therefore, that a portfolio analyst should not be faulted for warning an investor against choosing E, V efficient portfolios with higher E and V but smaller $E\log(1+R)$, perhaps not even presenting that part of the E, V curve which lies above the point with approximate maximum $E\log(1+R)$, on the grounds that such higher E, V combinations have greater variability in the short run and less "return in the long run".

Appendix

Using the notation of footnote 7, we will show that if $U_{\text{low}}(\alpha) = U^{\text{hi}}(\alpha)$ for all α then MEL is an optimum strategy quite generally; whereas, if $U^{\text{hi}}(\alpha) > U_{\text{low}}(\alpha)$ for some α_0, then a game can be constructed in which MEL is not optimum. $U_{\text{low}}(\alpha) = U^{\text{hi}}(\alpha)$ for all α is the "continuity condition" referred to in the text.

For $v = L$ or 0, indicating the MEL strategy or some other strategy, respectively, we define

$$y_t^v = L_t^v + u_t^v \tag{29}$$

where

$$L_t^v = E\{y_t^v \mid L_1^v, L_2^v, \ldots, L_{t-1}^v, u_1^v, u_2^v, \ldots, u_{t-1}^v\} \tag{30}$$

is the expected value of y_t^v given the events prior to time t. From this follows

$$E\{u_t^v \mid L_1^v, \ldots, u_{t:1}^v\} = 0. \tag{31}$$

The u_t^v (for a given v) are thus what Doob [4] refers to as "orthogonal" random variables, and Feller [5] calls "completely fair" random variables. Therefore, writing var for variance,

$$\sum_{n=1}^{\infty} \frac{\operatorname{var}(u_n^v)}{n^2} < \infty \tag{32}$$

implies

$$\frac{1}{n}\sum_{t=1}^{n} u_t^v \qquad \text{converges to 0 almost always.} \tag{33}$$

(In particular, (32) holds if the var(u_n^v) are bounded.) In addition to now assuming condition (32) we will also assume that the game is such that

$$\lim_{n\to\infty} \frac{1}{n}\sum_{i=1}^{n} L_i^L \qquad \text{exists almost always.} \tag{34}$$

This is the case, for example, when the same distributions are available each time, whether or not "the other" strategy uses a constant distribution. Since $L_i^L \geqslant L_i^0$ always, we have

$$\lim_n \frac{1}{n}\sum_{i=1}^{n} L_i^L = \limsup_n \frac{1}{n}\sum_{i=1}^{n} L_i^L \geqslant \limsup_n \frac{1}{n}\sum_{i=1}^{n} L_i^0 \qquad \text{always.} \tag{35}$$

Thus when (32) holds we have

$$\lim \frac{1}{n}\sum_{i=1}^{n} y_i^L \geqslant \limsup \frac{1}{n}\sum_{i=1}^{n} y_i^0 \qquad \text{almost always.} \tag{36}$$

In general,

$$\alpha = \limsup \frac{1}{n}\sum y_i^v \qquad \text{implies} \qquad U^{\text{hi}}(\alpha) \geqslant U(W_0, W_1^v, W_2^v, \ldots); \tag{37}$$

(since there always exists another series $y_1^*, y_2^*, \ldots$ such that

$$\alpha = \lim \frac{1}{n}\sum_{i=1}^{n} y_i^*$$

and

$$\frac{1}{n}\sum_{i=1}^{n} y_i^* \geqslant \frac{1}{n}\sum_{i=1}^{n} y_i^v \quad \cdot \text{ for all } n;$$

hence

$$U^{\text{hi}}(\alpha) \geqslant U(W_0, W_1^*, W_2^*, \ldots) \geqslant U(U_0, W_1^v, W_2^v, \ldots)).$$

If we now add to the assumptions expressed in equations (32) and (34), the assumption that $U^{\text{hi}}(\alpha) = U_{\text{low}}(\alpha)$ for all α, we get directly from (36) and (37) that

$$EU^L \geqslant EU^0.$$

Conversely, the following is an example in which $U^{\text{hi}}(\alpha_0) > U_{\text{low}}(\alpha_0)$ and which MEL is not optimum: let $W_0 = 1$ and suppose that for some fixed positive α we have $U(1, 0.5, 0.5(1+\alpha), 0.5(1+\alpha)^2, \ldots)$ equals

$$U\big(1, (1+\alpha), (1+\alpha)^2, (1+\alpha)^3, \ldots\big) < U\big(1, 1.5, 1.5(1+\alpha), 1.5(1+\alpha)^2, \ldots\big).$$

With such a U-function it would be better to take a 50-50 chance of $W_1 = 0.5$ or 1.5 followed by $W_t = (1+\alpha)W_{t-1}, t \geqslant 2$, rather than have $W_t = (1+\alpha)\cdot W_{t-1}$ with certainty for $t \geqslant 1, \ldots$.

While the above shows that MEL can fail to be optimal when $U^{\text{hi}}(\alpha) > U_{\text{low}}(\alpha)$ for some α, recall that we can have $U^{\text{hi}}(\alpha) > U_{\text{low}}(\alpha)$ for at most a countable number of values of α. Thus MEL is optimal in a game in which

$$\alpha = \lim_n \frac{1}{n} \sum_{i=1}^{n} L_i^L$$

has a continuous distribution, or in which α has a discrete or mixed distribution but in which none of the points of discontinuity of the cumulative probability distribution of α have $U^{\text{hi}}(\alpha) > U_{\text{low}}(\alpha)$.

REFERENCES

1. Leo Breiman. "Investment Policies for Expanding Businesses Optimal in a Long Run Sense," *Naval Research Logistics Quarterly*, 7:4, 1960, pp. 647–651.
2. ———. "Optimal Gambling Systems for Favorable Games," *Fourth Berkeley Symposium on Probability and Statistics, I*, 1961, pp. 65–78.
3. Daniel Bernoulli. "Exposition of a New Theory on the Measurement of Risk," *Econometrica*, XXII, January 1954, pp. 23–63. Translated by Louise Sommer—original 1738.
4. J. L. Doob. *Stochastic Processes*, John Wiley and Sons, New York, 1953.
5. William Feller. *An Introduction to Probabality Theory and Its Applications*, Volume II, John Wiley and Sons, New York, 1966.
6. M. B. Goldman. "A Negative Report on the 'Near Optimality' of the Max-Expected-Log Policy As Applied to Bounded Utilities for Long Lived Programs." *Journal of Financial Economics*, Vol. 1, No. 1, May 1974.
7. J. L. Kelly, Jr. "A New Interpretation of Information Rate," *Bell System Technical Journal*, pp. 917–926, 1956.
8. H. A. Latané. "Rational Decision Making in Portfolio Management," Ph.D. dissertation, University of North Carolina, 1957.
9. ———. "Criteria for Choice Among Risky Ventures," *Journal of Political Economy*, April 1959.
10. H. M. Markowitz. *Portfolio Selection: Efficient Diversification of Investments*, John Wiley and Sons, New York, 1959; Yale University Press, 1972.
11. ———. "Investment for the Long Run," Rodney L. White Center for Financial Research Working Paper no. 20-72 (University of Pennsylvania) 1972.

12. R. C. Merton and P. A. Samuelson. "Fallacy of the Log-Normal Approximation to Optimal Portfolio Decision-Making Over Many Periods," *Journal of Financial Economics*, Volume 1 No. 1, May 1974.
13. Karl Menger. "Das Unsicherheitsmoment in der Wertlehre. Betrachtungen im Anschluss an das sogenannte Petersburger Spiel," *Zeitschrift für Nationalökonomie*, Vol. 5, 1934. Translated in *Essays in Mathematical Economics in Honor of Oskar Morgenstern*, M. Shubik ed., Princeton University Press, 1967.
14. P. A. Samuelson. "Risk and Uncertainty: A Fallacy of Large Numbers," *Scientia*, 6th Series, 57th year, April–May 1963.
15. ———. "Lifetime Portfolio Selection by Dynamic Stochastic Programming," *Review of Economics and Statistics*, August 1969.
16. W. E. Yong and R. M. Trent. "Geometric Mean Approximation of Individual Security and Portfolio Performance," *Journal of Financial and Quantitative Analysis*, June 1969.
17. John von Neuman and Oskar Morgenstern. *Theory of Games and Economic Behavior*, Princeton University Press, 1944. John Wiley and Sons, 1967.

36

Understanding the Kelly Criterion*

Edward O. Thorp

In January 1961, I spoke at the annual meeting of the American Mathematical Society on "Fortune's Formula: The Game of Blackjack". This announced the discovery of favorable card counting systems for blackjack. My 1962 book *Beat the Dealer* explained the detailed theory and practice. The 'optimal' way to bet in favorable situations was an important feature. In *Beat the Dealer*, I called this, naturally enough, "The Kelly gambling system", since I learned about it from the 1956 paper by John L. Kelly (Claude Shannon, who refereed the Kelly paper, brought it to my attention in November of 1960). I have continued to use it successfully in gambling and in investing. Since 1966, I've called it "the Kelly Criterion". The rising tide of theory about and practical use of the Kelly Criterion by several leading money managers received further impetus from William Poundstone's readable book about the Kelly Criterion, *Fortune's Formula*. (As this title came from that of my 1961 talk, I was asked to approve the use of the title). At a value investor's conference held in Los Angeles in May, 2007, my son reported that 'everyone' said they were using the Kelly Criterion.

The Kelly Criterion is simple: bet or invest so as to maximize (after each bet) the expected growth rate of capital, which is equivalent to maximizing the expected value of the logarithm of wealth; but the details can be mathematically subtle. Since they're not covered in Poundstone (2005), you may wish to refer to my article, Thorp (2006), and other papers in this volume. Also some services such as Morningstar and Motley Fool have recommended it. These sources use the rule: "optimal Kelly bet equals edge/odds" that applies only to the very special case of a two-valued payoff.

Hedge fund manager, Mohnish Pabrai (2007), gives examples of the use of the Kelly Criterion for investment situations (Pabrai won the bidding for the 2008 lunch with Warren Buffett, paying over $600,000). Consider his investment in Stewart Enterprises (Pabrai, 2007: 108-115), his analysis gave what he believed to be a list of worst case scenarios and payoffs over the next 24 months which I summarize in Table 1.

The expected growth rate of capital $g(f)$ if we bet a fraction f of our net

*Reprinted revised from two columns from the series *A Mathematician on Wall Street* in *Wilmott* Magazine, May and September 2008. Edited by Bill Ziemba.

Table 1 Stewart enterprises, payoff within 24 months.

Probability	Return
$p_1 = 8.80$	$R_1 > 100\%$
$p_2 = 0.19$	$R_2 > 0\%$
$p_3 = 0.01$	$R_3 = -100\%$

Sum = 1.00.

worth is

$$g(f) = \sum_{i=1}^{3} p_i \ln(1 + R_i f) \tag{1}$$

where ln means the logarithm to the base e. When we use Table 1 to insert the p_i values, replacing the R_i by their lower bounds gives the conservative estimate

$$g(f) = 0.80 \ln(1 + f) + 0.01 \ln(1 - f)\,. \tag{2}$$

Setting $g'(f) = 0$ and solving gives the optimal Kelly fraction $f^* = 0.975$ noted by Pabrai. Not having heard of the Kelly Criterion in 2000, Pabrai only bet 10% of his fund on Stewart. Would he have bet more, or less, if he had then known about Kelly's Criterion? Would I have? Not necessarily. Here are some of the many reasons why:

(1) **Opportunity costs.** A simplistic example illustrates the idea. Suppose Pabrai's portfolio already had one investment which was statistically independent of Stewart and with the same payoff probabilities. Then, by symmetry, an optimal strategy is to invest in both equally. Call the optimal Kelly fraction for each f^*, then $2f^* < 1$ since $2f^* = 1$ has a positive probability of total loss, which Kelly always avoids, so $f^* < 0.50$. The same reasoning for n such investments gives $f^* < 1/n$. Hence, we need to know the other investments currently in the portfolio, any candidates for new investments, and their (joint) properties, in order to find the Kelly optimal fraction for each new investment, along with possible revisions for existing investments. Formally, we solve the nonlinear programming problem: maximize the expected logarithm of final wealth subject to the various constraints on the asset weights (see the papers in Section 6 of this volume for examples).

Pabrai's discussion (e.g. pp. 78–81) of Buffett's concentrated bets gives considerable evidence that Buffet thinks like a Kelly investor, citing Buffett bets of 25% to 40% of his net worth on single situations. Since $f^* < 1$ is necessary to avoid total loss, Buffett must be betting more than 0.25 to 0.40 of f^8 in these cases. The opportunity cost principle suggests it must be higher, perhaps much higher. Here's what Buffett himself says, as reported in http://undergroundvalue.blogspot.com/2008/02/notes-from-buffett-meeting-2152008_23.html, notes from a Q & A session with business students:

Emory:

With the popularity of "Fortune's Formula" and the Kelly Criterion, there seems to be a lot of debate in the value community regarding diversification vs. concentration. I know where you side in that discussion, but was curious if you could tell us more about your process for position sizing or averaging down.

Buffett:

I have 2 views on diversification. **If you are a professional and have confidence, then I would advocate lots of concentration.** *For everyone else, if it's not your game, participate in total diversification. So this means that professionals use Kelly and amateurs better off with index funds following the capital asset pricing model.*

If it's your game, diversification doesn't make sense. It's crazy to put money in your 20th choice rather than your 1st choice. If you have LeBron James on your team, don't take him out of the game just to make room for some else.

Charlie and I operated mostly with 5 positions. *If I were running 50, 100, 200 million, I would have 80% in 5 positions, with 25% for the largest. In 1964, I found a position I was willing to go heavier into, up to 40%. I told investors they could pull their money out. None did. The position was American Express after the Salad Oil Scandal. In 1951 I put the bulk of my net worth into GEICO. With the spread between the on-the-run versus off-the-run 30 year Treasury bonds, I would have been willing to put 75% of my portfolio into it. There were various times I would have gone up to 75%, even in the past few years. If it's your game and you really know your business, you can load up.*

This supports the assertion in Rachel and Bill Ziemba's 2007 book, that Buffett thinks like a Kelly investor when choosing the size of an investment. They discuss Kelly and investment scenarios at length.

Computing f^* without considering the available alternative investments is one of the most common oversights I've seen in the use of the Kelly Criterion. It is a dangerous error because it generally overestimates f^*.

(2) **Risk tolerance.** As discussed at length in Thorp (2006), "full Kelly" is too risky for the tastes of many, perhaps most, investors and using instead an $f = cf^*$, with fraction c where $0 < c < 1$ or "fractional Kelly" is much more to their liking. Full Kelly is characterized by drawdowns which are too large for the comfort of many investors.[1]

[1]Several papers in Section 3 in this volume, as do the following two papers in this section, discuss fractional Kelly strategies.

(3) **The "true" scenario** is worse than the supposedly conservative lower bound estimate. Then we are inadvertently betting more than f^* and, as discussed in Thorp (2006), we get more risk and less return, a strongly suboptimal result. Betting $f = cd^*$, $0 < c < 1$ gives some protection against this (see the graphs in Section 3, (MacLean, Ziemba and Blazenko (1992)).

(4) **Black swans.** As fellow Wilmott columnist Nassim Nicholas (Taleb 2007) has pointed out so eloquently in his bestseller *The Black Swan*, humans tend not to appreciate the effect of relatively infrequent unexpected high impact events. Failing to allow for these "black swans", scenarios often don't adequately consider the probabilities of large losses. These large loss probabilities may substantially reduce f^*. One approach to successfully model such black swans is to use a scenario optimization stochastic programming model.[2] For Kelly bets that simply means that you include such extreme scenarios and their consequences in the nonlinear programming optimization to compute the optimal asset weights. The f^* will be reduced by these negative events.

(5) **The "long run".** The Kelly Criterion's superior properties are asymptotic, appearing with increasing probability as time increases. For instance:

As time t tends to infinity the Kelly bettor's fortune will, with probability tending to 1, permanently surpass that of any bettor following an "essentially different" strategy.

The notion of "essentially different" has confounded some well known quants so I'll take time here to explore some of its subtleties. Consider for simplicity repeated tosses of a favorable coin, the outcome of the nth trial is X_n where $P(X_n = 1) = p > 1/2$ and $P(X_n = -1)$ is $1 - p = q > 0$. The $\{X_n\}$ are independent identically distributed random variables. The Kelly fraction is $f^* = p - q = E(X_n) > 0$. The Kelly strategy is to bet a fraction $f_n = f^*$ at each trial $n = 1, 2, \ldots$. Now consider a strategy which bets g_n, $n = 1, 2, \ldots$ at each trial with $g_n \neq f^*$ for some $n \leq N$ and $g_n = f^*$ thereafter. The $\{g_n\}$ strategy differs from Kelly on at least one of the first N trials but copies it thereafter, but it does not differ infinitely often. There is a positive probability that $\{g_n\}$ is ahead of Kelly at time N, hence ahead for all $n \geq N$. For example consider the sequence of the first N outcomes such that $X_n = 1$ if $g_n > f^*$ and $X_n = -1$ if $g_n \leq f^*$. Then for this specific sequence, which has probability $\geq q^N$, $\{g_n\}$ gains more than Kelly for each $n \leq N$ where $g_n \neq f^*$, hence exceeds Kelly for all $n \geq N$.

What if instead in this coin tossing example we require that $g_n \neq f^*$ for infinitely many n? This question arose indirectly about 15 years ago in the newsletter *Blackjack Forum* when a well known anti Kellyite, John Leib, challenged a well known blackjack expert with (approximately) this proposition bet: Leib would produce a strategy which differed from Kelly at every trial but would (with probability as

[2]There you assume the possibility of an event, specifying its consequences but not what it is. See Geyer and Ziemba (2008) for the application to the Siemens Austria Pension Fund. Correlations change as the scenario sets move from normal conditions to volatile to crash which include the black swans. See also Ziemba (2003) for additional applications of this approach.

close to 1 as you wish), after a finite number of trials, get ahead of Kelly and stay ahead forever. When I read the challenge I immediately saw how Leib could win the bet.

Leib's Paradox: Assuming capital is infinitely divisible,footnoteThe infinite divisibility of capital is a minor assumption and can be dealt with as needed in examples where there is a minimum monetary unit by choosing a sufficiently large starting capital. then given $\varepsilon > 0$ there is an $N > 0$ and a sequence $\{f_n\}$ with $f_n \neq f^*$ for all n, such that $P(V_n^* < V_n \text{ for all } n \geq N) > 1 - varepsilon$ where $V_n = \prod_{i=1}^n (1 + f_i X_i)$ and $V_n^* = \prod_{i=1}^n (1 + f^* X_i)$. Furthermore, there is a $b > 1$ such that $P(V_n/V_n^* \geq b, n \geq N) > 1 - varepsilon$ and $P(V_n - V_n^* \to \infty) > 1 - varepsilon$. That is, for some N there is a non Kelly sequence that beats Kelly "infinitely badly" with probability $1 - \varepsilon$ for all $n \geq N$.

Proof. The proof has two parts. First we want to establish the assertion for $n = N$. Second we show that once we have an $\{f_n, n \leq N\}$ that is ahead of Kelly at $n = N$, we can construct $\{f_n \neq f^*, n > N\}$ to stay ahead.

To see the second part, suppose $V_M > V_N^*$. Then $V_N \geq a + bV_N^*$ for some $a > 0$, $b > 1$. For instance, $V_N >_N^* \geq C > 0$ since there are only a finite number of sequences of outcomes in the first N trials, hence, only a finite number with $V_N > V_N^*$. So:

$$V_N \geq c + V_N^* \geq c/2 + [(c/2) + V_N^*] = c/2 + [d \operatorname{Max} V_N^* + V_N^*] \geq c/2 + (d+1)V_N^*$$

where $d \operatorname{Max} V_N^* = c/2$ defines $d > 0$ and $\operatorname{Max} V_N^*$ is over all sequences of the first N trials such that $V_N > V_N^*$. Setting $c/2 = a > 0$ and $d + 1 = b > 1$ suffices. Once we have $V_N \geq a + bV_N^*$ we can, for bookkeeping purposes, partition our capital into two parts: a and bV_N^*. For $n > N$ we bet $f_n = f^*$ from bV_N^* and an additional amount $a/2^n$ from the a part, for a total which is generally unequal to f^* of our capital. If by chance for some n the total equals f^* of our total capital we simply revise $a/2^n$ to $a/3^n$ for that n. The portion bV_N^* will become bV_N^* for $n > N$ and the portion a will never be exhausted so we have $V_n > bV_n^*$ for all $n > N$. Hence, since $P(V_N^* \to \infty) = 1$, we have $P(V_n/V_N^* \geq b) = 1$ from which it follows that $P(V_n - V_N^* \to \infty) = 1$.

To prove the first part, we show how to get ahead of Kelly with probability $1 - \varepsilon$ within a finite number of trials. The idea is to begin by betting less than Kelly by a very small amount. If the first outcome is a loss, then we have more than Kelly and use the strategy from the proof of the second part to stay ahead. If the first outcome is a win, we're behind Kelly and now underbet on the second trial by enough so that a loss on the second trial will put us ahead of Kelly. We continue this strategy until either there is a loss and we are ahead of Kelly or until even betting 0 is not enough to surpass Kelly after a loss. Given any N, if our initial underbet is small enough, we can continue this strategy for up to N trials. The probability of the strategy failing is p^N, $1/2 < p < 1$ Hence, given $\varepsilon > 0$, we can choose N such that $p^N < \varepsilon$ and the strategy therefore succeeds on or before trial N with probability $1 - p^N > 1 - \varepsilon$.

More precisely: suppose the first n trials are wins and we have bet a fraction $f^* - a_i$ with $a_i > 0$, $i = 1, \ldots, n$, on the ith trial. Then:

$$\frac{V_n}{V_N^*} = \frac{(1 + f^* - a_1) \cdots (1 + f^* - a_n)}{(1 + f^*) \cdots (1 + f^*)}$$

$$= \left(1 - \frac{a_1}{1 + f^*}\right) \cdots \left(1 - \frac{a_n}{1 + f^*}\right) > (1 - a_1) \cdots (1 - a_n) > 1 - (a_1 + \cdots + a_n)$$

where the last inequality is proven easily by induction. Letting $a_1 + \cdots + a_n = a$, so $V_n/V_n^* > 1 - a$, what betting fraction $f^* - b$ will put us ahead of Kelly if the next trial is a loss? A sufficient condition is

$$\frac{V_{n+1}}{V_{n+1}^*} = \frac{V_n(1 - f^* + b)}{V_n^*(1 - f^*)} > (1-a)\left\{1 + \frac{b}{1 - f^*}\right\} > (1-a)(1+b) \geq 1 \quad \text{or} \quad b \geq \frac{a}{1 - a}$$

provided $b \leq f^*$ and $0 < a < 1$. If $a \leq 1/2$ then $b = 2a$ suffices. Proceeding recursively, we have these conditions on the a_i: choose $a_1 > 0$. Then $a_{n+1} = 2(a_1 + \cdots + a_n)$, $n = 1, 2, \ldots$ provided all the $a_n \leq 1/2$. Letting $f(x) = a_1 x + a_2 x^2 + \cdots$ we get the equation

$$\begin{aligned} f(x) - a_1 x &= 2x f(x)(1 + x + x^2 + \cdots) \\ &= 2x f(x)/(1 - x) \end{aligned}$$

whose solution is $f(x) = a_1\{x + 2\sum_{n=2}^{\infty} 3^{n-2} x^n\}$ from which $a_n = 2a_1 3^{n-2}$ if $n \geq 2$. Then given $\varepsilon > 0$ and an N such that $p^N < \varepsilon$ it suffices to choose a_1 so that $a_N = 2a_1 3^{N-2} \leq \min(f^*, 1/2)$. □

Although Leib did not have the mathematical background to give such a proof he understood the idea and indicated this sort of procedure.

So far we've seen that all sequences which differ from Kelly for only a finite number of trials, and some sequences which differ infinitely often (even always), are not essentially different. How can we tell, then, if a betting sequence is essentially different than Kelly? Going to a more general setting than coin tossing, assume now for simplicity that the payoff random variables X_i are independent and bounded below but not necessarily identically distributed.

At this point we come to an important distinction. In financial applications, one commonly assumes that the f_i are constants that are dependent only on the current period payoff random variable (or variables). Such "myopic strategies" might arise for instance, by selecting a utility function and maximizing expected utility to determine the amount to bet. However, for gambling systems, the amount depend on previous outcomes, i.e., $f_n = f_n(X_1, X_2, \ldots, X_{n-1})$, just as it does in the Leib example. As Professor Stewart Ethier pointed out, our discussion of "essentially different" is for the constant f_i case. For a more general case, including the Leib example and many of the classical gambling systems, I recommend Ethier's 2010 book on the mathematics of gambling.

We assume $R(X_i) > 0$ for all X_i from which it follows that $f_i^* > 0$ for all i. As before, $V_n = \prod_{i=1}^{n}(1 + f_i X_i)$ and $V_n^* = \prod_{i=1}^{n}(1 + f_i^* X_i)$ from which $\ln V_n = \sum_{i=1}^{n} \ln(1 + f_i X_i)$ and $\ln V_n^* = \sum_{i=1}^{n} \ln(1 + f_i^* X_i)$. Note from the definition f^* that $E \ln(1 + f_i^* X_i) \geq E \ln(1 + f_i X_i)$, where E denotes the expected value, with equality if and only if $f_i^* = f_i$. Hence:

$$E \ln(V_n^*/V_n) = E \sum_{i=1}^{n} \{\ln(1 + f_i^* X_i) - \ln(1 + f_i X_i)\} = \sum_{i=1}^{n} a_i$$

where $a_i \geq 0$ and $a_i = 0$ if and only if $f_i^* = f_i$. This series of non-negative terms either increases to infinity or to a positive limit M. We say $\{f_i\}$ is essentially different from $\{f_i^*\}$ if and only if $\sum_{i=1}^{n} a_i$ tends to infinity as n increases. Otherwise, $\{f_i\}$ is not essentially different from $\{f_i^*\}$. The basic idea here can be applied to more general settings.

(6) **Given a large fixed goal**, e.g., to multiply your capital by 100, or 1000, the expected time for the Kelly investor to get there tends to be least.

Is a wealth multiple of 100 or 1000 realistic? Indeed. In the 51½ years from 1956 to mid 2007, Warren Buffett has increased his wealth to about $\$5 \times 10^{10}$. If he had $\$2.5 \times 10^4$ in 1956, that's a multiple of 2×10^6. We know he had about $\$2.5 \times 10^7$ in 1969 so his multiple over these 38 years is about 2×10^3. Even my own efforts, as a late starter on a much smaller scale, have multiplied capital by more than 2×10^4 over the 41 years from 1967 to early 2007. I know many investors and hedge fund managers who have achieved such multiples. One of the best is Jim Simons, who recently retired from running the Renaissance Medallion Fund. His record to 2005 is analyzed in Section 6 of this book.

The caveat here is that an investor or bettor many not choose to make, or be able to make, enough Kelly bets for the probability to be "high enough" for these asymptotic properties to prevail, i.e., he doesn't have enough opportunities to make it into this "long run". Below I explore investors for which Kelly or fractional Kelly may be a more or less appropriate approach. An important consideration will be the investor's expected future wealth multiple.

Using Kelly Optimization at PIMCO

During a recent interview in the *Wall Street Journal* (March 22–23, 2008), Bill Gross and I discussed turbulence in the markets, hedge funds, and risk management. Bill considered the question of risk management after he read *Beat the Dealer* in 1966. That summer he was off to Las Vegas to beat blackjack. Just as I did some years earlier, he sized his bets in proportion to his advantage, following the Kelly Criterion as described in *Beat the Dealer*, and ran his \$200 bankroll up to \$10,000 over the summer. Bill has gone from managing risk for his tiny bankroll to managing risk for Pacific Investment Management Company's (PIMCO) investment pool of almost \$1 trillion.[3] He still applies lessons he learned from the Kelly Criterion. As Bill said: "Here at PIMCO it doesn't matter how much you have, whether it's \$200 or \$1 trillion. ... Professional blackjack is being played in this trading room from the standpoint of risk management and that's a big part of our success".

The Kelly Criterion applies to multiperiod investing and we can get some insights by comparing it with Markowitzs standard portfolio theory for single period investing.

Compound Growth and Mean-Variance Optimality

Nobel Prize winner Harry Markowitz introduced the idea of mean-variance optimal portfolios. This class is defined by the property that, among the set of admissible portfolios, no other portfolio has both higher mean return and lower variance. The set of such portfolios as you vary return or variance is known as the efficient frontier. The concept is a cornerstone of modern portfolio theory, and the mean and variance refer to one period arithmetic returns.[4] In contrast, the Kelly Criterion is used to maximize the long term compound rate of growth, a multiperiod problem. It seems natural, then to ask the question: is there an analog to the Markowitz efficient frontier for multiperiod growth rates, i.e., are there portfolios such that no other portfolio has both a higher expected growth rate and a lower variance in the growth rate? We'll call the set of such portfolios the compound growth mean-variance efficient frontier.

Let's explore this in the simple setting of repeated independent identically distributed returns per unit invested, where the payoff random variables are $\{X_i : i = 1, \ldots, n\}$ with $E(X_i) > 0$ so the "game" is favorable, and where the non-negative fractions bet at each trial, specified in advance, are $\{f_i : i = 1, \ldots, n\}$. To keep the math simpler, we also assume that the X_i have a finite number of distinct values. After n trials the compound, growth rate per period is $G(\{f_i\}) = \frac{1}{n}\sum_{i=1}^{n} \log(1+f_iX_i)$ and the expected growth rate $g(\{g_i\}) = E[G(\{f_i\})] = \frac{1}{n}\sum_{i=1}^{n} E\log(1+f_iX_i) = \frac{1}{n}\sum_{i=1}^{n} E\log(1+f_iX) \leq E\log(1+\bar{f}X)$. The last step follows from the (strict) concavity of the log function, where as X has the common distribution of the X_i, we define $\bar{f} = \frac{1}{n}\sum_{i=1}^{N} f_i$ and we have equality if and only if $f_i = \bar{f}$ for all i. Therefore, if some f_i differ from $\bar{f}$, we have $g(\{f_i\}) < g(\{\bar{f}\})$. This tells us that betting the same fixed fraction always produces a higher expected growth rate than betting a varying fraction with the same average value. Note that whatever $\bar{f}$ turns out to be, it can always be written as $\bar{f} = cf^*$, a fraction c of the Kelly fraction.

[3]PIMCO is widely regarded as the top bond trading operation in the world.

[4]A comprehensive survey of mean-variance theory is in Markowitz and Van Dijk (2006).

Now consider the variance of $G(\{f_i\})$. If is a random variable with:

$$P(X = a) = p, \quad P(x = -1) = q \quad \text{and} \quad a > 0$$

$$\text{then} \quad \text{Var}[\ln(1 + fX)] = pq\left[\ln\left(\frac{1+af}{1-f}\right)\right]^2.$$

(Compare Thorp, 2006, Section 3.1).

Note: the change of variable $f = bh$, $b > 0$, shows the results apply to any two valued random variable. We chose $b = 1$ for convenience.)

A calculation shows that the second derivative with respect to f is strictly positive for $0 < f < 1$ so $\text{Var}[\ln(1 + fX)]$ is strictly convex in f. It follows that:

$$\text{Var}[G(\{f_i\})] = \frac{1}{n}\sum_{i=1}^{n}\text{Var}[\log(1+f_iX_i)] = \frac{1}{n}\sum_{i=1}^{n}\text{Var}[\log(1+f_iX)] \geq \text{Var}[\log(1+\bar{f}X)]$$

with equality if and only if $f_i = \bar{f}$ for all i. Since every admissible strategy is therefore "dominated" by a fractional Kelly strategy, it follows that the mean-variance efficient frontier for compound growth is a subset of the fractional Kelly strategies. If we now examine the set of fractional Kelly strategies $\{f\} = \{cf^*\}$, we see that for $0 \leq c \leq 1$, both the mean and the variance increase as c increases but for $c \geq 1$, the mean decreases and the variance increases as c increases. Consequently $\{f^*\}$ dominates the strategies for which $c > 1$ and they are not part of the efficient frontier. No fractional Kelly strategy is dominated for $0 \leq c \leq 1$. We have established *in this limited* setting:

Theorem. *For repeated independent trials of a two valued random variable, the mean-variance efficient frontier for compound growth* over a finite number of trials *consists precisely of the fractional Kelly strategies* $\{cf^* : 0 \leq c \leq 1\}$.

So, given any admissible strategy, there is a fractional Kelly strategy with $0 \leq c \leq 1$, which has a growth rate that is no lower and a variance of the growth rate that is no higher. The fractional Kelly strategies in this instance are preferable in this sense to all the other admissible strategies, regardless of any utility function upon which they may be based. This deals with yet another objection to the fractional Kelly strategies, namely that there is a wide spread in the distribution of wealth levels as the number of periods increases. In fact, this eventually enormous dispersion is simply the magnifying effect of compound growth on small differences in growth rate and we have shown in the theorem that in the two outcome setting this dispersion is minimized by the fractional Kelly strategies. Note that in this simple setting, a one-period utility function will choose a constant $f_i = cf$ which will either be a fractional Kelly with $c \leq 1$ in the efficient frontier or will be too risky, with $c > 1$, and not be in the efficient frontier.

As a second example, suppose we have a lognormal diffusion process with instantaneous drift rate m and variance s^2 where as before the admissible strategies are to specify a set of fixed fractions $\{f_i\}$ for each of n unit time periods, $i = 1, \ldots, n$. Then, for a given f and unit time period $\operatorname{Var} G(f) = s^2 f^2$ as noted in (Thorp, 2006, eq. (7.3)). Over n periods $\operatorname{Var} G(\{f_i\}) = s^2 \sum_{i=1}^{n} f_i^2 \geq s^2 \sum_{i=1}^{n} \bar{f}^2$ with equality if and only if $f_i = \bar{f}$ for all i. This follows from the strict convexity of the function $h(x) = x^2$. So the theorem also is true in this setting. I don't currently know how generally the convexity of $\operatorname{Var}[\ln(1 + fX)]$ is true but whenever it is, and we also have $\operatorname{Var}[\ln(1 + fX)]$ increasing in f, then the compound growth mean variance efficient frontier is once again the set of fractional Kelly strategies with $0 \leq c \leq 1$. In email correspondence, Stewart Ethier subsequently showed that $\operatorname{Var}[\ln(1+fX)]$ need not be convex. Example (Ethier):

> Let X assume values -1, 0 and 100 with probabilities 0.5, 0.49 and 0.01, respectively. Then, on approximately the interval [0.019, 0.180] the second derivative of the variance is negative, hence the variance is strictly concave on that interval. The first derivative of $\operatorname{Var}[\ln(1+fX)]$ equals $2\operatorname{Cov}(\ln(1+fX), X/(1+fX))$, which is always nonnegative because the two functions of X in the covariance are increasing in X. Thus $\operatorname{Var}[\ln(1+fX)]$ is always increasing in f. The second derivative of $\operatorname{Var}[\ln(1+fX)]$ equals $2\operatorname{Var}(X/(1+fX))$-$2\operatorname{Cov}(\ln(1+fX)-1, X^2/(1+fX)^2)$. However, the covariance term sometimes exceeds the variance term.

Samuelson's Criticisms

The best known "opponent" of the Kelly Criterion is Nobel Prize winning economist, Paul Samuelson, who has written numerous polemics, both published and private, over the last 40 years. William Poundstone's book *Fortune's Formula* gives an extensive account with references. The gist of it seems to be:

(1) Some authors once made the error of claiming, or seeming to claim, that acting to maximize the expected growth rate (i.e., logarithmic utility) would approximately maximize the expected value of any other continuous concave utility (the "false corollary").

Response: Samuelson's point was correct but, to others as well as me, obvious the first time I saw the false claim. However, the fact that some writers made mistakes has no bearing on an objective evaluation of the merits of the criterion. So this is of no further relevance.

(2) In private correspondence to numerous people Samuelson has offered examples and calculations in which he demonstrates, with a two valued X ("stock") and three utilities, $H(W) = -1/W$, $K(W) = \log W$, and $T(W) = W^{1/2}$, that if any one who values his wealth with one of these utilities uses one of the other utilities to choose how much to invest then he will suffer a loss as measured with his own utility in each period and the sum of these losses will tend to infinity as the number of periods increases.

Response: Samuelson's computations are simply instances of the following general fact proven 30 years earlier by Thorp and Whitley (1972, 1974).[5]

Theorem 1. *Let U and V be utilities defined and differentiable on $(0, \infty)$ with $U'(x)$ and $V'(x)$ positive and strictly decreasing as x increases. Then if U and V are inequivalent, there is a one period investment setting such that U and V have distinct sets of optimal strategies. Furthermore, the investment setting may be chosen to consist only of cash and a two-valued random investment, in which case the optimal strategies are unique.*

Corollary 2. *If the utilities U and V have the same (sets of) optimal strategies for each finite sequence of investment settings, then U and V are equivalent.*

Two utilities U_1 and U_2 are equivalent if and only if there are constants a and b such that $U_2(x) = aU_1(x) + b(a > 0)$, otherwise $U - 1$ and U_2 are inequivalent.

Thus, no utility in the class described in the theorem either dominates or is dominated by any other member of the class.

Samuelson offers us utilities without any indication as to how we ought to choose among them, except perhaps for this hint. He says that he and an apparent majority of the investment community believe that maximizing $U(x) = -1/x$ explains the data better than maximizing $U(x) = \log x$. How is it related to fractional Kelly? Does this matter? Here are two examples showing that this utility can choose cf^* for any $0 < c < 1$, $c \neq 1/2$, depending on the setting:

For a favorable coin toss and $U(x) = -1/x$, we have $f^* = \mu/(\sqrt{p} + \sqrt{q})^2$ which increases from $\mu/2$ or half Kelly to μ or full Kelly as p increases from $1/2$ to 1, giving us the set $1/2 < c1$. On the other hand, if $P(X = A) = P(X = -1) = 1/2$ describe the returns and $A > 1$ so $\mu > 0$, $\mu = (A-1)/2$ and the Kelly $f^* = \mu/A$. For $U(x) = -1/x$ we find U maximized for $f = \{-2A \pm (4A^2 + A(A-1)^2)^{1/2}\}/A(A-1)$, which is asymptotic to $A^{-1/2}$ as A increases, compared to the Kelly f^*, which is asymptotic to $1/2$ as A increases, giving us the set $0 < c < 1/2$.

In the continuous case, the relation between c, $g(f)$ and $\sigma(G(f))$ is simple and the tradeoff between growth and spread in growth rate as we adjust between 0 and 1 is easy to compute and it's easy to visualize the correspondence between fractional Kelly and the compound growth mean-variance efficient frontier. This is not the case for these two examples so the fact that $U(x) = -1/x$ can choose any c, $0 < c < 1$, $c \neq 1/2$ doesn't necessarily make it undesirable.[6] I suggest that

[5]The first Thorp and Whitley paper is reprinted in this book in Section 4 where three of Samuelson's papers are reprinted and discussed in the introduction to that part of this book.

[6]MacLean, Ziemba and Li (2005) reprinted in Section 4 of this book, show that for lognormally distributed assets, a fractional Kelly strategy is uniquely related to the coefficient $\alpha < 0$ in the negative power utility function αw^α via the formula $c = 1/(1 - \alpha)$ so $1/2$ Kelly is $-1/w$. However, when assets are lognormal this is only an approximation and, as shown here, it can be a poor approximation.

a useful way to look at the problem for any specific example involving n period compound growth is to map the admissible portfolios into the $(\sigma(G\{f_i\}), g(\{f_i\}))$ plane, analogous to the Markowitz one period mapping into the (standard deviation, return) plane. Then examine the efficient frontier and decide what tradeoff of growth versus variability of growth you like. Professor Tom Cover points out that there is no need to invoke utilities. Adopting this point of view, we're simply interested in portfolios on the compound growth efficient frontier whether or not any of them happen to be generated by utilities. The Samuelson's preoccupation with utilities becomes irrelevant. The Kelly or maximum growth portfolio, which as it happens can be computed using the utility $U(x) = \log x$, has the distinction of being at the extreme high end of the efficient frontier.

For another perspective on Samuelson's objections, consider the three concepts: normative, descriptive and prescriptive. A normative utility or other recipe tells us what portfolio we "ought" to choose, such as "bet according to log utility to maximize your own good". Samuelson has indicated that he wants to stop people from being deceived by such a pitch. I completely agree with him on this point. My view is instead prescriptive: how to achieve an objective. If you know future payoff for certain and want to maximize your long term growth rate then Kelly does it. If, as is usually the case, you only have estimates of future payoffs and want to come close to maximizing your long term growth rate, then to avoid damage from inadvertently betting more than Kelly you need to back off from your estimate of full Kelly and consider a fractional Kelly strategy. In any case, you may not like the large drawdowns that occur with Kelly fractions over $1/2$ and may be well advised to choose lower values. The long term growth investor can construct the compound growth efficient frontier and choose his most desirable geometric growth Markowitz type combinations.

Samuelson also says that $U(x) = -1/x$ seems roughly consistent with the data. That is descriptive, i.e., an assertion about what people actually do. We don't argue with that claim — it's something to be determined by experimental economists and its correctness or lack thereof has no bearing on the prescriptive recipe for growth maximizing.

I met the economist Oscar Morgenstern (1902–1977), coauthor with John von Neumann of the great book, *The Theory of Games and Economic Behavior*, at his company, Mathematica; in Princeton, New Jersey, in November of 1967 and, when I outlined these views on normative, prescriptive and descriptive, he liked them so much that he asked if he could incorporate them into an article he was writing at the time. He also gave me an autographed copy of his book, *On the Accuracy of Economic Observations*, which has an honored place in my library today and which remains timely. (For instance, think about how the government has made successive revisions in the method of calculating inflation so as to produce lower numbers, thereby gaining political and budgetary benefits).

Proebsting's Paradox

Next, we look at a curious paradox. Recall that one property of the Kelly Criterion is that if capital is infinitely divisible, arbitrarily small bets are allowed, and the bettor can choose to bet only on favorable situations, then the Kelly bettor can never be ruined absolutely (capital equals zero) or asymptotically (capital tends to zero with positive probability). Here's an example that seems to flatly contradict this property. The Kelly bettor can make a series of favorable bets yet be (asymptotically) ruined! Here's the email discussion through which I learned of this.

> **From:** Todd Proebsting
> **Subject:** FW: incremental Kelly Criterion
>
> Dear Dr. Thorp,
>
> I have tried to digest much of your writings on applying the Kelly Criterion to gambling but I have found a simple question that is unaddressed. I hope you find it interesting:
>
> Suppose that you believe an event will occur with 50% probability and somebody offers you 2:1 odds. Kelly would tell you to bet 25% of your capital. Similarly, if you were offered 5:1 odds, Kelly would tell you to bet 40%. Now, suppose that these events occur in sequence. You are offered 2:1 odds, and you place a 25% bet. Then another party offers you 5:1 odds. I assume you should place an additional bet, but for what amount?
>
> If you have any guidance or references on this question, I would appreciate it.
>
> Thank you.

> **From:** Ed Thorp
> **To:** Todd Proebsting
> **Subject:** Fw: incremental Kelly Criterion Interesting.
>
> After the first bet the situation is:
>
> A win gives a wealth relative of $1 + 0.25 * 2$
> A loss gives a wealth relative of $1 - 0.25$
> Now bet an additional fraction f at 5:1 odds and we have:
>
> A win gives a wealth relative of $1 + 0.25 * 2 + 5f$
> A loss gives a wealth relative of $1 - 0.25 - f$
>
> The exponential rate of growth $g(f) = 0.5*\ln(1.5+5f)+0.5*\ln(0.75-f)$
> Solving $g'(f) = 0$ yields $f = 0.225$ which was a bit of a surprise until I thought about it for a while and looked at other related situations.

> **From:** Todd Proebsting
> **To:** Ed Thorp
> **Subject:** RE: incremental Kelly Criterion
>
> Thank you very much for the reply.

I, too, came to this result, but I thought it must be wrong since this tells me to bet a total of 0.475 (0.25 + 0.225) at odds that are on average worse than 5:1, and yet at 5:1, Kelly would say to bet only 0.400.

Do you have an intuitive explanation for this paradox?

From: Ed Thorp
To: Todd Proebsting
Subject: Re: incremental Kelly Criterion

I don't know if this helps, but consider the example:

A fair coin will be tossed (Pr Heads = Pr Tails = 0.5). You place a bet which gives a wealth relative of $1 + u$ if you win and $1 - d$ if you lose (u and d are both nonnegative). (No assumption about whether you should have made the bet.) Then you are offered odds of 5:1 on any additional bet you care to make. Now the wealth relatives are, each with Pr 0.5, $1 + u + 5f$ and $1 - d - f$. The Kelly fraction is $f = (4 - u - 5d)/10$. It seems strange that increasing either u or d reduces f. To see why it happens, look at the $\ln(1 + x)$ function. This odd behavior follows from its concave shape.

From: Todd Proebsting
To: Ed Thorp
Subject: RE: incremental Kelly Criterion

Yes, this helps. Thank you.

It is interesting to note that Kelly is often thought to avoid ruin. For instance, no matter how high the offered odds, Kelly would never have you bet more than 0.5 of bankroll on a fair coin with one single bet. Things change, however, when given these string bets. If I keep offering you better and better odds and you keep applying Kelly, then I can get you to bet an amount arbitrarily close to your bankroll.

Thus, string bets can seduce people to risking ruin using Kelly. (Granted at the risk of potentially giant losses by the seductress.)

From: Ed Thorp
To: Todd Proebsting
Subject: Re: incremental Kelly Criterion

Thanks. I hadn't noticed this feature of Kelly (not having looked at string bets). To check your point with an example I chose consecutive odds to one of A_n : 1 where $A_n = 2^n$, $n = 1, 2, \ldots$ and showed by induction that the amount bet at each n was $f_n = 3^{(n-1)}/4^n$ (where ^ is exponentiation and is done before division or multiplication) and that $\text{sum}\{f_n : n = 1, 2, \ldots\} = 1$.

A feature (virtue?) of fractional Kelly strategies, with the multiplier less than 1, e.g. $f = c^* f(\text{kelly})$, $0 < c < 1$, is that it (presumably) avoids this.

In contrast to Proebsting's example, the property that betting Kelly or any fixed fraction thereof less than one leads to exponential growth is typically derived by assuming a series of independent bets or, more generally, with limitations on the degree of dependence between successive bets. For example, in blackjack there is weak dependence between the outcomes of successive deals from the same unreshuffled pack of cards but zero dependence between different packs of cards, or equivalently between different shufflings of the same pack. Thus the paradox is a surprise but doesn't contradict the Kelly optimal growth property.

References

Ethier, S. (2010). *The Doctrine of Chances*. Berlin: Springer-Verlag.

Geyer, A. and W. T. Ziemba (2008). The innovest Austrian pension fund financial planning model InnoALM. *Operations Research*, 56(4), 797–810.

MacLean, L. C., W. T. Ziemba and G. Blazenko (1992). Growth versus security in dynamic investment analysis. *Management Science*, 38, 1562–1585.

Markowitz, H. M. and E. van Dijk (2006). Risk return analysis, in S. A. Zenios and W. T. Ziemba (eds.), *Handbook of Asset and Liability Management, Vol. I: Theory and Methodology*. Amsterdam: North Holland, 139–197.

Pabrai, M. (2007). *The Dhandho Investor*. New York: Wiley.

Poundstone, W. (2005). *Fortune's Formula*. US: Hill and Wang.

Taleb, N. N. (2007). *The Black Swan: The Impact of the Highly Improbable*. US: Barnes and Noble.

Thorp, E. O. (2006). The Kelly Criterion in blackjack, sports betting and the stock market, in S. A. Zenios and W. T. Ziemba (eds.), *Handbook of Asset and Liability Management, Vol. I: Theory and Methodology*. Amsterdam: North Holland, 385–428.

Thorp, E. O. and R. Whitley (1972). Concave utilities are distinguished by their optimal strategies. *Colloquia Mathematica Societatis Janos Bolyai*, 9.

Thorp, E. O. and R. Whitley (1974). Progress in statistics, in *Proceedings of the European Meeting of Statisticians*, Budapest. North Holland, pp. 813–830.

Ziemba, R. E. S. and W. T. Ziemba (2007). *Scenarios for Risk Management and Global Investment Strategies*. New York: Wiley.

Ziemba, W. T. (2003). *The Stochastic Programming Approach to Asset Liability Management*. AIMR.

Ziemba, W. T. and R. G. Vickson, eds. (2006). *Stochastic Optimization Models in Finance, 2nd Edition*. Singapore: World Scientific.

COLLOQUIA MATHEMATICA SOCIETATIS JÁNOS BOLYAI
9. EUROPEAN MEETING OF STATISTICIANS, BUDAPEST (HUNGARY), 1972.

37

CONCAVE UTILITIES ARE DISTINGUISHED BY THEIR OPTIMAL STRATEGIES

E. THORP – R. WHITLEY

1. INTRODUCTION

Mossin [5], Thorp [7], and Samuelson [6] showed for specific pairs of utility functions that different utilities can lead to different optimal strategies. In particular the optimal investment strategy for the utility $\log x$ is not necessarily the optimal strategy for the utility $\frac{1}{\gamma} x^\gamma$ $(\gamma \neq 0)$.

These examples suggest the following generalization, of obvious importance to general utility theory.

Consider a T stage investment process. At each stage allocate resources among the available investments. Each chosen sequence A of allocations ("strategy") yields a corresponding terminal probability distribution F_T^A of assets at the completion of stage T. For each utility function $U(\cdot)$, consider those strategies $A^*(U)$ which maximize the expected value $\int U(x) dF_T^A(x)$ of terminal utility. Assume sufficient hypotheses on U and the set of F_T^A so that the integral is defined and that furthermore the maximizing strategy $A^*(U)$ exists. Then is it true in general that $A^*(U_1)$ is not $A^*(U_2)$ for "distinct" utilities U_1 and U_2?

As we now show, the answer is yes: the Mossin – Thorp – Samuelson results for specific utility pairs generalizes to the principal class of interest in modern utility theory.

2. THE MAIN THEOREM

We prove this for the class of "interesting" concave utilities. We begin with more special hypotheses.

Theorem 1. *Let U and V be utilities defined and differentiable on $(0, \infty)$ with $U'(x)$ and $V'(x)$ positive and strictly decreasing as x increases. Then if U and V are inequivalent, there is a one period investment setting such that U and V have distinct sets of optimal strategies. Furthermore, the investment setting may be chosen to consist only of cash and a two-valued random investment, in which case the optimal strategies are unique.*

Corollary 2. *If the utilities U and V have the same (sets of) optimal strategies for each finite sequence of investment settings, then U and V are equivalent.*

Two utilities U_1 and U_2 are equivalent if and only if there are constants a and b such that $U_2(x) = aU_1(x) + b$ $(a > 0)$, otherwise U_1 and U_2 are inequivalent.

Let X_i $(1 \leqslant i \leqslant k)$ be the (random) outcome per unit invested in the ith "security". We call $(X_1, \ldots, X_k)$ the investment setting. We assume X_i is independent of the amount invested. Let the initial capital be Z_0 and let the final capital be Z_1. A strategy is an allocation $W = (w_1, \ldots, w_k)$ where w_i is the fraction of Z_0 allocated to security i. We assume $w_i \geqslant 0$ for all i, that $\sum_i w_i = 1$, and that wealth is infinitely divisible. Thus the w_i may assume any real values consistent with the constraints and with the requirement that $\sum_i w_i X_i$ is in the domain of the utility function U.

Given a particular U satisfying the hypotheses of the theorem, suppose $\mathrm{E}U(Z_1(W))$ is maximized by some strategy W^*. Then W^* is an *optimal* (or *best*) *strategy* for U relative to the given investment setting.

Proof of Theorem. Suppose that U and V have the same optimal strategies for every one period investment setting consisting of cash and a two-valued random investment. It will be shown that U and V are equivalent, which will establish the logical contrapositive to the theorem and hence the theorem itself.

In the proof of theorems we shall assume for technical simplicity that the initial capital $Z_0 = 1$. When theorems have been established for this case, consideration of the transformation $U_0(s) = U(Z_0 s) = U(t)$ gives the theorems for arbitrary $Z_0 > 0$. We shall therefore state the general results without further comment after proving the $Z_0 = 1$ case.

Let the only investment (besides cash) be X where $\mathrm{P}(X = 1 - b) = q = 1 - p$ and $\mathrm{P}(X = 1 + a) = p$, where $a > 0$ and $0 < p, b < 1$. The choice $0 < b < 1$, rather than simply $b = 1$, has been made because for $b = 1$ and $w = 1$, the expression $U(0)$ would arise and 0 is not necessarily in the domain of U (e.g., $U(x) = \log x$). The available strategies are to allocate the fraction w of recources to X and $1 - w$ to cash, with $0 \leqslant w \leqslant 1$.

At the end of the period, we have

$$\mathrm{E}U(Z_1(w)) = pU(1 + aw) + qU(1 - bw) = f(w)\,. \tag{2.1}$$

To find the maximum, consider $f'(w) = apU'(1 + aw) - bqU'(1 - bw)$. Since $U'(t)$ strictly decreases as t increases, we have $f'(w)$ decreasing strictly as w increases. Thus there is a unique maximum. If $f'(w^*) = 0$ for some w^* with $0 \leqslant w^* \leqslant 1$, then the maximum is at this unique w^*. If $f'(w) > 0$ for all w with $0 \leqslant w \leqslant 1$, then the unique maximum is at $w = 1$. If instead $f'(w) < 0$ for $0 \leqslant w \leqslant 1$, then the unique maximum is at $w = 0$.

If $f'(w) = 0$ we have $\frac{U'(1 + aw)}{U'(1 - bw)} = \frac{bq}{ap}$. Suppose $a > 0$ and $\frac{1}{2} < b < 1$ are given and we wish $f'\left(\frac{1}{2b}\right) = 0$. Letting $\lambda = \frac{U'\left(1 + \frac{a}{2b}\right)}{U'\left(\frac{1}{2}\right)}$, we can solve $\lambda = \frac{bq}{ap}$ for p, with $0 < p < 1$. Thus for each $a > 0$ there is a choice of p, hence an X, such that $w^* = \frac{1}{2b}$ is optimal for U.

Now suppose that U and V have the same optimal strategies for all such investment settings. Then $w^* = \frac{1}{2b}$ for V also and we have $\frac{U'\left(1 + \frac{a}{2b}\right)}{U'\left(\frac{1}{2}\right)} = \frac{V'\left(1 + \frac{a}{2b}\right)}{V'\left(\frac{1}{2}\right)}$ for all $a > 0$. Letting $V'\left(\frac{1}{2}\right) = \alpha U'\left(\frac{1}{2}\right)$ we find $V'(t) = \alpha U'(t)$ $(t > 1)$ whence $V(t) = \alpha U(t) + \beta$ $(t > 1)$.

When $t < 1$, we proceed similarly. Choose X so that $\mathsf{P}(X = 2) = p$ and $\mathsf{P}(X = 1 - b) = q$, where $0 < b < 1$. Then

$$\mathsf{E}U(Z_1(w)) = pU(1 + w) + qU(1 - bw) = f(w)$$

$$f'(w) = pU'(1 + w) - bqU'(1 - bw)$$

and the maximum is unique and located as before.

If $f'(w) = 0$ we have $\lambda = \frac{U'(1 - aw)}{U'(1 + w)} = \frac{p}{aq}$ and given $w = b$, $0 < b < 1$, we can choose p with $0 < p < 1$ such that $\lambda = \frac{p}{aq}$. Then as before we find $V'(1 - ab) = \gamma U'(1 - ab)$ and since a and b can be any numbers such that $0 < a$, $b < 1$, then $V'(t) = \gamma U'(t)$ $(0 < t < 1)$ where $\gamma = \frac{V'(1 + b)}{U'(1 + b)}$. But γ was shown to be α.

Thus $V(t) = \alpha U(t) + \delta$ $(0 < t < 1)$. Also $V(1) = \alpha U(1) + \epsilon$. Hence $V(t) - \alpha U(t) = \beta$ if $t > 1$, δ if $t < 1$ and ϵ if $t = 1$. But $V(t) - \alpha U(t)$ is continuous so $\beta = \delta = \epsilon$ so $V(t) = \alpha U(t) + \beta$. Thus U and

V are equivalent under the assumption that they have the same optimal strategies for all one period investment settings containing only (cash and) a two-valued random investment. The logical contrapositive assertion is the Theorem. This completes the proof. The Corollary follows a fortiori.

Note that a single investment setting of the type in the proof will not in general distinguish inequivalent utility functions. For instance, if $\mathsf{E}(X) \leqslant 0$ then $w = 0$ is the unique optimal strategy for all the utilities of Theorem 1 (more generally, for all strictly concave utilities, as defined below) so such X distinguish between none of these utilities. It may be of interest to characterize each investment setting by the pairs of utility functions it distinguishes between or "separates", and to similarly characterize collections of investment settings.

For a security X, let $m(X)$ and $M(X)$ be the greatest and least numbers, respectively, such that $\mathsf{P}(m(X) \leqslant X \leqslant M(X)) = 1$. Then for a collection C of investment settings whose securities are $\{X_\alpha : \alpha \in A\}$, where A is some index set, let $m_A = \inf\{m(X_\alpha): \alpha \in A\}$ and $M_A = \sup\{M(X_\alpha): \alpha \in A\}$. Evidently, if $U(t) = V(t)$ for $m_A \leqslant t \leqslant M_A$, the collection C will not separate U and V. Thus a collection with $m_A = 0$ and $M_A = \infty$ will be needed in general to prove the conclusion of Theorem 1.

Next we generalize Theorem 1 to concave non-decreasing utilities defined on $(0, \infty)$. We do not make the common assumption that first or even second derivatives exist. A function f is *concave* on an interval I if for each pair of points $x_1 \neq x_2$ in I and each number s with $0 < s < 1$, then $f(sx_1 + (1-s)x_2) \geqslant sf(x_1) + (1-s)f(x_2)$. If $f(sx_1 + (1-s)x_2) > sf(x_1) + (1-s)f(x_2)$ always, then f is *strictly concave*. (We use "concave" to mean "concave from below".)

The more general definition includes such computationally and empirically natural functions as the "polygonal" utilities. In these, the utility is a sequence of linear segments. The vertices are such that the function lies on or below each segment extended, and the ordinates of the vertices increase as the abscissas increase.

First, recall some facts from the elementary theory of concave functions. (Most texts give results for convex functions. But f is concave exactly when $-f$ is convex so the theories of concave and convex functions are equivalent.) A concave function is either continuous in the interior of its domain or non-measurable. An increasing function is always measurable so our utilities are continuous. A continuous concave function f defined on an open interval has a left derivative f'_- and a right derivative f'_+ defined everywhere. (If the left endpoint a is included in the interval of definition, then $f'_-(a)$ is not defined and $f'_+(a)$ may or may not be defined. Similarly, if the right endpoint b is included in the interval of definition, then $f'_+(b)$ is not defined and $f'_-(b)$ may or may not be defined.) Furthermore, $f'_-(t) \geqslant f'_+(t)$ for all t except the endpoints in the domain of f and whenever $t_1 < t_2$ then $f'_-(t_1) \geqslant f'_-(t_2)$ and $f'_+(t_1) \geqslant f'_+(t_2)$. There are at most countably many points where $f'_-(t) >$ $> f'_+(t)$; otherwise $f'_-(t) = f'_+(t) = f'(t)$ and f is differentiable. Proofs of these assertions and further theorems on concave functions are given for instance in Hardy, Littlewood, Polya [3].

Theorem 3. *Let U and V be concave utilities defined on $(0, \infty)$, one of which is strictly increasing on $(0, 1 + e)$ for some $e > 0$. If U and V are inequivalent then there is a one period investment setting such that the sets of optimal strategies for U and for V are distinct. The investment setting may be chosen to consist only of cash and a two-valued random investment. If U and V are each strictly concave on the same one of the sets $(0, Z_0]$ or $[Z_0, \infty)$, then the optimal strategies are unique and U and V therefore have distinct optimal strategies.*

Proof. We proceed as in the proof of Theorem 1 until we obtain equiation (2.1).

Note that f is concave and that if U is strictly concave on either $(0, 1]$ or $[1, \infty)$ then f is strictly concave. Now $f(w)$ is a continuous function defined on the closed bounded set $\{w\colon 0 \leqslant w \leqslant 1\}$ hence f has an absolute maximum. Let w^* be a point where f attains its maximum. It follows from the continuity of f that the set of all such w^* is closed.

From the concavity of f, the set of points w^* where f attains its maximum is also convex, hence it is a closed interval in $[0, 1]$. If f is strictly concave the maximum is unique.

For any w^* with $0 < w^* < 1$, f is a maximum if and only if $f'_-(w^*) \geqslant 0 \geqslant f'_+(w^*)$. A maximum occurs at $w^* = 0$ if and only if $f'_+(0) \leqslant 0$. A maximum occurs at $w^* = 1$ if and only if $f'_-(1) \geqslant 0$. If the maxima occur on an interval $[a, b]$ with $0 \leqslant a < b \leqslant 1$, then $f'_-(a) \geqslant 0$ and $f'_+(a) = 0$, $f'_-(b) = 0$ and $f'_+(b) \leqslant 0$, and $f'(w^*)$ exists and is zero for $a < w^* < b$.

Equation (2.1) yields

$$\begin{aligned} f'_-(w) &= apU'_-(1 + aw) - bqU'_-(1 - bw) \geqslant \\ &\geqslant apU'_+(1 + aw) - bqU'_+(1 - bw) = f'_+(w) . \end{aligned} \tag{2.2}$$

Since $U'_-(t)$ and $U'_+(t)$ are non-increasing as t increases, it follows from equation (2.2) that $f'_-(w)$ and $f'_+(w)$ are non-increasing as w increases.

Let c be such that $0 < c < b$ and $U'(1 - c)$ and $V'(1 - c)$ are defined. This is possible because U' and V' are both defined except at countably many points hence there are uncountably many points in $(0, 1)$ where both U' and V' exist. With a and b already given, choose $w = \frac{c}{b}$. Consider now the case where $U'_-\left(1 + \frac{ac}{b}\right) > 0$. Then we may choose p with $0 < p < 1$ in equation (2.2) so that $f'_-\left(\frac{c}{b}\right) = 0$. This means $f'_+\left(\frac{c}{b}\right) \leqslant 0$ and since $w = \frac{c}{b}$ is not an endpoint of $[0, 1]$ this means f attains its maximum at $\frac{c}{b}$, thus $\frac{c}{b}$ is optimal for U in the given investment setting.

Since U and V have the same optimal strategies, $w = \frac{c}{b}$ is optimal for V hence V attains its maximum there so for $w = \frac{c}{b}$, $g'_-(w) = apV'_-(1 + aw) - bqV'_-(1 - bw) \geqslant 0$ and $apV'_+(1 + aw) -$

$-bqV'_+(1-bw) = g'_+(w) \leqslant 0$. Note that $g'_-\left(\frac{c}{b}\right) \geqslant 0$ and the fact $V'_-(1-c) > 0$ implies that $V'_-\left(1+\frac{ac}{b}\right) > 0$. We may show similarly that if $V'_-\left(1+\frac{ac}{b}\right) > 0$ then $U'_-\left(1+\frac{ac}{b}\right) > 0$. Since a is chosen independently of b and c this means that for each $t > 1$, $U'_-(t) > 0$ if and only if $V'_-(t) > 0$. But this is readily shown to be equivalent to the statement that $\{t\colon U(t) = \sup U(t)\} = \{t\colon V(t) = \sup V(t)\}$, i.e. that if either U or V become horizontal for $t \geqslant e > 1$ then they both become horizontal for $t \geqslant e > 1$. For $t > e$, we have of course $U'(t) = V'(t) = 0$. For $t < e$, the argument continues as follows.

From $f'_-(w) = 0$, $apU'_-\left(1+\frac{ac}{b}\right) = bqU'(1-c)$, noting that $U'_-(1-c) = U'(1-c)$. Thus $\dfrac{U'_-\left(1+\frac{ac}{b}\right)}{U'(1-c)} = \dfrac{bq}{ap}$. From $g'_-(w) \geqslant 0$, it follows similarly that $\dfrac{V'_-\left(1+\frac{ac}{b}\right)}{V'(1-c)} \geqslant \dfrac{bq}{ap}$. Letting $\alpha = \dfrac{V'(1-c)}{U'(1-c)}$ yields

$V'_-\left(1+\frac{ac}{b}\right) \geqslant \alpha U'_-\left(1+\frac{ac}{b}\right)$. Since the choices of b and c were independent of that a, the result holds for all $a > 0$, therefore $V'_-(t) \geqslant \alpha U'_-(t)$ for all $t > 1$.

A similar argument shows that $V'_+(t) \leqslant \alpha U'_+(t)$ for all $t > 1$. Thus, except for at most countably many points, $V'(t) = \alpha U'(t)$ for $t > 1$. Now U and V are readily shown to be absolutely continuous on any closed subinterval of $(1, \infty)$, as a consequence of the fact they are continuous, concave, and non-decreasing, thus $V - \alpha U$ is absolutely continuous. The absolute continuity of $U - \alpha V$ and the fact that $(V - \alpha U)' = 0$ almost everywhere implies that $V - \alpha U = \beta$, a constant (Goffman [2], p. 242, Prop. 12).

A similar argument shows that $V(t) = \alpha U(t) + \gamma$ for $t < 1$. The role of 2 in the proof of Theorem 1 is played by any number c such that $1 < c < e$ and $U'(c)$ and $V'(c)$ are both defined. One then shows as in the proof of Theorem 1 that $V(t) = \alpha U(t) + \beta$ for $0 < t < \infty$. We have established the contrapositive assertion as in the proof of Theorem 1. This completes the proof.

The hypothesis that either U or V (hence both, from the proof) is strictly increasing for a positive distance to the right of 1 is required. If instead U and V are merely concave and non-decreasing, the conclusion of Theorem 3 need not hold. For instance, let $U(t) = V(t) = 0$ if $t \geqslant d$, where $0 < d \leqslant 1$. Let $U(t)$ and $V(t)$ each be extended to $(0, d)$ so that they are continuous, concave, and strictly increasing on $(0, d)$. Then all such utilities have the same optimal strategies, yet many pairs are inequivalent.

To obtain an inequivalent pair, let $U(t) = t - d$ if $0 < t < d$ and let $V(t) = -(t-d)^2$. If for some constants α and β, $V(t) = \alpha U(t) + \beta$ then $V'(t) = \alpha U'(t)$. But $V'(t) = -2(t-d) \not\equiv \alpha = \alpha U'(t)$.

To see that all such utilities U have the same optimal strategies, note that $W = (w_1, \ldots, w_k)$ is optimal for the investment setting $(X_1, \ldots$ $\ldots, X_k)$ if and only if $\mathsf{P}\big(\sum_i w_i X_i \geqslant d\big) = 1$, in which case $\mathsf{E}U(Z_1(W)) =$ $= 0$. If instead $\mathsf{P}\big(\sum_i w_i X_i < d\big) > 0$ then for some $\epsilon > 0$, $\mathsf{P}\big(\sum_i w_i X_i \leqslant$ $\leqslant d - e\big) = \delta > 0$. Then $\mathsf{E}U(Z_1(W)) \leqslant \delta U(d - \epsilon) < 0$ so W is not optimal.

3. OTHER SEPARATING FAMILIES

We next establish the conclusion of Theorem 1 using investment settings with n points in their range. We determine the effect of varying the payoffs $(x_1, \ldots, x_n)$ and their probabilities $(p_1, \ldots, p_n)$ separately. One surprising conclusion (part (b)) can be stated in terms of an example. Suppose X consists of betting on a wheel of fortune divided into red, white and blue sectors, with payoffs of $\frac{1}{2}, \frac{3}{2}$, and $\frac{3}{4}$ respectively. Then if U and V are inequivalent on $\left[\frac{1}{2}, \frac{3}{2}\right]$ the areas of the sectors may be chosen so U and V have distinct optimal strategies. But if the wheel is divided into just red and blue sectors, with payoffs of $\frac{1}{2}$ and $\frac{3}{2}$, then there are two inequivalent utilities on $\left[\frac{1}{2}, \frac{3}{2}\right]$ which have the same optimal strategies for every choice of areas for the two sectors.

Theorem 4. *Suppose* U *and* V *are increasing strictly concave utilities on* $(0, \infty)$. *Let* X *be a random variable with outcomes* $0 \leqslant x_1 \leqslant \leqslant x_2 \leqslant \ldots \leqslant x_n$ *with* $x_1 < 1$ *and* $x_n > 1$. *Suppose* $\mathsf{P}(X = x_i) = p_i > 0$, $\sum_{i=1}^{n} p_i = 1$.

(a) *Let* n *and the* p_i *be given. If* U *and* V *have the same optimal strategies for each* X *(i.e.* $x_1, \ldots, x_n$ *vary), then* U *and* V *are equivalent.*

(b) *Let* n *and the* x_i *be given. Suppose* U' *and* V' *exist and are continuous at* 1. *If* U *and* V *have the same optimal strategies for each* X *(i.e.* $p_1, \ldots, p_n$ *vary) and at least three* x_i *are unequal to* 1, *then* U *and* V *are equivalent on* $[Z_0 x_1, Z_0 x_n]$. *If exactly two of the* x_i*'s are unequal to one, there are utilities* U *and* V *which are not equivalent on* $[Z_0 x_1, Z_0 x_n]$, *but which have the same optimal strategy for each* X.

Proof. Assume $Z_0 = 1$. Let $R = X - 1$ and $r_i = x_i - 1$. Then investing w in X gives an expected return (with respect to U) of $\mathsf{E}(U(wR + 1)) = \sum_{i=1}^{n} p_i U(wr_i + 1)$. Each function $U(wr_i + 1)$ is differentiable except at a countable set C_i of points, so except for w in the countable set $C_1 \cup \ldots \cup C_n$ the expectation $\mathsf{E}(U(wR + 1))$ is differentiable at w with $\frac{d\mathsf{E}(U(wR + 1)}{dw} = \sum_{i=1}^{n} p_i r_i U'(wr_i + 1)$. Similarly each function $V(wr_i + 1)$ is differentiable except at a countable set. Thus, except at a countable set D of points in $[0, \infty]$ both $\mathsf{E}(U(wR + 1))$ and $\mathsf{E}(V(wR + 1))$ are differentiable functions of w. They are also strictly concave functions of w.

For part (a) let $p_1, \ldots, p_n$ be given and choose w_0 in $(0, 1) - D$. Consider the vectors $\alpha = (U'(w_0 r_1 + 1), \ldots, U'(w_0 r_n + 1))$ and $\beta = (V'(w_0 r_1 + 1), \ldots, V'(w_0 r_n + 1))$. Suppose that the non-zero vector

$\gamma = (c_1, c_2, \ldots, c_n)$ is perpendicular to α, i.e., the inner product $(\alpha, \gamma) = 0$. Choose $r_i = \dfrac{c_i}{p_i\left\{\epsilon \sum_{j=1}^{n} |c_j|\right\}}$ with $\epsilon = \max \dfrac{1}{p_i}$, $1 \leqslant i \leqslant n$.

Since each component of α is positive, some $c_i > 0$ and some $c_j < 0$, hence some $x_i > 1$ and some $x_j < 1$. Also $r_i + 1 = x_i > 0$. Then $\dfrac{d\mathsf{E}(U(wR+1))}{dw}\Big|_{w=w_0} = \sum_{i=1}^{n} p_i r_i U'(w_0 r_i + 1) = 0$ and $\mathsf{E}(U(wR+1))$ has a maximum at w_0. By hypothesis $\mathsf{E}(V(wR+1))$ has a maximum at w_0 and, since it is differentiable there, $\dfrac{d\mathsf{E}(V(wR+1))}{dw}\Big|_{w=w_0} =$ $= \sum_{i=1}^{n} p_i r_i V'(w_0 r_i + 1) = 0$ i.e., $(\beta, \gamma) = 0$. Hence the set of vectors perpendicular to α is also perpendicular to β which implies that $\beta = a\alpha$. Since the components of α and β are non-negative, $a \geqslant 0$. Equating components

$$U'(w_0 r_i + 1) = aV'(w_0 r_i + 1) \tag{3.1}$$

where a is a non-negative function of $r_1, \ldots, r_n$ and w_0. Since U and V are strictly concave there is a point t_0 not in D, $w_0 < t_0 < 1$, with $V'(t_0) > 0$ and $U'(t_0) > 0$. Choose r_1 so that $w_0 r_1 + 1 = t_0$, choose $r_2 > 0$ with $t = w_0 r_2 + 1$ not in D, and choose $r_3 < \ldots < r_n$ so they are not in D. Then $U'(t_0) = a(r_1, \ldots, r_n, w_0)V'(t_0)$ and $U'(w_2 r_2 + 1) = a(r_1, \ldots, r_n, w_0)V'(w_0 r_2 + 1)$. Thus $a = \dfrac{U'(t_0)}{V'(t_0)} > 0$ is constant. So $V'(t) = aU'(t)$ for any $t > 1$ not in D. Since V and U are absolutely continuous on any closed subinterval, $V(t) = aU(t) + b$ for all $t > 1$. A similar argument shows that $V(t) = cU(t) + d$ for $t < 1$ with $c = \dfrac{V'(t_0)}{U'(t_0)} = a$. The equivalence of U and V now follows (as in the proof of Theorem 1) from their continuity.

For part (b) suppose that the x_i are given, with $0 < x_1 < x_2 < \ldots \ldots < x_p \leqslant 1 \leqslant x_{p+1} < \ldots < x_n$. We proceed as before, but now consider, for $0 < w_0 < 1$ and U, V differentiable at $w_0 r_j + 1$, $1 \leqslant j \leqslant n$, the vectors $\tilde{\alpha} = (r_1 U'(w_0 r_1 + 1), \ldots, r_n U'(w_0 r_n + 1))$ and $\tilde{\beta} = (r_1 V'(w_0 r_1 + 1), \ldots, r_n V'(w_0 r_n + 1))$. Since $\tilde{\alpha}$ has both positive and negative components there is a vector $(d_1, d_2, \ldots, d_n)$ perpendicular to $\tilde{\alpha}$ with each $d_i > 1$. Choose $p_i = \frac{d_i}{\sum_{i=1}^{n} d_j}$, thus $p_i > 0$ and $\sum_i p_i = 1$, and define X by $\mathrm{P}(X = x_i) = p_i$. Thus

$$0 = (\tilde{\alpha}, (d_1, d_2, \ldots, d_n)) = \sum_{i=1}^{n} \frac{d_i}{\sum_{i=1}^{n} d_j} r_i U'(w_0 r_i + 1) =$$

$$= \frac{d\mathrm{E}(U(wR + 1))}{dw}\bigg|_{w = w_0}$$

By hypothesis $\frac{d\mathrm{E}(U(wR + 1))}{dw}\bigg|_{w = w_0} = \frac{(\tilde{\beta}, (d_1, \ldots, d_n))}{\sum_i d_j} = 0$ so $(\tilde{\beta}, (d_1, \ldots, d_n)) = 0$. Suppose that $\tilde{\gamma} = (e_1, e_2, \ldots, e_n)$ is perpendicular to $\tilde{\alpha}$. Let $d_0 > \max|e_i|$ and choose $p_i = \frac{e_i + d_0 d_i}{\sum_{i=1}^{n} e_j + d_0 d_j}$. Note that

$p_i > 0$ and $\sum_{i=1}^{n} p_i = 1$. Thus

$$\frac{d\mathrm{E}(U(wR + 1))}{dw}\bigg|_{w = w_0} = \sum_{i=1}^{n} p_i [r_i U'(w_0 r_i + 1)]$$

and letting $D = \sum_{j=1}^{n} (e_j + d_0 d_j)$ gives

$$\frac{1}{D} \sum_{j=1}^{n} e_i r_i U'(w_0 r_i + 1) + \frac{1}{D} d_0 \sum_i d_i r_i U'(w_0 r_i + 1) =$$

$$= \frac{1}{D} (\tilde{\alpha}, \tilde{\gamma}) + \frac{1}{D} d_0 (\tilde{\alpha}, (d_1, \ldots, d_n)) = 0\,.$$

Hence, $\frac{dE(V(wR+1))}{dw}\Big|_{w=w_0} = 0$. This yields $(\widetilde{\beta}|\widetilde{\gamma}) + d_0(\widetilde{\beta},(d_1,\dots$ $\dots, d_n)) = (\widetilde{\beta},\widetilde{\gamma}) = 0$. Thus

$$U'(w_0 r_i + 1) = a(p_1, p_2, \dots, p_n, w_0)V'(w_0 r_i + 1) \quad (1 \leqslant i \leqslant n). \tag{3/2}$$

For w in $(0, 1)$ with U, V differentiable at $wr_i + 1$ $(1 \leqslant i \leqslant n)$ we have (3.2) with $w_0 = w$. Consider the quotient

$$h(w) = a(p_1, \dots, p_n, w) = \frac{U'(wr_i + 1)}{V'(wr_i + 1)} \qquad (1 \leqslant i \leqslant n). \tag{3.3}$$

First look at the case where at least three x_i's are unequal to one. Suppose that $x_1 < 1 < x_{n-1} < x_n$; the proof where two or more points fall to the left of 1 is similar.

Let $\varphi_i(w) = wr_i + 1$. The countable collection of functions $\{U, V, U \circ \varphi_n^{-1}, V \circ \varphi_n^{-1}, U \circ \varphi_n^{-1} \circ \varphi_{n-1} \circ \varphi_n^{-1}, V \circ \varphi_n^{-1} \circ \varphi_{n-1} \circ \varphi_n^{-1}, U \circ \varphi_n^{-1} \circ \varphi_{n-1} \circ \varphi_n^{-1} \circ \varphi_{n-1} \circ \varphi_n^{-1}, V \circ \varphi_n^{-1} \circ \varphi_{n-1} \circ \varphi_n^{-1} \circ \varphi_{n-1} \circ$ $\circ\, \varphi_n^{-1}, \dots\}$ is simultaneously differentiable except at a countable set of points D_0 in $(0, 1)$.

Choose t in $(1, x_n) - D_0$ and write $t = w_1 r_n + 1$, so $w_1 =$ $= \varphi_n^{-1}(t)$, and set $t_1 = w_1 r_{n-1} + 1 = \varphi_{n-1}[\varphi_n^{-1}(t)]$. We can also write $t_1 = w_2 r_n + 1$; $w_2 = \varphi_n^{-1}(t_1)$. By (3.3) we have $h(w_1) = h(w_2)$, since U and V are differentiable at w_1 and w_2. Note that $w_2 < w_1$, in fact, $w_2 = \lambda w_1$ with $\lambda = \frac{r_{n-1}}{r_n}$. Setting $t_2 = w_2 r_{n-1} + 1 = \varphi_{n-1}(w_2)$, $t_2 = w_3 r_n + 1$ and $w_3 = \varphi_n^{-1}(t_2)$. Then $h(w_2) = h(w_3)$ since U and V are differentiable at $w_2 = \varphi_n^{-1} \circ \varphi_{n-1} \circ \varphi_n^{-1}(t)$ and at $w_3 = \varphi_n^{-1} \circ$ $\circ\, \varphi_{n-1} \circ \varphi_n^{-1} \circ \varphi_{n-1} \circ \varphi_n^{-1}(t)$. Continuing inductively $t_j = w_j r_{n-1} + 1 =$ $= w_{j+1} r_n + 1$ and $w_{j+1} = \lambda w_j$. Iterating this equation $w_{j+1} = \lambda^j w_1 \to 0$ as $j \to \infty$, thus $h(w_1) = \dots = h(w_n) \to h(1)$ since U' and V' are continuous at 1. Hence the equation $\frac{U'(t)}{V'(t)} = h(1)$ holds except for countably many t in $(1, x_n)$ and thus, since U and V are absolutely continuous on any closed subinterval, $U(t) = h(1)V(t) + c$ for all t in $[1, x_n)$.

Let t belong to $(x_1, 1)$ with U and V differentiable at $wr_j + 1$, $1 \leqslant j \leqslant n$. Then $t = wr_1 + 1$ and from equation (3.3) $\frac{U'(t)}{V'(t)} = \frac{U'(wr_n + 1)}{V'(wr_n + 1)} = h(1)$. Since U and V are absolutely continuous on closed subintervals of $(x_1, 1)$, $U(t) = h(1)V(t) + d$. The continuity of U and V at 1 implies that $c = d$ and thus U and V are equivalent on $[x_1, x_n]$.

To complete the proof we must consider the case where there are only two x_i's distinct from one, say, $0 < x_1 < 1 < x_2$. Let g_0 be any non-constant positive function on $[1, x_2]$ with a continuous derivative which is zero at 1. Define g on $[x_1, 1]$ by $g(wr_1 + 1) = g(wr_2 + 1)$ for $0 \leqslant w \leqslant 1$. Choose a so that $\max_{x_1 \leqslant t \leqslant x_2} |g'(t)| - a \cdot \min_{x_1 \leqslant t \leqslant x_2} |g(t)| < < 0$ and define $U(t) = \int_0^t e^{-at}dt = \frac{1 - e^{-at}}{a}$ and $V(t) = \int_0^t e^{-at}g(t)dt$. Because $U''(t) = -ae^{-at} < 0$ and $V''(t) = e^{-at}(g'(t) - ag(t)) < 0$, U and V are strictly concave. Also $U'(t) = e^{-at}$ and $V'(t) = e^{-at}g(t)$ are positive so U and V are strictly increasing. Clearly U and V are not equivalent on $[x_1, x_2]$.

For these two functions U and V and $0 < w < 1$,

$$\frac{dE(V(wR + 1))}{dw} = r_1 p_1 V'(wr_1 + 1) + r_2 p_2 V'(wr_2 + 1) =$$

$$= r_1 p_1 g(wr_1 + 1)U'(wr_1 + 1) + r_2 p_2 g(wr_2 + 1)U'(wr_2 + 1) =$$

$$= g(wr_1 + 1)\frac{dE(U(wR + 1))}{dw}.$$

Hence $\frac{dE(U(wR + 1))}{dw} = 0$ if and only if $\frac{dE(V(wR + 1))}{dw} = 0$, and so w_0, $(0 < w_0 < 1)$, is an optimal strategy for U (with respect to X) if and only if it is an optimal strategy for V. If the derivative $\frac{dE(U(wR + 1))}{dw}$ is never 0, the equation above shows that it has the same sign as $\frac{dE(V(wR + 1))}{dw}$; so 0 (or 1) is an optimal strategy for U if and only if it is an optimal strategy for V.

We have seen that U and V are two utilities on $[x_1, x_2]$ which are not equivalent, but which have the same optimal strategies for all random variables with outcomes x_1 and x_2.

Remark. Our proofs may be modified readily to prove the theorems when U and V are defined on the *closed* interval $[0, \infty)$ and also when the interval is (c, ∞) or $[c, \infty)$, with $c < Z_0$. Presumably $c > 0$. (Alternately, the $[c, \infty)$ result implies the (c, ∞) result: if $U(x) = V(x)$ on every interval $[c + \epsilon, \infty)$ $(0 < \epsilon < Z_0 - c)$ then $U(x) = V(x)$ on (c, ∞).)

4. QUESTIONS FOR FURTHER INVESTIGATION

Friedman – Savage [1] and Markowitz [4] have shown that utilities which are not everywhere concave are of interest. This leads us to a question which we have not been able to answer yet:

Is the class of utilities which are continuous and strictly increasing (and differentiable everywhere, bounded, and even strictly positive derivative, if you like) distinguished by their optimal strategies?

In the real world factors such as human error, the discreteness of assets and monetary units, etc. make it in general not possible to choose the optimal allocation $W^* = (w_1^*, \ldots, w_k^*)$. The continuity of the utility in conjunction with boundedness of the *attainable* utilities implies that "sufficiently small" deviations from W^* will ensure that the realized utility is "close" to the optimum.

One feels as well that in the real world, the exact values of the utility function should not be critical. In other words, if two utility functions are somehow "close," the consequences of choosing one rather than the other should be "close."

What should it mean for two utility functions to be "close? " First, observe that we must define closeness not for functions, but for equivalence classes of functions. Let U be a utility. The equivalence class of U, written $[U]$, is the set $\{V: V = \alpha U + \beta, \alpha > 0\}$. For the class β of bounded utilities, i.e., $M(U) \equiv \sup U(t) < \infty$, $m(U) \equiv \inf U(t) > -\infty$, we suggest that each $[U]$ equivalence class be represented by $\widetilde{U} = \frac{U - M}{M - m} + 1$. Note that $M(\widetilde{U}) = 1$ and $m(\widetilde{U}) = 0$. Then the "closeness" of U and V, i.e., of $[U]$ and $[V]$, is defined to be $\sup[\widetilde{U}(t) - \widetilde{V}(t)]$ and written either $d(U, V)$ or $d([U], [V])$ or $d(\widetilde{U}, \widetilde{V})$.

We now show that U and V can be "close" yet the optimal strategies for U and V need not be. For $n \geqslant 2$, let $\widetilde{U}_n$ and $\widetilde{V}_n$ be defined as follows:

$$\widetilde{U}_n(t) = \frac{2nt}{n+1} - 1 \text{ if } 0 \leqslant t \leqslant 1 + \frac{1}{n} \text{ and } 1 \text{ if } t > 1 + \frac{1}{n};$$

$$\widetilde{V}_n(t) = \frac{2n-1}{n+1}t - 1 \text{ if } 0 \leqslant t \leqslant 1 + \frac{1}{n},$$

$$\frac{t+n-3}{n-1} \text{ if } 1 + \frac{1}{n} < t \leqslant 2, \text{ and } 1 \text{ if } t > 2.$$

Then $d(\widetilde{U}_n, \widetilde{V}_n) = \frac{1}{n}$. Now choose an investment setting consisting only of cash and the security X, where $\mathrm{P}(X = 1 - \epsilon) = q$, $\mathrm{P}(X = 1 + a) = p$, $\frac{1}{n} < a < 1$, and $0 < \epsilon, p, q < 1$. Assume $Z_0 = 1$. A calculation shows that if $ap > q\epsilon \frac{(2n-1)(n-1)}{n+1}$, then the unique optimal strategy for U_n is $w^* = \frac{1}{an}$ and for V_n the unique optimal strategy is $w^* = 1$.

Thus for any $\delta > 0$ we can construct sequences $\widetilde{U}_n$ and $\widetilde{V}_n$ such that $d(\widetilde{U}_n, \widetilde{V}_n) \to 0$ as $n \to \infty$ and $|w^*(\widetilde{V}_n) - w^*(\widetilde{U}_n)| \geqslant 1 - \delta$, where $w^*(\widetilde{U})$ means an optimal strategy for $\widetilde{U}$.

Even though a small "error" in the utility function can lead to a large change in optimal strategy, it can only lead to a small change in consequences, in the following sense. (We use the abbreviation $U(W)$ for $\mathrm{E}U\left(Z_0 \sum_i w_i X_i\right)$. Thus for each W, $U(W)$ is a number and $U\left(Z_0 \sum_i w_i X_i\right)$ is a random variable.)

Lemma. *If* $d(\tilde{U}, \tilde{V})$ *is "small," then* $\tilde{U}(W^*(\tilde{V})) \doteq \tilde{U}(W^*(\tilde{U}))$ *and* $\tilde{V}(W^*(\tilde{U})) \doteq \tilde{V}(W^*(\tilde{V}))$, *i.e., if* U *and* V *are "close," an optimal strategy for one is "nearly optimal" for the other.*

Proof. Let $d(\tilde{U}, \tilde{V}) \leqslant \epsilon$ so $\tilde{V}(t) + \epsilon \geqslant \tilde{U}(t)$. Then for any allocation W, $\tilde{V}\left(Z_0 \sum_i w_i X_i\right) + \epsilon \geqslant \tilde{U}\left(Z_0 \sum_i w_i X_i\right)$ and $\mathrm{E}\left(\tilde{V}\left(Z_0 \sum_i w_i X_i\right) + \right.$ $\left. + \epsilon\right) = \mathrm{E}\left(\tilde{V}\left(Z_0 \sum_i w_i X_i\right)\right) + \epsilon \geqslant \mathrm{E}\tilde{U}\left(Z_0 \sum_i w_i X_i\right)$, or $\tilde{V}(W) + \epsilon \geqslant \tilde{U}(W)$. Interchanging $\tilde{U}$ and $\tilde{V}$ in the argument yields $\tilde{U}(W) + \epsilon \geqslant \tilde{V}(W)$ so $|\tilde{U}(W) - \tilde{V}(W)| \leqslant \epsilon$. The choices for W of $W^*(\tilde{U})$ and $W^*(\tilde{V})$ yield the conclusion of the lemma.

The lemma and the example show us what may happen if we replace a U by a nearby V which may have more desirable properties, such as differentiability (of various orders), strictly increasing, etc.: The optimal strategies may change drastically but the maximum utility over all strategies changes only slightly.

Note added in proof: The authors have since extended the central result of the paper, Theorem 3, as follows.

Theorem. *Let* U *and* V *be continuous non-decreasing functions defined on an arbitrary interval* I *of the real line. Then if* U *and* V *are inequivalent, there is a one-period two security investment setting such that* U *and* V *have distinct optimal strategies if either* (a) U *and* V *are in the class of all functions which are either concave or convex, or* (b) U *and* V *are in the class of all functions with a second derivative which exists and is continuous, except perhaps for a set of isolated points.*

Thus the Theorem includes the utility functions generally encountered.

REFERENCES

[1] M. Friedman – L.J. Savage, The Utility Analysis of Choices Involving Risk, *Journal of Political Economy*, 56 (1948), 279-304.

[2] C. Goffman, *Real Functions,* Holt, Rinehart and Winston Inc., New York, 1953.

[3] G. Hardy – J. Littlewood – G. Polya, *Inequalities,* Cambridge University Press, 1959.

[4] H. Markowitz, The Utility of Wealth, *Journal of Political Economy,* (1952), 151-158.

[5] J. Mossin, Optimal Multiperiod Portfolio Policies, *Journal of Business,* (April 1968).

[6] P.A. Samuelson, The 'Fallacy' of Maximizing the Geometric Mean in Long Sequences of Investing or Gambling, unpublished preliminary preprint, 1971.

[7] E.O. Thorp, Optimal Gambling Systems for Favorable Games, *Review of the International Statistical Institute,* 37 (1969), 273-293.

38

Medium Term Simulations of The Full Kelly and Fractional Kelly Investment Strategies

Leonard C. MacLean[*] Edward O. Thorp[†] Yonggan Zhao[‡] and William T. Ziemba[§]

Abstract

Using three simple investment situations, we simulate the behavior of the Kelly and fractional Kelly proportional betting strategies over medium term horizons using a large number of scenarios. We extend the work of Bicksler and Thorp (1973) and Ziemba and Hausch (1986) to more scenarios and decision periods. The results show:

(1) the great superiority of full Kelly and close to full Kelly strategies over longer horizons with very large gains a large fraction of the time;

(2) that the short term performance of Kelly and high fractional Kelly strategies is very risky;

(3) that there is a consistent tradeoff of growth versus security as a function of the bet size determined by the various strategies; and

(4) that no matter how favorable the investment opportunities are or how long the finite horizon is, a sequence of bad results can lead to poor final wealth outcomes, with a loss of most of the investor's initial capital.

1 Introduction

The Kelly optimal capital growth investment strategy has many long term positive theoretical properties (MacLean, Thorp and Ziemba 2009). It has been dubbed " fortunes formula" by Thorp (see Poundstone, 2005). However, properties that hold in the long run may be countered by negative short to medium term behavior because of the low risk aversion of log utility. In this paper, three well known experiments are revisited. The objectives are: (i) to compare the Bicksler - Thorp (1973) and Ziemba - Hausch (1986) experiments in the same setting; and (ii) to study them using an expanded range of scenarios and investment strategies. The class of investment strategies generated by varying the fraction of investment capital allocated to the Kelly portfolio are applied to simulated returns from the experimental models, and the distribution of accumulated capital is described. The conclusions from the expanded experiments are compared to the original results.

[*]Herbert Lamb Chair, School of Business Administration, Dalhousie University, Halifax, Canada B3H 3J5 l.c.maclean@dal.ca

[†]E.O. Thorp and Associates, Newport Beach, CA, Professor Emeritus, University of Califirnia , Irvine, CA

[‡]Canada Research Chair, School of Business Administration, Dalhousie University, Halifax, Canada B3H 3J5 yonggan.zhao@dal.ca

[§]Alumni Professor of Financial Modeling and Stochastic Optimization (Emeritus), University of British Columbia, Vanvouver, Canada, Visiting Professor, Mathematical Institute, Oxford University, UK, ICMA Centre, University of Reading, UK, and University of Bergamo, Italy wtzimi@mac.com

2 Fractional Kelly Strategies: The Ziemba and Hausch (1986) example

We begin with an investment situation with five possible independent investments where one wagers \$1 and either loses it with probability $1-p$ or wins $\$(O+1)$ with probability p, where O is the odds. The five wagers with odds of $O = 1, 2, 3, 4$ *and* 5 to one all have expected value of 1.14. The optimal Kelly wagers are the expected value edge of 14% over the odds. So the wagers run from 14%, down to 2.8% of initial and current wealth at each decision point. Table 1 describes these investments. The value 1.14 was chosen as it is the recommended cutoff for profitable place and show racing bets using the system described in Ziemba and Hausch (1986).

Win Probability	Odds	Prob of Selection in Simulation	Kelly Bets
0.570	1-1	0.1	0.140
0.380	2-1	0.3	0.070
0.285	3-1	0.3	0.047
0.228	4-1	0.2	0.035
0.190	5-1	0.1	0.028

Table 1: The Investment Opportunities

Ziemba-Hausch (1986) used 700 decision points and 1000 scenarios and compared full with half Kelly strategies. We use the same 700 decision points and 2000 scenarios and calculate more attributes of the various strategies. We use full, 3/4, 1/2, 1/4, and 1/8 Kelly strategies and compute the maximum, mean, minimum, standard deviation, skewness, excess kurtosis and the number out of the 2000 scenarios that the final wealth starting from an initial wealth of \$1000 is more than \$50, \$100, \$500 (lose less than half), \$1000 (breakeven), \$10,000 (more than 10-fold), \$100,000 (more than 100-fold), and \$1 million (more than a thousand-fold). Table 2 shows these results and illustrates the conclusions stated in the abstract. The final wealth levels are much higher on average, the higher the Kelly fraction. With 1/8 Kelly, the average final wealth is \$2072, starting with \$1000. Its \$4339 with 1/4 Kelly, \$19,005 with half Kelly, \$70,991 with 3/4 Kelly and \$524,195 with full Kelly. So as you approach full Kelly, the typical final wealth escalates dramatically. This is shown also in the maximum wealth levels which for full Kelly is \$318,854,673 versus \$6330 for 1/8 Kelly.

			Kelly Fraction		
Statistic	1.0k	0.75k	0.50k	0.25k	0.125k
Max	318854673	4370619	1117424	27067	6330
Mean	524195	70991	19005	4339	2072
Min	4	56	111	513	587
St. Dev.	8033178	242313	41289	2951	650
Skewness	35	11	13	2	1
Kurtosis	1299	155	278	9	2
$> 5 \times 10$	1981	2000	2000	2000	2000
10^2	1965	1996	2000	2000	2000
$> 5 \times 10^2$	1854	1936	1985	2000	2000
$> 10^3$	1752	1855	1930	1957	1978
$> 10^4$	1175	1185	912	104	0
$> 10^5$	479	284	50	0	0
$> 10^6$	111	17	1	0	0

Table 2: Final Wealth Statistics by Kelly Fraction: Ziemba-Hausch (1986) Model

Figure 1 shows the wealth paths of these maximum final wealth levels. Most of the gain is in the last 100 of the 700 decision points. Even with these maximum graphs, there is much volatility in the final wealth with the amount of volatility generally higher with higher Kelly fractions. Indeed with 3/4 Kelly, there were losses from about decision point 610 to 670.

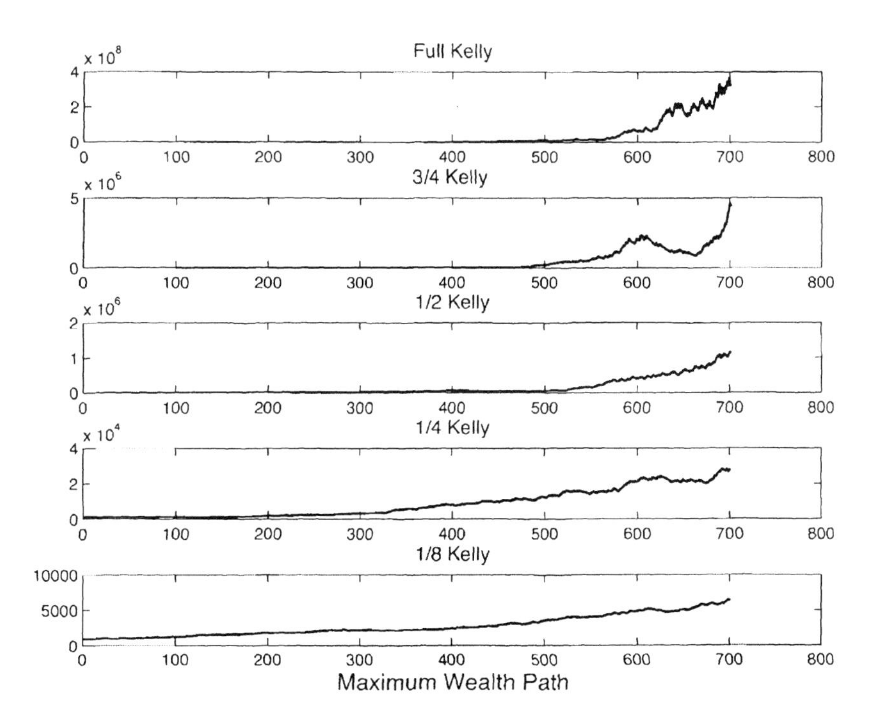

Figure 1: Highest Final Wealth Trajectory: Ziemba-Hausch (1986) Model

Looking at the chance of losses (final wealth is less than the initial $1000) in all cases, even with

1/8 Kelly with 1.1% and 1/4 Kelly with 2.15%, there are losses even with 700 independent bets each with an edge of 14%. For full Kelly, it is fully 12.4% losses, and it is 7.25% with 3/4 Kelly and 3.5% with half Kelly. These are just the percent of losses. But the size of the losses can be large as shown in the >50, >100, and >500 and columns of Table 2. The minimum final wealth levels were 587 for 1/8 and 513 for 1/4 Kelly so you never lose more than half your initial wealth with these lower risk betting strategies. But with 1/2, 3/4 and full Kelly, the minimums were 111, 56, and only $4. Figure 2 shows these minimum wealth paths. With full Kelly, and by inference 1/8, 1/4, 1/2, and 3/4 Kelly, the investor can actually never go fully bankrupt because of the proportional nature of Kelly betting.

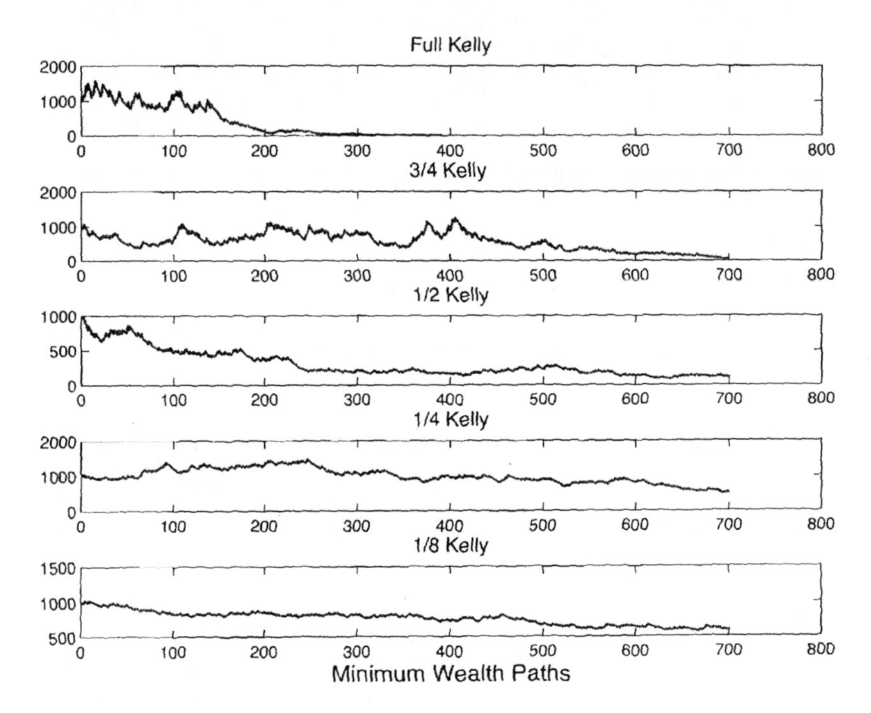

Figure 2: Lowest Final Wealth Trajectory: Ziemba-Hausch (1986) Model

If capital is infinitely divisible and there is no leveraging than the Kelly bettor cannot go bankrupt since one never bets everything (unless the probability of losing anything at all is zero and the probability of winning is positive). If capital is discrete, then presumably Kelly bets are rounded down to avoid overbetting, in which case, at least one unit is never bet. Hence, the worst case with Kelly is to be reduced to one unit, at which point betting stops. Since fractional Kelly bets less, the result follows for all such strategies. For levered wagers, that is, betting more than one's wealth with borrowed money, the investor can lose more than their initial wealth and become bankrupt.

3 Proportional Investment Strategies: Alternative Experiments

The growth and risk characteristics for proportional investment strategies such as the Kelly depend upon the returns on risky investments. In this section we consider some alternative investment experiments where the distributions on returns are quite different. The mean return is similar: 14% for Ziemba-Hausch, 12.5%for Bicksler-Thorp I, and 10.2% for Bicksler-Thorp II. However, the variation around the mean is not similar and this produces much different Kelly strategies and corresponding wealth trajectories for scenarios.

3.1 The Ziemba and Hausch (1986) Model

The first experiment is a repeat of the Ziemba - Hausch model in Section 2. A simulation was performed of 3000 scenarios over $T = 40$ decision points with the five types of independent investments for various investment strategies. The Kelly fractions and the proportion of wealth invested are reported in Table 3. Here, $1.0k$ is full Kelly, the strategy which maximizes the expected logarithm of wealth. Values below 1.0 are fractional Kelly and coincide in this setting with the decision from using a negative power utility function. Values above 1.0 coincide with those from some positive power utility function. This is overbetting according to MacLean, Ziemba and Blazenko (1992), because long run growth rate falls and security (measured by the chance of reaching a specific positive goal before falling to a negative growth level) also falls.

				Kelly Fraction: f			
Opportunity	*1.75k*	*1.5k*	*1.25k*	*1.0k*	*0.75k*	*0.50k*	*0.25k*
A	0.245	0.210	0.175	0.140	0.105	0.070	0.035
B	0.1225	0.105	0.0875	0.070	0.0525	0.035	0.0175
C	0.08225	0.0705	0.05875	0.047	0.03525	0.0235	0.01175
D	0.06125	0.0525	0.04375	0.035	0.02625	0.0175	0.00875
E	0.049	0.042	0.035	0.028	0.021	0.014	0.007

Table 3: The Investment Proportions (λ) and Kelly Fractions

The initial wealth for investment was 1000. Table 4 reports statistics on the final wealth for $T = 40$ with the various strategies.

				Fraction			
Statistic	1.75k	1.5k	1.25k	1.0k	0.75k	0.50k	0.25k
Max	50364.73	25093.12	21730.90	8256.97	6632.08	3044.34	1854.53
Mean	1738.11	1625.63	1527.20	1386.80	1279.32	1172.74	1085.07
Min	42.77	80.79	83.55	193.07	281.25	456.29	664.31
St. Dev.	2360.73	1851.10	1296.72	849.73	587.16	359.94	160.76
Skewness	6.42	4.72	3.49	1.94	1.61	1.12	0.49
Kurtosis	85.30	38.22	27.94	6.66	5.17	2.17	0.47
$> 5 \times 10$	2998	3000	3000	3000	3000	3000	3000
10^2	2980	2995	2998	3000	3000	3000	3000
$> 5 \times 10^2$	2338	2454	2634	2815	2939	2994	3000
$> 10^3$	1556	11606	1762	1836	1899	1938	2055
$> 10^4$	43	24	4	0	0	0	0
$> 10^5$	0	0	0	0	0	0	0
$> 10^6$	0	0	0	0	0	0	0

Table 4: Wealth Statistics by Kelly Fraction: Ziemba-Hausch Model (1986)

Since the Kelly bets are small, the proportion of current wealth invested is not high for any of the fractions. The upside and down side are not dramatic in this example, although there is a substantial gap between the maximum and minimum wealth with the highest fraction. Figure 3 shows the trajectories which have the highest and lowest final wealth for a selection of fractions. The log-wealth is displayed to show the rate of growth at each decision point. The lowest trajectories are almost a reflection of the highest ones.

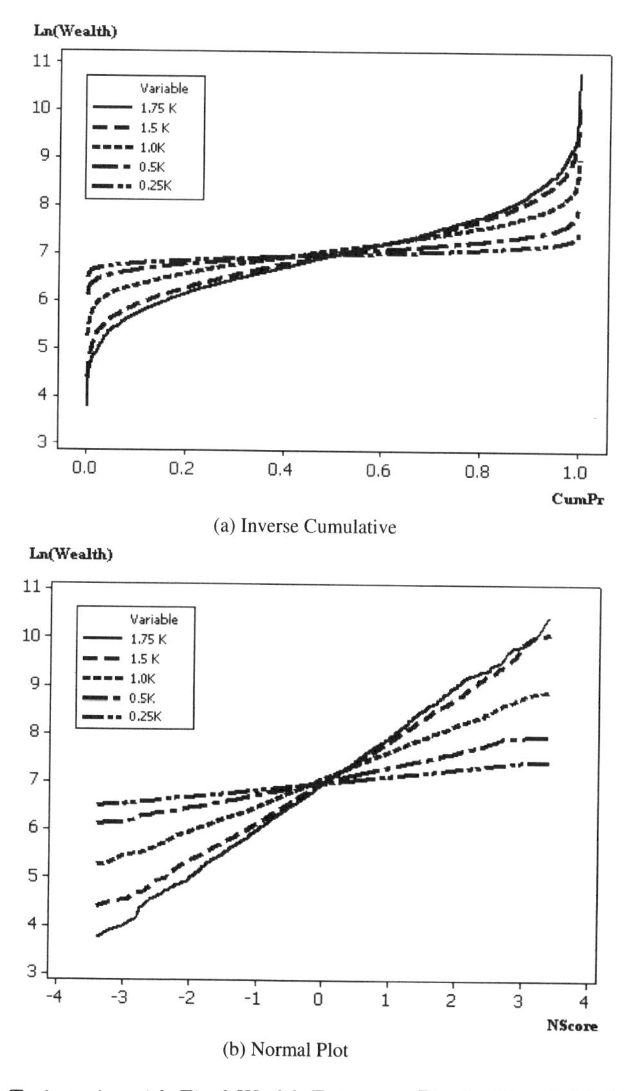

(a) Inverse Cumulative

(b) Normal Plot

Figure 3: Trajectories with Final Wealth Extremes: Ziemba-Hausch Model (1986)

The skewness and kurtosis indicate that final wealth is not normally distributed. This is expected since the geometric growth process suggests a log-normal wealth. Figure 4 displays the simulated log-wealth for selected fractions at the horizon $T = 40$. The normal probability plot will be linear if terminal wealth is distributed log-normally. The slope of the plot captures the shape of the log-wealth distribution. In this case the final wealth distribution is close to log-normal. As the Kelly fraction increases the slope increases, showing the longer right tail but also the increase in downside risk in the wealth distribution.

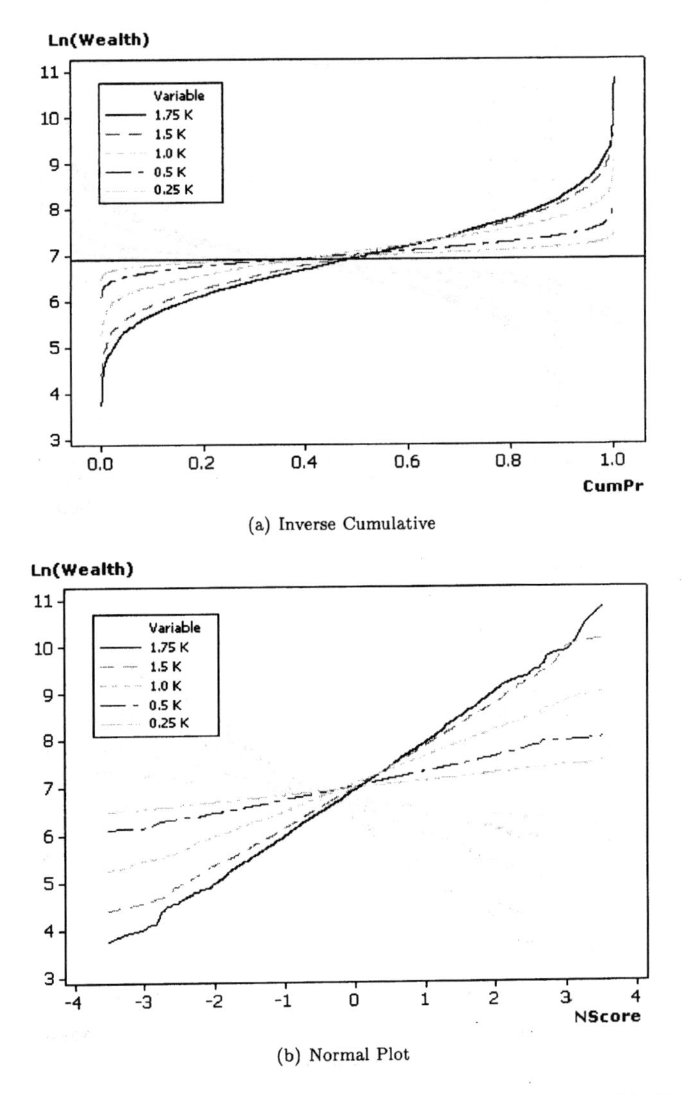

(a) Inverse Cumulative

(b) Normal Plot

Figure 4: Final Ln(Wealth) Distributions by Fraction: Ziemba-Hausch Model (1986)

On the inverse cumulative distribution plot, the initial wealth $\ln(1000) = 6.91$ is indicated to show the chance of losses. The inverse cumulative distribution of log-wealth is the basis of comparisons of accumulated wealth at the horizon. In particular, if the plots intersect then first order stochastic dominance by a wealth distribution does not exist (Hanoch and Levy, 1969). The mean and standard deviation of log-wealth are considered in Figure 5, where the trade-off as the Kelly fraction varies can be understood. Observe that the mean log-wealth peaks at the full Kelly strategy whereas the standard deviation is monotone increasing. Fractional strategies greater than full Kelly are inefficient in log-wealth, since the growth rate decreases and the the standard deviation of log-wealth increases.

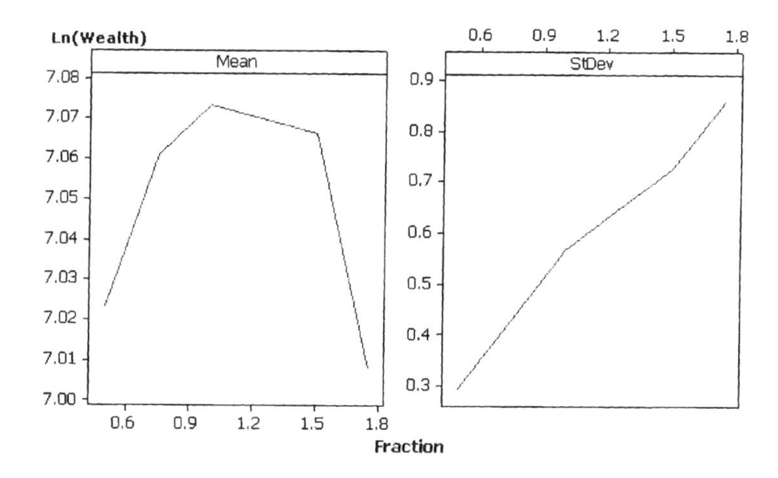

Figure 5: Mean-Std Tradeoff: Ziemba-Hausch Model (1986)

The results in Table 4 and Figures 3 - 5 support the following conclusions for Experiment 1.

1. The statistics describing end of horizon ($T = 40$) wealth are all monotone in the fraction of wealth invested in the Kelly portfolio. Specifically the maximum terminal wealth and the mean terminal wealth increase in the Kelly fraction. In contrast the minimum wealth decreases as the fraction increases and the standard deviation grows as the fraction increases. There is a trade-off between wealth growth and risk. The cumulative distribution in Figure 4 supports the theory for fractional strategies, as there is no dominance, and the distribution plots all intersect.

2. The maximum and minimum final wealth trajectories clearly show the wealth growth - risk trade-off of the strategies. The worst scenario is the same for all Kelly fractions so that the wealth decay is greater with higher fractions. The best scenario differs for the low fraction strategies, but the growth path is almost monotone in the fraction. The mean-standard deviation trade-off demonstrates the inefficiency of levered strategies (greater than full Kelly).

3.2 Bicksler - Thorp (1973) Case I - Uniform Returns

There is one risky asset R having mean return of $+12.5\%$, with the return uniformly distributed between 0.75 and 1.50 for each dollar invested. Assume we can lend or borrow capital at a risk free rate $r = 0.0$. Let $\lambda =$ the proportion of capital invested in the risky asset, where λ ranges from 0.4 to 2.4 . So $\lambda = 2.4$ means \$1.4 is borrowed for each \$1 of current wealth. The Kelly optimal growth investment in the risky asset for $r = 0.0$ is $x = 2.8655$. The Kelly fractions for the different values of λ are shown in Table 3. (The formula relating λ and f for this expiriment is in the Appendix.) In their simulation, Bicksler and Thorp use 10 and 20 yearly decision periods, and 50 simulated scenarios. We use 40 yearly decision periods, with 3000 scenarios.

Proportion: λ	0.4	0.8	1.2	1.6	2.0	2.4
Fraction: f	0.140	0.279	0.419	0.558	0.698	0.838

Table 5: The Investment Proportions and Kelly Fractions for Bicksler-Thorp (1973) Case I

The numerical results from the simulation with $T = 40$ are in Table 6 and Figures 7 - 9. Although the Kelly investment is levered, the fractions in this case are less than 1.

			Fraction			
Statistic	0.14k	0.28k	0.42k	0.56k	0.70k	0.84k
Max	34435.74	743361.14	11155417.33	124068469.50	1070576212.0	7399787898
Mean	7045.27	45675.75	275262.93	1538429.88	7877534.72	36387516.18
Min	728.45	425.57	197.43	70.97	18.91	3.46
St. Dev.	4016.18	60890.61	674415.54	6047844.60	44547205.57	272356844.8
Skewness	1.90	4.57	7.78	10.80	13.39	15.63
Kurtosis	6.00	31.54	83.19	150.51	223.70	301.38
$> 5 \times 10$	3000	3000	3000	3000	2999	2998
10^2	3000	3000	3000	2999	2999	2998
$> 5 \times 10^2$	3000	2999	2999	2997	2991	2976
$> 10^3$	2998	2997	2995	2991	2980	2965
$> 10^4$	529	2524	2808	2851	2847	2803
$> 10^5$	0	293	1414	2025	2243	2290
$> 10^6$	0	0	161	696	1165	1407

Table 6: Final Wealth Statistics by Kelly Fraction for Bicksler-Thorp Case I

In this experiment the Kelly proportion is high, based on the attractiveness of the investment in stock. The largest fraction (0.838k) shows strong returns, although in the worst scenario most of the wealth is lost. The trajectories for the highest and lowest terminal wealth scenarios are displayed in Figures 6. The highest rate of growth is for the highest fraction, and correspondingly it has the largest wealth fallback.

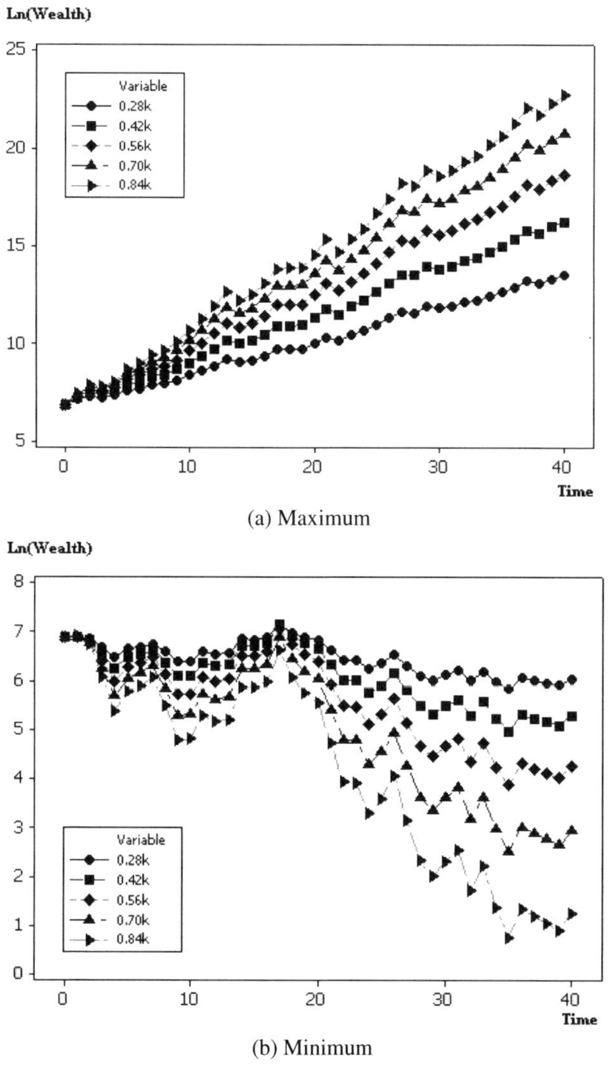

(a) Maximum

(b) Minimum

Figure 6: Trajectories with Final Wealth Extremes: Bicksler-Thorp (1973) Case I

The distribution of terminal wealth in Figure 7 illustrates the growth of the $f = 0.838k$ strategy. It intersects the normal probability plot for other strategies very early and increases its advantage. The linearity of the plots for all strategies is evidence of the log-normality of final wealth. The inverse cumulative distribution plot indicates that the chance of losses is small - the horizontal line indicates log of initial wealth.

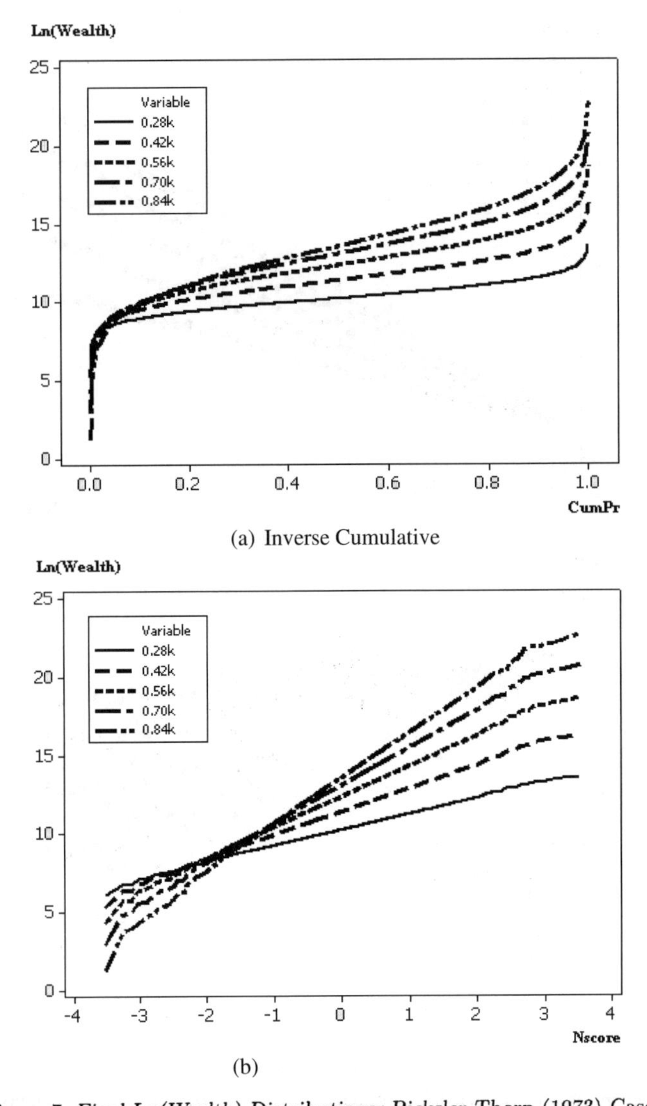

(a) Inverse Cumulative

(b)

Figure 7: Final Ln(Wealth) Distributions: Bicksler-Thorp (1973) Case I

As further evidence of the superiority of the $f = 0.838k$ strategy consider the mean and standard deviation of log-wealth in Figure 8. The growth rate (mean ln(Wealth)) continues to increase since the fractional strategies are less then full Kelly.

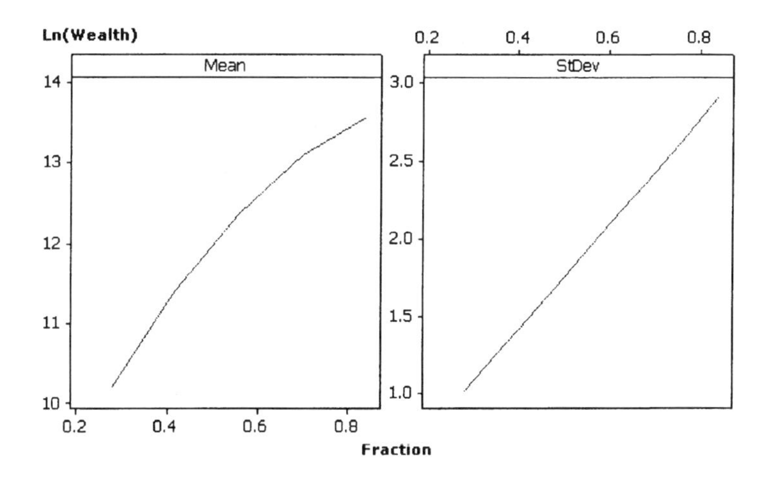

Figure 8: Mean-Std Trade-off: Bicksler-Thorp (1973) Case I

From the results of this experiment we can make the following statements.

1. The statistics describing end of horizon ($T = 40$) wealth are again monotone in the fraction of wealth invested in the Kelly portfolio. Specifically the maximum terminal wealth and the mean terminal wealth increase in the Kelly fraction. In contrast the minimum wealth decreases as the fraction increases and the standard deviation grows as the fraction increases. The growth and decay are much more pronounced than was the case in experiment 1. The minimum still remains above 0 since the fraction of Kelly is less than 1. There is a trade-off between wealth growth and risk, but the advantage of leveraged investment is clear. As illustrated with the cumulative distributions in Figure 7, the log-normality holds and the upside growth is more pronounced than the downside loss. Of course, the fractions are less than 1 so improved growth is expected.

2. The maximum and minimum final wealth trajectories clearly show the wealth growth - risk of various strategies. The mean-standard deviation trade-off favors the largest fraction, even though it is highly levered.

3.3 Bicksler - Thorp (1973) Case II - Equity Market Returns

In the third experiment there are two assets: US equities and US T-bills. According to Siegel (2002), during 1926-2001 US equities returned of 10.2% with a yearly standard deviation of 20.3%, and the mean return was 3.9% for short term government T-bills with zero standard deviation. We assume the choice is between these two assets in each period. The Kelly strategy is to invest a proportion of wealth $x = 1.5288$ in equities and sell short the T-bill at $1 - x = -0.5228$ of current wealth. With the short selling and levered strategies, there is a chance of substantial losses. For the simulations, the proportion: λ of wealth invested in equities and the corresponding Kelly fraction f are provided in Table 7. (The formula relating λ and f for this expiriment is in the Appendix.)

In their simulation, Bicksler and Thorp used 10 and 20 yearly decision periods, and 50 simulated scenarios. We use 40 yearly decision periods, with 3000 scenarios.

λ	0.4	0.8	1.2	1.6	2.0	2.4
f	0.26	0.52	0.78	1.05	1.31	1.57

Table 7: Kelly Fractions for Bicksler-Thorp (1973) Case II

The results from the simulations with experiment 3 are contained in Table 8 and Figures 9, 10, and 11. This experiment is based on actual market returns. The striking aspects of the statistics in Table 8 are the sizable gains and losses. For the the most aggressive strategy ($1.57k$), it is possible to lose 10,000 times the initial wealth. This assumes that the shortselling is permissable through to the horizon.

Table 8: Final Wealth Statistics by Kelly Fraction for Bicksler-Thorp (1973) Case II

			Fraction			
Statistic	0.26k	0.52k	0.78k	1.05k	1.31k	1.57k
Max	65842.09	673058.45	5283234.28	33314627.67	174061071.4	769753090
Mean	12110.34	30937.03	76573.69	182645.07	416382.80	895952.14
Min	2367.92	701.28	-4969.78	-133456.35	-6862762.81	-102513723.8
St. Dev.	6147.30	35980.17	174683.09	815091.13	3634459.82	15004915.61
Skewness	1.54	4.88	13.01	25.92	38.22	45.45
Kurtosis	4.90	51.85	305.66	950.96	1755.18	2303.38
$> 5 \times 10$	3000	3000	2998	2970	2713	2184
10^2	3000	3000	2998	2955	2671	2129
$> 5 \times 10^2$	3000	3000	2986	2866	2520	1960
$> 10^3$	3000	2996	2954	2779	2409	1875
$> 10^4$	1698	2276	2273	2112	1794	1375
$> 10^5$	0	132	575	838	877	751
$> 10^6$	0	0	9	116	216	270

The highest and lowest final wealth trajectories are presented in Figures 9. In the worst case, the trajectory is terminated to indicate the timing of vanishing wealth. There is quick bankruptcy for the aggressive strategies.

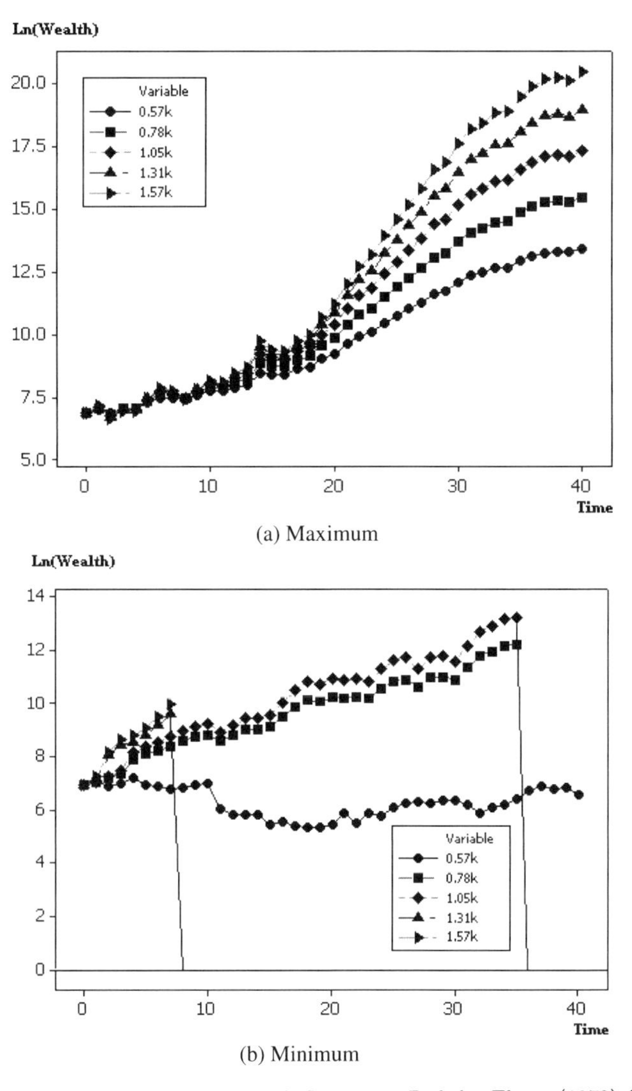

(a) Maximum

(b) Minimum

Figure 9: Trajectories with Final Wealth Extremes: Bicksler-Thorp (1973) Case II

The strong downside is further illustrated in the distribution of final wealth plot in Figure 10. The normal probability plots are almost linear on the upside (log-normality), but the downside is much more extreme than log-normal for all strategies except for $0.52k$. Even the full Kelly is risky in this case. The inverse cumulative distribution shows a high probability of large losses with the most aggressive strategies. In constructing these plots the negative growth was incorporated with the formula $growth = [signW_T]\, ln(|W_T|)$.

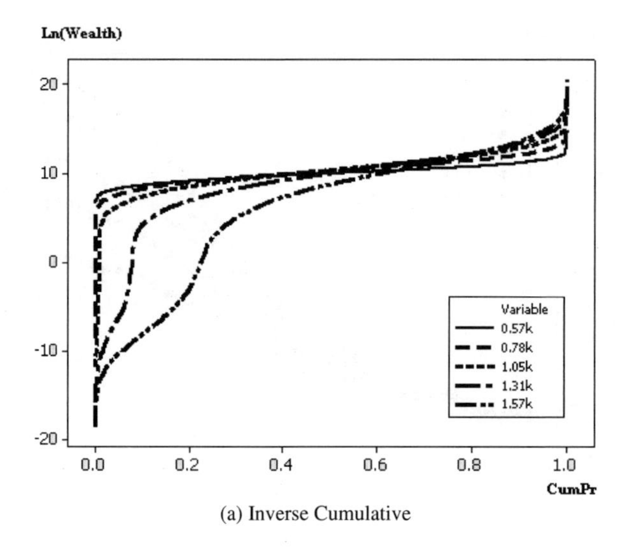

(a) Inverse Cumulative

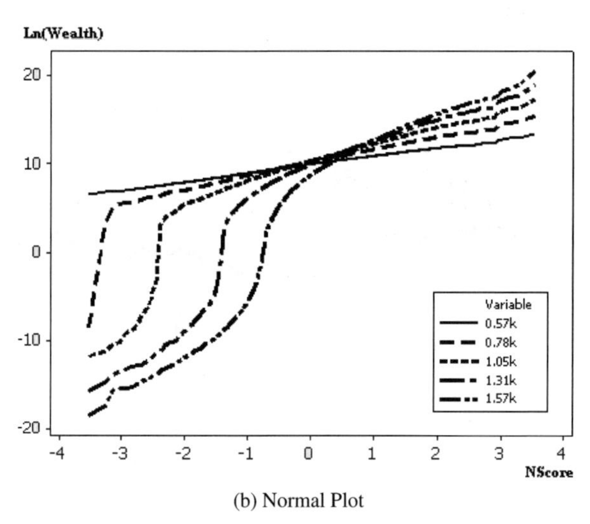

(b) Normal Plot

Figure 10: Final Ln(Wealth) Distributions: Bicksler-Thorp (1973) Case II

The mean-standard deviation trade-off in Figure 11 provides more evidence to the riskyness of the high proportion strategies. When the fraction exceeds the full Kelly, the drop-off in growth rate is sharp, and that is matched by a sharp increase in standard deviation.

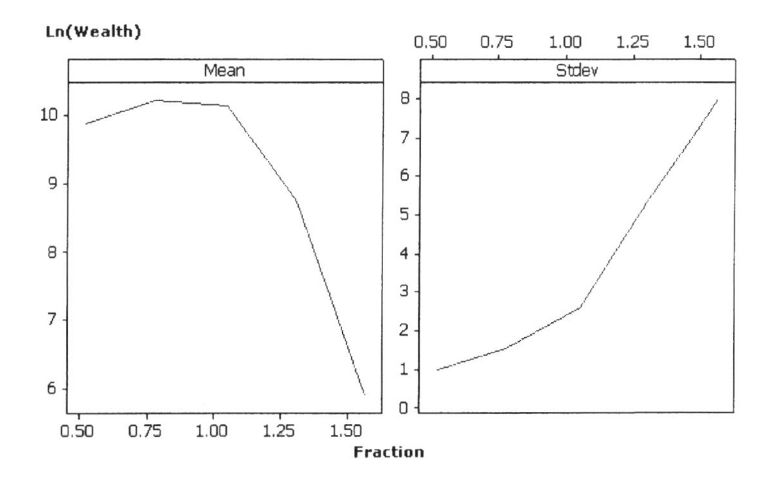

Figure 11: Mean-Std Tradeoff: Bicksler-Thorp (1973) Case II

The results in experiment 3 lead to the following conclusions.

1. The statistics describing the end of the horizon ($T = 40$) wealth are again monotone in the fraction of wealth invested in the Kelly portfolio. Specifically (i) the maximum terminal wealth and the mean terminal wealth increase in the Kelly fraction; and (ii) the minimum wealth decreases as the fraction increases and the standard deviation grows as the fraction increases. The growth and decay are pronounced and it is possible to have large losses. The fraction of the Kelly optimal growth strategy exceeds 1 in the most levered strategies and this is very risky. There is a trade-off between return and risk, but the mean for the levered strategies is growing far less than the standard deviation. The disadvantage of leveraged investment is clearly illustrated with the cumulative distribution in Figure 10. The log-normality of final wealth does not hold for the levered strategies.

2. The maximum and minimum final wealth trajectories clearly show the return - risk of levered strategies. The worst and best scenarios are the not same for all Kelly fractions. The worst scenario for the most levered strategy shows the rapid decline in wealth. The mean-standard deviation trade-off confirms the riskyness/folly of the aggressive strategies.

4 Discussion

The Kelly optimal capital growth investment strategy is an attractive approach to wealth creation. In addition to maximizing the rate of growth of capital, it avoids bankruptcy and overwhelms any essentially different investment strategy in the long run (MacLean, Thorp and Ziemba, 2009). However, automatic use of the Kelly strategy in any investment situation is risky. It requires some adaptation to the investment environment: rates of return, volatilities, correlation of alternative assets, estimation error, risk aversion preferences, and planning horizon. The experiments in this paper represent some of the diversity in the investment environment. By considering the Kelly and its variants we get a concrete look at the plusses and minusses of the capital growth model. The main points from the Bicksler and Thorp (1973) and Ziemba and Hausch (1986) studies are confirmed.

- The wealth accumulated from the full Kelly strategy does not stochastically dominate fractional Kelly wealth. The downside is often much more favorable with a fraction less than one.
- There is a tradeoff of risk and return with the fraction invested in the Kelly portfolio. In cases of large uncertainty, either from intrinsic volatility or estimation error, security is gained by reducing the Kelly investment fraction.
- The full Kelly strategy can be highly levered. While the use of borrowing can be effective in generating large returns on investment, increased leveraging beyond the full Kelly is not warranted. The returns from over-levered investment are offset by a growing probability of bankruptcy.
- The Kelly strategy is not merely a long term approach. Proper use in the short and medium run can achieve wealth goals while protecting against drawdowns.

References

[1] Bicksler, J.L. and E.O. Thorp (1973). The capital growth model: an empirical investigation. Journal of Financial and Quantitative Analysis 8\2, 273–287.

[2] Hanoch, G. and Levy, H. (1969). The Efficiency Analysis of Choices Involving Risk. The Review of Economic Studies 36, 335-346.

[3] MacLean, L.C., Thorp, E.O., and Ziemba, W.T. (2010). Good and bad properties of the Kelly criterion. in The Kelly Capital Growth Investment Criterion: Theory and Practice. Scientific Press, Singapore.

[4] MacLean, L.C., Ziemba, W.T. and Blazenko, G. (1992). Growth versus Security in Dynamic Investment Analysis. Management Science 38, 1562-85.

[5] Merton, Robert C. (1990). Continuous Time Finance. Malden, MA Blackwell Publishers Inc..

[6] Poundstone, W. (2005). Fortunes Formula: The Untold Story of the Scientific Betting System That Beat the Casinos and Wall Street. Farrar Straus & Giroux, New York, NY. Paperback version (2005) from Hill and Wang, New York.

[7] Siegel, J.J. (2002). Stocks for the long run. Wiley.

[8] Ziemba, W.T. and D.B. Hausch (1986). Betting at the Racetrack. Dr. Z. Investments Inc., San Luis Obispo, CA.

5 Appendix

The proportional investment strategies in the experiments of Bicksler and Thorp (1973) have fractional Kelly equivalents. The Kelly investment proportion for the experiments are deveolped in this appendix.

5.1 Kelly Strategy with Uniform Returns

Consider the problem

$$Max_x \{E(ln(1 + r + x(R - r)\},$$

where R is uniform on $[a, b]$ and r =the risk free rate.

We have the first order condition

$$\int_a^b \frac{R - r}{1 + r + x(R - r)} \times \frac{1}{b - a} dR = 0,$$

which reduces to

$$x(b - a) = (1 + r)ln\left(\frac{1 + r + x(b - r)}{1 + r + x(a - r)}\right) \Longleftrightarrow \left[\frac{1 + r + x(b - r)}{1 + r + x(a - r)}\right]^{\frac{1}{x}} = e^{\frac{b-a}{1+r}}.$$

In the case considered in Experiment II, $a = -0.25, b = 0.5, r = 0$. The equation becomes $\left[\frac{1+0.5x}{1-0.25x}\right]^{\frac{1}{x}} = e^{0.75}$, with a solution $x = 2.8655$. So the Kelly strategy is to invest 286.55% of wealth in the risky asset.

5.2 Kelly Strategy with Normal Returns

Consider the problem

$$Max_x \{E(ln(1 + r + x(R - r)\},$$

where R is Gaussian with mean μ_R and standard deviation σ_R, and r =the risk free rate. The solution is given by Merton (1990) as

$$x = \frac{\mu_R - r}{\sigma_R}.$$

The values in Experiment III are $\mu_R = 0.102, \sigma_R = 0.203, r = 0.039$, so the Kelly strategy is $x = 1.5288$.

39

Good and Bad Properties of the Kelly Criterion*

Leonard C. MacLean
School of Business, Dalhousie University, Halifax, NS

Edward O. Thorp
E. O. Thorp and Associates, Newport Beach, CA
Professor Emeritus, University of California, Irvine

William T. Ziemba
Professor Emeritus, University of British Columbia, Vancouver, BC
Visiting Professor, Mathematical Institute, Oxford University, UK
ICMA Centre, University of Reading, UK
University of Bergamo, Italy

Abstract

We summarize what we regard as the good and bad properties of the Kelly criterion and its variants. Additional properties are discussed as observations.

The main advantage of the Kelly criterion, which maximizes the expected value of the logarithm of wealth period by period, is that it maximizes the limiting exponential growth rate of wealth. The main disadvantage of the Kelly criterion is that its suggested wagers may be very large. Hence, the Kelly criterion can be very risky in the short term.

In the one asset two valued payoff case, the optimal Kelly wager is the edge (expected return) divided by the odds. Chopra and Ziemba (1993), reprinted in Part II of this volume, following earlier studies by Kallberg and Ziemba (1981, 1984) showed for any asset allocation problem that the mean is much more important than the variances and co-variances. Errors in means versus errors in variances were about 20 : 2 : 1 in importance as measured by the cash equivalent value of final wealth. Table 1 and Figure 1 show this and illustrate that the relative importance depends on the degree of risk aversion. The lower is the Arrow-Pratt risk aversion, $R_A = -u''(w)/u'(w)$, the higher are the relative errors from incorrect means. Chopra (1993) further shows that portfolio turnover is larger for errors in means

*Special thanks go to Tom Cover and John Mulvey for helpful comments on an earlier draft of this paper.

Table 1 Average Ratio of Certainty Equivalent Loss for Errors in Means, Variances and Covariances. [Source: Chopra and Ziemba (1993)]

Risk Tolerance*	Errors in Means vs Covariances	Errors in Means vs Variances	Errors in Variances vs Covariances
25	5.38	3.22	1.67
50	22.50	10.98	2.05
75	56.84	21.42	2.68
	↓	↓	↓
	20	10	2
	Error Mean	Error Var	Error Covar
	20	2	1

*Risk tolerance=$R_T(w) = \frac{100}{\frac{1}{2}R_A(w)}$ where $R_A(w) = -\frac{u''(w)}{u'(w)}$

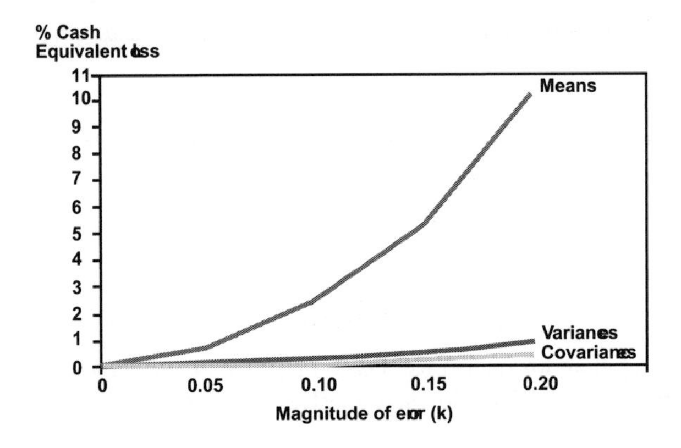

Figure 1 Mean Percentage Cash Equivalent Loss Due to Errors in Inputs

than for variances and for co-variances but the degree of difference in the size of the errors is much less than the performance as shown in Figure 2.

Since log has $R_A(w) = 1/w$, which is close to zero, the Kelly bets may be exceedingly large and risky for favorable bets. In MacLean *et al.* (2009), we present simulations of medium term Kelly, fractional Kelly, and proportional betting strategies. The results show that, with favorable investment opportunities, Kelly bettors attain large final wealth most of the time. However, because a long sequence of bad scenario outcomes is possible, any strategy can lose substantially even if there are many independent investment opportunities and the chance of losing at each investment decision point is small. The Kelly and fractional Kelly rules, like all other rules, are never a sure way of winning for a finite sequence.

In Part VI of this volume, we describe the use of the Kelly criterion in many applications and by many great investors. Two of them, Keynes and Buffett, were long term investors whose wealth paths were quite rocky but with good long term

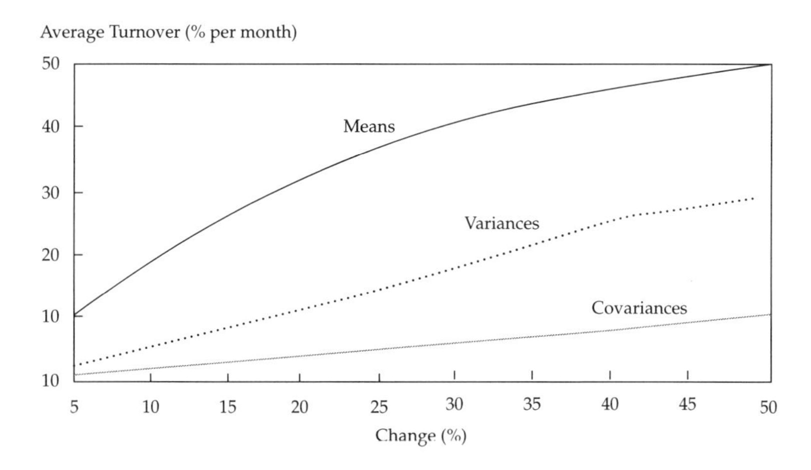

Figure 2 Average turnover for different percentage changes in means, variances and covariances. [Source: Based on data from Chopra (1993)]

outcomes. Our analyses suggest that Buffett seems to act similar to a fully Kelly bettor (subject to the constraint of no borrowing) and Keynes like a 80% Kelly bettor with a negative power utility function $-w^{-0.25}$, see Ziemba (2003) and the wealth graphs reprinted in Part VI from Ziemba (2005).

Graphs such as Figure 3 show that growth is traded off for security with the use of fractional Kelly strategies and negative power utility functions. Log maximizes the long run growth rate. Utility functions such as positive power that bet more than Kelly have more risk and lower growth. One of the properties shown below that is illustrated in the graph is that for processes, which are well approximated by continuous time, the growth rate becomes zero plus the risk free rate when one bets exactly twice the Kelly wager.

Hence, it never pays to bet more than the Kelly strategy because then risk increases (lower security) and growth decreases, so Kelly dominates all these strategies in geometric risk-return or mean-variance space. See Ziemba (2009) in this volume. As you exceed the Kelly bets more and more, risk increases and long term growth falls, eventually becoming more and more negative. Long Term Capital is one of many real world instances in which overbetting led to disaster. See Ziemba and Ziemba (2007) for additional examples.

Thus, long term growth maximizing investors should bet Kelly or less. We call betting less than Kelly "fractional Kelly", which is simply a blend of Kelly and cash. Consider the negative power utility function $\delta\omega^{\delta}$ for $\delta < 0$. This utility function is concave and when $\delta \to 0$, it converges to log utility. As δ becomes more negative, the investor is less aggressive since his absolute Arrow-Pratt risk aversion index is also higher. For the case of a stationary lognormal process and a given δ for utility function δw^{δ} and $\alpha = 1/(1-\delta)$ between 0 and 1, they both will

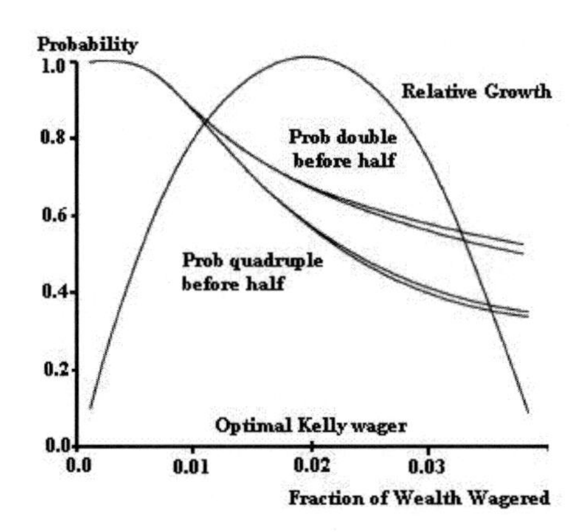

Figure 3 Probability of doubling and quadrupling before halving and relative growth rates versus fraction of wealth wagered for Blackjack (2% advantage, $p = 0.51$ and $q = 0.49$). [Source: MacLean and Ziemba (1999)]

provide the same optimal portfolio when α is invested in the Kelly portfolio and $1-\alpha$ is invested in cash. This handy formula relating the coefficient of the negative power utility function to the Kelly fraction is correct for lognormal investments and approximately correct for other distributed assets; see MacLean, Ziemba and Li (2005). For example, half Kelly is $\delta = -1$ and quarter Kelly is $\delta = -3$. So if you want a less aggressive path than Kelly, pick an appropriate δ. This formula does not apply more generally. For example, for coin tossing, where $Pr(X = 1) = p$, $Pr(X - 1) = q$, $p + q = 1$,

$$f_\delta^* = \frac{p^{\frac{1}{1-\delta}} - q^{\frac{1}{1-\delta}}}{p^{\frac{1}{1-\delta}} + q^{\frac{1}{1-\delta}}} = \frac{p^\alpha - q^\alpha}{p^\alpha + q^\alpha}$$

which is not αf^*, where $f^* = p - q \geqslant 0$ is the Kelly bet.

We now list these and other important Kelly criterion properties, updated from MacLean, Ziemba, and Blazenko (1992), MacLean and Ziemba (1999), and Ziemba and Ziemba (2007). See also Cover and Thomas (2006, Chapter 16).

The Good Properties

Good Maximizing ElogX asymptotically maximizes the rate of asset growth. See Breiman (1961), Algoet and Cover (1988).

Good The expected time to reach a preassigned goal A is asymptotically least as A increases without limit with a strategy maximizing ElogX_N. See Breiman (1961), Algoet and Cover (1988), Browne (1997a).

Good Under fairly general conditions, maximizing ElogX also asymptotically maximizes median logX. See Ethier (1987, 2004, 2010).

Good The ElogX bettor never risks ruin. See Hakansson and Miller (1975).

Good The absolute amount bet is monotone increasing in wealth.

Good The ElogX bettor has an optimal myopic policy. He does not have to consider prior nor subsequent investment opportunities. This is a crucially important result for practical use. Hakansson (1971) proved that the myopic policy obtains for dependent investments with the log utility function. For independent investments and any power utility, a myopic policy is optimal, see Mossin (1968). In fact, past outcomes can be taken into account by maximizing the conditional expected logarithm given the past (Algoet and Cover, 1988).

Good Simulation studies show that the ElogX bettor's fortune pulls ahead of other "essentially different" strategies' wealth for most reasonable-sized samples. "Essentially different" has a limited meaning. For example, $g^* \geqslant g$ but $g^* - g = \epsilon$ will not lead to rapid separation if ϵ is small enough. The key again is risk. See Bicksler and Thorp (1973), Ziemba and Hausch (1986), and MacLean, Thorp, Zhao and Ziemba (2009) in this volume. General formulas are in Aucamp (1993).

Good If you wish to have higher security by trading it off for lower growth, then use a negative power utility function, δw^δ, or fractional Kelly strategy. See MacLean, Sanegre, Zhao and Ziemba (2004) reprinted in Part III, who show how to compute the coefficent to stay above a growth path at discrete points in time with given probability or to be above a given drawdown with a certain confidence limit. MacLean, Zhao and Ziemba (2009) add the feature that path violations are penalized with a convex cost function. See also Stutzer (2009) for a related but different model of such security.

Good Competitive optimality. Kelly gambling yields wealth X^* such that $E(\frac{X}{X^*}) \leqslant 1$, for all other strategies X. This follows from the Kuhn Tucker conditions. Thus, by Markov's inequality, $Pr\left[X \geqslant tX^*\right] \leqslant \frac{1}{t}$, for $t \geqslant 1$, and for all other induced wealths x. Thus, an opponent cannot outperform X^* by a factor t with probability greater than $\frac{1}{t}$. This inequality can be improved when $t = 1$ by allowing fair randomization U. Let U be drawn according to a uniform distribution over the interval [0,2], and let U be independent of X^*. Then the result improves to $Pr\left[X \geqslant UX^*\right] \leqslant \frac{1}{2}$ for all portfolios X. Thus, fairly randomizing one's initial wealth and then investing it according to the Kelly criterion, one obtains a wealth UX^* that can only be beaten half the time. Since a competing investor can use the same strategy, probability $\frac{1}{2}$ is the best competitive performance one can expect. We see that Kelly gambling is the heart of the solution for this two-person zero sum game of who ends up with the most money. So we see that X^* (actually UX^*) is competitively optimal in a single investment period (Bell and Cover 1980, 1988).

Good If X^* is the wealth induced by the log optimal (Kelly) portfolio, then the expected wealth ratio is no greater than one, i.e., $E(\frac{X}{X^*}) \leqslant 1$, for the wealth X induced by any other portfolio (Bell and Cover, 1980, 1988).

Good Super St. Petersburg. Any cost c for the St. Petersburg random variable X, $Pr\left[X = 2^k\right] = 2^{-k}$, is acceptable. But the larger the cost c, the less wealth one should invest. The growth rate G^* of wealth resulting from repeated such investments is

$$G^* = \max_{0 \leqslant f \leqslant 1} E \ln \left(1 - f + \frac{f}{c} X\right)$$

where f is the fraction of wealth invested. The maximizing f^* is the Kelly proportion (Cover and Bell, 1980). The Kelly fraction f^* can be computed even for a super St. Petersburg random variable $Pr\left[Y = 2^{2^k}\right] = 2^{-k}$, $k = 1, 2, \ldots$, where $E \ln Y = \infty$, by maximizing the relative growth rate

$$\max_{0 \leqslant f \leqslant 1} E \ln \frac{1 - f + \frac{f}{c} Y}{\frac{1}{2} + \frac{1}{2c} Y}$$

This is bounded for all f in [0,1].

Now, although the exponential growth rate of wealth is infinite for all proportions f and it seems that all $f \in [0, 1]$ are equally good, the maximizing f^* in the previous equation guarantees that the f^* portfolio will asymptotically exponentially outperform any other portfolio $f \in [0, 1]$. Both investors' wealth have super exponential growth, but the f^* investor will exponentially outperform any other essentially different investor.

The Bad Properties

Bad The bets may be a large fraction of current wealth when the wager is favorable and the risk of loss is very small. For one such example, see Ziemba and Hausch (1986; 159–160). There, in the inaugural 1984 Breeders Cup Classic \$3 million race, the optimal fractional wager, using the Dr. Z place and show system using the win odds as the probability of winning, on the 3–5 shot Slew of Gold was 64% (see also the 74% future bet on the January effect in MacLean, Ziemba and Blazenko (1992) reprinted in this volume). Thorp and Ziemba actually made this place, show bet and won with a low fractional Kelly wager. Slew finished third but the second place horse Gate Dancer was disqualified and placed third. Wild Again won this race and it was the first great victory of the masterful jockey Pat Day.

Bad For coin tossing, any fixed fraction strategy has the property that if the number of wins equals the number of losses, then the bettor is behind. For n wins and n losses and initial wealth W_0, we have $W_{2n} = W_0(1 - f^2)^n$.

Bad The unweighted average rate of return converges to half the arithmetic rate of return. Thus, you may regularly win less than you expect. This is a consequence of weighting equally rather than by size of the wager. See Ethier and Tavaré (1983) and Griffin (1985).

Some Observations

- For an iid process and a myopic policy, which results from maximizing expected utility in case the utility function is log or a negative power, the result is fixed fraction betting, hence fractional Kelly includes all these policies.
- A betting strategy is "essentially different" from Kelly if $S_n \equiv \sum_{i=1}^{n} Elog(1+f_i^* X_i) - \sum_{i=1}^{n} Elog(1+f_i X_i)$ tends to infinity as n increases. The sequence $\{f_i^*\}$ denotes the Kelly betting fractions and the sequence $\{f_i\}$ denotes the corresponding betting fractions for the essentially different strategy.
- The Kelly portfolio does not necessarily lie on the efficient frontier in a mean-variance model (Thorp, 1971).
- Despite its superior long-run growth properties, Kelly, like any other strategy, can have a poor return outcome. For example, making 700 wagers, all of which have a 14% advantage, the least of which has a 19% chance of winning, can turn $1000 into $18. But with full Kelly 16.6% of the time, $1000 turns into at least $100,000, see Ziemba and Hausch (1996). Half Kelly does not help much as $1000 can become $145 and the growth is much lower with only $100,000 plus final wealth 0.1% of the time. For more such calculations, see Bicksler and Thorp (1973) and MacLean, Thorp, Zhao and Ziemba (2009) in this volume.
- Fallacy: If maximizing $E \log X_N$ almost certainly leads to a better outcome, then the expected utility of its outcome exceeds that of any other rule provided N is sufficiently large. *Counter-example:* $u(x) = x, 1/2 < p < 1$, Bernoulli trials $f = 1$ maximizes $EU(x)$ but $f = 2p - 1 < 1$ maximizes $ElogX_N$. See Samuelson (1971) and Thorp (1971, 2006).
- It can take a long time for any strategy, including Kelly, to dominate an essentially different strategy. For instance, in continuous time with a geometric Wiener process, suppose $\mu_\alpha = 20\%$, $\mu_\beta = 10\%$, $\sigma_\alpha = \sigma_\beta = 10\%$. Then in five years, A is ahead of B with 95% confidence. But if $\sigma_\alpha = 20\%, \sigma_\beta = 10\%$ with the same means, it takes 157 years for A to beat B with 95% confidence. As another example, in coin tossing, suppose game A has an edge of 1.0% and game B 1.1%. It takes two million trials to have an 84% chance that game A dominates game B, see Thorp (2006).

The theory and practical application of the Kelly criterion is straightforward when the underlying probability distributions are fairly accurately known. However, in investment applications, this is usually not the case. Realized future equity returns may be very different from what one would expect using estimates based on historical returns. Consequently, practitioners who wish to protect capital above all, sharply reduce risk as their drawdown increases.

Prospective users of the Kelly Criterion can check our list of good properties, bad properties and observations to test whether Kelly is well suited to their intended application. Given the extreme sensitivity of $E\log$ calculations to errors in mean estimates, these estimates must be accurate and to be on the safe side, the size of the wagers should be reduced.

For long term compounders, the good properties dominate the bad properties of the Kelly criterion, but the bad properties may dampen the enthusiasm of naive prospective users of the Kelly criterion. The Kelly and fractional Kelly strategies are very useful if applied carefully with good data input and proper financial engineering risk control.

Appendix

In continuous time, with a geometric Wiener process, betting exactly double the Kelly criterion amount leads to a growth rate equal to the risk free rate. This result is due to Thorp (1997), Stutzer (1998), Janacek (1998), and possibly others. The following simple proof, under the further assumption of the Capital Asset Pricing Model, is due to Harry Markowitz and appears in Ziemba (2003).

In continuous time

$$g_p = E_p - \frac{1}{2}V_p$$

E_p, V_p, g_p are the portfolio expected return, variance and expected log, respectively. In the CAPM

$$E_p = r_o + (E_M - r_0)X$$

$$V_p = \sigma_M^2 X^2$$

where X is the portfolio weight and r_0 is the risk free rate. Collecting terms and setting the derivative of g_p to zero yields

$$X = (E_M - r_0)/\sigma_M^2$$

which is the optimal Kelly bet with optimal growth rate

$$\begin{aligned} g^* &= r_0 + (E_M - r_0)^2 - \frac{1}{2}[(E_M r_0)/\sigma_M^2]^2 \sigma_M^2 \\ &= r_0 + (E_M - r_0)^2/\sigma_M^2 - \frac{1}{2}(E_M - r_0)^2/\sigma_M^2 \\ &= r_0 + \frac{1}{2}[(E - M - r))/\sigma_M]^2 . \end{aligned}$$

Substituting double Kelly, namely $Y = 2X$ for X above into

$$g_p = r_0 + (E_M - r_0)Y - \frac{1}{2}\sigma_M^2 Y^2$$

and simplifying yields

$$g_0 - r_0 = 2(E_M - r_0)^2/\sigma_M^2 - \frac{4}{2}(E_M - r_0)^2/\sigma_M^2 = 0.$$

Hence $g_0 = r_0$ when $Y = 2S$.

The CAPM assumption is not needed. For a more general proof and illustration, see Thorp (2006).

References

Algoet, P. H. and T. Cover (1988). Asymptotic optimality and asymptotic equipartition properties of log-optimum investment. *Annals of Probability*, 16(2), 876–898.

Aucamp, D. (1993). On the extensive number of plays to achieve superior performance with the geometric mean strategy. *Management Science*, 39, 1163–1172.

Bell, R. M. and T. M. Cover (1980). Competitive optimality of logarithmic investment. *Math of Operations Research*, 5, 161–166.

Bell, R. M. and T. M. Cover (1988). Game-theoretic optimal portfolios. *Management Science*, 34(6), 724–733.

Bicksler, J. L. and E. O. Thorp (1973). The capital growth model: an empirical investigation. *Journal of Financial and Quantitative Analysis*, 8(2), 273–287.

Breiman, L. (1961). Optimal gambling system for favorable games. *Proceedings of the 4th Berkeley Symposium on Mathematical Statistics and Probability*, 1, 63–8.

Browne, S. (1997). Survival and growth with a fixed liability: optimal portfolios in continuous time. *Math of Operations Research*, 22, 468–493.

Chopra, V. K. (1993). Improving optimization. *Journal of Investing*, 2(3), 51–59.

Chopra, V. K. and W. T. Ziemba (1993). The effect of errors in mean, variance and covariance estimates on optimal portfolio choice. *Journal of Portfolio Management*, 19, 6–11.

Cover, T. and J. A. Thomas (2006). *Elements of Information Theory* (2 ed.).

Ethier, S. (1987). The proportional bettor's fortune. Proceedings 7th International Conference on Gambling and Risk Taking, Department of Economics, University of Nevada, Reno.

Ethier, S. (2004). The Kelly system maximizes median fortune. *Journal of Applied Probability*, 41, 1230–1236.

Ethier, S. (2010). *The Doctrine of Chances: Probabilistic Aspects of Gambling.* New York: Springer.

Ethier, S. and S. Tavaré (1983). The proportional bettor's return on investment. *Journal of Applied Probability*, 20, 563–573.

Finkelstein, M. and R. Whitley (1981). Optimal strategies for repeated games. *Advanced Applied Probability*, 13, 415–428.

Griffin, P. (1985). Different measures of win rates for optimal proportional betting. *Management Science*, 30, 1540–1547.

Hakansson, N. H. (1971). On optimal myopic portfolio policies with and without serial correlation. *Journal of Business*, 44, 324–334.

Hakansson, N. H. and B. Miller (1975). Compound-return mean-variance efficient portfolios never risk ruin. *Management Science*, 22, 391–400.

Janacek, K. (1998). Optimal growth in gambling and investing. M.Sc. Thesis, Charles University, Prague.

Kallberg, J. G. and W. T. Ziemba (1981). Remarks on optimal portfolio selection. In G. Bamberg and O. Opitz (Eds.), *Methods of Operations Research*, pp. 507–520. Cambridge, MA: Oelgeschlager.

Kallberg, J. G. and W. T. Ziemba (1984). Mis-specifications in portfolio selection problems. In G. Bamberg and K. Spremann (Eds.), *Risk and Capital*, pp. 74–87. New York: Springer Verlag.

MacLean, L., R. Sanegre, Y. Zhao, and W. T. Ziemba (2004). Capital growth with security. *Journal of Economic Dynamics and Control*, 28(4), 937–954.

MacLean, L., E. O. Thorp, Y. Zhao, and W. T. Ziemba (2009). Medium term simulations of Kelly and fractional Kelly strategies.

MacLean, L. and W. T. Ziemba (1999). Growth versus security tradeoffs in dynamic investment analysis. *Annals of Operations Research*, 85, 193–227.

MacLean, L., W. T. Ziemba, and G. Blazenko (1992). Growth versus security in dynamic investment analysis. *Management Science*, 38, 1562–85.

MacLean, L., W. T. Ziemba, and Li (2005). Time to wealth goals in capital accumulation and the optimal trade-off of growth versus security. *Quantitative Finance*, 5(4), 343–357.

MacLean, L. C., Y. Zhao, and W. T. Ziemba (2009). Optimal capital growth with convex loss penalties. Working Paper, Dalhousie University.

Mossin, J. (1968). Optimal multiperiod portfolio policies. *Journal of Business*, 41, 215–229.

Samuelson, P. A. (1971). The fallacy of maximizing the geometric mean in long sequences of investing or gambling. *Proceedings, National Academy of Science*, 68, 2493–2496.

Stutzer, M. (2009). On growth optimality versus security against underperformance.

Thorp, E. O. (1971). Portfolio choice and the Kelly criterion. *Proceedings of the Business and Economics Section of the American Statistical Association*, 215–224.

Thorp, E. O. (2006). The Kelly criterion in blackjack, sports betting and the stock market. In S. A. Zenios and W. T. Ziemba (Eds.), *Handbook of Asset and Liability Management*, Handbooks in Finance, pp. 385–428. Amsterdam: North Holland.

Ziemba, R. E. S. and W. T. Ziemba (2007). *Scenarios for Risk Management and Global Investment Strategies.* NY: Wiley.

Ziemba, W. T. (2003). *The Stochastic Programming Approach to Asset Liability and Wealth Management.* AIMR, Charlottesville, VA.

Ziemba, W. T. (2005). The symmetric downside risk Sharpe ratio and the evaluation of great investors and speculators. *Journal of Portfolio Management*, Fall, 108–122.

Ziemba, W. T. (2009). Utility theory for growth versus security. Working Paper.

Ziemba, W. T. and D. B. Hausch (1986). *Betting at the Racetrack.* Dr. Z. Investments, San Luis Obispo, CA.

Part V

Utility foundations

40

Introduction to the Utility Foundations of Kelly

The criticisms of the Kelly strategy based on utility are general. If preferences are identified with wealth characteristics such as growth and tail values, then there is a utility foundation for Kelly and fractional Kelly strategies.

Hakansson and Ziemba (1995) survey their and others' works on capital growth theory. They discuss Hakansson's consumption-investment model over time in the expected log context and MacLean and Ziemba's growth and security tradeoffs and its relationship with fractional Kelly strategies. They discuss the properties of these models and relate capital growth to expected utility.

Luenberger (1993) investigates the long term behavior of $E\log$ strategies. Merton and Samuelson (1974) show that

$$\lim_{n\to\infty} E\left[u(W_n)\right]$$

is not an expected utility so the idea of using utility at a terminal time and taking the limit is not valid. Hence, the log mean-variance approach is not consistent with expected utility theory. Luenberger provides a new preference based approach to the problem assuming independent non-negative identically distributed returns. He establishes preferences on infinite sequences of wealth rather than wealth at a fixed (but later taken to the limit) terminal time. This approach avoids the limit problems and allows use of probability limit theorems.

If the tail preference is defined by a simple utility function, then the utility must be the expected log of wealth. Tail preferences are appropriate if the goal of investment is long term wealth maximization. If preferences are represented by a compound utility function, then that function will be a function of the expected logarithm and the variance of the logarithm of wealth. So in that case, there is a Markowitz type efficient tradeoff frontier of $E\log W$ and $Var\log W$ that was originally suggested by Williams (1936).

Stutzer (2003) provides an alternative behavioral foundation for an investor's use of power utility and its defining risk aversion parameter. This involves an investor's desire to minimize the probability that the wealth growth rate will not exceed an exogenous target growth rate. Stutzer uses the Gärtner-Ellis large deviations theory to show that this leads to an equivalent power utility function. This means that the investor's risk aversion parameter depends on the investment opportunity set contrary to typical model assumptions. Stutzer's formulation uses a wealth variable in the utility function that is the ratio of current wealth to the level of wealth growing at the constant target rate. Also the power coefficient is not specified in advance but rather is an output of the optimization. For ease of estimation, an independent identically distributed asset return assumption is made.

Stutzer (2010) in a followup paper discusses further how to increase security against under-performance in dynamic investment analysis over finite horizons against a specified exogenous benchmark. Motivation is provided because it takes a long time to be fairly sure that Kelly strategies dominate, and the Kelly strategy has considerable risk of under-performing or having very low returns. He uses a Bayesian formulation of the Occam's Razor Principle to illustrate the unavoidable reduction of statistical testability (namely the ability to more easily falsify) inherent in objective functions that have non-directly observable parameters. The coefficient of the power utility function is endogenously determined. The setting used is the familiar blackjack example previously discussed by Thorp and by MacLean, Ziemba, and Blazenko (1992).

R. Jarrow et al., Eds., *Handbooks in OR & MS, Vol. 9*

Chapter 3 **41**

Capital Growth Theory

Nils H. Hakansson
Walter A. Haas School of Business, University of California, Berkeley, CA 94720, U.S.A.

William T. Ziemba
Faculty of Commerce and Business Administration, University of British Columbia, Vancouver, B.C. V6T 1Y8, Canada

1. Introduction

Even casual observation strongly suggests that capital growth is not just a catch-phrase but something which many actively strive to achieve. It is therefore rather surprising that capital growth *theory* is a relatively obscure subject. For example, the great bulk of today's MBA's have had little or no exposure to the subject, having had their attention focussed almost exclusively on the single-period mean–variance model of portfolio choice. The purpose of this essay is to review the theory of capital growth, in particular the so-called growth-optimal investment strategy, its properties, its uses, and its links to betting and other investment models. We also discuss several applications that have tended to refine the basic theory.

The central feature of the growth-optimal investment strategy, also known as the geometric mean model and the Kelly criterion, is the logarithmic shape of the objective function. But the power and durability of the model is due to a remarkable set of properties. Some of these are unique to the growth-optimal strategy and the others are shared by all the members of the (remarkable) small family to which the growth optimal strategy belongs.

Investment over time is multiplicative, not additive, due to the compounding nature of the process itself. This makes a number of results in dynamic investment theory appear nonintuitive. For example, in the single-period portfolio problem, the optimal investment policy is very sensitive to the utility function being used; the set of policies that are inadmissible or dominated across all utility functions is relatively small. The same observation holds in the dynamic case when the number of periods is not large. But as the number of periods does become large, the set of investment policies that are optimal for *current* investment tends to shrink drastically, at least in the basic reinvestment case without transaction costs. As we will see, many strikingly different investors will, in essence, invest the same way when the horizon is distant and will only begin to part company as their horizons near.

It is tempting to conjecture that all long-run investment policies to which risk-averse investors with monotone increasing utility functions will flock, under a favorable return structure, insure growth of capital with a very high probability. Such a conjecture is false; many investors will, even in this case, converge on investment policies which almost surely risk ruin in the long run, in effect ignoring feasible policies which almost surely lead to capital growth. Similarly, the relationship between the behavior of capital over time and the behavior of the expected utility of that same capital over time often appears strikingly nonintuitive.

Section 2 reviews the origins of the capital growth model while Section 3 contains a derivation and identifies its key properties. The conditions for capital growth are examined in Section 4. The model's relationship to other long-run investment models is studied in Section 5 and Section 6 contains its role in intertemporal investment/consumption models. Section 7 adds various constraints for accomplishing tradeoffs between growth and security, while Section 8 reviews various applications. A concluding summary is given in Section 9.

2. Origins of the model

The approach to investment commonly known as the growth-optimal investment strategy has a number of apparently independent origins. In particular, Williams [1936], Kelly [1956], Latane [1959], and Breiman [1960, 1961] seem to have been unaware of each other's papers. But one can also argue that Bernoulli (1738) unwittingly stumbled on it in 1738 in his resolution of the St. Petersburg Paradox — see the 1954 translation — and Samuelson's survey [1977].

Samuelson [1971] appears to be the earliest to have related the geometric mean criterion to utility theory — and to find it wanting. The growth optimal strategy's inviolability in the larger consumption–investment context when preferences for consumption are logarithmic was first noted by Hakansson [1970]. Finally, models considering tradeoffs between capital growth and security appear to have been pioneered by MacLean & Ziemba [1986].

3. The model and its basic properties

The following notation and basic assumptions will be employed:

w_t = amount of investment capital at decision point t (the end of the tth period);

M_t = the number of investment opportunities available in period t, where $M_t \leq M$;

S_t = the subset of investment opportunities which it is possible to sell short in period t;

r_{1t} = rate of interest in period t;

r_{it} = return per unit of capital invested in opportunity i, where $i = 2, \ldots, M_t$, in the tth period (random variable). That is, if we invest an amount θ in i at the beginning of the period, we will obtain $(1 + r_{it})\theta$ at the end of that period;

z_{1t} = amount lent in period t (negative z_{1t} indicate borrowing) (decision variable);

z_{it} = amount invested in opportunity i, $i = 2, \ldots, M_t$ at the beginning of the tth period (decision variable);

$F_t(y_2, y_3, \ldots, y_{M_t}) \equiv \Pr\{r_{2t} \le y_2, r_{3t} \le y_3, \ldots, r_{M_t t} \le y_{M_t}\}$;

$z_t \equiv (z_{2t}, \ldots, z_{M_t t})$;

$x_{it} \equiv \dfrac{z_{it}}{w_{t-1}}, \qquad i = 1, \ldots, M_t$;

$x_t \equiv x_{2t}, \ldots, x_{M_t t}$;

$\langle x_t \rangle \equiv x_1, \ldots, x_t$.

The capital market will generally be assumed to be perfect, i.e. that there are no transaction costs or taxes, that the investor has no influence on prices or returns, that the amount invested can be any real number, and that the investor has full use of the proceeds from any short sale.

The following basic properties of returns will be assumed:

$$r_{1t} \ge 0, \qquad t = 1, 2, \ldots \tag{1}$$

$$E[r_{it}] \ge \delta + r_{1t}, \qquad \delta > 0, \qquad \text{some } i, t = 1, 2, \ldots \tag{2}$$

$$E[r_{it}] \le K, \qquad \text{all } i, t. \tag{3}$$

These assumptions imply that the financial market provides a ‘favorable game.’

We also assume that the (nonstationary) return distributions F_t are either independent from period to period or obey a Markov process and they also satisfy the ‘no-easy-money condition’

$$\mathrm{P}\left\{\sum_{i=2}^{M_t}(r_{it} - r_{1t})\theta_i < \delta_1\right\} > \delta_2 \text{ for all } t \text{ and all } \theta_i \text{ such that } \sum_{i=2}^{M_t}|\theta_i| = 1, \text{ and } \theta_i \ge 0 \text{ for all } i \notin S_t, \tag{4}$$

where $\delta_1 < 0$, $\delta_2 > 0$.

Condition (4) is equivalent to what is often referred to as the no-arbitrage condition. It is generally a necessary condition for the portfolio problem to have a solution.

We also assume that the investor must remain solvent in each period, i.e., that he or she must satisfy the solvency constraints

$$\Pr\{w_t \ge 0\} = 1, \qquad t = 1, 2, \ldots. \tag{5}$$

The amount invested at time $t-1$ is

$$\sum_{i=1}^{M_t} z_{it} = w_{t-1}$$

and the value of the investment at time t, broken down between its risky and riskfree components, is

$$w_t = \sum_{i=2}^{M_t}(1+r_{it})z_{it} + (1+r_{1t})\Big(1 - \sum_{i=2}^{M_t} z_{it}\Big),$$

which together yield the basic difference equation

$$\begin{aligned} w_t &= \sum_{i=2}^{M_t}(r_{it} - r_{1t})z_{it} + w_{t-1}(1+r_{1t}), \qquad t = 1, 2, \ldots \\ &= w_{t-1}R_t(x_t) = w_0 R_1(x_1)\ldots R_t(x_t), \qquad t = 1, 2, \ldots \end{aligned} \tag{6}$$

where

$$R_t(x_t) \equiv \sum_{i=2}^{M_t}(r_{it} - r_{1t})x_{it} + 1 + r_{1t}. \tag{7}$$

Let us now turn to the basic reinvestment problem which (ignores capital infusions and distributions and) simply revises the portfolio at discrete points in time. In view of (5), (6) may be written

$$w_t = w_0 \exp\Big\{\sum_{n=1}^{t} \ln R_n(x_n)\Big\}, \qquad t = 1, 2, \ldots. \tag{8}$$

Defining

$$G_t(\langle x_t\rangle) \equiv \frac{\Big[\sum_{n=1}^{t} \ln R_n(x_n)\Big]}{t}, \tag{9}$$

(8) becomes

$$\begin{aligned} w_t &= w_0[\exp\{G_t(\langle x_t\rangle)\}]^t \\ &= w_0(1+g_t)^t, \end{aligned} \tag{10}$$

where $g_t = \exp G_t(\langle x_t\rangle) - 1$ is the compound growth rate of capital over the first t periods.

By the law of large numbers,

$$G_t(\langle x_t\rangle) \rightarrow E[G_t(\langle x_t\rangle)]$$

under mild conditions. Thus, it is evident that for large T,

$$w_t \rightarrow 0 \text{ if } E[G_t] \leq \delta < 0, \quad t \geq T, \tag{11}$$

$$w_t \to \infty \text{ if } E[G_t] \geq \delta > 0, \quad t \geq T \tag{12}$$

$$g_t \to \exp E[G_t] - 1. \tag{13}$$

Under stationary returns and policies $\langle x_t \rangle$, (11) and (12) simplify to

$$\left.\begin{aligned} w_t &\to 0 \text{ if } E[\ln R_n(x_n)] < 0 \\ w_t &\to \infty \text{ if } E[\ln R_n(x_n)] > 0 \end{aligned}\right\} \text{any } n.$$

There is nothing intuitive that would suggest that the sign of $E[\ln R_n(x_n)]$ is the determinant of whether our capital will decline or grow in the (stationary) simple reinvestment problem. What *is* evident is that the *expected* return on capital, $E[R_n] - 1$, is *not* what matters. As (6) reminds us, capital growth (positive or negative) is a multiplicative, not an additive process.

To illustrate the point, consider the case of only two assets, one riskfree yielding 5% per period, and the other returning either −60% or +100% with equal probabilities in each period. Always putting all of our capital in the riskfree asset clearly gives a 5% growth rate of capital. The *expected* return on the risky asset is 20% per period. Yet placing all of our funds in the risky asset at the beginning of each period results in a capital growth rate that converges to −10.55%! It is easy to see this. We will double our money to 200% roughly half of the time. But we will also lose 60% (bringing the 200% to 80%) of our beginning-of-period capital about half the time, for a 'two-period return' of −20% on average, or −10.55% per period. *Expected* capital $E[w_t]$, on the other hand, has a growth rate of 20% per period.

What this simple example demonstrates is that there are many investment strategies for which, as $t \to \infty$,

$$E[w_t] \to \infty$$

$$\text{Median}\,[w_t] \to 0$$

$$\text{Mode}\,[w_t] \to 0$$

$$\Pr\{w_t < \$1\} \to 1.$$

The coexistence of the above four measures results when $E[G_t] \leq \delta < 0$ for $t \geq T$ and a long (but thin) upper tail is generated as w_t moves forward in time.

In view of (7), (9) and (10), we observe that to 'maximize' the long-run growth rate g_t, it is necessary and sufficient to maximize $E[G_t(\langle x_t \rangle)]$, or

$$\text{Max}\,\{E[\ln R_1(x_1)] + E[\ln R_2(x_2)] + \ldots\} \tag{14}$$

Whenever returns are independent from period to period or the economy obeys a Markov process [1], it is necessary and sufficient to accomplish (14) to

$$\underset{x_t}{\text{Max}}\ E[\ln R_t(x_t)] \text{ sequentially at each } t-1. \tag{15}$$

[1] Algoet & Cover [1988] show formally that the growth-optimal strategy maintains its basic properties under arbitrary returns processes.

Since the geometric mean of $R_t(x_t) = \exp\{E[\ln R_t(x_t)]\}$, we observe that (15) is also equivalent to maximizing the geometric mean of principal plus return at each point in time.

3.1. Properties of the growth-optimal investment strategy

Since the solution $\langle x_t^* \rangle$ to (15), in view of (10) and (13), *almost surely* leads to more capital in the long run than any other investment policy which does not converge to it, $\langle x_t^* \rangle$ is referred to as the growth-optimal investment strategy. Existence is assured by the no-easy-money condition (4), the bounds on expected returns (1)–(3), and the solvency constraint (5). The strict concavity of the objective function in (15) implies that the optimal payoff distribution $R_t(x_t^*)$ is unique; the optimal policy x_t^* itself will be unique only if, for any security i, there is no portfolio of the other assets which can replicate the return pattern r_{it}.

It is probably not surprising that the growth-optimal strategy never risks ruin, i.e.

$$\Pr\{R_t(x_t^*) = 0\} = 0$$

— because to grow you have to survive. But this need not mean that the solvency constraint is not binding: $E[\ln R_t(x_t)]$ may exist even when R_t touches 0 as long as the lower tail is very thin. The conditions (1)–(3) imply that positive growth is feasible. Another dimension of the consistency between short-term and long-term performance was observed by Bell & Cover [1988].

As shown by Breiman [1961], the growth-optimal strategy also has the property that it asymptotically minimizes the expected time to reach a given level of capital. This is not surprising in view of the characteristics noted in the previous two paragraphs.

It is also evident from (15) that the growth-optimal strategy is *myopic* even when returns obey a Markov process (Hakansson 1971c). This property is clearly of great practical significance since it means that the investor only needs to estimate the coming period's (joint) return structure in order to behave optimally in a long-run sense; future periods' return structures have no influence on the current period's optimal decision. *No* other dynamic investment model has this property in a Markov economy; only a small set of other families have it when returns are independent from period to period (see Section 5).

The growth-optimal strategy implies, and is implied by, logarithmic utility of wealth at the end of each period. This is because at each $t-1$

$$\begin{aligned} &\operatorname*{Max}_{x_t} \; E[\ln R_t(x_t)] \\ &\sim \operatorname*{Max}_{x_t} \{E[\ln R_t(x_t) + \ln w_{t-1}]\} \\ &= \operatorname*{Max}_{x_t} \; E[\ln(w_{t-1} R_t(x_t))] = \operatorname*{Max}_{z_t} \; E[\ln w_t(z_t)]. \end{aligned}$$

Since every utility function is unique (up to a positive linear transformation), it also follows that the growth-optimal strategy is *not* consistent with any other end-of-period utility function (more on this in the next subsection).

The relative risk aversion function

$$q(w) \equiv -\frac{wu''(w)}{u'(w)}$$

equals 1 when $u(w) = \ln(w)$ (it is 0 for a risk-neutral investor). Thus, we observe that to do 'the best' in the long run in terms of capital growth, it is not only necessary to be risk averse in each period. We must also display the 'right' amount of risk aversion. The long-run growth rate of capital will be lower either if one invests in a way which is *more* risk averse than the logarithmic function or relies on an objective function which is *less* risk averse.

The growth-optimal investment strategy is not only linear in beginning-of-period wealth but proportional as well since definitionally

$$z_t^* = w_{t-1}x_t^*.$$

Both of these properties are shared by only a small family of investment models.

Since the growth-optimal strategy is consistent with a logarithmic end-of-period utility function only, it is clearly *not* consistent with the mean–variance approach to portfolio choice — which in turn is consistent with quadratic utility for arbitrary security return structures, and, for normally distributed returns, with those utility functions whose expected utilities exist when integrated with the normal distribution, plus a few other cases, as shown by Ziemba & Vickson [1975] and Chamberlain [1983]. This incompatibility is easy to understand; in solving for the growth-optimal strategy, *all* of the moments of the return distributions matter, with positive skewness being particularly favored. When the returns on the risky assets are normally distributed, no matter how favorable the means and variances are, the growth-optimal strategy cooly places 100% of the investable funds in the riskfree asset.

The preceding does not imply that the growth-optimal portfolio necessarily is far from the mean–variance efficient frontier (although this *may* be the case [see e.g. Hakansson, 1971a]). It will generally be close to the MV-efficient frontier, especially when returns are fairly symmetric. And as shown in Section 8, the mean–variance model can in some cases be used to (sequentially) generate a close approximation to the growth-optimal portfolios.

Other properties of the Kelly criterion can be found in MacLean, Ziemba & Blazenko [1992, table 1].

3.2. Capital growth vs. expected utility

Based on (10), the uniqueness properties implied by (15), and the law of large numbers, it is undisputable, as noted in the previous subsection, that the growth-optimal strategy almost surely generates more capital (under basic reinvestment) in the long run than any other strategy which does not converge to it. At the same time, however, we observed that the growth-optimal strategy is consistent with logarithmic end-of-period utility of wealth *only*. This clearly implies that there must be 'reasonable' utility functions which value almost surely less capital in the long run more than they value the distribution generated by the Kelly criterion.

Consider the family

$$u(w) = \frac{1}{\gamma} w^{\gamma}, \qquad \gamma < 1, \tag{16}$$

to which $u(w) = \ln(w)$ belongs via $\gamma = 0$, and let $\langle x_t(\gamma) \rangle$ be the optimal portfolio sequence generated by solving

$$\operatorname*{Max}_{x_t} E\left[\frac{1}{\gamma} w_t^{\gamma}\right] \text{ at each } t-1.$$

For simplicity, consider the case of stationary returns. Since $x_t(\gamma) \neq x_t(0) = x_t^*$, it is evident that

$$\text{Max } E\left[\frac{1}{\gamma} w_t(\langle x_t(\gamma)\rangle)^{\gamma}\right] > E\left[\frac{1}{\gamma} w_t(\langle x_t^* \rangle)^{\gamma}\right], \quad \gamma \neq 0 \tag{17}$$

even though there exist numbers $a > 1$ and $T(\epsilon)$ such that

$$\Pr\{w_t(\langle x_t(\gamma)\rangle) < w_0 a^t < w_t(\langle x_t^* \rangle)\} \geq 1 - \epsilon, \quad t > T(\epsilon) \tag{18}$$

for every $(1 >)\ \epsilon > 0$.

Many a student of investment has stubbed his toe by interpreting (18) to mean that $\langle x_t^* \rangle$ generates higher expected utility than, say, $\langle x_t(\gamma) \rangle$. (17) and (18) may seem like a paradox but clearly implies that the geometric mean criterion *does not* give rise to a 'universally best' investment strategy.

The intuition behind this truth is as follows. For $\gamma < 0$ in (16), (17) and (18) occur because, despite the fact that the wealth distribution for $\langle x_t(\gamma) \rangle$ lies almost entirely to the *left* of the wealth distribution for $\langle x_t^* \rangle$, the lower tail of the distribution for $\langle x_t(\gamma) \rangle$ is shorter and (imperceptibly) thinner than the (bounded) left tail of the growth-optimal distribution. Thus, for negative powers, very small adverse changes in the lower tail overpower the value of almost surely ending up with a higher compound return. Conversely, for $\gamma > 0$, it is the *longer* (though admittedly very thin) upper tail that gives rise to (17) in the presence of (18) even though, again, the wealth distribution for $\langle x_t(\gamma) \rangle$ lies almost entirely to the *left* of the wealth distribution for $\langle x_t^* \rangle$.

4. Conditions for capital growth

As already noted, the determinants of whether capital will grow or decline (almost surely) in the long run are given by (12) and (11). Conditions (1)–(2) insure that (12) is feasible; in the absence of (1)–(2), positive growth may be infeasible. If a positive long-run growth rate (bounded away from zero) is achievable, then the growth-optimal strategy will find it. Thus we can state:

Theorem. *In the absence of (1) and (2), a necessary and sufficient condition for long-run capital growth to be feasible is that the growth-optimal strategy achieves a positive growth rate, i.e. that for some $\epsilon > 0$ and large T*

$$E[\ln R_t(x_t^*)] \geq \epsilon, \quad t \geq T \tag{19}$$

For $\gamma < 0$, the objective functions in (16) attain long-run growth rates of capital between those of the risk-free asset and of the growth-optimal strategy. But for $\gamma > 0$, the long run growth-rate may be negative. Consider for a moment the utility function $u(w) = w^{1/2}$, one of the most frequently cited examples of 'substantial' risk aversion since Bernoulli's time. Even this venerable function may, however, lead to (almost sure) ruin in the long-run: suppose, for example, that the riskfree asset yields 2% per period and that there is only one risky asset, which gives either a loss of 8.2% with probability 0.9, or a gain of 206% with probability 0.1. The optimal policy then calls for investing the fraction 1.5792 in the risky asset (by borrowing the fraction 0.5792 of current wealth to complete the financing) in each period. But the average compound growth rate g_t in (10) will now tend to -0.00756, or $-3/4\%$. Thus, expected utility 'grows' as capital itself almost surely vanishes.

What this example illustrates is that risk aversion plus a favorable return structure [see (1)–(3)] are *not* sufficient to insure capital growth in the basic reinvestment case.

5. Relationship to other long-run investment models

As shown in Section 3, the growth-optimal investment strategy has its traditional origin in arguments concerning capital growth and the law of large numbers. But it can also be derived strictly from an expected utility perspective — but only as a member of a small family.

Let n be the number of periods left to a terminal horizon point at time 0. Assume that wealth at that point, w_0, has utility $U_0(w_0)$, where $U_0' > 0$ everywhere and $U_0'' < 0$ for large w_0. Then, with one period to go, we have the single-period portfolio problem

$$U_1(w_1) \equiv \operatorname*{Max}_{z_1 | w_1} \; E[U_0(w_0(z_1))]$$

where $U_1(w_1)$ is the induced, or derived, utility of wealth w_1 at time 1 and the difference equation (6) has been trivially modified to

$$w_{n-1} = \sum_{i=2}^{M_n} (1 + r_{in}) z_{in} + w_n(1 + r_{1n}), \quad n = 1, 2, \ldots. \tag{20}$$

Thus, with n periods to go, we obtain

$$U_n(w_n) \equiv \operatorname*{Max}_{z_n | w_n} \; E[U_{n-1}(w_{n-1}(z_n))], \quad n = 1, 2, \ldots \tag{21}$$

where (21) is a standard recursive equation.

The induced utility of current wealth, $U_n(w_n)$, of course, generally depends on all the inputs to the problem, that is the utility of terminal wealth U_0, the joint distribution functions of future returns $F_n, \ldots, F_1$, and the future interest rates $r_{1n}, \ldots, r_{11}$. But there are two rather interesting special cases. The first is the case in which the induced utility functions $U_n(w_n)$ depend *only* on the terminal utility function U_0. This occurs when the returns are independent from period to period and $U_0(w_0)$ is isoelastic, i.e.

$$U_0(w_0) = \frac{1}{\gamma} w_0^{\gamma}, \text{ some } \gamma < 1.$$

As first shown by Mossin [1968], (21) now gives

$$\begin{aligned} U_n(w_n) &= a_n U_0(w_n) + b_n \\ &\sim U_0(w_n) \end{aligned}$$

(where $\sim$ means equivalent to) since a_n and b_n are constraints with a_n positive. The optimal investment policy is both myopic and proportional, i.e.

$$z_{in}^*(w_n) = x_{in}(\gamma) w_n, \text{ all } i$$

where the $x_{in}(\gamma)$ are constants.

The second special case obtains when returns are independent from period to period, interest rates are deterministic, and the terminal utility function reflects hyperbolic absolute risk aversion, that is (Hakansson 1971c)

$$U_0(w_0) = \begin{cases} \frac{1}{\gamma}(w_0 + \phi)^{\gamma}, & \gamma < 1, \\ \text{or} & \\ (\phi - w_0)^{\gamma}, & \gamma > 1, \ \phi \text{ large}; \\ \text{or} & \\ -\exp\{-\phi w_0\} & \phi > 0. \end{cases} \tag{22}$$

In the first subcase

$$U_n(w_n) = \frac{1}{\gamma}\left(w_n + \frac{\phi}{(1 + r_1) \ldots (1 + r_{1n})}\right)^{\gamma} \tag{23}$$

where (23) holds globally for $\phi \leq 0$ and locally for $\phi > 0$, i.e. for $w_n \geq L_n > 0$. The optimal investment policy is

$$z_{in}^*(w_n) = x_{in}(\gamma)\left(w_n + \frac{\phi}{(1 + r_{11}) \ldots (1 + r_{1n})}\right), \quad i \geq 2.$$

In the other two subcases, a closed form solution holds only locally.

But the most interesting result associated with (21) is surprisingly general. Under mild conditions on $U_0(w_0)$, and independent (but nonstationary) returns from period to period, we obtain [Hakansson, 1974; see also Leland, 1972; Ross,

1974; Huberman and Ross 1983]:

$$u_n(w_n) \to \frac{1}{\gamma} w_n^\gamma \tag{24}$$

and, if returns are stationary,

$$z_{in}^*(w_n) \to x_{in}^*(\gamma) w_n. \tag{25}$$

Thus, the class of utility functions

$$u(w) = \frac{1}{\gamma} w^\gamma, \quad \gamma < 1, \tag{16}$$

the only family with constant relative risk aversion (ranging from 0 to infinity) and exhibiting myopic and proportional investment policies, is evidently applicable to a *large* class of long-run investors. The optimal policies above are not mean–variance efficient, but for reasonably symmetric return distributions, they come close to MV efficiency.

Since $\gamma = 0$ in (24) corresponds to logarithmic utility of wealth, the growth-optimal strategy is clearly a member of this elite family of long-run oriented investors. In other words, the geometric mean investment strategy has a solid foundation in utility theory as well.

6. Relationship to intertemporal consumption–investment models

Up to this point, we have examined the basic dynamic investment problem, i.e. without reference to cash inflows or outflows. Under some conditions, the inclusion of these factors is straightforward and does not materially affect the optimal investment policy. But a realistic model incorporating noncapital in- and outflows typically complicates the model substantially.

The basic dynamic consumption–investment model incorporates consumption and a labor income into the dynamic reinvestment model. Following Fisher [1936], wealth is viewed as a means to an end, namely consumption.

The basic difference equation (6) now becomes

$$w_t = \sum_{i=2}^{M_t} (r_{it} - r_{1t}) z_{it} + (1 + r_{1t})(w_{t-1} - c_t) + y_t, \quad t = 1, \ldots, T, \tag{26}$$

where c_t is the amount consumed in period t (set aside at the beginning of the period) and y_t is the labor income received at the end of period t.

Consistent with the foregoing, the individual's objective becomes

$$\text{Max } E[U(c_1, \ldots, c_T)]$$

subject to

$$c_t \geq 0, \text{ all } t$$

where U is assumed to be monotone, strictly concave, and to reflect impatience, i.e. considering the two consumption streams

$$(a,\ b,\ c_3,\dots,c_T)$$
$$(b,\ a,\ c_3,\dots,c_T),\quad a>b$$

the first is preferred to the second.

In order to attain tractability, several strong assumptions are usually imposed:

1) the individual's lifetime (horizon) is known,
2) interest rates are viewed as deterministic,
3) the labor income y_t is deterministic; its present value is thus

$$Y_{t-1}\equiv\frac{y_t}{r_{1t}}+\dots+\frac{y_T}{(1+r_{1t})\dots(1+r_{1T})},$$

4) the utility function is assumed to be additive, i.e.

$$U(c_1,\dots,c_T)=u_1(c_1)+\alpha_1u_2(c_2)+\dots+\alpha_1\dots\alpha_{T-1}u_T(c_T),\tag{27}$$

where $u_t'>0$, $u_t''<0$, and typically $\alpha_t<1$, for all t, which implies that preferences are independent of past consumption.

Let

$$f_{t-1}(w_{t-1})\ =\ \text{maximum expected utility at } t-1 \text{ given } w_{t-1}.$$

This gives

$$f_{t-1}(w_{t-1})=\max_{c_t,z_t}\{u_t(c_t)+\alpha_tE[f_t(w_t)]\},\quad t=1,\dots,T,\tag{28}$$

where $f_T(w_T)\equiv 0$ or $b_T(w_T)$

$$\text{subject to}\quad c_t\ge 0\tag{29}$$
$$\Pr\{w_t\ge -Y_t\}=1\tag{30}$$
$$z_{it}\ge 0,\quad i\notin S_t\tag{31}$$

for each t, where $b_T(w_T)$ represents a possible bequest motive. It is apparent that $f_{t-1}(w_{t-1})$ represents the utility of wealth and that it is *induced* or *derived*; it clearly depends on everything in the model. Solving (28) recursively, it is evident that, under our assumptions concerning labor income and interest rates, Y_t can be exchanged for cash in the solution.

Suppose that in (27)

$$u_t(c_t)=\frac{1}{\gamma}c_t^{\gamma},\quad \gamma<1,\ t=1,\dots,T.\tag{32}$$

Then [Hakansson, 1970]

$$f_{t-1}(w_{t-1})=A_{t-1}(w_{t-1}+Y_{t-1})^{\gamma}+B_{t-1},$$

$$c_t^*(w_{t-1})=C_t(w_{t-1}+Y_{t-1}),$$

$$z^*_{it}(w_{t-1}) = (1 - C_t)x^*_{it\gamma}(w_{t-1} + Y_{t-1}), \quad i \geq 2,$$

and

$$z^*_{1t}(w_{t-1}) = w_{t-1} - c^*_t - \sum_{i=2}^{M_t} z^*_{it}(w_{t-1}),$$

where the A_t, B_t, and C_t are constants. Thus, the optimal consumption and investment policies are again proportional, not to w_{t-1} but to $w_{t-1} + Y_{t-1}$. The latter quantity is sometimes referred to as permanent income.

Note that when $\gamma = 0$ in (32), the consumer-investor does indeed employ the growth-optimal strategy to invested funds.

Finally, thc model (28)–(31) has been extended in a number of directions, to incorporate a random lifetime, life insurance, a subsistence level constraint on consumption, a Markov process for the economy, and an uncertain income stream from labor — with limited success [see Hakansson 1969, 1971b, 1972; Miller, 1974]. In general, closed-form solutions do not exist when income streams, payment obligations, and interest rates are stochastic. In such cases, multi-stage stochastic programming models are helpful [see e.g. Mulvey & Ziemba, 1995].

7. Growth vs. security

Empirical evidence suggests that the average investor is more risk averse than the growth-optimal investor, with a risk-tolerance corresponding to $\gamma \approx -3$ in (16) [see e.g. Blume & Friend, 1975]. While real-world investors exhibit a wide range of attitudes towards risk, this means that the majority of investors are in effect willing to sacrifice a certain amount of growth in favor of less variability, or greater 'security'.

7.1 The discrete-time case

In view of the convergence results (24) and (25), it is evident that repeated employment of (16) for any $\gamma < 0$ attains an efficient tradeoff between growth and security, as defined above, for the long-run investor. The concept of 'efficiency' is thus employed in a sense analogous to that used in mean–variance analysis.

A number of more direct measures of the sacrifice of growth for security have also been examined. In particular, MacLean, Ziemba & Blazenko [1992] analyzed the tradeoffs based on three growth and three security measures. The three growth measures are:

1. $E(w_t(\langle x_t \rangle))$], the expected wealth level after t periods;
2. $E[g_t]$, the mean compound growth rate over the first t periods;
3. $E[t : w_t(\langle x_t \rangle) \geq y]$, the mean first passage time to reach wealth level y;

while the three security measures are:

4. $\Pr\{w_t(\langle x_t \rangle) \geq y\}$, the probability that wealth level y will be reached in t periods;

5. $\Pr\{w_t(\langle x_t \rangle) \geq b_t,\ t = 1, 2, \ldots\}$, the probability that the investor's wealth is on or above a specified path;

6. $\Pr\{w_t(\langle x_t \rangle) \geq y$ before $w_t(\langle x_t \rangle) \leq b$, where $b < w_0 < y\}$, which includes the probability of doubling before halving.

Tradeoffs were generated via fractional Kelly strategies, i.e. strategies involving (stationary) mixtures of cash and the growth-optimal investment portfolio. Applied to a stationary environment, these strategies were shown to produce effective tradeoffs in that as growth declines, security increases. However, these tradeoffs, while easily computable, are generally not efficient, i.e. do not maximize security for a given (minimum) level of growth. Other comparisons involving the growth-optimal strategy and half Kelly or other strategies may be found in Ziemba & Hausch [1986], Rubinstein [1991], and Aucamp [1993].

7.2. The continuous-time case

Since transaction costs are zero under the perfect market assumption, it is natural to consider shorter and shorter periods between reinvestment decisions. In the limit, reinvestment takes place continuously. Assuming that the returns on risky assets can be described by diffusion processes, we obtain that optimal portfolios are mean–variance efficient in that the instantaneous variance is minimized for a given instantaneous expected return. The intuitive reason for this is that as the trading interval is shortened, the first two moments of the security's return become more and more dominant [see Samuelson, 1970]. The optimal portfolios also exhibit the separation property — as if returns over very short periods were normally distributed. Over any fixed interval, however, payoff distributions are, due to the compounding effect, usually lognormal. In other words, all investors with the same probability assessments, but regardless of risk attitude, invest in only two mutual funds, one of which is riskfree [Merton, 1971]. See also Karatzas, Lehoczky, Sethi and Shreve [1986] and Sethi and Taksar [1988].

In view of the above, it is evident that the tradeoff between growth and security generated by the fractional Kelly strategies in the continuous-time model when the wealth process is lognormal is efficient in a mean–variance sense. Li [1993] has addressed the growth vs. security question for the two asset case while Li & Ziemba [1992] and Dohi, Tanaka, Kaio & Osaki [1994] have done so when there are n risky assets that are jointly lognormally distributed.

8. Applications

8.1. Asset allocation

In view of the myopic property of the optimal investment policy in the dynamic reinvestment problem [see (24) and (25)], it is natural to apply (15) for different values of γ to the problem of choosing investment portfolios over time. In particular, the choice of broad asset categories, also known as the asset allocation problem, lends itself especially well to such treatment. Thus, to implement the growth-optimal strategy, for example, we merely solve (15) subject to relevant constraints (on borrowing when available and on short positions) at the beginning of each period.

To implement the model, it is necessary to estimate the joint distribution function for next period's returns. Since all moments and comoments matter, one way to do this is to employ the joint empirical distribution for the previous n periods. This approach provides a simple and realistic means of generating nonstationary scenarios of the possible outcomes over time. The raw distribution may of course may be modified in any number of ways, for example via Stein estimators [Jorion, 1985, 1986, 1991; Grauer & Hakansson, 1995], an inflation adapter [Hakansson, 1989], or some other method.

Grauer and Hakansson applied the dynamic reinvestment model in a number of settings with up to 16 different risk attitudes γ under both quarterly and annual portfolio revision. In the domestic setting [Grauer & Hakansson, 1982, 1985, 1986], the model was employed to construct and rebalance portfolios composed of U.S. stocks, corporate bonds, government bonds, and a riskfree asset. Borrowing was ruled out in the first article while margin purchases were permitted in the other two. The third article also included small stocks as a separate investment vehicle. On the whole, the growth-optimal strategy lived up to its reputation. On the basis of the empirical probability assessment approach, quarterly rebalancing, and a 32-quarter estimating period applied to 1934–1992, the growth-optimal strategy outperformed all the others — with borrowing permitted, it earned an average annual compound return of nearly 15%.

In Grauer & Hakansson [1987], the model was applied to a global environment by including in the universe the four principal U.S. asset categories and up to fourteen non-U.S. equity and bond categories. The results showed that the gains from including non-U.S. asset classes in the universe were remarkably large (in some cases statistically significant), especially for the highly risk-averse strategies. With leverage permitted and quarterly rebalancing, the geometric mean strategy again came out on top, generating an annual compound return of 27% over the 1970–1986 period. A different study examined the impact from adding three separate real estate investment categories to the universe of available categories [Grauer & Hakansson, 1994b]. Finally, Grauer, Hakansson & Shen [1990] examined the asset allocation problem when the universe of risky assets was composed of twelve equal- and value-weighted industry components of the U.S. stock market.

Mulvey [1993] developed a multi-period model of asset allocation which incorporates transaction costs, including price impact. The objective function is a general concave utility function. A computational version developed by Mulvey & Vladimirou [1992] focused on the isoelastic class of functions in which the objective was to maximize the expected utility of wealth at the end of the planning horizon. This model, like those based on the empirical distribution approach, can handle assets possessing skewed returns, such as options and other derivatives, and can be extended to include liabilities [see Mulvey & Ziemba, 1995]. Based on historical data over the period 1979 to 1988, this research, based on multi-stage stochastic programming, showed that efficiencies could be gained vis-à-vis myopic models in the presence of transaction costs by taking advantage of the network or linear structure of the problem.

Mean–variance approximations. A number of authors have argued that, in the single period case, power function policies can be well approximated by MV policies, e.g. Levy & Markowitz [1979], Pulley [1981, 1983], Kallberg & Ziemba [1979, 1983], and Kroll, Levy & Markowitz [1984]. However, there is an opposing intuition which suggests that the power functions' strong aversion to low returns and bankruptcy will lead them to select portfolios that are not MV-efficient, e.g. Hakansson [1971a] and Grauer [1981, 1986]. It is therefore of interest to know whether the power policies differ from the corresponding MV and quadratic policies when returns are compounded over many periods.

Let μ_{it} be the expected rate of return on security i at time t and σ_{ijt} be the covariance between the returns on securities i and j at time t. Then the MV investment problem is

$$\operatorname*{Max}_{x_t} \{T(1+\mu_t) - \tfrac{1}{2}\sigma_t^2\},$$

subject to the usual constraints. The MV approximation to the power functions in (16) are obtained [Ohlson, 1975; Pulley, 1981] when

$$T = \frac{1}{1-\gamma}.$$

Under certain conditions this result holds exactly in continuous time [see Merton, 1973, 1980].

With quarterly revision, the MV model was found to approximate the exact power function model very well [Grauer & Hakansson, 1993]. But with annual revision, the portfolio compositions and returns earned by the more risk averse power function strategies bore little resemblance to those of the corresponding MV approximations. Quadratic approximations proved even less satisfactory in this case. These results contrast somewhat with those of Kallberg & Ziemba [1983], who in the quadratic case with smaller variances obtained good approximations for horizons up to a whole year [see also MacLean, Ziemba & Blazenko, 1992].

8.2. Growth–security tradeoffs

The growth vs. security model has been applied to four well-known gambling-investment problems: blackjack, horse race wagering, lotto games, and commodity trading with stock index futures. In at least the first three cases, the basic investment situation is unfavorable for the average player. However, systems have been developed that yield a positive expected return. The various applications use a variety of growth and security measures that appear to model each situation well. The size of the optimal investment gamble also varies greatly, from over half to less than one millionth of one's fortune.

Blackjack. By wagering more in favorable situations and less or nothing when the deck is unfavorable, an average weighted edge is about 2%. An approximation to provide insight into the long-run behavior of a player's fortune is to assume that the game is a Bernoulli trial with a probability of success equal to 0.51. With a 2% edge, the optimal wager is also 2% of one's fortune. Professional blackjack teams often use a fractional Kelly wagering strategy with the fraction drawn from the interval 0.2 to 0.8. For further discussion, see Gottlieb [1985] and Maclean, Ziemba & Blazenko [1992].

Horseracing. There is considerable evidence supporting the proposition that it is possible to identify races where there is a substantial edge in the bettor's favor (see the survey by Hausch & Ziemba [1995] in this volume). At thoroughbred racetracks, one can find about 2–4 profitable wagers with an edge of 10% or more on an average day. These opportunities arise because (1) the public has a distaste for the high probability–low payoff wagers, and (2) the public is unable to properly evaluate the worth of multiple horse place and show and exotic wagers because of their complexity; for example, in a ten-horse race there are 120 possible show finishes, each with a different payoff and chance of occurrence. In this situation, interesting tradeoffs between growth and security arise as well.

The Kentucky Derby represents an interesting special case because of the long distance (1 1/4 miles), the fact that the horses have not previously run this distance, and the fame of the race. Hausch, Bain & Ziemba [1995] tabulated the results from Kelly and half Kelly wagers using the system in Ziemba & Hausch [1987] over the 61-year period 1934–1994. They also report the results from using a filter rule based on the horse's breeding.

Lotto games. Lotteries tend to have very low expected payoffs, typically on the order of 40 to 50%. One way to 'beat' parimutuel games is to wager on unpopular numbers — see Hausch & Ziemba [1995] for a survey. But even when the odds are 'turned' favorable, the optimal Kelly wagers are extremely small and it may take a very long time to reach substantial profits with high probability. Often an initial wealth level in the seven figures is required to justify the purchase of even a single $1 ticket. Comparisons between fractional and full Kelly strategies can be found in MacLean, Ziemba & Blazenko [1992].

Commodity trading. Repeated investments in commodity trades can be modeled as a capital growth problem via suitable modifications for margin requirements, daily mark-to-the-market procedures, and other practical details. An interesting example is the turn-of-the-year effect exhibited by U.S. small stocks in January.

One way to benefit from this anomaly is to take long positions in a small stock index and short positions in large stock indices, because the transaction costs (commissions plus market impact) are less than a tenth of what they would be by transacting in the corresponding basket of securities. Using data from 1976 through January 1987, Clark & Ziemba [1987] calculated that the growth-optimal strategy would invest 74% of one's capital in this opportunity. Hence fractional Kelly strategies are suggested. See also Ziemba [1994].

9. Summary

Capital growth theory is useful in the analysis of many dynamic investment situations, with many attractive properties. In the basic reinvestment case, the growth-optimal investment strategy, also known as the Kelly criterion, almost surely leads to more capital in the long run than any other investment policy which does not converge to it. It never risks ruin, and also has the appealing property that it asymptotically minimizes the expected time to reach a given level of capital. The Kelly criterion implies, and is implied by, logarithmic utility of wealth (only) at the end of each period; thus, its relative risk aversion equals 1, which makes it more risk-tolerant than the average investor. As a result, tradeoffs between growth and security have found application in a rich set of circumstances.

The fact that the growth-optimal investment strategy is proportional to beginning-of-period wealth is of great practical value. But perhaps the most significant property of the Kelly criterion is that it is myopic not only when returns are nonstationary and independent but also when they obey a Markov process. In the dynamic investment model with a given terminal objective function, the growth-optimal strategy is a member of the set to which the optimal policy converges as the horizon becomes more distant. Finally, the Kelly criterion is optimal in many environments in which consumption, noncapital income, and payment obligations are present.

References

Algoet, P.H., and T.M. Cover (1988). Asymptotic optimality and asymtotic equipartition properties at log-optimum investment. *Ann. Prob.*, 16, 876–898.

Aucamp, D. (1993). On the extensive number of plays to achieve superior performance with the geometric mean strategy. *Manage. Sci.* 39, 1163–1172.

Bell, R.M., and T.M. Cover (1980). Competitive optimality of logarithmic investment. *Math. Oper. Res.* 5, 161–166.

Bell, R., and T.M. Cover (1988). Game-theoretic optimal portfolios. *Manage. Sci.* 34, 724–733.

Bernoulli, D. (1738/1954). Exposition of a new theory on the measurement of risk (translation Louise Summer). *Econometrica*, 22, 23–36.

Blume, M.E., and I. Friend (1975). The asset structure of individual portfolios and some implications for utility functions. *J. Finance* 30, 585–603.

Breiman, L. (1960). Investment policies for expanding business optimal in a long-run sense. *Nav. Res. Logist. Q.* 7, 647–651.

Breiman, L. (1961). Optimal gambling system for favorable games, in: *Proc. 4th Berkeley Symp. on Mathematics, Statistics and Probability* 1, 63–68.

Chamberlain, G. (1983). A characterization of the distributions that imply mean–variance utility functions. *J. Econ. Theory* 29, 185–201.

Chernoff, H. (1980/1981). An analysis of the Massachusetts Numbers Game, Tech. Rep. No. 23, MIT Department of Mathematics, Massachusetts Institute of Technology, Cambridge, MA.,

1980; shortened version published in *Math. Intell.* 3, 166–172.
Clark, R., and W.T. Ziemba (1987). Playing the turn of the year with index futures. *Oper. Res.* 35, 799–813.
Dohi, T., H. Tanaka, N. Kaio and S. Osaki (1994). Alternative Growth Versus Security in Continuous Dynamic Trading. *Eur. J. Oper. Res.*, in press.
Efron, B., and C. Morris (1973). Stein's estimation rule and its competitors — An empirical Bayes approach. *J. Am. Stat. Assoc.* 68, 117–130.
Efron, B., and C. Morris (1975). Data analysis using Stein's estimator and its generalizations. *J. Am. Stat. Assoc.* 70, 311–319.
Efron, B., and C. Morris (1977). Stein's paradox in statistics. *Sci. Am.* 236, 119–127.
Epstein, R.A. (1977). *The Theory of Gambling and Statistical Logic*, 2nd edition, Academic Press, New York, NY.
Ethier, S.N. (1987). The Proportional Bettor's Fortune, in: *Proc. 7th Int. Conf. on Gambling and Risk Taking*, Department of Economics, University of Nevada, Reno, NV.
Ethier, S.N., and S. Tavare (1983). The proportional bettor's return on investment. *J. Appl. Probab.* 20, 563–573.
Feller, W. (1962). *An Introduction to Probability Theory and Its Applications*, 1, 2nd edition, John Wiley & Sons, New York, NY.
Ferguson, T.S. (1965). Betting systems which minimize the probability of ruin. *J. Soc. Appl. Math.* 13, 795–818.
Finkelstein, M., and R. Whitley (1981). Optimal strategies for repeated games. *Adv. Appl. Prob.* 13, 415–428.
Fisher, I. (1930). *The Theory of Interest*, New York, MacMillan, reprinted Augustus Kelley, 1965.
Friedman, J. (1982). Using the Kelly criterion to select optimal blackjack bets, Mimeo, Stanford University.
Goldman, B. (1974). A negative report on the 'near optimality' of the max-expected log policy as applied to bounded utilities for long-lived programs. *J. Financ. Econ.* 1, 97–103.
Gottlieb, G. (1984). An optimal betting strategy for repeated games, Mimeo, New York University.
Gottlieb, G. (1985). An analystic derivation of blackjack win rates. *Oper. Res.* 33, 971–988.
Grauer, R.R. (1981). A comparison of growth optimal and mean variance investment policies. *J. Financ. Quant. Anal.* 16, 1–21.
Grauer, R.R. (1986). Normality, Solvency and Portfolio Choice, *J. Financ. Quant. Anal.* 21, 265–278.
Grauer, R.R., and N.H. Hakansson (1982). Higher return, lower risk: Historical returns on long-run, actively managed portfolios of stocks, bonds and bills, 1936–1978. *Financ. Anal. J.* 38, 39–53.
Grauer, R.R., and N.H. Hakansson (1985). Returns on levered, actively managed long-run portfolios of stocks, bonds and bills, 1934–1984. *Financ. Anal. J.* 41, 24–43.
Grauer, R.R., and N.H. Hakansson (1986). A half-century of returns on levered and unlevered portfolios of stocks, bonds, and bills, with and without small stocks. *J. Bus.* 59, 287–318.
Grauer, R.R., and N.H. Hakansson (1987). Gains from international diversification: 1968–85 returns on portfolios of stocks and bonds. *J. Finance* 42, 721–739.
Grauer, R.R., and N.H. Hakansson (1993). On the use of mean–variance and quadratic approximations in implementing dynamic investment strategies: A comparison of returns and investment policies. *Manage. Sci.* 39, 856–871.
Grauer, R.R., and N.H. Hakansson (1994a). On timing the market: The empirical probability approach with an inflation adapter, Manuscript.
Grauer, R.R., and N.H. Hakansson (1994b). Gains from diversifying into real estate: Three decades of portfolio returns based on the dynamic investment model, *Real Estate Economics* 23, 119–159.
Grauer, R.R., and N.H. Hakansson (1995). Stein and CAPM estimators of the means in asset allocation, Working Paper (forthcoming)
Grauer, R.R., N.H. Hakansson and F.C. Shen (1990). Industry rotation in the U.S. stock market: 1934-1986 returns on passive, semi-passive, and active strategies. *J. Banking Finance* 14, 513–535.

Griffin, P. (1985). Different measures of win rate for optimal proportional betting. *Manage. Sci.* 30, 1540–1547.

Hakansson, N. (1969). Optimal investment and consumption strategies under risk, an uncertain lifetime, and insurance. *Int. Econ. Rev.* 10, 443–466.

Hakansson, N. (1970). Optimal investment and consumption strategies under risk for a class of utility functions. *Econometrica* 38, 587–607.

Hakansson, N. (1971a). Capital growth and the mean–variance approach to portfolio selection. *J. Financ. Quant. Anal.* 6, 517–557.

Hakansson, N. (1971b). Optimal entrepreneurial decisions in a completely stochastic environment. *Manage. Sci., Theory* 17, 427–449.

Hakansson, N. (1971c). On optimal myopic portfolio policies, with and without serial correlation of yields. *J. Bus.* 44, 324–234.

Hakansson, N. (1972). Sequential investment–consumption strategies for individuals and endowment funds with lexicographic preferences, in: J. Bicksler (ed.), *Methodology in Finance — Investments*, D.C. Heath & Company, Lexington, MA, pp. 175–203.

Hakansson, N. (1974). Convergence to isoelastic utility and policy in multiperiod portfolio choice. *J. Financ. Econ.* 1, 201–224.

Hakansson, N. (1979). A characterization of optimal multiperiod portfolio policies, in: E. Elton and M. Gruber (eds.), *Portfolio Theory, 25 Years Later*, Amsterdam, North Holland, pp. 169–177.

Hakansson, N. (1989). On the value of adapting to inflation in sequential portfolio decisions, in: B. Fridman and L. Ostman (eds.), *Accounting Development — Some Perspectives*, The Economic Research Institute, Stockholm School of Economics, pp. 151–185.

Hakansson, N., and B. Miller (1975). Compound-return mean–variance efficient portfolios never risk ruin. *Manage. Sci.* 22, 391–400.

Hausch, D., and W.T. Ziemba (1985). Transactions costs, extent of inefficiencies, entries and multiple wagers in a racetrack betting model. *Manage. Sci.* 31, 381–392.

Hausch, D., and W.T. Ziemba (1990). Arbitrage strategies for cross-track betting on major horse races. *J. Bus.* 63, 61–78.

Hausch, D., and W.T. Ziemba (1995). Efficiency of sports and lottery betting markets, in: R. Jarrow, V. Maksimovic and W.T. Ziemba (eds.), *Finance*, Handbooks in Operations Research and Management Science, Vol 9, Elsevier, Amsterdam, pp. 545–580 (this volume).

Hausch, D., W.T. Ziemba and M. Rubinstein (1981). Efficiency of the market for racetrack betting. *Manage. Sci.* 27, 1435–1452.

Hausch, D., R. Bain and W.T. Ziemba (1995). Wagering on the Kentucky Derby, 1934–1994, Mimeo, University of British Columbia.

Huberman, G., and S. Ross (1983). Portfolio turnpike theorems, risk aversion and regularly varying utility functions. *Econometrica* 51, 1345–1361.

Ibbotson Associates, Inc. (1986). *Stocks, Bonds, Bills and Inflation: Market Results for 1926–1985*, Ibbotson Associates, Inc., Chicago.

Ibbotson Associates, Inc. (1988). *Stocks, Bonds, Bills and Inflation: 1987 Yearbook*, Ibbotson Associates, Inc., Chicago.

James, W., and C. Stein (1961). Estimation with quadratic loss, in: *Proc. 4th Berkeley Symp. on Probability and Statistics I*, Berkeley, University of California Press, pp. 361–379.

Jobson, J.D., and B. Korkie (1981). Putting Markowitz theory to work. *J. Portfolio Manag.* 7, 70–74.

Jobson, J.D., B. Korkie and V. Ratti (1979). Improved estimation for Markowitz portfolios using James–Stein type estimators. in: *Proc. American Statistical Association*, Business and Economics Statistics Section 41, 279–284.

Jorion, P. (1985). International portfolio diversification with estimation risk. *J. Bus.* 58, 259–278.

Jorion, P. (1986). Bayes–Stein estimation for portfolio analysis. *J. Financ. Quant. Anal.* 21, 279–292.

Jorion, P (1991). Bayesian and CAPM estimators of the means: Implications for portfolio selection. *J. Banking Finance* 15, 717–727.

Kalymon, B. (1971). Estimation risk in the portfolio selection model. *J. Financ. Quant. Anal.* 6, 559–582.

Kallberg, J.G., and W.T. Ziemba (1979). On the robustness of the Arrow–Pratt risk aversion measure. *Econ. Lett.* 2, 21–26.

Kallberg, J.G., and W.T. Ziemba (1983). Comparison of alternative utility functions in portfolio selection problems. *Manage. Sci.* 9, 1257–1276.

Karatzas, I., J. Lehoczky, S.P. Sethi and S.F. Shreve (1986). Explicit Solution of a General Consumption/Investment Problem. *Math. Oper. Res.* 11, 261–294.

Kelly, J.L., Jr. (1956). A new interpretation of information rate. *Bell Syst. Tech. J.* 35 917–926.

Kroll, Y., H. Levy and H. Markowitz (1984). Mean–variance versus direct utility maximization. *J. Finance* 39, 47–75.

Latane, H. (1959). Criteria for choice among risky ventures. *J. Polit. Econ.* 67, 144–145.

Leland, H. (1972). On turnpike portfolios, in: K. Shell and G.P. Szego (eds.), *Mathematical Methods in Investment and Finance*, Amsterdam, North-Holland.

Levy, H., and H. Markowitz (1979). Approximating expected utility by a function of mean and variance. *Am. Econ. Rev.* 69, 308–317.

Li, Y. (1993). Growth–security investment strategy for long and short runs. *Manage. Sci.* 39, 915–934.

Li, Y., and W.T. Ziemba (1992). Security aspects of optimal growth models with minimum expected time criteria, Mimeo, University of British Columbia, Canada.

Loistl, O. (1976). The erroneous approximation of expected utility by means of a Taylor's series expansion: Analytic and computational results. *Am. Econ. Rev.* 66, 904–910.

MacLean, L.C., and W.T. Ziemba (1986). Growth versus security in a risky investment model, in: F. Archetti, G. DiPillo and M. Lucertini (eds.), *Stochastic Programming*, Springer Verlag, pp. 78–87.

MacLean, L.C., and W.T. Ziemba (1990). Growth–security profiles in capital accumulation under risk. *Ann. Oper. Res.* 31, 501–509.

MacLean, L.C., and W.T. Ziemba (1994). Capital growth and proportional investment strategies, Mimeo, Dalhousie University.

MacLean, L.C., W.T. Ziemba and G. Blazenko (1992). Growth versus security in dynamic investment analysis. *Manage. Sci.* 38, 1562–1585.

Markowitz, H.M. (1959). *Portfolio Selection: Efficient Diversification of Investments*. John Wiley & Sons, Inc., New York, NY.

Markowitz, H. (1976). Investment for the long run: New evidence for an old rule. *J. Finance* 31, 1273–1286.

Merton, R.C. (1971). Optimal consumption and portfolio rules in a continuous-time model. *J. Econ. Theory* 3, 373–413.

Merton, R.C. (1973). An intertemporal capital asset pricing model. *Econometrica* 41, 867–887.

Merton, R.C. (1980). On estimating the expected return on the market: An exploratory investigation. *J. Financ. Econ.* 8, 323–361.

Miller, B.L. (1974). Optimal consumption with a stochastic income stream. *Econometrica* 42, 253–266.

Mossin, J. (1968). Optimal multiperiod portfolio policies. *J. Bus.* 41, 215–229.

Mulvey, J.M. (1993). Incorporating transaction costs in models for asset allocation, in: *Financial Optimization*, S. Zenios (ed.) Cambridge University Press.

Mulvey, J.M., and H. Vladimirou (1992). Stochastic network programming for financial planning problems. *Manage. Sci.* 38, 1642–1664.

Mulvey, J.M., and W.T. Ziemba (1995). Asset and liability allocation in a global environment. in: R. Jarrow, V. Maksimovic and W.T. Ziemba (eds.), *Finance*, Handbooks in Operations Research and Management Science, Vol 9, Elsevier, Amsterdam, pp. 435–464 (this volume).

Ohlson, J.A. (1975). The asymptotic validity of quadratic utility as the trading interval approaches zero, in: W.T. Ziemba and R.G. Vickson (eds.), *Stochastic Optimization Models in Finance*, Academic Press, New York, NY.

Pulley, L.B. (1981). A general mean–variance approximation to expected utility for short holding periods. *J. Financ. Quant. Anal.* 16, 361–373.

Pulley, L.B. (1983). Mean–variance approximation to expected logarithmic utility. *Oper. Res.* 31, 685–696.

Ritter, J.R. (1988). The buying and selling behavior of individual investors at the turn of the year: Evidence of price pressure effects. *J. Finance* 43, 701–719.

Roll, R. (1983). Was ist das? The turn of the year effect and the return premia of small firms. *J. Portfolio Manag.* 10, 18–28.

Ross, S. (1974). Portfolio turnpike theorems for constant policies. *J. Financ. Econ.* 1, 171–198.

Rotando, L.M., and E.O. Thorp (1992). The Kelly criterion and the stock market. *Am. Math. Mon.* December, 922–931.

Rubinstein, M. (1977). The strong case for log as the premier model for financial modeling, in: H. Levy and M. Sarnat (eds.), *Financial Decisions Under Uncertainty*, Academic Press, New York, NY.

Rubinstein, M. (1991). Continuously rebalanced investment strategies. *J. Portfolio Manage.* 17, 78–81.

Samuelson, P.A. (1970). The fundamental approximation theorem of portfolio analysis in terms of means, variances, and higher moments. *Rev. Econ. Studies* 36, 537–542.

Samuelson, P.A. (1971). The 'fallacy' of maximizing the geometric mean in long sequences of investing or gambling. *Proc. Nat. Acad. Sci.* 68, 2493–2496.

Samuelson, P.A. (1977). St. Petersburg paradoxes: Defanged, dissected, and historically described. *J. Econ. Lit.* XV, 24–55.

Sethi, S.P. and M.I. Taksar (1988). A Note on Merton's Optimum Consumption and Portfolio Rules in a Continuous-Time Model. *J. Econ. Theory* 46, 395–401.

Stein, C. (1955). Inadmissibility of the usual estimator for the mean of a multivariate normal distribution, in: *Proc. 3rd Berkeley Symp. on Probability and Statistics I*, University of California Press, Berkeley, CA, pp. 197–206.

Thorp, E.O. (1966). *Beat the Dealer*, 2nd edition, Random House, New York, NY.

Thorp, E.O. (1975). Portfolio choice and the Kelly criterion, in: W.T. Ziemba and R.G. Vickson (eds.), *Stochastic Optimization Models in Finance*, Academic Press, New York, NY.

Williams, J. (1936). Speculation and carryover. *Q. J. Econ.* L, 436–455.

Wu, M.G.H., and W.T. Ziemba (1990). Growth versus security tradeoffs in dynamic investment analysis, Mimeo, University of British Columbia, B.C.

Ziemba, W.T. (1994). Investing in the turn of the year effect in the U.S. futures markets. *Interfaces* 24, 46–61.

Ziemba, W.T., and D.B. Hausch (1986). *Betting at the Racetrack*, Dr. Z Investments, Inc., Los Angeles and Vancouver.

Ziemba, W.T., and D.B. Hausch (1987). *Dr. Z's Beat the Racetrack*, William Morrow, New York (revised and expanded second edition of Ziemba–Hausch, *Beat the Racetrack*, Harcourt, Brace and Jovanovich, 1984).

Ziemba, W.T., C. Parkan and R. Brooks-Hill (1974). Calculation of investment portfolios with risk-free borrowing and lending. *Manage. Sci.* 21, 209–222.

Ziemba, W.T., S.L. Brumelle, A. Gautier and S.L. Schwartz (1986). *Dr. Z's 6/49 Lotto Guidebook*, Dr. Z Investments, Inc., Los Angeles and Vancouver.

Ziemba, W.T. and R.G. Vickson (eds.) (1975). Stochastic Optimization Models in Finance, Academic Press, New York, N.Y.

Journal of Economic Dynamics and Control 17 (1993) 887–906. North-Holland

42

A preference foundation for log mean–variance criteria in portfolio choice problems

David G. Luenberger
Stanford University, Stanford, CA 94305, USA

Received October 1991, final version received June 1992

The appropriate criterion for evaluating, and hence also for properly constructing, investment portfolios whose performance is governed by an infinite sequence of stochastic returns has long been a subject of controversy and fascination. A criterion based on the expected logarithm of one-period return is known to lead to exponential growth with the greatest exponent, almost surely; and hence this criterion is frequently proposed. A refinement has been to include the variance of the logarithm of return as well, but this has had no substantial theoretical justification.

This paper shows that log mean–variance criteria follow naturally from elementary assumptions on an individual's preference relation for deterministic wealth sequences. As a first and fundamental step, it is shown that if a preference relation involves only the tail of a sequence, then that relation can be extended to stochastic wealth sequences by almost sure equality. It is not necessary to introduce a von Neumann–Morgenstern utility function or the associated axioms.

It is then shown that if tail preferences can be described by a 'simple' utility function, one that is of the form $\overline{\lim}_{n\to\infty}\bar{\rho}(W_n, n)$ where W_n is wealth at period n, this utility must under suitable conditions be a function of the expected logarithm of return, independent of the functional form of $\bar{\rho}$. Finally, 'compound' utility functions are introduced; and they are shown under suitable conditions to be functions of the expected value and variance of the logarithm of one-period return, again independently of the specific form of the underlying function. The infinite repetitions of the dynamic process essentially 'hammer' all utility functions into a log mean–variance form.

1. Introduction

Consider an investment situation wherein a fund is initially endowed with capital and, through returns from investment, grows (or decreases) with time. After the initial endowment, no additions or withdrawals are made. The initial endowment is simply transformed by repeated chance operations of investment as controlled by the portfolio structure. The management objective in this

Correspondence to: David G. Luenberger, Department of Engineering-Economic Systems, Stanford University, Stanford, CA 94305–4025, USA.

*The author wishes to thank David R. Cariño for many helpful suggestions on this paper.

situation, roughly, is to cause the investment value to increase as rapidly as possible, subject perhaps to considerations of security. Beyond this general statement it is, at least initially, difficult to define a concrete investment criterion. Yet this situation, even though highly idealized, closely approximates many real money management situations, such as those involving university endowment funds, mutual funds, gambling, expansion of a business enterprise, and personal wealth-building. It is a situation that has received a great deal of attention in the literature.

To make the situation tractable, it is usually assumed that the investment environment is stationary in the sense that there is a fixed set of investment opportunities each of which are available in every period and that the returns of these investments are independent and identically distributed between periods (although there is dependency among the different investments). With this framework, the academic literature has traditionally taken three main approaches to this idealized management situation. One approach is to argue, on the basis of very strong limit properties, that the best policy is the one that maximizes $\mathrm{E}\log W_n$, where W_n denotes the wealth in the fund at some terminal period n and E denotes the expectation operator [Kelly (1956), Brieman (1961), Latané (1959)]. Because of the stationarity assumption, this policy reduces to that of successively repeating the single-period policy of maximizing $\mathrm{E}\log W_1$. The second approach is based on adherence to the expected utility framework of von Neumann and Morgenstern and suggests that one should maximize $\mathrm{E}U(W_n)$ for some utility function U. This includes $U(W_n) = \log W_n$, but there is no particular rationale for selecting that over any other utility function [see Samuelson (1971)]. A third approach, also based on the expected utility framework, suggests that one maximize $\mathrm{E}\sum_{i=1}^{n}(\beta)^i u(c_i)$, where c_i is (a monetary equivalent of) consumption drawn out of the wealth at period i, and β, $0 < \beta < 1$, is a discount factor [Samuelson (1969)]. This last approach has good economic justification within the von Neumann–Morgenstern framework, although the special form of additive discounted utility is quite specialized and does not really capture the spirit of the original (vague) objective of managing money which is not to be drawn down until the (distant) future. Our objective is not to argue that one approach is superior to another, but simply to provide a new preference foundation for the $\mathrm{E}\log W_1$ approach (or a generalization of it), so that it can be addressed by economic principles.

There has been considerable interest in considering the limiting behavior of the $\mathrm{E}U(W_n)$ criterion as $n \to \infty$, to obtain approximate and simple results. It has been hoped that this might provide some economic justification for the expected log approach. The principal technique in such investigations has been to impose restrictions on the utility function (usually through boundedness assumptions) in order to deduce that the investment policy derived from the $\mathrm{E}\log W_n$ criterion also maximizes $\mathrm{E}U(W_n)$ as $n \to \infty$ [cf. Markowitz (1959), Leland (1972)]. Such attempts have been of limited success, as elucidated in

Samuelson (1971) and Goldman (1974). No convincing argument has yet been made that reduces the general utility function approach (for a wide class of useful utility functions) to the $\mathrm{E}\log W_n$ (or equivalently $\mathrm{E}\log W_1$) approach.

The $\mathrm{E}\log W_n$ approach has been brought closer to standard economic theory by generalizing it to a criterion based on a trade-off between $\mathrm{E}\log W_n$ and $\mathrm{var}(\log W_n)$ [Williams (1936)]. Specifically, one defines an efficient frontier for fixed n, where $\mathrm{var}(\log W_n)$ is minimized subject to $\mathrm{E}\log W_n \geq m$, for various values of m. An efficient policy corresponds to a point on this frontier. This procedure parallels the familiar mean–variance approach to single-period investment [Markowitz (1952), Sharpe (1970), Tobin (1965), Lintner (1965)] and has the appeal of simplicity of that familiar procedure. Of course, this itself may not eliminate the conceptual gap between the asymptotic approach and classical utility theory, since even single-period mean–variance theory is not consistent with expected utility maximization except in special circumstances [Borch (1963)]. However, it has been shown that the log mean–variance approach[1] serves as an approximation to the terminal utility approach when uncertainty is 'small' [Samuelson (1970)].

Because of its intuitive appeal and simplicity, it is natural to look for stronger justification for log mean–variance criteria, and the central limit theorem might seem to provide the basis for such a justification. The distribution of the sum of the logarithms of n independent and identically distributed returns, when its mean is subtracted and the result is normalized by dividing by $n^{1/2}$, converges to a normal distribution. Ignoring the required normalization, one might argue that the distribution of the sum itself is close to a normal distribution for large n. Then since this normal distribution is completely characterized by its mean and variance, it is plausible to use these two parameters as determinants for an investment criterion. This idea has been explored in Hakansson (1971a, b), and it was found that the corresponding efficient frontier degenerates to a single point as $n \to \infty$, since the ratio of standard deviation to mean goes to zero.

If the normalization is accounted for, the situation is far more complex, since there seems to be no valid limit argument. This was forcefully pointed out in Merton and Samuelson (1974) and observation of this limit fallacy led them once again to dismiss log mean–variance criteria as too simplistic. (Additionally, the very idea of using utility at a terminal time and taking the limit also implicitly involves a fallacy since $\lim_{n\to\infty} \mathrm{E}\{U(W_n)\}$ is not an expected utility.)

So where does all of this leave us? The log mean criterion, or more generally, the log mean–variance approach is intuitively appealing, but apparently not consistent with standard expected utility theory.

[1] The phrase 'log mean–variance criteria' is used throughout as shorthand for 'criteria involving only the mean and variance of the logarithm of the return'.

In section 3 we set the stage for a new preference-based approach to the problem. We begin by establishing preferences on infinite sequences of wealth rather than wealth at a fixed (but later taken to the limit) terminal time. This approach circumvents the limit problems encountered by the earlier approaches, and in fact provides a path whereby powerful limit theorems of probability theory can be applied. The preference relations on infinite deterministic sequences are extended to stochastic sequences, not by imposition of a von Neumann–Morgenstern expected value criterion, but instead quite naturally by an 'almost sure' criterion, which is applicable when the original preference relation depends entirely on asymptotic properties, as is appropriate for this problem. It is then shown that a wide spectrum of these asymptotic preferences reduce to log mean–variance criteria. Thus, under quite broad conditions the log mean–variance approach is identical with an asymptotic preference approach. The justification for the variance comes about, by the way, not from the central limit theorem, but from its stronger cousin, the law of the iterated logarithm.

2. The investment environment

We begin by more explicitly defining the investment environment. There is an underlying vector random process $Z = \{Z_k\}$ defined on a probability space $(\Omega, \mathcal{F}, P)$. Each $Z_k = (Z_{1k}, Z_{2k}, \ldots, Z_{Sk})$ is a vector of S random variables. An investment policy $(\alpha_1, \alpha_2, \ldots)$ is a sequence of mappings α_k: $R^s \to R$. This policy defines a new process $X = \{X_k\}$ of random variables by $X_k = \alpha_k(Z_k)$. The X process is the single-period return process. The X process in turn generates a wealth process W according to

$$W_k = W_{k-1} X_k ,$$

where W_0 is the (fixed) initial wealth. The investment problem, in this general setting, is that of selecting the policy α within a class of feasible policies that leads to the most desirable W process.

It should be noted that systematic withdrawals from the accumulated wealth can be accommodated in this framework by incorporating them into the definition of α_k. For instance, a withdrawal of a fraction β_k of the wealth at period k using investment policy α_k is treated by setting $X_k = \alpha'_k(Z_k) = (1 - \beta_k)\alpha_k(Z_k)$.

A concrete example of the general framework is where there are S securities, each of which may be purchased in any amount at the beginning of each period. If one unit of wealth (one dollar, say) is invested in security j, $j = 1, 2, \ldots, S$, during period k, its value at the end of the period is Z_{jk}, where Z_{jk} is a non-negative random variable. The random vector $Z_k = (Z_{1k}, Z_{2k}, \ldots, Z_{Sk})$ thus describes the single-period returns for period k. At the beginning of any period k, the investment manager apportions the current wealth W_{k-1} among the S securities by choosing a weighting vector $\alpha_k = (\alpha_{1k}, \alpha_{2k}, \ldots, \alpha_{nk})$ such that $\sum_{i=1}^{n} \alpha_{ik} = 1$, and, accordingly, invests $W_{k-1}\alpha_{ik}$ in security i. The composite single-period return is

$$X_k = \alpha_{1k}Z_{1k} + \alpha_{2k}Z_{2k} + \cdots + \alpha_{Sk}Z_{Sk} = \alpha_k \cdot Z_k .$$

Wealth at the end of the period is then

$$W_k = W_{k-1}X_k .$$

A choice of a sequence of vectors $\alpha_1, \alpha_2, \ldots$ is an investment policy.[2]

The analysis in this paper is based on the following:

(A.1) Basic assumptions

(a) The Z_k are independent, identically distributed, and nonnegative on $(\Omega, \mathscr{F}, P)$.
(b) The Z_k are bounded above.
(c) Policies $(\alpha_1, \alpha_2, \ldots)$ must be constant (that is, $\alpha_k = \alpha$ for all $k = 1, 2, \ldots$), and α: $R_+^S \to R_+$ (no short sales).
(d) Each α is measurable and bounded. In particular,[3] there is a $C > 0$ such that $\alpha(Z_k) \leq C$ for all α and k.

Primarily we shall be concerned with the processes X and W derived from Z and a policy α. For this purpose we only require the following assumptions (the first two of which follow directly from above):

(A.2) Properties of X and W

(a) The X_k are independent, identically distributed, and nonnegative random variables on the probability space $\{\Omega, \mathscr{F}, P\}$.

For all k:

(b) $X_k \leq C$.
(c) $\log X_k$ has finite first and second moments.
(d) $W_k = W_{k-1}X_k$, with $W_0 = 1$.[4]

[2] In this example α_k is linear. Nonlinear mappings can accommodate options or other nonlinear functions of a security's return without introducing additional securities.

[3] This uniform upper bound requirement is for convenience. It can be relaxed.

[4] $W_0 = 1$ is a normalization that is used for convenience only.

The max E log policy

The wealth W_n can be written

$$W_n = X_1 X_2 \dots X_n .$$

Taking the logarithm of this equation we obtain

$$\log W_n = \log X_1 + \log X_2 + \cdots + \log X_n ,$$

or equivalently,

$$\log [W_n]^{1/n} = \frac{1}{n}(\log X_1 + \log X_2 + \cdots + \log X_n) . \tag{1}$$

Under our assumptions, as n approaches infinity the right-hand side of (1) approaches (almost surely) $\mathrm{E} \log X_1$ (or $\mathrm{E} \log X_k$ for any $k = 1, 2, \dots$ since all are equal). Hence for large n

$$\log [W_n]^{1/n} \simeq \mathrm{E} \log X_1 ,$$

and, in a formal manner, we write

$$W_n \simeq \exp n(\mathrm{E} \log X_1) . \tag{2}$$

Roughly speaking (2) says that wealth grows approximately exponentially with a coefficient in the exponent equal to $\mathrm{E} \log X_1$. This is an indication that maximization of $\mathrm{E} \log X_1$ might be desirable. This reasoning can in fact be made rigorous. [See Breiman (1961), Algoet and Cover (1985), and Bell and Cover (1980).] However, this argument is not based on a definitive concept of optimality. Nevertheless, as shown in section 4 of this paper, the optimal log mean policy does in fact correspond to optimization of any 'simple' asymptotic utility function.

3. Preferences on sequences

Our approach is based on recognizing from the outset that one is selecting from a set of competing infinite sequences of random variables. Hence the preference relation should be constructed on these sequences directly.

Consider first deterministic sequences. Let Γ be the set of all such sequences $w_1, w_2, \dots$ having the following properties:

(i) $w_k \geq 0$ for all k,

(ii) $w_k \leq C^k$ for all k,

where C is the bound in assumptions (A.1-d) and (A.2-b).

Assume that on $\Gamma \times \Gamma$ there is a preference relation $\succsim$. That is, $\succsim$ defines a preference relation on a pair of sequences. (If w, $v \in \Gamma$ and $w \succsim v$, we say w is *preferred* to v.) Such a preference relation satisfies the following properties:

(P.1) *Completeness*: For every pair $v, w \in \Gamma$, either $w \succsim v$ or $v \succsim w$.

(P.2) *Reflexivity*: For every $w \in \Gamma$, $w \succsim w$.

(P.3) *Transitivity*: For every $u, v, w \in \Gamma$, if $u \succsim v$ and $v \succsim w$, then $u \succsim w$.

Examples of preference relations can be easily constructed by restricting attention to a finite set of indices $k = 1, 2, \ldots, n$ and using any standard finite-dimensional preference relation. In particular, one might define the relation in terms of wealth at a fixed (surrogate terminal) time n; i.e., $w \succsim v$ if $w_n \geq v_n$. More interesting examples defining $w \succsim v$ include[5]

$$\overline{\lim_{n\to\infty}} \frac{1}{n} \log w_n \geq \overline{\lim_{n\to\infty}} \frac{1}{n} \log v_n , \tag{3}$$

$$\overline{\lim_{n\to\infty}} \frac{1}{n^2} \sum_{k=1}^{n} \log w_k \geq \overline{\lim_{n\to\infty}} \frac{1}{n^2} \sum_{k=1}^{n} \log v_k , \tag{4}$$

$$\sum_{n=1}^{\infty} \frac{w_n}{(2C)^n} \geq \sum_{n=1}^{\infty} \frac{v_n}{(2C)^n} , \tag{5}$$

$$\overline{\lim_{n\to\infty}} \frac{1}{n} \log w_n w_{n+1} \geq \overline{\lim_{n\to\infty}} \frac{1}{n} \log v_n v_{n+1} . \tag{6}$$

We assume an additional property that is really the heart of our analysis.

(P.4) *Tail Property*: The preference relation depends only on the tail of the sequences in Γ. That is if $w, v \in \Gamma$ and $w \succsim v$, then $\bar{w} \succsim \bar{v}$ for any $\bar{w}, \bar{v} \in \Gamma$ that differ from w and v in at most a finite number of elements.

The examples (3), (4), and (6) are tail preferences.

The motivation for this assumption stems from the original proposed setting of the investment problem. The investor is assumed to care only about the long-term behavior of the investment; no finite portion is of concern. From a technical perspective the introduction of tail preferences is precisely the concept required to introduce the limit idea early, so that it need not be

[5] The notation $\overline{\lim}$ indicates lim sup (that is, limit superior).

introduced at a later stage of the logical development and then interchanged with some other operation.

Preferences on stochastic processes

Now let $\mathscr{W}$ be the family of stochastic wealth processes derived from the investment procedure of section 2. We wish to establish a preference relation on $\mathscr{W} \times \mathscr{W}$. That is, we wish to be able to compare two different stochastic processes. This new preference relation should be an extension of the preference relation on deterministic sequences defined above. $\Gamma \times \Gamma$ is a subset of $R^{\infty} \times R^{\infty}$ and we let $\mathscr{B}$ be the corresponding subset of the Borel field on $R^{\infty} \times R^{\infty}$. This leads to an additional technical requirement on the preference relation.

(P.5) *Measurability*: The subset $\{(w, v): w \succsim v\}$ of $\Gamma \times \Gamma$ is in $\mathscr{B}$.

We now extend the preference relation to a preference relation $\succsim_S$ on stochastic wealth processes. The extension is by almost sure association.

Axiom. *Let $W = (W_1, W_2, \ldots)$, $V = (V_1, V_2, \ldots)$ be stochastic processes in $\mathscr{W}$. Then $W \succsim_S V$ if $W \succsim V$ a.s. That is*

$$W \succsim_S V \quad \text{if} \quad P\{W \succsim V\} = 1\,. \tag{7}$$

In other words, if for almost every realization of the processes the sequence generated by W is deterministically preferred to the sequence generated by V, then the stochastic process W is preferred to the stochastic process V.

This is a very natural (and very weak) axiom. Notice that the usual axioms required to define preferences over stochastic events (through an expected utility criterion) are not imposed. Our stochastic preference criterion is not an 'expected' criterion; it is an 'almost sure' criterion. Stochastic preference is induced by almost sure deterministic preference. In a very real sense then this preference structure is substantially weaker than that employed in traditional preference approaches to this problem. (To be sure, it is in another sense stronger because of the tail assumption.)

A potential difficulty with this axiom is that it may not make sense. It might be that $P\{W \succsim V\} = \frac{1}{2}$ in which case neither $W \succsim_S V$ nor $V \succsim_S W$ would hold. A zero–one law insures that this is never a problem.

Theorem 1. *If (P.1)–(P.5) hold, the relation $\succsim_S$ is complete, reflexive, and transitive.*

Proof. Given W and V in $\mathscr{W}$, the sequence of pairs $\{(W_i, V_i)\}$ is a process on $(\Omega, \mathscr{F}, P)$. Let $X = (X_1, X_2, \ldots)$ and $Y = (Y_1, Y_2, \ldots)$ be the corresponding single-period return processes. Under our assumptions (A.1) and (A.2) the pairs (X_k, Y_k) and (X_l, Y_l) are independent for $k \neq \ell$. Because X_i and Y_i are uniformly bounded, all realizations of W and V are in Γ. The event $\{W \succsim V\} \subset \Omega$ is measurable by (P.5) and is a tail event on the process of pairs $\{(W_i, V_i)\}$. However, any tail event in this process is a symmetric event [see Breiman (1968) or Shiryayev (1984)] for the process $\{(X_i, Y_i)\}$ since such an event is not influenced by a permutation of any finite number of elements of the processes. This latter process is independent, and hence by the Hewitt–Savage Zero–One Law [see Breiman (1968)], the probability of the event $\{W \succsim V\}$ is either zero or one.

If $P\{W \succsim V\} = 1$, then $W \succsim_S V$. Suppose, however, that $P\{W \succsim V\} = 0$. We have

$$1 = P\{\{W \succsim V\} \cup \{V \succsim W\}\} \leq P\{W \succsim V\} + P\{V \succsim W\},$$

where the first equality follows from the completeness of the order $\succsim$. Hence, in this case $P\{V \succsim W\} = 1$, and $V \succsim_S W$. Therefore in any case either $W \succsim_S V$, or $V \succsim_S W$.

Reflexivity follows immediately from the reflexivity of $\succsim$. Transitivity is fairly immediate also: Let $U \succsim_S V$, $V \succsim_S W$. Then $U \succsim V$ and $V \succsim W$ except on subsets M_1 and M_2, respectively, of Ω, each of measure zero. Thus by transitivity of $\succsim$, $U \succsim W$ except on $M_1 \cup M_2$. ∎

The key ingredient of this result is that measurable tail events on wealth processes have probability of either zero or one. It is for that reason we can define preference for stochastic wealth processes by almost sure association – the relation is always almost sure one way or the other. We exploit this property throughout the remainder of the paper.

4. Simple utility functions

A particularly useful way to describe a large class of preferences is through a utility function. Again, a utility function first can be introduced on deterministic sequences, that is, on Γ. Then this utility, if it defines a tail preference relation, can be extended by almost sure association to a utility function on the set $\mathscr{W}$ of stochastic processes.

A *utility function* on Γ is a measurable real-valued function $U: \Gamma \to R$. It defines a preference relation through the definition $w \succsim v$ if $U(w) \geq U(v)$. We say that a utility function is a *tail utility function* if $U(w)$ depends only on the tail of $w = (w_1, w_2, \ldots)$ [that is, $U(w) = U(\bar{w})$ if w and $\bar{w}$ differ in at most a finite number of elements].

We extend a tail utility function U from Γ to $\mathscr{W}$ by use of the fact that, by the zero–one law, if $W \in \mathscr{W}$, then if $U(W)$ is finite-valued, it is a degenerate random variable; that is, it is a constant almost surely. The value of this constant is taken to be the value of $U(W)$.

Although there is an infinite variety of possible tail utility functions, we focus first on what we call *simple* utility functions. These are defined in terms of limiting operations on functions of wealth at individual periods, with no interaction between periods. Specifically, simple utility functions have the form

$$U(w) = \overline{\lim_{n \to \infty}} \ \bar{\rho}(w_n, n),$$

or

$$U(w) = \underline{\lim_{n \to \infty}} \ \bar{\rho}(w_n, n),$$

or algebraic combinations of these forms. In these expressions, $\bar{\rho}$ is a continuous and increasing function of w_n for each n. This appears to be a very general class of tail utility functions; and if we accept the idea of basing preferences on tail events, this form of utility is very natural. We shall show, however, that this form (with some minor restrictions) leads inevitably to the expected log return criterion.

We note first that without loss of generality we may write $\bar{\rho}(w, n) = \rho(\log w, n)$ by applying the one-to-one transformation $\log w$, which maps between $(0, \infty)$ and $(-\infty, \infty)$. The resulting $\rho(\cdot, n)$ is continuous and increasing since $\bar{\rho}(\cdot, n)$ is. Finally, by adding a constant to U, if necessary, we may assume $\bar{\rho}(1, n) = 0$ or equivalently $\rho(0, n) = 0$. Thus we define a simple utility function as having the form

$$U(w) = \overline{\lim_{n \to \infty}} \ \rho(\log w_n, n), \tag{8}$$

or

$$U(w) = \lim_{n \to \infty} \ \rho(\log w_n, n), \tag{9}$$

or combinations of the forms (8) and (9), where ρ has the properties stated above.

We extend these to the stochastic wealth process W by use of the fact, from the proof of Theorem 1, that tail events on W have probability of either zero or one. This means, in particular, that if $f: \Gamma \to R$ is measurable and $f(w)$ depends only on the tail of the sequence w, then the random variable $f(W)$ is constant almost surely. Using this fact we define

$$U(W) = \overline{\lim_{n \to \infty}} \ \rho(\log W_n, n), \tag{10}$$

or

$$U(W) = \varliminf_{n \to \infty} \rho(\log W_n, n). \tag{11}$$

The random variables $U(W)$ defined above are measurable as extended random variables – see Shiryayev (1984, pp. 171, 358). If $U(W)$ is finite it is constant a.s., and the utility is taken to be this constant.

Not all functions $\rho(\cdot,\cdot)$ will lead to meaningful utility functions. For example, if $\rho(z, n) = z$, the process $W_n = c^n$ with $c > 1$ will lead to $U(W) = \infty$. In general, in order to obtain finite utility for feasible wealth processes, it is necessary that $\rho(z, n)$ satisfy appropriate growth-rate conditions with respect to n. In particular, ρ must decrease with n to counterbalance possible increases in z. The following theorem gives suitable conditions.

Theorem 2. For each n, let $\rho(z, n)$ be continuous and increasing in z with $\rho(0, n) = 0$. Let X and W be processes satisfying (A.2). Assume $\mathrm{E} \log X_1 = m$, *and define $\underline{U}$ by (10).*

(a) If $\varlimsup_{n \to \infty} \rho(nz, n) = \operatorname{sgn}(z) \cdot \infty$, *then*

$$U(W) = \begin{cases} +\infty, & m > 0, \\ -\infty, & m < 0. \end{cases}$$

(b) If $\varlimsup_{n \to \infty} \rho(nz, n) = g(z)$ *where $g(\cdot)$ is continuous, then $U(W) = g(m)$.*

Proof. We have $(1/n) \log W_n \to m$ a.s. Let $z_n = (1/n) \log w_n$, $n = 1, 2, \ldots$, correspond to a particular realization of the process where convergence of $z_n \to m$ holds. Given $\varepsilon > 0$, there is an N such that $|z_n - m| < \varepsilon$ for all $n > N$. By monotonicity of ρ it follows that

$$\rho(n(m - \varepsilon), n) \leq \rho(nz_n, n) \leq \rho(n(m + \varepsilon), n),$$

for all $n > N$. Hence

$$\varlimsup_{n \to \infty} \rho(n(m - \varepsilon), n) \leq \varlimsup_{n \to \infty} \rho(nz_n, n) \leq \varlimsup_{n \to \infty} \rho(n(m + \varepsilon), n).$$

If $m > 0$, then for sufficiently small $\varepsilon > 0$ both $m - \varepsilon > 0$ and $m + \varepsilon > 0$. Therefore, if the hypothesis (a) holds, it follows that $\varlimsup_{n \to \infty} \rho(nz_n; n) = \infty$. A similar argument shows that for $m < 0$ there holds $\varlimsup_{n \to \infty} \rho(nz_n, n) = -\infty$. If the hypothesis in (b) holds, then

$$\varlimsup_{n \to \infty} \rho(nz_n, n) = g(m).$$

Finally, the conclusion follows since

$$\frac{1}{n}\log W_n = \frac{1}{n}\sum_{k=1}^{n} \log X_k \to m \quad \text{a.s.}$$

by the strong law of large numbers, and hence the above values hold for

$$\overline{\lim_{n\to\infty}}\ \rho(\log W_n, n) = \overline{\lim_{n\to\infty}}\ \rho\left(n\frac{1}{n}\log W_n, n\right)$$

almost surely. ▮

The parallel of Theorem 2 holds for the form (10), with $\overline{\lim}_{n\to\infty}$ replaced everywhere by $\underline{\lim}_{n\to\infty}$. It is clear that in either case $g(m)$ is an increasing function of m.

Theorem 2, and the above remarks, say roughly that, to within a monotone transformation, the only real-valued tail simple utility function is $m = \mathrm{E}\log X_1$. We remark again, however, that although this 'appears' to be an expected value criterion, it is actually an 'almost sure' criterion. The expected value arises from the strong law of large numbers, rather than from an assumption of expected value utility.

It is perhaps useful to relate the theorem more directly to our earlier discussion of other utility functions. For example, consideration of the popular utility function defined as w_n^{γ} on terminal wealth leads one to set $\rho(\log w_n, n) = \exp(\gamma \log w_n)$ in (8). However, clearly $\rho(nz, n) \to \operatorname{sgn}(z)\cdot\infty$, leading to the infinite result of part (a) of Theorem 2. The idea can be salvaged by introduction of a $1/n$ in the power, i.e., $w_n^{\gamma/n}$ or $\rho(z, n) = \mathrm{e}^{\gamma z/n}$. This leads to $U(W) = \mathrm{e}^{\gamma m}$, which again gives the expected logarithm criterion.

5. Compound utility functions

It is apparent that there is a slight gap in Theorem 2. Part (a) covers the cases $m > 0$ and $m < 0$ but not $m = 0$. This is no real detraction from the theorem, for part (a) says that under its conditions the corresponding utility function is degenerate unless it is applied to a process with $m = 0$. However, this gap motivates the introduction of compound utility functions; since by subtracting out the mean m from the process, there is room for other possible forms.

Another motivation for compound utility functions is purely structural. We let a compound utility function have a simple utility function as an argument.

However, by Theorem 2, we can take this simple utility function to be $U_S(W) = m$. This motivates the specific compound forms

$$U(W) = \overline{\lim_{n\to\infty}}\ \psi(\log W_n - nm, m, n) \quad \text{a.s.}, \tag{12}$$

$$U(W) = \underline{\lim_{n\to\infty}}\ \psi(\log W_n - nm, m, n) \quad \text{a.s.}, \tag{13}$$

where ψ is continuous and increasing in its first two arguments. As before, the equalities in (12) and (13) are to be interpreted in the almost sure sense. Such a utility, depending on m, is clearly still a tail function.

Once we recognize the role of the mean log return m in compound utility functions, it is not necessary to carry it along in our analysis. Instead, we consider

$$U(W) = \overline{\lim_{n\to\infty}}\ \phi(\log W_n, n) \quad \text{a.s.}, \tag{14}$$

$$U(W) = \underline{\lim_{n\to\infty}}\ \phi(\log W_n, n) \quad \text{a.s.}, \tag{15}$$

where $\phi(z, n)$ is increasing in z, for the special case where $m = \mathrm{E}\log X_1 = 0$. Then, later, we can simply put m back into the utility as an additional variable, as in (12) or (13). This restriction to the study of utility functions (14) and (15) amounts, really, to a further examination of simple utility functions when it is known that $m = 0$.

A special situation is the degenerate one where $W_n = 1$ for all n. We wish this to have finite utility, and hence without loss of generality we may take $\phi(0, n) = 0$ for all n.

As before the behavior of ϕ as a function of n is critical. To characterize the appropriate behavior of ϕ it turns out that the asymptotic behavior of $\phi(\sqrt{n}z, n)$ is critical.

The following proposition and its corollary rule out a large class of functions.

Lemma 1. Suppose $\phi(z, n)$ is increasing in z for each n with $\phi(0, n) = 0$ and

$$\lim_{z\to\infty} \underline{\lim_{n\to\infty}}\ \phi(\sqrt{n}z, n) = A\,.$$

Let X and W be processes satisfying (A.2) and assume $\mathrm{E}\log X_1 = 0$, $\mathrm{var}(\log X_1) = \sigma^2 > 0$. *Then, for U defined by (14), $U(W) \geq A$.*

Proof. Since $\phi(0, n) = 0$, it follows that $A \geq 0$. In order to treat the case $A = \infty$ as well as $A < \infty$, we prove that if

$$\lim_{z \to \infty} \varliminf_{n \to \infty} \phi(\sqrt{n}z, n) \geq A',$$

with $A' \leq \infty$, then $U(W) \geq A'$. The lemma follows from this.

Let $\rho(z) = \varliminf_{n \to \infty} \phi(\sqrt{n}z, n)$. Select $\varepsilon > 0$. Then there is a $\bar{z}$ such that $\rho(\bar{z}) > A' - \varepsilon$. Then, by the definition of ρ, there is N such that for all $n > N$, $\phi(\sqrt{n}\bar{z}, n) > A' - 2\varepsilon$. By the monotonicity of $\phi(\cdot, n)$ we have

$$\phi(\sqrt{n}z, n) \geq \phi(\sqrt{n}\bar{z}, n) > A' - 2\varepsilon,$$

for all $z \geq \bar{z}$ and all $n > N$.

Thus,[6]

$$P\left\{\phi\left(\sqrt{n}\frac{\log W_n}{\sqrt{n}}, n\right) > A' - 2\varepsilon \quad \text{i.o.}\right\} \geq P\left\{\frac{\log W_n}{\sqrt{n}} > \bar{z} \quad \text{i.o.}\right\} = 1,$$

where the last equality follows from the law of the iterated logarithm,

$$\varlimsup_{n \to \infty} \frac{\log W_n}{(2\sigma^2 n \log\log n)^{1/2}} = 1 \quad \text{a.s.},$$

which shows that $\varlimsup_{n \to \infty} (\log W_n)/\sqrt{n} = \infty$ a.s. Therefore $U(W) > A' - 2\varepsilon$. Since this is true for all $\varepsilon > 0$, it follows that $U(W) \geq A'$. ▮

Corollary. If for all $z > 0$,

$$\lim_{n \to \infty} \phi(\sqrt{n}z, n) = \infty,$$

then, for U defined by (14), $U(W) = \infty$.

Proof. We have $A = \infty$ in the above lemma. ▮

Let us consider some examples. The function $\phi(z, n) = z/n$ corresponds, according to Theorem 2 of the previous section, to the simple utility function with value $U(W) = m$; and hence in this context, with $m = 0$, $U(W) = 0$. $A = 0$ for this function, so Lemma 1 gives $U(W) \geq 0$. The reverse inequality can be established by a symmetric argument (see Lemma 3), and together these agree with what is known from Theorem 2.

[6] i.o. denotes 'infinitely often'.

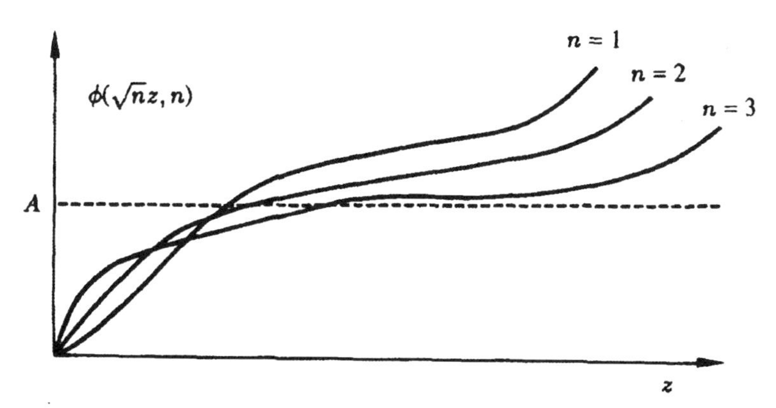

Fig. 1. An unusual function that is decreasing in n above A.

In general, if $\phi(z, n) = z/q(n)$, then clearly $q(n)$ must go to infinity faster than $\sqrt{n}$ in order that A be finite and hence for the corresponding compound utility function to be real-valued. For example, the function $\phi(z, n) = z/\sqrt{n}$ has $A = \infty$.

Definition. Let A be a constant. The function of two variables $\phi(\sqrt{n}z, n)$ is said to be *decreasing with respect to n above A* if $\phi(\sqrt{n}z, n) > A$ implies $\phi(\sqrt{n+1}z, n+1) \leq \phi(\sqrt{n}z, n)$.

The monotonicity property with respect to n is quite reasonable even if $A > 0$. An example is shown in fig. 1.

Corresponding to a function $\phi(z, n)$ that is continuous and strictly increasing in z for each n, let $f(\cdot, n)$: $D_n \to R$ [where D_n is the range of $\phi(\cdot, n)$] be the inverse function of $\phi(\cdot, n)$; that is,

$$f(\phi(z, n), n) = z,$$

for every z. This inverse exists and is also strictly increasing. We need the following result which shows that if $\phi(\sqrt{n}z, n)$ is decreasing with respect to n, then $f/\sqrt{n}$ is increasing with respect to n.

Lemma 2. Suppose $\phi(\sqrt{n+1}\,z, n+1) \leq \phi(\sqrt{n}z, n) = r$. *Then,*

$$f(r, n)/\sqrt{n} \leq f(r, n+1)/\sqrt{n+1}.$$

Proof. We have

$$\phi(\sqrt{n}z, n) = r,$$

$$\phi(\sqrt{n+1}z, n+1) = s,$$

with $s \leq r$. By applying f to these we have

$$z = f(r, n)/\sqrt{n},$$

$$z = f(s, n+1)/\sqrt{n+1} \leq f(r, n+1)/\sqrt{n+1},$$

where the inequality follows from the monotonicity of $f(\cdot, n+1)$. Therefore $f(r, n)/\sqrt{n} \leq f(r, n+1)/\sqrt{n+1}$. ∎

We are now prepared to state the basic result of this section. We show, essentially, that any compound utility function of wealth (14), derived from a process with zero mean (of the logarithm), is a function only of $\sigma^2 = \operatorname{var}(\log X_1)$ (or, equivalently, of σ).

Proposition 1. Let U be defined by (14) and assume that $\phi(z, n)$ is continuous and strictly increasing with respect to z for each n, and that $\phi(0, n) = 0$. Suppose further that $\phi(\sqrt{n}z, n)$ is decreasing with respect to n above A as defined in Lemma 1. Then there is a function h on R_+ possibly taking values $-\infty$ or $+\infty$ such that

(i) *$h(\sigma)$ is nondecreasing with respect to σ;*

(ii) *for any process X and W satisfying (A.2) with $m = \mathrm{E}(\log X_1) = 0$, $\operatorname{var}(\log X_1) = \sigma^2 > 0$, there holds*

$$U(W) = h(\sigma).$$

Proof. Defining A as in Lemma 1, we know that $U(W) \geq A$. For any $r > A$, we have $U(W) > r$ if and only if

$$P\{\phi(\log W_n, n) > r \quad \text{i.o.}\} = 1.$$

This is equivalent to

$$P\left\{\frac{\log W_n}{\sigma\sqrt{n}} > \frac{f(r, n)}{\sigma\sqrt{n}} \quad \text{i.o.}\right\} = 1.$$

According to the assumptions on ϕ and the above lemma, $y_n = f(r, n)/(\sigma\sqrt{n})$ is increasing for $r > A$. Furthermore, $f(r, n) > 0$ for $r > A \geq 0$. Therefore the extended law of the iterated logarithm [Feller (1943), Breiman (1968, p. 297)] states that the probability above is either zero or one, depending on whether the sum

$$\sum_{n=1}^{\infty} \frac{y_n}{n} \mathrm{e}^{-y_n^2/2}$$

converges or diverges, respectively. If the sum converges for a given sequence $\{y_n\}$, then it will converge also for $\{y'_n\}$ with $y'_n > y_n$. This is because $(\mathrm{d}/\mathrm{d}x) \times [x\mathrm{e}^{-x^2/2}] = [1 - x^3]\mathrm{e}^{-x^2/2} < 0$ for $x > 1$. Thus, for sufficiently large n, each term in the sum for $\{y'_n\}$ is smaller than the corresponding term in the sum for $\{y_n\}$. Therefore, if the sum converges for a given value of r, it also converges for $r' > r$. It follows that

$$U(W) = \inf\left\{r\colon \sum_{n=1}^{\infty} (y_n/n)\mathrm{e}^{-y_n^2/2} \text{ converges}\right\}.$$

In a similar way, if the series converges for a given σ, it will also converge for $\sigma' < \sigma$. Hence $U(W) = h(\sigma)$ where $h(\cdot)$ is nondecreasing. ▮

A useful ϕ for consideration is

$$\phi(z, n) = \frac{z}{(2n \log\log n)^{1/2}}.$$

In this case the standard law of the iterated logarithm states that

$$U(W) = \overline{\lim_{n\to\infty}} \frac{\log W_n}{(2n \log\log n)^{1/2}} = \sigma \quad \text{a.s.},$$

and hence this ϕ gives $h(\sigma) = \sigma$.

Proposition 1 shows that an investor with a utility function of the general form (14), with the $\overline{\lim}$ operation, *prefers* increased variance. Increased variance increases the extent of upward variations, and presumably the investor values these peaks.

The comparison to Proposition 1 treats utilities of form (15), with the $\underline{\lim}$ operation. In this case the investor wants to increase the lowest deviations, and hence such an investor values low variances more than high variances.

Specifically, we have the following results.

Lemma 3. *Suppose $\phi(z, n)$ is increasing in z for each n with $\phi(0, n) = 0$ and*

$$\lim_{z \to -\infty} \overline{\lim_{n \to \infty}} \phi(\sqrt{n}z, n) = B .$$

Let X and W be processes satisfying (2) and assume $\mathrm{E} \log X_1 = 0$, $\mathrm{var}(\log X_1) = \sigma^2 > 0$. *Then, for U defined by (15), $U(W) \leq B$.*

Proposition 2. *Let U be defined by (15) and assume that $\phi(z, n)$ is continuous and strictly increasing with respect to z for each n, and that $\phi(0, n) = 0$. Suppose further that $\phi(\sqrt{n}z, n)$ is increasing with respect to n below B as defined in Lemma 3. Then there is a function g on R_+ possibly taking values $-\infty$ or $+\infty$ such that*

(*i*) *$g(\sigma)$ is nonincreasing with respect to σ;*

(*ii*) *for any processes X and W satisfying (A.2) with* $m = \mathrm{E}(\log X_1) = 0$, $\mathrm{var}(\log X_1) = \sigma^2 > 0$, *there holds*

$$U(W) = g(\sigma) .$$

Finally, we may combine Propositions 1 and 2 and insert the explicit dependence of a general compound utility function on the mean m. We let $\psi(z, m, n)$ be continuous and strictly increasing with respect to z and increasing with respect to m.

Theorem 3. *Let U be defined by either*

(*a*) $U(W) = \overline{\lim}_{n \to \infty} \psi(\log W_n - nm, m, n)$ *a.s., or*

(*b*) $U(W) = \underline{\lim}_{n \to \infty} \psi(\log W_n - nm, m, n)$ *a.s.*

Correspondingly, assume that for each m, $\phi_m(z, n) = \psi(z, m, n)$ satisfies the conditions of Propositions 1 or 2. Then there is a function f on $R \times R_+$ possibly taking values $-\infty$ or $+\infty$ such that

(*i*) *$f(m, \sigma)$ is increasing with respect to m and either increasing or decreasing with respect to σ corresponding to (a) or (b);*

(*ii*) *for any processes X and W satisfying (A.2) with* $m = \mathrm{E} \log X_1$, $\mathrm{var}(\log X_1) = \sigma^2 > 0$, *there holds $U(W) = f(m, \sigma)$.*

For a given process Z and a given family of feasible investment policies, one can define the set S of attainable (m, σ) pairs (achieved as the policy ranges over all possibilities). If the set S is compact, the *right frontier* of the set S is the right-most boundary of this set; that is, the points (m, σ) having the largest σ for a given m. Then, according to the above theorem, an investor with a utility function of the form (a) will select a policy corresponding to a point on this frontier. Similarly, an investor with a utility function of the form (b) will select a policy corresponding to a point on the *left frontier*.

6. Conclusion

We have approached infinite-horizon investment situations by considering preference orders on infinite sequences of wealth, and we suggested that tail preferences are appropriate if the goal of investment is long-term 'wealth building'. We showed that if the tail preference takes the form of a simple utility function, then utility must be equivalent to the expected logarithm of return. If preferences are represented by a compound utility function, that function will be equivalent to a function of the expected logarithm and variance of the logarithm.

Although our results require that certain technical assumptions be satisfied, they are robust enough to support, at least roughly, the statement that: a tail utility function involving the limits of total return must be equivalent to a log mean–variance criterion. We do not anticipate, however, that the controversy over the use of such a criterion will subside. Indeed it should not. Our result is based on the tail preference concept, which is an idealization of what is generally a complex economic problem. However, since the log mean–variance approach is so intuitively appealing, it is nice to have a framework in which it is validated.

References

Algoet, P.H. and T.M. Cover, 1985, Asymptotic optimality and asymptotic equipartition properties of log-optimum investment, Technical report 57 (Department of Statistics, Stanford University, Stanford, CA).

Bell, R. and T.M. Cover, 1980, Competitive optimality of logarithm investment, Mathematics of Operations Research 5, 161–166.

Borch, K., 1963, A note on utility and attitudes to risk, Management Science 9, 697–701.

Breiman, L., 1961, Optimal gambling systems for favourable games, Proceedings of the Fourth Berkeley Symposium 1, 65–78.

Breiman, L., 1968, Probability (Addison-Wesley, Reading, MA).

Feller, W., 1943, The general form of the so-called law of the iterated logarithm, American Mathematical Society Transactions 54, 373–402.

Goldman, M.B., 1974, A negative report on the 'near optimality' of the max-expected-log policy as applied to bounded utilities for long-lived programs, Journal of Financial Economics 1, 97–103.

Hakansson, N.H., 1971a, Capital growth and the mean–variance approach to portfolio selection, Journal of Financial and Quantitative Analysis 6, 517–557.

Hakansson, N.H., 1971b, Multi-period mean–variance analysis: Toward a general theory of portfolio choice, Journal of Finance 26, 857–884.

Kelley, J.L., Jr., 1956, A new interpretation of information rate, Bell System Technical Journal 35, 917–926.

Latané, H., 1959, Criteria for choice among risky ventures, Journal of Political Economy 67, 144–155.
Leland, H., 1972, On turnpike portfolios, in: G. Szegö and K. Shell, eds., Mathematical methods in investment and finance (North-Holland, Amsterdam).
Lintner, J., 1965, The valuation of risk assets and the selection of risky investments in stock portfolios and capital budgets, Review of Economics and Statistics 48, 13–37.
Markowitz, H., 1952, Portfolio selection, Journal of Finance 12, 77–91.
Markowitz, H., 1959, Portfolio selection: Efficient diversification of investment (Wiley, New York, NY).
Merton, R.C. and P.A. Samuelson, 1974, Fallacy of the log-normal approximation to optimal portfolio decision-making over many periods, Journal of Financial Economics 1, 67–94.
Samuelson, P.A., 1969, Lifetime portfolio selection by dynamic stochastic programming, Review of Economics and Statistics 51, 239–246.
Samuelson, P.A., 1970, The fundamental approximation theorem of portfolio analysis in terms of means, variances, and higher moments, Review of Economic Studies 37, 537–542.
Samuelson, P.A., 1971, The 'fallacy' of maximizing the geometric mean in long sequences of investing or gambling, Proceedings of the National Academy of Sciences 68, 2493–2496.
Sharpe, W., 1970, Portfolio theory and capital markets (McGraw-Hill, New York, NY).
Shiryayev, A.N., 1984, Probability (Springer-Verlag, New York, NY).
Tobin, J., 1965, The theory of portfolio selection, in: I.H. Hahn and F.P.R. Brechling, eds., The theory of interest rates (Macmillan, New York, NY).
Williams, J.B., 1936, Speculation and the carryover, Quarterly Journal of Economics 50, 436.

ELSEVIER

Available online at www.sciencedirect.com

Journal of Econometrics 116 (2003) 365–386

JOURNAL OF Econometrics

www.elsevier.com/locate/econbase

43

Portfolio choice with endogenous utility: a large deviations approach

Michael Stutzer*

Burridge Center for Securities Analysis and Valuation, Leeds School of Business, University of Colorado, Boulder, CO 20309-0419, USA

Abstract

This paper provides an alternative behavioral foundation for an investor's use of power utility in the objective function and its particular risk aversion parameter. The foundation is grounded in an investor's desire to minimize the objective probability that the growth rate of invested wealth will not exceed an investor-selected target growth rate. Large deviations theory is used to show that this is equivalent to using power utility, with an argument that depends on the investor's target, and a risk aversion parameter determined by maximization. As a result, an investor's risk aversion parameter is not independent of the investment opportunity set, contrary to the standard model assumption.

JEL classification: C4; D8; G0

Keywords: Portfolio theory; Large deviations; Safety-first; Risk aversion

1. Introduction

What criterion function should be used to guide personal investment decisions? Perhaps the earliest contribution was Bernoulli's critique of expected wealth maximization, which led him to advocate maximization of the expected log wealth as a resolution of the St. Petersburg Paradox. This was resurrected as a long-term investment strategy in the 1950s, and is now synonymously described as either the log optimal, growth-maximal, geometric mean, or Kelly investment strategy. As also noted in their

* Tel.: +1-319-335-1239.

E-mail address: michael-stutzer@colorado.edu (M. Stutzer).

0304-4076/03/$ - see front matter © 2003 Elsevier B.V. All rights reserved.
doi:10.1016/S0304-4076(03)00112-X

excellent survey on this portfolio selection rule, Hakansson and Ziemba (1995, pp. 65–70) argue that "...the power and durability of the model is due to a remarkable set of properties", e.g. that it "*almost surely* leads to more capital in the long run than any other investment policy which does not converge to it". [1]

But even as a long-term investment strategy, the log optimal portfolio is problematic. It often invests very heavily in risky assets, which has led several researchers to highlight the possibilities that invested wealth will fall short of investor goals, even over the multi-decade horizons typical of young workers saving for retirement. For example, MacLean et al. (1992, p. 1564) note that "the Kelly strategy never risks ruin, but in general it entails a considerable risk of losing a substantial portion of wealth". Findings like these motivated Browne (1995, 1999a) to develop a variety of alternative, shortfall probability-based criteria, in specific continuous-time portfolio choice problems. Browne (1999b) considers these ideas in the context of the simplest possible investment decision, which will also be utilized to illustrate the criterion developed herein. Further discussion of his work is included in Section 2.3. Another similarly motivated criterion for continuous time portfolio choice is developed in Bielecki et al. (2000), which will be discussed further in Section 2.2.

The problem is exacerbated when investors have specific, short to medium term values for their respective investment horizons. If so, some criteria will lead to horizon-dependent optimal asset allocations, but others will not. For example, Samuelson (1969) proposes the criterion of intertemporal maximization of expected discounted, time-additive constant relative risk aversion (CRRA) power utility of consumption. He proves that when asset returns are IID, portfolio weights are independent of the horizon length. So in that case, long horizon investors should not invest more heavily in stocks than do short horizon investors. Samuelson (1994) provided caveats to this investor advice, citing six modifications of this specification that will result in horizon dependencies. [2]

But an investment advisor, hired to help an investor formulate asset allocation advice, may have difficulty determining a specific value for the investor's horizon. The advisor may be unable to determine an investor's exact horizon length when it exists, while other investors may not have a specific investment horizon length at all. A considerable simplification results when an infinite horizon is assumed, as has also been done when deriving many, but not all, consumption-based asset pricing models. [3] An exception to the infinite horizon formulations is found in Detemple and Zapatero (1991). Of course, the cost of this simplification is the inability to model horizon dependencies.

While the time horizon parameter is irrelevant for Samuelson's intertemporal power utility investor with IID returns, the optimal asset allocation is still very sensitive to the specific risk aversion parameter adopted, so an advisor would have to determine it with

[1] See the analysis of Algoet and Cover (1988) and the lucid exposition of Cover and Thomas (1991, Chapter 15) for more information on the growth maximal portfolio problem. For a spirited normative defense of the growth maximal portfolio criterion, see Thorp (1975).

[2] However not all of these modifications would support the oft-repeated advice to invest more heavily in stocks when the investor's horizon is longer.

[3] For a survey, see Kocherlakota (1996).

precision. An even more basic consideration is specification of the utility functional form and its argument. Should it be a power function, or an exponential function, or perhaps some function outside the HARA class? Should the argument be a function of current wealth, current consumption, or some function of the consumption path (as in habit formation models)? As a first step toward answering these questions, Section 2 of this paper develops a new criterion of investor behavior. It starts from the observation that the realized growth rate of investor wealth is a random variable, dependent on the returns to invested wealth and the time that it is left invested (i.e. the investment horizon). To obviate the need to specify a value for the latter, first assume that an investor acts as-if she wants to ensure that the (horizon-dependent) realized growth rate of her invested wealth will exceed a numerical target that she has, e.g. 8% per year. By choosing a portfolio that results in a higher expected growth rate of wealth than the target rate, the investor can ensure that the probability of not exceeding the target growth rate decays to zero asymptotically, as the time horizon $T \to \infty$. But the probability that the realized growth rate of wealth at finite time T will not exceed the target might vary from portfolio to portfolio. Which portfolio should be chosen? Without adopting a specific value of T, a sensible strategy is to choose a portfolio that makes this probability decay to zero as fast as possible as $T \to \infty$. This will ensure that the probability will be minimized for all but the relatively small values of T. In other words, the decay rate maximizing portfolio will maximize the probability that the realized growth rate will exceed the target growth rate at time T, for all but relatively small values of T. In fact, this turns out to be true for all values of T in the special IID cases studied in Sections 2.1 and 3.

Calculation of the decay rate maximizing portfolio is enabled by use of a simply stated, yet powerful result from large deviations theory, known as the Gärtner–Ellis Theorem. Straightforward application of it in Section 2.2 provides an expected power utility formulation of the decay rate criterion. But there are two important differences between this formulation and the standard expected power utility problem. First, the argument of the utility function is the ratio of invested wealth to a level of wealth growing at the constant target rate. Second, the value of the power, i.e. the risk aversion parameter, is also determined by maximization. As a result, a decay rate maximizing investor's degree of relative risk aversion will depend on the investment opportunity set, an effect absent in extant uses of power utility.

Because this endogenous degree of risk aversion is greater than 1, the third derivative of the utility is positive, so there is also an endogenous degree of skewness preference. This is fortunate, as some have argued that skewness preference helps explain expected asset returns. To see why, note that in the standard CAPM, investor aversion to variance makes an asset return's covariance with the market return a risk factor, so it is positively related to an asset's expected return. Kraus and Litzenberger (1976) argue that investor preference for positively skewed wealth distributions (ceteris paribus) should make market coskewness an additional factor, that should be negatively related to an asset's expected return. They thus generalized the standard CAPM to incorporate a market coskewness factor. The estimated model supports this implication of investor skewness preference. Harvey and Siddique (2000) extended this approach by incorporating conditional coskewness, concluding that "a model incorporating coskewness is helpful in explaining the cross-sectional variation of asset returns".[4]

The decay rate maximization criterion also nests Bernoulli's expected log maximization (a.k.a. growth optimal) criterion. An investor who has a target growth rate suitably close to the maximum feasible expected growth rate has an endogenous degree of risk aversion slightly greater than 1. As a result, the associated decay rate maximizing portfolio approaches the expected log maximizing portfolio. If the investor's target growth rate is lower, the investor uses a higher degree of risk aversion, and the associated decay rate maximizing portfolio is more conservative, with a lower expected growth rate, but a higher decay rate for the probability of underperforming that target growth rate (and hence a higher probability of realizing a growth rate of wealth in excess of that target). The (perhaps unlikely) presence of an unconditionally riskless asset, i.e. one with an intertemporally constant return, provides a floor on the attainable target growth rates. When the target growth rate is sufficiently near that floor, the investor's risk aversion will be quite high, and the associated decay rate maximizing portfolio will be close to full investment in the unconditionally riskless asset. The relationship between the target growth rate and the associated (maximum) decay rate of the probability that it will not be exceeded quantifies the tradeoff between growth and shortfall risk that has concerned analysts studying the expected log criterion.

Exact calculation of the decay rate (or equivalently, the expected power utility) requires the exact portfolio return process. In practice, the distribution is not exactly known. Even if its functional form is known, its parameters must still be estimated. To cope with this lack of exact knowledge, Section 3 adopts the assumption that portfolio log returns are IID with an unknown distribution, and follows Kroll et al. (1984) in estimating expected utility by substitution of a time average for the expectation operator. The estimated optimal portfolio and endogenous risk aversion parameter are those that jointly maximize the estimated expected power utility. An illustrative application of this estimator is included, contrasting decay rate maximization to both Sharpe Ratio and expected log maximization when allocating funds among domestic industry sectors. In it, decay rate maximization selects portfolios with higher skewness than Sharpe ratio maximization does. The IID assumption that underlies the estimator also permits the use of both a relative entropy minimizing, Esscher transformed log return distribution and a cumulant expansion to help interpret the empirical findings.

Section 4 summarizes the most important results, and concludes with some good topics for future research.

2. Porfolio analysis

Following Hakansson and Ziemba (1995, p. 68), the wealth at time T resulting from investment in a portfolio is $W_T = W_0 \prod_{t=1}^{T} R_{pt}$, where R_{pt} is the gross (hence positive)

[4] Hence, it is possible that an asset pricing model incorporating decay rate maximizing investors could outperform the CAPM, which incorporates Sharpe ratio maximizing investors. This topic is left for another paper.

rate of return between times $t-1$ and t from a portfolio p. Note that W_T does not depend on the length of the time interval between return measurements, but only on the product of the returns between those intervals. Dividing by W_0, taking the log of both sides, multiplying and dividing the right-hand side by T and exponentiating both sides produces the alterative expression

$$W_T = W_0 \left[e^{\sum_{t=1}^{T} \log R_{pt}/T} \right]^{T} = W_0 [e^{\overline{\log R_p}}]^{T}. \tag{1}$$

From (1), we see that W_T is a monotone increasing function of the realized time average of the log gross return, denoted $\overline{\log R_p}$, which is the realized growth rate of wealth through time T. When the log return process is ergodic in the mean, this will converge to a number denoted $\mathrm{E}[\log R_p]$, as $T \to \infty$. Accordingly, there was early (and still continuing) interest in the portfolio choice that maximizes this expected growth rate, i.e. selects the portfolio $\arg\max_p \mathrm{E}[\log R_p]$, also known as the "growth optimal" or "Kelly" criterion. As noted by Hakansson and Ziemba (1995, p. 65) "...the power and durability of the model is due to a remarkable set of properties", e.g. that it "*almost surely* leads to more capital in the long run than any other investment policy which does not converge to it".[5]

But maximizing the expected log return often invests very heavily in assets with volatile returns, which has led several researchers to highlight its substantial downside performance risks. Specifically, we will now examine the probability of the event that the realized growth rate of wealth $\overline{\log R_p}$ will not exceed a target growth rate $\log r$ specified by the investor or analyst. This is an event that will cause W_T in (1) to fail to exceed an amount equal to that earned by an account growing at a constant rate $\log r$. The following subsection uses a simple and widely analyzed portfolio problem to calculate this downside performance risk for the growth optimal portfolio and a portfolio chosen to minimize it.

2.1. The normal case

A simple portfolio choice problem, used in Browne (1999b), requires choice of a proportion of wealth p to invest in single stock, whose price is lognormally distributed at all times, with the rest invested in a riskless asset with continuously compounded constant return i. In this case, $\log R_{pt} \sim IID\ \mathcal{N}(\mathrm{E}[\log R_p], Var[\log R_p])$. We now compute the probability that $\overline{\log R_p} \leqslant \log r$. Because the returns are independent, $\overline{\log R_p} \sim \mathcal{N}(\mathrm{E}[\log R_p], Var[\log R_p]/T)$. The elementary transformation to the standard normal variate Z shows that the desired probability is

$$Prob[\overline{\log R_p} \leqslant \log r] = Prob\left[Z \leqslant \frac{\log r - \mathrm{E}[\log R_p]}{\sqrt{Var[R_p]/T}} \right]. \tag{2}$$

[5] See the analysis of Algoet and Cover (1988) and the lucid exposition of Cover and Thomas (1991, Chapter 15) for more information on the expected log criterion. For a spirited normative defense of this criterion, see Thorp (1975).

In order to minimize (2), i.e. to maximize the complementary probability that $\overline{\log R_p} > \log r$, one must choose the proportion of wealth p to minimize the expression on the right-hand side of (2). This is equivalent to maximizing -1 times this expression. Independent of the specific value of T, this portfolio stock weight is

$$\arg\max_p \frac{\mathrm{E}[\log R_p] - \log r}{\sqrt{Var[\log R_p]}}. \tag{3}$$

Portfolio (3) will differ considerably from the following growth optimal portfolio

$$\arg\max_p \mathrm{E}[\log R_p] \tag{4}$$

because of the presence of the target $\log r$ in the numerator of (3) and the standard deviation of the log portfolio return in its denominator. Portfolio (3) will also differ from the following Sharpe Ratio maximizing portfolio:

$$\arg\max_p \frac{E[R_p] - i}{\sqrt{Var[R_p]}} \tag{5}$$

because of the presence of the presence of the target $\log r$ in (3) in place of the riskless rate i in (5), and because of the presence of log gross returns in (3) in place of the net returns used in (5).

It will soon prove useful to reformulate the rule (3) in the following way. Note that $Prob[\overline{\log R_p} \leqslant \log r]$ will not decay asymptotically to zero unless the numerator of (3) is positive, so we need only consider portfolios p for which

$$\mathrm{E}[\log R_p] > \log r, \tag{6}$$

in which case the objective in problem (3) can be equivalently reformulated by squaring, and dividing by 2. The result is the following criterion:

$$\arg\max_p D_p(\log r) \equiv \arg\max_p \frac{1}{2}\left(\frac{\mathrm{E}[\log R_p] - \log r}{\sqrt{Var[\log R_p]}}\right)^2. \tag{7}$$

In order to quantitatively compare criteria (4), (5), and (7), it is useful to follow Browne (1999b) in using a parametric stochastic stock price process that results in the stock price being lognormally distributed at all times t, so that $\log R_{pt} \sim IID$ $\mathcal{N}(\mathrm{E}[\log R_p], Var[\log R_p])$ as assumed above. Specifically, the stock price S follows the geometric brownian motion with drift $\mathrm{d}S/S = m\,\mathrm{d}t + v\,\mathrm{d}W$, where m denotes the instantaneous mean parameter, v denotes the instantaneous volatility parameter, and W denotes a standard Wiener process. The bond price B follows $\mathrm{d}B/B = i\,\mathrm{d}t$. Denoting the period length between times t and $t+1$ by Δt, Hull (1993, p. 210) shows that

$$\mathrm{E}[\log R_p] = (pm + (1-p)i - p^2 v^2/2)\Delta t, \tag{8}$$

$$Var[\log R_p] = p^2 v^2 \Delta t. \tag{9}$$

Now substitute Eqs. (8) and (9) into (7), and write down the first-order condition for the maximizing stock weight p. You can verify by substitution that the following

p solves it:

$$\arg\max_p D_p(\log r) = \sqrt{\frac{2(\log r - i)}{v^2}}. \tag{10}$$

Using (8), the growth optimal criterion (4) yields the portfolio

$$\arg\max_p E[\log R_p] = \frac{(m - i)}{v^2}. \tag{11}$$

Using Browne's (1999b, p. 77) parameter values $m = 15\%$, $v = 30\%$, $i = 7\%$, and a target growth rate $\log r = 8\%$, the outperformance probability maximizing rule (10) advocates investing a constant $p=47\%$ of wealth in the stock, while the growth optimal rule (11) advocates $p=89\%$. Of course, (10)'s $p=47\%$ minimizes the probability that the realized growth rate $\overline{\log R_p} \leqslant 8\%$. Fig. 1 illustrates the phenomena, by graphing $Prob[\overline{\log R_p} \leqslant 8\%]$ for the two portfolios, and a third portfolio with just 33% invested in the stock. It shows that $Prob[\overline{\log R_p} \leqslant 8\%]$ decays to zero for all three portfolios, but decays at the fastest rate when (10)'s $p=47\%$ is used. Section 2.2 will show that the rate of probability decay rate in Fig. 1 is $D_p(\log r)$ *in* (7). Fig. 1 also shows that even though investors can invest in a riskless asset earning 7%, and can try to beat the modest 8% target growth rate by also investing in a stock with an instantaneous expected return of 15%, there is still almost a 20% probability that the investor's realized growth rate of wealth after 50 years will be less than 8%!

Table 1 contrasts performance statistics for the outperformance probability maximizing portfolios and the growth optimal portfolio $p = 89\%$ over the feasible range of target growth rates $\log r$. Because the riskless rate of interest is only 7%, the probability of earning more than a target rate $\log r > 7\%$ is always less than one. If the target rate $\log r \leqslant 7\%$, the investor could always ensure outperforming that rate by investing solely in the riskless asset. Hence the lower limit of the feasible target growth rates is the 7% riskless rate.[6] Line 1 in Table 1 shows that in order to maximize the probability of outperforming a target growth rate one basis point higher than this, i.e. $\log r = 7.01\%$, the investor need invest only $p = 5\%$ of wealth in the stock. As a result of this conservative portfolio, this investor will have a relatively low probability of not exceeding this target; the decay rate of the underperformance probability is $\max_p D_p(7.01) = 3.19\%$. But by investing 89% of wealth in the stock, the growth optimal investor will have a higher probability of not exceeding this 7.01% target, because its associated decay rate is just 0.88%. This occurs despite its much higher expected growth rate $E[\log R_p]$ (10.6% vs. 7.4%) and higher expected net return $\mu = pm + (1-p)i$ (14.1% vs. 7.4%). Of course, the major reason for this is its higher volatility $\sigma = pv$ (26.7% vs. 1.5%), which increases the probability that a bad series of returns will drive the growth optimal portfolio's realized growth rate below $\log r = 7.01\%$. Also note in line 1 that in order to maximize the probability of outperforming the 7.01% target, the investor must choose a portfolio with a higher expected growth rate (7.4%) than the target, as explained earlier.

[6] Of course, if a riskless rate does not exist, it would not provide a floor on the feasible target rates.

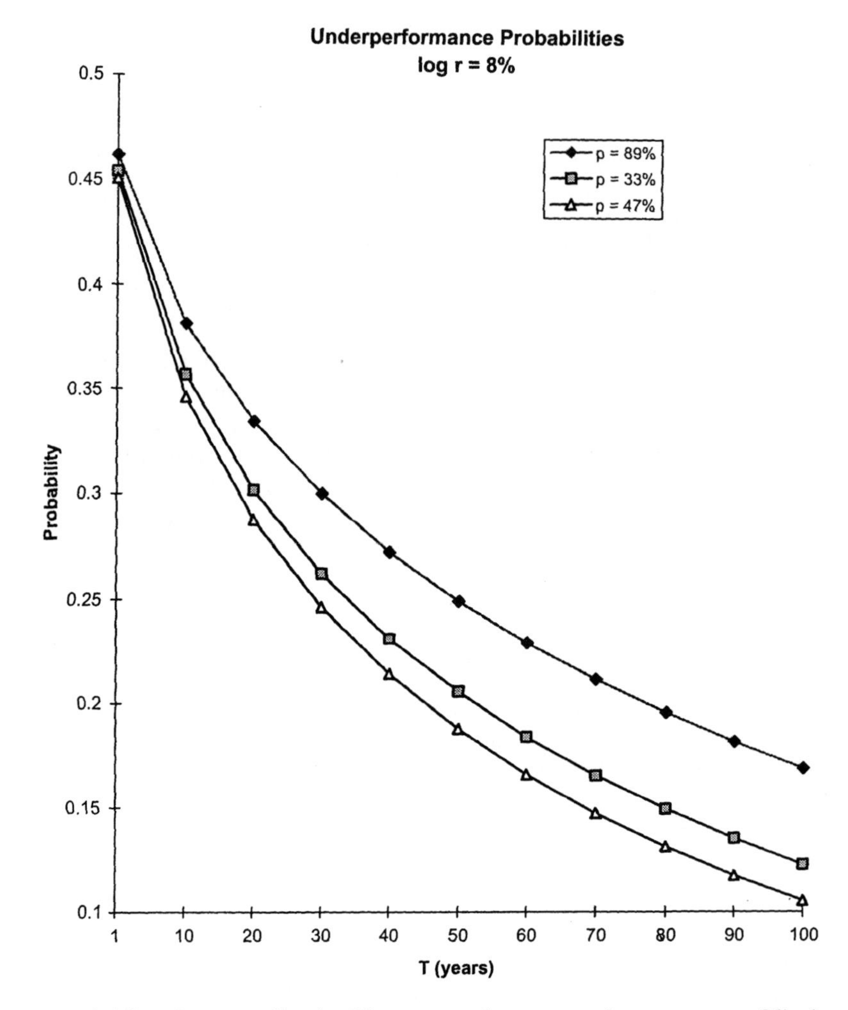

Fig. 1. The probability of not exceeding the 8% target growth rate approaches zero at a portfolio dependent rate of decay. The rate of decay is highest for the portfolio with $p = 47\%$ invested in the stock.

There is an important tradeoff present in Table 1. Note from columns 1 and 3 that investors with successively higher growth targets $\log r$ have successively lower underperformance probability decay rates $\max_p D_p(\log r)$. This implies that investors with higher targets will be exposed to a higher probability of realizing growth rates of wealth that do not exceed their respective targets. This occurs despite the fact that they did the best they could to minimize the probability of that happening. This is a consequence of the successively more aggressive portfolios needed to ensure asymptotic outperformance of their successively higher targets.

This tradeoff is analogous to the tradeoff between mean and standard deviation associated with the efficiency criterion that selects the portfolio with the smallest standard

Table 1
Performance statistics for the maximum expected log portfolio and maximum decay rate portfolios associated with feasible target growth rates $\log r$, when portfolios are formed from a lognormally distributed stock with $m=15\%$ instantaneous mean return and $v=30\%$ instantaneous volatility, and a riskless asset with instantaneous riskless rate $i = 7\%$

	Performance of portfolio (11) vs. portfolio (10)				
$\log r\%$	Stock weight $p\%$	$D_p(\log r)\%$	$E[\log R_p]\%$	$\mu\%$	$\sigma\%$
7.01	89 (5)	0.88 (3.19)	10.6 (7.4)	14.1 (7.4)	26.7 (1.5)
7.5	89 (33)	0.66 (1.39)	10.6 (9.2)	14.1 (9.6)	26.7 (9.9)
8.0	89 (47)	0.46 (0.78)	10.6 (9.8)	14.1 (10.8)	26.7 (14.1)
8.5	89 (58)	0.30 (0.44)	10.6 (10.1)	14.1 (11.6)	26.7 (17.4)
9.0	89 (67)	0.17 (0.22)	10.6 (10.3)	14.1 (12.4)	26.7 (20.1)
9.5	89 (75)	0.08 (0.09)	10.6 (10.5)	14.1 (13.0)	26.7 (22.5)
10.0	89 (82)	0.02 (0.02)	10.6 (10.5)	14.1 (13.6)	26.7 (24.6)
10.6	89 (89)	0.00 (0.00)	10.6 (10.6)	14.1 (14.1)	26.7 (26.7)

deviation of return, once the investor fixes a mean return. Here, the criterion selects the portfolio with the highest underperformance probability decay rate, once the investor fixes a target growth rate. In this way, the tradeoff between $\log r$ and $\max_p D_p(\log r)$ can be thought of as an alternative efficiency frontier, which yields the growth optimal portfolio on one extreme and full investment in the constant interest rate (when it exists) on the other. The efficiency frontier is graphed in Fig. 2, which shows it to be a convex curve in this example. In Section 2.2, we will see that this is true more generally, i.e. with multiple risky assets, whose log returns are not necessarily normal nor IID.

Finally, there is just one risky asset used to form the optimal portfolios in Table 1, so the mean-standard deviation efficiency frontier is just swept out by varying the stock weight and calculating the mean μ and standard deviation σ of the net returns. For comparison purposes, this is reported in the last two columns of Table 1. Reading down the last three columns of the table, note that the difference between μ and the expected growth rate $E[\log R_p]$ of wealth grows wider as the standard deviation of portfolio returns σ gets larger. The mean return increasingly overstates the expected growth rate of wealth as portfolio volatility increases. This is due to (1); as Hakansson and Ziemba (1995, p. 69) note, "...capital growth (positive or negative) is a multiplicative, not an additive process".[7] Here, due to lognormality, there is a precise relationship between the two: $E[\log R_p] = \mu - \sigma^2/2$ (see Hull, 1993, p. 212).

The following section will show that a simple, yet powerful result from large deviations theory permits us to rigorously characterize $D_p(\log r)$ in Table 1 as the decay rate of the portfolios' underperformance probabilities graphed in Fig. 1. More importantly, the result also shows how to correctly calculate the decay rate and associated decay rate maximizing portfolios when portfolio returns are not lognormally distributed.

[7] In this regard, see Stutzer (2000) for a simpler model of fund managers who use arithmetic average net returns rather than average log gross returns, under the assumption that net returns are IID.

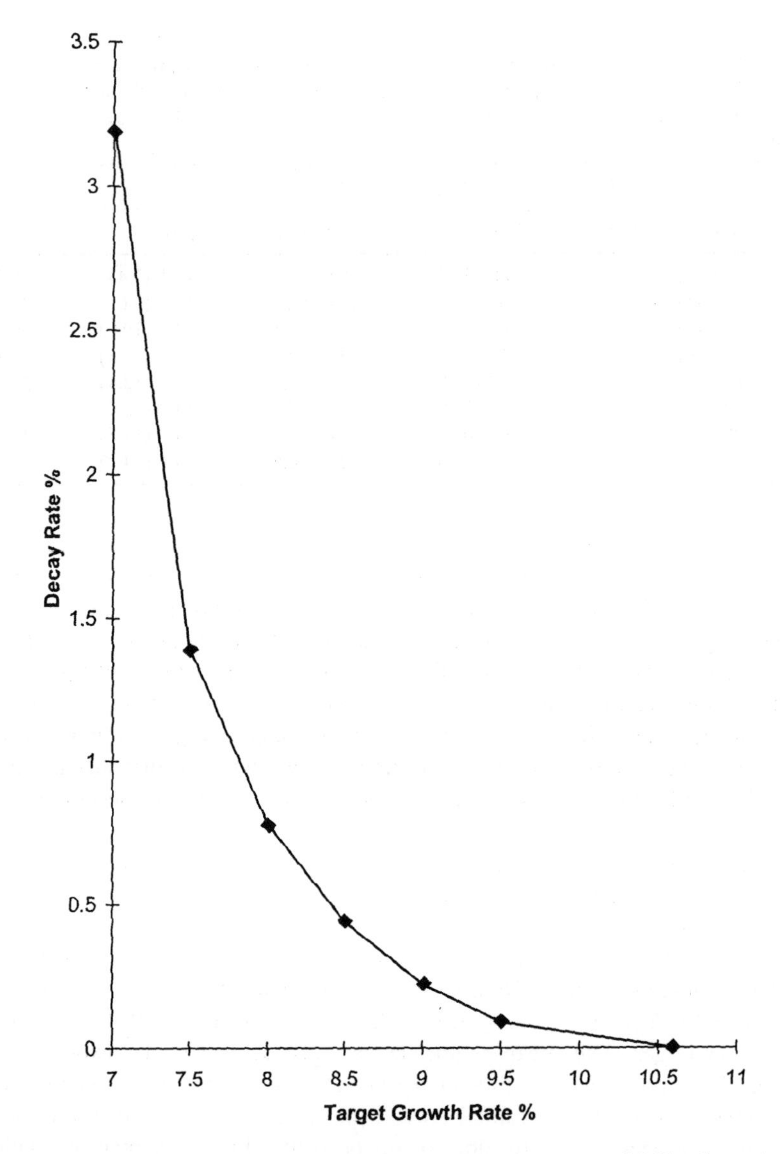

Fig. 2. The convex tradeoff between the target growth and underperformance probability decay rates for the optimal portfolios in Table 1. The convexity is generic.

2.2. *The general case*

As shown in the last section, when a portfolio's log returns were IID normally distributed, exact underperformance probabilities of the realized growth rate could be easily calculated using (2). But it is widely accepted that stock returns are often skewed and leptokurtotic. Even if they weren't, the skewed returns of derivative securities like stock options are inherently non-normally distributed. Hence there is an important need to rank portfolios according to their underperformance probabilities $Prob[\overline{\log R_p} \leqslant \log r]$ in non-IID, non-normal circumstances. It is now shown how to calculate the decay rate of this probability in more general cases. We will then apply the general result to prove that $D_p(\log r)$ in (7) is indeed the correct decay rate for the IID normal case.

As in the previous section, we seek to rank portfolios p for which the underperformance probability $Prob[\overline{\log R_p} \leqslant \log r] \to 0$ as $T \to \infty$. Calculation of this probability's decay rate $D_p(\log r)$ is facilitated by use of the powerful, yet simple to apply, Gärtner–Ellis Large Deviations Theorem, e.g. see Bucklew (1990, pp. 14–20). For a log portfolio return process with random log return $\log R_{pt}$ at time t, consider the following time average of the partial sums' log moment generating functions, i.e.:

$$\phi(\theta) \equiv \lim_{T\to\infty} \frac{1}{T} \log \mathrm{E}[\mathrm{e}^{\theta \sum_{t=1}^{T} \log R_{pt}}] = \lim_{T\to\infty} \frac{1}{T} \log \mathrm{E}\left[\left(\frac{W_T}{W_0}\right)^{\theta}\right], \tag{12}$$

where the last expression is found by using (1) to compute $W_T/W_0 = \prod_t R_{pt}$, substituting $\log(W_T/W_0)$ for the sum of the logs in (12), and simplifying. Hence (12) depends on the value of the random W_T, and so does not depend on the particular discrete time intervals between the log returns $\log R_{pt}$. We maintain the assumptions that the limit in (12) exists for all θ, possibly as the extended real number $+\infty$, and is differentiable at any θ yielding a finite limit. From the last expression in (12), these assumptions must apply to the asymptotic growth rate of the expected power of W_T/W_0. Some well-analyzed log return processes will satisfy these hypotheses, as will be demonstrated shortly by example. However, these assumptions do rule out some proposed stock return processes, e.g. the stable Levy processes with characteristic exponent $\alpha < 2$ and hence infinite variance, used in Fama and Miller (1972, pp. 261–274).

The calculation of the decay rate $D_p(\log r)$ is the following Legendre–Fenchel transform of (12):

$$D_p(\log r) \equiv \max_{\theta} \theta \log r - \phi(\theta). \tag{13}$$

When log returns are independent, but not identically distributed, (12) specializes to

$$\phi(\theta) = \lim_{T\to\infty} \frac{1}{T} \sum_{t=1}^{T} \log \mathrm{E}[\mathrm{e}^{\theta \log R_{pt}}] = \lim_{T\to\infty} \frac{1}{T} \sum_{t=1}^{T} \log \mathrm{E}[R_{pt}^{\theta}]. \tag{14}$$

When log returns are additionally identically distributed (IID), (12) simplifies to

$$\phi(\theta) = \log \mathrm{E}[\mathrm{e}^{\theta \log R_p}] = \log \mathrm{E}[R_p^{\theta}], \tag{15}$$

which when substituted into (13) yields the decay rate calculation for the IID case. This result will form the basis for the empirical application in Section 3. It is known as *Cramer's Theorem* (Bucklew, 1990, pp. 7–9).

To illustrate these calculations, let us return to the widely analyzed case where the log portfolio return $\log R_{pt}$ is a covariance-stationary normal process with absolutely summable autocovariances. Then the partial sum of log returns in (12) is also normally distributed. The mean and variance of it can be easily calculated by adapting Hamilton's (1994, p. 279) calculations for the distribution of the sample mean, i.e. the partial sum divided by T. One immediately obtains

$$\log(W_T/W_0) \equiv \sum_{t=1}^{T} \log R_{pt} \sim \mathcal{N}\left(T\mathrm{E}[\log R_p],\ T\,Cov_0 + \sum_{\tau=1}^{T-1}(T-\tau)(Cov_\tau + Cov_{-\tau})\right), \tag{16}$$

where $\mathrm{E}[\log R_p]$ denotes the log return process' common mean and Cov_τ denotes its τ-lagged autocovariance. Formula (12) is the limiting time average of the log moment generating functions of these normal distributions. Now remember from elementary statistics that a normal distribution's log moment generating function is linear in its mean and quadratic in its variance. As a result, use (12) to calculate

$$\begin{aligned}\phi(\theta) &= \lim_{T\to\infty}\frac{1}{T}\left(T\mathrm{E}[\log R_p]\theta + \left(T\,Cov_0 + \sum_{\tau=0}^{T-1}(T-\tau)(Cov_\tau + Cov_{-\tau})\right)\theta^2/2\right)\\ &= \mathrm{E}[\log R_p]\theta + \sum_{\tau=-\infty}^{\tau=+\infty} Cov_\tau \theta^2/2.\end{aligned} \tag{17}$$

Now substitute (17) into (13) and set its first derivative with respect to θ equal to zero to find that the maximum in (13) is attained by the following maximizer:

$$\theta_{\max} = (\log r - \mathrm{E}[\log R_p]) \Big/ \sum_{\tau=-\infty}^{\tau=+\infty} Cov_\tau\ . \tag{18}$$

Substituting (18) back into (17) and rearranging yields the underperformance probability decay rate

$$D_p(\log r) = \frac{1}{2}\left(\frac{\mathrm{E}[\log R_p] - \log r}{\sqrt{\sum_{\tau=-\infty}^{\tau=+\infty} Cov_\tau}}\right)^2. \tag{19}$$

Note that maximization of the decay rate (19) rewards portfolios with a high expected growth rate $E[\log R_p]$ (in its numerator) and a low asymptotic variance $\sum_{\tau=-\infty}^{\tau=+\infty} Cov_\tau \equiv \lim_{T\to\infty} Var[\log(W_T/W_0)]/T$ (in its denominator). This differs from the criterion in Bielecki et al. (2000), which is approximately the asymptotic expected growth rate minus a multiple of the asymptotic variance. For the IID case used in Section 2.1, all covariance terms in (19) are zero except $Cov_0 \equiv Var[\log R_p]$, so the decay rate function (19) reduces to the expression (7) used in Section 2.1 and Table 1. Fig. 3 depicts this decay rate function over a range of $\log r$, for each of the three portfolios whose underperformance probabilities are graphed in Fig. 1. There, we see that the portfolio $p = 47\%$ from (10) does indeed have the highest decay rate when $\log r = 8\%$.

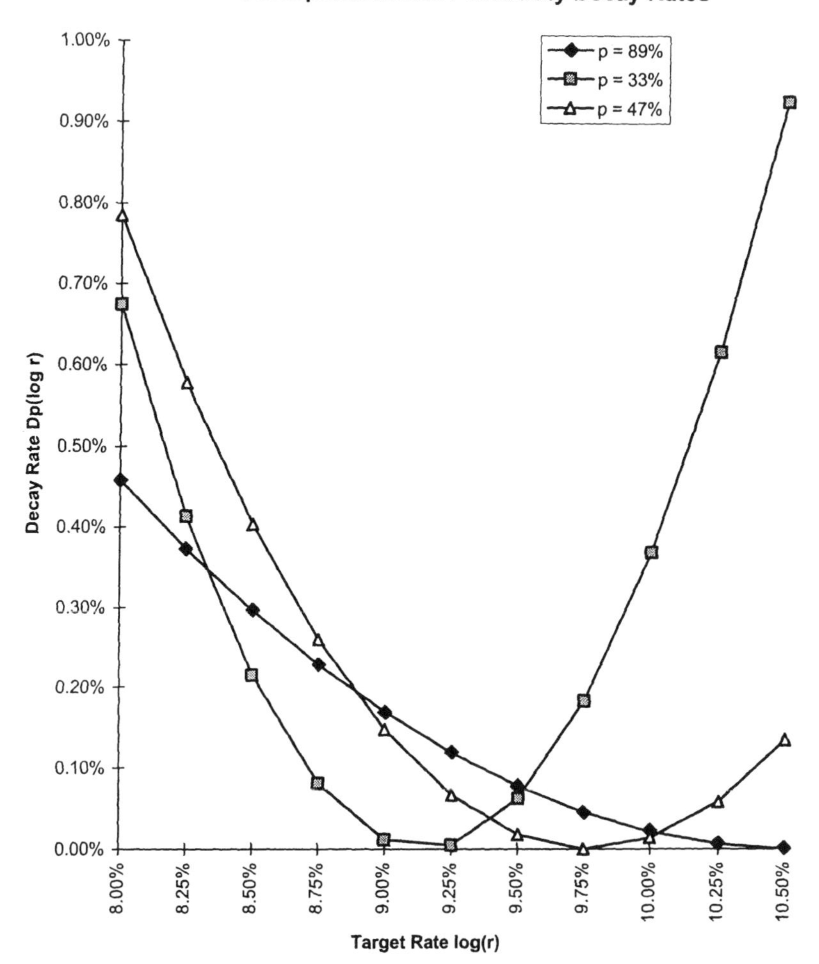

Fig. 3. The decay rate function $D_p(\log r)$ is convex, with a minimum at $\log r = E[\log R_p]$. The portfolio $p = 47\%$ attains the highest decay rate when $\log r = 8\%$.

Note that the decay rate function $D_p(\log r)$ in (19) for a covariance stationary Gaussian portfolio log return process is non-negative, and is a strictly convex function of $\log r$, achieving its global minimum of zero at the value $\log r = \mathrm{E}[\log R_p]$. These properties are true for more general processes (for a discussion, see Bucklew, 1990). As a result, remember from (6) that the decay rate criterion ranks portfolios with $\mathrm{E}[\log R_p] > \log r$, and apply the envelope theorem to the general rate function (13) to yield

$$\frac{\mathrm{d}D_p(\log r)}{\mathrm{d}\log r} = \frac{\partial}{\partial \log r} \max_{\theta} \theta \log r - \phi(\theta) = \theta_{\max} < 0 \tag{20}$$

as seen in the special case (18). Now differentiate (20) to find

$$\frac{\mathrm{d}D_p^2(\log r)}{\mathrm{d}\log r^2} = \frac{\mathrm{d}\theta_{\max}}{\mathrm{d}\log r} > 0 \tag{21}$$

due to convexity of $D_p(\log r)$. Again due to the envelope theorem, (20) and (21) continue to hold for $\max_p D_p(\log r)$ as well. Fig. 2 depicts the convexity of $\max_p D_p(\log r)$ over the relevant range of $\log r$ in the example of Table 1.

2.3. Analogy with power utility

The general decay rate criterion is a generalization of the expected power utility criterion. To uncover the generalization, substitute the right-hand side of (12) into (13), to derive

$$\begin{aligned}\max_p D_p(\log r) &\equiv \max_p \max_\theta \theta \log r - \lim_{T\to\infty} \frac{1}{T} \log \mathrm{E}\left[\left(\frac{W_T}{W_0}\right)^\theta\right] \\ &= \max_p \max_\theta \log r^\theta - \lim_{T\to\infty} \frac{1}{T} \log \mathrm{E}\left[\left(\frac{W_T}{W_0}\right)^\theta\right] \\ &= \max_p \max_\theta - \lim_{T\to\infty} \frac{1}{T} \log \mathrm{E}\left[\left(\frac{W_T}{W_0 r^T}\right)^\theta\right],\end{aligned} \tag{22}$$

which yields the following large T approximation:

$$-\mathrm{e}^{-\max_p D_p(\log r)T} \approx \max_p \mathrm{E}\left[-\left(\frac{W_T}{W_0 r^T}\right)^{\theta_{\max}(p)}\right], \tag{23}$$

where we write $\theta_{\max}(p)$ in (23) to stress dependence (through the joint maximization (22)) of θ on the portfolio p. The left-hand side of (23) increases with D_p, so a large T approximation of the portfolio ranking is produced by use of the expected power utility on the right-hand side of (23).

There are both similarities and differences between the right-hand side of (23) and a conventional expected power utility $\mathrm{E}[-(W_T)^\theta]$. From (20), $\theta_{\max}(p) < 0$. Evaluating it at the investor's decay rate maximizing portfolio p, note that the power function in (23) with the form $U = -(\bullet)^{\theta_{\max}(p)}$ increases toward zero as its argument grows to infinity, is strictly concave, and has a constant degree of relative risk aversion $\gamma \equiv 1 - \theta_{\max}(p) > 1$. Furthermore, $\theta_{\max}(p) < 0$ implies that the third derivative of U is positive, so the criterion exhibits positive skewness preference. But there are two important differences between the concepts. First, the argument of the power function in (23) is altered; it is the *ratio* of invested wealth to a "benchmark" level of wealth accruing in an account that grows at the geometric rate r. While absent from traditional criteria, this ratio is also present in other non-standard criteria, such as Browne's (1999a, p. 276) criterion to "maximize the probability of beating the benchmark by some predetermined percentage, before going below it by some other predetermined percentage". Browne (1999a, p. 277) notes that "...the relevant state variable is the *ratio* of the investor's wealth to the benchmark".[8] Second, conventional portfolio theory assumes that the risk aversion parameter θ is a preference parameter that is independent of the investment opportunity set. But in (23), $\theta = \theta_{max}(p)$ is determined by maximization, and hence is *not* independent of the investment opportunity set. Investors could utilize different investment opportunity sets, either because of differential regulatory constraints, such as hedge funds' greater ability to short sell, or because of different opinions about the parameters of portfolios' log return processes. *When this happens, investors will have different decay rate maximizing portfolios p, and different degrees of risk aversion* $\gamma = 1 - \theta_{max}(p)$, *even if they have the same target growth rate* $\log r$.

Assuming that asset returns are generated by a continuous time, correlated geometric Brownian process, Browne (1999a, p. 290) compares the formula for the optimal portfolio weights resulting from his criterion, to the formula resulting from conventional maximization of expected power utility at a fixed terminal time T. In this special case, he finds that the two formulae are isomorphic, i.e. there is a mapping between the models' parameters that equates the two formulae. He concludes that "there is a connection between maximizing the expected utility of terminal wealth for a power utility function, and the objective criteria of maximizing the probability of reaching a goal, or maximizing or minimizing the expected discounted reward of reaching certain goals". Connection (23) between decay rate maximization and expected power utility is quite specific, yet does not depend on a specific parametric model of the assets' joint return process.

Critics such as Bodie (1995, p. 19) have argued that "the probability of a shortfall is a flawed measure of risk because it completely ignores how large the potential shortfall might be". It is possible that this is a fair assessment of expected power utility maximization of wealth at a fixed horizon date T, subject to a "Value-At-Risk" (VaR) constraint that fixes a low probability for the event that terminal wealth could fall below a fixed floor. This problem was intensively studied by Basak and Shapiro (2001, p. 385), who concluded that "The shortcomings...stem from the fact that the VaR agent is concerned with controlling the probability of a loss rather than its magnitude". They proposed replacing the VaR constraint with an ad hoc expected loss constraint, resulting in fewer shortcomings. The investor's target growth rate serves a similar function in the horizon-free, unconstrained criterion (22).

3. Non-parametric implementation

In the IID case, there is a simple, distribution-free way to estimate $D_p(\log r)$ for a portfolio p. Following the comparative portfolio study of Kroll et al. (1984), we replace the expectation operator in (15) by an historical time average operator, substitute into (13), and numerically maximize that.[9] This estimator eliminates the need for prior

[8] While Browne considers a stochastic benchmark, the constant growth benchmark here can be modified to consider an arbitrary stochastic benchmark, at the cost of fewer concrete expository results.

[9] It is important to remember that the log moment generating function of the log return distribution necessarily has to exist near θ_{max} in order for this technique to work here.

knowledge of the log return distribution's functional form and parameters. Specifically, let $R_p(t)=\sum_{j=0}^{n} p_jR_j(t)$ denote the historical return at time t of a portfolio comprised of $n+1$ assets with respective returns $R_j(t)$, with constantly rebalanced portfolio weights $\sum_j p_j = 1$. The estimator is

$$\hat{D_p}(\log r) = \max_{\theta} \theta \log r - \log\left[\frac{1}{T}\sum_{t=1}^{T}\left(\sum_{j=0}^{n} p_jR_j(t)\right)^{\theta}\right] \tag{24}$$

and the optimal portfolio weights are estimated to be

$$\hat{p} = \arg\max_{p_1,\ldots,p_n}\max_{\theta} \theta \log r - \log\left[\frac{1}{T}\sum_{t=1}^{T}\left(\sum_{j=1}^{n} p_jR_j(t) + \left(1-\sum_{j=1}^{n} p_j\right)R_0(t)\right)^{\theta}\right]. \tag{25}$$

The maximum expected log portfolio was similarly estimated, by numerically finding the weights that maximize the time average of $\log R_p(t)$.

Let us now contrast the estimated decay rate maximizing portfolio (25) to both the expected log and Sharpe ratio maximizing, constantly rebalanced portfolios formed from Fama and French's 10 domestic industry, value-weighted assets,[10] whose annual returns run from 1927 through 2000. The sample cross-correlations of the 10 industries' gross returns range from 0.32 to 0.86, suggesting that diversified portfolios of them will provide significant investor benefits. The sample covariance matrix is invertible, permitting estimation of the Sharpe ratio maximizing "tangency" portfolio, by multiplying this inverse by the vector of sample mean excess returns over a riskless rate, and then normalizing the result. We assume that it was possible to costlessly store money between 1927–2000, with no positive *constant* nominal rate riskless asset available.[11] Hence we assume a zero constant riskless rate when computing the Sharpe ratio maximizing tangency portfolio of the 10 industry assets.

The results are seen in Table 2.

The performance statistics in Table 2 show that the Sharpe ratio maximizing portfolio has almost no skewness. But the decay rate maximizing portfolios all have a skewness of about 1, as does the expected log maximizing portfolio. This reflects the skewness preference inherent in the generalized expected power utilities with degrees of risk aversion greater than (in the log case, equal to) one.[12] In fact, these investors prefer all odd order moments and are averse to all even order moments. To see this, note that (15) is the cumulant generating function for the (assumed) IID log portfolio return

[10] The data are currently available for download from a website maintained by Kenneth French at MIT.

[11] Treasury Bills are not a *constant* rate riskless asset, like the one used to form portfolios in Section 2.1. A fixed percentage of wealth invested in Treasury Bills is just like any other risky asset.

[12] See Kraus and Litzenberger (1976) and Harvey and Siddique (2000) for evidence that investors prefer skewness.

Table 2
Comparison of estimated Sharpe ratio, expected log, and decay rate maximizing portfolios from Fama-French 10 industry indices, 1927–2000

Industries	Asset moments			Portfolio weights				
	μ	σ	Skewness	Max Sharpe	$\log r$ 5%	$\log r$ 10%	$\log r$ 15%	Max Log
NoDur	0.130	0.198	−0.12	0.80	0.92	1.0	1.11	1.15
Durbl	0.166	0.328	0.86	−0.01	0.27	0.52	0.97	1.22
Oil	0.137	0.220	0.01	0.75	0.77	0.96	1.24	1.36
Chems	0.146	0.225	0.63	0.14	0.35	0.52	0.89	1.15
Manuf	0.136	0.254	0.21	0.03	−0.10	0	0.11	0.20
Telcm	0.123	0.200	0.07	0.35	0.48	0.38	0.30	0.28
Utils	0.118	0.225	0.25	0.05	−0.20	−0.34	−0.61	−0.76
Shops	0.141	0.256	−0.25	−0.13	−0.44	−0.60	−0.96	−1.2
Money	0.142	0.245	−0.43	−0.23	−0.20	0.07	0.48	0.70
Other	0.106	0.242	−0.04	−0.76	−0.86	−1.5	−2.52	−3.09
Performance statistics								
Mean				0.148	0.162	0.195	0.248	0.278
Std. dev.				0.153	0.181	0.240	0.368	0.446
Skewness				−0.02	1.05	1.07	1.06	1.05
Decay rate $D_p(\log r)$					0.18	0.04	0.004	0
Risk aversion $1-\theta_{\max}(p)$					5.3	2.5	1.3	1

distribution. Substituting it into (13) and evaluating it at $\theta_{\max}(p)$ yields the following cumulant expansion:

$$D_p(\log r) = (\log r - \mathrm{E}[\log R_p])\theta_{\max}(p) - \frac{Var[\log R_p]}{2}\theta_{\max}(p)^2 - \sum_{i=3}^{\infty}\frac{\kappa_i}{i!}\theta_{\max}(p)^i, \tag{26}$$

which uses the facts that $\mathrm{E}[\log R_p]$ is the first cumulant of the log return distribution and that $Var[\log R_p]$ is its second cumulant, while κ_i denotes its ith order cumulant. Because $\theta_{\max}(p) < 0$, we see that the decay rate increases in odd-order cumulants and decreases in even-order cumulants. With normally distributed log returns, all the cumulants in the infinite sum are zero. But with non-normally distributed returns, increased skewness will increase the decay rate (due to κ_3). The relative weighting of the mean, variance and skewness in (26) is determined by their sizes, the sizes of the higher order cumulants, the target growth rate $\log r$, and the value of $\theta_{\max}(p) < 0$ associated with $\log r$.

The top panel of Table 2 contains the 10 industry weights in each portfolio. As is typical of estimated Sharpe ratio maximizing portfolios with more than a few assets, it is heavily long in just three industries (Non-durables, Oil, and Telecommunications). The decay rate maximizing portfolio for the target growth rate $\log r = 0.10$ is also heavily invested in these industries, but in addition it has considerable long positions in the two most positively skewed industries (Durables and Chemicals). The Sharpe ratio maximizing portfolio is heavily short in one industry (Other). The decay rate maximizing portfolios are heavily short in both this industry and as well as two others (Shops and Utilities). The differences between Sharpe ratio and decay rate maximizing portfolios are due to the presence of the target growth rate in decay rate maximization, its use of log gross returns rather than net returns when calculating portfolio means and variances, and the presence of higher order moments. It is difficult to assess the impact of higher order moments on the differences in portfolio weights. Bekaert et al. (1998, p. 113) were able to produce only a two percentage point difference in an asset weight, when simulating the effects of its return's skewness over the range -1 to 2.0, on the portfolio chosen by an expected power utility maximizing agent whose degree of risk aversion was close to 10. This suggests that the use of a target growth rate, and the use of log gross returns rather than arithmetic net returns, account for most of the differences between the decay rate and Sharpe ratio maximizing portfolios' weights.

The convergence of decay rate maximizing portfolios to the expected log maximizing portfolio is seen when reading across the last four columns of Table 2. The last two rows in the bottom panel of Table 2 show the relationship between the target growth rates, their respective efficient portfolios' maximum decay rates, and their respective endogenous degrees of risk aversion. Despite the fact that $\theta_{\max}(p)$ is determined by maximization in (25), we see that the degree of risk aversion $1-\theta_{\max}(p)$ is not unusually large in any of the decay rate maximizing portfolios tabled,[13] and converges toward 1 as $\log r \to \max_p \mathrm{E}[\log R_p]$. An alternative interpretation of this is enabled by computing the first order condition for $\theta_{\max}(p)$ in the IID case. To do so, substitute (15) into (13) and differentiate to find

$$\mathrm{E}\left[\log R_p \frac{\mathrm{d}Q}{\mathrm{d}P}\right] = \log r, \tag{27}$$

where the *Esscher* transformed probability density

$$\frac{\mathrm{d}Q}{\mathrm{d}P} = \frac{R_p^{\theta_{\max}(p)}}{\mathrm{E}[R_p^{\theta_{\max}(p)}]} \tag{28}$$

is used to compute the expected log return (i.e. growth rate) in (27).[14] Furthermore, a result known as Kullback's Lemma (1990) shows that the Esscher transformed density (28) is the solution to the following constrained minimization of relative entropy, whose minimized value is the decay rate, i.e.

$$D_p(\log r) = \min \mathrm{E}\left[\frac{\mathrm{d}Q}{\mathrm{d}P} \log \frac{\mathrm{d}Q}{\mathrm{d}P}\right] \text{ s.t. (27).} \tag{29}$$

From (27), an efficient portfolio has the highest decay rate among those with a fixed *transformed* expected growth rate equal to $\log r$. As $\log r \to \max_p \mathrm{E}[\log R_p]$,

[13] Of course, it can get unusually large when the target growth rate is unusually low, i.e. when the investor is unusually conservative.

[14] See Gerber and Shiu (1994) for option pricing formula derivations that use the Esscher transform to calculate the risk-neutral density required for option pricing.

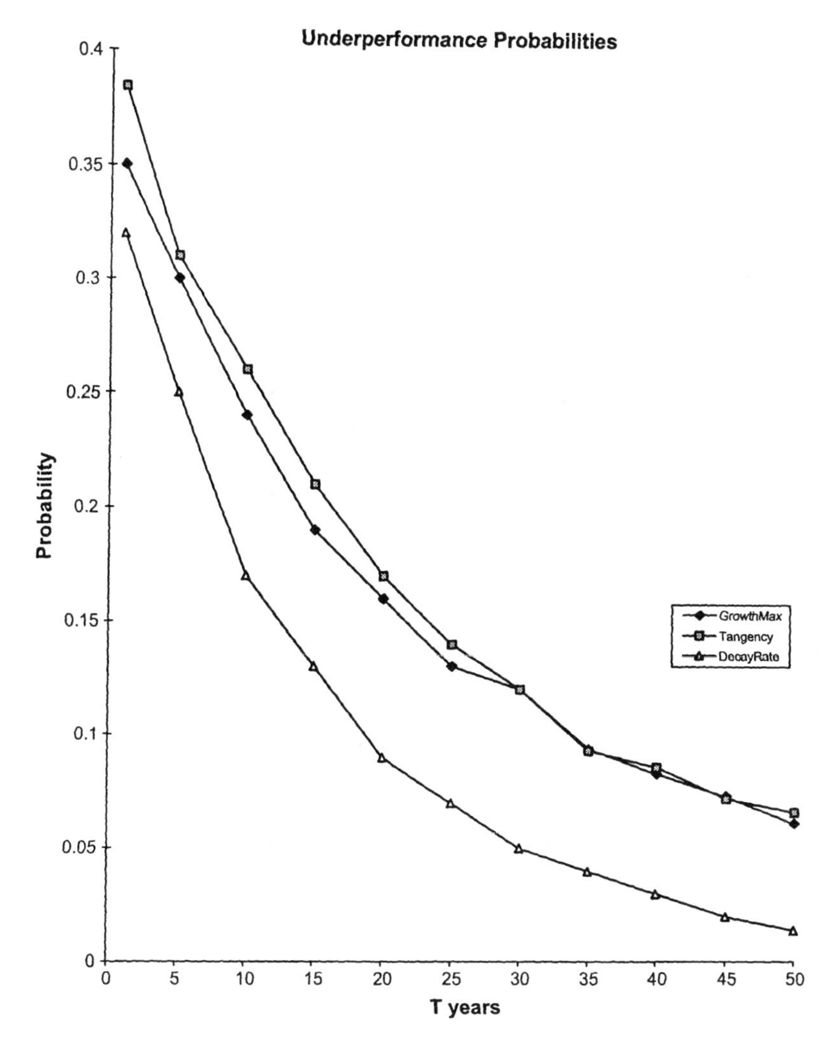

Fig. 4. Bootstrap estimated underperformance probabilities for portfolios in Table 2, when $\log r = 10\%$.

$\theta_{\max}(p) \to 0$, density (28) concentrates at unity and the minimal relative entropy in (29) approaches zero, i.e. the transformed probabilities approach the actual probabilities. As a result, the transformed expected log return in (27) approaches the actual expected log return, so constraint (27) collapses the portfolio constraint set onto the log optimal portfolio.

In order to determine if a decay rate maximizing portfolio in Table 2 will have lower underperformance probabilities than the Sharpe ratio and expected log maximizing portfolios do, the probabilities were estimated by resampling the portfolios' log returns 5000 times for each investment horizon length T, and then tabulating the empirical frequency of underperformance for each T. The results for the target decay rate

$\log r = 10\%$ are graphed in Fig. 4. Fig. 4 shows that the estimated decay rate maximizing portfolio in Table 2 had lower underperformance probabilities for all values of T.

3.1. More general estimators

The empirical estimates above were made under the assumption of IID returns. There is little evidence of serially correlated log returns in many equity portfolios, and what evidence there is finds low serial correlation. Hence there is little benefit in using an efficient estimator for the covariance stationary Gaussian rate function (20), e.g. using a Newey–West estimator of its denominator. But the presence of significant GARCH (perhaps with multiple components) effects (see Bollerslev, 1986) in log returns, as described in Tauchen (2001, p. 58), motivates the need for additional research into efficient estimation of (12) and (13) under specific parametric process assumptions. Alternatively, it may be possible to find an efficient nonparametric estimator for (12) and (13) by utilizing the smoothing technique exposited in Kitamura and Stutzer (1997, 2002) to estimate the expectation in (12).

4. Conclusions and future directions

A simple large deviations result was used to show that an investor desiring to maximize the probability of realizing invested wealth that grows faster than a target growth rate should choose a portfolio that makes the complimentary probability, i.e. of wealth growing no faster than the target rate, decay to zero at the maximum possible rate. A simple result in large deviations theory was used to show that this *decay rate maximization* criterion is equivalent to maximizing an expected power utility of the ratio of invested wealth to a "benchmark" wealth accruing at the target growth rate. The risk aversion parameter that determines the required power utility, and the investor's degree of risk aversion, is also determined by maximization and is hence endogenously dependent on the investment opportunity set. Yet it was not seen to be unusually large in the applications developed here.

The highest feasible target growth rate of wealth is that attained by the portfolio maximizing the expected log utility, i.e. that with the maximum expected growth rate of wealth. Investors with lower target growth rates choose decay rate maximizing portfolios that are more conservative, corresponding to degrees of risk aversion that exceed 1. As the target growth rate falls, it is easier to exceed it, so the decay rate of the probability of underperforming it goes up. The relationship between possible target growth rates and their corresponding maximal decay rates form an efficiency frontier that replaces the familiar mean-variance frontier. An investor's specific target growth rate determines the specific decay rate maximizing portfolio chosen by her. A decay rate maximizing investor does not choose a portfolio attaining an expected growth rate of wealth equal to her target growth rate (instead it is higher than her target). But there is an *Esscher* transformation of probabilities, under which the *transformed* expected growth rate of wealth *is* the target growth rate.

Researchers choosing to work in this area may select from several interesting topics. First, it is easy to generalize the analysis to incorporate a stochastic benchmark. This would be helpful in modelling an investor who wants to rank the probabilities that a group of similarly styled mutual funds will outperform their common style benchmark. Second, one could calculate the theoretical decay rate function using a multivariate GARCH model for the asset return processes, and then estimate the resulting function. Third, one could extend the decay rate maximizing investment problem to the joint consumption/portfolio choice problem, enabling the derivation of consumption-based asset pricing model with a decay rate maximizing representative agent. If it is possible to construct a model like this, the representative agent's degree of risk aversion will depend on the investment opportunity set—an effect heretofore unconsidered in the equity premium puzzle.

Acknowledgements

Thanks are extended to Eric Jacquier, the editors, and other participants at the Duke University Conference on Risk Neutral and Objective Probability Measures, to seminar participants at NYU, Tulane University, University of Illinois-Chicago, Chicago Loyola University, Georgia State University, University of Alberta, Bachelier Finance Conference, Eurandom Institute, Morningstar, Inc., and Goldman Sachs Asset Management, and to Edward O. Thorp, David Bates, Ashish Tiwari, Georgios Skoulakis, Paul Kaplan, and John Cochrane for their timely and useful comments on the analysis.

References

Algoet, P., Cover, T.M., 1988. Asymptotic optimality and asymptotic equipartition property of log-optimal investment. Annals of Probability 16, 876–898.

Basak, S., Shapiro, A., 2001. Value-at-risk based risk management: optimal policies and asset prices. Review of Financial Studies 14, 371–405.

Bekaert, G., Erb, C., Harvey, C., Viskanta, T., 1998. Distributional characteristics of emerging market returns and asset allocation. Journal of Portfolio Management 24, 102–116.

Bielecki, T.R., Pliska, S.R., Sherris, M., 2000. Risk sensitive asset allocation. Journal of Economic Dynamics and Control 24, 1145–1177.

Bodie, Z., 1995. On the risk of stocks in the long run. Financial Analysts Journal 51, 18–22.

Bollerslev, T., 1986. Generalized autoregressive conditional heteroskedasticity. Journal of Econometrics 31, 307–327.

Browne, S., 1995. Optimal investment policies for a firm with a random risk process: exponential utility and minimizing the probability of ruin. Mathematics of Operations Research 20, 937–958.

Browne, S., 1999a. Beating a moving target: optimal portfolio strategies for outperforming a stochastic benchmark. Finance and Stochastics 3, 275–294.

Browne, S., 1999b. The risk and rewards of minimizing shortfall probability. Journal of Portfolio Management 25, 76–85.

Bucklew, J.A., 1990. Large Deviation Techniques in Decision, Simulation, and Estimation. Wiley, New York.

Cover, T.M., Thomas, J.A., 1991. Elements of Information Theory. Wiley, New York.

Detemple, J., Zapatero, F., 1991. Asset prices in an exchange economy with habit formation. Econometrica 59, 1633–1657.

Fama, E., Miller, M., 1972. The Theory of Finance. Holt, Rhinehart and Whinston, New York.

Gerber, H., Shiu, E., 1994. Option pricing by Esscher Transforms. Transactions of the Society of Actuaries 46, 99–140.

Hamilton, J.D., 1994. Time Series Analysis. Princeton University Press, Princeton, NJ.

Hakansson, N.H., Ziemba, W.T., 1995. Capital growth theory. In: Jarrow, R.A., Maksimovic, V., Ziemba, W.T. (Eds.), Handbooks in Operations Research and Management Science: Finance, Vol. 9. North-Holland, Amsterdam.

Harvey, C., Siddique, A., 2000. Conditional skewness in asset pricing tests. Journal of Finance 40, 1263–1293.

Hull, J., 1993. Options, Futures, and Other Derivative Securties. Prentice-Hall, Englewood Cliffs, NJ.

Kitamura, Y., Stutzer, M., 1997. An information-theoretic alternative to generalized method of moments estimation. Econometrica 65, 861–874.

Kitamura, Y., Stutzer, M., 2002. Connections between entropic and linear projections in asset pricing estimation. Journal of Econometrics 107, 159–174.

Kocherlakota, N.R., 1996. The equity premium: Its still a puzzle. Journal of Economic Literature 34, 42–71.

Kraus, A., Litzenberger, R.H., 1976. Skewness preference and the valuation of risk assets. Journal of Finance 31, 1085–1100.

Kroll, Y., Levy, H., Markowitz, H., 1984. Mean-variance versus direct utility maximization. Journal of Finance 39, 47–61.

MacLean, L.C., Ziemba, W.T., Blazenko, G., 1992. Growth versus security in dynamic investment analysis. Management Science 38, 1562–1585.

Samuelson, P.A., 1969. Lifetime portfolio selection by dynamic programming. Review of Economics and Statistics 51, 239–246.

Samuelson, P.A., 1994. The long-term case for equities. Journal of Portfolio Management 21, 15–24.

Stutzer, M., 2000. A portfolio performance index. Financial Analysts Journal 56, 52–61.

Tauchen, G., 2001. Notes on financial economics. Journal of Econometrics 100, 57–64.

Thorp, E.O., 1975. Portfolio choice and the Kelly criterion. In: Ziemba, W.T., Vickson, R.G. (Eds.), Stochastic Optimization Models in Finance. Academic Press, New York.

44

On Growth-Optimality vs. Security Against Underperformance

Michael Stutzer
Professor of Finance and Director,
Burridge Center for Securities Analysis and Valuation,
University of Colorado, Boulder, CO
michael.stutzer@colorado.edu

Abstract

The expected log utility of wealth (i.e., the growth-optimal or Kelly) criterion has been oft-studied in the management science literature. It leads to the highest asymptotic growth rate of wealth, and has no adjustable "preference parameters" that would otherwise need to be precisely "adjusted" to a specific individual's needs. But risk-control concerns led to alternative criteria that stress security against underperformance over finite horizons. Large deviations theory enables a straightforward generalization of log utility's asymptotic analysis that incorporates these security concerns. The result is a power utility criterion that (like log utility) is free of an adjustable risk aversion parameter, because the latter is endogenously determined by expected utility maximization itself! A Bayesian formulation of the Occam's Razor Principle is used to illustrate the unavoidable reduction of scientific testability (i.e., the ability to more easily falsify) inherent in criterion functions that introduce additional adjustable parameters that are not directly observable.

1 A Simple Repeated Betting Problem

All concepts and results are illustrated using the simplest repeated betting problem analyzed in the management science literature. This is the popular "Blackjack" example, discussed by Thorp (1984), MacLean, Ziemba, and Blazenko (1992), and MacLean and Ziemba (1999). The agent must choose a fraction p of wealth to bet on an IID Bernoulli process that has a gross return per bet $R_p = 1 + p$ with probability $\pi > 1/2$, and $R_p = 1 - p$ with probability $1 - \pi$. In other words, the bettor either wins or loses the fraction p of accumulated wealth each try. For example, if the bettor's initial wealth is $W_0 = \$500$ and chooses $p = 5\%$, the bettor will either win or lose \$25 on the first try. If the bettor wins, the second try will either win or lose $\$525 * .05 = \$26.25$. But if the better lost the first time, the second try will either win or lose $475 * .05 = \$23.75$. This is isomorphic to a binomially distributed stock portfolio initially worth \$500, and that returns $\pm 5\%$ in each subsequent time period.

The accumulated wealth after T bets is $W_T = W_0 \prod_{t=1}^{T} R_{pt}$. The bettor is interested in the security against underperforming some benchmark wealth path $W_{bT} = W_0 \prod_{t=1}^{T} R_{bt}$. Taking the log of both establishes that:

$$Prob[W_T \leq W_{bT}] = Prob\left[\frac{\sum_{t=1}^{T} \log R_{pt}}{T} - \frac{\sum_{t=1}^{T} \log R_{bt}}{T} \leq 0\right] \tag{1}$$

while taking the exponentials and rearranging shows:

$$W_T = W_0 e^{\frac{\sum_{t=1}^{T} \log R_{pt}}{T} T} \tag{2}$$

By laws of large numbers valid for this binomial and more realistic processes, $\frac{\sum_{t=1}^{T} \log R_{pt}}{T} \overset{T\to\infty}{\longrightarrow} E[\log R_p]$. So (2) shows that the *expected* growth rate $E[\log R_p]$ resulting from a bet p is *realized only asymptotically.* Hence $\arg\max_p E[\log R_p]$ is the bet yielding the highest *asymptotic* growth rate of wealth. *For illustrative purposes only,* let us assume that the bettor's "edge" is $60 - 40 = 20\%$. It is easy to verify that betting $\arg\max_p E[\log R_p] = .20$ yields the maximum asymptotic growth rate $E[\log R_{.20}] = 0.60 \log 1.20 + 0.40 \log 0.80 \approx 0.0201$.

As Hakansson and Ziemba (1995) noted, this maximum expected log, a.k.a. growth-optimal or "Kelly", criterion "*almost surely* leads to more capital in the long run than any other investment policy which does not converge to it", and that "the growth-optimal strategy also has the property that it asymptotically minimizes the expected time to reach a given level of capital". Other interesting characterizations of the criterion are found in Bell and Cover (1988) as well as in the textbook of Cover and Thomas (1991).

But some other researchers, e.g., Aucamp (1993), argued that in practice it is quite possible that the asymptotic outperformance won't be realized with high probability until after a very, very long time – perhaps hundreds of years. Over finite horizons, MacLean, Ziemba and Blazenko (1992) note that "the Kelly strategy never risks ruin, but in general it entails a considerable risk of losing a substantial portion of wealth", and define three measures of security against underperforming a benchmark. Figure 1 graphs the probability (1) of underperforming a benchmark wealth path growing at a constant *target growth rate* denoted $\log r = 1\%$ per bet, i.e., substitute $\log R_{bt} \equiv 1\%$ for all t in (1). Money growing at a constant rate of $\log r = 1\%$ will double every $\log 2/.01 \approx 70$ bets. Figure 1 shows that the probability of underperformance achieved by each of the three betting fractions p approaches zero as $T \to \infty$, as the law of large numbers guarantees (because $E[\log R_p] > 1\%$ for all three values of p). But Figure 1 shows that the convergence to zero is quite slow, and that betting either too high a fraction of wealth (e.g., the growth-optimal $p = 20\%$ of wealth) or too low a fraction (e.g., $p = 8\%$) results in lower security against underperformance than betting $p = 14.1\%$ of wealth per bet. This following section will show how that fraction is found.

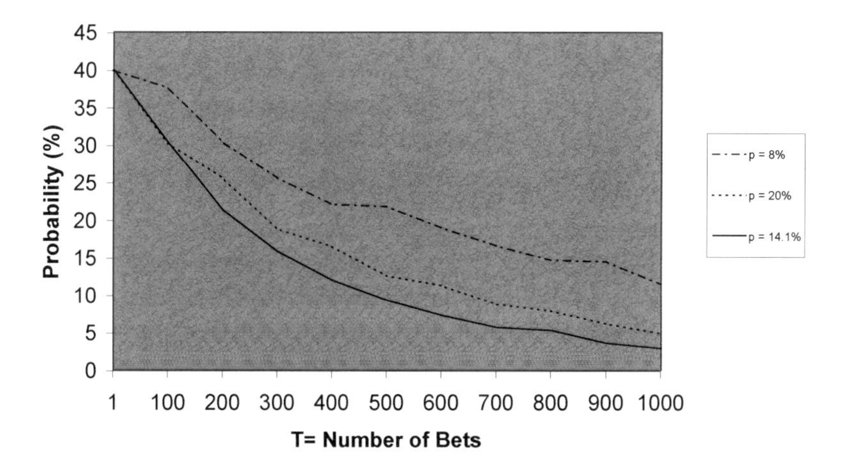

Figure 1 Probability of underperforming $\operatorname{Log} r = 1\%$ benchmark.

2 Asymptotics of Benchmark Underperformance Probabilities

The underperformance probabilities (1) graphed in Figure 1 decay to zero *exponentially* as $T \to \infty$. Our emphasis on the security after suitably large T differs from the security notions in MacLean, Ziemba, and Blazenko's (1992), who define security at either a *fixed* T (their definition S1) or at *all* T both small and large (their definition S2). But Figure 1 also suggests that the betting fraction with the relatively best large-T security will also have relatively good security when T is smaller, which is usually the case in other problems studied by the author. *For each betting fraction p, large deviations theory can be used to calculate the exponential decay rate of the underperformance probability curve associated with it. The optimal betting fraction is the one associated with the most rapidly decaying curve.* This criterion focuses attention on the asymptotics of the realized growth rate (i.e., $\sum_t \log R_{pt}/T$) relative to a benchmark's realized growth rate ($\sum_t \log R_{bt}/T$), rather than just the asymptotics of the realized growth rate itself (i.e., $E[\log R_p)$. When the benchmark *is* the growth-optimal policy (i.e., $\arg\max_p E[\log R_p]$), the two criteria coincide. *Hence the criterion exposited herein is a natural generalization of the asymptotic reasoning used by growth-optimal policy advocates, that additionally incorporates the objective fear of underperforming a benchmark (fixed or stochastic) over finite horizons.*

To find the optimal policy in our example, one must first determine the betting fractions p that make those probabilities decay to zero asymptotically (i.e., $E[\log R_p] - \log r > 0$), like the three fractions in Figure 1 do when $\log r = 1\%$. This restriction defines both *minimum* (denoted $\underline{p}$) and maximum (denoted $\overline{p}$) acceptable betting fractions. In our example, $\underline{p} \approx 5.9\% < p < \overline{p} \approx 33.6\%$.

Within this range of p, it is a consequence of the Gärtner-Ellis Large Deviation Theorem (e.g., see Bucklew (1990)) that under fairly general conditions, the underperformance probabilities approach zero at an asymptotic exponential *rate* of decay denoted D_p, computed below:

$$D_p \equiv \max_{\gamma} - \lim_{T\to\infty} \frac{1}{T} \log E\left[e^{\theta \sum_{t=1}^{T}(\log R_{pt} - \log R_{bt})}\right] \tag{3}$$

But there are more insightful ways of restating this criterion. It turns out that when $\underline{p} < p < \overline{p}$, the maximizing $\theta < 0$ in (3) (Stutzer, 2003). So without loss of generality we can replace θ in (3) by $\gamma \equiv -\theta > 0$, yielding the equivalent formulation

$$D_p \equiv \max_{\gamma>0} - \lim_{T\to\infty} \frac{1}{T} \log E\left[e^{-\gamma \sum_{t=1}^{T}(\log R_{pt} - \log R_{bt})}\right] \tag{4}$$

Using the right hand side of (1) to calculate the summed exponent in (4), a bit of algebra (Stutzer, 2003) yields the following restatements of the optimal choice problem:

$$p_{opt} \equiv \arg\max_p D_p \equiv \arg\max_p \max_{\gamma>0} \lim_{T\to\infty} -\frac{1}{T} \log E\left[\left(\frac{W_T}{W_{bT}}\right)^{-\gamma}\right] \tag{5}$$

$$\stackrel{IID}{\equiv} \arg\max_p \max_{\gamma>0} - \log E\left[\left(\frac{R_p}{r}\right)^{-\gamma}\right] \tag{6}$$

$$\equiv \arg\max_p \max_{\gamma>0} E\left[-\left(\frac{R_p}{r}\right)^{-\gamma}\right] \tag{7}$$

$$\stackrel{Blackjack}{\equiv} \arg\max_p \max_{\gamma>0} - \left[0.60\left(\frac{1+p}{r}\right)^{-\gamma} + 0.40\left(\frac{1-p}{r}\right)^{-\gamma}\right] \tag{8}$$

where $W_{bT} \equiv W_0 \prod_{t=1}^{T} R_{bt} = W_0 r^T$ when the benchmark is just a constant target growth rate.

Fixing the choice vector or scalar (in our example) denoted p, the inner maximization over γ in (5) is the most general expression for the asymptotic decay rate of its underperformance probability curve (see Figure 1), valid in both IID or non-IID situations with either constant or stochastic benchmark log returns. The inner maximization over γ in (6) is the decay rate in IID situations when the benchmark is a constant target growth rate.

There are two major differences between (4) or (5) or the IID special case (7), and conventional maximization of the following expected power utility with constant relative risk aversion $1 + \gamma$:

$$\max_p E[-R_p^{-\gamma}] \tag{9}$$

First, the argument in our criterion (7) is the ratio of the bet's return to the benchmark return, rather than just the former in (9). The second major difference is that γ is *not* adjustable by the analyst to "fit" a bettor's supposedly fixed, exogenous degree of relative risk aversion $(1 + \gamma)$. Instead, for a fixed p, γ is determined by the inner maximization, and thus varies across the values of p considered. The endogenous degree of risk aversion *exhibited* by the agent is $1+\gamma(p_{opt})$, which arises from the subsequent outer maximization over p that determines the agent's optimal choice p_{opt}. This will depend on both the opportunity set of alternatives evaluated by the agent, and the constant target growth rate (or stochastic benchmark) that the agent wants to beat. In other words, the bettor *jointly* maximizes expected power utility by jointly varying both the feasible betting fraction *and* the utility's curvature or risk aversion "coefficient".

This agent's rank ordering of each betting fraction in the range of p whose underperformance probability curves decay to zero (i.e., $5.9\% = \underline{p} < p < \overline{p} = 33.6\%$) is determined by numerical solution of (8). A summary of the optimization is seen in Table 1 below:

Table 1 The best security against underperforming a $\log r = 1\%$ target growth rate of wealth per bet is achieved by betting $p_{opt} = 14.1\%$, resulting in the fastest decay rate (0.18% per bet) of underperformance probabilities.

Bettor With $\log r = 1.0\%$ Per Bet Growth Target				
$p\%$	Value of (8)	D_p % from (6)	$E[\log R_p]$ %	Risk Aversion $1+\gamma$
5.9	≈ -1	≈ 0	≈ 1.0	≈ 1
8.0	-.9994	0.06	1.28	1.45
14.1	**-.9982**	**0.18**	**1.83**	**1.43**
20.0	-.9987	0.13	2.01	1.25
33.6	≈ -1	≈ 0	≈ 1.0	≈ 1

Table 1 shows that the highest attainable value of the decay rate is found by solving (8), producing $p_{opt} = 14.1\%$ and $\gamma(14.1\%) = .43$. The asymptotic growth rate of wealth associated with the optimal bet is $E[\log R_{14.1\%}] = 1.83\%$. This is higher than the agent's target growth rate of $\log r = 1\%$, as it must be in order to maximize the probability of outperforming that target (equivalently, to minimize the probability of underperforming it) over finite horizons T, as depicted in Figure 1, and is of course lower than the 2.01% asymptotic growth rate associated with the growth-optimal $p = 20\%$. The bettor *exhibits* an endogenous degree of relative risk aversion $1+\gamma(14.1\%) = 1.43$, but uses different values of γ to evaluate the expected utility of different p values, listed in the last column of Table 1.

The optimization problem can be reformulated as a constrained entropy problem of the sort analyzed in the management science literature, e.g., Dinkel and Kochenberger (1979). Let $D(Q \mid \pi) \equiv E_Q\left[\log \frac{Q}{\pi}\right]$ denote the familiar Kullback-Leibler discrimination statistic (a.k.a. *relative entropy*, see Cover and Thomas, 1991) measuring the difference between a variable probability measure Q and the fixed measure π. Here, E_Q is used to denote the expectation taken with respect to the measure Q. There is a close connection between entropy and the large deviations rate function, so it is not surprising that the optimal betting problem can be reformulated as a constrained entropy problem. That connection (e.g., Kitamura and Stutzer, 2002) shows that the optimal solution to the

Table 2 A bettor with a lower target growth rate of wealth will use different p-dependent coefficients of risk aversion to evaluate the same bets, will choose to bet less, and will exhibit a higher degree of endogenous risk aversion.

Bettor With $\log r = 0.1\%$ Per Bet Growth Target				
$p\%$	Value of (9)	D_p % from (6)	$E[\log R_p]$ %	Risk Aversion $1+\gamma$
0.507	≈ -1	≈ 0	≈ 0.1	≈ 1
1.0	-.9954	0.46	0.20	10.73
4.5	**-.9877**	**1.23**	**0.79**	**4.53**
14.1	-.9924	0.77	1.83	1.88
20.0	-.9954	0.46	2.01	1.48
33.0	-.9996	0.04	1.09	1.09
38.5	≈ -1	≈ 0	≈ 0.1	≈ 1

betting problem can also be computed by solving the constrained entropy problem: $\max_p \min_Q D(Q \mid \pi)$ *s.t.* $E_Q[\log R_p] \leq \log r$. In the above example, this problem reduces to finding scalar values for $0 \leq Q \leq 1$ and $0 \leq p \leq 1$ solving

$$\max_p \min_Q Q\log(Q/0.60) + (1-Q)\log((1-Q)/0.40)$$
$$s.t. \quad Q\log(1+p) + (1-Q)\log(1-p) \leq 0.01$$

It is easy to verify that a numerical solution to this problem is $p_{opt} = 14.1\%$ and $Q = 57\%$, so that the relative entropy $D_{p_{opt}} = 0.18\%$ at the saddlepoint. Furthermore, use of the notation $D_{p_{opt}}$ for both the optimal relative entropy and the underperformance probability decay rate is not an accident: The bold faced row in Table 1 shows that the latter (0.18% per bet) is the same as the former.

The following Table 2 summarizes the decision process for a more conservative agent, who would be satisfied by beating a lower, and hence more easily beaten target growth rate of $\log r = 0.1\%$ (i.e., wealth doubles only every 700 bets or so), over the (now slightly different) range of bets $0.507\% = \underline{p} < p < \overline{p} = 38.5\%$ that yield underperformance probabilities decaying to zero:

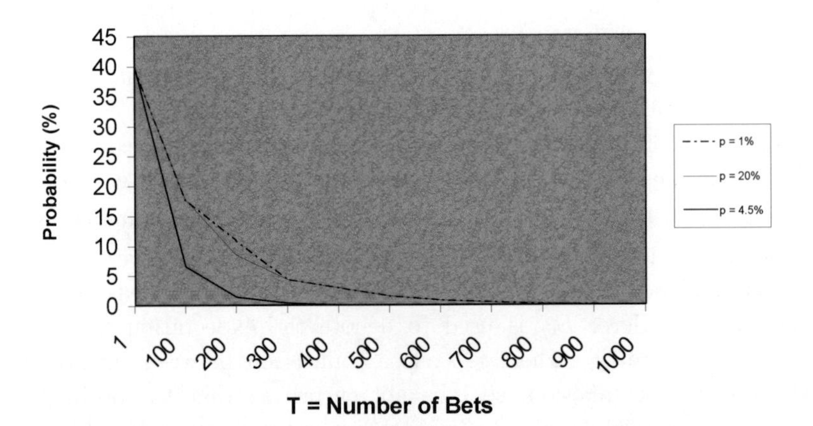

Figure 2 Probability of underperforming $\text{Log}\, r = 0.1\%$ benchmark.

Table 2 shows that this agent should bet only $p_{opt} = 4.5\%$ of wealth per bet to maximize (minimize) the probability of outperforming (underperforming) the more conservative (and hence more easily beaten) growth target $\log r = 0.1\%$ per bet. Comparing $D_{p_{opt}}$ in the bold faced row of Tables 2 and 1 shows that the lower target enables more rapid decay of the optimal bet's underperformance probability curve, seen by comparing the lowest curves in Figures 2 and 1. Table 2 shows that this bettor exhibits a higher (but still plausible) endogenous degree of relative risk aversion equal to 4.53.

The following Table 3 shows the optimal betting fractions and degrees of relative risk aversion (solving (8)) exhibited by bettors satisfied with beating lower target growth rates than the growth-optimal 2.01%.

Finally, it is easy to show that the relationship between the first and third columns in Table 3 is convex, as it will be in other betting or investment problems. This relationship shows the tradeoff between an index of desired growth (i.e., the target growth rate) and an index of security against underperformance (i.e., the decay rate). This efficiency frontier will shift out when the betting opportunity set is more favorable (i.e., a higher probability of winning π), as depicted in Figure 3.

Table 3 A Bettor who adopts a lower (than growth-optimal) target $\log r < 2.01\%$ to attain more security against underperforming it (i.e., a higher decay rate $D_p > 0$) should choose a smaller betting fraction $p_{opt} < 20\%$ than growth-optimal investors do, and will exhibit a higher endogenous degree of relative risk aversion $1 + \gamma(p_{opt}) > 1$ than growth-optimal investors do.

Bettors With Other Growth Targets				
$\log r\%$	$p_{opt}\%$	$D_{p_{opt}}(\log r)\%$	$E[\log R_{p_{opt}}]\%$	Risk Aversion $1 + \gamma(p_{opt})$
2.01	20.0	**0**	2.01	1
1.0	14.1	**0.18**	1.83	1.43
0.5	10.0	**0.51**	1.50	2.02
0.1	4.5	**1.23**	0.79	4.53

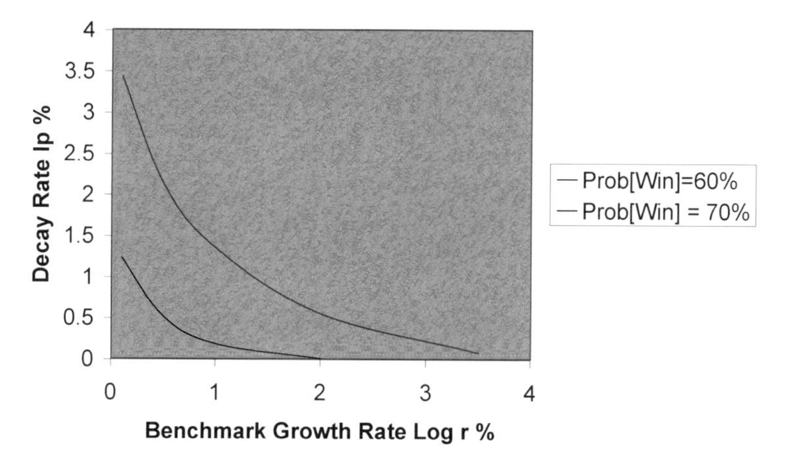

Figure 3 Efficiency frontiers.

3 A Brief Discussion of Closely Related Literature

There is a management science literature advocating that the growth-optimal policy be combined with a riskless asset to strike an agent's desired tradeoff of growth for security against underperformance. Such *fractional Kelly strategies* have been studied by MacLean, Ziemba, and Blazenko (1992), MacLean and Ziemba (1999), Li (1993), and others. The criterion embodied in (5) permits the policy p to be any adapted process. For example, in portfolio choice over N-assets, the policy p would be a (generally time-varying) N-vector of portfolio weights varying over time. The only restriction on p is that the underperformance probabilities decay to zero as $T \to \infty$, in accord with the motive to minimize such probabilities. There is (as yet) no basis for restricting the strategy space to fractional Kelly strategies.

Samuelson (1974, 1979) challenged the accuracy and/or relevance of *some* of the outperformance probability-based asymptotic reasoning advanced by those who advocated the use of *log* utility, i.e., the limiting case of exogenous $\gamma \downarrow 0$ in (9). But an analysis in Stutzer (2004) showed that Samuelson's critiques do not apply to the criterion developed in the previous section. Furthermore, an important risk-scaling paradox highlighted by Rabin (2000) applies to the expected concave utility function maximization (like (9)) that Samuelson preferred to outperformance probability-based criteria. The Blackjack problem can be used to illustrate the nature of Rabin's critique. The optimal betting fraction when *conventionally* maximizing the expected power utility $E[W_T] = E[-W_T^{-\gamma}]$ with exogenous γ does not depend on the number of bets T. Suppose you were offered the opportunity to bet $p = 10.8\%$ of your initial wealth $T = 10,000$ times, the end result of which is determined in a split second by computer simulation. The same calculations used to produce Figures 1 and 2 (see footnote 3) show that after betting $p = 10.8\%$ of accumulated wealth $T = 10,000$ times, there is almost no chance that you could underperform a $\log r = 0.6\%$ growth rate per bet. That is, there is near-certainty that your wealth would be multiplied by a factor of at least $e^{.006*10000} \approx \$10^{26}$ in the split second it takes for the computer-determined outcome. Yet conventional use of expected constant relative risk aversion utility (9) – arguably the most commonly made behavioral assumption in the economics literature – implies that a bettor with an exogenous degree of risk aversion greater than just $1 + \gamma = 3.76$ would rather not play at all than have to bet $p = 10.8\%$, no matter the size of T nor the initial wealth W_0, even though it is *always impossible* to lose more than W_0 in the Blackjack problem. Nonetheless maintaining the assumption that each individual acts as-if he/she *does have* an exogenous constant degree of relative risk aversion, Barsky *et al.* (1997) designed a questionnaire to measure it, which was given to thousands of individuals in person by Federal interviewers. About 2/3 of those surveyed had a degree of relative risk aversion higher than 3.76, so their maintained assumption implies that all of them would have refused to bet $p = 10.8\%$ once, twice, ten thousand, or ten million times. Would you? Fortunately, a bettor using the criterion in Section 2, and who has the relatively modest growth target used above (i.e., $\log r = 0.6\%$), would indeed be willing to bet this much.

Most importantly, Rabin (2000) also showed that analogous expected utility paradoxes are constructable when concave utility functions other than (9) are assumed. His concerns about expected concave utility of wealth criteria are not easily dismissed.

Finally, there is of course a large literature proposing other ways to integrate shortfall probability into portfolio analysis, e.g., Browne (2000) and the other papers cited by him, or maximization of non-log expected utility subject to a "Value-At-Risk" constraint (2001). A full exploration of the connection between these alternatives and the criterion in Section 2 is too lengthy to be attempted in this note on the growth-optimal literature.

4 Occam's Razor Critique of Alternative Decision Criteria

One long-held disideratum was expressed by Albert Einstein: "*Everything should be made as simple as possible, but not simpler*". This is the principle of parsimonious parameterization that has historically been termed *Occam's (or Ockham's) Razor.* There are other ways to generalize the standard CRRA (9) and its limiting log utility criteria. But they have more adjustable parameters than the criterion described in Section 2. Each of the alternatives introduced additional adjustable parameters when modifying a conventional CRRA power utility (such as log utility), gaining flexibility that is helpful when confronting individual behavioral observations. But by impeding the ability to make as sharp a prediction about what could *potentially* be observed, the flexibility enabled by the additional adjustable parameters comes at a cost. This tradeoff is quantified in the Bayesian analysis of Jefferys and Berger (1992), summarized as follows:

> Bayesian analysis can shed new light on what the notion of the "simplest" hypothesis consistent with the data actually means... By choosing prior probabilities of hypotheses, one can quantify the scientific judgment that simpler hypotheses are more likely to be correct. Bayesian analysis also shows that a hypothesis with fewer adjustable parameters automatically has an enhanced posterior probability, because the predictions it makes are sharp.

To formalize this, suppose we are trying to assess the plausibility (a.k.a. subjective probability, as defined and interpreted by Bayesians) of one hypothesized decision criterion H_0 with fewer adjustable parameters, e.g., that described in Section 2, relative to another criterion H_A. A Bayesian assesses prior probabilities $\mu(H_0)$ and $\mu(H_A)$, finds the likelihoods L of each *potential* individual decision under the two hypotheses, and finally calculates the following ratio of posterior probabilities (i.e., the posterior odds in favor of H_0):

$$\frac{Prob[H_0]}{Prob[H_A]} = \frac{L(D_{obs} \mid H_0)}{L(D_{obs} \mid H_A)} \frac{\mu(H_0)}{\mu(H_A)} \stackrel{Def.}{=} B * Prior\ Odds \tag{10}$$

where D_{obs} is the individual decision that is actually observed and B is the *Bayes factor*, i.e., the ratio of the actual observation's likelihood under each hypothesis. Because H_0 has fewer adjustable parameters, it is possible that the likelihood of a large subset of *potential* observations will be smaller under H_0 than under H_A. But if so, the likelihood of the complementary (possibly smaller) set of potential observations will be higher under H_0. If the actually observed individual decision D_{obs} lies in this complementary set, the Bayes factor in (10) will favor H_0, even though its general fitting ability is lower than an alternative with more adjustable parameters. Unless there is some reason to place higher prior probability on the more profligately parameterized alternative H_A, the posterior odds ratio (10) will thus favor the "simpler", sharper hypothesis H_0 over the more flexible H_A.

The crucial claim that the H_0 developed in section 2 makes a sharper prediction than the conventional (without benchmark) CRRA hypothesis H_A hinges on the ability to more accurately determine an agent's benchmark than direct measurements can accurately determine the hypothesized exogenous curvature parameter γ (e.g., the aforementioned problematic questionnaire used by Barsky, *et al.* (1997)). It is fair to question this ability in the context of our betting example, where the analyst would have had to more accurately determine a value for the bettor's target growth rate $\log r$ than a value for γ, before observing the fraction of wealth bet. But the ability to more accurately determine the benchmark is easy to envision in the more important context of investment behavior. A typical fund manager, and an investor who uses the manager, expressly try to outperform an identified benchmark portfolio tailored to the fund's style. For example, a recent advertisement by Standard and Poors cited independently gathered data indicating that \$3.5 trillion of funds are professionally invested in attempts to beat the S&P 500 index benchmark. Hence when considering fund managers (and their voluntary clients), the benchmark required for the criterion in Section 2 is a directly observable portfolio (e.g., the S&P 500), in which case the criterion in Section 2 has no adjustable parameters, and hence makes the sharpest conceivable predictions.

Notes

[1]Ergodic processes that are not IID still have limiting infinite time averages, that would need to be denoted differently than $E[\log R_p]$.

[2]MacLean, Ziemba, and Blazenko (1992) assumed that a value of $\pi = 51\%$ might be achievable by card counting techniques.

[3]For each value of T and p, the event of underperformance will occur when there are x or fewer wins, where $[x \log(1+p)+(T-x)\log(1-p)]/T \leq 0.01$. Hence the probability of this event is computed by evaluating the cumulative binomial distribution (with probability of winning equal to $\pi = 0.60$) at the highest integer x satisfying this inequality.

[4]In brief, define the partial sum $Y_T \equiv \sum_{t=1}^{T} \log R_{pt} - \sum_{t=1}^{T} \log R_{bt}$. Define the time-averaged log moment generating function $\phi_T(\theta) \equiv \frac{1}{T} \log E[e^{\theta Y_T}]$. Bucklew (1990) lists the following two implicit restrictions on the process needed to ensure that (3) is the aforementioned decay rate: (i) $\lim_{T \to \infty} \phi_T(\theta)$ exists for all θ – but where ∞ is allowed both as a limit value as well as member of the sequence and (ii) this limit function is differentiable at all values of θ for which it is finite. As shown in Bucklew (1990), these restrictions hold for a wide variety of both independent and dependent stochastic processes used in practice to model Y_T, including our example's IID Bernoulli return process with return R_p and constant benchmark return $\log r$. Stutzer (1995) used a vector version of the result to characterize an asset pricing model error diagnostic. I recently discovered that Sornette (1998) applied the IID version (Cramer's Theorem) in several portfolio choice settings. With specific parametric processes, Pham (2003) found solutions for related continous time, dynamic portfolio choice problems. Dembo, Deuschel, and Duffie (2004) also used Cramer's Theorem to approximate loss probabilities for suitably large numbers of loans.

[5]Probably the earliest hint of a result like this is in Ferguson (1965). He studied a class of betting problems, and noted that *if* "the probability of ruin goes to zero exponentially as the fortune tends to infinity", maximization of expected *exponential* utility would be "approximately valid" for some value of its θ coefficient "when the fortune is large". A precise finite T result of this nature was developed by Browne (1995), for a problem where wealth is generated by a specific parametric controlled stochastic differential equation. The more general, nonparametric results (5)–(7) were enabled by formulating shortfall probabilities using (1), and then applying the powerful Gärtner-Ellis Theorem to simply compute the asymptotics.

[6]Perhaps the first use of constrained entropy in finance was Cozzolino and Zahner (1973).

[7]The constraint is binding at the solution, so (in this scalar p example!) the constraint equation can be used to write Q as a function of p in closed form. Substitute this for Q in $D(Q \mid \pi)$ and numerically maximize over p to find $p_{opt} = 14.1\%$.

[8]See Stutzer (2004) for a proof applicable to this and more general situations.

[9]For example, consider the following "habit formation" modification of the CRRA utility (9): divide R_p by the exogenous parameter r as in (7), but leave γ as an additional exogenous parameter.

[10]The likelihood *functions* must be calculated *before* the individual's decision p_{obs}) is observed. This is to avoid the following ersatz calculation: observe an individual's p_{obs}, then "fit" the adjustable parameters in H_A by working backwards to find adjustable parameter values that exactly "explain" p_{obs}, and then assert that $L(p_{obs} \mid H_A) = 1$ in (10)! This procedure would make it impossible to reject this or any other hypothesis that is flexible enough to generate all possible data that could be observed (and hence any particular p_{obs}. According to the widely accepted view of Popper (1977), an hypotheses that can't be rejected (in his terms, falsified) is not a *scientific* hypothesis.

References

Donald, C. A. (1993). On the extensive number of plays to achieve superior performance with the geometric mean strategy. *Management Science*, 39(9), 1163–1172.

Barsky, R. B., F. T. Juster, M. S. Kimball, and M. D. Shapiro (1997). Preference parameters and behavioral heterogeneity: An experimental approach in the health and retirement study. *Quarterly Journal of Economics*, 112(2), 537–579.

Suleiman, B. and A. Shapiro (2001). Value-at-risk based risk management: Optimal policies and asset prices. *Review of Financial Studies*, 14(2), 371–405.
Robert, M. B. and T. M. Cover (1988). Game-theoretic optimal portfolios. *Management Science*, 34(6), 724–733.
Browne, S. (1995). Optimal investment policies for a firm with a random risk process: Exponential utility and minimizing the probability of ruin. *Mathematics of Operations Research*, 20, 937–958.
Browne, S. (2000). Risk-constrained dynamic active portfolio management. *Management Science*, 46(9), 1188–1199.
Bucklew, J. A. (1990). *Large Deviation Techniques in Decision, Simulation, and Estimation.* New York: Wiley.
Cover, T. M. and J. A. Thomas (1991). *Elements of Information Theory.* New York: Wiley.
Cozzolino, J. M. and M. J. Zahner (1973). The maximum entropy distribution of the future market price of a stock. *Operations Research*, 23, 1200–1211.
Dembo, A., J.-D. Deuschel, and D. Duffie (2004). Large portfolio losses. *Finance and Stochastics*, 8(1), 3–16.
Dinkel, J. J. and G. A. Kochenberger (1979). Constrained entropy models: Solvability and sensitivity. *Management Science*, 25(6), 555–564.
Ferguson, T. S. (1965). Betting systems which minimize the probability of ruin. *SIAM Journal*, 13(3), 795–818.
Hakansson, N. H. and W. T. Ziemba (1995). Capital growth theory. In R. A. Jarrow, V. Maksimovic, and W. T. Ziemba, editors, *Handbooks in Operations Research and Management Science: Finance*, Volume 9, pp. 65–86. Amsterdam: North Holland.
Jefferys, W. H. and J. O. Berger (1992). An application of robust Bayesian analysis to hypothesis testing and Occam's razor. *Journal of the Italian Statistical Society*, 1, 17–32.
Kitamura, Y. and M. Stutzer (2002). Connections between entropic and linear projections in asset pricing estimation. *Journal of Econometrics*, 107, 159–174.
Li, Y. (1993). Growth-security investment strategy for long and short runs. *Management Science*, 39, 915–924.
MacLean, L. C. and W. T. Ziemba (1999). Growth versus security: Tradeoffs in dynamic investment analysis. *Annals of Operations Research*, 85, 193–225.
MacLean, L. C., W. T. Ziemba, and G. Blazenko (1992). Growth versus security in dynamic investment analysis. *Management Science*, 38, 1562–1585.
Merton, R. and P. Samuelson (1974). Fallacy of the log-normal approximation to optimal portfolio decision-making over many periods. *Journal of Financial Economics*, 1(1), 67–94.
Pham, H. (2003). A large deviations approach to optimal long term investment. *Finance and Stochastics*, 7, 169–195.
Popper, K. (1977). *The Logic of Scientific Discovery.* US: Routledge (14th Printing).
Rabin, M. (2000). Risk aversion and expected-utility theory: A calibration theorem. *Econometrica*, 68(5), 1281–1292.
Samuelson, P. A. (1979). Why we should not make mean log of wealth big though years to act are long. *Journal of Banking and Finance*, 3, 305–307.
Sornette, D. (1998). Large deviations and portfolio optimization. *Physica A*, 256, 251–283.
Stutzer, M. (1995). A Bayesian approach to diagnosis of asset pricing models. *Journal of Econometrics*, 68, 367–397.
Stutzer, M. (2003). Portfolio choice with endogenous utility: A large deviations approach. *Journal of Econometrics*, 116, 365–386.

Stutzer, M. (2004). Asset allocation without unobservable parameters. *Financial Analysts Journal*, 60(5), 38–51.

Thorp, E. O. (1984). *The Mathematics of Gambling*. New Jersey: Lyle Stuart.

Part VI

Evidence of the use of Kelly type strategies by the great investors and others

45

Introduction to the Evidence of the Use of Kelly Type Strategies by the Great Investors and Others

We know that in the long run, full Kelly strategies dominate other strategies, but they are very risky short term. Hence, practical applications to long sequences of wagers are especially appropriate. Current hedge fund trading that enters and exits in a few seconds is such an application. Thorp (1960) coined the term *Fortunes's Formula* in his application of Kelly to the game of blackjack using his card counting system. The count provides an estimate of the mean so that with more favorable counts the player should wager more. Typically, blackjack players can play about 60+ hands per hour. So this application works well for full Kelly. But the risk is high that even after a few hours, with say a 1% edge, a player may be behind. Hence, blackjack teams typically use fractional Kelly strategies, with fractions of 0.2 to 0.8 being common. Gottlieb (1984, 1985) describes the early use of these fractional Kelly strategies.

It is no coincidence that many of the greatest hedge fund and portfolio managers are former blackjack players. Billionaires like trend follower Harold McPike, bond guru Bill Gross, options trader Blair Hull, and racetrack guru Bill Benter are some examples. Other top investors like Renaissance Medallion hedge fund trader Jim Simons and Warren Buffett, the legendary greatest investor in the world, use strategies that employ Kelly-type ideas to attain long term wealth growth. So did the great economist John Maynard Keynes in managing the King's College Cambridge endowment. As we have seen the size of Kelly wagers is largely dependent upon the probability of losing and it is important not to overbet. Since, the effect of errors in means are risk aversion dependent, these wagers for Kelly bettors are very sensitive to good mean estimates. Otherwise, it is easy to overbet.

The book *Fortune's Formula* written by William Poundstone (2005) brought the Kelly ideas to a wider audience. So now many investment newsletters such as *Morningstar* and *The Motley Fool* suggest Kelly strategies (see Fuller (2006) and Lee, 2006). Unfortunately, most of these sources do not fully understand the computation of Kelly strategies in the multivariate case and when there are portfolio management imperfections (such as slippage). As well, they often miss the subtleties and danger of these Kelly applications. In this final part of the book, we present applications of Kelly and fractional Kelly modeling.

Hausch, Ziemba, and Rubinstein (HRZ) (1981) present a weak market efficiency anomaly in the horserace place and show markets and how it can be exploited and wagered on using the Kelly criterion. The idea is to use probabilities derived in a simple market (the win pool involving n horses) in a market that is more complex — the place pool involving $n(n-1)$ combinations and/or the show pool involving $n(n-1)(n-2)$ combinations. Previous studies (see Hausch and Ziemba, 2008, for an up to date survey of results), indicate that the win market at racetracks is efficient except for a favorite-longshot bias. Favorites are underbet and longshots are overbet. This bias has been apparent for well over 50 years, although recent changes in the betting to allow rebates for large bettors and head to head wagering on a betting exchange have changed this shape to some extent; see graphs in Hausch and Ziemba (2008). This biases for win is in the opposite direction of the bias for place and show where favorites are overbet and longshots underbet so in the HZR analysis that was the probability of finishing one-two for place or one-two-three for show, the biases tend to cancel. So HZR assume that the probability that horse i wins is simply its betting fraction W_i/W, and the probability that i is first and j is second is the chance that i wins and j wins a race that does not contain i, namely

$$(W_i/W)\left\{\frac{W_j/W}{1-W_i/W}\right\}$$

where W_i are the wagers to win on horse i and $W=\sum W_i$. While its known that these Harville formulas are biased, the bias cancels in the HZR application, again see Hausch and Ziemba (2008) for current research on this bias. HZR compute the expected value of place and show bets and note that there frequently are favorable bets with expected values well above 1.0 per dollar bet. Applying the Kelly criterion leads to a complex non-concave non-linear programming model to compute optimal wagers that include the transactions costs or slippage that comes from the effect of our investor's bets on the odds. They show that the system yields positive profits at several racetracks. To make the system operational, they solved thousands of Kelly models and regressed the optimal bets and expected values on the ratio of the place and show pool bet on horse i. Then at the track, one only needs to look at W_i and W and P_i and P or S_i and S, where P_i are the wagers to place on horse i and S_i are the wagers to show on horse i. A quick scan can indicate whether a bet is good or not.

Hausch and Ziemba (1985) added further refinements to the basic system. These included equations for the optimal place and show bets for varying initial wealth and place or show betting pool size. These were subsequently used in a calculator that allows players to punch in the four parameters from above and compute quickly the expected value and Kelly bet. This includes the track take adjustment, and coupled entry adjustments if needed that Hausch and Ziemba worked out. They also looked into the effects of the track take and breakage (the rounding to 2.20 or 2.10 for example from 2.25 and 2.18 payoffs to the nearest dime per \$2 or \$1 bet, respectively. Finally, they computed how many players can be using the system at the same time before the market will become efficient. Ziemba and Hausch (1984, 1986, 1987) wrote two books that made the system popular and famous. Now twenty-five years late, the system is still in use although the conditions are rougher in 2010 than in the 1980s because of the two factors mentioned above (rebates and betting exchanges) and the fact that there is so much cross track betting these days that about half the betting pool has not arrived when the horses begin running. But the clever punter can estimate this odds movement that occurs after the bets have to be made.

Ziemba and Hausch (1994, 2008) show how to modify the place and show system developed in the HZR and HZ papers to a different betting system that is used in some other countries such as England. In the UK, there is no show betting, only place, but the number of horses that place is not necessarily two as in North America. Rather, it depends on the number of horses running in the race. There can be one, two, three or four horses that place when the number of horses running is at most four, 5–7, 8–15 or 16+, respectively. In the UK there are bookies offering fixed odds and the parimutuel tote as discussed here. In North America place and show bets pay back the stakes of the horses that place or show and then the net profit is divided equally among these two or three qualifiers. The the payoffs depend on the number of winning tickets on each qualifier. But in the UK, the entire pool, after the take, is divided into 1, 2, 3 or 4 equal portions which is then split into the individual payoffs. So you can see that with a large amount bet on one horse, the payoffs for a £1 bet can be as low as £1. With this minimum payoff there are arbitrages possible, see Hausch and Ziemba (1990) for the US and Jackson and Waldron (2003) for the UK and other locales. Ziemba and Hausch develop the expected value equation and Kelly criterion bet sizing. These equations yield simple graphs to apply the system in the UK.

In all of these applications, the theory indictes that the calculations should be based, as Thorp has argued, on total liquid wealth not just the track wealth. Wilcox (2003ab, 2005) has argued for a related theory where decisions are made on excess wealth after liabilities are deducted. This is a fractional Kelly strategy.

Grauer and Hakansson (1986) and in a number of other papers apply the Kelly modeling approach to asset-allocation problems. They use a simple asset return distribution procedure: it is assumed that the past will repeat itself. At the beginning of each period t, the investor chooses a portfolio on the basis of a utility function for gross returns, R, given by

$$\frac{1}{\gamma}\{R\}^{\gamma}, \text{ for } \quad \gamma \leqslant 1\,.$$

The data used are the total monthly and yearly returns including interest and dividends on stocks, long-term government bonds, long-term corporate bonds and small capitalized stocks for 1926–1983 in the Ibbotson and Sinquefield CRSP file. They compute results for the no-leverage and leveraged cases with and without small capitalized stocks for $\gamma = 1$ (positive power), 3/4, 1/2, 1/4 and 0 (full Kelly) and negative powers $-1, -2, \ldots, -75$. They did quarterly and yearly revisions. The simple historical probability approach gives good results. Small stocks generally add value.

Mulvey, Bilgili and Vural (2010) apply Kelly type models to deal with various practical issues involving imperfections such as transaction costs, operational constraints and path dependencies, etc for use by investors such as defined benefit pension plans. Their model is modified to also provide downside protection in turbulent markets. They apply the theory to recent US stock market behavior including the 2008 crash period. They show how large endowments can modify their typical three step process: select benchmarks, revise portfolio allocation targets and select active managers to deal with these issues. Fixed mix portfolio rates along with exchange traded ETFs provide volatility induced wealth growth. They show that gains are larger in upswings and losses lower in crash periods than the underlying indices. The increasing correlation of equity assets during crash periods is dealt with by developing a portfolio of low correlation investment strategies.

Rudolf and Ziemba (2004) present a continuous time model for pension or insurance company surplus management over time. Such lifetime intertemporal portfolio investment models date to Ramsey (1928), Phelps (1962) and Samuelson (1969) in discrete time and Merton (1969) in continuous time. Rudolf and Ziemba use an extension of the Merton (1973, 1990) model that maximizes the intertemporal expected utility of the surplus of assets net of liabilities using liabilities as a new variable. They assume that both the asset and the liability return follow Itô processes as functions of a risky state variable. The optimum occurs for investors holding four funds: the market portfolio, the hedge portfolio for the state variable, the hedge portfolio for the liabilities, and the riskless asset. This is a four fund CAPM, while Merton has a three fund CAPM, and the ordinary Sharpe-Lintner-Mossin CAPM has two funds. The hedge portfolio provides maximum correlation to the state variable, that is, it provides the best possible hedge against the variance of the state variable. In contrast to Merton's result in the assets only case, the liability hedge is independent of preferences and depends only on the funding ratio. With hyperbolic risk aversion utility, which includes negative exponential, power and log, the investments in the state variable hedge portfolio are also preference independent, and with Kelly type log utility, the market portfolio investment depends only on the current funding ratio.

Having surplus over time is what life and other insurance companies, pension funds and other organizations try to achieve. With both life insurance companies and pension funds, parts of the surplus are distributed to the clients usually once every year. Hence, optimizing their investment strategy is well represented by maximizing the expected lifetime utility of the surplus.

A case study involves a US based investor investing in the stock and bond markets of the US, UK, Japan, the EMU, Canada and Switzerland. The investor faces an exposure to five foreign currencies euro, British pound, Japanese yen, Canadian dollar and Swiss franc, and can invest in eight funds. Five of them are hedge portfolios for the state variables, which are assumed to be currency returns and the others by portfolio separation are the market portfolios, the riskless asset and the liability hedge portfolio.

The holdings of the various funds depend only on the funding ratio and on the currency betas of the distinct markets. For a funding ratio of one, there is no investment in the market portfolio and only diminishing investments in the currency hedge portfolios. The portfolio betas against the five currencies are close to zero for all funding ratios. The higher the funding ratio is, the higher is the investment in the market portfolio, the lower are the investments in the liability and the state variable hedge portfolios, and the lower is the investment in the riskless fund. Negative currency hedge portfolios imply an increase of the currency exposure instead of a hedge against it. The increase of the market portfolio holdings and the reduction of the hedge portfolio holdings and the riskless fund for increasing funding ratios shows that the funding ratio is directly related to the ability to bear risk. Rather than risk aversion coefficients, the funding ratio provides an objective measure to quantify attitudes towards risk.

Ziemba (2005) investigates the records of great investors and proposes a simple modification of the ordinary Sharpe ratio to evaluate them more fairly. In most cases these investors have mostly monthly or quarterly gains with few losses so the ordinary Sharpe ratio penalizes them for gains by increasing the portfolio standard deviation. The funds discussed are Windsor of George Neff, the Ford Foundation, the Harvard Endowment, Tiger of Julian Robertson, Quantum of George Soros and Berkshire Hathaway of Warren Buffett. The modified Sharpe ratio simply replaces the gains by the mirror image of the losses. So gains are not penalized but only losses.

With the ordinary Sharpe ratio the Ford Foundation beats Berkshire Hathaway even though the geometric mean return is much less. So the question is does Berkshire dominate with the modified DSSR downside symmetric Sharpe ratio. Only Berkshire improves with the DSSR but the Ford Foundation and the Harvard endowment are still slightly better. This is because Berkshire has the most large gains of all the funds studied but also has a lot of large losses on a monthly or quarterly basis and more than the other funds. So Berkshire is oriented towards a full Kelly long term valuation where intermediate monthly, quarterly and even yearly returns are not the goal, rather long terms growth is the goal. So we believe Buffett acts like a full Kelly bettor in his investment strategies. These funds have DSSRs about 1.0.

Princeton-Newport, run by Edward O. Thorp, a co-editor of this book, had a DSSR of 13.8 with a very smooth wealth path with only three monthly losses, and no quarterly or yearly losses, over the twenty year period 1969–1988. Thorp had similar good results in other funds run later. Renaissance Medallion, run by James Simon, arguably the current top hedge fund in the world, had, as reported by Ziemba and Ziemba (2007), an even higher 26.4 DSSR with very few monthly or quarterly losses. no yearly losses, and very high mean returns during the period January 1993 to April 2005.

Thorp (2006) presents a comprehensive survey of his considerable use over almost fifty years of Kelly strategies in blackjack, sports betting and the stock market along with theoretical results he has found useful in practice. He traces the early first serious application of Kelly methods to blackjack bet sizing, to various sports betting applications and his great success as a hedge fund trader with the Princeton-Newport and later hedge funds.

46

EFFICIENCY OF THE MARKET FOR RACETRACK BETTING*

DONALD B. HAUSCH,† WILLIAM T. ZIEMBA‡ AND MARK RUBINSTEIN§

Many racetrack bettors have systems. Since the track is a market similar in many ways to the stock market one would expect that the basic strategies would be either fundamental or technical in nature. Fundamental strategies utilize past data available from racing forms, special sources, etc. to "handicap" races. The investor then wagers on one or more horses whose probability of winning exceeds that determined by the odds by an amount sufficient to overcome the track take. Technical systems require less information and only utilize current betting data. They attempt to find inefficiencies in the "market" and bet on such "overlays" when they have positive expected value. Previous studies and our data confirm that for win bets these inefficiencies, which exist for underbet favorites and overbet longshots, are not sufficiently great to result in positive profits. This paper describes a technical system for place and show betting for which it appears to be possible to make substantial positive profits and thus to demonstrate market inefficiency in a weak form sense. Estimated theoretical probabilities of all possible finishes are compared with the actual amounts bet to determine profitable betting situations. Since the amount bet influences the odds and theory suggests that to maximize long run growth a logarithmic utility function is appropriate the resulting model is a nonlinear program. Side calculations generally reduce the number of possible bets in any one race to three or less hence the actual optimization is quite simple. The system was tested on data from Santa Anita and Exhibition Park using exact and approximate solutions (that make the system operational at the track given the limited time available for placing bets) and found to produce substantial positive profits. A model is developed to demonstrate that the profits are not due to chance but rather to proper identification of market inefficiencies.
(FINANCE-PORTFOLIO; GAMES-GAMBLING)

1. The Racetrack Market

For the most part,[1] academic research on security markets has bypassed an interesting and accessible market—the racetrack for thoroughbred horses—with its highly standardized form of security—the tote ticket. The racetrack shares many of the characteristics of the archtypical securities market in listed common stocks. Moreover the racetrack gains further interest from its significant differences, and more importantly, because it is inherently a more elementary market context, lacking many of the dynamic features which complicate analysis of the stock market.

The "market" at the track in North America convenes for about 20 minutes, during which participants place bets on any number of the six to twelve horses in the following race. In a typical race, participants can bet on each horse, either to win, place or show.[2] The horses that finish the race first, second or third are said to finish

*Accepted by Vijay S. Bawa, former Departmental Editor; received April 3, 1980. This paper has been with the authors 2 months for 1 revision.

†Northwestern University.
‡University of British Columbia.
§University of California, Berkeley.

[1] For surveys, see Copeland and Weston [6], Fama [9], [10] and Rubinstein [22].

[2] Other bets such as the daily double (pick the winners in the first and second race), the quinella (pick the first two finishers regardless of order in a given race) and the exacta (pick the first two finishers in exact order in a given race) as well as various combinations are possible as well. Such bets are utilized by the public to construct low probability high payoff bets. For discussion of some of the implications of such bets see Rosett [21].

"in-the-money." All participants who have bet a horse to win, realize a positive return on that bet only if the horse is first, while a place bet realizes a positive return if the horse is first or second, and a show bet realizes a positive return if the horse is first, second or third. Regardless of the outcome, all bets have limited liability. Unlike most casino games such as roulette, but like the stock market, security prices (i.e. the "odds") are jointly determined by all the participants and a rule governing transaction costs (i.e. the track "take"). To take the simplest case, for win, all bets across all horses to win are aggregated to form the win pool. If W_i represents the total amount bet by all participants on horse i to win, then $W = \sum_i W_i$ is the win pool and WQ/W_i is the payoff per dollar bet on horse l to win if and only if horse i wins, where $1 - Q$ is the percentage transaction costs.[3] If horse i does not win, the payoff per dollar bet is zero.

The track is "competitive" in the sense that every dollar bet on horse i to win, regardless of the identification of the bettor, has the same payoff. The "state contingent" price of a dollar received if and only if horse i wins is $\rho_i = (W_i/WQ)$ and the one plus riskless return is $1/\sum_i \rho_i = Q$, a number less than one. The interest rate at the track is thus negative and solely determined by the level of transactions costs. Thus, apart from an exogeneously set riskless rate, the participants at the track jointly determine the security prices. For example, by betting more on one horse than another, the state-contingent price of the first horse increases relative to the second, or alternatively the payoff per dollar bet decreases on the first relative to the second.

The rules for division of the place and show pools are slightly more complex than the win pool. Let P_j be the amount bet on horse j to place and $P \equiv \sum_j P_j$ be the place pool. Similarly S_k is the amount bet on horse k to show and the show pool is $S \equiv \sum_k S_k$. The payoff per dollar bet on horse j to place is

$$1 + [PQ - P_i - P_j]/(2P_j) \quad \text{if} \quad \begin{cases} i \text{ is first and } j \text{ is second or} \\ j \text{ is first and } i \text{ is second} \end{cases}$$
$$0 \qquad \text{otherwise.}$$

Thus if horses i and j are first and second each bettor on j (and also i) to place first receives the amount of his bet back. The remaining amount in the place pool, after the track take, is then split evenly between the place bettors on i and j. The payoff to horse j to place is independent of whether j finishes first or second, but it is dependent on which horse finishes with it. A bettor on horse j to place hopes that a longshot, not a favorite, will finish with it.

The payoff per dollar bet on show is analogous

$$1 + [SQ - S_i - S_j - S_k]/(3S_k) \quad \text{if} \quad \begin{cases} k \text{ is first, second or third} \\ \text{and finishes with } i \text{ and } j \end{cases}$$
$$0 \qquad \text{otherwise.}$$

In many ways the racetrack is like the stock market. A technical strategy based on discrepancies between the amounts bet on the same horses to win, place and show, is examined in this paper. Since a short position has a perfectly negative correlated outcome to the result of normal bet a given horse can be "shorted" by buying tickets on all the other horses in a race.

The racetrack also differs from the stock market in important ways. In the stock market, an investor's profit depends not only on the initial price he pays for a security,

[3]The actual transactions cost is more complicated and is described below.

but also on what some other investor is willing to pay him for it when he decides to sell. Thus his profit depends not only on how well the underlying firm does in terms of earnings over the time he holds its stock (i.e. supply uncertainty), but also on how other investors value that stock in the future (i.e. demand uncertainty). Given the initial price, both the nature and behavior of other market participants determine his profit. Thus current stock prices might depend not only on "fundamental" factors but also on market "psychology"—the tastes, beliefs, and endowments of other investors, etc. In contrast once all bets are placed at the track prior to a given race (i.e. initial security prices are given), the result of the race and the corresponding payoffs depend only on nature. There is no demand uncertainty at the track.

2. Previous Work on Racetrack Efficiency

A market is efficient if current security prices fully reflect all available relevant information. If this is the case, experts should not be able to achieve higher than average returns with regularity. A number of investigators have demonstrated that the New York Stock Exchange and other major security markets are efficient and so-called experts in fact achieve returns when adjusted for risk that are no higher than those that would be received from random investments (see Copeland and Weston [6], Fama [9], [10] and Rubinstein [22] for discussion, terminology, and relevant references). For an exception see Downes and Dyckman [7].

Snyder [24] provided an investigation of the efficiency of the market for racetrack bets to win. The question Snyder poses is whether or not bets at different odds levels yield the same average return. The rate of return for odds group i is

$$R_i = \frac{N_i^*(O_i + 1) - N_i}{N_i}$$

where N_i and N_i^* are the number of horses, and the number who won, respectively, at odds $O_i = (WQ/W_i - 1)$. A weakly-efficient market in Snyder's sense would set $R_i = Q$ for all i, where $1 - Q$ is the percentage track take. His results as well as those of Fabricant [8], Griffith [12], [13], McClothlin [18], Seligman [23] and Weitzman [27] suggest that there are "strong and stable biases but these are not large enough to make it possible to earn a positive profit" [24; 1110]. In particular, extreme favorites tend to be under bet and longshots overbet. The combined results of several studies comparing over 30,000 races are summarized in Table 1.

TABLE 1 (Snyder [24])

Rates of Return on Bets to Win by Grouped Odds, Take Added Back

	Midpoint of grouped odds							
Study	0.75	1.25	2.5	5.0	7.5	10.0	15.0	33.0
Fabricant	11.1[a]	9.0[a]	4.6[a]	– 1.4	– 3.3	– 3.7	– 8.1	– 39.5[a]
Griffith	8.0	4.9	3.1	– 3.1	– 34.6[a]	– 34.1[a]	– 10.5	– 65.5[a]
McGlothlin	8.0[b]	8.0[a]	8.0[a]	– 0.8	– 4.6	– 7.0[b]	– 9.7	– 11.0
Seligman	14.0	4.0	– 1.0	1.0	– 2.0	– 4.0	– 7.8	– 24.2
Snyder	5.5	5.5	4.0	– 1.2	3.4	2.9	2.4	– 15.8
Weitzman	9.0[a]	3.2	6.8[a]	– 1.3	– 4.2	– 5.1	– 8.2[b]	– 18.0[a]
Combined	9.1[a]	6.4[a]	6.1[a]	– 1.2	– 5.2[a]	– 5.2[a]	– 10.2[a]	– 23.7[a]

[a] Significantly different from zero at 1% level or better.
[b] Significantly different from zero at 5% level or better.

Since the track take averages about 18%, the net rate of return for any strategy which consistently bets within a single odds category is −9% or less. For horses with odds averaging 33 the net rate of return is about −42%.

Conventional financial theory does not explain these biases because it is usual to assume that as nondiversifiable risk (e.g. variance) rises expected return rises as well. In the win pool expected return declines as risk increases. An explanation consistent with the expected utility hypothesis is that investors (as a composite) are risk lovers and behave as if the betting opportunities are limited to a single race. Weitzman [27] and Ali [1] have estimated such utility functions. Ali's estimated utility function over wealth w is the convex function

$$u(x) = 1.91 w^{1.1784} \qquad (R^2 = .9981),$$

which has increasing absolute risk aversion. Thus by the Arrow-Pratt [3], [19] theory investors will take more risk as their wealth declines. This explains the common phenomenon that bettors, when losing, tend to bet more and more on longer odds horses in a desperate attempt to recoup earlier losses. Moreover, since u is nearly linear for large w investors are nearly risk neutral at such wealth levels.

A second explanation is that gamblers simply prefer low probability high prize combinations (i.e. longshots) to high probability low prize combinations. Besides the possible gains involved, gamblers have egos associated with analyzing racing forms and pitting one's predictions against others. Luck and entertainment as well are largely absent in betting favorites. The thrill is to successfully detect a moderate or long odds winner and thus confirm one's ability to outperform the other bettors. Such a scenario is consistent with the data and leads to the biases. Rosett [21] provides an analysis of ways to construct low probability high prize bets through parlays and other combinations that can be used to avoid the longshot tail bias problem and to take advantage of the favorite bias. In an effort to capitalize on this market many tracks feature such bets in the form of the daily double, the exacta and the quinella. However, none of these schemes appear to yield bets with positive net returns.[4]

Other studies of racetrack efficiency have been conducted by Ali [2], Figlewski [11] and Snyder [24]. Ali shows that the win market is efficient in the sense that independently derived bets with identical probabilities of winning do in fact have the same odds in a statistical sense. His analysis utilizes daily double bets and the corresponding parlays, i.e. bet the proceeds if the chosen horse wins the first race on the chosen horse in the second race, for 1089 races. Figlewski, using a multinomial logit probability model to measure the information content of the forecasts of professional handicappers and data from 189 races at Belmont in 1977, found that these forecasts do contain considerable information but the track odds generated by bettors discount almost all of it. Snyder provided strong form efficiency tests of the form: are there persons with special information that would allow them to outperform the general public? He found using data on 846 races at Arlington Park in Chicago that forecasts from three leading newspapers, the daily racing form and the official track handicapper did not lead to bets that outperformed the general public.

[4] For an entertaining account of an "expert" who was able to achieve positive net returns over a full racing season, see Beyer [5]. See also Vergin [26].

3. Proposed Test

The studies described in §2 examine racetrack efficiency with respect to the win pool only. In this paper our concern is with the efficiency of the place and show pools relative to the win pool and with the development of procedures to best capitalize on potential inefficiencies between these three "markets." We utilize the following definition of weak-form efficiency: the market is weakly-efficient if no individual can earn positive profits using trading rules based on historical price information. In Baumol's [4; 46] words ... "all opportunities for profit by systematic betting are eliminated". Our analysis utilizes the following two data sets: 1) Data Set 1: all dollar bets to win, place and show for the 627 races over 75 days involving 5895 horses running in the 1973/74 winter season at Santa Anita Racetrack in Arcadia, California, collected by Mark Rubinstein; and 2) Data Set 2: all dollar bets to win, place and show for the 1065 races over 110 days involving 9037 horses running in the 1978 summer season at Exhibition Park, Vancouver, British Columbia, collected by Donald B. Hausch and William T. Ziemba.

In the analysis of the efficiency of the win pool one may compare the actual frequency of winning with the theoretical probability of winning as reflected through the odds. Similar analyses are possible for the place and show pools once an estimate of the theoretical probabilities of placing and showing for all horses is available. There is no unique way to obtain these estimates. However, very reasonable estimates obtain from the natural generalization of the following simple procedure. Suppose three horses have probabilities to win of 0.5, 0.3, and 0.2, respectively. Now if horse 2 wins, a Bayesian would naturally expect that the probabilities that horses 1 and 3 place (i.e. win second place) are 0.5/0.7 and 0.2/0.7, respectively. In general if q_i $(i = 1, \ldots, n)$ is the probability horse i wins, then the probability that i is first and j is second is

$$(q_i q_j)/(1 - q_i) \tag{1}$$

and the probability that i is first, j is second and k is third is

$$q_i q_j q_k/(1 - q_i)(1 - q_i - q_j). \tag{2}$$

Harville [14] gives an analysis of these formulas. Despite their apparent reasonableness they suffer from at least two flaws:

1) no account is made of the possibility of the "Silky Sullivan" problem; that is, some horses generally either win or finish out-of-the-money—for example, see footnote 5 in [15]; for these horses the formulas greatly over-estimate the true probability of finishing second or third; and

2) the formulas are not derivable from first principles involving individual horses running times; even independence of these random variables $T_1, \ldots, T_n$ is neither necessary nor sufficient to imply the formulas.

In addition to assuming (1) and (2) we assume that

$$q_i = W_i \Big/ \sum_{i=1}^{n} W_i. \tag{3}$$

That is, the win pool is efficient. The discussion above, of course, indicates that this assumption is suspect in the tails and this is discussed below. Table 2a–c compares the actual versus theoretical probability of winning, placing and showing for data set 2.

Similar tables for data set 1 appear in King [17]; see also Harville [14]. The usual tail biases appear in Table 2a (although they are not significant at the 5% confidence level). One has reverse tail biases in the finishing second and third probabilities, see Tables 2b, c in [15]. This occurs because if the probability to win is overestimated (underestimated) and probabilities sum to one it is likely that the probabilities of finishing second and third would be underestimated (overestimated). Tables 2b, c, indicate how these biases tend to cancel when they are aggregated to form the theoretical probabilities and frequencies of placing and showing.[5]

TABLE 2

Actual vs. Theoretical Probability of Winning, Placing and Showing: Exhibition Park 1978

(a)	Theoretical Probability of Winning	Number of Horses	Average Theoretical Probability	Actual Frequency of Winning	Estimated Standard Error
	0.000 –0.025	540	0.019	0.016	0.005
	0.026 –0.050	1498	0.037	0.036	0.005
	0.051 –0.100	2658	0.073	0.079	0.005
	0.101 –0.150	1772	0.123	0.126	0.008
	0.151 –0.200	1199	0.172	0.156	0.010
	0.201 –0.250	646	0.223	0.227	0.016
	0.251 –0.300	341	0.272	0.263	0.024
	0.301 –0.350	199	0.323	0.306	0.033
	0.351 –0.400	101	0.373	0.415	0.049
	0.401 +	83	0.450	0.469	0.055
		9037			

(b)	Theoretical Probability of Placing	Number of Horses	Average Theoretical Probability	Actual Frequency of Placing	Estimated Standard Error
	0.000 –0.025	21	0.022	0.000*	0.000
	0.026 –0.050	391	0.040	0.030	0.009
	0.051 –0.100	1394	0.075	0.080	0.007
	0.101 –0.150	1335	0.124	0.152*	0.010
	0.151 –0.200	1295	0.174	0.174	0.011
	0.201 –0.250	1057	0.223	0.243	0.013
	0.251 –0.300	871	0.274	0.304	0.016
	0.301 –0.350	772	0.323	0.314	0.017
	0.351 –0.400	580	0.373	0.313*	0.019
	0.401 –0.450	420	0.424	0.395	0.024
	0.451 –0.500	321	0.472	0.457	0.028
	0.501 –0.550	202	0.523	0.415*	0.035
	0.551 –0.600	149	0.573	0.483*	0.041
	0.601 –0.650	114	0.623	0.570	0.046
	0.651 –0.700	51	0.672	0.627	0.068
	0.701 –0.750	41	0.721	0.731	0.069
	0.751 +	23	0.792	0.782	0.086
		9037			

[5] Since it is the accuracy of these probabilities rather than the q_i's that is of crucial importance in the calculations and model below this canceling provides some justification for omitting the tail biases in (3). In practice, bets are only made on horses with expected returns considerably above 1, e.g., 1.16 at Santa Anita. Modification of the q_i to include these biases might change the 1.16 to 1.14, for example. See also the discussion below and in §§4 and 5.

TABLE 2 (continued)

Actual vs. Theoretical Probability of Winning, Placing and Showing: Exhibition Park 1978

(c) Theoretical Probability of Showing	Number of Horses	Average Theoretical Probability	Actual Frequency of Showing	Estimated Standard Error
0.000 -0.025	0	—	—	—
0.026 -0.050	55	0.043	0.036	0.025
0.051 -0.100	592	0.078	0.081	0.011
0.101 -0.150	895	0.125	0.165*	0.012
0.151 -0.200	909	0.175	0.195	0.013
0.201 -0.250	799	0.224	0.292*	0.016
0.251 -0.300	885	0.275	0.289	0.015
0.301 -0.350	794	0.324	0.346	0.017
0.351 -0.400	703	0.374	0.398	0.018
0.401 -0.450	655	0.425	0.433	0.019
0.451 -0.500	617	0.475	0.452	0.020
0.501 -0.550	542	0.524	0.477*	0.021
0.551 -0.600	396	0.573	0.477*	0.025
0.601 -0.650	375	0.624	0.599	0.025
0.651 -0.700	264	0.672	0.609*	0.030
0.701 -0.750	206	0.722	0.582*	0.034
0.751 -0.800	154	0.773	0.655*	0.038
0.801 -0.850	113	0.821	0.752	0.041
0.851 -0.900	53	0.873	0.830	0.052
0.901 +	30	0.925	0.833	0.068
	9037			

[a]Categories when the theoretical probability and the actual frequency are different at the 5% significance level are denoted by *'s. The estimated standard error is $(s^2/N)^{\frac{1}{2}}$ where the actual frequency sample variance $s^2 = N(E(X^2) - (EX)^2)/(N-1)$. Since the X_i are either 0 or 1, $E(X^2) = EX$ and $s^2 = N(EX - (EX)^2)/(N-1)$.

Formulas (1)–(3) can be used to develop procedures that yield net rates of return for place and show betting that are higher than expected (i.e.–18%) and indeed make positive profits. As a first step towards development of a "system" we present the results on the two data sets of \$1 bets when the theoretical expected return is α for varying α. The expected return from a \$1 bet to place on horse l is[6,7]

$$EX_l^P \equiv \sum_{\substack{j=1\\ j\neq l}}^{n}\left(\frac{q_l q_j}{1-q_l}\right)\left[1+\frac{1}{20}\,\mathrm{INT}\left\{\left(\frac{Q(P+1)-(1+P_l+P_j)}{2}\right)\left(\frac{1}{1+P_l}\right)\times 20\right\}\right]$$

$$+\sum_{\substack{i=1\\ i\neq l}}^{n}\left(\frac{q_i q_l}{1-q_i}\right)\left[1+\frac{1}{20}\,\mathrm{INT}\left\{\left(\frac{Q(P+1)-(1+P_i+P_l)}{2}\right)\left(\frac{1}{1+P_l}\right)\times 20\right\}\right], \tag{4}$$

[6]The expressions (4) and (5) give the marginal expected return for an additional \$1 bet to place or show on horse l. To obtain the average expected rates of return one simply replaces $(1 + p_l)$ and $(1 + s_l)$ in these expressions by p_l and s_l, respectively. From a practical point of view with usual track data these quantities are virtually identical.

[7]A further complication, not reflected in (4) and (5) below, is that a \$2 winning bet must return at least \$2.10. Hence in these "minus pools" involving an extreme favorite the track's take is less than $1 - Q$.

where Q is 1 minus the track take, $P \equiv \sum P_i$ is the place pool, P_i is bet on horse i to place and INT(Y) means the largest integer not exceeding Y. In (4) the quantities P and P_i are the amounts bet before an additional \$1 is bet; similarly with S and S_i in (5), below. The expressions involving INT take into account the fact that \$2 bets return payoffs rounded down to the nearest \$0.10. The two expressions represent the expected payoffs if l is first or second, respectively. Similarly the expected payoff from a \$1 bet to show on horse l is

$$
\begin{aligned}
EX_l^s \equiv & \sum_{\substack{j=1 \\ j \neq l}}^{n} \sum_{\substack{k=1 \\ k \neq l,j}}^{n} \frac{q_l q_j q_k}{(1-q_l)(1-q_l-q_j)} \\
& \times \left[1 + \frac{1}{20} \mathrm{INT} \left\{ \left(\frac{Q(S+1) - (1 + S_l + S_j + S_k)}{3} \right) \left(\frac{1}{1+S_l} \right) \times 20 \right\} \right] \\
& + \sum_{\substack{i=1 \\ i \neq l}}^{n} \sum_{\substack{k=1 \\ k \neq i,l}}^{n} \frac{q_i q_l q_k}{(1-q_i)(1-q_i-q_l)} \\
& \times \left[1 + \frac{1}{20} \mathrm{INT} \left\{ \left(\frac{Q(S+1) - (1 + S_i + S_l + S_k)}{3} \right) \left(\frac{1}{1+S_l} \right) \times 20 \right\} \right] \\
& + \sum_{\substack{i=1 \\ i \neq l}}^{n} \sum_{\substack{j=1 \\ j \neq l,i}}^{n} \frac{q_i q_j q_l}{(1-q_i)(1-q_i-q_j)} \\
& \times \left[1 + \frac{1}{20} \mathrm{INT} \left\{ \left(\frac{Q(S+1) - (1 + S_i + S_j + S_l)}{3} \right) \left(\frac{1}{1+S_l} \right) \times 20 \right\} \right], \qquad (5)
\end{aligned}
$$

where $S \equiv \sum S_i$ is the show pool, S_i is bet on horse i to show and the three expressions represent the expected payoffs if l is first, second and third, respectively.

Naturally one would expect that positive profits would not be obtained, given the inherent inaccuracies in assumptions (1)–(3), unless the theoretical expected return α was significantly greater than 1. However we might hope that the actual rate of return would at least increase with α and be somewhat near α. Table 3 indicates this is true for both data sets. The perverse behavior for high α in the place pool is presumably a small sample phenomenon. Additional calculations along these lines appear in Harville [14].

4. A Betting Model

The results in Table 3 give a strong indication that there are significant inefficiencies in the place and show pools and that it is possible not only to achieve above average returns but to make substantial profits. In this section we develop a model indicating not only which horses should be bet but how much should be bet taking into account investor preferences and wealth levels and the effect of bet size on the odds.

We consider an investor having initial wealth w_0 contemplating a series of bets. It is

TABLE 3

Results of Betting \$1 to Place or Show on Horses with a Theoretical Expected Return of at Least α

Exhibition Park

	Place			Show		
α	Number of Bets	Total Net Profit (\$)	Net Rate of Return (%)	Number of Bets	Total Net Profit (\$)	Net Rate of Return (%)
1.04	225	5.10	2.3	612	33.20	5.4
1.08	126	−10.10	−8.0	386	53.50	13.9
1.12	69	11.10	16.1	223	40.80	18.3
1.16	40	5.10	12.8	143	26.30	18.4
1.20	18	5.30	29.4	95	21.70	22.8
1.25	11	−2.70	−24.5	44	11.20	25.5
1.30	3	−3	−100.0	27	10.80	40.0
1.50	0	0	—	3	6	200.0

Santa Anita

	Place			Show		
α	Number of Bets	Total Net Profit (\$)	Net Rate of Return (%)	Number of Bets	Total Net Profit (\$)	Net Rate of Return (%)
1.04	103	12.30	11.9	307	−18.00	−5.9
1.08	52	12.80	24.6	162	6.90	4.3
1.12	22	9.20	41.8	89	3.00	3.4
1.16	7	2.30	32.9	46	12.40	27.0
1.20	3	−1.30	−43.3	27	6.20	23.0
1.25	0	0	—	9	6.00	66.7
1.30	0	0	—	5	5.10	102.0
1.50	0	0	—	0	0	—

natural to suppose that the investor would wish to maximize the long run rate of asset growth and thus employ the so-called Bernoulli capital growth model; see Ziemba and Vickson [28] for references and discussion of various assumptions and results. We use the following result: if in each time period $t = 1, 2, \ldots$ there are I investment opportunities with returns per unit invested denoted by the random variables $x_{t1}, \ldots, x_{tI}$, where the x_{ti} have finitely many distinct values and for distinct t the families are independent, then maximizing $E \log \sum \lambda_{ti} x_{ti}$, s.t. $\sum \lambda_{ti} \leq w_t$, $\lambda_{ti} \geq 0$ maximizes the asymptotic rate of asset growth. The assumptions are quite reasonable in a horseracing context because there are a finite number of return possibilities and the race by race returns are likely to be nearly independent since different horses will be running (although the jockeys and trainers may not be).

The second key feature of the model is that it considers an investor's ability to influence the odds by the size of his bets.[8] This yields the following model to calculate

[8]The first model to include this feature seems to be Isaacs [16]. Only win bets are considered with linear utility, and he is able to determine the exact solution in closed form. His model may be useful in situations where the perfect market assumption (3) is violated or where special expertise leads one to believe their estimates of the q_i are better than those of the other bettors.

the optimal amounts to bet for place and show.

$$\underset{\{p_l\}\{s_l\}}{\text{Maximize}} \sum_{i=1}^{n} \sum_{\substack{j=1 \\ j\neq i}}^{n} \sum_{\substack{k=1 \\ k\neq i,j}}^{n} \frac{q_i q_j q_k}{(1-q_i)(1-q_i-q_j)} \log \left[\begin{array}{l} \dfrac{Q(P+\sum_{l=1}^{n} p_l) - (p_i+p_j+P_i+P_j)}{2} \\ \quad \times \left(\dfrac{p_i}{p_i+P_i} + \dfrac{p_j}{p_j+P_j} \right) \\ + \dfrac{Q(S+\sum_{l=1}^{n} s_l) - (s_i+s_j+s_k+S_i+S_j+S_k)}{3} \\ \quad \times \left(\dfrac{s_i}{s_i+S_i} + \dfrac{s_j}{s_j+S_j} + \dfrac{s_k}{s_k+S_k} \right) \\ \quad + w_0 - \sum_{\substack{l=1 \\ l\neq i,j,k}}^{n} s_l - \sum_{\substack{l=1 \\ l\neq i,j}}^{n} p_l \end{array} \right] \tag{6}$$

$$\text{s.t.} \sum_{l=1}^{n} (p_l + s_l) \leqslant w_0, \qquad p_l \geqslant 0, s_l \geqslant 0, l = 1, \ldots, n,$$

where $Q = 1 -$ the track take, W_i, P_j and S_k are the total dollar amounts bet to win, place and show on the indicated horses by the crowd, respectively, $W \equiv \sum W_i$, $P \equiv \sum P_j$ and $S \equiv \sum S_k$ are the win, place and show pools, respectively, $q_i \equiv W_i / W$ is the theoretical probability that horse i wins, w_0 is initial wealth and p_l and s_l are the investor's bets to place and show on horse l, respectively.

The formulation (6) maximizes the expected logarithm of final wealth considering the probabilities and payoffs from all possible horserace finishes. It is exact except for the minor adjustment made that the rounding down to the nearest \$0.10 for a two dollar bet, see (4) and (5), is omitted.[9] For the values of $\alpha \geqslant 1.16$ it was observed that in a given race at most three p_l and three s_l were nonzero. When (6) is then simplified it can be solved in less than 1 second of CPU time.[10] A discussion of the generalized concavity properties of (6) will appear in a forthcoming paper by Kallberg and Ziemba.

The results are illustrated by function 1 in Figures 1 and 2 for the two data sets using an initial wealth of \$10,000. In both cases the bets produced from (6) lead to well above average returns and to positive profits.[11] These results may be contrasted with random betting; function 2 in Figures 1 and 2. Intuition suggests that Santa Anita with its larger betting pools would have more accurate estimates of the q_i than would be obtained at Exhibition Park. Hence positive profits would result from lower values of α. The results bear this out and only horses with $\alpha \geqslant 1.20$ for Exhibition Park were considered for possible bets. Generally speaking the bets are usually favorites and almost always on those horses with maximum $(W_i/W)/(P_i/P)$ and $(W_i/W)/(S_i/S)$

[9] It is possible to include this feature in (6) but it greatly complicates the solution procedure (e.g. differentiability is lost) with little added gain in accuracy.

[10] All calculations were made on UBC's AMDAHL 470V6 Model II computer using a code for the generalized reduced gradient algorithm.

[11] The procedure was to calculate the optimal bets to place and show in each race using (6) with the present wealth level. The results of the race and the actual payoffs that reflect the track's take and breakage are known. The payoff for our investor's bets were calculated using all the bets of the crowd plus the bets of our investor taking into account the track's take and breakage, i.e. the payoffs are thus those that would have occurred had our investor actually made his bets. The wealth level was then adjusted to reflect the race's gain or loss and the procedure continued for all races.

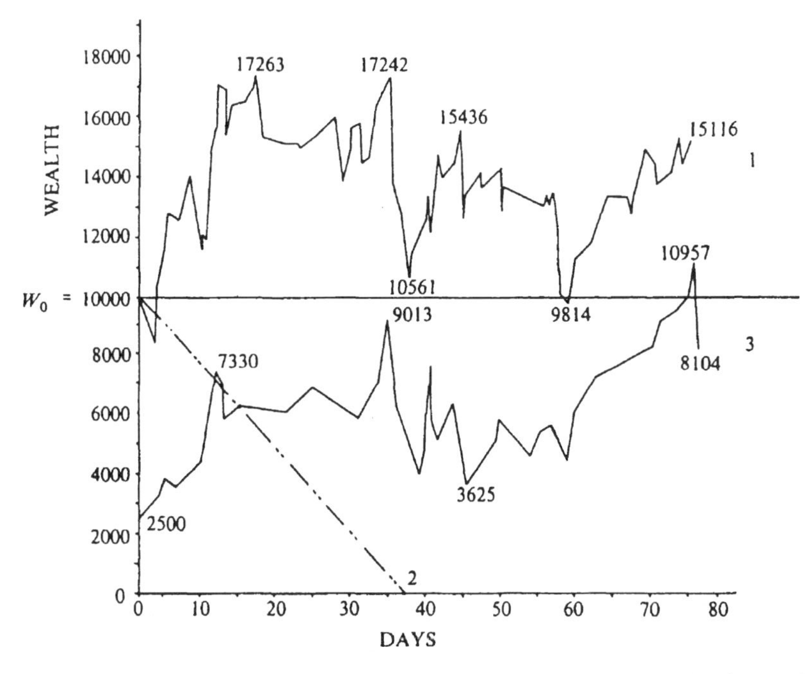

[1] Results from expected log betting to place and show when expected returns are 1.16 or better with initial wealth $10,000.

[2] Approximate wealth level history for random horse betting. Total dollars bet is as in system 1 ($116,074). Track payback is 82.5%, therefore final wealth level is $10,000 – 0.175($116.074) = –$10,313 (Note: breakage is not taken into consideration)

[3] Results from using the Exhibition Park approximate regression scheme (with initial wealth $2,500) at Santa Anita.

FIGURE 1. Wealth Level Histories for Alternative Betting Schemes; Santa Anita: 1973/74 Season.

ratios. These ratios of the theoretical probability of winning to the track take unadjusted odds to place or show form a type of cost-benefit ratio that provides a first approximation to α. This is discussed further in §5, below. Most of the bets are to show and one tends to bet only about once per day. The numbers of bets and their size distribution are presented in Table 4. As expected, the influence on the odds made by our investor's bets is much greater at Exhibition Park hence the bets there tend to be much smaller than at Santa Anita. However, even there about 10% of the bets exceed $1000.

The log formulation has absolute risk aversion $1/w$, which for wealth around $10,000 is virtually zero. Zero absolute risk aversion is, of course, achieved by linear utility. One then will bet on the horse (or horses) with the highest α until the influence on the odds drops this horse (or horses) below another horse's α, etc., continuing until there are no favorable bets or the betting wealth has been fully utilized. The results of such linear utility betting with $w_0 = \$10,000$ are: at Exhibition Park with $\alpha \geqslant 1.20$ final wealth is $14,818; at Santa Anita with $\alpha \geqslant 1.16$ final wealth is $10,910. Such a strategy is a very risky one and leads to some very large bets. The log function has the distinct advantage that it implies negative infinite utility at zero wealth hence bets having any significant probability of yielding final wealth near zero are avoided.

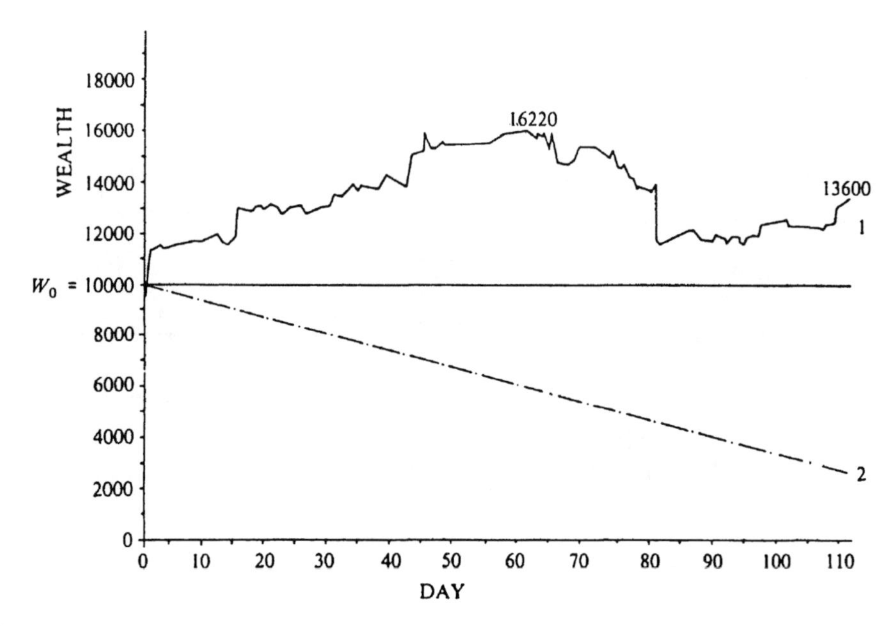

[1]Results from expected log betting to place and show when expected returns are 1.20 or better with initial wealth $10,000.

[2]Approximate wealth level history for random horse betting. Total dollars bet is as in system 2 ($8461). Track payback is 81.9% therefore final wealth level is $10,000 – 0.181($38461) = $3, 035. (Note: Breakage is not taken into consideration.)

FIGURE 2. Wealth Level Histories for Alternative Betting Schemes; Exhibition Park: 1978 Season.

TABLE 4

Size Distribution of Bets, w_0 = $10,000, Log Utility

	Santa Anita				Exhibition Park			
	Place		Show		Place		Show	
Size	% of Bets	% of $Bet	% of Bets	% of $Bet	% of Bets	% of $Bet	% of Bets	% of $Bet
0–50	7.1	0.3	2.6	0.1	29.4	3.6	17.0	1.0
51–100	0	0	1.3	0.1	23.5	6.7	13.8	2.8
101–200	0	0	3.9	0.4	5.9	2.9	22.3	9.3
201–300	21.4	6.1	5.2	1.0	5.9	4.4	14.9	10.3
301–500	7.1	3.2	9.1	2.6	23.5	30.6	8.5	9.1
501–700	14.3	10.3	13.0	5.7	0	0	6.4	10.5
701–1000	14.3	14.4	7.8	4.8	0	0	7.4	17.5
> 1001	35.8	65.7	57.1	85.3	11.8	51.8	9.7	39.5
	n = 14	$11,932	n = 77	$104, 142	n = 17	$4, 954	n = 94[a]	$33, 507

Total Place Bets; Total Show Bets; Total Santa Anita Bettings = $116, 074

Total Place Bets; Total Show Bets; Total Exhibition Park Betting = $38, 461

[a]Two of these bets had $EX_s^l > 1.20$ and $s_l^* = 0$.

5. Making the System Operational

The calculations reported in Figures 1 and 2 were made under the assumption that the investor is free to bet once all other bettors have placed their bets. In practice one can only attempt to be one of the last bettors.[12] There is a natural tradeoff between placing a bet too early and increasing the inaccuracies and running the possibility of arriving too late at the betting window to place a bet.[13] The time just prior to the beginning of a race is crucial since many bets are typically made then including the so called "smart money" bets made very close to the beginning of the race so their impact on other bettors is minimized. It is thus extremely important that the investor be able to perform all calculations necessary to place the bet(s) very quickly. Typically, since tracks have neither public phones nor electricity, calculations can at most utilize battery operated calculators or possibly a battery operated special purpose computer. Even if computing times were negligible the very act of punching in the data needed for an exact calculation is too time-consuming since it takes more than one minute. Therefore, in practice, approximations that utilize a limited number of input data elements are required. Several types of approximations are possible such as the tabular rules of thumb developed by King [17] or the regression procedures suggested here. Our procedure indicates whether or not a bet to place or show is warranted and at what level using the following eight data inputs: w_0, W_{i*}, P_{i*}, Q_{j*}, S_{j*}, W, P and S, where i^* is an i for which $(W_i/W)/(P_i/P)$ is maximized and j^* is a j for which $(W_j/W)/(S_j/S)$ is maximized. It is easy to determine i^* and j^*, particularly since i^* often equals j^*, by inspection of the tote board. The approximation supposes that the only possible bets are i^* to place and j^* to show. The regressions, as given below, must be calibrated to a given track and initial wealth level. As a prelude to actual betting the regressions were calibrated at Exhibition Park for $w_0 = \$2500$. For calculated i^* the expected return on a \$1 bet to place is approximated by

$$E\tilde{X}^P_{i*} = 0.39445 + 0.51338 \frac{W_{i*}/W}{P_{i*}/P}, \qquad R^2 = 0.776. \tag{7}$$

If $E\tilde{X}^P_{i*} \geqslant 1.20$ then the optimal bet to place is approximated by

$$p_{i*} = -459.32 + 1715.6 q_{i*} - 0.042518 q_{i*} P - 7440.1 q_{i*}^2$$
$$+ 13791 q_{i*}^3 + 0.10247 P_{i*} + 49.572 \ln w_0, R^2 = 0.954. \tag{8}$$

Similarly for j^* the expected return on a \$1 bet to show is approximated by

$$E\tilde{X}^s_{j*} = 0.64514 + 0.32806 \frac{W_{j*}/W}{S_{j*}/S}, \qquad R^2 = 0.650. \tag{9}$$

[12] The model as developed in this paper utilizes an inefficiency in the place and show pools to yield positive profits. An investor's bets are determined by his wealth level as well as the profitability of one or more such bets. There may or may not exist "enough" inefficiency to provide positive profits for additional investors using a system of this nature. In the context of the Isaac's model, for win bets with linear utility, Thrall [25] showed that if there were positive profits to be made and each new investor was aware of all previous investor's bets then the profits for these various bettors are shared and in the limit become zero. It is likely that a similar result obtains for the model discussed here.

[13] At some tracks, such as Santa Anita, betting ends precisely at post time when the electric totalizator machines are shut off. At other tracks including Exhibition Park, betting ends when the horses enter the starting gate which may be 2 or even 3 minutes past the post time.

If $E\tilde{X}_j^s \geqslant 1.20$ then the optimal bet to show is approximated by

$$s_{j*} = -660.97 - 867.69q_{j*} + 0.25933q_{j*}S + 3715.2q_{j*}^2 - 0.19572S_{j*} + 77.014 \ln w_0, \qquad R^2 = 0.970. \tag{10}$$

All the coefficients in (7)–(10) are highly significant at levels not exceeding 0.05. Utilizing the equations (7)–(10) results in a scheme in which the data input and execution time on a modern hand held programmable calculator is about 35 seconds.

The results of utilizing this method on the Exhibition Park data with initial wealth of \$2500 are shown in Figure 3, functions 1 and 2. Using the exact calculations yields a final wealth of \$5197 while the approximation scheme has an even higher final wealth of \$7698. The approximation scheme leads to 63 more bets (174 versus 111) than the exact calculation. Since these bets had a positive net return the total profit of the approximation scheme exceeds that of the exact calculation. Thus, it is clear that one would maximize profits with a cutoff rule below the conservative level of 1.20. The size distribution of bets from the exact and approximate solutions are remarkably similar; see Table 6 in [15].

For place there were 17 bets where EX^p exceeded or equalled 1.20 of which 7 (41%) were in the money. Only 2 of these 17 bets were not chosen by the regression. However 14 horses were chosen for betting by the regressions even though their "true" EX^p was less than 1.20. Most of these had "true" EX^p values close to 1.20 and were favorites. Four of these horses (29%) were in the money. For show there were 94 bets where EX^s exceeded or equalled 1.20 of which 52 (55%) were in the money. Only 8 of these were

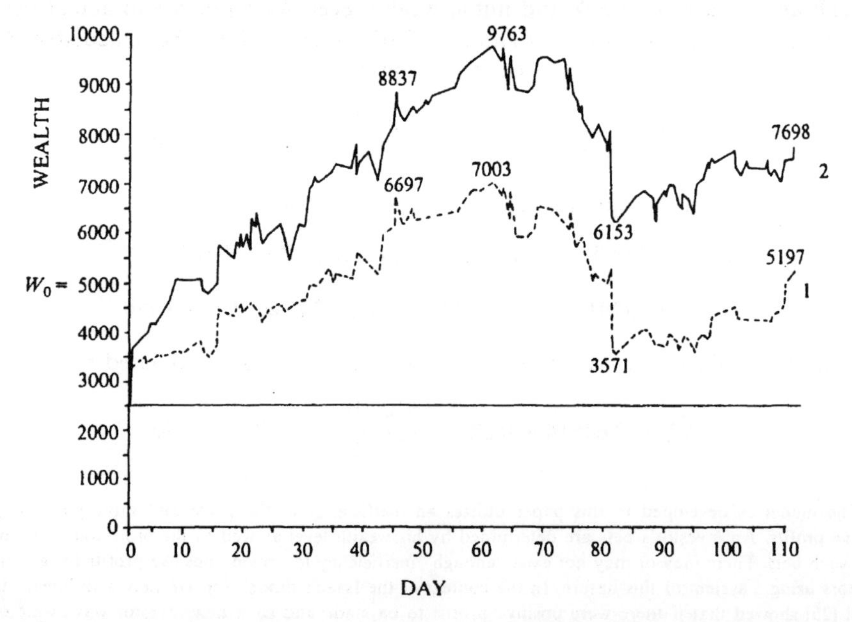

[1]Results from expected log betting to place and show when expected returns are 1.20 or better with initial wealth \$2,500.

[2]Results from using the approximate regression scheme with initial wealth \$2,500.

FIGURE 3. Wealth Level Histories for Exact and Approximate Regression Betting Schemes; Exhibition Park: 1978 Season.

overlooked by the regressions while 59 bets with "true" EX^s values less than 1.20 were chosen by the regressions. Of these 39 (66%) were in the money.

The regression method is a simple procedure that seems to work well. For example using it on the Santa Anita data (without reestimating the coefficients) indicates that initial wealth of \$2500 would yield a final wealth of \$8104; see function 3 in Figure 1. Conceivably, there are many possible refinements using fundamental information that could be added. Many of these refinements as well as discussion of the results of actual betting will appear in a forthcoming book by Ziemba and Hausch. One such refinement is the supposition that the win market is more efficient under normal fast track conditions and less efficient when the track is slow, muddy, heavy, wet, sloppy, etc. Bets were made on 87 of the 110 days at Exhibition Park; of these 57 days had a fast track. Using the regressions yields "fast track" bets of \$32,501 returning \$38,364 for a net profit of \$5863. On the nonfast days the regressions suggest bets of \$17,180 returning \$16,515 for a loss of \$665. Hence this refinement decreases the time spent at the track and increases the net profit from \$5198 to \$5863.

6. Implementation and Reliability of the System

The model presented in this paper assumed that one can utilize the betting data that prevails at the end of the betting period, say t, to calculate optimal bets. In practice, however, even with the approximations given by (7)–(10), one requires about 1–1.5 minutes to physically calculate the optimal bets and place them. Thus one can only utilize betting information from τ ($\approx t - 1.5$). Hence bets that were optimal at τ may not be as profitable at t. Some evidence by Ritter [20] seems to indicate that the odds on the expost favorite at t often decrease from τ to t. In which case an optimal bet at τ may be a poor bet at t. Ritter investigated the systems: bet on i^* if $(P_{i*}/P/(W_{i*}/W) \leqslant 0.7$ and on j^* if $(S_{j*}/S)/(W_{j*}/W) \leqslant 0.7$. (For comparison equations (7) and (9) indicate the more restrictive constraints 0.6373 and 0.5912, respectively, instead of 0.7. The less restrictive Ritter constraints yield about three times as many bets as (7) and (9) indicate.) Using a sample of 229 harness races at Sportsman's Park and Hawthorne Park in Chicago he found that with \$2 bets these systems gained 24% and 16%, respectively. But a 15% advantage shrunk to -1% if one uses the τ bets with a random shock over the τ to t period (for the show bets for the 95 races at Hawthorne with a 0.65 cutoff). There are some difficulties with the design of Ritter's experiment, such as the small sample, the inclusion of only \$2 bets and the use of a random shock rather than an estimate of usual trends from τ to t, etc., however, his results point to the difficulty of actually making substantial profits in a racetrack setting.

An attempt was made during the summer 1980 racing season at Exhibition Park to implement the proposed system and observe the "end of betting problem." Nine racing days were attended during which 90 races were run. At $t - 2$ minutes the win, place, and show data were recorded on any horse that, through equations (7) and (9), warranted a bet. Equations (8) and (10) were then used on this data to determine the regression estimates of the optimal bets. Updated toteboard data on these horses were recorded until the end of betting to determine if the horse remained a system bet. Results of this experiment are shown in Table 5 where: 1) an initial wealth of \$2500 is assumed; 2) size of bet calculations are done on data at $t - 2$ minutes; and 3) returns are based on final data. Twenty-two bets were made that yielded final wealth of \$3716 for a profit of \$1216. Actual betting was not done but by making the calculations two minutes before the end of betting allows 1.5 minutes to place the bet, more than

TABLE 5

Results from Summer 1980 Exhibition Park Betting

Date	Race	Regression estimate of expected return per dollar, 2 minutes before end of betting	Regression estimate of expected return per dollar, at the end of betting	Regression estimate of optimal bet 2 minutes before end of betting	Finish	Net return based on final data with consideration of our bets affecting odds	Final wealth
							$2500
July 2	9	120	122	$19,SHOW ON 4	5-6-7	− $19	2481
"	10	120	123	72,SHOW ON 8	8-1-2	72	2553
July 9	7	121	110	292,SHOW ON 1	2-7-1	131	2684
"	10	135	122	248,PLACE ON 1	1-6-2	260	2944
July 16	6	131	122	487,PLACE ON 9 }	9-8-6	536 } 682	3626
		139	117	292,SHOW ON 9 }		146 }	
"	7	125	127	7,SHOW ON 1	5-2-8	− 7	3619
July 23	3	149	149	30,SHOW ON 2	2-10-7	92	3711
"	4	139	134	573,SHOW ON 10	6-10-4	201	3912
July 30	8	121	111	215,PLACE ON 4	4-1-5	129	4041
"	9	123	125	591,SHOW ON 6	8-1-5	− 591	3450
Aug 6	6	128	112	39,SHOW ON 4	4-3-1	59	3509
"	9	124	103	51,SHOW ON 2	4-1-3	− 51	3458
Aug 8	1	121	132	87,SHOW ON 1	1-10-4	139	3597
"	3	127	111	635,SHOW ON 3	3-4-7	127	3724
"	4	126	113	126,SHOW ON 2	2-7-1	82	3806
Aug 11	8	121	112	94,SHOW ON 8	8-6-2	113	3919
"	9	131	130	688,SHOW ON 5	5-3-4	138	4057
Aug 13	3	128	106	33,SHOW ON 2	1-6-7	− 33	4024
"	6	131	122	205,SHOW ON 5	5-8-4	144	4168
"	7	134	133	511,SHOW ON 6	8-5-9	− 511	3657
"	10	123	109	108,SHOW ON 5	3-5-1	59	3716

enough time. Note in Table 5 that many systems bets at $t-2$ were not system bets at t, however all had regression expected returns of more than one.

An important question concerns the reliability of the results: are the results true exploitations of market inefficiencies or could they be obtained simply by chance? This question is investigated utilizing a simple model which was suggested to us by an anonymous referee. The first application is concerned with an estimate of the probability that the system's theory is vacuous and indeed the observations conform to specific favorable samples from a random betting population. The second application estimates the probability of not making a positive profit. The calculations utilize the 1980 Exhibition Park data; see Table 5.

Let π be the probability of winning a bet in each trial and

$$X_i = \begin{cases} 1 + w & \text{if the bet is won,} \\ 0 & \text{otherwise,} \end{cases}$$

be the return from a $1 bet in trial i. In n trials, the probability of winning at least $100y\%$ of the total bet is

$$\Pr\left[\frac{1}{n}\left(\sum_{i=1}^{n} X_i\right) - 1 > y\right]. \tag{11}$$

Assume that the trials are independent. Since the X_i are binomially distributed (11) can be approximated by a normal probability distribution as

$$1 - \Phi\left[\frac{\sqrt{n}\,\{y - (1 + w)\pi + 1\}}{(1 + w)\pi\sqrt{(1 - \pi)/\pi}}\right] \tag{12}$$

where Φ is the cumulative distribution function of a standard $N(0, 1)$ variable. The observed probability of winning a bet, weighted by size of bet made, yields 0.771 as an estimate of π. If the systems theory was vacuous and random betting was actually being made then $(1 + w)\pi = 0.83$ since the track's payback is approximately 83%. The 22 bets made totalled $5304 and resulted in a profit of $1216 for a rate of return of 22.9%. Using equation (12) with $n = 22$, gives 3×10^{-5} as the probability of making 22.9% through random betting.

Suppose that the 1980 Exhibition Park results represent typical system behavior so $\pi = 0.771$ and $(1 + w)\pi = 1.229$. In n trials the probability of making a non-positive net return is

$$\Pr\left[\frac{1}{n}\left(\sum_{i=1}^{n} X_i\right) - 1 < 0\right] \tag{13}$$

which can be approximated as

$$\Phi\left[\frac{\sqrt{n}\,(1 - (1 + w)\pi)}{(1 - w)\pi\sqrt{(1 - \pi)/\pi}}\right] = \Phi(-0.342\sqrt{n}). \tag{14}$$

For $n = 22$ this probability is only 0.054 and for $n = 50$ and 100 this probability is 0.008 and 0.0003, respectively. Thus it is reasonable to suppose that the results from the 1980 Exhibition Park data (as well as the 1978 Exhibition Park and 1973/1974 Santa Anita data with their larger samples) represent true exploitation of a market inefficiency.[14]

[14] Without implicating them we would like to thank Michael Alhadeff of Longacres Racetrack in Seattle, the American Totalizator Corporation and the staff of the Jockey Club at Exhibition Park in Vancouver for helping us obtain the data used in this study, and M. J. Brennan and E. U. Choo for helpful discussions. Thanks are also due to V. S. Bawa, J. R. Ritter, and two anonymous referees for helpful comments on an earlier draft of this paper. This paper is a condensation of [15].

References

1. Ali, M. M., "Probability and Utility Estimates for Racetrack Bettors," *J. Political Economy*, Vol. 85 (1977), pp. 803–815.
2. ———, "Some Evidence of the Efficiency of a Speculative Market," *Econometrica*, Vol. 47, No. 2 (1979), pp. 387–392.
3. Arrow, K. J., *Aspects of the Theory of Risk Bearing*, Yrjö Jahnsson Foundation, Helsinki, 1965.
4. Baumol, W. J., *The Stock Market and Economic Efficiency*, Fordham Univ. Press, New York, 1965.
5. Beyer, A., *My $50,000 Year at the Races*, Harcourt, Brace, Jovanovitch, New York, 1978.
6. Copeland, J. E. and Weston, J. F., *Financial Theory and Corporate Policy*, Addison-Wesley, Reading, Mass., 1979.
7. Downes, D. and Dyckman, T. R., "A Critical Look at the Efficient Market Empirical Research Literature as it Relates to Accounting Information," *Accounting Rev.* (April 1973), pp. 300–317.
8. Fabricant, B. F., *Horse Sense*, David McKay, New York, 1965.

9. FAMA, E. F., "Efficient Capital Markets: A Review of Theory and Empirical Work," *J. Finance*, Vol. 25 (1970), pp. 383–417.
10. ———, *Foundations of Finance*, Basic Books, New York, 1976.
11. FIGLEWSKI, S., "Subjective Information and Market Efficiency in a Betting Model," *J. Political Economy*, Vol. 87 (1979), pp. 75–88.
12. GRIFFITH, R. M., "Odds Adjustments by American Horse Race Bettors," *Amer. J. Psychology*, Vol. 62 (1949), pp. 290–294.
13. ———, "A Footnote on Horse Race Betting," *Trans. Kentucky Acad. Sci.*, Vol. 22 (1961), pp. 78–81.
14. HARVILLE, D. A., "Assigning Probabilities to the Outcomes of Multi-Entry Competitions," *J. Amer. Statist. Assoc.*, Vol. 68 (1973), pp. 312–316.
15. HAUSCH, D. B., ZIEMBA, W. T. AND RUBINSTEIN, M. E., "Efficiency of the Market for Racetrack Betting," U.B.C. Faculty of Commerce, W. P. No. 712, September 1980.
16. ISAACS, R., "Optimal Horse Race Bets," *Amer. Math. Monthly* (1953), pp. 310–315.
17. KING, A. P., "Market Efficiency of a Multi-Entry Competition," MBA essay, Graduate School of Business, University of California, Berkeley, June 1978.
18. MCGLOTHLIN, W. H., "Stability of Choices Among Uncertain Alternatives," *Amer. J. Psychology*, Vol. 63 (1956), pp. 604–615.
19. PRATT, J., "Risk Aversion in the Small and in the Large," *Econometrica*, Vol. 32 (1964), pp. 122–136.
20. RITTER, J. R., "Racetrack Betting: An Example of a Market with Efficient Arbitrage," mimeo, Department of Economics, University of Chicago, March 1978.
21. ROSETT, R. H., "Gambling and Rationality," *J. Political Economy*, Vol. 73 (1965), pp. 595–607.
22. RUBINSTEIN, M., "Securities Market Efficiency in an Arrow-Depreu Market," *Amer. Econom. Rev.*, Vol. 65, No. 5 (1975), pp. 812–824.
23. SELIGMAN, D., "A Thinking Man's Guide to Losing at the Track," *Fortune*, Vol. 92 (1975), pp. 81–87.
24. SNYDER, W. W., "Horse Racing: Testing the Efficient Markets Model," *J. Finance*, Vol. 33 (1978), pp. 1109–1118.
25. THRALL, R. M., "Some Results in Non-Linear Programming," *Proc. Second Sympos. in Linear Programming*, Vol. 2, National Bureau of Standards, Washington, D. C., January 27–29, 1955, pp. 471–493.
26. VERGIN, R. C., "An Investigation of Decision Rules for Thoroughbred Race Horse Wagering," *Interfaces*, Vol. 8, No. 1 (1977), pp. 34–45.
27. WEITZMAN, M., "Utility Analysis and Group Behaviour: An Empirical Study," *J. Political Economy*, Vol. 73 (1965), pp. 18–26.
28. ZIEMBA, W. T. AND VICKSON, R. G., eds., *Stochastic Optimization Models in Finance*, Academic Press, New York, 1975.

47

TRANSACTIONS COSTS, EXTENT OF INEFFICIENCIES, ENTRIES AND MULTIPLE WAGERS IN A RACETRACK BETTING MODEL*

DONALD B. HAUSCH AND WILLIAM T. ZIEMBA

School of Business, University of Wisconsin, Madison, Wisconsin 53706

Faculty of Commerce, University of British Columbia, Vancouver, British Columbia, Canada V6T 1W5

In a previous paper (*Management Science*, December 1981) Hausch, Ziemba and Rubinstein (HZR) developed a system that demonstrated the existence of a weak market inefficiency in racetrack place and show betting pools. The system appeared to make possible substantial positive profits. To make the system operational, given the limited time available for placing bets, an approximate regression scheme was developed for the Exhibition Park Racetrack in Vancouver for initial betting wealth between $2500 and $7500 and a track take of 17.1%. This paper: (1) extends this scheme to virtually any track and initial wealth level; (2) develops a modified system for multiple horse entries; (3) allows for multiple bets; (4) analyzes the effects of the track take and breakage on profits; (5) presents recent results using this system; and (6) considers the extent of the inefficiency, i.e., how much can be bet before the market becomes efficient?

(FINANCE–PORTFOLIO; GAMES—GAMBLING)

1. The Racetrack Market

The "market" at the track in North America convenes for about 20 minutes, during which participants make bets on any number of the six to twelve horses in the following race. In a typical race, participants can bet on each horse, either to win, place or show. All participants who have bet a horse to win realize a positive return on that bet only if the horse is first, while a place bet realizes a positive return if the horse is first or second, and a show bet realizes a positive return if the horse is first, second or third. Regardless of the outcome, all bets have limited liability. Unlike casino games such as roulette, but like the stock market, security prices (i.e. the "odds") are jointly determined by all the participants and the rules governing transactions costs (i.e. the track "take" and "breakage"). To take the simplest case, all bets across all horses to win are aggregated to form the win pool. If W_i represents the total amount bet by all participants on horse i to win, then $W = \sum_i W_i$ is the win pool and WQ/W_i is the payoff per dollar bet on horse i to win if and only if horse i wins, where Q is the track payback proportion.

The rules for division of the place and show pools are as follows. Let P_j be the amount bet on horse j to place and $P \equiv \sum_j P_j$ be the place pool. The payoff per dollar bet on horse j to place is

$$\begin{array}{ll} 1 + [PQ - P_i - P_j]/(2P_j) & \text{if } \begin{cases} i \text{ is first and } j \text{ is second} \quad \text{or} \\ j \text{ is first and } i \text{ is second,} \\ \text{otherwise.} \end{cases} \\ 0 & \end{array} \quad (1)$$

*Accepted by Donald G. Morrison as Special Departmental Editor; received December 13, 1983. This paper has been with the authors 2 months for 2 revisions.

Thus if horses i and j are first and second each bettor on j (and also i) to place first receives the amount of his bet back. The remaining amount in the place pool, after the track take, is then split evenly between the place bettors and i and j. The payoff to horse j to place is independent of whether j finishes first or second, but it is dependent on which horse i finishes with it. A bettor on horse j to place hopes that a longshot with a small P_i not a favorite will finish with it.

The payoff per dollar bet on horse k to show is analogous

$$\begin{matrix} 1+[SQ-S_i-S_j-S_k]/(3S_k) \\ 0 \end{matrix} \quad \text{if} \quad \begin{cases} k \text{ is first, second or third} \\ \text{and finishes with } i \text{ and } j, \\ \text{otherwise,} \end{cases} \tag{2}$$

where S_k is the amount bet on horse k to show and the show pool is $S \equiv \sum_k S_k$.

Equations (1) and (2) are not quite correct as they do not account for "breakage". Breakage is discussed in §6; it is an extra commission resulting from payoffs being rounded down to the nearest 10¢ or 20¢ on a $2 bet.

2. Racetrack Efficiency

A market is efficient, see Fama (1970), if current prices fully reflect all available relevant information. In this case experts should not be able to achieve higher than average returns with regularity.

To investigate the efficiency of the racetrack's win market Snyder (1978) tested whether or not bets at different odds levels yielded the same average return. A weakly-efficient market in Snyder's sense would have the average rate of return for each odds level equal to Q, the track payback ratio which currently varies in North America from 0.852 in Ontario to 0.779 in Saskatchewan. His results suggest there are "strong and stable biases but these are not large enough to make it possible to earn a positive profit" (Snyder 1978, p. 1101). In particular, favorites tend to be underbet and longshots overbet. See Ziemba and Hausch (1984), hereafter referred to as ZH, for a survey of the literature on this bias.

To test the efficiency of the racetrack's place and show market HZR assumed:

(1) If q_i ($i = 1, \ldots, n$ horses) is the probability that i wins, then the probability that i is first and j is second is

$$\frac{q_i q_j}{1-q_i}, \tag{3}$$

and the probability that i is first, j is second and k is third is

$$\frac{q_i q_j q_k}{(1-q_i)(1-q_i-q_j)}. \tag{4}$$

Harville (1973) developed and analyzed these formulas.

(2) If W_i is the total amount bet on horse i to win and $W \equiv \sum_{i=1}^{n} W_i$ then $q_i = W_i/W$, i.e., the win market is efficient. While this assumption ignores the bias for favorites and longshots mentioned above there are reverse tail biases in the probabilities of finishing second and third. They occur because if the probability to win is overestimated (underestimated) and probabilities sum to one it is likely that the probabilities of finishing second and third are underestimated (overestimated). Tables 2b,c in HZR indicate how these biases tend to cancel when they are aggregated to form the theoretical probabilities of placing and showing which are used in the HZR model.[1]

[1] See HZR, §§4–6, for a more in-depth discussion of the model.

Using equations (1)–(4), the Bernoulli capital growth model (see Ziemba and Vickson 1975), and assuming initial wealth is w_0, the HZR model to calculate optimal amounts to bet for place (p_l, $l = 1, 2, \ldots, n$) and show (s_l, $l = 1, 2, \ldots, n$) is

$$\underset{\{p_l\}\{s_l\}}{\text{Maximize}} \sum_{i=1}^{n} \sum_{\substack{j=1 \\ j \neq i}}^{n} \sum_{\substack{k=1 \\ k \neq i,j}}^{n} \frac{q_i q_j q_k}{(1-q_i)(1-q_i-q_j)} \log \left[\begin{array}{l} \dfrac{Q(P + \sum_{l=1}^{n} p_l) - (p_i + p_j + P_{ij})}{2} \\ \quad \times \left[\dfrac{p_i}{p_i + P_i} + \dfrac{p_j}{p_j + P_j} \right] \\ + \dfrac{Q(S + \sum_{l=1}^{n} s_l) - (s_i + s_j + s_k + S_{ijk})}{3} \\ \quad \times \left[\dfrac{s_i}{s_i + S_i} + \dfrac{s_j}{s_j + S_j} + \dfrac{s_k}{s_k + S_k} \right] \\ + w_0 - \sum\limits_{\substack{l=1 \\ l \neq i,j,k}}^{n} s_l - \sum\limits_{\substack{l=1 \\ l \neq i,j}}^{n} p_l \end{array} \right] \tag{5}$$

$$\text{s.t.} \quad \sum_{l=1}^{n} (p_l + s_l) \leqslant w_0, \quad p_l \geqslant 0, \quad s_l \geqslant 0, \quad l = 1, \ldots, n.$$

The formulation (5) maximizes the expected logarithm of final wealth considering the probabilities and payoffs from the possible horserace finishes. For notational simplicity $P_{ij} \equiv P_i + P_j$ and $S_{ijk} \equiv S_i + S_j + S_k$.

The generalized concavity properties of (5) are discussed in Kallberg and Ziemba (1981).

Using equations (1) and (3), the expected return on an additional \$1 bet to place on horse 1 is:

$$\mathrm{EX}_1^p \equiv \sum_{j=2}^{n} \left[\frac{q_1 q_j}{1-q_1} + \frac{q_j q_1}{1-q_j} \right] \left[1 + \frac{1}{20} \mathrm{INT} \left[\frac{Q(P+1) - (1 + P_1 + P_j)}{2(P_1 + 1)} \times 20 \right] \right]. \tag{6}$$

INT[Y] means the largest integer not exceeding Y. The INT and multiplying and dividing by 20 accounts for breakage, i.e. the payoffs on a \$2 bet being rounded down to the nearest 10¢ in this instance. The $q_1 q_j/(1 - q_1)$ term is the probability that 1 is first and j is second while the $q_j q_1/(1 - q_j)$ term is the probability that j is first and 1 is second. A similar equation is available for show bets (see HZR).

One might hope that when EX_1^p or EX_1^s equals say, α, the actual average rate of return would be near α or at least increasing in α. Despite the inherent inaccuracies in assumptions (1) and (2) this is indeed the case as is borne out in Table 3 in HZR and Table 5.1 in ZH. In fact for cases of α above about 1.02, positive profits seem realizable. These profits are maximized when α is about 1.16.

An ideal model is then:

(i) for each i check EX_i^p and EX_i^s to decide whether or not to bet on horse i, and then

(ii) solve (5) to determine the optimal bet size after setting p_i and s_j to zero for horses you definitely do not want to bet on.

This analysis presumes one can bet after all other bettors have placed their bets. In practice one can only attempt to be one of the last bettors. Thus it is extremely important that one be able to perform all calculations necessary to determine the bet(s) quickly so that the bets can be placed as close to the end of the betting period as possible.

The approximations developed in HZR are regression schemes with the minimal data inputs: w_0, W_{i^*}, P_{i^*}, W_{j^*}, S_{j^*}, W, P and S, where $i^* = \mathrm{argmax}_i((W_i/W)/(P_i/P))$

and $j^* = \text{argmax}_j((W_j/W)/(S_j/S))$. The ratios $(W_i/W)/(P_i/P)$ and $(W_j/W)/(S_j/S)$ may be thought of as simple measures of the inefficiency to place on i and show on j respectively. The regressions were calibrated for an initial wealth w_0 between $2500 and $7500 and a track about the size of Exhibition Park in Vancouver, B.C. (daily handle about $1.2 million) with $Q = 0.829$. Using this regression scheme on the Exhibition Part data with an initial wealth of $2500 and updating wealth over time resulted in a final wealth of $7698 at the end of the 110-day season (see Figure 3 in HZR).

We now extend equations (7)–(10) in HZR to account for different wealth levels, different size tracks, different Q's, and coupled entries.

3. Track Size and Wealth Level

Track size and wealth level do not affect the expected return per dollar bet to place or show. However both are important factors in determining the optimal amounts to bet to place and show because the larger the betting pools at the track the less our bet

TABLE 1

The Optimal Place Bet for Various Betting Wealth Levels and Place Pools Sizes

	$W_0 = \$50$	$W_0 = \$500$	$W_0 = \$2{,}500$	$W_0 = \$10{,}000$
Place Pool = $2,000		$261q + 256q^2 + 180q^3 - \left(\frac{199qP_i}{qP_0 - 0.70P_i}\right)$ [P2]	$426q + 802q^2 - \left(\frac{459qP_i}{qP - 0.60P_i}\right)$ [P5]	$487q + 901q^2 - \left(\frac{521qP_i}{qP - 0.60P_i}\right)$ [
Place Pool = $10,000	$39q + 52q^2 - \left(\frac{25qP_i}{qP - 0.75P_i}\right)$ [P1]	$375q + 525q^2 - \left(\frac{271qP_i}{qP - 0.70P_i}\right)$ [P3]	$1{,}307q + 1{,}280q^2 + 902q^3 - \left(\frac{993qP_i}{qP - 0.70P_i}\right)$ [P6]	$2{,}497q + 1{,}806q^2 + 2{,}073q^3 - \left(\frac{2{,}199qP_i}{qP - 0.60P_i}\right)$ [
Place Pool = $150,000		$505q + 527q^2 - \left(\frac{386qP_i}{qP - 0.60P_i}\right)$ [P4]	$2{,}386q + 2{,}668q^2 - \left(\frac{1{,}877qP_i}{qP - 0.60P_i}\right)$ [P7]	$7{,}072q + 10{,}470q^{\cdot} - \left(\frac{5{,}273qP_i}{qP - 0.70P_i}\right)$ [I

TABLE 2

The Optimal Show Bet for Various Betting Wealth Levels and Show Pool Sizes

	$W_0 = \$50$	$W_0 = \$500$	$W_0 = \$2{,}500$	$W_0 = \$10{,}000$
Show Pool = $1,200		$9 + 994q^2 - 464q^3 - \left(\frac{150qS_i}{qS - 0.80S_i}\right)$ [S2]	$13 + 1{,}549q^2 - 901q^3 - \left(\frac{303qS_i}{qS - 0.60S_i}\right)$ [S5]	
Show Pool = $6,000	$10 + 183q^2 - 135q^3 - \left(\frac{11S_i}{qS - 0.80S_i}\right)$ [S1]	$86 + 1{,}516q^2 - 968q^3 - \left(\frac{90.7S_i}{qS - 0.85S_i}\right)$ [S3]	$53 + 5{,}219q^2 - 2{,}513q^3 - \left(\frac{934qS_i}{qS - 0.70S_i}\right)$ [S6]	$58 + 7{,}406q^2 - 4{,}211q^3 - \left(\frac{1{,}359qS_i}{qS - 0.65S_i}\right)$ [
Show Pool = $100,000		$131 + 2{,}150q^2 - 1{,}778q^3 - \left(\frac{150S_i}{qS - 0.70S_i}\right)$ [S4]	$533 + 9{,}862q^2 - 7{,}696q^3 - \left(\frac{571S_i}{qS - 0.80S_i}\right)$ [S7]	$1{,}682 + 28{,}200q^2 - 16{,}880q^3 - \left(\frac{1{,}769S_i}{qS - 0.85S_i}\right)$ [

influences the odds and as our wealth increases we tend to wager more. Therefore new regressions were calculated for most reasonable track sizes and wealth levels. The data for these regressions were the true optimal bets from the NLP model (5) over a broad range of wealth levels and track sizes. The ideal data to represent a broad range of track sizes would be a season's data from many different tracks. A more practical alternative was to multiply the Exhibition Park data by varying constants to simulate data from smaller and larger tracks. Nineteen different regressions which appear in Tables 1 and 2 were determined for place and show depending upon wealth and track size. These regressions labelled P1–P10 and S1–S9 give the optimal place or show bet for specific values of initial wealth and size of pool. For intermediate values of these variables one may determine accurate betting amounts by taking convex combinations of these basic regressions. For example, for a place bet with initial wealth \$1000 and place pool of \$5000 the optimal bet is α_2 (equation $[P_2]$) + α_3 (equation $[P_3]$) + α_5 (equation $[P_5]$) + α_6 (equation $[P_6]$), where

$$\alpha_2 = \tfrac{3}{4}\tfrac{5}{8} = \tfrac{15}{32}, \qquad \alpha_3 = \tfrac{3}{4}\tfrac{3}{8} = \tfrac{9}{32}, \qquad \alpha_5 = \tfrac{1}{4}\tfrac{5}{8} = \tfrac{5}{32}, \qquad \alpha_6 = \tfrac{1}{4}\tfrac{3}{8} = \tfrac{3}{32}.$$

4. Track Payback

Both the expected return per dollar bet and the optimal bet size are increasing functions of Q, the track's payback. Equations (7) to (10) in HZR were calculated with $Q = 0.829$. Modification of these equations for use at tracks with $Q \neq 0.829$ are now developed.

4.1. *Adjustment of* $\widetilde{\mathrm{EX}}_i^p$ *and* $\widetilde{\mathrm{EX}}_i^s$ *for* Q

The regression equivalent for EX_i^p when $Q = 0.829$ is:[2]

$$\widetilde{\mathrm{EX}}_i^p = 0.319 + 0.559\,\frac{W_i/W}{P_i/P}. \tag{7}$$

How can this equation be adjusted for a track payback different from 0.829? The "true" expected return on a one dollar place bet on horse i is

$$\mathrm{EX}_i^p = \sum_{\substack{j=1\\ j\neq i}}^{n}\left(\frac{q_i q_j}{1-q_i} + \frac{q_i q_j}{1-q_j}\right)\left(1 + \frac{QP - (P_i + P_j)}{2P_i}\right). \tag{8}$$

Equation (7) is linear in Q with

$$\frac{\partial\,\mathrm{EX}_i^p}{\partial Q} = \frac{q_i P}{2P_i}\left[1 + \sum_{\substack{j=1\\ j\neq i}}\left(\frac{q_j}{1-q_j}\right)\right].$$

Using 124 Exhibition Park races with EX_i^p in the range 1.10 and greater, the true $\partial\,\mathrm{EX}_i^p/\partial Q$ was regressed against q_i to give $\partial\,\mathrm{EX}_i^p/\partial Q \approx 2.22 - 1.29q_i$ ($R^2 = 0.86$, $SE = 0.055$, both coefficients highly significant). Therefore when the track payback is Q, the expected return on a \$1 place bet can be approximated by adjusting (7) to

$$\begin{aligned}\widehat{\mathrm{EX}}_i^p &\equiv \widetilde{\mathrm{EX}}_i^p + (2.22 - 1.29q_i)(Q - 0.829)\\ &= 0.319 + 0.559\left(\frac{W_i/W}{P_i/P}\right) + \left(2.22 - 1.29\left(\frac{W_i}{W}\right)\right)(Q - 0.829).\end{aligned} \tag{9}$$

[2]Note that the coefficients of $\widetilde{\mathrm{EX}}_i^p$ here are different from those of equation (7) in HZR. In HZR only cases of expected return greater than 1.16 were considered. Here we may wish to bet on horses with expected returns as low as 1.10 to reflect a high quality track and high quality horses. Thus the $\widetilde{\mathrm{EX}}_i^p$ had to be recalculated to be accurate in the range 1.10 to 1.16. A similar change will be noted for $\widetilde{\mathrm{EX}}_i^s$.

A similar analysis for show yields

$$\hat{EX}_i^s \equiv \tilde{EX}_i^s + (3.60 - 2.13q_i)(Q - 0.829)$$

$(R^2 = 0.565,\ SE = 0.198$, both coefficients highly significant)

$$= 0.543 + 0.369\left(\frac{W_i/W}{S_i/S}\right) + \left(3.60 - 2.13\left(\frac{W_i}{W}\right)\right)(Q - 0.829). \qquad (10)$$

4.2. *Adjustment of $\tilde{p}^*$ and $\tilde{s}^*$ for Q*

Equations (8) and (10) in HZR were calibrated for $Q = 0.829$. Since the true p^* and s^* are nondecreasing in Q which varies from track to track we must adjust $\tilde{p}^*$ and $\tilde{s}^*$ at tracks with $Q \neq 0.829$.

The exact NLP model was used on a number of Exhibition Park examples to compute the optimal place or show bets at different initial wealths, track sizes and different Q's (from 0.809 to 0.859). The results indicated that $\Delta p^*/\Delta Q$ and $\Delta s^*/\Delta Q$ are independent of Q in this range. Therefore with a ΔQ of 0.01, $\Delta p^*/\Delta Q$ and $\Delta s^*/\Delta Q$ were regressed on p^*, w_0, P_i, P and s^*, w_0, S_i, S, respectively. The analysis showed for $\Delta p^*/\Delta Q$ that p^* and w_0 were very significant independent variables but neither P_i and P were significant; similar results for $\Delta s^*/\Delta Q$ were observed. Then the $(\Delta p^*/\Delta Q, p^*, w_0)$ and $(\Delta s^*/\Delta Q, s^*, w_0)$ were aggregated (due to a small number of place data points) to give:

$$\begin{bmatrix} \Delta p^*/\Delta Q \\ \Delta s^*/\Delta Q \end{bmatrix} = [0.0316]\begin{bmatrix} \tilde{p}^* \\ \tilde{s}^* \end{bmatrix} + 0.000351 w_0$$

$(R^2 = 0.948,\ SE = 2.23,\ n = 56$, both coefficients highly significant).

Therefore $\hat{p}^*$ and $\hat{s}^*$ (i.e. $\tilde{p}^*$ and $\tilde{s}^*$ adjusted for Q) are:

$$\hat{p}^* = \tilde{p}^* + (Q - 0.829)(3.16\tilde{p}^* + 0.0351 w_0) \quad \text{and} \qquad (11)$$

$$\hat{s}^* = \tilde{s}^* + (Q - 0.829)(3.16\tilde{s}^* + 0.0351 w_0). \qquad (12)$$

4.3. *Example Involving $Q \neq 0.829$*

May 7, 1983–Kentucky Derby at Churchill Downs, Louisville, Kentucky. The final win and show pools and the win and show bets on #8, Sunny's Halo, were:

$$W = \$3{,}143{,}669, \qquad W_8 = \$745{,}524,$$
$$S = \$1{,}099{,}990, \qquad S_8 = \$179{,}758.$$

Using just the $\tilde{EX}_i^s$ portion of equation (10) gives $\tilde{EX}_8^s = 1.08$, i.e. not enough to consider a bet if one is using a typical cutoff of 1.10 as recommended in ZH for a race like the Kentucky Derby. But Kentucky has $Q = 0.85$ and therefore it is more accurate to use equation (10) resulting in $\hat{EX}_8^s = 1.14$. Hence a show bet should be made. With an initial wealth of \$1000, Table 2 gives the optimal show bet as $\tilde{s}^* = \$48$. The correction for $Q = .85$ using equation (12) yields $\hat{s}^* = \$52$. Sunny's Halo won the Derby and paid \$4.00 per \$2.00 bet to show so the \$52 bet returned \$104 for a \$52 profit. Full details on this race appear in ZH.

5. Coupled Entries

Occasionally two or more horses are run as a single "coupled entry" or simply "entry" because (1) an owner or a trainer has two or more horses in the same race, or

(2) there are more horses than the toteboard can accommodate (commonly called a field). The entry wins, places or shows if just one of the horses wins, places or shows. If any two of the horses in the entry come first and second all the place pool goes to the place tickets on the entry. If two of three of the 'in-the-money' horses are the entry then typically two thirds of the show pool goes to the draw tickets on the entry (rather than the usual third).

Suppose the coupled entry has number 1 and let $q_1 = W_1/W$. Then q_1 estimates the probability that one of the horses in the entry will win the race. Suppose further that q_{1A} and q_{1B} (with $q_{1A} + q_{1B} = q_1$) are the correct winning probability estimates of the two horses in the entry. Using q_1 (i.e. thinking of the entry as a single horse) and equation (3) to calculate the probability of the entry placing gives

$$\Pr(\text{entry 1 is 1st or 2nd}) = q_1 + \sum_{i=2}^{n} \frac{q_1 q_i}{1 - q_i}. \tag{13}$$

Using q_{1A} and q_{1B} (i.e. thinking of the entry as two horses) and equation (3) to calculate the probability of the entry placing gives

$$\begin{aligned}
\Pr(\text{entry 1 is 1st and/or 2nd}) &= \Pr(1A \text{ is 1st and any horse but } 1B \text{ is 2nd}) \\
&+ \Pr(1B \text{ is 1st and any horse but } 1A \text{ is 2nd}) \\
&+ \Pr(1A \text{ is 2nd and any horse but } 1B \text{ is 1st}) \\
&+ \Pr(1B \text{ is 2nd and any horse but } 1A \text{ is 1st}) \\
&+ \Pr(1A \text{ and } 1B \text{ are 1st and 2nd in either order}) \\
&= \sum_{i=2}^{n} \frac{q_{1A} q_i}{1 - q_{1A}} + \sum_{i=2}^{n} \frac{q_{1B} q_i}{1 - q_{1B}} + \sum_{i=2}^{n} \frac{q_{1A} q_i}{1 - q_i} \\
&+ \sum_{i=2}^{n} \frac{q_{1B} q_i}{1 - q_i} + \frac{q_{1A} q_{1B}}{1 - q_{1A}} + \frac{q_{1A} q_{1B}}{1 - q_{1B}},
\end{aligned}$$

which equals

$$q_{1A} + q_{1B} + \sum_{i=2}^{n} \frac{(q_{1A} + q_{1B}) q_i}{1 - q_i},$$

which is equation (13). Hence considering the entry as two horses does not affect our estimate of the entry's probability of placing. It does, however, affect our estimate of the expected return on a dollar bet to place since the possibility of a $1A$-$1B$ or $1B$-$1A$ finish exists and for those finishes the place payoff will be high since the whole place pool (net of the track take and breakage) goes to the holders of place tickets on the entry 1. Thus equation (9) will underestimate $\hat{EX}_1^p$. For the same reason p^*, from Table 1, will also be underestimated. We now consider the use of Tables 1 and 2 and equations (9), (11), (10) and (12) on $\hat{EX}^p$, $\hat{p}^*$, $\hat{EX}^s$ and $\hat{s}^*$, respectively, to account for an entry.

5.1. *Adjustments of* EX_1^p *and* EX_1^s *for Coupled Entries*

Equation (6) gives the expected return on a dollar bet to place on a horse, considering the entry 1 as a single horse. Ignoring breakage, this expected return is

$$EX_1^p = \sum_{i=2}^{n} \left(\frac{q_1 q_i}{1 - q_1} + \frac{q_1 q_i}{1 - q_i} \right) \left(1 + \frac{QP - (P_1 + P_i)}{2P_1} \right).$$

More properly considering entry 1 as two horses, $1A$ and $1B$, the expected return on a one dollar bet to place is

$$\mathrm{EX}^{p}_{1A,1B} = \sum_{i=2}^{n}\left(\frac{q_{1A}q_i}{1-q_{1A}} + \frac{q_{1A}q_i}{1-q_i}\right)\left(1 + \frac{QP-(P_1+P_i)}{2P_1}\right)\left\{\text{horses } 1A \text{ and } i \text{ place}\right.$$

$$+ \sum_{i=2}^{n}\left(\frac{q_{1B}q_i}{1-q_{1B}} + \frac{q_{1B}q_i}{1-q_i}\right)\left(1 + \frac{QP-(P_1+P_i)}{2P_1}\right)\left\{\text{horses } 1B \text{ and } i \text{ place}\right.$$

$$+ \left(\frac{q_{1A}q_{1B}}{1-q_{1A}} + \frac{q_{1A}q_{1B}}{1-q_{1B}}\right)\left(1 + \frac{QP-P_1}{P_1}\right)\left\{\text{horses } 1A \text{ and } 1B \text{ place.}\right.$$

Let $\Delta^p \equiv \mathrm{EX}^p_{1A,1B} - \mathrm{EX}^p_1$. It is generally the case that the two horses in the coupled entry are not of equal ability. We assume that $q_{1A} = \frac{2}{3}q_1$ and $q_{1B} = \frac{1}{3}q_1$.

Using the Exhibition Park data,[3] Δ^p was regressed on W_1/W and P_1/P, giving

$$\tilde{\Delta}^p = 0.867\frac{W_1}{W} - 0.857\frac{P_1}{P} \tag{14}$$

($R^2 = 0.996$, $SE = 0.00267$, and both coefficients highly significant).

Then using equations (7) and (18) the regression approximation for expected return on a one dollar bet on the coupled entry 1 is

$$\tilde{\mathrm{EX}}^p_{1A,1B} \equiv \tilde{\mathrm{EX}}^p_1 + \tilde{\Delta}^p = 0.319 + 0.559\left(\frac{W_1/W}{P_1/P}\right) + 0.867\frac{W_1}{W} - 0.857\frac{P_1}{P}. \tag{15}$$

The same procedure for the expected return on a one dollar show bet on the coupled entry 1 yields

$$\tilde{\mathrm{EX}}^s_{1A,1B} \equiv \tilde{\mathrm{EX}}^s_1 + \tilde{\Delta}^p = 0.543 + 0.369\left[\frac{W_1/W}{S_1/S}\right] + 0.842\frac{W_1}{W} - 0.810\frac{S_1}{S}. \tag{16}$$

5.2. *Adjustments of p^* and s^* for Coupled Entries*

When the possible bet is on an entry, equations (7) or (9) underestimate the expected return on an additional dollar bet to place. Therefore the optimal bet from the NLP (5) will underestimate the true optimal coupled entry place bet. To understand this phenomenon many Exhibition Park examples were solved using the exact NLP (5) assuming the entry was one horse. Then the same examples were solved supposing the entry was two horses (the formulation of the NLP was adjusted to consider the possibility of the two horses finishing first and second and then receiving a high place payoff) but the win bet on the entry was lowered, using an iterative scheme, until the optimal bet was the same as the optimal bet assuming the entry was one horse. This procedure gave pairs of $\tilde{q}^p_i$ and q_i, where $\tilde{q}^p_i$ was the probability of entry i winning in part a (thinking of the entry as one horse); and q_i was the adjusted probability that gave the same optimal bet when thinking of the entry as two horses as was observed when treating it as one horse.

Since the regression formula (13) gives the approximate optimal place bet when the horse's probability of winning is q_i, then using $\tilde{q}^p_i$ in that formula gives the approximate optimal place bet when the coupled entry's probablity of winning is q_i.

[3]The data used were cases where the expected return (equation (7)) was > 1.16, i.e., the cases of interest. For instances with a low expected return the correction factor is meaningless.

The regression relating $\tilde{q}_i^p$ and q_i is

$$\tilde{q}_i^p = 0.991 q_i + 0.137 q_i^2 + 3.47 \times 10^{-7} W_0 \tag{17}$$

($R^2 = 0.9998$, $SE = 0.00161$, all coefficients highly significant).

The examples from which the data were derived spanned many wealth levels and pool sizes. While the wealth level was a very significant independent variable the pool size was found to be statistically insignificant. Therefore to compute the optimal place bet on coupled entry i the procedure is:

(0) determine that EX^p is large enough to consider a place bet,
(1) set $q_i = W_i / W$,
(2) determine $\tilde{q}_i^p$, and
(3) substitute $\tilde{q}_i^p$, w_0, P, P_i in Table 1.

The same procedure was carried out for the optimal show bet on coupled entry i:

(0) determine that EX^s is large enough to consider a show bet,
(1) set $q_i = W_i / W$,
(2) determine

$$\tilde{q}_i^s = 1.07 q_i + 4.13 \times 10^{-7} W_0 - 0.00663, \tag{18}$$

and ($R^2 = 0.999$, $SE = 0.00298$, all coefficients highly significant), and
(3) substitute $\tilde{q}_i^s$, w_0, S, and S_i, in Table 2.

Three additional questions to be considered are: (1) Will the results be different with a weighting other than 1/3 and 2/3 on the two horses in the entry? (2) Should there be larger adjustments for entries of three or more horses? and (3) Should there be adjustments of the equations for a single horse running against a coupled entry? The answer to all three questions is yes, but how important is it to account for these possibilities? (1) The coupled entry adjustment appears to be fairly robust to the weighting. Also each race would require handicapping to determine its more accurate weighting. Since that goes beyond the scope of this research (in fact the system described in ZH requires absolutely no handicapping), we choose to opt for no adjustments for different weightings. In the extreme case where $q_{1A} \approx q_1$ and $q_{1B} \approx 0$ it may be better to treat the entry as a single horse and ignore the entry adjustments. (2) The additional benefits of three or more horses in an entry are small beyond accounting for the entry as two horses. Thus we suggest the simplification of no further adjustment. (3) Generally the expected return equations on a single horse running against an entry will overestimate the true expected return. In most cases though the bias is small and again we suggest no further adjustment.

6. Multiple Bets

Occasionally there is more than one system bet in a given race. Since the optimal bet equations in Tables 1 and 2 were calibrated assuming only one place bet or one show bet in a race it is not correct, in a multiple betting situation, to calculate each bet individually from the tables and then wager those amounts. Often that would result in overbetting but there are also times where, for diversification reasons, that would actually result in underbetting.

We have attempted to deal with the most common multiple betting situation—a place and show bet on the same horse. Ninety-eight cases of place and show system bets on the same horse were analyzed covering a wide range of track handles, q_i's, and w_0's. Using the optimization model (5) resulted in the quadruples $(p_T^*, s_T^*, p_A^*, s_A^*)$. The p_A^* is the optimal place bet supposing it is the only good bet in the race, s_A^* is the optimal show bet supposing it is the only good bet in the race, and (p_T^*, s_T^*) is the

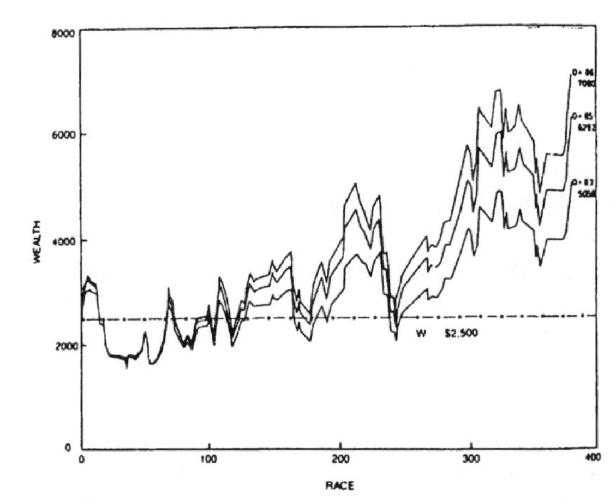

FIGURE 1. Betting Wealth Level Histories for System Bets at the 1981–82 Aqueduct Winter Meeting Using an Expected Value Cutoff of 1.14 for Track Takes of 14%, 15% and 17%.

optimal pair of place and show bets when they are considered together. The p_A^* and s_A^* are the values we should calculate from Tables 1 and 2. Then p_T^* was regressed on p_A^*, s_A^*, w_0, P_i, P and q_i. The only statistically significant independent variables were p_A^* and s_A^* leading to the regression equation

$$\tilde{p}_T^* = 1.59p_A^* - 0.639s_A^* \tag{19}$$

($R^2 = 0.967$, $SE = 73.7$, both coefficients highly significant).

A similar procedure for s_T^* yields

$$\tilde{s}_T^* = 0.907s_A^* - 0.134p_A^* \tag{20}$$

($R^2 = 0.992$, $SE = 72.6$, both coefficients highly significant).

7. Effect of the Track Take

The track take is a commission of 14–22% on every dollar wagered. As mentioned in §4 a change in Q, the track payback proportion, can have a substantial effect on EX_i^p and EX_i^s. A change in Q can also have a surprisingly dramatic effect on long run profit. This is illustrated in Figure 1 using data from the 1981–82 Winter Season at Aqueduct, New York[4] and supposing three different track takes: 14%, 15% and 17%. In 1981–82 the track take was 15%, earlier it had been 14% and just recently it has increased to 17%. Assuming an initial wealth of \$2,500 the final wealths are \$7,090, \$6,292 and \$5,058 at track's takes of 14%, 15% and 17% respectively. The track take increasing from 14% to 15% dropped profits by \$798 (17.4%). The track take increasing from 15% to 17% dropped profits by \$1,234 (32.5%).

8. Effect of Breakage

In addition to the track take, bettors must also pay an additional commission called breakage. This commission refers to the funds not returned to the betting public because the payoffs are rounded down to the nearest 10¢ or 20¢ on a \$2 bet. For example, a payoff net of the track take of \$6.39 would pay \$6.30 or \$6.20, respectively.

[4]The data consisted of the final win, place and show mutuels for the 43-day period December 27, 1981—March 27, 1982. During this period 3,470 horses ran in 380 races. Thanks go to Dr. Richard Van Slyke for collecting these data for us.

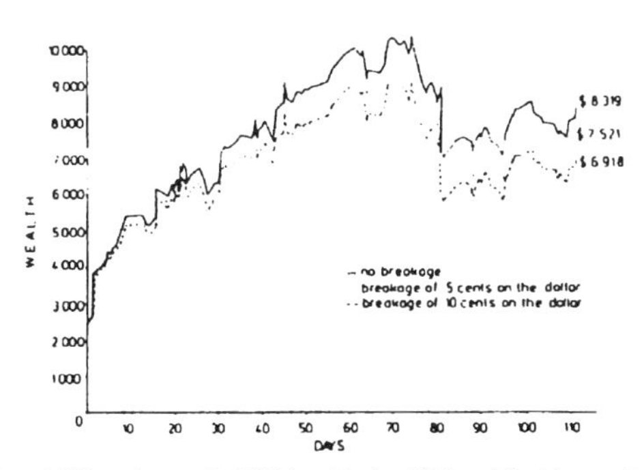

FIGURE 2. Wealth Level Histories at Exhibition Park (1978) with Alternative Breakage Schemes.

Breakage occurs in the win, place and show as well as other pools. We refer to rounding down to the nearest 10¢ on a $2 bet as "5¢ breakage" that is 5¢ per dollar and rounding down to the nearest 20¢ on a $2 bet as "10¢ breakage". Initially most tracks utilized a 5¢ breakage procedure. However, in recent years more and more tracks have switched to 10¢ breakage. The 10¢ breakage is never less than 5¢ breakage and usually is considerably more. As a percentage of the payoff breakage usually increases as the payoff becomes smaller unless the payoff is close to the breakage roundoff amount. An exception is the "minus pool" when $2.10[5] must be paid on a winning $2 ticket even if the payoff before breakage is only $2.08, or even $1.73.

On average bettors lose about 1.79% of the total payoff to 5¢ breakage and 3.14% to 10¢ breakage on bets using the system in ZH. Adding these amounts to the track take gives the total commission. For example, at Churchill Downs in Louisville, Kentucky the 15% track take becomes about 18.1% with their 10¢ breakage, and at Exhibition Park in Vancouver, British Columbia the 15.8% track take becomes about 17.6% with their 5¢ breakage. Thus to determine the full commission at a given racetrack one must take into account the breakage as well as the track take.

The full extent of the effect of breakage is shown in its effect on profits. Using the 1978 system bets for Exhibition Park with an initial wealth of $2500 one would have $8319 at the end of the year without breakage. With 5¢ breakage one would have $7521 and $6918 with 10¢ breakage. The effect of breakage throughout the 1978 season is shown in Figure 2. In addition to taking money away from total wealth the breakage has the effect of lowering the bet size (because of this lower wealth) thus resulting in lower future profits. These calculations indicate that 5¢ breakage averages 13.7% of profits and 10¢ breakage averages 24.1% of profits on system bets.

These calculations indicate that breakage (especially the very common 10¢ variety) is a very substantial cost. The costs are highest when one is placing bets on short odds horses. This is, unfortunately, an unavoidable aspect of the system presented in ZH.

9. Results Using the Betting System

In ZH many examples are presented showing precisely how to use the system. In Table 3 we present summary statistics on system bets made at several tracks over different seasons. The data sets for Aqueduct, Santa Anita and Exhibition Park 1978 were collected after their seasons finished. The Exhibition Park 1980 and Kentucky Derby Days data were collected race by race at the track. In all cases initial betting wealth is assumed to be $2,500. The different expected cutoff levels reflect the quality of the horses at the different tracks.

The most common system bet is to show on a favorite. Show and place bets occur about 85% and 15% of the time, respectively. The percent of bets won is about 59%

[5]In Kentucky it is $2.20.

TABLE 3

Summary Statistics on System Bets Made at Aqueduct in 1981/82, *Santa Anita in* 1973/74, *Exhibition Park in* 1978 *and* 1980, *and at the Kentucky Derby Days* 1981/82/83 *with an Initial Betting Wealth of* $2500

Track and Season	Number of Days	Number of Races	Track Take	Expected Value Cutoff	Number of System Bets	Number of Bets Won	Percent of Bets Won	Percent of Bets Won Weighted by Size of Bet	Total Money Wagered	Track Take	Total Profits	Average Payout Per $2 Bet	Rate of Return on Bets Made
Aqueduct 1981/82	43	380	15%	1.14	124	68	55%	65%	$42.686	$6.403	$3,792	$3.33	8.9%
Santa Anita 1973/74	75	627	15%	1.14	192	114	59%	69%	$51.631	$7.745	$2,837	$3.16	5.5%
Exhibition Park 1978	110	1.065	18.1%	1.20	174	97	56%	72%	$49.991	$9.048	$5.198	$3.08	10.4%
Exhibition Park 1980	10	90	17.1%	1.20	22	16	73%	77%	$ 5.403	$ 924	$1.216	$3.18	22.5%
Derby Days* 1981/82/83	3	30	15%	1.10	19	17	89%	96%	$12.766	$1.915	$5,462	$2.97	42.8%
Totals and Weighted Averages	241	2,192	—	—	531	312	59%	71%	$162,477	$26,035	$18,505	$3.12	11.4%

while the percent of bets won weighted by the size of the bet is 71%. This difference is because the bets are on shorter odds horses which finish in-the-money more often. At a track such as Santa Anita with large betting pools, our bets do not affect the odds very much and the average bet is about 7% of the betting pool. Over all these thousands of races and hundreds of system bets the total amount wagered was $162,477. The track take was $26,035. Our profit was $18,505, for an 11.4% rate of return on dollars wagered. The higher rates of return were on the races where we were at the track, so we were able to skip rainy days and reject certain horses on the basis of very simple handicapping rules. The lower rates of return, as expected, were at the tracks where we had no information other than the win, place and show mutuel pools. A simple correction which has a surprisingly large effect is removing from the Exhibition Park 1978 data the days when the track was not a fast track, i.e. rainy days when the track was slow, muddy, heavy, sloppy, etc. Doing so decreases the total money wagered from $49,991 to $32,811 but increases the profit from $5,198 to $5,863. Thus the rate of return on the "fast track" days is 17.9%, up from 10.4% over all days. It also increases the rate of return over all the racetracks from 11.4% to 13.2%.

Since the bets usually have a high EX_i^p or EX_i^s we might expect a rate of return around 16%–20%. Remember that EX_i^p and EX_i^s are on the first dollar bet. These values drop as we bet large amounts due to our bets affecting the odds.

These profits do not consider several minor "entertainment type" costs that one must incur by actual attendance at the track to apply the system. Parking, gasoline, racing program, racing form, track admission and food amount to $3–10 or more. For example, at $10 per day Aqueduct's profits of $3792 over 43 days become $3362.

Finally the average payout per $2 bet ranged from $2.97 to $3.33 at the various tracks, with an average of $3.12. This value is actually a high show return when one considers the heavy favorites the system often picks.

10. Will the Market Become Efficient?

As more and more individuals use this system the markets for place and show betting will tend to become efficient. Two important questions are: (1) How many people can play this system and still have it provide a return of 10–20%? and (2) How many people can play this system before the market becomes efficient enough that expected profits are zero?

To consider the first question we can determine how much additional money can be wagered on a particular horse to place or show before the expected value per dollar bet drops to the suggested cutoff for good betting opportunities. As an illustration Figure 3 indicates this amount to show for a cutoff of 1.14. These figures are based on a track take of 17.1% so for lower track takes more can be bet and for higher track takes less can be bet.

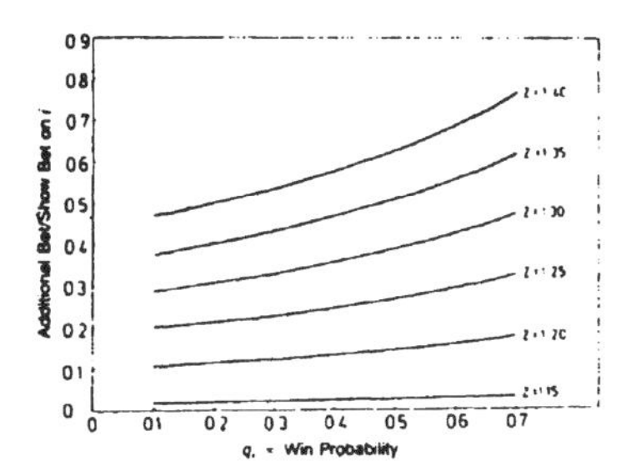

FIGURE 3. How Much Can Be Bet, B, by System Bettors Relative to the Crowd's Show Bet, S_i, on Horse i to Lower the Expected Value to Show on the Horse i from Z to 1.14, When the Track Take is 17.1%.

An example that more directly answers questions 1 and 2 is provided below and is based on the data given in §4.3 on Sunny's Halo, the 1983 Kentucky Derby winner. Additional examples appear in ZH.

Sunny's Halo

	Betting Wealth w_0			
	$200	$500	$1000	$2000
Optimal System Bets (Using the Actual Data One Minute Before Post Time) Assuming a Betting Wealth of w_0	$11	$31	$52	$96

α	Total Amount That Can be Bet Before the Expected Return per Dollar Bet Drops to α	Number of System Bettors Needed to Drop the Expected Return per Dollar Bet to α Assuming a Betting Wealth of $200	$500	$1000	$2000
1.10	$19,323	1,757	623	372	201
1.06	$41,175	3,743	1,328	792	429
1.02	$68,409	6,219	2,207	1,316	713

Our results show that $\alpha = 1.02$ is a breakeven cutoff and $\alpha = 1.06$ yields a rate of return of about 5–6%.[6]

References

FAMA, E. F., "Efficient Capital Markets: A Review of Theory and Empirical Work," *J. Finance*, 25 (1970), 383–417.

HAUSCH, D. B., W. T. ZIEMBA AND M. RUBINSTEIN, "Efficiency of the Market for Racetrack Betting," *Management Sci.*, 27 (1981), 1435–1452.

HARVILLE, D. A., "Assigning Probabilities to the Outcome of Multi-Entry Competitions," *J. Amer. Statist. Assoc.*, 68 (1973), 312–316.

KALLBERG, J. G. AND W. T. ZIEMBA, "Generalized Concave Functions in Stochastic Programming and Portfolio Theory," in *Generalized Concavity in Optimization and Economics*, S. Schaible and W. T. Ziemba (Eds.), Academic Press, New York, 1981, 719–767.

SNYDER, W. W., "Horse Racing: Testing the Efficient Markets Model," *J. Finance*, 33 (1978), 1109–1118.

ZIEMBA, W. T. AND D. B. HAUSCH, *Beat the Racetrack*, Harcourt, Brace and Jovanovich, San Diego, 1984.

——— AND R. G. VICKSON, EDS., *Stochastic Optimization Models in Finance*, Academic Press, New York, 1975.

48

The Dr.Z Betting System in England[1]

William T. Ziemba and Donald B. Hausch

Management Science Division, Faculty of Commerce,
University of British Columbia, Vancouver BC, Canada V6T 1Z2

School of Business, University of Wisconsin,
Madison, Wisconsin 53706

Abstract

The betting strategy proposed in Hausch, Ziemba and Rubinstein (1981) and Ziemba and Hausch (1984,1987) has had considerable some success in North American place and show pools. The place pool is England is very different. This paper applies a similar strategy with appropriate modifications for places bet at British racetracks. The system or minor modifications also applies in a number of other countries such as Singapore with similar betting rules. The system appears to provide positive expectation wagers. However, with the higher track take it is not known how often profitable wagers will exist or what the long run performance might be.

I. Introduction

At North American racetracks, the parimutuel system of betting utilizing electric totalizator boards in the dominant method of betting. Las Vegas and the other legal sports books may set odds on particular betting situations, but these fixed odds are not available at racetracks. In England and in other Commonwealth countries, such as Italy and France, odds betting against bookies is the dominant betting scheme. This fixed-odds system is introduced in Hausch, Lo and Ziemba (1994). They also indicate a tendency of higher (lower) returns for lower (higher) odds ranges. Thus, the favorite-longshot bias appears to exist in England (see e.g. Ali (1977), Busche and Hall (1988)).

This paper applies the betting system proposed by Hausch, Ziemba and Rubinstein (1981) and Hausch and Ziemba (1985) in England. This strategy is also called the Dr.Z system in the trade books Ziemba and Hausch (1984,1987) who discuss it more fully. The strategy utilizes the Kelly criterion (Kelly (1956)) which maximizes the expected logarithm of wealth. The Kelly criterion has several advantages. First, it maximizes the capital growth asymptotically. Second, it prevents bankruptcy. Third, the expected time to reach a specified goal is minimum when the goal increases. Fourth, it is superior to any different strategy in the long run. These properties were proved in Breiman (1961). See McLean, Ziemba and Blazenko (1992) for discussion of these properties.

II. The System

Instead of the North American parimutuel system of win, place, and show, the bets in England are to win and place. By "place" the British mean "finish in the money." This is what North Americans call show except for one important difference. The number of horses that can place in a particular race is dependent on the number of starters.

[1] Modified from an Appendix in Ziemba and Hausch (1987).

Efficiency of Racetrack Betting Markets

Table 1. Relationship between number of horses that place and number of starters

Number of Horses that Place	Number of Starters
one: the winner	four or less
two: winner and second	five, six, or seven
three: winner, second, and third	eight to fifteen
four: winner, second, third, and fourth	sixteen or more

The place pools are not shown on the tote board, but the current payoffs for place bets for each horse are flashed on the screen. Bookies, on the other hand, simply pay a percentage of the win odds, as shown in Table 2.

There are many types of exotic bets as well. The tote jackpot corresponds to what North Americans call the pick six or sweep six. The tote placepot bet has no analogue in North America. The average rates of return on various bets on and off course against a bookmaker or the tote are listed in Table 3. The track take is 5% larger in the place pool than in the win pool. The tote take is larger than what the bookies make on average, and on-course betting takes are much less than off-course takes.

The races in England are on the turf for distances of generally at least a mile, except for some shorter races for two-year-olds. The season in southern England is unique in that races are run for about three days at each race course. The jockeys, trainers, and so forth then move on to a new course. After a month or so they return to the same course. Handicapping is very sophisticated in England. It has to be, with little information easily accessible (they have no analogue of the *Daily Racing Form*, although some past performances are available in newspapers) and all that moving from course to course.

The method of computing the place payoffs in England differs from that used in North America. In both locales, the net pool is the total amount wagered minus the track take. In North America, the cost of the winning in-the-money tickets is first subtracted to form the profit. This profit is then shared equally among the in-the-money horses. Holders of winning tickets receive a payoff consisting of the original stake plus their proportionate share of the horse's profits. This means that the amount of money wagered on the other horses in the money greatly affects the payoff. In England, the total net pool is divided equally among the horses that finish in the money. This means that the payoff on a particular horse depends upon how much is bet on this horse to place but not on how much is bet on the other horses. Since the minimum payoff is £1 per £1 wager, management is able to keep a control on betting for particular favorites. Once this minimum level is reached, it does not pay to wager on a given horse. This occurs whenever the percentage of the place pool that is bet on a given horse becomes as large as Q_p, which is the track take for place, divided by m, which is the number of in-the-money horses. In a race with 8-15 starters, if Q_p is about 0.735, and $m=3$, the just-get-your-money-back point is reached when the bet on a particular horse to place becomes 24.5% of the total place pool: $0.735/3 = 24.5\%$. Hence in England you will often see horses whose place payoffs are £1 or just slightly higher. This method of sharing the place pool tends to favor longer-priced horses at the expense of the favorites.

Table 2. Bookmakers payoff for place bets

Number of Runners	Type of Race	Fraction of Win Odds Paid on Place Element	Horses Regarded
Two to five		No place betting	
Six or seven	Any	$\frac{1}{4}$	First and second
Eight or more	Any except handicaps involving twelve or more runners	$\frac{1}{5}$	First, second, and third
Twelve to fifteen	Handicaps	$\frac{1}{4}$	First, second, and third
Sixteen to twenty-one	Handicaps	$\frac{1}{5}$	First, second, third, and fourth
Twenty-two or more	Handicaps	$\frac{1}{4}$	First, second, third, and fourth

Source: Rothschild (1978).

Table 3. Rates of return on different types of bets in England on thoroughbred and greyhound racing

Type of Bet	Rate of Return (%)
On-course bookmaker	90
Off-course bookmaker	81
Single bet to win with off-course bookmaker	85
Double bet to win with off-course bookmaker	78
Treble bet to win with off-course bookmaker	72
ITV Seven bet to win with off-course bookmaker	70–75
Computer straight forecast with off-course bookmaker	65
Greyhound forecast with off-course bookmaker	76
Greyhound forecast double with off-course bookmaker	58
Place element of each-way with off-course bookmaker	80
Ante-post betting with off-course bookmaker	96
Horse race tote win pool (on course)	80
Horse race tote win pool (off course)	77
Horse race tote place pool (on course)	75
Horse race tote place pool (off course)	72
Horse race tote daily double pool (on course)	74
Horse race tote daily double pool (off course)	71
Horse race tote daily treble pool (on course)	70
Horse race tote daily treble pool (off course)	67
Horse race tote daily forecast pool (on course)	70
Horse race tote daily forecast pool (off course)	67
Horse race tote jackpot pool (on course)	70
Horse race tote jackpot pool (off course)	67
Horse race tote placepot pool (on course)	70
Horse race tote placepot pool (off course)	67
Greyhound tote pool betting, average	83.5

Source: Rothschild (1978).

The current track take to win is about 20.6% and to place is 26.5%, and the breakage is of 10¢ variety, or more properly 10p, for pence[2]. These track takes are much higher than those in North America. Since the track paybacks to win and place are different, we call the former, $Q_w=0.794$, and the latter, $Q_p=0.735$.

It is easy to apply the Dr.Z system in Great Britain, although with its much higher track takes, there may not be many Dr.Z system bets, see Mordin (1992) for a discussion of this. We utilize the substitution that $q_i=Q_w/O_i$, where O_i are the odds to win on the horse under consideration.

The expected value per pound bet to place on horse i is

$$EX\text{Place} = (\text{probability of placing})\ (\text{place odds}) = (\text{Prob})\ (PO_i). \tag{1}$$

In (1), PO_i refers to the odds to place on horse i. Prob, the probability of placing is determined as follows.[3] [4]

---

[2]We can calculate these track takes as follows: The payoff on horse i if it wins is $Q_w W_i/W$, where Q_w is the track payback to win, and W_i and W are the bet amount of horse i and the total bet amount, respectively. So let $q_i = W_i/W$, the efficient-market assumption. Let B be the average breakage, namely, 4.5p. Since breakage can be 0,1,2,...,,9 pence, its average is 4.5p. Then the payoff on i is $Q_w/q_i\text{-}B$, which equals the odds O_i, since the odds are based on total return (not return plus original stake as in North America). So $q_i = Q_w/(B+O_i)$. Summing over all n horses gives

$$\sum_{i=1}^{n} q_i = 1 = Q_w \sum_{i=1}^{n} \left(\frac{1}{B+O_i}\right),$$

since some horse must win. Hence

$$Q_w = \frac{1}{\sum_{i=1}^{n}(B+O_i)}.$$

For place, there are one, two, three, or four horses that are in the money, depending upon the number of starters. So

$$Q_p = \frac{m}{\sum_{i=1}^{n}(B+O_i)},$$

where $m=1,2,3$ or 4.

With the above formulas, $Q_w \simeq 0.794$ and $Q_p \simeq 0.745$.

[3]These equations were developed using the 1981-1982 Aqueduct data to relate probability of in-the-money finishes to q, the probability of winning and n, the number of horses. Equations (2), (3) and (4) had R^2 of 0.991, 0.993 and 0.998, respectively. These equations are valid when q ranges from 0 to 0.6

With $n=5$ to 7 horses, the first 2 horses place and

$$\text{Prob} = 0.0667 + 2.37q - 1.61q^2 - 0.0097n. \quad (2)$$

With n=8 to 15 horses, the first 3 horses place and

$$\text{Prob} = 0.0665 + 3.44q - 3.47q^2 - 0.0049n. \quad (3)$$

With n=16 or more horses, the first 4 horses place and

$$\text{Prob} = 0.0371 + 4.47q - 6.29q^2 - 0.00164n. \quad (4)$$

Figures 1,2, and 3 determine Prob directly using only O_i, the win odds on the horse in question. Figure 1 applies when there are five, six or seven horses. Figure 2 corresponds to equation (3) and applies when there are eight to fifteen horses. Finally, Figure 3 corresponds to equation (4) and applies when there are sixteen or more horses.

The optimal Kelly criterion bet is to wager (Prob PO_i-1)/(PO_i-1) percent of your betting wealth[5]. We can determine the optimal fraction of your wealth to bet indicated by equation (5) using Figure 4.

---

for (2), from 0 to 0.45 for (3), and 0 to 0.3 for (4), which should be the case in most instances. However, Figures 1,2 and 3 are valid for any q.

[4]In a race with n=2,3, or 4 horses, only one horse places, the winner. Such races are rare. Also, it is unlikely that the win and place pools would then become so unbalanced as to yield a Dr.Z system bet. However, one would occur when PO_i/O_i was at least 1.44, for a track payback of 0.794 and an expected-value cutoff of 1.14, since 1.14/0.794 is 1.44. In such a case, one would have a good bet.

[5]We have assumed that your bets will be small and hence will not affect the odds very much. Thus to determine the optimal bet b for betting wealth w_0, you maximize Prob log[w_0 + (PO_i-1)b] + (1 - Prob)log(w_0-b), whose solution is equation (5).

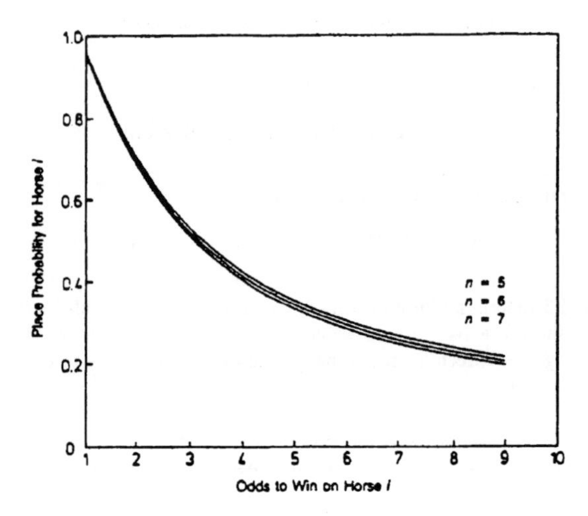

Figure 1. Probabilities of placing for different odds horses when the race has five to seven starters

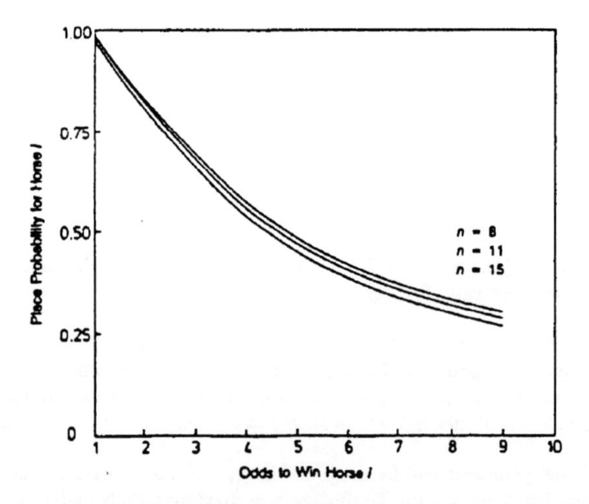

Figure 3. Probabilities of placing for different odds horses when the race has eight to fifteen starters

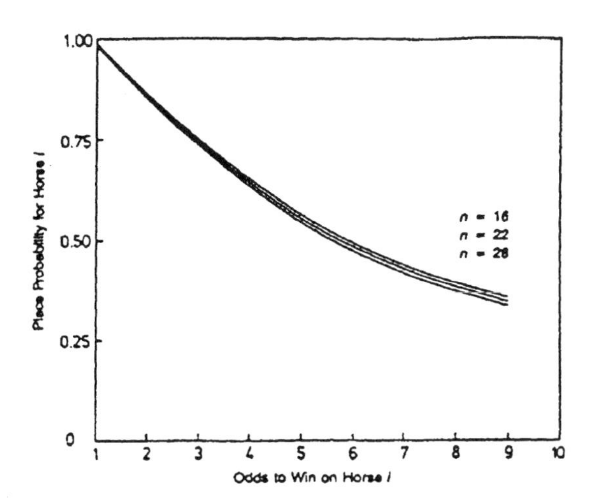

Figure 3. Probabilities of placing for different odds horses when the race has sixteen or more starters

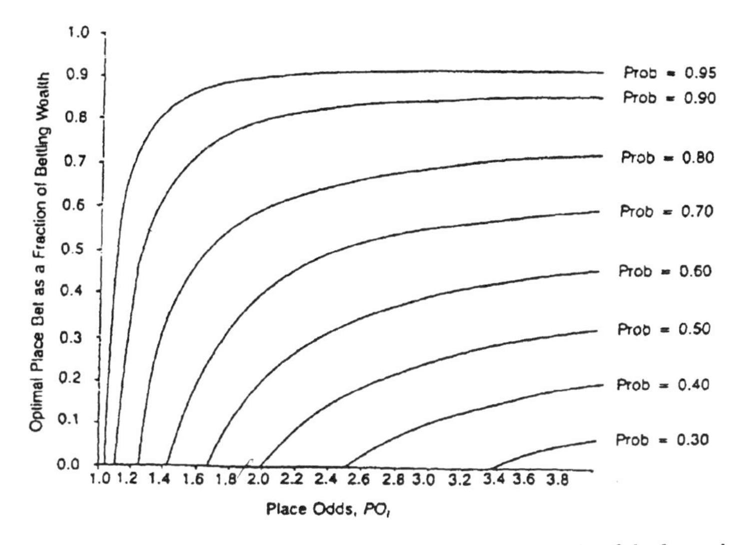

;ure 4. Optimal bets when the probability of placing is Prob and the place odds of the horse in question : PO_i

References

Breiman,L. (1961) "Optimal gambling systems for favorable games." in Proceedings of the Fourth Berkeley Symposium. Mathematical Statistics and Probability 1, 65-78. University of California Press.

Figgis,E.L. (1974) "Rates of return from flat race betting in England in 1973." *Sporting Life* 11 (March).

Hausch,D.B., Ziemba,W.T. and Rubinstein,M. (1981) "Efficiency of the market for racetrack betting." *Management Science* 27, 1435-1452.

Hausch,D.B. and Ziemba,W.T. (1985) "Transactions costs, extent of inefficiencies, entries and multiple wagers in a racetrack betting model." *Management Science* 31, 381-394.

Kelly,J.L. (1956) "A new interpretation of information rate." *Bell System Technical Journal* 35, 917-926.

MacLean,L.C., Ziemba,W.T. and Blazenko,G. (1992) "Growth versus security in dynamic investment analysis." *Management Science* 38, 1562-1585.

Mordin,N. (1992) "Grab your place in the pool!" *Sporting Life*.

Rothschild, Lord (1978) Royal commission on gambling, Vols I and II. Presented to parliament by Command of Her Majesty (July).

Ziemba,W.T. and Hausch,D.B. (1984) Beat the racetrack. Harcourt Brace Jovanovich, San Diego.

Ziemba,W.T. and Hausch,D.B. (1987) Dr.Z's beat the racetrack. Revised edition. William Morrow and Co. Inc. New York.

49

Robert R. Grauer
Simon Fraser University

Nils H. Hakansson
University of California, Berkeley

A Half Century of Returns on Levered and Unlevered Portfolios of Stocks, Bonds, and Bills, with and without Small Stocks*

This paper applies multiperiod portfolio theory to the construction and rebalancing of portfolios composed of U.S. stocks, corporate bonds, government bonds, and a risk-free asset, with small stocks included as a separate investment vehicle. Probability assessments are based on the past, joint empirical distribution. Our principal findings are (1) small stocks, while totally ignored at times, entered even the most risk-averse portfolios most of the time and (2) small stocks, when chosen, tended to replace common stocks, except in the 1970s and early 1980s, when they were primarily held in lieu of the risk-free asset.

I. Introduction

In two earlier papers (Grauer and Hakansson 1982; Grauer and Hakansson 1984), we applied the multiperiod portfolio theory of Mossin (1968), Hakansson (1971, 1974), Leland (1972), Ross (1974), and Huberman and Ross (1983) to the construction and rebalancing of portfolios composed of U.S. stocks, corporate bonds, government bonds, and a risk-free asset. Borrowing was ruled out in the first article, while margin purchases were permitted in the second. The probability distributions used were naively estimated from past realized returns in Ibbotson

* Presented at the Australian Graduate School of Management and at Macquarie University, both in Sydney, Southern Methodist University, Duke University, the University of Southern California, the London Graduate School of Business, and the annual meeting of the Western Financo Association in Scottsdale, Arizona. The authors thank the participants of these seminars, especially Michael Brennan and Ehud Ronn, for helpful comments. Financial support from the Social Sciences and Humanities Research Council of Canada and the Financial Research Foundation of Canada is gratefully acknowledged. The authors also wish to thank Changwoo Lee for research assistance and are greatly indebted to Frederick Shen for computational assistance.

(*Journal of Business*, 1986, vol. 59, no. 2, pt. 1)

0021-9398/86/5902-0005$01.50

and Sinquefield (1982) and in their Center for Research in Security Prices (CRSP) 1926–83 data base, and both annual and quarterly holding periods were employed from the mid-1930s forward. The results of both papers revealed that the gains from active diversification among the major asset categories were substantial, especially for the highly risk averse strategies. In addition, we found evidence of substantial use of, and gains from, margin purchases for the more risk-tolerant strategies from the mid-1930s to the mid-1960s.

In the present paper, small stocks are included as a separate investment vehicle. Thus the opportunity set is expanded to five categories: common stocks, government bonds, corporate bonds, small stocks, and a risk-free asset. The resulting sequential portfolio problem is solved both with and without leverage. When used, all borrowing is assumed to bear interest at the call money rate +1% and to be limited in size by applicable initial margin requirements. Our principal findings are (1) small stocks, while totally ignored at times, entered even the most risk-averse portfolios most of the time, (2) small stocks, when chosen, tended to replace common stocks, except in the 1970s and early 1980s, when they were primarily held in lieu of the risk-free asset, and (3) small stocks had a salutary effect on the geometric means of realized returns, especially from the mid-1930s to the early 1950s in the presence of margin purchases. In the 1934–47 subperiod with leverage, for example, small stocks raised the geometric mean return for the power .5 strategy from 9.09% to 25.95% while reducing variability.

II. Theory

Despite explosive development over three decades and extensive application to the construction of equity portfolios, modern portfolio theory has found relatively little use in the larger portfolio context, namely, the choice of the proportions to be held in the major categories of common stocks and in different types of bonds, money market instruments, real estate, and foreign securities. There are several reasons for this. First, extant portfolio theory, being principally based on the mean-variance model, is fundamentally single-period in nature, whereas the larger problem accents the multiperiod, sequential nature of investment decisions. On top of this, since the universe of interest extends well beyond common stocks, extant betas are too narrowly defined to be useful, and the appropriate betas are not easily estimated because of data problems concerning the market weights of bonds, for example. At the other extreme, continuous-time portfolio theory is somewhat intractable in a world of nontrivial transaction costs. Finally, extant models rely heavily on narrow classes of theoretical (stationary) return distributions, with limited ability to capture the richness of joint, real-world stochastic processes.

There is, however, a middle category of investment models, usually classified under the heading of discrete-time multiperiod portfolio theory, which has been largely ignored in portfolio selection applications. This is so despite the fact that these models have a strong foundation in theory and lend themselves naturally to the problem of rebalancing portfolios over many periods (200 quarters in the present study). An additional virtue of these models is that they can handle return distributions of every conceivable shape and are unfazed by such things as nonstationary returns or distributions that are just plain unknown beyond the period just ahead.

To review, consider the simplest reinvestment problem, in which the market is perfect and returns are independent over time but otherwise arbitrary and not necessarily stationary. The investor has a preference function U_0 (with $U_0' > 0$, $U_0'' < 0$) defined on wealth w_0 at some terminal point (time 0). Let w_n denote the investor's wealth with n periods to go, r_{in} the return on asset i in period n, z_{in} the amount invested in asset i in period n (with $i = 1$ being the safe asset), and $U_n(w_n)$ the relevant (unknown) utility of wealth with n periods to go. At the end of period n (time $n - 1$), the investor's wealth is

$$w_{n-1}(z_n) = \sum_{i=2}^{M} (r_{in} - r_{1n})z_{in} + w_n(1 + r_{1n}),$$

where $z = (z_{2n}, \ldots, z_{Mn})$ and M is the number of securities.

Consider the portfolio problem with one period to go. The investor, with w_1 to invest, must now solve

$$\max_{z_1|w_1} E[U_0(w_0(z_1))] \equiv U_1(w_1).$$

Clearly, $U_1(w_1)$ represents the highest attainable expected utility level from capital level w_1 at time 1 and thus the "derived" utility of w_1. Employing the induced utility function $U_1(w_1)$, the portfolio problem with two periods to go becomes

$$U_2(w_2) \equiv \max_{z_2|w_2} E[U_1(w_1(z_2))].$$

Thus with n periods to go we obtain (the recursive equation)

$$U_n(w_n) = \max_{z_n|w_n} E[U_{n-1}(w_{n-1}(z_n))], \; n = 1, 2, \ldots$$

Examining the above system, it is evident that the induced utility of current wealth, $U_n(w_n)$, generally depends on "everything," namely, the terminal utility function U_0, the joint distribution functions of future returns, and future interest rates. There is, however, a special case in which $U_n(w_n)$ depends only on U_0: this occurs (Mossin 1968) if and only if $U_0(w_0)$ is isoelastic; that is, if and only if

$$U_0(w_0) = \frac{1}{\gamma} w^{\gamma}, \quad \gamma < 1.$$

(Note that for $\gamma = 0$, $U_0[w_0] = \ln w_0$.) The variable $U_n(w_n)$ is now a positive linear transformation of $U_0(w_n)$; that is, we can write

$$U_n(w_n) = \frac{1}{\gamma} w_n^\gamma. \tag{1}$$

For these preferences, the optimal investment policy $z^*_{n_\gamma}$ is proportional to wealth; that is,

$$z^*_{in_\gamma}(w_n) = x^*_{in_\gamma} w_n, \quad \text{all } i, \tag{2}$$

where the $x^*_{in_\gamma}$ are constants. It is also completely myopic since it only depends on U_0 and the current period's return structure and not on returns beyond the current period. Both of these properties hold only for family (1), which is also the only class of preferences exhibiting constant relative risk aversion.[1] Finally, (2) also implies that the utility of wealth relatives, $V_n(1 + r_n)$, is of the same form only for this family; that is,

$$U_n(w_n) = \frac{1}{\gamma} w_n^\gamma <=> V_n(1 + r_n) = \frac{1}{\gamma}(1 + r_n)^\gamma.$$

While the above properties are interesting, they are clearly rather special. However, the isoelastic family's influence extends far beyond its numbers. As shown by Leland (1972), Hakansson (1974), Ross (1974), and Huberman and Ross (1983), there is a very broad class of terminal utility functions $U_0(w_0)$ for which the induced utility functions U_n converge to an isoelastic function, that is, for which

$$U_n(w_n) \rightarrow \frac{1}{\gamma} w_n^\gamma \quad \text{for some } \gamma < 1. \tag{3}$$

Hakansson (1974) has also shown that (3) is usually accompanied by convergence in policy; that is,

$$z^*_n \rightarrow x^*_{n_\gamma} w_n.$$

Thus the objectives given by (1) are quite robust and encompass a broad variety of different goal formulations for investors with intermediate- to long-term investment horizons.[2] In particular, class (1) spans a continuum of risk attitudes all the way from risk neutrality ($\gamma = 1$) to infinite risk aversion ($\gamma = -\infty$).[3]

Having selected our model, we turn next to what we need to operate it. The major input to the model is an estimate of next period's joint return distribution for the various asset categories.[4] As in our previous

1. This measure is defined as $-wU''_n(w)/U'_n(w)$ and equals $1 - \gamma$ for the class (1).

2. The simple reinvestment formulation does ignore consumption of the course.

3. A plot of the functions $(1/\gamma)(1 + r)^\gamma$ for several values of γ was given in Grauer and Hakansson (1982, p. 42).

4. For a comprehensive overview of the issues and problems associated with the estimation of return distributions, see Bawa, Brown, and Klein (1979).

studies, we based this estimate on the so-called simple probability assessment approach. In this approach, the realized returns of the most recent n periods are recorded; each of the n joint realizations is then assumed to have probability $1/n$ of occurring in the coming period. Thus estimates are obtained on a moving basis and used in raw form without adjustment of any kind. Since the whole joint distribution is specified and used, there is no information loss; all moments—marginal and conditional, for example—and every last shred of correlation are implicitly taken into account. It may be noted that the empirical distribution of the past n periods is optimal if the investor has no information about the form and parameters of the true distribution but believes that this distribution went into effect n periods ago.[5] Consequently, we have, as a starting point, resisted all temptations to parameterize the input distributions.

III. Calculations

The model used can be summarized as follows. At the beginning of each period t, the investor chooses a portfolio, x_t, on the basis of some member, γ, of the family of utility functions for returns r given by

$$V(1 + r) = \frac{1}{\gamma}(1 + r)^{\gamma}. \tag{4}$$

This is equivalent to solving the following problem in each period t:

$$\begin{aligned} &\max_{x_t} E\left[\frac{1}{\gamma}(1 + \sum_i x_{it}r_{it} + x_{Lt}r_{Lt} + x_{Bt}r^d_{Bt})^{\gamma}\right] \\ &= \max_{x_t} \sum_s \pi_{ts}\frac{1}{\gamma}(1 + \sum_i {}_{it}r_{its} + x_{Lt}r_{Lt} + x_{Bt}r^d_{Bt})^{\gamma} \end{aligned} \tag{5a}$$

subject to

$$x_{it} \geq 0, \quad x_{Lt} \geq 0, \quad x_{Bt} \leq 0, \qquad \text{all } i, t, \tag{6}$$

$$\sum_i x_{it} + x_{Lt} + x_{Bt} = 1, \quad \text{all } t, \tag{7}$$

$$\sum_i m_{it}x_{it} \leq 1, \quad \text{all } t, \tag{8}$$

$$\text{pr}(1 + \sum_i x_i r_{it} + x_{Lt}r_{Lt} + x_{Bt}r^d_{Bt} \geq 0) = 1, \tag{9}$$

where

$\gamma \leq 1$ = a parameter that remains fixed over time;

$x_t \equiv (x_{1t}, \ldots, x_{nt}, x_{Lt}, x_{Bt})$;

x_{it} = the amount invested in risky asset category i in period t as a fraction of own capital;

5. See Bawa et al. 1979, p. 100.

x_{Lt} = the amount invested in the risk-free asset in period t as a fraction of own capital;
x_{Bt} = the amount borrowed in period t as a fraction of own capital;
r_{it} = the anticipated return on asset category i in period t;
r_{Lt} = the return on the risk-free asset of period t;
r_{Bt}^{d} = the borrowing rate at the time of the decision at the beginning of period t;
m_{it} = the initial margin requirement for asset category i in period t expressed as a fraction; and
π_{ts} = the probability of event s in period t, in which case the random return r_{it} will assume the value r_{its}.

As noted, when γ equals zero, (4) reduces to the logarithmic utility function:

$$V = \ln(1 + r).$$

The equivalent of (5a) is then

$$\max_{x_t} E[\ln(1 + \sum_i x_{it}r_{it} + x_{Lt}r_{Lt} + x_{Bt}r_{Bt}^{d})]. \tag{5b}$$

Constraint (6) rules out short positions, and (7) is the budget constraint. Constraint (8) serves to limit borrowing (when desired) to the maximum permissible under the margin requirements that apply to the various asset categories. Finally, since

$$\sum_i x_{it}r_{it} + x_{Lt}r_{Lt} + x_{Bt}r_{Bt}^{d}$$

is the (ex ante) return distribution for portfolio x_t, the expressions in parentheses in (5a) and (5b) represent the (ex ante) wealth relative in period t.

On the basis of the probability estimation method described earlier, the (sequential) solution of the portfolio problem may be pictured as follows. Suppose quarterly revision is used. Then, at the beginning of quarter t, the portfolio problem (5a) or (5b) for that quarter uses the following inputs—the (observable) risk-free return for quarter t; the (observable) call money rate +1% at the beginning of quarter t; and the (observable) realized returns for common stocks, government bonds, corporate bonds, and small stocks for the previous n quarters. Each joint realization in quarters $t - n$ through $t - 1$ is given probability $1/n$ of occurring in quarter t.

With these inputs in place, the portfolio weights for the various asset categories and the proportion of assets borrowed are calculated by solving system (5a)–(9) (or [5b]–[9]) via nonlinear programming methods.[6] At the end of quarter t, the realized returns on stocks, gov-

6. The nonlinear programming algorithm employed is described in Best (1975).

ernment bonds, corporate bonds, and small stocks are observed, along with the realized borrowing rate r^r_{Bt} (which may differ from the decision borrowing rate r^d_{Bt}).[7] Then, using the weights selected at the beginning of the quarter, the realized return on the portfolio chosen for quarter t is recorded. The cycle is then repeated in all subsequent quarters.[8]

All reported returns are gross of transaction costs and taxes and assume that the investor in question had no influence on prices. There are several reasons for this approach. First, since this is one of the first studies in this area, we wish to keep the complications to a minimum. Second, the Ibbotson and Sinquefield series used as inputs and for comparisons also exclude transaction costs (for reinvestment of interest and dividends) and taxes. Third, many investors are tax-exempt, and various techniques are available for keeping transaction costs low. Finally, since the proper treatment of these items is nontrivial, they are better left to a later study.

IV. Data

The data used to estimate the probabilities of next period's returns and to calculate each period's realized returns on risky assets are the total monthly and annual returns on common stocks, long-term government bonds, long-term corporate bonds, and small stocks for 1926–83 in the Ibbotson and Sinquefield CRSP data file; both dividends and capital appreciation are taken into account. The risk-free asset used for quarterly revision was assumed to be 90-day U.S. Treasury bills maturing at the end of the quarter; we used the *Survey of Current Business* and the *Wall Street Journal* as sources. In the annual portfolio revision case, the risk-free return was obtained from the yield, as of the beginning of the year, on that U.S. government obligation (note or bond) that matured on the date closest to the end of the year in question; we obtained the 1926–76 data privately from Roger Ibbotson and the remainder from the *Wall Street Journal*.

Margin requirements for stocks were obtained from the *Federal Reserve Bulletin*. Initial margins were set at 10% for government bonds and at 35% for corporate bonds. These levels are on the conservative side and are designed to compensate for the absence of maintenance requirements.[9]

7. The realized borrowing rate r^r_{Bt} was calculated as a monthly average.

8. Note that, if $n = 32$ under quarterly revision, then the first quarter for which a portfolio can be selected is the first quarter of 1934 since the period 1926–33 is required to develop the estimated return distributions used for that quarter's portfolio choice.

9. There was no practical way to take maintenance margins into account in our programs. In any case, it is evident from the results that they would come into play only for the more risk tolerant strategies, and even for them only occasionally, and that the net effect would be relatively neutral.

TABLE 1 Composition of Fixed Weight Portfolios

	Proportion in:				
	Stocks	Government Bonds	Corporate Bonds	Small Stocks	Risk-free Asset
Portfolio A	...	.40	.40	...	.20
Portfolio B	.20	.30	.30	...	.20
Portfolio C	.45	.20	.25	...	.10
Portfolio D	.55	.15	.20	...	.10
Portfolio E	.65	.10	.15	...	.10
Portfolio F	.75	...	.15	...	.10
Portfolio G	.90	.05	.05	...	...
Portfolio H	.70	.05	.05	.20	...
Portfolio I	.50	.05	.05	.40	...
Portfolio J	.30	.05	.05	.60	...

As noted, the borrowing rate was assumed to be the call money rate +1%; for decision purposes (but not for rate of return calculations), the applicable beginning of period rate, r^d_{Bt}, was viewed as persisting throughout the period and thus as risk free.[10] For 1934–76, the call money rates were obtained from the *Survey of Current Business;* for later periods, the *Wall Street Journal* was the source.

V. Results

Because of space limitations, only a portion of the results can be reported here. However, tables 2–6 below and figures 1 and 2 provide a fairly representative sample of our findings.

For comparison, we have calculated and included the returns for 10 fixed-weight portfolios. The compositions of these fixed-weight portfolios are shown in table 1.

Portfolio Returns: The No-Leverage Case

Table 2 compares the geometric means of the annual returns, along with the standard deviations of $\ln(1 + r_t)$, with and without small stocks for the subperiods 1936–47, 1948–65, and 1966–83 and for the full 1936–83 period, under quarterly portfolio revision (with a 40-quarter estimating period) in the absence of leverage. Note that the realized returns without small stocks differed substantially between the three subperiods. For example, the highest geometric mean was 7.75% for 1936–47, 15.69% for 1948–65, 7.38% for 1966–83, and 10.11% for

10. It may be noted that the minimum differential between the borrowing and the lending rate was 1.4% (1935), while the maximum was 5.0% (1981). The differential was 2% or less through 1952 and again in 1964–65. For most of the 1950s, 1960s, and 1970s, and again in 1983, the differential was between 2% and 3%; the exceptions were 1974 and 1980, when it was 4.5%, and, as already mentioned, 1981.

the whole 1936–83 period. In contrast, the realized returns were much more uniform with small stocks included: the highest geometric means were then 15.79% for 1936–47, 15.27% for 1948–65, 14.97% for 1966–83, and 14.54% for the full period.

For the 1936–47 subperiod, table 2 shows that the presence of the small stock category increased both the standard deviation and the geometric mean for each of the 16 strategies. These increases are relatively small for powers −75 through −10 but quite substantial for the others, especially powers −1 and up. The situation for 1966–83 is similar but less dramatic. In the 1948–65 subperiod, on the other hand, the presence of the small stock category again led to uniformly higher standard deviations but also to uniformly lower geometric means. The differences are quite small, however. Standard deviations also increased across the board for the full period, as did the geometric means.

Portfolio Returns: The Leverage Case

Table 3 shows the geometric means of the annual returns, and the standard deviations of $\ln(1 + r_t)$, with and without small stocks for the leverage case. This table is based on quarterly portfolio revision and a 32-quarter estimating period. Again it is evident that the presence of small stocks tended to increase both the geometric means and the standard deviations of returns. This was uniformly so in the 1966–83 subperiod as well as for the full 1934–83 period. This pattern was broken in only two cases: in the 1934–47 subperiod, when powers 0, .25, and .50 achieved lower standard deviations in the presence of the small stock category, and in the 1948–65 subperiod, when powers −3 through 1 attained higher geometric means in the absence of small stocks.

Table 3 also shows that small stocks had a rather uneven effect across the three subperiods. In the 1934–47 subperiod, they provided an enormous lift to most geometric means with very little effect on standard deviations. On the other hand, their effect in the 1948–65 subperiod was minimal. The effect in the last subperiod was fairly modest, while the effect of small stocks on returns over the full 50 years was rather substantial.

Portfolio Compositions

An examination of the strategies used reveals some interesting patterns. In the no-borrowing case, small stocks were ignored by all the active strategies, including the risk-neutral one, from the third quarter of 1955 through the first quarter of 1962, in the third quarter of 1963, and again in the first quarter of 1979. The rest of the time, however, small stocks tended to be an important investment outlet, mostly at the expense of common stocks. Even the ultraconservative power −75

TABLE 2 **Comparison of Geometric Means and Standard Deviations of Annual Returns with and without Small Stocks, 1936–83: No-Leverage Case (Quarterly Portfolio Revision, 40-Quarter Estimating Period)**

A. 1936–47

	Without Small Stocks		With Small Stocks	
Portfolio	Geometric Mean	Standard Deviation*	Geometric Mean	Standard Deviation*
Common stocks	6.71	22.28	6.71	22.28
Government bonds	3.46	3.62	3.46	3.62
Corporate bonds	3.25	2.26	3.25	2.26
Small stocks	...	...	14.65	41.68
Risk free	.24	.18	.24	.18
Power 1	4.78	23.74	14.65	41.68
Power .75	5.19	20.98	12.64	41.72
Power .50	6.95	18.21	14.29	35.77
Power .25	7.75	14.92	15.79	30.19
Power 0	7.72	12.92	15.48	26.11
Power −1	6.18	8.31	9.88	14.29
Power −2	5.27	6.04	7.67	9.87
Power −3	4.79	4.96	6.54	7.71
Power −5	4.31	3.96	5.43	5.63
Power −7	4.03	3.39	4.94	4.66
Power −10	3.73	2.99	4.48	3.81
Power −15	3.53	2.68	4.03	3.24
Power −20	3.11	2.49	3.50	2.92
Power −30	2.52	2.13	2.78	2.51
Power −50	2.09	1.94	2.28	2.27
Power −75	1.74	1.65	2.07	2.14
Portfolio A	2.74	2.19	2.74	2.19
Portfolio B	3.78	5.45	3.78	5.45
Portfolio C	5.13	10.58	5.13	10.58
Portfolio D	5.47	12.63	5.47	12.63
Portfolio E	5.75	14.71	5.75	14.71
Portfolio F	5.97	16.76	5.97	16.76
Portfolio G	6.59	20.13	6.59	20.13
Portfolio H	...	...	8.75	23.70
Portfolio I	...	...	10.68	27.45
Portfolio J	...	...	12.36	31.32

TABLE 2 *(Continued)*

B. 1948–65

Portfolio	Without Small Stocks: Geometric Mean	Without Small Stocks: Standard Deviation*	With Small Stocks: Geometric Mean	With Small Stocks: Standard Deviation*
Common stocks	15.67	14.54	15.67	14.54
Government bonds	1.94	4.96	1.94	4.96
Corporate bonds	2.52	4.11	2.52	4.11
Small stocks	...	...	15.61	19.36
Risk free	2.29	.94	2.29	.94
Power 1	15.67	14.54	14.03	17.12
Power .75	15.67	14.54	14.64	17.14
Power .50	15.67	14.54	14.70	17.16
Power .25	15.67	14.54	14.83	17.04
Power 0	15.67	14.54	14.90	16.98
Power −1	15.69	14.54	15.27	16.21
Power −2	15.52	14.48	15.25	15.54
Power −3	15.18	14.42	14.93	14.72
Power −5	13.65	14.45	13.24	14.60
Power −7	11.85	13.58	11.59	13.79
Power −10	9.48	10.91	9.35	11.09
Power −15	7.27	7.60	7.15	7.81
Power −20	6.12	5.68	6.06	5.86
Power −30	4.98	3.73	4.95	3.87
Power −50	3.92	2.19	3.91	2.31
Power −75	3.39	1.46	3.38	1.56
Portfolio A	2.26	3.56	2.26	3.56
Portfolio B	5.00	3.02	5.00	3.02
Portfolio C	8.41	5.99	8.41	5.99
Portfolio D	9.75	7.46	9.75	7.46
Portfolio E	11.09	9.01	11.09	9.01
Portfolio F	12.44	10.63	12.44	10.63
Portfolio G	14.37	12.90	14.37	12.90
Portfolio H	...	...	14.44	13.55
Portfolio I	...	...	14.47	14.41
Portfolio J	...	...	14.46	15.46

TABLE 2 *(Continued)*

C. 1966–83

Portfolio	Without Small Stocks		With Small Stocks	
	Geometric Mean	Standard Deviation*	Geometric Mean	Standard Deviation*
Common stocks	7.62	17.03	7.62	17.03
Government bonds	4.13	9.91	4.13	9.91
Corporate bonds	4.90	10.82	4.90	10.82
Small stocks	. . .	. . .	16.11	27.78
Risk free	7.37	2.86	7.37	2.86
Power 1	7.38	12.03	14.97	27.29
Power .75	6.65	11.33	13.06	25.72
Power .50	6.40	11.09	12.85	24.50
Power .25	6.34	10.85	11.86	24.30
Power 0	6.35	10.39	11.48	23.98
Power −1	6.84	9.07	10.24	22.37
Power −2	7.08	7.79	10.01	19.65
Power −3	7.18	6.22	9.89	15.98
Power −5	7.28	4.58	9.23	10.88
Power −7	7.31	3.86	8.84	8.35
Power −10	7.33	3.37	8.49	6.33
Power −15	7.35	3.05	8.17	4.74
Power −20	7.36	2.94	7.99	4.00
Power −30	7.36	2.86	7.80	3.36
Power −50	7.37	2.84	7.63	2.99
Power −75	7.37	2.83	7.55	2.88
Portfolio A	5.20	8.27	5.20	8.27
Portfolio B	6.03	7.96	6.03	7.96
Portfolio C	6.61	10.08	6.61	10.08
Portfolio D	6.92	10.99	6.92	10.99
Portfolio E	7.20	12.06	7.20	12.06
Portfolio F	7.48	13.38	7.48	13.38
Portfolio G	7.44	15.62	7.44	15.62
Portfolio H	. . .	. . .	9.41	16.86
Portfolio I	. . .	. . .	11.25	18.73
Portfolio J	. . .	. . .	12.95	21.06

TABLE 2 ***(Continued)***

D. 1936–83

	Without Small Stocks		With Small Stocks	
Portfolio	Geometric Mean	Standard Deviation*	Geometric Mean	Standard Deviation*
Common stocks	10.34	17.64	10.34	17.64
Government bonds	3.14	6.96	3.14	6.96
Corporate bonds	3.59	7.12	3.59	7.12
Small stocks	...	...	15.56	28.66
Risk free	3.64	3.40	3.64	3.40
Power 1	9.74	16.69	14.54	27.97
Power .75	9.57	15.63	13.54	27.45
Power .50	9.93	14.65	13.90	24.97
Power .25	10.11	13.65	13.94	23.11
Power 0	10.11	13.03	13.75	21.78
Power −1	9.91	11.77	12.01	18.14
Power −2	9.70	11.10	11.35	16.06
Power −3	9.50	10.56	10.89	13.93
Power −5	8.86	9.97	9.74	11.63
Power −7	8.15	9.12	8.87	10.24
Power −10	7.21	7.33	7.79	8.11
Power −15	6.35	5.33	6.74	5.92
Power −20	5.82	4.33	6.13	4.79
Power −30	5.24	3.52	5.46	3.82
Power −50	4.73	3.13	4.88	3.30
Power −75	4.44	3.06	4.59	3.16
Portfolio A	3.48	5.67	3.48	5.67
Portfolio B	5.08	5.82	5.08	5.82
Portfolio C	6.91	8.80	6.91	8.80
Portfolio D	7.60	10.19	7.60	10.19
Portfolio E	8.27	11.70	8.27	11.70
Portfolio F	8.93	13.34	8.93	13.34
Portfolio G	9.77	15.93	9.77	15.93
Portfolio H	...	...	11.10	17.50
Portfolio I	...	...	12.30	19.51
Portfolio J	...	...	13.37	21.84

* Standard deviation is for the variables $\ln(1 + r_t)$.

TABLE 3 **Comparison of Geometric Means and Standard Deviations of Annual Returns with and without Small Stocks, 1934–83: Leverage Case (Quarterly Portfolio Revision, 32-Quarter Estimating Period)**

A. 1934–47

	Without Small Stocks		With Small Stocks	
Portfolio	Geometric Mean	Standard Deviation*	Geometric Mean	Standard Deviation*
Common stocks	8.59	22.42	8.59	22.42
Government bonds	4.03	3.72	4.03	3.72
Corporate bonds	4.42	3.61	4.42	3.61
Small stocks	...	...	16.98	38.76
Risk free	.24	.17	.24	.17
Power 1	−5.86	56.91	11.57	88.59
Power .75	5.30	40.76	19.95	48.99
Power .50	9.09	38.63	25.95	35.10
Power .25	9.19	34.75	23.43	31.01
Power 0	10.67	31.30	22.58	28.25
Power −1	10.19	22.95	17.47	24.83
Power −2	9.32	17.86	13.94	20.89
Power −3	8.83	14.91	12.33	17.41
Power −5	7.17	10.36	9.44	11.87
Power −7	6.24	7.67	7.85	8.67
Power −10	5.53	5.59	6.58	6.36
Power −15	4.85	4.43	5.69	5.06
Power −20	3.92	3.60	4.55	4.18
Power −30	2.99	2.64	3.39	3.02
Power −50	2.42	2.09	2.63	2.31
Power −75	1.86	1.64	2.29	1.98
Portfolio A	3.44	2.72	3.44	2.72
Portfolio B	4.68	5.60	4.68	5.60
Portfolio C	6.40	10.74	6.40	10.74
Portfolio D	6.83	12.80	6.83	12.80
Portfolio E	7.22	14.88	7.22	14.88
Portfolio F	7.57	16.96	7.57	16.96
Portfolio G	8.38	20.28	8.38	20.28
Portfolio H	...	...	10.63	23.01
Portfolio I	...	...	12.64	26.09
Portfolio J	...	...	14.40	29.40

TABLE 3 ***(Continued)***

B. 1948–65

	Without Small Stocks		With Small Stocks	
Portfolio	Geometric Mean	Standard Deviation*	Geometric Mean	Standard Deviation*
Common stocks	15.67	14.54	15.67	14.54
Government bonds	1.94	4.96	1.94	4.96
Corporate bonds	2.52	4.11	2.52	4.11
Small stocks	...	...	15.61	19.36
Risk free	2.29	.94	2.29	.94
Power 1	24.70	24.79	23.40	28.61
Power .75	24.70	24.79	22.92	28.01
Power .50	24.70	24.79	23.31	27.60
Power .25	24.70	24.79	23.54	27.49
Power 0	24.70	24.79	23.75	27.41
Power −1	22.79	25.18	21.93	27.25
Power −2	20.17	23.67	19.49	24.30
Power −3	17.97	22.31	17.79	22.82
Power −5	15.04	19.35	15.08	19.73
Power −7	12.38	16.07	12.49	16.28
Power −10	9.96	12.82	10.09	13.00
Power −15	8.00	9.45	8.17	9.62
Power −20	6.73	7.16	6.81	7.37
Power −30	5.36	4.71	5.39	4.88
Power −50	4.17	2.73	4.19	2.87
Power −75	3.56	1.77	3.57	1.90
Portfolio A	2.26	3.56	2.26	3.56
Portfolio B	5.00	3.02	5.00	3.02
Portfolio C	8.41	5.99	8.41	5.99
Portfolio D	9.75	7.46	9.75	7.46
Portfolio E	11.09	9.01	11.09	9.01
Portfolio F	12.44	10.63	12.44	10.63
Portfolio G	14.37	12.90	14.37	12.90
Portfolio H	...	...	14.44	13.55
Portfolio I	...	...	14.47	14.41
Portfolio J	...	...	14.46	15.46

TABLE 3 (*Continued*)

C. 1966–83

Portfolio	Without Small Stocks		With Small Stocks	
	Geometric Mean	Standard Deviation*	Geometric Mean	Standard Deviation*
Common stocks	7.62	17.03	7.62	17.03
Government bonds	4.13	9.91	4.13	9.91
Corporate bonds	4.90	10.82	4.90	10.82
Small stocks	...	...	16.11	27.78
Risk free	7.37	2.86	7.37	2.86
Power 1	9.22	15.82	13.26	33.56
Power .75	10.16	14.98	11.80	34.55
Power .50	9.91	13.58	10.11	32.42
Power .25	9.74	11.93	11.54	31.12
Power 0	9.25	11.44	12.17	30.16
Power −1	9.32	10.42	10.87	25.07
Power −2	8.96	8.68	10.51	19.69
Power −3	8.41	7.24	10.33	16.11
Power −5	8.05	5.46	9.54	11.13
Power −7	7.89	4.58	9.08	8.61
Power −10	7.75	3.92	8.66	6.58
Power −15	7.64	3.45	8.29	4.97
Power −20	7.58	3.25	8.08	4.21
Power −30	7.51	3.07	7.86	3.53
Power −50	7.46	2.96	7.67	3.11
Power −75	7.43	2.92	7.58	2.97
Portfolio A	5.20	8.27	5.20	8.27
Portfolio B	6.03	7.96	6.03	7.96
Portfolio C	6.61	10.08	6.61	10.08
Portfolio D	6.92	10.99	6.92	10.99
Portfolio E	7.20	12.06	7.20	12.06
Portfolio F	7.48	13.38	7.48	13.38
Portfolio G	7.44	15.62	7.44	15.62
Portfolio H	...	...	9.41	16.86
Portfolio I	...	...	11.25	18.73
Portfolio J	...	...	12.95	21.06

TABLE 3 ***(Continued)***

D. 1934–83

Portfolio	Without Small Stocks		With Small Stocks	
	Geometric Mean	Standard Deviation*	Geometric Mean	Standard Deviation*
Common stocks	10.73	17.84	10.73	17.84
Government bonds	3.31	6.88	3.31	6.88
Corporate bonds	3.90	7.14	3.90	7.14
Small stocks	...	...	16.17	28.22
Risk free	3.51	3.40	3.51	3.40
Power 1	9.89	35.87	16.32	52.70
Power .75	13.74	28.00	17.98	36.62
Power .50	14.78	26.70	19.09	31.49
Power .25	14.75	24.96	19.05	29.63
Power 0	14.99	23.59	19.13	28.44
Power −1	14.25	20.67	16.61	25.62
Power −2	12.97	18.09	14.64	21.60
Power −3	11.88	16.31	13.53	18.95
Power −5	10.26	13.39	11.47	14.88
Power −7	9.01	10.87	9.95	11.87
Power −10	7.91	8.57	8.58	9.28
Power −15	6.98	6.48	7.51	6.97
Power −20	6.24	5.19	6.63	5.61
Power −30	5.45	3.97	5.71	4.23
Power −50	4.84	3.27	4.99	3.41
Power −75	4.45	3.12	4.63	3.16
Portfolio A	3.64	5.62	3.64	5.62
Portfolio B	5.28	5.81	5.28	5.81
Portfolio C	7.19	8.89	7.19	8.89
Portfolio D	7.91	10.31	7.91	10.31
Portfolio E	8.59	11.85	8.59	11.85
Portfolio F	9.27	13.50	9.27	13.50
Portfolio G	10.16	16.12	10.16	16.12
Portfolio H	...	...	11.54	17.51
Portfolio I	...	...	12.79	19.39
Portfolio J	...	...	13.90	21.60

* Standard deviation is for the variables $\ln(1 + r_t)$.

strategy, for example, had an uninterrupted presence in small stocks from the beginning of 1942 through the first quarter of 1952, from the fourth quarter of 1963 through the second quarter of 1974, and again from the third quarter of 1980 on.

Table 4 gives a comparison of the quarter-by-quarter portfolio compositions and returns for powers 0 and −15 with and without small stocks in the leverage case. For each power, the first five columns show the proportions invested in common stocks, government bonds, corporate bonds, small stocks, and the risk-free asset. The sixth column reports the fraction in borrowing and the last column the portfolio's return for the quarter. This is then repeated with small stocks excluded.

Turning first to the logarithmic investor, we note that he stayed away from small stocks in the first quarter of 1934, from the beginning of 1953 through mid-1960, during the second quarter of 1961, and from the beginning of 1974 through mid-1977 with the exception of three quarters. During these periods, portfolio holdings and returns were thus unaffected by the opportunity to invest in small stocks. In the 1930s, small stocks tended to replace holdings in governments and, to a lesser extent, common stocks. In the 1940s and early 1950s, and again in the 1960s and early 1970s, it was principally common stocks that were being crowded out by small stocks for the logarithmic investor. In the late 1970s through the third quarter of 1982, it was primarily the risk-free asset that gave way for small stocks, followed by common stocks one more time.

The logarithmic investor's use of leverage was remarkably similar with and without the small stock category. Its presence did not always increase leverage; while they markedly did so in 1982 and 1983, small stocks sharply reduced the use of leverage in 1946 and 1947.

Turning now to the rather conservative power −15 strategy, we first observe that this investor was a heavy and repeated user of the risk-free asset beginning with the third quarter of 1962. Prior to the second quarter of 1951, however, the risk-free asset was practically ignored. In fact, from the beginning of 1940 through the third quarter of 1946, the power −15 strategy was a consistent user of leverage, with holdings of government bonds on margin.

Small stocks first entered the power −15 portfolio in the last quarter of 1939, growing gradually in importance to a 19% allocation in the first quarter of 1950, then declining and disappearing as of mid-1951, 47 quarters later. As table 4 shows, these holdings were primarily at the expense of common stocks. After a nearly total hiatus during the rest of the 1950s, small stocks reappeared for 55 consecutive quarters beginning with the second quarter of 1960, reaching a maximum allocation of 30% in the second quarter of 1962, in a streak that lasted through 1973. These positions were also essentially at the expense of common stocks, except in the second half of 1973, when they replaced the risk-free asset. When the power −15 investor returned to small stocks at the beginning of 1978, they first replaced principally investments in the risk-free asset and, beginning in the last quarter of 1982, holdings in common stocks.

Many investigators have noted that small stocks have, over long periods, produced what is usually referred to as excess risk-adjusted returns (e.g., Banz 1981; and Reinganum 1981). This suggests that the empirical (past) distribution used in our model should, at a minimum, have found small stocks a consistently attractive outlet. While this was obviously the case most of the time, small stocks were, as noted, totally ignored in the middle and late 1950s and for intervals in the middle 1970s even by the more risk-tolerant strategies.

Other Results

We also examined the case in which portfolios were revised only once a year rather than quarterly, with 8- and 10-year estimating periods. As in the previous studies, the differences in returns and portfolio compositions were fairly small. The use of 24-quarter and 40-quarter estimating periods also led to only minor differences in the results.

VI. Tests

The returns under quarterly reinvestment with leverage in the presence of small stocks over the 1934–47 subperiod are shown as round dots in figure 1, while the returns over the full 1934–83 period are similarly depicted in figure 2. The same figures also plot the returns for the higher powers in the absence of small stocks (see diamonds) as well as for the fixed-weight portfolios in table 1 (see triangles). The geometric means of the annual returns are measured on the vertical axis and the standard deviations of $\ln(1 + r_t)$ on the horizontal.

The attained returns reflect clearly the benefits from diversification among the five (or four) asset categories (represented in the figures by the square points RL [risk-free asset], GB [government bonds], CB [corporate bonds], CS [common stocks], and SS [small stocks]). In other words, the portfolios selected by the model have enabled investors to travel in a distinctly "northwesterly" direction, confirming the results of the previous studies.

Recall that terminal wealth w_0 in terms of beginning wealth w_n is given by

$$w_0 = w_n(1 + r_n)(1 + r_{n-1}) \ldots (1 + r_1)$$
$$= w_n \exp\left[\sum_{t=1}^{n} \ln(1 + r_t)\right].$$

TABLE 4 **Portfolio Composition and Realized Returns for Powers 0 and − 15 with and without Small Stocks, 1934–83 (Quarterly Revision, with Leverage, 32-Quarter Estimating Period)**

	Power 0												
	With Small Stocks							Without Small Stocks					
	Inv. Fractions							Inv. Fractions					
Period	CS	GB	CB	SS	RL	B	r_t	CS	GB	CB	RL	B	r_t
1934:1	.29	4.74	1.34	...	...	−5.37	33.21	.29	4.74	1.34	...	−5.37	33.21
1934:2	.32	5.23	1.16	.04	...	−5.75	15.46	.38	5.29	1.13	...	−5.80	15.52
1934:3	.04	5.78	1.10	.14	...	−6.07	−17.50	.25	5.98	1.01	...	−6.23	−17.65
1934:4	...	4.39	1.53	.13	...	−5.05	18.89	.14	4.61	1.46	...	−5.21	18.56
1935:1	...	4.26	1.50	.11	...	−4.87	14.94	.03	4.48	1.54	...	−5.05	17.56
1935:2	...	4.82	1.42	.05	...	−5.29	7.95	...	4.92	1.45	...	−5.37	7.62
1935:3	...	6.28	1.03	.03	...	−6.34	−5.75	...	6.34	1.05	...	−6.39	−6.27
1935:4	...	5.06	1.31	.08	...	−5.44	10.30	...	5.23	1.36	...	−5.59	7.91
1936:1	...	2.73	1.85	.18	...	−3.76	15.90	.10	3.02	1.87	...	−3.98	10.90
1936:2	...	3.34	1.59	.24	...	−4.18	−.71	.11	3.79	1.64	...	−4.53	4.03
1936:3	...	3.09	1.75	.17	...	−4.02	7.84	.08	3.42	1.77	...	−4.28	5.98
1936:4	...	4.90	1.25	.16	...	−5.31	15.04	.04	5.29	1.30	...	−5.62	12.61
1937:1	...	5.27	1.09	.17	...	−5.52	−18.69	...	5.69	1.23	...	−5.92	−24.08
1937:2	...	5.84	.80	.25	...	−5.88	−4.75	...	6.32	1.05	...	−6.37	3.19
1937:3	...	4.98	1.15	.18	...	−5.31	−2.24	...	5.33	1.33	...	−5.66	1.95
1937:4	...	5.60	1.03	.14	...	−5.77	6.40	...	5.88	1.18	...	−6.06	12.40
1938:1	...	4.03	1.50	.18	...	−4.71	−5.63	...	4.54	1.56	...	−5.10	.11
1938:2	...	4.47	1.53	.04	...	−5.05	15.33	...	4.59	1.54	...	−5.14	13.11
1938:3	...	4.35	1.35	.24	...	−4.93	2.90	...	5.07	1.41	...	−5.48	2.75
1938:4	...	4.13	1.35	.28	...	−4.77	12.11	...	4.98	1.43	...	−5.42	7.97
1939:1	...	4.16	1.22	.39	...	−4.77	−1.89	.15	5.15	1.22	...	−5.52	9.92
1939:2	...	6.23	.86	.19	...	−6.28	14.67	...	6.80	.92	...	−6.71	15.88
1939:3	...	6.32	.78	.24	...	−6.34	−30.65	.04	7.02	.80	...	−6.86	−48.79
1939:4	...	5.59	.47	.69	...	−5.75	32.70	.79	5.67	.33	...	−5.80	37.90
1940:1	...	7.28	...	.68	...	−6.96	20.44	.67	7.32	...	...	−6.99	9.62
1940:2	...	7.04	...	.74	...	−6.78	−28.20	.75	7.01	...	...	−6.76	−21.87
1940:3	...	6.84	...	.79	...	−6.63	15.40	1.03	5.89	...	...	−5.92	16.87
1940:4	...	7.21	.03	.67	...	−6.91	20.70	.75	6.25	.21	...	−6.22	16.92
1941:1	...	6.84	...	.79	...	−6.63	−8.89	.88	6.48	...	...	−6.36	−12.80
1941:2	...	6.74	...	.82	...	−6.55	12.38	1.00	6.00	...	...	−6.00	11.50
1941:3	...	7.88	...	.53	...	−7.41	6.58	.48	6.71	.39	...	−6.58	1.65
1941:4	...	7.45	...	.64	...	−7.09	−22.65	.75	7.01	...	...	−6.76	−17.81
1942:1	...	7.98	...	.50	...	−7.49	14.96	.45	8.21	...	...	−7.66	7.46
1942:2	...	8.11	...	.47	...	−7.58	−.10	.41	8.13	.06	...	−7.61	2.65
1942:3	...	7.86	...	.54	...	−7.39	12.18	.60	7.60	...	...	−7.20	5.74
1942:4	...	7.56	...	.61	...	−7.17	5.12	.61	7.56	...	...	−7.17	6.86
1943:1	...	7.43	...	.64	...	−7.07	41.32	.75	6.99	...	...	−6.75	14.04
1943:2	...	6.28	...	.93	...	−6.21	23.69	1.12	5.53	...	...	−5.65	12.80
1943:3	...	6.16	...	.96	...	−6.12	−7.91	1.00	5.98	...	...	−5.99	−2.25
1943:4	...	6.61	...	.85	...	−6.46	−.82	.81	6.77	...	...	−6.58	−3.29
1944:1	...	6.93	...	.77	...	−6.69	15.42	.69	7.25	...	...	−6.93	4.74
1944:2	...	7.04	...	.74	...	−6.78	11.72	.68	7.27	...	...	−6.95	6.68
1944:3	...	6.45	...	.89	...	−6.34	1.72	.78	6.88	...	...	−6.66	1.60
1944:4	...	6.66	...	.84	...	−6.49	12.71	.72	7.12	...	...	−6.84	5.99
1945:1	...	6.68	...	.83	...	−6.51	16.37	.67	6.39	.27	...	−6.32	14.14
1945:2	...	7.08	...	.58	...	−6.66	40.15	.39	8.07	...	...	−7.45	32.01
1945:3	...	5.79	...	.56	...	−5.36	.54	.27	7.99	...	...	−7.26	−1.76
1945:4	...	4.76	...	.70	...	−4.46	33.27	.38	7.16	...	...	−6.54	30.68
1946:1	...	2.83	...	.96	...	−2.79	11.22	.47	6.44	...	...	−5.91	3.77
1946:2	...	1.31	...	.87	...	−1.18	5.38	.37	6.26	...	...	−5.64	−6.59
1946:3	...	1.49	...	.85	...	−1.34	−26.30	.39	6.14	...	...	−5.53	−19.88
1946:4	...	.75	...	.93	...	−.67	1.81	.50	5.00	...	...	−4.50	7.38

TABLE 4 ***(Continued)***

Power −15												
With Small Stocks							Without Small Stocks					
Inv. Fractions							Inv. Fractions					
CS	GB	CB	SS	RL	B	r_t	CS	GB	CB	RL	B	r_t
...	.15	.83	...	.02	...	5.83	...	.15	.83	.02	...	5.83
...	.15	.85	...	...	...	3.52	...	.15	.85	...	...	3.52
...	.14	.86	...	...	...	−.04	...	.14	.86	...	...	−.04
...	.08	.92	...	...	...	3.36	...	.08	.92	...	...	3.36
...	.05	.95	...	...	...	3.95	...	.05	.95	...	...	3.95
...	...	1.00	...	...	...	2.68	...	...	1.00	...	...	2.68
...	...	1.00	...	...	...	.69	...	...	1.00	...	...	.69
...	...	1.00	...	...	...	1.95	...	...	1.00	...	...	1.95
...	...	1.00	...	...	...	2.20	...	...	1.00	...	...	2.20
...	...	1.00	...	...	...	1.49	...	...	1.00	...	...	1.49
...	...	1.00	...	...	...	1.46	...	...	1.00	...	...	1.46
...	...	1.00	...	...	...	1.44	...	...	1.00	...	...	1.44
...	...	1.00	...	...	...	−1.36	...	...	1.00	...	...	−1.36
...	.05	.95	...	...	...	1.58	...	.05	.95	...	...	1.58
...	...	1.00	...	...	...	.47	...	...	1.00	...	...	.47
...	...	1.00	...	...	...	2.02	...	...	1.00	...	...	2.02
...	...	1.00	...	...	...	−.39	...	...	1.00	...	...	−.39
...	...	1.00	...	...	...	2.45	...	...	1.00	...	...	2.45
...	.01	.99	...	...	...	1.55	...	.01	.99	...	...	1.55
...	.03	.97	...	...	...	2.38	...	.03	.97	...	...	2.38
...	...	1.00	...	...	...	1.08	...	...	1.00	...	...	1.08
...	.11	.89	...	...	...	1.61	...	.11	.89	...	...	1.61
...	.15	.85	...	...	...	−3.11	...	.15	.85	...	...	−3.11
...	.14	.81	.02	.03	...	4.07	...	.03	.95	.02	...	3.98
...	...	2.64	.04	...	−1.68	2.92	...	...	2.42	...	−1.42	2.17
...	...	2.70	.05	...	−1.75	−1.84	...	...	2.46	...	−1.46	−.60
...	...	2.79	.06	...	−1.85	2.83	.06	...	2.69	...	−1.74	2.83
...	...	2.79	.06	...	−1.85	1.74	.06	...	2.72	...	−1.78	1.59
...	...	2.79	.06	...	−1.85	−1.17	.07	...	2.67	...	−1.74	−1.44
...	...	2.78	.06	...	−1.85	4.44	.08	...	2.63	...	−1.71	4.26
...	...	2.79	.06	...	−1.85	4.10	.08	...	2.68	...	−1.76	3.44
...	...	2.78	.07	...	−1.85	−3.92	.09	...	2.67	...	−1.75	−3.46
...	...	2.72	.06	...	−1.77	1.31	.05	...	2.43	...	−1.47	.45
...	...	2.69	.06	...	−1.74	.71	.03	...	2.35	...	−1.39	.92
...	...	2.58	.06	...	−1.64	2.28	.04	...	2.27	...	−1.31	1.39
...	...	2.70	.07	...	−1.77	1.40	.06	...	2.34	...	−1.40	1.49
...	...	2.61	.07	...	−1.68	5.53	.07	...	2.25	...	−1.33	2.43
...	...	2.54	.08	...	−1.62	4.52	.09	...	2.16	...	−1.25	3.28
...	...	2.51	.08	...	−1.58	−.27	.09	...	2.13	...	−1.22	.21
...	...	2.43	.07	...	−1.50	−.26	.07	...	2.09	...	−1.16	−.36
...	...	2.26	.07	...	−1.32	2.80	.05	...	1.92	...	−.98	1.69
...	...	2.21	.06	...	−1.27	1.67	.05	...	1.85	...	−.90	1.11
...	...	2.13	.07	...	−1.21	1.25	.06	...	1.78	...	−.83	1.10
...	...	2.07	.07	...	−1.14	4.80	.05	...	1.71	...	−.76	3.59
...	...	2.11	.07	...	−1.17	2.75	.04	...	1.75	...	−.80	2.24
...	...	2.67	.06	...	−1.73	1.88	.05	...	2.29	...	−1.33	.73
...	.21	2.24	.08	...	−1.53	.31	.07	...	2.20	...	−1.27	.56
...	.24	2.22	.10	...	−1.56	6.73	.10	...	2.20	...	−1.30	4.56
...	.75	1.25	.13	...	−1.13	3.82	.14	.26	1.68	...	−1.09	3.69
...	.33	2.31	.16	...	−1.80	...	.21	.05	2.25	...	−1.51	−.29
...	.21	2.35	.16	...	−1.71	−8.51	.20	...	2.28	...	−1.48	−7.35
...	.59	.29	.12	...	...	1.41	.11	.20	.78	...	−.09	1.53

TABLE 4 *(Continued)*

	Power 0												
	With Small Stocks							Without Small Stocks					
	Inv. Fractions							Inv. Fractions					
Period	CS	GB	CB	SS	RL	RB	RP	CS	GB	CB	RL	RB	RP
1947:1	...	1.47	...	.85	...	−1.32	−.03	.47	5.30	...	...	−4.77	−.89
1947:2	...	...	...	1.33	...	−.33	−14.08	.96	2.78	...	...	−2.74	.30
1947:3	...	...	...	1.33	...	−.33	11.44	1.17	1.19	...	...	−1.37	1.04
1947:4	...	.97	...	1.20	...	−1.18	−.60	...	10.00	...	...	−9.00	−45.20
1948:1	...	...	...	1.33	...	−.33	−0.61	1.33	...	...	...	−.33	−.46
1948:2	...	...	...	1.33	...	−.33	20.06	1.33	...	...	...	−.33	16.55
1948:3	...	...	...	1.33	...	−.33	−14.46	1.33	...	...	...	−.33	−8.54
1948:4	...	...	...	1.33	...	−.33	−6.34	1.33	...	...	...	−.33	−.01
1949:1	...	...	...	1.33	...	−.33	3.81	1.33	...	...	...	−.33	.59
1949:2	...	...	...	2.00	...	−1.00	−20.04	2.00	...	...	...	−1.00	−9.04
1949:3	...	...	...	2.00	...	−1.00	28.94	2.00	...	...	...	−1.00	22.72
1949:4	...	...	...	2.00	...	−1.00	23.58	2.00	...	...	...	−1.00	19.98
1950:1	...	...	...	2.00	...	−1.00	13.00	2.00	...	...	...	−1.00	8.80
1950:2	...	...	...	2.00	...	−1.00	−3.72	2.00	...	...	...	−1.00	7.66
1950:3	...	...	...	2.00	...	−1.00	34.02	2.00	...	...	...	−1.00	23.20
1950:4	...	...	...	2.00	...	−1.00	24.14	2.00	...	...	...	−1.00	15.14
1951:1	...	...	...	2.00	...	−1.00	6.77	2.00	...	...	...	−1.00	11.95
1951:2	...	...	...	1.33	...	−.33	−7.01	1.33	...	...	...	−.33	−.77
1951:3	...	...	...	1.33	...	−.33	16.22	1.33	...	...	...	−.33	16.24
1951:4	...	...	...	1.33	...	−.33	−3.75	1.33	...	...	...	−.33	5.27
1952:1	...	...	...	1.33	...	−.33	.50	1.33	...	...	...	−.33	4.94
1952:2	.19	...	...	1.14	...	−.33	−2.14	1.33	...	...	...	−.33	5.24
1952:3	1.25	...	...	.09	...	−.33	−1.03	1.33	...	...	...	−.33	−1.03
1952:4	1.28	...	...	.05	...	−.33	12.76	1.33	...	...	...	−.33	12.99
1953:1	1.33	...	...	...	...	−.33	−5.14	1.33	...	...	...	−.33	−5.14
1953:2	2.00	...	...	...	...	−1.00	−6.91	2.00	...	...	...	−1.00	−6.91
1953:3	2.00	...	...	...	...	−1.00	−5.24	2.00	...	...	...	−1.00	−5.24
1953:4	2.00	...	...	...	...	−1.00	15.18	2.00	...	...	...	−1.00	15.18
1954:1	2.00	...	...	...	...	−1.00	18.93	2.00	...	...	...	−1.00	18.93
1954:2	2.00	...	...	...	...	−1.00	18.80	2.00	...	...	...	−1.00	18.80
1954:3	2.00	...	...	...	...	−1.00	22.48	2.00	...	...	...	−1.00	22.48
1954:4	2.00	...	...	...	...	−1.00	25.00	2.00	...	...	...	−1.00	25.00
1955:1	1.67	...	...	...	...	−.67	3.77	1.67	...	...	...	−.67	3.77
1955:2	1.67	...	...	...	...	−.67	21.53	1.67	...	...	...	−.67	21.53
1955:3	1.43	...	...	...	...	−.43	10.00	1.43	...	...	...	−.43	10.00
1955:4	1.43	...	...	...	...	−.43	7.15	1.43	...	...	...	−.43	7.15
1956:1	1.43	...	...	...	...	−.43	10.43	1.43	...	...	...	−.43	10.43
1956:2	1.43	...	...	...	...	−.43	−3.56	1.43	...	...	...	−.43	−3.56
1956:3	1.43	...	...	...	...	−.43	−4.32	1.43	...	...	...	−.43	−4.32
1956:4	1.43	...	...	...	...	−.43	4.94	1.43	...	...	...	−.43	4.94
1957:1	1.43	...	...	...	...	−.43	−7.06	1.43	...	...	...	−.43	−7.06
1957:2	1.43	...	...	...	...	−.43	11.49	1.43	...	...	...	−.43	11.49
1957:3	1.43	...	...	...	...	−.43	−14.31	1.43	...	...	...	−.43	−14.31
1957:4	1.43	...	...	...	...	−.43	−7.31	1.43	...	...	...	−.43	−7.31
1958:1	1.43	...	...	...	...	−.43	8.54	1.43	...	...	...	−.43	8.54
1958:2	2.00	...	...	...	...	−1.00	15.87	2.00	...	...	...	−1.00	15.87
1958:3	2.00	...	...	...	...	−1.00	22.18	2.00	...	...	...	−1.00	22.18
1958:4	1.43	...	...	...	...	−.43	15.59	1.43	...	...	...	−.43	15.59
1959:1	1.11	...	...	...	...	−.11	1.22	1.11	...	...	...	−.11	1.22
1959:2	1.11	...	...	...	...	−.11	6.84	1.11	...	...	...	−.11	6.84
1959:3	1.11	...	...	...	...	−.11	−2.34	1.11	...	...	...	−.11	−2.34
1959:4	1.11	...	...	...	...	−.11	6.71	1.11	...	...	...	−.11	6.71
1960:1	1.11	...	...	...	...	−.11	−7.72	1.11	...	...	...	−.11	−7.72
1960:2	1.11	...	...	...	...	−.11	3.98	1.11	...	...	...	−.11	3.98

TABLE 4 ***(Continued)***

Power − 15												
With Small Stocks							Without Small Stocks					
Inv. Fractions							Inv. Fractions					
CS	GB	CB	SS	RL	B	r_t	CS	GB	CB	RL	B	r_t
...	.74	.15	.12	...	...	.40	.12	.38	.51	...	...	.55
...	.34	.52	.14	...	...	−1.24	.15	.15	.70	...	...	.60
...	.08	.78	.13	...	...	−.17	.16	...	.84	...	...	−1.44
...	.89	...	.11	...	...	−3.21	.07	1.08	...	...	−.15	−4.16
...	...	.87	.13	...	...	1.51	.13	.30	.58	...	...	1.30
...	...	.87	.13	...	...	1.58	.13	.11	.76	...	...	1.48
...	.12	.69	.18	...	...	−1.79	.25	.40	.35	...	...	−1.49
...	...	.83	.17	...	...	1.25	.21	.29	.50	...	...	1.66
...	...	.84	.16	...	...	1.18	.21	...	.79	...	...	.78
...	...	.84	.16	...	...	−.31	.22	.19	.59	...	...	.30
...	...	.85	.15	...	...	3.53	.20	.24	.55	...	...	3.59
...	.08	.78	.15	...	...	1.39	.21	.37	.42	...	...	2.29
.03	.17	.61	.19	...	...	1.78	.31	.52	.17	...	...	1.43
.20	...	.70	.11	...	...	.70	.37	.04	.59	...	...	1.58
.19	...	.70	.11	...	...	4.61	.36	.10	.53	...	...	4.70
.25	...	.68	.07	...	...	3.36	.37	.05	.58	...	...	3.31
.18	...	.71	.11	...	...	−.27	.37	...	.63	...	...	.67
.38	.01	.03	.01	.57	...	−.05	.40	.03	.02	.56	...	−.01
.38	...	...	...	.62	...	4.93	.38	...	...	.62	...	4.93
.40	...	...	...	.60	...	1.91	.40	...	...	.60	...	1.91
.42	...	...	...	.58	...	1.88	.42	...	...	.58	...	1.88
.42	...	...	...	.58	...	1.97	.42	...	...	.58	...	1.97
.41	...	...	...	.59	...	.05	.41	...	...	.59	...	.05
.41	...	...	...	.59	...	4.33	.41	...	...	.59	...	4.33
.41	...	...	...	.59	...	−1.18	.41	...	...	.59	...	−1.18
.38	...	...	...	.62	...	−.78	.38	...	...	.62	...	−.78
.35	...	...	...	.65	...	−.38	.35	...	...	.65	...	−.38
.34	...	...	...	.66	...	2.96	.34	...	...	.66	...	2.96
.34	...	...	...	.66	...	3.61	.34	...	...	.66	...	3.61
.35	...	.09	...	.57	...	3.57	.35	...	.09	.57	...	3.57
.35	...	.25	...	.39	...	4.48	.35	...	.25	.39	...	4.48
.92	...	.08	...	...	...	11.98	.92	...	.08	...	...	11.98
.92	...	.08	...	...	...	2.40	.92	...	.08	...	...	2.40
.93	...	...	...	.07	...	12.35	.93	...	...	.07	...	12.35
.94	...	...	...	.06	...	6.87	.94	...	...	.06	...	6.87
.92	...	...	...	.08	...	4.99	.92	...	...	.08	...	4.99
.92	...	...	...	.08	...	7.09	.92	...	...	.08	...	7.09
.94	...	...	...	.06	...	−1.97	.94	...	...	.06	...	−1.97
.90	...	...	...	.10	...	−2.31	.90	...	...	.10	...	−2.31
1.00	...	...	...	...	...	3.86	1.00	...	...	...	...	3.86
1.00	...	...	...	...	...	−4.53	1.00	...	...	...	...	−4.53
.90	...	...	...	.10	...	7.72	.90	...	...	.10	...	7.72
.98	...	...	.02	...	...	−9.58	1.00	...	...	...	...	−9.60
.65	...	...	...	.35	...	−2.75	.65	...	...	.35	...	−2.75
.61	...	...	...	.39	...	4.13	.61	...	...	.39	...	4.13
.69	...	.26	...	.05	...	6.29	.69	...	.26	.05	...	6.29
.70	...	.30	...	...	...	6.47	.70	...	.30	...	...	6.47
.61	...	...	...	.39	...	7.18	.61	...	...	.39	...	7.18
.61	...	...	...	.39	...	1.02	.61	...	...	.39	...	1.02
.59	...	...	...	.41	...	4.02	.59	...	...	.41	...	4.02
.61	...	...	...	.39	...	−.88	.61	...	...	.39	...	−.88
.53	...	...	...	.47	...	3.74	.53	...	...	.47	...	3.74
.52	...	...	...	.48	...	−2.99	.52	...	...	.48	...	−2.99
.43	...	...	.04	.52	...	2.20	.47	...	...	.53	...	2.20

TABLE 4 (*Continued*)

	Power 0												
	With Small Stocks							Without Small Stocks					
	Inv. Fractions							Inv. Fractions					
Period	CS	GB	CB	SS	RL	B	r_t	CS	GB	CB	RL	B	r_t
1960:3	.79	...	...	.32	...	−.11	−5.66	1.11	...	...	...	−.11	−5.93
1960:4	.90	...	...	.52	...	−.43	9.92	1.43	...	...	...	−.43	13.11
1961:1	1.13	...	...	.30	...	−.43	20.71	1.43	...	...	...	−.43	17.71
1961:2	1.43	...	...	...	...	−.43	−.48	1.43	...	...	...	−.43	−.48
1961:3	.88	...	...	.55	...	−.43	1.92	1.43	...	...	...	−.43	5.09
1961:4	.89	...	...	.54	...	−.43	11.86	1.43	...	...	...	−.43	10.95
1962:1	...	...	...	1.43	...	−.43	4.91	1.43	...	...	...	−.43	−3.59
1962:2	...	...	...	1.43	...	−.43	−34.28	1.43	...	...	...	−.43	−30.05
1962:3	...	...	...	1.43	...	−.43	4.34	1.43	...	...	...	−.43	4.67
1962:4	...	...	...	2.00	...	−1.00	13.26	2.00	...	...	...	−1.00	25.18
1963:1	1.64	...	...	.36	...	−1.00	12.98	2.00	...	...	...	−1.00	11.30
1963:2	1.76	...	...	.24	...	−1.00	8.96	2.00	...	...	...	−1.00	8.64
1963:3	.34	...	...	1.66	...	−1.00	6.32	2.00	...	...	...	−1.00	6.82
1963:4	...	...	...	2.00	...	−1.00	.20	2.00	...	...	...	−1.00	9.84
1964:1	...	...	...	1.43	...	−.43	12.01	1.43	...	...	...	−.43	8.07
1964:2	...	...	...	1.43	...	−.43	5.38	1.43	...	...	...	−.43	5.41
1964:3	...	...	...	1.43	...	−.43	10.62	1.43	...	...	...	−.43	4.81
1964:4	...	...	...	1.43	...	−.43	.87	1.43	...	...	...	−.43	1.67
1965:1	...	...	...	1.43	...	−.43	16.55	1.43	...	...	...	−.43	2.75
1965:2	...	...	...	1.43	...	−.43	−7.92	1.43	...	...	...	−.43	−2.94
1965:3	...	...	...	1.43	...	−.43	20.17	1.43	...	...	...	−.43	10.40
1965:4	...	...	...	1.43	...	−.43	22.89	1.43	...	...	...	−.43	4.60
1966:1	...	...	...	1.43	...	−.43	11.86	1.43	...	...	...	−.43	−4.57
1966:2	...	...	...	1.43	...	−.43	−10.16	1.43	...	...	...	−.43	−6.77
1966:3	...	...	...	1.43	...	−.43	−18.42	1.43	...	...	...	−.43	−13.41
1966:4	...	...	...	1.43	...	−.43	5.61	1.17	...	...	...	−.17	6.68
1967:1	...	...	...	1.43	...	−.43	43.99	1.00	...	...	...	...	13.21
1967:2	...	...	...	1.43	...	−.43	16.72	1.43	...	...	...	−.43	1.13
1967:3	...	...	...	1.43	...	−.43	22.06	1.43	...	...	...	−.43	10.02
1967:4	...	...	...	1.43	...	−.43	9.98	1.43	...	...	...	−.43	.12
1968:1	...	...	...	1.43	...	−.43	−10.31	1.43	...	...	...	−.43	−8.92
1968:2	...	...	...	1.43	...	−.43	36.96	1.43	...	...	...	−.43	15.26
1968:3	...	...	...	1.25	...	−.25	7.14	1.25	...	...	...	−.25	4.39
1968:4	...	...	...	1.25	...	−.25	10.34	1.25	...	...	...	−.25	1.99
1969:1	...	...	...	1.25	...	−.25	−10.37	1.25	...	...	...	−.25	−2.38
1969:2	...	...	...	1.25	...	−.25	−8.76	1.00	...	...	...	...	−3.00
1969:3	...	...	...	1.25	...	−.25	−8.92	1.00	...	...	...	...	−3.92
1969:4	...	...	...	1.25	...	−.25	−8.97	.72	...	...	.28	...	.27
1970:1	...	...	...	1.25	...	−.25	−7.12	...	...	...	1.00	...	1.98
1970:2	...	...	...	1.25	...	−.25	−41.44	.50	...	...	.50	...	−8.20
1970:3	...	...	...	1.05	...	−.05	29.38	.75	...	...	.25	...	13.01
1970:4	...	...	...	1.37	...	−.37	.66	1.00	...	...	...	...	10.43
1971:1	...	...	...	1.45	...	−.45	37.53	1.00	...	...	...	...	9.69
1971:2	...	...	...	1.54	...	−.54	−11.34	1.54	...	...	...	−.54	−.61
1971:3	...	...	...	1.44	...	−.44	−4.21	1.01	...	...	...	−.01	−.61
1971:4	...	...	...	1.35	...	−.35	1.27	1.00	...	...	...	...	4.66
1972:1	...	...	...	1.47	...	−.47	18.42	1.19	...	...	...	−.19	6.54
1972:2	...	...	...	1.51	...	−.51	−6.34	1.17	...	...	...	−.17	.53
1972:3	...	...	...	1.37	...	−.37	−8.46	1.00	...	...	...	...	3.91
1972:4	...	...	...	1.12	...	−.12	1.87	1.00	...	...	...	...	7.56
1973:1	...	...	...	1.09	...	−.09	−15.27	1.00	...	...	...	...	−4.89
1973:2	...	...	...	.75	.25	...	−11.86	.62	...	...	.38	...	−2.98
1973:3	...	...	...	.32	.68	...	7.21	...	...	...	1.00	...	2.00
1973:4	...	...	...	.48	.52	...	−8.20	...	...	...	1.00	...	1.79

TABLE 4 ***(Continued)***

Power − 15												
With Small Stocks							Without Small Stocks					
Inv. Fractions							Inv. Fractions					
CS	GB	CB	SS	RL	B	r_t	CS	GB	CB	RL	B	r_t
.40	...	.21	.15	.25	...	−1.90	.52	...	.15	.33	...	−2.03
.36	.39	...	.19	.06	...	4.86	.51	.30	...	.19	...	5.58
.37	.43	...	.19	.01	...	9.29	.52	.34	...	.14	...	6.95
.53	.42	...	.06	...	...	...	.58	.42	...	...	...	.02
.47	.40	...	.13	...	...	2.14	.60	.40	...	...	...	2.88
.47	.37	...	.16	...	...	5.05	.62	.38	...	...	...	4.75
.41	.37	...	.23	...	...	1.27	.63	.36	...	.01	...	−.07
.31	.39	...	.30	...	...	−13.30	.60	.39	...	.01	...	−12.18
.10	...	.07	.13	.70	...	1.49	.24	...	.07	.70	...	1.54
.10	...	.13	.13	.64	...	2.94	.24	...	.13	.64	...	3.76
.20	...	.11	.03	.66	...	2.21	.23	...	.11	.66	...	2.06
.22	...	.16	.02	.60	...	1.74	.24	...	.16	.60	...	1.71
.12	...	.09	.10	.69	...	1.43	.22	...	.10	.69	...	1.48
.03	...	.04	.17	.76	...	.96	.20	...	.04	.76	...	1.80
.08	...	...	.11	.80	...	2.19	.20	...	...	.80	...	1.90
.04	...	.03	.15	.78	...	1.52	.20	...	.03	.78	...	1.54
.05	...	.11	.16	.68	...	2.18	.22	...	.11	.67	...	1.54
.06	...	.23	.18	.53	...	1.06	.25	...	.25	.50	...	1.18
.03	...	.30	.20	.46	...	3.24	.25	...	.32	.44	...	1.32
.03	...	.23	.21	.53	...	−.57	.26	...	.22	.52	...	.13
...	...	.42	.22	.36	...	3.58	.24	...	.39	.37	...	2.20
.03	...	.46	.25	.26	...	3.73	.31	...	.44	.25	...	.67
...	...	...	.29	.71	...	3.35	.29	...	...	.71	...	.02
...	...	...	.28	.72	...	−1.06	.27	...	...	.73	...	−.29
...	...	...	.25	.75	...	−2.15	.22	...	...	.78	...	−1.02
...	...	...	.18	.82	...	1.90	.13	...	...	.87	...	1.96
...	...	...	.17	.83	...	6.43	.14	...	...	.86	...	2.84
...	.01	...	.20	.79	...	3.15	.19	...	...	.81	...	1.02
...	...	...	.21	.79	...	4.19	.17	...	...	.83	...	2.15
...	...	...	.23	.77	...	2.60	.19	...	...	.81	...	1.04
...	...	...	.22	.78	...	−.50	.16	...	...	.84	...	.16
...	...	...	.22	.78	...	6.74	.16	...	...	.84	...	2.89
...	...	...	.23	.77	...	2.44	.17	...	...	.83	...	1.79
...	...	...	.25	.75	...	3.15	.21	...	...	.79	...	1.46
...	...	...	.24	.76	...	−.77	.17	...	...	.83	...	1.03
...	...	...	.21	.79	...	−.13	.12	...	...	.88	...	.98
...	...	...	.18	.82	...	.23	.08	...	...	.92	...	1.28
...	...	...	.17	.83	...	.33	.05	...	...	.95	...	1.66
...	...	...	.13	.87	...	1.05	...	...	...	1.00	...	1.98
...	...	...	.13	.87	...	−2.87	.03	...	...	.97	...	1.00
...	...	...	.10	.90	...	4.15	.05	...	...	.95	...	2.35
...	...	...	.12	.88	...	1.43	.11	...	...	.89	...	2.50
...	...	...	.13	.87	...	4.28	.15	...	...	.85	...	2.44
...	...	...	.14	.86	...	−.13	.18	...	...	.82	...	.81
...	...	...	.11	.89	...	.94	.12	...	...	.88	...	1.13
...	...	...	.11	.89	...	1.13	.13	...	...	.87	...	1.56
...	...	...	.12	.88	...	2.31	.16	...	...	.84	...	1.62
...	...	...	.12	.88	...	.37	.15	...	...	.85	...	.89
...	...	...	.11	.89	...	.27	.12	...	...	.88	...	1.36
...	...	...	.09	.91	...	1.24	.10	...	...	.90	...	1.82
...	...	...	.08	.92	...	.05	.10	...	...	.90	...	.69
...	...	...	.05	.95	...	.72	.04	...	...	.96	...	1.28
...	...	...	.02	.98	...	2.33	...	...	...	1.00	...	2.00
...	...	...	.03	.97	...	1.16	...	...	...	1.00	...	1.79

TABLE 4 ***(Continued)***

	Power 0												
	With Small Stocks							Without Small Stocks					
	Inv. Fractions							Inv. Fractions					
Period	CS	GB	CB	SS	RL	RB	RP	CS	GB	CB	RL	RB	RP
1974:1	...	...	...	...	1.00	...	1.94	...	...	...	1.00	...	1.94
1974:2	...	...	...	...	1.00	...	2.06	...	...	...	1.00	...	2.06
1974:3	...	...	...	...	1.00	...	1.94	...	...	...	1.00	...	1.94
1974:4	...	...	...	...	1.00	...	1.81	...	...	...	1.00	...	1.81
1975:1	...	...	...	...	1.00	...	1.62	...	...	...	1.00	...	1.62
1975:2	...	...	...	.02	.98	...	1.72	...	...	...	1.00	...	1.42
1975:3	...	...	...	.08	.92	...	.68	.03	...	...	.97	...	1.17
1975:4	...	...	...	...	1.00	...	1.52	...	...	...	1.00	...	1.52
1976:1	...	...	.62	...	.38	...	3.07	...	...	.62	.38	...	3.07
1976:2	.23	...	.35	.23	.19	...	.27	.49	...	.35	.16	...	1.52
1976:3	.27	...	...	...	.73	...	1.48	.27	...	...	.73	...	1.48
1976:4	.14	...	.64	...	.22	...	5.55	.14	...	.64	.22	...	5.55
1977:1	...	...	1.00	...	...	...	−2.31	...	...	1.00	...	...	−2.31
1977:2	...	...	1.01	...	...	−.01	2.94	...	...	1.01	...	−.01	2.94
1977:3	...	...	.90	.10	...	...	.99	...	...	1.00	...	...	1.09
1977:4	...	...	.85	.15	...	...	.51	...	...	1.00	...	...	−.82
1978:1	...	...	.69	.31	...	...	3.75	...	...	1.00	...	...	.03
1978:2	...	...	.36	.64	...	...	8.90	...	...	1.00	...	...	−1.08
1978:3	...	...	...	1.28	...	−.28	20.47	.36	...	.64	...	...	5.10
1978:4	...	...	...	1.00	...	...	−17.37	.14	...	...	.86	...	1.05
1979:1	...	...	...	.73	.27	...	16.92	...	...	...	1.00	...	2.43
1979:2	...	...	...	.71	.29	...	7.19	...	...	...	1.00	...	2.37
1979:3	...	...	...	1.00	...	...	5.64	...	...	...	1.00	...	2.15
1979:4	...	...	...	1.00	...	...	1.70	...	...	...	1.00	...	2.52
1980:1	...	...	...	.81	.19	...	−10.24	...	...	...	1.00	...	2.96
1980:2	...	...	...	.05	.95	...	4.58	...	...	...	1.00	...	3.68
1980:3	...	...	...	1.00	...	...	25.10	...	...	...	1.00	...	1.97
1980:4	...	...	...	1.00	...	...	7.49	...	...	...	1.00	...	2.85
1981:1	...	...	...	.85	.15	...	13.13	...	...	...	1.00	...	3.73
1981:2	...	...	...	1.08	...	−.08	8.72	...	...	...	1.00	...	3.24
1981:3	...	...	...	1.06	...	−.06	−18.98	...	...	...	1.00	...	3.73
1981:4	...	...	...	1.00	...	...	11.29	...	...	...	1.00	...	3.77
1982:1	...	...	...	1.74	...	−.74	−12.76	...	...	...	1.00	...	2.86
1982:2	...	...	...	1.11	...	−.11	−.86	...	...	...	1.00	...	3.46
1982:3	...	...	...	1.51	...	−.51	13.78	...	...	...	1.00	...	3.33
1982:4	...	...	...	2.00	...	−1.00	44.04	1.00	...	....	...	...	18.14
1983:1	...	...	...	2.00	...	−1.00	36.76	1.00	...	...	...	...	10.05
1983:2	...	...	...	2.00	...	−1.00	39.57	1.25	...	...	...	−.25	13.25
1983:3	...	...	...	2.00	...	−1.00	−5.90	1.00	...	...	...	...	−.14
1983:4	...	...	...	2.00	...	−1.00	−7.32	1.78	...	...	...	−.78	−1.59

NOTE.—CS = common stocks, GB = government bonds, CB = corporate bonds, SS = small stocks, RL = risk-free asset, B = the fraction in borrowing, and r_t = return for the quarter.

TABLE 4 ***(Continued)***

Power − 15												
With Small Stocks							Without Small Stocks					
Inv. Fractions							Inv. Fractions					
CS	GB	CB	SS	RL	RB	RP	CS	GB	CB	RL	RB	RP
...	...	...	...	1.00	...	1.94	...	...	...	1.00	...	1.94
...	...	...	...	1.00	...	2.06	...	...	...	1.00	...	2.06
...	...	...	...	1.00	...	1.94	...	...	...	1.00	...	1.94
...	...	...	...	1.00	...	1.81	...	...	...	1.00	...	1.81
...	...	...	...	1.00	...	1.62	...	...	...	1.00	...	1.62
...	...	...	...	1.00	...	1.44	...	...	...	1.00	...	1.42
...	...	...	...	1.00	...	1.49	...	...	...	1.00	...	1.52
...	...	...	...	1.00	...	1.52	...	...	...	1.00	...	1.52
...	...	.04	...	.96	...	1.35	...	...	.04	.96	...	1.35
.02	...	.02	.01	.95	...	1.17	.03	...	.02	.95	...	1.24
.02	...	...	...	.98	...	1.33	.02	...	...	.98	...	1.33
.01	...	.04	...	.95	...	1.50	.01	...	.04	.95	...	1.50
...	.07	.20	...	.73	...	−.05	...	.07	.20	.73	...	−.05
...	...	.24	...	.76	...	1.56	...	...	.24	.76	...	1.56
...	...	.22	...	.78	...	1.21	...	...	.22	.78	...	1.21
...	...	.18	...	.82	...	1.05	...	...	.18	.82	...	1.05
...	...	.21	.01	.78	...	1.29	...	...	.23	.77	...	1.18
...	...	.06	.04	.91	...	1.93	...	...	.13	.87	...	1.26
...	...	...	.09	.91	...	3.15	.02	...	.09	.89	...	1.99
...	...	...	.08	.92	...	.44	.01	...	...	.99	...	1.93
...	...	...	.04	.96	...	3.31	...	...	...	1.00	...	2.43
...	...	...	.04	.96	...	2.66	...	...	...	1.00	...	2.37
...	...	...	.07	.93	...	2.38	...	...	...	1.00	...	2.15
...	...	...	.06	.94	...	2.47	...	...	...	1.00	...	2.52
...	...	...	.05	.95	...	2.15	...	...	...	1.00	...	2.96
...	...	...	...	1.00	...	3.74	...	...	...	1.00	...	3.68
...	...	...	.07	.93	...	3.64	...	...	...	1.00	...	1.97
...	...	...	.07	.93	...	3.18	...	...	...	1.00	...	2.85
...	...	...	.05	.95	...	4.31	...	...	...	1.00	...	3.73
...	...	...	.09	.91	...	3.73	...	...	...	1.00	...	3.24
...	...	...	.11	.89	...	1.36	...	...	...	1.00	...	3.73
...	...	...	.07	.93	...	4.30	...	...	...	1.00	...	3.77
...	...	...	.14	.86	...	1.70	...	...	...	1.00	...	2.86
...	...	...	.10	.90	...	3.08	...	...	...	1.00	...	3.46
...	...	...	.12	.88	...	4.19	...	...	...	1.00	...	3.33
...	...	...	.21	.79	...	6.56	.17	...	...	.83	...	4.70
...	...	...	.23	.77	...	6.16	.18	...	...	.82	...	3.44
...	...	...	.23	.77	...	6.45	.14	...	...	.86	...	3.44
...	...	...	.22	.78	...	1.42	.12	...	...	.88	...	1.98
...	...	...	.24	.76	...	1.16	.18	...	...	.82	...	1.90

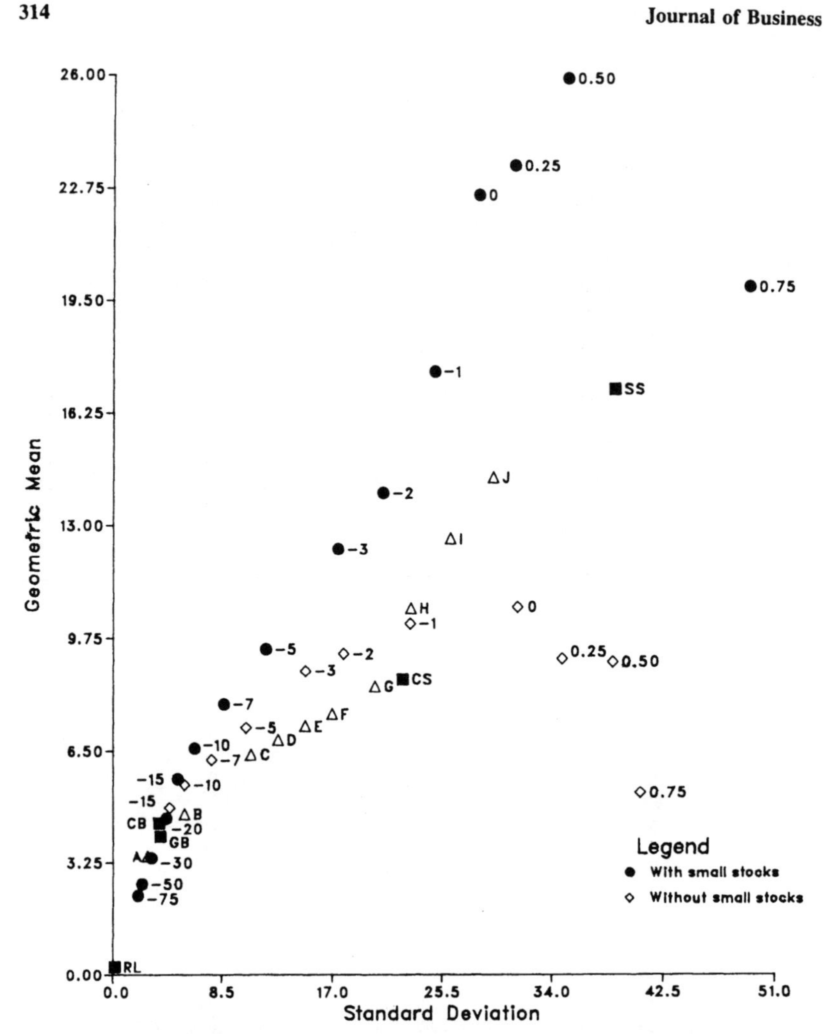

FIG. 1.—Geometric means and standard deviations of annual returns with and without small stocks, 1934–47 (quarterly revision, with leverage, 32-quarter estimating period).

Since the returns themselves are not additive but compound multiplicatively, we employed the paired t-test for dependent observations to the quarterly (and additive) variables $\ln(1 + r_t)$ to test whether the presence of small stocks improved returns significantly.[11] Thus to compare the return series $r_1^1, \ldots, r_n^1$ with the return series $r_1^2, \ldots, r_n^2$ for two different strategies, we calculate the statistic

$$t = \frac{\bar{d}}{\sigma(d)/\sqrt{n}},$$

11. This test was also employed by Fama and MacBeth (1974).

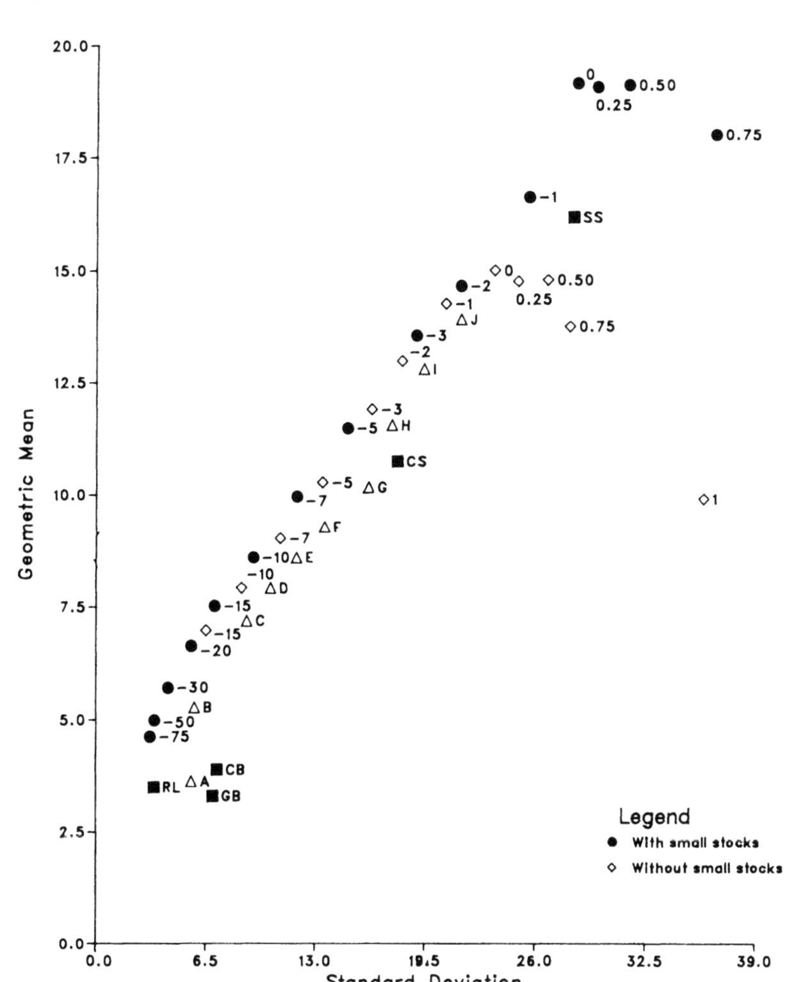

FIG. 2.—Geometric means and standard deviations of annual returns with and without small stocks, 1934–83 (quarterly revision, with leverage, 32-quarter estimating period).

where

$$\bar{d} = \sum_{t=1}^{n} \frac{\ln(1 + r_t^1) - \ln(1 + r_t^2)}{n}$$

and $\sigma(d)$ is the standard deviation of the differences between $\ln(1 + r_t^1)$ and $\ln(1 + r_t^2)$. In each case, the null hypothesis is that

$$E[\ln(1 + r_t^1)] = E[\ln(1 + r_t^2)]$$

and the alternative hypothesis is that

$$E[\ln(1 + r_t^1)] > E[\ln(1 + r_t^2)].$$

TABLE 5 **Paired t-Tests, 1934–47 Subperiod, with Leverage**

Comparison	$\bar{d}$	$\sigma(d)$	t
Power −1 vs. common stocks	.0196	.1267	1.16
Power 0 vs. portfolio I	.0378	.1438	1.97*
Power .5 with small stocks vs. power .5 without small stocks	.0360	.1813	1.48
Power .25 with small stocks vs. power .25 without small stocks	.0306	.1333	1.72*
Power .25 with small stocks vs. power 0 without small stocks	.0273	.1157	1.76*
Power 0 with small stocks vs. power 0 without small stocks	.0256	.1068	1.79*
Power −1 with small stocks vs. power −1 without small stocks	.0160	.0591	2.03**
Power −2 with small stocks vs. power −2 without small stocks	.0103	.0389	1.99**
Power −3 with small stocks vs. power −3 without small stocks	.0079	.0311	1.90*
Power −5 with small stocks vs. power −5 without small stocks	.0052	.0211	1.85*
Power −7 with small stocks vs. power −7 without small stocks	.0038	.0156	1.81*
Power −10 with small stocks vs. power −10 without small stocks	.0025	.0104	1.79*
Power −15 with small stocks vs. power −15 without small stocks	.0020	.0073	2.06**

* Significant at the 10% level.
** Significant at the 5% level.

The results for selected pairs with comparable standard deviations in the 1934–47 period (56 quarters) are given in table 5, while table 6 gives the results for the full 200-quarter 1934–83 period.

Table 5 suggests that the presence of small stocks was especially significant in improving returns for the "middle" strategies (powers −15–.25) from the early 1930s to the late 1940s. Statistically, the improvements were less impressive for the full period (fig. 2 and table 6). The active strategies performed extremely well when compared to bonds and bills, attaining highly significant improvements, through diversification, in the geometric means without increasing variability. Figures 1 and 2 and tables 5 and 6 also reveal that the active strategies performed surprisingly well when compared to the 10 fixed-weight portfolio strategies.

VII. Concluding Remarks

By way of summary, several conclusions emerge. First, the present study confirms that the simple probability assessment approach, which uses only the past to (naively) forecast the future, is apparently not without merit when combined with multiperiod investment theory. Not

TABLE 6 **Paired *t*-Tests, 1934–83 Period, with Leverage**

Comparison	$\bar{d}$	$\sigma(d)$	t
Power −3 vs. common stocks	.0062	.0758	1.16
Power −15 vs. corporate bonds	.0085	.0452	2.67***
Power −15 vs. government bonds	.0100	.0458	3.08***
Power −50 vs. risk-free asset	.0036	.0102	4.94***
Power −10 vs. portfolio C	.0032	.0419	1.09
Power −20 vs. portfolio B	.0032	.0311	1.45
Power −20 vs. portfolio A	.0071	.0356	2.82***
Power 0 with small stocks vs. power .75 without small stocks	.0116	.1073	1.52

*** Significant at the 1% level.

only did it produce respectable results, but the logarithmic policy also always came out at or near the top in the geometric mean dimension, suggesting an absence of clear biases in the estimating method; the exception occurs in the last subperiod, when the approach appears to have been somewhat too conservative.[12] Second, small stocks entered even the most risk-averse portfolios most of the time. On the other hand, they were ignored by all the strategies some of the time (especially in the mid- and late 1950s and mid-1970s). Third, small stocks, when chosen, tended to replace common stocks, except in the 1970s and early 1980s, when they were primarily held in lieu of the risk-free asset. This was true even when (additional) margin purchases would have been possible. Consequently, the presence of small stocks had relatively little effect on the use of leverage; in fact, the effect was not always in the positive direction. Thus the use of, and gains from, margin purchases occurred, as in our previous paper (Grauer and Hakansson 1982), primarily for the more risk-tolerant strategies from the mid-1930s to the mid-1960s.

Fourth, small stocks had a notable positive effect on returns, not only for the risk-tolerant strategies but also for conservative investors. This effect was especially significant from the mid-1930s to the early 1950s in the presence of margin purchases. Fifth, the strong gains from diversification among the major asset categories reported in the previous papers are, if anything, strengthened when small stocks are included.

Sixth, the performances of the active strategies, when measured against those of the fixed-weight strategies, were surprisingly strong. Could it be that the naive, simple-minded empirical distribution contains the kind of information money managers usually pay good money for?

12. Under perfectly valid distributions, power $\gamma = 1/T$, where T is the number of periods, has the highest asymptotic probability of attaining the largest geometric mean as T increases.

Finally, one cannot escape the conclusion that U.S. financial markets offered a generous environment during the half-century we studied. As a point of reference, consider the following. The maximum employee-portion of the social security contribution for an individual in 1984 was $2,532.60. The purchasing power of this amount (in January 1, 1984, dollars) at the beginning of 1934 was $328.78. Under a naive logarithmic policy in an environment with access to small stocks and leverage, this initial contribution alone would have grown to $2,078,925 as of the end of 1983.

Of course, the reader should also be reminded of the limitations of the study. The model used is focused on sequential reinvestments only, without concern for intermediate consumption; even though its birth occurred in the mid-1970s, it was applied as far back as 1934. The latter statement also applies at least partially to the data base used. The joint probability estimates were based on periods ranging from the most recent 5–10 years only. All investors were assumed to be strict price takers. Transactions costs and taxes were ignored (as in the underlying returns series). Finally, maintenance margins were ignored whenever leverage was used.

References

Banz, Rolf. 1981. The relationship between return and market value of common stocks. *Journal of Financial Economics* 9 (March): 3–18.

Bawa, Vijay; Brown, Stephen; and Klein, Roger. 1979. *Estimation Risk and Optimal Portfolio Choice*. Amsterdam: North-Holland.

Best, Michael. 1975. A feasible conjugate direction method to solve linearly constrained optimization problems. *Journal of Optimization Theory and Applications* 16 (July): 25–38.

Fama, Eugene, and MacBeth, James. 1974. Long-term growth in a short-term market. *Journal of Finance* 29 (June): 857–85.

Grauer, Robert, and Hakansson, Nils. 1982. Higher return, lower risk: Historical returns on long-run, actively managed portfolios of stock, bonds and bills, 1936–1978. *Financial Analysts Journal* 38 (March–April): 39–53.

Grauer, Robert, and Hakansson, Nils. 1985. 1934–1984 returns on levered, actively managed long-run portfolios of stocks, bonds and bills. *Financial Analysts Journal* 41 (September–October): 24–43.

Hakansson, Nils. 1971. On optimal myopic portfolio policies, with and without serial correlation of yields. *Journal of Business* 44 (July): 324–34.

Hakansson, Nils. 1974. Convergence to isoelastic utility and policy in multiperiod choice. *Journal of Financial Economics* 1 (September): 201–24.

Huberman, Gur, and Ross, Stephen. 1983. Portfolio turnpike theorems, risk aversion and regularly varying utility functions. *Econometrica* 51 (September): 1104–19.

Ibbotson, Roger, and Sinquefield, Rex. 1982. *Stocks, Bonds, Bills, and Inflation: The Past and the Future*. Charlottesville, Va.: Financial Analysts Research Foundation.

Leland, Hayne. 1972. On turnpike portfolios. In Karl Shell and G. P. Szego (eds.), *Mathematical Methods in Investment Finance*. Amsterdam: North-Holland.

Mossin, Jan. 1968. Optimal multiperiod portfolio policies. *Journal of Business* 41 (April): 215–29.

Reinganum, Marc. 1981. Misspecification of capital asset pricing: Empirical anomalies based on earnings yields and market values. *Journal of Financial Economics* 9 (March): 19–46.

Ross, Stephen. 1974. Portfolio turnpike theorems for constant policies. *Journal of Financial Economics* 1 (July): 171–98.

50

A Dynamic Portfolio of Investment Strategies: Applying Capital Growth with Drawdown Penalties

Professor John M. Mulvey, Mehmet Bilgili and Taha M. Vural
Department of Operations Research and Financial Engineering, Princeton University

Abstract

The growth optimal investment strategy has been shown to be highly effective for structured decision problems such as blackjack, sports betting, and high frequency trading. For securities markets, these strategies are more difficult to apply due to a variety of practical issues: structural changes in market behavior due to varying risk premium and related factors, transaction costs, operational constraints, and path dependent risk measures for many investors, including surplus risks for a defined-benefit pension plan. In addition, the standard three step approach for institutional money management does not allow for rapid changes in asset allocation — especially needed during highly turbulent periods. We modify the growth models to address downside protection, along with applying a portfolio of investment strategies — to improve diversification of the portfolio. Empirical results show the benefits of the concepts during normal and crash (2008) periods.

1 Introduction

The growth optimal model and its siblings have been applied successfully in a wide variety of decision problems. As an early example, Thorp (1969) implemented card counting approximations in the area of gambling situations such as blackjack, as well as extensions in the investment domain including option pricing models. Grauer and Hakkasson (1986) evaluated asset allocation models with the iso-elastic expected utility model (including log-utility); these authors applied this framework for many years in a series of related papers. More recently, Stutzer (2003, 2010), Ziemba (2005), and MacLean *et al.* (2004, 2009) extended the growth models to address downside protection. See Bell and Cover (1980), Kelly (1956), and Samuelson (1971, 1979) and the remainder of this book for details of growth optimal strategies.

There are a number of practical issues to address when implementing growth models. For example, individual investors rarely are able to monitor their affairs on a short term basis and take proper corrective actions. In fact, individuals are often accused of rendering poor judgment during market turning points —

increasing risk assets at market tops, and vice versa. Individuals can be unaware of their asset allocation or surplus capital, except during occasional visits to a financial planner.

In contrast, institutional investors often apply well established risk management tools via a three step process. First, a set of generic asset categories is defined, such as large-cap U.S. equities, small-cap value equities, long duration government bonds, real estate, and so on. Standard asset categories along with tracking benchmarks are provided by firms including, Russell, Standard and Poor's, MSCI, and Dow Jones. In most cases, these asset categories can be passively managed by means of index funds or exchange traded funds (ETFs); they serve as benchmarks for active portfolio managers. Second, on a semi-regular basis, the investor conducts an asset allocation or asset-liability management study to ascertain their asset proportions to best meet their goals, liabilities, and risk tolerances. For example, many university endowment or pension plan administrators carry out this task every 2 to 3 years. As a third step, the investor selects active managers to beat their respective benchmarks, or to invest in a low-cost passive index (which can be simply defined such as the usual cap-weighted index or a variant such as fundamental weighted index). The three step process has evolved to provide an implementable process for large institutions; otherwise, there can be great difficulty achieving diversification benefits and simply managing a large pool of capital.

The three step process has worked reasonably well over the past twenty plus years. It is durable and has certain benefits in terms of ease of implementation and adding judgment by an advisory firm or investment management organization. Unfortunately, several issues have arisen over the past few years that have slowly undermined the approach. First, the correlation among many of the traditional style segmentation has increased over time, and reached almost unity in 2008. See Figure 1 for the average correlation among six segments along two dimensions: market capitalization (large, medium, and small), and fundamental valuation (value, growth) over the past 15 years. This situation became critical in 2008 when the contagion increased even further. The great majority of stocks performed quite poorly in 2008 — and diversification strategies failed in many cases.

Second, many institutional investors have turned to alternative private markets to improve their performance. This shift has been championed by David Swensen at Yale University endowment (with his book Pioneering Portfolio Management (2000) and superb performance up until early 2008). Private markets are much harder to track since the securities are rarely traded. Also, private markets possess substantial spreads between the return of the top managers (say top decile) and the average manager in these segments. Thus, it is difficult to apply a passive index. And importantly, the return patterns can be hard to model since there are few detailed data available for analysis and issues such as leverage are not always evident to outside researchers. Conducting an asset allocation study with a majority of alternative investments is a treacherous assignment. Importantly, a portfolio of

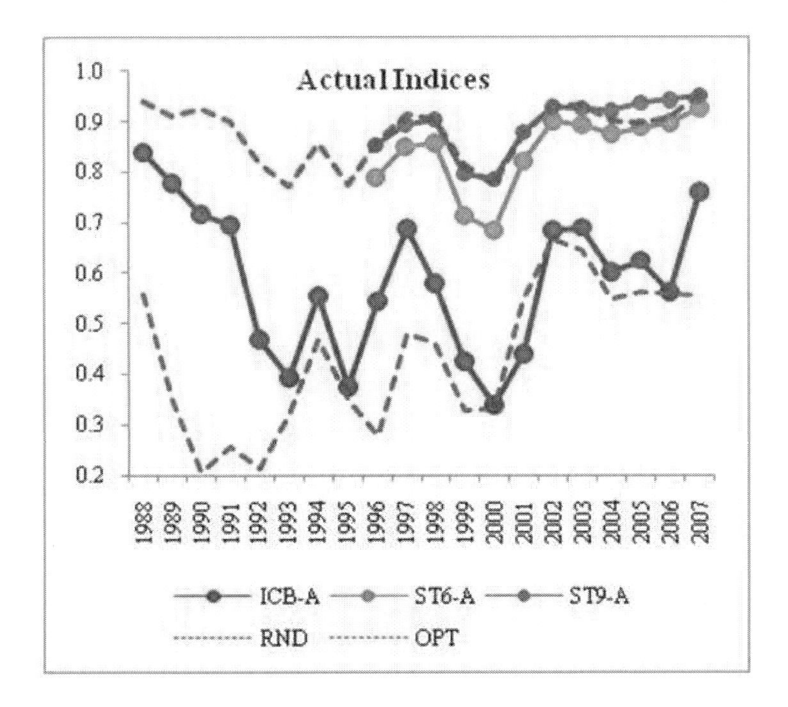

Figure 1 Rolling correlation of traditional equity style classification

private market assets is quite difficult to modify as conditions warrant due to their illiquidity.

Third, the emergence of ETFs and passive mutual funds in particular segments, along with more fundamental causes, has changed the relationships of securities to each other. For example, institutional investors will trade a basket of commodities (e.g., Goldman's GSCI, or Deutsche Bank's commodity index) in large volumes — again altering the diversification benefits. Thus, in late 2008, we saw very high correlation in returns for a very wide variety of securities — equities, corporate bonds of all types, commodities, mortgage-backed securities, and so on. Even many alternative investors saw substantial losses — 20, to 30 to 40% in 2008. There was a very rapid and substantial increase in risk premium — with commensurate increases in volatility and correlation.

The correlation has increased to close to one for the classical segments: Value/growth and Large/mid/small (ST6-A = style segmentation with six categories, ST9-A = style segmentation with nine categories). Industry segments (ICB = Industry Classification Benchmark) and an optimal classification scheme (OPT) give better diversification. The Traditional segmentations are close to a randomly selected classification (RND).

Given our knowledge about these turbulent events, investor could have benefited by applying capital growth models. A critical issue involves ensuring a consistent relationship between the investor's wealth and the anticipatory risks in the market. The investor must avoid taking risks (bets) above the log-optimal solution, in order to optimize the long-term consequence of their decisions. Critically, as the investor's wealth decreases, they must lower their risk threshold — by reducing committed capital. This rule is quite well known; however, it has been difficult or almost impossible for many investors to take the appropriate action (due to a number of factors including those mentioned above).

In the next section, we propose an alternative to the usual three-step approach to address the abovementioned issues. In this regard, we link traditional growth models with strong downside protection. We emphasize capital and risk allocation across investment strategies — rather than individual assets (or asset categories).

2 Dynamic Portfolio Tactics

This section describes the fundamental tenets of dynamic portfolio tactics (DPT). Two of the primary concepts are: protect the investor's capital (or goal capital) as a first priority — prevent large losses at all costs; and construct and revise a portfolio of investment strategies — rather than a portfolio of assets — based on dynamic risk analysis. To achieve the first concept, we employ highly liquid investments such as futures markets and high-volume exchange traded funds. In both cases, the transaction costs are quite low for most investors (except the largest institutions such as CALPERs and the Canadian Pension System). For investors possessing illiquid assets, we advocate a variation of the DPT approach for tactical asset allocation based on replication strategies (see Mulvey and Ling, 2009). In this paper, we focus on single-period, asset only investments; see Mulvey *et al.* (2003a, 2008, 2009), Zenios and Ziemba (2006, 2007), and Ziemba and Mulvey (1998) and it references for applications of asset-liability and multi-stage models. The developed approach can be extended in the natural manner.

To start, we define the traditional growth model as a sequence of nonlinear stochastic optimization models. First, we are interested in a relatively long sequence of decisions occurring at equal time steps (for simplicity), $t = \{1, 2, \ldots, T\}$ where the horizon equals time T. For our purposes in the empirical section, we will render decisions every five trading days (a balance between high frequency and mid-term models) in order to capture relative rapid changes in the investor's wealth. Roughly, the investor will makes fifty asset allocation decisions per year, or 500 per decade.

Next, at time point t, we define uncertainties via a set of stochastic scenarios $\{S\}$, in which the returns for the next five days will depend upon a number of factors known at the time of the decision — including recent volatility, recent momentum returns, current risk premium, and so on. Importantly, we do not assume a stationary process; instead the risk premium and volatility/correlation will encounter bursts of high turbulence (as we have seen displayed in 2008 and in several previous episodes).

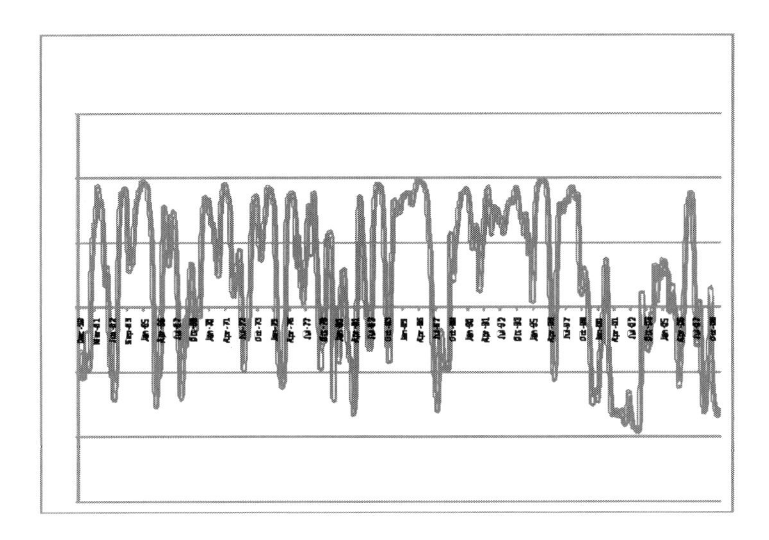

Figure 2 Rolling correlation of equity returns and government long bond returns

We give a straightforward example of a changing relationship in market behavior (Figure 2). Here we display the rolling correlation between the return of large U.S. cap stocks (S&P 500) and the return on long-duration U.S. government bonds. Many portfolio models assume that this correlation remains constant. However, during sharp downturns, the correlation generally becomes negative — due to fundamental factors — mostly due to a flight to quality during a recession. It is clear that the scenario generator must take these changing regimes into account (more on multiple regimes later).

As compared with traditional asset allocation models, we will focus on a set of investment strategies $\{J\}$, rather than subsets of securities. Several example investment strategies will be described in the next section. We distinguish two types of investment strategies: core strategies taking capital $\{J1\}$, and overlay strategies $\{J2\}$ that employ futures markets and swaps, taking risk capital. We separate these strategies in order to emphasize the linkage of overlay strategies with the core strategies, and when the core strategies cannot be easily traded, such as private equity or venture capital. We are interested in the optimal combination of core and overlay strategies that best suits the investor at a given time.

The traditional growth models employ the following stochastic optimization model [GO].

$$[\text{GO}] \qquad \text{Max}\, EU(w_2)$$

$$\text{Subject to} \qquad w_1 = \sum_{j \in j1} x_j \qquad x_j \geq 0, \quad \text{and} \quad x \in X$$

$$\text{and} \qquad w_{2,s} = \sum_{J1} x_j * (1 + r_{j,s}) \quad \text{for} \quad s \in \{S\}$$

The decision variables, x_j, are defined over set $\{J1\}$, in our case investment strategies rather than securities or groups of securities, except when indicated otherwise. We define the investor's initial wealth equal to w_1, to reference the start of the planning period $t = [1 \text{ to } 2]$. For this initial model, we greatly simplify the investor's situation by ignoring transaction costs, by assuming a static, single period planning period, by assuming no cashflow considerations, by focusing on the assets without reference to liabilities or goals, defining a long-only perspective, and so on. In a discrete-time model such as [GO], there is an implicit assumption that the model will be executed over a short enough horizon so that the investor's wealth is not greatly affected by intermediate changes in returns. Next, we assume the standard growth-optimal utility function $U(w_2) = Ln(w_2)$ within the family of iso-elastic functions $U(w) = (1/\gamma) \times w^\gamma$ for a single risk aversion parameter gamma. Returns are designated over a finite set of stochastic scenarios $\{S\}$. All policy and legal constraints are contained in set $\{X\}$.

This model is a straightforward nonlinear program, which is concave for any risk adverse decision maker, and thus can be readily solved with a standard package (such as the solver in Excel). For the optimization, we employ versions of the iso-elastic utility function. However, the model can be readily adapted for other convex reward and risk measures.

To extend the model, we allow for a borrowing variable x_b (with upper limit u_b) and a set of overlay variables — set $\{J2\}$. The first equation above is modified as:

$$w_1 + x_b = \sum_{j \in J1} x_j \quad \text{and} \quad 0 \leq x_b \leq u_b$$

The revised second equation, ending wealth w_2 is:

$$w_{2,s} = \sum_j x_j * (1 + r_{j,s}) - y * (r_b) + \sum_{j \in J2} x_j * r_j$$

The overlay variables do not require borrowing (implicit) and do not take direct capital. Rather, the core assets $\{J1\}$ serve as the margin requirements for the overlay strategies (Mulvey *et al.*, 2006, 2007). We assure that the total level of overlay risk capital is small enough to prevent margin calls and to maintain the investor's overall risk profile (limits in constraints X). The cost of borrow is defined by $r_{b,s}$. This value is uncertain due to two causes: since the cost of borrowing may depend upon interest rates during the planning period, and secondly due to the variable amount of leverage that may occur during the planning period as specified by the investment strategies.

The growth strategy [GO] (and variants) can be applied in a straightforward manner given a reliable and robust model for generating/forecasting the return scenarios embedded in set $\{S\}$. Unfortunately, due to the changing structure of security markets (e.g., DeBondt and Thaler, 1985, 1987) and the limited number of time junctures for most investors, there is a need for a carefully designated risk management system. A typical theoretical approach would be to increase the risk

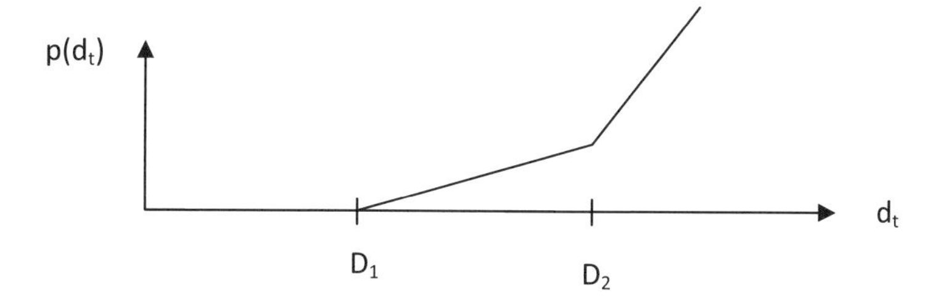

Figure 3 A penalty function to protect downside losses

aversion parameter γ within the iso-elastic family — lowering risks throughout the entire planning period. Instead, we design an alternative approach in which the investor's risk aversion changes as a function of her wealth path. Accordingly, the utility function is modified to be conditional on the wealth path $\{w_t\}$ and especially drawdown values $\{d_t\}$ at time t. The following function reflects this concept:

$$[\text{GO} - \text{DRAW}] \quad U(w_2) = (1/\gamma) * w_2^{\gamma} - p(d_2) * w_t$$

here the function $p(\cdot)$ penalizes investment decisions with outcomes falling below specified drawdown values. The drawdown variable is a function of the distance between the high water mark of the wealth path and the current wealth.

$$d_t = w_t^{high} - w_t \quad \text{where} \quad w_t^{high} = \max_{\tau=1,\dots,t-1} \{w_\tau\}$$

For the empirical tests, we have found that two thresholds values, $D_1 = 12\%$, and $D_2 = 16\%$, in conjunction with a piecewise approximation have proven to be a robust approach. Thus, when drawdown exceeds these thresholds, additional risk aversion is obtained by assessing penalties over the iso-elastic function. This approach is similar in spirit to Ziemba (2005) and MacLean *et al.* (2004, 2009). Our approach allows for changing risk premium as a function of the investor's wealth and market conditions.

The downside penalty function is particularly important during periods of high turbulence for assets possessing returns with a positive function of risk premium, such as equities or corporate bonds. To this point, we refer to the fall 2008 months in which drawdown became quite excessive, and risk protection was important at that time.

A critical aspect of any investment model involves the forecasting system for projecting the returns for the assets and investment strategies over the upcoming period. We advocate a projection system based on multiple regimes and an embedded Markov switching matrix (Guidolin and Timmermann, 2007; Hamilton, 1989; Kim, 1994; and Mulvey and Bilgili, 2009). In other words, at each time period, we estimate the current regime (for example, normal, bubble, crash) by means of current conditions such as volatility and correlation over the past 5 to 10 days. Several algorithms can be applied to this problem (Ding and He, 2004; Mulvey and

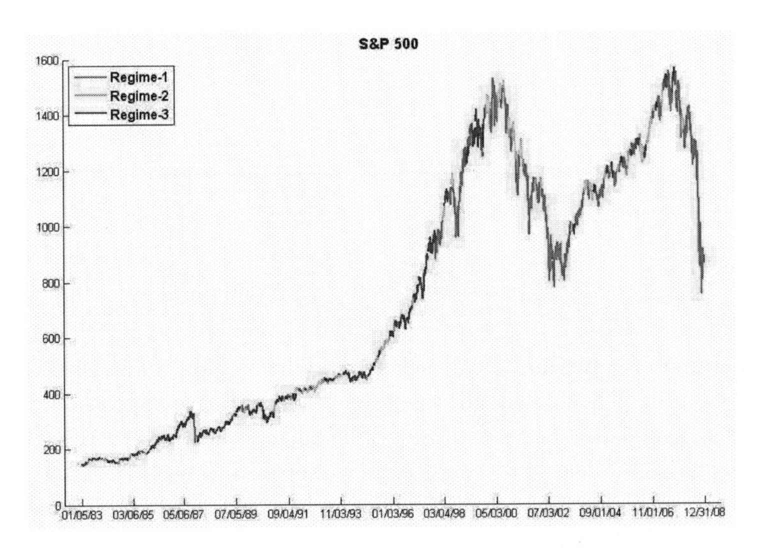

Figure 4 A time series of a triple regime analysis for U.S. equities (1/1/1983 to 1/1/2009)

Table 1 Transition probability matrix for three regime analysis

	Regime 1	Regime 2	Regime 3	Number of Periods
Regime 1	0.65	0.22	0.33	101
Regime 2	0.21	0.35	0.44	81
Regime 3	0.12	0.19	0.63	158

Bilgili, 2009). Once a regime is estimated, we can calculate a subset of internally consistent scenarios for each regime. As an example, suppose that we estimate 3 regimes for equities. Then, we form three sets of scenarios $\{S1\}$, $\{S2\}$, and $\{S3\}$, with probabilities, π_1, π_2, and π_3, respectively. The number of scenarios in each set depicts the relative probabilities. And the probability of moving between scenarios is indicated by the Markov transition matrix. Mulvey and Bilgili (2009) provide further details. Figure 4 and Table 1 show the path and the transition probability matrix of a three regime analysis over the period 1983 to 2009 for the U.S. equity market, respectively. Note that the down cycles correspond mostly to the third regime, and the persistence of these regimes over time. Regimes one and three are the most stable (60–65% probability of remaining in that regime for the subsequent period), whereas regime two is the least stable (35% probability).

3 Investment Strategies and Empirical Results

This section takes up the proposed growth model with downside protection. The first step is to define several investment strategies that possess good long term

Table 2 Performance of equity strategies for emerging markets (Benchmark EEM, rebalanced = equal weighted, and 130/20 strategies)

Period: January 2005-June 2009			
Period Length: 4.42 years	*EEM*	*Rebalanced*	*130/20*
Annual Geometric Return	10.80%	12.80%	13.90%
Annual Standard Deviation	36.90%	33.30%	36.10%
Best Month Return	30.60%	29.60%	32.30%
Sharpe Ratio	0.18	0.27	0.27
Worst Month Return	-26.10%	-25.70%	-28.20%
Max Drawdown	63.80%	63.30%	67.10%
Return/MaxDrawdown	16.92%	20.28%	20.66%
Correlation with EEM Strategy	1.00	0.98	0.96

Period: January 1999-June 2009			
Period Length: 14.42 years	*EEM*	*Rebalanced*	*130/20*
Annual Geometric Return	6.61%	11.70%	13.61%
Annual Standard Deviation	25.86%	25.11%	27.71%
Best Month Return	30.60%	29.60%	32.30%
Sharpe Ratio	0.10	0.31	0.35
Worst Month Return	-26.10%	-25.70%	-28.20%
Max Drawdown	63.80%	63.30%	67.10%
Return/MaxDrawdown	10.40%	18.50%	20.30%
Correlation with EEM Strategy	1.00	0.95	0.93

characteristics. We illustrate the modeling principles by reference to several strategies that have proven effective over long time periods.

The core investment strategy employs three equity markets: (a) countries in the emerging market segment (symbol = EEM), (b) sectors of the S&P 500 index (symbol = SPY), and (c) developed countries outside the U.S (symbol = EFA). In each case, there are cap-weighted indices that can be readily purchased (EEM, SPY, and ETA, respectively) via high volume exchange traded funds (ETFs). These instruments are among the most highly traded in the world (over $500 million per day in mid-2009). To develop an investment strategy, we form a portfolio of ten ETFs that depict each of the three underlying markets. For example, the emerging markets can be largely covered by purchasing single country ETF (China, Brazil, India, Korea, etc.). The long portfolio will be set up by equally weighting the 10 countries (10% each), with rebalancing back to the equal weight target on a short-term basis (5 days in our tests). We can show that in many cases, a fixed-mix rebalanced portfolio will outperform a traditional buy-and-hold cap-weighted index (next subsection). Table 2 shows a comparison of the equal-weighted index versus the cap-weighted index from the emerging markets, over two periods 1/1/1995 to 7/1/2009 and 1/1/2005 to 7/1/2009. Similar results occur for EAFE and S&P 500 sectors. In fact, equal weighting is the optimal solution to a dynamic portfolio model if the returns of the individual components are equal and independent.

This section illustrates the advantages of applying a multi-period policy rule. We begin with the well-known fixed-mix investment rule due to its simplicity and profitability. This policy rule serves as a benchmark both for other types of rules and for the recommendations of a stochastic programming model.

3.1 *Fixed mix policy rules*

First, we describe the performance advantages of the fixed-mix rule over a static, buy-and-hold perspective. This rule generates greater return than the static model by means of rebalancing. The topic of re-balancing gains (also called excess growth or volatility pumping) as derived from the fixed-mix decision rule is well understood for a theoretical perspective. The fundamental solutions were developed by Merton (1969) and Samuelson (1969) for long-term investors. Further work was done by

Fernholz and Shay (1982). Luenberger (1998) presents a clear discussion. We illustrate how rebalancing the portfolio to a fixed-mix creates excess growth (Mulvey and Kim, 2008). Suppose that a stock price process P_t is lognormal so it can be represented by the equation

$$dP_t = \alpha P_t dt + \sigma P_t dz_t \tag{1}$$

where α is the rate of return of P_t and σ^2 is its variance, z_t is Brownian motion with mean 0 and variance t.

The risk-free asset follows the same price process with rate of return equal to r and standard deviation equal to 0. We represent the price process of risk-free asset by B_t:

$$dB_t = rB_t dt \tag{2}$$

When we integrate the equation (1), the resulting stock price process is

$$P_t = P_0 e^{(\alpha-\sigma^2/2)t+\sigma z_t} \tag{3}$$

It is well documented that the growth rate $\gamma = \alpha - \sigma^2/2$ is the most relevant measure for long-run performance. For simplicity, we assume equality of growth rates across all assets. This assumption is not required for generating excess growth, but it makes the illustration easier to understand.

Next, let's assume that the market consists of n stocks, each with stock price processes $P_{1,t}, \ldots, P_{n,t}$ following the lognormal process. A fixed-mix portfolio has a wealth process W_t that can be represented by the equation

$$dW_t/W_t = \eta_1 dP_{1,t}/P_{1,t} + \cdots + \eta_n dP_{n,t}/P_{n,t} \tag{4}$$

where $\eta_1, \ldots, \eta_n$ are the fixed weights given to each stock (proportion of capital allocated to each stock). In this case, the weights sum up to one

$$\sum_{i=1}^{n} \eta_i = 1 \tag{5}$$

The fixed-mix strategy in continuous time always applies the same weights to stocks over time. The instantaneous rate of return of the fixed-mix portfolio at anytime is the weighted average of the instantaneous rates of returns of the stocks in the portfolio.

In contrast, a buy-and-hold portfolio is one where there is no rebalancing and therefore the number of shares for each stock remains constant over time. This portfolio can be represented by the wealth process W_t:

$$dW_t = m_1 dP_{1,t} + \cdots + m_n dP_{n,t} \tag{6}$$

where $m_1, \ldots, m_n$ depicts the number of shares for each stock.

Again for simplicity, let's assume that there is one stock and a risk-free instrument in the market. This case is sufficient to demonstrate the concept of excess growth in a fixed-mix portfolio as originally presented in Fernholz and Shay (1982).

Assume that we invest η portion of our wealth in the stock and the rest $(1-\eta)$ in the risk-free asset. Then the wealth process W_t with these constant weights over time can be expressed as

$$dW_t/W_t = \eta dP_t/P_t + (1-\eta)dB_t/B_t \tag{7}$$

where P_t is the stock price process and B_t is the risk-free asset.

When we substitute the dynamic equations for P_t and B_t, we get

$$dW_t/W_t = (r + \eta(\alpha - r))dt + \eta\sigma dz_t \tag{8}$$

For simplicity, assume the growth rate of all assets in the ideal market should be the same over long-time periods so that the growth rate of the stock and the risk-free asset are equal. Hence

$$\alpha - \sigma^2/2 = r \tag{9}$$

From equation (8), we can see that the rate of return of the portfolio, α_ω, is

$$\alpha_w = r + \eta(\alpha - r) \tag{10}$$

By using (9), this rate of return is equal to

$$\alpha_w = r + \eta\sigma^2/2 \tag{11}$$

The variance of the resulting portfolio is

$$\sigma_w^2 = \eta^2\sigma^2 \tag{12}$$

Hence, the growth rate of the fixed-mix portfolio becomes

$$\gamma_w = \alpha_w - \sigma_w^2/2 = r + (\eta - \eta^2)\sigma^2/2 \tag{13}$$

This quantity is greater than r for $0 < \eta < 1$. As it is greater than r, which is the growth rate of individual assets, the portfolio growth rate has an excess component, which is $(\eta - \eta^2)\sigma^2/2$. Excess growth is due to rebalancing the portfolio constantly to the target fixed-mix. The strategy moves capital out of stock when it performs well and moves capital into stock when it performs poorly. By moving capital between the two assets in the portfolio, a higher growth rate than each individual asset is achievable. It can be shown that the buy-and-hold investor with equal returning assets lacks the excess growth component. Therefore, buy-and-hold portfolios under-perform fixed-mix portfolios in various cases. We can easily see that the excess growth component is larger when σ takes a higher value.

Next, we design a simple long/short strategy based on the observation that an equal weighted portfolio will generally outperform a capitalized weighted portfolio such as the S&P 500 or the MSCI emerging market index EEM. A good compromise is 130/20 — wherein the equal weighted index is set at 130% long, and the cap-weighted index is 20% short. The portfolio is slightly levered at 110% with a commensurate cost of borrowing the extra 10%. The statistics for this variant is listed in Table 2, along with the associated statistics for the benchmarks: capital weighted index, equal weighted/5 trading days, and the 130/20 index.

Table 3 Performance of equity strategies with drawdown constraints (January 1999 to June 2009) (Benchmarks = EEM, EAFE, and S&P500; equal weighted; and Dynamic Portfolio Tactics DPT)

Emerging Markets	EEM	Equal Weight	DPT	Emerging Markets	EEM	Equal Weight	DPT
Annual Geometric Return	10.80%	12.80%	31.00%	Annual Geometric Return	6.61%	11.70%	20.38%
Annual Standard Deviation	36.90%	33.30%	21.60%	Annual Standard Deviation	25.86%	25.11%	19.02%
Best Month Return	30.60%	29.60%	7.80%	Best Month Return	30.60%	29.60%	8.80%
Sharpe Ratio	0.18	0.27	1.25	Sharpe Ratio	0.10	0.31	0.86
Worst Month Return	-26.10%	-25.70%	-8.70%	Worst Month Return	-26.10%	-25.70%	-10.60%
Max Drawdown	63.80%	63.30%	30.80%	Max Drawdown	63.80%	63.30%	33.90%
Return/MaxDrawdown	16.92%	20.28%	100.80%	Return/MaxDrawdown	10.40%	18.50%	60.10%
Correlation with EEM Strategy	1.00	0.98	0.74	Correlation with EEM Strategy	1.00	0.95	0.74

Europe, Australia, Far East	EAFE	EW	DPT	Europe, Australia, Far East	EAFE	EW	DPT
Annual Geometric Return	4.10%	7.00%	11.00%	Annual Geometric Return	4.37%	9.04%	16.55%
Annual Standard Deviation	23.50%	25.30%	12.10%	Annual Standard Deviation	18.95%	20.07%	13.94%
Best Month Return	17.30%	20.60%	6.50%	Best Month Return	17.30%	20.60%	9.20%
Sharpe Ratio	0.01	0.12	0.58	Sharpe Ratio	0.02	0.25	0.90
Worst Month Return	-12.80%	-13.50%	-8.50%	Worst Month Return	-12.80%	-13.50%	-8.50%
Max Drawdown	57.70%	59.80%	24.10%	Max Drawdown	57.70%	59.80%	24.10%
Return/MaxDrawdown	7.11%	11.71%	45.64%	Return/MaxDrawdown	7.57%	15.12%	68.67%
Correlation with EAFE Strategy	1.00	0.99	0.59	Correlation with EAFE Strategy	1.00	0.95	0.75

S&P 500	S&P 500	EW	DPT	S&P 500	S&P 500	EW	DPT
Annual Geometric Return	-4.70%	-1.70%	9.40%	Annual Geometric Return	6.29%	9.29%	16.50%
Annual Standard Deviation	20.60%	21.80%	13.70%	Annual Standard Deviation	18.75%	18.09%	14.43%
Best Month Return	15.70%	16.30%	5.50%	Best Month Return	15.70%	16.30%	9.60%
Sharpe Ratio	-0.42	-0.26	0.40	Sharpe Ratio	0.12	0.29	0.87
Worst Month Return	-15.00%	-15.80%	-5.90%	Worst Month Return	-15.00%	-15.80%	-6.00%
Max Drawdown	55.20%	54.80%	24.00%	Max Drawdown	55.20%	54.80%	24.00%
Return/MaxDrawdown	-8.51%	-3.10%	39.17%	Return/MaxDrawdown	11.39%	16.95%	68.75%
Correlation with S&P 500 Strategy	1.00	0.99	0.69	Correlation with S&P 500 Strategy	1.00	0.97	0.76

As mentioned in the previous section, capital growth theory requires a consistent relationship between the investor's capital at time t and their risks at the same time. Thus, as capital is lost and drawdown increases, the investor must lower risks — otherwise the size of the bets will be too large in most cases (roughly speaking). The DPT model takes this into account. As mentioned, we designate two breakpoints for approximating the nonlinear objective function, D_1 and D_2 to reduce risks by reducing capital in a complementary fashion.

The results of applying the approximation are shown in Table 3. Here we see that the overall performance has been increased, especially with regard to the worst case losses. Especially, note the worst and best case time periods. In the case of the cap-weighted index, we see that the upside periods are much larger than the upside periods for DPT. Conversely, the worst downside periods are much better with DPT than with the cap-weighted indices. As mentioned, we are willing to give up a substantial portion of the best periods, if we are able to protect the investor's capital during drawdown periods. This condition is an important criterion for achieving the capital growth path.

To improve upon the core equity results (core assets), we add two strategies. These strategies are designated as overlays, i.e., they do not require any dedicated capital for their purchase (only as marginable assets). See Mulvey *et al.* (2004) for a well known example of a successful overlay strategy based on trend following. Table 4 shows the historical performance of the two developed overlay strategies.

Table 4 Performance of two overlay strategies (DEO and PUI) at 100

Period: July 1999-June 2009	***100% DEO***	***100% PUI***
Annual Geometric Return	8.69%	7.32%
Annual Standard Deviation	16.39%	7.73%
Max Drawdown	14.29%	8.13%
Return/Volatility	0.53	0.95
Return/MaxDrawdown	0.61	0.90
Worst Month Return	-8.58%	-6.37%
Best Month Return	35.59%	5.56%

The first overlay strategy is called Duration Enhancing Overlay — DEO. See Mulvey *et al.* (2006) and Zhang (2006). This strategy increases the duration of a portfolio by entering into a swap agreement (or a long future purchase of long government bonds). DEO, in general, takes a long position on the long term bond and a short position on T-bill. This strategy has a particular benefit for a defined benefit pension plan to help them in matching duration of assets and liabilities (Mulvey *et al.*, 2006).

Dynamic DEO is a strategy that uses certain signals to improve performance vis a vis returns and reduced correlation with equity markets. The signals that the strategy uses are from (1) interest rates, (2) equity market returns, and (3) the strategy's performance over recent time periods. The basic idea is to use the trend property of interest rates and the volatility level of the equity market returns. When equity market volatility increases, the strategy increases its capital commitment. In addition to these, past performance of the strategy itself is used; it is assumed that in the DEO strategy a bad day precedes a cluster of bad days.

The second overlay strategy applies momentum and futures curve factors to a collection of futures markets — extending upon the traditional trend following rules developed by Dr. Frank Vannerson at Commodity Corp and later at Mt. Lucas Management, and many others (e.g., Bodie and Rosansky, 1980; Chan *et al.*, 1996; Erb and Harvey, 2006; Mulvey *et al.*, 2004; and Rouwenhorst, 1998). We call our strategy the Princeton University Index (PUI). Herein, the goal is to improve performance by capitalizing on patterns occurring in commodity markets.

The index is based on two ideas: the expected return of commodities futures depends upon the degree of hedging by producers or consumers — as evidenced by the shape of the futures curve for each commodity (Brennan *et al.*, 1997). Thus, a curve with sharp contango signifies that consumers are the primary hedgers, whereas backwardation signifies that the producers are the primary hedgers. We assume, and with much evidence to support the supposition, that hedgers will pay an expected costs for their behavior (and conversely that speculators will possess a positive expected cost) over long time periods and on average. Accordingly, we take positions that correspond to positive expected values. A rebalanced portfolio of commodity futures is constructed to greatly reduce risks for the PUI strategy.

Table 5 Performance of portfolio of strategies — July 1, 1999 to June 30, 2009

Period: July 1999-June 2009	100% SP500	100% EAFE	100% Tbond (30 year)	AIG commodity Index	GS commodity Index	DPT Combined Strategies
Annual Geometric Return	-2.26%	1.88%	6.95%	4.02%	10.67%	25.25%
Annual Standard Deviation	15.92%	18.08%	13.59%	17.46%	25.05%	14.94%
Max Drawdown	50.80%	54.24%	23.77%	54.50%	61.03%	12.99%
Return/Volatility	-0.14	0.10	0.51	0.23	0.43	1.69
Return/MaxDrawdown	-0.04	0.03	0.29	0.07	0.17	1.94
Worst Month Return	-16.52%	-19.05%	-14.56%	-21.34%	-27.77%	-7.68%
Best Month Return	9.93%	13.19%	16.05%	12.99%	21.10%	22.79%

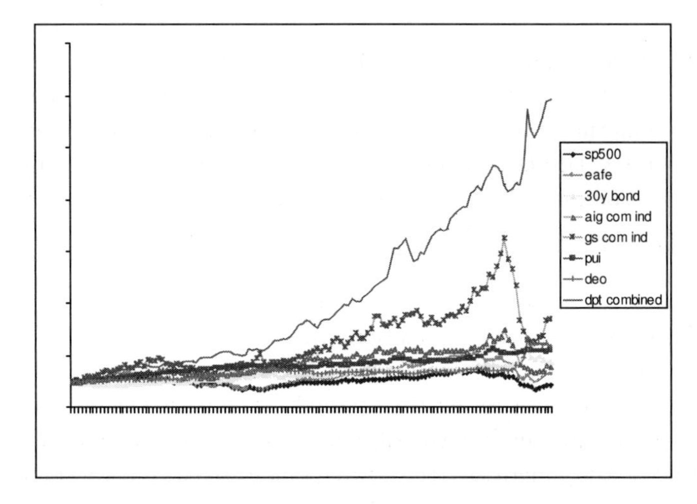

Figure 5 Time series performance for equity and bond benchmarks and combined DPT strategy (July 1, 1999 to June 30, 2009)

As a second factor, we measure the momentum of return for each of the commodities over the past six to twelve months. If momentum is positive, we increase our exposure to that commodity in the index. The overall commodity portfolio is rebalanced twice per month.

We next combine the three strategies: core equities as discussed in the previous section, the commodity overlay via PUI, and the interest rate overlay via DEO in an integrated portfolio (called DPT). Note that we are employing the core equity strategy as traditional investable assets, along with the two overlay strategies DEO and PUI. The percentage of traditional assets assigned for the overlays is a small portion of capital — around 10 to 15%. This small percentage prevents margin calls and remains within the target level of risks. Also, from the standpoint of risk allocation, we designate the proper proportion of the overlays, in this case — 100% for PUI, and 50% for DEO. These proportions are targets for a fixed-mix rebalancing rule on a monthly basis.

Turning to Table 5 and Figure 5, we see that the overall performance is enhanced by combining the three strategies in a rebalanced portfolio. In particular, the Sharpe ratio is improved since the strategies provide their best returns at different time conjunctures. In a similar fashion and more importantly, the return per drawdown ratios are superior for the integrated system. This performance is robust with respect to the modeling parameters since the individual strategies are designed to provide wide diversification. Also, the rebalancing gains can be significant and can be readily implemented due to the liquidity of ETFs and futures markets.

4 Conclusions

This paper extends capital growth models with downside protection. To achieve the twin goals of protecting the downside while generating good returns during upswings, we penalize drawdown values in a sequential fashion, first via a stochastic nonlinear program, and second by implementing relatively simple policy rules. We show that a long-term investor can perform well by avoiding large losses, as is well known. As a corollary, we demonstrate that a portion of the upside gains can be forsaken in our quest to improve the downside protection. This latter observation is somewhat controversial since some have argued that an investor must stay fully invested in order to improve on long term performance. Our objective function has an asymmetric relationship between the upside and the downside. But the changing market structure necessitates a more conservative strategy than is the case with traditional capital growth models. We focus on the drawdown values to implement this primary concept.

A second feature involves combining a set of investment strategies via dynamic portfolio tactics. Since sharp market corrections take place within the context of rising volatility and correlation, we cannot depend upon diversification benefits among asset securities. There is simply too much contagion during periods of high turbulence. Instead we design a portfolio of strategies in which the performance of one strategy is relatively uncorrelated with the performance of neighboring strategies. Several examples of these phenomena are presented.

The primary damage of the 2008 market crash was done in a quick amount of time — three months September toNovember (along with earlier losses in February). It is evident, therefore, that investors must remain vigilant and nimble if they wish to avoid large losses. The traditional asset allocation procedure possessing relatively long periods between revisions of policy have not served the investor well; as we mentioned, many investor experienced large losses (25–30% or more). A more dynamic framework is needed, with much shorter time intervals, in order to achieve the goals of capital growth. The investor must be careful to not only grow their capital during positive markets, but they must protect their capital during turbulent periods if they are to achieve the goals of capital growth theory.

References

Bell, R. M. and T. M. Cover (1980). Competitive optimality of logarithmic investment. *Math of Operations Research*, 5, 161–166.

Brennan, D., J. Williams, and B. D. Wright (1997). Convenience yield without the convenience: A spatial-temporal interpretation of storage under backwardation. *Economic Journal*, 107, 1009–1022.

Bodie, Z. and V. I. Rosansky (1980). Risk and return in commodity futures. *Financial Analysts Journal*, 36, 27–39.

Chan, L. K. C., N. Jegadeesh, and J. Lakonishok (1996). Momentum strategies. *The Journal of Finance*, 51(5), 1681–1713.

De Bondt, W. F. M. and R. Thaler (1985). Does stock market overreact? *The Journal of Finance*, 40(3), 793–805.

De Bondt, W. F. M. and R. Thaler (1987). Further evidence on investor overreaction and stock market seasonality. *The Journal of Finance*, 42(3), 557–581.

Ding, C., X. He (2004). K-Means clustering via principal component analysis. *ACM International Conference Proceeding Series.* Vol 69.

Erb, C. B., CFA, and C. R. Harvey (2006). The strategic and tactical value of commodity futures. *Financial Analysts Journal.* Volume 62, Number 2. CFA Institute.

Fernholz, R. and B. Shay (1982). Stochastic portfolio theory and stock market equilibrium. *The Journal of Finance*, 37, 615–624.

George, T. J. and C. Hwang (2004). The 52-week high and momentum investing. *The Journal of Finance*, 59(5), 2145–2176.

Grauer, R. R. and N. H. Hakansson (1986). A half century of returns on levered and unlevered portfolios of stocks, bonds and bills with and without small stocks. *Journal of Business*, 59(2), 287–318.

Guidolin, M., A. Timmermann (2007). Asset allocation under multivariate regime switching. *Journal of Economic Dynamics and Control*, 31, 3503–3544.

Hamilton, J. D. (1989). A new approach to the economic analysis of nonstationary time series and business cycles. *Econometrica*, 57, 357–384.

Jegadeesh, N. (1990). Evidence of predictable behavior of security returns. *The Journal of Finance*, 45(3), 881–898.

Kelly, Jr., J. R. (1956). A new interpretation of the information rate. *Bell System Technical Journal*, 35, 917–926.

Kim, C. (1994). Dynamic linear models with Markov switching. *Journal of Econometric*, 60, 1–22.

Luenberger, D. (1998). *Investment Science.* New York: Oxford University Press.

MacLean, L. C., R. Sanegre, Y. Zhao, and W. T. Ziemba (2004). Capital Growth with Security. *Journal of Economic Dynamics and Control*, 28, 937–954.

MacLean, L. C., E. Thorp, and W. T. Ziemba (2009). Optimal Capital Growth with Convex Loss Penalties, Working paper.

Merton, R. C. (1969). Lifetime portfolio selection under uncertainty: The continuous-time case. *Review Economics Statistics*, 51, 247–257.

Merton, R. C. (1973). An intertemporal capital asset pricing model. *Econometrica*, 41, 867–887.

Merton, R. C. and P. A. Samuelson (1974). Fallacy of the log-normal approximation to optimal portfolio decision-making over many periods. *Journal of Financial Economics*, 1, 67–94.

Mulvey, J. M., M. Bilgili, and W. C. Kim (2007/2008). Dynamic investment strategies and rebalancing gains. *The Euromoney Algorithmic Trading Handbook.* Chapter 13.

Mulvey, J. M. and M. Bilgili (2007). Discovering economic regimes and their application in stochastic optimization. Working paper, Princeton University.

Mulvey, J. M. and M. Bilgili (2009). Persistent clustering and its application to financial data. Working paper, Princeton University.

Mulvey, J. M., S. S. N. Kaul, and K. D. Simsek (2004). Evaluating a trend-following commodity index for multi-period asset allocation. *Journal of Alternative Investments*, 7, 54–69.
Mulvey, J. M. and W. C. Kim (Fall 2009). Evaluating style investment: Does a fund market defined along equity styles add value? *Quantitative Finance*, 9, 637–651.
Mulvey, J. M. and W. C. Kim (2009). Multi-stage financial planning: Integrating stochastic programs and policy simulators. In G. Infanger (Ed.). *Stochastic Programming: The State of the Art.*
Mulvey, J. M. and W. C. Kim (2008). Constantly rebalanced portfolio — Is mean reversion necessary? *Encyclopedia of Quantitative Finance*, 2, 714–717.
Mulvey, J. M., B. Pauling, and R. E. Madey (2003a). Advantages of multiperiod portfolio models. *Journal of Portfolio Management*, 29, 35–45.
Mulvey, J. M., K. Simsek, Z. Zhang, and F. Fabozzi (2008). Assisting defined-benefit pension plans. *Operations Research*, 56, 1066–1078.
Mulvey, J. M., K. Simsek, and Z. Zhang (2006). Improving investment performance for pension plans. *Journal of Asset Management*, 7, 93–108.
Mulvey, J. M., C. Ural, and Z. Zhang (2007). Improving performance for long-term investors: Wide diversification, leverage and overlay strategies. *Quantitative Finance*, 7(2), 1–13.
Mulvey, J. M. and S. Ling (2009). Replicating private equity returns by means of sector ETFs. Working paper, Princeton University.
Rouwenhorst, K. G. (1998). International momentum strategies. *The Journal of Finance*, 53(1), 267–284.
Samuelson, P. A. (1969). Lifetime portfolio selection by dynamic stochastic programming. *Review Economics Statistics*, 51, 239–246.
Samuelson, P. A. (1971). The fallacy of maximizing the geometric mean in long sequences of investing or gambling. *Proceedings National Academy of Science*, 68, 2493–2496.
Samuelson, P. A. (1979). Why we should not make mean log of wealth big though years to act are long. *Journal of Banking and Finance*, 3, 305–307.
Stutzer, M. (2003). Portfolio choice with endogenous utility: A large deviation approach. *Journal of Econometrics*, 116, 365–386.
Swensen, D. (2000). *Pioneering Portfolio Management.* US: Free Press.
Thorp, E. O. (1969). Optimal gambling systems for favorable games. *Review of the International Statistical Institute*, 37(3), 273–293.
Zenios, S. A. and W. T. Ziemba (2006), eds. Handbook of Asset and Liability Management. *Series of Handbooks in Finance*, Vol. 1. Amsterdam: Elsevier.
Zenios, S. A. and W. T. Ziemba (2007), eds. Handbook of Asset and Liability Management. *Series of Handbooks in Finance*, Vol. 2. Amsterdam: Elsevier.
Zhang, Z. (2006). Stochastic optimization for enterprise risk management. Doctoral Dissertation. Department of Operations Research and Financial Engineering, Princeton University, Princeton, NJ.
Ziemba, W. T. and J. M. Mulvey (1998), eds. *Worldwide Asset and Liability Modeling.* UK: Cambridge University Press.
Ziemba, W. T. (2005). The symmetric downside risk Sharpe ratio and the evaluation of great investors and speculators. *Journal of Portfolio Management*, Fall, 108–122.

ELSEVIER

Available online at www.sciencedirect.com
SCIENCE @ DIRECT•

Journal of Economic Dynamics & Control 28 (2004) 975–990

JOURNAL OF
Economic
Dynamics
& Control

www.elsevier.com/locate/econbase

51

Intertemporal surplus management

Markus Rudolf[a,*], William T. Ziemba[b]

[a] WHU-Otto Beisheim Graduate School of Management, Dresdner Bank chair of Finance, Burgplatz 2, 56179 Vallendar, Germany

[b] Faculty of Commerce, University of British Columbia, 2053 Main Mall, Vancouver, BC, Canada V6T 1Z2

Abstract

This paper presents an intertemporal portfolio selection model for pension funds or life insurance funds that maximizes the intertemporal expected utility of the surplus of assets net of liabilities. Following Merton (Econometrica 41 (1973) 867), it is assumed that both the asset and the liability return follow Itô processes as functions of a state variable. The optimum occurs for investors holding four funds: the market portfolio, the hedge portfolio for the state variable, the hedge portfolio for the liabilities, and the riskless asset. In contrast to Merton's result in the assets only case, the liability hedge is independent of preferences and only depends on the funding ratio. With HARA utility the investments in the state variable hedge portfolio are also preference independent. Finally, with log utility the market portfolio investment depends only on the current funding ratio.

JEL classification: G23; G11

Keywords: Asset; Beta; Funding ratio; HARA utility function; Hedge portfolio; Intertemporal capital asset pricing model; Itô process; *J*-function; Liabilities; Log utility function; Safety first; Shortfall risk; State variable; Surplus management

1. Introduction

Intertemporal asset allocation models date to Merton (1969, 1971, 1973) (see also the summary in Merton, 1990) and Samuelson (1969). Merton presents a continuous time intertemporal model whereas Samuelson discusses the discrete time case. Both models

* Corresponding author. WHU-Otto Beisheim Hochschule, Burgplatz 2, 56179 Vallendar, Germany. Tel.: +49-261-6509421; fax: +49-261-6509409.

E-mail addresses: markus.rudolf@unisg.ch, mrudolf@whu.edu (M. Rudolf), ziemba@interchange.ubc.ca (W.T. Ziemba).

URL: http://www.whu.edu/banking

0165-1889/03/$ - see front matter
doi:10.1016/S0165-1889(03)00058-7

976 M. Rudolf, W.T. Ziemba / Journal of Economic Dynamics & Control 28 (2004) 975–990

formulate a lifetime portfolio selection problem. Merton's (1973) intertemporal capital asset pricing model derives equilibrium asset premia. In contrast to Sharpe's (1964) CAPM, Merton's CAPM is based on a three-fund theorem. Each rational investor holds the riskless asset, the market portfolio, and a hedge portfolio for a so-called state variable in order to maximize his lifetime expected utility. The state variable is a stochastic term, which affects the asset price processes. The hedge portfolio provides maximum correlation to the state variable, i.e. it provides the best possible hedge against the state variable variance. Continuous time models can be applied to surplus optimization problems, since surplus optimizers, such as pension funds or life insurers, distribute the surplus to those who are insured. Pension funds or life insurance funds are frequently broadly internationally diversified. Exchange rate movements are considered as state variables.

Merton, (1993, reprinted in Ziemba and Mulvey (1998), see also Constantinides' (1993) comments on Merton's paper) addresses a problem similar to surplus optimization. He advocates the view that University endowment funds can be managed by using Merton's (1969) intertemporal portfolio selection model. His objective function is to maximize the lifetime expected utility of University activities with specific costs. The University activity portfolio includes education, training, research, storage of knowledge, etc. Since the activities of a University have to be optimized with respect to their costs, the activity costs are "liabilities" for universities. In contrast to our approach, Merton (1993) specifies the University's non-endowment cash flows as Itô processes dependent on the activity costs. This is the classical setting of Merton's (1973) intertemporal CAPM, where the state variables are given by the activity costs. The result is that each utility maximizing University holds three funds, the market portfolio, the riskless portfolio and a hedge portfolio against fluctuations of the activity costs. The composition across those three funds depends on the risk preferences towards market and activity cost volatility.

This paper provides a synthesis of the surplus management and the continuous time finance literature. Both the assets and liabilities of a pension fund are modeled as stochastic processes dependent on stochastic state variables. Only the asset mix can be influenced by the fund manager's decision. The paper provides a solution to his decision problem. In comparison to Merton's (1993) setting, the advance made by the paper is that one (or several) state variable(s) are allowed to be distinct from the liabilities. I.e. in contrast to Merton (1993), this paper explicitly examines the impact of long-term liabilities. Three important results developed are: first, if only one state variable is considered, Merton's three-fund theorem is extended to a four-fund theorem. The four distinct mutual funds are: the market portfolio, the state variable hedge portfolio, the liability hedge portfolio, and the riskless asset. Secondly and most important, the investment in the liability hedge portfolio depends only on the current funding ratio of a pension or life insurance fund and is independent of the utility function. This result differs from Merton's (1993) findings. The preference independence of the liability hedge portfolio has major implications for monitoring pension funds and life insurance companies. Thirdly, the state variable hedging policy is preference independent if hyperbolic absolute risk aversion (HARA) utility functions are assumed. With log utility, even the market portfolio investment is independent of the risk aversion

M. Rudolf, W.T. Ziemba / Journal of Economic Dynamics & Control 28 (2004) 975–990 977

coefficient which is caused by the fact that log utility is equivalent to HARA utility with a risk tolerance coefficient of $\alpha = 0$.

In Section 2, the intertemporal surplus management model and a four-fund theorem are derived. In Section 3 we show that with HARA utility, the risk tolerance coefficients of the model are related to the funding ratio and to the currency betas of the portfolio. A k state variable case is derived in Section 4. Section 5 describes a case study for a surplus optimizer who is diversified across the stock, the bond markets, and cash equivalents of four countries (EMU countries are treated as a single country). Section 6 concludes the paper.

2. An intertemporal surplus management model—a four-fund theorem

Having surplus over time is what life insurance companies and pension funds try to achieve. In both life insurance companies and pension funds, parts of the surplus are distributed to the clients usually once every year. Hence, optimizing the investment strategy of a life insurance or a pension fund is equivalent to maximizing the expected lifetime utility of the surplus.

For $t \geqslant 0$ we have the stochastic processes $A(t), L(t), Y(t)$, representing assets, liabilities, and a state variable Y. An extension of the setting to k state variables is possible without difficulties (see e.g. Richard (1979) and Adler and Dumas (1983)). Adler and Dumas (1983) take purchasing power parities as state variables. The surplus $S(t)$ and the funding ratio $F(t)$ are defined by

$$S(t) := A(t) - L(t), \qquad F(t) := A(t)/L(t). \tag{1}$$

According to Merton (1973), the state variable Y follows a geometric Brownian motion (i.e. log-normally distributed) where μ_Y and σ_Y are constants representing the drift and volatility, and $Z_Y(t)$ is a standard Wiener process. The state variable as well as the assets and the liabilities are assumed to follow the stochastic processes:

$$\begin{aligned}
\mathrm{d}Y(t) &= Y(t)[\mu_Y \,\mathrm{d}t + \sigma_Y \,\mathrm{d}Z_Y(t)],\\
\mathrm{d}A(t) &= A(t)[\mu_A(t, Y(t))\,\mathrm{d}t + \sigma_A(t, Y(t))\,\mathrm{d}Z_A(t)],\\
\mathrm{d}L(t) &= L(t)[\mu_L(t, Y(t))\,\mathrm{d}t + \sigma_L(t, Y(t))\,\mathrm{d}Z_L(t)],
\end{aligned} \tag{2}$$

where the drift and volatility parameters for the assets and liabilities are allowed to depend on both time and state variable, and where $Z_A(t), Z_L(t)$ are Wiener processes correlated with $Z_Y(t)$ and also with each other. Hence, we have cross variations

$$\begin{aligned}
\mathrm{d}Z_A(t)\,\mathrm{d}Z_Y(t) &= \rho_{AY}\,\mathrm{d}t,\\
\mathrm{d}Z_L(t)\,\mathrm{d}Z_Y(t) &= \rho_{LY}\,\mathrm{d}t,\\
\mathrm{d}Z_A(t)\,\mathrm{d}Z_L(t) &= \rho_{AL}\,\mathrm{d}t,
\end{aligned}$$

where ρ_{AY} represents the instantaneous correlation between the Wiener processes $Z_A(t)$ and $Z_Y(t)$, ρ_{LY} is the instantaneous correlation between $Z_L(t)$ and $Z_Y(t)$, and ρ_{AL} refers to the instantaneous correlation between $Z_A(t)$ and $Z_L(t)$. The return processes $R_Y(t), R_A(t), R_L(t)$ for the state variable, assets and liabilities are defined by

$$R_Y(t) := \mathrm{d}Y(t)/Y(t),$$

$$R_A(t) := \mathrm{d}A(t)/A(t),$$

$$R_L(t) := \mathrm{d}L(t)/L(t).$$

Following Sharpe and Tint (1990) and according to Eq. (1), the surplus return process is defined as

$$R_S(t) := \frac{\mathrm{d}S(t)}{A(t)} = R_A(t) - \frac{R_L(t)}{F(t)}$$

$$= \left[\mu_A - \frac{\mu_L}{F(t)}\right] \mathrm{d}t + \sigma_A \,\mathrm{d}Z_A(t) - \frac{\sigma_L}{F(t)} \mathrm{d}Z_L(t). \tag{3}$$

Since $\mathrm{d}S(t) = R_S(t)A(t)$, using Eq. (3) we have

$$\mathrm{E}[\mathrm{d}S(t)] = A(t)\left(\mu_A - \frac{\mu_L}{F(t)}\right) \mathrm{d}t,$$

$$\mathrm{d}S(t)\,\mathrm{d}S(t) = A^2(t)\left(\sigma_A^2 + \frac{\sigma_L^2}{F^2(t)} - 2\frac{\sigma_{AL}}{F(t)}\right) \mathrm{d}t,$$

$$\mathrm{d}S(t)\,\mathrm{d}Y(t) = A(t)Y(t)\left(\sigma_{AY} - \frac{\sigma_{LY}}{F(t)}\right) \mathrm{d}t,$$

$$\mathrm{d}Y(t)\,\mathrm{d}Y(t) = Y^2(t)\sigma_Y^2 \,\mathrm{d}t, \tag{4}$$

where $\sigma_{AL} := \rho_{AL}\sigma_A\sigma_L, \sigma_{AY} := \rho_{AY}\sigma_A\sigma_Y, \sigma_{LY} := \rho_{LY}\sigma_L\sigma_Y$ are the covariances of assets with liabilities and of the state variable with assets and liabilities, respectively.

The objective is to maximize the expected lifetime utility of surplus, which implies identifying an optimum surplus strategy.[1] For an analysis of a situation with convex penalties for underfunding, see Cariño and Ziemba (1998). The expected utility is positively related to the surplus in each period. This is because positive surpluses improve the wealth position of the insurants of a pension or a life insurance fund, even if the yearly retirement benefits are over-covered by the surplus. Hence, in this interpretation, the insurants are like shareholders of the fund. Then the following equation is the fund's optimization problem, where U is an additively separable, twice differentiable,

[1] To avoid underfunding in particular periods we assume a steady state implying that the dollar value of employees (insurance policies) entering equals those leaving the pension fund (insurance company) dollar value.

concave utility function and T is the end of the fund's existence:

$$J(S,Y,t) := \max_{\mathbf{w}} \mathrm{E}_t \left(\int_t^T U(S,Y,\tau)\,\mathrm{d}\tau \right)$$

$$= \max_{\mathbf{w}} \mathrm{E}_t \left(\int_t^{t+\mathrm{d}t} U(S,Y,\tau)\,\mathrm{d}\tau + \int_{t+\mathrm{d}t}^T U(S,Y,\tau)\,\mathrm{d}\tau \right). \tag{5}$$

E_t denotes expectation with respect to the information set at time t and the J-function is the maximum of the life insurance or pension fund's expected lifetime utility. The maximum in (5) is taken with respect to $\mathbf{w}$, the vector of portfolio fractions of the risky assets. Let n be the number of portfolio assets and ω_i $(1 \leqslant i \leqslant n)$ be the portfolio fraction of asset i, then $\mathbf{w}' := (\omega_1, \ldots, \omega_n)$. Applying the Bellman principle (see e.g. Dixit and Pindyck (1994, Chapter 4)), according to Merton (1969) and Merton (1990, p. 102) and according to the mean value theorem for integrals yields

$$J(S,Y,t) = U(S,Y,t)\,\mathrm{d}t + \max_{\mathbf{w}} \mathrm{E}_t[J(S+\mathrm{d}S, Y+\mathrm{d}Y, t+\mathrm{d}t)]. \tag{6}$$

Let J_S be the first partial derivative of J with respect to S, J_{SS} the second partial derivative of J with respect to S, J_Y and J_{YY} the first and second partial derivatives with respect to Y, and J_{SY} the derivative of J with respect to S and Y. Applying Itô's lemma and Eq. (4) yields (see the appendix for proof),

$$0 = U(t,S,Y)$$

$$+ \max_{\mathbf{w}} \begin{bmatrix} J_t + J_S A(t) \left(\mu_A - \dfrac{\mu_L}{F(t)} \right) \\ + J_Y Y(t)\mu_Y + \dfrac{1}{2} J_{SS} A^2(t) \left(\sigma_A^2 + \dfrac{\sigma_L^2}{F^2(t)} - 2 \cdot \dfrac{\sigma_{AL}}{F(t)} \right) \\ + \dfrac{1}{2} J_{YY} Y^2(t) \sigma_Y^2 + J_{SY} A(t) Y(t) \left(\sigma_{AY} - \dfrac{\sigma_{LY}}{F(t)} \right) \end{bmatrix}. \tag{7}$$

Let $\mathbf{m}_A$ be the vector of expected asset returns of the risky assets of dimension n, $\mathbf{e}$ be the n-dimensional vector of ones, $\mathbf{V}$ be the $n \times n$ matrix of covariances between the n risky assets, $\mathbf{v}_{AL}$ and $\mathbf{v}_{AY}$ be the vectors of covariances between the n assets and the liabilities, respectively, with the state variable. It is assumed that a riskless asset with return r exists. Then rearranging (7) yields the following equation (see the appendix for proof and definitions of the matrix and the vectors):

$$0 = U(t,S,Y)$$

$$+ \max_{\mathbf{w}} \begin{bmatrix} J_t + J_S A(t) \left(\mathbf{w}'(\mathbf{m}_A - r\mathbf{e}) + r - \dfrac{\mu_L}{F(t)} \right) \\ + J_Y Y(t)\mu_Y + \dfrac{1}{2} J_{SS} A^2(t) \left(\mathbf{w}'\mathbf{V}\mathbf{w} + \dfrac{\sigma_L^2}{F^2(t)} - 2 \dfrac{\mathbf{w}'\mathbf{v}_{AL}}{F(t)} \right) \\ + \dfrac{1}{2} J_{YY} Y^2(t) \sigma_Y^2 + J_{SY} A(t) Y(t) \left(\mathbf{w}'\mathbf{v}_{AY} - \dfrac{\sigma_{LY}}{F(t)} \right) \end{bmatrix}. \tag{8}$$

980 M. Rudolf, W.T. Ziemba / Journal of Economic Dynamics & Control 28 (2004) 975–990

Differentiating (8) with respect to $\mathbf{w}$ yields[2]

$$\mathbf{w} = -\frac{J_S}{A(t)J_{SS}}\mathbf{V}^{-1}(\mathbf{m}_A - r\mathbf{e}) - \frac{Y(t)J_{SY}}{A(t)J_{SS}}\mathbf{V}^{-1}\mathbf{v}_{AY} + \frac{\mathbf{V}^{-1}\mathbf{v}_{AL}}{F(t)}$$

$$=: -a\frac{J_S}{A(t)J_{SS}} \cdot \mathbf{w}_M - b\frac{Y(t)J_{SY}}{A(t)J_{SS}} \cdot \mathbf{w}_Y + \frac{c}{F(t)} \cdot \mathbf{w}_L, \tag{9}$$

where

$$\mathbf{w}_M := \frac{\mathbf{V}^{-1}(\mathbf{m}_A - r\mathbf{e})}{\mathbf{e}'\mathbf{V}^{-1}(\mathbf{m}_A - r\mathbf{e})}, \quad \mathbf{w}_Y := \frac{\mathbf{V}^{-1}\mathbf{v}_{AY}}{\mathbf{e}'\mathbf{V}^{-1}\mathbf{v}_{AY}}, \quad \mathbf{w}_L := \frac{\mathbf{V}^{-1}\mathbf{v}_{AL}}{\mathbf{e}'\mathbf{V}^{-1}\mathbf{v}_{AL}},$$

$$a := \mathbf{e}'\mathbf{V}^{-1}(\mathbf{m}_A - r\mathbf{e}), \quad b := \mathbf{e}'\mathbf{V}^{-1}\mathbf{v}_{AY}, \quad c := \mathbf{e}'\mathbf{V}^{-1}\mathbf{v}_{AL}.$$

The vectors $\mathbf{w}_M$, $\mathbf{w}_Y$, and $\mathbf{w}_L$ are of dimension n with elements that sum to 1; a, b, and c are real constants. The optimum portfolio consists of four single portfolios: the market portfolio $\mathbf{w}_M$, the hedge portfolio for the state variable $\mathbf{w}_Y$, which is Merton's (1973) state variable hedge portfolio, the hedge portfolio for the liabilities $\mathbf{w}_L$, and the riskless asset. The state variable hedge portfolio $\mathbf{w}_Y$ reveals the maximum correlation with the state variable Y (see the appendix). A perfect hedge for the state variable could be achieved if the universe of n risky assets contains forward contracts on the state variable. Then the state variable hedge portfolio would consist of a single asset, which is the forward contract. The third portfolio $\mathbf{w}_L$ is interesting. For the liabilities there exist no hedging opportunities at the financial markets (i.e. a portfolio which hedges wage increases or inflation rates). Eq. (9) shows how a liability hedge can be constructed. This is related to the problem addressed by Ezra (1991) and Black (1989) and solves it intertemporally. In the four-fund theorem, life insurance and pension funds invest in the following four funds, if they maximize their expected lifetime utility (5) based on the stochastic differential equations (2) and (3):

1. The market portfolio $\mathbf{w}_M$ with level $-a(J_S/A(t)J_{SS})$.
2. The state variable hedge portfolio $\mathbf{w}_Y$ with level $-b(Y(t)J_{SY}/A(t)J_{SS})$.
3. The riskless asset with level $1 + a(J_S/A(t)J_{SS}) + b(Y(t)J_{SY}/A(t)J_{SS}) - c/F(t)$.
4. Finally, the liability hedge portfolio $\mathbf{w}_L$ with level $c/F(t)$.

Thus, the holdings of the liability hedge portfolio are independent of preferences. The most interesting result in the portfolio selection equation (9) is that the liability hedge portfolio holdings depend only on the current funding ratio and not on the form of the utility function. In order to maximize its lifetime expected utility, each life insurance or pension fund should hedge the liabilities according to the financial endowment.

The percentages of each of the three other funds differ according to the risk preferences of the investors. For example, $-a(J_S/A(t)J_{SS})$ is the percentage invested in

[2] The derivative of (8) with respect to $\mathbf{w}$ is: $0 = J_S(\mathbf{m}_A - r\mathbf{e}) + A(t)J_{SS}\mathbf{V}\mathbf{w} - [A(t)/F(t)]J_{SS}\mathbf{v}_{AL} + Y(t)J_{SY}\mathbf{v}_{AY}$.

the market portfolio. Since J is a "derived" utility function, this ratio is a times the Arrow/Pratt relative risk tolerance with respect to changes in the surplus. That is, the higher the risk tolerance towards market risk, the higher the fraction of the market portfolio holdings. The percentage of the state variable hedge portfolio is $-b(Y(t)J_{SY}/A(t)J_{SS})$. Merton (1973) showed that this ratio is b times the Arrow/Pratt relative risk tolerance with respect to changes in the state variable. The percentage of the liability hedge portfolio is $c/F(t)$. Surprisingly, and in contrast to Merton's results, this portfolio does not depend on preferences nor on a specific utility function, but only on the funding ratio of the pension fund. The lower the funding ratio, the higher the percentage of the liability hedge portfolio.

This allows for a simple technique to monitor life insurance and pension funds, which extends Merton's (1993) approach. In most funding systems, funds are legally obliged to invest subject to a deterministic threshold return. Since payments of life insurance or pension funds depend on the growth and the volatility of wage rates, this is not appropriate. For instance, if the threshold return is 4% p.a. and the wages grow by more than this, the liabilities cannot be covered by the assets. Our model suggests instead that a portfolio manager of a pension fund should invest in a portfolio which smoothes the fluctuation of the surplus returns caused by wage volatility, i.e. in a liability hedge portfolio. Since the liability hedge portfolio depends only on the funding ratio, preferences of the insurants have not to be specified.

3. Risk preference, funding ratio, and currency betas

Assume that the utility function is from the HARA class. Merton (1971 and 1990, p. 140) shows that this is equivalent to assuming that the J-function (6) belongs to the HARA class as well. Let $\alpha < 1$ be the risk tolerance coefficient. Then

$$U(S,Y,t) \subset HARA \Leftrightarrow J(S,Y,t) \subset HARA,$$

$$J(S,Y,t) = J[S(A(Y)),t] = \frac{1-\alpha}{e^{\rho t}\alpha}\left(\frac{\kappa S}{1-\alpha}+\eta\right)^{\alpha}, \tag{10}$$

where κ and $\eta > 0$ are real constants and ρ is the utility deflator. Observe that (10) implies linear absolute risk tolerance since $-J_S/J_{SS} = S/(1-\alpha) + \eta/\kappa$. The HARA class of utility functions as defined in (10) is commonly used; it implies the negative exponential utility functions $J = -e^{-aS}$ when α approaches $-\infty$, $\eta = 1$, and $\rho = 0$. Moreover, the isoelastic power utility (see Ingersoll (1987, p. 39)) $J = S^{\alpha}/\alpha$ is obtained for $\eta = 0$, $\rho = 0$, and $\kappa = (1-\alpha)^{(\alpha-1)/\alpha}$. Log utility is that member of isoelastic power utility when α approaches 0 ($\eta = \rho = 0$), since by applying l'Hôpital's rule to the equivalent utility function,

$$\lim_{\alpha\to 0} J(S,Y,t) = \lim_{\alpha\to 0}\frac{S^{\alpha}-1}{\alpha} = \lim_{\alpha\to 0}(S^{\alpha}\ln S) = \ln S. \tag{11}$$

982 M. Rudolf, W.T. Ziemba / Journal of Economic Dynamics & Control 28 (2004) 975–990

Under the HARA utility function assumption,

$$J_S = \frac{\kappa}{e^{\rho t}}\left(\frac{\kappa S}{1-\alpha}+\eta\right)^{\alpha-1}, \quad J_{SS} = -\frac{\kappa^2}{e^{\rho t}}\left(\frac{\kappa S}{1-\alpha}+\eta\right)^{\alpha-2},$$

$$J_{SY} = \frac{dJ_S}{dY} = \frac{dJ_S}{dS}\cdot\frac{dS}{dA}\cdot\frac{dA}{dY} = J_{SS}\cdot\frac{dA}{dY}. \tag{12}$$

The percentage holdings of the market portfolio may be re-expressed as

$$-\frac{J_S}{AJ_{SS}} = \left(\frac{S}{A(1-\alpha)}+\frac{\eta}{A\kappa}\right) = \frac{1}{1-\alpha}\left(1-\frac{1}{F}\right)+\frac{\eta}{A\kappa}. \tag{13}$$

The market portfolio holdings thus depend on the funding ratio and on the risk aversion. The higher F, the higher the investment in the market portfolio, and the higher α, the lower the market portfolio investment for funding ratios smaller than 1. If a sufficient funding is observed (i.e. $F > 1$), then there is a positive relationship between α and F. If α approaches 0 (log utility case), the coefficient for the market portfolio investment becomes $1-1/F$ plus the constant $\eta/(A\kappa)$. Thus, for log utility pension or life insurance funds, risk aversion does not matter. Indeed only the funding ratio matters to determine the market portfolio investment. For either case of utility functions, if the funding ratio is 100%, there will be no investment in the risky market portfolio. Thus, the funding ratio of a fund does not only determine the capability to bear risk but also the willingness to take risk.

We now consider the percentage holding of the state variable hedge portfolio. Suppose the state variable Y is an exchange rate fluctuation, which affects the surplus of a life insurance or pension fund. Given (12) it follows that

$$-\frac{YJ_{SY}}{AJ_{SS}} = -\frac{dA/A}{dY/Y} = -\frac{R_A}{R_Y}. \tag{14}$$

Hence, the weight of the state variable hedge portfolio (14) is preference independent if HARA utility is assumed. Assume the regression model $R_A = \beta(R_A, R_Y)R_Y$ (asset returns are linear functions of currency returns). Hence, $-R_A/R_Y$ is the negative beta of the portfolio with respect to the state variable. Using (14) it follows that

$$-\frac{YJ_{SY}}{AJ_{SS}} = -\beta(R_A, R_Y). \tag{15}$$

If Y is an exchange rate, then $-\beta$ equals the minimum variance hedge ratio for the foreign currency position. The holdings of the state variable hedge portfolio are independent of preferences; they only depend on the foreign currency exposure of the portfolio. The higher the exchange rate risk in the portfolio, the higher the currency hedging.

Hence, for the HARA utility case, only the investment in the market portfolio depends on the risk aversion α. The investments in all other funds are preference independent. They depend only on the funding ratio and on the exposure of the asset portfolio to the state variable.

M. Rudolf, W.T. Ziemba / Journal of Economic Dynamics & Control 28 (2004) 975–990 983

4. The multiple state variable case

Let k be the number of foreign currencies contained in a life insurance or pension fund's portfolio, $Y_1,\ldots,Y_k$ be the exchange rates in terms of the domestic currency, and $R_{Y_1},\ldots,R_{Y_k}$ the exchange rate returns of the k currencies. Then

$$-\frac{Y_1 J_{SY_1}}{AJ_{SS}} = -\beta(R_A,R_{Y_1}),\ldots,-\frac{Y_k J_{SY_k}}{AJ_{SS}} = -\beta(R_A,R_{Y_k}),$$

which are the Arrow/Pratt relative risk tolerances with respect to changes in the exchange rates for the HARA case. Let $\mathbf{v}_{AY_1},\ldots,\mathbf{v}_{AY_k}$ be the covariance vectors of the returns of the asset portfolio with the k exchange rate returns,

$$\mathbf{w}_{Y_1} := \frac{\mathbf{V}^{-1}\mathbf{v}_{AY_1}}{\mathbf{e}'\mathbf{V}^{-1}\mathbf{v}_{AY_1}},\ldots,\mathbf{w}_{Y_k} := \frac{\mathbf{V}^{-1}\mathbf{v}_{AY_k}}{\mathbf{e}'\mathbf{V}^{-1}\mathbf{v}_{AY_k}},$$

which are the state variable hedge portfolios 1–k, and $b_1 := \mathbf{e}'\mathbf{V}^{-1}\mathbf{v}_{AY_1},\ldots,b_k := \mathbf{e}'\mathbf{V}^{-1}\mathbf{v}_{AY_k}$ are the coefficients of the state variable hedge portfolios. A pension or a life insurance fund with HARA utility function facing k state variables has the following investment strategy ($R_l := \mathrm{d}A_l/A_l$: return on risky asset l, $l = 1,\ldots,n$):

$$\mathbf{w} = \left[\frac{a}{1-\alpha}\left(1-\frac{1}{F(t)}\right)+\frac{a\eta}{A\kappa}\right]\mathbf{w}_M - \sum_{i=1}^{k} b_i\beta(R_A,R_{Y_i})\mathbf{w}_{Y_i} + \frac{c}{F(t)}\mathbf{w}_L, \tag{16}$$

where

$$\beta(R_A,R_{Y_i}) = \sum_{l=1}^{n} \omega_l \beta(R_l,R_{Y_i}).$$

Since the right-hand side of Eq. (16) depends on the portfolio allocation $\mathbf{w}$, it is not possible to solve (16) analytically for $\mathbf{w}$. However, numerical solutions can be applied. For k state variables a $(k+3)$-fund theorem thus follows.

5. Case study

The following case study is based on a USD-based surplus optimizer investing in the stock and bond markets of the US, UK, Japan, the EMU countries, Canada, and Switzerland. Monthly MSCI data between January 1987 and July 2000 (163 observations) are used for the stock markets. The monthly JP Morgan indices are used for the bond markets in this period (Salomon Brothers for Switzerland). The stochastic benchmark for a surplus optimizer is the quarterly Thomson Financial Datastream index for US wages and salaries. Quarterly data are linearly interpolated in order to obtain monthly wages and salaries data. The average growth rate of wages and salaries in the U.S. between January 1987 and July 2000 was 5.7% p.a. (see Table 1) with annualized volatility of 4.0%. Table 1 also contains the stock and bond market descriptive statistics in USD, and the currency betas of the indices.

All foreign currencies except CAD, i.e. GBP, JPY, EUR, and CHF, have volatility of about 12% p.a., and all currencies except JPY depreciated against the USD by a little

Table 1
Descriptive statistics

		Mean return	Volatility	Beta GBP	Beta JPY	Beta EUR	Beta CAD	Beta CHF
Stocks	USA	13.47	14.74	0.18	0.05	0.35	−0.59	0.35
	UK	9.97	17.96	−0.47	−0.36	−0.29	−0.48	−0.14
	Japan	3.42	25.99	−0.61	−1.11	−0.45	−0.44	−0.43
	EMU countries	10.48	15.80	−0.32	−0.27	−0.26	−0.45	−0.11
	Canada	5.52	18.07	0.05	−0.02	0.27	−1.44	0.34
	Switzerland	11.56	18.17	−0.14	−0.32	−0.26	0.13	−0.32
Bonds	USA	5.04	4.50	−0.03	0.00	−0.06	−0.04	−0.06
	UK	6.86	12.51	−0.92	−0.44	−0.80	−0.39	−0.59
	Japan	3.77	14.46	−0.53	−1.04	−0.75	0.09	−0.72
	EMU countries	7.78	10.57	−0.69	−0.42	−0.93	−0.06	−0.74
	Canada	5.16	8.44	−0.09	0.03	−0.02	−1.14	0.02
	Switzerland	3.56	12.09	−0.67	−0.53	−1.03	0.19	−0.99
Exchange rates in USD	GBP	0.11	11.13	1	0.37	0.86	0.42	0.64
	JPY	−2.75	12.54	0.48	1	0.65	0.04	0.61
	EUR[a]	1.13	10.08	0.7	0.42	1	0.1	0.79
	CAD	0.79	4.73	0.08	0	0.02	1	−0.02
	CHF	0.32	11.57	0.69	0.52	1.04	−0.13	1
	Wages and salaries	5.71	4.0	0	0.01	0	−0.01	0

The stock market data are based on MSC indices and the bond data on JP Morgan indices (Switzerland on Salomon Brothers data). The wage and salary growth rate is from datastream. Monthly data between January 1987 and July 2000 (163 observations) is used. All coefficients are in USD. The average returns and volatilities are in percent per annum.

[a]ECU before January 1999.

more than 0 to 1.13% per year. From a USD viewpoint, for GBP-beta is especially high (absolute value) for the UK bond market. The Japanese stock and bond market reveal a JPY-beta of −1.11 and −1.04, respectively, and the EMU bond market has a EUR-beta of −0.93. Furthermore, the CAD-beta of the Canadian bond market is −1.14, the CHF-beta of the Swiss bond market is −0.99. All other countries have substantially lower currency betas. Since the betas are close to zero, the wages and salaries in the U.S. obviously do not depend on currency movements.

The investor faces an exposure against five foreign currencies (GBP, JPY, EUR, CAD, CHF), and has to invest into eight funds. Five of them are hedge portfolios for the state variables, which are assumed to be currency returns. The next step is to calculate the compositions of the eight funds. The results appear in Table 2. The major holdings in the market portfolio are investments in the US stock market and the EMU bond market. Substantial short positions for the tangency portfolio are obtained for the Canadian stock and the Swiss bond market. The US bond portfolio is a major part of a portfolio providing the best hedge against fluctuations in wages and salaries. More than 126% of the liability hedge portfolio are invested in the US bond market. The bond markets of the respective currencies dominate the currency hedge portfolios.

Table 2
Optimum portfolios of an internationally diversified pension fund from a USD perspective

		Market portfolio	Liability hedge portfolio	Hedge portfolio GBP	Hedge portfolio JPY	Hedge portfolio EUR	Hedge portfolio CAD	Hedge portfolio CHF
Stocks	USA	83.9	−30.4	3.9	−4.5	−7.5	60.9	−5.1
	UK	−14.8	60.6	−31.0	−1.1	−8.6	−85.3	1.9
	Japan	−6.7	2.7	6.1	8.9	−2.4	−4.0	−0.7
	EMU	−19.2	−68.3	35.3	12.8	23.5	138.3	−3.8
	Canada	−39.2	5.1	14.6	13.6	5.5	−2.6	−0.9
	Switzerl.	21.6	1.5	−24.8	−14.6	−14.7	−100.7	1.0
Bonds	USA	14.8	126.0	−126.4	−28.6	−41.2	−627.3	−35.9
	UK	−9.7	−56.8	189.7	7.9	−3.4	97.5	−8.5
	Japan	0.6	39.1	−31.2	133.9	−3.6	−30.1	6.4
	EMU	138.5	9.6	−16.4	−38.0	97.3	−170.1	34.0
	Canada	7.9	5.0	−11.8	−32.6	−7.9	679.6	0.9
	Switzerl.	−77.7	5.9	91.9	42.4	62.9	143.8	110.7

The portfolio holdings are based on Eq. (9). All portfolio fractions are percentages. A riskless rate of interest of 2% per annum is assumed.

Table 3
Weightings of the funds due to different funding ratios

Funding ratio	0.9	1	1.1	1.2	1.3	1.5
Market portfolio (%)	−11.5	0.0	6.3	12.3	18.2	24.6
Liability hedge portfolio (%)	14.8	13.3	12.1	11.1	10.2	8.9
Hedge portfolio GBP (%)	−0.6	−0.5	−0.5	−0.5	−0.4	−0.4
Hedge portfolio JPY (%)	1.2	1.1	1.0	0.9	0.9	0.8
Hedge portfolio EUR (%)	0.3	0.2	0.1	0.0	0.0	−0.1
Hedge portfolio CAD (%)	−0.1	−0.1	−0.1	−0.1	−0.1	−0.1
Hedge portfolio CHF (%)	1.1	1.0	0.9	0.8	0.8	0.7
Riskless assets (%)	94.9	85.1	80.2	75.3	70.4	65.6
Portfolio beta against GBP	−0.01	−0.01	−0.01	−0.01	−0.01	−0.01
Portfolio beta against JPY	0.02	0.02	0.02	0.02	0.02	0.02
Portfolio beta against EUR	0.01	0.00	0.00	0.00	0.00	0.00
Portfolio beta against CAD	−0.02	−0.02	−0.01	−0.01	−0.01	−0.01
Portfolio beta against CHF	0.02	0.01	0.01	0.01	0.01	0.01

The weightings of the portfolios according to Eq. (16), where $\alpha = 0$, i.e. log utility, is assumed.

As derived in Section 3, the holdings of the six funds depend only on the funding ratio and on the currency betas of the distinct markets. This is shown in Table 3. For a funding ratio of one, there is no investment in the market portfolio and only diminishing investments in the currency hedge portfolios. The portfolio betas against the five currencies are close to zero for all funding ratios. The higher the funding ratio, the higher the investment in the market portfolio, the lower the investments in the liability and the state variable hedge portfolios, and the lower the investment in the riskless fund. Negative currency hedge portfolios imply an increase of the currency exposure instead of a hedge against it. The increase of the market portfolio holdings and the reduction of the hedge portfolio holdings and the riskless fund for increasing funding ratios shows that the funding ratio is directly related to the ability to bear risk. Rather than risk aversion coefficients, the funding ratio provides an objective measure to quantify attitudes towards risk.

6. Conclusions

This paper derives a four-fund theorem for intertemporal surplus optimizers such as life insurance and pension funds. In addition to the three funds identified by Merton (1973), the expected utility maximizing portfolio contains a liability hedge portfolio, which is preference independent. Its holdings depend only on the funding ratio of a pension fund. The higher the funding ratio, the lower the necessity for liabilities hedging. As a practical consequence, the hedging policy of pension or life insurance funds could very easily be monitored by authorities. This is due to the fact that in the optimum, only the funding ratio is decisive for the liabilities hedge, and not the utility function, which hardly can be determined by law. Today's pension fund laws do not contain any rules for the treatment of stochastic wage growths. Although Merton (1993) addresses a similar problem of the optimal investment strategy for University endowment funds, his setting is different. He describes the "liabilities" of universities, i.e. the costs for their activities, as state variables. He obtains a preference-dependent hedge portfolio for the activity costs. In contrast, here the liability returns are specified as a part of the surplus return. In contrast to Merton's results, the hedge portfolio for the liabilities is exclusively dependent on the funding ratio (and not on risk preferences) of a life insurance or pension fund. The funding ratio is an objective measure, whereas risk preferences are hard to determine.

The model provides an intertemporal portfolio selection approach for surplus optimizers. The intertemporal surplus management approach holds for investors who have to cover liabilities by their assets in each moment of time. Since the investment strategy of all investors is consumption orientated, it is reasonable to assume that all investors invest in order to cover their liabilities. If the growth rate of individual consumption is related to the growth of wages and salaries, the relevant benchmark for surplus optimizers refers to the growth rate and the volatility of wages and salaries. Finally, this model suggests a new type of product for investment banks. Pension and life insurance funds need to protect themselves against unanticipated changes in the growth rates of wages and salaries. Investment banks could offer liability protection and hedge those positions by purchasing a liability hedge portfolio as is given by Eq. (9).

Acknowledgements

Two anonymous referees of this journal have contributed valuable comments. Without implicating them, the authors thank the participants in a workshop at the ETH in

Zürich, May 1996, the Annual Conference of the Deutsche Gesellschaft für Finanzwirtschaft (DGF) in Berlin, September 1996, the EURO INFORMS Meeting in Barcelona, July 1997, the Eighth International Conference on Stochastic Programming, Vancouver, August 1998, the INQUIRE, Venice Meeting, October 2000, Günter Franke, Christian Hipp, Astrid Eisenberg, Karl Keiber, Alexander Kempf, André Kronimus, Matthias Muck, Stanley Pliska, Bernd Schips, and Heinz Zimmermann for helpful comments on an earlier draft. All remaining errors are ours.

Appendix A

A.1. Derivation of (7)

The arguments of the *J-function* are dropped for simplicity. $\mathrm{o}(\mathrm{d}t)$ summarizes all terms of higher order than 1 of $\mathrm{d}t$. For $\mathrm{o}(\mathrm{d}t)$: $\lim_{\mathrm{d}t\to 0}\mathrm{o}(\mathrm{d}t)/\mathrm{d}t = 0$. Applying Itôs lemma yields

$$\begin{aligned} J(S,Y,t) &= U(S,Y,t)\,\mathrm{d}t + \max_{\mathbf{w}} \mathrm{E}_t[J(S+\mathrm{d}S, Y+\mathrm{d}Y, t+\mathrm{d}t)] \\ &= U\,\mathrm{d}t + \max_{\mathbf{w}} \mathrm{E}_t[J + J_t\,\mathrm{d}t + J_S\,\mathrm{d}S + J_Y\,\mathrm{d}Y + \tfrac{1}{2}J_{SS}\,\mathrm{d}S^2 + \tfrac{1}{2}J_{YY}\,\mathrm{d}Y^2 \\ &\quad + J_{SY}\,\mathrm{d}S\,\mathrm{d}Y + \mathrm{o}(\mathrm{d}t)]. \end{aligned}$$

Transforming the expression above yields

$$\begin{aligned} 0 &= U\,\mathrm{d}t + \max_{\mathbf{w}} \mathrm{E}_t[J_t\,\mathrm{d}t + J_S\,\mathrm{d}S + J_Y\,\mathrm{d}Y + \tfrac{1}{2}J_{SS}\,\mathrm{d}S^2 + \tfrac{1}{2}J_{YY}\,\mathrm{d}Y^2 \\ &\quad + J_{SY}\,\mathrm{d}S\,\mathrm{d}Y + \mathrm{o}(\mathrm{d}t)], \\ 0 &= U\,\mathrm{d}t + \max_{\mathbf{w}}[J_t\,\mathrm{d}t + J_S\mathrm{E}_t(\mathrm{d}S) + J_Y\mathrm{E}_t(\mathrm{d}Y) + \tfrac{1}{2}J_{SS}\,\mathrm{d}S\,\mathrm{d}S \\ &\quad + \tfrac{1}{2}J_{YY}\,\mathrm{d}Y\,\mathrm{d}Y + J_{SY}\,\mathrm{d}S\,\mathrm{d}Y], \end{aligned}$$

and substituting in the expressions in (4) yields (7).

A.2. Derivation of (8)

Denote the prices of the n risky assets by $A_i(t)$, $i=1,\ldots,n$, and suppose they follow the processes

$$\mathrm{d}A_i(t) = A_i(t)[\mu_i\,\mathrm{d}t + \sigma_i\,\mathrm{d}Z_i(t)], \quad i = 1,\ldots n,$$

988 M. Rudolf, W.T. Ziemba / Journal of Economic Dynamics & Control 28 (2004) 975–990

which defines the vector of risky expected asset returns $\mathbf{m}'_A := (\mu_i, \ldots, \mu_n)$. The Wiener processes $Z_i(t)$, $i = 1, \ldots n$, are correlated according to

$$dZ_i(t)\, dZ_j(t) = \rho_{ij}\, dt, \quad i \neq j,$$

where ρ_{ij} is the instantaneous correlation between $Z_i(t)$ and $Z_j(t)$. If the covariance between assets i and j is denoted by σ_{ij}, then we have $\sigma_{ij} = \sigma_i \sigma_j \rho_{ij}$ and the covariance matrix $\mathbf{V}$ has entries σ_{ij}, i.e.

$$\mathbf{V} = \begin{bmatrix} \sigma_{11} & \cdots & \sigma_{1n} \\ \vdots & \vdots & \vdots \\ \sigma_{n1} & \cdots & \sigma_{nn} \end{bmatrix}.$$

Suppose the fund holds $x_i(t)$ shares of asset i at time t, and that the cash in the asset portfolio at time t is $B(t)$. Then the asset portfolio value $A(t)$ is given by ($A_i(t)$: price of asset i)

$$A(t) = \sum_{i=1}^{n} x_i(t) A_i(t) + B(t).$$

In the next infinitesimal time interval dt, this evolves according to

$$dA(t) = \sum_{i=1}^{n} x_i(t)\, dA_i(t) + dB(t) = \sum_{i=1}^{n} x_i(t)\, dA_i(t) + rB(t)\, dt$$

where r is the riskless interest rate. Define the vector of portfolio fractions $\mathbf{w}' = (\omega_1, \ldots, \omega_n)$ by

$$\omega_i(t) := \frac{x_i(t) A_i(t)}{A(t)}, \quad i = 1, \ldots, n.$$

The sum of risky and riskless investments is assumed to be 1. Hence, the fraction invested in the riskless asset is $B(t)/A(t) = 1 - \sum_{i=1}^{n} \omega_i$. Therefore, the expression above for $dA(t)$ becomes

$$dA(t) = A(t) \left[\sum_{i=1}^{n} [\omega_i(\mu_i - r) + r]\, dt + \sum_{i=1}^{n} \omega_i \sigma_i\, dZ_i(t) \right].$$

Comparing the above expression with Eq. (2), i.e. $dA(t) = A(t)[\mu_A\, dt + \sigma_A\, dZ_A(t)]$, and assuming that $\mathbf{e}' := (1, \ldots, 1)$ is the n-dimensional vector of ones, gives

$$\mu_A = \mathbf{w}'(\mathbf{m}_A - r\mathbf{e}) + r,$$

$$\sigma_A\, dZ_A(t) = \sum_{i=1}^{n} \omega_i \sigma_i\, dZ_i(t).$$

From this follows that

$$\sigma_A^2 = \mathbf{w}'\mathbf{V}\mathbf{w},$$

where $\mathbf{V}$ is the covariance matrix given earlier. Furthermore, define $\sigma_{iL}(\sigma_{iY})$ as the covariance between the ith risky asset and the liabilities (state variable). Denote the vector of these covariances by

$$\mathbf{v}'_{AL} := (\sigma_{1L}, \ldots, \sigma_{nL}), \quad \mathbf{v}'_{AY} := (\sigma_{1Y}, \ldots, \sigma_{nY}).$$

Then the relationship $\sigma_{AL} = \mathbf{w}'\mathbf{v}_{AL}$ and $\sigma_{AY} = \mathbf{w}'\mathbf{v}_{AY}$ is obvious. From the last four representations of expected returns, variances, and covariances Eq. (8) follows.

A.3. Discussing Eq. (9)

Maximizing the covariance between the asset portfolio and the state variable,

$$Cov\left(\mathbf{w}'\begin{pmatrix} R_1 \\ \vdots \\ R_n \end{pmatrix}, \frac{\mathrm{d}Y}{Y}\right) = \mathbf{w}'\mathbf{v}_{AY} \underset{\mathbf{w}}{\rightarrow} \max \quad \text{s.t. } \mathbf{w}'\mathbf{V}\mathbf{w} = \sigma_A^2,$$

where $R_1, \ldots, R_n$ refer to the n asset returns. Let λ be a Lagrange multiplier:

$$\begin{aligned} & L = \mathbf{w}'\mathbf{v}_{AY} - \lambda(\mathbf{w}'\mathbf{V}\mathbf{w} - \sigma_A^2) \\ \Rightarrow \quad & \partial L/\partial \mathbf{w} = \mathbf{v}_{AY} - 2\lambda \cdots \mathbf{V}\mathbf{w} = 0 \\ \Rightarrow \quad & \mathbf{w} = \mathbf{V}^{-1}\mathbf{v}_{AY}/(2\lambda). \\ & \partial^2 L/(\partial \mathbf{w})^2 < 0 \text{ if and only if } \mathbf{V} \text{ is positive definite.} \end{aligned}$$

Hence, $\mathbf{w}_Y$ (the state variable hedge portfolio) maximizes the correlation between the asset portfolio and the state variable. Furthermore,

$$\begin{aligned} & 2\lambda \cdot \mathbf{w}'\mathbf{V}\mathbf{w} = \mathbf{w}'\mathbf{v}_{AY} \Leftrightarrow 2\lambda\sigma_A^2 = \sigma_{AY} \\ \Leftrightarrow \quad & 2\lambda = \frac{\sigma_{AY}}{\sigma_A^2} = \beta_{AY} \Rightarrow \mathbf{V}^{-1}\mathbf{v}_{AY} = \beta_{AY}\mathbf{w}. \end{aligned}$$

Multiplying the fractions of the asset portfolio by the regression coefficient β_{AY} provides the fractions of the hedge portfolio.

References

Adler, M., Dumas, B., 1983. International portfolio choice and corporation finance: a synthesis. The Journal of Finance 38 (3), 925–984.

Black, F., 1989. Should you use stocks to hedge your pension liability? Financial Analysts Journal 22, 10–12.

Cariño, D.R., Ziemba, W.T., 1998. Formulation of the Russell Yasuda-Kasai financial planning model. Operations Research 46 (4), 433–449.

Constantinides, G.M., 1993. Comment. In: Clotfelter, C.T., Rothschild, M. (Eds.), The Economics of Higher Education. University of Chicago Press, National Bureau of Economic Research, Chicago, pp. 236–242.

Dixit, A.K., Pindyck, R.S., 1994. Investment under uncertainty. Princeton University Press, New Jersey.

Ezra, D.D., 1991. Asset allocation by surplus optimization. Financial Analysts Journal 24, 51–57.

Ingersoll Jr., J.E., 1987. Theory of Financial Decision Making. Rowman & Littlefield, New York.

Merton, R.C., 1969. Lifetime portfolio selection under uncertainty: the continuous-time case. The Review of Economics and Statistics 51, 247–257.

Merton, R.C., 1971. Optimal consumption and portfolio rules in a continuous-time model. Journal of Economic Theory 3, 373–413.

Merton, R.C., 1973. An intertemporal capital asset pricing model. Econometrica 41, 867–887.

Merton, R.C., 1990. Continuous Time Finance. Blackwell Publishers Inc., Cambridge, MA (reprinted 1995).

Merton, R.C., 1993. Optimal investment strategies for university endowment funds. In: Clotfelter, C.T., Rothschild, M. (Eds.), The Economics of Higher Education. University of Chicago Press, National Bureau of Economic Research, Chicago (reprinted in: Ziemba,W.T., Mulvey, J.M. (Eds.), (1998), Worldwide Asset and Liability Modeling, Cambridge University Press, Cambridge, MA).

Richard, S.F., 1979. A generalized capital asset pricing model. In: Gruber, M.J., Elton, E.J. (Eds.), Portfolio Theory, 25 Years After. North-Holland, New York, pp. 215–232.

Samuelson, P.A., 1969. Lifetime portfolio selection by dynamic stochastic programming. The Review of Economics and Statistics 51, 239–246.

Sharpe, W.F., 1964. Capital asset prices: a theory of market equilibriums under conditions of risk. The Journal of Finance 19 (3), 425–442.

Sharpe, W.F., Tint, L.G., 1990. Liabilities—a new approach. The Journal of Portfolio Management 17, 5–10.

Ziemba, W.T., Mulvey, J.M. (Eds.) 1998. World Wide Asset and Liability Modeling. Cambridge University Press, Cambridge, MA.

52

The Symmetric Downside-Risk Sharpe Ratio

And the evaluation of great investors and speculators.

William T. Ziemba

The Sharpe ratio is a very useful measure of investment performance. Because it is based on mean-variance theory, and thus is basically valid only for quadratic preferences or normal distributions, skewed investment returns can lead to misleading conclusions. This is especially true for superior investors with many high returns. Superior investors may use capital growth wagering ideas to implement their strategies, which produces higher growth rates but also higher variability of wealth.

My simple modification of the Sharpe ratio to assume that the upside deviation is identical to the downside risk provides a useful modification and gives more realistic results.

Exhibit 1 plots wealth levels using monthly data from December 1985 through March 2000 for the Windsor Fund of George Neff, the Ford Foundation, the Tiger Fund of Julian Robertson, the Quantum Fund of George Soros, and Berkshire Hathaway, the fund run by Warren Buffett, as well as the S&P 500 total return index, U.S. Treasuries, and T-bills. Yearly data are shown in Exhibit 2.

The means, standard deviations, and Sharpe [1966, 1994] ratios of these six funds, based on monthly, quarterly, and yearly net arithmetic and geometric total return data, are shown in Exhibit 3. Shown as well are data on the Harvard endowment (quarterly) plus U.S. Treasuries, T-bills, and U.S. inflation, and number of negative months and quarters.

The first panel in Exhibit 3 shows the data behind

William T. Ziemba
is the alumni professor of financial modeling and stochastic optimization emeritus at the Sauder School of Business of the University of British Columbia in Vancouver, Canada, and a visiting professor of finance at the Sloan School of Management of the Massachusetts Institute of Technology in Cambridge, MA.

PERFORMANCE EVALUATION AND RISK ANALYSIS

The Symmetric Downside-Risk Sharpe Ratio 108

WILLIAM T. ZIEMBA

The Sharpe ratio, a most useful measure of investment performance, has the disadvantage that it is based on mean-variance theory and thus is valid basically only for quadratic preferences or normal distributions. Hence skewed investment returns can engender misleading conclusions. This is especially true for superior investors with a number of high returns. Many of these superior investors use capital growth wagering ideas to implement their strategies, which means higher growth rates but also higher variability of wealth. A simple modification of the Sharpe ratio to assume that the upside deviation is identical to the downside risk gives more realistic results.

EXHIBIT 1

Growth of Assets—Monthly Data December 1985–April 2000

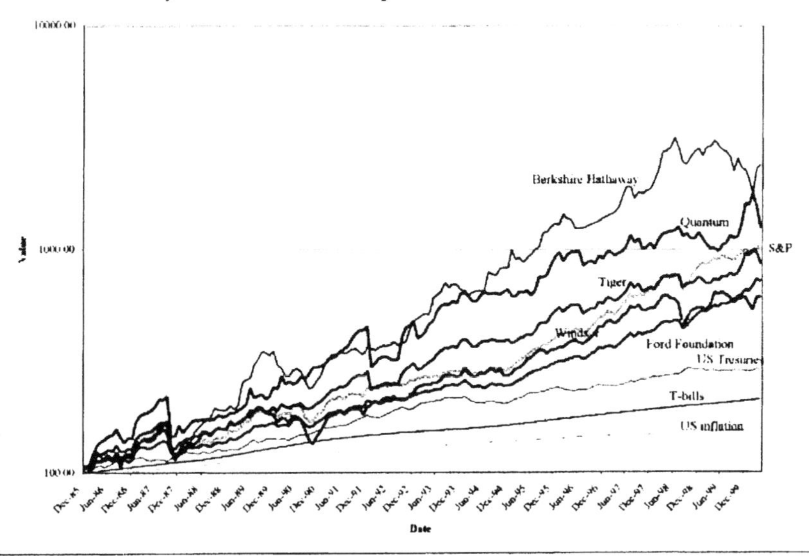

Exhibit 1, which illustrates the high mean returns of Berkshire Hathaway and the Tiger Fund as well as that the Ford Foundation's standard deviation was about a third of Berkshire's. The Ford Foundation actually trailed the S&P 500 mean return. We can see the much lower monthly and quarterly Sharpe ratios compared to the annualized values based on monthly, quarterly, and yearly data.

By the Sharpe ratio, the Harvard endowment and the Ford Foundation had the best performance, followed by the Tiger Fund, the S&P 500 total return index, Berkshire Hathaway, Quantum, Windsor, and U.S. Treasuries. The basic conclusions remain the same according to monthly or quarterly data and arithmetic or geometric means. Because of data smoothing, the Sharpe ratios with yearly data usually exceed those with quarterly data, which in turn exceed the monthly calculations.

The reason for this ranking is that the Ford Foundation and the Harvard endowment, with less growth, also had much less variability. These endowments have different purposes, different investors, different portfolio managers, and different fees, so such differences are not surprising.

Clifford, Kroner, and Siegel [2001] give us similar calculations for a larger group of funds. They also show that over July 1977–March 2000, Berkshire Hathaway's Sharpe ratio was 0.850, Ford's 0.765, and the S&P 500's 0.676. The geometric mean returns were: 32.07% (Buffett), 14.88% (Ford), and 16.71% (S&P 500).

Exhibit 4 shows Buffett's returns on a yearly basis in terms of increase in per share book value of Berkshire Hathaway versus the S&P 500 total return yearly index values for the 40 years 1965–2004. Buffett's geometric and arithmetic means were 22.02% and 22.84%, respectively, versus 10.47% and 11.83% for the S&P 500. This measure does not fully reflect the trading prices of Berkshire Hathaway shares and thus yearly net returns, but it does indicate that Buffett has easily beaten the S&P 500 over these 40 years with a 286,841% increase versus the S&P 500's 5,371% increase.

Typically the Sharpe ratio is computed using arithmetic returns. This is because the basic static theories of portfolio investment management such as mean-variance analysis and the capital asset pricing model are based on arithmetic means. These are static, one-period theories. For asset returns over time, however, the geometric mean is a more accurate measure of average performance because the arithmetic mean is biased upward.

The geometric mean helps mitigate the autocorrelated and time-varying mean and other statistical proper-

EXHIBIT 2
Yearly Return Data (%)

Date	Windsor	Berkshire Hathaway	Quantum	Tiger	Ford	Harvard	S&P 500 Total	US Trea	US T-bills	US Infl
Yearly data, 14 years										
Neg years	2	2	1	0	1	1	1	2	0	0
Dec - 86	20.27	14.17	42.12	26.83	18.09	22.16	18.47	15.14	6.16	1.13
Dec - 87	1.23	4.61	14.13	7.28	5.20	12.46	5.23	2.90	5.47	4.41
Dec - 88	28.69	59.32	10.13	15.76	10.42	12.68	16.81	6.10	6.35	4.42
Dec - 89	15.02	84.57	35.21	24.72	22.15	15.99	31.49	13.29	8.37	4.65
Dec - 90	-15.50	-23.05	23.80	5.57	1.96	-1.01	-3.17	9.73	7.81	6.11
Dec - 91	28.55	35.58	50.58	37.59	22.92	15.73	30.55	15.46	5.60	3.06
Dec - 92	16.50	29.83	6.37	8.42	5.26	4.88	7.67	7.19	3.51	2.90
Dec - 93	19.37	38.94	33.03	24.91	13.07	21.73	9.99	11.24	2.90	2.75
Dec - 94	-0.15	24.96	3.94	1.71	-1.96	3.71	1.31	-5.14	3.90	2.67
Dec - 95	30.15	57.35	38.98	34.34	26.47	24.99	37.43	16.80	5.60	2.54
Dec - 96	26.36	6.23	-1.50	8.03	15.39	26.47	23.07	2.10	5.21	3.32
Dec - 97	21.98	34.90	17.09	18.79	19.11	20.91	33.36	8.38	5.26	1.70
Dec - 98	0.81	52.17	12.46	11.21	21.39	12.14	28.58	10.21	4.86	1.61
Dec - 99	11.57	-19.86	34.68	27.44	27.59	23.78	21.04	-1.77	4.68	2.68

ties of returns that are not independent and identically distributed. If one has returns of +50% and –50% in two periods, for example, the arithmetic mean is zero, which does not reflect the fact that 100 became 150 and then 75. The geometric mean, which is –13.7%, is the correct measure to use.

For investment returns in the 10%-15% range, the arithmetic returns are about 2 percentage points above the geometric returns. But for higher returns, this approximation is not accurate. Hence, I use geometric means as well as more typical arithmetic means in this article.

Lo [2002] points out that we must take care in making Sharpe ratio estimations when the investment returns are not iid, which they are for the investors I discuss here. For dependent but stationary returns Lo derives a correction of the Sharpe ratios that deflates artificially high values to correct values using an estimation of the correlation of serial returns.

Miller and Gehr [1978] and Knight and Satchell [2005] derive exact statistical properties of the Sharpe ratio with normal and lognormal assets, respectively. The Sharpe ratios are almost always lower when geometric means are used rather than arithmetic means; the difference between these two measures is a function of return volatility. The basic conclusions in this research, such as the relative ranking of the various funds, are the same for the arithmetic and geometric means.

Exhibit 5 shows that the Harvard Investment Company, that great school's endowment, had essentially the same wealth record over time as the Ford Foundation. This conclusion is based on quarterly data, which are all I have on Harvard. Harvard beats Ford by the ordinary Sharpe ratio, but Ford is better by the symmetric downside risk measure I develop later.

Before evaluating positive and negative returns performances of these various funds using the Sharpe ratio and my modified version, it is useful to discuss how these funds got their outstanding but sometimes volatile records.

SOME GREAT INVESTORS

Ideally, we would want to penalize only losses such as those shown in Exhibit 6, while rewarding positive returns. The Sharpe ratio penalizes high-return but volatile records.

In the theory of optimal investment over time, it is not quadratic (the utility function behind the Sharpe ratio) but log that yields the most long-term growth. But the elegant results in the Kelly [1956] criterion, as it is known in the gambling literature, and the capital growth theory, as it is known in the investments literature, as proven rigorously by Breiman [1961] and generalized by Algoet and Cover [1988], are long-run asymptotic results (see Hakansson and Ziemba [1995], Ziemba [2003], and MacLean and Ziemba [2005]).

The Arrow-Pratt absolute risk aversion of the log utility criterion:

$$R_A(w) = \frac{-u''(w)}{u'(w)} = 1/w$$

is essentially zero, where u is the utility function of wealth w, and the primes denote differentiation. Hence, in the

EXHIBIT 3

Fund Return Data—December 1985–April 2000

	Windsor	Berkshire Hathaway	Quantum	Tiger	Ford Found	Harvard	S&P500 Total	US Trea	US T-bills	US Infl
Monthly data, 172 months										
Neg months	61	58	53	56	44	na	56	54	0	13
arith mean, mon	1.17	2.15	1.77	2.02	1.19	na	1.45	0.63	0.44	0.26
st dev, mon	4.70	7.66	7.42	6.24	2.68	na	4.41	1.32	0.12	0.21
Sharpe, mon	0.157	0.223	0.180	0.54	0.80	na	0.230	0.14	50.000	-0.827
arith mean	14.10	25.77	21.25	24.27	14.29	na	17.44	7.57	5.27	3.14
st dev	16.27	26.54	25.70	21.62	9.30	na	15.28	4.58	0.43	0.74
Sharpe, yr	0.543	0.773	0.622	0.879	0.970	na	0.797	0.50	40.000	-2.865
geomean, mon	1.06	1.87	1.48	1.83	1.16	na	1.35	0.62	0.44	0.26
geo st dev,mon	4.70	7.67	7.42	6.25	2.69	na	4.41	1.32	0.12	0.21
Sharpe, mon	0.133	0.186	0.140	0.222	0.267	na	0.208	0.13	90.000	-0.828
geo mean, yr	12.76	22.38	17.76	21.92	13.86	na	16.25	7.47	5.27	3.14
geo st dev, yr	16.27	26.56	25.72	21.63	9.30	na	15.28	4.58	0.43	0.74
Sharpe, yr	0.460	0.644	0.486	0.770	0.924	na	0.719	0.48	20.000	-2.868
Quarterly data, 57 quarters										
Neg quarters	14	15	16	11	11	11	10	15	0	1
mean, qtly	3.55	6.70	5.70	4.35	3.68	3.86	4.48	1.93	1.32	0.79
st dev, qtly	8.01	14.75	12.67	7.70	4.72	4.72	7.52	2.67	0.36	0.49
mean, yr	14.20	26.81	22.79	17.42	14.71	15.44	17.91	7.73	5.29	3.16
st dev, yr	16.03	29.50	25.33	15.40	9.43	9.45	15.05	5.34	0.73	0.97
Sharpe, yr	0.556	0.729	0.691	0.788	0.999	1.074	0.839	0.45	60.000	-2.188
geomean, qtly	3.23	5.67	4.94	4.07	3.57	3.75	4.20	1.90	1.32	0.79
geo st dev,qtly	8.02	14.79	12.69	7.70	4.72	4.73	7.53	2.67	0.36	0.49
geo mean, yr	12.90	22.67	19.78	16.28	14.29	15.01	16.80	7.59	5.29	3.16
geo st dev, yr	16.04	29.58	25.38	15.41	9.43	9.45	15.06	5.34	0.73	0.97
Sharpe, yr	0.475	0.588	0.571	0.713	0.954	1.029	0.764	0.43	10.000	-2.190
Yearly Data, 14 years										
Neg years	2	2	1	0	1	1	1	2	0	0
mean,	14.63	28.55	22.93	18.04	14.79	15.47	18.70	7.97	5.40	3.14
st dev,	13.55	30.34	16.17	11.40	9.38	8.52	12.88	6.59	1.50	1.35
Sharpe, yrly	0.681	0.763	1.084	1.109	1.001	1.181	1.033	0.39	00.000	-1.673
geom mean	13.83	24.99	21.94	17.54	14.43	15.17	18.04	7.78	5.39	3.13
st dev	13.58	30.57	16.20	11.41	9.39	8.53	12.90	6.59	1.50	1.35
Sharpe	0.621	0.641	1.022	1.064	0.962	1.146	0.981	0.36	20.000	-1.672

EXHIBIT 4

Increase in Per Share Book Value of Berkshire Hathaway versus S&P 500 (dividends included) 1965-2004 (%)

Year	BH	S&P 500	Diff	Year	BH	S&P 500	Diff
1965	23.8	10.0	13.8	1985	48.2	31.6	16.6
1966	20.3	(11.7)	32.0	1986	26.1	18.6	7.5
1967	11.0	30.9	(19.9)	1987	19.5	5.1	14.4
1968	19.0	11.0	8.0	1988	20.1	16.6	3.5
1969	16.2	(8.4)	24.6	1989	44.4	31.7	12.7
1970	12.0	3.9	8.1	1990	7.4	(3.1)	10.5
1971	16.4	14.6	1.8	1991	39.6	30.5	9.1
1972	21.7	18.9	2.8	1992	20.3	7.6	12.7
1973	4.7	(14.8)	19.5	1993	14.3	10.1	4.2
1974	5.5	(26.4)	31.9	1994	13.9	1.3	12.6
1975	21.9	37.2	(15.3)	1995	43.1	37.6	5.5
1976	59.3	23.6	35.7	1996	31.8	23.0	8.8
1977	31.9	(7.4)	39.3	1997	34.1	33.4	.7
1978	24.0	6.4	17.6	1998	48.3	28.6	19.7
1979	35.7	18.2	17.5	1999	.5	21.0	(20.5)
1980	19.3	32.3	(13.0)	2000	6.5	(9.1)	15.6
1981	31.4	(5.0)	36.4	2001	(6.2)	(11.9)	5.7
1982	40.0	21.4	18.6	2002	10.0	(22.1)	32.1
1983	32.3	22.4	9.9	2003	21.0	28.7	(7.7)
1984	13.6	6.1	7.5	2004	10.5	10.9	(0.04)
Overall Gain	286,841	5,371					
Arithmetic Mean	22.84	11.83	11.01				
Geometric Mean	22.02	10.47	10.07				

Sources: Berkshire Hathaway 2004 annual report, Hagstrom [2004].

EXHIBIT 5

Ford Foundation and Harvard Investment Corporation Returns—Quarterly Data June 1977–March 2000

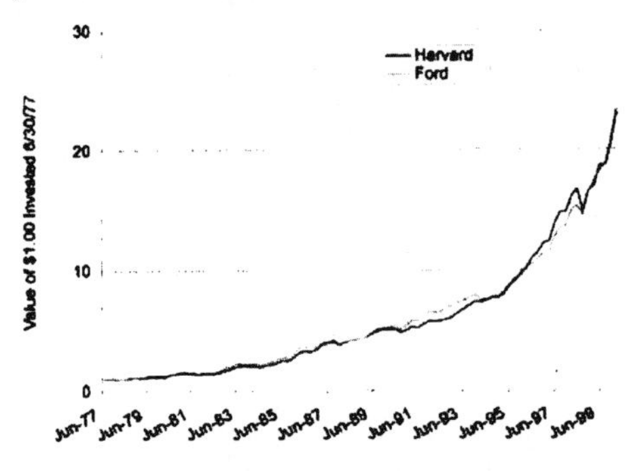

short run, log can be an exceedingly risky utility function with wide swings in wealth values because the optimal bets can be so large.

Long-run exponential growth is equivalent to maximizing the one-period expected log of that period's returns. To illustrate how large Kelly (expected log) bets are, consider the simplest case with Bernoulli trials where you win with probability p and lose with probability $q = 1 - p$.

Log utility is related to negative power utility, namely, for αw^α for $\alpha < 0$ since negative power converges to log when $\alpha \rightarrow 0$. Kelly [1956] discovered that log utility investors had the best utility function, provided they were very long-run investors. The asymptotic rate of asset growth is

$$G = \lim_{N \to \infty} log\left(\frac{w_N}{w_0}\right)^{\frac{1}{N}}$$

where w_N is period N's wealth and w_0 is initial wealth.

Consider Bernoulli trials that win +1 with probability p and lose −1 with probability $1 - p$. If we win M out of N of these independent trials, the wealth after period N is:

$$w_N = w_0(1 + f)^M(1 - f)^{N-M}$$

where f is the fraction of our wealth bet in each period. Then:

$$G(f) = \lim_{N \to \infty}\left[\frac{M}{N}log(1 + f) + \frac{N - M}{N}log(1 - f)\right]$$

which by the strong law of large numbers is

$$G(f) = plog(1 + f) + qlog(1 - f) = E(logw)$$

Hence, the criterion of maximizing the long-run exponential rate of asset growth is equivalent to maximizing the one-period expected logarithm of wealth. Hence, to maximize long-run (asymptotic) wealth, maximizing expected log is the way to do it period by period.

The optimal fractional bet, obtained by setting the derivative of $G(f)$ to zero, is $f^* = p - q$, which is simply the investor's edge or expected gain on the bet. If the bets are win $O + 1$ or lose 1—that is, the odds are O to 1 to win—the optimal Kelly bet is $f^* = \frac{p-q}{O}$ or the $\frac{edge}{odds}$. Since edge is a mean concept and odds is a risk concept, you wager more with higher mean and less with higher risk.[1]

In continuous time:

$$f^* = \frac{\mu - r}{\sigma^2} = \frac{edge}{risk(odds)}$$

with optimal growth rate

EXHIBIT 6

Summary—Negative Observations and Arithmetic and Geometric Means

Number of negative	Windsor	Berkshire Hathaway	Quantum	Tiger	Ford Found	Harvard	S&P Total	US Treas
...months out of 172	*61*	58	53	56	44	na	56	54
...quarters out of 57	14	15	*16*	11	10	11	10	15
...years out of 14	*2*	*2*	1	**0**	1	1	1	*2*

small-cap stocks under Democrats had 24.5 times as much wealth as a 60-40 large-cap/bond investor. That's the idea, more or less.

Keynes and Buffett are essentially Kelly bettors. Kelly bettors will have bumpy investment paths, but most of the time, in the end, accumulate more money than other investors.

$$G^* = \frac{1}{2}\left(\frac{\mu - r}{\sigma}\right)^2 + r = \frac{1}{2}(\text{Sharpe Ratio})^2 + \text{Risk-Free Asset}$$

where μ is the mean portfolio return, r is the risk-free return, and σ^2 is the portfolio return variance. So the ordinary Sharpe ratio determines the optimal growth rate.

Kelly bets can be large. Consider Bernoulli trials where you win 1 or lose 1 with probabilities p and $1 - p$, respectively. Then:

p .5 .51 .6 .8 .9 .99
$1 - p$ .5 .49 .4 .2 .1 .01
f^* 0 .02 .2 .6 .8 .98

So, if the edge is 98%, the optimal bet is 98% of one's fortune. With longer-odds bets, the wagers are lower.

The Kelly bettor is sure to win in the end if the horizon is long enough. Breiman [1960, 1961] was the first to clean up the math in Kelly's [1956] and Latané's [1959] heuristic analyses. He proves that:

$$\lim_{N\to\infty} \frac{w_{KB}(N)}{w_B(N)} \to \infty$$

where w_{KB} (N) and w_B (N) are the wealth levels of the Kelly bettor and another essentially different bettor after N play. That is, the Kelly bettor wins infinitely more than bettor B and moves farther and farther ahead as the long time horizon becomes more distant.[2]

Breiman also shows that the expected time to reach a preassigned goal is asymptotically the shortest with an expected log strategy. Moreover, the ratio of the expected log bettor's fortune to that of any other essentially different investor goes to infinity. The log investor gets all the money in the end if one plays forever.

Hensel and Ziemba [2000] calculate that over 1942 through 1997 a 100% long investor solely in large-cap stocks under Republican administrations and solely in

Ziemba and Hausch [1986] perform a simulation to show *medium*-run properties of log utility and half-Kelly betting (using the $-w^{-1}$ utility function). Starting with an initial wealth of $1,000 and considering 700 independent wagers with probability of winning 0.19 to 0.57, all with expected values of $1.14 per dollar wagered, they compute the final wealth profiles over 1,000 simulations. These wagers correspond to odds of 1-1, 2-1, 3-1, 4-1, and 5-1.

The results show that the log bettor has more than 100 times initial wealth 16.6% of the time and more than 50 times as much 30.2% of the time. This demonstrates the great power of log betting, as the half-Kelly strategy has very few such high outcomes. Yet this high return comes at a price. Despite making 700 independent bets, each with a 14% success rate, the investor could have lost more than 98% of the initial wealth of $1,000.

Exhibit 7 provides the simulation results. The Ziemba-Hausch simulation used the data in Exhibit 8. The edge over odds gives f^* equal to between 0.14 and 0.028 for the optimal Kelly wagers for 1-1 versus 5-1 odds bets.

The 18 in the first column in Exhibit 7 shows it is possible for a Kelly bettor to make 700 independent wagers, all with a 14% edge, having 19% to 57% chance of winning each wager, and still lose over 98% of one's wealth. Even with half-Kelly, the minimum starting with $1,000 was $145, or a 85.5% loss. This shows the effect of a sequence of very bad scenarios that may be unlikely but are certainly possible.

The last column in Exhibit 7 shows that 16.6% of the time the Kelly bettor increases initial wealth more than 100-fold. The half-Kelly strategy is much safer, as the chance of being ahead after the 700 wagers is 95.4% versus 87.0% for full-Kelly. But the growth rate is much lower, since the 16.6% chance of making 100 times initial wealth is only 0.1% for half-Kelly wagerers. The Kelly bettor accumulates more wealth but with a much riskier time path of wealth accumulation. The Kelly bettor can take a long time to get ahead of another bettor.

EXHIBIT 7
Distributions of Final Wealth—Kelly and Half-Kelly Wagers

Final Wealth					Number of Times Final Wealth out of 1000 Trials Was				
Strategy	Min	Max	Mean	Median	>500	>1000	>10,000	>50,000	>100,000
Kelly	18	483,883	48,135	17,269	916	870	598	302	166
Half-Kelly	145	111,770	13,069	8,043	990	954	480	30	1

EXHIBIT 8
Value of Odds on Wagers

Probability of Winning	Odds	Probability of Being Chosen in the Simulation at Each Decision Point	Optimal Kelly Bets Fraction of Current Wealth
0.57	1-1	0.1	0.14
0.38	2-1	0.3	0.07
0.285	3-1	0.3	0.047
0.228	4-1	0.2	0.035
0.19	5-1	0.1	0.028

EXHIBIT 9
Place-and-Show Betting on Kentucky Derby 1934-1994

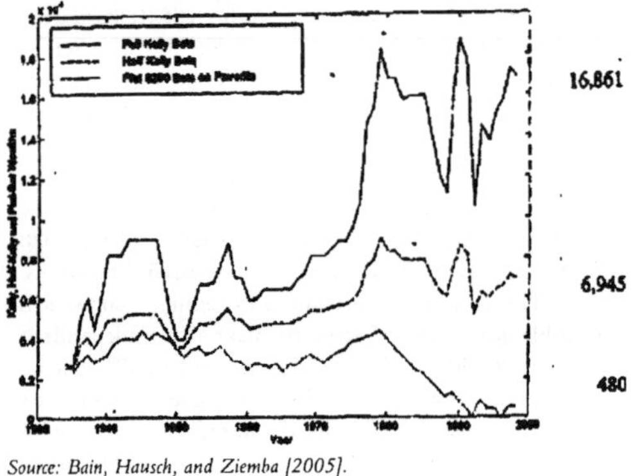

Source: Bain, Hausch, and Ziemba [2005].

What one wants is to have those swings in the right direction. And who better, it seems, at predicting these swings than Warren Buffett. Chopra and Ziemba [1993] show that, in portfolio problems, errors in estimating means, variances, and covariances influence investment performance roughly in the ratio 20:2:1. When risk aversion is lower, as it is with log, the errors are even greater. In this case, the effect on portfolio performance of mean errors can be 100 times the errors in covariances.

Who then would use a log utility function if it is so risky, or use a toned-down log utility function by mixing the log utility investment fraction of one's wealth with cash? These so-called fractional Kelly strategies are actually mathematically equivalent to negative power utility when the assets are lognormally distributed and approximately equivalent otherwise; see MacLean, Ziemba, and Li [2005].

The difference in wealth paths between Kelly and half-Kelly strategies is illustrated in Exhibit 9, which shows the wealth level histories from place-and-show betting on the Kentucky Derby over 1934-1994 following the DrZ system. This system uses a 4.00 dosage index breeding filter rule with full- and half-Kelly wagering and $200 flat bets on the favorite, with initial wealth of $2,500 (*dosage* is a measure of a horse's speed over stamina ratio). Starting with $2,500, full-Kelly yields a final wealth of $16,861, while half-Kelly, with a much smoother path, has final wealth of $6,945. These two strategies for the Kentucky Derby are winning ones compared to the losing strategy of betting on the favorite, which turns the $2,500 into $480.

Some rather good investors, including four I have worked with or consulted with, have used the Kelly and fractional Kelly approach to turn humble starts into fortunes of hundreds of millions. One was the world's most successful racetrack bettor; see Benter [1994].

Another was a trend-following futures trader in the Caribbean for whom I designed a Kelly betting system for the 90 liquid futures markets he traded. The Kelly system added $9 million extra profits per year to his already good betting system based on an ad hoc but sound probability-of-success approach. Another was an options trader eking out nickels and dimes with slightly mispriced options in Chicago.

The fourth was the popularizer of the Kelly approach in sports betting, Edward O. Thorp. Thorp's Princeton-Newport Fund had a net mean return of 15.1%, when the S&P 500's was 10.2% and T-bills returned 8.1%. Interest rates were very high in 1968-1988 while Thorp was running his fund. Thorp had no losing quarters, only three losing months, and a most impressive yearly standard deviation of 4%.

To show the differences between the gamblers and other investors, Exhibit 10 shows Benter's racetrack bet-

EXHIBIT 10
Gamblers Like Smooth Wealth Paths

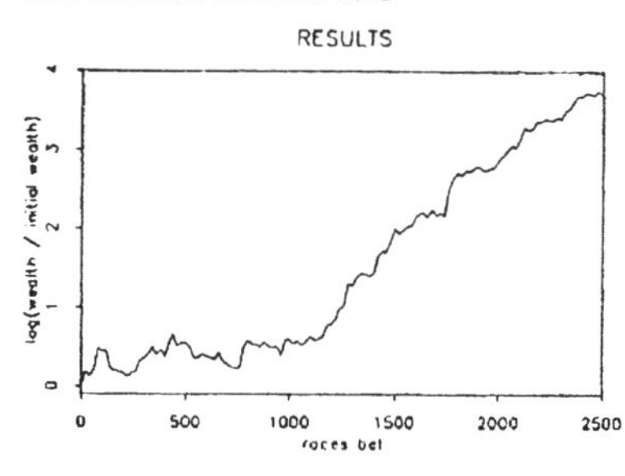

(a) Hong Kong racing syndicate 1989 to 1994. See Benter [1994].

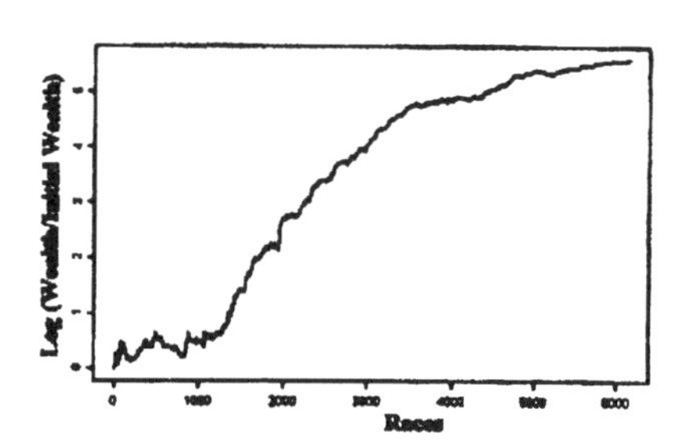

(b) Hong Kong racing syndicate 1989 to 2001 See Benter [2001].

EXHIBIT 11
Princeton-Newport Partners—Cumulative Results November 1968-December 1988

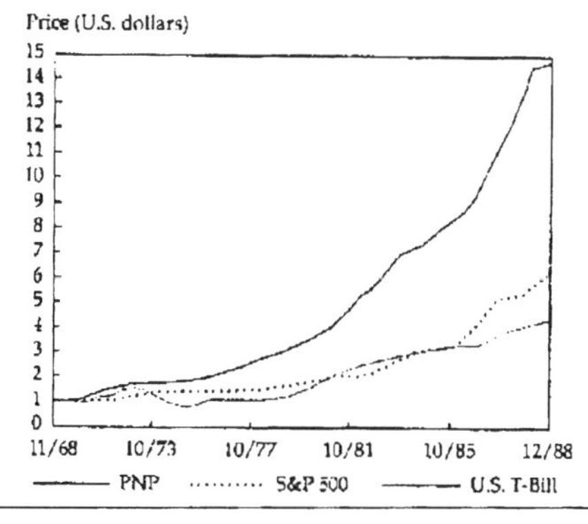

ting record over 1989–1994 and 1989–2001, and Exhibit 11 shows Thorp's record over 1968–1988. Compare these rather smooth graphs with the brilliant but quite volatile record in Exhibit 12 of the eminent economist, John Maynard Keynes, who ran the King's College Chest Fund, the college's endowment, from 1927 until his death in 1945.

Keynes lost more than 50% of his fortune during the difficult years of the depression around the 1929 crash. This bad start was followed by many years of outperformance on relative and absolute terms, so that by 1945 Keynes's geometric mean return was 9.12% versus the U.K. index of -0.89%. Keynes's Sharpe index was 0.385.

The gamblers have several common characteristics:

- They carefully developed anomaly systems with positive means.
- They carefully developed computerized betting systems that automated the betting process.
- They constantly updated their research.
- They were more focused on not losing than on winning in their careful risk control.

Thorp [2006] shows that the Buffett trades are actually very similar to those of a Kelly trader. In Ziemba [2003] I find Keynes is well approximated as a fractional Kelly bettor with 80% Kelly and 20% cash; this is equivalent to the negative power utility function $-w^{-0.25}$. Recall log is when the power coefficient goes to zero, and that log is the most risky utility function that should ever be considered. Positive power utility has less growth and more risk, so is not an acceptable utility function.

SYMMETRIC DOWNSIDE SHARPE RATIO PERFORMANCE MEASURE

Now back to the records of the funds in Exhibit 1. How do we compare these various investors? And can we determine if Warren Buffett really is a better investor than the rather good funds, especially the Ford Foundation and the Harvard endowment?

Exhibit 13 plots the Berkshire Hathaway and Ford Foundation monthly returns as a histogram and shows the losing months and the winning months in a smooth curve. We want to penalize Buffett for losing but not for winning. We define the downside risk as:

EXHIBIT 12

King's College Chest Fund 1927-1945

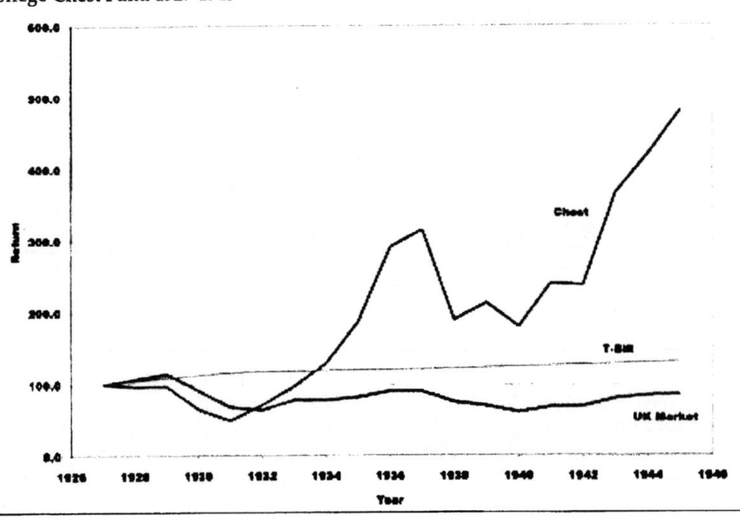

EXHIBIT 13

Berkshire Hathaway versus Ford Foundation—Monthly Returns Distribution January 1977–April 2000

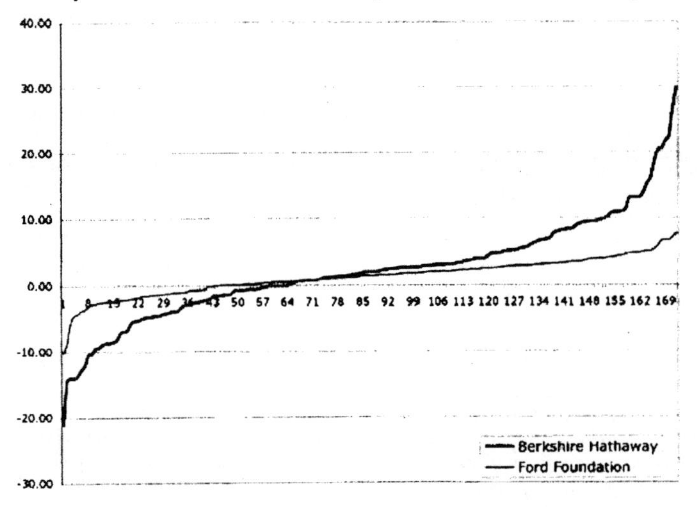

EXHIBIT 14
Comparison of Ordinary and Symmetric Downside Sharpe Yearly Performance Measures

	Ordinary	Downside
Ford Foundation	0.970	0.920
Tiger Fund	0.879	0.865
S&P500	0.797	0.696
Berkshire Hathaway	0.773	0.917
Quantum	0.622	0.458
Windsor	0.543	0.495

$$\sigma_{x_-}^2 = \frac{\sum_{i=1}^{n}(x_i - \bar{x})_-^2}{n-1}$$

where the benchmark $\bar{x}$ is zero, i is the index on the n months in the sample, and the x_i taken are those below $\bar{x}$, namely, those m of the n months with losses. This is the downside variance measured from zero, not the mean, so it is more precisely the downside risk. To get the total variance, we use twice the downside variance ($2\sigma_{x-}^2$), so that Buffett gets the symmetric gains added, not his actual gains. Using $2\sigma_{x}^2$, the usual Sharpe ratio with monthly data and arithmetic returns is:

$$S_- = \frac{\bar{R} - R_F}{\sqrt{2}\sigma_{x-}}$$

Exhibit 14 provides the results for the ordinary and symmetric downside Sharpe ratios using monthly data and arithmetic means.

My measure moves Warren Buffett higher, to 0.917, but not up to the Ford Foundation and not higher because of his high monthly losses. He does gain in the switch from ordinary Sharpe to downside symmetric Sharpe, while all the other funds drop. Ford is 0.920, and Tiger 0.865. The Berkshire Hathaway monthly losses when annualized are over 64% versus under 27% for the Ford Foundation.

Exhibit 14 shows these rather fat tails on the up side and down side of Berkshire Hathaway versus the much less volatile Ford Foundation returns. When Berkshire Hathaway had a losing month, it averaged –5.36% versus +2.15% for all months. Meanwhile, Ford lost 2.44% and won, on average, 1.19%.

Exhibit 15 shows the histogram of quarterly returns for all funds including Harvard (for which monthly data are not available). The plots show that the distributions of all the funds lie between those of Berkshire Hathaway, Harvard, and Ford.

By the quarterly data, the Harvard endowment has a record almost as good as the Ford Foundation's; see Exhibits 16 and 17. Berkshire Hathaway made the most money but also took more risk, and by either the Sharpe or the downside Sharpe measure the Ford Foundation and the Harvard endowment had superior rewards.

I first used this measure in Ziemba and Schwartz [1991] to compare the results of superior investment in Japanese small-cap stocks during the late 1980s. The choice of $\bar{x} = 0$ is convenient and has a good interpretation. But other $\bar{x}$ are possible and might be useful in other applications. This measure is closely related to the Sortino ratio (see Sortino and van der Meer [1991] and Sortino

EXHIBIT 15
Quarterly Returns Distributions—December 1985–March 2000

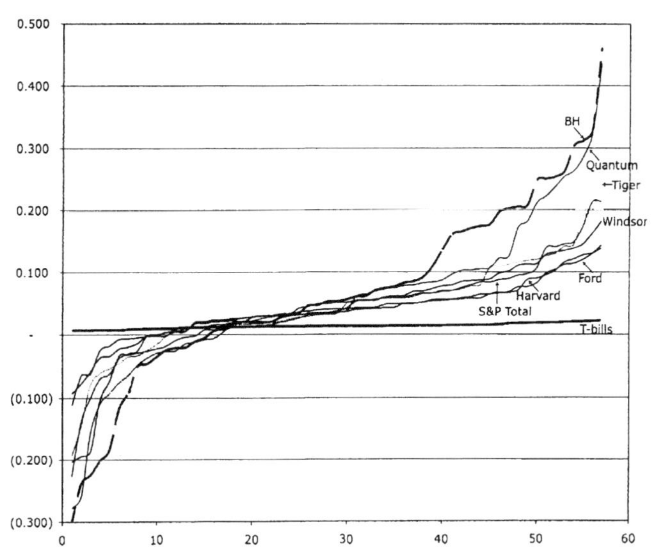

EXHIBIT 16

Yearly Sharpe and Symmetric Downside Sharpe Ratios December 1985–April 2000

	Windsor	Berkshire Hathaway	Quantum	Tiger	Ford Found	Harvard	S&P Total	Trea
qtly data, 57 quarters								
neg qts	14	15	16	11	10	11	10	15
mean, neg	-6.69	-10.50	-7.77	-6.26	-3.59	-2.81	-6.92	-1.35
mean, qtly	3.55	6.70	5.70	4.35	3.68	3.86	4.48	1.93
ds st dev, qtly	10.52	14.00	12.09	8.18	4.44	5.12	9.89	1.35
mean, yr	14.20	26.81	22.79	17.42	14.71	15.44	17.91	7.73
ds st dev, yr	21.04	28.00	24.17	16.35	8.89	10.24	19.78	2.70
ds Sharpe	0.424	0.769	0.724	0.742	1.060	0.991	0.638	0.903
geomean, qtly	3.23	5.67	4.94	4.07	3.57	3.75	4.20	1.90
ds geo st dev,qtly	10.20	13.49	11.80	7.71	3.91	4.99	9.34	1.28
geo mean, yr	12.90	22.67	19.78	16.28	14.29	15.01	16.80	7.59
ds geo st dev, yr	20.41	26.97	23.60	15.43	7.83	9.97	18.68	2.56
ds Sharpe	0.373	0.644	0.614	0.712	1.150	0.975	0.616	0.900

and Price [1994]), which considers downside risk only. That measure does not have the two-sided interpretation of my measure, and the $\sqrt{2}$ does not appear.

The notion of focusing on downside risk is popular these days as it represents real risk better. I started using it in asset-liability models in the 1970s; see Kallberg, White, and Ziemba [1982] and Kusy and Ziemba [1986] for early applications. In those models we measure risk as the downside non-attainment of investment target goals that can be deterministic such as wealth growth over time, or stochastic such as the non-attainment of a portfolio of weighted benchmark returns. See Geyer et al. [2003] for an application of this to the Siemens Austria pension fund.[3]

Calculating the Sharpe ratio and the downside-symmetric Sharpe ratio using quarterly or yearly data does not change the results much, even though it smooths the data because individual monthly losses are combined with gains for lower volatility. The yearly data move us closer to normally distributed returns so the symmetric downside and ordinary Sharpe measures will yield similar rankings.

In Exhibit 2, I showed the yearly returns for the various funds and their Sharpe ratios computed using arithmetic and geometric means and the yearly data. There are insufficient data to compute the downside Sharpe ratios based on yearly data. The Ford Foundation had only one losing year, and that loss was only 1.96%. Berkshire Hathaway had two losing years with losses of 23.1% and 19.9%. The Ford Foundation had a higher Sharpe ratio than Berkshire Hathaway, but that was exceeded by the Tiger and Quantum funds and the S&P 500.

Exhibits 17 and 18 summarize the annualized results using monthly, quarterly, and yearly data. The Ford Foundation had the highest Sharpe ratio, followed by the Harvard endowment, and both of these exceeded Berkshire Hathaway. The Ford Foundation had the highest symmetric-downside Sharpe ratio, followed by Harvard, and both exceeded Berkshire Hathaway and the other funds.

EXHIBIT 17

Summary of Means (%) and Sharpe and Symmetric Downside Sharpe Ratios Annualized

	Windsor	Berkshire Hathaway	Quantum	Tiger	Ford Found	Harvard	S&P Total	Trea
mean (arith, mon)	14.10	25.77	21.25	24.27	14.29	na	17.44	7.57
Sharpe (arith, mon)	0.543	0.773	0.622	0.879	0.970	na	0.797	0.504
ds Sharpe (arith,mon)	0.495	0.917	0.458	0.865	0.920	na	0.696	0.631
mean (geom, mon)	12.76	22.38	17.76	21.92	13.86	na	16.25	7.47
Sharpe (geom, mon)	0.460	0.644	0.486	0.770	0.924	na	0.719	0.482
ds Sharpe (geom, mon)	0.420	0.765	0.358	0.758	0.876	na	0.628	1.053
mean (arith, qtly)	14.20	26.81	22.79	17.42	14.71	15.44	17.91	7.73
Sharpe (arith, qtly)	0.556	0.729	0.691	0.788	0.999	1.074	0.839	0.456
ds Sharpe (arith,qtly)	0.424	0.769	0.724	0.742	1.060	0.991	0.638	0.903
mean (geom, qtly)	12.90	22.67	19.78	16.28	14.29	15.01	16.80	7.59
Sharpe (geom, qlty)	0.475	0.588	0.571	0.713	0.954	1.029	0.764	0.431
ds Sharpe (geom, qtly)	0.373	0.644	0.614	0.712	1.150	0.975	0.616	0.900
mean (arith, yrly)	14.63	28.55	22.93	18.04	14.79	15.47	18.70	7.97
Sharpe (arith, yrly)	0.681	0.763	1.084	1.109	1.001	1.181	1.033	0.390
mean (geom, yrly)	13.83	24.99	21.94	17.54	14.43	15.17	18.04	7.78
Sharpe (geom, yrly)	0.621	0.641	1.022	1.064	0.962	1.146	0.981	0.362

EXHIBIT 18

Summary of Means and Sharpe and Downside Sharpe—Monthly, Quarterly, and Yearly Data Annualized December 1985–March 2000

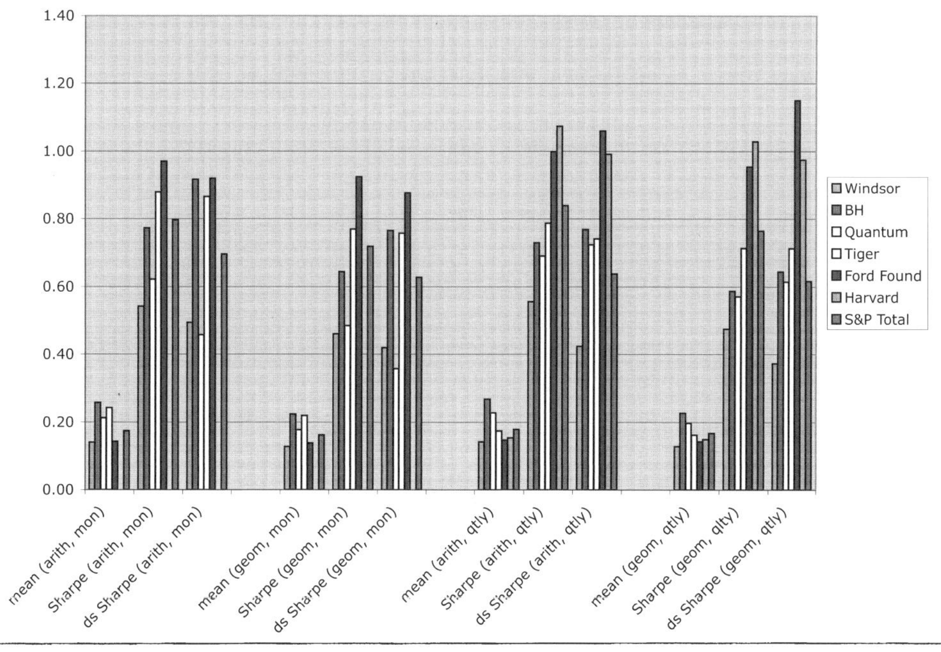

EXHIBIT 19

Asset Allocation of Harvard Endowment—June 2004

Harvard's Holdings By Sector	Annual Returns 1-Year, %	10-Year, %	Weight In Endowment, %
Domestic Equities	22.8	17.8	15
Foreign Equities	36.1	8.5	10
Emerging Markets	6.6	9.7	5
Private Equity	20.8	31.5	13
Hedge Funds	15.7	n.a	12
High Yield	12.4	9.7	5
Commodities	19.7	10.9	13
Real Estate	16.0	15.0	10
Domestic Bonds	9.2	14.9	11
Foreign Bonds	17.4	16.9	5
Inflation-Indexed Bonds	4.2	n.a.	6
Total Endowment	21.1	15.9	105%*

*includes slight leverage

Source: *Barrons*, February 1, 2005.

As Lawrence Siegel of the Ford Foundation privately acknowledges, some of the Ford Foundation's high Sharpe ratio results from dividing by an artificially smoothed standard deviation, due to that institution's high private equity allocation whose market prices do not reflect actual volatility. This is also true of Harvard.

Exhibit 19 shows the Harvard endowment's asset allocation. Its ten-year 15.9% performance on a $22.6 billion portfolio was 3.1 percentage points better than the median of the 25 largest university endowments through June 2004. Harvard has continued its good performance, with extensive use of private equity and other non-traditional investments. Ford's performance in 2001-2002 was poor, and Berkshire Hathaway has doubled in price since the 2000 lows, so the rankings based on more data may well have changed.

Exhibits 6, 16-17, and 20 give a short overview of key data. The numbers in italics indicate the worst outcomes and those in bold the best. Windsor had the most negative months and negative years (tied with Berkshire Hathaway) and the lowest means. Berkshire Hathaway had the highest annual mean returns.

EXHIBIT 20

Yearly Sharpe and Symmetric Downside Sharpe Ratios December 1985–April 2000 (%)

	Windsor	Berkshire Hathaway	Quantum	Tiger	Ford Found	Harvard	S&P Total	US Trea	T-bills	US Infl
Neg months	61	58	53	56	44	na	56	54	0	13
arith mean, neg	-3.54	-5.36	-5.65	-4.41	-2.24	na	-3.31	-0.87	na	-0.14
arith mean, mon	1.17	2.15	1.77	2.02	1.19	na	1.45	0.63	0.44	0.26
ds st dev, mon	5.15	6.46	10.08	6.34	2.83	na	5.05	1.06	0.00	0.19
arith mean, yr	14.10	25.77	21.25	24.27	14.29	na	17.44	7.57	5.27	3.14
ds st dev, yr	17.85	22.36	34.91	21.97	9.81	na	17.49	3.66	0.00	0.64
ds Sharpe	0.495	0.917	0.458	0.865	0.920	na	0.696	0.631	na	-3.320
geomean, neg	-3.62	-5.48	-5.95	-4.53	-2.26	na	-3.38			
geomean, mon	1.06	1.87	1.48	1.83	1.16	na	1.35	0.62	0.44	0.26
ds geo st dev,mon	5.15	6.46	10.09	6.34	2.83	na	5.05	0.60	0.00	0.00
geo mean, yr	12.76	22.38	17.76	21.92	13.86	na	16.25	7.47	5.27	3.14
ds geo st dev, yr	17.86	22.37	34.94	21.98	9.81	na	17.49	2.09	0.00	0.00
ds Sharpe	0.420	0.765	0.358	0.758	0.876	na	0.628	1.053		

CONCLUSIONS AND CAVEATS

My downside-risk Sharpe ratio measure is ad hoc, as all performance measures are, and adds to but does not close the debate on this subject.

Hodges [1998] proposes a generalized Sharpe measure that eliminates some of the paradoxes the Sharpe measure leads to. It uses a constant, absolute risk-return exponential utility function and general return distributions. When the returns are normally distributed, this is the usual Sharpe ratio, as that portfolio problem is equivalent to a mean-variance model. A better utility function is the constant relative risk aversion negative power.

Leland [1999] shows how to modify βs when there are fat tails into more correct βs in a CAPM framework.

Goetzmann et al. [2002] and Spurgin [2000] show how the Sharpe ratio may be manipulated using option strategies to obtain what looks like a superior record to obtain more funds to manage. Managers sell calls to cut off upside variance, and use the proceeds to buy puts to cut off downside variance, leading to higher Sharpe ratios because of the reduced portfolio variance. These options transactions may actually lead to poorer investment performance in final wealth terms even with their higher Sharpe ratios. Tompkins, Ziemba, and Hodges [2003], for example, show how on average the calls sold and the puts purchased on the S&P 500 both had negative expected values over 1985-2002.

I have not tried to establish when the symmetric downside-risk Sharpe ratio might give misleading results in real investment situations or to establish its mathematical and statistical properties. I note only that it is consistent with an investor whose utility is based on the negative of the disutility of losses. The technique does seem to provide a simple way to avoid penalizing superior performance in order to more fairly evaluate performance.

We will have to find another way to measure and establish the superiority of Warren Buffett. One likely candidate is related to the Kelly approach to evaluate investments, which looks at compounded wealth over a long period of time, which we know at the limit is attained by the log bettor (which Buffet seems to be). After 40 years, most of us believe Buffett is in the skill, not luck, category. After all, $15 a share in 1965 became $87,000 in June 2005, but since he is a log bettor only more time will tell.

ENDNOTES

[1]If there are two independent wagers, and the size of the bets does not influence the odds, an analytic expression can be derived; see Thorp [1997, pp. 19-20]. In general, to solve for the optimal wagers when the bets influence the odds, there is dependence. In the case of three or more wagers, one must solve a non-convex nonlinear program; see Ziemba and Hausch [1984, 1987] for technique. This gives the optimal wager, taking into account the effect of our bets on the odds (prices).

[2]Bettor B must use an essentially different strategy from our Kelly bettor for this to be true. This means that the strategies differ infinitely often. For example, they are the same for the first ten years, and then every second trial is different. This is a technical point to get proofs correct, but nothing much to worry about in practice since non-log strategies will differ infinitely often.

[3]Roy [1952], Markowitz [1959], Mao [1970], Bawa [1975, 1978], Bawa and Lindenberg [1977], Fishburn [1977], Harlow and Rao [1989], and Harlow [1991] have used downside-risk measures in portfolio theories other than those based on mean-variance and related analyses.

REFERENCES

Algoet, P., and T. Cover. "Asymptotic Optimality and Asymptotic Equipartition Properties of Log-Optimum Investment." *Annals of Probability*, 16 (1988), pp. 876-898.

Bain, R., D.B. Hausch, and W.T. Ziemba. "An Application of Expert Information to Win Betting on the Kentucky Derby, 1981-2005." Working paper, University of British Columbia, 2005.

Bawa, V. "Optimal Rules for Ordering Uncertain Prospects." *Journal of Financial Economics*, 2 (March 1975), pp. 95-121.

——. "Safety First, Stochastic Dominance and Optimal Portfolio Choice." *Journal of Financial and Quantitative Analysis* 13 (June 1978), pp. 255-271.

Bawa, V., and E. Lindenberg. "Capital Market Equilibrium in a Mean, Lower Partial Moment Framework." *Journal of Financial Economics*, 5 (November 1977), pp. 189-200.

Benter, W. "Computer-Based Horse Race Handicapping and Wagering Systems: A Report." In D.B. Hausch, V. Lo, and W.T. Ziemba, eds., *Efficiency of Racetrack Betting Markets*. San Diego: Academic Press, 1994.

——. "Development of a Mathematical Model for Successful Horse Race Wagering." Presented at University of Nevada, Las Vegas, 2001.

Breiman, L. "Investment Policies for Expanding Business Optimal in a Long-Run Sense." *Naval Research Logistics Quarterly*, 7 (1960), pp. 647-651.

——. "Optimal Gambling System for Favorable Games." *Proceedings of the 4th Berkeley Symposium on Mathematical Statistics and Probability*, 1 (1961), pp. 63-68.

Chopra, V.K., and W.T. Ziemba. "The Effect of Errors in Mean, Variance and Co-Variance Estimates on Optimal Portfolio Choice." *The Journal of Portfolio Management*, 19 (1993), pp. 6-11.

Clifford, S.W., K.F. Kroner, and L.B. Siegel. "In Pursuit of Performance: The Greatest Return Stories Ever Told." *Investment Insights*, Barclays Global Investors, 4 (1) (2001), pp. 1-25.

Fishburn, P.C. "Mean-Risk Analysis with Risk Associated with Below Target Returns." *American Economic Review*, 76 (March 1977), pp. 116-126.

Geyer, A., W. Herold, K. Kontriner, and W.T. Ziemba. "The Innovest Austrian Pension Fund Financial Planning Model InnoALM." Working paper, UBC, 2003.

Goetzmann, W., J. Ingersoll, M. Spiegel, and I. Welch. "Sharpening Sharpe Ratios." Working paper, Yale School of Management, 2002.

Hagstrom, R.G. *The Warren Buffett Way*. New York: Wiley, 2004.

Hakansson, N.H., and W.T. Ziemba. "Capital Growth Theory." In R.A. Jarrow, V. Maksimovic, and W.T. Ziemba, eds., *Finance Handbook*. Amsterdam: North-Holland, 1995, pp. 123-144.

Harlow, W.V. "Asset Allocation in a Downside-Risk Framework." *Financial Analysts Journal*, September/October 1991, pp. 28-40.

Harlow, W.V., and R.K.S. Rao. "Asset Pricing in a Generalized Mean-Lower Partial Moment Framework: Theory and Evidence." *Journal of Financial and Quantitative Analysis*, 24 (3) (1989), pp. 285-311.

Hausch, D.B., V. Lo, and W.T. Ziemba, eds. *Efficiency of Racetrack Betting Markets*. San Diego: Academic Press, 1994.

Hensel, C.R., and W.T. Ziemba. "How Did Clinton Stand Up to History? US Stock Market Returns and Presidential Party Affiliations." In D.B. Keim and W.T. Ziemba, eds., *Security Market Imperfections in World Wide Equity Markets*. Cambridge: Cambridge University Press, pp. 203-217.

Hodges, S. "A Generalization of the Sharpe Ratio and its Applications to Valuation Bounds and Risk Measures." Warwick Business School, 1998.

Kallberg, J., R. White, and W. Ziemba. "Short Term Financial Planning under Uncertainty." *Management Science*, XXVIII (1982), pp. 670-682.

Kelly, J. "A New Interpretation of Information Rate." *Bell System Technology Journal*, 35 (1956), pp. 917-926.

Knight, J., and S. Satchell. "A Re-examination of Sharpe's Ratio for Log-Normal Prices." *Applied Mathematical Finance* 12 (1) (2005), pp. 87-100.

Kusy, M., and W. Ziemba. "A Bank Asset and Liability Management Model." *Operations Research*, XXXIV (1986), pp. 356-376.

Latané, H. "Criteria for Choice among Risky Ventures." *Journal of Political Economy*, 38 (April 1959), pp. 144-155.

Leland, H. "Beyond Mean-Variance: Performance Measurement in a Nonsymmetrical World." *Financial Analysts Journal*, January/February 1999, pp. 27-36.

Lo, A.W. "The Statistics of Sharpe Ratios." *Financial Analysts Journal*, 56 (2002), pp. 36-52.

MacLean, L.C., and W.T. Ziemba. "Capital Growth: Theory and Practice." In S.A. Zenios and W.T. Ziemba, eds., *Handbook of Asset-Liability Management, Volume 1: Theory and Methodology*. Amsterdam: North-Holland, 2005.

MacLean, L.C., W.T. Ziemba, and Y. Li. "Time to Wealth Goals in Capital Accumulation and the Optimal Trade-Off of Growth versus Security." *Quantitative Finance*, 5 (2005), pp. 343-357.

Mao, J. "Models of Capital Budgeting, e-v vs. e-s." *Journal of Financial and Quantitative Analysis*, 5 (January 1970), pp. 657-675.

Markowitz, H.M. *Portfolio Selection*. New York: Wiley & Sons, 1959.

Miller, R., and A. Gehr. "Sample Bias and Sharpe's Performance Measure: A Note." *Journal of Financial and Quantitative Analysis*, 13 (2005), pp. 943-946.

Roy, A. "Safety First and the Holding of Assets." *Econometrica*, 20 (July 1952), pp. 431-449.

Sharpe, W. "Mutual Fund Performance." *Journal of Business*, 39 (1966), pp. 119-138.

Sharpe, W.F. "The Sharpe Ratio." *The Journal of Portfolio Management*, 21 (1) (1994), pp. 49-58.

Sortino, F.A., and L.N. Price. "Performance Measurement in a Downside Risk Framework." *The Journal of Investing*, Fall 1994.

Sortino, F.A., and R. van der Meer. "Downside Risk." *The Journal of Portfolio Management*, Summer 1991.

Spurgin, R.B. "How to Game Your Sharpe Ratio." *The Journal of Alternative Investments* 4 (3) (2000), pp. 38-46.

Thorp, E.O. "The Kelly Criterion in Blackjack, Sports Betting and the Stock Market." In S.A. Zenios and W.T. Ziemba, eds., *Handbook of Asset Liability Management, Volume 1: Theory and Methodology*. Amsterdam: North Holland Elsevier, 2006.

Tompkins, R., W. Ziemba, and S. Hodges. "The Favorite-Longshot Bias in S&P 500 Futures Options: The Return to Bets and the Cost of Insurance." Working paper, Sauder School of Business, UBC, 2003.

Ziemba, W.T. "The Stochastic Programing Approach to Asset Liability and Wealth Management." AIMR, 2003.

Ziemba, W.T., and D.B. Hausch. *Beat the Racetrack*, 1st ed. San Diego: Harcourt, Brace, Jovanovich, 1984.

——. *DrZ's Beat the Racetrack*, 2nd ed. New York: William Morrow, 1987.

——. *Betting at the Racetrack*. San Luis Obispo, CA: DrZ Investments, Inc., 1986.

Ziemba, W.T., and S.L. Schwartz. *Invest Japan*. Chicago: Probus, 1991.

53

POSTSCRIPT: THE RENAISSANCE MEDALLION FUND

The Medallion Fund uses mathematical ideas such as the Kelly criterion to run a superior hedge fund.[1] The staff of technical researchers and traders, working under mathematician James Simons, is constantly devising edges that they use to generate successful trades of various durations including many short term trades that enter and exit in seconds. The fund, whose size is in the $5 billion area, has very large fees (5% management and 44% incentive). Despite these fees and the large size of the fund, the net returns have been consistently outstanding, with a few small monthly losses and high positive monthly returns; see the histogram in Figure A1. Table A1 shows the monthly net returns from January 1993 to April 2005. There were only 17 monthly losses in 148 months and 3 losses in 46 quarters and no yearly losses in these 12+ years of trading in our data sample. The mean monthly, quarterly and yearly net returns, Sharpe and Symmetric Downside Sharpe ratios are shown in Table A2.[2]

We calculated the quarterly standard deviation for the DSSR by multiplying the monthly standard deviation by sqrt(3) and multiplied it by sqrt(12) for the annual standard deviation. All calculations use arithmetic means. We know from Ziemba (2005) that the results using geometric means will have essentially the same conclusions.

In Figure A3 we assumed that the fund had initial wealth of 100 dollars on Dec 31, 1992. Figures A2 and A3 show the rates of return over time and the wealth graph over time assuming an initial wealth of 100 on December 31, 2002.

Medallion's outstanding yearly DSSR of 26.4 is the best we have seen even higher than Princeton Newport's 13.8 during 1969–1988. Jim Simon's Medallion fund is near or at the top of the worlds hedge funds. Indeed Simons' $1.4 billion in 2005 was the highest in

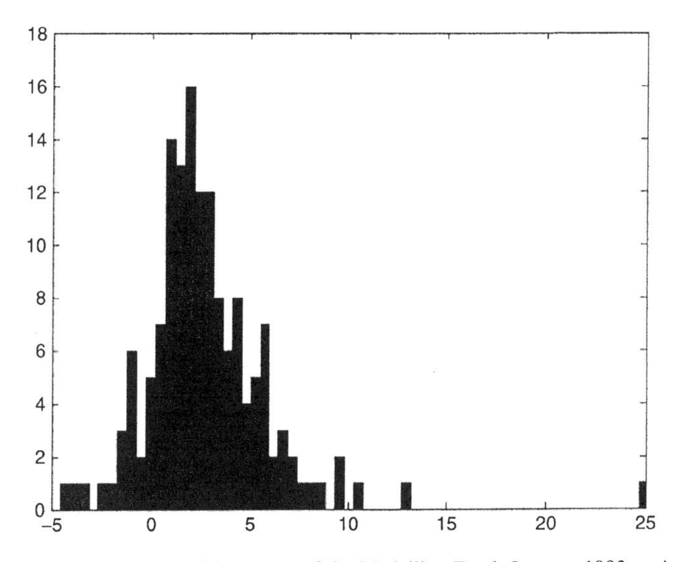

Figure A1 Histogram of monthly returns of the Medallion Fund, January 1993 to April 2005.

[1] WTZ is pleased to have had a minor role in teaching Simons about the Kelly criterion in 1992.
[2] Thanks to Ilkay Boduroglu for making these calculations.

Table A1 Net returns in percent of the Medallion Fund, January 1993 to April 2005, Yearly, Quarterly and Monthly

	1993	1994	1995	1996	1997	1998	1999	2000	2001	2002	2003	2004	2005
Annual													
	39.06	70.69	38.33	31.49	21.21	41.50	24.54	98.53	31.12	29.14	25.28	27.77	
Quarterly													
Q1	7.81	14.69	22.06	7.88	3.51	7.30	(0.25)	25.44	12.62	5.90	4.29	9.03	8.30
Q2	25.06	35.48	4.84	1.40	6.60	7.60	6.70	20.51	5.64	7.20	6.59	3.88	
Q3	4.04	11.19	3.62	10.82	8.37	9.69	6.88	8.58	7.60	8.91	8.77	5.71	
Q4	(0.86)	(1.20)	4.31	8.44	1.41	11.73	9.48	20.93	2.42	4.44	3.62	6.72	
Monthly													
January	1.27	4.68	7.4	3.25	1.16	5.02	3.79	10.5	4.67	1.65	2.07	3.76	2.26
February	3.08	5.16	7.54	1.67	2.03	1.96	−2.44	9.37	2.13	3.03	2.53	1.97	2.86
March	3.28	4.19	5.68	2.77	0.29	0.21	−1.49	3.8	5.36	1.12	−0.35	3.05	2.96
April	6.89	2.42	4.1	0.44	1.01	0.61	3.22	9.78	2.97	3.81	1.78	0.86	0.95
May	3.74	5.66	5.53	0.22	4.08	4.56	1.64	7.24	2.44	1.11	3.44	2.61	
June	12.78	25.19	−4.57	0.73	1.36	2.28	1.71	2.37	0.15	2.13	1.24	0.37	
July	3.15	6.59	−1.28	4.24	5.45	−1.1	4.39	5.97	1	5.92	1.98	2.2	
August	−0.67	7.96	5.91	2.97	1.9	4.31	1.22	3.52	3.05	1.68	2.38	2.08	
September	1.54	−3.38	−0.89	3.25	0.85	6.33	1.15	−1.02	3.38	1.13	4.18	1.33	
October	1.88	−2.05	0.3	6.37	−1.11	5.33	2.76	6.71	1.89	1.15	0.35	2.39	
November	−1.51	−0.74	2.45	5.93	−0.22	2.26	5.42	8.66	0.17	1.42	1.42	3.03	
December	−1.2	1.62	1.52	−3.74	2.77	3.73	1.06	4.3	0.35	1.81	1.81	1.16	

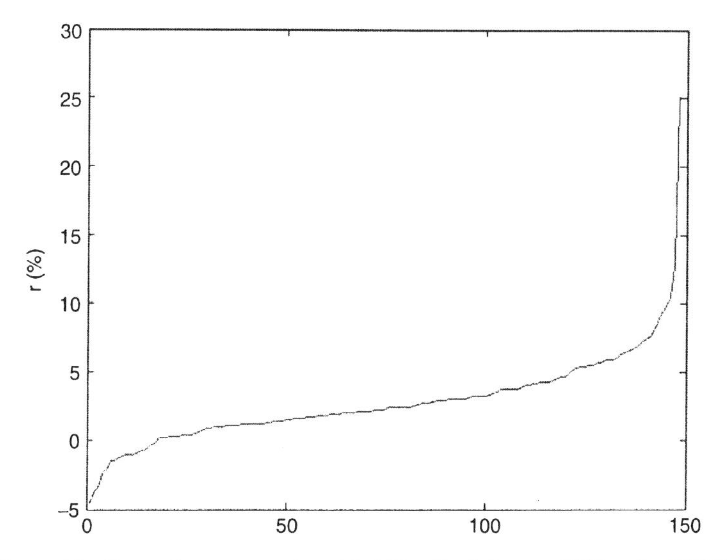

Figure A2 Rates of return over time, Medallion Fund, January 1993 to April 2005

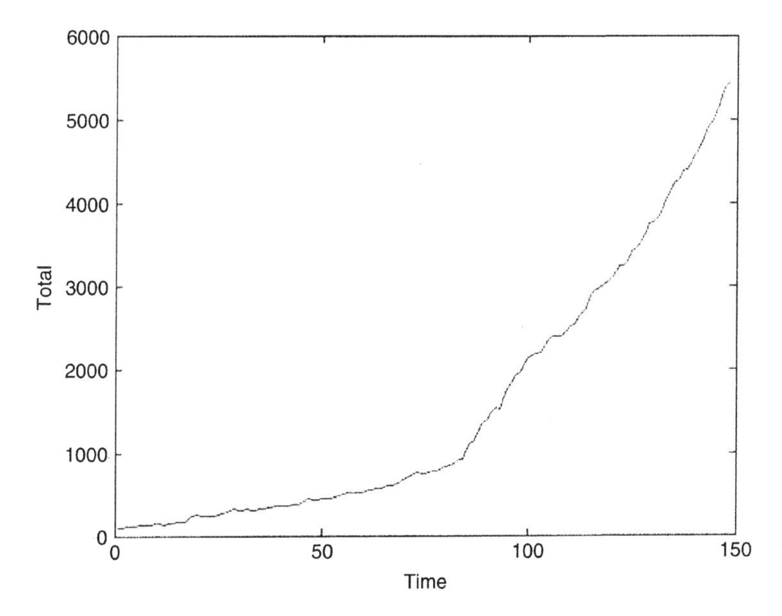

Figure A3 Wealth over time, Medallion Fund, January 1993 to April 2005

Table A2 Sharpe and Downside Symmetric Sharpe Ratios for the Medallion Fund, January 1993 to April 2005

	Yearly	Quarterly	Monthly
SR	1.68	1.09	0.76
DSSR	26.4	11.6	2.20
Mean Risk Free Rate			

Table A3 T-bill interest rates in percent, January 1993 to April 2005, Yearly, Quarterly and Monthly

	1993	1994	1995	1996	1997	1998	1999	2000	2001	2002	2003	2004	2005
Annual													
	3.33	4.98	5.69	5.23	5.36	4.85	4.76	5.86	3.36	1.68	1.05	1.56	3.39
Quarterly													
Q1	2.99	3.27	5.77	4.94	5.06	5.07	4.41	5.53	4.85	1.71	1.14	0.90	2.56
Q2	2.98	4.05	5.61	5.04	5.08	5.00	4.43	5.75	3.70	1.71	1.05	1.06	
Q3	3.02	4.52	5.38	5.13	5.06	4.89	4.67	6.00	3.25	1.63	0.92	1.48	
Q4	3.08	5.31	5.28	4.97	5.08	2.30	5.05	6.03	1.93	1.36	0.91	2.01	
Monthly	1993	1994	1995	1996	1997	1998	1999	2000	2001	2002	2003	2004	2005
January	2.99	3.14	5.46	5.16	5.00	5.07	4.33	5.21	5.63	1.85	1.28	0.91	2.18
February	2.99	3.21	5.62	5.05	5.03	5.07	4.37	5.37	5.24	1.78	1.21	0.91	2.36
March	2.99	3.27	5.77	4.94	5.06	5.07	4.41	5.53	4.85	1.71	1.14	0.90	2.54
April	2.99	3.53	5.72	4.97	5.07	5.05	4.41	5.61	4.45	1.71	1.11	0.95	2.65
May	2.98	3.79	5.67	5.01	5.07	5.02	4.42	5.68	4.05	1.71	1.08	1.01	
June	2.98	4.05	5.61	5.04	5.08	5.00	4.43	5.75	3.66	1.71	1.05	1.06	
July	2.99	4.21	5.54	5.07	5.07	4.96	4.51	5.83	3.52	1.68	1.00	1.20	
August	3.00	4.36	5.46	5.10	5.07	4.92	4.59	5.92	3.39	1.66	0.96	1.34	
September	3.02	4.52	5.38	5.13	5.06	4.89	4.67	6.00	3.25	1.63	0.92	1.48	
October	3.04	4.78	5.34	5.08	5.07	4.69	4.80	6.01	2.81	1.54	0.92	1.66	
November	3.06	5.05	5.31	5.03	5.07	4.49	4.93	6.02	2.37	1.45	0.91	1.83	
December	3.08	5.31	5.28	4.97	5.08	4.30	5.05	6.03	1.93	1.36	0.91	2.01	

the world for hedge fund managers and his $1.6 billion in 2006 was second best. Since the fund is closed to all but about six outside investors plus employees we watch with envy but Renaissance's new $100 billion fund accepts qualified investors.

Chapter 9

54

THE KELLY CRITERION IN BLACKJACK SPORTS BETTING, AND THE STOCK MARKET*

EDWARD O. THORP

Edward O. Thorp and Associates, Newport Beach, CA 92660, USA

Contents

* Paper presented at: The 10th International Conference on Gambling and Risk Taking, Montreal, June 1997, published in: Finding the Edge: Mathematical Analysis of Casino Games, edited by O. Vancura, J.A. Cornelius, W.R. Eadington, 2000. Corrections added April 20, 2005.

Handbook of Asset and Liability Management, Volume 1
Edited by S.A. Zenios and W.T. Ziemba

DOI: 10.1016/S1872-0978(06)01009-X

Abstract

The central problem for gamblers is to find positive expectation bets. But the gambler also needs to know how to manage his money, i.e., how much to bet. In the stock market (more inclusively, the securities markets) the problem is similar but more complex. The gambler, who is now an "investor", looks for "excess risk adjusted return". In both these settings, we explore the use of the Kelly criterion, which is to maximize the expected value of the logarithm of wealth ("maximize expected logarithmic utility"). The criterion is known to economists and financial theorists by names such as the "geometric mean maximizing portfolio strategy", maximizing logarithmic utility, the growth-optimal strategy, the capital growth criterion, etc. The author initiated the practical application of the Kelly criterion by using it for card counting in blackjack. We will present some useful formulas and methods to answer various natural questions about it that arise in blackjack and other gambling games. Then we illustrate its recent use in a successful casino sports betting system. Finally, we discuss its application to the securities markets where it has helped the author to make a thirty year total of 80 billion dollars worth of "bets".

Keywords

Kelly criterion, betting, long run investing, portfolio allocation, logarithmic utility, capital growth

JEL classification: C61, D81, G1

1. Introduction

The fundamental problem in gambling is to find positive expectation betting opportunities. The analogous problem in investing is to find investments with excess risk-adjusted expected rates of return. Once these favorable opportunities have been identified, the gambler or investor must decide how much of his capital to bet. This is the problem which we consider here. It has been of interest at least since the eighteenth century discussion of the St. Petersburg Paradox (Feller, 1966) by Daniel Bernoulli.

One approach is to choose a goal, such as to minimize the probability of total loss within a specified number of trials, N. Another example would be to maximize the probability of reaching a fixed goal on or before N trials (Browne, 1996).

A different approach, much studied by economists and others, is to value money using a utility function. These are typically defined for all non-negative real numbers, have extended real number values, and are non-decreasing (more money is at least as good as less money). Some examples are $U(x) = x^a$, $0 \leqslant a < \infty$, and $U(x) = \log x$, where log means $\log_e$, and $\log 0 = -\infty$. Once a utility function is specified, the object is to maximize the expected value of the utility of wealth.

Daniel Bernoulli used the utility function $\log x$ to "solve" the St. Petersburg Paradox. (But his solution does not eliminate the paradox because every utility function which is unbounded above, including log, has a modified version of the St. Petersburg Paradox.) The utility function $\log x$ was revisited by Kelly (1956) where he showed that it had some remarkable properties. These were elaborated and generalized in an important paper by Breiman (1961). Markowitz (1959) illustrates the application to securities. For a discussion of the Kelly criterion (the "geometric mean criterion") from a finance point of view, see McEnally (1986). He also includes additional history and references.

I was introduced to the Kelly paper by Claude Shannon at M.I.T. in 1960, shortly after I had created the mathematical theory of card counting at casino blackjack. Kelly's criterion was a bet on each trial so as to maximize $E \log X$, the expected value of the logarithm of the (random variable) capital X. I used it in actual play and introduced it to the gambling community in the first edition of Beat the Dealer (Thorp, 1962). If all blackjack bets paid even money, had positive expectation and were independent, the resulting Kelly betting recipe when playing one hand at a time would be extremely simple: bet a fraction of your current capital equal to your expectation. This is modified somewhat in practice (generally down) to allow for having to make some negative expectation "waiting bets", for the higher variance due to the occurrence of payoffs greater than one to one, and when more than one hand is played at a time.

Here are the properties that made the Kelly criterion so appealing. For ease of understanding, we illustrate using the simplest case, coin tossing, but the concepts and conclusions generalize greatly.

2. Coin tossing

Imagine that we are faced with an infinitely wealthy opponent who will wager even money bets made on repeated independent trials of a biased coin. Further, suppose that on each trial our win probability is $p > 1/2$ and the probability of losing is $q = 1 - p$. Our initial capital is X_0. Suppose we choose the goal of maximizing the expected value $E(X_n)$ after n trials. How much should we bet, B_k, on the kth trial? Letting $T_k = 1$ if the kth trial is a win and $T_k = -1$ if it is a loss, then $X_k = X_{k-1} + T_k B_k$ for $k = 1, 2, 3, \ldots$, and $X_n = X_0 + \sum_{k=1}^{n} T_k B_k$. Then

$$E(X_n) = X_0 + \sum_{k=1}^{n} E(B_k T_k) = X_0 + \sum_{k=1}^{n} (p - q) E(B_k).$$

Since the game has a positive expectation, i.e., $p - q > 0$ in this even payoff situation, then in order to maximize $E(X_n)$ we would want to maximize $E(B_k)$ at each trial. Thus, to maximize expected gain we should bet *all of our resources* at each trial. Thus $B_1 = X_0$ and if we win the first bet, $B_2 = 2X_0$, etc. However, the probability of ruin is given by $1 - p^n$ and with $p < 1$, $\lim_{n\to\infty}[1 - p^n] = 1$ so ruin is almost sure. Thus the "bold" criterion of betting to maximize expected gain is usually undesirable.

Likewise, if we play to minimize the probability of eventual ruin (i.e., "ruin" occurs if $X_k = 0$ on the kth outcome) the well-known gambler's ruin formula in Feller (1966) shows that we minimize ruin by making a *minimum* bet on each trial, but this unfortunately also minimizes the expected gain. Thus "timid" betting is also unattractive.

This suggests an intermediate strategy which is somewhere between maximizing $E(X_n)$ (and assuring ruin) and minimizing the probability of ruin (and minimizing $E(X_n)$). An asymptotically optimal strategy was first proposed by Kelly (1956).

In the coin-tossing game just described, since the probabilities and payoffs for each bet are the same, it seems plausible that an "optimal" strategy will involve always wagering the same fraction f of your bankroll. To make this possible we shall assume from here on that capital is infinitely divisible. This assumption usually does not matter much in the interesting practical applications.

If we bet according to $B_i = f X_{i-1}$, where $0 \leqslant f \leqslant 1$, this is sometimes called "fixed fraction" betting. Where S and F are the number of successes and failures, respectively, in n trials, then our capital after n trials is $X_n = X_0(1+f)^S(1-f)^F$, where $S+F = n$. With f in the interval $0 < f < 1$, $\Pr(X_n = 0) = 0$. Thus "ruin" in the technical sense of the gambler's ruin problem cannot occur. "Ruin" shall henceforth be reinterpreted to mean that for arbitrarily small positive ε, $\lim_{n\to\infty}[\Pr(X_n \leqslant \varepsilon)] = 1$. Even in this sense, as we shall see, ruin *can* occur under certain circumstances.

We note that since

$$e^{n \log\left[\frac{X_n}{X_0}\right]^{1/n}} = \frac{X_n}{X_0},$$

the quantity

$$G_n(f) = \log\left[\frac{X_n}{X_0}\right]^{1/n} = \frac{S}{n}\log(1+f) + \frac{F}{n}\log(1-f)$$

measures the exponential rate of increase per trial. Kelly chose to maximize the expected value of the growth rate coefficient, $g(f)$, where

$$g(f) = E\left\{\log\left[\frac{X_n}{X_0}\right]^{1/n}\right\} = E\left\{\frac{S}{n}\log(1+f) + \frac{F}{n}\log(1-f)\right\}$$
$$= p\log(1+f) + q\log(1-f).$$

Note that $g(f) = (1/n)E(\log X_n) - (1/n)\log X_0$ so for n fixed, maximizing $g(f)$ is the same as maximizing $E\log X_n$. We usually will talk about maximizing $g(f)$ in the discussion below. Note that

$$g'(f) = \frac{p}{1+f} - \frac{q}{1-f} = \frac{p-q-f}{(1+f)(1-f)} = 0$$

when $f = f^* = p - q$.

Now

$$g''(f) = -p/(1+f)^2 - q/(1-f)^2 < 0$$

so that $g'(f)$ is monotone strictly decreasing on $[0, 1)$. Also $g'(0) = p - q > 0$ and $\lim_{f\to 1^-} g'(f) = -\infty$. Therefore by the continuity of $g'(f)$, $g(f)$ has a unique maximum at $f = f^*$, where $g(f^*) = p\log p + q\log q + \log 2 > 0$. Moreover, $g(0) = 0$ and $\lim_{f\to q^-} g(f) = -\infty$ so there is a unique number $f_c > 0$, where $0 < f^* < f_c < 1$, such that $g(f_c) = 0$. The nature of the function $g(f)$ is now apparent and a graph of $g(f)$ versus f appears as shown in Figure 1.

The following theorem recounts the important advantages of maximizing $g(f)$. The details are omitted here but proofs of (i)–(iii), and (vi) for the simple binomial case can be found in Thorp (1969); more general proofs of these and of (iv) and (v) are in Breiman (1961).

Theorem 1. (i) *If* $g(f) > 0$, *then* $\lim_{n\to\infty} X_n = \infty$ *almost surely, i.e., for each* M, $\Pr[\liminf_{n\to\infty} X_n > M] = 1$;

(ii) *If* $g(f) < 0$, *then* $\lim_{n\to\infty} X_n = 0$ *almost surely*; *i.e., for each* $\varepsilon > 0$, $\Pr[\limsup_{n\to\infty} X_n < \varepsilon] = 1$;

(iii) *If* $g(f) = 0$, *then* $\limsup_{n\to\infty} X_n = \infty$ *a.s. and* $\liminf_{n\to\infty} X_n = 0$ *a.s.*

(iv) *Given a strategy* Φ^* *which maximizes* $E\log X_n$ *and any other "essentially different" strategy* Φ *(not necessarily a fixed fractional betting strategy), then* $\lim_{n\to\infty} X_n(\Phi^*)/X_n(\Phi) = \infty$ *a.s.*

(v) *The expected time for the current capital* X_n *to reach any fixed preassigned goal* C *is, asymptotically, least with a strategy which maximizes* $E\log X_n$.

(vi) *Suppose the return on one unit bet on the* ith *trial is the binomial random variable* U_i; *further, suppose that the probability of success is* p_i, *where* $1/2 < p_i < 1$.

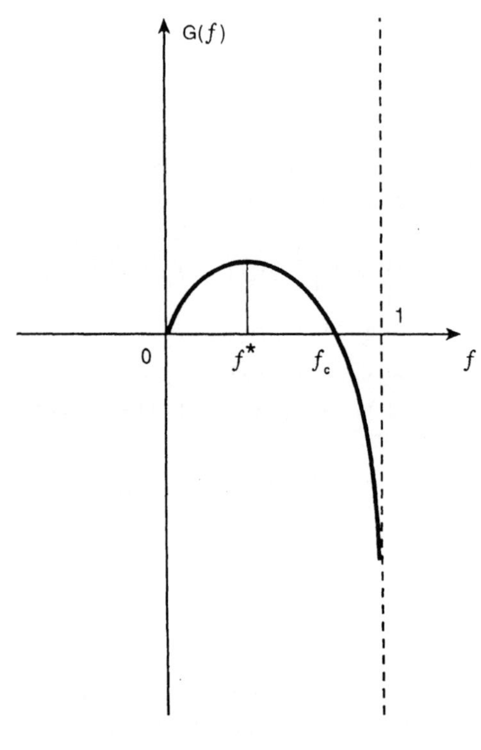

Fig. 1.

Then $E \log X_n$ *is maximized by choosing on each trial the fraction* $f_i^* = p_i - q_i$ *which maximizes* $E \log(1 + f_i U_i)$.

Part (i) shows that, except for a finite number of terms, the player's fortune X_n will exceed any fixed bound M when f is chosen in the interval $(0, f_c)$. But, if $f > f_c$, part (ii) shows that ruin is almost sure. Part (iii) demonstrates that if $f = f_c$, X_n will (almost surely) oscillate randomly between 0 and $+\infty$. Thus, one author's statement that $X_n \to X_0$ as $n \to \infty$, when $f = f_e$, is clearly contradicted. Parts (iv) and (v) show that the Kelly strategy of maximizing $E \log X_n$ is asymptotically optimal by two important criteria. An "essentially different" strategy is one such that the difference $E \ln X_n^* - E \ln X_n$ between the Kelly strategy and the other strategy grows faster than the standard deviation of $\ln X_n^* - \ln X_n$, ensuring $P(\ln X_n^* - \ln X_n > 0) \to 1$. Part (vi) establishes the validity of utilizing the Kelly method of choosing f_i^* on each trial (even if the probabilities change from one trial to the next) in order to maximize $E \log X_n$.

Example 2.1. Player A plays against an infinitely wealthy adversary. Player A wins even money on successive independent flips of a biased coin with a win probability of $p = .53$ (no ties). Player A has an initial capital of X_0 and *capital is infinitely divisible*.

Applying Theorem 1(vi), $f^* = p - q = .53 - .47 = .06$. Thus 6% of current capital should be wagered on each play in order to cause X_n to grow at the fastest rate possible consistent with zero probability of ever going broke. If Player A continually bets a fraction smaller than 6%, X_n will also grow to infinity but the rate will be slower.

If Player A repeatedly bets a fraction larger than 6%, up to the value f_c, the same thing applies. Solving the equation $g(f) = .53 \log(1 + f) + .47 \log(1 - f) = 0$ numerically on a computer yields $f_c = .11973^-$. So, if the fraction wagered is more than about 12%, then even though Player A may temporarily experience the pleasure of a faster win rate, eventual downward fluctuations will inexorably drive the values of X_n toward zero. Calculation yields a growth coefficient of $g(f^*) = f(.06) = .001801$ so that after n successive bets the log of Player A's average bankroll will tend to $.001801n$ times as much money as he started with. Setting $.001801n = \log 2$ gives an expected time of about $n = 385$ to double the bankroll.

The Kelly criterion can easily be extended to uneven payoff games. Suppose Player A wins b units for every unit wager. Further, suppose that on each trial the win probability is $p > 0$ and $pb - q > 0$ so the game is advantageous to Player A. Methods similar to those already described can be used to maximize

$$g(f) = E \log(X_n/X_0) = p \log(1 + bf) + q \log(1 - f).$$

Arguments using calculus yield $f^* = (bp - q)/b$, the optimal fraction of current capital which should be wagered on each play in order to maximize the growth coefficient $g(f)$.

This formula for f^* appeared in Thorp (1984) and was the subject of an April 1997 discussion on the Internet at Stanford Wong's website, http://bj21.com (miscellaneous free pages section). One claim was that one can only lose the amount bet so there was no reason to consider the (simple) generalization of this formula to the situation where a unit wager wins b with probability $p > 0$ and loses a with probability q. Then if the expectation $m \equiv bp - aq > 0$, $f^* > 0$ and $f^* = m/ab$. The generalization does stand up to the objection. One can buy on credit in the financial markets and lose much more than the amount bet. Consider buying commodity futures or selling short a security (where the loss is potentially unlimited). See, e.g., Thorp and Kassouf (1967) for an account of the E.L. Bruce short squeeze.

For purists who insist that these payoffs are not binary, consider selling short a binary digital option. These options are described in Browne (1996).

A criticism sometimes applied to the Kelly strategy is that capital is not, in fact, infinitely divisible. In the real world, bets are multiples of a minimum unit, such as \$1 or \$.01 (penny "slots"). In the securities markets, with computerized records, the minimum unit can be as small as desired. With a minimum allowed bet, "ruin" in the standard sense is always possible. It is not difficult to show, however (see Thorp and Walden, 1966) that if the minimum bet allowed is small relative to the gambler's initial capital, then the probability of ruin in the standard sense is "negligible" and also that the theory herein described is a useful approximation. This section follows Rotando and Thorp (1992).

3. Optimal growth: Kelly criterion formulas for practitioners

Since the Kelly criterion asymptotically maximizes the expected growth rate of wealth, it is often called the optimal growth strategy. It is interesting to compare it with the other fixed fraction strategies. I will present some results that I have found useful in practice. My object is to do so in a way that is simple and easily understood. These results have come mostly from sitting and thinking about "interesting questions". I have not made a thorough literature search but I know that some of these results have been previously published and in greater mathematical generality. See, e.g., Browne (1996, 1997) and the references therein.

3.1. The probability of reaching a fixed goal on or before n trials

We first assume coin tossing. We begin by noting a related result for standard Brownian motion. Howard Tucker showed me this in 1974 and it is probably the most useful single fact I know for dealing with diverse problems in gambling and in the theory of financial derivatives.

For standard Brownian motion $X(t)$, we have

$$P\big(\sup\big[X(t) - (at + b)\big] \geqslant 0,\ 0 \leqslant t \leqslant T\big) = N(-\alpha - \beta) + e^{-2ab} N(\alpha - \beta) \tag{3.1}$$

where $\alpha = a\sqrt{T}$ and $\beta = b/\sqrt{T}$. See Figure 2. See Appendix B for Tucker's derivation of (3.1).

In our application $a < 0, b > 0$ so we expect $\lim_{T\to\infty} P(X(t) \geqslant at+b,\ 0 \leqslant t \leqslant T) = 1$.

Let f be the fraction bet. Assume independent identically distributed (i.d.d.) trials Y_i, $i = 1, \ldots, n$, with $P(Y_i = 1) = p > 1/2$, $P(Y_i = -1) = q < 1/2$; also assume $p < 1$ to avoid the trivial case $p = 1$.

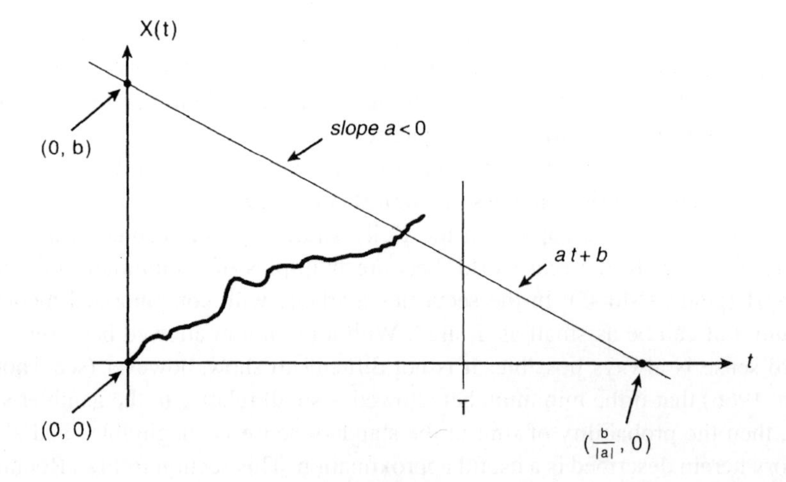

Fig. 2.

Bet a fixed fraction f, $0 < f < 1$, at each trial. Let V_k be the value of the gambler or investor's bankroll after k trials, with initial value V_0. Choose initial stake $V_0 = 1$ (without loss of generality); number of trials n; goal $C > 1$.

What is the probability that $V_k \geqslant C$ for some k, $1 \leqslant k \leqslant n$? This is the same as the probability that $\log V_k \geqslant \log C$ for some k, $1 \leqslant k \leqslant n$. Letting $\ln = \log_e$ we have:

$$V_k = \prod_{i-1}^{k} (1 + Y_i f) \quad \text{and}$$

$$\ln V_k = \sum_{i=1}^{k} \ln(1 + Y_i f),$$

$$E \ln V_k = \sum_{i=1}^{k} E \ln(1 + Y_i f),$$

$$\operatorname{Var}(\ln V_k) = \sum_{i=1}^{k} \operatorname{Var}\big(\ln(1 + Y_i f)\big),$$

$$E \ln(1 + Y_i f) = p \ln(1 + f) + q \ln(1 - f) = m = g(f),$$

$$\begin{aligned}
\operatorname{Var}\big[\ln(1 + Y_i f)\big] &= p\big[\ln(1 + f)\big]^2 + q\big[\ln(1 - f)\big]^2 - m^2 \\
&= \big(p - p^2\big)\big[\ln(1 + f)\big]^2 + \big(q - q^2\big)\big[\ln(1 - f)\big]^2 \\
&\quad - 2pq \ln(1 + f) \ln(1 - f) \\
&= pq\Big\{\big[\ln(1 + f)\big]^2 - 2 \ln(1 + f) \ln(1 - f) + \big[\ln(1 - f)\big]^2\Big\} \\
&= pq\Big\{\ln\big[(1 + f)/(1 - f)\big]\Big\}^2 \equiv s^2.
\end{aligned}$$

Drift in n trials: mn.
Variance in n trials: $s^2 n$.

$$\ln V_k \geqslant \ln C, \quad 1 \leqslant k \leqslant n, \quad \text{iff}$$

$$\sum_{i=1}^{k} \ln(1 + Y_i f) \geqslant \ln C, \quad 1 \leqslant k \leqslant n, \quad \text{iff}$$

$$S_k \equiv \sum_{i=1}^{k} \big[\ln(1 + Y_i f) - m\big] \geqslant \ln C - mk, \quad 1 \leqslant k \leqslant n,$$

$$E(S_k) = 0, \quad \operatorname{Var}(S_k) = s^2 k.$$

We want $\operatorname{Prob}(S_k \geqslant \ln C - mk,\ 1 \leqslant k \leqslant n)$.

Now we use our Brownian motion formula to approximate S_n by $\operatorname{Prob}(X(t) \geqslant \ln C - mt/s^2,\ 1 \leqslant t \leqslant s^2 n)$ where each term of S_n is approximated by an $X(t)$, drift 0 and variance s^2 ($0 \leqslant t \leqslant s^2$, $s^2 \leqslant t \leqslant 2s^2$, ..., $(n-1)s^2 \leqslant t \leqslant ns^2$). Note: the approximation is only "good" for "large" n.

Then in the original formula (3.1):

$$
\begin{aligned}
T &= s^2 n, \\
b &= \ln C, \\
a &= -m/s^2, \\
\alpha &= a\sqrt{T} = -m\sqrt{n}/s, \\
\beta &= b/\sqrt{T} = \ln C/s\sqrt{n}.
\end{aligned}
$$

Example 3.1.

$$
\begin{aligned}
C &= 2, \\
n &= 10^4, \\
p &= .51, \\
q &= .49, \\
f &= .0117, \\
m &= .000165561, \\
s^2 &= .000136848.
\end{aligned}
$$

Then

$$P(\cdot) = .9142.$$

Example 3.2. Repeat with

$$f = .02,$$

then

$$m = .000200013, \quad s^2 = .000399947 \quad \text{and} \quad P(\cdot) = .9214.$$

3.2. The probability of ever being reduced to a fraction x of this initial bankroll

This is a question that is of great concern to gamblers and investors. It is readily answered, approximately, by our previous methods.

Using the notation of the previous section, we want $P(V_k \leqslant x$ for some k, $1 \leqslant k \leqslant \infty)$. Similar methods yield the (much simpler) continuous approximation formula:

$$\text{Prob}(\cdot) = e^{2ab} \quad \text{where } a = -m/s^2 \text{ and } b = -\ln x$$

which can be rewritten as

$$\text{Prob}(\cdot) = x^\wedge(2m/s^2) \quad \text{where } \wedge \text{ means exponentiation.} \tag{3.2}$$

Example 3.3.

$$p = .51, \quad f = f^* = .02,$$
$$2m/s = 1.0002,$$
$$\text{Prob}(\cdot) \doteq x.$$

We will see in Section 7 that for the limiting continuous approximation and the Kelly optimal fraction f^*, $P(V_k(f^*) \leqslant x \text{ for some } k \geqslant 1) = x$.

My experience has been that most cautious gamblers or investors who use Kelly find the frequency of substantial bankroll reduction to be uncomfortably large. We can see why now. To reduce this, they tend to prefer somewhat less than the full betting fraction f^*. This also offers a margin of safety in case the betting situations are less favorable than believed. The penalty in reduced growth rate is not severe for moderate underbetting. We discuss this further in Section 7.

3.3. The probability of being at or above a specified value at the end of a specified number of trials

Hecht (1995) suggested setting this probability as the goal and used a computerized search method to determine optimal (by this criterion) fixed fractions for $p - q = .02$ and various c, n and specified success probabilities.

This is a much easier problem than the similar sounding in Section 3.1. We have for the probability that $X(T)$ at the end exceeds the goal:

$$\begin{aligned} P\big(X(T) \geqslant aT + b\big) &= \frac{1}{\sqrt{2\pi T}} \int_{aT+b}^{\infty} \exp\{-x^2/2T\}\, dx \\ &= \frac{1}{\sqrt{2\pi T}} \int_{aT^{1/2}+bT^{-1/2}}^{\infty} \exp\{-u^2/2\}\, du \end{aligned}$$

where $u = x/\sqrt{T}$ so $x = aT + b$ gives $u\sqrt{T} = aT + b$ and $U = aT^{1/2} + bT^{-1/2}$. The integral equals

$$\begin{aligned} 1 - N\big(aT^{1/2} + bT^{-1/2}\big) &= N\big(-\big(aT^{1/2} + bT^{-1/2}\big)\big) \\ &= 1 - N(\alpha + \beta) = N(-\alpha - \beta). \end{aligned} \tag{3.3}$$

For example (3.1) $f = .0117$ and $P = .7947$. For example (3.2) $P = .7433$. Example (3.1) is for the Hecht optimal fraction and example (3.2) is for the Kelly optimal fraction. Note the difference in P values.

Our numerical results are consistent with Hecht's simulations in the instances we have checked.

Browne (1996) has given an elegant continuous approximation solution to the problem: What is the strategy which maximizes the probability of reaching a fixed goal C on or before a specified time n and what is the corresponding probability of success? Note that the optimal strategy will in general involve betting varying fractions, depending on the time remaining and the distance to the goal.

As an extreme example, just to make the point, suppose $n = 1$ and $C = 2$. If $X_0 < 1$ then no strategy works and the probability of success is 0. But if $1 \leqslant X_0 < 2$ one should bet at least $2 - X_0$, thus any fraction $f \geqslant (2 - X_0)/X_0$, for a success probability of p. Another extreme example: $n = 10$, $C = 2^{10} = 1024$, $X_0 = 1$. Then the only strategy which can succeed is to bet $f = 1$ on every trial. The probability of success is p^{10} for this strategy and 0 for all others (if $p < 1$), including Kelly.

3.4. Continuous approximation of expected time to reach a goal

According to Theorem 1(v), the optimal growth strategy asymptotically minimizes the expected time to reach a goal. Here is what this means. Suppose for goal C that $m(C)$ is the greatest lower bound over all strategies for the expected time to reach C. Suppose $t^*(C)$ is the expected time using the Kelly strategy. Then $\lim_{C\to\infty}(t^*(c)/m(c)) = 1$.

The continuous approximation to the expected number of trials to reach the goal $C > X_0 = 1$ is

$$n(C, f) = (\ln C)/g(f)$$

where f is any fixed fraction strategy. Appendix C has the derivation. Now $g(f)$ has a unique maximum at $g(f^*)$ so $n(C, f)$ has a unique minimum at $f = f^*$. Moreover, we can see how much longer it takes, on average, to reach C if one deviates from f^*.

3.5. Comparing fixed fraction strategies: the probability that one strategy leads another after n trials

Theorem 1(iv) says that wealth using the Kelly strategy will tend, in the long run, to an infinitely large multiple of wealth using any "essentially different" strategy. It can be shown that any fixed $f \neq f^*$ is an "essentially different" strategy. This leads to the question of how fast the Kelly strategy gets ahead of another fixed fraction strategy, and more generally, how fast one fixed fraction strategy gets ahead of (or behind) another.

If W_n is the number of wins in n trials and $n - W_n$ is the number of losses,

$$G(f) = (W_n/n)\ln(l + f) + (1 - W_n/n)\ln(1 - f)$$

is the actual (random variable) growth coefficient.

As we saw, its expectation is

$$g(f) = E\big(G(f)\big) = p\log(1 + f) + q\log(1 - f) \tag{3.4}$$

and the variance of $G(f)$ is

$$\operatorname{Var} G(f) = \big((pq)/n\big)\big\{\ln\big((1 + f)/(1 - f)\big)\big\}^2 \tag{3.5}$$

and it follows that $G(f)$, which has the form $G(f) = a(\sum T_k)/n + b$, is approximately normally distributed with mean $g(f)$ and variance $\operatorname{Var} G(f)$. This enables us to give the distribution of X_n and once again answer the question of Section 3.3. We illustrate this with an example.

Example 3.4.

$$p = .51, \quad q = .49, \quad f^* = .02, \quad N = 10{,}000 \quad \text{and}$$
$$s = \text{standard deviation of } G(f)$$

g/s	f	g	s	$\Pr(G(f) \leqslant 0)$
1.5	.01	.000150004	.0001	.067
1.0	.02	.000200013	.0002	.159
.5	.03	.000149977	.0003	.309

Continuing, we find the distribution of $G(f_2) - G(f_1)$. We consider two cases.

Case 1. The *same* game

Here we assume both players are betting on the same trials, e.g., betting on the same coin tosses, or on the same series of hands at blackjack, or on the same games with the same odds at the same sports book. In the stock market, both players could invest in the same "security" at the same time, e.g., a no-load S&P 500 index mutual fund.

We find

$$E\big(G(f_2) - G(f_1)\big) = p \log\big((1 + f_2)/(1 + f_1)\big) + q \log\big((1 - f_2)/(1 - f_1)\big)$$

and

$$\operatorname{Var}\big(G(f_2) - G(f_1)\big) = (pq/n)\left\{\log\left[\left(\frac{1 + f_2}{1 - f_2}\right)\left(\frac{1 - f_1}{1 + f_1}\right)\right]\right\}^2$$

where $G(f_2) - G(f_1)$ is approximately normally distributed with this mean and variance.

Case 2. Identically distributed independent games

This corresponds to betting on two different series of tosses with the same coin. $E(G(f_2) - G(f_1))$ is as before. But now $\operatorname{Var}(G(f_2) - G(f_1)) = \operatorname{Var}(G(f_2)) + \operatorname{Var}(G(f_1))$ because $G(f_2)$ and $G(f_1)$ are now independent. Thus

$$\operatorname{Var}\big(G(f_2) - G(f_1)\big) = (pq/n)\left\{\left[\log\left(\frac{1 + f_2}{1 - f_2}\right)\right]^2 + \left[\log\left(\frac{1 + f_1}{1 - f_1}\right)^2\right]\right\}.$$

Let

$$a = \log\left(\frac{1 + f_1}{1 - f_1}\right), \qquad b = \log\left(\frac{1 + f_2}{1 - f_2}\right).$$

Then in Case 1, $V_1 = (pq/n)(a - b)^2$ and in Case 2, $V_2 = (pq/n)(a^2 + b^2)$ and since $a, b > 0$, $V_1 < V_2$ as expected. We can now compare the Kelly strategy with other fixed fractions to determine the probability that Kelly leads after n trials. Note that this probability is always greater than 1/2 (within the accuracy limits of the continuous approximation, which is the approximation of the binomial distribution by the normal, with its well known and thoroughly studied properties) because $g(f^*) - g(f) > 0$ where $f^* = p - q$ and $f \neq f^*$ is some alternative. This can fail to be true for small n, where the approximation is poor. As an extreme example to make the point, if $n = 1$, any $f > f^*$ beats Kelly with probability $p > 1/2$. If instead $n = 2$, $f > f^*$ wins with probability p^2 and $p^2 > 1/2$ if $p > 1/\sqrt{2} \doteq .7071$. Also, $f < f^*$ wins with probability $1 - p^2$ and $1 - p^2 > 1/2$ if $p^2 < 1/2$, i.e., $p < 1/\sqrt{2} = .7071$. So when $n = 2$, Kelly always loses more than half the time to some other f unless $p = 1/\sqrt{2}$.

We now have the formulas we need to explore many practical applications of the Kelly criterion.

4. The long run: when will the Kelly strategy "dominate"?

The late John Leib wrote several articles for Blackjack Forum which were critical of the Kelly criterion. He was much bemused by "the long run". What is it and when, if ever, does it happen?

We begin with an example.

Example 4.1.

$$p = .51, \quad n = 10{,}000,$$

V_i and s_i, $i = 1, 2$, are the variance and standard deviation, respectively, for Section 3.5 Cases 1 and 2, and $R = V_2/V_1 = (a^2 + b^2)/(a - b)^2$ so $s_2 = s_1\sqrt{R}$. Table 1 summarizes some results. We can also approximate $\sqrt{R}$ with a power series estimate using only the first term of a and of b: $a \doteq 2f_1$, $b \doteq 2f_2$ so $\sqrt{R} \doteq \sqrt{f_1^2 + f_2^2}/|f_1 - f_2|$. The approximate results, which agree extremely well, are 2.236, 3.606 and 1.581, respectively.

The first two rows show how nearly symmetric the behavior is on each side of the optimal $f^* = .02$. The column $(g_2 - g_1)/s_1$ shows us that $f^* = .02$ only has a .5 standard deviation advantage over its neighbors $f = .01$ and $f = .03$ after $n = 10{,}000$

Table 1
Comparing strategies

f_1	f_2	$g_2 - g_1$	s_1	$(g_2 - g_1)/s_1$	$\sqrt{R}$
.01	.02	.00005001	.00010000	.50	2.236
.03	.02	.00005004	.00010004	.50	3.604
.03	.01	.00000003	.00020005	.00013346	1.581

Table 2
The long run: $(g_2 - g_1)/s$ after n trials

f_1	f_2	$n = 10^4$	$n = 4 \times 10^4$	$n = 16 \times 10^4$	$n = 10^6$
.01	.02	.5	1.0	2.0	5.0
.03	.02	.5	1.0	2.0	5.0
.03	.01	.000133	.000267	.000534	.001335

trials. Since this advantage is proportional to $\sqrt{n}$, the column $(g_2 - g_1)/s_1$ from Table 1 gives the results of Table 2.

The factor $\sqrt{R}$ in Table 1 shows how much more slowly f_2 dominates f_1 in Case 2 versus Case 1. The ratio $(g_2 - g_1)/s_2$ is $\sqrt{R}$ times as large so the same level of dominance takes R times as long. When the real world comparisons of strategies for practical reasons often use Case 2 comparisons rather than the more appropriate Case 1 comparisons, the dominance of f^* is further obscured. An example is players with different betting fractions at blackjack. Case 1 corresponds to both betting on the same sequence of hands. Case 2 corresponds to them playing at different tables (not the same table, because Case 2 assumes independence). (Because of the positive correlation between payoffs on hands played at the same table, this is intermediate between Cases 1 and 2.)

It is important to understand that "the long run", i.e., the time it takes for f^* to dominate a specified neighbor by a specified probability, can vary without limit. Each application requires a separate analysis. In cases such as Example 4.1, where dominance is "slow", one might argue that using f^* is not important. As an argument against this, consider two coin-tossing games. In game 1 your edge is 1.0%. In game 2 your edge is 1.1%. With one unit bets, after n trials the difference in expected gain is $E_2 - E_1 = .001n$ with standard deviation s of about $\sqrt{2n}$ hence $(E_2 - E_1)/s \doteq .001\sqrt{n}/\sqrt{2}$ which is 1 when $n = 2 \times 10^6$. So it takes two million trials to have an 84% chance of the game 2 results being better than the game 1 results. Does that mean it's unimportant to select the higher expectation game?

5. Blackjack

For a general discussion of blackjack, see Thorp (1962, 1966), Wong (1994) and Griffin (1979). The Kelly criterion was introduced for blackjack by Thorp (1962). The analysis is more complicated than that of coin tossing because the payoffs are not simply one to one. In particular the variance is generally more than 1 and the Kelly fraction tends to be less than for coin tossing with the same expectation. Moreover, the distribution of various payoffs depends on the player advantage. For instance the frequency of pair splitting, doubling down, and blackjacks all vary as the advantage changes. By binning the probability of payoff types according to ex ante expectation, and solving the Kelly equations on a computer, a strategy can be found which is as close to optimal as desired.

There are some conceptual subtleties which are noteworthy. To illustrate them we'll simplify to the coin toss model.

At each trial, we have with probability .5 a "favorable situation" with gain or loss X per unit bet such that $P(X = 1) = .51$, $P(X = -1) = .49$ and with probability .5 an unfavorable situation with gain or loss Y per unit bet such that $P(Y = 1) = .49$ and $P(Y = -1) = .51$. We know before we bet whether X or Y applies.

Suppose the player must make small "waiting" bets on the unfavorable situations in order to be able to exploit the favorable situations. On these he will place "large" bets. We consider two cases.

Case 1. Bet f_0 on unfavorable situations and find the optimal f^* for favorable situations. We have

$$g(f) = .5\big(.51\log(1+f) + .49\log(1-f)\big) + .5\big(.49\log(1+f_0) + .51\log(1-f_0)\big). \tag{5.1}$$

Since the second expression in (5.1) is constant, f maximizes $g(f)$ if it maximizes the first expression, so $f^* = p - q = .02$, as usual. It is easy to verify that when there is a spectrum of favorable situations the same recipe, $f_i^* = p_i - q_i$ for the ith situation, holds. Again, in actual blackjack f_i^* would be adjusted down somewhat for the greater variance. With an additional constraint such as $f_i \leqslant kf_0$, where k is typically some integral multiple of f_0, representing the betting spread adopted by a prudent player, then the solution is just $f_i \leqslant \min(f_i^*, kf_0)$.

Curiously, a seemingly similar formulation of the betting problem leads to rather different results.

Case 2. Bet f in favorable situations and af in unfavorable situations, $0 \leqslant a \leqslant 1$.

Now the bet sizes in the two situations are linked and both the analysis and results are more complex. We have a Kelly growth rate of

$$g(f) = .5\big(.51\log(1+f) + .49\log(1-f)\big) + .5\big(.49\log(1+af) + .51\log(1-af)\big). \tag{5.2}$$

If we choose $a = 0$ (no bet in unfavorable situations) then the maximum value for $g(f)$ is at $f^* = .02$, the usual Kelly fraction.

If we make "waiting bets", corresponding to some value of $a > 0$, this will shift the value of f^* down, perhaps even to 0. The expected gain divided by the expected bet is $.02(1-a)/(1+a)$, $a \geqslant 0$. If $a = 0$ we get .02, as expected. If $a = 1$, we get 0, as expected: this is a fair game and the Kelly fraction is $f^* = 0$. As a increases from 0 to 1 the (optimal) Kelly fraction f^* decreases from .02 to 0. Thus the Kelly fraction for favorable situations is less *in this case* when bets on unfavorable situations reduce the overall advantage of the game.

Arnold Snyder called to my attention the fact that Winston Yamashita had (also) made this point (March 18, 1997) on the "free" pages, miscellaneous section, of Stanford Wong's web site.

Table 3
f^* versus a

a	f^*	a	f^*	a	f^*
0	.0200	1/3	.0120	.7	.0040
.1	.0178	.4	.0103	.8	.0024
.2	.0154	.5	.0080	.9	.0011
.3	.0128	.6	.0059	1.0	.0000

For this example, we find the new f^* for a given value of a, $0 < a < 1$, by solving $g'(f) = 0$. A value of $a = 1/3$, for instance, corresponds to a bet of $1/3$ unit on Y and 1 unit on X, a betting range of 3 to 1. The overall expectation is .01. Calculation shows $f^* = .012001$. Table 3 shows how f^* varies with a.

To understand why Cases 1 and 2 have different f^*, look first at Equation (5.1). The part of $g(f)$ corresponding to the unfavorable situations is fixed when f_0 is fixed. Only the part of $g(f)$ corresponding to the favorable situations is affected by varying f. Thus we maximize $g(f)$ by maximizing it over just the favorable situations. Whatever the result, it is then reduced by a fixed quantity, the part of g containing f_0. On the other hand, in Equation (5.2) both parts of $g(f)$ are affected when f varies, because the fraction af used for unfavorable situations bears the constant ratio a to the fraction f used in favorable situations. Now the first term, for the favorable situations, has a maximum at $f = .02$, and is approximately "flat" nearby. But the second term, for the unfavorable situations, is negative and decreasing moderately rapidly at $f = .02$. Therefore, it we reduce f somewhat, this term increases somewhat, while the first term decreases only very slightly. There is a net gain so we find $f^* < .02$. The greater a is, the more important is the effect of this term so the more we have to reduce f to get f^*, as Table 3 clearly shows. When there is a spectrum of favorable situations the solution is more complex and can be found through standard multivariable optimization techniques.

The more complex Case 2 corresponds to what the serious blackjack player is likely to need to do in practice. He will have to limit his current maximum bet to some multiple of his current minimum bet. As his bankroll increases or decreases, the corresponding bet sizes will increase or decrease proportionately.

6. Sports betting

In 1993 an outstanding young computer science Ph.D. told me about a successful sports betting system that he had developed. Upon review I was convinced. I made suggestions for minor simplifications and improvements. Then we agreed on a field test. We found a person who was extremely likely to always be regarded by the other sports bettors as a novice. I put up a test bankroll of $50,000 and we used the Kelly system to estimate our bet size.

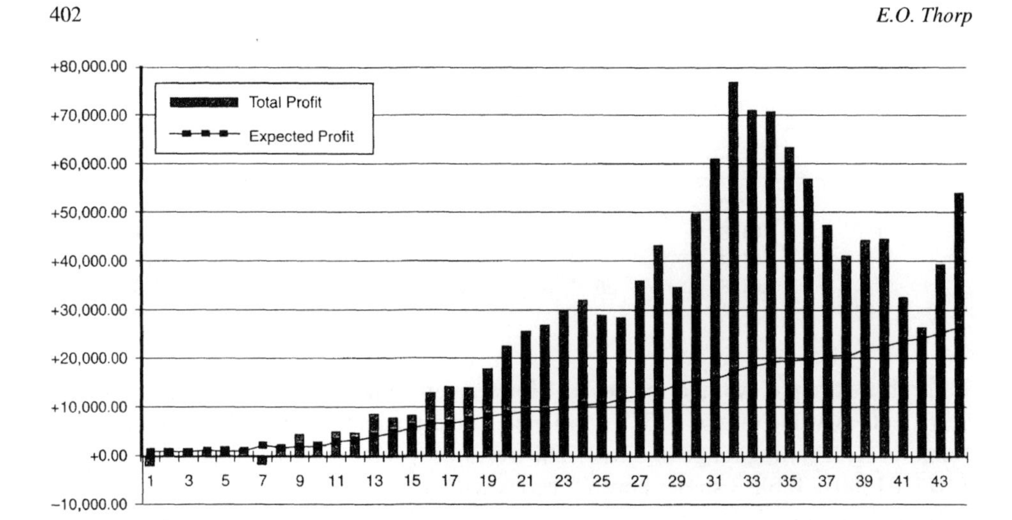

Fig. 3. Betting log Type 2 sports.

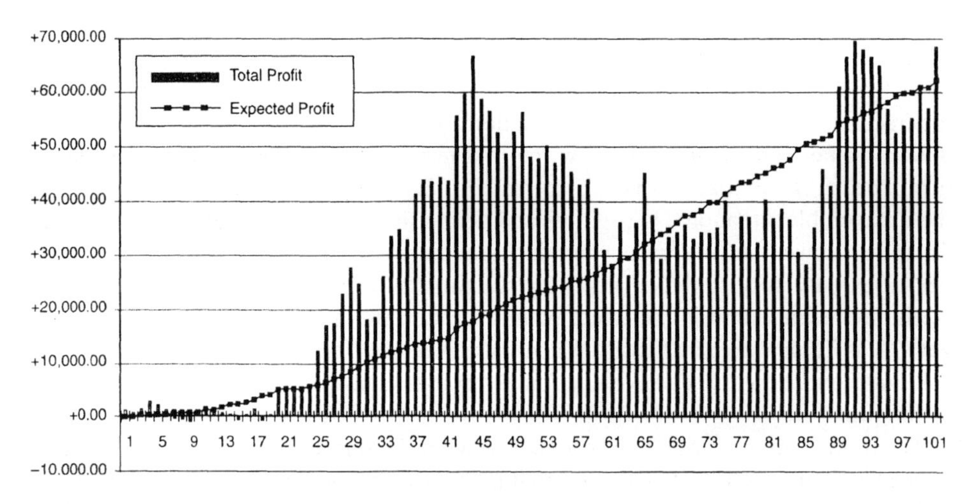

Fig. 4. Betting log Type 1 sports.

We bet on 101 days in the first four and a half months of 1994. The system works for various sports. The results appear in Figures 3 and 4. After 101 days of bets, our \$50,000 bankroll had a profit of \$123,000, about \$68,000 from Type 1 sports and about \$55,000 from Type 2 sports. The expected returns are shown as about \$62,000 for Type 1 and about \$27,000 for Type 2. One might assign the additional \$34,000 actually won to luck. But this is likely to be at most partly true because our expectation estimates from the model were deliberately chosen to be conservative. The reason is that using too large an f^* and overbetting is much more severely penalized than using too small an f^* and underbetting.

Though \$123,000 is a modest sum for some, and insignificant by Wall Street standards, the system performed as predicted and passed its test. We were never more than a few thousand behind. The farthest we had to invade our bankroll to place bets was about \$10,000.

Our typical expectation was about 6% so our total bets ("action") were about \$2,000,000 or about \$20,000 per day. We typically placed from five to fifteen bets a day and bets ranged from a few hundred dollars to several thousand each, increasing as our bankroll grew.

Though we had a net win, the net results by casino varied by chance from a substantial loss to a large win. Particularly hard hit was the "sawdust joint" Little Caesar's. It "died" towards the end of our test and I suspect that sports book losses to us may have expedited its departure.

One feature of sports betting which is of interest to Kelly users is the prospect of betting on several games at once. This also arises in blackjack when (a) a player bets on multiple hands or (b) two or more players share a common bankroll. The standard techniques readily solve such problems. We illustrate with:

Example 6.1. Suppose we bet simultaneously on two independent favorable coins with betting fractions f_1 and f_2 and with success probabilities p_1 and p_2, respectively. Then the expected growth rate is given by

$$\begin{aligned} g(f_1, f_2) = {} & p_1 p_2 \ln(1 + f_1 + f_2) + p_1 q_2 \ln(1 + f_1 - f_2) \\ & + q_1 p_2 \ln(1 - f_1 + f_2) + q_1 q_2 \ln(1 - f_1 - f_2). \end{aligned}$$

To find the optimal f_1^* and f_2^* we solve the simultaneous equations $\partial g / \partial f_1 = 0$ and $\partial g / \partial f_2 = 0$. The result is

$$\begin{aligned} f_1 + f_2 &= \frac{p_1 p_2 - q_1 q_2}{p_1 p_2 + q_1 q_2} \equiv c, \\ f_1 - f_2 &= \frac{p_1 q_2 - q_1 p_2}{p_1 q_2 + q_1 p_2} \equiv d, \\ f_1^* &= (c + d)/2, \qquad f_2^* = (c - d)/2. \end{aligned} \tag{6.1}$$

These equations pass the symmetry check: interchanging 1 and 2 throughout maps the equation set into itself.

An alternate form is instructive. Let $m_i = p_i - q_i$, $i = 1, 2$ so $p_i = (1 + m_i)/2$ and $q_i = (1 - m_i)/2$. Substituting in (6.1) and simplifying leads to:

$$\begin{aligned} c &= \frac{m_1 + m_2}{1 + m_1 m_2}, \qquad d = \frac{m_1 - m_2}{1 - m_1 m_2}, \\ f_1^* &= \frac{m_1(1 - m_2^2)}{1 - m_1^2 m_2^2}, \qquad f_2^* = \frac{m_2(1 - m_1^2)}{1 - m_1^2 m_2^2} \end{aligned} \tag{6.2}$$

which shows clearly the factors by which the f_i^* are each reduced from m_i^*. Since the m_i are typically a few percent, the reduction factors are typically very close to 1.

In the special case $p_1 = p_2 = p$, $d = 0$ and $f^* = f_1^* = f_2^* = c/2 = (p-q)/(2(p^2+q^2))$. Letting $m = p - q$ this may be written $f^* = m/(1+m^2)$ as the optimal fraction to bet on each coin simultaneously, compared to $f^* = m$ to bet on each coin sequentially.

Our simultaneous sports bets were generally on different games and typically not numerous so they were approximately independent and the appropriate fractions were only moderately less than the corresponding single bet fractions. Question: Is this always true for independent simultaneous bets? Simultaneous bets on blackjack hands at different tables are independent but at the same table they have a pairwise correlation that has been estimated at .5 (Griffin, 1979, p. 142). This should substantially reduce the Kelly fraction per hand. The blackjack literature discusses approximations to these problems. On the other hand, correlations between the returns on securities can range from nearly -1 to nearly 1. An extreme correlation often can be exploited to great advantage through the techniques of "hedging". The risk averse investor may be able to acquire combinations of securities where the expectations add and the risks tend to cancel. The optimal betting fraction may be very large.

The next example is a simple illustration of the important effect of covariance on the optimal betting fraction.

Example 6.2. We have two favorable coins as in the previous example but now their outcomes need not be independent. For simplicity assume the special case where the two bets have the same payoff distributions, but with a joint distribution as in Table 4.

Now $c+m+b = (1+m)/2$ so $b = (1-m)/2 - c$ and therefore $0 \leqslant c \leqslant (1-m)/2$.

Calculation shows $\mathrm{Var}(X_i) = 1 - m^2$, $\mathrm{Cor}(X_1, X_2) = 4c - (1-m)^2$ and $\mathrm{Cor}(X_1, X_2) = [4c - (1-m)^2]/(1-m^2)$. The symmetry of the distribution shows that $g(f_1, f_2)$ will have its maximum at $f_1 = f_2 = f$ so we simply need to maximize $g(f) = (c+m)\ln(1+2f) + c\ln(1-2f)$. The result is $f^* = m/(2(2c+m))$. We see that for m fixed, as c decreases from $(1-m)/2$ and $\mathrm{cor}(X_1, X_2) = 1$, to 0 and $\mathrm{cor}(X_1, X_2) = -(1-m)/(1+m)$, f^* for each bet increases from $m/2$ to $1/2$, as in Table 5.

Table 4
Joint distribution of two "identical" favorable coins with correlated outcomes

X_1 :	X_2 : 1	-1
1	$c+m$	b
-1	b	c

Table 5
f^* increases as $\mathrm{Cor}(X_1, X_2)$ decreases

$\mathrm{Cor}(X_1, X_2)$	c	f^*
1	$(1-m)/2$	$m/2$
0	$(1-m^2)/4$	$m/(1+m^2)$
$-(1-m)/(1+m)$	0	$1/2$

It is important to note that for an exact solution or an arbitrarily accurate numerical approximation to the simultaneous bet problem, covariance or correlation information is not enough. We need to use the entire joint distribution to construct the g function.

We stopped sports betting after our successful test for reasons including:

(1) It required a person on site in Nevada.

(2) Large amounts of cash and winning tickets had to be transported between casinos. We believed this was very risky. To the sorrow of others, subsequent events confirmed this.

(3) It was not economically competitive with our other operations.

If it becomes possible to place bets telephonically from out of state and to transfer the corresponding funds electronically, we may be back.

7. Wall street: the biggest game

To illustrate both the Kelly criterion and the size of the securities markets, we return to the study of the effects of correlation as in Example 6.2. Consider the more symmetric and esthetically pleasing pair of bets U_1 and U_2, with joint distribution given in Table 6.

Clearly $0 \leqslant a \leqslant 1/2$ and $\mathrm{Cor}(U_1, U_2) = \mathrm{Cor}(U_1, U_2) = 4a - 1$ increases from -1 to 1 as a increases from 0 to 1/2. Finding a general solution for (f_1^*, f_2^*) appears algebraically complicated (but specific solutions are easy to find numerically), which is why we chose Example 6.2 instead. Even with reduction to the special case $m_1 = m_2 = m$ and the use of symmetry to reduce the problem to finding $f^* = f_1^* = f_2^*$, a general solution is still much less simple. But consider the instance when $a = 0$ so $\mathrm{Cor}(U_1, U_2) = -1$. Then $g(f) = \ln(1 + 2mf)$ which increases without limit as f increases. This pair of bets is a "sure thing" and one should bet as much as possible.

This is a simplified version of the classic arbitrage of securities markets: find a pair of securities which are identical or "equivalent" and trade at disparate prices. Buy the relatively underpriced security and sell short the relatively overpriced security, achieving a correlation of -1 and "locking in" a riskless profit. An example occurred in 1983. My investment partnership bought $ 330 million worth of "old" AT&T and sold short $332.5 million worth of when-issued "new" AT&T plus the new "seven sisters" regional telephone companies. Much of this was done in a single trade as part of what was then the largest dollar value block trade ever done on the New York Stock Exchange (December 1, 1983).

In applying the Kelly criterion to the securities markets, we meet new analytic problems. A bet on a security typically has many outcomes rather than just a few, as in

Table 6
Joint distribution of U_1 and U_2

U_1 :	$U_2 : m_2 + 1$	$m_2 - 1$
$m_1 + 1$	a	$1/2 - a$
$m_1 - 1$	$1/2 - a$	a

most gambling situations. This leads to the use of continuous instead of discrete probability distributions. We are led to find f to maximize $g(f) = E\ln(1 + fX) = \int \ln(1 + fx)\,\mathrm{d}P(x)$ where $P(x)$ is a probability measure describing the outcomes. Frequently the problem is to find an optimum portfolio from among n securities, where n may be a "large" number. In this case x and f are n-dimension vectors and fx is their scalar product. We also have constraints. We always need $1 + fx > 0$ so $\ln(\cdot)$ is defined, and $\sum f_i = 1$ (or some $c > 0$) to normalize to a unit (or to a $c > 0$) investment. The maximization problem is generally solvable because $g(f)$ is concave. There may be other constraints as well for some or all i such as $f_i \geqslant 0$ (no short selling), or $f_i \leqslant M_i$ or $f_i \geqslant m_i$ (limits amount invested in ith security), or $\sum |f_i| \leqslant M$ (limits total leverage to meet margin regulations or capital requirements). Note that in some instances there is not enough of a good bet or investment to allow betting the full f^*, so one is forced to underbet, reducing somewhat both the overall growth rate and the risk. This is more a problem in the gaming world than in the much larger securities markets. More on these problems and techniques may be found in the literature.

7.1. Continuous approximation

There is one technique which leads rapidly to interesting results. Let X be a random variable with $P(X = m + s) = P(X = m - s) = .5$. Then $E(X) = m$, $\mathrm{Var}(X) = s^2$. With initial capital V_0, betting fraction f, and return per unit of X, the result is

$$V(f) = V_0\big(1 + (1 - f)r + fX\big) = V_0\big(1 + r + f(X - r)\big),$$

where r is the rate of return on the remaining capital, invested in, e.g., Treasury bills. Then

$$\begin{aligned} g(f) &= E\big(G(f)\big) = E\big(\ln\big(V(f)/V_0\big)\big) = E\ln\big(1 + r + f(X - r)\big) \\ &= .5\ln\big(1 + r + f(m - r + s)\big) + .5\ln\big(1 + r + f(m - r - s)\big). \end{aligned}$$

Now subdivide the time interval into n equal independent steps, keeping the same drift and the same total variance. Thus m, s^2 and r are replaced by m/n, s^2/n and r/n, respectively. We have n independent X_i, $i = 1, \ldots, n$, with

$$P\big(X_i = m/n + sn^{-1/2}\big) = P\big(X_i = m/n - sn^{-1/2}\big) = .5.$$

Then

$$V_n(f)/V_0 = \prod_{i=1}^{n}\big(1 + (1 - f)r + fX_i\big).$$

Taking $E(\log(\cdot))$ of both sides gives $g(f)$. Expanding the result in a power series leads to

$$g(f) = r + f(m - r) - s^2f^2/2 + \mathrm{O}\big(n^{-1/2}\big) \tag{7.1}$$

where $\mathrm{O}(n^{-1/2})$ has the property $n^{1/2}\mathrm{O}(n^{-1/2})$ is bounded as $n \to \infty$. Letting $n \to \infty$ in (7.1) we have

$$g_\infty(f) \equiv r + f(m - r) - s^2f^2/2. \tag{7.2}$$

The limit $V \equiv V_\infty(f)$ of $V_n(f)$ as $n \to \infty$ corresponds to a log normal diffusion process, which is a well-known model for securities prices. The "security" here has instantaneous drift rate m, variance rate s^2, and the riskless investment of "cash" earns at an instantaneous rate r. Then $g_\infty(f)$ in (7.2) is the (instantaneous) growth rate of capital with investment or betting fraction f. There is nothing special about our choice of the random variable X. Any bounded random variable with mean $E(X) = m$ and variance $\text{Var}(X) = s^2$ will lead to the same result. Note that f no longer needs to be less than or equal to 1. The usual problems, with $\log(\cdot)$ being undefined for negative arguments, have disappeared. Also, $f < 0$ causes no problems. This simply corresponds to selling the security short. If $m < r$ this could be advantageous. Note further that the investor who follows the policy f must now adjust his investment "instantaneously". In practice this means adjusting in tiny increments whenever there is a small change in V. This idealization appears in option theory. It is well known and does not prevent the practical application of the theory (Black and Scholes, 1973). Our previous growth functions for finite sized betting steps were approximately parabolic in a neighborhood of f^* and often in a range up to $0 \leqslant f \leqslant 2f^*$, where also often $2f^* \doteq f_c$. Now with the limiting case (7.2), $g_\infty(f)$ is exactly parabolic and very easy to study.

Lognormality of $V(f)/V_0$ means $\log(V(f)/V_0)$ is $N(M, S^2)$ distributed, with mean $M = g_\infty(f)t$ and variance $S^2 = \text{Var}(G_\infty(f))t$ for any time t. From this we can determine, for instance, the expected capital growth and the time t_k required for $V(f)$ to be at least k standard deviations above V_0. First, we can show by our previous methods that $\text{Var}(G_\infty(f)) = s^2 f^2$, hence $\text{Sdev}(G_\infty(f)) = sf$. Solving $t_k g_\infty = k t_k^{1/2}\,\text{Sdev}(G_\infty(f))$ gives $t_k g_\infty^2$ hence the expected capital growth $t_k g_\infty$, from which we find t_k. The results are summarized in Equations (7.3).

$$
\begin{aligned}
&f^* = (m-r)/s^2, \qquad g_\infty(f) = r + f(m-r) - s^2 f^2/2,\\
&g_\infty(f^*) = (m-r)^2/2s^2 + r,\\
&\text{Var}\big(G_\infty(f)\big) = s^2 f^2, \qquad \text{Sdev}\big(G_\infty(f)\big) = sf,\\
&t_k g_\infty(f) = k^2 s^2 f^2 / g_\infty,\\
&t_k = k^2 s^2 f^2 / g_\infty^2.
\end{aligned}
\tag{7.3}
$$

Examination of the expressions for $t_k g_\infty(f)$ and t_k show that each one increases as f increases, for $0 \leqslant f < f_+$ where f_+ is the positive root of $s^2 f^2/2 - (m-r)f - r = 0$ and $f_+ > 2f^*$.

Comment: The capital asset pricing model (CAPM) says that the market portfolio lies on the Markowitz efficient frontier E in the (s, m) plane at a (generally) unique point $P = (s_0, m_0)$ such that the line determined by P and $(s = 0, m = r)$ is tangent to E (at P). The slope of this line is the Sharpe ratio $S = (m_0 - r_0)/s_0$ and from (7.3) $g_\infty(f^*) = S^2/2 + r$ so the maximum growth rate $g_\infty(f^*)$ depends, for fixed r, only on the Sharpe ratio. (See Quaife (1995).) Again from (7.3), $f^* = 1$ when $m = r + s^2$ in which case the Kelly investor will select the market portfolio without borrowing or lending. If $m > r + s^2$ the Kelly investor will use leverage and if $m < r + s^2$ he will invest partly in T-bills and partly in the market portfolio. Thus the Kelly investor will dynamically reallocate as f^* changes over time because of fluctuations in the forecast m, r and s^2, as well as in the prices of the portfolio securities.

From (7.3), $g_\infty(1) = m - s^2/2$ so the portfolios in the (s, m) plane satisfying $m - s^2/2 = C$, where C is a constant, all have the same growth rate. In the continuous approximation, the Kelly investor appears to have the utility function $U(s, m) = m - s^2/2$. Thus, for any (closed, bounded) set of portfolios, the best portfolios are exactly those in the subset that maximizes the one parameter family $m - s^2/2 = C$. See Kritzman (1998), for an elementary introduction to related ideas.

Example 7.1. The long run revisited. For this example let $r = 0$. Then the basic equations (7.3) simplify to

$$\begin{aligned}
&r = 0\colon\quad f^* = m/s^2, \quad g_\infty(f) = mf - s^2 f^2/2,\\
&g_\infty(f^*) = m^2/2s^2,\\
&\mathrm{Var}\big(G_\infty(f)\big) = s^2 f^2, \qquad \mathrm{Sdev}\big(G_\infty(f)\big) = sf.
\end{aligned} \tag{7.4}$$

How long will it take for $V(f^*) \geqslant V_0$ with a specified probability? How about $V(f^*/2)$? To find the time t needed for $V(f) \geqslant V_0$ at the k standard deviations level of significance ($k = 1$, $P = 84\%$; $k = 2$, $P = 98\%$, etc.) we solve for $t \equiv t_k$:

$$t g_\infty(f) = k t^{1/2}\,\mathrm{Sdev}\big(G_\infty(f)\big). \tag{7.5}$$

We get more insight by normalizing all f with f^*. Setting $f = cf^*$ throughout, we find when $r = 0$

$$\begin{aligned}
&r = 0\colon\quad f^* \doteq m/s^2, \quad f = cm/s^2,\\
&g_\infty(cf^*) = m^2(c - c^2/2)/s^2,\\
&\mathrm{Sdev}\big(G_\infty(cf^*)\big) = cm/s,\\
&t g_\infty(cf^*) = k^2 c/(1 - c/2),\\
&t(k, cf^*) = k^2 s^2/\big(m^2(1 - c/2)^2\big).
\end{aligned} \tag{7.6}$$

Equations (7.6) contain a remarkable result: $V(f) \geqslant V_0$ at the k standard deviation level of significance occurs when expected capital growth $t g_\infty = k^2 c/(1 - c/2)$ and this result is *independent of m and s*. For $f = f^*$ ($c = 1$ in (7.6)), this happens for $k = 1$ at $t g_\infty = 2$ corresponding to $V = V_0 e^2$ and at $k = 2$ for $t g_\infty = 8$ corresponding to $V = V_0 e^8$. Now $e^8 \doteq 2981$ and at a 10% annual (instantaneous) growth rate, it takes 80 years to have a probability of 98% for $V \geqslant V_0$. At a 20% annual instantaneous rate it takes 40 years. However, for $f = f^*/2$, the number for $k = 1$ and 2 are $t g_\infty = 2/3$ and $8/3$, respectively, just $1/3$ as large. So the waiting times for Prob($V \geqslant V_0$) to exceed 84% and 98% become 6.7 years and 26.7 years, respectively, and the expected growth rate is reduced to $3/4$ of that for f^*.

Comment: Fractional Kelly versus Kelly when $r = 0$

From Equations (7.6) we see that $g_\infty(cf^*)/g_\infty(f^*) = c(2-c)$, $0 \leqslant c < \infty$, showing how the growth rate relative to the maximum varies with c. The relative risk $\mathrm{Sdev}(G_\infty(cf^*))/\mathrm{Sdev}(G_\infty(f^*)) = c$ and the relative time to achieve the same expected total growth is $1/c(2-c)$, $0 < c < 2$. Thus the relative "spread" for the same expected total growth is $1/(2-c)$, $0 < c < 2$. Thus, even by choosing c very small, the spread around a given expected growth cannot be reduced by $1/2$. The corresponding results are not quite as simple when $r > 0$.

7.2. The (almost) real world

Assume that prices change "continuously" (no "jumps"), that portfolios may be revised "continuously", and that there are no transactions costs (market impact, commissions, "overhead"), or taxes (Federal, State, city, exchange, etc.). Then our previous model applies.

Example 7.2. The S&P 500 Index. Using historical data we make the rough estimates $m = .11$, $s = .15$, $r = .06$. The equations we need for $r \neq 0$ are the generalizations of (7.6) to $r \neq 0$ and $f = cf^*$, which follow from (7.3):

$$
\begin{aligned}
&cf^* = c(m-r)/s^2,\\
&g_\infty(cf^*) = \big((m-r)^2(c-c^2/2)\big)/s^2 + r,\\
&\mathrm{Sdev}\big(G_\infty(cf^*)\big) = c(m-r)/s,\\
&tg_\infty(cf^*) = k^2c^2/\big(c - c^2/2 + rs^2/(m-r)^2\big),\\
&t(k, cf^*) = k^2c^2\big((m-r)^2/s^2\big)/\big(((m-r)^2/s^2)(c-c^2/2)+r\big)^2.
\end{aligned} \tag{7.7}
$$

If we define $\widetilde{m} = m - r$, $\widetilde{G}_\infty = G_\infty - r$, $\tilde{g}_\infty = g_\infty - r$, then substitution into Equations (7.7) give Equations (7.6), showing the relation between the two sets. It also shows that examples and conclusions about $P(V_n > V_0)$ in the $r = 0$ case are equivalent to those about $P(\ln(V(t)/V_0) > rt)$ in the $r \neq 0$ case. Thus we can compare various strategies versus an investment compounding at a constant riskless rate r such as zero coupon U.S. Treasury bonds.

From Equations (7.7) and $c = 1$, we find

$$
\begin{aligned}
&f^* = 2.2\bar{2}, \qquad g_\infty(f^*) = .11\bar{5}, \qquad \mathrm{Sdev}\big(G_\infty(f^*)\big) = .3\bar{3},\\
&tg_\infty(f^*) = .96k^2, \qquad t = 8.32k^2 \text{ years.}
\end{aligned}
$$

Thus, with $f^* = 2.2\bar{2}$, after 8.32 years the probability is 84% that $V_n > V_0$ and the expected value of $\log(V_n/V_0) = .96$ so the median value of V_n/V_0 will be about $e^{.96} = 2.61$.

With the usual unlevered $f = 1$, and $c = .45$, we find using (7.3)

$$
\begin{aligned}
&g_\infty(1) = m - s^2/2 = .09875, \qquad \mathrm{Sdev}\big(G_\infty(1)\big) = .15,\\
&tg_\infty(1) = .23k^2, \qquad t(k, .45f^*) = 2.31k^2 \text{ years.}
\end{aligned}
$$

Writing $tg_\infty = h(c)$ in (7.7) as

$$h(c) = k^2/\big(1/c + rs^2/\big((m-r)^2c^2\big) - 1/2\big)$$

we see that the measure of riskiness, $h(c)$, increases as c increases, at least up to the point $c = 2$, corresponding to $2f^*$ (and actually beyond, up to $1 + \sqrt{1 + \frac{rs^2}{(m-r)^2}}$).

Writing $t(k, cf^*) = t(c)$ as

$$t(c) = k^2\big((m-r)^2/s^2\big)/\big((m-r)^2/s^2\big)(1 - c/2) + r/c^2$$

shows that $t(c)$ also increases as c increases, at least up to the point $c = 2$. Thus for smaller (more conservative) $f = cf^*$, $c \leqslant 2$, specified levels of $P(V_n > V_0)$ are reached earlier. For $c < 1$, this comes with a reduction in growth rate, which reduction is relatively small for f near f^*.

Note: During the period 1975–1997 the short term T-bill total return for the year, a proxy for r if the investor lends (i.e., $f < 1$), varied from a low of 2.90% (1993) to a high of 14.71% (1981). For details, see Ibbotson Associates, 1998 (or any later) Yearbook.

A large well connected investor might be able to borrow at broker's call plus about 1%, which might be approximated by T-bills plus 1%. This might be a reasonable estimate for the investor who borrows ($f > 1$). For others the rates are likely to be higher. For instance the prime rate from 1975–1997 varied from a low of 6% (1993) to a high of 19% (1981), according to Associates First Capital Corporation (1998).

As r fluctuates, we expect m to tend to fluctuate inversely (high interest rates tend to depress stock prices for well known reasons). Accordingly, f^* and g_∞ will also fluctuate so the long term S&P index fund investor needs a procedure for periodically re-estimating and revising f^* and his desired level of leverage or cash.

To illustrate the impact of $r_b > r$, where r_b is the investor's borrowing rate, suppose r_b in example (7.2) is $r + 2\%$ or .08, a choice based on the above cited historical values for r, which is intermediate between "good" $r_b \doteq r + 1\%$, and "poor" $r_b \doteq$ the prime rate $\doteq r + 3\%$. We replace r by r_b in Equations (7.7) and, if $f^* > 1$, $f^* = 1.3\bar{3}$, $g_\infty(f^*) = .100$, $\mathrm{Sdev}(G_\infty(f^*)) = .20$, $tg_\infty(f^*) = .4k^2$, $t = 4k^2$ years. Note how greatly f^* is reduced.

Comment: Taxes

Suppose for simplicity that all gains are subject to a constant continuous tax rate T and that all losses are subject to a constant continuous tax refund at the same rate T. Think of the taxing entities, collectively, as a partner that shares a fraction T of all gains and losses. Then Equations (7.7) become:

$$cf^* = c(m-r)/s^2(1-T),$$

$$g_\infty(cf^*) = \big((m-r)^2\big(c - c^2/2\big)\big)/s^2 + r(1-T),$$

$$\mathrm{Sdev}\big(G_\infty(cf^*)\big) = c(m-r)/s,$$

$$tg_\infty(cf^*) = k^2c^2/(c - c^2/2 + r(1-T)s^2/(m - r^2)),$$
$$t(k, cf^*) = k^2c^2((m-r)^2/s^2)/(((m-r)^2/s^2)(c - c^2/2) + r(1-T))^2. \tag{7.7T}$$

It is interesting to see that cf^* increases by the factor $1/(1-T)$. For a high income California resident, the combined state and federal marginal tax rate is 45% so this factor is $1/.55 = 1.82$. The amplification of cf^* leads to the same growth rate as before except for a reduction by rT. The Sdev is unchanged and $t(k, cf^*)$ is increased slightly. However, as a practical matter, the much higher leverage needed with a high tax rate is typically not allowed under the margin regulation or is not advisable because the inability to continuously adjust in the real world creates dangers that increase rapidly with the degree of leverage.

7.3. The case for "fractional Kelly"

Figure 5 shows three g curves for the true m : $m_t = .5m_e$, $1.0m_e$ and $1.5m_e$, where m_e is the estimated value of m. The vertical lines and the slanting arrows illustrate the reduction in g for the three choices of: $f = .5f_e^*$, f_e^* and $1.5f_e^*$. For example with $f = .5f_e^*$ or "half Kelly", we have no loss and achieve the maximum $g = .25$, in case $m_t = .5m_e$. But if $m_t = m_e$ then $g = .75$, a loss of .25 and if $m_t = 1.5m_e$ then $g = 1.25$, a loss of 1.0, where all g are in units of $m_e^2/2s^2$. This is indicated both by LOSS_1 and LOSS_2 on the vertical line above $f/f_e^* = .5$, and by the two corresponding arrows which lead upward, and in this case to the right, from this line. A disaster occurs when $m_t = .5m_e$ but we choose $f = 1.5f_e^*$. This combines overbetting f_e^* by 50% with the overestimate of $m_e = 2m_t$. Then $g = -.75$ and we will be ruined. It is still bad to choose $f = f_e^*$ when $m_t = .5m_e$ for then $g = 0$ and we suffer increasingly wild oscillations, both up and down, around our initial capital. During a large downward oscillation experience shows that bettors will generally either quit or be eliminated by a minimum bet size requirement.

Some lessons here are:

(1) To the extent m_e is an uncertain estimate of m_t, it is wise to assume $m_t < m_e$ and to choose $f < f_e^*$ by enough to prevent $g \leqslant 0$.

Estimates of m_e in the stock market have many uncertainties and, in cases of forecast excess return, are more likely to be too high than too low. The tendency is to regress towards the mean. Securities prices follow a "non-stationary process" where m and s vary somewhat unpredictably over time. The economic situation can change for companies, industries, or the economy as a whole. Systems that worked may be partly or entirely based on data mining so m_t may be substantially less than m_e. Changes in the "rules" such as commissions, tax laws, margin regulations, insider trading laws, etc., can also affect m_t. Systems that do work attract capital, which tends to push exceptional m_t down towards average values. The drift down means $m_e > m_t$ is likely.

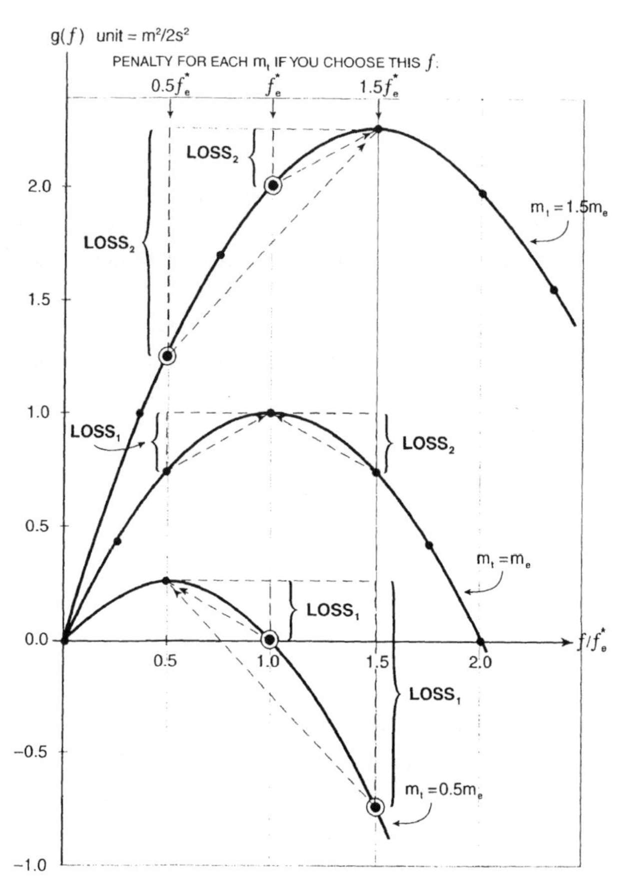

Fig. 5. Penalties for choosing $f = f_e \neq f^* = f_t$.

Sports betting has much the same caveats as the securities markets, with its own differences in detail. Rules changes, for instance, might include: adding expansion teams; the three point rule in basketball; playing overtime sessions to break a tie; changing types of bats, balls, gloves, racquets or surfaces.

Blackjack differs from the securities and sports betting markets in that the probabilities of outcomes can in principle generally be either calculated or simulated to any desired degree of accuracy. But even here m_t is likely to be at least somewhat less than m_e. Consider player fatigue and errors, calculational errors and mistakes in applying either blackjack theory or Kelly theory (e.g., calculating f^* correctly, for which some of the issues have been discussed above), effects of a fixed shuffle point, non-random shuffling, preferential shuffling, cheating, etc.

(2) Subject to (1), choosing f in the range $.5 f_e^* \leqslant f < f_e^*$ offers protection against $g \leqslant 0$ with a reduction of g that is likely to be no more than 25%.

Example 7.3. The great compounder. In 1964 a young hedge fund manager acquired a substantial interest in a small New England textile company called Berkshire Hathaway. The stock traded then at 20. In 1998 it traded at 70,000, a multiple of 3500, and an annualized compound growth rate of about 27%, or an instantaneous rate of 24%. The once young hedge fund manager Warren Buffett is now acknowledged as the greatest investor of our time, and the world's second richest man. You may read about Buffett in Buffett and Clark (1997), Hagstrom (1994, 2004), Kilpatrick (1994) and Lowenstein (1995). If, as I was, you were fortunate enough to meet Buffett and identify the Berkshire opportunity, what strategy does our method suggest? Assume (the somewhat smaller drift rate) $m = .20$, $s = .15$, $r = .06$. Note: Plausible arguments for a smaller future drift rate include regression towards the mean, the increasing size of Berkshire, and risk from the aging of management. A counter-argument is that Berkshire's compounding rate has been as high in its later years as in its earlier years. However, the S&P 500 Index has performed much better in recent years so the spread between the growth rates of the Index and of Berkshire has been somewhat less. So, if we expect the Index growth rate to revert towards the historical mean, then we expect Berkshire to do so even more. From Equations (7.3) or (7.7),

$$f^* = 6.2\bar{2}, \qquad g_\infty(f^*) = .49\bar{5}, \qquad \mathrm{Sdev}\big(G_\infty(f^*)\big) = .9\bar{3},$$
$$t g_\infty(f^*) = 1.76k^2, \qquad t = 3.54k^2 \text{ years}.$$

Compare this to the unlevered portfolio, where $f = 1$ and $c = 1/6.2\bar{2} \doteq .1607$. We find:

$$f = 1, \qquad g_\infty(f) = .189, \qquad \mathrm{Sdev}\big(G_\infty(f)\big) = .15,$$
$$t_k g_\infty(f) = .119k^2, \qquad t_k = .63k^2 \text{ years}.$$

Leverage to the level $6.2\bar{2}$ would be inadvisable here in the real world because securities prices may change suddenly and discontinuously. In the crash of October, 1987, the S&P 500 index dropped 23% in a single day. If this happened at leverage of 2.0, the new leverage would suddenly be $77/27 = 2.85$ before readjustment by selling part of the portfolio. In the case of Berkshire, which is a large well-diversified portfolio, suppose we chose the conservative $f = 2.0$. Note that this is the maximum initial leverage allowed "customers" under current regulations. Then $g_\infty(2) = .295$. The values in 30 years for median V_∞/V_0 are approximately: $f = 1$, $V_\infty/V_0 = 288$; $f = 2$, $V_\infty/V_0 = 6,974$; $f = 6.2\bar{2}$, $V_\infty/V_0 = 2.86 \times 10^6$. So the differences in results of leveraging are enormous in a generation. (Note: Art Quaife reports $s = .24$ for 1980–1997. The reader is invited to explore the example with this change.)

The results of Section 3 apply directly to this continuous approximation model of a (possibly) leveraged securities portfolio. The reason is that both involve the same "dynamics", namely $\log G_n(f)$ is approximated as (scaled) Brownian motion with drift. So we can answer the same questions here for our portfolio that were answered in Section 3 for casino betting. For instance (3.2) becomes

$$\mathrm{Prob}\big(V(t)/V_0 \leqslant x \text{ for some } t\big) = x^\wedge\big(2g_\infty/\mathrm{Var}(G_\infty)\big) \qquad (7.8)$$

where $\wedge$ means exponentiation and $0 < x < 1$. Using (7.4), for $r = 0$ and $f = f^*$, $2g_\infty/\mathrm{Var}(G_\infty) = 1$ so this simplifies to

$$\mathrm{Prob}(\cdot) = x. \tag{7.9}$$

Compare with Example 3.3 For $0 < r < m$ and $f = f^*$ the exponent of x in (7.9) becomes $1 + 2rs^2/(m-r)^2$ and has a positive first derivative so, as r increases, $P(\cdot)$ decreases since $0 < x < 1$, tending to 0 as r tends to m, which is what we expect.

7.4. *A remarkable formula*

In earlier versions of this chapter the exponent in Equations (3.2), (7.8) and (7.9) were off by a factor of 2, which I had inadvertently dropped during my derivation. Subsequently Don Schlesinger posted (without details) two more general continuous approximation formulas for the $r = 0$ case on the Internet at www.bjmath.com dated June 19, 1997.

If V_0 is the initial investment and $y > 1 > x > 0$ then for f^* the probability that $V(t)$ reaches yV_0 before xV_0 is

$$\mathrm{Prob}\big(V(t, f^*) \text{ reaches } yV_0 \text{ before } xV_0\big) = (1-x)/\big(1-(x/y)\big) \tag{7.10}$$

and more generally, for $f = cf^*$, $0 < c < 2$,

$$\mathrm{Prob}\big(V(t, cf^*) \text{ reaches } yV_0 \text{ before } xV_0\big) = \big[1 - x\wedge(2/c-1)\big]/\big[1-(x/y)\wedge(2/c-1)\big] \tag{7.11}$$

where $\wedge$ means exponentiation.

Clearly (7.10) follows from (7.11) by choosing $c = 1$. The $r = 0$ case of our Equation (7.8) follows from (7.11) and the $r = 0$ case of our Equation (7.9) follows from (7.10). We can derive a generalization of (7.11) by using the classical gambler's ruin formula (Cox and Miller, 1965, p. 31, Equation (2.0)) and passing to the limit as step size tends to zero (Cox and Miller, 1965, pp. 205–206), where we think of $\log(V(t, f)/V_0)$ as following a diffusion process with mean g_∞ and variance $v(G_\infty)$, initial value 0, and absorbing barriers at $\log y$ and $\log x$. The result is

$$\mathrm{Prob}\big(V(t, cf^*) \text{ reaches } yV_0 \text{ before } xV_0\big) = [1 - x\wedge a]/\big[1-(x/y)\wedge a\big] \tag{7.12}$$

where $a = 2g_\infty/V(G_\infty) = 2M/V$ where M and V are the drift and variance, respectively, of the diffusion process per unit time. Alternatively, (7.12) is a simple restatement of the known solution for the Wiener process with two absorbing barriers (Cox and Miller, 1965, Example 5.5).

As Schlesinger notes, choosing $x = 1/2$ and $y = 2$ in (7.10) gives $\mathrm{Prob}(V(t, f^*)$ doubles before halving) $= 2/3$. Now consider a gambler or investor who focuses only on values $V_n = 2^n V_0$, $n = 0, \pm 1, \pm 2, \ldots$, multiples of his initial capital. In log space, $\log(V_n/V_0) = n \log 2$ so we have a random walk on the integer multiples of $\log 2$, where the probability of an increase is $p = 2/3$ and of a decrease, $q = 1/3$. This gives us a convenient compact visualization of the Kelly strategy's level of risk.

If instead we choose $c = 1/2$ ("half Kelly"), Equation (7.11) gives Prob($V(t, f^*/2)$ doubles before halving) $= 8/9$ yet the growth rate $g_\infty(f^*/2) = .75 g_\infty(f^*)$ so "half Kelly" has 3/4 the growth rate but much less chance of a big loss.

A second useful visualization of comparative risk comes from Equation (7.8) which gives

$$\text{Prob}\big(V(t, cf^*)/V_0 \leqslant x \text{ for some } t\big) = x^\wedge(2/c - 1). \tag{7.13}$$

For $c = 1$ we had Prob$(\cdot) = x$ and for $c = 1/2$ we get Prob$(\cdot) = x^3$. Thus "half Kelly" has a much lessened likelihood of severe capital loss. The chance of ever losing half the starting capital is 1/2 for $f = f^*$ but only 1/8 for $f = f^*/2$. My gambling and investment experience, as well as reports from numerous blackjack players and teams, suggests that most people strongly prefer the increased safety and psychological comfort of "half Kelly" (or some nearby value), in exchange for giving up 1/4 of their growth rate.

8. A case study

In the summer of 1997 the XYZ Corporation (pseudonym) received a substantial amount of cash. This prompted a review of its portfolio, which is shown in Table 7 in the column 8/17/97. The portfolio was 54% in Biotime, ticker BTIM, a NASDAQ biotechnology company. This was due to existing and historical relationships between people in XYZ Corp. and in BTIM. XYZ's officers and directors were very knowledge-

Table 7
Statistics for logs of monthly wealth relatives, 3/31/92 through 6/30/97

		Berkshire	BioTime	SP500	T-bills
Monthly	Mean	.0264	.0186	.0146	.0035
	Standard deviation	.0582	.2237	.0268	.0008
Annual	Mean	.3167	.2227	.1753	.0426
	Standard deviation	.2016	.7748	.0929	.0028
Monthly	Covariance	.0034	−.0021	.0005	1.2E−06
			.0500	−.0001	3.2E−05
				.0007	5.7E−06
					6.7E−07
Monthly	Correlation	1.0000	−.1581	.2954	.0257
			1.0000	−.0237	.1773
				1.0000	.2610
					1.0000

able about BTIM and felt they were especially qualified to evaluate it as an investment. They wished to retain a substantial position in BTIM.

The portfolio held Berkshire Hathaway, ticker BRK, having first purchased it in 1991.

8.1. The constraints

Dr. Quaife determined the Kelly optimal portfolio for XYZ Corp. subject to certain constraints. The list of allowable securities was limited to BTIM, BRK, the Vanguard 500 (S&P 500) Index Fund, and T-bills. Being short T-bills was used as a proxy for margin debt. The XYZ broker actually charges about 2% more, which has been ignored in this analysis. The simple CAPM (capital asset pricing model) suggests that the investor only need consider the market portfolio (for which the S&P 500 is being substituted here, with well known caveats) and borrowing or lending. Both Quaife and the author were convinced that BRK was and is a superior alternative and their knowledge about and long experience with BRK supported this.

XYZ Corp. was subject to margin requirements of 50% initially and 30% maintenance, meaning for a portfolio of securities purchased that initial margin debt (money lent by the broker) was limited to 50% of the value of the securities, and that whenever the value of the account net of margin debt was less than 30% of the value of the securities owned, then securities would have to be sold until the 30% figure was restored.

In addition XYZ Corp. wished to continue with a "significant" part of its portfolio in BTIM.

8.2. The analysis and results

Using monthly data from 3/31/92 through 6/30/97, a total of 63 months, Quaife finds the means, covariances, etc. given in Table 7.

Note from Table 7 that BRK has a higher mean and a lower standard deviation than BTIM, hence we expect it to be favored by the analysis. But note also the negative correlation with BTIM, which suggests that adding some BTIM to BRK may prove advantageous.

Using the statistics from Table 7, Quaife finds the following optimal portfolios, under various assumptions about borrowing.

As expected, BRK is important and favored over BTIM but some BTIM added to the BRK is better than none.

If unrestricted borrowing were allowed it would be foolish to choose the corresponding portfolio in Table 8. The various underlying assumptions are only approximations with varying degrees of validity: Stock prices do not change continuously; portfolios can't be adjusted continuously; transactions are not costless; the borrowing rate is greater than the T-bill rate; the after tax return, if different, needs to be used; the process which generates securities returns is not stationary and our point estimates of the statistics in Table 7 are uncertain. We have also noted earlier that because "over-

Table 8
Optimal portfolio allocations with various assumptions about borrowing

Security fraction			
Security	No borrowing	50% margin	Unrestricted borrowing
Berkshire	.63	1.50	6.26
BioTime	.37	.50	1.18
S&P 500	.00	.00	12.61
T-bills	.00	−1.00	−19.04
Portfolio growth rate			
Mean	.36	.62	2.10
Standard deviation	.29	.45	2.03

betting" is much more harmful than underbetting, "fractional Kelly" is prudent to the extent the results of the Kelly calculations reflect uncertainties.

In fact, the data used comes from part of the period 1982–1997, the greatest bull market in history. We would expect returns in the future to regress towards the mean so the means in Table 7 are likely to be overestimates of the near future. The data set is necessarily short, which introduces more uncertainty, because it is limited by the amount of BTIM data. As a sensitivity test, Quaife used conservative (mean, std. dev.) values for the price relatives (not their logs) for BRK of (1.15, .20), BTIM of (1.15, 1.0) and the S&P 500 from 1926–1995 from Ibbotson (1998) of (1.125, .204) and the correlations from Table 7. The result was fractions of 1.65, .17, .18 and −1.00 respectively for BRK, BTIM, S&P 500 and T-bills. The mean growth rate was .19 and its standard deviation was .30.

8.3. *The recommendation and the result*

The 50% margin portfolio reallocations of Table 8 were recommended to XYZ Corp.'s board on 8/17/97 and could have been implemented at once. The board elected to do nothing. On 10/9/97 (in hindsight, a good sale at a good price) it sold some BTIM and left the proceeds in cash (not good). Finally on 2/9/98 after a discussion with both Quaife and the author, it purchased 10 BRK (thereby gaining almost $140,000 by 3/31/98, as it happened). The actual policy, led to an increase of 73.5%. What would have happened with the recommended policy with no rebalance and with one rebalance on 10/6/97? The gains would have been 117.6% and 199.4%, respectively. The gains over the suboptimal board policy were an additional $475,935 and $1,359,826, respectively.

The optimal policy displays three important features in this example: the use of leverage, the initial allocation of the portfolio, and possible rebalancing (reallocation) of the portfolio over time. Each of these was potentially important in determining the final result. The potential impact of continuously rebalancing to maintain maximum margin is illustrated in Thorp and Kassouf (1967), Appendix A, The Avalanche Effort.

The large loss from the suboptimal policy was much more than what would have been expected because BRK and BTIM appreciated remarkably. In .62 years, BRK was up 60.4% and BTIM was up 62.9%. This tells us that—atypically—in the absence of rebalancing, the relative initial proportions of BRK and BTIM did not matter much over the actual time period. However, rebalancing to adjust the relative proportions of BRK and BTIM was important, as the actual policy's sale of some BTIM on 10/9/97 illustrated. Also, rebalancing was important for adjusting the margin level as prices, in this instance, rose rapidly.

Table 8 illustrates what we might have normally expected to gain by using 50% margin, rather than no margin. We expect the difference in the medians of the portfolio distributions to be $\$1{,}080{,}736[\exp(.62 \times .62) - \exp(.36 \times .62)] = \$236{,}316$ or 21.9% which is still large.

8.4. *The theory for a portfolio of securities*

Consider first the unconstrained case with a riskless security (T-bills) with portfolio fraction f_0 and n securities with portfolio fractions $f_1, \ldots, f_n$. Suppose the rate of return on the riskless security is r and, to simplify the discussion, that this is also the rate for borrowing, lending, and the rate paid on short sale proceeds. Let $C = (s_{ij})$ be the matrix such that s_{ij}, $i, j = 1, \ldots, n$, is the covariance of the ith and jth securities and $M = (m_1, m_2, \ldots, m_n)^T$ be the row vector such that m_i, $i = 1, \ldots, n$, is the drift rate of the ith security. Then the portfolio satisfies

$$
\begin{aligned}
&f_0 + \cdots + f_n = 1, \\
&m = f_0 r + f_1 m_1 + \cdots + f_n m_n = r + f_1(m_1 - r) + \cdots + f_n(m_n - r) \\
&\quad = r + F^T(M - R), \\
&s^2 = F^T C F
\end{aligned}
\tag{8.1}
$$

where $F^T = (f_1, \ldots, f_n)$ and ^{T} means "transpose", and R is the column vector $(r, r, \ldots, r)^T$ of length n.

Then our previous formulas and results for one security plus a riskless security apply to $g_\infty(f_1, \ldots, f_n) = m - s^2/2$. This is a standard quadratic maximization problem. Using (8.1) and solving the simultaneous equations $\partial g_\infty / \partial f_i = 0$, $i = 1, \ldots, n$, we get

$$
\begin{aligned}
&F^* = C^{-1}[M - R], \\
&g_\infty(f_1^*, \ldots, f_n^*) = r + (F^*)^T C F^*/2
\end{aligned}
\tag{8.2}
$$

where for a unique solution we require C^{-1} to exist, i.e., $\det C \neq 0$. When all the securities are uncorrelated, C is diagonal and we have $f_i^* = (m_i - r)/s_{ii}$ or $f_i^* = (m_i - r)/s_i^2$, which agrees with Equation (7.3) when $n =, 1$.

Note: BRK issued a new class of common, ticker symbol BRK.B, with the old common changing its symbol to BRK.A. One share of BRK.A can be converted to 30 shares of BRK.B at any time, but not the reverse. BRK.B has lesser voting rights and no right to assign a portion of the annual quota of charitable contributions. Both we and the market consider these differences insignificant and the A has consistently traded at about 30 times the price of the B.

If the price ratio were always exactly 30 to 1 and both these securities were included in an analysis, they would each have the same covariances with other securities, so $\det C = 0$ and C^{-1} does not exist.

If there is an initial margin constraint of q, $0 \leqslant q \leqslant 1$, then we have the additional restriction

$$|f_1| + \cdots + |f_n| \leqslant 1/q. \tag{8.3}$$

The n-dimensional subset in (8.3) is closed and bounded.

If the rate for borrowing to finance the portfolio is $r_b = r + e_b$, $e_b \geqslant 0$, and the rate paid on the short sale proceeds is $r_s = r - e_s$, $e_s \geqslant 0$, then the m in Equation (8.1) is altered. Let $x^+ = \max(x, 0)$ and $x^- = \max(0, -x)$ so $x = x^+ - x^-$ for all x. Define $f^+ = f_1^+ + \cdots + f_n^+$, the fraction of the portfolio held long. Let $f^- = f_1^- + \cdots + f_n^-$, the fraction of the portfolio held short.

Case 1. $f^+ \leqslant 1$

$$m = r + f_1(m_1 - r) + \cdots + f_n(m_n - r) - e_s f^-. \tag{8.4.1}$$

Case 2. $f^+ > 1$

$$m = r + f_1(m_1 - r) + \cdots + f_n(m_n - r) - e_b(f^+ - 1) - e_s f^-. \tag{8.4.2}$$

9. My experience with the Kelly approach

How does the Kelly-optimal approach do in practice in the securities markets? In a little-known paper (Thorp, 1971) I discussed the use of the Kelly criterion for portfolio management. Page 220 mentions that "On November 3, 1969, a private institutional investor decided to ... use the Kelly criterion to allocate its assets". This was actually a private limited partnership, specializing in convertible hedging, which I managed. A notable competitor at the time (see Institutional Investor (1998)) was future Nobel prize winner Harry Markowitz. After 20 months, our record as cited was a gain of 39.9% versus a gain for the Dow Jones Industrial Average of +4.2%. Markowitz dropped out after a couple of years, but we liked our results and persisted. What would the future bring?

Up to May 1998, twenty eight and a half years since the investment program began. The partnership and its continuations have compounded at approximately 20% annually with a standard deviation of about 6% and approximately zero correlation with the market ("market neutral"). Ten thousand dollars would, tax exempt, now be worth 18 million dollars. To help persuade you that this may not be luck, I estimate that during this period I have made about $80 billion worth of purchases and sales ("action", in casino language) for my investors. This breaks down into something like one and a quarter million individual "bets" averaging about $65,000 each, with on average hundreds of "positions" in place at any one time. Over all, it would seem to be a moderately "long run" with a high probability that the excess performance is more than chance.

10. Conclusion

Those individuals or institutions who are long term compounders should consider the possibility of using the Kelly criterion to asymptotically maximize the expected compound growth rate of their wealth. Investors with less tolerance for intermediate term risk may prefer to use a lesser function. Long term compounders ought to avoid using a greater fraction ("overbetting"). Therefore, to the extent that future probabilities are uncertain, long term compounders should further limit their investment fraction enough to prevent a significant risk of overbetting.

Acknowledgements

I thank Dr. Jerry Baesel, Professor Sid Browne, Professor Peter Griffin, Dr. Art Quaife, and Don Schlesinger for comments and corrections and to Richard Reid for posting this chapter on his website. I am also indebted to Dr. Art Quaife for allowing me to use his analysis in the case study.

This chapter has been revised and expanded since its presentation at the 10th International Conference on Gambling and Risk Taking.

Appendix A. Integrals for deriving moments of E_∞

$$I_0(a^2, b^2) = \int_0^\infty \exp\left[-\left(a^2x^2 + b^2/x^2\right)\right] \mathrm{d}x,$$

$$I_n(a^2, b^2) = \int_0^\infty x^n \exp\left[-\left(a^2x^2 + b^2/x^2\right)\right] \mathrm{d}x.$$

Given I_0 find I_2

$$\begin{aligned} I_0(a^2, b^2) &= \int_0^\infty \exp\left[-\left(a^2x^2 + b^2/x^2\right)\right] \mathrm{d}x \\ &= -\int_\infty^0 \exp\left[-\left(a^2/u^2 + b^2u^2\right)\right]\left(-\mathrm{d}u/u^2\right) \end{aligned}$$

where $x = 1/u$ and $\mathrm{d}x = -\mathrm{d}u/u^2$ so

$$I_0(a^2, b^2) = \int_0^\infty x^{-2} \exp[-(b^2x^2 + a^2/x^2)] = I_{-2}(b^2, a^2),$$

hence

$$I_{-2}(a^2, b^2) = I_0(b^2, a^2) = \frac{\sqrt{\pi}}{2 \mid b \mid} \mathrm{e}^{-2|ab|},$$

$$I_0 = \int_0^\infty \exp[-(a^2x^2 + b^2/x^2)]\,\mathrm{d}x = U \cdot V|_0^\infty - \int_0^\infty V\,\mathrm{d}U \tag{A.1}$$

where $U = \exp[\,\cdot\,]$, $\mathrm{d}V = \mathrm{d}x$, $\mathrm{d}U = (\exp[\,\cdot\,])(-2a^2x + 2b^2x^{-3})$ and $V = x$ so

$$\begin{aligned} I_0 &= \exp[-(a^2x^2 + b^2/x^2)] \cdot x|_0^\infty \\ &\quad - \int_0^\infty (-2a^2x^2 + 2b^2/x^2) \exp[-(a^2x^2 + b^2/x^2)]\,\mathrm{d}x \\ &= 2a^2 I_2(a^2, b^2) - 2b^2 I_{-2}(a^2, b^2). \end{aligned}$$

Hence:

$$I_0(a^2, b^2) = 2a^2 I_2(a^2, b^2) - 2b^2 I_{-2}(a^2, b^2)$$

and $I_{-2}(a^2, b^2) = I_0(b^2, a^2)$ so substituting and solving for I_2 gives

$$I_2(a^2, b^2) = \frac{1}{2a^2}\{I_0(a^2, b^2) + 2b^2 I_0(b^2, a^2)\}.$$

Comments.

(1) We can solve for all even n by using I_0, I_{-2} and I_2, and integration by parts.

(2) We can use the indefinite integral J_0 corresponding to I_0, and the previous methods, to solve for J_{-2}, J_2, and then for all even n. Since

$$I_0(a^2, b^2) = \frac{\sqrt{\pi}}{2|a|}\mathrm{e}^{-2|ab|}$$

then

$$I_{-2}(a^2, b^2) = \frac{\sqrt{\pi}}{2|b|}\mathrm{e}^{-2|ab|} \quad \text{and}$$

$$I_2(a^2, b^2) = \frac{1}{2a^2}\left\{\frac{\sqrt{\pi}}{2|a|} + 2b^2\frac{\sqrt{\pi}}{2|b|}\right\}\mathrm{e}^{-2|ab|} = \frac{\sqrt{\pi}}{4a^2}\mathrm{e}^{-2|ab|}\{1/|a| + 2|b|\}.$$

Appendix B. Derivation of formula (3.1)

This is based on a note from Howard Tucker. Any errors are mine.

From the paper by Paranjape and Park, if $x(t)$ is standard Brownian motion, if $a \neq 0$, $b > 0$,

$$P\big(X(t) \leqslant at + b,\ 0 \leqslant t \leqslant T \mid X(T) = s\big)$$
$$= \begin{cases} 1 - \exp\left\{-\dfrac{2b}{T}(aT + B - s)\right\} & \text{if } s \leqslant aT + b, \\ 0 & \text{if } s > aT + B. \end{cases}$$

Write this as:

$$P\big(X(t) \leqslant at + b,\ 0 \leqslant t \leqslant T \mid X(T)\big)$$
$$\overset{\text{a.s.}}{=} 1 - \exp\left\{-2b\big(aT + b - X(T)\big)\frac{1}{T}\right\} \quad \text{if } X(T) \leqslant aT + b.$$

Taking expectations of both sides of the above, we get

$$P\big(X(t) \leqslant at + b,\ 0 \leqslant t \leqslant T\big)$$
$$= \int_{-\infty}^{aT+b} \big(1 - e^{-2b(aT+b-s)1/T}\big)\frac{1}{\sqrt{2\pi T}} e^{-s^2/2T}\,ds$$
$$= \frac{1}{\sqrt{2\pi T}} \int_{-\infty}^{aT+b} e^{-s^2/2T}\,ds - \frac{e^{-2ab}}{\sqrt{2\pi T}} \int_{-\infty}^{aT+b} e^{-(s-2b)^2/2T}\,ds.$$

Hence

$$P(X \text{ going above line } at + b \text{ during } [0, T]) = 1 - \text{previous probability}$$
$$= \frac{1}{\sqrt{2\pi T}} \int_{aT+b}^{\infty} e^{-s^2/2T}\,ds + e^{-2ab} \cdot \frac{1}{\sqrt{2\pi T}} \int_{-\infty}^{aT-b} e^{-u^2/2T}\,du,$$
$$\text{where } u = s - 2b. \tag{B.1}$$

Now, when $a = 0$, $b > 0$,

$$P\Big[\sup_{0 \leqslant t \leqslant T} X(t) \geqslant b\Big] = \sqrt{\frac{2}{\pi T}} \int_{b}^{\infty} e^{-v^2/2T}\,dv,$$

which agrees with a known formula (see, e.g., p. 261 of Tucker (1967)). In the case $a > 0$, when $T \to \infty$, since $\sqrt{T}/T \to 0$ and $\sqrt{T} = s.d.$ of $X(T)$, the first integral $\to 0$, the second integral $\to 1$, and $P(X \text{ ever rises above line } at + b) = e^{-2ab}$. Similarly, in the case $a < 0$, $P(\text{ever rises above line } at + b) = 1$.

The theorem it comes from is due to Sten Malmquist, On certain confidence contours for distribution functions, Ann. Math. Stat. 25 (1954), pp. 523–533. This theorem is stated in S.R. Paranjape and C. Park, Distribution of the supremum of the two-parameter Yeh–Wiener process on the boundary, J. Appl. Prob. 10 (1973).

Letting $\alpha = a\sqrt{T}$, $\beta = b/\sqrt{T}$, formula (B.1) becomes

$$P(\cdot) = N(-\alpha - \beta) + e^{-2\alpha\beta} N(\alpha - \beta) \text{ where } \alpha, \beta > 0 \quad \text{or}$$

$$P\big(X(t) \leqslant at + b,\ 0 \leqslant t \leqslant T\big) = 1 - P(\cdot) = N(\alpha + \beta) - e^{-2\alpha\beta} N(\alpha - \beta)$$

for the probability the line is never surpassed. This follows from:

$$\frac{1}{\sqrt{2\pi T}} \int_{aT+b}^{\infty} e^{-s^2/2T}\, ds = \frac{1}{\sqrt{2\pi}} \int_{a\sqrt{T}+b/\sqrt{T}}^{\infty} e^{-x^2/2}\, dx = N(-\alpha - \beta) \quad \text{and}$$

$$\frac{1}{\sqrt{2\pi T}} \int_{-\infty}^{aT-b} e^{-u^2/2T}\, du = N(\alpha - \beta)$$

where $s = aT + b$, $x = s/\sqrt{T} = a\sqrt{T} + b/\sqrt{T}$, $\alpha = a\sqrt{T}$ and $\beta = b/\sqrt{T}$.

The formula becomes:

$$\begin{aligned}
&P\big(\sup\big[X(t) - (at + b)\big] \geqslant 0\colon\ 0 \leqslant t \leqslant T\big) \\
&\quad = N(-\alpha - \beta) + e^{-2ab} N(\alpha - \beta) \\
&\quad = N(-\alpha - \beta) + e^{-2\alpha\beta} N(\alpha - \beta), \qquad \alpha, \beta > 0.
\end{aligned}$$

Observe that

$$\begin{aligned}
P(\cdot) &< N(-\alpha - \beta) + N(\alpha - \beta) = \big\{1 - N(\alpha + \beta)\big\} + N(\alpha - \beta) \\
&= \int_{-\infty}^{\alpha-\beta} \alpha(x)\, dx + \int_{\alpha+\beta}^{\infty} \alpha(x)\, dx < 1
\end{aligned}$$

as it should be.

Appendix C. Expected time to reach goal

Reference: Handbook of Mathematical Functions, Abramowitz and Stegun, Editors, N.B.S. Applied Math. Series 55, June 1964.

P. 304, 7.4.33 gives with $\operatorname{erf} z \equiv \frac{2}{\sqrt{\pi}} \int_0^z e^{-t^2}\, dt$ the integral:

$$\begin{aligned}
&\int \exp\big\{-\big(a^2x^2 + b^2/x^2\big)\big\}\, dx \\
&\quad = \frac{\sqrt{\pi}}{4a}\big[e^{2ab} \operatorname{erf}(ax + b/x) + e^{-2ab} \operatorname{erf}(ax - b/x)\big] + C, \quad a \neq 0. \qquad \text{(C.1)}
\end{aligned}$$

Now the left side is >0 so for real a, we require $a > 0$ otherwise the right side is <0, a contradiction.

We also note that p. 302, 7.4.3. gives

$$\int_0^{\infty} \exp\big\{-\big(at^2 + b/t^2\big)\big\}\, dt = \frac{1}{2}\sqrt{\frac{\pi}{a}}\, e^{-2\sqrt{ab}} \qquad \text{(C.2)}$$

with $\Re a > 0$, $\Re b > 0$.

To check (C.2) v. (C.1), suppose in (C.1) $a > 0, b > 0$ and find $\lim_{x\to 0}$ and $\lim_{x\to\infty}$ of $\text{erf}(ax + b/x)$ and $\text{erf}(ax - b/x)$,

$$\lim_{x\downarrow 0+}(ax + b/x) = +\infty, \qquad \lim_{x\downarrow 0+}(ax - b/x) = -\infty,$$
$$\lim_{x\to\infty}(ax + b/x) = +\infty, \qquad \lim_{x\to\infty}(ax - b/x) = +\infty.$$

Equation (C.1) becomes

$$\frac{\sqrt{\pi}}{4a}\mathrm{e}^{-2ab}\big[\text{erf}(\infty) - \text{erf}(-\infty)\big] = \frac{\sqrt{\pi}}{4a}\mathrm{e}^{-2ab}2\,\text{erf}(\infty) = \frac{\sqrt{\pi}}{2a}\mathrm{e}^{-2ab}$$

since we know $\text{erf}(\infty) = 1$.

In (C.2) replace a by a^2, b by b^2 to get

$$I_0\big(a^2, b^2\big) \equiv \int_0^\infty \exp\big\{-\big(a^2t^2 + b^2/t^2\big)\big\}\,\mathrm{d}t = \frac{1}{2}\frac{\sqrt{\pi}}{|a|}\mathrm{e}^{-2|ab|}$$

which is the same.

Note: if we choose the lower limit of integration to be 0 in (C.1), then we can find C:

$$0 = \int_0^{0^+} \exp\big\{-\big(a^2x^2 + b^2/x^2\big)\big\}\,\mathrm{d}x = \frac{\sqrt{\pi}}{4a}\big[\mathrm{e}^{2ab}\text{erf}(\infty) + \mathrm{e}^{-2ab}\,\text{erf}(-\infty)\big] + C$$
$$= \frac{\sqrt{\pi}}{4a}\big[\mathrm{e}^{2ab} - \mathrm{e}^{-2ab}\big] + C.$$

Whence

$$F(x) \equiv \int_0^x \exp\big\{-\big(a^2x^2 + b^2/x^2\big)\big\}\,\mathrm{d}x$$
$$= \frac{\sqrt{\pi}}{4a}\big\{\mathrm{e}^{2ab}\big[\text{erf}(ax + b/x) - 1\big] + \mathrm{e}^{-2ab}\big[\text{erf}(ax - b/x) + 1\big]\big\}. \tag{C.3}$$

To see how (C.3) might have been discovered, differentiate:

$$F'(x) = \exp\big\{-\big(a^2x^2 + b^2/x^2\big)\big\}$$
$$= \frac{\sqrt{\pi}}{4a}\big\{\mathrm{e}^{2ab}\big(a - b/x^2\big)\,\text{erf}'(ax + b/x)$$
$$+\,\mathrm{e}^{-2ab}\big(a + b/x^2\big)\,\text{erf}'(ax - b/x)\big\}.$$

Now $\text{erf}'(z) = \frac{2}{\sqrt{\pi}}\exp(-z^2)$ so

$$\text{erf}'(ax + b/x) = \frac{2}{\sqrt{\pi}}\exp\big[-(ax + b/x)^2\big] = \frac{2}{\sqrt{\pi}}\exp\big\{-\big(a^2x^2 + b^2/x^2 + 2ab\big)\big\}$$
$$= \frac{2}{\sqrt{\pi}}\mathrm{e}^{-2ab}\exp\big\{-\big(a^2x^2 + b^2/x^2\big)\big\}\ ,$$

and, setting $b \leftarrow -b$,

$$\mathrm{erf}'(ax - b/x) = \frac{2}{\sqrt{\pi}} \mathrm{e}^{2ab} \exp\{-(a^2x^2 + b^2/x^2)\}$$

whence

$$\begin{aligned} F'(x) &= \frac{\sqrt{\pi}}{4a} \left\{ \frac{2}{\sqrt{\pi}}(a - b/x^2) + \frac{2}{\sqrt{\pi}}(a + b/x^2) \right\} \exp\{-(a^2x^2 + b^2/x^2)\} \\ &= \frac{1}{2a}\{2a\} \exp\{-(a^2x^2 + b^2/x^2)\} = \exp\{-(a^2x^2 + b^2/x^2)\}. \end{aligned}$$

Case of interest: $a < 0$, $b > 0$.
Expect:

$$\begin{aligned} b > 0,\ a \leqslant 0 \quad &\Rightarrow \quad F(T) \uparrow 1 \text{ as } T \to \infty, \\ b > 0,\ a > 0 \quad &\Rightarrow \quad F(T) \uparrow c < 1 \text{ as } T \to \infty. \end{aligned}$$

If $b > 0, a = 0$:

$$F(T) = N(-\beta) + N(-\beta) = 2N(-b/\sqrt{T}) \uparrow 2N(0) = 1 \quad \text{as } T \uparrow \infty.$$

Also, as expected $F(T) \uparrow 1$ as $b \downarrow 0$.
If $b > 0, a < 0$: See below.
If $b > 0, a > 0$:

$$\begin{aligned} F(T) &= N(-a\sqrt{T} - b/\sqrt{T}) + \mathrm{e}^{-2ab} N(a\sqrt{T} - b/\sqrt{T}) \\ &\to N(-\infty) + \mathrm{e}^{-2ab} N(\infty) \\ &= \mathrm{e}^{-2ab} < 1 \quad \text{as } T \uparrow \infty. \end{aligned}$$

This is correct.

If $b = 0$: $F(T) = N(-a\sqrt{T}) + N(a\sqrt{T}) = 1$. This is correct.

Let $F(T) = P(X(t) \geqslant at + b$ for some $t, 0 \leqslant t \leqslant T)$ which equals $N(-\alpha - \beta) + \mathrm{e}^{-2ab} N(\alpha - \beta)$ where $\alpha = a\sqrt{T}$ and $\beta = b/\sqrt{T}$ so $ab = \alpha\beta$; we assume $b > 0$ and $a < 0$ in which case $0 \leqslant F(T) \leqslant 1$ and $\lim_{T\to\infty} F(T) = 1$, $\lim_{T\to 0} F(T) = 0$; F is a probability distribution function:

$$\begin{aligned} \lim_{T\to 0} F(T) &= N(-\infty) + \mathrm{e}^{-2ab} N(-\infty) = 0, \\ \lim_{T\to\infty} F(T) &= N(+\infty) + \mathrm{e}^{-2ab} N(-\infty) = 1. \end{aligned}$$

The density function is

$$f(T) = F'(T) = \frac{\partial}{\partial T}(-\alpha - \beta) N'(-\alpha - \beta) + \mathrm{e}^{-2ab} \frac{\partial}{\partial T}(\alpha - \beta) N'(\alpha - \beta)$$

where

$$\frac{\partial \alpha}{\partial T} = \frac{1}{2} a T^{-1/2}, \qquad \frac{\partial \beta}{\partial T} = -\frac{1}{2} b T^{-3/2},$$

$$N'(-\alpha-\beta) = \frac{1}{\sqrt{2\pi}}\mathrm{e}^{-(\alpha+\beta)^2/2} = \frac{1}{\sqrt{2\pi}}\exp\left\{-\frac{(a^2T + b^2/T + 2ab)}{2}\right\},$$

$$N'(\alpha-\beta) = \frac{1}{\sqrt{2\pi}}\mathrm{e}^{-(\alpha-\beta)^2/2} = \frac{1}{\sqrt{2\pi}}\exp\left\{-\frac{(a^2T + b^2/T - 2ab)}{2}\right\},$$

$$\begin{aligned}
Tf(T) &= T\left(-\frac{1}{2}aT^{-1/2} + \frac{1}{2}bT^{-3/2}\right)\frac{1}{\sqrt{2\pi}}\mathrm{e}^{-ab}\exp\left\{-\frac{(a^2T + b^2/T)}{2}\right\} \\
&\quad + T\mathrm{e}^{-2ab}\left(\frac{1}{2}aT^{-1/2} + \frac{1}{2}bT^{-3/2}\right)\frac{1}{\sqrt{2\pi}}\mathrm{e}^{ab}\exp\left\{-\frac{(a^2T + b^2/T)}{2}\right\} \\
&= \frac{\mathrm{e}^{-ab}}{2\sqrt{2\pi}}\left[(-aT^{+1/2} + bT^{-1/2})\exp\left\{-\frac{(a^2T + b^2/T)}{2}\right\}\right. \\
&\quad \left. + \left(aT^{+1/2} + bT^{-1/2}\right)\exp\left\{-\frac{(a^2T + b^2/T)}{2}\right\}\right] \\
&= \frac{b\mathrm{e}^{-ab}}{\sqrt{2\pi}}T^{1/2}\exp\left\{\frac{-(a^2T + b^2/T)}{2}\right\}.
\end{aligned}$$

The expected time to the goal is

$$E_\infty = \int_0^\infty Tf(T)\,\mathrm{d}T = \frac{b\mathrm{e}^{-ab}}{\sqrt{2\pi}}\int_0^\infty T^{-1/2}\exp\left\{-\frac{(a^2T + b^2/T)}{2}\right\}\mathrm{d}T,$$

$$\left.\begin{aligned} T^{1/2} &= x \\ T &= x^2 \\ \mathrm{d}T &= 2x\,\mathrm{d}x \end{aligned}\right\} = \frac{2b\mathrm{e}^{-ab}}{\sqrt{2\pi}}\int_0^\infty \exp\left\{-\left[\left(\frac{a}{\sqrt{2}}\right)^2 x^2 + \left(\frac{b}{\sqrt{2}}\right)^2 x^{-2}\right]\right\}\mathrm{d}x$$

$$= \frac{2b\mathrm{e}^{-ab}}{\sqrt{2\pi}}I_0\left(\left(\frac{a}{\sqrt{2}}\right)^2, \left(\frac{b}{\sqrt{2}}\right)^2\right).$$

Now

$$I_0\left(a^2, b^2\right) = \frac{\sqrt{\pi}}{2|a|}\mathrm{e}^{-2|ab|} \quad \text{so}$$

$$I_0\left(\left(\frac{a}{\sqrt{2}}\right)^2, \left(\frac{b}{\sqrt{2}}\right)^2\right) = \frac{\sqrt{\pi}}{\sqrt{2}|a|}\mathrm{e}^{-|ab|} \quad \text{whence}$$

$$E_\infty = \frac{2b\mathrm{e}^{-ab}}{\sqrt{2\pi}}\frac{\sqrt{\pi}}{\sqrt{2}|a|}\mathrm{e}^{-|ab|} = \frac{b}{|a|}, \quad a < 0,\ b > 0.$$

Note:

$$f(T) \equiv F'(T) = \frac{b\mathrm{e}^{-ab}}{\sqrt{2\pi}}T^{-3/2}\exp\left\{\frac{-(a^2T + b^2/T)}{2}\right\} > 0$$

for all a, e.g., $a < 0$, so $F(T)$ is monotone increasing. Hence, since $\lim_{T\to\infty} F(T) = 1$ for $a < 0$ and $\lim_{T\to\infty} F(T) < 1$ for $a > 0$, $0 \leqslant F(T) \leqslant 1$ for all T so we have more confidence in using the formula for $a < 0$ too.

$$\begin{array}{lll} \text{Check:} & E_\infty(a,b) \downarrow 0 \text{ as } \downarrow -\infty & \text{yes,} \\ & E_\infty(a,b) \uparrow \text{ as } b \uparrow & \text{yes,} \\ & E_\infty(a,b) \uparrow \text{ as} |a| \downarrow & \text{yes,} \\ \text{note} & \lim_{a \downarrow 0^+} E_\infty(a,b) = +\infty & \text{as suspected.} \end{array}$$

This leads us to believe that in a fair coin toss (fair means no drift) and a gambler with finite capital, the expected time to ruin is infinite.

This is correct. Feller gives $D = z(a-z)$ as the duration of the game, where z is initial capital, ruin is at 0, and a is the goal. Then $\lim_{a\to\infty} D(a) = +\infty$.

Note: $E_\infty = b/|a|$ means the expected time is the same as the point where $aT + b$ crosses $X(t) = 0$. See Figure 2.

$$\begin{aligned} &E_\infty = b/|a|, \quad a = -m/s^2, \quad b = \ln\lambda, \\ &\lambda = C/X_0 = \text{normalized goal}, \\ &m = p\ln(1+f) + q\ln(1-f) \equiv g(f), \\ &s^2 = pq\{\ln[(1+f)/(1-f)]\}^2, \\ &\text{Kelly fraction } f^* = p - q, \quad g(f^*) = p\ln 2p + q\ln 2q, \\ &\text{For } m > 0, \quad E_\infty = (\ln\lambda)s^2/g(f). \end{aligned}$$

Now this is the expected time in variance units. However s^2 variance units $= 1$ trial so

$$n(\lambda, f) \equiv \frac{E_\infty}{s^2} = \frac{\ln\lambda}{g(f)} = \frac{\ln\lambda}{m}$$

is the expected number of trials.

$$\begin{array}{lll} \text{Check:} & n(\lambda, f) \uparrow & \text{as } \lambda \uparrow, \\ & n(\lambda, f) \to \infty & \text{as } \lambda \to \infty, \\ & n(\lambda, f) \uparrow & \text{as } m \downarrow 0, \\ & n(\lambda, f) \to \infty & \text{as } m \to 0. \end{array}$$

Now $g(f)$ has unique maximum at $g(f^*)$ where $f^* = p - q$, the "Kelly fraction", therefore $n(\lambda, f)$ has a unique minimum for $f = f^*$. Hence f^* reaches a fixed goal in least expected time in this, the continuous case, so we must be asymptotically close to least expected time in the discrete case, which this approximates increasing by well in the sense of the CLT (Central Limit Theorem) and its special case, the normal approximation to the binomial distribution. The difference here is the trials are asymmetric. The positive and negative step sizes are unequal.

References

Associates First Capital Corporation, 1998. 1997 Annual Report. Associates First Capital Corporation, Dallas, TX.

Black, F., Scholes, M., 1973. The pricing of options and corporate liabilities. Journal of Political Economy 81, 637–659.

Breiman, L., 1961. Optimal gambling systems for favorable games. In: Fourth Berkeley Symposium on Probability and Statistics, vol. I, pp. 65–78.

Browne, S., 1996. Reaching Goals by a Deadline: Continuous-Time Active Portfolio Management. Columbia University, New York.

Browne, S., 1997. The return on investment from proportional investment strategies. Advances in Applied Probability 30 (1), 216–238.

Buffett, M., Clark, D., 1997. Buffettology. Rawson Associates, Simon and Schuster, New York.

Cox, D.R., Miller, H.D., 1965. The Theory of Stochastic Processes. Wiley, New York.

Feller, W., 1966. An Introduction to Probability Theory and Its Applications, vol. I, Revised. Wiley, New York.

Griffin, P.A., 1979. The Theory of Blackjack. Huntington Press, Las Vegas. Revised 1995.

Hagstrom, R.G. Jr., 1994. The Warren Buffett Way. Wiley, New York.

Hagstrom, R.G. Jr., 2004. The Warren Buffett Way, second ed. Wiley, New York.

Hecht, R., 1995. Private correspondence.

Ibbotson Associates, 1998. Yearbook: Stocks, Bonds, Bills and Inflation (or any later edition). Ibbotson Associates, Chicago.

Institutional Investor, 1998. Ivory Tower Investing, pp. 43–55 (see p. 44), March.

Kelly, J.L., 1956. A new interpretation of information rate. Bell System Technical Journal 35, 917–926.

Kilpatrick, A., 1994. Of Permanent Value, the Story of Warren Buffet. Distributed by Southern Publishers Group, Birmingham, AL.

Kritzman, M., 1998. Risk and utility: basics. In: Bernstein, Damodaran (Eds.), Investment Management. Wiley, New York. Chapter 2.

Lowenstein, R., 1995. The Making of an American Capitalist. Random House, New York.

Markowitz, H., 1959. Portfolio Selection. Cowles Monograph, vol. 16. Wiley, New York.

McEnally, R.W., 1986. Latané's bequest: the best of portfolio strategies. Journal of Portfolio Management 12 (2), 21–30, Winter.

Quaife, A., 1995. Using the Sharpe ratio to evaluate investments. The Trans Times 4 (1), February. Trans Time Inc., Oakland, CA.

Rotando, L.M., Thorp, E.O., 1992. The Kelly criterion and the stock market. American Mathematical Monthly 99, 922–931, December.

Thorp, E.O., 1962. Beat the Dealer. Random House, New York.

Thorp, E.O., 1966. Beat the Dealer, second ed. Vintage, New York.

Thorp, E.O., 1969. Optimal gambling systems for favorable games. Review of the International Statistical Institute 37, 273–293.

Thorp, E.O., 1971. Portfolio choice and the Kelly criterion. In: Proceedings of the 1971 Business and Economics Section of the American Statistical Association, pp. 215–224.

Thorp, E.O., 1984. The Mathematics of Gambling. Lyle Stuart, Secaucus, NJ.

Thorp, E.O., Kassouf, S.T., 1967. Beat the Market. Random House, New York.

Thorp, E.O., Walden, W., 1966. A winning bet in Nevada baccarat, part I. Journal of the American Statistical Association 61, 313–328.

Tucker, H., 1967. A Graduate Course in Probability. Academic Press, San Diego, CA.

Wong, S., 1994. Professional Blackjack. Pi Yee Press, La Jolla, CA.

Bibliography

Aase, K. K. (2001). On the St. Petersburg Paradox. *Scandinavian Actuarial Journal*, 3(1), 69–78.

Algoet, P. H. and T. M. Cover (1988). Asymptotic optimality and asymptotic equipartition properties of log-optimum investment. *Annals of Probability*, 16(2), 876–898.

Aucamp, D. (1977). An investment strategy with overshoot rebates which minimizes the time to attain a specified goal. *Management Science*, 23(11), 1234–1241.

Aucamp, D. (1993). On the extensive number of plays to achieve superior performance with the geometric mean strategy. *Management Science*, 39, 1163–1172.

Barron, A. R. and T. M. Cover (1988). A bound on the financial value of information. *IEEE Transactions of Information Theory*, 34(5), 1097–1100.

Bell, R. M. and T. M. Cover (1980). Competitive optimality of logarithmic investment. *Math of Operations Research*, 5, 161–166.

Bell, R. M. and T. M. Cover (1988). Game-theoretic optimal portfolios. *Management Science*, 34(6), 724–733.

Bellman, R. and R. Kalaba (1957). On the role of dynamic programming in statistical communication theory. *IRE Transactions of the Professional Group on Information Theory, IT*, 3(3), 197–203.

Bernoulli, D. (1954). Exposition of a new theory on the measurement of risk (translated by Louise Sommer). *Econometrica*, 22, 23–36.

Bicksler, J. L. and E. O. Thorp (1973). The capital growth model: An empirical investigation. *Journal of Financial and Quantitative Analysis*, 8(2), 273–287.

Breiman, L. (1960). Investment policies for expanding businesses optimal in a long run sense. *Naval Research Logistics Querterly*, 4(4), 647–651.

Breiman, L. (1961). Optimal gambling system for favorable games. *Proceedings of the 4th Berkeley Symposium on Mathematical Statistics and Probability*, 1, 63–8.

Browne, S. (1997). Survival and growth with a liability: Optimal portfolios in continuous time. *Math of Operations Research*, 22, 468–493.

Browne, S. (2000). Risk-constrained dynamic active portfolio management. *Management Science*, 46(9), 1188–1199.

Chapman, S. J. (2007). The Kelly criterion for spread bets. *IMA Journal of Applied Mathematics*, 72, 43–51.

Chopra, V. K. and W. T. Ziemba (1993). The effect of errors in means, variances and co-variances on optimal portfolio choice. *Journal of Portfolio Management*, 19, 6–11.

Christensen, M. M. (2005). On the history of the growth optimal portfolio. Working Paper, University of South Denmark.

Clark, R. and W. T. Ziemba (1987). Playing the turn-of-the-year effect with index futures. *Operations Research*, 35, 799–813.

Cover, T. M. (1984). An algorithm for maximizing expected log investment return. *IEEE Transactions on Information Theory*, 30(2), 369–373.

Cover, T. M. (1987). Log optimal portfolios. In W. Eadington (Ed.), *Research: Gambling and Risk Taking, Seventh International Conference, Vol 4: Quantitative Analysis and Gambling, Reno, NV.*

Cover, T. M. (1991). Universal portfolios. *Math Finance*, 1, 1–29.
Cover, T. M. and D. H. Gluss (1986). Empirical Bayes stock market portfolios. *Advances in Applied Mathematics*, 7, 170–181.
Cover, T. M. and E. Ordentlich (1996). Universal portfolios with side information. *IEEE Transactions on Information Theory*, 42(2), 348–363.
Cover, T. M. and J. Thomas (2006). *Elements of Information Theory, 2nd Edition*. New York: John Wiley & Sons.
Davis, M. and S. Lleo (2010). On benchmarks and fractional Kelly strategies.
Davis, M. H. A. and S. Lleo (2008a). A risk sensitive asset and liability management model. Working Paper.
Davis, M. H. A. and S. Lleo (2008b). Risk-sensitive benchmarked asset management. *Quantitative Finance*, 8(4), 415426.
Dempster, M. A. H., I. V. Evstigneev, and K. R. Schenk-Hoppé (2003). Exponential growth of fixed mix assets in stationary markets. *Finance and Stochastics*, 7(2), 263–276.
Dempster, M. A. H., I. V. Evstigneev, and K. R. Schenk-Hoppé (2007). Volatility-induced financial growth. *Quantitative Finance*, 7(2), 151–160.
Dempster, M. A. H., I. V. Evstigneev, and K. R. Schenk-Hoppé (2010). Gaining wealth with fixed-mix strategies.
Dubins, L. and L. Savage (1965). *How to Gamble If You Must*. New York: McGraw-Hill.
Epstein, R. A. (1977). *The Theory of Gambling and Statistical Logic*. London: Academic Press.
Erkip, E. and T. M. Cover (1998). The efficiency of investment information. *IEEE Transactions on Information Theory*, 4(3), 1026–1040.
Ethier, S. and N. Tavaré (1983). The proportional bettor's return on investment. *Journal of Applied Probability*, 20, 563–573.
Ethier, S. N. (2004). The Kelly system maximizes median wealth. *Journal of Applied Probability*, 41(4), 1230–1236.
Evstigneev, I. V., T. Hens, and K. R. Schenk-Hoppé (2008). Globally evolutionarily stable portfolio rules. *Journal of Economic Theory*, 140, 197–228.
Evstigneev, I. V., T. Hens, and K. R. Schenk-Hoppé (2010). Survival and evolutionary stability of the Kelly rule.
Feller, W. (1962). *An Introduction to Probability Theory and Its Applications* (2 ed.). New York: John Wiley & Sons.
Ferguson, T. S. (1965). Betting systems which minimize the probability of ruin. *Journal of the Society for Industrial and Applied Mathematics*, 13(3), 795–818.
Fernholz, R. and B. Shay (1982). Stochastic portfolio theory and stock market equilibrium. *Journal of Finance*, 37, 615–624.
Finkelstein, M. and R. Whitley (1981). Optimal strategies for repeated games. *Advanced Applied Probability*, 13, 415–428.
Fitt, A. D., C. J. Howls, and M. Kabelka (2006). The valuation of soccer spread bets. *Journal of the Operational Research Society*, 77, 975–985.
Fuller, J. (2006). Optimize your portfolio with the Kelly formula. www.morningstar.com, October 6.
Geyer, A. and W. T. Ziemba (2008). The Innovest Austrian pension fund planning model InnoALM. *Operations Research*, 56(4), 797–810.
Goldman, M. B. (1974). A negative report on the "near optimality" of the max-expected log policy and appled to bounded utilities for long-lived programs. *Journal of Financial Economics*, 1, 97–103.
Gottlieb, G. (1984). An optimal betting strategy for repeated games. Working paper, New York University.

Gottlieb, G. (1985). An analytic derivation of blackjack win rates. *Operations Research*, 33, 971–988.

Grauer, R. R. and N. H. Hakansson (1986). A half century of returns on levered and unlevered portfolios of stocks, bonds and bills, with and without small stocks. *Journal of Business*, 592, 287–318.

Haigh, J. The Kelly criterion and bet comparisons in spread betting. *Statistician*, 49, 531–539.

Hakansson, N. (1970). Optimal investment and consumption strategies under risk for a class of utility functions. *Econometrica*, 38, 587–607.

Hakansson, N. H. (1971). On optimal myopic portfolio policies with and without serial correlation. *Journal of Business*, 44, 324–334.

Hakansson, N. H. and W. T. Ziemba (1995). Capital growth theory. In R. A. Jarrow, V. Maksimovic, and W. T. Ziemba (Eds.), *Finance, Handbooks in OR & MS*, 65–86. Amsterdam: North Holland.

Hausch, D. B. and W. T. Ziemba (1985). Transaction costs, extent of inefficiencies, entries and multiple wagers in a racetrack betting model. *Management Science*, 31, 381–394.

Hausch, D. B., W. T. Ziemba, and M. E. Rubinstein (1981). Efficiency of the market for racetrack betting. *Management Science*, 27, 1435–1452.

Hens, T. and K. Schenk-Hoppe (2005). Evolutionary stability of portfolio rules in incomplete markets. *J. of Mathematical Economics*, 41, 43–66.

Iyengar, G. N. and T. M. Cover (2000). Growth optimal investment in horse race markets with costs. *IEEE Transactions on Information Theory*, 46(7), 2675–2683.

Jackson, D. and P. Waldron (2003). Parimutuel place betting in Great Britain and Ireland. In L. V. Williams (Ed.), *The Economics of Gambling*, 18–29.

Kelly, Jr., J. R. (1956). A new interpretation of the information rate. *Bell System Technical Journal*, 35, 917–926.

Latané, H. (1959). Criteria for choice among risky ventures. *Journal of Political Economy*, 67, 144–155.

Latané, H. (1978). The geometric-mean principle revisited – a reply. *Journal of Banking and Finance*, 2(4), 395–398.

Lee, E. (2006). How to calculate the Kelly formula. www.fool.com, October 31.

Leland, H. (1968). Dynamic portfolio theory. PhD thesis, Economics Department, Harvard University.

Long, Jr, J. B. (1990). The numeraire portfolio. *Journal of Financial Economics*, 26(1), 29–69.

Luenberger, D. G. (1993). A preference foundation for log mean-variance criteria in portfolio choice problems. *Journal of Economic Dynamics and Control*, 17, 887–906.

Lv, Y. and B. K. Meister (2009). Application of the Kelly criterion to Ornstein-Uhlenbeck processes. Lecture Notes of the Institute for Computer Sciences, 4, 1051–1062, Springer.

MacLean, L. C., R. Sanegre, Y. Zhao, and W. T. Ziemba (2004). Capital growth with security. *Journal of Economic Dynamics and Control*, 28(4), 937–954.

MacLean, L. C., E. O. Thorp, Y. Zhao, and W. T. Ziemba (2010). Medium term simulations of Kelly and fractional Kelly and proportional betting strategies.

MacLean, L. C., E. O. Thorp, and W. T. Ziemba (2010). Good and bad properties of the Kelly criterion.

MacLean, L. C., Y. Zhao, and W. T. Ziemba (2009). Optimal capital growth with convex loss penalties. Working paper, Dalhousie University.

MacLean, L. C., W. T. Ziemba, and G. Blazenko (1992). Growth versus security in dynamic investment analysis. *Management Science*, 38, 1562–85.

MacLean, L. C., W. T. Ziemba, and Li (2005). Time to wealth goals in capital accumulation and the optimal trade-off of growth versus security. *Quantitative Finance*, 5(4), 343–357.

Markowitz, H. M. (1976). Investment for the long run: New evidence for an old rule. *Journal of Finance*, 31(5), 1273–1286.

Mathis, C. and T. M. Cover (2005). A statistic for measuring the influence of side information in investment. *Proceedings of the IEEE International Symposium on Information Theory*, 1156–1157.

McEnally, R. W. (1986). Latané's bequest: The best of portfolio strategies. *Journal of Portfolio Management*, 12(2), 21–30.

Menger, K. (1967). The role of uncertainty in economics. In *Essays in Mathematical Economics in Honor of Oskar Morgenstern*. New Jersey: Princeton University Press.

Merton, R. C. (1969). Lifetime portfolio selection under uncertainty: The continuous time case. *Review of Economics and Statistics*, 51, 247–259.

Merton, R. C. (1973). An intertemporal capital asset pricing model. *Econometrica*, 41, 867–887.

Merton, R. C. (1990). *Continuous-Time Finance*. Cambridge, MA: Blackwell Publishers.

Merton, R. C. and P. A. Samuelson (1974). Fallacy of the log-normal approximation to optimal portfolio decision-making over many periods. *Journal of Financial Economics*, 1, 67–94.

Mossin, J. (1968). Optimal multiperiod portfolio policies. *Journal of Business*, 41, 215–229.

Mulvey, J. M., M. Bilgili, and T. M. Vural (2010). A dynamic portfolio of investment strategies: Applying capital growth with drawdown penalties.

Ordentlich, E. and T. M. Cover (1998). The cost of achieving the best portfolio in hindsight. *Mathematics of Operations Research*, 23(4), 960–982.

Osorio, R. (2008). A prospect theory approach to the Kelly criterion for fat-tail portfolios: The case of the Student t-Distribution. Evnine and Associates Report.

Phelps, E. S. (1962). The accumulation of risky capital: A sequential utility analysis. *Econometrica*, 30, 729–743.

Platen, E. (2010). A benchmark approach to investing and pricing.

Poundstone, W. (2005). *Fortune's Formula: The Untold Story of the Scientific System that Beat the Casinos and Wall Street*. New York: Hill and Wang.

Pulley, L. B. (1983). Mean-variance approximation to expected logarithmic utility. *Operations Research*, 31, 685–697.

Ramsey, F. P. (1928). A mathematical theory of saving. *Economic Journal*, 38, 543–559.

Roll, R. (1973). Evidence on the growth optimum model. *The Journal of Finance*, 28(3), 551–566.

Rotando, L. M. and E. Thorp (1992). The Kelly criterion and the stock market. *The American Mathematical Monthly*, 99(10), 922–931.

Rubinstein, M. (1976). The strong case for the generalized logarithmic utility model as the premier model of nancial markets. *Journal of Finance*, 31(2), 551–571.

Rubinstein, M. (1991). Continuously rebalanced portfolio strategies. *Journal of Portfolio Management*, 18(1), 78–81.

Rudolf, M. and W. T. Ziemba (2004). Intertemporal surplus management. *Journal of Economic Dynamics and Control*, 28, 975–990.

Samuelson, P. A. (1969). Lifetime portfolio selection by dynamic stochastic programming. *Review of Economics and Statistics*, 51, 239–246.

Samuelson, P. A. (1971). The "fallacy" of maximizing the geometric mean in long sequences of investing or gambling. *Proceedings National Academy of Science*, 68, 2493–2496.

Samuelson, P. A. (1977). St. Petersburg paradoxes: Defanged, dissected and historically described. *Journal of Economic Literature*, 15(1), 24–55.

Samuelson, P. A. (1979). Why we should not make mean log of wealth big though years to act are long. *Journal of Banking and Finance*, 3, 305–307.

Samuelson, P. A. (1991). Long-run risk tolerance when equity returns are mean regressing: Pseudoparadoxes and vindication of businessmens risk. In W. C. Brainard, W. D. Nordhaus, and H. W. Watts (Eds.), *Money, Macroeconomics and Economic Policy*, 181–200. Cambridge, MA: MIT Press.

Stutzer, M. (2003). Portfolio choice with endogenous utility: A large deviations approach. *Journal of Econometrics*, 116, 365–386.

Stutzer, M. (2004). Asset allocation without unobservable parameters. *Financial Analysts Journal*, 60(5), 38–51.

Stutzer, M. (2010). On growth optimality security against underperformance.

Thorp, E. O. (1966). A favorable side bet in Nevada Baccarat.

Thorp, E. O. (1969). Optimal gambling systems for favorable games. *Review of the International Statistical Institute*, 37(3), 273–293.

Thorp, E. O. (1971). Portfolio choice and the Kelly criterion. *Proceedings of the Business and Economics Section of the American Statistical Association*, 215–224.

Thorp, E. O. (2006). The Kelly criterion in blackjack sports betting and the stock market. In S. A. Zenios and W. T. Ziemba (Eds.), *Handbook of Asset and Liability Management, Volume 1*, 387–428. Amsterdam: North Holland.

Thorp, E. O. (2008). Understanding the Kelly criterion. *Wilmott*, May and September.

Thorp, E. O. and R. Whitley (1972). Concave utilities are distinguished by their optimal strategies. *Colloquia Mathematica Societatis Janos Bolyai*, 813–830.

Wilcox, J. (2003a). Harry Markowitz and the discretionary wealth hypothesis. *Journal of Portfolio Management*, 29 (Spring), 58–65.

Wilcox, J. (2003b). Risk management: Survival of the fittest. Wilcox Investment Inc.

Wilcox, J. (2005). A better paradigm for finance. *Finance Letters*, 3(1), 5–11.

Williams, J. B. (1936). Speculation and the carryover. *Quarterly Journal of Economics*, 50(3), 436–455.

Ziemba, R. E. S. and W. T. Ziemba (2007). *Scenarios for Risk Management and Global Investment Strategies*. New York: Wiley.

Ziemba, W. T. (2005). The symmetric downside risk Sharpe ratio and the evaluation of great investors and speculators. *Journal of Portfolio Management Fall*, 108–122.

Ziemba, W. T. and D. B. Hausch (1984). *Beat the Racetrack*. Harcourt, Brace and Jovanovich.

Ziemba, W. T. and D. B. Hausch (1986). *Betting at the Racetrack*. Dr. Z Investments.

Ziemba, W. T. and D. B. Hausch (1987). *Dr. Z's Beat the Racetrack*. Dr. Z Investments.

Ziemba, W. T. and D. B. Hausch (2008). The Dr. Z betting system in England. In *Efficiency of Racetrack Betting Markets*, 567–574. Singapore: World Scientific.

Ziemba, W. T. and R. G. Vickson (2010). Models of optimal capital accumulation and portfolio selection.

Author Index

Subject Index

H

I

J

K

L